THE INSECTS AND ARACHNIDS OF CANADA

PART 14

The Grasshoppers, Crickets, and Related Insects of Canada and Adjacent Regions

Ulonata: Dermaptera, Cheleutoptera, Notoptera, Dictuoptera, Grylloptera, and Orthoptera

Vernon R. Vickery[1] and D. Keith McE. Kevan[1]

Charles D. Dondale, Scientific Adviser

Biosystematics Research Institute
Ottawa, Ontario

Research Branch
Agriculture Canada

Publication 1777 1985

[1] Lyman Entomological Museum and Research Laboratory, Macdonald College of McGill University, Sainte-Anne-de-Bellevue, Québec

Available in Canada through

Associated Bookstores
and other booksellers

or by mail from

Canadian Government Publishing Centre
Supply and Services Canada
Ottawa, Canada K1A 0S9

Catalogue No. A42.42/1985-14E
ISBN 0-660-11749-5

Canada: $36.95 Other Countries: $44.35

Price subject to change without notice

Canadian Cataloguing in Publication Data

Vickery, V. R. (Vernon Randolph)

The grasshoppers, crickets, and related insects of Canada and adjacent regions

(The Insects and arachnids of Canada,
ISSN 0706-7313 ; pt. 14)
(Publication ; 1777)

Includes bibliographical references and index.

1. Locusts — Canada. 2. Crickets — Canada.
I. Kevan, D. Keith McE. (Douglas Keith McEwan)
II. Canada. Agriculture Canada. Research
Branch. III. Title. IV. Series. V. Series:
Publication (Canada. Agriculture Canada).
English ; 1777.

QL507.1.C3V5 1985 595.7'26 C85-097201-9

Staff editor: Sheilah V. Balchin

The Insects and Arachnids of Canada

Part 1. Collecting, Preparing, and Preserving Insects, Mites, and Spiders, compiled by J. E. H. Martin, Biosystematics Research Institute, Ottawa, 1977. 182 p. Price: Canada $3.50, other countries $4.20 (Canadian funds). Cat. No. A42-42/1977-1.

Partie 1. Récolte, préparation et conservation des Insectes, des Acariens et des Araignées, compilé par J.E.H. Martin, Institut de recherches biosystématiques, Ottawa, 1983. 205 p. Prix: $3.50 (Canada), à l'étranger $4.20 (en devises canadiennes). Cat. No. A42-42-1977-1F.

Part 2. The Bark Beetles of Canada and Alaska (Coleoptera: Scolytidae), by D. E. Bright, Jr., Biosystematics Research Institute, Ottawa, 1976. 241 p. Price: Canada $11.95, other countries $14.35 (Canadian funds). Cat. No. A42-42/1976-2.

Part 3. The Aradidae of Canada (Hemiptera: Aradidae), by R. Matsuda, Biosystematics Research Institute, Ottawa, 1977. 116 p. Price: Canada $4.00, other countries $4.80 (Canadian funds). Cat. No. A42-42/1977-3.

Part 4. The Anthocoridae of Canada and Alaska (Heteroptera: Anthocoridae), by L. A. Kelton, Biosystematics Research Institute, Ottawa, 1978. 101 p. Price: Canada $4.00, other countries $4.80 (Canadian funds). Cat. No. A42-42/1977-4.

Part 5. The Crab Spiders of Canada and Alaska (Araneae: Philodromidae and Thomisidae), by C. D. Dondale and J. H. Redner, Biosystematics Research Institute, Ottawa, 1978. 255 p. Price: Canada $7.50, other countries $9.00 (Canadian funds). Cat. No. A42-42/1978-5.

Part 6. The Mosquitoes of Canada (Diptera: Culicidae), by D. M. Wood, P. T. Dang, and R. A. Ellis, Biosystematics Research Institute, Ottawa, 1979. 390 p. Price: Canada $8.00, other countries $9.60 (Canadian funds). Cat. No. A42-42/1979-6.

Partie 7. Genera des Trichoptères du Canada et des États adjacents, par F. Schmid, Institut de recherches biosystématiques, Ottawa, 1980. 296 p. Prix: $7.75 (Canada), à l'étranger $9.30 (en devises canadiennes). Cat. No. A42-42/1980-7F. (Available in French only.)

Part 8. The Plant Bugs of the Prairie Provinces of Canada (Heteroptera: Miridae), by L. A. Kelton, Biosystematics Research Institute, Ottawa, 1980. 408 p. Price: Canada $9.95, other countries $11.95 (Canadian funds). Cat. No. A42-42/1980-8.

Part 9. The Sac Spiders of Canada and Alaska (Araneae: Clubionidae and Anyphaenidae), by C. D. Dondale and J. H. Redner, Biosystematics Research Institute, Ottawa, 1982. 194 p. Price: Canada $8.95, other countries $10.75 (Canadian funds). Cat. No. A42-42/1982-9E.

Part 10. The Spittlebugs of Canada (Homoptera: Cercopidae), by K. G. A. Hamilton, Biosystematics Research Institute, Ottawa, 1982. 102 p. Price: Canada $7.95, other countries $9.55 (Canadian funds). Cat. No. A42-42/1982-10E.

Part 11. The Genera of Larval Midges of Canada (Diptera: Chironomidae), by D. R. Oliver and M. E. Roussel, Biosystematics Research Institute, Ottawa, 1983. 263 p. Price: Canada $11.95, other countries $14.35 (Canadian funds). Cat. No. A42-42-1983-11E.

Part 12. The Families and Subfamilies of Canadian Chalcidoid Wasps (Hymenoptera: Chalcidoidea), by C. M. Yoshimoto, Biosystematics Research Institute, Ottawa, 1984. 149 p. Price: Canada $5.95, other countries $7.15 (Canadian funds). Cat. No. A42-42/1983-12E.

Part 13. The Carrion Beetles of Canada and Alaska (Coleoptera: Silphidae and Agyrtidae), by R. S. Anderson and S. B. Peck, Biosystematics Research Institute, Ottawa, 1985. 121 p. Price: Canada $7.00, other countries $8.40 (Canadian funds). Cat. No. A42-42-1985-13E.

(Prices subject to change without notice.)

Contents

Acknowledgments

Many people have been involved with this project. We are indebted to Dr. C.-C. Hsiung and Miss Diane E. Johnstone for many of the illustrations. Dr. S. M. Ulagaraj assisted in the literature search. Mr. A. T. Finnamore, Mrs. S. Wolter, and Mr. D. N. Duffy, employed in the Lyman Museum, searched out the localities and prepared the range maps.

Dr. T. J. Walker, University of Florida, Gainesville; Dr. D. Otte, Academy of Natural Sciences, Philadelphia, PA; Dr. G. K. Morris, Erindale College, University of Toronto, Toronto, Ont.; and Dr. G. E. Kerr, formerly of the same institution, were particularly generous in supplying locality data.

We also acknowledge the assistance of the following who allowed us to borrow or to examine specimens at their institutions: Mr. J. E. H. Martin, Canadian National Collection of Insects and Arachnids, Ottawa, Ont., who was particularly helpful; Dr. A. B. Gurney (now retired), U.S. National Museum of Natural History, Washington, DC; Dr. I. J. Cantrall and Dr. T. H. Hubbell (both now retired), Museum of Zoology, University of Michigan, Ann Arbor; the late Dr. H. J. Grant, the late Mr. J. A. G. Rehn, and Dr. H. R. Roberts (now retired), Academy of Natural Sciences, Philadelphia, PA; Dr. P. J. Darlington (now retired), Museum of Comparative Zoology, Harvard University, Cambridge, MA; Dr. D. R. Ragge and Mrs. J. Marshall, British Museum (Natural History), London; Dr. J. R. Beaudry, Dr. J.-G. Pilon, and Mme Colloudon, Université de Montréal, Montréal, Qué.; Dr. G. G. Wiggins, Royal Ontario Museum, Toronto; Dr. A. G. Robinson, University of Manitoba, Winnipeg; Dr. H. McDonald, Agriculture Canada, Winnipeg, Man.; the late Mr. A. R. Brooks, Agriculture Canada, Saskatoon, Sask.; the late Dr. B. Hocking and Dr. G. E. Ball, University of Alberta, Edmonton; Dr. R. H. Handford (now retired), Agriculture Canada, Kamloops, B.C.; Dr. G. G. E. Scudder, University of British Columbia, Vancouver; and the staff (collectively) of the British Columbia Provincial Museum, Victoria.

Many people assisted by collecting specimens for the Lyman Entomological Museum collection. These include Peter and Susan Arntfield, Colin Boyle, Wayne Boyle, Alan Campbell, Lorne Crozier, Bert Finnamore, P. Giorgio Fontana, Michel Guibord, Wyb Hoek, Glen Jamieson, Kaye Kevan, Rowan Lalonde, Joe Lovrity, Dave Marsh, Pak Haup Ng, Bob Randell, A. C. Sheppard, Elsie Smythe, John Stainer, Bill Vickery, Edwin Vickery, and Sandra Wolter. Without their contributions the work would have been much more incomplete.

Numerous others have assisted by donating or lending material to the Lyman Entomological Museum. Particular thanks are due Dr. Luc Jobin, Environment Canada, Forestry Service, Sainte-Foy, Qué.; Dr. W. Edmonds, Toronto, Ont.; and Dr. C. Chantal, formerly University of Waterloo, Waterloo, Ont. (now Agriculture Canada, Ottawa, Ont.).

Acknowledgment must be made to the following for permission to reproduce published illustrations: the trustees of the British Museum (Natural History); the editors of Danmarks Fauna; the editors of the California Insect Survey; Masson S. A. Paris, the publisher of Chopard's Biologie des Orthoptères; and the Fédération française des Sociétés de Sciences Naturelles, Paris.

The research was partially carried out under contract for the Research Branch, Agriculture Canada; the report was prepared under the same contract. The views, conclusions, and recommendations expressed herein are those of the authors and not necessarily those of the Department.

Introduction

The orthopteroid insects include the grasshoppers, locusts, crickets, and their kind as well as cockroaches, mantids, termites, rock-crawlers (or grylloblattids), earwigs, and stick-insects. Unlike such assemblages as "beetles" (Coleoptera), "bugs" (Hemiptera), or "flies" (Diptera), the orthoperoid insects have no single vernacular name in any language that is applicable to them all. A few languages do, nevertheless, possess single words that are used to include most orthopteroid insects in that they refer to those forms which leap, and which are commonly called by the Latin name "Saltatoria." These leaping insects are the true locusts, grasshoppers, and their immediate relatives, and English is one of those few languages in which there is a single, succinct, expressive noun that covers them all. The word is "grigs." We may, therefore, include under the umbrella of this almost obsolete name all insects belonging to orders herein treated collectively.

Even the term "insect" has changed its meaning several times over the years, although all insects in the modern sense are included in the superphylum Arthropoda. Arthropoda is the name used to include the greater part of the animal kingdom. It replaces the ambiguous "Insecta," used by Linnaeus (1761) for all animals having a chitinous exoskeleton and segmented appendages. Within this superphylum falls the phylum Entoma, whose name was used by Aristotle about 330 B.C. for the arthropods exclusive of crustaceans, and which was resurrected by Latreille (1804) in 1796 (although in a wider sense than it is now used, for he included arachnids and some other animals). Among the Entoma we recognize subphyla, the largest of which is the Uniramia of Manton. This includes all arthropods that have primarily unbranched appendages and whose respiratory mechanism is basically a ramifying and usually anastomosing system of tubules, or tracheae. They are all land-based or derived from terrestrial ancestors (unlike the arachnids which, although now mainly terrestrial, have a marine origin). The great majority of the Uniramia constitute the infraphylum Dicera, or Atelocerata, of Latreille. These include the superclass Myriapoda, e.g., centipedes and millipedes, as well as the six-legged arthropods, or superclass Hexapoda. It should, however, be remembered that there are some arthropods (mainly in the immature stages) that are not Hexapoda, although they have six legs, just as there are many Hexapoda (again, mainly in the immature stages) that are not six-legged.

Orthopteroid insects belong to what we may call supercohort Neopterygota, i.e., they are insects in which the wings (if present) are capable of being folded backward to lie along the dorsum of the body. The Neopterygota may be subdivided into cohorts, the orthopteroids falling into what other early workers called the Polynephria, or Polyneoptera. The Polynephria are Neopterygota in which the excretory Malpighian tubules discharging into the posterior part of the alimentary canal are numerous, the metamorphosis is incomplete, i.e., the wing buds develop externally, the mouthparts are of a biting, or chewing, type (more generally determinable by inspection), and the abdomen bears a pair of caudal appendages called cerci. Some authors regard all such insects as being orthopteroid and call them the Paurometabola.

All the orthopteroid orders treated in this work fall within a subcohort constituting one of the major groups of insects established long ago by Fabricius (1793), namely the Ulonata. The orthopteroid insects, or Ulonata, take their place within the hierarchical classification system of the superphylum Arthropoda, as follows:

Superphylum ARTHROPODA
  Phylum ENTOMA
    Subphylum UNIRAMIA
      Infraphylum DICERA, or ATELOCERATA
        Superclass HEXAPODA
          Class INSECTA
            Subclass DICONDYLIA
              Infraclass PTILOTA, or ALATA
                Supercohort NEOPTERYGOTA
                  Cohort POLYNEPHRIA, or POLYNEOPTERA
                    Subcohort ULONATA (Orthopteroid insects)

The principal characteristics of the Ulonata will be given later when the major groups within the subcohort are indicated. Brief preliminary accounts of these insects, insofar as they are found in Canada, are given by Kevan (1979) and, for earwigs, by Lamb (1979). The tables accompanying these articles indicated the approximate numbers of species that were known to occur in Canada, together with the numbers of additional species that might eventually be found there. The latter figures were, in effect, based on the numbers occurring in the northern United States, and both sets need minor revision in the light of more recent studies.

As indicated by Vickery et al. (1974), grigs are excellent study material for both professional biologists and amateur entomologists. They are not so numerous in species that the whole fauna of an area may not be studied by a single person. Many species are easily reared in the laboratory, or even in the home, making them popular experimental animals, although, admittedly, a large number present considerable problems in this respect. Many grigs are aesthetically pleasing, either in their appearance or in the sounds they produce; some excite curiosity on account of their bizarre aspect or unusual lives; others teach much about past (and present) geographical and ecological conditions and about evolutionary principles. Many grigs have the advantage for the student in that they "can be studied in the field without undue discomfort, although a fair amount of energy may need to be expended in their capture." The complex behavior patterns, both auditory and, in some cases, visual, of any orthopteroids, especially among the saltatorial true grigs, provide rewarding fields for study.

Cytologists and physiologists have been well served by these insects. From an economic standpoint there is a very important minority of orthopteroid species that are serious pests of crops and some that attack stored produce or structural materials. Scarcely a single field of human endeavor on land is not in some way influenced directly or indirectly by these insects—and not always adversely. There has, also, been a place for grigs in folklore and culture from the earliest times until the present, and in the simplest to the most sophisticated forms (see, for example, Kevan 1974*b*, 1978*b*).

With the exception of the few known species of Notoptera, all orthopteroid orders are predominantly tropical. As a result, people who live in countries that have temperate climates, particularly those who live in higher latitudes, obtain not only a distorted impression of them but also, despite the prevalence of a few species that gain public attention on account of their economic significance, an unrealistic idea of their abundance and diversity. In order to put matters into perspective, therefore, we present, in tabular form (Table 1), the approximate numbers of living families, genera, and species known for each orthopteroid order for the world as a whole, compared with those for the region covered by the present work. Subspecies are not considered separately. About a score of alien species that have become naturalized, even if but temporarily, either in the field or under artificial conditions, are included in the numbers given for Canada and adjacent regions. It will be immediately apparent that Canada and its immediate neighbors are inhabited by only a minute fraction of the world's orthopteroid fauna—less than 0.01%! This does not, however, mean that they are either insignificant or unimportant.

Approximate numbers of non-fossil Ulonota known for the world and for Canada

Order	Number of families[1]		Number of genera		Number of species	
	World	Canada	World	Canada	World	Canada[2]
Dictuoptera	37	9	1 000	15	7 500	24
Notoptera	1	1	3	1	23	4
Dermaptera	9	3	132	6	1 850	7
Dermodermaptera	1	0	1	0	8	0
Cheleutoptera	11	1	357	1	2 500	3
Grylloptera	32	13	1 500	37	8 500	104
Orthoptera	32	5	2 000	64	11 000	155
Total	123	32	4 993	124	31 381	297[3]

[1] The number of families varies with different authors. The figures given here are for those families recognized by the present authors.

[2] Excluding nonbreeding adventives but including aliens that have been established, at least temporarily, even if under artificial conditions.

[3] Of this total, almost 250 species together with 15 or 16 subspecies are native or established in Canada.

Review of literature to 1850

Over the centuries, and before European settlement in Canada, numerous visits were made to Canada's shores by different expeditions and for different purposes. Until Giovanni Caboto's visit in 1497, however, none left contemporary written records, and, so far as can be determined, only one, that of John Davys in 1586, mentions any insect—and then only mosquitoes, apparently in southern Greenland (Davys 1589).

The earliest Europeans to settle in the northern part of North America were, of course, those who came to Newfoundland (the first colony in 1583), New France, and New England, but if (even among those who professed an interest in natural history) they had anything to say about insects at all, it rarely concerned grigs.

The so-called "Jesuit Relations" (Thwaites 1896–1901), spanning a period of 180 years, from 1610 to 1791, contain a few scattered early references to insects in New France, one or two of them referring to grigs.

Nearly a quarter of a century later, Pierre Boucher, who came to Canada when he was only 12 years old and who became governor of Trois-Rivières, published a book that included observations on natural history. He mentioned that he had a collection of several hundred insect specimens (Boucher 1664). He gave no detail, but presumably some grasshoppers would have been included. The fate of his collection is unknown (Comeau 1965).

When, in a search for early records of orthopteroid insects, we turn to the history of New England, we fare little better.

We may notice what seems to be the first, brief mention in English (and the first to be published) of a few orthopteroid insects specifically from northeastern North America in John Josselyn's account of his second visit to New England begun in 1663 (Josselyn 1674; see also Weiss 1936). He went first to Boston, Massachusetts, and later lived in Black Point, Scarborough, Maine, for more than 8 years. He notes crickets (presumably *Gryllus pennsylvanicus* Burmeister), grasshoppers (presumably Acrididae of various kinds), and "a stinking black and red Bug called a Cacarooch or Cockroach."

The first grasshoppers to be reported during the eighteenth century from our area were definitely not popular. The occurrence was in 1743, and again in 1746. In Maine, grasshoppers were said to have "covered the whole country" and to have "threatened to devour every thing green." In 1749 and 1754, grasshoppers were again reported from the same region as being "very numerous and voracious; no vegetables escaped these greedy troops; they even devoured the potato tops . . . Indeed, so great was the alarm they occasioned among the people, that days of fasting and prayer were appointed" (Williamson 1832:102–103; see also T. W. Harris 1841; Scudder 1872; Riley 1875, 1877*b*).

The overland explorers and those who visited the coasts of the Pacific northwest during the eighteenth century were singularly uncommunicative about insects in general and about grigs in particular. In the East, however, it would be appropriate to mention the Finnish professor Pehr Kalm of Abo (now Turku), sent by the Royal Swedish Academy to report on conditions in North America. He visited parts of Pennsylvania, New Jersey, New York, and southern Canada during 1748–1751, making various observations in his diaries concerning the natural history of the regions that he visited (Kalm 1756*b*, 1761). Although there is no record of the content of his collection, or of the fate of his specimens, it is not improbable that there were some insects, doubtless including grasshoppers.

On 9 December 1748, Kalm himself (1756*b*) made reference to damage done by grasshoppers to pasture. He said that when the grasshoppers appeared, the grass-grubs and caterpillars become scarce. Various people, he said, assured him that these three kinds of insects followed one another, so that the grasshoppers appeared in the first year, the caterpillars in the second year, and the grass-grubs in the third year. In his own experience this was partly correct.

With regard to orthopteroids other than true grasshoppers, Kalm (op. cit.) mentions that cockroaches, or mill beetles, were present in almost every house in the city of New York, and, furthermore, that they were common "everywhere," even in woods and wilderness, where he correctly regarded them as native. In the former, at the time, they would probably have been virtually all *Blatta orientalis* Linnaeus; in the latter, *Parcoblatta* spp., although one interesting observation of large numbers of cockroaches in a rotten log being used for fuel, suggests that he may have collected *Cryptocercus punctulatus* Scudder also.

With regard to the house cricket (*Acheta domesticus* (Linnaeus), Kalm 1756*b*) did not observe it in any part of Pennsylvania or New Jersey, and others that he asked were unable to say that they had ever seen any. He notes, however, that in summer, black crickets, which made the same chirping sound, were found in the fields, but that they were silent as soon as winter or cold weather came. Anticipating his account of his later excursion to the north (cf. Kalm 1761), he notes that in some parts of New York, and in Canada, every farmhouse and most of the houses in the towns, swarmed with crickets, which continued with their "music" there throughout the whole winter and summer. He states furthermore that there they had, likewise, not put aside their customary bad habits (in Europe), namely, of getting among and cutting to pieces, clothes. These crickets were, at least for the most part, surely *Acheta domesticus*.

Field crickets, according to Kalm, were plentiful in summer in all parts of North America that he visited, and were sometimes so numerous and vociferous that they were painful to the ears. Most of those that he observed would have been either *Gryllus pennsylvanicus* Burmeister or *G. veletis* (Alexander & Bigelow), according to whether they were observed later or earlier in the season. Kalm (1761) further notes that crickets passed the winter in the soil and emerged again in March to recommence their singing. (This would apply to *G. veletis*, which hibernates in the nymphal stage and becomes adult in spring, but not to *G. pennsylvanicus*, which passes the winter in the egg stage and does not become mature until later in the year; the same individual cricket does not sing in both spring and autumn.) So early a date would apply only to the southernmost parts of the region and not to that here covered, which Kalm visited only briefly in the summer of 1749. In his entry for 13 February 1749, Kalm records that crickets were found at a depth of about 25 cm and that they were quite torpid, but as soon as they were brought into a warm place, they revived and were quite lively. (The crickets in this instance may have been overwintering nymphs of *G. veletis* that had passed the critical point of their diapause period.) Kalm also remarks on the fact that, when crickets were abundant, they could prove dangerous where rattlesnakes abounded because the warning rattle of the serpents could be rendered inaudible by their chirping.

On 7 September 1762, the Lieutenant-Governor of Nova Scotia, Jonathan Belcher, sent a dispatch from Halifax to the Lords of Trade in London, referring to the loss of crops that year brought about by a combination of drought and grasshoppers. Which species was (or were) concerned is uncertain, but Piers (1918), who cites the report, strongly suspects "*Melanoplus atlantis* (Riley)," i.e., *M. sanguinipes sanguinipes* (Fabricius), but it may well have been *M. femurrubrum femurrubrum* (De Geer).

Up until the end of the eighteenth century, we know of no record of orthopteroid insects from the north central or northwestern parts of North America. Pritchett (1942), however, places the first observation by a European of swarms of the Rocky Mountain locust, *Melanoplus spretus* (Walsh), in the Canadian West at 13 years before the establishment of the first Red River Settlement in what is now Manitoba. As this latter occurred in 1812, the record, if correct, would be for 1799. It is probable that there would have been swarms of locusts in the general area in 1799 to give rise to those that were actually reported in 1800, but there is no written record of them. Nevertheless, grasshoppers in general, and locust swarms in particular, were surely familiar to the aboriginal Amerindians from time immemorial. As Bird (1961) says, "These outbreaks must have undoubtedly occurred periodically for centuries." He supports this by reference to the vast numbers of grasshoppers frozen in the glaciers of Montana (Gurney 1953*a*; Gurney and Brooks 1959).

The early part of the nineteenth century saw a slow growth in knowledge of the orthopteroid insects of Canada and adjacent regions in two different ways. On one hand, the spread westward of European settlers (and the accompanying introduction of agricultural and horticultural systems that had not previously existed in the West) led to reports on swarms of locusts, which had become of great economic significance, rather than mere matters of curiosity. On the other, a beginning was made, mainly in the East, on the collection and scientific description of the insects.

The first swarms of locusts (acridid grasshoppers) to be reported from the middle of the North American continent were not, as might have been anticipated, in territory now forming part of the United States (although they were doubtless present there at the time), but in the Red River valley of what is now Manitoba and Saskatchewan. Alexander Henry, who was eventually drowned in 1814 in British Columbia, recorded in his diary for 17 August 1800 that, on the "east shore of Lake Winnipeg, six leagues from the mouth of the Red River. . . The beach was covered with grasshoppers which had been thrown up by the waves and formed one continuous line as far as the eye could reach; in some places they lay from six to nine inches deep, and in a state of putrefaction which occasioned a horrid stench" (from Ms., James Campbell, Librarian of the Library of Parliament, Ottawa, *in* Packard 1878, and Coues 1897; see also Morton 1937; Mitchener 1954*a*, 1954*b*; Bird 1961; Riegert 1980).

Although we cannot be certain, the above records, and presumably most reports of swarms in the early days in the Prairie regions, doubtless referred to the Rocky Mountain locust, *Melanoplus spretus* (Walsh). In the East, although several species were undoubtedly involved, the main cause of damage and comment was often *M. femurrubrum femurrubrum* (De Geer). *M. sanguinipes* (Fabricius), which is nowadays the most important species in the West, was probably not the most injurious grasshopper pest in the East, any more than it is now.

In the years 1800–1801, infestations were said to have caused great damage on Ile d'Orléans in the St. Lawrence River by Québec City. Lambert (1810, 1813; see also Vickery et al. 1974) gives a colorful version of the incident.

Two years after Henry's experiences in the West, and a year after the Québec incident, Daniel William Harmon (born 1778, died 1845), entered in his journal for 23 July 1802, that locusts were (still or again) present in great numbers on the Prairies. Harmon (1820; see also Morton 1937; Riegert 1968) reports that, at Fort Alexandria on the upper Assiniboine River, about 16 km southeast of the present town of Sturgis, in eastern Saskatchewan, there were "grasshoppers, in such prodigious numbers, as I never saw in any place. In fair weather, between eight and ten o'clock A.M., which [was] the only part of the day when many of them [left] the ground, they [were] flying in such numbers that they obscure[d] the sun, like a light cloud passing over it. They also devour[ed] everything before them, leaving scarcely a leaf on the trees, or a blade of grass on the prairies, and our potato tops escape[d] not their ravages." This was the first record of damage to cultivated crops in the West.

Ten years after Henry began his migration westward, locusts returned to the Prairies. In 1812–1815, Lord Douglas, Earl of Selkirk, established his Red River Settlement of Assiniboia. History shows that things did not go smoothly for them. In 1817, when all looked well, the wheat crop was destroyed by frost, but in 1818, things were "looking up" and reasonable crops were produced. Then disaster struck again, for the locusts came!

Numerous authors have referred, at greater or lesser length, to the locust plagues of the Red River Settlement during 1818–1821. In addition to those already cited, they include the following: Hind (1857); Hargrave (1871); Huyshe (1871); G. M. Dawson (1875*a*, 1875*b*); Bethune (1875); Riley (1875, 1877*b*), who also discusses the contemporary invasions of Minnesota in 1818 and 1819, which are recorded by Neill (1858), and reports on grasshopper damage elsewhere in 1820 and 1821; Packard (1878), who also refers to grasshopper damage elsewhere in 1820–1821; Gunn (1878); Macoun (1882); and others later. The disastrous years are commemorated by Pratt (1952) in Section 5 of "The Gathering," a poem in his saga of the building of the transcontinental Canadian Pacific Railway.

At about the same time as the devastating swarms of locusts visited the unfortunate Red River settlers, Eastern Canada and New England were also experiencing unusually severe attacks by grasshoppers. Talbot (1824), who came from Ireland in 1818 and who seems to have been even more prone to exaggerate than his predecessor Lambert (1810, 1813), wrote that "locusts and grass-hoppers which infest the whole country . . . are often as destructive to the corn crops in Canada as Sampson's foxes were to the wheat of the Philistines." He also stated that some of the grasshoppers in what is now eastern Ontario and Québec, were "as large as a field mouse" and so numerous "that a single person with a waggoner's whip might drive ten thousand of them before him with as great ease as a shepherd can drive a flock of sheep. The whole face of the earth appears so thickly covered with them, that crops of every description seemed destined to immediate destruction."

All interest in insects in the early nineteenth century was not of an exclusively pragmatic nature, nor did it relate entirely to economic considerations. Captain William Edward Parry's Northwest Passage expedition of 1819–1820 brought back a few northern insects to Great Britain (W. Kirby 1824), although it is scarcely surprising that no orthopteroid was among them. Some were, however, collected (presumably as a pastime), about 1821, by an army officer, Lt. R. S. Redman, in the vicinity of Halifax, N.S. Such orthopteroids collected by Redman as we know of, are listed in the catalogs of F. Walker (1869*a*, 1869*b*, 1870, 1871). The species, under their current names, are *Ceuthophilus maculatus* (Harris), *Scudderia curvicauda* (De Geer), *S. furcata furcata* Brunner von Wattenwyl, *Conocephalus fasciatus* (De Geer), *Allonemobius fasciatus* (De Geer), *Melanoplus femurrubrum femurrubrum* (De Geer), *M. bivittatus* (Say), *Dissosteira carolina* (Linnaeus), *Arphia sulphurea* (Fabricius), *Pardalophora apiculata* (Harris), *Chorthippus curtipennis curtipennis* (Harris), and *Tetrix ornata* (Say). F. Walker (1872), in his list of orthopteroids known from north of the United States border (omitting reference to anything west of Ontario), bases his Nova Scotia records entirely on Redman's material, all or most of which is still preserved in the British Museum (Natural History), London. These seem to be the oldest Canadian specimens still in existence, for they must have been collected not later than the beginning of 1822, although not before the end of 1820.

Dr. John Jeremiah Bigsby, a keen naturalist (born 1792, died 1881), was a member of the Canadian Boundary Commission and, at an early date in his career, in the early 1820s, he visited, among other places, southern Lake Huron and the surrounding regions (Bigsby 1824). He apparently collected various natural history specimens at the time, for much later (Bigsby 1850) he published a list of insects that he had collected, "principally in [*sic*] Lake Huron." Although one cannot say for certain that these belonged to the early period, it may be assumed that some, at least, did so. The names were *Gryllus bipunctatus* (not of De Geer; other than that this was a cricket of some kind, it cannot be identified); *Acrida coeca* and *A. stenoptera* (in earlier times, Kirby used *Acrida* Linnaeus for *Tettigonia* Linnaeus; that these "species" were tettigonioids with relatively broader or narrower wings, respectively, is all we can say of them); *Locusta borealis* (Kirby used *Locusta* for anything we should now place in the Acridoidea); and *Acrydium 4-punctatum* (Kirby used *Acrydium* specifically for Tetrigoidea). This last species we can tentatively identify because a specimen collected by Bigsby from Lake Huron is listed as *Tettix granulata* (Kirby) (= *Tetrix subulata* (Linnaeus)) by F. Walker (1871).

Captain George G. Back's Arctic Land Expedition of 1833–1835 (G. Back 1836) returned to the United Kingdom with a few insects. Among these was a solitary grasshopper, which is listed by Children (1836) as *Acridium sulphureum* Palisot de Beauvois (i.e., *Arphia sulphurea* (Fabricius)), but the specimen almost certainly belonged to *Trimerotropis verruculata verruculata* (Kirby). W. Kirby (1837) notes the reference and F. Walker (1870) refers to the actual specimen (in the genus *Oedipoda* from "Arctic Regions, Capt. Back." It probably came from the general region of Lake Garry, northwest Keewatin.

From the higher latitudes, we should now return to the lower ones. In the 1830s, Thaddeus William Harris, whose father was also a keen naturalist, began a serious study of the insects of Massachusetts. He published a list of insects for the state, which included a dozen or so species of orthopteroids (T. W. Harris 1833), about half of which he regarded as new species, for which he provided names but no description. Two years later, his revised list (T. W. Harris 1835) included some 35 names, 25 of them new, but again there was no description. The only one of these "new" names that is now valid as of that date, was *Locusta* (now *Pardalophora*) *apiculata*, by virtue of its being indicated as a replacement name for what had previously been recorded under the name of an Old World species (*tuberculata* Palisot de Beauvois) (J. A. G. Rehn and Hebard 1911*b*). Most of them were, however, validly described by him, usually (but not always) under the same names, later (T. W. Harris 1841). This last list contained about 32 or 33 nominal species, of which only 9 or 10 were not regarded as being new. Those which are still valid, with their current names, are as follows:

Harris's name	Current name
Rhaphidophora maculata	*Ceuthophilus maculatus* (Harris)
Conocephalus ensiger	*Neoconocephalus ensiger* (Harris)
Orchelimum vulgare	*Orchelimum vulgare* Harris
Acrydium alutaceum	*Schistocerca alutacea* (Harris)
Locusta maritima	*Trimerotropis maritima maritima* (Harris)
Locusta marmorata	*Scirtetica marmorata* (Harris)
Locusta (*Chloealtis*) *conspersa*	*Chloealtis conspersa* Harris
Locusta (*Chloealtis*) *curtipennis*	*Chorthippus curtipennis curtipennis* (Harris)

It may also be noted here that T. W. Harris (1841) mentioned various natural enemies of grigs, including endoparasitic worms (several in one host, which he referred to as *Gordius*, although they may have been mermithid nematodes) and ectoparasitic red mites (which he called *Ocypete*, although they would now presumably be referable to larval Erythraeidae), both from "locusts" (i.e., Acrididae). "A kind of sand-wasp" is recorded as preying upon "grasshoppers" (i.e., Tettigonioidea). Many birds, including domestic fowls, he said, will consume "locusts, grasshoppers and crickets," and turkeys, if permitted, will, during the summer, "derive nearly the whole of their sustenance from these insects." In respect of the injurious nature of Acrididae, Harris notes the references by Dwight (1821) and Williamson (1832) and says, "The southern and western parts of New Hampshire, the northern and eastern parts of Massachusetts, and the southern part of Vermont have been overrun by swarms of these miscalled grasshoppers (mainly *Melanoplus femurrubrum femurrubrum* (De Geer), and have suffered more or less from their depredations" (see also Riley 1875, 1877*b*). As regards the local importance in Massachusetts, of "*Acrydium femur-rubrum*" (probably both *Melanoplus femurrubrum femurrubrum* and *M. sanguinipes sanguinipes* (Fabricius) were involved), he says, "These, in certain seasons, almost entirely consume the grass of these [salt] marshes from whenceth ey [*sic*] then take their course to the uplands, devouring in their way, grass, corn, and vegetables, until checked by the early frosts. . . It is stated that, in Maine, during dry seasons, they often appear in great multitudes, and are greedy destroyers of the half-parched herbage."

About the time that T. W. Harris was becoming involved with entomology, so was a young English immigrant to Newfoundland, Philip Henry Gosse. At the age of 17, in 1827, he obtained a post there at Carbonear. As recorded by Bruton (1930) and Weiss (1936), his interest in natural history received added stimulus in 1832, and he painted insects, including orthopteroids, that he collected there. He left Newfoundland in 1835 and spent 3 years, until 1838, at Compton in the Eastern Townships of Québec, before departing for the United States and then returning to England. While at Compton, he began a book on the entomology of Newfoundland, which, however, was never published. After his departure from Québec, Gosse (1840) published the first book exclusively on Canadian natural history, based on his experience at Compton, in which he mentioned various tettigonioid grigs, two of which he vaguely named: "*Phyllopterus Myrtifolius* ?" (probably *Scudderia furcata furcata* Brunner von Wattenwyl) and "*Locusta* ?" (*Conocephalus* sp., probably *C. fasciatus* (De Geer)).

About this time the first record of the lesser earwig, *Labia minor* (Linnaeus), from the northeastern United States was published, when Doubleday (1838) reported it from Wanborough, New York. Further records of orthopteroids were not published for New York until some years later, when Fitch (1847) listed about 10 saltatorial species from the state. There was apparently an outbreak of grasshoppers or locusts in Minnesota in 1842 (Packard 1878; Lugger 1898) but it does not seem to have affected Canada. In 1845, Léon (later l'Abbé) Provancher began collecting insects at Bécancourt, Québec, but his calling put a stop to this in 1846, and he did not take up entomology again for another 11 years (Comeau 1965). There is no reason to believe that his earliest collecting involved orthopteroids.

Toward the end of the fifth decade of the nineteenth century two further expeditions to the Canadian Arctic took place following the disappearance of the ill-fated Franklin expedition of 1845. A narrative of the first of these, in 1846 and 1847, is given by Rae (1850), and that for the second, in 1848 and 1849, by Richardson (1851). Both Rae and Richardson brought back insect specimens to the United Kingdom, and some, at least, of these are still preserved in the British Museum (Natural History), London. A. White (1851) lists material from both collectors, but only two species of orthopteroids are named, both from the same two localities, the "Borders (i.e., margins) of Mackenzie and Slave Rivers," and Fort Simpson (61° 30′N). In addition, four unnamed species of "*Locusta*" are recorded from the first of these localities. The named species were "*Acrydium granulatum*" (= *Tetrix subulata* (Linnaeus)) and "*Locusta tuberculata*" (= *Pardalophora apiculata* (Harris)). F. Walker (1871) mentions (as *Tettix granulata*) what may be one of the same specimens from "Arctic America" presented by Sir J. Richardson, and (1870) records "*Oedipoda phoenicoptera*" (= *Pardalophora apiculata* (Harris)) with similar data, which may be any one of the "*Locusta*" specimens recorded by White (1851). Specimens collected by Rae and recorded by F. Walker (1870) for Arctic America are *Caloptenus femur-rubrum* (= *Melanoplus femurrubrum femurrubrum* (De Geer)), which E. M. Walker (1920) believes (probably correctly) to have been *M. borealis* (Fieber), and a "new" species, *Caloptenus arcticus* (= *Melanoplus bilituratus bilituratus* (F. Walker) or, according to Gurney and Brooks (1959), *M. sanguinipes sanguinipes* (Fabricius)).

Richardson's material included, in addition to the two aforementioned species, "*Caloptenus femur-rubrum*"; the "new" species *C. extremus* (= *Melanoplus borealis*); *Oedipoda* (= *Arphia*) *sulphurea* (Fabricius); *O.* (= Xanthippus) *corallipes* Haldeman); *Stenobothrus curtipennis* (Harris) and *S. longipennis* Scudder (both = *Chorthippus curtipennis curtipennis* (Harris)); and *S. maculipennis* Scudder (= *Orphulella pelidna pelidna* (Burmeister)). These species are the basis for all the records for "Arctic America" given by F. Walker (1872), which omits "*S. longipennis.*" It should be realized, however, that the expression "Arctic" is now misleading as many of the specimens would have come from (at the most) subarctic environments in the western part of the continent and probably from lower than 60°N.

F. Walker (1869*a*, 1869*b*, 1870, 1871) makes several references to other early Canadian specimens of orthopteroids, but most of these may date from after 1850. On the positive side, however, two specimens of *Oedipoda* (= *Arphia sulphurea* (Fabricius)), dated 1845, were collected in the Rocky Mountains and presented to the British Museum (Natural History) by the Earl of Derby. The dubiously pre-1850 material included *Parcoblatta pennsylvanica* (De Geer) from "Canada" (presumably eastern Ontario), collected by W. S. M. D'Urban, who lived in Montréal for some years before returning to England; *Scudderia furcata furcata* Brunner von Wattenwyl (as *Phylloptera myrtifolia* from "Canada" (i.e., Québec or Ontario; no collector, but probably D'Urban or Barnston; *Sphagniana sphagnorum* (Walker) (in genus *Decticus*), type described from "St. Martin's Falls, Hudson's Bay" (= Martin Falls, Albany River, northern Ontario), collected by George Barnston, one of the co-founders of the *Canadian Naturalist and Geologist*); *Gryllus pennsylvanicus* Burmeister (as *G. luctuosus*) from "Canada," collected by both D'Urban and Barnston; *Melanoplus bivittatus* (Say), from "Canada" and Albany River (Martin Falls), collected by Barnston; *M. fasciatus* (Walker), types, Albany River (Martin Falls); *M. sanguinipes sanguinipes* (Fabricius) (types of *Caloptenus bilituratus* Walker and *C. scriptus* Walker) from "Vancouver's Island" collected by a Dr. Lyall; *M. femurrubrum femurrubrum* (De Geer) (type of *Caloptenus repletus* (Walker)) from the same locality but collected by Lieut.-Col. Hawkins; *Dissosteira carolina* (Linnaeus) from "W. Canada" (meaning Canada West = Ontario), collected by a Mr. Bush (? Richard James Bush, active in the 1860s) and from "Vancouver's Island," collected by Lieut.-Col. Hawkins; *Arphia sulphurea* (Fabricius) from the same locality but collected by Dr. Lyall; *Pardalophora apiculata* (Harris) collected from "Hudson's Bay" (i.e., Martin Falls) (as "*Oedipoda phoenicoptera*") and by Lieut.-Col. Hawkins from "Vancouver's Island" (as "*O. rugosa*"); *Chorthippus curtipennis curtipennis* (Harris) from "Hudson's Bay" (Martin Falls) collected by Barnston, and from Newfoundland, presented by a Prof. Shephard; and *Tetrix subulata* (Linnaeus) (as "*Tettix granulata*") and *T. ornata* (Say) from "Hudson's Bay" (= Martin Falls).

One of the pioneer entomologists of Canada was William Couper, who came to Canada from England in 1842, and who first collected Canadian insects of "all orders," no later than 1843 (Baillie 1929), but if any orthopteroids were among them (and there probably were), there is no way of telling this.

Review of literature after 1850

Locust and grasshopper outbreaks

Since the trials and tribulations of 1818–1821, the settlers in the Red River valley of what is now Manitoba had no serious locust problem for many years, although they had other troubles. The Canadian West had, nevertheless, by no means seen the last of the Rocky Mountain locust and other major grasshopper pests, and Seamans (1956) notes, with some inaccuracy, most of the "locust" and "grasshopper" years from 1818 until 1951. The first signs of serious trouble ahead seem to have been in 1855, when, as reported by Taylor (1859), severe outbreaks occurred in Oregon and Washington, the "entire territories" being covered by grasshoppers. There were also, according to Taylor (op. cit.), swarms in Minnesota that year—although this has been denied by Whitman (*in* Packard 1877)—and again in 1856 (Walsh 1866, quoting O. H. Kelley in *Country Gentleman*; Scudder 1872; Riley 1875, 1877*b*; Packard 1877, 1878; Lugger 1898). In 1857, the swarms became more numerous in various parts of the northwestern United States, including Minnesota (references as aforementioned), and many entered Manitoba. Riegert (1980) quotes earlier authors who described swarms in the Pembina River district. There was not much serious crop loss as a result of the visitations because the swarms arrived too late in the year (Scudder 1872, with reference to a letter from the Honourable Donald Gunn; G. M. Dawson 1875*b*, cf. Mitchener 1954*b*). Nevertheless, they deposited vast numbers of eggs that gave rise to abundant progeny in the following year, 1858 (Hind 1859:40, 43; Gunn 1878). Hind (1859:44) gives a graphic account of a devastating flight of *Melanoplus spretus* to the west of what is now Souris, Man., on 2 July 1858. He also refers to other occurrences of locusts in Saskatchewan, and records observations on phenology.

The seriousness of the locust plague of 1858 is stressed by many later authors. For the following few years, however, locusts were not prevalent until they appeared in Dakota territory in 1863 (Whitman, *in* Packard 1877) and Minnesota (Lugger 1898). They reappeared in force in 1864 and 1865, causing (according to some authors) widespread damage, particularly in the former year. Morton (1938), following Hargrave (1871), notes that the 1864 swarms of grasshoppers had been preceded by "a severe drought, such as had not hitherto been known in the Settlement," and that "the depredations of the grasshoppers continued in 1865, but were largely confined to the part of the Settlement occupied by the Scots 'who were probably better able to support the loss than any other portion of the community.' "

The following year, 1866, although listed for Minnesota by Lugger (1898), is not generally regarded as having been a "locust" year in Manitoba. For 1867, the same source reports that the harvest was considerably injured by "a vast cloud" of grasshoppers. This was said to be the largest swarm ever known (Bethune 1875). In 1868, the progeny of these locusts completely destroyed the crops in Manitoba in that year also (Huyshe 1871; Bethune 1875*b*; Gunn 1878). They "devoured everything, causing a famine" (Dawson 1875*b*; Macoun 1882; Morton 1939; Mitchener 1954*a*, 1954*b*).

According to Huyshe (1871), the locusts of the 1868 outbreak "came in such countless myriads that they were lying piled up against the angles of the bastions and walls of Fort Garry [near Winnipeg, Man.] to a depth of three feet. The stench from their dead bodies was almost unsupportable, and they had to be carted away and thrown into the river." The following year, 1869, was another marked by heavy locust invasion of Manitoba, but the swarms arrived too late to affect very much the bountiful crops of that year (Dawson 1875*b*). The progeny of these swarms, however, did much localized damage in 1870 (Dawson 1875*b*; Bethune 1875; Gunn 1878; Macoun 1882; Mitchener 1954*a*; Riegert 1980). Some Manitoba farmers did not sow crops in 1870 (Gunn 1878). In Minnesota, locust damage was apparently more restricted in 1869, but severe again in 1870 (Lugger 1898).

In 1871, the Canadian Prairie Provinces seem to have been free from serious injury by grasshoppers or locusts. It was in the eastern part of our region that the most serious damage by Acrididae (presumably mainly *Melanoplus femurrubrum femurrubrum* (De Geer)) occurred. Riley (1875, 1877*b*; cited also by Packard 1877) quotes a letter from one L. Colby and from reports in the *Country Gentleman, American Agriculturist*, and the *Monthly Reports of the U.S. Department of Agriculture* for 1871, which indicate that damage was severe and widespread in Vermont, New Hampshire, Massachusetts, and Maine.

In 1872, apparently a bad locust year in Minnesota also (Lugger 1898), fresh swarms of *Melanoplus spretus* appeared in Manitoba, and thereafter the locust situation again became serious (see Dawson 1875*b*, quoted also by Packard 1878, Macoun 1882, Mitchener 1954*a*, Riegert 1980). The locusts deposited an abundance of eggs, particularly in the "northern" (now central) part of the province. Some of the resulting hoppers marched, in 1873, up the Red River valley, but over considerable areas, notably around Winnipeg and Stone Fort, the farmers cut their losses by sowing no crop.

The following year, 1874, saw immigration of flying swarms from the West earlier than was considered usual, and much injury to crops was inflicted in some districts. Dawson (1875*b*) says that, according to a Mr. Solbey, five million bushels of grain were destroyed by locusts in Minnesota alone. Packard (1877) quotes Dawson's (1876) report regarding the Canadian locust outbreak of 1875, during which little or no cultivation was attempted in Manitoba (see also Riley 1876; Packard 1878), or as Macoun (1882) puts it: "The grasshopper plague [of 1873–1875] was upon the whole province and no wheat or next to none was raised in 1875. Many got a crop of potatoes, but the country was on the verge of starvation and all the seed wheat had to be brought in from Minnesota." This did, however, result eventually in a long-term benefit. The utter reduction of seed stocks in Manitoba led to the replacement of inferior varieties by Red Fife, the wheat that placed Canada "on the wheat map of the world" (Morton 1937). A note on the Manitoba "Grasshopper Seed Grain Mortgages" of 1876 is given by Painter (1941). Nevertheless, at the time, the locust outbreaks of 1873–1875 were directly responsible for a great reduction in immigration by settlers during 1876 to the Canadian prairies (Morton 1937, 1938), which was an embarrassing set-back to the youthful federal government who had just expended a great deal of effort and treasure in attempting to encourage settlement of the West. "The great tide of immigrants" that was expected "refused to come to a land cursed with one of the plagues of Egypt" (Morton 1937).

The Manitobans for the most part remained resolute despite adversity, and the cornerstone for the old Winnipeg City Hall was laid on 17 August 1875. Beside the stone was buried a bronze casket (it would be called a "time capsule" nowadays) containing various contemporary articles regarded of interest to posterity. Included were "three photographs of Manitoba scourge, 'grasshoppers,' 1875" (one actually dating from 1874, is shown in Plate I, *A*), "a [glass "Preston Salts"] bottle containing samples of the scourge of Manitoba 'grasshoppers' in spirits" and sealed with red sealing-wax (Plate I, *B*, *C*), and "a box containing heads of wheat from a field partially destroyed by grasshoppers, 1875" (Begg and Nursey 1879:122–123). When recent excavations were being undertaken for the extension of the City Hall, the casket was exhumed 15 February 1962 (together with another buried 19 July 1884, when the cornerstone of the "new" City Hall was laid) and the objects recovered (Anonymous 1962*a*, 1962*b*). The "grasshoppers," although the alcohol had, of course, dried out, were in an excellent state of preservation. Most of them were *Melanoplus spretus*, although other species were represented.[1]

The year 1875 saw the last of the really serious crop losses to the Rocky Mountain locust in the region treated here. There was an exodus of locusts from Manitoba in 1876 and no problem there in 1877. Packard (1878) shows 1877 as being an unimportant locust year in "British North America," as well as in the adjacent United States from Minnesota westward. Locust flights are shown for 1877 in Manitoba, present Saskatchewan, the Dakotas, and Montana by Packard (1881), but, for 1878, a few in the last two named locations only (see also Packard and Riley 1881). By 1879, activity seems to have become reduced to a small amount in the western Dakotas and in

[1] The bottle (capacity 38 mL, height 53 mm, greatest circumference 135 mm) eventually made its way to Saskatoon, where it was examined by our former colleague Dr. R. L. Randell. We are grateful to him for the photograph and for information regarding its contents. Besides a male and a female *Melanoplus spretus* there were one male *M. sanguinipes sanguinipes* (Fabricius), one each of second-, third-, and fourth-instar nymphs of one or other of these, one female *Cratypedes neglectus* (Thomas), and one female "Mormon cricket," *Anabrus simplex* Haldeman. The bottle also contained some newspaper clippings. Bottle and contents were returned to the City of Winnipeg at the end of August 1970, and were placed in the city archives (W. A. Quayle, City Clerk, Winnipeg, *in litt.* to R. L. Randell, 1.IX.1970).

western Montana (which had otherwise become virtually free from the Rocky Mountain locust for the first time since settlement in 1861).

There has been much speculation over the decline (and eventual demise) of the Rocky Mountain locust. Bird (1961) believes that, in former times, ecological conditions were made particularly favorable for *Melanoplus spretus* by a combination of dry seasons and an abundance of bison (buffalo) that grazed and trampled the arid prairie and made dust wallows. Increase in settlement and plowing, the elimination of the bison and their replacement by domestic cattle, which do not make wallows, together with other environmental changes induced by man in succeeding decades, led, Bird maintains, to the decline and eventual disappearance of the Rocky Mountain locust—and to the rise of other injurious species in its place. So far as the northern United States herds of bison were concerned, the greatest hunting seasons were from 1878 to 1883, by which time the peak outbreaks of Rocky Mountain locust had passed. It would seem safe to say, therefore, that if the elimination of bison did play a part in the decline of *M. spretus*, it would have been a partial, and probably a minor one. The Rocky Mountain locust, in fact, briefly returned in force to the Canadian prairies, but not until the end of the century. There was, however, in Minnesota, a build-up from very small beginnings in 1886 and 1887 and a serious local outbreak in 1888 (Lugger 1898). Few of the 1889 eggs hatched, possibly partly owing to an energetic plowing campaign.

During the summer of 1891, "myriads" of *Melanoplus sanguinipes sanguinipes* (Fabricius) made a sudden appearance on Sable Island about 180 km off the coast of Nova Scotia. Whence they came is unknown. The species was said to be uncommon at the time in the vicinity of Halifax, N.S., although *M. femurrubrum femurrubrum* (De Geer) was numerous. From that year until 1896 they rapidly devastated the island (Piers 1896, 1918 [as *M. atlanis*]). In 1894, the driest season on record, only one load of hay was cut where 14 had been harvested previously. In 1895, their depredations were worse than ever. Gardens were destroyed, houses invaded, and furnishings damaged; there was insufficient fodder for the wild ponies, so that hay had to be imported to feed some, while others were shipped to the mainland. There was considerable anxiety in 1894 and 1895 that, since the grasshoppers ate the grass down to near the roots the sands would shift and possibly result in the disappearance of the island beneath the sea, rendering it an even greater hazard to shipping than it already was. In 1896, cold weather occurred and the grasshoppers suddenly disappeared. No specimen was reported again until 1969 (Vickery et al. 1974).

A second noteworthy grasshopper outbreak of 1891 has only recently come to our attention in the journal of a remarkable Acadian-born lighthouse keeper, Placide Vigneau, 1842–1926 (Gallienne 1969). In his diary, written at "Pointe-aux-Esquimaux" (Eskimo Point, now Havre-Saint-Pierre), on the remote coast of Québec's "Lower North Shore," opposite Anticosti Island, under 25 July 1891, he mentions a "prodigious quantity of *grasshoppers*" that devoured almost all the grass and completely ravaged gardens (of which, however, there cannot have been many), even consuming rhubarb leaves and angelica; the sky was darkened by them during the day; in the evening, they covered everything including the walls of houses.

The species involved cannot be determined with certainty, but it was probably the same as that which appeared on Sable Island in the same year. Unlike the latter, it does not seem that the North Shore grasshoppers persisted, as Vigneau does not mention any again until 1899. Of this invasion, he says only that they were as abundant as in 1891 and that they did much damage to garden plants. In neither case do we have any idea where they came from.

The return of *Melanoplus spretus* to the West in 1898 through 1902 was naturally a cause for much anxiety, but, although this species was present, it seems that the principal injurious Acrididae (at least in Manitoba) were *M. sanguinipes sanguinipes* (Fabricius), *M. packardii* Scudder, and *Camnula pellucida* (Scudder) (Fletcher 1899–1902*a*; Mitchener 1954*a*, 1954*b*, 1956). Losses in 1900 and 1901 were considerable and greater than they had been in the previous two years, but it was in 1901 that the successful, world-famous grasshopper bait known as "Criddle mixture" was first used (N. Criddle *in* Fletcher 1902*a*; Seamans 1956). It was developed by a budding entomologist, Norman Criddle, and his half-brother, Harry Vane (A. Criddle 1973:230). Losses were also heavy in 1902, but not appreciable in 1903, after which infestations subsided (Mitchener 1954*a*, 1954*b*). Grasshopper outbreaks occurred in the twentieth century, but the Rocky Mountain locust had completely disappeared everywhere by 1903—although N. Criddle (1920) gives 1904 (possibly referring to other species). The last known living specimens of *M. spretus* to be collected and which are still in existence are, by strange coincidence, from southern Manitoba, whence swarms of the species were first reported in 1800. The last known specimen was taken by Norman Criddle at Aweme, near Treesbank, in 1902 (N. Criddle 1903; Gurney and Brooks 1959; Riegert 1980). The same year, also by coincidence, saw another grasshopper outbreak in Eastern Canada (Seamans 1956).

After the turn of the century some trouble with grasshoppers was experienced in one part of Canada or another, particularly in the West, but even in Newfoundland (R. F. Morris and Morry 1978, 1979), in almost every year until the present. The intensity of the outbreaks varied from minor, local nuisances to grave regional problems. There was an outbreak in Manitoba 1911–1912 (N. Criddle 1913; Bird 1961), although Mitchener (1954*a*, 1954*b*, 1956) and Seamans (1956) do not mention it, and there were others both in British Columbia (T. Wilson 1915) and in the East (A. Gibson 1914, 1915*a*, 1915*b*, 1916*a*, 1916*b*, 1916*c*; Petch 1915; Seamans 1956) during 1913–1915. The first major upsurge in Western Canada did not, however, occur until 1919 (N. Criddle 1920; Mitchener 1954*b*). It became alarming in certain regions in 1920 (Mitchener 1954*b*) and declined in 1923. In 1920, grasshoppers were again a problem in the East (Seamans 1956). These might be termed local skirmishes in comparison with what was to come on the prairies, where the most serious recurrence of the grasshopper problem began in 1929 and continued with varying intensity until 1952 (Mitchener 1954*b*), after which grasshopper populations remained low for some time. Riegert (1980) devotes a chapter to the outbreaks of the 1930s. During this period there were also serious grasshopper outbreaks in Ontario in the middle 1930s (Gilbert 1936; Gilbert and Thompson 1937), in British Columbia, notably in 1944 (Buckell 1945), and in Eastern Canada, particularly in 1949 (Seamans 1956). Seamans declares that the depredations by grasshoppers in Canada during 1930–1951 were "the worst of all," including, it seems to be implied, the dreadful times of the Rocky Mountain locust plagues of the nineteenth century. Indeed they may well have been so in terms of total volume of crops lost, since the area under cultivation and the potential yield per plant had increased considerably. It was during this period that practical cooperation between Canada and the northern United States in the matter of grasshopper "control" programs and exchange of relevant information began in earnest.

It is not the intention here to discuss all these grasshopper outbreaks, but the reader is referred to Mitchener (1954*a*, 1956) for a résumé, with references to the literature, up to 1953, particularly in relation to Manitoba, to Riegert (1968, 1980) in relation to Saskatchewan, and to D. S. Smith and Holmes (1977) in relation to Alberta since 1918. The species involved varied in importance from year to year and from place to place, but they included principally *Melanoplus sanguinipes sanguinipes* (Fabricius), *M. femurrubrum femurrubrum* (De Geer), *M. packardii* Scudder, *M. bivittatus* (Say) (which does not seem to have been a problem in the West until the 1920s), and *Camnula pellucida* (Scudder). Bird (1961) comments on the man-induced ecological changes in the aspen parkland areas of Western Canada and how, among other things, the "new" grasshopper pests came to the fore as a result. The worst "grasshopper years" since the heyday of the Rocky Mountain locust remain the 1930s ("The Dirty Thirties"), both in Canada and in the northern United States (Parker 1934; Mitchener 1954*a*; Riegert 1980). There have been serious crop losses in later times, up until the present, although the development of modern insecticides and other methods of treatment and prevention, and, more recently, the growth in understanding of ecological principles, have alleviated the problem to a considerable extent.

Systematic and related literature

While information on the economic aspects of the orthopteroid insects of Canada and the adjacent United States was being published, and as attempts were being made to devise methods of controlling pest species, knowledge of the orthopteroid fauna of the region was gradually built up. It was rather spasmodic, the number of workers engaged in the study of these insects (in contrast to others, such as Coleoptera and Lepidoptera) always being small.

No comprehensive review of the progress made specifically in the region here treated has before been attempted, although some of the references in the introductory works of Scudder (1868*c*, 1897*b*) on the North American orthopteroids, and in his later catalog for the continent (Scudder 1900*a*), are applicable. His extensive bibliography (Scudder 1901*a*) contained entries for virtually all the relevant available literature up to that time, and is still of great value as a reference tool. It is, however, perhaps indicative of the poor state of early knowledge and the general lack of interest in orthopteroid insects, other than injurious species, that Bethune (1898*b*), in his account of "the rise and progress of entomology in Canada," had almost nothing to say about them. The world catalogs of W. F. Kirby (1904–1910) also contain specific references to our region. Piers (1918) gave an abridged summary of some of the pertinent publications dating from 1869 to 1915, mainly, but not exclusively, for Eastern Canada and the northeastern United States, and, recently, Vickery et al. (1974) have noted the relevant literature for Québec and the Atlantic Provinces of Canada from the early nineteenth century until 1973. As we have already reviewed the literature up until 1850—which is approximately when Glen (1956) reckoned that scientific entomology began in Canada—we may largely confine our attention here to works published after that date.

The first Canadian reference is to the stick-insect now known to have been *Diapheromera femorata* (Say) (Gibb 1859). The paper was, in fact, mainly concerned with the sounds produced by local insects, including grigs. Three years later, Scudder (1862) listed a number of species from the West and described a new North American grasshopper, now *Stethophyma gracile* (Scudder), from Manitoba and Maine. In January of the following year, the same author (Scudder 1863*a*) published his first treatise on the orthopteroid insects of North America. In the same year, also, Scudder (1863*b*) described *Pezotettix* [now *Booneacris*] *glacialis* from New Hampshire. Ritchie (1867) again referred to the stick-insect *Diapheromera femorata* (Say) (as *Spectrum femoratum*) in Eastern Canada from Montréal in 1865. Scudder's (1868*b*) paper on the stridulation of New England grigs was the first of several in this field. S. I. Smith's (1868) list of 38 nominal species of orthopteroids for Maine was an important landmark.

The catalogs of F. Walker (1869*a*, 1869*b*, 1870, 1871) included various references to early records of Canadian orthopteroids, as noted previously, but also other information, including descriptions of several new species, some of which still remain valid. A few records for the northwestern United States and Canada are noted by C. Thomas (1873) in his general work on the "Acrididae of North America."

Although it is impossible here to attempt to itemize every subsequent work that had a direct bearing upon the study of the orthopteroid insects of the northern United States and Canada, we may at least draw attention to the more significant of these, beginning, perhaps, with the survey for New Hampshire, given by Scudder (1874). Together with S. I. Smith's (1868) account for Maine, this provided the impetus for launching various other regional studies in the northern United States during the next 30 years or so. These included the works of Fernald (1888) and the earlier writings of Morse (1894–1906) for New England, of Beutenmüller (1894*c*) for New York, and, farther west, of Lugger (1898) for Minnesota. McNeill (1896/1897, 1901) wrote two revisional studies of different groups of Acrididae which were of great importance. Scudder (1900*b*) made a further important contribution to the knowledge of New England species. In the early years of the twentieth century, we may also note the contributions of Walden (1911) for Connecticut, of F. L. Washburn (1912*b*) and Somes (1914) for Minnesota, of Pettit and McDaniel (1918) for Michigan, of Morse (1919*a*, 1919*b*, 1920, 1921) for New England and Maine, and of Britton (1920) for Connecticut. The book by Blatchley (1920) on the orthopteroids of eastern North America, naturally included the species occurring in much of the present region, and, although now out of date, remains the most comprehensive work for the eastern United States.

In referring more particularly to Canada, Scudder (1875*b*) provided the first account to deal exclusively with the western provinces, but soon afterward, real progress was made, particularly in Québec, notably by Provancher (1876, 1877, 1883) and Caulfield (1886, 1887, 1888), the latter author also covering eastern Ontario. Before Caulfield began publishing, however, there appeared a label list of the species of insects known from Canada (Brodie and White 1883) in which, on a single page [67], were set out, with their classification, the names of 87 species of orthopteroids. For some time, the publications of Provancher and Caulfield remained the only standard references to the orthopteroid insects of Québec, Ontario, and the neighboring provinces. Soon, however, Piers (1895, 1896) began taking an interest in the Nova Scotian species, Lochhead (1898) gave a general account of crickets (in Ontario), and E. M. Walker (1898*a*) published the first of his many papers on the orthopteroid insects of Canada.

If any one Canadian worker can be cited as producing the most important early contributions to our knowledge of the Canadian orthopteroids, the credit must go to Edmund M. Walker. He wrote mainly on the species found in Ontario, but also on those from Western Canada, Prince Edward Island, and the Magdalen Islands of Québec. Walker, although he became much more widely known as a student of dragonflies, maintained his interest in orthopteroids until he died. Quite late in life he was joined by F. A. Urquhart in a publication on the group in Ontario (E. M. Walker and Urquhart 1940). In the course of time, he described eight species or subspecies of orthopteroids that are still regarded as valid. He also recognized the first known grylloblattids (Notoptera) and described the first species in this unusual order of insects, which the Entomological Society of Canada has adopted as its emblem.

Among American authors who contributed most significantly to our knowledge of the Canadian orthopteroid fauna in the earlier years of the present century were A. N. Caudell, James A. G. Rehn, and Morgan Hebard. Caudell reported mainly upon species from western and northern localities. Rehn and Hebard, either individually or together, produced an abundance of information on the orthopteroid orders as they occurred in North America generally, but also with reference to particular regions. Some of their publications dealt directly with Canada (e.g., Rehn 1910; Hebard 1915*a*, 1931*a*, 1934*b*, 1935*a*); others added greatly to the information available on species occurring in Minnesota, the Dakotas, and Montana (J. A. G. Rehn and Hebard 1906; Hebard 1925*a*, 1928, 1932*a*, 1932*c*, 1936*a*, 1936*b*, 1936*c*, 1937*a*). Many other papers by these authors were also, in part, of direct or indirect importance to our understanding of the Canadian fauna. Significant contributions to the understanding of North American Tetrigodea, including those of Canada and the northern United States, were also made by Rehn in collaboration with Harold J. Grant, Jr. (J. A. G. Rehn and Grant 1956*a*, 1956*b*, 1958, 1961). Other particularly notable United States authors who contributed to our understanding of the fauna of the region here treated were Theodore H. Hubbell, whose monographic revision of the North American camel-crickets (Hubbell 1936) was especially valuable, Ashley B. Gurney, who worked on grasshoppers, cockroaches, and grylloblattids (Gurney 1937*b*, 1939, 1940*a*, 1948, 1949, 1953*a*, 1953*b*, 1961, 1971), and

Irving J. Cantrall, who produced one of the best ecological studies of orthopteroids, based upon his observations in Michigan (Cantrall 1943). One should also mention the faunistic papers of Mills and Pepper (1938) on the grasshoppers of Montana, and of J. W. H. Rehn (1939*a*, 1939*b*) on the orthopteroids of Newfoundland, Saint Pierre and Miquelon, and Nova Scotia. The important pioneer study by Tuck and Smith (1940) paved the way for the study of eggs by Onsager and Mulkern (1963).

Returning now to consider Canadian authors, we should first note the contributions of Gooderham (1917, 1918) to the study of the orthopteroids of Nova Scotia, and the monograph of Piers (1918) for the same province. The latter was not superseded until that of Vickery (1961) was published. In Québec, DuPorte (1919) began publishing on *Gryllus pennsylvanicus* Burmeister. On the opposite side of Canada, Ernest R. Buckell (later associated with R. C. Treherne) began about 1918 to take a particular interest in the orthopteroids of British Columbia (Buckell 1920–1937; Treherne and Buckell 1924), while, in the middle of the country, Norman Criddle had already made and was continuing to make valuable observations upon the behavior, habitats, and life histories of prairie species in Manitoba. His observations, too few of which were actually published, have provided models for subsequent workers. His pioneer studies have seldom received due credit, at least beyond the limits of the Canadian prairies. Much regarding Norman Criddle will be found in the history of his remarkable family (A. Criddle 1973). Of his published papers, which included several relating more particularly to injurious species, we will mention here only that on the field crickets and those on the immature stages of grasshoppers (N. Criddle 1924, 1925, 1926*b*, 1931). It was not until Handford (1946) made detailed studies of immature stages of *Melanoplus* species in the West that further significant work was published in this field.

In the middle years of the present century, E. M. Walker and Urquhart (1940) and Urquhart (1938, 1940*a*, 1941*a*, 1941*b*, 1941*c*, 1941*d*, 1942) made further valuable contributions to our knowledge of the distribution and ecology of the orthopteroids of southern Ontario, and Chagnon (1944*a*, 1944*b*, 1947, 1948) and Beaudry (1952, 1954) wrote on Québec species. Urquhart and Beaudry (1953) reported the discovery of the European "bush-cricket," *Metrioptera roeselii* (Hagenbach) in the Montréal area, and, in the West, Brooks (1958) produced his important monograph of the acridid grasshoppers of the southern parts of the Canadian Prairie Provinces.

It was during the middle 1950s, also, that R. S. Bigelow, at the authors' institution, began his studies in experimental taxonomy related to field crickets, particularly of the genus *Gryllus* in North America, which continued after he had left Canada. These investigations, together with the studies of R. D. Alexander at the University of Michigan, Ann Arbor, clarified the taxonomic status of the Canadian and northern United States species, long since questioned by N. Criddle (1925). A new species of field cricket, *Acheta* [now *Gryllus*] *veletis*, which matures in spring and early summer, instead of in late summer and early autumn, was recognized (R. D. Alexander and Bigelow 1960). Of greater significance than the formal recognition of this species, however, was the hypothesis of "allochronic speciation" which resulted from the work and which became of considerable importance in evolutionary theory.

Also in 1960, two small faunistic studies of orthopteroids in Eastern Canada appeared, one for Newfoundland (Ander 1960) and one for Mont Tremblant Park, Qué. (Robert 1960). In 1961, D. K. McE. Kevan and V. R. Vickery began publishing on Canadian orthopteroids. Their publications have continued to the present time, together with those of a third contributor to this work, Miss Diane E. Johnstone, who began independent publication in 1970. This group's most significant contributions were Vickery's (1967*a*) records for Alaska, Yukon, and MacKenzie District, Vickery and Kevan's (1967) annotated list for Ontario, and Vickery et al.'s (1974) monograph for Québec and the Atlantic Provinces of Canada.

Other recent Canadian work on orthopteroids is that of G. K. Morris (1970–1972) and his co-workers (Morris and Pipher 1967, 1972; Pipher and Morris 1974; Kerr 1974, 1978; Morris et al. 1975*a*, 1975*b*, 1978; Gwynne 1977; Morris and Gwynne 1979) on the analysis of stridulatory and other behavior of Ontario and northern New York grigs. Morris and Walker (1976), in work also associated with these studies, have published the first part of a critical review of the North American katydids of the genus *Orchelimum*. In British Columbia, Lamb and Wellington (1975) and Lamb (1975–1976*b*) have studied the bioecology of earwigs (Dermaptera). Lamb (1979), also, gave a brief account of what is known of this order so far as Canada is concerned. Larochelle (1978) issued a summary of the Québec orthopteroids based upon the monograph of Vickery et al. (1974). He and his associates have been publishing frequent records for the province over the past few years. Kamp's (1979) work on *Grylloblatta* should also be noted.

In conclusion, we may also draw attention to one or two of the publications by American authors that involve species found in the region covered by this book. First and foremost we should mention Helfer's (1963) useful "pictured keys" to the genera and most of the species of orthopteroids of all orders that occur in North America, north of Mexico. Range of each species by state and province (so far as was known at the time) is given. The "tree crickets" (Oecanthidae, mainly the genus *Oecanthus*) have long been in a confused state taxonomically, so that the several species living in the northern United States and Canada were often incorrectly identified. T. J. Walker (1962, 1963) has clarified the situation so far as most of North America is concerned, so that identification of our species is now possible. Other recent works having a definite bearing on the Canadian fauna include those of Hewitt and Barr (1967) and Scoggan and Brusven (1972) (Idaho), and of Cantrall (1968) and R. D. Alexander et al. (1972) (Michigan). We have also had occasion in the present work to refer frequently to the behavioral studies of Robert Willey and his associates and of Daniel Otte, for, although the work of these authors was done in various of the United States, mainly outside our region, it is relevant in the case of a number of species that occur in our area (see Willey and Willey 1967, 1969, 1971; Otte 1970; Steinberg and Willey 1974; Willey 1975). The reader is also referred to Weissman and Rentz (1980) for cytological data on the species in three genera of Locustini, and to Tyrkus et al. (1983) for several papers in many different fields (including economic) that are applicable to the grigs of Canada and adjacent regions.

Anatomy

The following discussion is intended to facilitate use of the keys and descriptions, especially for those who are not familiar with this group of insects. A glossary, which explains the terms used in this work, will be found at the end of the book. The basic structure of all orthopteroids is generally similar, although each order has its own peculiar characteristics involving important structural and other features that have evolved as a result of their specialized behavior. Fig. 6 illustrates the basic structure of an ulonate insect of a familiar form, that of a short-horned grasshopper.

The body is made up of three regions: the head (*h*), the thorax (*t*), and the abdomen (*ab*). On the head are a pair of antennae (*a*), three simple eyes, or ocelli (*o*), although these may be reduced or absent, a pair of compound eyes (*e*), and the mouthparts, i.e., structures associated with feeding. The antennae vary considerably in length. The top of the head is the vertex (*v*); the sides, or cheeks, are the genae (*g*); the front of the head, or face, is the frons (*f*), often with a vertical ridge, or frontal costa (*fc*), whereas the posterior part of the head is the occiput (*oc*). The ocelli are arranged, basically, in a triangle, one ocellus near each antenna and the third in the middle of the frons. The two compound eyes are generally large and usually prominent, each being composed of a large number of facets, each of which is the cornea of an individual component, or ommatidium, of the eye.

The mouthparts in all Ulonata are of a typical biting or chewing type, consisting of two pairs of movable jaws, the mandibles (*md*), which are large and usually heavy, and the maxillae (*mx*), which are located behind the mandibles, together with the labium (*li*), which lies behind the maxillae. Attached to the lower part of the frons, and above the mouthparts proper, is the clypeus (*c*), to which is attached the labrum (*lr*), which functions as an upper, or outer, lip. The labium acts as a lower lip and prevents loss of food while the insect is chewing; it bears a pair of three-segmented appendages, the labial palps. Similar palps, each with five segments, are attached to the maxillae. The palps have a sensory function.

The thorax, i.e., the body region which has locomotion as its main function, is made up of three segments: the prothorax (*pr*), the mesothorax (*ms*), and the metathorax (*mt*). Each of these segments bears ventrally a pair of legs. The uppermost part of the prothorax, which is generally bent down over the sides of the rest of prothorax, is the pronotum, normally a large and prominent structure. The mesonota and metanota, i.e., the uppermost parts of the mesothorax and metathorax, respectively, each bear a pair of wings (except in wingless species) and laterally a pair of spiracles (respiratory pores). The fore wings, or tegmina (*teg*), are typically thickened, usually close-veined and tough, serving to protect the more delicate membranous hind wings, which are folded, fanlike, beneath the tegmina when not in use. In some groups the pronotum is projected backward over the bases of the wings. In the Orthoptera, *sensu stricto*, the pronotum is usually divided transversely by a major furrow, or "typical" sulcus, into an anterior area, the prozona (*prz*), and a posterior area, the metazona (*mez*). (Sometimes a less prominent sulcus between the anterior margin of the pronotum and the "typical" sulcus is sufficiently obvious for the region immediately behind it to merit the term mesozona, but this is not in general use.)

The legs are basically similar throughout the orthopteroid orders, although many modifications and adaptions occur between and within different groups. In almost all Orthoptera and Grylloptera the hind legs are large and modified for jumping; the fore legs may be equipped for grasping prey (mantids) or for digging (some crickets). The segments of the legs from the base to the apex are coxa (*cx*), trochanter (*tr*), femur (*fe*), tibia (*ti*), tarsus (*ta*) (usually three- or four-segmented, but five-segmented in cockroaches, in a few termites, and in almost all Cheleutoptera), and pretarsus (*pt*). The trochanter often is reduced or fused with the coxa. The pretarsus nearly always bears a pair of claws, with or without an arolium between them.

The abdomen, in most groups, is relatively unspecialized. It consists of successive similar segments (except for the terminal ones), with a clearly definable dorsal tergum (*ter*) and ventral sternum (*st*), which are heavily sclerotized plates, and, on each side, a softer pleuron (*pl*). Spiracles (*sp*) are located near the bases of the tergal plates on each side of the first six abdominal segments. These are the breathing pores used in respiration and are connected to tracheal trunks, the principal breathing tubes, inside the body.

The terminal part of the abdomen is specialized, bearing the genital structures. The external appearance of these varies widely between the orders, but cerci (*ce*) of various types are always present (even if occasionally they are very small). In the males of some forms, these serve as copulatory claspers; they are often sensory; in earwigs they form forceps. A female has an egg-laying device, or ovipositor (Fig. 6, inset, *ov*), which varies considerably in form and size in different groups. It is absent in most termites, in many cockroaches, in earwigs, and in a few members of other groups. The principal copulatory, or phallic, structures of a male, which are usually complex, are, strictly speaking, external but, except when carrying out their function, they are retracted and concealed within the abdominal apex by the supra-anal plate (the terminal part of which is called the epiproct) above, the subgenital plate below, and the paraprocts on either side. The form of the male copulatory structures is of great taxonomic importance. The female also has subgenital and supra-anal plates, and a pair of paraprocts.

Collection, preparation, and preservation

It is impossible to present a full exposition on collecting, preparing, and preserving grigs, but we may give some guidelines, partly taken, with a little modification, from Martin (1977).

True, or saltatorial, grigs may be found in trees, shrubs, herbs, grasses, swamps, and bogs, or beneath stones, logs, and debris. Collecting methods vary according to the kinds being collected: sweeping low vegetation; beating trees and shrubs; and using fingers or forceps to collect specimens under stones or logs. Quick-flying or jumping species may be captured by covering them quickly with a standard net while they are at rest on the ground. When they are covered, the net handle is moved slightly and the insects usually hop upward into the net bag; then a rapid sweep and twist of the net ensures they do not escape. A conventional large long-handled net may not always be the most satisfactory tool, especially for collecting crickets. These insects may best be stalked and swatted with an ordinary small aquarium fishnet. This technique, developed many years ago by one of us (D.K.K.), does not seem to have been mentioned in the literature. It has certain advantages which may outweigh the lack of reach: the insect's position (if secured!) is immediately obvious (because of the small search area and lack of folds formed by a conventional net); the net, because of its lightness, may rapidly be used a second (or third) time with great precision, if the quarry is missed or crawls from under the edge; the small size of the net enables it to be stowed in a pocket when not in use or when both hands need to be free; and such nets are inexpensive and readily obtainable.

In areas of rank low herbage, dense grass, or sedges, grigs are sometimes difficult to locate or capture. By heavily trampling the vegetation around an ever-diminishing perimeter (i.e., roughly in a spiral), starting with an area of, say 50 m^2 and reducing it to a little over 1 m^2, one can bring about a concentration of insects such that the remaining standing vegetation may profitably be worked (Cantrall 1939, 1941).

Grouse grasshoppers (Tetrigodea) and pygmy hoppers (Tridactylodea) may often be obtained by sweeping a net "blindly" over damp mud and sand. These small insects may not be readily seen, but the disturbance causes them to jump vigorously, and they may be found in the net.

Specialized methods may have to be used to collect adequate numbers of specimens of some groups. Camel, or cave, crickets (Rhaphidophoridae) may be collected by using "bait-jars" (Hubbell 1936; Beaudry 1954): diluted molasses coated on the inside of jars 8–13 cm in diameter and 13–15 cm tall. The bottom of the jar is covered with dead leaves and the jar buried to its rim in the ground. The method is most productive in rocky, wooded, or forested areas. The combined use of a trail of dry oatmeal bait and a miner's headlight is also very effective for obtaining these insects (Hubbell 1956).

Many species of Grylloptera stridulate at night, and the males may be discovered by following their sound to the source. Then, with the aid of a flashlight or a headlight and a net, a specimen may be caught before it is able to fly away. The songs of the insects are different for nearly every species and are useful for identification. Songs may be recorded, before the specimen is captured, by using a good-quality tape-recorder with a microphone attached to a parabola.

For each specimen collected, pertinent information such as host plant, location, and date of capture should be recorded either on a label or in a notebook with a cross-reference number on the data label.

Kill specimens promptly with potassium cyanide in standard entomological killing bottles. Grasshoppers do not die as quickly as some other insects do in cyanide bottles. They may appear to be dead, but then, later, they may revive and damage other specimens in the bottle by thrashing around. Specimens should, however, not be left too long in cyanide, because it affects the color, particularly that of green specimens. To kill grasshoppers as quickly as possible, use fresh cyanide bottles.

Some species, such as those of camel crickets, tend to shrivel when they dry out. Such species are best stored in alcohol or other fluid, or preserved by the following process: pass specimens through two changes of 95% alcohol and two changes of absolute alcohol, then transfer to xylene. Specimens become translucent when they are completely dehydrated; they can, if so desired, be pinned directly from the xylene; the appendages, although somewhat stiff, can be arranged carefully while the specimens are still wet. Specimens that have been preserved in alcohol and then become dried and shriveled, or specimens of soft-bodied insects, like termites, that have become desiccated naturally, may be restored to some semblance of their former shape by immersing them for about 24 hours in a 1% solution of trisodium phosphate before putting them in fresh preservative. A surfactant, Decon 90, is also recommended by McNutt (1976); the process is less direct, but the result superior. Another method is to soak the dried specimens, for a few minutes only, in a mixture of warm water (three parts) and household detergent (one part). Instead of using standard ethyl alcohol as a fluid preservative, we recommend as a general fixative and preserving fluid, what has been termed CFAA preservative (Frings and Frings 1971). This consists of a mixture of chloroform, isopropyl alcohol, formalin (40% formaldehyde solution), glacial acetic acid, and water (preferably distilled) in the following proportions by volume: 8:45:10:2:35. Although this mixture does not preserve color well, it is generally useful for most purposes where fluid preservation is preferred (especially if subsequent dissection or histological study is the objective). It has the advantage of being quick-acting, rapidly killing insects dipped into it, thus dispensing with the cyanide bottle; it is easy to make up; the ingredients are relatively easily obtainable; and it is inexpensive.

If orthopteroids are to be pinned, this should be done within 24 hours after their death, and they should be dried as quickly as possible. Insert the pin through the right side of the pronotum near its posterior edge. Take care not to damage the wings, particularly in crickets and other singing species, where examination of the stridulatory apparatus on the tegmina may be necessary for positive identification. Ensure that the long axis of the body is nearly at right angles to the pin, with the head of the insect directed slightly downward. Bend the legs beneath the body to minimize the possibility of breakage and to occupy the least amount of storage space. Set the abdomen so that it droops below the wings and is not obscured by the hind legs, because several taxonomic characters are found on the terminal end. Until the specimen is thoroughly dry, support its body parts with extra pins so that it dries in the desired position. Mount very small specimens, such as grouse grasshoppers, pygmy hoppers, or small crickets, either on card points or on small pins in *Polyporus*, cork, or balsa supports, and then place these on larger pins.

Large specimens (about 3 cm, or more) should be eviscerated when they are fresh, because the internal organs and food in the gut tend to decompose before drying can be completed, and color changes occur. This is particularly important with large green specimens. The most satisfactory method is to cut the membranous dorsal region between the head and pronotum, remove the internal contents with fine curved forceps, swab the inside with cotton, and insert a mixture of one part boric acid to three parts talc. The specimen is then shaken gently to distribute the talc. A drop of adhesive is finally used to hold the head close to the thorax. This method promotes quick internal drying and preserves color better than other known methods. Evisceration can also be done by making an incision through the first three abdominal sterna along the midline (or, in certain cases, where the abdomen is robust, along the pleural region) with fine sharp scissors, removing the thoracic contents and those of the abdomen with fine straight forceps, and then proceeding as previously described. For very large specimens, the body cavity can be packed lightly with absorbent cotton. The abdomen should be molded back to as near its original form as possible when still moist. It will thus retain a natural appearance. Kaltenbach (1958) and Rentz (1962*b*) describe more elaborate although superior methods for dry preservation of soft-bodied grigs.

If specimens cannot be pinned soon after they are killed, they should be eviscerated (unless they are small), and placed in layers between either absorbent cotton or absorbent paper in small wooden or plastic (not metal) boxes. Later they can be relaxed by standard means, pinned, placed in the desired position, and allowed to dry. Since the tegmina and/or hind wings of many species are important in identification because of their venation or coloration, it is often desirable to set specimens with their wings extended. This is particularly true of bandedwinged (locustine) grasshoppers. The method of spreading is standard for most insects, but it is conventional in the case of grigs to spread the wings of only one side. Unless there is a good reason to the contrary (as with a damaged specimen), the left side (away from the insertion of the pin) is selected. Uniform treatment leads to conservation of storage space as well as being aesthetically more pleasing.

The foregoing remarks are directed mainly toward the study of saltatorial forms, but cockroaches (Blattodea) and mantids (Mantodea) may be treated in the same way, except that it is often preferable to pin bulky specimens through the right tegmen and abdomen (like beetles). Although some species occur outdoors, amongst litter, for example, cockroaches are usually found in buildings, where they may be collected by means of "bait-jars" similar to those mentioned for camel crickets. Place the jars in out-of-the-way corners and make sure their outsides are not too clean or smooth (any labels should be left on them), otherwise the cockroaches may have difficulty in climbing the jar.

In the East and in southern British Columbia, adult specimens of introduced mantids are usually found in late summer on rank vegetation, including bushes. Females, especially green ones, should be eviscerated after a slit is made in the basal abdominal pleura on one side. The small native western mantid is normally found running on the ground or on low vegetation. Because this mantid is fragile in a dried condition, it is best to mount it like a stick-insect.

The termites (Termitodea) of our region are nearly always (although not exclusively) associated with buildings, except in British Columbia, where they are also found in natural rotting wood and in underground nests. Because they are soft-bodied, they are not usually pinned, but are preserved in fluid. To do this, use either ordinary 70–75% ethyl alcohol (or isopropyl alcohol), to which a drop of glycerol has been added in order to ensure that the insects do not become brittle, or CFAA preservative. It is desirable to collect members of all castes. Workers, which are by far the most numerous, are often difficult or impossible to identify to species with certainty; soldiers, however, are readily distinguishable, but it is the alate forms that indicate the group of termites to which a species belongs, and thus assist the novice to place the other castes correctly. With few native species and little geographical overlap there is usually no difficulty in our region, but imported material or occurrences in unexpected places may cause problems of identification.

Rock-crawlers (Notoptera, Grylloblattidae) are found only in the western mountainous regions and mainly at high elevations. They should be sought in cracks and fissures, either in the ground or in old damp logs (especially of Douglas fir showing cuboid rot). In alpine areas they are sometimes found foraging on the surface of snow, and freshly dead specimens that have died of exposure overnight by venturing too far from shelter may be the first evidence that they are present. They should be preserved in fluid, as for termites.

Earwigs (Dermaptera) are most frequently found under sticks or stones or similar objects, often on beaches, although the most generally distributed species is found everywhere and often likes to hide in flowers. Earwigs may be pinned in the same way as other small orthopteroids, or they may be preserved in fluid.

Stick-insects (Cheleutoptera, Phasmatodea) are usually taken by beating trees, although some outside Canada occur on lower vegetation. They are best left unpinned, stored singly in cellophane envelopes into which heavy white cards of the same dimensions have been inserted to prevent breakage of the specimen. Pertinent data may be written on the card instead of on a normal data label. If it is preferred to pin the specimens, this should be done as soon as possible after killing, immediately arranging appendages as desired (but not too widely spread), supporting them, until they are dry, by means of a large stiff card pinned immediately below the specimen. The support is removed altogether, or cut much smaller and returned to its original position before labeling and storing.

In taxonomic work on grigs, much recourse is often had to the study of the concealed genitalic structures. Since little use has had to be made of these in the present work, we need here say only that, when they must be examined, the general methods by H. R. Roberts (1941*b*), for males, and by Randell (1963), for females, may be followed. These methods have, of course, been variously modified over the years and for different kinds of grigs, but the basic methods used are still much the same. The most important modification is undoubtedly that of Cohn and Cantrall (1974). These authors recommend a chloral hydrate macerating fluid, originally developed for studying water mites (formula: 9 g chloral hydrate, 6 mL glacial acetic acid, and 9 mL distilled water) in place of the more drastic potassium hydroxide solution of Roberts (op. cit.) or even of the sodium hydroxide of Gurney et al. (1964).

We prefer to store genitalic structures in glycerol in microvials pinned through their stoppers immediately below the specimens from which they have been taken (Gurney et al. 1964). The microvials are placed above, rather than below, the data labels, although this is a matter of personal preference. Cork stoppers, if they come into contact with glycerol, eventually blacken and the glycerol becomes stained, but this is not usually serious. It may be largely avoided if the proper precautions are taken (Gurney et al. 1964; Deitz 1979). Although neoprene or *Polyporus* stoppers tend to be difficult to insert properly and are expensive, they are generally preferable to cork stoppers, which are weak and tend to break. Some workers prefer to store in alcohol, because the structures are rendered less transparent thereby, but the greater risk of desiccation due to evaporation of the preservative necessitates the use of larger vials (in which minute specimens may be difficult to locate) and demands that the vials be themselves stored in greater quantities of fluid in larger, well-sealed vessels. This, in turn, results in the separation of the genitalia from their owners and a time-consuming system of cross-references and additional labeling. A more compact method even than the pinned microvial is the use of the card microslide pinned below the specimen (Kevan 1952*b*), but there are several drawbacks to this, most important being some flattening of the structures and the impossibility of viewing them from all angles.

Some orthopteroid species, such as various kinds of cockroaches, crickets, and camel crickets, are easy to rear at room temperature in almost any kind of closed container that is provided with a metal gauze ventilation panel, crumpled paper toweling for concealment and absorption of excess humidity, dry food (e.g., bran, commercially available pellets such as are used for feeding small rodents and young rabbits, or broken dog biscuits), sometimes with the addition of discarded fruit, and a vial of drinking water plugged with cotton. The bottom of the container may profitably be covered with coarse wood-shavings or dry sterilized natural leaf-litter. Crickets and other species also require containers of moist sand in which to lay their eggs. Apart from the cockroaches, however, some species undergo an obligatory winter diapause and will not develop unless the appropriate stage is chilled for a suitable length of time.

Although many species of grasshopper also have this last limitation, others do not and may be reared fairly readily and continuously in large airy cages. They should be provided with twigs and sticks upon which they can climb or roost, a water vial, a fairly deep container of moist sand for oviposition, and fresh green food. Food preferences vary with the species, but most will accept lettuce leaves, although grass-feeders are best supplied with young wheat or corn grown for the purpose in a greenhouse. In season, the insects may be supplied with natural vegetation cut locally outdoors, but this sometimes may introduce disease or parasites, notably mermithid nematodes, into the culture. Species that have an obligatory diapause and those that lay their eggs in plant tissues (many katydids, tree crickets, and others) are more difficult to rear. Some species may have a prolonged life cycle lasting more than 1 year; thus impatience at slow hatching of eggs should not lead to the immediate conclusion of failure.

Although we cannot deal with the subject more fully here, we recommend that those interested in rearing a particular species consult the literature listed in the synonymies of the appropriate and related species. Often the publications (if any) describing methods of culturing the insect in question will be cited, or following the appropriate lead may guide the reader to such information. Only a few of the species treated in this work have, in fact, been reared in the laboratory, but many more could be cultured without much difficulty.

Zacher (1928) gives a lengthy review of rearing methods for various orthopteroids, although many are not of this region and he does not treat the termites, for which Delong and Keagy (1949) give a brief introduction.

A useful brief simplified guide to the rearing of saltatorial grigs was circulated in the early 1930s by N. Criddle (1930*b*), but a more up-to-date résumé is that of Frings and Frings (1962). More detail is given by Gangwere (1960*a*). For designs for rearing cages of proven reliability, we suggest those described by Hunter-Jones (1961), although the variations on these are almost endless. For a more general guide to the rearing of insects (including grigs) and other technical matters, the reader should consult Nicholls (1963) and Peterson (1964).

The orthopteroid orders, or Ulonata

The Ulonata, which form the subject of this work, are Polynephria in which the immature stages are not aquatic and lack gills. (Only rarely are they even subaquatic, in which case the adults have similar habitats.) The wings of the Ulonata, when present, typically have reticular venation or numerous crossveins (whereas other Polynephria rarely do so), and the anterior pair are usually stiffened, parchmentlike, or leathery, forming tegmina which may, like the hind wings, be fully developed or abbreviated to varying degrees; the hind wings, also, are normally broad and characteristically fanlike, typically with enlarged posterior, or anal, lobes. (Plecoptera often have a somewhat similar type of hind wing.) Even when wings are present, however, there are exceptions to the foregoing. This is particularly true of the termites, in which the fore wing is not stiffened, the hind wing is rarely broad or with an anal lobe, and reticulation or cross-venation is usually much reduced.

Key to orders of Ulonata

1. Cerci strong, movable, forcepslike (in immature forms sometimes elongate-conical or rodlike and close together) .... **Dermaptera** (p. 39)

   Cerci of various forms, but, even if inwardly curved, not forcepslike or close together 2

2(1). Body greatly elongate, sticklike, wingless. Tarsi 5-segmented; legs all similar, with coxae small, wide apart; fore pair not raptorial. Abdomen terminating in a pair of inwardly curved, unsegmented cerci .... **Cheleutoptera** (p. 54)

   Without the above combination of characters 3

3(1). Body somewhat elongate, wingless. Eyes greatly reduced. Antennae not whiskerlike. Thoracic segments short, subequal, independent of each other, even in adult. Legs subequal, with coxae broad; tarsi 5-segmented. Cerci generally 8-segmented. Male abdominal terminalia asymmetrical. Female abdomen terminating in long exserted ovipositor .... **Notoptera** (p. 60)

   Without the above combination of characters 4

4(3). Hind legs with femora not thickened or modified for jumping; coxae broad, close together; tarsi usually 5-segmented, but, if 4-segmented, then sometimes with basal segment subdivided ventrally. Wings, when present, held more or less horizontally over dorsum when at rest (anterior pair sometimes not thickened or posterior pair broad). Cerci rarely unsegmented, even if very short .... **Dictuoptera** (p. 66)

Hind legs with femora thickened and modified for jumping; coxae small, widely separated; tarsi with 5, or fewer, segments (usually 3 or 4). Fore wings, when present, always in form of thickened tegmina held rooflike over body or angled or curved downward along sides of abdomen when at rest. Cerci usually unsegmented, even if long 5

5(4). Antennae whiskerlike, usually longer than body, with many (well over 30) indistinct segments (rarely bristlelike and much shorter than body, when hind legs are not adapted for jumping). Anterior tibiae usually with basal, oval, or slitlike tympanal organs. Male tegmina, when present (even if reduced), with wing venation generally modified, at least at wing base, to form a tegmen-to-tegmen stridulating organ. Ovipositor long, sword-, dagger-, sickle-, or needle-like, the valves forming a single unit .... **Grylloptera** (p. 117)

Antennae not whiskerlike (or bristlelike), shorter than body, with less than 30 distinct segments. Anterior tibiae without, but sides of base of abdomen generally with, rounded tympanal organs. Male tegmina without tegmen-to-tegmen stridulating organ, although tegmina may be modified for this function (but not at base), and hind femur also frequently involved. External part of ovipositor consisting of two pairs of rather short sharp triangular plates, or valves, with divergent tips .... **Orthoptera** (p. 296)

Clé des ordres d'Ulonata

1. Cerques forts, mobiles, en forme de pinces (chez les formes immatures, parfois allongés et coniques ou de forme cylindrique et rapprochés l'un de l'autre) .... **Dermaptera** (p. 39)

Cerques de formes diverses, mais jamais en forme de pinces ni rapprochés l'un de l'autre même si tournés vers l'intérieur 2

2(1). Corps très long, en forme de bâton et aptère. Tarses à 5 segments; pattes toutes semblables à petites coxas clairement séparées l'une de l'autre; pattes antérieures non ravisseuses. Abdomen se terminant par une paire de cerques non segmentés et incurvés vers l'intérieur .... **Cheleutoptera** (p. 54)

Ne présente pas la combinaison de caractères donnée ci-dessus 3

3(1). Corps assez long et aptère. Yeux très petits. Antennes non filiformes. Segments thoraciques courts, subégaux et indépendants l'un de l'autre, même chez l'adulte. Pattes subégales à coxas larges; tarses à 5 segments. Cerques comptant en général 8 segments. Terminalia asymétriques chez le mâle. Femelle pourvue, au bout de l'abdomen, d'un long ovipositeur saillant .... **Notoptera** (p. 60)

Ne présente pas la combinaison de caractères donnée ci-dessus 4

4(3). Pattes postérieures ayant un fémur non adapté au saut; coxas larges, rapprochées l'une de l'autre; tarses comptant d'habitude 5 segments mais, s'il n'y en a que 4, le segment basal est parfois subdivisé ventralement. Ailes, lorsque présentes, posées plus ou moins à plat sur le dos au repos (paire antérieure parfois non épaissie ou paire postérieure large). Cerques rarement non segmentés, même s'ils sont très courts .... **Dictuoptera** (p. 66)

Pattes postérieures ayant un fémur adapté au saut; coxas petites, clairement séparées l'une de l'autre; tarses comptant 5 segments ou moins (habituellement 3 ou 4). Ailes antérieures, lorsque présentes, prenant toujours la forme de tegmina épaissis et posés comme un toit sur le dos ou décrivant un angle ou une courbe vers le bas, de part et d'autre de l'abdomen, au repos. Cerques d'habitude non segmentés, même s'ils sont longs ...5

5(4). Antennes filiformes, en général plus longues que le corps, comptant beaucoup plus que 30 segments indistincts (plus rarement, antennes sétiformes et beaucoup plus courtes que le corps, lorsque les pattes postérieures ne sont pas adaptées au saut). Tibias antérieurs d'habitude pourvus à la base d'un tympan de forme ovale ou allongée. Tegmina du mâle, lorsque présents (même de taille réduite), présentant des nervures généralement modifiées, au moins à la base, pour former l'une contre l'autre un organe stridulant. Ovipositeur long, en forme de sabre, de poignard, de faux ou d'aiguille, et aux valves soudées pour ne donner qu'une seule pièce .... **Grylloptera** (p. 117)

Antennes ni filiformes ni sétiformes, plus courtes que le corps, comptant moins de 30 segments distincts. Tibias antérieurs dépourvus d'un tympan. Un tympan de forme arrondie existe toutefois généralement, de part et d'autre de l'abdomen, à sa base. Tegmina des mâles ne formant pas un organe stridulant (ils peuvent cependant être modifiés de manière à jouer ce rôle, mais pas à la base), mais fémurs postérieurs souvent adaptés pour la stridulation. Partie externe de l'ovipositeur faite de deux paires de plaques triangulaires tranchantes, ou valves, à pointes divergentes .... **Orthoptera** (p. 296)

Order Dermaptera—earwigs

Members of this order are characterized as follows: body usually small to moderate (about 10–30 mm long), depressed, generally elongate. Head small, more or less prognathous, with distinct gular sclerite. Antennae filiform or slightly moniliform, seldom more than half as long as body. Compound eyes usually well-developed, sometimes reduced; ocelli absent. Cranial sutures distinct. Pronotum somewhat larger than mesonota and metanota, usually more or less subquadrate and flat dorsally, not much, if at all, wider than head. Legs similar in size and form, rather short, with tarsi 3-segmented. Wings frequently absent; anterior pair, if present, modified to form short subquadrate veinless elytronlike tegmina which meet at mid-dorsal line and which barely reach first abdominal tergum; hind wings (which may be absent, even if tegmina are present) variously developed, if present composed almost entirely of greatly enlarged semicircular anal lobe with a number of radiating veins, but, except for a submarginal connecting row, lacking crossveins; base strongly sclerotized, incorporating two reduced longitudinal veins only, with the wing, at rest, folded first longitudinally in a fanlike manner, and then transversely along two folds so that they are stowed as packages under the tegmina, only the small sclerotized basal part projecting for a short distance over the base of the abdomen. Abdomen 11-segmented, with first tergum fused to metanotum; first sternum reduced or absent; in females 8th and 9th terga nearly always small and invisible (without dissection); 11th segment represented only by epiproct (supra-anal plate) and lateral paraprocts; 9th sternum in male largely overlying 10th (which is represented in both sexes by a pair of plates at bases of cerci); 7th sternum in females concealing 8th and 9th. Epiproct divided transversely into opisthomeres (3 in the most completely divided forms—the so-called "pygidium," "metapygidium," and "telson." Cerci well-developed, unsegmented (except in the immature stages of some species), typically in the form of forceps whose bases are close together (they vary considerably in length, often in a single species), those of the males generally larger, heavier, and more strongly curved, often asymmetrical (in immature forms they may be almost straight and, in some, are held pressed together so as to appear almost like a tail; cerci of some nymphal forms long, segmented. Ovipositor usually lacking, but present (although reduced) in some primitive forms and represented by two pairs of short valves developed from the 8th and 9th sterna. Male genitalia basically incorporating a pair of phallic structures (penes), one member of which is usually reduced, sometimes directed forward, but, in most species, absent.

Remarks. Earwigs are generally cryptic insects, concealing themselves, at least during the day, under stones, debris, logs, or bark, in crevices or leaf sheaths, or deep in flowers. They emerge at night, however, and may forage widely. Most species, even if able to fly, seem generally reluctant to do so, although sometimes (particularly in the case of smaller species) they may fly to lights. On rare occasions, earwigs have been known to fly by day. A few species are cavernicolous or burrow actively in soil.

Most species of earwigs are more or less omnivorous, although they probably prefer animal food, whether it be dead or alive. Some species are known to be exclusively predaceous, such as those that feed on worker ants in Africa. That some, at least, consume vegetable matter is, however, well known in the case of *Forficula auricularia* Linnaeus, which often damages garden blooms in addition to preying upon other insects. Earwigs are said to capture their prey by means of their forceps (cerci)—although these are probably as much defensive as offensive weapons. (They are also used in courtship.) The Dermaptera are all but unique among the Ulonata in that they include epizoic (if not truly parasitic) species, which spend much of their life on the bodies of Malaysian, Indonesian, and Philippine free-tailed bats.

Distribution. The 135 or so genera and over 1350 known species of Dermaptera are largely tropical in distribution, occurring widely throughout the warmer parts of the world, with some species inhabiting temperate climates. A few can tolerate cold winters, but these are exceptions. Because of their retiring habits and omnivorous diet, earwigs, like cockroaches, are frequently transported successfully by commerce. Indeed some species seem to have accompanied primitive man on some of his earliest migrations, so that their native homelands may be only surmised. As with cockroaches, however, relatively few of the total number of species are synanthropic. Unlike cockroaches, those few that have managed to adapt to new environments seldom become serious domiciliary nuisances or major pests of commercial premises—although there are localized exceptions. Most successful synanthropic species have, in fact, established themselves out-of-doors in their adopted countries. Of these, only *Forficula auricularia* Linnaeus has become a serious crop pest, and then really only of prized or commercially valuable flowers. For the most part, these earwigs probably do more good than harm from a human standpoint, for they feed largely on other insects. If abundant, like any other insect, they may incur the ire of the gardener or the wrath of the householder, more as a matter of prejudice than of practicality.

There is much folklore associated with earwigs, and traditionally, of course, they are supposed to enter the human ear with dire results. As they seek out crevices, they may indeed, on rare occasions, enter ears if these are placed on the ground when the owner is recumbent, but probably no more so than do other similarly retiring insects.

An interesting sidelight on Canadian folklore is that, at Sandspit, Queen Charlotte Islands, B.C., earwig races are an annual gambling event, attended with much enthusiasm by locals and visitors alike (Canadian Broadcasting Corporation 1970). The species involved is *Forficula auricularia* Linnaeus.

Biology. The life histories of few species of Dermaptera have been studied, but the eggs always seem to be rather small, smooth, and whitish, laid in batches but not cemented together or in capsules or oothecae. In general, there are four or five nymphal instars, according to species, but there are occasionally more. The number of antennal segments (and, in species in which the nymphal cerci—which may be much longer than the body—are segmented, of these too) increases with each instar. The females of several species, including *Forficula auricularia* Linnaeus and at least some others, are known to brood over their eggs and to show a degree of maternal care for the emerging offspring. The young earwigs may remain associated in family groups until they are nearly mature, so that the insects are in some measure semisocial. This is, however, not general among Dermaptera. Earwig biology is well discussed by Chopard (1938). The most recent comprehensive accounts of the Dermaptera are those of Beier (1959) and Günther and Herter (1974). For suprageneric classifications, the reader is referred to Popham (1963, 1965), to Günther and Herter (op. cit.), and to Steinmann (1975). For a checklist of world species, see Steinmann (1973).

Four superfamilies of living earwigs may be recognized. Only two of these concern us here: Spongiphoroidea (= Labiodea) and Forficuloidea. They differ from other superfamilies by not being hairy, by possessing a collarlike structure in the neck region, and by the anterior ventral cervical sclerite being smaller than the posterior one.

Key to superfamilies of Dermaptera

1. Second tarsal segment typically subcylindrical, not projecting beneath third; third segment, in lateral view, arising more or less terminally from second toward ventral part (Figs. 90, 94) .... **Spongiphoroidea** (p. 42)

   Second tarsal segment typically bilobed, projecting distally beneath third (Figs. 98, 109), or if not, then third (in lateral view) arising subterminally toward dorsal part of second segment ..... **Forficuloidea** (p. 49)

Clé des superfamilles de Dermaptera

1. Deuxième segment des tarses typiquement subcylindrique, ne se prolongeant pas sous le troisième segment; troisième segment, vu de côté, s'articulant plus ou moins vers l'extrémité à partir du deuxième segment vers la partie ventrale (fig. 90 et 94) .... **Spongiphoroidea** (p. 42)
   Deuxième segment des tarses typiquement bilobé, à extrémité faisant saillie sous le troisième (fig. 98 et 109) ou sinon, troisième segment (vu de côté) s'articulant avant l'extrémité et vers le côté dorsal du deuxième segment .... **Forficuloidea** (p. 49)

Superfamily Spongiphoroidea

This superfamily comprises two families, if the Arixeniidae be excluded. These are the Anisolabididae and the Spongiphoridae, which include, between them, 48 genera and over 350 species. Both are represented by established alien species in the region with which we are here concerned, there being no native species.

Key to families of Spongiphoroidea

1. Wings lacking. Pygidium (11th abdominal segment) distinguishable but concealed by being bent downward between cerci (Figs. 83, 89). Male with two "penis lobes" .... **Anisolabididae** (p. 42)
   Wings usually present. Pygidium distinct (Figs. 91, 93), and other opisthomeres reduced. Male with single "penis lobe" .... **Spongiphoridae** (p. 46)

Clé des familles de Spongiphoroidea

1. Ailes absentes. Pygidium (onzième segment abdominal) distinct, mais replié vers le bas et dissimulé entre les cerques (fig. 83 et 89). Mâles pourvus de deux lobes «péniens» .... **Anisolabididae** (p. 42)
   Ailes habituellement présentes. Pygidium distinct (fig. 91 et 93) et autres opisthomères réduits. Mâles munis d'un seul lobe «pénien» .... **Spongiphoridae** (p. 46)

Family Anisolabididae

This family is widely distributed naturally throughout the warmer parts of the world. There are 18 genera and about 115 known species.

Key to genera of Anisolabididae

1. Antennae with 20–24 segments, with basal segment longer than combined segments 4–6. Legs (and body) entirely black (Fig. 83) .... ***Anisolabis*** (p. 43)

Antennae with 14–16 segments, with basal segment about equal to combined segments 4–6. Legs pale, with femora and tibiae usually distinctly banded with black (Fig. 84) .... ***Euborellia*** (p. 45)

Clé des genres d'Anisolabididae

1. Antennes comptant de 20 à 24 segments, à segment basal plus long que les segments 4 à 6 inclusivement. Pattes (et corps) d'un noir uniforme (fig. 83) .... ***Anisolabis*** (p. 43)

   Antennes comptant de 14 à 16 segments, à segment basal de longueur approximativement égale aux segments 4 à 6 inclusivement. Pattes de couleur pâle, fémurs et tibias d'habitude marqués de bandes noires (fig. 84) .... ***Euborellia*** (p. 45)

Genus *Anisolabis* Fieber

Description. Body apterous, long, slender. Antennae about half as long as body. Head and pronotum about equal in width; pronotum nearly square, with posterior margin and angles evenly rounded; metanotum with posterior margin conspicuously concave. Legs relatively short. Abdomen large, somewhat enlarged at middle. Male forceps well-separated at base, strongly curved, asymmetrical; right forceps (dorsal aspect) more incurved than left one (Figs. 85, 86); female forceps short, nearly straight, symmetrical (Fig. 87).

Anisolabis maritima (Bonelli)

Figs. 83, 85–87; Map 1

Forficula maritima Bonelli in Gené, 1832:221.
Anisolabis maritima; Helfer 1963:15.

Diagnosis. Body large, shiny black or brown. Legs yellow. Found only at or near seashores.

Description. Body large (length, male 18.0–23.5, forceps 3–4.5; female 15.5–25.0, forceps 4.5–5.4 mm), shiny black or brown. Antennae with 20–24 segments, uniformly dark. Legs pale yellow. Immatures identical in shape and color but with fewer antennal segments, and forceps straight in both sexes.

Range. British Columbia to California; Ontario and Québec to Florida.

Behavior and habitats. Bennett (1904) reported watching caged *A. maritima* employ their forceps to capture crickets (species unspecified), sand fleas, and smaller earwigs. They devoured them while still holding them with the forceps. This earwig is able to twist and bend its abdomen so that the forceps are brought around nearly in front of the head.

Guppy (1950) found *A. maritima* at Departure Bay, Vancouver Island, and observed its behavior. Adults were found in trash at the high-water level and also well above this mark, and under the bark of logs on the shore. Members of the species are nocturnal, remaining active until temperatures drop to near freezing. Then they retreat to well above the high-tide mark,

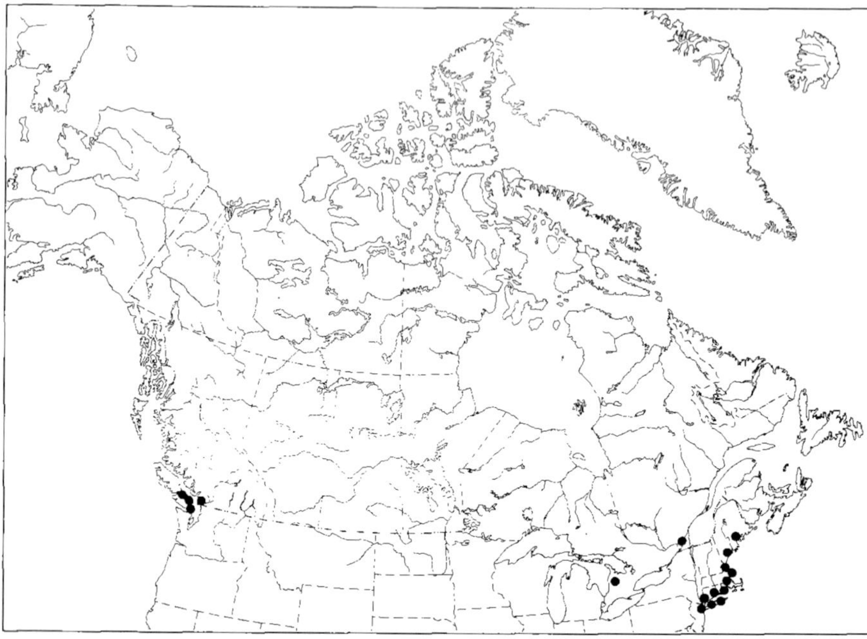

Map 1. Collection localities for *Anisolabis maritima*.

seek shelter under logs, boards, or other debris, and become completely dormant. Burrows are dug in sand, using only the mandibles. Eggs are hatched and young are fed in the burrows. The food supply usually consists of small crustaceans and drowned insects that wash up on the shore. Eggs and larvae of Diptera which breed in decaying seaweed were thought to provide a great part of the food supply.

Life history. The life cycle takes 2 years on Vancouver Island. Eggs are laid in burrows, usually in July, although warm weather in May may induce egg-laying at an earlier date. Eggs hatch in 30–45 days, and immatures pass the first winter in the first or second instar. They may reach adulthood during the second summer or pass the second winter as fifth instar nymphs, maturing early during the following spring. Females can produce eggs within 6 weeks of becoming adult. Burrows are dug in areas that remain moist, if these are constructed in sand, often under driftwood or other debris. However, the brooding chamber may be under bark of rotten logs. Females are aggressive in protecting their eggs, using their forceps and striking forward over their backs with such force as to throw themselves on end. If eggs have not hatched by the time cold weather arrives, they are eaten by their parents. The rate of growth of nymphs is strongly influenced by temperature (Guppy 1950).

Genus *Euborellia* Burr

Description. Size as in genus *Anisolabis*. Tegmina normal, reduced, or absent, and wings present or absent. Pronotum nearly square to slightly broader posteriorly. Abdomen slightly expanded at middle. Male forceps stout, separated at base, triangular in cross-section basally, with apices curved inward, asymmetrical to nearly symmetrical (Figs. 88, 89). Female forceps more slender, symmetrical (Fig. 84*a*). Body black to dark brown, with legs yellow, often with darker bands. Antennae often with 1-several segments near apex (but not apical segment) white.

Euborellia annulipes (Lucas)

Figs. 84, 88-90; Map 2

Forficesila annulipes Lucas, 1847:IXXXIV.

Forcinella azteca Dohrn, 1862:226.

Euborellia annulipes; Hebard 1917*b*:312; Helfer 1963:163; Vickery et al. 1974:127.

Common name. Ringlegged earwig.

Diagnosis. Legs yellow with darker rings.

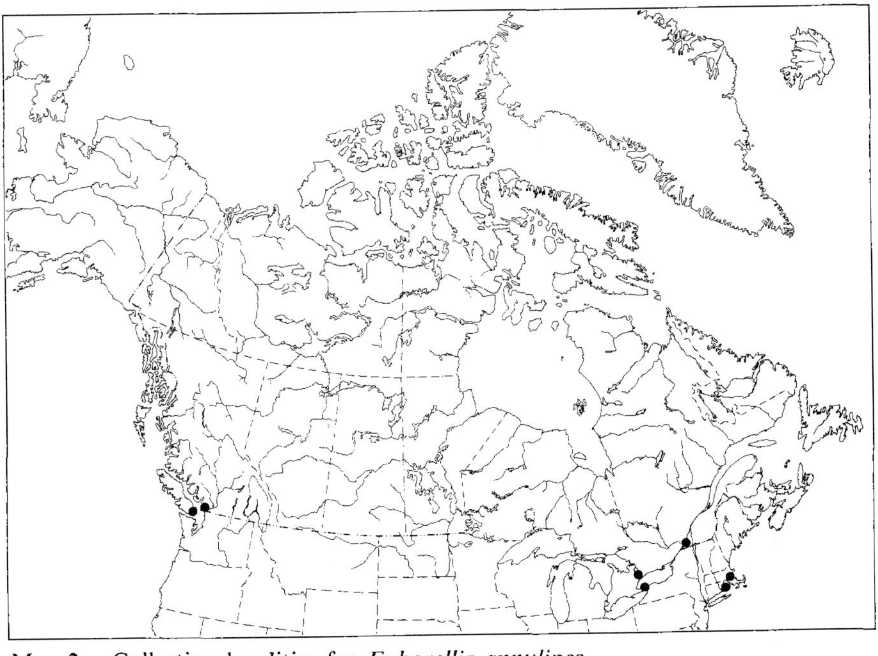

Map 2. Collection localities for *Euborellia annulipes*.

Description. Body apterous, dark brown. Legs pale yellow with dark rings (which may, however, be faint). Third and fourth (and sometimes fifth) subapical antennal segments pale.

Range. British Columbia; Ontario and Québec; cosmopolitan.

Behavior and habitats. *E. annulipes* is found under debris in both damp and dry locations. It is not restricted to seashores like *A. maritima*, although it is commonest in maritime regions. It is a general feeder and may be found in greenhouses and warehouses. It has been reported damaging stored potatoes, flour, meat products, and roots of vegetables grown in greenhouses. It is also predaceous on other insects, and, like other Dermaptera, probably attacks products of plant origin only when insect prey is scarce. Bohart (1947), as reported also by Langston and Powell (1975), found *E. annulipes* to be an effective predator on larvae of *Crambus* species (Crambidae: Lepidoptera) in lawns. Bharadwaj (1966) found females guarding their eggs, which they handled and rearranged until the eggs hatched. Any damaged or infertile egg was eaten by the mother. Maternal care ceased when hatching was completed.

Life history. Bharadwaj (1966) found egg clutches to average 53 eggs. Incubation lasted from 6 to 17 days at 20–29°C. Five nymphal instars are usual, but some individuals pass through a sixth instar before becoming mature. An average life cycle requires 120 days. Under good conditions there may be two generations each year.

Family Spongiphoridae

This family is also widely distributed throughout the world. There are 30 genera and about 240 known species, which are placed in 8 subfamilies that mainly occur in the tropics and subtropics. The Labiinae are widely distributed, reaching high temperate latitudes, and comprise 11 genera and more than 100 species. It includes a number of synanthropic species, particularly in the genus *Labia* Leach, one of which has been present in Canada and the northeastern United States for a long time. This species has, indeed, become so well adapted to North American conditions as to give the impression that it is a native. The Labiinae may be distinguished from other Spongiphoridae as follows: size small (less than 12 mm long); body not greatly flattened; head narrow with small eyes that are not as wide as first antennal segment is long; tegmina without lateral keels.

Key to genera of Spongiphoridae

1. Body small (4–5 mm), pubescent. Hind wings visible when folded, twice as long as tegmina. Antennae filiform, with fourth segment cylindrical, as long as third segment (Fig. 91) .... ***Labia*** (p. 47)
   Body larger (5–8 mm), not pubescent. Tegmina short, with hind wings absent or concealed. Antennae clavate, with fourth segment conical, shorter than third segment (Fig. 95) .... ***Marava*** (p. 48)

Clé des genres de Spongiphoridae

1. Corps petit (de 4 à 5 mm) et pubescent. Ailes postérieures visibles lorsque repliées, deux fois plus longues que les tegmina. Antennes filiformes ayant le quatrième segment cylindrique et aussi long que le troisième (fig. 91) .... ***Labia*** (p. 47)

   Corps plus gros (de 5 à 8 mm) et glabre. Tegmina courts, et ailes postérieures absentes ou dissimulées. Antennes claviformes ayant le quatrième segment conique et plus court que le troisième (fig. 95) .... ***Marava*** (p. 48)

Genus *Labia* Leach

Description. Body flattened, slender, pubescent, with abdomen slightly dilated at middle. Eyes small, shorter than basal antennal segment is long. Antennae about half as long as body, 10–12-segmented; each segment somewhat conical, about three times longer than broad. Tegmina and wings present. Basal and third tarsal segments equal in size, the second small and compressed. Forceps about half as long as abdomen, of male slightly to strongly curved (Fig. 92), of female straight and incurved apically (Fig. 93).

Labia minor (Linnaeus)

Figs. 91–94; Map 3

Forficula minor Linnaeus, 1758:423.

Labia minor; Doubleday 1838:279; Helfer 1963:17; Vickery et al. 1974:126.

Labia minuta Scudder, 1863*a*:415.

Diagnosis. Body small, alate. Male pygidium rectangular, narrow.

Description. Body small (Fig. 91) (length 4–5 mm, length of tegmina 1.5 mm, male forceps 1.5–2 mm, female forceps 1.2 mm). Pronotum narrower than head, only slightly longer than broad. Tegmina about twice pronotal length; hind wings, when folded, twice as long as and extending beyond tegmina. Male forceps simple, minutely toothed on inner edge (Fig. 92). Penultimate sternum of male with rectangular projection; pygidium visible from above (Fig. 91). Forceps of female short, rather stout, straight, with narrowly curved tips (Fig. 93). Color usually dull yellowish brown to brown, with head, forceps, and apical abdominal segment darker. Body, tegmina, and wings with fine yellowish pubescence.

Range. British Columbia to Nova Scotia, south to California; Europe.

Behavior and habitats. The species occurs in a wide variety of habitats. Morse (1920) reported it from stables, manure heaps, and fungi. It is mainly nocturnal, flies well, and often is attracted to lights at night. It is primarily a scavenger, although it may feed upon vegetation (floral parts) when other food is scarce.

Life history. Morse (1920) recorded adults from 25 May to 4 November in the New England states.

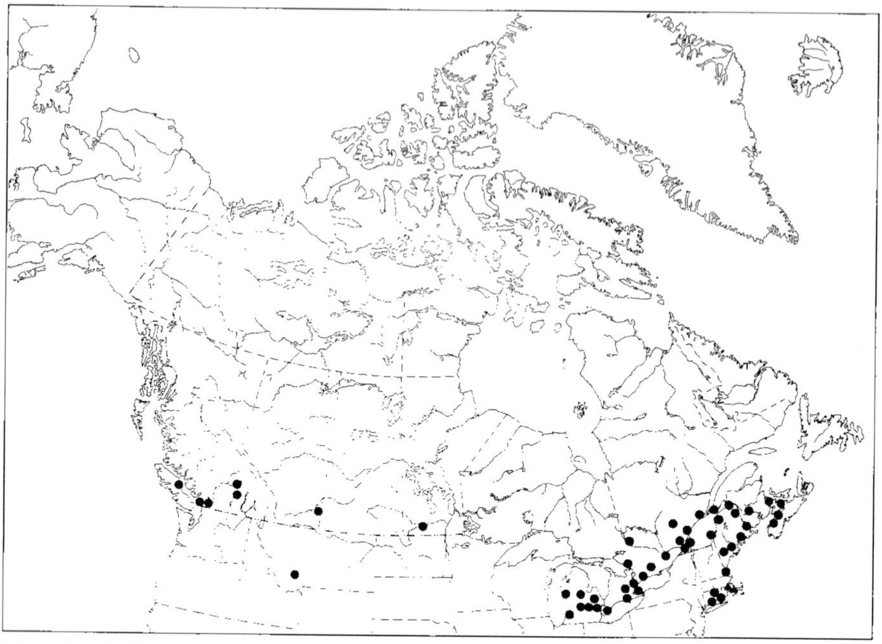

Map 3. Collection localities for *Labia minor* (●) and *Marava arachidis* (▲).

Genus *Marava* Burr

Description. Antennae with basal segments shorter than distance between antennal bases; second segment short, squarish; third segment long, three times, or more, as long as fourth and fifth segments combined. Pronotum nearly square, with posterior margin convex. Tegmina present, well-developed or abbreviated; wings present or absent. Femora, especially fore femora, broad. Male forceps well-separated at base, with small elongate projection (pygidium) between them (Fig. 95). Female forceps short (Fig. 95*a*).

Marava arachidis (Yersin)

Fig. 95; Map 3

Forficula arachidis Yersin, 1860:509; pl. 10, figs. 33–35.
Marava arachidis; Hincks 1954:163.
Prolabia arachidia [*sic*]; Helfer 1963:17.

Diagnosis. Body small. Wings generally absent or concealed, with tegmina short. Male pygidium broad.

Description. Body small (length 5–9 mm; forceps, male 1.5–2.75, female 0.75–1.25 mm). Tegmina short; hind wings generally absent or concealed. Abdomen broad. Male forceps arcuate, cylindrical, widely separated at base, each with small tooth on inner side near apex; pygidium broad. Forceps of female closer together, only slightly arcuate. Color dark brown to reddish brown or yellowish brown, shiny in appearance. Antennae yellow or brown, usually with basal segments paler.

Range. Massachusetts; cosmopolitan in tropics and subtropics.

Behavior and habitats. Although a tropical and subtropical species, *M. arachidis* can survive under artificial conditions in temperate countries. It is often transported from place to place in stored produce such as groundnuts (as its specific name suggests), and it seems particularly prone to infest materials of animal origin. In Europe, it was formerly known to breed occasionally in large numbers in such places as bone-meal plants and glue factories. In Massachusetts, large numbers were found in a slaughterhouse as well as in a sugar refinery (Morse 1919*b*, 1920). With more modern hygienic conditions in such places, it is not likely to establish itself so readily.

Life history. Not a great deal is known of this. Under artificial conditions, the species seems to breed more or less continuously, with overlapping generations, but probably not usually producing more than a single brood per year.

Superfamily Forficuloidea

This superfamily is made up of three families: Labiduridae, Chelisochidae, and Forficulidae. Only the last of these is of direct concern here, but both of the others include cosmopolitan synanthropic species that may well be found as adventives in our region and which have become established in other parts of North America.

Family Forficulidae

This family is one of the two largest of the Dermaptera, being only slightly bigger, although with many more genera, than the Spongiphoridae. Some 57 genera and nearly 250 species are included, coming from all parts of the world where earwigs are found. Of the eight subfamilies, only the Forficulinae is represented in the region here covered, where it is represented by the only indigenous species of earwig and by a thoroughly established alien. Additional adventive species have also been found.

Key to genera of Forficulidae

1. Body brown or black with yellow longitudinal stripes. Antennae with 11–13 segments; fourth segment from base twice as long as broad. Tarsi with basal segment as long as but no longer than 1.5 times length of third segment. Male pygidium spinelike (Figs. 100–105) .... ***Doru*** (p. 50)

   Body brown and pale brownish yellow, not striped. Antennae with 14 or 15 segments; fourth segment from base less than twice as long as broad. Tarsi with basal segment twice as long as third segment. Male pygidium not spinelike (Figs. 106–109) .... ***Forficula*** (p. 51)

Clé des genres de Forficulidae

1. Corps brun ou noir ayant des bandes longitudinales jaunes. Antennes comptant de 11 à 13 segments, le quatrième étant à partir de la base deux fois plus long que large. Tarses pourvus d'un segment basal au moins aussi long, mais de longueur ne dépassant pas une fois et demie celle du troisième segment. Pygidium du mâle pointu (fig. 100 à 105) .. .. ***Doru*** (p. 50)

   Corps brun et jaune brunâtre pâle, mais sans rayure. Antennes comptant 14 ou 15 segments, le quatrième étant à partir de la base moins de deux fois plus long que large. Tarse à segment basal deux fois plus long que le troisième segment. Pygidium du mâle obtus (fig. 106 à 109) .... ***Forficula*** (p. 51)

Genus *Doru* Burr

Description. Body small to medium (length 8–15 mm). Head broad, smooth. Eyes small. Basal antennal segment shorter than distance between antennal bases; antennal segments cylindrical but not elongated. Pronotal shape variable. Tegmina normal or abbreviated, usually meeting at median line; hind wings present or absent. Abdomen with sides parallel or widening posteriorly; last tergum before pygidium usually with tubercle above base of each forceps or with numerous small tubercles between bases; male forceps widely separated at base, arcuate to nearly straight, with or without inner subapical tooth; pygidium spatulate or pointed. Female forceps slender, rather straight.

Doru aculeatum (Scudder)

Figs. 100-102; Map 4

Forficula aculeata Scudder, 18[illegible]:262.
Doru aculeatum; J. A. G. Rehn and Hebard 1914*a*:93; Helfer 1963:13.
Doru aculeatum aculeatum; Urquhart 1941*b*:9; Cantrall 1943:67.
Doru lineare; Helfer 1963:13 (partim, not fig. 10).

Diagnosis. Body brown and yellow. Pronotum almost as long as wide. Male pygidial spine much longer than wide, acute.

Description. Robust species (body length 7–12, male forceps 3.5–6.8, female forceps 3–3.6 mm). Pronotum rather large, of about equal length and width, with posterior margin convex. Tegmina short; hind wings concealed (rarely, if ever, fully developed). Male pygidial spine (between forceps) long, aciculate; each male forceps usually with distinct subapical internal tooth. Head and pronotum dark brown, and pronotum with yellow lateral margins; tegmina yellowish brown to brown, with sutures darker; legs yellowish brown; abdomen and forceps black.

Range. Nebraska to Ontario; southeastern United States.

Behavior and habitats. Cantrall (1943) has given a detailed account of *D. aculeatum* as it occurs in the northern part of its range, in southern Michigan. It persists only along margins of sedgy (*Carex*) marshes where it

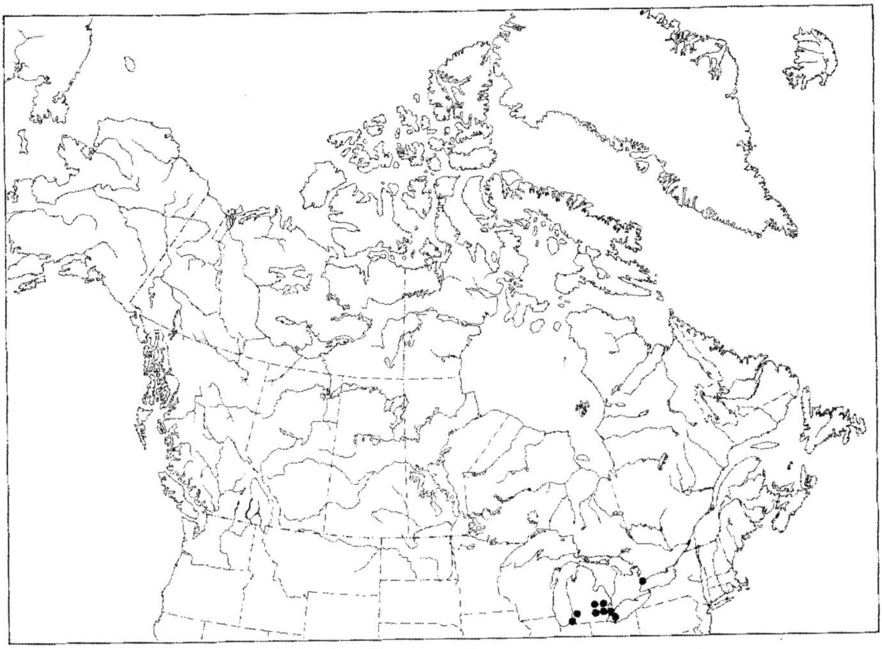

Map 4. Collection localities for *Doru aculeatum*.

can find suitable winter protection. He found it only in association with *Carex riparia* var. *lacustris*, although farther south it is found with other species. It spends much of its life in leaf axils. It usually chooses the greenest plants, is never found on dead ones, and remains head downward in the leaf axils during daytime. It is nocturnal, roaming freely on the foliage. If disturbed by a light it attempts to hide.

Life history. Adults hibernate during the winter. They emerge in late May and lay eggs until late June. These begin hatching in late June, and the insects mature by fall. Earliest and latest adults were reported by Cantrall (1968) to occur on 27 May and 30 September, although no adult was found during midsummer.

Genus *Forficula* Linnaeus

Description. Body flattened. Pronotum squarish with posterior angles rounded. Abdomen not expanded at middle. Hind wings, when folded, slightly longer than tegmina. First tarsal segment slightly longer than third, the second short, dilated, and lobed at apex, held beneath the third. Antennae with 14 or 15 segments, with third segment twice as long as fourth. Terminal abdominal segments beyond the sixth cylindrical, more than four times as long as broad.

Forficula auricularia Linnaeus

Figs. 106–109; Map 5

Forficula auricularia Linnaeus, 1758:423.
Forficula auricularia; Glaser 1914:157; Vickery et al. 1974:128.

Common name. European earwig.

Diagnosis. Antennae with 14 or 15 segments. Male forceps as long as abdomen, curved. Female forceps crossed.

Description. Medium to small species (body length 10–12, male forceps 4–7, female forceps 3, tegmina 2 mm). Male forceps varying from about half as long to as long as abdomen (sometimes longer); flat and somewhat broadened basally, then curved, semicircular, with crenulate teeth basally and on beginning of curvature of inner margin (Fig. 107). Female forceps slightly curved, crenulate on inner margins, usually crossed apically (Fig. 108). Color generally dark reddish brown with tegmina and wings dull yellow; legs, forceps, basal antennal segments, lateral areas of pronotum, and posterior margins of abdominal segments pale.

Range. British Columbia to Newfoundland; cosmopolitan in cooler parts of the world.

Behavior and habitats. *F. auricularia* is often of concern to the human population in areas where it has recently become established. New colonies tend to build up to very high population levels with consequent competition

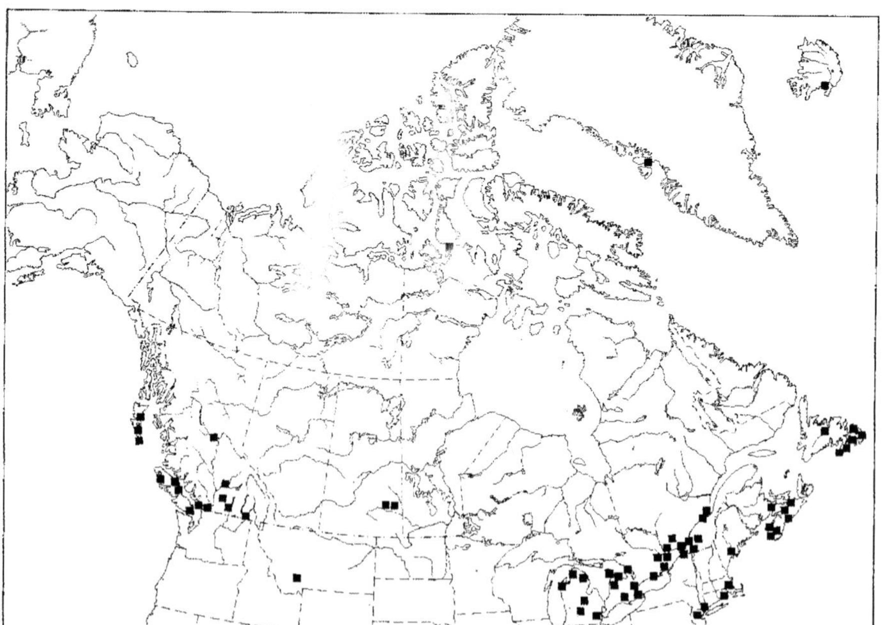

Map 5. Collection localities for *Forficula auricularia*.

for food and shelter. This occurred on the western part of Montréal Island during the mid-1960s (Vickery et al. 1974) and continues there at high level, causing much undue concern to householders. The earlier infestation in Vancouver exhibited the same characteristics, but that population has now declined greatly in numbers, and the species is no longer treated as a pest of consequence there (Lamb and Wellington 1975). Initial high levels of population, followed by gradual decline, is characteristic of many species introduced into new areas. *F. auricularia* is a prime example of this type of population behavior.

Earwigs in general tend to shun light and are mainly nocturnal. Also when crowded they take advantage of shelter wherever this can be found, often in buildings. Although they are primarily omnivorous, feeding on both animal and plant materials, they are largely beneficial as other insects make up a large part of the food supply. When population pressure is great, and insect food then becomes scarce, they do attack plants and may cause severe damage to blossoms in flower gardens (they are partial to ray florets of *Dahlia*, and were reported as a pest of these flowers in Ottawa, Ont., in 1978–1979). *F. auricularia*, although fully winged, is an infrequent flier (Brindle 1977). W. A. Wilson (1971) reports on parasitism by nematodes.

Life history. Lamb and Wellington (1975) and Lamb (1975) studied the behavior and life history of *F. auricularia* in Vancouver, where it spends the winter in the adult stage, living in pairs in underground nests or in aggregations on the surface. It is doubtful that surface aggregations could survive the winter in Ontario or Québec, although, with an early snow cover, this is possible. Females oviposit in the nests near the end of winter or early in spring. They then drive the males out of the nests. In May, some females and immatures appear on the surface. Some of these females may produce a second brood in spring. Most males die early in the spring after being driven from the nests. Females remain with the eggs and tend them and the newly hatched immatures. Then for a short period both females and immatures leave the nests to forage, returning before daylight. Finally the immatures cease to return, but seek shelter under surface debris elsewhere. The females die before midsummer. The immatures of both broods become adults by the end of August, and, by early October, fresh pairs have entered the soil and constructed winter nests (Lamb and Wellington 1975). Observations in Québec indicate a life history similar to this, but emergence from winter quarters is later in the season. There is a lower frequency of second brood, and retiral to winter quarters occurs earlier in the fall.

Economic importance. When populations of *F. auricularia* are dense, significant damage to garden blooms may occur, as previously noted, but the main economic effect is the amount of time wasted in worrying about the insects and in the expense incurred in attempts to eradicate them when they enter buildings to seek shelter! The principal natural enemy of *auricularia* is the tachinid fly *Bigonicheta spinipennis* (Meigen); McLeod (1962) summarized biological control attempts up to 1959.

Order Cheleutoptera—stick-insects

Members of this order have been given various vernacular names such as spectres, phasmids, walking sticks, walkingsticks, or stick-insects. They also include some leaf-insects and a few other Ulonata having, with some exceptions, an elongate twiglike sometimes dorsoventrally flattened and leaflike appearance. The relationships of the order are not clear, and those with the Orthoptera are probably not so close as was previously supposed. There are, indeed, indications of a nearer affinity with the Dermaptera.

Cheleutoptera are of moderate to large size, adults being rarely less than 3 cm long, and they include the longest of all living insects (some exceed 30 cm long). They may be characterized as follows: body form usually elongate, cylindrical, occasionally shorter, depressed and leaflike, rarely of other form. Antennae simple, usually filiform, of variable length. Prothorax short; mesothorax and metathorax usually elongate, the tergum of the latter being closely associated with (often fused to) that of the first abdominal segment. Legs all more or less similar, with hind pair not modified for jumping; coxae small and wide apart; tarsi 5-segmented. Wings often greatly reduced or absent; fore wings (when present) usually rather short, in the form of leathery or parchmentlike tegmina, overlapping each other slightly, with costal vein submarginal; hind wings (when present) fanlike, usually considerably surpassing fore wings; hind wing buds of nymph not overlapping front wing buds during development. Abdomen with distinct, rather short, symmetrical, unsegmented, typically curved cerci (sometimes expanded in female). Male external genitalia variable, asymmetrical, and concealed by enlarged 9th and 10th abdominal terga and 9th sternum (subgenital plate), behind which the 10th sternum may be clearly visible as a hooklike vomer, or reduced. Ovipositor of female (belonging to 8th and 9th abdominal segments) present, but short, rather complex.

Somewhat in excess of 2500 species have been described, but the real number of valid species is uncertain because males and females of the same species may often have been given different names. The majority occur in the humid tropics and a large proportion of these are from the Oriental-Australasian region. They do, however, occur in drier and in temperate regions. Most are arboreal or arbusticolous, but a few live on the ground or among grasses and herbs. None is aquatic, although some early reports suggested that one or two species were so. So far as known, all are phytophagous and most are nocturnal. Eggs are nearly always laid singly in frequently elaborate, seedlike oothecae, usually scattered at random on the ground, although a few species lay eggs in rows on twigs or foliage, or oviposit in debris such as rotten wood. Many species are parthenogenetic, often continuously so, although occasional males occur in such species.

The sexes usually differ greatly in size and often in form also. There seem to be usually about 5–8 nymphal stages, metamorphosis being gradual and usually rather slight, especially in wingless forms. Cheleutoptera are unusual for the powers possessed by many (if not all) to regenerate, to a considerable degree, damaged or lost appendages in the instar following an injury by autotomy (voluntary shedding of appendages). Adults cannot regenerate, as they do not molt, but they retain the facility for autotomy as an escape mechanism. The whole order is remarkable for the powers of camouflage that result from extraordinary mimicry of plant stems of many kinds and occasionally of foliage and the like. Associated with this is the state of catalepsy, whereby the insect becomes rigid and feigns death. Most species can change color from instar to instar so as to match their background; some can even change color, like chameleons, without molting. Some discharge defensive irritant fluids. Cheleutoptera are usually said not to stridulate, but female leaf-insects (Phylliidae), at least, have what are regarded as stridulatory teeth on the inner surfaces of the bases of the antennae, and some species in other families strike hind and fore wings together when alarmed.

Relatively few species are economically important, but some may suddenly become abundant and seriously defoliate trees. There is a keen interest among amateur naturalists in rearing these insects and a fairly large international exchange between them. One or two species (notably *Carausius morosus* (Brunner von Wattenwyl)) have become favorite experimental animals with European insect physiologists.

For comprehensive accounts of the Cheleutoptera, see Chopard (1938, 1949), and Beier (1957, 1968). Systematics and zoogeography are also covered by these authors (except Chopard 1938), by Günther (1953), by Clark (1975), who reproduces the key of Beier (1968) in English, and, in a more recent interpretation, by Bradley and Galil (1977). The most recent modifications are those of Kevan (1976*b*, 1977*b*). Clark (1979) gives the most comprehensive account to date of the eggs in the suborder Phasmatodea. The eggs differ from those of Orthoptera *sensu stricto* in lacking a distinct localized hydropyle at the posterior pole.

Family Heteronemiidae

This family may be characterized as follows: body greatly elongate, thin, sticklike. Antennae slender, indistinctly segmented, at least beyond the middle, longer than front femora, but not as long as body. Wings lacking. Thoracic mesonotum nearly always shorter than metanotum. Ventral keels of middle and hind femora unarmed, or virtually so. Penultimate (9th) abdominal tergum often with lateral margins turned downward and overlapping ventrally so as to form, at least apically, a sort of tube. Terminal abdominal segment of male with tergum not split and bilobed, nor with pair of fingerlike posterior processes, of female not beaklike.

Two subfamilies (Heteronemiinae and Libethrinae) and about 19 genera are known, which together probably include more than 100 species. All but one genus of Libethrinae, which occurs in southern Africa, are American, mainly tropical, but a few species of Heteronemiinae extend into temperate regions, north and south. A single genus is represented in Canada.

Genus *Diapheromera* Gray

Description: Body slender, long, nearly cylindrical; antennae inserted in front of eyes; pronotum about equal in length to head; mesonotum four times as long as pronotum; metanotum about three times as long as pronotum; apical abdominal segment oblong, twice as long as wide in male; legs very long, slender, with middle femur swollen in male; middle and hind femora armed on upper side near apex with curved acute spine; male cerci long, incurved, usually crossed, and female cerci short, straight.

Key to species and subspecies of *Diapheromera*

1. Hind femur unarmed apically beneath ...... ***blatchleyi* (Caudell)** ... 2
   Hind femur with apical spine beneath (Fig. 110) 3

2(1). Head as seen from above with subparallel sides. Midwestern United States .... ***blatchleyi blatchleyi* (Caudell)** (p. 57)
   Head broader across eyes than behind, then tapering slightly backward. Eastern coastal United States . ***blatchleyi atlantica* (Davis)** (p. 58)

3(1). Middle femur of male banded green and brown. Apical spine of hind femur strong. Cerci of male armed internally with blunt angular tubercle (Fig. 111), of female short, less than half length of terminal abdominal segment (Fig. 114) .... ***femorata* (Say)** (p. 58)
   Middle femur of male not, or at most feebly, banded. Coloration generally rather uniform. Apical spine of hind femur rather weak. Cerci of male armed internally at base with slender spine (Fig. 113), of female longer, nearly equal in length to terminal abdominal segment (Fig. 115)..... .... ***velii velii* Walsh** (p. 59)

Clé des espèces et des sous-espèces de Diapheromera

1. Fémur postérieur inerme sous son extrémité apicale .... ***blatchleyi* (Caudell)** 2
   Fémur postérieur portant une épine sous l'extrémité apicale (fig. 110) 3

2(1). Tête, vue du dessus, à côtés subparallèles. Midwest des États-Unis .... ***blatchleyi blatchleyi* (Caudell)** (p. 57)
   Tête plus large au niveau des yeux, puis allant se rétrécissant vers l'arrière. Côte est des États-Unis ..... ***blatchelyi atlantica* (Davis)** (p. 58)

3(1). Fémurs médians des mâles marqués de bandes vertes et brunes. Épine apicale des fémurs postérieurs forte. Cerques des mâles pourvus, vers l'intérieur, d'un tubercule arrondi (fig. 111); cerques des femelles courts, de longueur inférieure à la moitié de la longueur du dernier segment abdominal (fig. 114) .. .. ***femorata* (Say)** (p. 58)
   Fémurs médians des mâles de couleur uniforme ou faiblement marqués de bandes. Coloration généralement uniforme. Épine apicale des fémurs postérieurs plutôt petite. Cerques des mâles pourvus vers l'intérieur, à la base, d'une fine épine (fig. 113); cerques des femelles plus longs, de longueur presque égale à celle du dernier segment abdominal (fig. 115) .... ***velii velii* Walsh** (p. 59)

Diapheromera blatchleyi blatchleyi (Caudell)

Figs. 113, 116; Map 6

Bacunculus blatchleyi Caudell, 1905*b*:212.
Diapheromera persimilis blatchleyi; Hebard 1931*b*:130.
Diapheromera blatchleyi blatchleyi; Hebard 1934*a*:160.

Diagnosis. Hind femur without apical ventral spine. Head, seen from above, with subparallel sides.

Description. Middle and hind femora lacking subapical spines. Head slightly longer than pronotum, with more or less subparallel sides when viewed from above. Female with ninth abdominal segment broadly but angularly emarginate (Fig. 116). Cercus of male with blunt inner basal tooth (Fig. 113), of female about as long as that of male, subequal to terminal abdominal segment (Fig. 116). Color green or brownish green; male with pale lateral stripe from head to base of hind femur.

Range. Nebraska to Ohio, south to Oklahoma.

Behavior and habitats. Arboreal, according to Froeschner (1954), but host trees were not mentioned.

Life history. As adults have been found in July and August (Froeschner 1954) and undoubtedly occur in September, the life history appears to be univoltine, with overwintering in the egg stage.

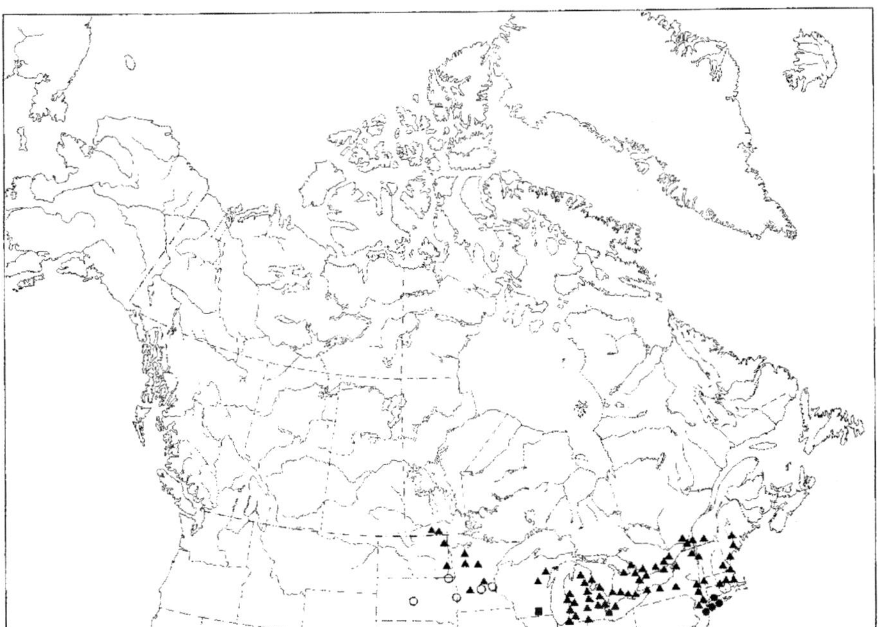

Map 6. Collection localities for *Diapheromera blatchleyi blatchleyi* (■), *D. blatchleyi atlantica* (●), *D. femorata* (▲), and *D. velii velii* (○).

Diapheromera blatchleyi atlantica (Davis)

Map 6

Manomera blatchleyi; Morse 1919*a*:16; Helfer 1963:24 (partim).
Manomera blatchleyi atlantica W. T. Davis, 1923:53, pl. X, figs. 1–4.
Diapheromera blatchleyi atlantica; Hebard 1931*b*:130.

Diagnosis and description. Like *D. blatchleyi blatchleyi*, but with head wider in front than behind. Female cerci shorter than last abdominal segment.

Range. New York to Virginia.

Behavior and habitats. Found on *Solidago* (goldenrod) and associated plants (Blatchley 1920; W. T. Davis 1923).

Life history. In New York, mature females occur from end of July to beginning of October, according to W. T. Davis (1923), who discovered no male.

Diapheromera femorata (Say)

Figs. 110, 111, 114; Map 6

Spectrum femoratum Say, 1824*b*:297.
Diapheromera Sayi Gray, 1835:18.
Diapheromera femorata; T. W. Harris 1841:119; Cantrall 1943:75; Helfer 1963:23; Vickery et al. 1974:130, fig. 5.

Common name. Walkingstick.

Diagnosis. Body long, slender. Hind femur with apical ventral spine. Middle femur of male banded green and brown. Female cerci less than half as long as terminal abdominal segment.

Description. General appearance as in Fig. 110. Body length 55–68 mm, occasional females longer; antennae slender, nearly as long as body. Cerci of male with short hairs, with cylindrical to oval-flattened apex, with blunt basal tooth or tubercle (Fig. 111), strongly incurved, usually crossing at or near middle. Cerci of female (Fig. 114) straight, stout, blunt, pubescent. Apex of ninth abdominal segment with shallow triangular emargination, with apex acute. Color variable, green, brown, or greenish brown; males usually brownish; both sexes brownish late in season; head yellowish with three vague fuscous stripes; in dark males middle femora with pale bands.

Range. Manitoba to Québec, south to Arizona and Florida.

Behavior and habitats. *D. femorata* is seldom found except where its principal host plants, several species of oak trees, occur. Black oak (*Quercus velutina* Lam.), northern red oak (*Q. rubra* L.), white oak (*Q. alba* L.), and others may serve as hosts. Small nymphs often feed upon understory plants as well as white oak. Their numbers on these hosts become progressively less as the nymphs move into the oak trees. Giese and Knauer (1977) found nymphal mortality to be very high when black oak was the only food supplied, although this species is heavily favored by adults. Pubescence on the oak leaves is thought to be a barrier to feeding by the early-instar nymphs. Occasionally other genera of trees may support this species (C. E. Rogers

1975), and, at one time, a lone *Robinia* (locust tree) on the grounds of McGill University, Montréal, was one of the few places in Québec where the species was known. (The tree has now gone, and so, from that locality, has *D. femorata*.)

Population density in tree crowns increases as the season progresses, but crowding and wind dislodge many individuals, which then climb back to the crowns, producing a vertical cycling of the population (Giese and Knauer 1977). *D. femorata* exhibits a well-defined feeding period from 21:00 to 03:00 hours, but feeding activity is less at lower than at higher temperatures (Giese and Knauer 1977). *D. femorata* may become numerous in areas where oak becomes predominant following logging and burning and subsequent fire prevention.

Life history. Giese and Knauer (1977) gave a detailed analysis of the life history of *D. femorata*. Eggs are deposited early in September, mainly between 15:00 and 03:00 hours with a peak at 18:00 hours. Eggs fall into the leaf litter beneath the trees and winter there, hatching during mid-June to mid-July. The number of nymphal instars is usually five, but some males require only four and some females do not mature until after the sixth instar. Maturity is reached in August and early September.

Economic importance. *D. femorata* can cause severe defoliation of its host trees (C. E. Rogers 1975) and such injury has been reported by Williams (1906) and by Brown (1940) in southern Ontario. Giese and Knauer (1977) stated that severe defoliation by this insect significantly reduced radial growth of black oak.

Diapheromera velii velii Walsh

Figs. 112, 115; Map 6

Diapheromera velii Walsh, 1864:410.
Diapheromera velii velii; Hebard 1934*a*:162.

Diagnosis. Body long, slender. Hind femur with ventral spine; middle femur of male not banded. Female cercus nearly as long as terminal abdominal segment.

Description. Middle femur of male not banded. Male cercus with slender curved spine on inner side near base (Fig. 112). Female with cerci a little shorter than in male, but nearly as long as terminal abdominal segment and with deep broadly rounded emargination at apex of ninth abdominal segment (Fig. 115). Spines of middle and hind femora small.

Range. Colorado to Minnesota, south to Texas.

Behavior and habitats. Unlike *D. femorata*, this species is not arboreal but is found in clumps of prairie grasses. It appears to prefer *Andropogon*.

Life history. Probably similar to that of *D. femorata*. Adults appear earlier than those of other species of *Diapheromera*. Froeschner (1954) reported adults from 9 July to 19 September in Iowa, and Hebard (1932*a*) recorded adults from 28 June to 27 August in Minnesota.

Order Notoptera—rock-crawlers

The placing of this order has presented considerable difficulty since the time of the discovery of its best-known species, *Grylloblatta campodeiformis*, by E. M. Walker and the description of a new orthopteroid family, Grylloblattidae, to contain it. Recent work on these insects, however, particularly that of Kamp (1973*a*, 1973*b*), Matsuda (1976), Pritchard and Scholefield (1978), and J. S. Edwards (unpublished)[1] strongly suggests that their affinities are more with the Dermaptera than with any other living order. Matsuda believes, probably correctly, that they may be neotenic (i.e., carrying juvenile characters into the adult stage). We tentatively regard the living forms as suborder Grylloblattodea, with a single family characterized by the features indicated here.

Family Grylloblattidae

The few known species of this family, which have been given the vernacular name of rock-crawlers, may be characterized as follows: Ulonata of modest size (adults about 1.5–2.5 cm long). Body elongate, subcylindrical but slightly flattened. Head short, rather flat, prognathous (although jaws often held pointing downward), with all principal sutures well-defined. Eyes small, reduced, or absent; ocelli lacking. Antennae moderately long, filiform, with about 28–40 clearly defined segments. Thorax with all three segments independent, similar, more or less subquadrate, and somewhat flattened. Legs similar, rather short, cursorial, with large coxae and 5-segmented tarsi; latter with more or less divided pulvilli; pretarsi without arolia between claws. Wings entirely lacking. Abdomen with 10 clearly defined segments, ending in supra-anal plate or epiproct, and with pair of well-defined paraprocts; cerci long, flexible, 8-segmented. Females with eighth sternum somewhat reduced and with pair of long (anterior) ovipositor valves; ninth sternum much reduced and bearing another two pairs of long ovipositor valves. Males with ninth sternum enlarged, bearing pair of enlarged asymmetrical coxites, each ending in small style; phallic structures complex, asymmetrical.

These insects are often regarded as being relict "living fossils." More than a score of living species are known, although a few of these still remain undescribed. They are placed in three genera, the majority belonging to *Grylloblatta* E. M. Walker, which occurs in the mountainous regions of western North America from California to the southern border of Yukon Territory (and possibly farther north).

The importance of these insects in the study of Pleistocene zoogeography is considerable (Kamp 1963, 1979). Associated with this is the unusual physiological adaptation of these rock-crawlers to low temperatures. There is, in fact, a "popular" belief among entomologists (and possibly prospectors who *may* have called them "ice-worms") that they live only among the ice

[1] Dr. J. S. Edwards of the University of Washington, Seattle (*in litt.*, 27.IX.1979) writes: ". . .our ultrastructural and neuroanatomical studies on *Grylloblatta* gives a picture almost indistinguishable from Dermaptera. . ."

and snow on high mountains. Although some are found at elevations of over 2000 m, and do, in fact, forage on ice or snow, they are, nevertheless, killed by undue exposure to temperatures below freezing (Kamp 1979). Conversely, whereas they may be found on talus slopes at fairly low elevations (down to 386 m, or less), they are intolerant of even moderately warm temperatures, from which they escape by descending deep into crevices, emerging only at night to feed. Their temperature tolerance range is, in fact, very limited, which makes them difficult to keep alive in the laboratory. There is a valuable recent study by Morrissey and Edwards (1979) on the neurophysiological effects of temperature on an undescribed species of *Grylloblatta* from Mt. Rainier, Washington.

Rock-crawlers occur mainly under stones or in underground tunnels, when not foraging at night. At moderate elevations they may be found inhabiting old damp logs (Kevan 1962*b*) (e.g., of Douglas fir) with cuboid rot. Some species occur in shallow mountain caves (Kamp 1970). Their food consists mainly of other insects, both alive and dead (Pritchard and Scholefield 1978). Their eggs are dark in color and deposited singly in soil crevices or in moss. The life history, at least where this has been studied, seems to be prolonged (Ando 1979). Eggs take about a year to develop, and the seven or eight nymphal stages may take as long as 5 years to attain adulthood. Females do not, apparently, lay eggs for about a year thereafter.

On a world basis, the systematic reviews of Gurney (1948, 1961) may be referred to, although several species have been described since his papers were published. Dr. J. W. Kamp is preparing a revised key to all the known North American forms. Ando (1979) may also be consulted.

Genus *Grylloblatta* E. M. Walker

Description. Body elongate, slender, depressed (Fig. 78). Head flattened, with epicranial suture evident. Antennae shorter than body, comprised of 26–37 segments. Eyes small, widely separated, without ocelli. Thorax depressed, decreasing in width posteriorly. Wings absent. Terga and sterna of abdomen not overlapping but separated laterally by pleural membrane. Cerci long, about same length as hind tibia, cylindrical. Legs slender except for large fore coxae; tibia with 2 spurs, inner and outer. Male external genitalia markedly asymmetrical (Fig. 80). Ovipositor exserted (Figs. 78, 79), in form resembling that of the tettigonioid Grylloptera.

Key to species and subspecies of *Grylloblatta*

1. Antennae of adults with more than 30 segments. Southwestern British Columbia and northern Washington 2
   Antennae of adults with less than 30 segments. Northeast British Columbia, western Alberta, and western Montana 3

2(1). Antennae of adults with 36 or 37 segments. Whistler Mountain, Wedge Peak, Mt. Garibaldi, and Garibaldi Provincial Park, B.C. .... *scudderi* **Kamp** (p. 62)

	Antennae of adults with 30–32 segments. Northern Washington near British Columbia border, Mt. Baker, and Hanagen Peak .... ***occidentalis* Silvestri** (p. 64)
3(1).	Cassiar Range, northern British Columbia, near Yukon border 4
	Mountainous areas of southeastern British Columbia, southwestern Alberta, northern Idaho, and western Montana .... ***campodeiformis campodeiformis* E. M. Walker** (p. 64)
4(3).	Pronotum long (2.80 mm male, 3.05 mm female), with margins converging posteriorly. Cerci long (4.80 mm male, 5.00 mm female). Mt. McDane, above 1525 m in Cassiar Range .... ***campodeiformis nahanni* Kamp** (p. 65)
	Pronotum shorter (2.59 mm male, 2.72 mm female), nearly square in male, with lateral margins only slightly converging. Cerci shorter (4.50 mm male, 4.40 mm female). Mt. St. Paul, at 1500 m in Stone Mountain Provincial Park .... ***campodeiformis athapaska* Kamp** (p. 66)

Clé des espèces et des sous-espèces de *Grylloblatta*

1.	Antennes des adultes comptant plus de 30 segments. Sud-ouest de la Colombie-Britannique et nord de l'État de Washington 2
	Antennes des adultes comptant moins de 30 segments. Nord-est de la Colombie-Britannique, ouest de l'Alberta et ouest du Montana 3
2(1).	Antennes des adultes comptant 36 ou 37 segments. Mont Whistler, pic Wedge, mont Garibaldi et parc provincial Garibaldi (C.-B.) .... ***scudderi* Kamp** (p. 62)
	Antennes des adultes comptant de 30 à 32 segments. Nord de l'État de Washington, près de la frontière de la Colombie-Britannique, mont Baker et pic Hanagen .... ***occidentalis* Silvestri** (p. 64)
3(1).	Chaîne des Cassiars, nord de la Colombie-Britannique, près de la frontière du Yukon 4
	Régions montagneuses du sud-est de la Colombie-Britannique, sud-ouest de l'Alberta, nord de l'Idaho et ouest du Montana .... ***campodeiformis campodeiformis* E.M. Walker** (p. 64)
4(3).	Pronotum long (2,80 mm chez les mâles, 3,05 mm chez les femelles), à bords convergeant vers l'arrière. Cerques longs (4,80 mm chez les mâles, 5,00 mm chez les femelles). Mont McDame, altitudes supérieures à 1 525 m dans la chaîne des Cassiars .... ***campodeiformis nahanni* Kamp** (p. 65)
	Pronotum plus court (2,59 mm chez les mâles, 2,72 mm chez les femelles), presque carré chez les mâles, les bords latéraux ne convergeant que légèrement. Cerques plus courts (4,50 mm chez les mâles, 4,40 mm chez les femelles). Mont Saint-Paul, altitudes de 1 500 m dans le parc provincial de Stone Mountain .... ***campodeiformis athapaska* Kamp** (p. 66)

Grylloblatta scudderi Kamp

Fig. 81; Map 7

Grylloblatta scudderi Kamp, 1979:30.

Diagnosis. Adults with 36- or 37-segmented antennae.

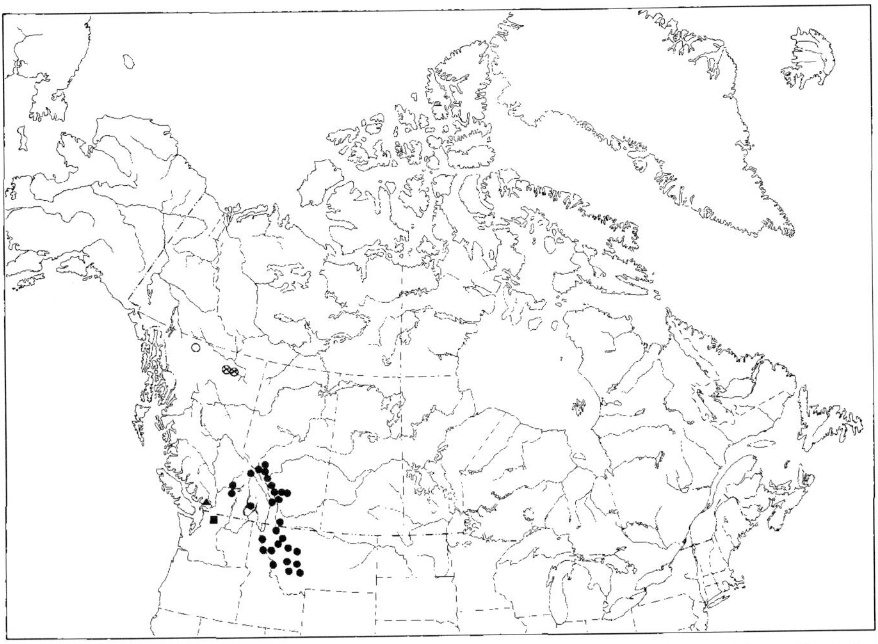

Map 7. Collection localities for *Grylloblatta scudderi* (▲), *G. occidentalis* (■), *G. campodeiformis campodeiformis* (●), *G. campodeiformis nahanni* (○), and *G. campodeiformis athapaska* (⊗).

Description. Male densely clothed with fine tan pubescence. Antennal segments 35–37 (holotype with 35 left, 37 right; allotype with 36 left and right). Pronotum with lateral margins slightly convergent and convex (male, length 2.67, width 2.22; female, length 2.90, width 2.32 mm). Legs long (male, fore tibia 2.82, fore femur 3.15, hind tibia 4.75, hind femur 3.82 mm; female, fore tibia 2.80, fore femur 3.15, hind tibia 4.85, hind femur 4.40 mm [note: hind femur length was given incorrectly by Kamp (1979) as 0.40 mm]. Ventral ovipositor valves moderately curved throughout. The genitalia are illustrated by Kamp (1979).

Range. Known only from the type locality.

Behavior and habitats. The "types were collected on a large persistent snowfield in a cirque approximately 1000 m below a glacier." In this area, cold air from higher elevations flowed off the glacier and was funneled into the cirque and over the snowfield. At the time of collection (13 July) the air temperature was 5°C. Nine immature specimens were found in rocky areas near the snowfield (Kamp 1979).

Life history. Not studied, but the life cycle is probably of several years duration.

Grylloblatta occidentalis Silvestri

Map 7

Grylloblatta campodeiformis var. *occidentalis* Silvestri, 1931:293.
Grylloblatta campodeiformis occidentalis; Helfer 1963:61.

Diagnosis. Antennae of adults with 30–32 segments.

Description. *G. occidentalis* was described from immature specimens. Little descriptive information is available and we have no adult specimen at the present time. Gurney (1953*b*:325) lists the following information in a key: dorsal valve of ovipositor reaching to middle of cercus, or at least to apical half of fifth segment. Stylus of male about four times as long as wide. Antennae of adult with more than 30 segments. Gurney (op. cit.) also apparently included the species now known as *G. scudderi* (see previous species) under *occidentalis*.

Range. Known only from the type locality, Mt. Baker, WA.

Behavior and habitats. Found under stones at base of hill near deep snow, and among stones along a roadside bank at lower elevations (ca. 1280 m).

Life history. Unknown.

Grylloblatta campodeiformis campodeiformis E. M. Walker

Figs. 78–80, 82; Map 7

Grylloblatta campodeiformis E. M. Walker, 1914*a*:94, pl. 6, figs. 1–7.
Grylloblatta campodeiformis; Crampton 1915:337–350; Helfer 1963:61.
Grylloblatta campodeiformis campodeiformis; Kamp 1979:27, 32–37, figs. 19–27.

Diagnosis. Body small, pale. Antennae with less than 30 segments. Occurring in mountainous areas of western Alberta, southeastern British Columbia, Montana, and Idaho.

Description. Body of female (including ovipositor) 30–34 mm, covered with fine pubescence (Figs. 78, 79). Eyes small, each with 70 slightly irregular facets. Antennae filiform, with 28 or 29 segments; basal segment much larger than others. Pronotum feebly convex, longer than broad, with anterior margin arcuate and with posterior margin truncate. Abdomen widening from base to fifth segment, which is wider than head, narrowing drastically in three apical segments. Cerci 8-segmented, with segments progressively longer toward apex. Ovipositor (ca. 3.0–3.3 mm) shorter than hind femur (Fig. 78). Supra-anal plate of male bluntly subtriangular, slightly wider than long (Fig. 80*a*); male subgenital plate bearing 2 hairy styli, scoop-shaped, enlarged, asymmetrical (Fig. 80*b*).

Range. British Columbia, Alberta, Idaho, and Montana.

Behavior and habitats. The optimum temperature for *G. campodeiformis campodeiformis* is 3.7°C (Mills and Pepper 1937). A temperature range of about 10 to −6°C is tolerated, but the insects cannot withstand temperatures outside this range, except for short periods, and they are susceptible to rapid temperature changes. The natural habitat is in moss, decaying wood, or damp crevices between stones. High humidity is necessary, close to 100% RH (Mills and Pepper 1937; Edwards and Nutting 1950). *G. campodeiformis campodeiformis* is usually found near the timberline on mountains, near margins of glacial bogs, and sometimes 1 m, or more, deep in rock scree. A population of *Grylloblatta* that occurs at low elevations and under dry conditions near Kamloops, B.C., and which has been referred to as *G. campodeiformis* by Gregson (1938, 1939) and Campbell (1949), probably represents a separate species.

Many authors regard members of this species as carnivorous, although E. M. Walker (1937) and Campbell (1949) mentioned that they also "nibble moss" or eat "decaying vegetable matter." Pritchard and Scholefield (1978) found gut contents to consist almost entirely of arthropods in adults, and arthropods and occasional vascular plant tissue in immatures. Collembola, Microcoryphia, oribatid mites, adult Diptera, aphids, and staphylinid beetles were all found as gut contents. However, the primary food source appears to be adults of the wingless cranefly, *Chionea obtusa* Byers. These insects live in the same habitats as *Grylloblatta*, and are found in late fall and winter. The living prey are usually smaller than the hunters. *G. campodeiformis campodeiformis* can exist for long periods without food. Small and medium-sized immature individuals have been kept alive without food for 18 months with no growth. Adults lose weight if not fed at least every 3 weeks. N. Ford (1926) gave an interesting account of the behavior of this species. E. M. Walker (1937) reported on mating in captivity. Although the sexes may live together for months, a female may suddenly decapitate or eat the male.

Life history. The total cycle has been calculated to take about 7 years (Henson 1957). Females become sexually mature, mate, and deposit eggs about 1 year after reaching the adult stage. Immatures take at least 5 years to grow to the adult stage. There are eight or so nymphal instars, three during the 1st year, four at 1-year intervals, and the last about 6 weeks after the preceding one (E. M. Walker 1937). Nymphs not only lack fully developed genitalic structures but have fewer segments to the antennae and cerci than do adults—see E. M. Walker (1919*a*, 1919*b*) and Fig. 82.

Grylloblatta campodeiformis nahanni Kamp

Map 7

Grylloblatta campodeiformis nahanni Kamp, 1979:28.

Diagnosis. Pronotum narrowing posteriorly. Known only from type locality, Mt. McDane, Cassier Range, or its vicinity.

Description. Legs long (male, fore tibia 2.40, fore femur 2.70, hind tibia 4.25, hind femur 3.43; female, fore tibia 2.80, fore femur 3.11, hind tibia 4.43, hind femur 4.00 mm). Pronotum wide in proportion to length (male 2.80 mm long, 2.48 mm wide; female 3.05 mm long, 2.70 mm wide). Pubescence on body conspicuous. Kamp (1979) illustrates genitalia.

Range. Known from type locality and from nearby Limestone Peak.

Behavior and habitats. Similar to *G. campodeiformis campodeiformis.*

Life history. Probably similar to *G. campodeiformis campodeiformis.*

Grylloblatta campodeiformis athapaska Kamp

Map 7

Grylloblatta campodeiformis athapaska Kamp, 1979:27.

Diagnosis. Pronotum nearly square in male. Known only from Mt. St. Paul, Stone Mountain Provincial Park, northern British Columbia.

Description. As for *G. campodeiformis*. Pronotum short, square, (male 2.59 mm long, 2.60 mm wide; female 2.72 mm long, 2.45 mm wide). Leg length (male, fore tibia 2.50, fore femur 2.73, hind tibia 4.00, hind femur 3.71; female, fore tibia 2.20, fore femur 2.89, hind tibia 3.00, hind femur 3.60 mm). Anterior margin of male lobiform clasper process smoothly curved, not obtuse. Ovipositor with ventral valves markedly curved in apical third; female cerci short, no longer than ovipositor, 6-segmented. Kamp (1979) illustrates the genitalia.

Range. Known only from Mt. St. Paul, Stone Mountain Provincial Park, northern British Columbia.

Behavior and habitats. Similar to those of *G. campodeiformis campodeiformis.*

Life history. Probably similar to that of *G. campodeiformis campodeiformis.*

Order Dictuoptera—termites, mantids, and cockroaches

The order Dictuoptera includes the "Dictyoptera" of many recent works (i.e., the cockroaches and mantids) together with the Isoptera, or termites, which, since they were removed from the old composite "Neuroptera" have long been recognized as a separate order. It is often argued that the general structure, and especially the social organization, of termites is sufficient to set them apart, but, in fact, when the most primitive termites, represented by the Australian genus *Mastotermes* and certain fossil forms (including North American species), are considered, basic anatomical differences from some cockroaches are slight (McKittrick 1965), and the level of social organization is equivalent (*Mastotermes* is only slightly more advanced than certain semisocial cockroaches). Certain tropical cockroaches, e.g., *Cardacopsis* spp., have virtually isopterous wings, the hind pair of which lack anal lobes. In addition, the characteristic manner of laying eggs in a package (ootheca), even if constructed of different material, is retained by *Mastotermes* (whereas in many ovoviviparous cockroaches it is not), and the peculiar flagellate protozoan gut fauna found in the same genus, and in Kalotermitidae, is remarkably similar to that of the wood-eating cockroach genus *Cryptocercus*. Boudreaux (1979), however, maintains that the Isoptera should continue to be recognized at the ordinal level and presents arguments in favor of this position, suggesting that the intestinal protozoa have been acquired from termites by the cockroaches in the remote past by virtue of their wood-eating behavior.

It has also been common practice, particularly in Continental Europe, but also among some North American authors, including Boudreaux (1979), to treat the mantids as a distinct order, but there is even less justification for this than for giving separate ordinal status to the termites. It is true that the raptorial front legs, the elongate rather than disclike pronotum, and the form of the tegmina of many mantids give them a different aspect from typical cockroaches, but, when certain short-bodied, primitive mantids with subquadrate pronota are compared with slenderly built cockroaches, the differences lose much of their significance. The primitive Neotropical mantid genus *Chaeteessa* is very cockroachlike. The fact that all mantids are predaceous, whereas cockroaches are not, does not warrant separate ordinal status.

We may now consider the general characteristics of the Dictuoptera as a whole. These insects are Ulonata with the following combination of characters: size small (a few millimetres long) to large (body length up to 16 cm and wingspan up to 18.5 cm, although not in the same insect); head either hypognathous or (in termites) prognathous; antennae long, whiskerlike, and annulated rather than distinctly segmented, or (termites and some mantids) comparatively short and multiarticulate; pronotum of diverse form, but not elevated; legs all similar or (mantids) front pair raptorial; coxae large and broad and rather close together; hind femora not thickened at their bases or modified for jumping; tarsi with 5 segments or 4 (in most termites); wings, if present, superposed more or less horizontally over the dorsum when at rest (somewhat bent downward along the sides of the abdomen, as in most mantids), typically reticulate or with numerous crossveins (except in a few cockroaches and many termites); fore wings, when developed, forming leathery tegmina (except in some mantids and cockroaches and in all termites), considerably overlapping when at rest; hind wings, when developed, broad and with large anal lobe (except in a few cockroaches and in nearly all termites); hind wing buds of immature stages (when present) not overlapping fore wing buds during development; cerci typically rather long and with numerous segments, although short to very short and with few segments in the majority of termites and in some cockroaches; ovipositor present but reduced and usually concealed by enlarged seventh abdominal sternum, or (most termites) not evident; ninth abdominal sternum of male with a pair of styles (in termites often inconspicuous but sometimes present in females and often in asexual forms and nymphs; in some cockroaches one or both reduced) and more or less concealing the copulatory structures, which are complex and asymmetrical (reduced or virtually lacking in most termites). Auditory and special stridulatory organs are not generally developed, although many species are capable of producing sounds—usually by striking the substrate or rustling the wings.

The number of known living species is about 7500, although only about a score inhabit the region here considered. They occur throughout most of the world, but they are predominantly tropical and subtropical, being infrequent at higher latitudes and at altitudes subject to lower temperatures. A few species of cockroaches, however, do reach such regions, even out-of-doors, and a number establish themselves readily there under artificial conditions. Within their general range of temperature tolerance, some Dictuoptera will be found in almost any terrestrial environment, from subterranean burrows to treetops, and from dense forest to savanna, prairie, and desert, although subaquatic forms (some cockroaches only) are infrequent. As the three suborders Blattodea, Mantodea, and Termitodea differ markedly in their biology, generalizations regarding this will be left until the suborders are referred to individually.

The economic importance of a few species of Dictuoptera that have become associated with human activities is great, but the vast majority of species are of little or no direct practical significance to man. Nevertheless, the importance of their roles in the general ecosystem may be great, differing according to the group concerned. Further comment will be made under the three individual suborders.

Key to suborders of Dictuoptera

1\. Body small, not distinctly flattened or elongate, usually rather pale. Head prognathous. Antennae multiarticulate, not whiskerlike or bristlelike. Pronotum not disclike or elongate. Legs similar, short, not markedly spinose. Wings, when present, long in relation to body length; fore wings not leathery; hind wings similar in shape to fore wings, without anal lobe. Species social, living in wood or soil, with winged forms feebly migrant .... **Termitodea** (p. 69)

Without the above combination of characters 2

2(1). Body elongate. Pronotum much longer than wide, not concealing head from above. Front legs raptorial; middle and hind legs slender .... **Mantodea** (p. 83)

Body broad, dorsoventrally flattened. Pronotum disclike, largely concealing head from above. Front legs not raptorial; middle and hind legs often stout .... **Blattodea** (p. 91)

Clé des sous-ordres de Dictuoptera

1\. Corps petit, ni aplati ni allongé, de couleur habituellement plutôt pâle. Tête prognathe. Antennes multiarticulées, ni filiformes ni sétiformes. Pronotum ni arrondi ni allongé. Pattes semblables, courtes, non clairement spinervées. Ailes, lorsque présentes, longues comparativement au corps; ailes antérieures non tannées; ailes postérieures de forme semblable aux ailes antérieures, sans lobe anal. Insectes sociaux, vivant dans le bois ou dans le sol, à formes ailées légèrement migratrices .... **Termitodea** (p. 69)

Ne présente pas la combinaison de caractères donnée ci-dessus 2

2(1). Corps allongé. Pronotum beaucoup plus long que large, ne surplombant pas la tête. Pattes antérieures ravisseuses; pattes médianes et postérieures grêles .... **Mantodea** (p. 83)

Corps large, aplati dorsoventralement. Pronotum arrondi, surplombant la tête. Pattes antérieures non ravisseuses; pattes médianes et postérieures souvent fortes .... **Blattodea** (p. 91)

Suborder Termitodea—termites

The suborder Termitodea comprises the termites, or white ants. As previously noted in the discussion of the order Dictuoptera, these have generally been regarded as constituting a separate order, Isoptera, and there are many who still adhere to this position. Most termites, indeed, are of different appearance and behavior from most other Dictuoptera (particularly the few that concern us here). They may be characterized as follows:

Dictuoptera of small size (except for the exaggeratedly swollen, gravid queens of some species), dimorphic or (most species) polymorphic (including sexual, soldier, and worker castes), found in colonies (except for sexual forms at time of migration, mating, and colony foundation). Body subcylindrical or slightly flattened, usually weakly sclerotized, pale (head and thorax often, rest of body seldom, more sclerotized and darker), with pigmentation limited to various shades of yellowish to brownish or blackish; head sometimes extremely heavily sclerotized or drawn out into conical process (some soldiers of exotic species). Antennae short, distinctly segmented, not greatly enlarged. Pronotum more or less quadrate or rectangular in dorsal view, not concealing head. Legs short, similar in form, without pronounced spines; tarsi typically 4-segmented, with basal segment in some indistinctly subdivided, with arolia between claws (two families). Wings developed only in primary "sexual" forms (and then only temporarily present, being shed after nuptial migratory flight), much longer than body, similar in shape, venation, and texture, with front pair not forming thickened tegmina (although the anterior veins are usually strongly developed, giving rigidity); wing membranes densely reticulate or not; hind wings narrow, without anal lobes. Abdominal cerci varying from long, 8-segmented, to short, tuberclelike, 2-segmented, or unsegmented. Females with seventh abdominal sternum much enlarged to form subgenital plate covering succeeding sterna but ovipositor usually more or less vestigal. Males with eighth and ninth abdominal sterna forming subgenital plate, which is often longitudinally divided; phallic structures usually reduced, membranous.

There are nearly 2000 species of termites in the world, distributed among 191 genera in 6 or 7 families. Most of them belong to a single superfamily, Termitoidea.

Termites are predominantly tropical, being poorly represented even in warmer temperate climates. In latitudes similar to those covered by the present work, very few species are found, and some of those mainly under artificial conditions. Termites are all social insects, but the complexity of their social structure, like that of their habitations, varies widely. Even in the most primitive termites, there are at least two castes of individuals (which, unlike those of Hymenoptera, may be of either sex). These are the primary reproductive forms (sexuales), which are fully winged in the adult stage, and the sterile forms (soldiers), which are wingless with enlarged mandibles or other specialized head features, notably in some "nasutiforms," which have a forward prolongation of the head by means of which a sticky poisonous substance is ejected.

Few genera of termites lack a soldier caste; many species have soldiers of more than one kind. In most species of termite there is a third caste (of which there may also be more than one kind). This is the worker caste, and when workers occur, they are the most numerous individuals in a colony. Soldierless species have workers; thus there are two castes here also. Strictly speaking, a worker caste is not found in the Mastotermitoidea or in the most primitive Termitoidea (the Termopsidae). Nor do they occur in the Kalotermitidae, in which their function is carried out by nymphal stages and "pseudergates." Pseudergates are wingless "subadults" that normally do not molt again, but which, under special circumstances, must do so before becoming capable of reproduction.

The most primitive form of termite habitation merely consists of galleries chewed through wood. The Kalotermitidae are "powder-post" termites, which attack sound wood and which maintain no direct contact with the soil. The most primitive of the Termitoidea (the Termopsidae) are the dampwood termites, and these live in moist decaying timber (usually in contact with the ground). These wood-feeding termites all have flagellate protozoan intestinal microfaunas, which enable them to digest cellulose. The great majority of termites, although they also consume wood and other products containing cellulose, have a bacterial intestinal microflora to achieve digestion. They also have more complex habitations than the wood-feeding species, and they are known either as mound-building termites (which are absent from the region here covered) or subterranean termites. This is not a very appropriate distinction (unless one is aware of what is meant by the terms), for all mound-building termites have a large part of their habitations underground and some build "nests" in trees and not mounds—or even hills or towers. The so-called subterranean termites do not, indeed, build massive structures aboveground, but they do construct covered galleries by means of which (like the mound-building termites) they traverse regions through which they cannot excavate. They also tunnel extensively in wood aboveground, often ascending to considerable heights in doing so.

In dull weather, or in deep forest, termites have been observed to march across open spaces in columns, but they rarely do so, and it is usually only the more strongly sclerotized kinds that proceed thus. Nearly all termites shun the light at all times even when flying, for nuptial flights mainly occur at night. A number of species do, however, make these flights by day. The habitations of termites almost always contain "guest" arthropods or inquilines, particularly various specialized beetles and flies, as well as representatives of many other groups. Some termite species have symbiotic relations with, and live in the habitations of, others, notably those few that lack a soldier caste. Some of the more advanced species of termites cultivate fungus gardens somewhat similar to those tended by certain kinds of ants.

The economic importance of many species of termites is great, particularly in respect of their ravaging structural timbers, but almost anything containing cellulose may be consumed if it is in contact with the ground, or sufficiently close to it for the termites to reach it by means of their galleries. Wood-feeding termites do not even need this provision as they can, and often do, establish themselves in timber aboveground. On the whole, termites are not usually thought of as major pests of growing plants, but far more damage of this kind is done than is generally believed, particularly to shrubs and trees, and also to such plants as sugarcane and pasture grasses.

In tropical and many subtropical countries, the role of termites in the ecology of the soil is great, particularly on account of the remarkable rapidity with which they consume wood. They tend to lock up the by-products of decomposition in such relatively small areas that the general fertility of the soil is not enhanced by their activity. However, their subterranean excavations may bring up enough minerals from the soil for the surface soil to become more fertile in their vicinity. Certain observant prospectors have also profited by geological information thus made available! As termites are not abundant in the region here covered, we refer the reader elsewhere for further information concerning them (see, for example, Hegh (1922), Kofoid (1934), Snyder (1935), Schmidt (1955), W. V. Harris (1961, 1971), United Nations (1962), Krishna and Weesner (1969, 1970), Howse (1970), Weidner (1970), and Lee and Wood (1971).

Key to families of Termitodea

1. Fontanelle absent (Figs. 54, 56–58). Wings with 3, or more, heavy oblique veins in anterior field (Figs. 51, 52). Soldiers with toothed jaws (Figs. 55, 59, 61, 64). True worker caste lacking 2

Fontanelle present in sexual forms (Figs. 65, 69, 70). Wings without anterior oblique veins (Fig. 53). Soldiers with virtually untoothed (even if serrated) jaws (Figs. 66, 68). True worker caste developed .... **Rhinotermitidae** (p. 72)

2(1). Cerci long (Figs. 54, 56–59), 3–8-segmented. Antennae usually with more than 21 segments. Sexuales without ocelli. Tarsi with basal segment partly divided below so that they are apparently 5-segmented from some angles; claws with an arolium. Pseudergates with left mandible possessing 3 marginal teeth .... **Termopsidae** (p. 77)

Cerci shorter (Fig. 63), 2-segmented. Antennae usually with fewer than 21 segments. Sexuales with pair of ocelli. Tarsi 4-segmented, with basal segments not divided; claws often without arolium. Pseudergates with left mandible possessing 2 marginal teeth ...... **Kalotermitidae** (p. 81)

Clé des familles de Termitodea

1. Fontanelle absente (fig. 54, 56 à 58). Ailes comptant trois nervures obliques épaisses ou plus dans la région costale (fig. 51 et 52). Chez les soldats, mandibules dentées (fig. 55, 59, 61 et 64). Pas de véritable caste d'ouvriers 2

Fontanelle présente chez les formes sexuées (fig. 65, 69 et 70). Pas de nervure oblique dans la région costale de l'aile (fig. 53). Chez les soldats, mandibules non dentées mais parfois denticulées (fig. 66 et 68). Existence d'une véritable caste d'ouvriers .... **Rhinotermitidae** (p. 72)

2(1). Cerques longs (fig. 54, 56 à 59), comptant de 3 à 8 segments. Antennes comptant habituellement plus de 21 segments. Formes sexuées sans ocelles. Tarses à segment basal partiellement divisé en-dessous, de sorte que vus d'un certain angle, ils semblent compter 5 segments; arolium présent. Chez les Pseudergates, mandibule gauche à bord muni de 3 dents marginales .... **Termopsidae** (p. 77)

Cerques plus courts (fig. 63) à 2 segments. Antennes comptant habituellement moins de 21 segments. Formes sexuées pourvues d'une paire d'ocelles. Tarses à 4 segments, à segment basal non divisé; arolium souvent absent. Chez les Pseudergates, mandibule gauche à bord muni de deux dents .... **Kalotermitidae** (p. 81)

Family Rhinotermitidae

Members of the Rhinotermitidae family are included among the subterranean termites. Their principal distinguishing features have already been given in the key to families. Their general biology is characteristic of smaller subterranean termites in general and may be typified by that of *Reticulitermes hesperus* Banks (Weesner 1956). There are 14 living genera of Rhinotermitidae and about 160 species. Except in the most southerly United States, only two genera, which belong to the subfamily Heterotermitinae, are represented in North America.

Genus *Reticulitermes* Holmgren

Diagnosis and description. Alates (Fig. 65) with wings not possessing hairs. Ocelli small but conspicuous. Antennae with less than 18 segments. Body yellow brown to black, with wings brown, grayish, or whitish (if whitish, veins are gray). Soldiers with short, heavy, bluntly pointed mandibles; mandibles usually shorter than head width (Figs. 67, 68).

Key to species of *Reticulitermes*

(The following key to species is based mainly upon alate sexual forms, for which it should be fairly reliable. For soldiers, it is less reliable, for these are highly variable, save that, in *R. hesperus*, the head is fully twice as long as wide, not about 1.5 times as long, as in other species. Workers are difficult or virtually impossible to tell apart.)

1. Body of alate, and head of soldier, yellowish brown to brown. Eastern species .... *arenicola* **Goellner** (p. 74)
   Body of alate brownish black to black; head of soldier darker 2

2(1). Alate with tibiae slightly darkened. Soldiers with head twice as long as wide (Fig. 68). Western, including coastal regions .... *hesperus* **Banks** (p. 74)
   Alate with tibiae either distinctly pale (eastern) or distinctly dark. Soldiers with head about 1.5 times as long as wide (Fig. 66). Eastern or western, but in west not coastal 3

3(2). Body of alate dark. Tibiae black (Fig. 65). Soldiers generally with gula less than twice as broad in front as in middle (Fig. 67). Eastern and western .... *tibialis* **Banks** (p. 76)
   Body of alate blackish brown to black. Tibiae pale (Fig. 69). Soldiers generally with gula fully twice as broad in front as in middle (Fig. 71). Eastern 4

4(3). Alate with wings almost colorless (anterior field may be slightly infumated). Overall length (including folded wings) about 8 mm (wings less than 6 mm). Ocelli close to eyes (distant by less than diameter of an ocellus). Soldiers usually 4.5–5.0 mm long ..... ***virginicus* (Banks)** (p. 76)

Alate with wings faintly to distinctly grayish or brownish (fore wings darker than hind wings). Overall length about 10 mm (wings usually more than 6 mm). Ocelli less close to eyes (distant by about diameter of an ocellus). Soldiers usually 6–7 mm long .... ***flavipes* (Kollar)** (p. 76)

Clé des espèces de *Réticulitermes*

(La présente clé d'identification des espèces porte principalement sur les formes ailées, pour lesquelles elle devrait être assez fiable. Pour les soldats, qui sont beaucoup plus variables, elle pourra s'avérer moins fiable à une exception près: chez le *R. hesperus*, la tête est nettement deux fois plus longue que large, comparativement à environ 1,5 fois plus longue que large chez les autres espèces. Par ailleurs, les ouvriers sont difficiles ou pratiquement impossibles à identifier.)

1. Corps des formes ailées et têtes des soldats de couleur brun jaunâtre à brun. Espèce orientale .... ***arenicola* Goellner** (p. 74)

Corps des formes ailées de couleur noir brunâtre à noir; tête des soldats plus foncée 2

2(1). Chez les formes ailées, tibias légèrement plus foncés. Chez les soldats, tête deux fois plus longue que large (fig. 68). Espèce vivant dans l'ouest, y compris les régions côtières .... ***hesperus* Banks** (p. 74)

Chez les formes ailées, tibias clairement plus pâles (est) ou nettement plus foncés. Chez les soldats, tête environ une fois et demie plus longue que large (fig. 66). Dans l'est ou dans l'ouest, sauf sur la côte ouest 3

3(2). Corps des formes ailées foncé. Tibias noirs (fig. 65). Chez les soldats, gula en général moins de deux fois plus large en avant qu'au milieu (fig. 67). Dans l'est et dans l'ouest .... ***tibialis* Banks** (p. 76)

Corps des formes ailées brun noirâtre à noir. Tibias pâles (fig. 69). Chez les soldats, gula en général au moins deux fois plus large en avant qu'au milieu (fig. 71). Espèces orientales 4

4(3). Chez les formes ailées, ailes presque incolores (la région costale peut être légèrement enfumée). Longueur totale (y compris les ailes repliées) d'environ 8 mm (ailes de longueur inférieure à 6 mm). Ocelles situés près des yeux (à une distance inférieure à celle de leur propre diamètre). Soldats mesurant d'habitude de 4,5 à 5 mm de long .... ***virginicus* (Banks)** (p. 76)

Chez les formes ailées, ailes légèrement ou nettement grisâtres ou brunâtres (ailes antérieures plus foncées que les ailes postérieures). Longueur totale d'environ 10 mm (ailes de longueur habituellement supérieure à 6 mm). Ocelles moins près des yeux (distance à peu près égale à leur propre diamètre). Soldats atteignant d'habitude de 6 à 7 mm de long .... ***flavipes* (Kollar)** (p. 76)

Reticulitermes arenicola Goellner

Map 8

Reticulitermes arenicola Goellner, 1931:227–234, figs. 1–6.

Diagnosis. Body of alate pale. Head of soldier brown.

Description. Small species. Alate with body yellowish brown to brown. Wings longer than 6 mm, narrow, not exceeding 2 mm in width, white to dusky. Soldier with head yellowish brown to brown, light in color.

Range. Illinois to Massachusetts.

Behavior and habitats. *R. arenicola* prefers dry sandy areas where it infests buried or partly buried pieces of wood. According to Snyder (1934), swarms issue late in May in the Chicago area, but, according to the same author, under the name of *R. lucifuga*, the species swarms in the Boston area a month later than does *R. flavipes*, namely in June and July.

Economic importance. Of minor, if any, significance.

Reticulitermes hesperus Banks

Fig. 68; Map 8

Reticulitermes hesperus Banks in Banks and Snyder, 1920:50.

Common name. Western subterranean termite.

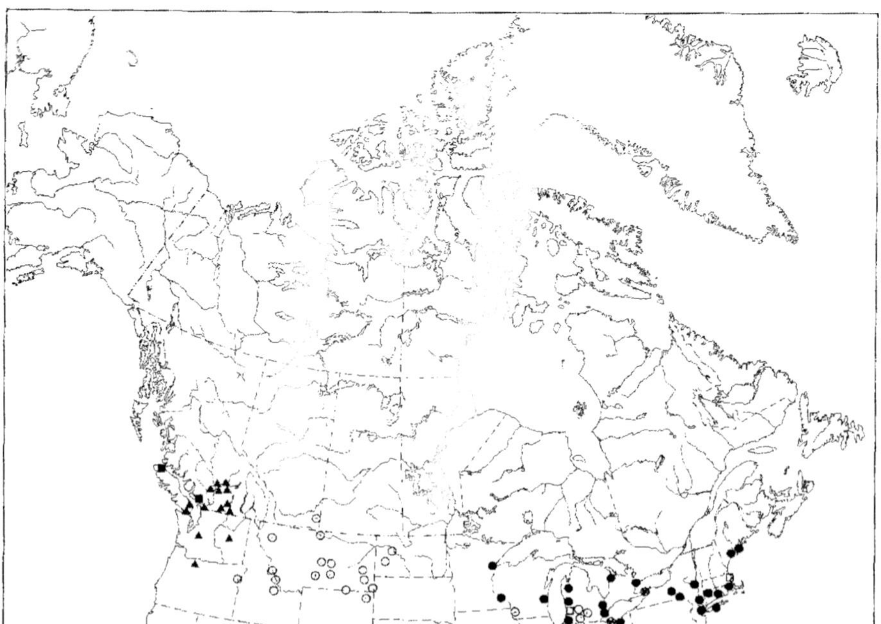

Map 8. Collection localities for *Reticulitermes arenicola* (■), *R. hesperus* (▲), *R. tibialis* (○), *R. flavipes* (●), *Incisitermes minor* (⊗), and *Cryptotermes brevis* (■).

Diagnosis. Alate dark brown. Soldier with long head.

Description. Alate dark brown to brownish black. Wings distinctly brownish gray (when seen against white background). Fore wings about 8.0–9.5 mm long. Tibiae slightly darkened but not black. Antennae and palpi dark. Soldier usually with dark head (yellowish brown to coal black), nearly to more than twice as long as wide (Fig. 68). Pronotum slightly emarginate anteriorly, narrowed posteriorly (Fig. 68). Gula broad anteriorly, more than twice as broad as narrowest part (often even broader than in Fig. 71).

Range. British Columbia south to California.

Behavior and habitats. This species is the common subterranean termite of the Pacific coast area. Although it (and other species of the genus) often burrow into soil, soil is necessary only to provide a source of moisture. *R. hesperus* colonizes wood that is in contact with soil, but can survive without soil contact if food, moisture, and temperature are satisfactory. It is the major economic termite species of the Pacific coastal area, although Spencer (1945) mentioned particularly that it did not appear to build up or to spread to any extent in interior British Columbia. Pickens (1934) noted that colonies consisted of three or four individuals up to several thousand, depending upon the age and location of the colony. Four thousand individuals can exist in a small piece of wood. Supplementary queens are commonly produced, enabling colonies to build up in population much more quickly than would be possible with only the primary queen.

Life history. As with other species, alate sexual forms are produced in large numbers and most of these perish without finding a suitable habitat in which to initiate a new colony. In the south, most swarms occur in late fall or even early winter, but in the northern part of the range spring swarms are usual. Flight is weak and direction is largely determined by prevailing wind, usually covering only 60–200 m. Almost immediately upon settling, an alate termite raises its abdomen, pushing against the wings, which break off at the natural sutures leaving only four small stumps. Females remain with their abdomens raised, emitting a pheromone which attracts males. When a male finds a female, she lowers her abdomen and, with the male following, searches for shelter, preferably in or under wood. The pair excavates an area and plugs all cracks. Copulation does not occur until the home is ready, usually not until the next day. Parental care occurs; the parents care for the eggs and the newly hatched nymphs. Colony growth is slow at first, since the nymphs require many months to become fully grown.

At the Lyman Entomological Museum's laboratory, a colony originating from a few individuals collected by Kevan at Spences Bridge, B.C., 20 April 1963, was maintained for 15 years. The original group did not contain primary reproductives, but secondary reproductives managed to build up a thriving colony.

Economic importance. Under natural conditions, subterranean termites play a valuable role in aerating and mixing soil. They accelerate the breakdown of cellulose. The activities of man in clearing, irrigating, building, and increasing the availability of dead wood (stumps, fallen logs, parts of trees, and so on) have made the environment in many localities much more favorable for these termites than they would be naturally. As urbanization and industrialization increase, so does termite damage. First to be attacked are structures of minor importance, such as poles and fences, then foundations, floors, walls, or any parts of buildings that are in contact with soil.

Effective control uses methods of construction that make infestation nearly impossible. Treatment of such timbers as will be below ground level, use of concrete where soil contact with a structure occurs, and allowing no direct contact between wood and soil, effectively prevent infestation. Control in an infested building usually requires the affected timbers to be replaced and the original problem to be corrected. This is often expensive.

Reticulitermes tibialis Banks

Figs. 65–67; Map 8

Reticulitermes tibialis Banks in Banks and Snyder, 1920:47.

Diagnosis. Alates with ocelli close to eyes. Tibiae black.

Description. Alate black, often shiny. Tibiae black (Fig. 65). Wings 7.5–9.0 mm long, nearly colorless. Ocelli distant from eyes by distance less than diameter of ocellus. Soldiers with dark heads about 1.5 times as long as wide (Fig. 66), with gula usually less than twice as wide in front as in middle (Fig. 67).

Range. Southern Alberta to Michigan, south to Mexico.

Behavior and habitats. There are records of flights of swarms in both spring and fall. In southern areas, *R. tibialis* frequently occurs in hard-baked and alkali soils. In Colorado, it is known to occur at elevations of up to 2100 m.

Economic importance. Although the feeding behavior is similar to that of other termites, the species is not classed as so troublesome as most, owing largely to its preference for arid conditions. It may, however, prove to be more troublesome in buildings in such situations than is generally appreciated.

Reticulitermes virginicus (Banks)

Termes virginicus Banks, 1907:392.
Reticulitermes virginicus; MacNay 1961*b*:206; Helfer 1963:74.

Diagnosis. Ocelli close to eyes. Tibiae yellowish.

Description. Alate brownish black to black; tibiae pale yellowish; wings almost colorless except along anterior margins. Wings less than 6 mm in length. Ocelli distant from eyes by distance less than diameter of ocellus. Soldiers similar to those of *R. flavipes* but smaller, about 4.5–5 mm long.

Range. Native to southeastern United States; established (once) in Ontario and eliminated.

Reticulitermes flavipes (Kollar)

Figs. 69–71; Map 8

Termes flavipes Kollar, 1837:411.
Reticulitermes flavipes; Banks and Snyder 1920:45; Helfer 1963:72.

Common name. Eastern subterranean termite.

Diagnosis. Alate with ocellus distant from eye by distance greater than width of ocellus.

Description. Alate with wings at least 6.5 mm long, usually 7.0–7.5 mm; fore wing membrane distinctly grayish to gray brown (when viewed against white background); hind wing paler. Body brownish to black; tibiae pale yellowish. Ocelli distant from eyes by distance more than width of an ocellus. Soldiers about 6–7 mm long, with dark heads about 1.5 times as long as wide; gula usually twice as wide in front as at middle (Fig. 71).

Range. Minnesota to Maine, south to Texas and Florida.

Behavior and habitats. *R. flavipes* attacks and destroys wood in buildings, and also wood-products such as cardboard and paper, and any wood that has contact with soil. Its natural spread is relatively slow, less than 100 m/year. It appears that most northern infestations have resulted from transport of infested soil or wood from one area to another. Like other species of *Reticulitermes* this species prefers sandy rather than heavier soils. Despite this, it is less tolerant of dry conditions than most other species of the genus. *R. flavipes* builds covered runways from the soil to suitable colony sites and these are often noticed on trunks of shade trees in infested areas such as Toronto.

Colonies are founded by adults from swarms. Build-up is slow at first, then accelerates. During cold weather, termites may migrate into cells in soil, thus escaping freezing, but specimens have been found during winter in debris at the soil surface, sluggish but not dead (Esenther 1969). Swarms issue early in the season, with records as early as late January and February in New York (Weesner 1965), and terminate by the end of April. It is probable that swarming flights are later than this, May or even June, in the northernmost part of the range.

Economic importance. Pickens (1934) classed this termite as destructive. It can seriously weaken foundations of buildings. The many pest-control operators who specialize in termite control in northern latitudes is mute testimony to its importance. *R. flavipes* is undoubtedly the most important termite species in the eastern United States and Canada. For a general account of the species as it affects Canada (specifically the Toronto region), see C. S. Kirby (1963).

Family Termopsidae

Members of this family are the dampwood termites, and they are among the most primitive of living species. Their principal characteristics have been given in the key to families. Their general mode of life has also been briefly noted. There are five genera, one of which is represented in Canada.

Genus *Zootermopsis* Emerson

Description. Alate forms with head lacking ocelli and fontanelle; clypeus short, 5–6 times broader than long; labrum large, truncate apically.

Basal wing scale of fore wings large, longer than pronotum, twice as large as those of hind wings. Tibiae with several spines before apex. Wings large, with 3 or 4 heavy veins near anterior margins. Cerci long, with 5 or 6 segments. Soldiers with head large; labrum broad, truncate apically; mandibles long, toothed, with number of teeth differing on opposite sides. Antennae with more than 20 segments. Tibiae spined. Cerci prominent, with 3, or more, segments.

Key to species of *Zootermopsis*

A. Alates

1. Body yellowish to light cinnamon brown or chestnut brown; wings grayish, usually measuring more than 20 mm from tip to basal suture .... ***angusticollis* (Hagen)** (p. 78)

   Body chocolate brown; wings darker, usually measuring less than 20 mm in length .... ***nevadensis* (Hagen)** (p. 80)

B. Soldiers

1. Sides of head almost parallel. Mandibles as long as or slightly longer than width of head (Fig. 55) .... ***nevadensis* (Hagen)** (p. 80)

   Sides of head slightly concave. Mandibles not as long as width of head (Fig. 59) .... ***angusticollis* (Hagen)** (p. 78)

Clé des espèces de *Zootermopsis*

A. Formes ailées.

1. Corps de couleur variant de jaunâtre à brun cannelle clair ou à brun noisette; ailes grisâtres, mesurant habituellement plus de 20 mm de l'extrémité à la suture basilaire .... ***angusticollis* (Hagen)** (p. 78)

   Corps de couleur brun chocolat; ailes plus foncées, mesurant habituellement moins de 20 mm de longueur ... ***nevadensis* (Hagen)** (p. 80)

B. Soldats

1. Côtés de la tête presque parallèles. Mandibules aussi longues ou un peu plus longues que la largeur de la tête (fig. 55) .... ***nevadensis* (Hagen)** (p. 80)

   Côtés de la tête légèrement concaves. Mandibules moins longues que la largeur de la tête (fig. 59) .... ***angusticollis* (Hagen)** (p. 78)

Zootermopsis angusticollis (Hagen)

Figs. 56–59; Map 9

Termopsis angusticollis Hagen, 1858:75.
Zootermopsis angusticollis; Emerson 1933:182; Helfer 1963:64.

Common name. Pacific dampwood termite.

Diagnosis. Alates yellowish brown to chestnut brown. Heads of soldiers narrow in front, concave at sides.

Description. As for genus *Zootermopsis* and in key. Alates up to 25 mm in length, including folded wings. Antennae with 25, or more, segments, longer than combined length of head and pronotum. Body yellowish brown to chestnut brown; wings grayish. Soldiers (Fig. 59) large,

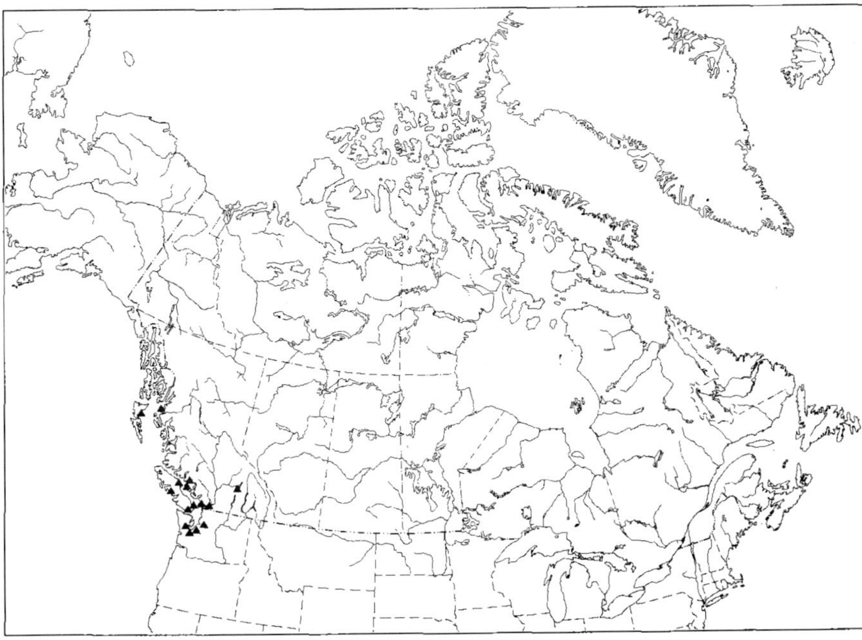

Map 9. Collection localities for *Zootermopsis angusticollis*.

up to 15–20 mm in length (size varies with age of colony, smaller in young colonies). Head with sides slightly concave, narrower in front.

Range. British Columbia to Baja California.

Behavior and habitats. Established colonies inhabit rotting wood; younger, establishing colonies may occur in sound wood, such as felled logs, dead trees still standing, and dead parts of living trees. Colonies occur most often in fallen pine or fir logs in contact with moisture.

Alates (sexual forms) swarm at dusk, pairing and beginning new colonies. Swarms may occur at any time during the year, but most swarms take place in late summer or early fall. One such swarm was observed at Miracle Beach, near Comox, Vancouver Island, by V. R. Vickery in August 1966. It appeared to contain many thousands of individuals. Castle (1934) provided a detailed account of the life of the colonies of this species.

Z. angusticollis, like all Termopsidae, has only two well-defined castes, soldiers and reproductives. The latter may be of three types, but usually only one or two of these occur in a colony at any time. Most of the members of a colony are nymphs (Fig. 57). The social system is primitive when compared with that of most other termites. The most common reproductive forms are the alate males and females. Following swarming, a pair of alates shed their wings and establish a new colony. Once a colony becomes well established, the abdomen of the female becomes greatly swollen (Fig. 56). A less common form of reproductive is a supplementary, or secondary, form, which has only short wing-pads (Fig. 58). The third type of reproductive is nearly apterous and usually develops only in colonies that have lost one or both of the primary reproductives or in groups that become isolated from the parent colony. Normally these forms do not molt to the adult stage and they remain as final-stage nymphs, carrying out a function similar to the workers (or ergates) in more advanced species of termites. Since they do not form a distinct, sterile worker caste, they are called pseudergates.

Life history. Continuous production of nymphs, except during cold months, occurs. During these months activity ceases, then resumes when spring arrives. Colonies grow very slowly. A year-old colony contains only the royal pair, one soldier, and 12–20 nymphs. A few years later it may contain up to 3500 individuals, with the same royal pair. Size of colony may be limited by the size of log that they inhabit.

Economic importance. *Z. angusticollis* is regarded as unimportant economically. However, it causes more damage in the northern part of its range than in the south. These termites are more likely to attack in shaded areas, where moisture is abundant. Wooden poles standing in soil (such as along telephone and telegraph lines) may be damaged severely. Wooden fence posts and sills of buildings such as garages may be attacked. Fence posts that have been treated below ground level may be infested as high as 60 cm above the ground in the untreated wood.

Zootermopsis nevadensis (Hagen)

Figs. 54, 55; Map 10

Termopsis angusticollis var. *nevadensis* Hagen, 1874:571.
Zootermopsis nevadensis; Emerson 1933:165.

Diagnosis. Alates chocolate brown. Soldiers with sides of head parallel.

Description. As for genus *Zootermopsis* and in key. Alates (Fig. 54) small, although sometimes exceeding 20 mm in length including folded wings. Body chocolate-colored, with wings dark. Soldiers with long jaws and with sides of head straight, parallel (Fig. 55).

Range. British Columbia to Idaho, south to northern California.

Behavior and habitats. These are much the same as for *Z. angusticollis* but *Z. nevadensis* can withstand lower temperatures and is found in drier areas and at altitudes up to 1[illegible]0 m. In coastal areas, swarming usually peaks in late summer and early fall, but at higher altitudes, the peak swarming period apparently occurs in the spring. Eggs are laid in spring following the autumn swarm. Composition of colonies at any given time is much the same as in *Z. angusticollis*.

Life history. More or less as for *Z. angusticollis*.

Economic importance. This species is regarded as relatively unimportant economically, but it can and does cause damage to untreated lumber in contact with the soil, such as house sills, poles, docks, and wooden bridges. Its importance is probably greater than is generally recognized. It is well controlled by use of construction methods designed to prevent damage by subterranean termites.

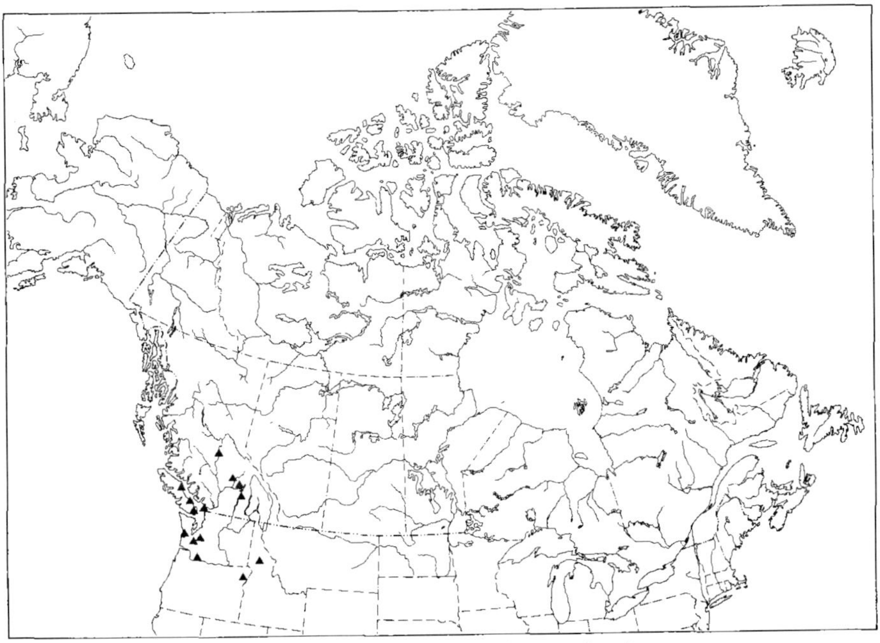

Map 10. Collection localities for *Zootermopsis nevadensis*.

Family Kalotermitidae

Members of this family are the drywood and powderpost termites. Their anatomical features have already been given in the key to families and need not be repeated. There are 21 living and 4 extinct genera, and it has been estimated that there are at least 343 species, although many are probably still undescribed (Weidner 1970). Only a single subfamily is recognized. Living species occur naturally throughout the warmer parts of the world, six genera being North American, but none is native to the region covered by the present work. Many species are of importance because of their destruction of such items as structural timber, containers, and furniture, and some species have been transported by commerce to many parts of the world, including the northern United States and southern Canada. Two of these have managed to establish themselves, for a time, in the region covered.

Key to genera of Kalotermitidae

1. Fore wing with median vein distinct, running about midway between adjacent long veins (radial sector and cubitus), reaching margin of wing near apex (Fig. 60). Soldier with head longer than wide, not boxlike or excavated in front. Mandibles long (Fig. 61) ..... ***Incisitermes*** (p. 82)

Fore wing with median vein running closer to vein in front (radial sector), uniting with it before reaching wing apex (Fig. 62). Soldier with short truncated boxlike head, which is strongly excavated in front. Mandibles short (Figs. 63, 64) .... *Cryptotermes* (p. 83)

Clé des genres de Kalotermitidae

1. Ailes antérieures à nervure médiane distincte, située environ à mi-distance entre les nervures adjacentes (secteur radial et nervure cubitale) et rejoignant le bord de l'aile près de l'apex (fig. 60). Chez les soldats, tête plus longue que large, ni cubique ni concave en avant. Mandibules longues (fig. 61) .... *Incisitermes* (p. 82)

   Ailes antérieures à nervure médiane plus rapprochée du secteur radial et se joignant à ce dernier avant d'atteindre l'apex (fig. 62). Chez les soldats, tête courte et cubique, fortement concave en avant. Mandibules courtes (fig. 63 et 64) .... *Cryptotermes* (p. 83)

Genus *Incisitermes* Krishna

Description. Alates lacking pigmented nodules on wing membrane. Third antennal segment slightly longer and usually darker than second segment, distinctly larger than fourth segment. Soldiers with head longer than broad, with sides straight, less than 4 mm wide. Eyes dark or hyaline. Third antennal segment as long as or longer than fourth and fifth segments combined.

Incisitermes minor (Hagen)

Figs. 60, 61; Map 8

Kalotermes marginipennis var. *minor* Hagen, 1858:47.
Kalotermes minor; Banks and Snyder 1920:20; Helfer 1963:66.

Common name. Western drywood termite.

Diagnosis. Winged forms nearly black. Antennae with less than 20 segments. Soldiers with third segment as long as next three segments.

Description. Alates (Fig. 60) nearly black, with wings grayish, lacking pigmented nodules. Ocelli present. Antennae with less than 20 segments. Soldiers with third antennal segment as long as segments four to six combined. Mandibles with conspicuous asymmetrically positioned teeth (Fig. 61).

Behavior and habitats. *I. minor* infests dead trees, dry wood, or dead wood of living trees in California, Arizona, and Mexico. Weesner (1970) reports its natural occurrence as far north as Washington State. It also infests buildings. It has high tolerance for dry conditions and high temperatures.

Snyder (1954) reports an infestation lasting at least 6 years (prior to eradication in 1951) at Niagara Falls, NY. Snyder (loc. cit., and Weesner (1970)) record introductions in Ohio, at Cleveland (eradicated 1948) and Columbus.

Genus *Cryptotermes* Banks

Description. Alates with median vein of wing not sclerotized, pigmented only on basal third; vein bending forward and joining radial sector (Fig. 62). Soldiers with head short, thick, concave in front. Mandibles recurved beneath head. Apex of tibiae with three spines but no spur.

Cryptotermes brevis (F. Walker)

Figs. 62–64; Map 8

Termes brevis F. Walker, 1853:524.
Cryptotermes brevis; Banks and Snyder 1920:36, 144; Helfer 1963:69.

Diagnosis. Winged forms brownish. Antennae with 16–18 segments. Soldiers with short dark rough heads with dorsal depressions.

Description. Alates (Fig. 62) 10–11 mm in total length. Antennae with 16–18 segments. Head width slightly more than 1 mm. Wing membrane colorless, with brownish veins. Soldiers with head black, rough in front and dorsally, with distinct "dished" depression (fontanelle) at top of head (Figs. 63, 64).

Behavior and habitats. *C. brevis* is known to be established from Florida to southern Texas and has been introduced into many other areas, including Hawaii and Bermuda. It does not occur in natural habitats in the United States or Canada. It infests furniture as well as wooden buildings. "This and other non-subterranean termites. . . have been found infesting furniture as far north as Belleville, Ontario . . . and as far west as Cleveland, Ohio" (Snyder 1954). Weesner (1970) records *C. brevis* from Columbis, OH, and Suffolk County, NY. What was thought to be the same species was found dead in a picture frame in Montréal in the early 1960s (the late Dr. F. O. Morrison, pers. com.). Subsequently, this proved to be *Incisitermes snyderi* (Light 1933). An established colony of *C. brevis* has recently been confirmed as occurring at Port McNeill, northwest Vancouver Island. A wooden case, imported from Peru early in 1978 by Mr. D. L. Kennedy, was found to be still infested with living termites of this species (all castes) 19 June 1979 and subsequently. Another colony occurs at Vancouver, within the woodwork of a cupboard in Dr. G. G. E. Scudder's laboratory.

Suborder Mantodea—mantids

The suborder Mantodea constitutes the praying mantids. Mantids are Dictuoptera having the following characteristics: Size moderate to large (body length of adult ranging from about 15 mm to 16 cm), often strongly sexually dimorphic, usually somewhat to very elongate. Head roundish or conical, hypognathous with prominent usually globular eyes which may, however, be acutely conical anteriorly or each may bear a prominent median thornlike or spinelike process; outer convexities of eyes projecting beyond lateral margins of pronotum, except when latter is foliarly expanded. Cervical region narrow, flexible, enabling the head to rotate. Antennae usually whiskerlike (not excessively long) or bristlelike, occasionally pectinate. Pronotum typically elongate, sometimes more or less quadrate, occasionally broadly rhomboid on account of the presence of wide lateral expansions. Front legs raptorial (Fig. 26), with coxae and femora elongate, the latter usually, but not invariably, heavily spined on ventral (posterior) margins, and tibiae strongly spined apically and usually ventrally, folding upward against femora and with laterally articulating tarsi; middle and hind legs ambulatory, long, slender. Wings fully developed, reduced, or absent, their form and/or degree of development often differing markedly between sexes; fore wings, if present, reticulate, normally in form of leathery tegmina, although sometimes, if fully developed, rather more membranous, typically with anterior field denser than rest, although main part of fore wings and the hind wings folded more or less horizontally over dorsum; latter pair, when developed, broad with wide anal lobe. Cerci long, with numerous segments, usually rather slender and pointed but occasionally flat and foliaceous. Ovipositor short, more or less concealed by enlarged seventh abdominal sternum (subgenital plate). Males with abdominal styles well-developed; copulatory structures distinctly asymmetrical and well-sclerotized. True stridulatory mechanism not developed, although a rustling sound may be produced by vibration of hind wings which strike against raised tegmina; some species, when in a "threatening" position, may raise the abdomen and bend it forward so that the cerci strike the tegminal veins and make an intermittent chirping sound.

The number of known species in the world is about 1800, the vast majority of them tropical. They are distributed among 3 superfamilies and 8 families, as currently recognized, and are placed in some 360 genera.

Mantids are all predatory, mainly on other insects, although some of the larger ones will take small reptiles, amphibia, and even nestlings of small birds. It is an old cliché that they should be called preying, not praying, mantids, the latter applying only to their supplicating attitude when awaiting their victims. Usually, prey is caught by waiting in ambush until the victim approaches, and then, as often as not, creeping up on it and suddenly seizing it with the formidable gin-traplike front legs. The mantids are often well camouflaged by resemblance to foliage (green or dried), such as dry sticks, stems, and bark. A few are brilliantly colored like the flowers in which they lurk, waiting for prey. Some ground-dwelling species are rapid runners and may pursue their prey actively. They, too, resemble their background closely. The females of some species of mantid make little, if any, initial distinction between a meal and a male, and, if the latter is to survive to fulfill its function, it has to be both adroit at avoiding attack and successful in courtship. Having done its duty, the male often pays the supreme penalty anyway!

Mantids mainly live among foliage, be it herbaceous (including grasses) or on shrubs, bushes, and trees. Some flattened species live on tree trunks and, as previously noted, a number live among blossoms; others live on the ground's surface, particularly in extremely arid desert situations. Some mantids inhabit savanna, steppe, or prairie, but most live in mesophyllic or humid situations. None is aquatic or subaquatic. Many are bizarre in form, the nymphs often being rather different from the adults in appearance. Most species adopt a threatening attitude when alarmed, and some with brightly colored or heavily pigmented hind wings appear to enhance their obvious ferocity by sudden display behavior; others adopt a grotesque posture. It is interesting that, in the Oriental martial art of kung fu, mantid posturing is overtly imitated. Unlike many other cultures, the Chinese correctly regard mantids as symbols of strength, ferocity, and courage, not of piety, prophecy, or pathfinding. In North America, and to some extent in Europe, attempts have been made to use mantids as "biological control" agents. They are even made available commercially with this in view, but they are insufficiently discriminating for the purpose. One or two introduced species have, however, managed to establish themselves.

Eggs are laid in typical papery or spongy packages (oothecae), which are usually attached to vegetation. Such packages usually contain a large number of eggs. Viviparity and ovoviviparity are not known to occur, and parthenogenesis is rare. The number of nymphal instars ranges from six to nine, there apparently being some variation within most individual species studied. The most recent general accounts of the mantids are those of Beier (1964, 1968). A useful account of North American species is that of Gurney (1951*a*).

Family Mantidae

This family includes the great majority of the known species of mantids (about 1480), which are divided among about 275 genera placed in 21 subfamilies. Of these last, only two concern us here.

Key to subfamilies of Mantidae

1. Small slender species (body of adult less than 35 mm). Antennae and costal (anterior) margins of male tegmina usually setose. Supra-anal plate of male usually enlarged .... **Amelinae** (p. 86)
   Larger species (body of adult more than 45 mm). Females robust. Antennae and costal margins of tegmina not setose. Males fully winged. Supra-anal plate of male transverse but not enlarged .... **Mantinae** (p. 87)

Clé des sous-familles de Mantidae

1. Espèces petites et grêles (corps des adultes mesurant moins de 35 mm). Antennes et bords antérieurs des tegmina des mâles habituellement sétifères. Plaque supra-anale des mâles en général développée .... **Amelinae** (p. 86)
   Espèces plus grosses (corps des adultes mesurant plus de 45 mm). Femelles robustes. Antennes et bords antérieurs des tegmina non sétifères. Mâles pourvus d'ailes. Plaque supra-anale des mâles transverse mais pas développée .... **Mantinae** (p. 87)

Subfamily Amelinae

This subfamily includes some 32 genera and more than 150 species widely distributed in the Ethiopian and Oriental regions, with a few representatives in the southern Palaearctic and Nearctic regions. The single Nearctic genus *Litaneutria* ranges from Mexico to southwestern Canada.

Genus *Litaneutria* Saussure

Description. Body small (less than 35 mm). Tegmina and wings of male fully developed, of female reduced and covering no more than one-third of abdomen. Color light buff to dark brown.

Litaneutria minor (Scudder)

Figs. 72, 73; Map 11

Stagmatoptera minor Scudder, 1872:251.
Litaneutria ocularis Saussure, 1892:123.
Litaneutria borealis Scudder, 1896*b*:209.
Litaneutria obscura Scudder, 1896*b*:209.
Litaneutria minor; Scudder 1896*b*:209; Gurney 1951*a*:358; Helfer 1963:28.

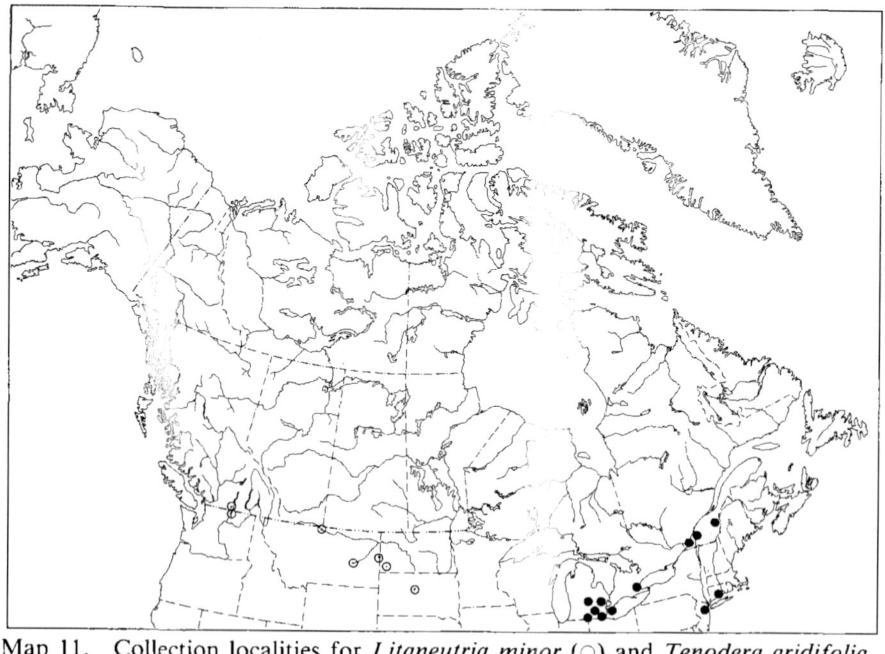

Map 11. Collection localities for *Litaneutria minor* (○) and *Tenodera aridifolia sinensis* (●).

Litaneutria pacifica Scudder, 1896*b*:210.
Litaneutria skinneri J. A. G. Rehn, 1907:26.
Litaneutria longipennis Beier, 1929:137.

Diagnosis. Body small. Insect ground-dwelling. Male usually with black spot on hind wing.

Description. As for genus *Litaneutria*. Males usually macropterous, with large black spot in center of hind wing; more rarely brachypterous or even micropterous (with tegmina covering no more than one-third of abdomen); females micropterous.

Range. British Columbia to North Dakota, south to Mexico.

Behavior and habitats. *L. minor* is most often found on the ground, less frequently on low vegetation. R. A. Roberts (1937) found specimens mainly on low rocky ridges that were sparsely covered with bunchgrass. The color is similar to that of the environment and specimens are not readily detected unless they move. Because *L. minor* usually runs on the ground it can be captured in pan traps set in the ground (Barnum 1964). R. A. Roberts (1937) observed mating and reported that copulation lasts 3–4 hours. The female may capture and devour the male following copulation, although it often escapes. Mated females produce small egg masses, about 7 mm long, nearly rectangular with rounded corners. Unmated females also produce egg masses but the eggs do not hatch.

Life history. The egg masses are deposited on stems of low-growing shrubs. A female may lay as many as 10 egg masses in its lifetime, each of which may produce about 16 nymphs. For overwintering eggs, the incubation period is 185–205 days. Nymphs mature about 93 days after the eggs hatch, females requiring one more molt than males. Adult longevity for males is about 47 days and for females about 156 days or until killed by frost. In the southern part of the range a partial second generation occurs, but, in northern latitudes, the life cycle is univoltine, with overwintering in the egg stage.

Subfamily Mantinae

This is by far the largest subfamily of Mantidae (and of the whole suborder), and it includes approximately 500 species in just over 60 genera. These are found throughout the whole range of the suborder, with the exception of southern South America and northwestern North America. Several tribes are recognized, but only the Mantini concern us here. Of these, but two genera breed in our region, both of them introduced by man.

Key to genera of Mantinae

1. Tegmina rather broad, with anterior margin convex and apex rounded. Middle and hind femora without apical spines. Body green, sometimes brown, with colored part of tegmina not sharply limited to anterior. Fore coxae each with large black-ringed spot near base .... ***Mantis*** (p. 88)

Tegmina narrower, with anterior margin rather straight and apex somewhat pointed. Middle and hind femora each with apical spine. Coloration of tegmina sharply limited to anterior area. Fore coxae without black-ringed spot .... ***Tenodera*** (p. 90)

Clé des genres de Mantinae

1. Tegmina plutôt larges, à bord antérieur convexe et à apex arrondi. Fémurs médians et postérieurs dépourvus d'épines apicales. Corps vert, parfois brun. Partie antérieure colorée des tegmina non clairement délimitée. Coxas antérieures marquées chacune d'une grande tache cerclée de noir près de la base .... ***Mantis*** (p. 88)

   Tegmina plus étroites, à bord antérieur plutôt rectiligne et à apex plutôt pointu. Fémurs médians et postérieurs portant chacun une épine apicale. Coloration des tegmina clairement limitée à la région antérieure. Coxas antérieures dépourvues de taches annelées ..... ***Tenodera*** (p. 90)

Genus *Mantis* Linnaeus

Description. Body size medium (male 47–51, female 52–56 mm long). Both sexes fully winged; tegmina extending to or surpassing apex of abdomen. Facial shield somewhat less than twice as broad as high. Pronotum rather long, narrow (13–18 mm), broadest well before middle, anterior to this with sides converging to rounded anterior margin, in female margins finely and densely serrated for more than half the length. Male subgenital plate small. Inner face of fore coxa with large black spot.

Mantis religiosa religiosa Linnaeus

Figs. 2*a*, 2*b*, 74, 75; Map 12

Gryllus Mantis religiosa Linnaeus, 1758:426.
Mantis religiosa; Scudder in Slingerland 1900:38; Helfer 1963:31.
Mantis religiosa religiosa; Bazyluk 1960:238, 239; Vickery et al. 1974:19.

Common name. European mantid.

Diagnosis. Body large (47–56 mm). Fore coxa with black spot on inner face.

Description. As for genus *Mantis* (Fig. 74). Female elongate (52–56 mm); male smaller and more slender (47–51 mm). Body and legs usually green or greenish yellow (Fig. 2*a*), some specimens entirely pale brown (as in Fig. 2*b*); in green form, female tegmina grass green; those of males paler green, sometimes hyaline with narrow brown borders; wings transparent or pale green. Pronotum of male finely serrated laterally anterior to the dorsal transverse sulcus.

Range. Michigan to Québec; interior British Columbia; northeastern United States.

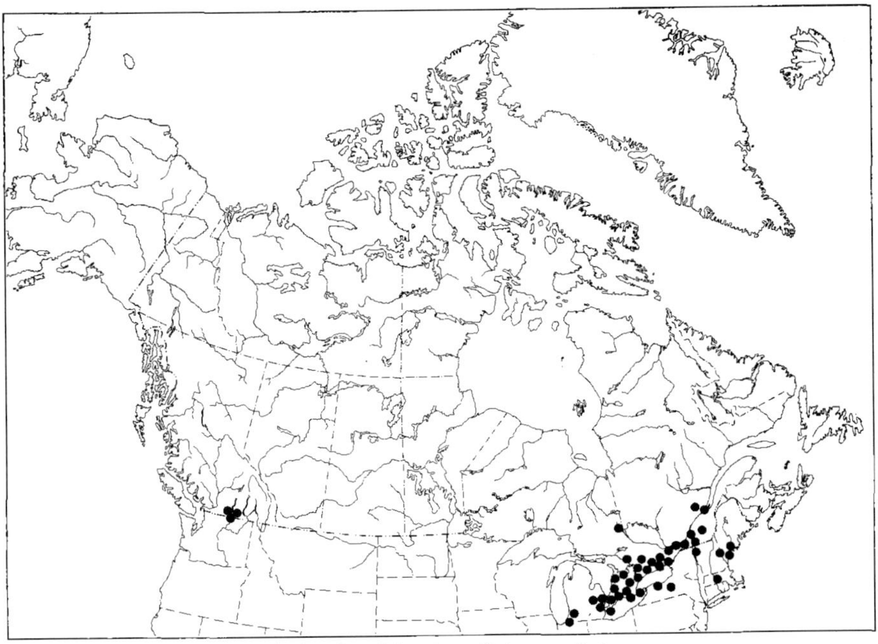

Map 12. Collection localities for *Mantis religiosa religiosa.*

Behavior and habitats. A general account of the biology of *M. religiosa religiosa* under western European conditions is found in Binet (1931). In Ontario and southwestern Québec, *M. religiosa religiosa* is found most often on vegetation bordering fields and roads, in uncultivated fields and pastures, or on tall species of *Solidago* (goldenrod).

Like all other mantids, this species is carnivorous, and, like most, it feeds upon a wide variety of other insects. Young immature individuals seize and devour small insects. The raptorial forelegs (Fig. 75) are ideally suited for this. The mantid usually waits motionless until its prey approaches, then strikes extremely rapidly. It may, however, stalk prey that is relatively immobile. As a mantid grows larger it attacks larger prey. Adults have been observed to kill fairly large female grasshoppers (*Melanoplus bivittatus*) that are bulkier than themselves. Cannibalism is common, particularly among young mantids in crowded conditions.

Female mantids commonly attack and devour males either before or after copulation. Males approach females only when they are in the range of vision. Approach is slow and hesitant with no display. If the female detects the male's presence it is attacked and, if caught, eaten. In many cases, copulation is effected and completed after the head of the male has been destroyed. The suboesophageal ganglion acts as an inhibitor of lateral movement and also of copulatory movements of the male abdomen. Destruction of the ganglion allows copulation to proceed (Roeder 1935). The male is not always destroyed by the female if it manages to get onto the female's back in a normal mating position. Copulation may last 4–5 hours. The male then withdraws, leaving a large white spermatophore attached to the female. The male releases its hold and attempts to drop from the female's back. The male may escape, but the female may turn and seize it.

Slingerland (1900) reported that, according to H. F. Atwood, *M. religiosa religiosa* produced sound by rubbing a leg against a tegmen—although it was presumably vice versa, as is known for other mantids (Chopard 1938). This appeared to be aggressive behavior and occurred only when two males were close to each other. Long ago, this species was also said to produce sound by rubbing the extremity of the body against the wings (DuFour 1841).

Females produce eggs in masses covered with a frothy substance which, when dry, acts as protection and insulation. Egg masses are attached to twigs, stems, or flat surfaces. Several times they have been seen in southwestern Québec attached to the sides of houses (Vickery and Kevan, unpublished). Vickery (1962) reported taking an egg mass that had wintered in Québec to Nova Scotia, but the eggs failed to hatch. The experiment was repeated on a larger scale about 10 years later. A number of egg masses (22) were collected in Québec in the fall and taken to Yarmouth County, N.S., where they were fastened in appropriate situations outdoors. The following summer most of the egg masses were intact, but the eggs had not hatched. Several other attempts, largely unsuccessful, have been made to introduce the species into various parts of Western and Atlantic Canada from Ontario (Baird 1938–1946; Buckell 1941; McLeod 1962; MacNay 1955*c*; G. D. Williamson 1966, 1973).

Life history. Univoltine, with overwintering in the egg stage. Adults are present from mid-August to the end of September or until killing frost occurs.

Genus *Tenodera* Burmeister

Description. Body large (length 70–104 mm), elongate, rather robust. Facial shield about twice as broad as high. Pronotum moderately long (13–18 mm), but quite stout in commonest species, only feebly broadened at transverse sulcus; lateral pronotal margins of females finely serrate. Tegmina and hind wings fully developed, with tips subacute. Forelegs stout, with lower margin of coxa strongly toothed.

Tenodera aridifolia sinensis Saussure

Fig. 76; Map 11

Tenodera aridifolia var. *sinensis* Saussure, 1871:419.

Paratenodera sinensis; Weiss 1914:279.

Tenodera aridifolia sinensis; Gurney 1951*a*:355; Vickery et al. 1974:20; Helfer 1963:32.

Common name. Chinese mantid.

Diagnosis. Body large (male 83–87, female 85–104 mm).

Description. As for genus *Tenodera*. Female with body and legs green to greenish yellow, with tegmina and wings greenish subhyaline; male same color as female, entirely brown, or brown with tegminal margin green. Pronotum of male subcarinate and of female strongly carinate behind transverse sulcus, bicarinate in front with carinae granulose. Tegmina with large marginal field.

Range. Southern Ontario and Québec, south to New Jersey.

Behavior and habitats. *T. aridifolia sinensis* is a voracious predator and is sold commercially as a biological control agent in the United States (Gurney 1951*a*), although its catholic tastes and slow rate of reproduction would render it very inefficient for the purpose. It seems to prefer taller vegetation than does *Mantis religiosa*, is often found on blackberry bushes, and avoids low wet ground and low-growing plants. Egg masses are attached to vegetation and solid surfaces aboveground; each mass contains 150–300 eggs.

Life history. Univoltine, with overwintering in the egg stage. An egg mass was dissected from a Pennsylvania female in October by Laurent and Saussure (1898). Weiss (1914) investigated the oothecae and eggs. In Michigan, Cantrall (1968) reported *T. aridifolia sinensis* as a late summer species that occurred in the adult stage from September to 13 October.

Suborder Blattodea—cockroaches

The insects in this suborder may be characterized as follows: size ranging from a few millimetres to 10 cm (including the projecting wings) or nearly 8 cm in body length alone, and with a wingspan of as much as 18.5 cm. Head hypognathous, with most sclerites and sutures well-defined. Antennae whiskerlike, usually rather long, often as long as or longer than body. Compound eyes normally well-developed; ocelli usually absent, never more than two, usually represented only by pair of pale frontolateral areas (fenestrae). Mouthparts of generalized biting and chewing type, with mandibles neither particularly large nor acute. Neck sclerites well-developed so that the head is unusually mobile (normally held tucked under pronotum), although not rotatable. Pronotum broad, typically disclike, usually rather flat, sometimes markedly convex. Legs similar, cursorial, with hind pair longer than others, typically with few to many strong ventral spines; coxae large, flat, platelike, close together, 5-segmented. Wings, when present, usually rather broad, held more or less flat over dorsum, usually shorter, sometimes much shorter, in females than in males of the same species, often reduced or lacking either in females or in both sexes; fore wings, if present, typically in form of leathery tegmina, more membranous in some small species, not bent downward along sides of abdomen when at rest; hind wings nearly always broad, with prominent anal lobe (reduced in a few small aberrant species). Abdomen with 10 terga, with 8th and 9th largely concealed beneath 7th; 11th (epiproct) reduced; male terga often with median glandular setose depression (licked by female when mating); 1st sternum small; 7th (in female) or 9th (in male) enlarged, forming distinct subgenital plate, which conceals most of the genitalia; 8th to 10th in females generally reduced and partly membranous; 9th sternum in males normally with pair of styles, although one or both occasionally lacking; abdominal cerci usually conspicuous, multiarticulate, tapering, sometimes blunt with few segments or, occasionally, even unsegmented. Ovipositor of female consisting of 3 pairs of valves, but not visible or only partly so. Male phallic structures strongly sclerotized, complex, asymmetrical.

A true stridulatory apparatus is not commonly present, although a number of sound-producing structures are known. Some species produce sounds by rustling the wings, sometimes against special ridges at the sides of the pronotum and on the costal veins, by means of which they produce a squeaking noise; a few species hiss by the expulsion of air from the tracheal system or by vibrating the abdominal segments, either against each other or against the undersides of the wings if these are present; some species drum on the substrate with the abdomen (Roth and Hartman 1967). Many species have integumental glands that produce repugnant fluids. Other defense mechanisms include the ability of some wingless species to roll up into a ball like some isopod crustaceans and short-bodied millipedes.

The popular concept of cockroaches is based upon a few synanthropic species that are by no means representative of the suborder. In the region covered by the present work, there are few other kinds.

The Blattodea are predominantly tropical and subtropical insects, although a few species have a natural distribution that extends to high latitudes, e.g., Lapland, central Ontario, or southern New Zealand. Associated with man, they can exist under artificial conditions wherever he goes, and one species, *Blattella germanica* (Linnaeus), has been found in heated buildings in Canada within 750 km of the North Pole. Most cockroaches, whether synanthropic or not, live under debris, stones, or logs or in cracks or crevices, and some species are cavernicolous or live in the nests of other animals such as ants or rodents. A number of species actively burrow in soil or bury themselves under sand. There are, however, many kinds of cockroach that spend their lives well above ground level, either on the trunks of trees and in epiphytic plants or amongst foliage and flowers. These latter are sometimes green or brightly colored, or spotted or gaily striped and not at all of the brownish, blackish, yellowish, or tawny coloration with which most people are familiar. Cockroaches are particularly abundant in warm humid climates, but many are quite at home under mesophilic or semiarid conditions. There are, in fact, true desert-living species. At the opposite extreme there are those, like *Dryadoblatta scotti* (Shelford) from Trinidad and Venezuela, that live in epiphytic bromeliads on tropical forest trees, where they spend most of their time under the water that accumulates there.

Cockroaches are typically but not always omnivorous. Many species, particularly those that are synanthropic, are largely detritus feeders and scavengers, feeding on decaying plant remains and the associated fungi and microorganisms. They are not, however, averse to eating animal matter or even living animal tissue, either in the form of small insects or, at the opposite extreme, the dermal products and indeed the skin of recumbent mammals (including man). Some species, however, feed directly on foliage or nectar. The small family Cryptocercidae is of special interest because the species feed on wood, igesting cellulose (but not lignin) by the aid of peculiar intestinal flagellate protozoa (Trichonymphidae and related families) that are virtually indistinguishable from those found in the guts of the "less advanced" groups of termites.

In their reproductive biology, cockroaches show a wide diversity. Whereas most species are bisexual, there are some, like *Pycnoscelus surinamensis* (Linnaeus), that are parthenogenetic. Cockroach eggs are produced in double rows in packages. Typically they are laid in capsules, which are often in the form of leathery or horny oothecae. These may be deposited almost at once or carried around for a period before deposition (when the eggs have a relatively long incubation period). Frequently, however, the oothecae are thin-walled, the eggs hatching within a relatively short time after deposition—sometimes almost at once. In fact, the hatching may be so advanced that the females are often ovoviviporous, the young emerging as (or even before) the oothecae are deposited. Roth and Willis (1954), McKittrick (1964), and Roth (1968, 1970) give the most comprehensive accounts of the various forms of reproduction. There may be any number between 4 and 12 nymphal instars according to species, with some variation, particularly when the number is greater. Females may sometimes pass through an extra instar as compared with males. Most cockroaches have at least one generation per year, but there are some in which the life history is longer.

Because of their catholic tastes and their ability (indeed preference) to hide in crevices, certain species of cockroach have been widely distributed by human commerce and have become unmitigated nuisances, if not serious pests or hazards to the health of humans and their livestock. General accounts of cockroaches in their various roles are given by Chopard (1938), Roth and Willis (1957*a*, 1960), Beier (1961, 1974), Guthrie and Tindall (1968), and Cornwell (1968, 1976).

Key to superfamilies of Blattodea

1. Middle and hind femora armed ventrally on both margins with rows of strong, more or less equally developed, spines along their lengths. Male subgenital plate symmetrical, with elongate slender symmetrical style (Figs. 18, 21). Female subgenital plate distinctly divided into valves by short longitudinal slit .... ***Blattoidea*** (p. 94)

   Middle and hind femora typically unarmed ventrally, or with distal spines only, or with 1-4 weak spines on posterior and/or anterior margins, those of the two series unequally developed; if leg spines as indicated above, then abdominal terminalia not as above 2

2(1). Male with subgenital plate symmetrical, rather narrowly subtriangular (with short symmetrical styles). Seventh abdominal tergum broadly triangular, somewhat rounded posteriorly, with heavy thickened margin (Fig. 7) .... **Cryptocercoidea** (p. 101)

   Male without the above characters .... **Ectobioidea** (p. 102)

Clé des super-familles de Blattodea

1. Fémurs médians et postérieurs munis, de part et d'autre de la portion ventrale, de rangées d'épines fortes et plus ou moins égales sur toute leur longueur. Plaque sous-génitale des mâles symétrique, munie de styles longs, grêles et symétriques (fig. 18 et 21). Plaque sous-génitale des femelles clairement divisée en valves par une courte incision longitudinale . . . . **Blattoidea** (p. 94)

   Fémurs médians et postérieurs typiquement dépourvus d'épines sur le ventre ou ne portant des épines que sur la partie distale ou ne portant que de 1 à 4 petites épines sur le bord postérieur ou sur le bord antérieur ou portant des épines de grosseur inégale sur le bord postérieur et sur le bord antérieur. Si les épines des fémurs répondent à la description donnée ci-haut, les terminalia sont différents 2

2(1). Plaque sous-génitale des mâles symétrique, pratiquement subtriangulaire (pourvue de styles courts et symétriques). Septième tergite de l'abdomen grossièrement triangulaire, plutôt arrondi postérieurement et muni d'un bord épaissi (fig. 7) . . . . **Cryptocercoidea** (p. 101)

   Mâle ne présentant pas la combinaison de caractères donnée ci-dessus . . . . **Ectobioidea** (p. 102)

Superfamily Blattoidea

The general characters distinguishing the members of Blattoidea have been given in the key to superfamilies. There are three included families, if the Archiblattidae be admitted. Blattidae, the only family of concern here, is a large family of mainly moderate-sized to large insects, with 47 genera and nearly 650 species distributed throughout the world, including a few that occur naturally in warmer temperate regions. A handful of species are of economic importance, some of them being among the most obnoxious of man's companions. The great majority are strongly pigmented, usually reddish brown, dark brown, or blackish species, often with pale lateral streaks on pronotum and tegmina. The tegmina and hind wings may be fully developed or reduced in both sexes, but, in either event, they are generally shorter in females than in males. Most species, under natural conditions, live under debris, stones, or logs, or under the loose bark of dead trees.

Family Blattidae

Apart from one or two adventives, the family Blattidae includes the largest of the cockroaches found in the region here covered. All are limited to artificial conditions, such as houses, warehouses, greenhouses, city sewers, and the like. They are commonly found on ships and in seaports. They are among the most unsavoury of the domiciliary species as they produce objectionable odors. Species of *Eurycotis*, in particular, are known as "stinking cockroaches," but, in the region here covered, these are only met with occasionally as adventives (or as laboratory insects) and do not breed "in the wild." Members of Blattidae are, however, not the most abundant or important pest species in Canada, although in some parts of the world, especially in tropical and subtropical countries, they may be of major significance, both indoors and in the open. As hazards to health they are second to none, and some people are actually allergic to them. Conversely, they have been widely used in folk medicine in different ways for curing various ailments. There is also a large body of folklore associated with them. Despite the numerous species belonging to this family, relatively few are pests.

Key to genera of Blattidae

1. Body uniformly dark pitchy brown. Tegmina of male somewhat shorter than abdomen, subtruncated apically, of female reduced to small lateral lobes (Figs. 8, 9) .... ***Blatta*** (p. 95)
   Body and tegmina not as above .... ***Periplaneta*** (p. 97)

Clé des genres de Blattidae

1. Corps de couleur brun foncé uniforme. Tegmina des mâles un peu plus courts que l'abdomen et à l'extrémité quelque peu tronquée. Chez les femelles, tegmina réduits à de petits lobes latéraux (fig. 8 et 9) .... ***Blatta*** (p. 95)
   Corps et tegmina différents de la description donnée ci-dessus .... ***Periplaneta*** (p. 97)

Genus *Blatta* Linnaeus

Description. Body size medium (18-24 mm long), robust. Antennae elongate, setaceous. Space between eyes of male slightly broader than that between antennal bases. Lower anterior margin of fore femur with numerous stout feebly curved spines that decrease in length toward apex with two longer apical spines (Fig. 10); other femora with fewer more widely separated spines; tarsi with segments one to four bearing small pulvilli.

Blatta orientalis Linnaeus

Figs. 8–10; Map 13

Blatta orientalis Linnaeus, 1758:424; Cantrall 1943:74; Vickery et al. 1974:21; R. J. Harrison 1980:10.

Common name. Oriental cockroach.

Diagnosis. Body size medium (18–24 mm long), shining dark chestnut and blackish brown. Male brachypterous, female micropterous.

Description. As for genus *Blatta*. Body size medium (pronotal length, male 5.7–6.3, female 6.1–7.1 mm), shining dark chestnut and blackish brown. Pronotum with broad impression on each side on anterior third of disc. Tegmina of male overlapping, covering about two-thirds of abdomen; tegmina of female reduced to small oval widely separated pads. Chromosome number, $2n$ male = 47 (Cohen and Roth 1970).

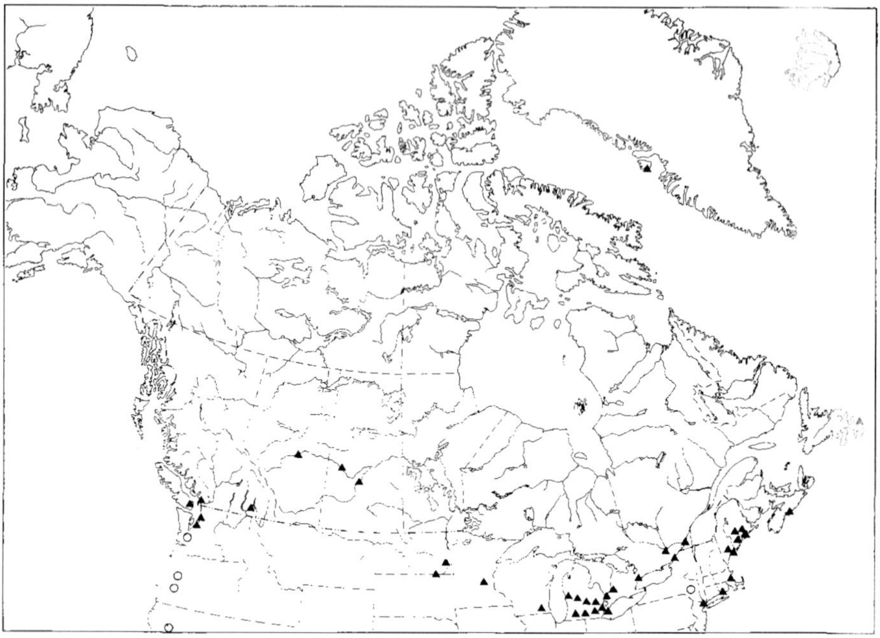

Map 13. Collection localities for *Blatta orientalis* (▲) and *Cryptocercus punctulatus* (○).

Range. British Columbia to Newfoundland; cosmopolitan.

Behavior and habitats. Cornwell (1968) has summarized the information on habitats of this insect. It is found in areas that are noticeably cooler than those preferred by the German cockroach, *Blattella germanica*, such as basements and cellars, service ducts, and crawl spaces. In kitchens, it may occur behind radiators, ovens, hot-water pipes, and under floor coverings. It is often found in toilets and behind bathtubs and sinks, where large numbers congregate near sources of water such as condensation on water- and waste-pipes. Although infestations are most frequently found below ground level or on ground floors of buildings, the insects may disperse via water-pipes to upper floors. Restaurants, bakeries, and candy factories are infested, as are ships and warehouses. This cockroach shuns light and is active in buildings at night. The species is gregarious; where one is seen many more are nearby. If a light is turned on while specimens are prowling, they instantly scramble for cover in a dark spot.

In Canada, most infestations in homes seem to be produced by the carrying home of specimens in grocery packages from infested retail outlets. In Great Britain and in the southern United States, *B. orientalis* occurs outside buildings and is able to survive winter conditions in dumps, refuse heaps, and even under dead leaves and stones. A few parasitic insects, Evaniidae and Eulophidae (Hymenoptera), are known as egg parasites of *B. orientalis* (Roth and Willis 1960). In domiciliary situations, parasitism does not generally occur.

Life history. Under optimum conditions a generation requires about 6 months to complete development (Willis et al. 1958), but it may take 2 years in unheated buildings (Cornwell 1968). Male nymphs develop more rapidly than females. Parthenogenesis is known in *B. orientalis* (Roth and Willis 1956) and a 1:1 sex ratio has been found in the progeny of both mated and virgin females. Copulation usually occurs within a few days of molting to the adult stage and lasts about 40 minutes. Females produce oothecae (egg cases), each containing up to 18 eggs. These are carried by the female for about 24 hours, then they are carefully hidden, glued to some surface. This apparently prevents their being eaten by other individuals of the species. The incubation period of the eggs is about 6 weeks at 30–36°C, longer at cooler temperatures. Adults live 1–6 months, and a single female may produce six oothecae during this time.

Economic importance. *B. orientalis* is an important pest species. The obnoxious smell associated with it makes it particularly annoying. It will feed upon meat, cheese, soiled woolen clothes, even old leather, but it prefers starchy foods.

Genus *Periplaneta* Burmeister

Description. Large stout species (23.5–34.5 mm long). Antennae long, slender. Pronotum convex, with lateral edges deflexed. Tegmina rather leathery, extending beyond apex of abdomen in both sexes, with lateral and apical margins rounded. Male subgenital plate symmetrical, with styles elongate, slender, located in sockets laterally near apex. Fore femur armed on anteroventral margin with row of stout spines; longest spine located near middle, and 3 apical spines unequal.

Key to species of *Periplaneta*

1. Pronotum yellowish with 2 large rather indistinct suffused chestnut brown blotches (Figs. 11, 12) 2
   Pronotum reddish brown, tinged with orange, with 2 large, sharply defined, usually confluent black blotches (Figs. 13, 14) .... ***australasiae* (Fabricius)** (p. 98)

2(1). Cerci long, slender (Figs. 16, 17). Male supra-anal plate nearly unsclerotized, extending half its length beyond subgenital plate, divided distally (Fig. 16). Male styles long (Fig. 18). Female supra-anal plate with deep acute median notch (Fig. 17) .... ***americana* (Linnaeus)** (p. 99)
   Cerci shorter and less slender (Figs. 19, 20). Male supra-anal plate nearly truncate, with shallow median emargination (Fig. 19). Male styles short (Fig. 21). Female supra-anal plate feebly bilobate at apex (Fig. 20) .... ***brunnea* Burmeister** (p. 101)

Clé des espèces de *Periplaneta*

1. Pronotum jaunâtre et marqué de deux taches plutôt diffuses, de couleur brun noisette (fig. 11 et 12) 2
   Pronotum de couleur brun rougeâtre, teinté d'orangé, marqué de 2 taches noires, claires et habituellement accolées (fig. 13 et 14) . . . . ***australasiae* (Fabricius)** (p. 98)

2(1). Cerques longs et grêles (fig. 16 et 17). Plaque supra-anale des mâles pratiquement non sclérifiée, s'étendant, sur la moitié de sa longueur, audelà de la plaque sous-génitale; divisée distalement (fig. 16). Styles des mâles longs (fig. 18). Plaque supra-anale des femelles pourvue d'une incision médiane profonde (fig. 17) . . . . ***americana* (Linnaeus)** (p. 99)
   Cerques courts et moins grêles (fig. 19 et 20). Plaque supra-anale des mâles presque tronquée et ayant une légère dépression médiane (fig. 19). Styles des mâles courts (fig. 21). Plaque supra-anale des femelles légèrement bilobée à l'apex (fig. 20) . . . . ***brunnea* Burmeister** (p. 101)

Periplaneta australasiae (Fabricius)

Figs. 13, 14; Map 14

[*Blatta*] *australasiae* Fabricius, 1775:271.
Blatta domingensis Palisot de Beauvois, 1817:182.
Periplaneta zonata Haan, 1842:49.
Periplaneta repanda F. Walker, 1869*a*:125.
Periplaneta subcincta F. Walker, 1869*a*:126.
Periplaneta inclusa F. Walker, 1869*a*:126.
Periplaneta emittens F. Walker, 1871:37.
Polyzosteria subornata F. Walker, 1871:35.
Periplaneta australasiae; Lugger 1898:184; Helfer 1963:51; Vickery et al. 1974:23; R. J. Harrison 1980:10.

Common name. Australian cockroach.

Diagnosis. Pronotum reddish brown to orange yellow, with 2 large sharply defined, usually confluent, black blotches.

Description. Body medium large (male 23.6–25.5, female 24.3–29 mm long). Pronotum without discal impressions, subelliptical, with all angles broadly rounded (pronotum length, male 6.2–7.2, female 7.8–8.8 mm), red brown to orange yellow, with 2 large black blotches (Fig. 14). Tegmina and wings fully developed. Male cerci short, acute. Chromosome number, $2n$ male = 27 (Cohen and Roth 1970).

Range. British Columbia to Newfoundland (in buildings); cosmopolitan.

Behavior and habitats. *P. australasiae* requires warmer temperatures than does *P. americana* and, in the present region, is found only in premises that are heated. In general, it is found in habitats similar to that species and feeds on the same types of food.

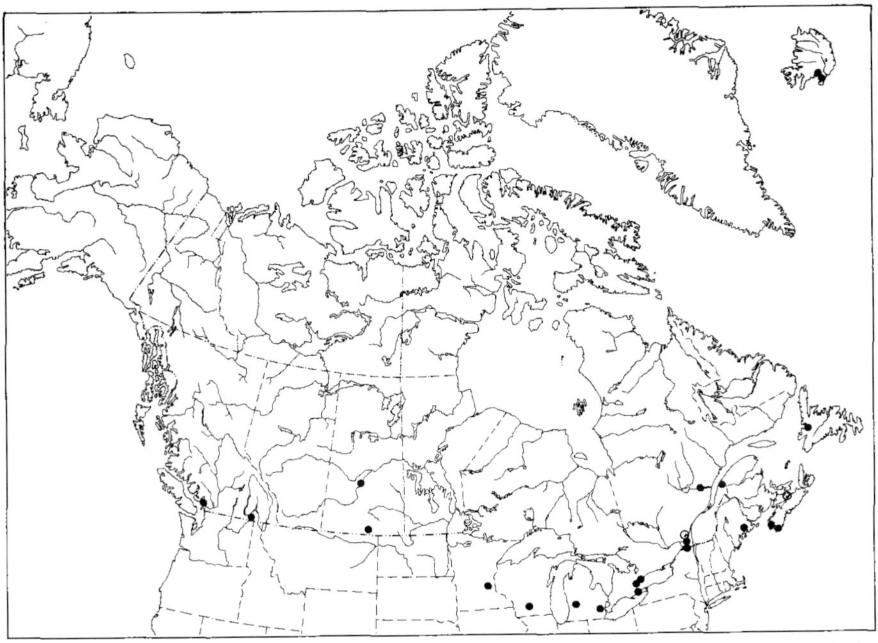

Map 14. Collection localities for *Periplaneta australasiae* (●) and *P. brunnea* (○).

Life history. Willis et al. (1958) reported that mating occurs about 5 days after molting to the adult stage. Females may produce 20–30 oothecae at about 10-day intervals. Each ootheca contains about 24 eggs. The incubation period for eggs is about 40 days but is dependent upon temperature. The period of nymphal development is not known but probably is of 4–12 months' duration.

Economic importance. In the present region, similar to but less than for *P. americana*.

Periplaneta americana (Linnaeus)

Figs. 11, 12, 16–18; Map 15

[*Blatta*] *americana* Linnaeus, 1758:424.

Blatta kakkerlac De Geer, 1773:535.

Periplaneta americana; Saussure 1864*b*:72; Vickery et al. 1974:22; R. J. Harrison 1980:10.

Periplaneta stolida F. Walker, 1869*a*:128.

Periplaneta americana colorata J. A. G. Rehn, 1901*a*:220.

Common name. American cockroach.

Diagnosis. Pronotum yellowish brown with 2 large suffused chestnut brown blotches.

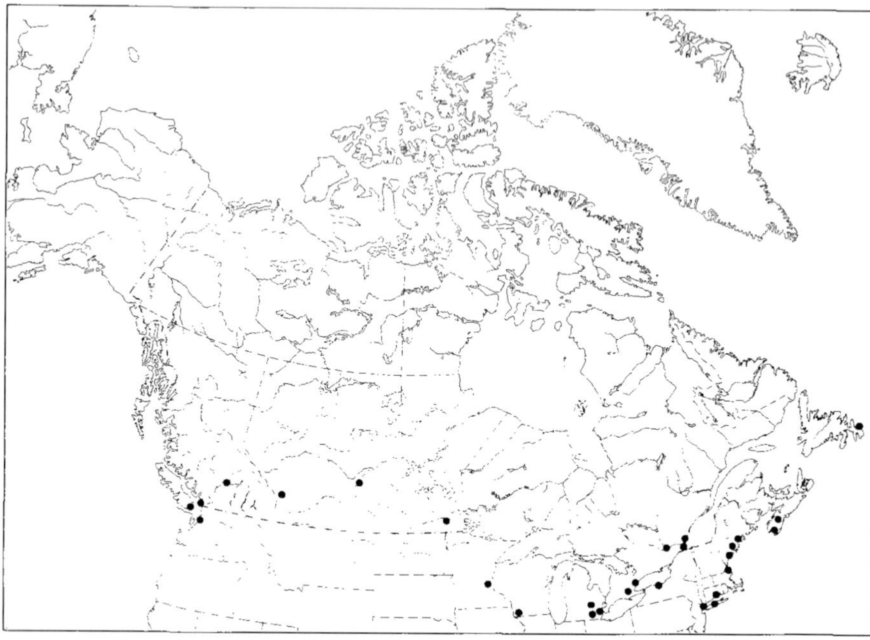

Map 15. Collection localities for *Periplaneta americana*.

Description. Body medium large (male 29.7–34.2, female 27.8–34 mm long), red brown. Pronotum 7.6 mm long (male) or 9.6 mm (female), subelliptical, widest behind middle, narrowed in front, with all angles broadly rounded (Fig. 12), yellow brown, with 2 large suffused chestnut brown blotches. Tegmina and wings fully developed in both sexes. Male cercus with 15 segments, long and strongly tapering. Supra-anal plates of male and female as in Figs. 16, 17. Subgenital plate of male as in Fig. 18. Chromosome number, $2n$ male = 33 (Cohen and Roth 1970).

Range. British Columbia to Newfoundland (in buildings); cosmopolitan.

Behavior and habitats. *P. americana* prefers a warm moist environment. In southern areas of the North American continent it is often abundant in dumps, wood piles, and outbuildings; it is found on palm trees on the Gulf coast of Texas, and in decaying debris, and so forth, almost universally in Bermuda. It also is known from caves and coal mines. In temperate regions, it cannot withstand winter temperatures except in heated buildings. It is a common pest of restaurants, bakeries, grocery stores, and premises where food is prepared, stored, or sold. It may also live in latrines, privies, and sewers and it has often been reported infesting cargo ships.

Roth and Willis (1956) reported sexual as well as parthenogenetic reproduction in this species. After mating, females produce oothecae which they carry for 24 hours then deposit in carefully selected locations. The ootheca is a strong ovate to quadrate capsule about 5.2–7.8 mm long. Each ootheca contains about 16 eggs in two parallel rows.

P. americana feeds upon a wide variety of products such as bread, pastry, cheese, and various fruits and vegetables. In greenhouses, they will feed upon flower petals and tender shoots. Willis and Lewis (1957) found this species able to survive without water for up to 40 days, but it was unable to use available dry food. Survival up to 90 days on water alone is possible. *P. americana* may disperse to new locations during warm weather due to population pressure or other factors, but normally it tends not to migrate (Cornwell 1968; Jackson and Maier 1961). New infestations usually result from accidental transfer by man.

Life history. Cornwell (1968) reported the incubation period for eggs as averaging 30–45 days with extremes of 24–100 days. Nymphs may take a year or more (depending upon conditions) to mature, 5–15 months during which time they molt 7–13 times. Under adverse conditions, nymphs cease growth, greatly prolonging the life cycle (Nigam 1933). Longevity of adults varies from 102 to 588 days, depending upon temperature.

Economic importance. This species often created severe problems during the years prior to 1945. With the advent of many newer insecticides, it is now somewhat easier to control, but the problem persists of introducing chemicals into the inaccessible places it inhabits, and of using chemicals in places where food stuffs may become contaminated.

Periplaneta brunnea Burmeister

Figs. 15, 19–21; Map 14

P[eriplaneta] brunnea Burmeister, 1838:503.

P[eriplaneta] truncata Krauss, 1892:165.

Periplaneta brunnea; Hebard 1917*c*:182; Helfer 1963:52; Vickery et al. 1974:24; R. J. Harrison 1980:10.

Common name. Brown cockroach.

Diagnosis. Supra-anal plate of male truncate, of female feebly bilobate.

Description. Body medium large (male 25.1–28.5, female 26.5–33.0 mm long), dark red brown. Pronotum with slight discal impressions (Fig. 15), with two indistinct dark blotches (Fig. 15). Tegmina fully developed, uniformly brown. Male cerci short, acute. Supra-anal plates of male and female as in Figs. 19 and 20. Subgenital plate of male as in Fig. 21. Chromosome number, $2n$ male = 27 (Cohen and Roth 1970).

Range. Native to South America; established (once) in Montréal and subsequently eliminated; adventive.

Superfamily Cryptocercoidea

The general characters of the Cryptocercoidea, or hooded cockroaches, have been given in the key to superfamilies of Blattodea and need not be repeated. These cockroaches are strongly sclerotized, and are of special interest because of their ability to feed on wood, digestion being achieved by the action of peculiar flagellate protozoa (Trichonymphidae and related families) similar to those found in the alimentary canals of certain wood-feeding termites. Whereas the Cryptocercoidea have certain termite features (McKittrick 1965), they do not much resemble in general appearance even the largest and most heavily sclerotized termites.

The superfamily contains a single genus, *Cryptocercus* Scudder, which has a disjunct relict geographical distribution. Two species occur in China and one in North America.

Family Cryptocercidae

Genus *Cryptocercus* Scudder

Description. Body size medium (23–29 mm long), robust, apterous. Eyes smaller than antennal sockets. Last dorsal and ventral abdominal segments large, concealing supra-anal and subgenital plates and genitalia. Pronotum convex, broader than long, with wide deep central depression, and with anterior margin broad, upturned, forming hood over head. Legs short, stout.

Cryptocercus punctulatus Scudder

Fig. 7; Map 13

Cryptocercus punctulatus Scudder, 1863*a*:420.

Diagnosis. Body apterous, dark. Eyes smaller than antennal sockets.

Description. As for genus *Cryptocercus*. Body shining dark brown to almost black, with reddish brown abdomen and venter, and with dark brown pronotum. Chromosome number, $2n$ male = 39 (Cohen and Roth 1970).

Range. Washington to California; New York to Northern Georgia.

Behavior and habitats. Hebard (1917*c*) reported two colonies in Oregon found in fir logs. The logs were soggy and the insects were found under the bark, burrowing into the softer decayed areas of sapwood. Hebard also stated that the species had always been found in similar situations in decaying trunks of a variety of species of trees. Blatchley (1920) reported it as most numerous in Virginia and North Carolina, in the decaying sapwood of pine or chestnut logs. As already noted *C. punctulatus* possesses intestinal flagellates, very like those of termites, that enable it to feed upon and derive nourishment from the wood which it excavates from its burrows.

Life history. Little is known to the seasonal history, but adults have been recorded only from June to August.

Superfamily Ectobioidea

Members of this superfamily are the "Blattellidae," *sensu lato*, of McKittrick (1964). Their characters are given in the key to superfamilies of Blattodea. Despite apparent evolutionary convergence, and despite similarities and overlapping of genitalic, proventricular, and biological characters, there would seem to be some practical advantage in continuing to recognize more than McKittrick's single family for this great array of species, which constitute about half the known species of cockroaches.

Key to families of Ectobioidea

1. Body small (seldom more than 10 mm long), pale. Hind wings with large apical intercalated triangle. Subgenital plate of male obliquely asymmetrical, widely triangular, with reduced styles .... **Ectobiidae** (p. 103)
   Body usually somewhat larger (10–20 mm, or more), generally darker, or pale with distinct dark marks. Hind wings without large apical triangle. Subgenital plate of male of various forms, but not as above .... **Blattellidae** (p. 105)

Clé des familles d'Ectobioidea

1. Corps petit (de longueur dépassant rarement 10 mm) et pâle. Ailes postérieures munies, dans la partie apicale, d'un large triangle intercalaire. Plaque sous-génitale des mâles obliquement asymétrique, formant un triangle large, munie de styles réduits .... **Ectobiidae** (p. 103)
   Corps habituellement plus grand (de 10 à 20 mm ou plus) et en général plus foncé ou pâle avec des marques foncées distinctes. Ailes postérieures dépourvues d'un large triangle apical. Plaque sous-génitale des mâles de formes diverses, mais différente de celle décrite plus haut .... **Blattellidae** (p. 105)

Family Ectobiidae

The main distinguishing features of this family have already been given in the key to families. There are 10 genera and 180 known species, the great majority of them being confined to the Old World. Some species are the most fully adapted of all cockroaches to life in cold climates.

Genus *Ectobius* Stephens

Description. Body small, rather delicate (less than 10 mm long), tawny or grayish, agreeing in general terms with the characteristics already given for the family. Apical triangle of hind wing traversed by single vein.

Ectobius pallidus (Olivier)

Fig. 50; Map 16

Blatta pallida Olivier, 1789:319.
Blatta livida [*nec* De Geer, 1773] Fabricius, 1793:10.
Blatta livens Turton, 1802:529.
Ectobius pallidus; Flint 1951:53 (first North American record).
Ectobius pallidus; Roth and Willis 1957*b*:31–37; Helfer 1963:58.

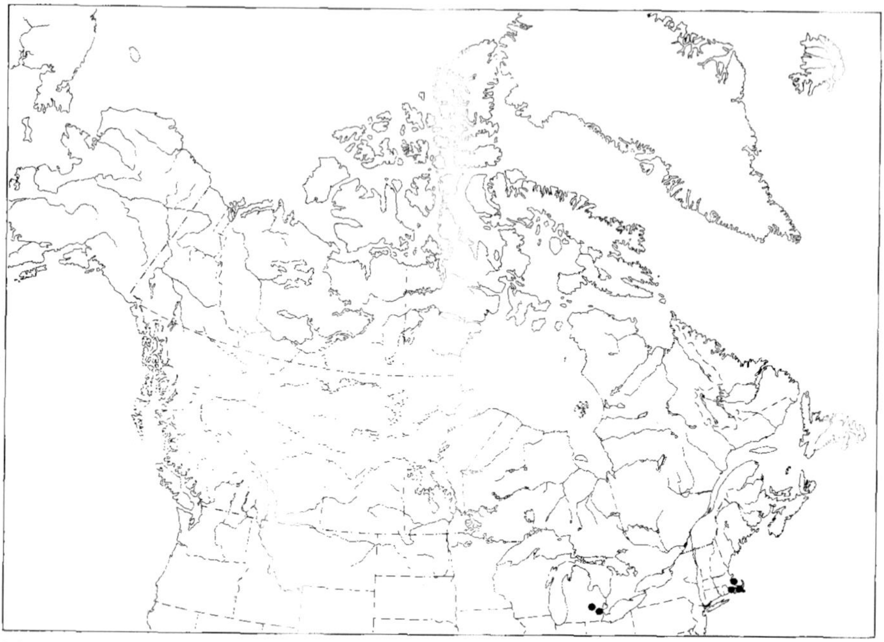

Map 16. Collection localities for *Ectobius pallidus*.

Common name. Spotted Mediterranean cockroach.

Diagnosis. Body small (9 mm, or less, long), rather broad, pale tawny with maculations.

Description. Body small (8–9 mm long, including tegmina), pale tawny yellow with numerous small dark brown and blackish dots, or maculae, particularly on the tegmina, but also on pronotum. Hind wing with large intercalated triangular area at apex, veinless except for single indistinct long vein. Tarsi each with distinct arolium, with claws of unequal length. Male subgenital plate with single small style. Chromosome number, $2n$ male = 21 (Cohen and Roth 1970).

Range. Michigan and Massachusetts; Europe.

Behavior and habitats. Individuals of *E. pallidus* are found among leaf litter in shady woods, under tree bark, and under debris. They are known to run on the surface of vegetation at night. They are fully alate, and both sexes can fly well. In both Massachusetts and Michigan they are known to invade domestic premises. Roth and Willis (1957*b*) give an account of the biology of the introduced populations of this species. They mention that two species of ant prey upon it and that there is a mermithid nematode parasite.

Life history. The species apparently passes the winter in the egg stage, although overwintering may take place under natural conditions in the nymphal stage (Roth and Willis 1957*b*), but it can probably survive in buildings as adults.

Economic importance. The species has no economic significance in Europe. Gurney (1968) states that it is not likely to become an important pest in North America but that the occasional infestation of buildings, particularly private houses, may occur.

Family Blattellidae

The family Blattellidae constitutes the largest single family of cockroaches. It occurs (at least as introductions) throughout almost the entire range known for the order Dictuoptera. On an individual species basis, one member of the family, *Blattella germanica* (Linnaeus), is among the most widely dispersed of all insects, occurring, as it does, on some of the remotest oceanic islands, and, under artificially heated conditions, within about 750 km of the North Pole. Blattellidae includes the only native Canadian cockroaches, and, with the exception of *Cryptocercus*, the only endemic species for the whole region here covered. The native species are, however, exclusively eastern; by contrast with the two introduced species, none can be regarded as a pest, although sensitive people object to their occasional presence in summer homes and camps.

Key to subfamilies of Blattellidae

1. Male with retractable hook on right phallomere; basal phallic sclerite long, unforked. Female carrying ootheca with keel uppermost.[1] Tenth tergum with deep terminal notch. First pair of ovipositor valves not embracing second and third valves laterally. Spermathecae in single pair, with openings at tip of genital papilla. Laterosternal shelf broad, with posterior border flat or slightly curved, rarely narrowly divided, but if not of this form, then first valvifer arms and spermathecal plate firmly joined anteriorly; vestibular sclerite small, broad .... **Pseudophyllodromiinae** (p. 106)

Male with retractable hook on left phallomere; basal phallic sclerite more or less forked. Female carrying ootheca with keel directed laterally (i.e., rotated through 90°).[1] Tenth tergum with slight notch or none. First pair of ovipositor valves embracing second and third valves laterally. Spermathecae usually in 2 (occasionally more) pairs (if 1 pair only, then first valvifer arms and spermathecal plate separated anteriorly by membrane), with openings on groove of genital papilla; vestibular sclerite small, slender, discrete, or absent .... **Blattellinae** (p. 108)

[1] The position of the ootheca in dried specimens or in species in which it is rotated, may differ from the positions indicated, either due to distortion or to the fact that rotation has not been completed. Caution should therefore be exercised in using this character.

Clé des sous-familles de Blattellidae

1. Chez les mâles, crochet rétractile sur le phallomère de droite; sclérite phallique basilaire long et non fourchu. Chez la femelle, l'oothèque est portée la partie inférieure retournée vers le haut[1]. Dixième tergite portant une indentation profonde. La première paire de valves de l'ovipositeur n'enveloppe pas la deuxième et la troisième paires latéralement. Une seule paire de spermathèques, s'ouvrant à l'extrémité de la papille génitale. Rebord latérosternal large, à bord postérieur plat ou légèrement incurvé et rarement divisé; sinon, premières valvifères et plaque de la spermathèque jointes solidement en avant; sclérite du vestibule petit et large .... **Pseudophyllodromiinae** (p. 106)

   Chez les mâles, crochet rétractile sur le phallomère de gauche; sclérite phallique basilaire plus ou moins fourchu. Chez la femelle, oothèque portée le dessous tourné vers le côté (rotation de 90°)[1]. Dixième tergite à indentation légère ou sans indentation. Première paire de valves de l'ovipositeur enveloppant la deuxième et la troisième paires latéralement. D'habitude deux paires de spermathèques (parfois plus). S'il n'y en a qu'une paire, les premières valvifères et la plaque de la spermathèque sont séparées antérieurement par une membrane. Les spermathèques s'ouvrent dans le sillon de la papille génitale. Sclérite du vestibule petit, grêle, dissimulé ou absent .... **Blattellinae** (p. 108)

[1] La position de l'oothèque chez les spécimens séchés ou chez les espèces où elle décrit un mouvement de rotation peut différer des positions décrites soit à cause d'une distorsion ou à cause d'une interruption du mouvement de rotation. Ce caractère ne doit donc être utilisé qu'avec précaution.

Subfamily Pseudophyllodromiinae

This is McKittrick's (1964) invalid subfamily Plectopterinae.

Genus *Supella* Shelford

Description. Body size medium (10.5–12 mm long), slender. Eyes well-separated, with interocular area flattened and raised to form rounded angulation with ocellar area. Tegmina of male narrow, delicate, of females somewhat reduced; wings hyaline. Male subgenital plate asymmetrical, elongate, triangular, fused with styles. Female subgenital plate large, broad. Legs slender; fore femur armed with weak spines that gradually diminish in size toward apex, followed by 2 large subequal spines near apex (Fig. 39).

Supella longipalpa (Fabricius)

Figs. 37–39; Map 17

Blatta longipalpa Fabricius, 1798:185.
Blatta supellectilium Audinet-Serville, 1838:114.
Ischnoptera vacillons F. Walker, 1868:114.
Supella supellectilium; Hebard 1917*c*:47; Helfer 1963:46.
Supella longipalpa; Princis 1960*b*:193; Vickery et al. 1974:30; R. J. Harrison 1980:10.

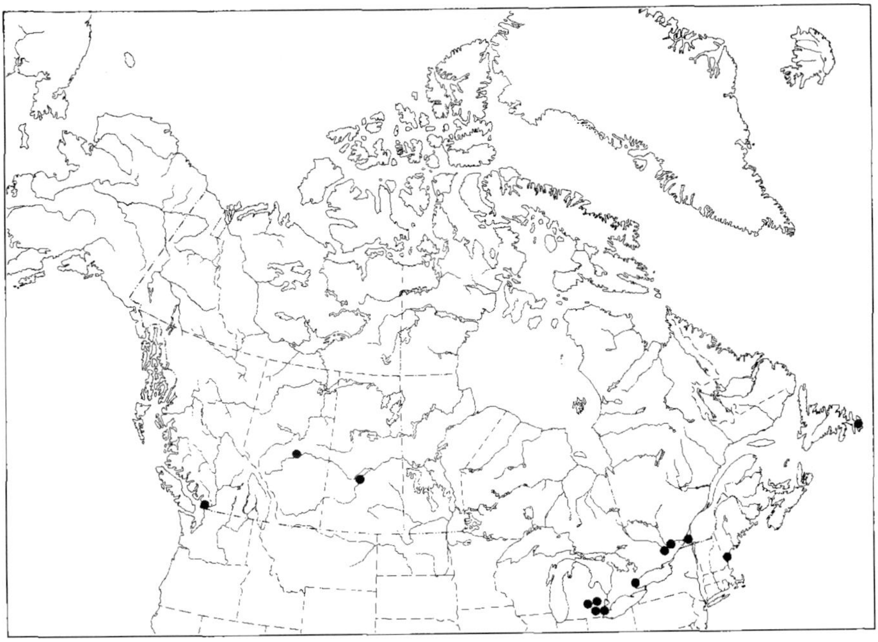

Map 17. Collection localities for *Supella longipalpa*.

Common name. Brownbanded cockroach.

Diagnosis. Tegmina pale brown, banded by 2 darker blotches.

Description. As for genus *Supella*. Pronotum one-third wider than long, narrowed from posterior third to anterior end, with disc feebly convex and with posterior margin weakly convex. Tegmina pale brown, with 2 darker blotches. Dorsum of sixth abdominal segment of male with circular deeply depressed area containing low conical tuft of closely adpressed hairs. Chromosome number, $2n$ male = 19 (Cohen and Roth 1970).

Range. British Columbia to Newfoundland (in buildings); cosmopolitan.

Behavior and habitats. In northern latitudes, the brownbanded cockroach is entirely domiciliary, surviving winter only in heated buildings. It ranks with *Blattella germanica* in importance as a household pest. In some instances it has displaced *B. germanica*. On account of its propensity to seek the extra warmth to be found in radio and television receivers, it has sometimes been dubbed the "television (cock)roach."

Life history. Gould and Deay (1940) and Hull and Davidson (1958) studied the life cycle of this species under laboratory conditions. The adult life span varies from 4 to 10 months at about 22°C. Each female deposits an average of 10 oothecae at about 10-day intervals. Nymphs mature in about 3 months at 28°C but require twice as long at 22°C. Under warm conditions, breeding is continuous.

Economic importance. *S. longipalpa* is increasing in importance and no doubt will continue to do so as it spreads within communities where it is established and to other cities and towns where it is not yet known. It is a very agile species. It tends to dominate in locations where it occurs, so that previously present species disappear. It does not seem to be so readily controlled by chemical means as are other domiciliary species.

Subfamily Blattellinae

In the restricted sense of McKittrick (1964), so far as the region here covered is concerned, this subfamily includes only the native genus *Parcoblatta* Hebard and the ubiquitous introduced domiciliary *Blattella germanica* (Linnaeus).

Key to genera of Blattellinae

1. Pronotum pale, with 2 conspicuous longitudinal dark bars (Fig. 41). Tegmina of female not distinctly abbreviated .... ***Blattella*** (p. 108)
   Pronotum dark with pale anterior and lateral margins. Tegmina of female usually abbreviated .... ***Parcoblatta*** (p. 110)

Clé des genres de Blattellinae

1. Pronotum pâle, marqué de 2 lignes longitudinales foncées (fig. 41). Tegmina des femelles non clairement tronqués .... ***Blattella*** (p. 108)
   Pronotum foncé, à bords antérieurs et latéraux pâles. Tegmina des femelles habituellement tronqués .... ***Parcoblatta*** (p. 110)

Genus *Blattella* Caudell

Description. Body small (male 10.5–11.5, female 11–12.8 mm long), rather narrow. Head long, nearly hidden by pronotum. Eyes well-separated, with interocular area not flattened. Tegmina narrow, surpassing apex of abdomen; hind wings hyaline. Male subgenital plate asymmetrical. Female subgenital plate large, convex, truncate. Anteroventral margin of fore femur armed on basal half with longer spines followed apically by row of much shorter spines, with 3 longer unequal spines near apex (Fig. 42); tarsal segments elongate, with arolia small.

Blattella germanica (Linnaeus)

Figs. 40–42; Map 18

Blatta transfuga Brunnich, 1763:697.
Blatta germanica Linnaeus, 1767:668.
Blatta bivittata Saussure, 1864*b*:102.
Blatella germanica; Caudell 1903*b*:239.

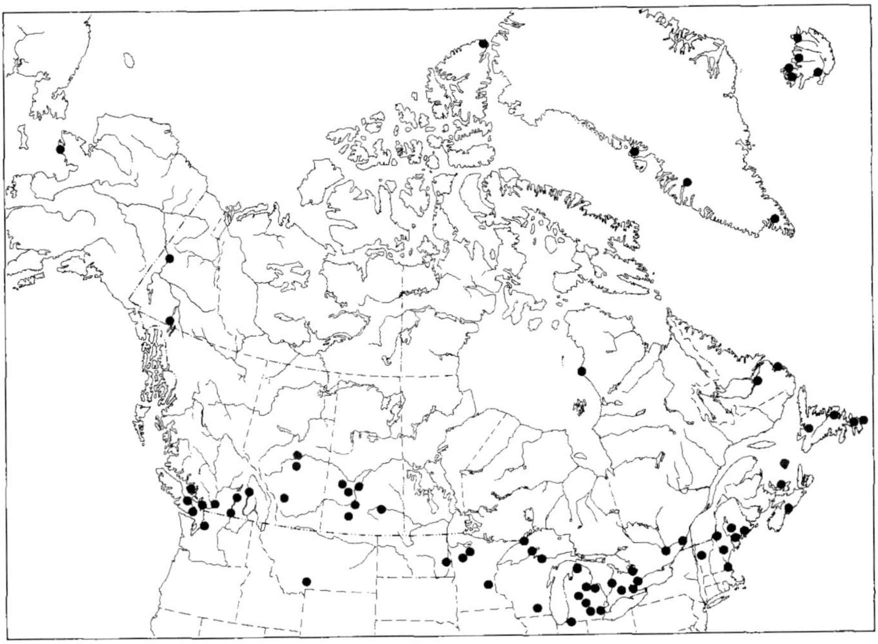

Map 18. Collection localities for *Blattella germanica*.

Common name. German cockroach.

Diagnosis. Pronotal disc with 2 dark longitudinal stripes (Fig. 41).

Description. As for genus *Blattella*. Pronotal disc with 2 blackish or dark brown longitudinal stripes separated by testaceous stripe (Fig. 41); general color dull brownish yellow, with legs paler; females often darker than males. Tegmina and wings of female extending to apex of abdomen, those of male longer. Chromosome number, $2n$ male = 23 (Cohen and Roth 1970).

Range. Alaska to Newfoundland; cosmopolitan.

Behavior and habitats. Requires food, warmth, and moisture and can establish itself anywhere if these requirements are met. It is reported from stores, warehouses, bakeries, food-processing and storage buildings, and dwellings, often in great numbers. It will displace other species of cockroach in places it infests, being an excellent competitor. It may itself be displaced by *Supella longipalpa*, which is, however, not so universally distributed.

Life history. Adults mate within a few days of reaching the adult stage. Females will copulate repeatedly. Willis et al. (1958) reported on development at 30°C. Each female produces 4–8 oothecae (average 7) with an average of 37 eggs per ootheca. Nymphs have 5–7 molts and, in isolation, males mature to adults in 38 days with 5 molts; females in 63 days with 7 molts. However, when reared in groups, both sexes mature in about 40 days. Adult longevity varies from 128 days (average for male) to 153 days (average for female). At lower temperatures, the life cycle is extended somewhat.

Economic importance. *B. germanica* is a common pest of houses, restaurants, and so forth, and is difficult to eradicate. Individuals move rapidly and are able to utilize very small spaces in which to hide. In many cities, this species is responsible for much of the livelihood of pest control operators; without it many would go out of business.

Genus *Parcoblatta* Hebard

Description. Body size medium (male 11.5–24.5, females 9.8–22 mm long), ovate–elongate in male (Fig. 43), broader and more compact in female. Pronotum feebly convex, in male elliptic, narrowed anteriorly, with posterior margin rounded, and with 2 oblique impressions on disc, in female semicircular, with posterior margin truncate, lacking impressions on disc. Tegmina membranous, translucent, extending beyond abdomen in males, not reaching apex in females (and often abbreviated) except in *P. caudelli.* Male styles slender, deflexed, cylindrical; right style slightly longer than left.

Key to species of *Parcoblatta*

(Tegmina and wings of males must be separated to show dorsal abdominal segments.)

1. Tegmina exceeding apex of abdomen (Fig. 43). Pronotum elliptic, narrowed anteriorly, with posterior margin rounded. Subgenital plate with right style slightly longer than left (males) 2

   Tegmina reduced (except in *caudelli*). Pronotum nearly semicircular, with posterior margin truncate (Figs. 44, 45). Subgenital plate with styles symmetrical (females) 7

2(1). Dorsum of median segment (apparent first abdominal[1]) alone specialized (Figs. 47, 48) 3

   Both median and true first abdominal segments (apparent first and second abdominals) with twin specialized areas (Fig. 46) 5

3(2). Median abdominal segment with single specialized area, this moderately large and with minute scattered hairs (Fig. 47) . . . . *virginica* **(Brunner von Wattenwyl)** (p. 112)

   Median abdominal segment with smaller twin specialized areas (Fig. 48) . . 4

4(3). Pronotum short, widest at middle. Tegmina much broader than pronotum. Subgenital plate when viewed from beneath, not elevated at base of "right" style[2] . . . . *uhleriana* **(Saussure)** (p. 113)

   Pronotum longer, widest behind middle. Tegmina little broader than pronotum. Subgenital plate elevated at base of "right" style . . . . *fulvescens* **(Saussure & Zehntner)** (p. 115)

5(2). Specialized areas on dorsal abdominal segments weakly elevated; anterior faces with heavy tuft of hairs 6

   Specialized areas on dorsal abdominal segments distinctly elevated; ventral face of overhanging extremities heavily supplied with hairs (Fig. 46) . . . . *pennsylvanica* **(De Geer)** (p. 115)

6(5). Body size medium, slender. Pronotum 2.9–3.4 mm long, 3.9–4.7 mm wide . . . . *caudelli* **Hebard** (p. 117)

---

[1] The median segment is actually the metathorax; it appears to be the first abdominal segment. The actual first abdominal segment thus appears to be the second.

[2] Meaning "right" only from the viewpoint of the observer; actually left in respect of the insect itself.

Body larger, robust. Pronotum 4.2–5.2 mm long, 5.4–6.7 mm wide . . . . *lata* **(Brunner von Wattenwyl)** (p. 117)

7(1). Tegmina strongly reduced 8

Tegmina moderately reduced, not confined to lateral position or apically truncate 11

8(7). Tegmina represented by rounded subtriangular lateral lobes (Fig. 45) . . . . *uhleriana* **(Saussure)** (p. 113)

Tegmina represented by subquadrate slightly overlapping lobes 9

9(8). Tegmina truncated at apex of anal field (Fig. 44). Pronotum 3.1–4.9 mm long, 4.6–6.6 mm wide 10

Tegmina truncated distal to apex of anal field. Pronotum 4.8–6.1 mm long, 6.8–8.2 mm wide . . . . *lata* **(Brunner von Wattenwyl)** (p. 117)

10(9). Body small, slender (pronotum 3.2–3.8 mm long, 4.0–4.9 mm wide), pale. Supra-anal plate with apex straight to sharply rounded . . . . *virginica* **(Brunner von Wattenwyl)** (p. 112)

Body medium in size (pronotum 3.1–4.9 mm long, 4.6–6.6 mm wide), brown to blackish brown. Supra-anal plate with apex feebly concave to bluntly rounded . . . . *fulvescens* **(Saussure & Zehntner)** (p. 115)

11(7). Apices of tegmina not reaching beyond apex of supra-anal plate. Body dark brown, with lateral pronotal margins pale . . . . *pennsylvanica* **(De Geer)** (p. 115)

Apices of tegmina exceeding apex of supra-anal plate. Body tawny . . . . *caudelli* **Hebard** (p. 117)

Clé des espèces de *Parcoblatta*

(Les tegmina et les ailes des mâles doivent être ouverts pour permettre d'observer les tergites abdominaux.)

1. Tegmina dépassant l'extrémité de l'abdomen (fig. 43). Pronotum elliptique, plus étroit en avant et à bord postérieur arrondi. Sur la plaque sous-génitale des mâles, le style droit est légèrement plus long que le style gauche 2

Tegmina réduits (sauf chez le *caudelli*). Pronotum presque semi-circulaire, à bord postérieur tronqué (fig. 44 et 45). Sur la plaque sous-génitale des femelles, les styles sont de longueur égale 7

2(1). Le tergite du segment médian (le premier segment abdominal apparent[1]) est le seul à porter des structures spécialisées (fig. 47 et 48) 3

Le segment médian et le premier véritable segment abdominal (premier et deuxième segments apparents) portent une paire de structures spécialisées (fig. 46) 5

3(2). Segment abdominal médian portant une seule structure spécialisée, de taille moyenne, parsemée de poils (fig. 47) . . . . *virginica* **(Brunner von Wattenwyl** (p. 112)

Segment abdominal médian portant une paire de structures spécialisées plus petites (fig. 48) 4

4(3). Pronotum court, plus large au milieu. Tegmina beaucoup plus larges que le pronotum. Plaque sous-génitale, vue du dessous, non relevée à la base du style droit[2] . . . . *uhleriana* **(Saussure)** (p. 113)

---

[1] Le segment médian est en réalité le métathorax, même s'il semble s'agir du premier segment abdominal. Ainsi, le véritable segment abdominal semble être le deuxième et non le premier.

[2] À droite par rapport à l'observateur; en réalité, il s'agit du côté gauche de l'insecte.

Pronotum plus long, plus large en arrière de la partie médiane. Tegmina un peu plus larges que le pronotum. Plaque sous-génitale relevée à la base du style droit ..... ***fulvescens* (Saussure & Zehntner)** (p. 115)

5(2). Structures spécialisées des segments abdominaux légèrement protubérantes; faces antérieures pourvues d'épaisses touffes de poils 6

Structures spécialisées des segments abdominaux clairement protubérantes; face ventrale des extrémités surplombantes abondamment recouvertes de poils (fig. 46) .... ***pennsylvanica* (De Geer)** (p. 115)

6(5). Corps de taille moyenne et grêle. Pronotum de 2,9 à 3,4 mm de longueur, de 3,9 à 4,7 mm de largeur .... ***caudelli* Hebard** (p. 117)

Corps plus gros et robuste. Pronotum de 4,2 à 5,2 mm de longueur, de 5,4 à 6,7 mm de largeur ...... ***lata* (Brunner von Wattenwyl)** (p. 117)

7(1). Tegmina de taille très réduite 8

Tegmina de taille légèrement réduite, non limités à une position latérale ni apicalement tronqués 11

8(7). Tegmina en forme de lobes latéraux subtriangulaires arrondis (fig. 45) . ... ***uhleriana* (Saussure)** (p. 113)

Tegmina en forme de lobes subquadrangulaires légèrement superposés ... 9

9(8). Tegmina tronqués à l'apex de la zone anale (fig. 44). Pronotum de 3,1 à 4,9 mm de longueur, de 4,6 à 6,6 mm de largeur 10

Tegmina tronqués en position distale de l'apex de la zone anale. Pronotum de 4,8 à 6,1 mm de longueur, de 6,8 à 8,2 mm de largeur .... ***lata* (Brunner von Wattenwyl)** (p. 117)

10(9). Corps petit, grêle (pronotum de 3,2 à 3,8 mm de longueur, de 4,0 à 4,9 mm de largeur) et pâle. Apex de la plaque supra-anale variant de rectiligne à fortement arrondi .... ***virginica* (Brunner von Wattenwyl)** (p. 112)

Corps de taille moyenne (pronotum de 3,1 à 4,9 mm de longueur, de 4,6 à 6,6 mm de largeur), brun à brun noirâtre. Plaque supra-anale ayant un apex variant de légèrement concave à légèrement convexe .... ***fulvescens* (Saussure & Zehntner)** (p. 115)

11(7). Apex des tegmina ne dépassant pas l'apex de la plaque supra-anale. Corps brun foncé, bords latéraux du pronotum pâles .... ***pennsylvanica* (De Geer)** (p. 115)

Apex des tegmina dépassant l'apex de la plaque supra-anale. Corps brun clair .... ***caudelli* Hebard** (p. 117)

Parcoblatta virginica (Brunner von Wattenwyl)

Figs. 44, 47; Map 19

Temnopteryx virginica Brunner von Wattenwyl, 1865:86.
I[schnoptera] borealis Brunner von Wattenwyl, 1865:153.
Parcoblatta virginica; Hebard 1917*c*:96; Cantrall 1943:72; Helfer 1963:41.

Diagnosis. Outdoor species. Male abdomen with median segment specialized. Female small, pale.

Description. Median abdominal segment possessing large specialized area with minute scattered hairs (Fig. 45). Male supra-anal plate four times as broad as long, truncate. Tegmina of females greatly reduced, subquadrate, slightly overlapping (Fig. 44). Supra-anal plate of female with lateral margins straight, converging posteriorly to sharply rounded apex.

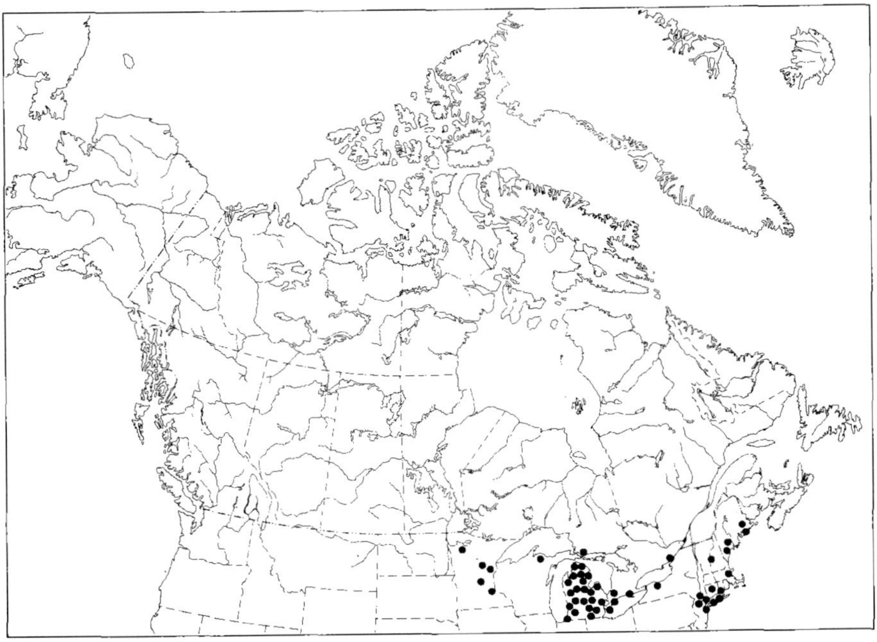

Map 19. Collection localities for *Parcoblatta virginica*.

Range. Minnesota to Maine, south to Alabama and North Carolina.

Behavior and habitats. This is a free-living species, found near woods and occasionally near cultivated fields, under logs, leaves, or other debris. Cantrall (1943) reported its favorite habitat as "deep, moist piles of dead and decaying leaves, humus, and the crevices beneath the bark of decaying logs." It appears to be confined to locations where moisture is abundant and continuous, and it apparently cannot survive under dry conditions.

Life history. Winters are passed as nymphs, which mature early in June; adults are present only until mid-August. Adults are most abundant about the middle of June. Males are seldom found after mid-July.

Parcoblatta uhleriana (Saussure)

Figs. 45, 48; Map 20

I[schnoptera] uhleriana Saussure, 1862:169.
P[latamodes] unicolor Scudder, 1863*a*:417.
E[ctobia] lithophila Scudder, 1863*a*:418.
Ischnoptera intricata Blatchley, 1903:186.
Parcoblatta uhleriana; Hebard 1917*c*:105; Cantrall 1943:72; Helfer 1963:43.

Diagnosis. Outdoor species. Pronotum of male short. Tegmina much broader than pronotum; female tegmina with rounded subtriangular lateral lobes.

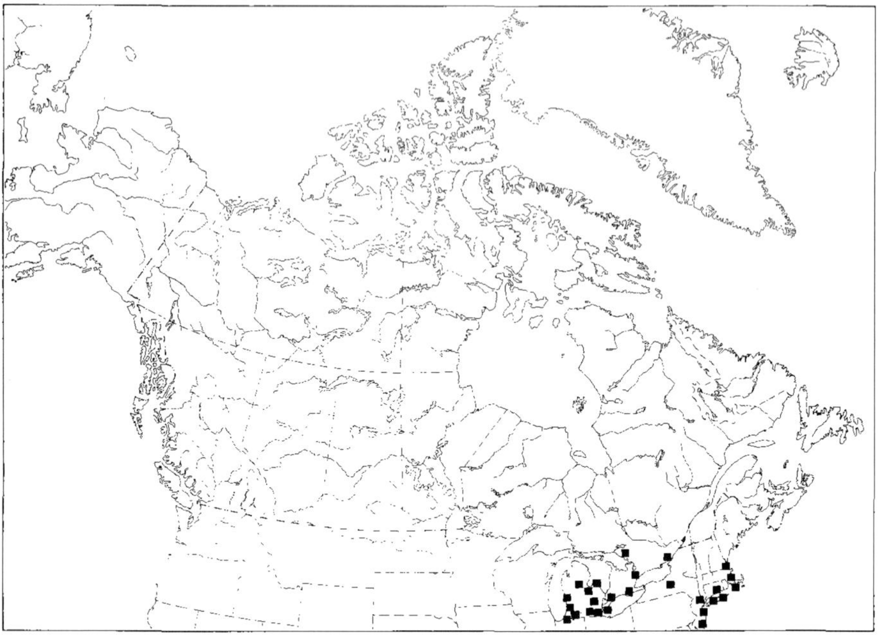

Map 20. Collection localities for *Parcoblatta uhleriana*.

Description. Body slender. Male with twin specialized areas on median abdominal segment (Fig. 48), each bearing tuft of hairs. Male supra-anal plate twice as broad as long, rounded at apex. Female tegmina represented by subtriangular lateral pads, which are separated by space equal to a quarter of width of one pad (Fig. 45). Female supra-anal plate with lateral margins convergent, weakly concave. Body pale brownish yellow, with tegmina often darker; head with brownish stripe from between eyes down two-thirds length of face.

Range. Iowa to southern Ontario, south to Mississippi and Florida.

Behavior and habitats. *P. uhleriana* is found in forested areas, in habitats similar to those of *P. virginica*, although *uhleriana* also inhabits areas that are much drier than those required by *virginica*. Cantrall (1943) found it on the George Reserve, southern Michigan, associated with rotten logs and large piles of dead leaves in oak-hickory forest. Hubbell (1922*b*) trapped specimens only in baited pit-traps. Cantrall (1943) found males in pit-traps as well as at light at night.

Life history. Like other species of *Parcoblatta, P. uhleriana* winters in the nymphal stage, maturing early in June. Adult males disappear early in July and females persist only until late August.

Parcoblatta fulvescens (Saussure & Zehntner)

[*Ischnoptera uhleriana* var.] *fulvescens* Saussure and Zehnter, 1893:36.
Temnopteryx texensis Saussure and Zehntner, 1893:52.
Parcoblatta fulvescens; Hebard 1917*c*:114; Helfer 1963:43.

Diagnosis and description. Body size medium (length 10.8–16.5 mm). Male with median abdominal segment specialized, this with twin weakly arcuate raised ridges that converge anteriorly; each anterior face bearing tuft of hairs. Female tegmina represented by subquadrate overlapping pads; pads truncate apically. Chromosome number, $2n$ male = 37 (Cohen and Roth 1970).

Range. Southern New York, south to Texas and Florida.

Parcoblatta pennsylvanica (De Geer)

Figs. 43, 46; Map 21

Blatta pensylvanica [*sic*] De Geer, 1773:537.
Blatta borealis Saussure, 1862:166.
[*Ischnoptera*] *couloniana* Saussure, 1862:169.
I[*schnoptera*] *nortoniana* Saussure, 1862:169.
Ectobia flavocincta Scudder, 1863*a*:419.
Ischnoptera translucida Saussure, 1864*b*:85.
Temnopteryx marginata Scudder, 1872:251.
Ischnoptera inaequalis Saussure and Zehntner, 1893:36.
Parcoblatta pensylvanica [*sic*]; Hebard 1917*c*:139.
Parcoblatta pennsylvanica; Blatchley 1920:86; Helfer 1963:42; Vickery et al. 1974:28; R. J. Harrison 1980:10.

Diagnosis. Found outdoors. Male long-winged with both median and true first abdominal segments possessing twin specialized areas. Female tegmina not reaching apex of abdomen. Body dark, with pronotal margins pale.

Description. Body large (pronotal length 4.2–5.8 mm). Dorsum of male abdomen specialized as in key (Fig. 46). Male subgenital plate 1.5 times broader than long, convex except near apex where it is narrowed, often concave near styles; left style (as seen from below) slightly stouter and more deflexed than right one. Female subgenital plate strongly convex, twice as wide as long, broadly rounded apically. Tegmina of male exceeding abdomen, of female variable, usually covering two-thirds, or more, of abdomen. Head brown, paler laterally, with pronotal disc and posterior margin brown, yellowish transparent laterally; tegmina pale reddish brown with yellowish transparent margin for one-third to one-half their length; abdomen dark brown; legs mainly yellowish brown on tibiae and tarsi. Chromosome number, $2n$ male = 37 (Cohen and Roth 1970).

Range. North Dakota to Québec, south to Texas and Georgia.

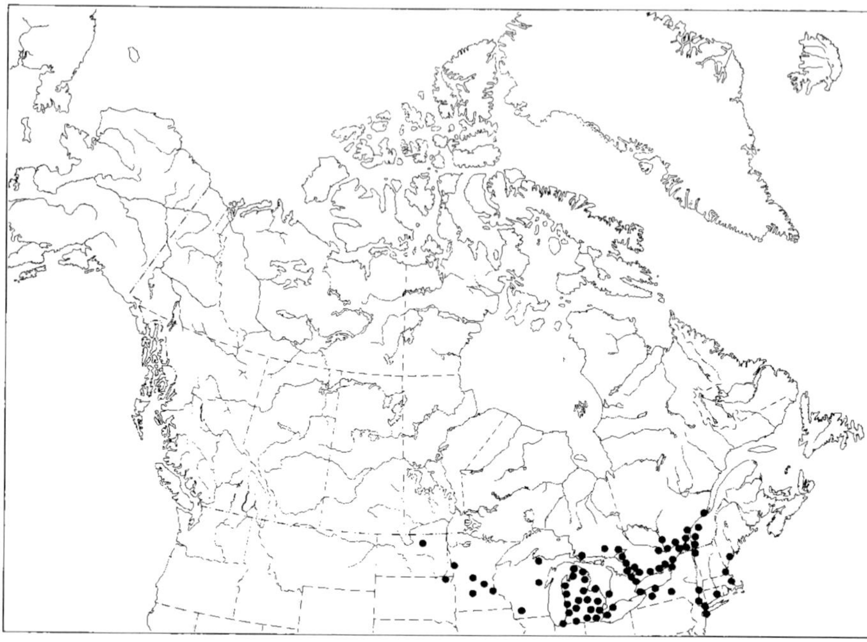

Map 21. Collection localities for *Parcoblatta pennsylvanica*.

Behavior and habitats. *P. pennsylvanica* is native to North America and is a free-living species. It is found usually among fallen leaves, or under rocks or logs, in deciduous forest. It is agile and often escapes detection. It can be trapped, using nearly buried jars or cans containing molasses. Blatchley (1920) reported this species feeding upon dead larvae of *Tenebrio*, so it is possible that other insects, as well as living and decaying vegetation, make up most of the food requirements. He also reported oothecae under tree bark. These are chestnut brown, 5–9 mm long and about 4 mm broad, with one edge slightly concave and the other minutely serrate.

Life history. Hebard (1917) stated that males probably winter as nearly full-grown nymphs, maturing early in the spring. Females mature somewhat later. Adults are common from late May to October. Nymphal stages are common from mid-August until they take shelter for the winter.

Economic importance. Although this species is not confined to domiciliary existence like most cockroach species in the region, it is reported frequently in the fall from summer cottages, motels, or other buildings in forested areas, where it apparently seeks shelter from winter conditions (MacNay 1959*a*, 1961*a*). It is sometimes attracted to light, now and then causing unwarranted alarm. It is not a pest, although occasionally it has damaged wallpaper, when other food was lacking, in getting at the flour paste beneath.

Parcoblatta caudelli Hebard

Parcoblatta caudelli Hebard, 1917*c*:122.
Parcoblatta caudelli; Blatchley 1920:89; Helfer 1963:45.

Diagnosis and description. Body small (10.7–16 mm long). Male with specialized areas on dorsum of both median and first abdominal segments, these weakly elevated with hair tufts on anterior faces. Females with tegmina long, exceeding apex of abdomen.

Range. Ontario (as adventive), Indiana to southern New York, south to Texas.

Parcoblatta lata (Brunner von Wattenwyl)

I[*schnoptera*] *lata* Brunner von Wattenwyl, 1865:135.
Parcoblatta lata; Hebard 1917*c*:126; Helfer 1963:44.

Diagnosis and description. Body large (15.7–22 mm long), robust. Male with specialized dorsal abdominal segments similar to those of *P. caudelli*. Female with tegmina represented by subquadrate slightly overlapping pads; pads truncate distal to anal field.

Range. Massachusetts (as adventive); Maryland, south to Texas and Florida.

Order Grylloptera—katydids and crickets

Members of this order are the insects most commonly called Ensifera in recent literature, where they are regarded as constituting a suborder of Orthoptera, *sensu lato*. They include the crickets, katydids, and long-horned grasshoppers. They share several features with the Orthoptera, *sensu stricto*, notably the generalized biting and chewing mouthparts, usually somewhat backwardly produced pronotum, enlarged and modified hind legs, basically similar wing venation, and a widespread ability, mainly in males, to stridulate loudly. They differ in many ways, however, and some of the shared features are doubtless due to their ancient, although not necessarily immediately common, lineage (mouthparts, wing venation) or to parallel or convergent evolution. (The jumping hind legs probably developed more than once as a "catapult" mechanism for launching early, aerodynamically rather awkward insects into the air; the stridulatory mechanisms are clearly independently evolved, as they are, for the most part, of quite a different nature; they vary throughout both groups, as they do in other orders of insects, notably Hemiptera.) The characters separating Grylloptera from Orthoptera, *sensu stricto*, include the following: antennae fine, threadlike, simple, nearly always at least as long as, and often longer than, body, and with numerous short ill-defined segments. Mandibles of biting type, more or less symmetrical, with rather acute teeth, occasionally greatly enlarged in males. Thoracic pleura largely concealed by lateral pronotal lobes, with visible parts small; prothoracic spiracle and associated tracheal system sometimes enlarged and modified as auditory organ (some Tettigonioidea). Legs with tarsi either

3- or 4-segmented; front tibiae often with proximal auditory organ; hind femur sometimes possessing longitudinal rasplike stridulatory ridge or a patch of fine, rather irregularly distributed teeth on inner faces (some Stenopelmatoidea). Tegmina frequently resembling broad-leaved (dicotyledonous) foliage (many Tettigonioidea), with venation of bases (at least) usually modified for stridulation, typically incorporating a file and enlarged cells, sound being produced by the rubbing of the two together; hind wing with "fan" not confined to anal region (i.e., apical margin normally continuously rounded, but if unequally biarcuate, the emargination is in front of, not behind, cubital veins). Base of abdomen without lateral auditory (tympanal) organs, although often with a smooth area, stridulatory teeth, or ridges. Cerci often long and flexible (not in Tettigonioidea). Ovipositor of female typically sickle-, dagger-, sword-, or needle-like, sometimes reduced, rarely absent, with valves tongued and grooved together so as to form virtually a single structure, and with all six well-developed except in Grylloidea (in which the inner pair are reduced) or when ovipositor is rudimentary or lacking. Male subgenital plate (ninth abdominal sternum) typically with pair of styles, except in Gryllodea; phallic structures lacking true intromittent organ. Salivary glands typically well-developed and having reservoirs; proventriculus (gizzard) normally well-developed, with sclerotized cuticular teeth or ridges; gut with one pair of large gastric caeca; intestine convoluted to some degree; Malpighian tubules coming together in groups and terminating in small common ampullae, or (Grylloidea) uniting in a pair of common ducts. Male ejaculatory duct with pair of associated "prostate" glands. Accessory glands of female reproductive system often opening at base of ovipositor independently of gonopore and spermathecal duct.

The number of chromosomes is variable, but commonly exceeds 12 pairs (female diploid) and may be more than double this. In some Grylloidea, however, there may be half this number, or even less.

Members of Grylloptera are largely, although by no means exclusively, nocturnal. Most of "the music of the night" is due to their stridulation. They are collectively found virtually throughout the habitat range of the Orthopteroida, but are more predominantly tropical and arboreal and less well represented at higher altitudes and latitudes than are Orthoptera, *sensu stricto*. Unlike the latter, many are subterranean burrowers or cave-dwellers, and subaquatic forms do not habitually swim. A large number are omnivorous or at least partially, and sometimes exclusively, carnivorous; many are mainly if not wholly phytophagous (probably a secondary evolutionary development). A few are commensals, living in the habitations of other animals (ants, termites, rodents, and man), but this relationship is not always obligatory. Although mimicry of foliage and other plant materials, and occasionally of other arthropods, is widespread, aposematic (warning) coloration is uncommon, as is the production of irritant and similar secretions.

Members of Grylloptera usually copulate with the female above the male (either the male comes from below or the female mounts the male). This is the opposite of the position in Orthoptera, *sensu stricto*. A spermatophore is produced and this is normally subspherical, flask-shaped, or pyriform, and remains partly extruded from the female, which commonly eats some of it. Eggs, although normally laid in batches of several at a time, are generally deposited separately, frequently in living plant tissues, and not held together in "egg pods."

Gregarious behavior is relatively uncommon, and sustained migration, particularly in flight, rare, although both occur, sometimes spectacularly. A number of species may become crop pests, mostly of a sporadic nature, and a few others may be nuisances in human habitations and in other ways, but, generally, the economic importance is not nearly as great as that of the Orthoptera, *sensu stricto*. There is a large body of folklore associated with some species.

Living Grylloptera number more than 1000 genera and something like 8500 species divided among 2 suborders. These suborders are the Tettigoniodea and Gryllodea.

Key to suborders of Grylloptera

1. Tarsi (at least middle pair) 4-segmented. Tegmina, when present, not bent angularly downward along sides of body when at rest. Cerci shorter and usually rigid, but if longer and flexible, sometimes segmented (although this is almost always weak and confined to the apices). Ovipositor not needlelike, virtually always present, even if small, and with 6 distinct valves .... **Tettigoniodea** (p. 119)

   Tarsi 3-segmented, although metatarsus may appear subdivided. Tegmina, when present and well-developed, bent angularly downward along sides of body when at rest. Cerci long, flexible, almost invariably unsegmented. Ovipositor typically needlelike, occasionally absent, sometimes flattened and/or curved, and with 4 valves well-developed .... **Gryllodea** (p. 245)

Clé des sous-ordres de Grylloptera

1. Tarses (au moins ceux des pattes médianes) à 4 segments. Tegmina, lorsque présents, non recourbés vers le bas sur les côtés du corps en position repliée. Cerques courts et habituellement rigides, mais parfois segmentés lorsqu'ils sont plus longs et flexibles (la segmentation est toutefois presque toujours légère et limitée aux extrémités). Ovipositeur non en forme d'aiguille, presque toujours présent, mais parfois petit et constitué de 6 valves distinctes .... **Tettigoniodea** (p. 119)

   Tarses à 3 segments (le métatarse peut parfois sembler subdivisé). Tegmina, lorsque présents et bien développés, incurvés vers le bas le long des côtés du corps en position repliée. Cerques longs, flexibles, presque jamais segmentés. Ovipositeur en forme typique d'aiguille, parfois absent, parfois aplati et ou incurvé et ne comptant que 4 valves bien développées. .... **Gryllodea** (p. 245)

Suborder Tettigoniodea

The suborder Tettigoniodea consists of the katydids and various other cricketlike insects other than true crickets and mole crickets. The two living infraorders are the Stenopelmatidea and the Tettigoniidea. The former includes about 170 genera and somewhere in the order of 1000 known species; the latter about 900 genera and 5000 species.

Key to infraorders of Tettigoniodea

1. General coloration brownish or grayish, often mottled, but never green. Head rather globular, sometimes appearing disproportionately large (eyes always small) (Fig. 117). Prothoracic spiracles small, simple. Legs often heavily spined and modified for digging; front and middle tibiae often with ventral articulated spines; front tibiae without auditory organs. Wings absent. Cerci usually rather long and flexible .... **Stenopelmatidea** (p. 120)

   General coloration usually largely or wholly green, although brownish, grayish, or mottled forms common. Head of various shapes but not disproportionately large. Prothoracic spiracles often enlarged. Legs, even if heavily spined, not modified for digging; front tibiae normally with basal auditory organs. Wings present, brachypterous, or sometimes micropterous, rarely completely apterous; veins of male tegmina (even if latter reduced) almost always modified at base as stridulatory organs; tegmina leathery unless reduced. Cerci shorter, rigid .... **Tettigoniidea** (p. 159)

Clé des infra-ordres de Tettigoniodea

1. Corps variant de brunâtre à grisâtre, souvent moucheté, mais jamais vert. Tête plutôt globuleuse, de taille parfois disproportionnée (yeux toujours petits) (fig. 117). Stigmates prothoraciques petits et simples. Pattes souvent couvertes de nombreuses épines et modifiées pour creuser; tibias antérieurs et médians souvent munis d'épines ventrales articulées; tibias antérieurs dépourvus d'organes auditifs. Ailes absentes. Cerques habituellement longs et flexibles .... **Stenopelmatidea** (p. 120)

   Corps d'habitude entièrement ou presque entièrement vert, parfois brunâtre, grisâtre ou moucheté. Tête de formes diverses, mais non disproportionnée. Stigmates prothoraciques souvent grands. Pattes jamais adaptées au creusage, même si elles sont couvertes de nombreuses épines; tibias antérieurs normalement pourvus à la base d'organes auditifs. Ailes présentes, insecte brachyptère, parfois microptère ou rarement complètement aptère; nervures des tegmina du mâle (même lorsqu'elles sont réduites) presque toujours modifiées à la base pour former des organes stridulants; tegmina tannés sinon réduits. Cerques plus courts, rigides .... **Tettigoniidea** (p. 159)

Infraorder Stenopelmatidea

Members of this infraorder include leaf-rolling crickets, stone crickets, sand crickets, cave crickets, camel crickets, and weta, characterized by the following combination of characters: general coloration brownish or grayish, sometimes variegated, never green. Head round and bulletlike, sometimes large in proportion to body. Eyes small, widely separated. Antennae occasionally short. Prothoracic spiracles small, simple. Legs often heavily spined and modified for digging, with tarsi 4-segmented (occasionally apparently 3-segmented by reduction of second segment); front and middle tibiae often with ventral articulated spines; lacking auditory organs. Wings absent. Cerci usually rather long and flexible, occasionally weakly segmented, at least toward apices. Subgenital plate (ninth abdominal sternum) of males with pair of styles. Ovipositor normally fairly long and sabrelike, with all 6 valves well-developed, occasionally reduced, rarely absent.

There are about 170 genera and rather more than 1000 living species, divided between two superfamilies as recognized here: Stenopelmatoidea and Rhaphidophoroidea. They are widely distributed throughout the world but are mainly tropical or from the Southern Hemisphere; they are poorly represented in the Palaearctic region. All fossil Stenopelmatidea (known from the early Triassic onward) seem to belong to these superfamilies.

A high proportion of the species are cavernicolous or burrowing forms, many others live on the surface of the ground, e.g., under stones or logs, but there are also many arboreal species, mainly in the Old World tropics. A few are commensals in rodent burrows, and so forth, and some are beetle mimics. Members of the infraorder are mainly nocturnal, largely omnivorous, or predatory. Most do not stridulate, although drumming sounds are made by some. A few species are sometimes minor pests of crops and household nuisances.

The most comprehensive work on the taxonomy of this group remains that of Karny (1937). Most North American species, however, are covered by the classic monograph of Hubbell (1936), which also includes much information other than that of a purely taxonomic nature.

Key to superfamilies of Stenopelmatidea

1. Head and mandibles disproportionately large. Antennae widely separated at bases. Hind femora neither greatly elongate nor unusually robust; tarsi depressed, with distinct pulvilli ... **Stenopelmatoidea** (p. 122)

   Head and mandibles not disproportionately large. Antennae arising fairly close together. Hind femora usually greatly elongate or greatly thickened; tarsi strongly compressed laterally, without pulvilli .... **Rhaphidophoroidea** (p. 126)

Clé des super-familles de Stenopelmatidea

1. Tête et mandibules disproportionnées. Antennes éloignées l'une de l'autre à la base. Fémurs postérieurs ni très allongés ni anormalement robustes; tarses aplatis et pourvus de pulvilles distincts. .... **Stenopelmatoidea** (p. 122)

   Tête et mandibules non disproportionnées. Antennes rapprochées l'une de l'autre à la base. Fémurs postérieurs habituellement très allongés ou très épaissis; tarses fortement aplatis latéralement et dépourvus de pulvilles .... **Rhaphidophoroidea** (p. 126)

Superfamily Stenopelmatoidea

The insects included in this superfamily are stone-crickets, sand-crickets, and leaf-rolling crickets. Although represented by only a few species in North America, they constitute the principal superfamily of Stenopelmatidea, including all winged forms, and all those with "outsize" heads, unusually short antennae or ovipositors, with auditory organs present at bases of anterior tibiae, with a (femoro-abdominal) stridulatory mechanism, or with hind legs that are neither unusually long nor stout. The antennae are usually rather cylindrical and depressed, and have distinct pulvilli, or, if compressed, they are only moderately so; the cerci are rarely segmented, but, if so, then throughout their length (Lezinidae).

Exclusive of the Rhaphidophoroidea, which are commonly regarded as an included family, there are about 125 genera and more than 750 species of Stenopelmatoidea.

Family Stenopelmatidae

The family Stenopelmatidae, *sensu stricto*, includes the stone-crickets, sand-crickets, ground-crickets, and Jerusalem crickets. The family consists of stout-bodied insects, differing from other members of the superfamily Stenopelmatoidea in having the following combination of characters: head disproportionately large. Antennae usually rather short. Pronotum distinctly widened anteriorly. Anterior coxae and tibiae above without spines, the latter without auditory organs; all tibiae flattened and with strong spines for digging; tarsi more or less cylindrical. Wings absent. Cerci rather short, unsegmented. Ovipositor short, but not entirely lacking.

About six genera and a little over 30 species are known. Most are Central American, whence a few extend into western North America. They clearly constitute a relict group, but their fossil record is apparently no earlier than the Miocene. The insects are mainly predators or scavengers that emerge, usually at night, from burrows dug in the earth. They will, however, gnaw at roots and parts of plants in contact with the ground, and thus may occasionally be minor local crop pests.

Genus *Stenopelmatus* Burmeister

Description. Tegmina and wings absent. Head large in proportion to body; head and pronotum glabrous or only indistinctly punctate; pronotum expanded anteriorly, concealing posterior part of head. Body brown. Median spur present on ventral face of fore tibia; hind tibia with prominent conical to subconical spines on apical margin, these forming an uneven ringlet; innermost teeth of ringlet usually longest.

Key to species of *Stenopelmatus*

1. Apical spurs of hind tibia cylindrical; innermost two spurs longer than others .... ***longispina*** **Brunner von Wattenwyl** (p. 123)

Apical spurs of hind tibia spatulate or trowel-shaped; innermost three spurs nearly equal and longer than outer spurs .... *fuscus* **Haldeman** (p. 124)

Clé des espèces de *Stenopelmatus*

1. Éperons apicaux des tibias postérieurs cylindriques; les deux éperons situés le plus à l'intérieur sont plus longs que les autres .... *longispina* **Brunner von Wattenwyl** (p. 123)

Éperons apicaux des tibias postérieurs en forme de spatule ou de truelle; les trois éperons situés le plus à l'intérieur presque égaux ou plus longs que ceux situés à l'extérieur .... *fuscus* **Haldeman** (p. 124)

Stenopelmatus longispina Brunner von Wattenwyl

Map 22

Stenopelmatus longispina Brunner von Wattenwyl, 1888:260, pl. V, fig. 1D.

Stenopelmatus californicus Brunner von Wattenwyl, 1888:261, pl. V, figs. 1A, 1B, 1C.

Stenopelmatus irregularis Brunner von Wattenwyl, 1888:261.

Stenopelmatus terrenus J. A. G. Rehn, 1902*b*:240.

Stenopelmatus longispina; Helfer 1963:305.

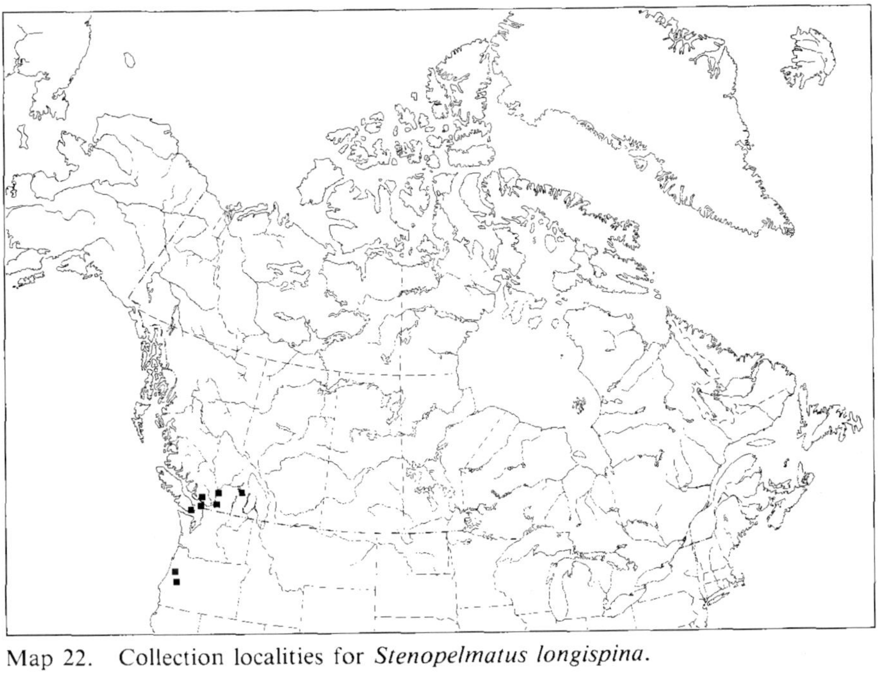

Map 22. Collection localities for *Stenopelmatus longispina*.

Diagnosis. Body large. Head large but not bulbous.

Description. Head large, but not with abnormally bulbous occiput. Hind tibiae weakly concave dorsally; hind tibial spurs and spines rather constant, with innermost two longer than others. Head, pronotum, and legs yellowish to brownish; females with triangular darker transverse markings dorsally on pronotum; abdomen shining dark brown with caudal margins paler; coloration of immature individuals less distinct and pronotal markings less apparent.

Range. British Columbia, south to northern Mexico.

Behavior, habitats, and life history. Presumably similar to those of *S. fuscus*.

Economic importance. Like *S. fuscus*, this species causes sporadic damage to roots of tubers (Spencer 1958).

Stenopelmatus fuscus Haldeman

Fig. 117; Map 23

Stenopelmatus fuscus Haldeman, 1852:372.
Stenopelmatus cephalotes F. Walker, 1869*a*:195.
Stenopelmatus fasciatus C. Thomas, 1872:434.
Stenopelmatus oculatus Scudder, 1876*a*:261.
Stenopelmatus hydrocephalus Brunner von Wattenwyl, 1888:261.
Stenopelmatus comanchus Saussure and Pictet, 1897:290.

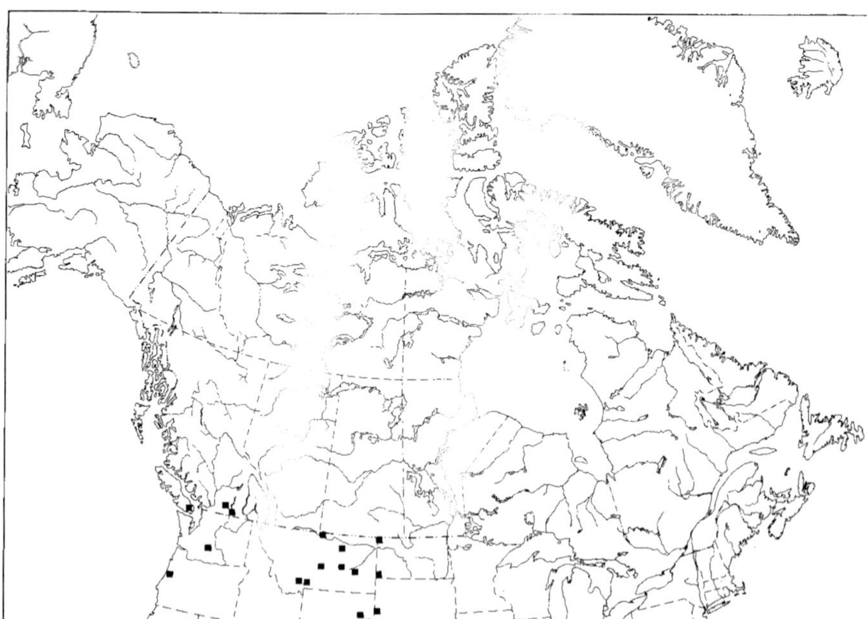

Map 23. Collection localities for *Stenopelmatus fuscus*.

Common name. Jerusalem cricket.

Diagnosis. Body large (up to 50 mm long), shining. Head large, with occiput bulbous in male.

Description. Dorsal surface shining. Head, especially in males, large in proportion to body, particularly in occipital region (Fig. 117). Body large (up to 50 mm long; pronotal length 6–9 mm, hind femur length 11–15 mm; hind tibia length 10–13.8 mm). Head, pronotum, legs, and underside of body buff to yellowish with dark brown markings (tending to orange red in southern areas); abdomen dark brown above with posterior margins of segments a much paler, buff color. Dorsolateral margins of subgenital plate of adult (and preadult) male armed with hooks.

Range. British Columbia to North Dakota, south to Mexico.

Behavior and habitats. Barnum (1964) reported *S. fuscus* as nocturnal and subterranean, found under rocks and debris. However, he also reported finding individuals wandering on the desert in Nevada during late afternoons when humidity was relatively high. The species occupies a wider variety of habitats than other species of the genus in California (Tinkham and Rentz 1969). Davis and Smith (1926) observed mating and reported that the female pursues the male prior to mating, whereas the reverse is true of *S. intermedius*. However, Tinkham and Rentz (1969) believe the specimens called *fuscus* by Davis and Smith may have been *longispina* rather than *fuscus*. Rentz (*in* Tinkham and Rentz 1969) reported on mating of *S. intermedius* and *S. nigrocapitatus*. When a male detected the presence of an acceptable mate, he drummed his abdomen against the substrate making an audible sound. Apparently the drumming is usually done within burrows. The next phase appeared like a battle, with the sexes grasping each other with their mandibles. If the opposite sex was not acceptable, a real battle ensued and the vanquished member was eaten. If the mate was acceptable, the male finally mounted the female with his hind legs placed dorsally between the middle and hind legs of the female at their bases. He then bent his abdomen to meet the genitalia of the female. At the instant of contact a spermatophore was ejected and attached to the female. The male with noticeable abdominal pulsations pumped fluid onto the back of the female until a good part of his abdominal contents seemed to have been emptied. The total act lasted about 5 minutes. It appears that the female attacks and devours the male, if she can, following copulation.

Life history. Hebard (1916*a*) suggested that the life cycle might extend over several years. Tinkham and Rentz (1969) reported that second- or third-instar nymphs which had been brought to the laboratory and reared took 4–5 years to mature.

The prolonged life cycle has resulted in collection of many immatures and few adults. The wingless condition has added to the confusion as late-instar nymphs are difficult to differentiate from adults. Adult males (and last-nymphal-instar males) of *S. fuscus* can be determined by the hooks on the dorsolateral margins of the subgenital plate. In earlier-instar males these appear as swellings or ridges.

Economic importance. In the southern part of its range, *S. fuscus* has been reported as damaging crops, particularly the roots and seedlings, as well as potatoes (Henderson 1931).

Superfamily Rhaphidophoroidea

All members of this superfamily, which includes camel crickets and cave crickets, are completely apterous, lack detectable stridulatory or auditory apparatus (although a few have been reported to make drumming sounds), have the antennae arising close together on the head, and usually have greatly elongate, or sometimes very stout hind legs, whose tarsi are strongly compressed laterally and almost or virtually without pulvilli, occasionally appearing to be 3-segmented by reduction of second segment. The cerci are typically unsegmented, but in some, particularly in males, they are feebly segmented toward their apices.

There are about 300 species and 45, or more, genera, mostly in the family Rhaphidophoridae, found more or less throughout the geographical range of the Stenopelmatidea. The great majority are cave dwellers, or live in crevices, or burrow in sand; a few live on bushes. They seem to feed mainly on dead and dying invertebrates, but some will also take plant material. A few species may be minor pests or nuisances to humans.

Family Rhaphidophoridae

Insects such as camel-crickets, sand-treader crickets, and false cave-crickets, constitute the principal family of Rhaphidophoroidea, in which the apices of the posterior metatarsi have only one dorsal spine or none. They typically have long hind legs, but in many North American digging forms these may be short and stout.

There are about 250 species and 30 genera, divided among 7 subfamilies. Four of these (Ceuthophilinae, Daihiniinae, Tropidischiinae, and Gammarotettiginae, the last two monogeneric) are exclusively North and/or Central American and many live lar[illegible]y in crevices or burrow in sand. The principal subfamily (Rhaphidophori[illegible] occurs mainly in the Indo-Australian region and in Polynesia, with som[illegible]rms reaching China and Japan. One member of this subfamily, *Tachy*[illegible] *asynamorus* Adelung, is now cosmopolitan in greenhouses, where, it i[illegible], it occasionally damages shoots. Most North American species belong [illegible]e Ceuthophilinae (see Hubbell 1936), and some members of the genus *C*[illegible]*ophilus* Scudder may inhabit basements and cellars of houses (caves of a[illegible]!), where they may cause consternation but no harm.

The oldest recognizable pi[illegible]e of an insect is one of *Troglophilus* sp. from a cave in southern Fran[illegible] where, incidentally, the genus no longer occurs (Magdalenian culture, [illegible] 16 000 B.C., or earlier).

Key to subfam[illegible]es of Rhaphidophoridae

1. Hind tibiae square in cr[illegible] section, armed along the 4 ridges with rows of small closely set te[illegible] (Fig. 123) .... **Tropidischiinae** (p. 127)
   Hind tibiae cylindrical, [illegible]t toothed as above 2

2(1). Hind tibiae armed above with pairs of rather large movable spines alternating with series of small teeth; latter sometimes reduced and replaced in distal part of tibia by large spines forming a "sand-basket." Fastigium of vertex not divided apically 3

Hind tibiae not so armed. Fastigium of vertex divided apically .... **Rhaphidophorinae** (p. 129)

3(2). All tarsi distinctly 4-segmented. Insects sometimes long-legged; short-legged species sometimes with "sand-baskets" at ends of hind tibiae .... **Ceuthophilinae** (p. 130)

At least anterior and middle tarsi, and sometimes hind tarsi also, apparently 3-segmented owing to reduction of second segments (Figs. 179, 180). Hind legs rather short, never with "sand-baskets" on tibiae . ... **Daihiniinae** (p. 158)

Clé des sous-familles de Rhaphidophoridae

1. Tibias postérieurs carrés vus en coupe transversale, les quatre arêtes étant munies de rangées de petites dents rapprochées (fig. 123) .... **Tropidischiinae** (p. 127)

Tibias postérieurs cylindriques, dépourvus de dents comme ci-dessus 2

2(1). Tibias postérieurs armés, en position proximale, de paires d'épines mobiles plutôt grandes alternant avec des séries de petites dents; ces dernières étant parfois réduites et remplacées, en position distale du tibia, par de grosses épines formant un panier. Fastigium du vertex non divisé apicalement 3

Tibias postérieurs dépourvus de telles épines. Fastigium du vertex divisé apicalement .... **Rhaphidophorinae** (p. 129)

3(2). Tous les tarses à 4 segments distincts. Insectes parfois pourvus de longues pattes; espèces à pattes courtes parfois pourvues de paniers aux extrémités des tibias postérieurs .... **Ceuthophilinae** (p. 130)

Tarses antérieurs et médians, et parfois tarses postérieurs, apparemment constitués de trois segments à cause d'une réduction du deuxième segment (fig. 179 et 180). Pattes postérieures plutôt courtes, jamais de panier sur les tibias .... **Daihiniinae** (p. 158)

Subfamily Tropidischiinae

All tibiae quadrangular in cross-section. All 4 carinae armed with many uniform overlapping spines (Fig. 123). Large insects with legs greatly elongated in proportion to body size.

Genus *Tropidischia* Scudder

Description. As for subfamily Tropidischiinae.

Tropidischia xanthostoma (Scudder)

Figs. 121–124; Map 24

Raphidophora xanthostoma Scudder, 1861:12.
Tropidischia xanthostoma; Scudder 1863*a*:441; Helfer 1963:307.

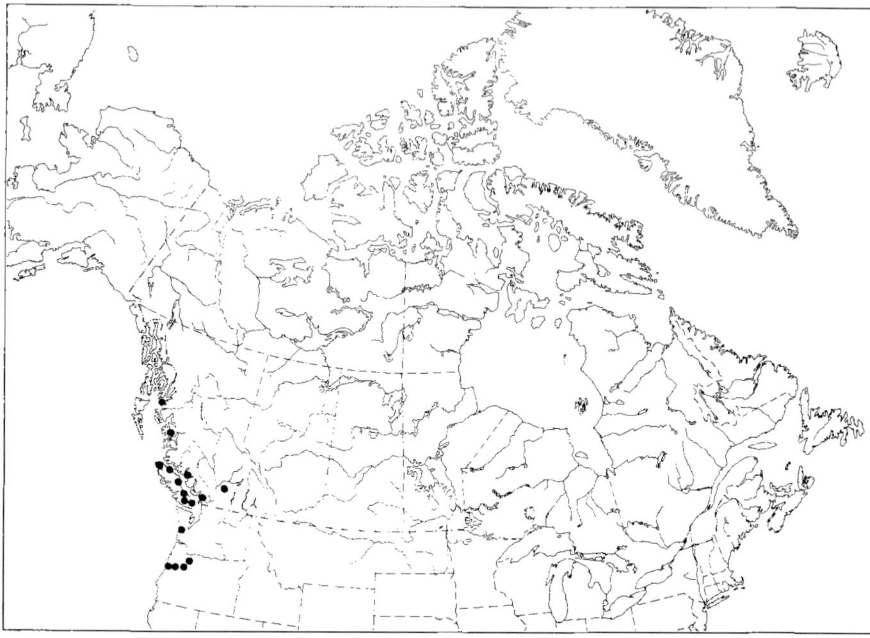

Map 24. Collection localities for *Tropidischia xanthostoma*.

Diagnosis. Legs long; all tibiae square in cross-section, spined.

Description. Head as broad in front as pronotum; fastigium between antennae formed as pair of rounded–triangular, flattened plates (Fig. 122). Antennae long, slender, with pale annulations. Pronotum broader posteriorly, with posterior margin truncate. Legs long and slender, with fore femora three times as long as pronotum; hind femora nearly 1.5 times body length (Fig. 124); all tibiae square in cross-section, spined. Abdomen broad. Male subgenital plate broad, with posterior margin rounded and with a pair of conical styles. Female subgenital plate broad, shallowly notched posteriorly. Ovipositor three times as long as pronotum, tapered, curved upward, without teeth at apex. General color dark chocolate brown, with head and legs paler; distal ends of fore and middle tibiae and tarsi nearly flesh-colored; mouthparts yellowish.

Range. British Columbia, south to California.

Behavior and habitats. Fulton (1928) found colonies of *T. xanthostoma* in dark areas under log bridges near Corvallis, OR; most of the individuals were immature in May.

Life history. The few adults examined in collections were collected in August and September, but maturation probably occurs earlier than this, and with some irregularity. Fulton (1928) captured one adult in May.

Subfamily Rhaphidophorinae

The characters of this subfamily are given in the foregoing key.

Genus *Tachycines* Adelung

Description. Head with paired prominent contiguous conical tubercles on vertex (Fig. 119); head twice as long as wide. Antennae about three times as long as body. Palpi long, slender, with fifth segment much longer than the others. Legs slender and long in proportion to body size. Male subgenital plate short, broad, truncate. Cerci long, slender, tapered. Ovipositor with apical half gently curved upward, and with inner valves crenulate beneath.

Tachycines asynamorus Adelung

Figs. 118–120; Map 25

Tachycines asynamorus Adelung, 1902:59.

Locusta marmorata Haan, 1843:217 (name preoccupied).

Diestrammena japanica [*sic*] Blatchley, 1920:611 (new name for *marmorata*).

Tachycines asynamorus; Henriksen 1939:11; Helfer 1963:308; Vickery et al. 1974:38.

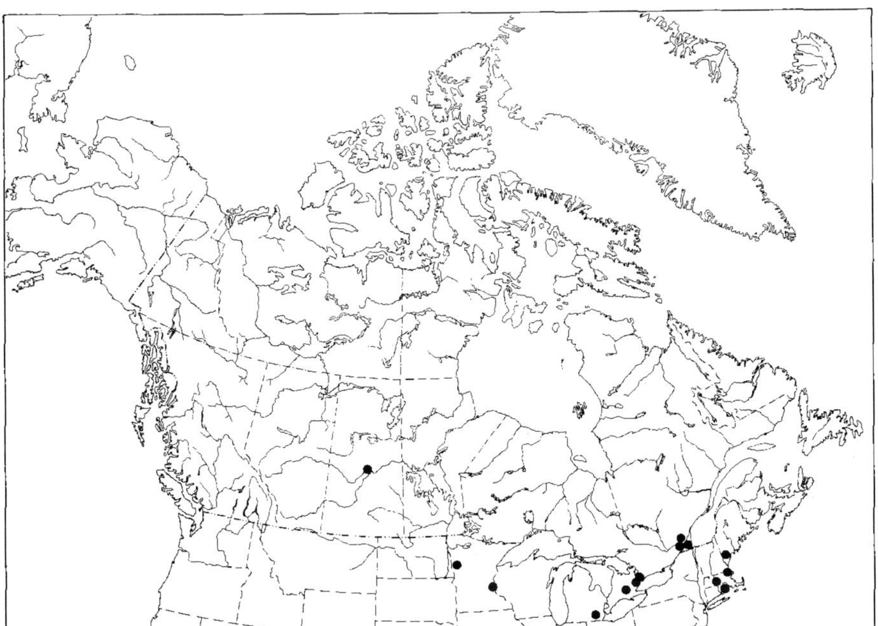

Map 25. Collection localities for *Tachycines asynamorus*.

Common name. Greenhouse stone cricket.

Diagnosis. Head with conical tubercles on vertex (Fig. 119).

Description. General characters as in key to subfamily. Legs long and slender; fore tibia with small median spur or spine between the paired disto-ventral spurs (Fig. 120). Ovipositor somewhat shorter than half length of hind tibia. Body brownish yellow, mottled with darker brown, covered with fine yellowish pubescence; hind femur with fuscous blotches. Sexes about equal in size (14–16 mm long).

Range. Saskatchewan to Québec, south to California; cosmopolitan.

Behavior and habitat. This species has become established only in greenhouses (Blatchley 1920; Vickery and Kevan 1967; Beirne 1972). It feeds on living plants, but also on dead plant material and on small living (and dead) insects. Boettger (1950) reported on reproductive behavior. Copulation takes place only in the dark. The male creeps backward beneath the female and attaches a spermatophore to her genital opening. When the sexes separate, the spermatophore sac is consumed by the female.

Life history. Breeding is continuous in heated greenhouses. Eggs are laid in soil and nymphs appear after 3–4 months. Following 11 molts, the adult stage is attained in about 7 months (Boettger 1950/1951).

Economic importance. Greenhouse plants may be damaged by feeding on seedlings, blossoms, seeds, or green leaves. Injury is usually not serious.

Subfamily Ceuthophilinae

Legs short and stout, with fore femora less than twice as long as pronotum; hind tarsus with 4 distinct segments.

Key to genera of Ceuthophilinae

1. Fastigium with conical or hornlike protuberance (Fig. 128). Tarsal claws with ventroproximal sensory seta (Fig. 127). Abdomen of male distinctly tuberculate dorsally (Figs. 129, 130). Male subgenital plate entire or emarginate distally. Ventral ovipositor valves with crenulations or numerous low serrulations .... ***Pristoceuthophilus*** (p. 131)

   Fastigium without hornlike protuberance. Tarsal claws without ventroproximal sensory seta. Abdomen of male smooth dorsally. Male subgenital plate often divided into lateral halves by median cleft, sometimes undivided. Ventral ovipositor valves with 4–8 triangular teeth ... 2

2(1). Dorsal surface of anterior tibia with stout spur near middle (Fig. 175). Body reddish brown to black .... ***Udeopsylla*** (p. 135)

   Dorsal surface of anterior tibia without spurs except at apex (Fig. 139) .... ***Ceuthophilus*** (p. 136)

Clé des genres de Ceuthophilinae

1. Fastigium surmonté d'une protubérance conique ou en forme de corne (fig. 128). Griffes des tarses munies d'une soie sensorielle ventro-proximale (fig. 127). Abdomen du mâle distinctivement tuberculé

dorsalement (fig. 129 et 130). Plaque sous-génitale du mâle entière ou marquée d'une encoche distale. Valves ventrales de l'ovipositeur crénelées ou serratulées .... ***Pristoceuthophilus*** (p. 131)

Fastigium dépourvu de protubérance en forme de corne. Griffes des tarses dépourvues d'une soie sensorielle ventro-proximale. Abdomen des mâles sans tubercule. Plaque sous-génitale du mâle souvent divisée en deux portions latérales par une incision médiane, mais parfois aussi entière. Valves ventrales de l'ovipositeur munies de 4 à 8 dents triangulaires ... 2

2(1). Surface dorsale des tibias antérieurs ornée, près du milieu, d'un éperon dur (fig. 175). Corps de couleur brun rougeâtre à noir .... ***Udeopsylla*** (p. 135)

Surface dorsale des tibias antérieurs dépourvue d'éperon sauf à l'apex (fig. 139) .... ***Ceuthophilus*** (p. 136)

Genus *Pristoceuthophilus* J. A. G. Rehn

Description. Vertex of head with conical projection. Hind tibiae with 3 pairs of spurs. Fore femur without spines. Apical segment of palps twice as long as preceding segment.

Key to species of *Pristoceuthophilus* (males)

Note: Keys are not reliable in distinguishing between females of this genus. Females should be associated with males from the same collection site; they do not have abdominal tubercles or curved or bowed hind tibiae (Fig. 126).

1. Dorsal abdominal terga with mesal raised smooth areas on first five (Fig. 129). Hind tibiae with slight curvature in basal fifth .... ***celatus* (Scudder)** (p. 132)

Dorsal abdominal terga without mesal raised smooth areas on first five. Hind tibiae either strongly bowed in basal half or showing no trace of curvature 2

2(1). Dorsal abdominal terga with numerous spines; fourth tergum with prominent protuberance (Fig. 130) .... ***gaigei* Hubbell** (p. 133)

Dorsal abdominal terga with numerous spines or tubercles; fourth tergum lacking protuberance 3

3(2). Dorsal abdominal terga with numerous tubercles. Cerci long, with acute deflexed tips (Fig. 135). Hind tibiae without curvature .... ***cercalis* Caudell** (p. 134)

Dorsal abdominal terga with lateral tubercles only, none on midline. Cerci tapered, unspecialized (Fig. 132). Hind tibiae strongly bowed in apical third to half. Hind femur with ventral spine adjacent to bow .... ***pacificus* (Thomas)** (p. 134)

Clé des espèces de *Pristoceuthophilus* (mâles)

Nota: Les clés d'identification ne permettent pas de distinguer l'une de l'autre les femelles appartenant à ce genre. Leur identification devrait passer par l'identification des mâles recueillis au même endroit; les femelles ne possèdent pas de tubercules abdominaux ni de tibias arqués (fig. 126).

1. Cinq premiers tergites dorsaux de l'abdomen munis d'une protubérance médiane non tuberculée (fig. 129). Le cinquième tergite proximal des tibias postérieurs légèrement incurvé .... *celatus* **(Scudder)** (p. 132)

Cinq premiers tergites dorsaux de l'abdomen dépourvus d'une protubérance médiane non tuberculée. La moitié proximale des tibias postérieurs est fortement arquée, sinon, il n'y a aucune trace de courbure 2

2(1). Tergites dorsaux de l'abdomen pourvus de nombreuses épines; quatrième tergite surmonté d'une forte protubérance (fig. 130) .... *gaigei* **Hubbell** (p. 133)

Tergites dorsaux de l'abdomen pourvus de nombreuses épines ou tubercules; quatrième tergite dépourvu d'une protubérance 3

3(2). Tergites dorsaux de l'abdomen pourvus de nombreux tubercules. Cerques longs, à extrémités fortement incurvées (fig. 135). Tibias postérieurs non courbés .... *cercalis* **Caudell** (p. 134)

Tergites dorsaux de l'abdomen munis de tubercules latéraux mais sans tubercule sur la ligne médiane. Cerques pointus et non spécialisés (fig. 132). Le tiers ou la moitié de l'apex des tibias postérieurs fortement arqué(e). Fémurs postérieurs munis d'une épine ventrale adjacente à l'arc .. .. *pacificus* **(Thomas)** (p. 134)

Pristoceuthophilus celatus (Scudder)

Figs. 128–130, 133, 137; Map 26

Ceuthophilus celatus Scudder, 1894*a*:26.
Ceuthophilus henshawi Scudder, 1894*a*:97.
Pristoceuthophilus celatus; Caudell 1916:671.

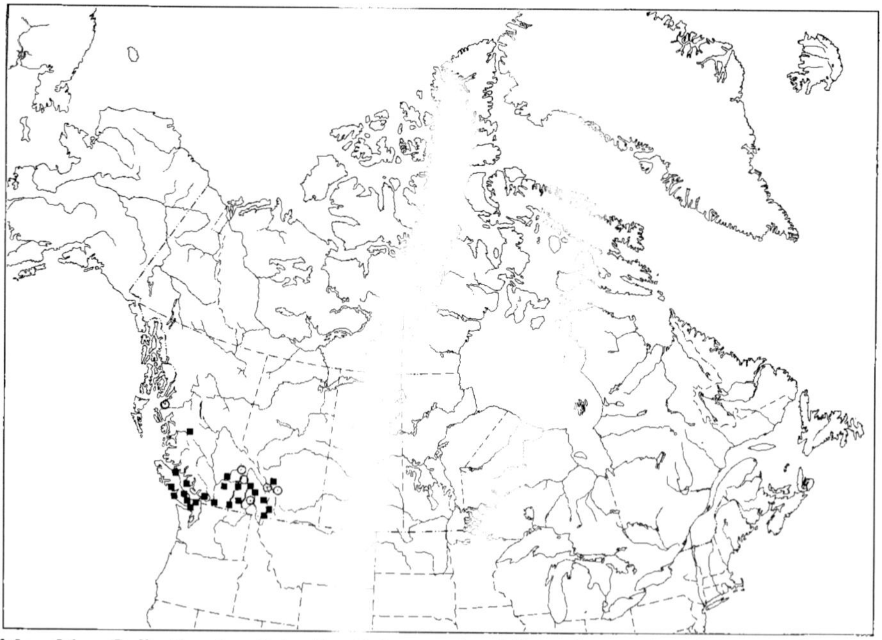

Map 26. Collection localities for *Pristoceuthophilus celatus* (■) and *P. cercalis* (○).

Diagnosis. Male abdomen with 4 or 5 large elevated smooth areas, and with remainder tuberculate.

Description. Body small (9–12 mm long). Dorsal surface of abdomen of male with numerous rounded tubercles and with 4 or 5 large elevated smooth areas (Fig. 130). Male cerci broad at base, each with blunt tooth on inner face, tapering to acute point (Fig. 133). Male subgenital plate as in Fig. 129. Fore femur longer than pronotum by one-third in males, by one-quarter in females. Ovipositor straight, slender, gradually attenuate at apex, slightly upcurved, with ventral valves crenulate (Fig. 137).

Range. British Columbia and Alberta, south to California.

Behavior and habitat. *P. celatus* is usually found under tree bark or in damaged decayed damp sapwood of trees.

Life history. Nymphs have been found in May, and adults from early July to late August.

Pristoceuthophilus gaigei Hubbell

Figs. 131, 134; Map 27

Pristoceuthophilus gaigei Hubbell, 1925:39–42.

Diagnosis. Fourth male abdominal tergum with large protuberance; other terga spined or tuberculate.

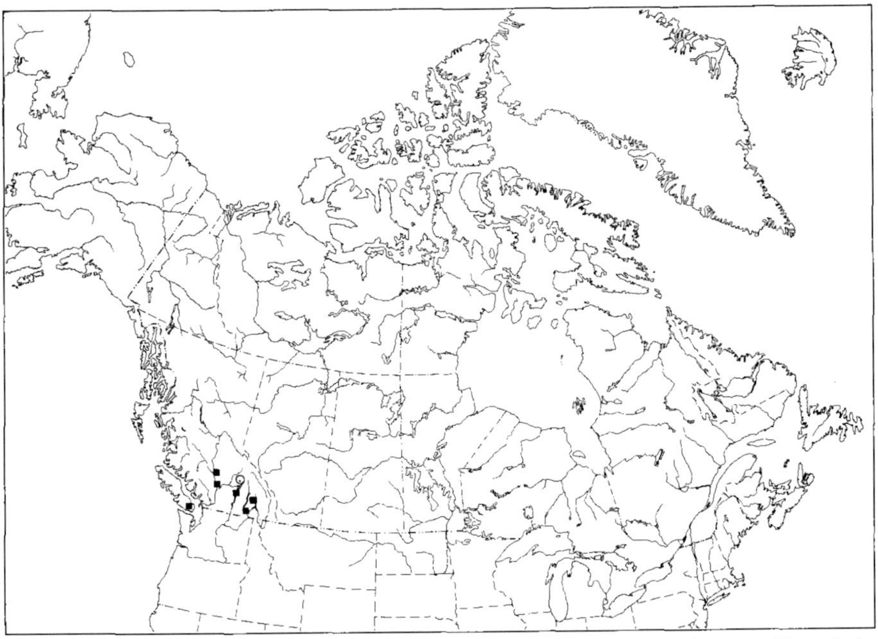

Map 27. Collection localities for *Pristoceuthophilus gaigei* (○) and *P. pacificus* (■).

Description. Body small. Dorsum of abdomen of male with large blunt spines or rounded or conical tubercles on first to eighth terga; fourth tergum with smooth protuberance about three times as long as width of third tergum (Fig. 131). Pronotum slightly longer than exposed parts of mesonotum and metanotum combined. Male cerci as in Fig. 134. General color yellowish with brown stripes.

Range. British Columbia and Washington.

Behavior and habitat. The type specimen was found in rotten logs in a clearing in maple–alder woods bordering a dense forest of conifers.

Pristoceuthophilus cercalis Caudell

Figs. 135, 138; Map 26

Pristoceuthophilus cercalis Caudell, 1916:673.

Diagnosis. Male abdomen with rounded tubercles on posterior half of all but two terminal segments.

Description. Dorsum of abdomen of male with all except terminal two segments covered on posterior half, or more, with large rounded tubercles. Male cerci slightly compressed basally, unsegmented to near tip, then bent inward and depressed, bearing short, weakly segmented apical part at right angles to base (Fig. 135). Abdomen of female smooth. Cerci of female simple. Ovipositor heavy, longer than pronotum, with ventral valves triangularly serrate beneath at tip (Fig. 138). General color brown with obscure lighter mottling.

Range. British Columbia and Alberta, south to Oregon.

Behavior and habitat. Caudell (1916) reported that specimens were found under boulders and logs in dense forests. Additional collecting would probably increase the known range of this species.

Life history. Adults have been collected in August and September.

Pristoceuthophilus pacificus (Thomas)

Figs. 125–127, 132, 136; Map 27

Ceuthophilus pacifica C. Thomas, 1872:436.
Pristoceuthophilus pacificus; Caudell 1916:675; Hubbell 1936:37; Helfer 1963:311.

Diagnosis. Male abdomen tuberculate dorsally on all segments.

Description. Body small (11–12.5 mm long). Dorsum of adult male tuberculate, the tubercles distributed on all segments. Hind femur usually armed beneath with heavy toothlike spine on each carina (lacking in some specimens); male hind tibia abruptly bowed before middle (Figs. 125, 127). Conical projection of vertex pronounced, even in immature specimens. Male cerci more or less straight, unspecialized (Fig. 132). Ovipositor as in Fig. 136. General color opaque yellowish brown with numerous brownish to purplish maculations.

Range. British Columbia, south to Nevada and California.

Behavior and habitats. *P. pacificus* is found under rocks where numerous fissures occur and where rocks have been loosened.

Life history. Barnum (1964) collected nymphs from 8 July to 2 November, and collected adults from 11 August to 10 January.

Genus *Udeopsylla* Scudder

Description. Head larger and body stouter than in *Ceuthophilus* (Fig. 174). Vertex of head flattened vertically. Antennae slender, hardly longer than body, with third segment longer than combined length of first and second. Pronotum 1.5 times as long as either mesonotum or metanotum. Fore tibia with stout spine dorsally near middle (Fig. 175); hind femur flattened, broad, thin, with outer ventral carina possessing 8–20 small teeth; inner carina with 8, or more, long stout teeth; hind tibia spinose on each dorsal margin with 4 long movable spines, spaced alternately, with 3 or 4 short teeth between each pair; apex with 3 pairs of stout spurs; tarsi 4-segmented; segments one and four longer than segments two and three combined.

Udeopsylla robusta (Haldeman)

Figs. 174–176; Map 28

Ph [*alangopsis*] [*Daihinia*] *robustus* Haldeman, 1850:346.
Daihinia robusta Scudder, 1862:284.
Udeopsylla nigra Scudder, 1862:284.
Ceuthophilus niger Scudder, 1863*a*:437.
Udeopsylla robusta; Scudder 1863*a*:442; Hebard 1936*a*:55; Helfer 1963:313.
Marsa arcuata F. Walker, 1869*b*:254 (Saskatchewan).
Daihinia gigantea Bruner, 1885*a*:127.
Udeopsylla compacta Bruner, 1891*a*:38.
Ceuthophilus devius Scudder, 1894*a*:30, 99.
Ceuthophilus politus Scudder in Scudder and Cockerell, 1902:56.
Ceuthophilus ater Scudder in Scudder and Cockerell, 1902:57.

Diagnosis. Fore tibia with stout dorsal spine near middle. Abdomen with nodules dorsally.

Description. As for genus. Body large (18–26 mm long). Dorsum of abdomen with numerous small nodules. Male subgenital plate deeply impressed but not cleft, with lobes swollen at middle and with apex feebly notched. Ovipositor short, stout at base, strongly tapering to middle, with apex of upper valves oblique, terminating in sharp spine; lower valves with large unequally spaced teeth and a slender decurved terminal hook (Fig. 176). Color extremely variable, typically dark brown to reddish brown, blotched with fuscous, or black with legs and dorsum of pronotum paler, or uniformly pale reddish brown.

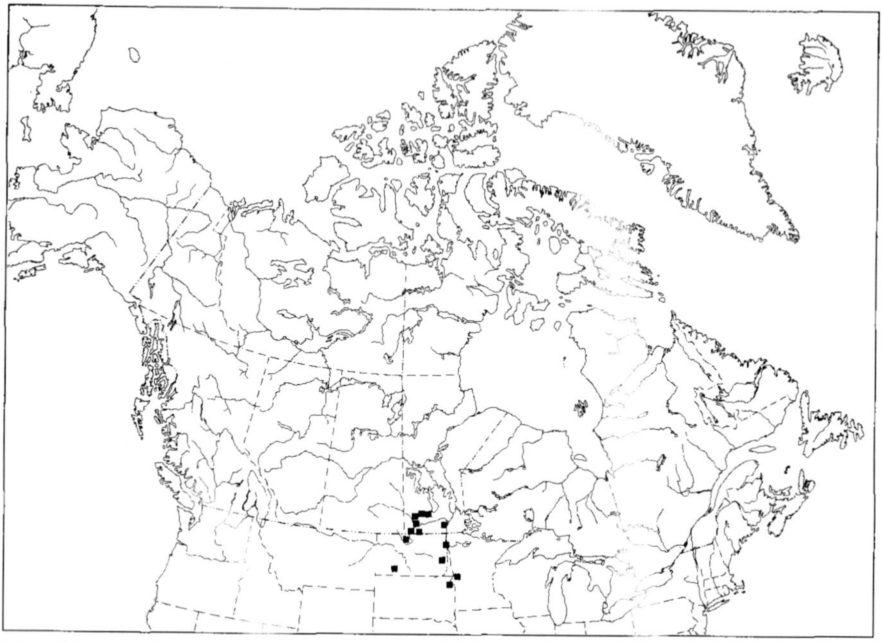

Map 28. Collection localities for *Udeopsylla robusta*.

Range. Idaho to Manitoba and Illinois, south to California and Texas.

Behavior and habitats. *U. robusta* is said to be nocturnal, living singly in burrows dug in loose soil. It has been found in plowed fields and on roadsides on cloudy days, early in the morning, or just before sunset. In Indiana, McCafferty and Stein (1976) found it in gardens under boards, and in old stumps.

Life history. Unknown. Six adult specimens at hand were all taken in September and October.

Genus *Ceuthophilus* Scudder

Description. Size medium. Body thick, dorsally arched (Fig. 139). Head oval, lacking anterior projection on vertex, bent downward and backward between fore legs. Antennae long, slender, cylindrical, tapered. Hind femur thick, usually spined beneath in males, feebly armed in females; all tarsi with 4 segments; fore tibia with spines only at apex. Ovipositor nearly straight, stout at base, tapering to middle, with dorsal valves upturned apically, and with ventral valves possessing 5 or 6 teeth near apex; apical tooth usually hooklike.

Key to species and subspecies of *Ceuthophilus*

1. Slender species with relatively long legs and antennae. Ninth tergum of male abdomen produced posteriorly (Figs. 144–147). Male subgenital plate usually divided vertically by median sulcus (Figs. 153–161). Ovipositor unspecialized (Fig. 151) (subgenus *Ceuthophilus*) 2

   Small robust compact species with relatively short stout legs. Ninth tergum of male abdomen not produced, and eighth tergum usually strongly produced and concealing ninth tergum (Figs. 148–150). Male subgenital plate scoop-shaped or conical, not divided by median sulcus (Figs. 162–166). Ovipositor moderately to strongly specialized (Fig. 152) (subgenus *Geotettix*) 12

2(1). Ninth tergum of male abdomen distinctly produced, subtruncate or deeply and broadly emarginate. Dorsal spurs of posterior tibiae small, not dorsally bicarinate 3

   Ninth tergum of male abdomen usually little produced, but, if distinctly produced, then not subquadrate nor emarginate. Dorsal spurs of posterior tibiae large, dorsally bicarinate 11

3(2). Male subgenital plate apically with distinct V-shaped or U-shaped notch, or with margins of mesal notch produced (Figs. 153, 154, 157). Posterior margin of ninth tergum of male convex or subtruncate (Fig. 144). Ovipositor relatively short 4

   Male subgenital plate with lateral margins produced but without distinct notch (Figs. 155, 156). Posterior margin of ninth tergum of male subtruncate or broadly emarginate, bilobate. Ovipositor stout .. 6

4(3). Ninth abdominal tergum of male broadly arcuate to subtruncate distally; margin of eighth tergum produced to cover much of ninth tergum (Fig. 144). Ovipositor with teeth of ventral valve close-spaced, crowded. Legs rather slender, moderately long; hind tibiae darkened, except spines and areas around spine bases. Dark eastern species .... ***brevipes*** **Scudder** (p. 142)

   Ninth abdominal tergum of male subquadrate, emarginate distally; eighth tergum not produced, not covering most of ninth tergum. Ovipositor with teeth of ventral valve short and widely separated. Legs not unusually long; hind tibiae pale. Pallid to pale western species .. 5

5(4). Anterior margins of abdominal terga darkened; ninth abdominal tergum of male deeply emarginate, forming paired diverging subtriangular lobes. Ovipositor slender, elongate, 2.0–2.4 times as long as length of pronotum. Found from Alberta to Manitoba and adjacent United States .... ***pallescens*** **Bruner** (p. 143)

   Anterior margins of abdominal terga not darkened; ninth abdominal tergum of male truncate or only slightly emarginate. Ovipositor shorter, 1.5–1.9 times as long as pronotal length. Found in British Columbia and Washington .... ***agassizii*** **(Scudder)** (p. 144)

6(3). Thoracic terga dull reddish brown, mottled with more or less indistinct weakly contrasted darker brown. Tibial spurs distinctly bicarinate dorsally. Found in Michigan and southern New York .... ***uhleri*** **Scudder** (p. 144)

   Thoracic terga polished, shining, with many distinct pale maculae (pronotum usually with pale bars outlining paler median stripe). Tibial spurs normally unicarinate dorsally 7

7(6). Hind tibiae of males with 1 or 2 subdistal ventral spurs (Fig. 140*b*). Dorsum without contrasting median color stripe, conspicuously maculate. Ninth abdominal tergum of male emarginate distally. Male subgenital plate without V-shaped notch, with dorsolateral angles prominent .... ***maculatus* (Harris)** (p. 146)

Hind tibiae of males with 2 pairs of subdistal ventral spurs (Fig. 142). Dorsum usually with contrasting median color stripe. Ninth abdominal tergum of male not emarginate distally. Male subgenital plate subtruncate distally 8

8(7). Hind tibiae with dorsal spines darkened at bases 10

Hind tibiae unicolorous, with dorsal surface sometimes darker than lateral surfaces, and with spines not darkened at bases 9

9(8). Posterior margins of abdominal terga darkened, appearing transverse-banded dorsally. Females with ratio of anterior femur length to pronotum length 1.4–1.9. Male subgenital plate rounded, mesally emarginate, with lobes triangulate (Fig. 155) .... ***gracilipes gracilipes* (Haldeman)** (p. 147)

Posterior margins of abdominal terga not darkened, and with dorsum infuscated on either side of broad orange stripe. Females with ratio of anterior femur length to pronotum length slightly less than 1.4. Male subgenital plate quadrate, sometimes concave distally but not emarginate (Fig. 156) .... ***meridionalis* Scudder** (p. 148)

10(8). Ninth abdominal tergum of male with distal margin more or less membranous mesally, with distal edge of heavily sclerotized part distinctly emarginate (Fig. 146). Ovipositor with intervals between teeth equal .... ***latens* Scudder** (p. 149)

Ninth abdominal tergum of male subquadrate, sclerotized throughout. Ovipositor teeth with intervals diminishing toward apex .... ***pallidipes* E. M. Walker** (p. 150)

11(2). Fastigium compressed, as prominent as mesal margins of antennal fossae. Dorsum of thorax glabrous, weakly polished, nearly always with distinct narrow pale median stripe .... ***divergens* Scudder** (p. 152)

Fastigium with rounded apex, not elevated, lower than or even with margins of antennal fossae. Dorsum of thorax densely setose, appearing dull, with contrasting light and dark blotches .... ***pallidus* Thomas** (p. 152)

12(1). Tarsi with ventral keels each bearing row of short stiff spiniform setae from base at least halfway to apex (Fig. 143) 13

Tarsi with ventral keels smooth, non-setose 14

13(12). Eighth tergum of male produced, truncate at apex .... ***fusiformis* Scudder** (p. 153)

Eighth tergum of male strongly produced, with posterior margin emarginate and forming paired lobes, which are bent downward and projected posteriorly past bases of cerci .... ***silvestris* Bruner** (p. 153)

14(12). Male subgenital plate scoop-shaped or conical, with apex semicircular, entire, or weakly notched. Dorsal spurs of hind tibiae short, stout. Dorsum, including lateral thoracic lobes, deeply infuscate 15

Male subgenital plate not scoop-shaped or conical, with apex entire. Dorsal spurs of hind tibiae usually elongate. Dorsum not infuscate ... 16

15(14). Dorsum, including lateral lobes, dark, often with contrasting broad median pale stripe. Distal part of abdomen of female often distinctly tectate-carinate, with margins of posterior 3 or 4 terga angulate or distinctly produced at midline. Male subgenital plate conical (Fig. 164), narrowly cleft at apex to form paired distal lobes .... ***guttulosus thomasi* Hubbell** (p. 154)

Dorsum brownish, sometimes infuscated, with lateral lobes usually paler, usually without median stripe. Abdomen of female seldom tectate posteriorly, with eighth tergum only produced moderately to spicate point. Male subgenital plate scoop-shaped, with free margin usually shallowly depressed at midline (Fig. 162) .... ***guttulosus guttulosus* F. Walker** (p. 155)

16(14). Body distinctly depressed. Eyes small. Hind tibia shorter than hind femur, swollen; subdistal ventral spurs of hind tibiae normally 2-2. Epiproct of male large, quadrate; male subgenital plate small, rounded, mesally cleft or emarginate (Fig. 167) .... ***alpinus* Scudder** (p. 156)

Body not depressed. Eyes large. Hind tibia not shorter than hind femur, not swollen; subdistal ventral spurs of hind tibiae normally 1-1. Epiproct of male elongate-triangular; male subgenital plate triangular, converging to narrow truncate apex .... ***vicinus* Hubbell** (p. 157)

Clé des espèces et des sous-espèces de *Ceuthophilus*

1. Espèces grêles, à pattes et à antennes relativement longues. Portion postérieure du neuvième tergite de l'abdomen des mâles développée (fig. 144 à 147). Plaque sous-génitale des mâles habituellement divisée verticalement par un sillon médian (fig. 153 à 161). Ovipositeur non spécialisé (fig. 151) (sous-genre *Ceuthophilus*) 2

Espèces petites et robustes, à pattes relativement courtes et fortes. Neuvième tergite de l'abdomen des mâles peu développé; huitième tergite habituellement très développé et dissimulant le neuvième tergite (fig. 148 à 150). Plaque sous-génitale des mâles en forme de cuillère ou de cône, non divisée par un sillon médian (fig. 162 à 166). Ovipositeur modérément à fortement spécialisé (fig. 152) (sous-genre *Geotettix*) ... 12

2(1). Neuvième tergite de l'abdomen des mâles bien développé, légèrement tronqué et marqué d'un sillon large et profond. Épines dorsales des tibias postérieurs petites, non bicarénées dorsalement 3

Neuvième tergite de l'abdomen des mâles habituellement peu développé mais dans le cas contraire, ni subquadrangulaire ni marqué d'un sillon. Épines dorsales des tibias postérieurs grandes et bicarénées dorsalement ... 11

3(2). Plaque sous-génitale du mâle marquée, à l'extrémité apicale, d'une encoche en forme de V ou de U ou aux bords faisant saillie (fig. 153, 154 et 157). Bord postérieur du neuvième tergite des mâles convexe ou d'apparence tronquée (fig. 144). Ovipositeur relativement court 4

Plaque sous-génitale du mâle à bords latéraux saillants mais dépourvus d'une véritable encoche (fig. 155 et 156). Bord postérieur du neuvième tergite du mâle d'apparence tronquée ou marqué d'une large incision, bilobé. Ovipositeur fort 6

4(3). Neuvième tergite de l'abdomen des mâles largement arqué ou d'apparence tronquée distalement; bord du huitième tergite développé, recouvrant en grande partie le neuvième tergite (fig. 144). Valve ventrale de l'ovipositeur pourvue de nombreuses dents rapprochées. Pattes plutôt minces, modérément longues; tibias postérieurs foncés à l'exception des épines et des zones à la base de ces épines. Espèce foncée de l'est .... ***brevipes* Scudder** (p. 142)

Neuvième tergite de l'abdomen du mâle subquadrangulaire et marqué d'une encoche distale; huitième tergite non développé et ne recouvrant pas la plupart du neuvième tergite. Valve ventrale de l'ovipositeur munie de dents courtes et largement séparées l'une de l'autre. Pattes pas trop longues; tibias postérieurs pâles. Espèce de l'ouest, décolorée ou pâle 5

5(4). Bords antérieurs des tergites de l'abdomen foncés; neuvième tergite de l'abdomen des mâles portant une profonde encoche et prenant la forme de deux lobes subtriangulaires divergents. Ovipositeur long et mince, de 2,0 à 2,4 fois plus long que le pronotum. Espèce vivant de l'Alberta au Manitoba ainsi que dans les États adjacents des États-Unis . . . . *pallescens* **Bruner** (p. 143)

Bords antérieurs des tergites de l'abdomen non foncés; neuvième tergite de l'abdomen du mâle tronqué ou ne portant qu'une légère encoche. Ovipositeur plus court, de 1,5 à 1,9 fois plus long que le pronotum. Espèce vivant en Colombie-Britannique et dans l'État de Washington . . . . *agassizii* **(Scudder)** (p. 144)

6(3). Tergite du thorax brun rougeâtre mat, moucheté de taches brun foncé plus ou moins distinctes. Épines des tibias distinctement bicarénées dorsalement. Espèce vivant au Michigan et dans le sud de l'État de New York . . . . *uhleri* **Scudder** (p. 144)

Tergite du thorax luisant, parsemé de nombreuses taches pâles distinctes (pronotum habituellement marqué de lignes pâles délimitant une bande médiane plus pâle). Épines des tibias normalement unicarénées dorsalement 7

7(6). Tibias postérieurs des mâles munis, près de l'extrémité distale, d'une ou deux épines ventrales (fig. 140*b*). Dos dépourvu d'une bande médiane de couleur contrastante, mais clairement mouchetée. Neuvième tergite de l'abdomen du mâle marqué d'une encoche distale. Plaque sous-génitale du mâle dépourvue d'une encoche en forme de V, mais à angles dorsolatéraux proéminents . . . . *maculatus* **(Harris)** (p. 146)

Tibias postérieurs des mâles armés, près de l'extrémité distale, de deux paires d'épines ventrales (fig. 142). Dos habituellement marqué d'une bande médiane de couleur contrastante. Neuvième tergite de l'abdomen des mâles dépourvu d'une encoche distale. Plaque sous-génitale du mâle d'apparence tronquée à l'extrémité distale 8

8(7). Tibias postérieurs armés d'épines dorsales foncées à la base 10

Tibias postérieurs de couleur uniforme, la surface dorsale étant parfois plus foncée que les côtés et les épines n'étant pas foncées à la base 9

9(8). Bords postérieurs des tergites de l'abdomen foncés, formant sur le dos une série de bandes transversales. Chez les femelles, fémurs antérieurs de 1,4 à 1,9 fois plus longs que le pronotum. Plaque sous-génitale du mâle arrondie qui porte une encoche médiane formant deux lobes triangulaires (fig. 155) . . . . *gracilipes gracilipes* **(Haldeman)** (p. 147)

Bords postérieurs des tergites de l'abdomen non foncés; dos plus foncé de part et d'autre d'une large bande orangée. Chez les femelles, fémurs antérieurs un peu moins de 1,4 fois plus longs que le pronotum. Plaque sous-génitale du mâle quadrangulaire, parfois concave à l'extrémité distale, mais jamais marquée d'un sillon (fig. 156) . . . . *meridionalis* **Scudder** (p. 148)

10(8). Neuvième tergite de l'abdomen des mâles à bord distal plus ou moins membraneux au milieu et fortement sclérifié de part et d'autre, présentant une concavité distincte (fig. 146). Ovipositeur muni de dents réparties également . . . . *latens* **Scudder** (p. 149)

Neuvième tergite de l'abdomen du mâle subquadrangulaire, entièrement sclérifié. Ovipositeur muni de dents plus rapprochées vers l'apex . . . . *pallidipes* **E.M. Walker** (p. 150)

11(2). Fastigium comprimé, aussi proéminent que les bords médians de la fovéa antennaire. Tergites du thorax glabres, légèrement luisants, presque toujours marqués d'une étroite bande médiane pâle et distincte ... . *divergens* **Scudder** (p. 152)

Fastigium à apex arrondi, non surélevé, de hauteur inférieure ou égale aux bords de la fovéa antennaire. Tergites du thorax fortement sétifères, mats, marqués de taches claires et foncées .... *pallidus* **Thomas** (p. 152)

12(1). Tarses à carènes ventrales munies, de la base jusqu'à mi-longueur, d'une rangée de soies courtes et spiniformes (fig. 143) 13

Tarses à carènes ventrales unies, non sétifère 14

13(12). Huitième tergite du mâle développé, tronqué à l'apex .... *fusiformis* **Scudder** (p. 153)

Huitième tergite du mâle très développé, à bord postérieur marqué d'un sillon formant deux lobes incurvés vers le bas et faisant saillie au-delà des cerques .... *silvestris* **Bruner** (p. 153)

14(12). Plaque sous-génitale du mâle en forme de cuillère ou de cône, à apex semi-circulaire, entier ou légèrement encoché. Épines dorsales des tibias postérieurs courtes et fortes. Dos et lobes latéraux du thorax très enfumés 15

Plaque sous-génitale du mâle ni en forme de cuillère, ni en forme de cône, à apex entier. Épines dorsales des tibias postérieurs habituellement longues. Dos non enfumé 16

15(14). Dos et lobes latéraux foncés, souvent marqués d'une large bande médiane pâle et contrastante. Portion distale de l'abdomen des femelles souvent clairement tectiforme ou carénée, les bords des 3 ou 4 tergites postérieurs formant un angle plus ou moins aigu au milieu. Plaque sous-génitale des mâles conique (fig. 164), marquée d'un sillon étroit à l'apex et formant deux lobes .... *guttulosus thomasi* **Hubbell** (p. 154)

Dos brunâtre, parfois enfumé, à lobes latéraux habituellement plus pâles, dépourvus d'habitude d'une bande médiane. Abdomen des femelles rarement tectiforme à l'extrémité postérieure; huitième tergite modérément développé, se terminant en pointe. Plaque sous-génitale du mâle en forme de cuillère, à bord distal en général légèrement concave au milieu (fig. 162) .... *guttulosus guttulosus* **F. Walker** (p. 155)

16(14). Corps clairement aplati dorso-ventralement. Yeux petits. Tibias postérieurs plus courts que les fémurs postérieurs et renflés; épines ventrales subdistales des tibias postérieurs habituellement 2–2. Épiprocte du mâle, grand et quadrangulaire; plaque sous-génitale du mâle petite, arrondie, encochée ou concave (fig. 167) ..... *alpinus* **Scudder** (p. 156)

Corps non aplati dorso-ventralement. Yeux grands. Tibias postérieurs ni plus courts que les fémurs postérieurs ni renflés; épines ventrales subdistales des tibias postérieurs habituellement 1–1. Épiprocte du mâle long et triangulaire; plaque sous-génitale du mâle triangulaire, se terminant en un apex étroit et tronqué. .... *vicinus* **Hubbell** (p. 157)

Ceuthophilus (Ceuthophilus) brevipes Scudder

Figs. 144, 151, 153, 169; Map 29

Ceuthophilus brevipes Scudder, 1863*a*:434.

Ceuthophilus brevipes; Hubbell 1936:114–126; Cantrall 1943:146; Helfer 1963:319; Vickery et al. 1974:43–44.

Ceuthophilus terrestris Scudder, 1894*a*:26, 46.

Diagnosis. Hind tibiae dark except for pale spurs and pale areas at spur bases.

Description. Body relatively small (pronotal length, male, 3.1–4.5, female, 3.5–4.7 mm). Hind margin of terminal dorsal segment of male rounded obtusely (Fig. 144). Fore femur one-third, or more, longer than pronotum; outer, lower carina of male with about 25 minute crowded teeth; female with even smaller teeth; hind tibia straight in both sexes. Subgenital plate of male and ovipositor as in Figs. 153 and 151, respectively; epiphallus as in Fig. 169. Color pale dull brown, thickly mottled with paler spots, usually without lighter dorsal stripe; ventrolateral margins of pronotum dark; hind tibia dark except for pale spurs and areas about spur bases.

Range. Ontario to Newfoundland, south to Virginia.

Behavior and habitats. *C. brevipes* is restricted to cool damp locations in or near swampy areas. It is found under fallen logs or in debris on the forest floor in deciduous, coniferous, or mixed forests. Occasional reports are received of infestations in damp basements where it may be a harmless nuisance.

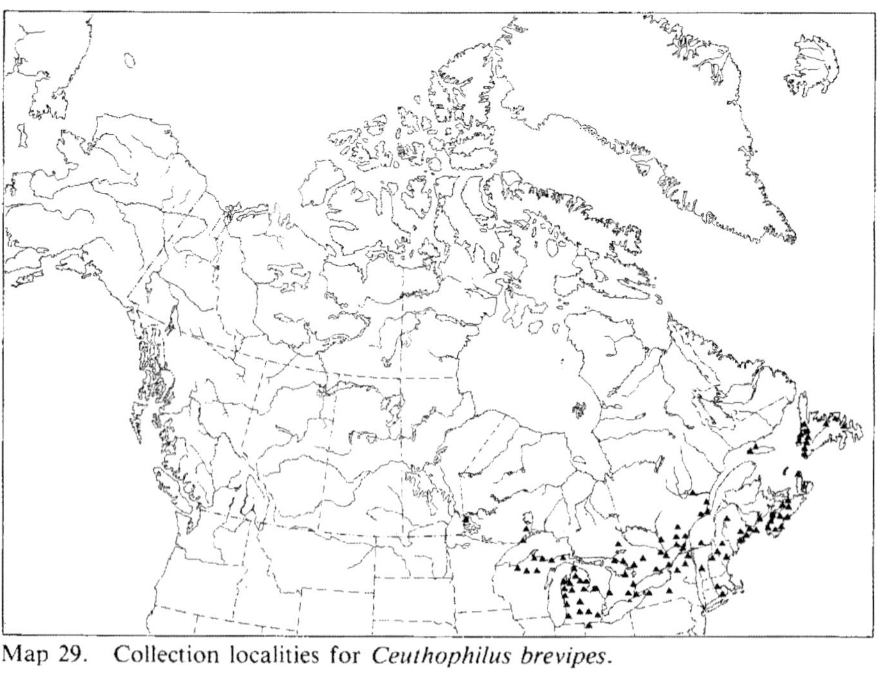

Map 29. Collection localities for *Ceuthophilus brevipes*.

Life history. According to Hubbell (1936) adults are found from April to September; most of the specimens examined during this study were taken June to August.

Ceuthophilus (Ceuthophilus) pallescens Bruner

Map 30

Ceuthophilus pallescens Bruner, 1891*a*:37.
Ceuthophilus pallescens; Hubbell 1926:14; Hubbell 1936:136–140.

Diagnosis. Male abdominal terga dark on anterior margins.

Description. Size small (pronotal length, male, 3.4–4.8, female, 4.1–5.2 mm), not robust. Ovipositor with 5 short slender aciculate teeth; apical teeth strongly decurved. Anterior margins of male abdominal terga strongly infuscated, these tending to disappear by normal telescoping of the segments.

Range. Alberta to Manitoba, south to Colorado and Nebraska.

Behavior and habitats. Little is known about *C. pallescens*. The type was found in a shallow well, and other specimens have been found in outside cellar stairways.

Life history. Probably univoltine. Hubbell (1936) recorded adults from July to October. E. M. Walker (1910) reported adults (as *C. aridus*) from Saskatchewan in September and nymphs in May, June, September, and November.

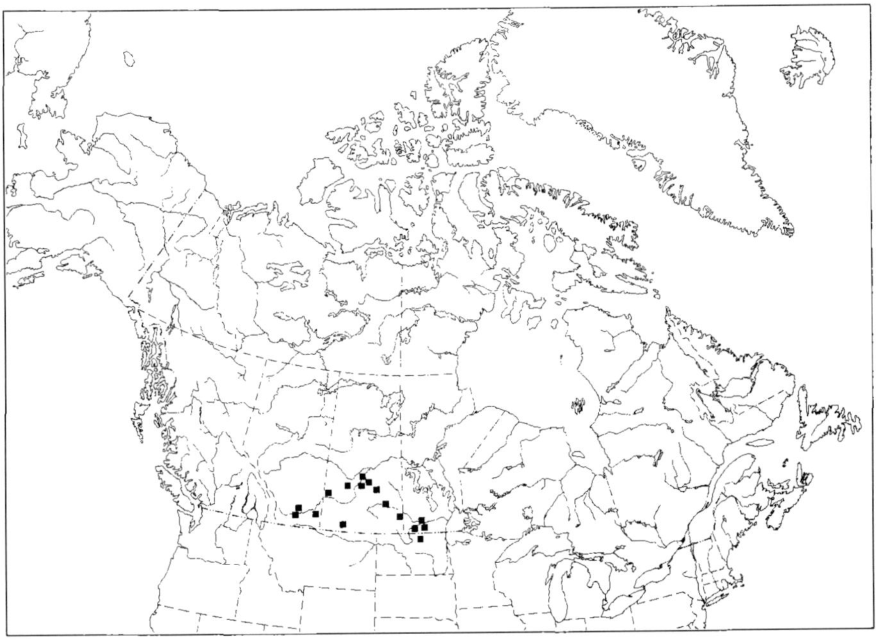

Map 30. Collection localities for *Ceuthophilus pallescens*.

Ceuthophilus (Ceuthophilus) agassizii (Scudder)

Fig. 154; Map 31

Rhaphidophora agassizii Scudder, 1861:11–12.
Ceuthophilus zonarius F. Walker, 1869*a*:203.
Ceuthophilus agassizii; Scudder 1894*a*:27, 81–82.
Ceuthophilus crassus Scudder, 1894*a*:30, 85–86.

Diagnosis. Pale western species.

Description. Pronotal length of males 3.7–5.5 mm; females, 3.9–5.8 mm. Hind femur with 20–60 tiny denticulations dorsally. Median sulcus of male subgenital plate bordered with membranous areas extending almost to base of plate (Fig. 154); male epiphallus with dorsolateral margins heavily sclerotized, rolled outward to form pair of auriculae with strongly tumid caudal surfaces. Body pale, varying from brown to brownish yellow.

Range. British Columbia and Alberta, south to Oregon.

Behavior and habitats. Nocturnal, but has been found during daylight hiding under stones or debris, pieces of board, or in garages or cellars (Buckell 1922).

Life history. Adults have been found in British Columbia from early June to early October.

Ceuthophilus (Ceuthophilus) uhleri Scudder

Fig. 161; Map 32

Ceuthophilus uhleri Scudder, 1863*a*:435–436.
Ceuthophilus blatchleyi Scudder, 1894*a*:26, 27.

Diagnosis. Legs stout. Body yellowish to reddish brown, glabrous.

Description. Size medium (pronotal length, male, 3.1–5.8, female, 3.8–5.3 mm). Legs stout. Male ninth tergum weakly produced; male subgenital plate persulcate with lateral lobes small, weakly sclerotized, appearing narrow with angular apices; distal area between lobes membranous, V-shaped (Fig. 161). Ovipositor short, straight, appearing heavy, with teeth of ventral valve short, triangular. Dorsum dull to weakly shining, glabrous, yellowish to reddish brown, usually with faintly indicated pattern of small maculation.

Range. Michigan and Ontario, south to Mississippi and Florida.

Behavior and habitats. *C. uhleri* is found in deciduous forests usually on heavier soils such as clay or clay loam. Blatchley (1920) reported small colonies of 3–6, or more, beneath logs or rails in dry locations. Hubbell (1936) stated that it hides by day but is active at night, and that, although it does not normally climb on vegetation, some specimens can be captured by sweeping. It has not been found in caves or on rocky slopes.

Life history. Over most of its range, *C. uhleri* is a late-maturing species, most numerous in the adult stage from late August to mid-October. It matures earlier in the northern part of the range, adults appearing in late June and July in Pennsylvania (Hubbell 1936).

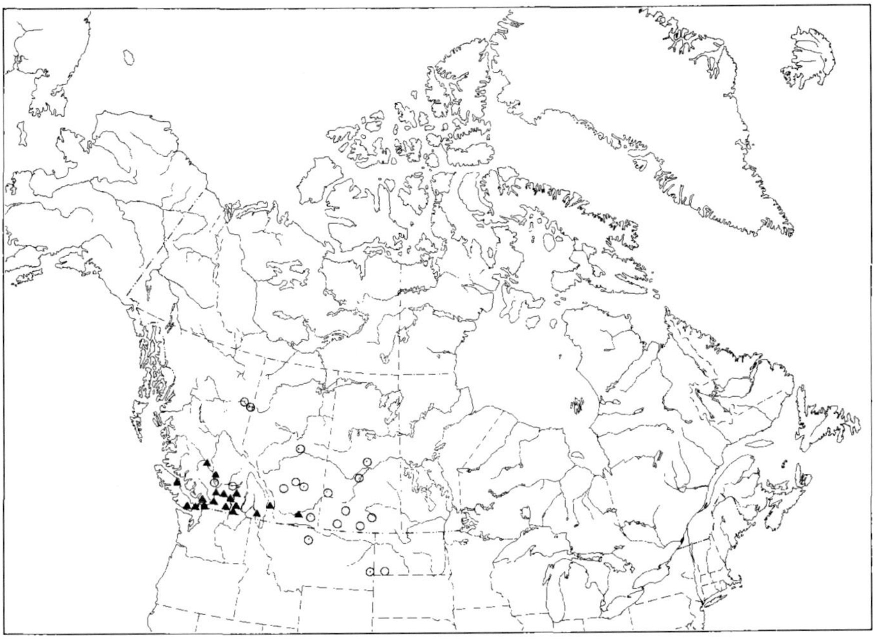

Map 31. Collection localities for *Ceuthophilus agassizii* (▲) and *C. alpinus* (○).

Map 32. Collection localities for *Ceuthophilus uhleri* (○) and *C. silvestris* (●).

Ceuthophilus (Ceuthophilus) maculatus (Harris)

Figs. 139, 140, 147, 159, 168; Map 33

Rhaphidophora maculata T. W. Harris, 1841:126.

Ceuthophilus maculatus; Scudder 1862:284 (invalid combination); 1863*a*:434; Hubbell 1936:197, 205; Helfer 1963:318; Vickery et al. 1974:41, 42.

Ceuthophilus scriptus F. Walker, 1869*a*:202.

Machamala armata F. Walker, 1869*a*:209.

Diagnosis. Ventrolateral margins of pronotum pale.

Description. Size medium (pronotal length, male, 3.6–5.0, female, 4.0–5.5 mm). Posterior margin of last evident tergum of male notched (Fig. 147). Male subgenital plate broad, more or less flattened, with lateral lobes more or less flattened and each consisting of two more or less distinct regions, namely, a heavily sclerotized distolateral lobe and a weakly sclerotized to membranous basal area (Fig. 159). Fore femur about one-third, or more, longer than pronotum; hind tibia and femur about equal in length, the latter broad and armed ventrally with 8–16 unequal rather coarse spines; hind tibia of male distinctly bowed or curved on basal third. Male epiphallus as in Fig. 168. Body blackish brown above, often with paler middorsal stripe; ventrolateral margins of pronotum pale; abdomen maculate with paler spots on darker background; hind femur with brownish bars.

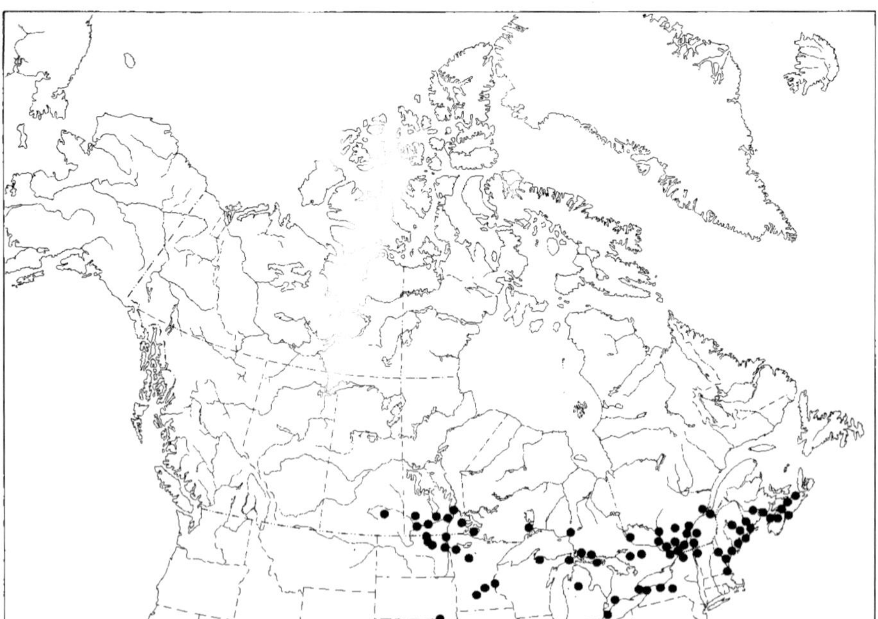

Map 33. Collection localities for *Ceuthophilus maculatus*.

Range. Saskatchewan to Nova Scotia, south to Arkansas.

Behavior and habitats. *C. maculatus* is commonly found under stones, logs, and soil surface debris in woods, particularly in drier areas.

Life history. Adults have been found from early May to the end of summer. They are most numerous during July and August. During winter, they are sometimes found in cellars. Nymphs are most common during early summer but may be found at any time during the frost-free period. In cellars and basements they may be found at almost any time of year.

Ceuthophilus (Ceuthophilus) gracilipes gracilipes (Haldeman)

Figs. 141, 155; Map 34

Phalangopsis gracilipes Haldeman, 1850:346.
Phalangopsis scabripes Haldeman, 1853:364.
Ceuthophilus gracilipes; Scudder 1863*a*:439.
Ceuthophilus grandis Scudder, 1894*a*:38–39.
Ceuthophilus heros Scudder, 1894*a*:54–56.

Diagnosis. Pronotal length 4.6–6.2 mm. Abdominal terga with posterior margins darkened.

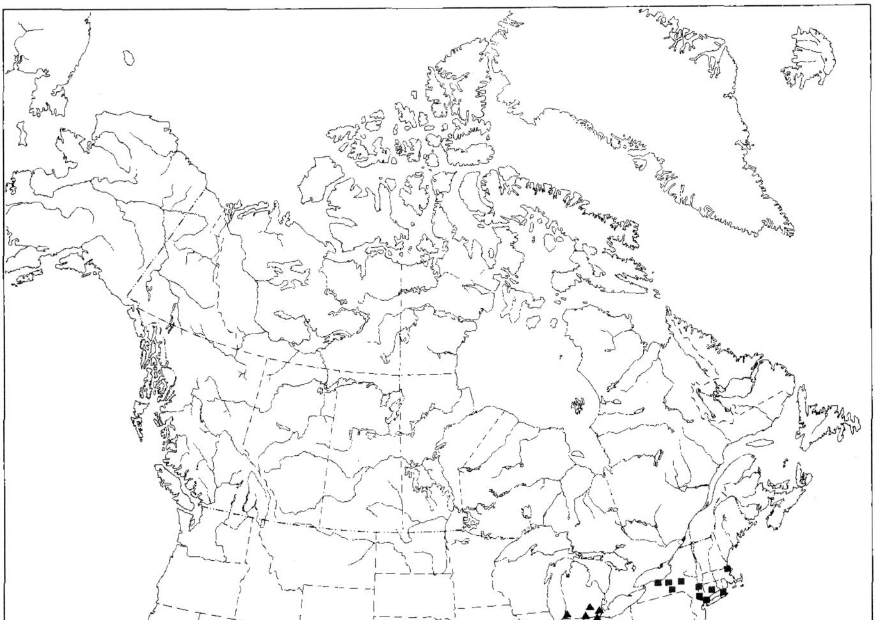

Map 34. Collection localities for *Ceuthophilus gracilipes gracilipes* (■) and *C. divergens* (▲).

Description. Body large (pronotal length, male, 4.6–6.2, female 5.1–6.2 mm), robust. Legs and antennae long; hind femur nearly as long as body, smooth above; hind femur of male with 6–14 (normally 8–11) robust spines with a few smaller spines interspersed between larger ones toward apical end (Fig. 141); hind femur of female with 5–14 (usually 8–10) small spines on outer ventral carina. Male subgenital plate divided by central sulcus, with lateral lobes narrow and ventral areas strongly convex; distal halves triangular with bluntly conical apices (Fig. 155). Ovipositor elongate; ventral valves with 5 stout triangular teeth, with proximal pair smaller and more widely separated than the others. Body brownish yellow to reddish brown, marked with darker brown above, this usually across hind margins of most segments, with or without median longitudinal pale area or row of lighter spots; lateral area sprinkled with pale yellow spots but these may be lacking in darker northern specimens.

Range. New York and Massachusetts, south to Mississippi.

Behavior and habitats. *C. gracilipes gracilipes* occurs widely in forested areas. It is not numerous at any locality, although it is both widespread and common throughout its range. It has been found in hollow trees, under loose bark of dead trees, under logs and stones, and in caves. In the southern part of the range, *C. gracilipes gracilipes* has been observed climbing trees, especially rough-barked trees. It is more active than most *Ceuthophilus* species. It can be attracted to molasses painted on tree trunks (Hubbell 1936).

Life history. Hubbell (1936) reported mature individuals at the northern part of the range in late June and early July and as late as October.

Ceuthophilus (*Ceuthophilus*) *meridionalis* Scudder

Figs. 142, 156; Map 35

Ceuthophilus meridionalis Scudder, 1894*a*:66–67.

Ceuthophilus meridionalis; J. A. G. Rehn and Hebard 1912*a*:70; Hubbell 1936:166–175.

Diagnosis. Dark, shiny, with dorsal median stripe orange yellow, conspicuous.

Description. Body size medium (pronotal length, male, 4.0–6.5, female, 4.6–7.0 mm), robust. Male subgenital plate similar to that of *gracilipes* but with apices of lobes not produced, angulate, with notch between distal parts of lobes broadly open (Fig. 156). Ovipositor elongate (9–13 mm). Female with 7–9 blunt teeth on mesal areas of posterior margins of terga. Dorsal surface dark, shiny, with conspicuous orange yellow median stripe on thorax and basal abdominal segments; abdomen dark with lighter spots.

Range. Michigan and Ontario, south to Pennsylvania.

Behavior and habitats. *C. meridionalis* is found in low moist areas (but not in swampy areas), in forests, in wooded margins of lakes and streams, and also in grassy pastured woodlots. It burrows beneath debris and probably emerges only at night. Some specimens appear to establish residence in certain burrows, returning to these after nocturnal forays (Hubbell 1936).

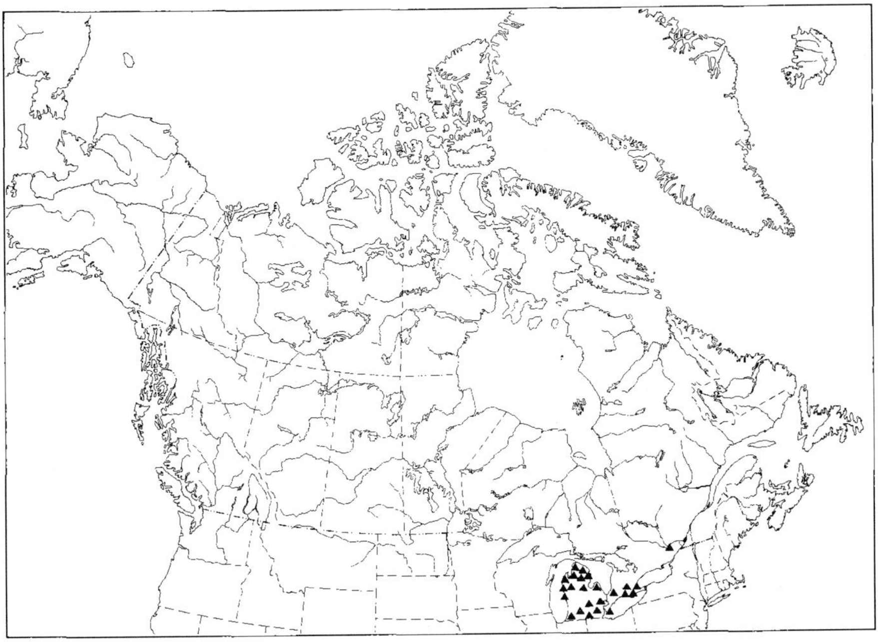

Map 35. Collection localities for *Ceuthophilus meridionalis*.

Life history. Adults have been taken in southern Michigan from 10 July to 11 November, but they are most numerous from late July to early September. Nymphs occur from early May to late July.

Ceuthophilus (*Ceuthophilus*) *latens* Scudder

Figs. 146, 158; Map 36

Ceuthophilus latens Scudder, 1863*a*:437.

Ceuthophilus latens; E. M. Walker in A. Gibson 1910*b*:126; Hubbell 1936:184.

Diagnosis. Hind femur nearly spineless. Body with broad orange band dorsally, darker laterally.

Description. Size medium (pronotum length, male, 3.6–5.0, female, 4.3–5.7 mm). Legs short, stout, with hind femur almost lacking spines. Male abdominal terminalia as in Fig. 146; male subgenital plate with lateral lobes widely separated dorsally and with apices strongly oblique (Fig. 158). Ovipositor with teeth nearly equally spaced. Dorsum of thorax with broad orange band; dorsolaterally much darker, reddish brown to shining intense black.

Range. Michigan to Maine, south to Tennessee and South Carolina.

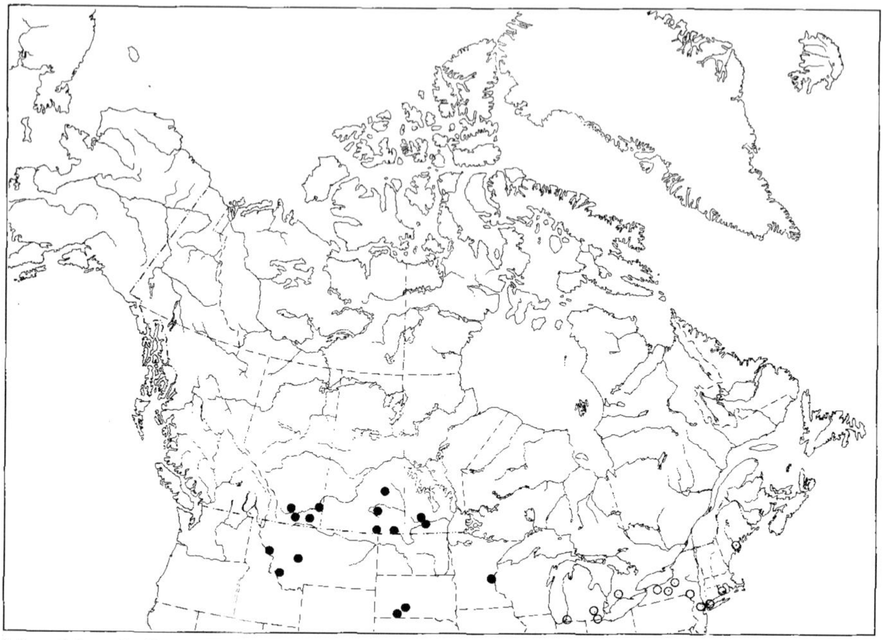

Map 36. Collection localities for *Ceuthophilus* latens (○) and *C. fusiformis* (●).

Behavior and habitats. Hubbell (1936) reported *C. latens* from a wide variety of deciduous forest environments to open grassy areas, particularly in drier areas. It does not appear to be associated with any particular soil type. It has also been found under surface debris, stones, and pieces of board, as well as in house cellars and near caves.

Life history. Eggs are laid in the fall. Some of these hatch in the autumn and winter as nymphs; others hatch in the spring. Adults appear in late June and July and decline in numbers, disappearing in October or early November.

Ceuthophilus (*Ceuthophilus*) *pallidipes* E. M. Walker

Figs. 145, 157, 170; Map 37

Ceuthophilus pallidipes E. M. Walker, 1905*b*:114–116, pl. IV, figs. 2, 2*a*, 2*b*, 2*c*.

Ceuthophilus pallidipes; E. M. Walker 1909:210; Vickery et al. 1974:44.

Diagnosis. Dorsal spurs of hind tibiae darkened at bases.

Description. Body small (pronotal length, male, 3.1–4.6, female, 3.4–4.9 mm). Dorsal spurs of hind tibiae darkened at base. Ninth abdominal tergum of male subquadrate, often upturned, heavily sclerotized (Fig. 145). Male subgenital plate with lateral lobes ovate; apices bluntly rounded, separated by shallow V-shaped notch (Fig. 157). Epiphallus as in Fig. 170.

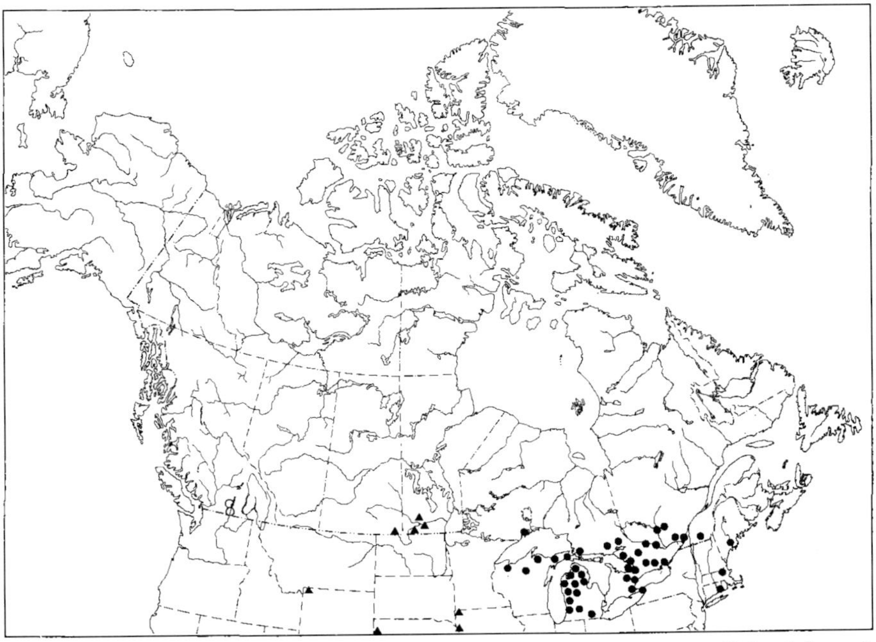

Map 37. Collection localities for *Ceuthophilus pallidipes* (●), *C. pallidus* (▲), and *C. vicinus* (○).

Ovipositor elongate, broad at base, tapering to apex, with 5 short closely spaced triangular teeth, the intervals between them diminishing toward apex. Body infuscate dorsally, with well-defined contrasting median pale stripe; dorsolateral thoracic bands not intensely black.

Range. Ontario to Maine, south to Georgia.

Behavior and habitats. *C. pallidipes* is nocturnal and is found in hollow logs, beneath tree bark, and under debris on the soil surface. Specimens have occasionally been taken up to 2 m above the ground on the sides of cabins and on tree trunks. Hubbell (1936) reported capture of specimens while sweeping bracken ferns and shrubbery at night in northern Michigan. He also mentioned that this species will visit molasses smeared on tree trunks. *C. pallidipes* is an active insect and a good climber.

Life history. Most of the information available on the life history of the species was provided by E. M. Walker (1905*b*). In southern Ontario, *pallidipes* was found to mature in the summer. A few specimens, which hatch from the eggs late in the season, pass the winter as young nymphs. Adults are usually found from early July until September, although captures in mid-October, in Massachusetts, and early November, in Michigan, were reported by Hubbell (1936).

Ceuthophilus (*Ceuthophilus*) *divergens* Scudder

Map 34

Ceuthophilus divergens Scudder, 1863*a*:436–437.
Ceuthophilus divergens; Hubbell 1926:14; 1936:355.
Ceuthophilus bicolor Scudder, 1894*a*:72.
Ceuthophilus sallei Scudder, 1894*a*:63.
Ceuthophilus caecus Scudder, 1894*a*:60.

Diagnosis. Body glabrous, with narrow longitudinal pale stripe bordered by darker areas.

Description. Body small (pronotal length of northern specimens, male, 2.8–3.8, female, 3.1–4.7 mm), not robust. Fastigium compressed, narrow, low, round, conelike. Male with posterior margin of ninth tergum subtruncate; male subgenital plate small, weakly sclerotized, broadest at base, with lateral lobes weakly convex at sides and converging to near apices; apices produced as short blunt lobes with open V-shaped notch between them. Ovipositor short, little longer than pronotum, stout at base, tapering only in basal half, with teeth of ventral valve slender and sharp. Body dark reddish brown dorsally, with narrow median pale stripe extending length of body; stripe bordered by broad infuscate areas that are broken by paler spots and dashes.

Range. South Dakota to Michigan, south to Louisiana and Georgia.

Behavior and habitats. *C. divergens* is generally found in low open-wooded areas but also in dense woods or grassy areas in woodland. It prefers dry conditions and is seldom found in marshy locations. Hubbell (1936) took large numbers in traps, possibly indicating high population densities to be present where the species occurs.

Life history. Hubbell (1936) recorded late June as the time of earliest appearance of adults in the northern part of the range, then abundant through July and August, declining in September. Late-instar nymphs were common in May and early June, indicating that winter is probably passed as small nymphs, although possibly also in the egg stage.

Ceuthophilus (*Ceuthophilus*) *pallidus* Thomas

Map 37

Ceuthophilus pallidus C. Thomas, 1872:434–435.
Ceuthophilus pallidus; Hubbell in Hebard 1928:301–302; Hubbell 1936:269–283.

Diagnosis. Thorax with minute setae (pilose) dorsally.

Description. Thorax minutely setose or pilose dorsally. Body small (pronotal length, male, 2.9–3.9, female, 3.0–4.1 mm), densely clothed on thoracic and abdominal terga with minute recumbent setae, appearing somewhat dull to unaided eye. Fastigium of head low, rounded, conelike, not elevated. Male subgenital plate shallow, stout at base, tapering only to middle. Color pattern consisting of light and darker blotches.

Four distinct geographical "phases" were characterized by Hubbell (1936). All the specimens occurring within the scope of the present work belong to the "Northern Great Plains phase." This "phase" is characterized by the dorsocephalic angle of middle coxa being strongly produced, forming an angle averaging from 60 to 70°; size small; spurs and calcar short; ovipositor elongate; epiphallus subquadrate.

Range. Saskatchewan and Manitoba, south to new Mexico and Texas.

Behavior and habitats. *C. pallidus* is generally in wooded areas but also in debris and under stones in short-grass plains or grassy hillsides. Specimens have been found in nests of bank swallows and kangaroo rats (Hubbell 1936). It is probable that *C. pallidus* is a useful food source for rodents. V. Bailey and Sperry (1929) reported that "camel-crickets" were eaten extensively by "grasshopper mice" of the genus *Onychomys*.

Life history. Earliest and latest dates for adults in the northern Great Plains are 2 July and 14 October, respectively. Maturity appears to be reached earlier in this region than farther south (Hubbell 1936).

Ceuthophilus (*Geotettix*) *fusiformis* Scudder

Figs. 143, 165; Map 36

Ceuthophilus fusiformis Scudder, 1894*a*:62–63.

Diagnosis. Legs pale. Prairie and plains species.

Description. Body small, compact, robust (pronotal length, male, 2.9–3.9, female, 3.0–3.9 mm). Legs short, stout. Ninth tergum of male completely concealed by eighth tergum, which is strongly produced with mesal part overhanging genital cavity; male subgenital plate small, entire, broadest at base, with lobes triangular, and with deep narrow cleft between lobes (Fig. 165). Ovipositor short, stout at base, tapering on basal half to slender apical half; teeth of ventral valve slender, curved. Body brownish yellow, banded with darker brown on posterior edges of terga but varying from completely pale to extensively infuscate; legs always pale.

Range. Alberta to Manitoba, south to New Mexico.

Behavior and habitats. Hubbell (1936) recorded *C. fusiformis* as a plains species, found under rocks, in nests of burrowing rodents, and among sedges in swales. He also stated that it excavates burrows when it cannot find a natural cavity or one prepared by another animal. Although it is primarily a grassland species, it is also found in open coniferous forests on mountain slopes. It has been found in Colorado at an elevation of 3140 m.

Life history. Adults have been found during every month of the year except January and February. Nymphs in various instars have been taken from April to October (Hubbell 1936).

Ceuthophilus (*Geotettix*) *silvestris* Bruner

Fig. 166; Map 32

Ceuthophilus silvestris Bruner, 1885*a*:126–127.
Ceuthophilus silvestris; Hubbell in Hebard 1932*a*:50; Cantrall 1943:149.

Diagnosis. Legs pale. Eastern forest-associated species. Male eighth tergum strongly produced, with bent lateral lobes at posterior margins.

Description. Body small, slender (pronotal length, male, 2.4–3.7, female, 2.9–3.5 mm). Eighth tergum of male in dorsal aspect strongly produced, with posterior margin emarginate and forming pair of lateral lobes bent downward over genital cavity, in lateral aspect projected past bases of cerci; subgenital plate with apices of lobes strongly rounded and prolonged, separated by cleft to middle (Fig. 166).

Range. Minnesota to Michigan, south to Arkansas.

Behavior and habitats. *C. silvestris* has been found in forests along stream margins, in roadside hedges bordering grassy fields, stone piles, and bushy clumps in grassy pastures.

Life history. Hubbell (1936) reported that adults had been collected in February and March and from August to November. The part of the range where these collections were made was not given. This may represent two generations but more likely is a representation of wintering as eggs and also as large nymphs.

Ceuthophilus (*Geotettix*) *guttulosus* F. Walker

Figs. 148, 150, 152, 162, 164, 171; Map 38

Ceuthophilus guttulosus F. Walker, 1869*a*:203.

Two subspecies are recognized in the area here covered. See discussion under the typical subspecies for general characters of the species as a whole.

Ceuthophilus (*Geotettix*) *guttulosus thomasi* Hubbell

Figs. 150, 164, 172; Map 38.

Ceuthophilus thomasi Hubbell, 1936:419–425.
Ceuthophilus thomasi; Cantrall 1943:148.
Ceuthophilus guttulosus thomasi; Eades 1962:147; Vickery et al. 1974:45.

Diagnosis. Body robust. Legs short, stout. Male subgenital plate entire, conical. Abdomen of both sexes somewhat tectate.

Description. Body small (pronotal length, male, 3.2–3.9, female, 3.5–4.2 mm), robust, with short stout legs. Male subgenital plate conical, entire, with a deep V-shaped notch posterodorsally (Fig. 164); eighth tergum of male strongly produced, nearly or entirely concealing ninth tergum (Fig. 150). Distal part of abdomen distinctly tectate to tectate–carinate dorsally (somewhat less apparent in males). Male epiphallus as in Fig. 172. Ovipositor as in Fig. 152. Body dark, usually with mediolongitudinal pale stripe (sometimes lacking); ventrolateral lobes of pronotum dark, rarely paler than dorsum.

Range. Michigan to Québec, south to Indiana and Ohio.

Behavior and habitats. Remarks under *C. guttulosus guttulosus* apply also to *C. guttulosus thomasi.*

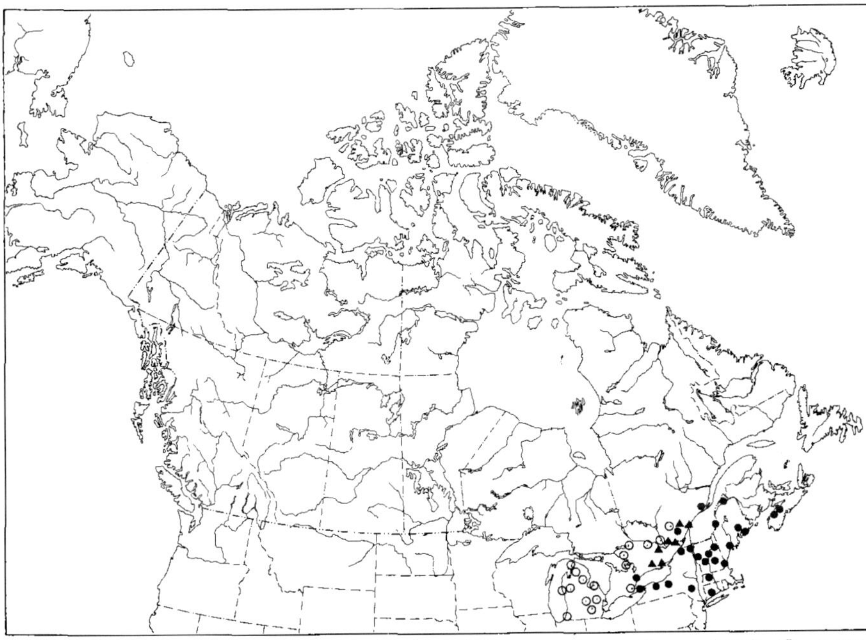

Map 38. Collection localities for *Ceuthophilus guttulosus thomasi* (○), *C. guttulosus guttulosus* (●), and intermediates (▲).

Life history. As for *C. guttulosus guttulosus.*

Ceuthophilus (Geotettix) guttulosus guttulosus F. Walker

Figs. 148, 152, 162, 171; Map 38

Ceuthophilus guttulosus F. Walker, 1869*a*:203.

Ceuthophilus guttulosus guttulosus; Eades 1962:147; Vickery et al. 1974:45.

Diagnosis. Body robust. Legs short, stout. Male subgenital plate entire, scoop-shaped.

Description. Body small, robust (pronotal length in northern part of range, male, 3.5–4.3, female, 3.7–4.4 mm). Legs short, stout. Ventral carina of hind femur with numerous close-spaced denticulations. Ninth tergum of male narrow, nearly concealed by eighth tergum (Fig. 148); male subgenital plate distinctive, scoop-shaped, undivided (Fig. 162); male epiphallus as in Fig. 171. Ovipositor short, heavy; ventral valve with crowded acute triangular teeth. Body buff to reddish brown, with broad irregular pale median stripe; dark areas interrupted laterally by pale maculae.

Range. Indiana to Nova Scotia, south to Virginia.

Behavior and habitats. At St. Hermas, Qué., specimens were found under stones in a wooded area where maple was the predominant tree (Vickery et al. 1974). Hubbell (1936) concluded that the (sub) species was associated with moist to medium dry forest areas. Eades (1964) stated that deep leaf litter seemed to be a key habitat requirement. He found greatest numbers in climax and subclimax forests dominated by sugar maple, basswood, beech, and yellow birch. The subspecies does not occur to any extent in oak–hickory forests.

C. guttulosus guttulosus does not climb like some other species, probably because the legs are relatively short. It is active at dusk and for about 3 hours afterward. At other times, these insects hide under logs or rocks amongst leaf litter. Eades (1964) reports that this subspecies is sensitive to drying and cannot survive under low humidity for more than a few hours.

Mating behavior was observed in the laboratory by Eades (1964). The sexes first touch antennae. The male then turns quickly and backs toward the female with abdomen twitching and cerci directed backward and in constant motion. When the male cerci contact the female abdomen, the male twists the tip of his abdomen through almost 90° and attempts to insert his genitalia beneath the ovipositor of the female. The male abdomen continues to twitch and cerci continue to move until insertion has been accomplished. The pair remain together for 15 minutes, or longer. Eades (1964) carried out experiments on crossbreeding the various subspecies of *C. guttulosus*, and found mating to be about random between them; no apparent barrier was found. That author also reported finding specimens heavily infested with mites in Québec.

Life history. Eades (1964) stated that nearly all individuals pass the winter as eggs or as large nymphs. There are two peaks of adult activity as a result of this, the first in late April, the second from August to October.

Ceuthophilus (*Geotettix*) *alpinus* Scudder

Figs. 167, 173; Map 31

Ceuthophilus alpinus Scudder, 1894a:30, 78–79.
Ceuthophilus vinculatus Scudder, 1894a:29, 91.
Ceuthophilus alpinus; Hubbell in Hubbard 1928:300.

Diagnosis. Body small, depressed, appearing broad. Male subgenital plate entire, rounded with shallow apical cleft.

Description. Form depressed, thus appearing unusually broad; size small (pronotal length, male, 3.1–3.9, female, 3.1–3.7 mm). Eyes small, widely separated. Male hind tibia shorter than hind femur, swollen, dorsally tumid, with dorsolateral carinae poorly defined; spurs short, stout, subcylindrical; subdistal ventral spurs of hind tibia 2-2. Eighth tergum of male completely covering ninth tergum, truncate posteriorly, bent downward mesally; subgenital plate rounded, entire, shallowly cleft (but appearing deeper in dried specimens) (Fig. 167); male epiphallus as in Fig. 173. Ovipositor noticeably curved, with apices of dorsal valves upturned, aciculate; ventral valves with 5 slender curved aciculate teeth and with apical 3 longer and more curved.

Range. British Columbia to Iowa, south to Colorado.

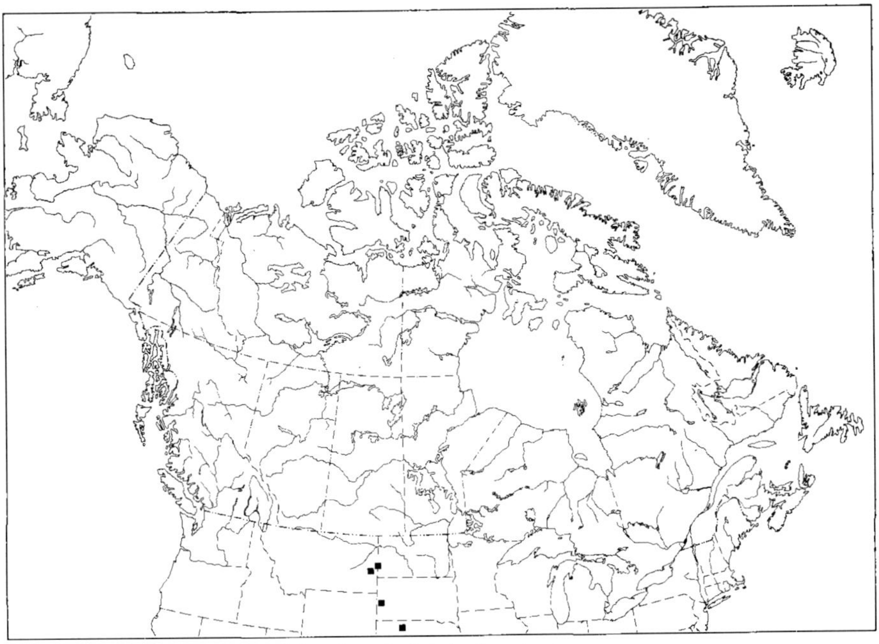

Map 39. Collection localities for *Daihinia brevipes*.

Behavior and habitats. Buckell (1928) reported this species as common under logs and stones in the Peace River area of British Columbia, near the Alberta border. Other specimens have been found under boards at the edge of a meadow in a grove of yellow pines in Wyoming. The structure of this species indicates that it is probably a burrowing species, seldom to be found aboveground (Hubbell 1936).

Life history. Poorly known; adults have been collected from April to October.

Ceuthophilus (*Geotettix*) *vicinus* Hubbell

Map 37

Ceuthophilus vicinus Hubbell, 1936:454–456.

Diagnosis. Western. Male subgenital plate elongate, triangular, truncate at apex.

Description. Pronotal length, male, 3.5–4.9, female, 4.4 mm; based upon 3 males, 2 females from Washington (Hubbell 1936). Body robust with short stout legs, nearly glabrous dorsally, moderately polished. Male with ninth tergum evaginated mesally; eighth tergum produced obtuse-triangulate, with apex extending over evaginated area of ninth tergum; male subgenital plate entire, elongate triangular, truncate at apex, not cleft or notched.

Ovipositor about 1.25 times longer than pronotum; ventral valves with 5 long attenuate teeth, of which apical 3 distinctly curved and longer than basal 2.

Range. British Columbia, Washington, and Idaho.

Behavior and habitats. Unknown.

Life history. Unknown; adults have been collected in April and August.

Subfamily Daihiniinae

Diagnosis. Hind legs short, fore and/or hind tarsi apparently 3-segmented.

Genus *Daihinia* Haldeman

Description. Fore and hind tarsi each with only 3 segments rather than the usual 4 (due to reduction of true second), the apparent second segment much smaller than the others (Fig. 179). Pronotum truncate anteriorly and posteriorly; mesonota and metanota together as long as pronotum. Fore tibia with large blunt spines; hind tibia with large claws (Fig. 179); hind femur armed ventrally; hind tibia armed above on both margins with about 6 large spines with smaller spines among the large ones on basal half (Fig. 180). Ovipositor stout, with ventral valves possessing 4 long teeth (Fig. 178) and an apical hook. Male subgenital plate with deep apical fissure.

Daihinia brevipes Haldeman

Figs. 177–180; Map 39

Phalangopsis (*Daihinia*) *brevipes* Haldeman, 1850:346.
Daihinia brevipes; Bruner 1893*b*:30.

Diagnosis and description. As for genus. Hind tibia with single spine on ventral face.

Range. North Dakota, South Dakota, Nebraska.

Behavior and habitats. Hebard (1931*b*) reported that *D. brevipes*, by preference, digs holes in hard soil, often along clay roads. The biology and behavior of the species have been extensively studied (in Oklahoma) by Whitehead and Miner (1944). The eggs are about 2 mm long and may be laid as deep underground as 1.25–1.5 m, although Hebard (1931*b*) reported a depth of only 45 cm, which is probably more normal. There are six nymphal instars. Feeding occurs at night, when the insects may emerge from the ground. Mating occurs in the burrows and several individuals of each sex may cohabit a single burrow system, although Hebard (1931*b*) states that only single males occur with the females.

Life history. According to Whitehead and Miner (1944), the winter, in Oklahoma, is spent in burrows, mostly as third- or fourth-instar nymphs. Rapid development occurs in spring and, by April or May, adults appear. Maturity is presumably later in the north. Oviposition takes place in the walls of burrows until the end of June, by which time most adults are dead.

Eggs hatch in the latter part of September, after an incubation period of 3.5–4 months. The young nymphs then dig their way to the surface from the collapsed adult burrows where they hatched, and excavate their own individual burrows. They remain active until the onset of cold weather. There is only one generation each year.

Economic importance. Hyslop (1934) first recorded this species as being injurious to truck crops and garden plants at Hollis, OK. During 1935 to 1938, in particular, it did considerable damage in various parts of that state, injuring seedlings of a wide variety of crops, according to Whitehead and Miner (1944). Injury does not seem to have been serious subsequently. The authors mentioned details concerning the kinds of plants attacked and the nature of the damage. They also discuss control measures, using poison baits.

Infraorder Tettigoniidea

Members of this infraorder are known by such vernacular names as bush-crickets (*not* bush crickets), katydids, long-horned grasshoppers, leaf-crickets, and hump-winged crickets. They may be distinguished from other Grylloptera (including the Stenopelmatidea) by the following combination of characters: general coloration most often predominantly green, although frequently brown, gray, or mottled. Head, if round, not bulletlike or disproportionately large, often conical or of other shape. Eyes often small, but generally not unusually so. Antennae seldom shorter than body. Prothoracic spiracles and associated ends of tracheal system sometimes enlarged and modified as auditory organ. Legs, if heavily spined, seldom modified for digging, with tarsi 4- or 5-segmented; front tibiae usually with basal auditory organs; hind femora not modified for stridulation. Wings usually fully developed or moderately abbreviated, sometimes greatly reduced or absent; tegmina usually held rooflike over abdomen, frequently resembling foliage, those of males in most species with stridulatory area of specialized veins at bases (even in species with greatly reduced wings, the stridulatory area remains). Cerci neither long nor flexible. Subgenital plate (ninth abdominal sternum) of males almost invariably with pair of styles. Ovipositor normally long, sickle-, sword-, or dagger-like, with all 6 valves well-developed, occasionally reduced.

There are almost 5000 species and probably somewhere in the order of 900 genera of living Tettigoniidea; all, except for four species of Hagloidea, belong to the superfamily Tettigonioidea, *sensu stricto*. There is also a small extinct group, the somewhat elongate-bodied Phasmomimidae of Sharov (1968), formerly known only from the Upper Jurassic to Lower Cretaceous deposits of Siberia. They are believed to have had a common ancestry with the Hagloidea, or to have been derived from these (Sharov 1968, 1971); they probably merit separate superfamily status (Phasmomimoidea). One of us (D. K. K.) examined one of the two fossil "Orthoptera" reported by P. Mitchell and Wighton (1979) from the Palaeocene Paskapoo deposits in the Red Deer River region of Alberta, and found that, of the single pair of wings (Figs. 181, 182), the tegmen bore a close resemblance, in both shape and venation, to similar structures in the genus *Phasmomimoides* Sharov.

What little could be seen of the hind wing venation agreed with this finding, and it seems that the Phasmomimidae did not become extinct in Cretaceous times, but persisted, at least in Canada, until the early Tertiary. This is a similar state of affairs to the recent discovery of Tertiary Locustopsidae in Montana and is not particularly surprising in view of the relict status of the few living members of the related Hagloidea.

Living Tettigoniidea are widely distributed throughout the world, but they are predominantly tropical, although many occur in temperate countries. They are uncommon, however, in regions with long cold winters. A high proportion of species are arboreal or arbusticolous, but there are also many that live in herbage, reeds, or grassland, and a considerable number live on the ground. A great many resemble broad-leaved foliage and many others mimic grass blades or stems. Procryptic resemblance to bark, lichen, and so forth is also exhibited by some; a few, in the immature stages, mimic other arthropods. There are species that exhibit aposematic (warning) coloration and some may produce irritant secretions, or eject streams of haemolymph, but their numbers are relatively few. The majority of species, although by no means all, are crepuscular or nocturnal. A great many are completely phytophagous, but it is probable that the majority have a diet that includes not only foliage, but also flowers, seeds, and insect prey. Some are largely or wholly carnivorous (probably the primitive condition), when they may be ferocious predators of large insects.

A few Tettigonioidea, under certain circumstances, may become strongly gregarious. This applies mainly to certain flightless species, but there are one or two that form migrating flying swarms somewhat like locusts (although not in the region here considered). Some serious, if spasmodic, crop pests are included in the Tettigonioidea, particularly among flightless, gregarious species. The flying swarms, in countries where they occur, tend to be welcomed as a source of human food. There are, of course, nongregarious species that are crop pests and a few ground-living species that attack stored produce, more or less in the open, but the general economic importance of the infraorder is not great in relation to its size.

Key to superfamilies of Tettigoniidea

1. Body brownish, usually mottled with green. Head short, with slightly concave fastigium and with antennae inserted near lower margins of eyes. Hind tibiae rather short and stout, with less than 10 dorsal spines in each row. Ovipositor short .... **Hagloidea** (p. 161)

   Body wholly or partly green (rarely with some pink). Head, if short, with fastigium not concave (often produced); antennae inserted near middle or upper margins of eyes. Hind tibiae long, slender, with numerous spines in each row. Ovipositor usually well-developed, sometimes conspicuously elongate .... **Tettigonioidea** (p. 165)

Clé des superfamilles de Tettigoniidea

1. Corps brunâtre, habituellement moucheté de vert. Tête courte, à fastigium légèrement concave; point d'insertion des antennes près du bord inférieur des yeux. Tibias postérieurs plutôt courts et forts, portant moins

de 10 épines dorsales par rangée. Ovipositeur court .... **Hagloidea** (p. 161)

Corps entièrement ou partiellement vert (plus rarement avec du rose). Tête, si elle est courte, munie d'un fastigium non concave (souvent convexe); point d'insertion des antennes près du milieu ou du dessus du bord des yeux. Tibias postérieurs longs, minces et munis de nombreuses épines dans chaque rangée. Ovipositeur en général bien développé, parfois long .... **Tettigonioidea** (p. 165)

Superfamily Hagloidea

Members of this superfamily have been called ambidextrous crickets and hump-winged crickets, and they were given superfamily rank by Chopard (1949).

Two families are recognized, namely, the extant Prophalangopsidae and the extinct Haglidae, *sensu stricto* (Triassic to Liassic or Lower Jurassic). Both living subfamilies of Prophalangopsidae are known from the Liassic until the present time, but most members of the family are also extinct. The living subfamily Prophalangopsinae is represented today by a single species from northeastern India, and the other living subfamily, Cyrtophyllitinae, is represented by one genus with three brachypterous species (all having normal, four-segmented tarsi) from western North America. These latter are all ground-dwelling, largely predatory, and mainly nocturnal. They stridulate loudly. Information concerning them is summarized by G. K. Morris and Gwynne (1979).

Family Prophalangopsidae

Characters are given for this family in the key to superfamilies.

Genus *Cyphoderris* Uhler

This genus is the sole North American representative of the family.

Description. Size medium (16–31 mm long), robust. Legs short; hind pair barely saltatorial (Figs. 183, 187); fore tibiae with spines on both ventral margins, and on inner dorsal margin; tympanal cavity broad and long on both faces. Tegmina reduced to small oval lobes in females, well-developed for stridulation (but not for flight) in males. Thorax of female cylindrical, of male dilated and inflated (''hump-backed'') posteriorly; posterior margin of metanotum typically with small denticles.

Only two of the three known species occur within the range of the present study, *C. monstrosa* Uhler and *C. buckelli* Hebard. The third species, *C. strepitans* Morris & Gwynne, is known from Wyoming and Colorado.

Key to species of *Cyphoderris*

1. Male with prominent process on subgenital plate projected ventrally at angle greater than 90° (Fig. 184*a*), usually terminally cleft and toothed (Fig. 184*b*); styli of subgenital plate depressed, sublanceolate (Fig. 184*a*). Ridged patch (Ander's organ) present dorsolaterally on first abdominal tergum just behind hind wing; adjacent edge of metanotum with strong teeth (Fig. 185) .... ***monstrosa* Uhler** (p. 162)

   Males lacking ventral process on subgenital plate; plate bulbous, with paired short stout somewhat incurved styles laterally (Figs. 186*a*, 187*a*, 187*b*). Ridged patch (Ander's organ) lacking; adjacent edge of metanotum sometimes with weakly developed teeth ..... ***buckelli* Hebard** (p. 164)

Clé des espèces de *Cyphoderris*

1. Plaque sous-génitale des mâles munie d'une protubérance ventrale formant un angle supérieur à 90° (fig. 184*a*), habituellement encochée à l'extrémité et munie de dents (fig. 184*b*); styles de la plaque sous-génitale aplatis, sublancéolés (fig. 184*a*). Plaque carénée (organe d'Ander) en position dorso-latérale, sur le premier tergite de l'abdomen, juste derrière les ailes postérieures; bord adjacent du métanotum muni de fortes dents (fig. 185) .... ***monstrosa* Uhler** (p. 162)

   Plaque sous-génitale des mâles dépourvue d'une protubérance ventrale, de forme bulbeuse et munie d'une paire de styles courts, forts et légèrement incurvés (fig. 186*a*, 187*a* et 187*b*). Plaque carénée (organe d'Ander) absente; bord adjacent du métanotum parfois muni de dents peu développées .... ***buckelli* Hebard** (p. 164)

Cyphoderris monstrosa Uhler

Figs. 183–185; Map 40

Cyphoderris monstrosus Uhler, 1864:552 (partim).
Cyphoderris monstrosa; Scudder 1899*a*:117.
Cyphoderris monstrosa var. *piperi* Caudell, 1904*b*:53; Helfer 1963:303.

Diagnosis. Process of male subgenital plate projected ventrally. Females with ridged patch dorsally on first abdominal tergum.

Description. Head subglobose, broad. Pronotum expanded and produced anteriorly, covering occipital region of head, subtruncate on anterior and posterior margins in males, somewhat more rounded posteriorly in females, deeply constricted before the middle and expanded posteriorly in males (lacking in females). Male genital structure as in Figs. 184*a*, 184*b*. Both sexes possessing Ander's organ, a ridged patch situated dorsolaterally on the first abdominal tergum (present also in many immature individuals) and associated with denticles on posterior margin of metanotum (Fig. 185).

Range. British Columbia to Montana, south to northern California.

Behavior and habitats. Hebard (1934*b*) reported finding specimens under stones and also that they are frequently found in buildings. G. K. Morris and Gwynne (1979) recorded observations in the Kananaskis Valley,

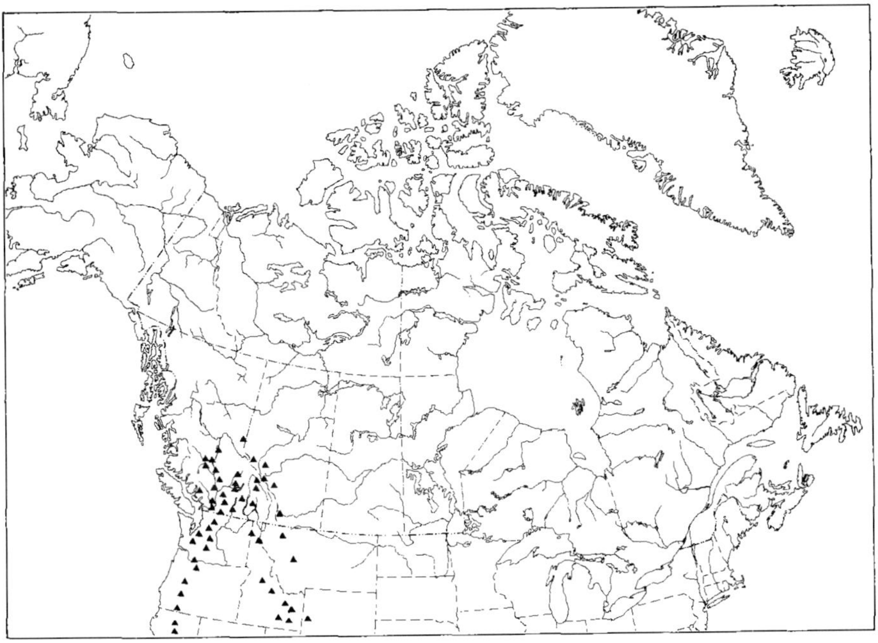

Map 40. Collection localities for *Cyphoderris monstrosa*.

Alta., where both adults and nymphs were observed in Lodgepole pine trees (*Pinus contorta*), feeding upon staminate cones prior to the "loose pollen" stage of these. A similar situation, for nymphs only, was observed by one of us (D. K. K.) in the same general locality 19 June 1959. *C. monstrosa* is the only member of the genus that is known to climb trees, the other species preferring to remain on the ground or on low structures. Several immature specimens were collected by one of us (V. R. V.), 20 August 1966, on Sulphur Mountain, Banff, Alta. The ridge is rocky and nearly devoid of vegetation.

The stridulation of *C. monstrosa* is a trill of sinusoidal pulses with dominant carrier frequency of 13 kHz and a pulse rate of 71.4 per second at 25°C. This species is unusual in that it switches the position of the tegmina during stridulation, so that first one, then the other, is uppermost. This causes variation in intensity and frequency of pulses (Spooner 1973). G. K. Morris and Gwynne (1979) found that, in the calling songs of *C. monstrosa*, tegminal switching does not occur; this apparently is normal only in "courtship" songs. Stridulation normally occurs in the evening and continues past midnight. Nearly all acoustic behavior by males of this species occurs in trees, often 5 m, or more, about ground level (G. K. Morris and Gwynne 1979).

Life history. It seems that this species passes the winter in concealment as a nymph and emerges early in summer, becoming adult and remaining active until the onset of cold weather.

Economic importance. Turley (1901) records extensive damage to fruit trees, which was attributed to this species in Idaho around the turn of the century.

Cyphoderris buckelli Hebard

Figs. 186, 187; Map 41

Cyphoderris buckelli Hebard, 1934*b*:371.
Cyphoderris buckelli; Helfer 1963:304; G. K. Morris and Gwynne 1979:147, 152–154.

Diagnosis. Process of male subgenital plate not directed ventrally. Female lacking ridged patches (Ander's organs) on first abdominal tergum.

Description. Male subgenital plate not directed ventrally at angle greater than 90°, bulbous, with paired short incurved styles dorsolaterally (Figs. 187*a*, 187*b*). Cerci not flattened.

Range. British Columbia to Montana, south to Oregon.

Behavior and habitats. Typically *C. buckelli* occurs in dry forest that is characterized by ponderosa pine (*Pinus ponderosa* Dougl. ex P. Laws. & C. Laws.) and, at higher elevations, by Douglas fir (*Pseudotsuga menziesii* (Mirb.) Franco). According to G. K. Morris and Gwynne (1979), it feeds during the spring upon blossoms of understory plants such as serviceberry (*Amelanchier*), arrow-leaf balsamroot (*Balsamorhiza*), and tall Oregon grape (*Berberis*). These authors state that they did not find *C. monstrosa* and *C. buckelli* together, although Buckell (1924) had stated that both species had been seen at Nicola, B.C., feeding on *Amelanchier* in May.

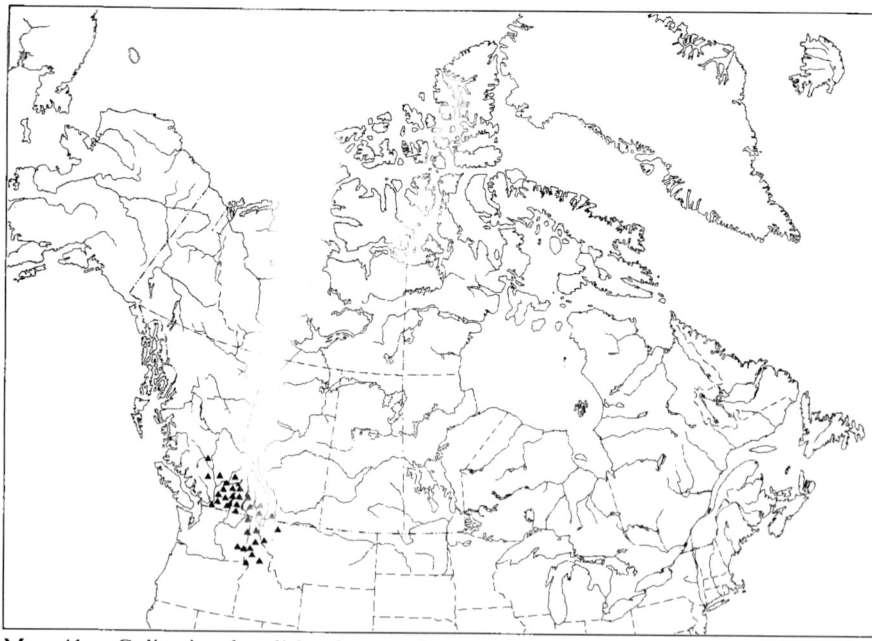

Map 41. Collection localities for *Cyphoderris buckelli*.

Stridulation occurs on the ground or on low shrubs or on tree trunks less than a metre from the ground. Unlike *C. monstrosa*, change of terminal overlap is infrequent in *C. buckelli* (less than 3%, according to G. K. Morris and Gwynne (1979)). The dominant carrier frequency is near 13.3 kHz, with a lesser harmonic peak from 28 to 30 kHz. Each pulse increases rapidly in amplitude, then decreases gradually to the end of the pulse. Stridulation of *C. buckelli* is apparently similar to that of *C. strepitans* Morris & Gwynne, but the geographical distributions of the two species do not overlap, the latter occurring only in the mountains of Wyoming and Colorado. This is fortunate, as there is, as yet, no known anatomical character to distinguish between females of the two species.

Life history. Presumably similar to that of *C. monstrosa*.

Superfamily Tettigonioidea

This is the principal superfamily of the infraorder Tettigoniidea and it includes the various insects known by such vernacular names as bush-crickets (*not* bush crickets), katydids, long-horned grasshoppers, leaf-crickets, and weed-insects. In French, the word "sauterelles" applies strictly to them, although it is often used as ambiguously as the English word "grasshopper" to include short-horned grasshoppers (the Acridoidea of the Orthoptera, *sensu stricto*, which are, strictly, "criquets" in French).

> Now do the katydids, leaf-crickets and weed-insects of the dusk,
> that stridulate the long night through,
> celebrate their Erotidia, or festivals of love,
> Or are they elves disguised, insect-like, in long close coats of green
> and gray. . . .
>
> (Cawein 1906)

All living Tettigoniidea, except for the four species of Hagloidea, belong to the Tettigonioidea.

Many of the 11 families that we recognize are unrepresented in Canada and adjacent regions. Only the Phaneropteridae, Pseudophyllidae, Tettigoniidae, *sensu stricto*, and Conocephalidae concern us here.

Key to families of Tettigonioidea

1. Head rounded, short, with fastigium not produced or pointed, and with face not flattened or slanting. Prothoracic spiracles small, not covered by pronotum; prosternum without pair of spines. Tegmina well-developed, usually rather broad, leaflike 2
 Not as above 3

2(1). Antennal insertions ("scrobes") weakly margined. Anterior tibia rather flattened dorsally and somewhat quadrate distally in cross-section; tarsi with first and second segments cylindrical, not grooved laterally. Tegmina lacking expanded area with prominent parallel crossveins. Ovipositor rather short, sickle-shaped, typically rather blunt, apically denticulate. . . . . **Phaneropteridae** (p. 167)

Antennal scrobes strongly margined. Anterior tibia more or less round or oval in cross-section throughout; tarsi with first and second segments grooved laterally. Tegmina (particularly of male) with expanded area having prominent, parallel crossveins. Ovipositor moderately long, gradually curved upward .... **Pseudophyllidae** (p. 186)

3(1). Head subconical to strongly pointed. Prosternum sometimes lacking spines. Anterior tibiae without terminal dorsal spines; tibiae and femora without ventral rows of strong spines. Wings sometimes short. Ovipositor long, typically rather slender, straight or only slightly curved .... **Conocephalidae** (p. 189)

Head short, not conical, with fastigium forming a short narrow rounded or truncated rostrum. Prosternum with pair of spines. Anterior tibiae with apicodorsal spine, or tibiae and femora with ventral rows of strong spines. Wings often strongly reduced. Ovipositor rather stout, swordlike or daggerlike, usually somewhat curved upward, sometimes fairly straight .... **Tettigoniidae** (p. 221)

Clé des familles de Tettigonioidea

1\. Tête ronde, courte, à fastigium ni convexe ni pointu, à face ni aplatie ni inclinée. Stigmates du prothorax petits, non recouverts par le pronotum; prosternum dépourvu d'épines. Tegmina bien développés, habituellement larges, ayant l'apparence d'une feuille 2

Combinaison de caractères différente de celle décrite ci-dessus 3

2(1). Points d'insertion des antennes (scrobes) faiblement délimités. Tibias antérieurs plutôt aplatis dorsalement et quelque peu quadrangulaires en coupe transversale, à l'extrémité distale; premier et deuxième segments des tarses cylindriques, dépourvus d'un sillon latéral. Tegmina dépourvus d'une zone étendue à nervures parallèles proéminentes. Ovipositeur plutôt court, en forme de faucille, d'habitude plutôt obtu, denticulé à l'extrémité apicale .... **Phaneropteridae** (p. 167)

Scrobes clairement délimités. Tibias antérieurs en coupe transversale plus ou moins ronde ou ovale sur toute leur longueur; premier et deuxième segments des tarses munis d'un sillon latéral. Tegmina (surtout chez les mâles) développés et munis de nervures parallèles proéminentes. Ovipositeur modérément long et recourbé vers le haut .... **Pseudophyllidae** (p. 186)

3(1). Tête de forme subconique à très pointue. Prosternum parfois dépourvu d'épines. Tibias antérieurs sans épines dorsales terminales; tibias et fémurs dépourvus de rangées ventrales d'épines fortes. Ailes parfois courtes. Ovipositeur long, habituellement plutôt mince, droit ou légèrement incurvé .... **Conocephalidae** (p. 189)

Tête courte, non conique, à fastigium formant un rostre court, étroit, arrondi ou tronqué. Prosternum muni d'une paire d'épines. Tibias antérieurs munis d'une épine dorsale terminale, ou tibias et fémurs munis de rangées ventrales de fortes épines. Ailes souvent très réduites. Ovipositeur plutôt fort, en forme de sabre ou de poignard, habituellement quelque peu incurvé vers le haut, parfois plutôt rectiligne .... **Tettigoniidae** (p. 221)

Family Phaneropteridae

Members of the family Phaneropteridae are the leaf-crickets, leaf-katydids, and lyre bush-crickets. They constitute the largest family of Tettigonioidea and are recognizable among other members of the superfamily by the following combination of characters: head rounded, short, with face not flattened or slanted. Antennae longer than body, inserted between eyes (which are typically small and round), their insertions ("scrobes") only weakly margined. Prothoracic spiracles small, not concealed by pronotum; prosternum without spines. Anterior tibiae flattened dorsally, distally rather square in cross-section, with auditory tympana generally covered; hind tibiae with apical spine on each side; tarsi with first and second segments subcylindrical, not laterally grooved. Wings often reduced (but not completely absent); although typically tegmina characteristically leaflike, occasionally resembling insect or fungal damage to plants, with basal stridulatory organ well-developed even when tegmina reduced; hind wings in fully alate species (all but one introduced species in region here covered) much longer than tegmina, projecting well beyond them at rest. Ovipositor rather short, sickle-shaped, flat, usually with apex rather obtusely pointed and often distinctly crenulated or serrated.

There are about 2000 species of Phaneropteridae, and probably more than 300 genera, distributed unevenly between two subfamilies: the Odonturinae (virtually confined to western Palaearctic region), and the Phaneropterinae, *sensu stricto*, which are nearly all fully winged and generally with spines on the front coxae (worldwide except for colder regions, predominantly tropical).

Phaneropterinae species are almost exclusively phytophagous and mainly nocturnal, most of them living in trees or on bushes, in the foliage or twigs of which they lay their eggs and among which they are extraordinarily well camouflaged. A number of species, in the immature stages, mimic ants, cicindeline ground-beetles, and even certain spiders. The Odonturinae species, although protectively colored, are not generally striking mimics; they are largely ground-living, often most active during the day. Some may occasionally eat small insects, and most lay their eggs in compact masses in the soil. Some species of this subfamily are prone to become gregarious and to form marching bands. Sporadically they may cause serious injury to crop plants. Most Phaneropteridae, however, are not of much economic significance, particularly among the Phaneropterinae, which are the only native members of the family represented in the region here covered.

Subfamily Phaneropterinae

Key to genera of Phaneropterinae

1. Tegmina elongate, not distinctly broadened in middle (Figs. 190, 191). Fastigium little, if any, wider than basal antennal segment (Fig. 192) .... ***Scudderia*** (p. 168)

Tegmina ovate to ovate-lanceolate, distinctly broadened in middle (Figs. 210, 217). Fastigium much wider than basal antennal segment (Fig. 211) 2

2(1). Hind femur long (Fig. 210). Fore and middle tibiae flat or sulcate above with raised margins. Tegmina oblong-elliptic or ovate (Fig. 210). Ovipositor long, curved upward, strongly serrate on both edges (Fig. 216) .... ***Amblycorypha*** (p. 179)

Hind femur rather short (Fig. 217). Fore and middle tibiae smooth, rounded, without angular margins. Tegmina ovate-lanceolate (Fig. 217). Ovipositor short, bent abruptly upward, finely crenulate, with apex obtuse (Fig. 219) .... ***Microcentrum*** (p. 184)

Clé des genres de Phaneropterinae

1. Tegmina longs, sans élargissement distinct au milieu (fig. 190 et 191). Fastigium parfois absent, sinon petit, plus large que le segment basilaire de l'antenne (fig. 192) .... ***Scudderia*** (p. 168)

Tegmina de forme ovée ou ovée-lancéolée, clairement élargis à mi-longueur (fig. 210 et 217). Fastigium beaucoup plus large que le segment basilaire de l'antenne (fig. 211) 2

2(1). Fémurs postérieurs longs (fig. 210). Tibias antérieurs et médians aplatis ou en forme de gouttière à bords relevés vers le haut. Tegmina oblongs, elliptiques ou ovés (fig. 210). Ovipositeur long, recourbé vers le haut, fortement denticulé des deux côtés (fig. 216) .... **Amblycorypha** (p. 179)

Fémurs postérieurs plutôt courts (fig. 217). Tibias antérieurs et médians lisses, arrondis, sans bord angulaire. Tegmina ovés-lancéolés (fig. 217). Ovipositeur court, recourbé brusquement vers le haut, finement crénelé et à apex obtu (fig. 219) .... **Microcentrum** (p. 184)

Genus *Scudderia* Stål

Description. Body size medium (15–28 mm long), having general form indicated in Fig. 191. Fastigium horizontal or bent slightly downward, not wider than basal antennal segment (Fig. 192). Pronotum flattened dorsally, with sides parallel or only slightly divergent posteriad (Fig. 193). Eyes rounded. Tegmina long, with margins subparallel but slightly widened toward apex. Male subgenital plate usually long, upcurved, with apical U- or V-shaped notch; male supra-anal plate greatly prolonged, decurved, forked, or notched to receive apical part of subgenital plate as in Figs. 201–204 (except in *S. septentrionalis*—Fig. 200). Ovipositor short, broad, curved sharply upward (Figs. 205–209), finely crenulate apically on both edges. Chromosome number, $2n$ male = 31 (except *S. pistillata*, which has $2n$ male = 29 (Beaudry 1973)).

Key to species of *Scudderia*

1. Male. Stridulatory apparatus present on tegmina. Subgenital and supra-anal plates elongate and/or elaborate (Figs. 195–204) 2

Female. Stridulatory apparatus absent. Ovipositor broad, curved (Figs. 205–209) 7

2(1). Supra-anal plate nearly triangular, without elongate curved process (Figs. 195, 200); subgenital plate not compressed toward apex. Cercus long, tapered (Fig. 195) .... *septentrionalis* **(Audinet-Serville)** (p. 170)

Supra-anal plate with elongate curved median process, notched or forked at apex (Figs. 196–199, 201–204); subgenital plate narrowed and compressed toward apex. Cercus short, strongly incurved (Figs. 196–199) 3

3(2). Apex of supra-anal plate distinctly and deeply forked, usually enclosing tip of subgenital plate (Fig. 203) 4

Apex of supra-anal plate not deeply forked, but with broad shallow notch (Figs. 201, 202, 204) 5

4(3). Tegmina with blackish streaks, one on dorsal (inner) margin from base of wing to apex and another on the costal vein .... *fasciata* **Beutenmüller** (p. 172)

Tegmina without blackish streaks .... *furcata furcata* **Brunner von Wattenwyl** (p. 173)

5(3). Notch of supra-anal plate broad, with median tooth (Fig. 204) .... *texensis* **Saussure & Pictet** (p. 174)

Notch of supra-anal plate narrower than subgenital plate (Figs. 201, 202), without median tooth 6

6(5). Tegmina broad, not more than four times as long as broad. Terminal lobes of supra-anal plate, in dorsal view, slightly tapering (Fig. 201) .... *pistillata* **Brunner von Wattenwyl** (p. 176)

Tegmina narrower and longer, nearly five times as long as broad. Terminal lobes of supra-anal plate, in dorsal view, subtruncate (Fig. 202) .... *curvicauda* (De Geer) (p. 177)

7(1). Tegmina broad, dull, with swollen veins. Ovipositor large, with both margins smoothly curved, not bent (Fig. 205), more than 1.5 times as long as pronotum .... *septentrionalis* **(Audinet-Serville)** (p. 170)

Tegmina narrow to moderately broad, with veins not swollen. Ovipositor shorter, rather abruptly bent upward (Figs. 206–209), distinctly shorter than 1.5 times as long as pronotum 8

8(7). Tegmina long, narrow. Ovipositor abruptly and strongly bent upward, broader at base than at middle (Fig. 209) .... *texensis* **Saussure & Pictet** (p. 174)

Tegmina shorter, broader. Ovipositor less abruptly bent upward, not broader at base than at middle (Figs. 206–208) 9

9(8). Disc of pronotum with sides nearly parallel .... *furcata furcata* **Brunner von Wattenwyl** (p. 173)

Disc of pronotum with sides distinctly diverging posteriad (Fig. 193)..... 10

10(9). Tegmina more than 8 mm wide, with ratio of width to length 1:3 .... *pistillata* **Brunner von Wattenwyl** (p. 176)

Tegmina less than 8 mm wide, with ratio of width to length 1:4.5 .... *curvicauda* (De Geer) (p. 177)

Clé des espèces et des sous-espèces de *Scudderia*

1. Mâles: organe stridulant présent sur les tegmina. Plaques sous-génitale et supra-anale de forme allongée ou élaborée (fig. 195 à 204) 2

Femelles. Organe stridulant absent. Ovipositeur large, courbé (fig. 205 à 209) 7

2(1). Plaque supra-anale presque triangulaire, sans prolongement long et incurvé (fig. 195 et 200); plaque sous-génitale non aplatie à l'apex. Cerques longs et pointus (fig. 195) .... *septentrionalis* **(Audinet-Serville)** (p. 170)

Plaque supra-anale munie d'un prolongement médian allongé et incurvé, encochée ou fourchue à l'apex (fig. 196 à 199 et 201 à 204); plaque sous-génitale rétrécie et aplatie vers l'apex. Cerques courts, très incurvés (fig. 196 à 199) 3

3(2). Apex de la plaque supra-anale clairement fourchu, dissimulant habituellement l'extrémité de la plaque sous-génitale (fig. 203) 4

Apex de la plaque supra-anale ne formant pas une fourche, mais muni d'une encoche large et peu profonde (fig. 201, 202 et 204) 5

4(3). Tegmina portant des rayures noirâtres, l'une sur le bord dorsal (intérieur), de la base de l'aile à l'apex, et l'autre sur la nervure costale .... *fasciata* **Beutenmüller** (p. 172)

Tegmina sans rayures noirâtres .... *furcata furcata* **Brunner von Wattenwyl** (p. 173)

5(3). Encoche de la plaque supra-anale large, munie d'une dent médiane (fig. 204) .... *texensis* **Saussure & Pictet** (p. 174)

Encoche de la plaque supra-anale plus étroite que la plaque sous-génitale (fig. 201 et 202) et dépourvue d'une dent médiane 6

6(5). Tegmina larges, pas plus de 4 fois plus longs que larges. Vus du dessus, bords des lobes terminaux de la plaque supra-anale légèrement convergents (fig. 201) .... *pistillata* **Brunner von Wattenwyl** (p. 176)

Tegmina plus étroits et plus longs, près de 5 fois plus longs que larges. Vus du dessus, lobes terminaux de la plaque supra-anale subtronqués (fig. 202) .... *curvicauda* **(De Geer)** (p. 177)

7(1). Tegmina larges, mats et à nervures proéminentes. Ovipositeur grand, aux bords décrivant une courbe régulière, non repliés (fig. 205) et plus de 1,5 fois plus long que le pronotum .... *septentrionalis* **(Audinet-Serville)** (p. 170)

Tegmina étroits à modérément larges, à nervures non proéminentes. Ovipositeur plus court, fortement incurvé vers le haut (fig. 206 à 209) et de longueur clairement inférieure à 1,5 fois celle du pronotum 8

8(7). Tegmina longs et étroits. Ovipositeur fortement incurvé vers le haut et plus large à la base qu'à mi-longueur (fig. 209) .... *texensis* **Saussure & Pictet** (p. 174)

Tegmina plus courts et plus larges. Ovipositeur à courbure vers le haut moins prononcée et pas plus large à la base qu'à mi-longueur (fig. 206 à 208) 9

9(8). Disque du pronotum à côtés presque parallèles .... *furcata furcata* **Brunner von Wattenwyl** (p. 173)

Disque du pronotum à côtés clairement divergents vers l'arrière (fig. 193) .. 10

10(9). Tegmina de largeur supérieure à 8 mm, rapport largeur/longueur de 1:3 .... *pistillata* **Brunner von Wattenwyl** (p. 176)

Tegmina de largeur inférieure à 8 mm, rapport largeur/longueur de 1:4,5 .... *curvicauda* **(De Geer)** (p. 177)

Scudderia septentrionalis (Audinet-Serville)

Figs. 195, 200, 205; Map 42

Phaneroptera septentrionalis Audinet-Serville, 1838:416.
Scudderia truncata Beutenmüller, 1894*b*:252.
Scudderia septentrionalis; Scudder 1898*b*:285; Cantrall 1943:124; Helfer 1963:228; Vickery et al. 1974:72.

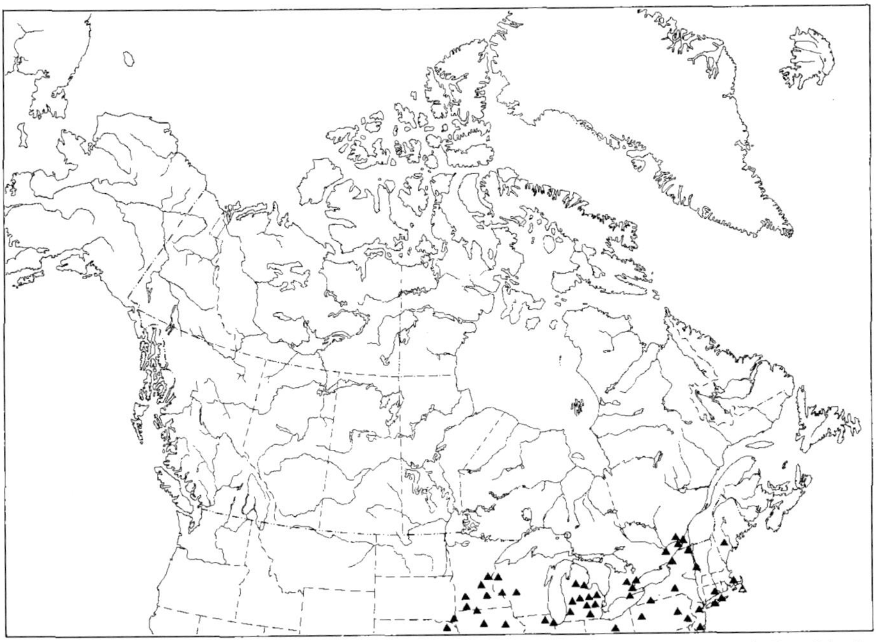

Map 42. Collection localities for *Leptophyes punctatissima* (△) and *Scudderia septentrionalis* (▲).

Diagnosis. Male supra-anal plate without elongate median process. Ovipositor directed more posteriorly than upward.

Description. Body small (pronotal length, male 4.7, female 5.0 mm), slender. Pronotal disc feebly divergent posteriorly, with lateral carinae only on posterior half. Tegmina short, broad, reticulate, rather coarse in texture. Supra-anal plate of male somewhat triangular, without median process (Fig. 200); subgenital plate not compressed at apex. Ovipositor with both margins smoothly curved, not abruptly bent, about 1.5 times as long as pronotum (Fig. 205). Color pale green tinged with yellow, with tegmina, posterior half of pronotum, posterior femora, and all tibiae dark green; antennae brownish on apical half.

Range. Minnesota to Québec, south to Tennessee and South Carolina.

Behavior and habitats. *S. septentrionalis* has been regarded as a rare species. Cantrall (1943) stated that this was due to ignorance of its behavior and habitat and the difficulty of capturing specimens. He found it to be common on the E. S. George Reserve, Livingston County, MI, where it was found characteristically in shady oak-hickory woodland. The only other plants on which it was observed were elm and dogwood, and these were regarded as incidental records. Beaudry (1952), although he did not say so in print, found the species in Québec in association with white pine, but the present authors have not been able to find it in any white pine stand in Québec. A few specimens, like that from St. Laurent, Qué., have been taken at light.

The stridulation is complex, a series of ticks followed by a series of clear brisk notes described by Cantrall (1943) as "dee-dee-dee-dee-dee," usually 6–8 in succession, the entire sequence lasting about 3 seconds. Each sequence increases in volume, producing a crescendo that ends abruptly.

Life history. Univoltine, presumably with wintering in the egg stage. Cantrall (1943) found adults from 8 July onward, and specimens in other areas have been taken in August. Vickery and Kerr (1975) mentioned that adults are found earlier than those of other species of *Scudderia*, and that they do not persist later than the end of August.

Scudderia fasciata Beutenmüller

Map 43

Scudderia fasciata Beutenmüller, 1894*b*:251.

Scudderia fasciata; Beutenmüller 1894*c*:276; R. D. Alexander et al. 1972:35, 43, 50.

Diagnosis. Tegmina with blackish longitudinal stripe.

Description. Tegmen with blackish stripe along upper edge, sometimes also with blackish to purplish elsewhere on wings and body, and rarely with "vivid red front femora" (Urquhart 1942). Male supra-anal plate slender.

Range. Michigan to Maine, south to Tennessee and North Carolina.

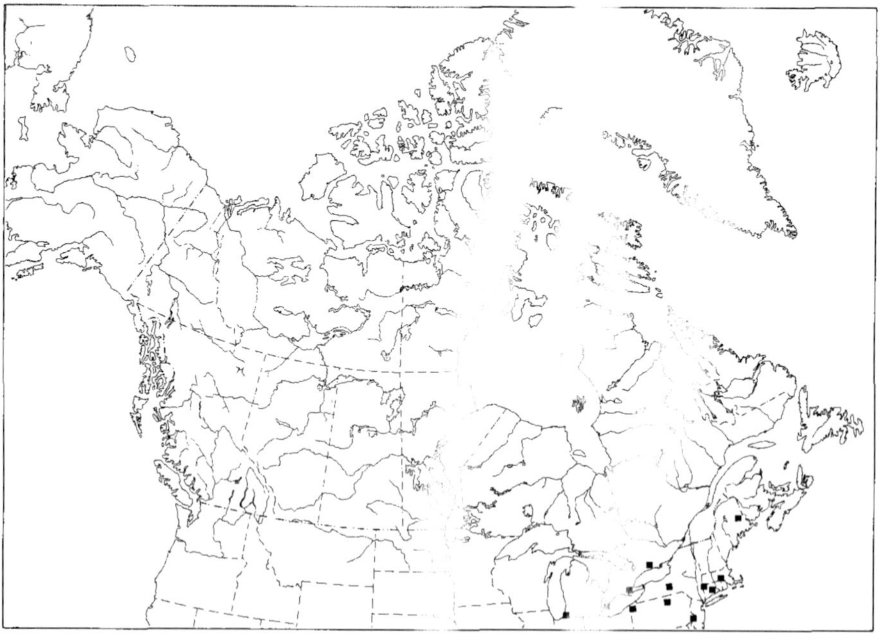

Map 43. Collection localities for *Scudderia fasciata*.

Behavior and habitats. The types were collected on pine trees and Urquhart (1942) found it on white pine (*Pinus strobus* L.). Hebard (1945) reported a female from the "edge of a sphagnum patch in the deciduous forest" in North Carolina. R. D. Alexander et al. (1972) also reported *S. fasciata* as an inhabitant of pines in Ohio, Michigan, and Ontario, but not on other vegetation. Stridulation of *S. fasciata* consists of rather soft lisps, repeated slowly, by day or at night, very much like that of its sibling partner, *S. furcata*. *S. fasciata*, however, occurs only in pine or hemlock trees (and possibly other conifers), whereas *S. furcata* sings in bushes or woods.

Life history. Probably univoltine, with wintering in the egg stage. Adults have been found in August and September.

Scudderia furcata furcata Brunner von Wattenwyl

Figs. 191, 198, 203, 208; Map 44

Scudderia furcata Brunner von Wattenwyl, 1878:239.

Scudderia furcata; Scudder 1898*b*:284, Helfer 1963:277; R. D. Alexander et al. 1972:43–50.

Scudderia furcata furcata; J. A. G. Rehn and Hebard 1914*b*:297; Cantrall 1943:129; Vickery et al. 1974:74.

Phaneroptera furcata furcata; Hebard 1925*a*:127.

Common name. Forktailed bush katydid.

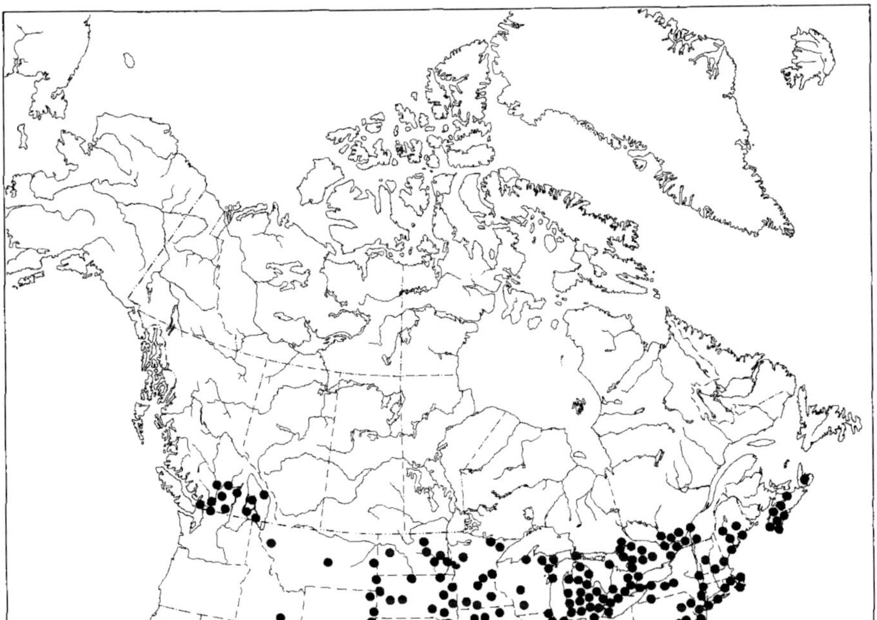

Map 44. Collection localities for *Scudderia furcata furcata*.

Diagnosis. Process of male supra-anal plate deeply forked. Disc of pronotum with lateral margins nearly parallel.

Description. Body small (males 14–18, females 18–22 mm, pronotal length, males and females, 4.6–5.0 mm). Tegmina narrow, with width to length ratio about 1:4.5–1:4.7. Pronotal disc with sides nearly parallel. Process of supra-anal plate of male deeply forked, with apical notch deep and U-shaped and with lateral processes decidedly swollen (Figs. 191, 198). Ovipositor as in Fig. 203. Color dark leaf green, occasionally somewhat suffused with brownish; head and pronotum paler; lateral margins of pronotal disc not outlined with paler color; hind tibiae and ovipositors of some specimens suffused with brownish purple or red.

Range. British Columbia to Nova Scotia, south to Mexico.

Behavior and habitats. Cantrall (1943) found *S. furcata furcata* in permanent marshes, but more numerous in or near semipermanent marsh habitats, and usually within no more than 50 m of them. E. M. Walker (1904*b*) stated that *S. furcata* occurred in trees and bushes at the edges of woods and thickets on both dry and marshy ground but most often near marshes.

Males stridulate in trees and bushes, often high off the ground, by day and at night. The song consists of rather soft lisps produced very slowly, a few seconds apart, in series of three or four. A lisp is defined as a pulse containing a large number of tooth strikes and has a consistent length that is typical for the species. Intervals between songs is erratic, from 1 to 30 minutes (Spooner 1968*b*). *S. furcata* may also produce a second sound, a short-pulsed phrase repeated at intervals of 4 or 5 seconds for up to 1 minute. Spooner (1968*b*) also observed that reports of increased diurnal stridulation could be the result of inhibition owing to low night temperatures. Males also produce "ticks" in the presence of responsive females. Spooner (1968*a*) timed ticks produced by females of several species in response to male stridulation. He found the mean time interval following male song to female response in *S. furcata* to be 1.37 seconds at 25°C. Nickle (1976) found the dominant carrier frequency of *S. furcata* stridulation to be 14.6–1.5 kHz.

Life history. Pair formation and mating occur 14–40 (with a mean of 20) days after molting to the adult stage. In the southeastern United States, *S. furcata* has two generations each year (Spooner 1968*b*), but in the north it is univoltine, wintering in the egg stage. Cantrall (1943) stated that it matures later than other species of the genus in southern Michigan, adults occurring from 20 July to 30 September. In Québec, adults have been collected from 13 August to 1 September (Larochelle 1978), although there is little doubt that it appears earlier and disappears later than these dates.

Scudderia texensis Saussure & Pictet

Figs. 199, 204, 209; Map 45

Scudderia texensis Saussure and Pictet, 1897:328.

Scudderia texensis; Scudder 1898*b*:277; Cantrall 1943:128; Helfer 1963:279; R. D. Alexander et al. 1972:44, 50; Vickery et al. 1974:75.

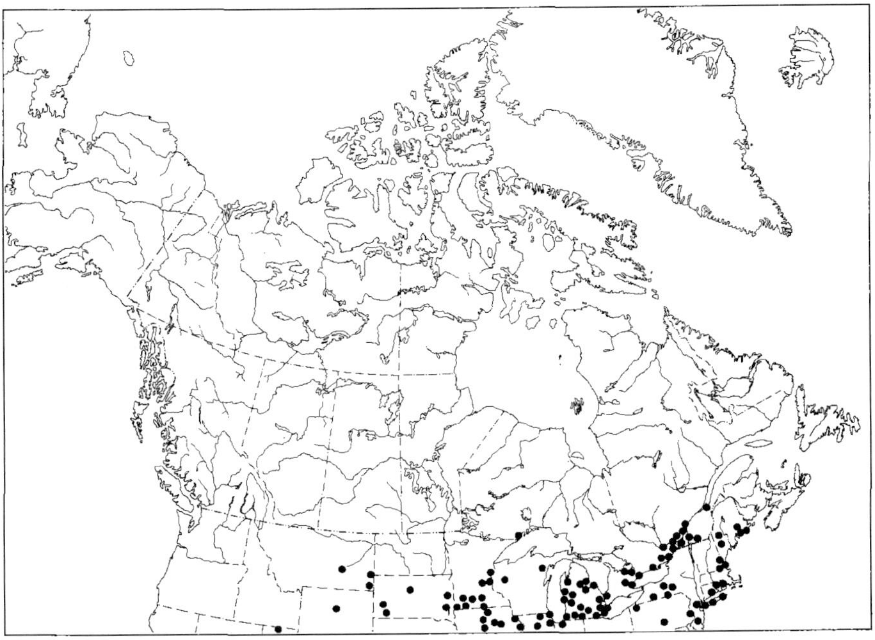

Map 45. Collection localities for *Scudderia texensis*.

Diagnosis. Male supra-anal process broad apically, emarginate with median tooth. Tegmina long, 5–5.5 times longer than broad.

Description. Body large (males 21–25, females 24–28 mm long). Pronotum much longer than broad (5.5–6.5 mm in both sexes), with margins of disc diverging posteriorly. Hind femur long, slender, with 3 or 4 minute spines on inner ventral carina. Tegmina long, narrow, more than five times as long as broad. Male supra-anal process broad apically, shallowly emarginate with small median tooth (Figs. 199, 204). Ovipositor suddenly and strongly bent upward, distinctly broader at base than at middle (Fig. 209). Color generally grass green with body and face tinged with yellow.

Range. Montana to Maine, south to Texas and Florida.

Behavior and habitats. J. A. G. Rehn and Hebard (1914*b*) observed *S. texensis* in salt marshes on the New Jersey coast where it was abundant. Males stridulated in the afternoon in taller vegetation. These authors state that it is "almost invariably found in or near marsh swamp or bog." Cantrall (1943) recorded it as "a characteristic species of the semipermanent marsh habitat" in Michigan. Spooner (1964) stated that *S. texensis* "frequents almost any open area where weeds and grasses abound such as in abandoned fields and along highway and railway embankments." It probably is not so restricted in habitat type as was indicated by J. A. G. Rehn and Hebard (1914*b*) and by Cantrall (1943). We have collected it in wasteland and in abandoned areas that were fairly extensively covered with regrowth scrub-brush vegetation. Eggs are laid in the edges of leaves between the upper and lower epidermis and, at first, are so thin that they are not noticeable (Blatchley 1920).

Stridulation is louder, and more harsh, and is produced more slowly than that of other species of *Scudderia*. Cantrall (1943) described the day song as "skee-deeck," the complete song lasting about 0.75 second, produced at irregular intervals. The nocturnal song is amplified and sounds like "skee-dee-dee-dee-deeck," with brief pauses, the entire song lasting about 1.25 seconds, emitted at irregular intervals. Spooner (1964) discovered that *S. texensis* males produce four sounds, three in solitary situations. Females produce lisping sounds in response to one of the male signals. In the afternoon, a fast-pulsed song is produced, consisting of "long series of irregularly spaced, short, lispy phrases." At twilight, males produce soft ticks, which are audible only a few metres away. After dark a slow-pulsed song is produced, composed of two phrases that are longer and are given more slowly than the fast-pulsed phrases. Then alternation of fast-pulsed and slow-pulsed song occurs. When males are situated close together, the stridulation of one produces intense ticks by the other. If the contact is very close, an erratic ticking, which Spooner calls "crackling," is elicited. These sounds serve to maintain territorial spacing of males. The slow-pulsed song is answered by the females 1.16 seconds later with 1–3 lisps (Spooner 1968*a*).

Life history. In the northern part of its range, *S. texensis* is univoltine, a single generation overwintering in the egg stage. In Georgia and Florida, there are two generations each year, adults appearing in June and September. In the region here considered, adults appear in July and persist until late September. Vickery and Kerr (1975) recorded adults in southern Ontario from 29 July to 17 September.

Economic importance. J. A. G. Rehn and Hebard (1914*b*) reported damage to cranberry crops by *S. texensis* in the eastern United States.

Scudderia pistillata Brunner von Wattenwyl

Figs. 192–194, 196, 201, 206; Map 46

Scudderia pistillata Brunner von Wattenwyl, 1878:240.

Scudderia pistillata; Piers 189[illegible]:11; Cantrall 1943:126; Helfer 1963:280; R. D. Alexander et al. 1972:44–[illegible]; Vickery et al. 1974:73.

Diagnosis. Tegmina broad, only three times as long as wide.

Description. Size medium (pronotal length, male and female 4.7–5.7 mm). Disc of pronotum distinctly broader posteriorly. Process of male supra-anal plate forked, with apical notch acute, shallow, narrower than upcurved subanal process (Figs. 196, 201). Ovipositor as in Fig. 206. Color pale apple green above, whitish green beneath; antennae brownish except greenish basally; vertex of head and dorsolateral angles of pronotum creamy white; abdomen green with annular stripes of brighter or darker green on posterior margins of segments; ventral surface of abdomen with 2 longitudinal raised white lines.

Range. British Columbia to Nova Scotia, south to Indiana and New York.

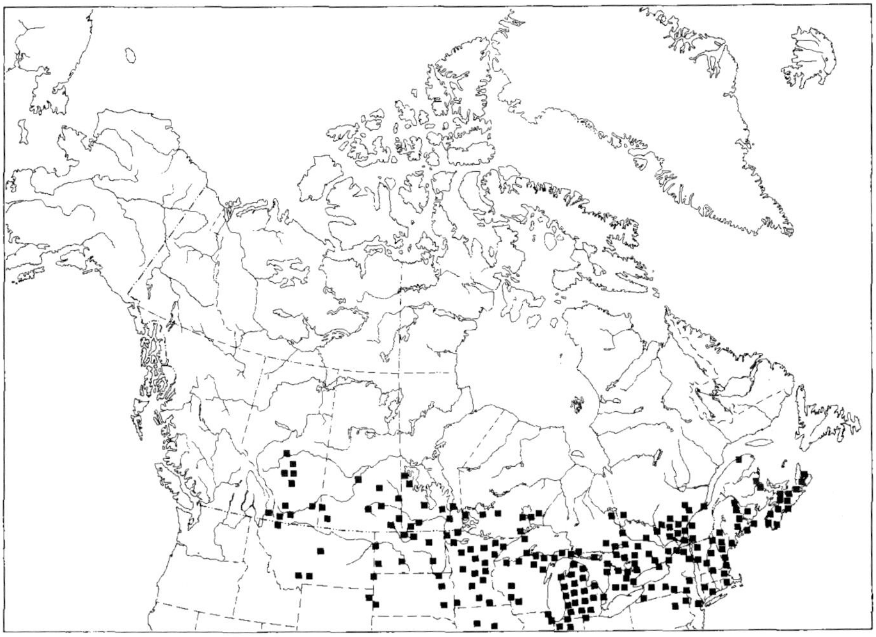

Map 46. Collection localities for *Scudderia pistillata*.

Behavior and habitats. *S. pistillata* seems to prefer damp locations and is usually found on bushes. It is slow-moving and not very active. Males stridulate rarely in daylight, but readily at dusk. The song is a harsh "tsick," consisting of 5–8 rapid pulses, increasing in loudness at the end. Intervals between songs vary from 2 minutes to more than 1 hour. Cantrall (1943) reported a frequency of about 20 minutes with favorable weather conditions. Males usually perch on the top of vegetation when stridulating. Flight is slow and clumsy.

Life history. Univoltine, with wintering in the egg stage. Vickery (1961) reported adults in Nova Scotia from 14 July to 21 October.

Scudderia curvicauda (De Geer)

Figs. 197, 202, 207; Map 47

Locusta curvicauda De Geer, 1773:446.

Phaneroptera angustifolia T. W. Harris, 1841:129.

Scudderia curvicauda; Scudder 1898*b*:278; Helfer 1963:279; R. D. Alexander et al. 1972:44–50; Vickery et al. 1974:73.

Scudderia curvicauda curvicauda; J. A. G. Rehn and Hebard 1914*b*:293; Cantrall 1943:127.

Scudderia curvicauda borealis; J. A. G. Rehn and Hebard 1914*b*:281.

Diagnosis. Tegmina long, narrow, 4.5 times as long as broad.

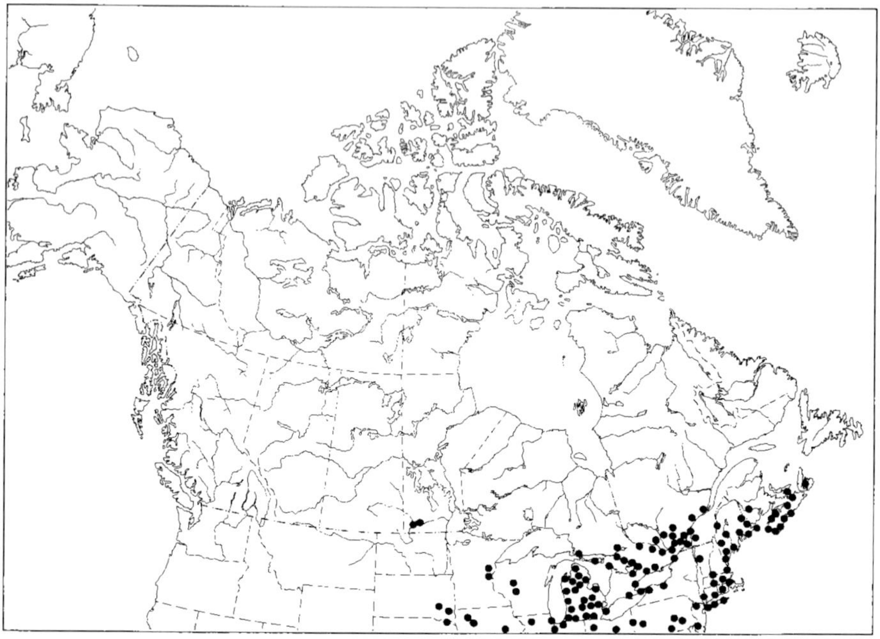

Map 47. Collection localities for *Scudderia curvicauda*.

Description. Body small (pronotal length, male and female, 5.0–6.0 mm). Tegmina narrow (width to length 1:4.5). Eyes large. Hind femora long; male tympanum small. Branches of fork of process of male supra-anal plate subequal in width (Fig. 202). Ovipositor as in Fig. 207. Color generally uniformly green with lateral pronotal angles weakly outlined in brownish white; some specimens reddish to purple on dorsal surfaces of hind tibiae, between spines; ovipositor sometimes brownish purple. In Québec, dorsal side of hind tibia often reddish purple and ovipositor often brownish purple.

Range. Manitoba to Nova Scotia, south to Texas and Florida.

Behavior and habitats. E. M. Walker (1904*b*) stated that all the specimens found by him near Toronto were from trees and bushes in more or less open, partly wooded country. J. A. G. Rehn and Hebard (1914*b*) found *S. curvicauda* to be common in undergrowth of woods in the New Jersey pine barrens. Specimens have been found in similar habitats in Nova Scotia and Québec. Cantrall (1943) found it in heavy marsh vegetation in southern Michigan. He also recorded that *S. curvicauda* is an inactive species, spending daylight hours in concealment. Eggs are laid in the stems or branches of plants.

Males of this species stridulate by day and also at night. The diurnal song is a single phrase sounding like "tsick," given at long intervals. At night, the phrase is similar but is repeated several times about 5 seconds apart. The phrases consist of 1–3 slow pulses (rate of two to three per second). An individual often begins with one phrase, followed by several two- and three-pulse phrases (Alexander et al. 1972). Spooner (1968*b*) reported the song as a series of phrases with each succeeding phrase having one more pulse than the previous phrase, usually 2–5, but occasionally with seven-pulsed phrases. Pulse rates at 25°C were 4.5–5.8 per second. Spooner (1968*a*) pointed out that the females of *Scudderia* answer the males with soft noises (0.84 second after the termination of male stridulation in the case of *S. curvicauda*), which produce variations in the male song, and also that the males then go to the females, rather than the reverse as in most singing Grylloptera.

Life history. Univoltine, with wintering in the egg stage. Vickery (1961) recorded adults from 15 August to 12 October in Nova Scotia. Cantrall (1943) reported adult activity from 13 July to 21 September in southern Michigan.

Genus *Amblycorypha* Stål

Description. Size medium to large (20–27 mm in body length). Fastigium rounded, strongly bent downward, more than twice as broad as basal antennal segment (Fig. 211). Eyes oval. Pronotum flattened dorsally, narrower in front, with hind margin broadly rounded. Tegmina broad, widest at middle. Male subgenital plate broad, slightly tapered, carinate beneath, with apex deeply notched and with paired short cylindrical lateral styles. Male cerci long, tapered, incurved, usually crossed. Female ovipositor broad, gradually curved upward, with apex rounded, serrate on both edges of apical half (Fig. 216).

Key to species and subspecies of *Amblycorypha*

1. Tegmina oblong-elliptic, 3.25–3.50 times as long as broad 3
   Tegmina broadly ovate, 2.25–2.75 times as long as wide .... ***rotundifolia*** **(Scudder)** ... 2

2(1). Hind wings extending beyond tegmina when wings folded (eastern subspecies) .... ***rotundifolia rotundifolia*** **(Scudder)** (p. 180)
   Hind wings not extending beyond tegmina when wings folded (Minnesota, South Dakota) .... ***rotundifolia iselyi*** **Caudell** (p. 181)

3(1). Stridulatory area of male small, not much larger than pronotal disc. Ratio of hind femur length to ovipositor length 2.16–2.25 .... ***carinata*** **Rehn & Hebard** (p. 181)
   Stridulatory area of male much larger than pronotal disc. Ratio of hind femur length to ovipositor length 2.38–2.61 .... ***oblongifolia*** (De Geer) (p. 182)

Clé des espèces et des sous-espèces d'*Amblycorypha*

1. Tegmina oblongs ou elliptiques, de 3,25 à 3,50 fois plus longs que larges 3
   Tegmina clairement ovés, de 2,25 à 2,75 fois plus longs que larges .... ***rotundifolia*** **(Scudder)** ... 2

2(1). Ailes postérieures dépassant les tegmina en position repliée (sous-espèce de l'est) .... ***rotundifolia rotundifolia*** **(Scudder)** (p. 180)

Ailes postérieures ne dépassant pas les tegmina en position repliée (Minnesota et Dakota sud) .... *rotundifolia iselyi* **Caudell** (p. 181)

3(1). Organe stridulant du mâle petit, pas beaucoup plus grand que le disque du pronotum. Fémurs postérieurs de 2,16 à 2,25 fois plus longs que l'ovipositeur .... *carinata* **Rehn & Hebard** (p. 181)

Organe stridulant du mâle beaucoup plus grand que le disque du pronotum. Fémurs postérieurs de 2,38 à 2,61 fois plus longs que l'ovipositeur .... *oblongifolia* **(De Geer)** (p. 182)

Amblycorypha rotundifolia (Scudder)

Phylloptera rotundifolia Scudder, 1863*a*:445.

There is some confusion regarding the exact limits of this species, but there is a general consensus that there are three subspecies, two of which occur in the region here covered.

Amblycorypha rotundifolia rotundifolia (Scudder)

Fig. 214; Plate IV*A*; Map 48

Phylloptera rotundifolia Scudder, 1863*a*:445.

Amblycorypha rotundifolia rotundifolia; J. A. G. Rehn and Hebard 1914*c*:335.

Amblycorypha rotundifolia; Pettit and McDaniel 1918:38; Cantrall 1943:130; R. D. Alexander et al. 1972:42, 50.

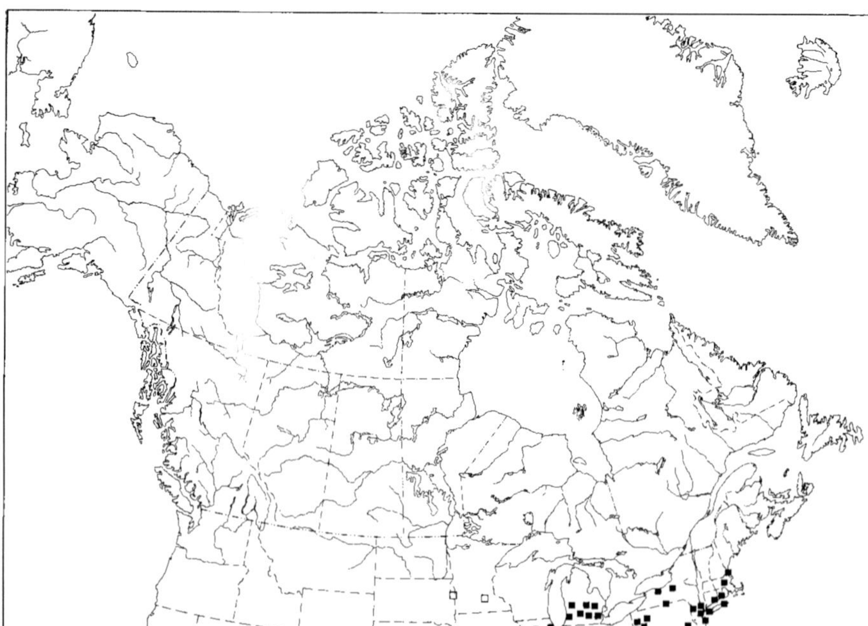

Map 48. Collection localities for *Amblycorypha rotundifolia rotundifolia* (■) and *A. rotundifolia iselyi* (□).

Diagnosis. Tegmina distinctly ovate; hind wings projecting beyond them.

Description. Tegmina ovate; wings slightly longer than tegmina. Metasternal lobe on each side broader than long with posterior margin subtruncate; humeral sinus of lateral pronotal lobe obsolete. Hind femur with 4 or 5 minute teeth on ventral carina. Ovipositor strongly curved, strongly serrate. Color bright green with abdomen and femora yellowish brown to brownish, occasionally pink.

Range. Iowa to Massachusetts, south to Arkansas and Georgia.

Behavior and habitats. *A. rotundifolia* is a common inhabitant of unimproved rough pastureland, on such plants as *Vaccinium, Kalmia*, and *Spiraea* (Allard 1911). Cantrall (1943) found it mainly in shady oak-hickory woods in southern Michigan. The species is inactive, has never been seen to fly, and jumps only slowly and feebly, or as Cantrall (1943) describes it, "more of a slow shove than a quick springing movement." The "leap" carries the insect only 5-8 cm. That author has given a detailed account of the cleaning and grooming activity of this species.

The song consists of notes that are dull and lisping, groups of rattling buzzes or phrases, 5-15 per group, with a long (5-second) phrase before termination of the song. Stridulation is nocturnal on vegetation about 1.5 m above the ground (R. D. Alexander et al. 1972).

Life history. Univoltine, with wintering in the egg stage. Cantrall (1943) recorded adults in southern Michigan from 18 July to 15 September, with a peak of adult population during the 2nd week of August. It appears to be susceptible to frost and does not survive beyond September.

Amblycorypha rotundifolia iselyi Caudell

Map 48

Amblycorypha iselyi Caudell, 1905*a*:50.

Amblycorypha rotundifolia; Lugger 1898:312.

Amblycorypha rotundifolia iselyi; J. A. G. Rehn and Hebard 1914*c*:340; Hebard 1932*a*:42; Helfer 1963:274.

Diagnosis and description. As for *A. rotundifolia*, but hind wings not projecting beyond tegmina when at rest.

Range. South Dakota and Minnesota, south to Missouri.

Behavior, habitats, and life history. Presumably as for *A. rotundifolia rotundifolia*.

Amblycorypha carinata Rehn & Hebard

Map 49

Amblycorypha floridana carinata J. A. G. Rehn and Hebard, 1914*c*:323.

Diagnosis. Male stridulatory area about same size as pronotal disc.

Description. Male stridulatory area small, about size of pronotal disc. Ovipositor unevenly curved, long in proportion to hind femur (ratio 2.06-2.25), heavy.

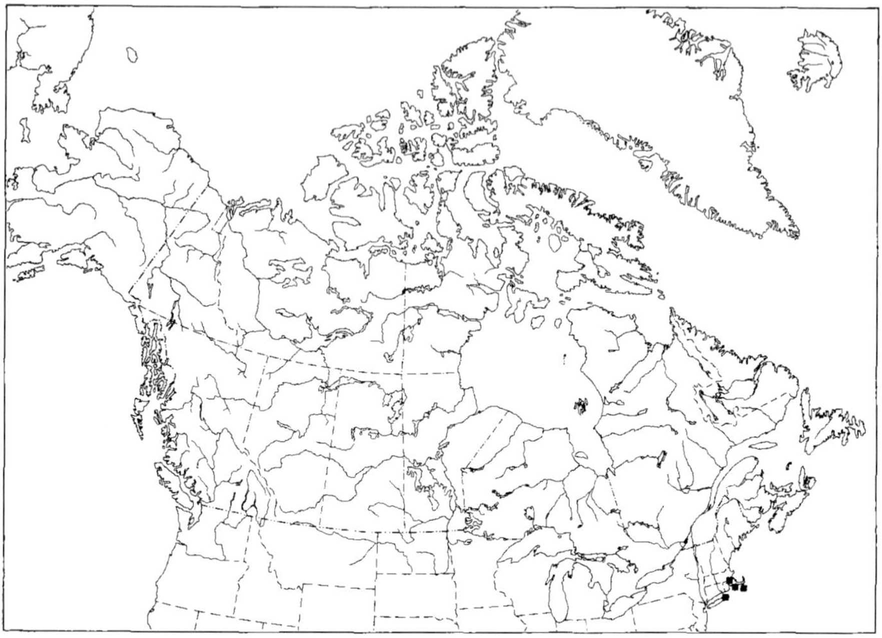

Map 49. Collection localities for *Amblycorypha carinata*.

Range. Pennsylvania to Massachusetts, south to Georgia.

Behavior and habitats. H. Fox (1917) stated that *A. carinata* is frequently found in tidewater areas in Virginia. Little else is known about the species.

Life history. Presumably univoltine, with wintering in the egg stage. The type specimen was collected 16 July.

Amblycorypha oblongifolia (De Geer)

Figs. 1, 210–213, 215, 216; Map 50

Locusta oblongifolia De Geer, 1773:445.

Amblyconypha [*sic*] *oblongifolia*; Caulfield 1886:212.

Amblycorypha scudderae Bruner, 1891*a*:73.

Amblycorypha oblongifolia; Lugger 1898:312; Cantrall 1943:130; Helfer 1963:274; R. D. Alexander et al. 1972:44, 50; Vickery et al. 1974:76.

Diagnosis. Male green with stridulatory area brown, larger than pronotal disc. Female entirely green, with ovipositor longer than hind femur.

Description. Body large, robust (length, male 21–23, female 22–25 mm); general form as in Figs. 1, 210. Fastigium rounded, more than twice as broad as basal antennal segment (Fig. 211). Pronotal disc flat, narrowed in front, broadly rounded behind (length, male 6–6.5, female 7 mm), with sides distinctly divergent on posterior two-thirds; lateral carinae of pronotum

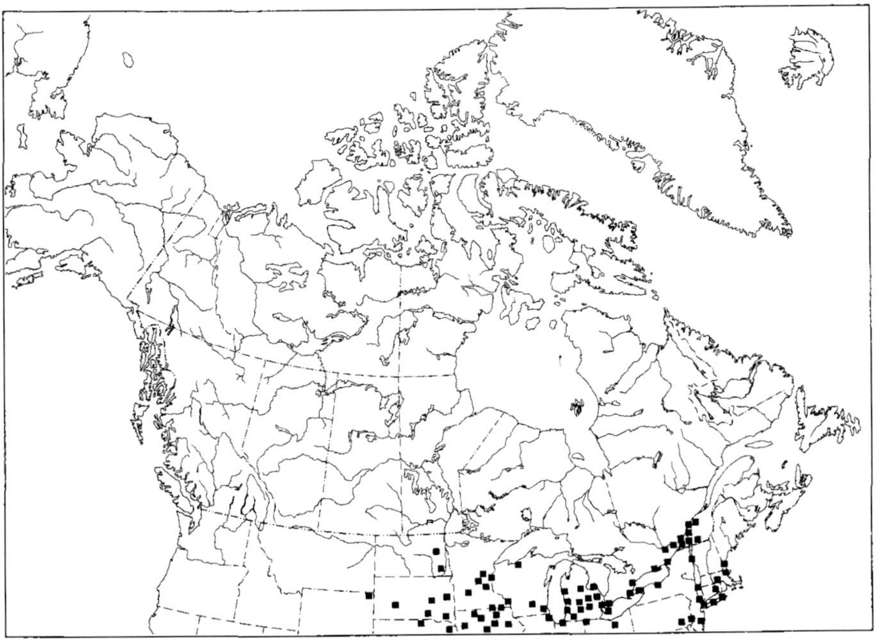

Map 50. Collection localities for *Amblycorypha oblongifolia*.

less distinct on anterior third; humeral sinus distinct. Tegmina elongate-elliptic, about 3.3 times longer than broad; stridulatory field of male large, much larger than pronotal disc; wings exceeding tegmina in length by about 6 mm. Posterior femur slender, not reaching tips of tegmina (shorter in female), with inner posterior carina possessing 6–12 strong teeth. Ovipositor evenly and strongly curved, strongly serrate, with ratio of hind femur length to ovipositor length 2.38–2.61. Color typically bright pea green, with abdomen and femora usually brown or yellowish brown (rarely pink); male stridulatory area brown with green crossvein; female usually entirely green (rarely pink), but some individuals with black marginal pronotal markings. Chromosome number, $2n$ male = 31 (Beaudry 1973).

Range. North Dakota to Québec, south to Texas and Florida.

Behavior and habitats. Cantrall (1943) found *A. oblongifolia* mainly in permanent and semipermanent marsh habitats. E. M. Walker (1904*b*) found it common on tall weeds and shrubs in western Ontario. We have found it in similar locations in southwestern Québec where it is not necessarily associated with marshes, but is usually found on vegetation where soil is moist or at least near bodies of water. It is annually present on bushes in both of the authors' gardens. Doyon (1962) recorded it feeding on milkweed (*Asclepias syriaca* L.) in Québec, but milkweed is not regarded as a primary food plant. Individuals are relatively inactive. They walk slowly and generally remain perched on top of a bush unless disturbed. Flight is leisurely and generally of short duration (Cantrall 1943). Eggs are deposited in moist soil.

The female, after locating a suitable spot, bends her abdomen forward under her body and grasps her ovipositor with her mandibles, then forces the ovipositor into the soil and deposits the eggs (Hancock 1916).

Stridulation is loud, sounding like "z-z-z-z-itzick," repeated every few seconds with some regularity. The phrase is not uniform, beginning with the long pulse and speeding up toward the end (R. D. Alexander et al. 1972). Females respond by ticking. The time interval between male stridulation and female response is very short. At 25°C, Spooner (1968*a*) found an attraction to artificial "ticks" of one-tenth of a second after the male song ceased. The same author (Spooner 1968*b*) recorded female response at 280 milliseconds, following the cessation of stridulation by the male. The male then (in every case) moved directly to the female. Males respond to other males by increased rate of stridulation, the two males producing sound alternately in rapid succession (Vickery, unpublished, 1980).

Life history. Although there is only a single generation each year, the cycle is somewhat complex. Eggs in the soil may pass through two or three winters before they hatch. Adults and nymphs do not hibernate. Adults have been taken in Québec and Ontario from 12 July to 10 September and have been heard in some years until the 1st week of October, but most have been found during August.

Genus *Microcentrum* Scudder

Description. Body large (male 25–28, female 28–30 mm; pronotal length, male and female 6.0–7.0 mm). Occiput of head convex, with fastigium obtuse, about as broad as basal antennal segment, slightly sulcate. Eyes prominent, broadly oval. Pronotal disc flat, with margins nearly parallel; posterior margin with lateral carinae distinct; lateral lobes deeper than broad; humeral sinus distinct. Tegmi[illegible]vate, lanceolate, broadest at middle, tapering to narrow rounded apices; [illegible]d wings projecting beyond tegmina. Femora with several small spines bene[illegible]; posterior spines short, slender. Male cerci long, slender, nearly cylindric[illegible] with apices usually incurved. Male subgenital plate bilobate, with long cy[illegible]drical styles. Ovipositor short, bent upward, with apex obtuse and finely [illegible]othed (Fig. 219). Female subgenital plate triangular, compressed, obtus[illegible]

Microcentrum rhombifolium (Saussure)

Figs. 217–219; Map 51

Phylloptera rhombifolia Saussure, 1859:204.

Microcentrum affiliatum Scudder, 1863*a*:447.

Microcentrum rhombifolium; J. A. G. Rehn and Hebard 1916*b*:256; Helfer 1963:274; R. D. Alexander et al. 1972:43, 50.

Common name. Broadwinged katydid.

Diagnosis. Fore and middle tibiae smooth. Tegmina ovate–lanceolate.

Description. As for genus. Anterior margin of pronotum sinuate with small median tubercle (Figs. 217, 218). Eyes subglobose. Male stridulatory

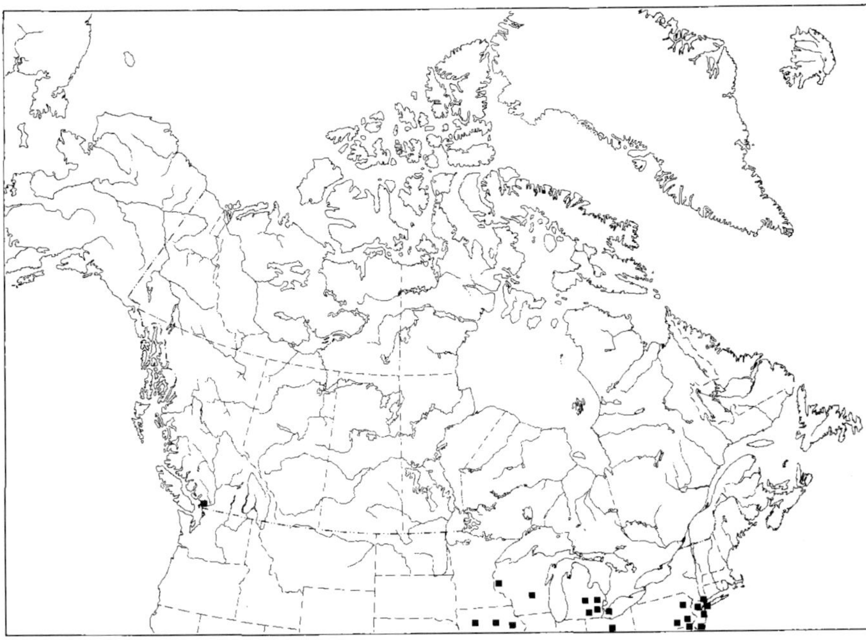

Map 51. Collection localities for *Microcentrum rhombifolium.*

field elongate-triangular, opaque, coarsely punctate. Female stridulatory area with 7 transverse veins bearing pointed spines. Hind femora short (22.5–25 mm), not reaching beyond apical third of tegmina. Ovipositor bent abruptly upward, with apex rounded-truncate, finely toothed (Fig. 219). Color rich deep green.

Range. Minnesota to New York; British Columbia (adventive).

Behavior and habitats. *M. rhombifolium* frequents bushes, shrubbery, and often crowns of deciduous trees (Cantrall 1968). Johnstone (1971) reported it from bushes growing in sandy soil at Point Pelee, Ont. Blatchley (1920) stated that it was found on low bushes and shrubbery and that it is attracted to light at night. Eggs are glued in double rows on the sides of slender twigs. A female first chews the area, thus roughening it, then deposits the eggs so that, in each row, they overlap for about one-quarter of their length.

Male stridulation was described by Spooner (1968*b*) as two distinct sounds, lisps and ticks, which are produced at any time of the day or night, although more often at night. The lisps and ticks are not usually given in sequence but this may occur after a male has had a prolonged silent period. Females are attracted to the lisp sequence at an intensity of 55 decibels, but not at higher intensities. Lisps may also produce reactions in other males and may serve to maintain territorial spacing. Females tick in response to a series of male ticks after a very short interval (0.16 second), according to Spooner (1968*a*), and, in such cases, males move to the responding females.

Life history. Univoltine, with wintering in the egg stage in the northern part of the range, but bivoltine in central and southern Florida. Specimens were collected at Point Pelee, Ont., on 23 August. Cantrall (1968) recorded 20 September to 2 November as the period of adult activity. Blatchley (1920) stated that, in central Indiana, nymphs were present from May until mid-August.

Family Pseudophyllidae

The family Pseudophyllidae includes the true katydids, "singing leaves," and bark-crickets, although only the first of these are native to Canada and the adjacent United States. The family, as a whole, is a large one, but it is poorly represented in North America. It is distinguishable from other members of the superfamily Tettigonioidea by the following combination of characters: size moderately large to large. Head usually short, rounded, with face not slanting or flattened. Antennae longer than body, inserted between eyes (which are typically small and round), the insertions ("scrobes") with characteristic, strongly raised margins, particularly mediodorsally. Prothoracic spiracles small, but not hidden by pronotum; prosternum often with pair of spines. Tibiae without apical spines, the front pair more or less cylindrical or oval throughout in cross-section and with tympanal auditory organs normally covered; tarsi with first and second segments grooved laterally. Wings seldom reduced, the tegmina strongly modified to resemble foliage, fruit, bark, and lichen; male tegmina with well-developed stridulatory organs and with crossveins often modified to form expanded resonating area; hind wings at rest not, or but slightly, projecting beyond tegmina. Ovipositor usually of moderate length, distinctly curved upward.

There are about 1000 species of Pseudophyllidae and nearly 250 genera. The number of subfamilies that should be recognized is uncertain, although there are about 20 recognizable tribes. We may regard the majority, which are American, as constituting the subfamily Cyrtophyllinae (= Pterophyllinae). Most of the rest, from the Old World tropics, are placed in the Pseudophyllinae, *sensu stricto*.

The Pseudophyllidae includes some large insects, with wings sometimes spanning more than 20 cm. Resemblance to broad leaves, often to decaying ones with transparent "holes" or irregular margins apparently bitten by insects, or with fungi or lichens growing on them, is often extraordinary (even more so than in the Phaneropteridae). So, too, is the procryptic form and coloration of species that rest on tree trunks or branches. Some species can be extraordinarily noisy at night in tropical forests, where most of them live. Both sexes may produce sounds and more than one mechanism may be used in sound production. Few species occur in temperate countries, species of *Pterophylla*, the katydids of eastern North America, being among the exceptions. Most, if not all, are phytophagous and most seem to oviposit in stems, although the biology of few is known. Very few species are known to have any economic significance. The family has been monographed and cataloged by Beier (1960, 1962, 1963).

Genus *Pterophylla* W. Kirby

Description. Body large (length 25–34 mm). Head broad, with fastigium short, triangular, acute, sulcate above. Eyes small, subglobose, widely separated. Pronotum saddle-shaped, with disc having faint lateral carinae; anterior margin truncate; posterior margin broadly rounded; prosternum with paired slender tapering spines. Tegmina broad, leaflike, strongly convex, entirely enclosing abdomen; wings shorter than tegmina. Femora sulcate, spined beneath; fore tibiae without apical spines; hind tibia quadrangular in section, with all margins bearing spines. Male cerci broad at base, widely forked, with apices of forks divergent, incurved. Male subgenital plate produced into long acute appendage, sulcate above. Ovipositor more than twice as long as pronotum, sickle-shaped, with apex acute, finely serrate on apical third of lower margin.

Pterophylla camellifolia (Fabricius)

Fig. 220; Map 52

Locusta camellifolia Fabricius, 1775:283.
Locusta perspicillata Fabricius, 1775:283 (male).
Platyphyllum concavum T. W. Harris, 1831:42.
Platyphyllus zimmermanni Saussure, 1859:206.
Cyrtophyllus elongatus Caudell, 1906:40.
Pterophylla camellifolia; Blatchley 1920:496; Helfer 1963:264; R. D. Alexander et al. 1972:33, 43, 49.
Pterophylla camellifolia camellifolia; Hebard 1941:205.

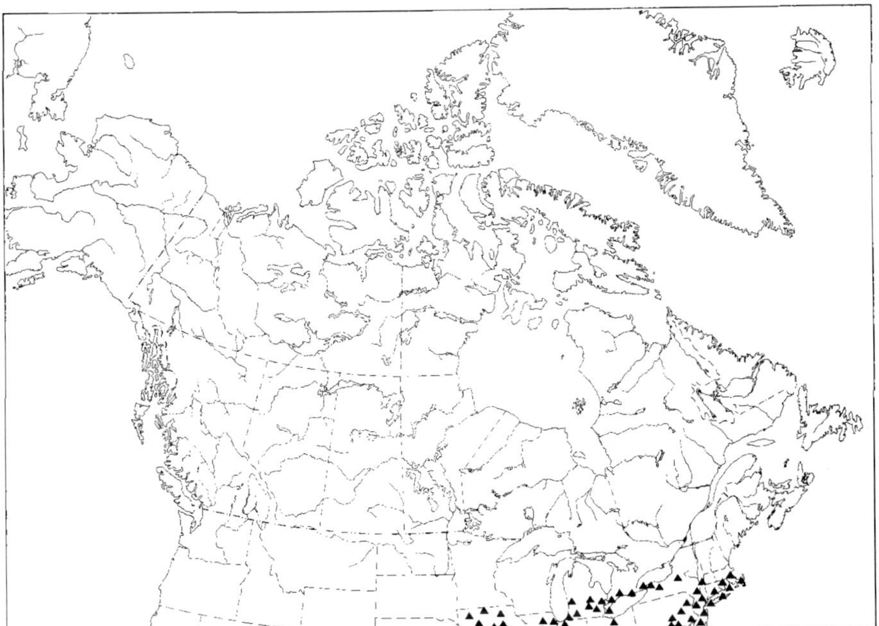

Map 52. Collection localities for *Pterophylla camellifolia*.

Diagnosis. Tegmina strongly convex, entirely enclosing abdomen.

Description. As for genus *Pterophylla*. Body large, robust, green (Fig. 220). Male cercus with lower branch of fork about one-quarter shorter than upper branch and with apices of both branches strongly incurved, partly clasping sides of elongate subgenital plate; upper branches crossing each other, each terminating in tiny black spine.

Range. Iowa to Ontario and Massachusetts, south to Texas and Florida.

Behavior and habitats. *P. camellifolia*, the "true" katydid, is more commonly heard than seen, as it often remains in dense foliage in high trees. Blatchley (1920) indicated that it prefers trees in the open rather than in forests, but this was disputed by Hebard (1941), who found it to be much more numerous in extensive areas of deciduous forests (Pennsylvania, New Jersey to Virginia), particularly where numerous high oak trees occurred. He was able to collect many specimens in oak scrub in New Jersey. Although it possesses tegmina and hind wings, *P. camellifolia* does not fly. It may glide from a high perch to a lower one, but, in regaining the former, it crawls up the trunk of a tree. Eggs are deposited in soft stems of woody plants and in tree bark (Caudell 1906).

The stridulation of the males is the loudest of any North American tettigonioid. R. D. Alexander et al. (1972) describe the song as "a very loud, harsh, 2-to-3-pulse phrase delivered steadily at about one phrase per second at night from trees." Hebard (1941), rather than describing the song as "harsh," found it to be "exceptionally pleasing." Cantrall (1968) called the species "a noisy inhabitant of the crowns of deciduous trees."

R. D. Alexander (1960) discovered that a male of this species in captivity would stridulate in response to an artificial stimulus and would alter speed of pulse production in response to such a stimulus. In this case, a typewriter nearby stimulated stridulation during the day (the species normally sings only at night) and would increase or decrease the rate of pulse production as the rate of typing was increased or decreased. In the field, males respond by stridulation to the calls of other males. Alternation of stridulation is apparently usual in this species if more than one male is present in a location. Shaw (1968, 1975) observed similar behavior. He also found that young males (1 to 2 weeks after sexual maturity) produce longer and more variable songs than older individuals. Stabilization of song length is probably a result of interaction with other males. Female stridulation is well known in this species (Caudell 1906). In response to the call of a male, a female produces a sharp scraping note that is audible to the human ear several metres away.

Life history. Univoltine, wintering in the egg stage. Adults are found from 10 July to 27 October in Indiana (Blatchley 1920). In Ontario and Michigan the active period is reduced, Cantrall (1968) giving the dates as 30 July to 10 October. At the end of the active season, adult females are sometimes found on the ground under trees or on tree trunks following frosty nights (Hebard 1941).

Family Conocephalidae

Members of this family include the cone-head bush-crickets, meadow katydids, and weed-insects. For the most part, they are easily recognized, particularly if only the present region is considered, although, when a world view is taken, they can be of rather diverse appearance. They are distinguishable from other members of the superfamily Tettigonioidea by the following combination of characters: body generally elongate, and size moderate to small. Coloration generally green or brownish, sometimes with darker stripes, particularly on pronotum. Head typically subconical to strongly pointed, with antennae longer than body, inserted between the eyes, the insertions ("scrobes") at most weakly margined. Prothoracic spiracle large, elongate, nearly covered by pronotum; prosternum generally with pair of spines. Anterior tibiae without terminal dorsal spines. Auditory organ usually covered. Hind tibiae with at least 1, and usually 2, terminal dorsal spines; first and second tarsal segments laterally grooved. Wings fully developed or reduced, but rarely, if ever, entirely lacking; tegmina, when developed, usually long and narrow, often resembling grass blades or stems or narrow leaves. Males with well-developed, but not exaggerated, basal stridulatory organs. Ovipositor long (sometimes longer than rest of body), straight or slightly curved upward.

There are probably about 1000 world species in perhaps 140 genera, the great majority placed in the subfamily Conocephalinae as here understood.

The males of Conocephalidae, like those of many Tettigoniidae, and unlike those of most Phaneropteridae, frequently stridulate by day, particularly in early evening. In general, the Conocephalini are more diurnal than the Copiphorini, which do not usually begin to stridulate much before sunset. The songs of these insects are typically in the form of a prolonged whine, which gives the Chinese name fan-shib-liang or lo-shib-liang, meaning "spinning-girl," to members of the genus *Euconocephalus*, particularly *E. pallidus* Redtenbacher (Copiphorini). This insect is often referred to in Chinese literature.

The Conocephalini and Copiphorini are mainly found in rank humid low vegetation such as grasses and reeds. Conocephalidae generally appear to favor a mixed diet of insects, seeds, or herbage, varying from species to species. Oviposition by most species, where known, seems to occur mainly in plant tissues (sometimes in galls). The family has little direct economic significance.

Key to tribes of Conocephalidae

1. Body large (length, without ovipositor, more than 24 mm). Vertex of head extending beyond basal antennal segment, produced to form tapering cone between antennal bases, notched beneath (Figs. 277, 278, 281). Fore and middle tibiae with spines beneath. Insect mainly nocturnal or crepuscular .... **Copiphorini** (p. 190)

   Body smaller (length, without ovipositor, 11–23 mm). Vertex of head not extending beyond basal antennal segment, produced to form rounded tubercle with concave sides, not notched beneath (Figs. 289, 309, 310). Fore and middle tibiae without spines beneath. Insect mainly diurnal and crepuscular .... **Conocephalini** (p. 196)

Clé des tribus de Conocephalidae

1. Corps grand (plus de 24 mm de long, ovipositeur exclu). Vertex s'étendant au-delà des segments basilaires des antennes, formant un cône effilé entre ces dernières et muni d'une encoche en dessous (fig. 277, 278 et 281). Tibias antérieurs et médians armés d'épines du côté ventral. Insectes plutôt nocturnes ou crépusculaires .... **Copiphorini** (p. 190)
   Corps petit (longueur de 11 à 23 mm, ovipositeur exclu). Vertex ne dépassant pas les segments basilaires des antennes, formant un tubercule rond aux côtés concaves, sans encoche dessous (fig. 289, 309 et 310). Tibias antérieurs et médians dépourvus d'épines du côté ventral. Insectes plutôt diurnes ou crépusculaires .... **Conocephalini** (p. 196)

Tribe Copiphorini

This tribe is easily recognized by the comparatively large size of the species and by the elongate fastigium of the vertex, which has a notch beneath it. One genus is represented in the region here covered.

Genus *Neoconocephalus* Karny

Description. Body large, elongate, moderately robust, characterized as indicated for the tribe Copiphorini, and, within it, by the conelike projection of the fastigium of the vertex (Figs. 278, 280) not being roughened above nor its apex being terminated in a strong sharp downwardly directed hook. Tegmina long, narrow, projecting beyond the well-developed hind wings when at rest. Male cerci of the general form shown in Fig. 279. Ovipositor long, straight (cf. Fig. 281).

Key to species of *Neoconocephalus*

1. Fastigium almost as wide as or wider than long, with apex broadly and evenly rounded, its ventral surface with narrow subterminal transverse arcuate black band (Figs. [illegible]6, 2[illegible]) .... ***retusus* (Scudder)** (p. 191)
   Fastigium distinctly longer than wide, with apex subacute or parabolic, not broadly and evenly rounded, its ventral surface either extensively or without black 2

2(1). Fastigium greatly elongate, slightly curved downward apically (Fig. 281) .... ***exiliscanorus* (Davis)** (p. 192)
   Fastigium less elongate, not downwardly curved apically (Fig. 278) .. 3

3(2). Underside of fastigium without black markings 4
   Underside of fastigium partly or entirely black 5

4(3). Large robust species. Shape of fastigium as in Fig. 284 .... ***robustus* (Scudder)** (p. 193)
   Smaller more slender species. Shape of fastigium as in Fig. 285 .... ***palustris* (Blatchley)** (p. 194)

5(3). Underside of fastigium almost entirely black (Fig. 283). Hind femur armed with spines on both margins ... ***lyristes* (Rehn & Hebard)** (p. 194)
   Underside of fastigium black only at tip and at sides (Fig. 282). Hind femur usually armed with spines only on inner lower carina .... ***ensiger* (Harris)** (p. 195)

Clé des espèces de *Neoconocephalus*

1. Fastigium presque aussi large ou plus large que long, à apex très régulièrement arrondi et marqué, du côté ventral, d'une étroite bande noire, arquée, subterminale (fig. 286 et 287) .. ***retusus* (Scudder)** (p. 191)
   Fastigium clairement plus long que large, à apex subaigu ou parabolique, ni très arrondi, ni régulièrement arrondi, à surface ventrale soit presque entièrement noire, soit dépourvue de noir 2

2(1). Fastigium très allongé et légèrement incurvé vers le bas à l'extrémité apicale (fig. 281) .... ***exiliscanorus* (Davis)** (p. 192)
   Fastigium moins allongé et non incurvé vers le bas à l'extrémité apicale (fig. 278) 3

3(2). Dessous du fastigium dépourvu de noir 4
   Dessous du fastigium partiellement ou entièrement noir 5

4(3). Espèce grande et robuste. Fastigium de la forme illustrée à la figure 284 .... ***robustus* (Scudder)** (p. 193)
   Espèce plus petite et plus grêle. Fastigium de la forme illustrée à la figure 285 .... ***palustris* (Blatchley)** (p. 194)

5(3). Dessous du fastigium presque entièrement noir (fig. 283). Fémurs postérieurs armés d'épines sur les deux côtés .... ***lyristes* (Rehn & Hebard)** (p. 194)
   Dessous du fastigium noir à l'extrémité et sur les côtés (fig. 282). Fémurs postérieurs ne portant habituellement des épines qu'au bas du côté intérieur .... ***ensiger* (Harris)** (p. 195)

Neoconocephalus retusus (Scudder)

Fig. 286; Map 53

Conocephalus retusus Scudder, 1879*c*:93.
Conocephalus atlanticus Bruner, 1899:38.
Conocephalus (*Neoconocephalus*) *retusus*; Karny 1907*c*:28.
Conocephalus (*Neoconocephalus*) *atlanticus*; Karny 1907*c*:29.
Neoconocephalus retesus [*sic*]; Helfer 1963:271.

Diagnosis. Rather small slender species. Fastigium about as wide as long; underside with subapical black band.

Description. Body small (pronotal length, male and female, 5.3–8 but usually less than 7 mm). Fastigium more or less as wide as long, broadly rounded apically and with narrow black subapical transverse band below (Fig. 286). Tegmina rather narrow, with male stridulatory area relatively small. Ovipositor much longer than hind femur. Color generally either green or brownish, the former usually more common.

Range. Indiana and Missouri to Connecticut, south to Mississippi and Florida.

Behavior and habitats. Walden (1911), referring to the species as *Conocephalus triops*, reported that he found it "in tall grass along a ditch which contained water only in very wet seasons." J. A. G. Rehn and Hebard (1915*a*) say that it is an inhabitant of grasses in "waste" fields and in marshland along the drier margins, where many are usually found together.

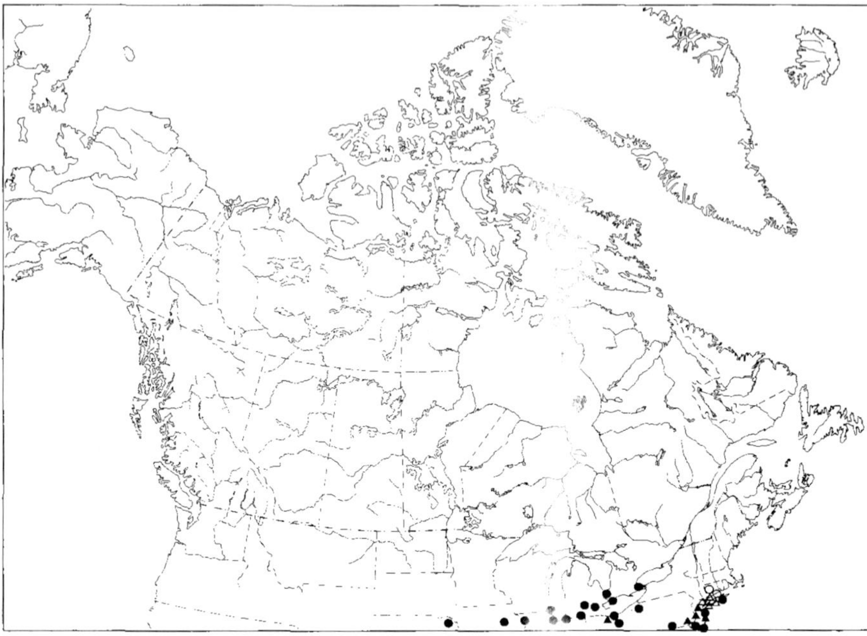

Map 53. Collection localities for *Neoconocephalus retusus* (○), *N. exiliscanorus* (△), *N. palustris* (▲), and *N. lyristes* (●).

Stridulation is described as being "of the exact pitch of that of *N. lyristes* but weaker . . . a continuous zeeeeeeeeee."

Life history. J. A. G. Rehn and Hebard (1915*a*) say that this species is the last of its genus to appear, in the season, and that in New Jersey it reaches the adult stage toward the end of August. It undoubtedly overwinters in the egg stage and has but one generation each year.

Neoconocephalus exiliscanorus (Davis)

Fig. 280; Map 53

Conocephalus exiliscanorus W. T. Davis, 1887:57.

Conocephalus bruneri Blatchley, 1903:267, fig. 90.

Neoconocephalus exiliscanorus; J. A. G. Rehn and Hebard 1915*a*:369, 371, 372, fig. 1*A*, pl. XV, figs. 1*B–E*; Helfer 1963:268.

Diagnosis. Fastigium greatly elongate, almost entirely black below.

Description. Body moderately short and robust (pronotal length, male 7.6–9.1, female 7.7–9.6 mm). Fastigium long, slender, acuminate, several times as long as wide, with apex slightly curved downward (Fig. 281), and with lower face in front of ventral notch almost entirely black. Tegmina of moderate length and width. Ovipositor about as long as tegmina, straight, pointed, similar to Fig. 280. Color green or brown, but, if former, then usually with yellow stripes along lateral margins of pronotal disc; if brown, often with minute black points scattered over tegmina.

Range. Indiana to Connecticut, south to Texas and Georgia.

Behavior and habitats. On the Atlantic seaboard the species occurs among tall reeds and cattails (*Typha*) in tidal marshes and, in inland localities, in thickets and cornfields. J. A. G. Rehn and Hebard (1915*a*) describe the stridulation as a rather loud vibrant rattling "ziit-ziit-ziit-ziit," about three to the second, rising and falling in intensity, often ceasing "as if from exhaustion." The song does not begin before sunset and is much more vigorous after dark.

Life history. Eggs are reported as being laid in grass blades (J. A. G. Rehn and Hebard 1915*a*). There is presumably only a single generation each year, at least in northern latitudes. Overwintering probably occurs in the egg stage. Adults occur most commonly in August, but they have been found from July until October.

Neoconocephalus robustus (Scudder)

Fig. 284; Map 54

Conocephalus robustus Scudder, 1863*a*:449.
Conocephalus crepitans Scudder, 1863*a*:450.
Neoconocephalus robustus; Karny 1912*b*:32; Helfer 1963:269.
Neoconocephalus robustus robustus; J. A. G. Rehn and Hebard 1915*a*:387.
Neoconocephalus robustus crepitans; J. A. G. Rehn and Hebard 1915*a*:391.

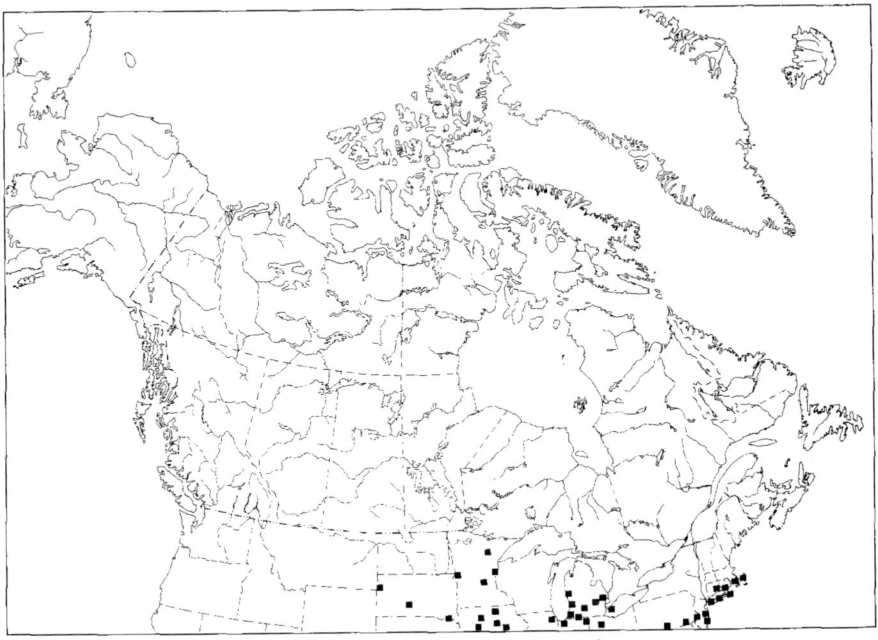

Map 54. Collection localities for *Neoconocephalus robustus*.

Diagnosis. Fastigium blunt, not marked with black beneath.

Description. Size large (pronotal length, male 8.2–10.8, female 7.3–9.3 mm). Fastigium with blunt apex and without black marking beneath (Fig. 284). Lateral carinae of pronotum distinctly divergent posteriorly; stridulatory area of male wide, greater than 4.9 mm (measured from median of yellow ridge to edge of left tegmen). Ovipositor long (16–37 mm) but less than 1.2 times length of hind femur. Color usually green, sometimes brown, greenish yellow beneath and on legs.

Range. South Dakota to southern Ontario and Massachusetts, south to Texas and Florida; California.

Behavior and habitats. Individuals are found generally in dry areas, often where ground is sandy. Johnstone (1971) reported specimens from Point Pelee, Ont., on grasses and bushes growing in sandy soil. Males are usually found in tall rank vegetation. When disturbed, these insects will dive downward. They may either run or remain motionless with the fastigial cone embedded in the soil. Males, when disturbed, may fly some distance.

Stridulation has been described as a continuous ear-splitting buzzing (Johnstone 1971). R. D. Alexander (1956) called this species a "whiner," because of the whining component of the stridulation, which is detectable to the human ear. T. J. Walker et al. (1973) described a new species, *N. bivocatus*, which is similar in anatomy to *robustus* but which differs somewhat in stridulation. In *N. robustus*, the pulse intervals are approximately of the same duration, whereas in *N. bivocatus*, consecutive intervals between pulses are unequal and alternate ones are equal; thus pulses are produced in pairs. The songs of *N. robustus* are usually intense and "whiney," but some individuals may sound "buzzy" and be confused with *N. bivocatus*. The latter species does not occur in the region covered here. The wing stroke rate of *N. robustus* at 30°C is 224 per second, the highest rate known for the Tettigonioidea (T. J. Walker 1975*b*).

Life history. Univoltine with wintering in the egg stage. Adults occur mainly in August and September but they have been reported from 16 June to 12 October in the United States (T. J. Walker et al. 1973).

Neoconocephalus palustris (Blatchley)

Fig. 285; Map 53

Conocephalus palustris Blatchley, 1893*c*:89.
Neoconocephalus palustris; Karny 1912*b*:32; Helfer 1963:269.

This species is somewhat similar to *N. robustus* but is smaller, more slender, and with a relatively less prominent fastigium (Fig. 285). It is known from Pennsylvania to Louisiana and North Carolina.

Neoconocephalus lyristes (Rehn & Hebard)

Fig. 283; Map 53

Conocephalus lyristes J. A. G. Rehn and Hebard, 1905:45.
Neoconocephalus lyristes; Karny 1912*b*:30.

Diagnosis. Fastigium jet black beneath.

Description. Size (pronotal length, male 7.6–8, female 7.2–8 mm), slender. Fastigium long, jet black beneath (Fig. 283). Male cercus stout, with internal tooth robust. Ovipositor long (23.3–32.5 mm), slender, slightly upcurved. General coloration either bright grassy green or yellowish tan to dark brown, with narrow yellowish lines on lateral carinae of pronotum.

Range. Nebraska to Ontario, and along Atlantic coast from New York to Virginia.

Behavior and habitats. J. A. G. Rehn and Hebard (1915*a*) reported *N. lyristes* as an inhabitant of freshwater and saltwater marshes and bogs, on *Scirpus* and other high marsh plants. E. M. Walker (1904*b*) recorded it from a large stretch of open grassy marshland. Stridulation is a high-pitched, continuous buzzing. R. D. Alexander (1956) recorded 130 pulses/second with dominant frequency of 7.5 kHz at 23°C from a specimen in Ohio.

Life history. Records of captures indicate a single generation each year, with adults occurring during late summer. The specimens reported by E. M. Walker (1904*b*) were taken 12 August.

Neoconocephalus ensiger (Harris)

Figs. 277–279, 281, 282; Map 55

Conocephalus ensiger T. W. Harris, 1841:131.
Conocephalus attenuatus Scudder, 1872:249.
Neoconocephalus ensiger; Karny 1907*c*:38; 1912*b*:32; Cantrall 1943:134; Helfer 1963:269; Vickery et al. 1974:63.

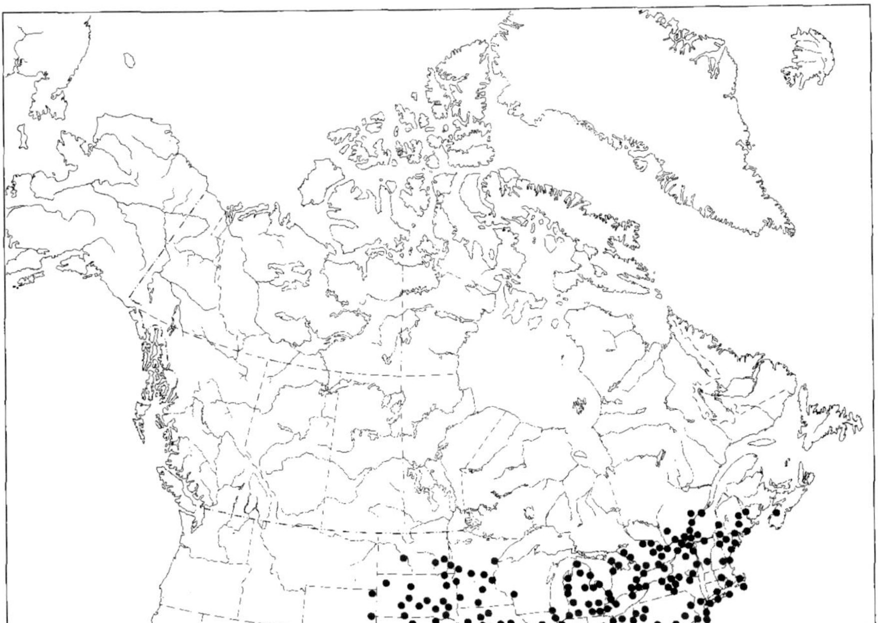

Map 55. Collection localities for *Neoconocephalus ensiger*.

Diagnosis. Margins and tip of lower face of fastigium black.

Description. Body elongate, slender (pronotal length, male and female, 7–7.5 mm). Fastigium narrow, slightly constricted in front of eyes, with margins convergent from middle forward and with tooth projecting ventrally from front of base (Fig. 278); margins and tip of lower face of fastigium black (Fig. 282), remainder green (or pale brown). Tegmina narrow, longer than body. Male cerci as in Fig. 279. Ovipositor long, straight, pointed (Fig. 281). Color generally grass green, occasionally pale brown.

Range. Colorado to Nova Scotia, south to New Mexico and North Carolina.

Behavior and habitats. *N. ensiger* is usually found in tall grasses bordering ditches, fields, or roadsides, often but by no means exclusively in damp locations. Sexually receptive conspecific females are attracted by male stridulation and move toward the source of the sound. Copulation occurs with the male positioned head downward on a plant stem. The female approaches the male and begins antennal contact. Both insects use the fore and middle legs to grasp the tegmina of the other. The female invariably faces upward. Following copulation, which lasts 27–68 minutes, there is no external evidence of a spermatophore on the female genitalia. After separation, the male grooms his genitalia. Females of *N. ensiger* have never been observed to do likewise, although this is common in some other species of *Neoconocephalus*.

Males stridulate in August and September, generally during evenings, often after dark, but also occasionally by daylight, usually above the surface of the ground on grasses, tall weeds, or even in trees. The stridulation sounds like a rapid succession of pulses, "tsip, tsip, tsip, tsip," at about 10 per second. T. J. Walker (1975*b*) and Gwynne (1977) have studied the stridulation and found it to consist of a wide band of frequencies with the most intense lying between 8 and 17 kHz. Opening of the tegmina is silent, but closure produces a train of rapid decay pulses ending with an intense prolonged pulse.

Life history. Winter is passed in the egg stage. Nymphs hatch in May and June and reach maturity in summer. Gwynne (1977) reported immature stages in Ontario during June and July. Adults are found during August and September. There is a single generation each year.

Tribe Conocephalini

Members of this tribe are small to medium-sized (11–23 mm long). Fastigium projected forward and upward as blunt rounded tubercle, concave at sides adjacent to antennal bases, never distinctly cone-shaped. Eyes relatively large, globose. Antennae slender, long. Pronotum with no more than one transverse sulcus; prosternum with spines long and slender or reduced to toothlike projections. Tegmina and hind wings well-developed, somewhat shortened or reduced to lateral pads (some species dimorphic in this respect). Color usually predominantly green, occasionally pale brown.

Key to genera of Conocephalini

1. Prosternal spines long, cylindrical, slender. Tegmina well-developed, longer than abdomen. Ovipositor stout, usually curved upward. Body robust, with length (excluding ovipositor) more than 18 mm (except *O. delicatum*) .... ***Orchelimum*** (p. 197)
   Prosternal spines short. Tegmina not surpassing abdomen, although variable even within species. Ovipositor slender, nearly straight. Body small, slender, with length (excluding ovipositor) less than 17 mm .... ***Conocephalus*** (p. 210)

Clé des genres de Conocephalini

1. Épines du prosternum longues, cylindriques et minces. Tegmina bien développés, plus longs que l'abdomen. Ovipositeur fort, habituellement incurvé vers le haut. Corps robuste, mesurant plus de 18 mm de long (ovipositeur exclu) (sauf *O. delicatum*) ...... ***Orchelimum*** (p. 197)
   Épines du prosternum courtes. Tegmina ne dépassant pas l'abdomen, ce caractère étant toutefois variable même à l'intérieur d'une même espèce. Ovipositeur mince, presque rectiligne. Corps petit, mince, mesurant moins de 17 mm de long (ovipositeur exclu) .... ***Conocephalus*** (p. 210)

Genus *Orchelimum* Audinet-Serville

Description. Size large (most species more than 18 mm long). Face short, moderately slanted; fastigial tubercle with ventral projection meeting dorsal part of frons. Pronotum short, smooth, somewhat saddle-shaped, with posterior margin broadly rounded; lateral pronotal lobes bulging posteroventrally to form convex callosity (Figs. 290–294); prosternal spines long, slender. Tegmina narrow, tapering; hind wings usually longer than tegmina. Male subgenital plate V- or U-emarginate. Male cercus stout, armed on inner face with tooth (Figs. 295–301). Ovipositor stout, rather broad, distinctly curved upward (Figs. 302–308). Pronotum with conspicuous brown stripe on dorsal midline.

Key to species of *Orchelimum*

1. Male (without ovipositor) 2
   Female (with ovipositor) 9
2(1). Cercus not elongate, with part beyond tooth not longer than basal part (Figs. 295–297) 3
   Cercus elongate, with part beyond tooth as long as or distinctly longer than basal part (Figs. 298–301) 6
3(2). Hind tibia blackish .... ***nigripes* Scudder** (p. 199)
   Hind tibia not blackish 4
4(3). Humeral sinus of lateral pronotal lobes distinct (Fig. 290) .... ***vulgare* Harris** (p. 201)
   Humeral sinus of lateral pronotal lobes scarcely evident (Fig. 291) .. 5

5(4). Tooth of cercus as long as apical part of shaft. Hind femora not spinose lateroventrally .... *gladiator* **Bruner** (p. 203)
Tooth of cercus longer than apical part of shaft. Hind femora with 3 or 4 spines lateroventrally .... *silvaticum* **McNeill** (p. 204)
6(2). Apical part of cercus distinctly tapering to subacute apex (Fig. 301) ... . *volantum* **McNeill** (p. 205)
Apical part of cercus not tapering to acute or subacute apex 7
7(6). Larger species (17 mm, or more, in length) 8
Smaller species (not greater than 16 mm in length) .... *delicatum* **Bruner** (p. 206)
8(7). Face with median brown stripe .... *concinnum* **Scudder** (p. 207)
Face without median stripe .... *campestre* **Blatchley** (p. 208)
9(1). Ovipositor longer than one-half length of hind femur (ratio 0.5, or greater) 10
Ovipositor less than one-half length of hind femur (ratio less than 0.5) 15
10(9). Ovipositor greater than two-thirds length of hind femur (ratio greater than 0.66) .... *delicatum* **Bruner** (p. 206)
Ovipositor greater than one-half but less than two-thirds length of hind femur (ratio 0.50–0.65) 11
11(10). Ovipositor with dorsal margin nearly straight 12
Ovipositor with dorsal (and ventral) margin curved 13
12(11). Robust species. Ovipositor long (ratio of ovipositor length to hind femur length 0.58–0.64) .... *gladiator* **Bruner** (p. 203)
Slender species. Ovipositor shorter (ratio of ovipositor length to hind femur length 0.53–0.55) .... *volantum* **McNeill** (p. 205)
13(10). Tibiae blackish .... *nigripes* **Scudder** (p. 199)
Tibiae green or brown 14
14(13). Lateral sinus of pronotum well-developed ..... *vulgare* **Harris** (p. 201)
Lateral sinus of pronotum scarcely evident .... *silvaticum* **McNeill** (p. 204)
15(9). Ovipositor strongly curved. Pronotum long (5.4–6.5 mm) .... *vulgare* **Harris** (p. 201)
Ovipositor gradually arcuate, not strongly curved. Pronotum shorter (4.0–5.0 mm) 16
16(15). Face without dark median stripe .... *campestre* **Blatchley** (p. 208)
Face with red brown median stripe ...... *concinnum* **Scudder** (p. 207)

Clé des espèces d'*Orchelimum*

1. Mâles (dépourvus d'ovipositeur) 2
Femelles (pourvues d'ovipositeur) 9
2(1). Cerques non allongés, portion située au-delà de la dent pas plus longue que la portion proximale (fig. 295 à 297) 3
Cerques allongés, portion située au-delà de la dent aussi longue ou clairement plus longue que la portion proximale (fig. 298 à 301) 6
3(2). Tibias postérieurs noirâtres .... *nigripes* **Scudder** (p. 199)
Tibias postérieurs non noirâtres 4
4(3). Sinus huméral des lobes latéraux du pronotum visible (fig. 290) .... *vulgare* **Harris** (p. 201)
Sinus huméral des lobes latéraux du pronotum à peine visible (fig. 291) ... 5
5(4). Dent des cerques aussi longue que la portion apicale de ces derniers (fig. 296). Fémurs postérieurs dépourvus d'épine latéro-ventrale .... *gladiator* **Bruner** (p. 203)

Dent des cerques plus longue que la portion apicale de ces derniers. Fémurs postérieurs armés de 3 ou 4 épines latéro-ventrales .... *silvaticum* **McNeill** (p. 204)

6(2). Portion apicale du cerque allant s'amenuisant pour donner un apex subaigu (fig. 301) .... *volantum* **McNeill** (p. 205)

Portion apicale du cerque ne se terminant pas en apex aigu ni subaigu ... 7

7(6). Espèces plus grosses (17 mm ou plus de longueur) 8

Espèces plus petites (de longueur ne dépassant pas 16 mm) .... *delicatum* **Bruner** (p. 206)

8(7). Face marquée d'une bande brune médiane ... *concinnum* **Scudder** (p. 207)

Face dépourvue d'une bande médiane .... *campestre* **Blatchley** (p. 208)

9(1). Longueur de l'ovipositeur égale à plus de la moitié de celle du fémur postérieur (rapport de 0,5 ou plus) 10

Longueur de l'ovipositeur inférieure à la moitié de celle du fémur postérieur (rapport inférieur à 0,5) 15

10(9). Longueur de l'ovipositeur supérieure aux deux tiers de celle du fémur postérieur (rapport supérieur à 0,66) ... *delicatum* **Bruner** (p. 206)

Longueur de l'ovipositeur supérieure à la moitié mais inférieure aux deux tiers de la longueur du fémur postérieur (rapport de 0,50 à 0,65) ... 11

11(10). Ovipositeur à bord dorsal presque rectiligne 12

Ovipositeur à bords dorsal et ventral incurvés 13

12(11). Espèce robuste. Ovipositeur long (rapport de la longueur de l'ovipositeur à celle du fémur postérieur variant de 0,58 à 0,64) .... *gladiator* **Bruner** (p. 203)

Espèce plus grêle. Ovipositeur plus court (rapport de la longueur de l'ovipositeur à celle du fémur postérieur variant de 0,53 à 0,55) .. .. *volantum* **McNeill** (p. 205)

13(10. Tibias noirâtres .... *nigripes* **Scudder** (p. 199)

Tibias vert ou brun 14

14(13). Sinus latéral du pronotum bien développé ..... *vulgare* **Harris** (p. 201)

Sinus latéral du pronotum à peine visible .... *silvaticum* **McNeill** (p. 204)

15(9). Ovipositeur fortement incurvé. Pronotum long (5,4 à 6,5 mm) .... *vulgare* **Harris** (p. 201)

Ovipositeur arqué, mais pas fortement incurvé. Pronotum plus court (4,0 à 5,0 mm) 16

16(15). Face sans bande médiane foncée .... *campestre* **Blatchley** (p. 208)

Face marquée d'une bande médiane brun rougeâtre .... *concinnum* **Scudder** (p. 207)

Orchelimum nigripes Scudder

Figs. 292, 297, 304; Plate II*D*; Map 56

Orchelimum nigripes Scudder, 1875*a*:459.

Xiphidium (Orchelimum) robustum Redtenbacher, 1891:494, 499.

Orchelimum nigripes; Lugger 1898:326; Cantrall 1943:139; R. D. Alexander et al. 1972:33; G. K. Morris and Walker 1976:785.

Diagnosis. Tibiae blackish. Male cercus with carina obtuse, sinuate.

Description. Body small, robust (pronotal length, male and female, 4.3–5.0 mm). Lateral lobes of pronotum as in Fig. 292. Stridulatory vein of male tegmen large, prominent. Male cercus slender, broadest at middle,

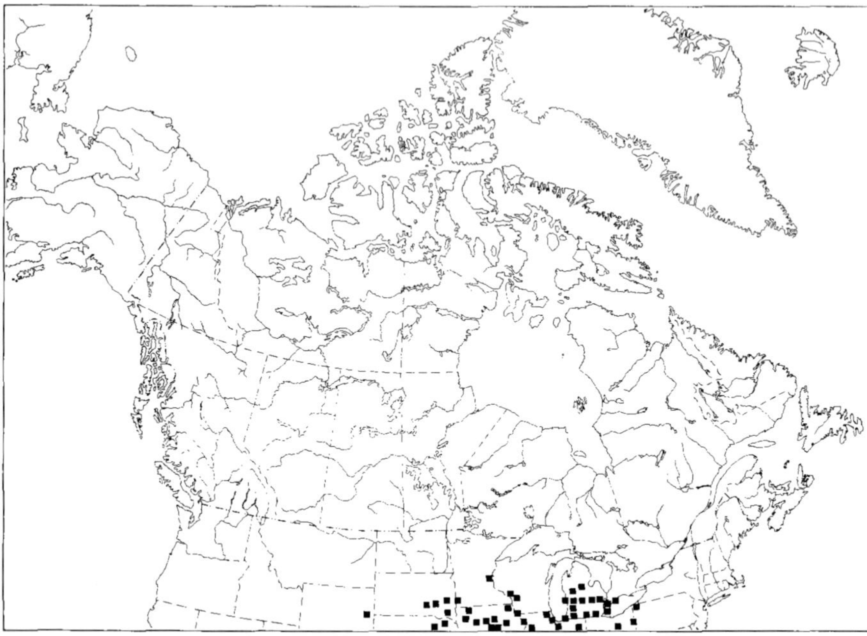

Map 56. Collection localities for *Orchelimum nigripes*.

tapering, concave on inner face, with apex obtuse; tooth long, acute, directed toward base (Fig. 297). Ovipositor slightly more than half as long as hind femur, strongly curved, broadest at middle, tapering to acute tip (Fig. 304). Males usually green, females usually reddish brown; antennae, all tarsi, and apical third of hind femur black or dark brown. Stridulatory file of "Type II" of G. K. Morris and Walker (1976), with basal teeth not separated by interval as wide as a tooth.

Range. Wyoming to southernmost Ontario, south to Texas and Alabama.

Behavior and habitats. This is mainly a lowland or hygrophilous species, found among sedges or weeds along roadside ditches, lake margins, or near ponds or marshes (Blatchley 1920). E. M. Walker (1905*a*) collected it by following the sound of stridulation to specimens that were on tall weeds and vines, and also to specimens that were up to 3–5 m above the ground in trees. Froeschner (1954) reported *O. nigripes* in Iowa from cornfields, where it was observed feeding upon pollen that had fallen from the tassels to the leaves or to the ground.

G. K. Morris and Walker (1976) described the song of *O. nigripes*. It produces both "buzz" and "tick" phonatomes (all the sound produced during one cycle of wing movement is described by them as a "phonatome"). During wing closure, the scraper apparently lodges behind the steep face of a tooth, creating a silent interval while the tegmina are held open. This causes energy storage, because force is being exerted without displacing the scraper, ultimately producing a major pulse train (grouping of two or more pulses preceded and followed by silent periods that are longer than the intervals between pulses).

Life history. Univoltine, with overwintering in the egg stage. Adults were collected in Iowa between 24 July and 18 October (Froeschner 1954).

Orchelimum vulgare Harris

Figs. 289–290, 295, 302; Plate II*A*; Map 57

Orchelimum vulgare T. W. Harris, 1841:130.

Orchelimum vulgare; Caulfield 1888:70; Cantrall 1943:137; Helfer 1963:265; R. D. Alexander et al. 1972:33; G. K. Morris and Walker 1976:785, 792, 798.

Orchelimum glaberrimum Burmeister, 1838:707. Name suppressed.

Orchelimum cuticulare Audinet-Serville, 1838:523. Name suppressed.

Diagnosis. Humeral sinus well-defined. Male cercus with tooth short. Upper margin of ovipositor curved.

Description. Body size medium (pronotal length, male 5–6, female 5.4–6.5 mm), rather robust. Pronotum long; lateral lobes nearly as long as deep; posteroventral angle obtusely rounded; convex callosity prominent, with humeral sinus broad and well-defined (Fig. 290). Tegmina usually extending

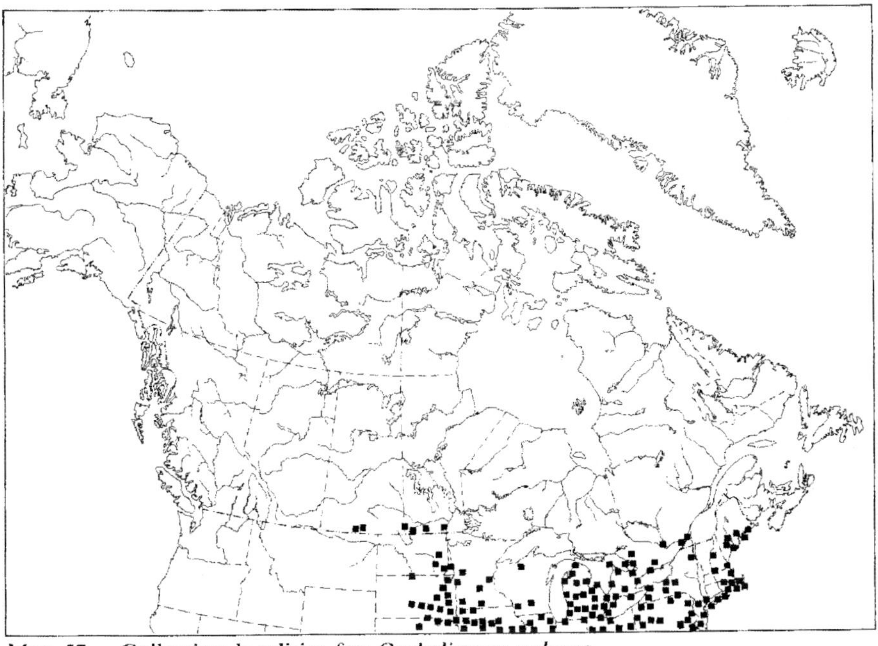

Map 57. Collection localities for *Orchelimum vulgare*.

to or slightly beyond apex of hind femur. Male cercus with tooth shorter than distal part, and with apex bluntly rounded (Fig. 295). Ovipositor usually less than half length of hind femur, with upper margin curved (Fig. 302). General color green to pale reddish brown; occiput and pronotal disc with reddish longitudinal band. Male with 2 short black marks on each tegmen; legs and tarsi buff.

Range. Saskatchewan to Maine, south to Texas and northern Georgia.

Behavior and habitats. According to E. M. Walker (1905*a*), *O. vulgare*, in Ontario, is found "in upland fields as well as low meadows and is fond of perching in clumps of tall grass." Blatchley (1920) reported it from "upland localities, along fence rows and in clover and timothy meadows." Cantrall (1968) found it in drier marshes and meadows and in mixed herbaceous upland habitats.

G. K. Morris and Walker (1976) reported the type of stridulation of *Orchelimum*. *O. vulgare* has a "Type I" stridulatory file, with widely spaced teeth in the basal half of the file elevated on a greatly swollen vein. Teeth are separated by spaces nearly as wide as a tooth. Their profile is asymmetrical, strongly buttressed basally. When the tegmina are opened, the scraper strikes a much steeper face. The most intense sounds are generated on tegminal closure, since the momentary trapping of the scraper ridge in the space between the teeth increases the energy stored in the bent scraper lobe. Pulses are of short duration. T. J. Walker (1975*b*) reported a wing-stroke rate of 68 per second at 25°C. He also demonstrated the effect of temperature on wing-stroke rate, i.e., that rate is a linear function of temperature. The low temperature base of *O. vulgare* is 10.5°C. Stridulation by males attracts conspecific females. G. K. Morris (1971, 1972) has shown that stridulation by *O. vulgare* also results in aggressive behavior between conspecific males. Males as well as females are attracted to singing males and the approach of male to male often produces aggressive behavior on the part of both males, presumably due to competition for females.

Blatchley (1920) reported *O. vulgare* to be carnivorous as well as phytophagous because he had seen it feeding upon moths and a beetle. It may also be cannibalistic in the field, since cannibalism has been noted in captivity. Eggs are deposited singly in the stems of plants, usually plants with pithy stems.

Life history. Univoltine, with wintering in the egg stage. Nymphs hatch in May or June. Adults appear in late July and August. Oviposition may continue through September into October, or until the insects are killed by frost.

Economic importance. Metcalf and Colby (1930) reported severe damage by this species to raspberry crops in Illinois. Sugar beets and leguminous plants have also been reported damaged by *O. vulgare*. Damage is caused by oviposition in the stems, the stems eventually breaking at an oviposition scar. Nymphs feed upon plant foliage but seldom cause noticeable injury.

Orchelimum gladiator Bruner

Figs. 291, 296, 303; Plate II*B*; Map 58

Orchelimum gladiator Bruner, 1891*a*:71.
Ochelimum gladiator; E. M. Walker 1905*a*:38; Cantrall 1943:137; Helfer 1963:266; Vickery et al. 1974:65; G. K. Morris and Walker 1976:785.
Orchelimum manitobense E. M. Walker, 1910:351.

Diagnosis. Humeral sinus scarcely evident. Upper margin of ovipositor nearly straight.

Description. Body rather robust (pronotal length, male 4.3–5, female 4.7–5.2 mm). Fastigium short, obtuse, shallowly sulcate at apex. Pronotum long; lateral lobes deep, with posteroventral angle nearly right-angled; humeral sinus scarcely evident; convex callosity poorly developed (Fig. 291). Tegmina usually slightly surpassing apex of hind femur in females, distinctly longer in males. Male cercus with tooth as long as apex beyond tooth; apex subacute (Fig. 296). Ovipositor about two-thirds length of hind femur, with upper margin nearly straight (Fig. 303). Color pale grass green with markings on occiput, and with 2 diverging narrow lines on pronotal disc; antennae, tarsi, and tip of ovipositor dark brown to reddish. Stridulatory file of male "Type II" of G. K. Morris and Walker (1976), with teeth on basal half close together, separated by an interval much narrower than width of a tooth; vein not greatly swollen.

Range. Washington and Alberta to Québec and Maine, south to California and Tennessee.

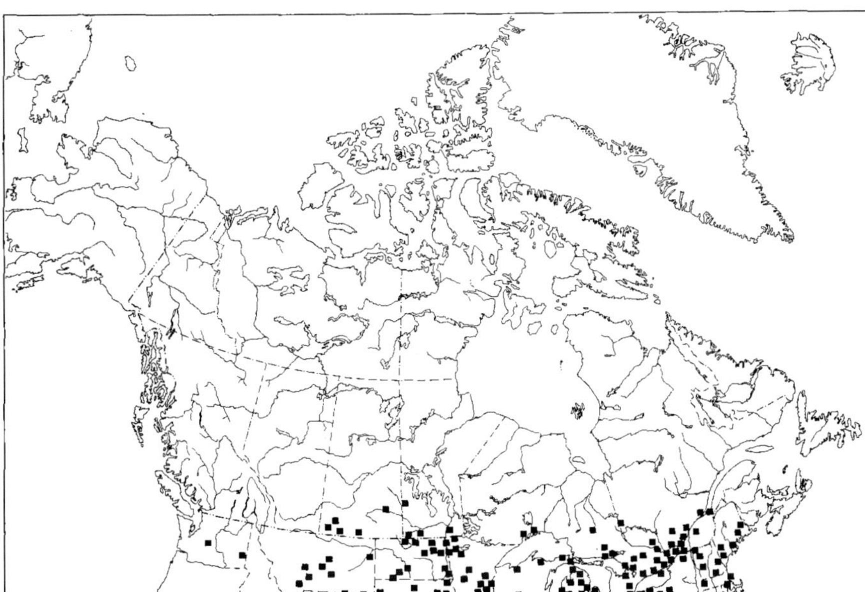

Map 58. Collection localities for *Orchelimum gladiator*.

Behavior and habitats. G. K. Morris (1972) found this species "in typical meadow habitats of waist-length sedge and grass." Blatchley (1920) reported it common in northern counties of Indiana in damp prairies, meadows, and marshes.

Stridulation does not occur when the temperature falls below 17°C. The song, a series of "buzzes" interspersed with "ticks," is similar to that of *O. vulgare* (E. S. Thomas and Alexander 1962; G. K. Morris 1971). Unlike *O. vulgare*, which produces double-pulse ticks, *O. gladiator* lacks the minor pulse train and produces single-pulse ticks. This species may increase the number of ticks in the tick phase of its song if closely approached by human observers (Cantrall 1943). The difference between single- and double-pulse ticks is not readily distinguishable by the human ear but is efficient in preventing attempted mismating where the two species occur together.

G. K. Morris et al. (1975*b*) conducted experiments with *O. gladiator* to determine female phonotaxis. Females developed responsiveness to playback of recorded conspecific male song 5 or 6 days after becoming adult. Responsiveness continued until mating, then was rapidly extinguished. Parasitized females were not mated; this indicated that a parasitized female may not be able to accomplish some step in the response chain, or that males can discriminate against parasitized females and thus avoid wasting reproductive potential. G. K. Morris (1971) showed that stridulating males spaced themselves regularly within the habitat.

Life history. Univoltine, with wintering in the egg stage. Adults are present in the field from mid-July to early September.

Orchelimum silvaticum McNeill

Map 59

Orchelimum silvaticum McNeill, 1891:26.

Orchelimum sylvaticum [*sic*]; Blatchley 1893*a*:136; G. K. Morris and Walker 1976:791, 792.

Diagnosis. Lateroventral margin of hind femur with 3 or 4 spines.

Description. Pronotum with humeral sinus scarcely evident (Fig. 291), with metazona two-thirds as long as prozona, upturned at posterior margin. Lateroventral margin of hind femur spinose in both sexes, bearing 3 or 4 spines. Tegmina long, reaching or exceeding apices of hind femora. Male cercus with apical part beyond tooth not longer than basal part (cf. Fig. 296); tooth longer than apical part. Ovipositor regularly and decidedly curved upward (cf. Fig. 302). Color pale green, usually with dark stripe on vertex and pronotum.

Range. South Dakota to Ontario, south to Ohio.

Behavior and habitats. McNeill (1891) found the species on corn, and later in open places in woods. He described the stridulation as "zip-zip-zip" repeated rapidly many times, followed by a "zee-e-e-e" phrase lasting about 8 seconds.

Life history. Presumably similar to that of other species of *Orchelimum*.

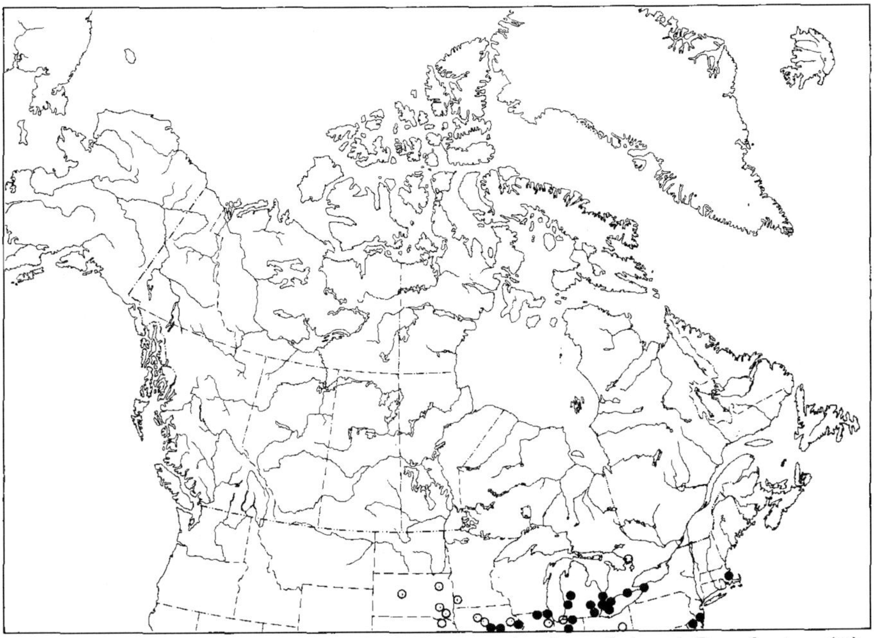

Map 59. Collection localities for *Orchelimum sylvaticum* (○) and *O. volantum* (●).

Orchelimum volantum McNeill

Figs. 294, 301, 308; Plate II*H*; Map 59

Orchelimum volantum McNeill, 1891:26.

Orchelimum volantum; McNeill 1900*b*:80, 83; R. D. Alexander et al. 1972:33; G. K. Morris and Walker 1976:785.

Orchelimum bruneri Blatchley, 1893*c*:92.

Diagnosis. Hind femur short, with 1–4 spines beneath.

Description. Pronotum short (male 3.7–4.5; female 4–5.0 mm); lateral lobes as long as deep (Fig. 294). Hind femur short, stout, armed beneath with 1–4 short spines. Male cercus with apical part distinctly and strongly tapering to subacute apex; tooth with base broad and flat (Fig. 301). Ovipositor (Fig. 308) about two-thirds as long as hind femur, with upper margin nearly straight. Color pale translucent brownish green, with lower halves of lateral pronotal lobes brighter green; occiput and prozona usually with 2 feebly divergent dark stripes; ovipositor and hind tibiae brown. Male stridulatory file "Type I" of G. K. Morris and Walker (1976); basal teeth widely spaced, the interval between teeth as wide or nearly as wide as a tooth.

Range. Kansas and Iowa to southern Michigan and Ohio.

Behavior and habitats. The type specimens were found by McNeill (1891) in a clump of rank-growing *Sagittaria variabilis* Engelm. E. M. Walker (1905*a*) found *O. volantum* in rushes and *Sagittaria* in open marshes bordering streams in southern Ontario. McNeill (1900*b*) says that this species "makes more use of its remarkably long wings than any other species of the genus" known to him. Blatchley (1920) found the species on stems and leaves of a tall broad-leaved knotweed, *Polygonum amphibium* L., growing in shallow water on margins of ponds and lakes.

The stridulation was reported by McNeill (1891) as "zip-zip, kr-ze-e-e, kr-ze-ee," the "ze-e-e" component lasting no more than 0.75 second. R. D. Alexander et al. (1972) described stridulation as buzzes preceded by ticks, 2–6 every 5 seconds, with each buzz slowing toward the end.

Life history. This species is univoltine, wintering in the egg stage. Blatchley (1920) reported *O. volantum* in the adult stage in August and September. E. M. Walker (1905*a*) collected this species during the same period in Kent and Essex counties, southern Ontario.

Orchelimum delicatum Bruner

Figs. 299, 306; Plate II*F*; Map 60

Orchelimum gracile Bruner, 1891*a*:70 (homonym).

Orchelimum delicatum Bruner, 1892:264 (replacement name for *gracile*); R. D. Alexander et al. 1972:33; G. K. Morris and Walker 1976:785.

Orchelimum concinnum delicatum; Blatchley 1920:556–557.

Diagnosis. Face green, lacking red stripe, occasionally with red flecks, and rarely with brown central area.

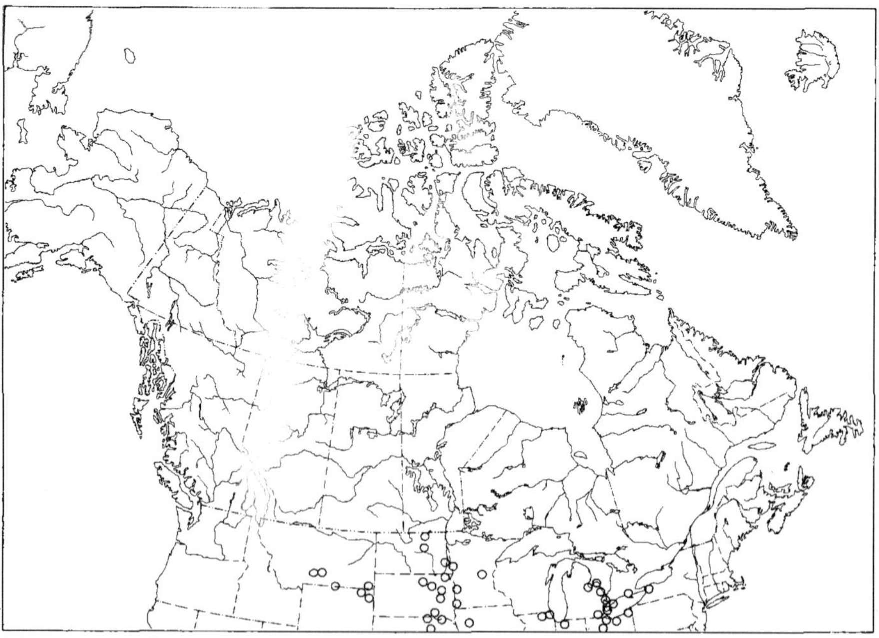

Map 60. Collection localities for *Orchelimum delicatum*.

Description. Lateral pronotal lobe much as in Fig. 293. Male stridulating mechanism nearly identical with that of *O. concinnum*; male cercus with tooth large, long (Fig. 299). Ovipositor (Fig. 306) with upper edge straight on apical two-thirds, more than half as long as hind femur.

Range. Montana to New York, south to California and Louisiana.

Behavior and habitats. "*O. delicatum* is largely restricted to swales adjacent to sand dunes or sand beaches, where it is often associated with blue joint grass, *Calamagrostis canadensis*" (E. S. Thomas and Alexander 1962).

Stridulation is hardly distinguishable to the human ear from that of *O. concinnum*, even though the wing-stroke rate is one-third faster at any given temperature. Like that species the song of *O. delicatum* consists of ticks and buzzes, but the ticks are stronger and are usually audibly double. The main frequencies lie between 7 and 16 kHz.

Life history. Univoltine, with wintering in the egg stage. Adults occur during late summer and early fall.

Orchelimum concinnum Scudder

Figs. 293, 298, 305; Plate II*E*; Map 61

Orchelimum concinnum Scudder, 1863*a*:452.

Orchelimum longipennis [*sic*, for *longipenne*] Scudder, 1863*a*:453.

X[*iphidium*] (*Orchelimum*) *inerme* Redtenbacher, 1891:495, 501.

Orchelimum indianense Blatchley, 1893*c*:90; E. M. Walker 1902*c*:85.

Orchelimum concinnum; Blatchley 1920:554; R. D. Alexander et al. 1972:36, 45, 51; G. K. Morris and Walker 1976:788–790, 792, 794, 795.

Orchelimum concinnum concinnum; Hebard 1931*b*:199.

Diagnosis. Face with median red to reddish brown stripe.

Description. Lateral pronotal lobe as in Fig. 293. Male cercus with apical part beyond tooth distinctly longer than basal part; tooth short, acute (Fig. 298). Ovipositor (Fig. 305) slender, with upper margin little curved in apical two-thirds, no longer than and usually less than one-half length of hind femur. Color usually green (glaucous in living specimens), with brown mid-dorsal stripe extending on supra-anal plate; cerci brown; specimens in Huron County, MI, suffused with purplish red over much of body, including tegmina and ovipositor. Male stridulatory file of "Type II" of G. K. Morris and Walker (1976, figs. 4*F*, 5*E*, 5*F*), with basal teeth separated by much less than the width of a tooth.

Range. Minnesota to southern Ontario and Ohio; Atlantic coast from Maine to Florida and Gulf Coast states; Bermudas.

Behavior and habitats. E. S. Thomas and Alexander (1962) found *O. concinnum* "in a wide variety of marshy areas, including brackish locations." They reported limited distribution and association with wetland plants, dominated by the sedge *Eleocharis rostellata*, or rushes, *Scirpus* spp. Cantrall (1968) reports it as "characteristic of alkaline situations such as northern relict marl bogs."

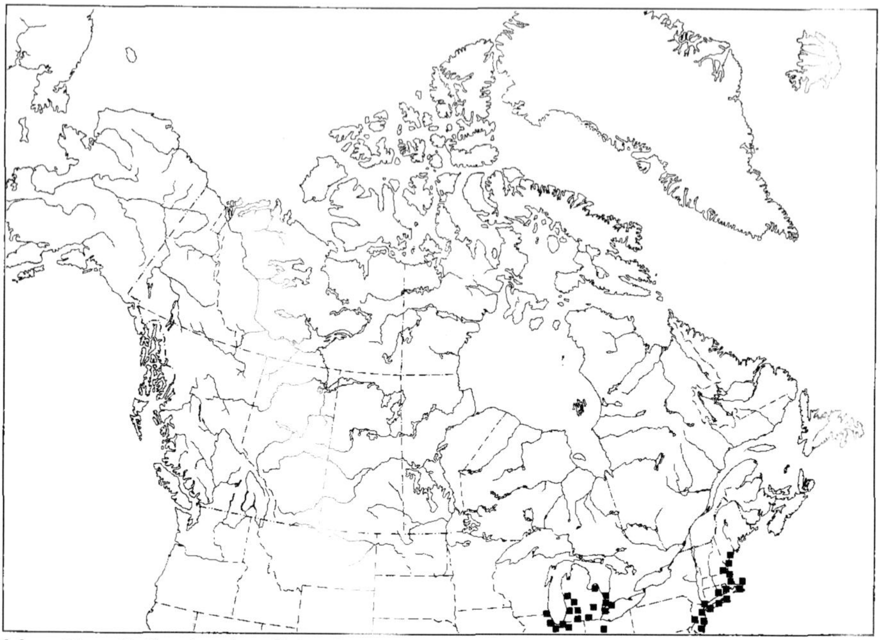

Map 61. Collection localities for *Orchelimum concinnum*.

Males stridulate at a height of about 0.6 m above the substrate. The song consists of 2–4 ticks, which can be distinguished, and of buzzes of less than 2 seconds duration, and complete phrases that are repeated, lasting slightly more than 2 seconds (R. D. Alexander et al. 1972).

Life history. Univoltine, with overwintering in the egg stage. Adults are found in Ontario in August and September.

Orchelimum campestre Blatchley

Figs. 300, 307; Plate II*G*; Map 62

Orchelimum campestre Blatchley, 1893*c*:91.

Orchelimum campestre; Lugger 1898:326; R. D. Alexander et al. 1972:45, 51; G. K. Morris and Walker 1976:789, 791.

Orchelimum concinnum campestre; Blatchley 1920:556.

Orchelimum concinnum concinnum; Cantrall 1943:140.

Diagnosis. Face uniformly dull amber or suffused with red.

Description. Color generally green. Lateral pronotal lobe as in Fig. 293. Male cercus with dorsal and ventral carinae weakly developed, with deep oval to subquadrate depression on inner face at base of tooth (Fig. 301). Ovipositor (Fig. 307) upwardly curved from middle to apex, shorter than half length of the hind femur, little acute at apex. Male stridulatory file of "Type II"

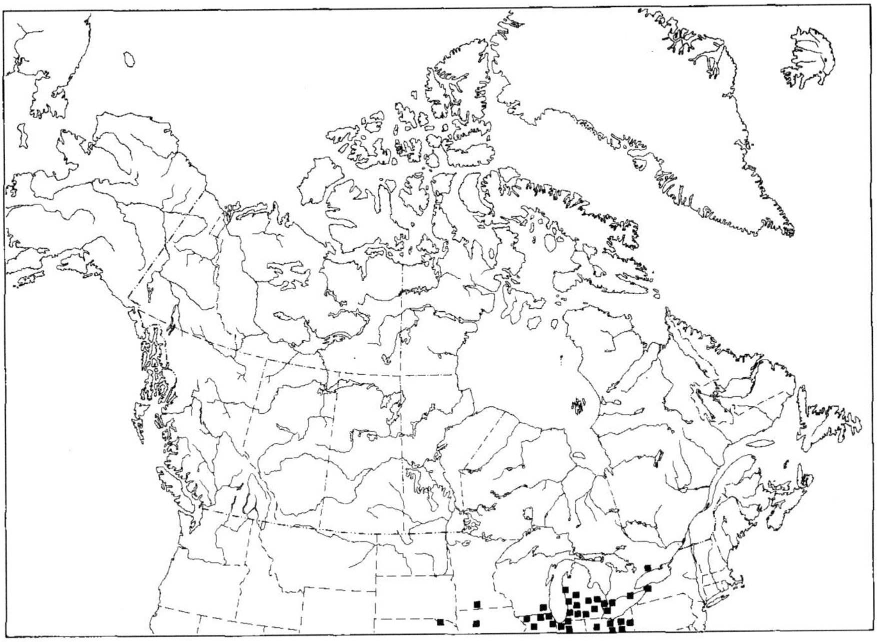

Map 62. Collection localities for *Orchelimum campestre*.

of G. K. Morris and Walker (1976), with about 66 teeth; about 1.67 mm in length, based on a single specimen (E. S. Thomas and Alexander 1962).

Range. Nebraska to southern Ontario, south to Louisiana and Tennessee.

Behavior and habitats. *O. campestre* is found generally in marshes. In Michigan and Ohio, it was found to be plentiful in vegetation over standing water, often in rice cutgrass (*Leersia oryzoides*), a common marsh grass of the area (E. S. Thomas and Alexander 1962).

Stridulation is a combination of ticks and buzzes, but differs from those of *delicatum* and *concinnum* in that it is not a "programmed" series of 2–4 ticks followed by a buzzing sequence. The stridulation of *O. campestre* may consist of as many as 150 ticks without buzzes, or continuous buzzing sequences for up to 3 minutes, without intervening ticks. Both single and double ticks are produced and pulses in buzzes are paired (E. S. Thomas and Alexander 1962). Stridulating males are often found on vertical stems usually more than 1.5 m above the substrate.

Life history. Univoltine, with wintering as eggs. Females probably oviposit in the stems of vegetation upon which they are found. It is probable that the stems are submerged or partly submerged in water for part of the time before the eggs hatch in spring. Adult specimens have been noted in the field during August in Ontario (Vickery and Kerr 1975).

Genus *Conocephalus* Thunberg

Description. Similar to *Orchelimum* but smaller (averaging less than 16 mm long) and more slender (Fig. 309). Prosternal spines short or lacking. Ovipositor usually straight or nearly so, occasionally slightly curved. Tegmina and hind wings fully developed or not; dimorphism for tegminal and hind wing length common in some species.

Key to species of *Conocephalus*

1. Hind tibia with 3 pairs of spurs apically (Fig. 311). Prosternum with paired spines. Tegmina distinctly longer than pronotum (except female of *strictus*). Male cercus with ventral tooth before or at middle (Figs. 313–318) 2

   Hind tibia with 1 pair of spurs apically (Fig. 312). Prosternum spineless. Tegmina padlike, hardly longer than pronotum. Male cercus with long inner spine situated well beyond middle (Fig. 319) . . . . ***saltans* (Scudder)** (p. 212)

2(1). Male cerci neither narrow and acuminate nor swollen about middle, armed on inner margin, near or before middle, with stout tooth plainly visible from above (Figs. 313, 314). Small slender species (females less than 14 mm long, excluding ovipositor) 3

   Male cerci either long and acuminate (Fig. 315), or swollen about the middle, flattened at the apex and with a subventral tooth not fully visible from above (Figs. 316–318). Larger species (females more than 15 mm long, excluding ovipositor) 4

3(2). Tegmina as long as or longer than abdomen. Male cercus slender, depressed at tip but not strongly flattened on inner face (Fig. 313). Ovipositor straight to slightly curved (Fig. 320). Slender species . . . . *fasciatus* (De Geer) (p. 213)

   Tegmina usually shorter than abdomen (macropterous specimens rare). Male cercus robust, strongly flattened on inner face (Fig. 314). Ovipositor usually straight (Fig. 321) or bent slightly at basal fifth. Robust species . . . . ***brevipennis* (Scudder)** (p. 215)

4(2). Male cercus narrow, tapered, not swollen at middle nor flattened apically, with apex acute and with strong inner tooth (Fig. 315). Head, sides of body, and femora green . . . . ***strictus* (Scudder)** (p. 216)

   Male cercus swollen at middle, flattened apically, with a subventral tooth (Figs. 316–318). Body not colored as above 5

5(4). Abdomen of male pale brown, with apical half pale orange. Male cercus with tooth small, situated about one-fourth distance from base (Fig. 318). Ovipositor short (less than 12 mm), slightly curved (Fig. 325) . . . . ***spartinae* (Fox)** (p. 217)

   Abdomen of male pale brown or black, but not pale orange apically. Male cercus with tooth large or small, situated about one-third distance from base. Ovipositor longer than 16 mm, more or less straight (Figs. 323, 324) 6

6(5). Sides of abdomen shining black. Male cercus slightly pointed, with strong tooth (Fig. 316). Ovipositor moderately long (Fig. 323) . . . . ***nigropleurum* (Bruner)** (p. 218)

   Sides of abdomen brownish to reddish brown. Male cercus blunter, with weaker tooth (Fig. 317). Ovipositor longer (Fig. 324) . . . . ***attenuatus* (Scudder)** (p. 220)

Clé des espèces de *Conocephalus*

1. Tibias postérieurs armés de trois paires d'épines apicales (fig. 311). Prosternum armé d'une paire d'épines. Tegmina clairement plus longs que le pronotum (sauf chez la femelle de *strictus*). Cerques des mâles munis d'une dent ventrale en position proximale ou médiane (fig. 313 à 318) 2

   Tibias postérieurs armés d'une paire d'épines apicales (fig. 312). Prosternum inerme. Tegmina en forme de tampon, à peine plus longs que le pronotum. Cerques des mâles armés d'une longue épine intérieure en position distale (fig. 319) .... ***saltans* (Scudder)** (p. 212)

2(1). Cerques des mâles ni étroits, ni aigus, ni renflés à mi-longueur; armés, sur le bord intérieur, en position proximale ou médiane, d'une forte dent clairement visible du dessus (fig. 313 et 314). Espèces petites et grêles (femelles de longueur inférieure à 14 mm, ovipositeur exclu) 3

   Cerques des mâles longs et pointus (fig. 315) ou renflés à mi-longueur, aplatis à l'apex et armés d'une dent subventrale non entièrement visible du dessus (fig. 316 à 318). Espèces plus grandes (femelles de longueur supérieure à 15 mm, ovipositeur exclu) 4

3(2). Tegmina aussi longs ou plus longs que l'abdomen. Cerques des mâles minces, aplatis à l'extrémité, mais à face interne non fortement aplatie (fig. 313). Ovipositeur rectiligne ou légèrement incurvé (fig. 320). Espèce grêle .... ***fasciatus* (De Geer)** (p. 213)

   Tegmina habituellement plus courts que l'abdomen (spécimens macroptères rares). Cerques des mâles robustes, à face interne fortement aplatie (fig. 314). Ovipositeur en général rectiligne (fig. 321) ou légèrement incurvé au cinquième proximal. Espèce robuste .... ***brevipennis* (Scudder)** (p. 215)

4(2). Cerques des mâles minces, allant s'amenuisant, non renflés à mi-longueur ni aplatis à l'apex; apex aigu; forte dent dirigée vers l'intérieur (fig. 315). Tête, côtés du corps et fémurs verts .... ***strictus* Scudder)** (p. 216)

   Cerques des mâles renflés à mi-longueur, aplatis à l'apex et armés d'une dent subventrale (fig. 316 à 318). Coloration du corps différente de celle décrite ci-dessus 5

5(4). Abdomen du mâle brun pâle et à moitié apicale orangé pâle. Cerques armés d'une petite dent, au quart de leur longueur à partir de la base (fig. 318). Ovipositeur court (moins de 12 mm) et légèrement incurvé (fig. 325) .... ***spartinae* (Fox)** (p. 217)

   Abdomen du mâle brun pâle ou noir, mais sans coloration orangé pâle sur la moitié apicale. Cerques des mâles munis d'une dent grande ou petite, située au tiers proximal de leur base. Ovipositeur de longueur supérieure à 16 mm, plus ou moins rectiligne (fig. 323 et 324) 6

6(5). Côtés de l'abdomen noir luisant. Cerques des mâles plus ou moins pointus, armés d'une forte dent (fig. 316). Ovipositeur modérément long (fig. 323) .... ***nigropleurum* (Bruner)** (p. 218)

   Côtés de l'abdomen brunâtre à brun rougeâtre. Cerques des mâles obtus, armés d'une dent plus faible (fig. 317). Ovipositeur plus long (fig. 324) .... ***attenuatus* (Scudder)** (p. 220)

Conocephalus saltans (Scudder)

Figs. 312, 319, 326; Map 63

Xiphidium saltans Scudder, 1872:249.
Xiphidium modestum Bruner, 1891*a*:56.
Xiphidium taeniatum Redtenbacher, 1891:498, 520.
Conocephalus saltans; J. A. G. Rehn and Hebard 1910*b*:9; R. D. Alexander et al. 1972:45, 52.
Conocephalus viridifrons Blatchley, 1920:583.

Diagnosis. Hind tibia with single pair of apical spurs (Fig. 312).

Description. Body small (pronotal length, male and female, 3–3.5 mm). Prosternum lacking spines; lateral pronotal lobes with anteroventral angle obsolete and with posteroventral angle obtusely rounded; lateral sinus evident but shallow; convex callosity prominent. Tegmina normally short, covering about one-quarter of abdomen, with tips rounded (rarely macropterous, slightly surpassing apices of hind femora; hind wings extending about 5 mm beyond apices of tegmina). Hind tibia with single pair of apical spurs. Male cercus with apical part longer than basal part, distinctly twisted, incurved, with inner face flattened near apex and with tooth long, sharp, bent inward and downward, inserted beyond middle (Fig. 319). Ovipositor (Fig. 326) longer than hind femur, straight, tapering abruptly in apical quarter. Color generally dull reddish brown; dorsal stripe dark brown bordered by pale yellow; sides of abdomen with narrow yellow stripes.

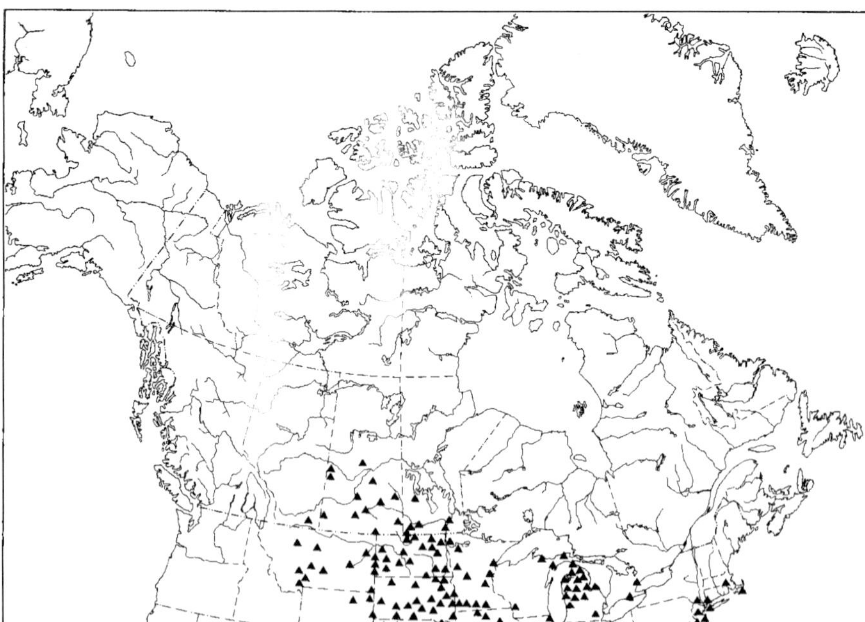

Map 63. Collection localities for *Conocephalus saltans*.

Range. Alberta to southern Ontario, south to New Mexico and Florida.

Behavior and habitats. *C. saltans* is often found near marshy areas but also in open grassy uplands on sandy soils. Specimens have been taken in tall prairie grasses and various other types of vegetation. E. M. Walker (1904*b*) found *C. saltans* to be numerous in open grassy uplands, on sandy soil at High Park, Toronto. Blatchley (1920) states that it appears to be xerophilous, occurring "mainly on dry upland prairies and sandy barrens." No information is available on stridulation.

Life history. Univoltine, with wintering in the egg stage. Adults are present from late July to early October.

Conocephalus fasciatus (De Geer)

Figs. 310, 311, 313, 320; Map 64

Locusta fasciata De Geer, 1773:458, pl. 40, fig. 4.

Orchelimum gracile T. W. Harris, 1841:131.

Xiphidium vicinum Morse, 1901:203.

Conocephalus fasciatus; J. A. G. Rehn and Hebard 1910*b*:9; Cantrall 1943:141; Helfer 1963:266; R. D. Alexander et al. 1972:33; Vickery et al. 1974:67.

Conocephalus fasciatus fasciatus; J. A. G. Rehn and Hebard 1915*c*:170–177; Cantrall 1943:141.

Conocephalus fasciatus vicinus; J. A. G. Rehn and Hebard 1915*c*:177, 180; Helfer 1963:267.

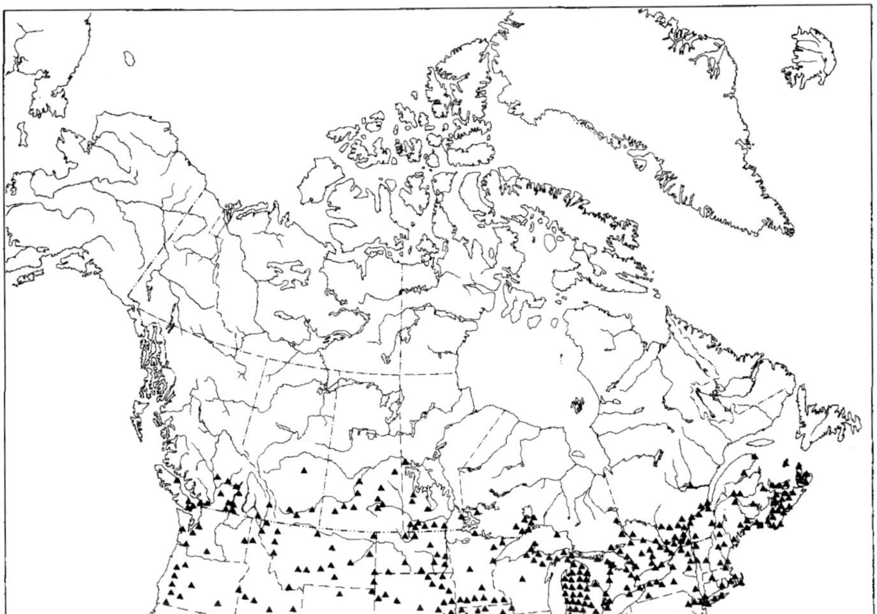

Map 64. Collection localities for *Conocephalus fasciatus*.

Diagnosis. Tegmina fully developed. Male cercus with stout tooth on inner face near base. Ovipositor long, straight.

Description. Body small, slender (pronotal length, male 3.0–3.5, female 2.8–3.2 mm). Lateral lobes of pronotum deeper than long; posterior margin strongly oblique, with humeral sinus not evident and with convex callosity low and narrow. Tegmina extending beyond hind femur; wings 2–3 mm longer than tegmina. Male cercus (Fig. 313) with stout tooth on inner face, slightly nearer base than apex; apex somewhat depressed, rounded. Ovipositor (Fig. 320) straight, about two-thirds length of hind femur. Occiput, pronotal disc, and dorsum of abdomen brown; face, pronotum laterally, legs, and basal third of ovipositor green; tegmina and apical two-thirds of ovipositor reddish brown.

No brachypterous specimen of *C. fasciatus* is known. The ovipositors of western specimens are proportionally longer than in those from the east. This is a clinal geographic effect, as shown by Stainer (1975), but was the basis for the subspecific name *vicinus* Morse, which persisted in the literature for many years.

Range. British Columbia to Nova Scotia, south to Mexico; Bermudas.

Behavior and habitats. Often abundant in dry upland fields as well as in damp locations near bodies of water, *C. fasciatus* is found in a wide variety of habitats; it shows, however, much less hygroscopic affinity than species of *Orchelimum*. It will also eat the flowers of Graminaceae, and Whitcomb (1967) reported predation by *C. fasciatus* on second-instar cotton bollworm larvae in Arkansas.

R. D. Alexander et al. (1972) described the calling song of *C. fasciatus* as "soft sound, noticeable only a few feet away. . .10 to 30 second buzzes, with 10 to 25 ticks between them." Allard (1911) used the term "staccato lisps" in referring to the ticks, which are very faint.

Briand and Rivard (1964) reported parasitism of this species in Québec by a mermithid nematode parasite, presumably *Agamermis decaudata* Cobb. Evans and Kurczewski (1966) reported that a sphecid wasp, *Tachytes crassus* Patton, captured large numbers of this and other species of *Conocephalus* and *Orchelimum*, as larval food. A female wasp straddled a grasshopper, which was held head forward and dorsum uppermost, grasped the bases of the antennae with her mandibles, then, in flight, grasped the body of the prey with all her legs.

Life history. Vickery (1961) reported adults from 18 July to 22 September in Nova Scotia. Vickery et al. (1974) found adults in Québec from mid-July until the insects were killed by frost. In the northern part of the range, *C. fasciatus* is univoltine, passing the winter in the egg stage.

Economic importance. A. Gibson (1918) reported injury to field corn by this species in 1917 near Norway Bay, Qué. The male flowers were favored. It cannot, however, be regarded as more than a sporadic and very minor pest.

Conocephalus brevipennis (Scudder)

Figs. 314, 321; Map 65

Xiphidium brevipenne Scudder, 1862:285.
Xiphidium ensifer Scudder, 1863*a*:451.
Xiphidium ensiferum F. Walker, 1869*b*:270.
Xiphidium gossypii Scudder, 1875*a*:462.
Xiphidium fasciatum var. *brevipennis*; Caulfield 1888:64, 70.
Conocephalus brevipennis; J. A. G. Rehn and Hebard 1915*c*:182; Cantrall 1943:142; R. D. Alexander et al. 1972:45, 52; Vickery et al. 1974:68.

Diagnosis. Tegmina short. Male cercus short, with tooth prominent, depressed on inner side. Ovipositor straight to slightly curved.

Description. Body size medium, rather robust. Fastigium scarcely ascending, with sides feebly divergent; apex narrower than basal antennal segment. Lateral pronotal lobes as deep as long, with lower margin oblique; posteroventral angle broadly rounded; humeral sinus nearly obsolete. Tegmina short, in male usually reaching bases of cerci, in female covering about two-thirds of abdomen. Male cercus rather short, with tooth prominent (Fig. 314). Ovipositor (Fig. 321) straight or slightly curved upward. Color pale reddish brown, with face and sides of pronotum usually green; occiput and pronotal disc with dark brown stripe having yellow margins; tegmina and wings pale reddish brown; ovipositor same but darker at apex. Macropterous specimens rare.

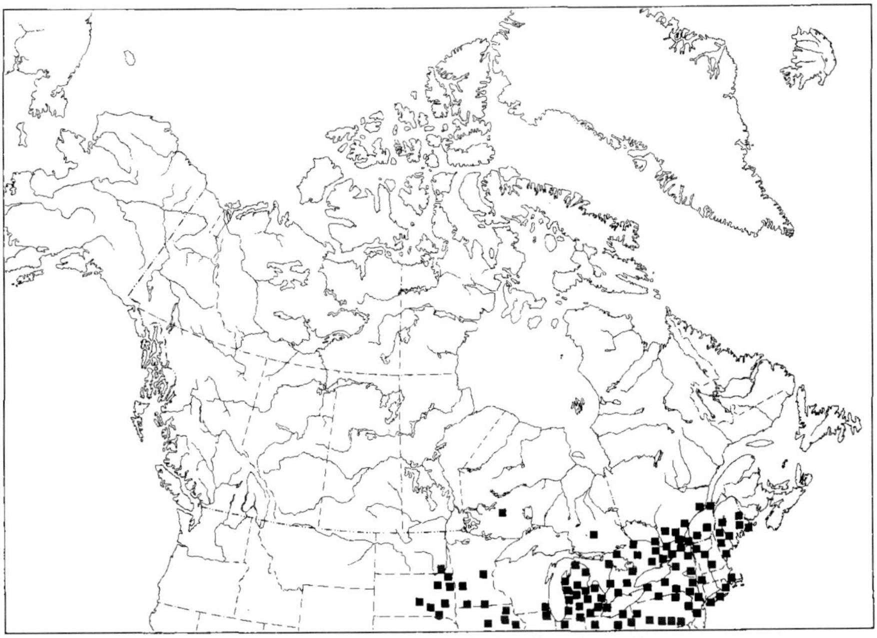

Map 65. Collection localities for *Conocephalus brevipennis*.

Range. North Dakota to Québec and Maine, south to Texas and Florida.

Behavior and habitats. Species of *Conocephalus* usually oviposit in stems of grasses. *C. brevipennis*, in addition to laying eggs in grass stems, often oviposits in willow galls, which have been formed by the plants as a result of oviposition by dipterous flies of the family Cecidomyiidae. The eggs are pushed between the scales on the surface of the galls (Blatchley 1920). Other species of the genus in Europe also oviposit in galls. E. M. Walker (1904*b*) found *C. brevipennis* in Algonquin Park, Ont., in rank herbs and bushes in more or less shady spots. The authors have found it to be common in similar habitats in the Morgan Arboretum, Macdonald College, Sainte-Anne-de-Bellevue, Qué. The species seems to inhabit locations of higher humidity, such as swamps and river margins, much more than does *C. fasciatus*.

The calling song of *C. brevipennis* consists of 2–4 ticks and buzz pulses, 3–5 complete phrases in 5 seconds; it is very faint. Males usually stridulate on vegetation about 0.25 m above the ground (R. D. Alexander et al. 1972).

Certain sphecid wasps provision their nests with *C. brevipennis. Tachytes validus* Cresson was reported to take large numbers in Kansas (Kurczewski and Ginsburg 1971). They found more males captured than females, so that the predators might have been using male stridulation to locate their prey.

Life history. Univoltine, with wintering in the egg stage. Adults are found from July to October. In Québec, Larochelle (1978) reported adults from 6 July to 14 October.

Conocephalus strictus (Scudder)

Figs. 315, 322; Map 66

Xiphidium strictum Scudder, 1875*a*:[illegible]0.

Conocephalus strictus; J. A. G. Rehn and Hebard 1915*c*:193; Cantrall 1943:143; Helfer 1963:266; R. D. Alexander et al. 1972:33.

Diagnosis. Tegmina short. Apical part of male cercus long, tapered (Fig. 315). Ovipositor (Fig. 322) slender, long (more than 14 mm). Hind femur lacking spines.

Description. Body large (pronotal length, male 3.5–4, female 4–4.5 mm). Lateral pronotal lobes deeper than long, with anterior and lower margins straight; humeral sinus shallow but evident. Tegmina normally short, in male covering about half of abdomen, in female padlike, covering about one-third of abdomen. Male cercus with apical part elongate (1.5 times to twice as long as tooth), slender, tapered, subacute, somewhat flattened (Fig. 315). Ovipositor (Fig. 322) slender, straight, (more than 14 mm long). General coloration green with dorsal longitudinal reddish brown stripe having narrow whitish border on occiput; tegmina reddish brown; cerci and ovipositor dark brown. Macropterous form rare, constituting less than 2% of a population.

Range. Montana to southern Ontario, south to Arizona and Georgia.

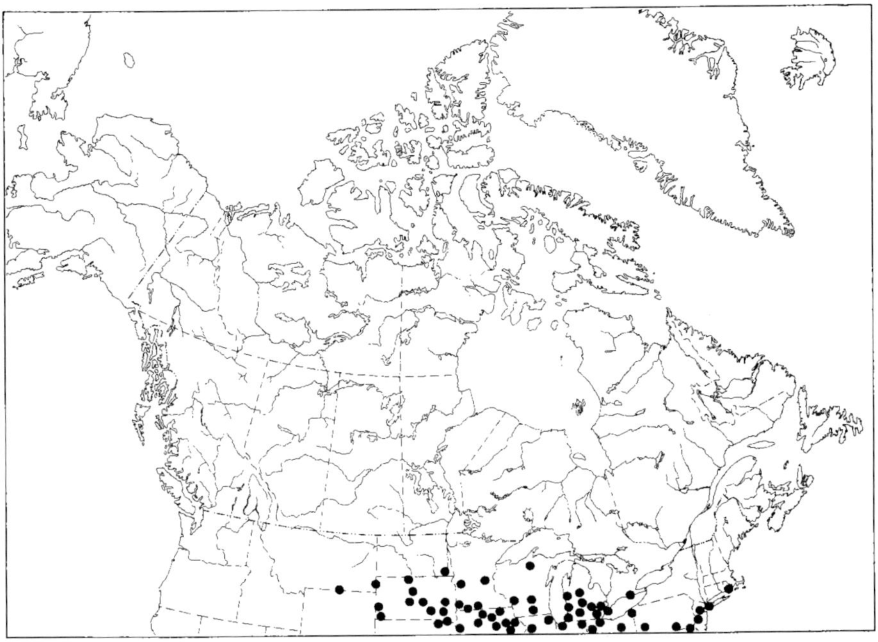

Map 66. Collection localities for *Conocephalus strictus*.

Behavior and habitats. E. M. Walker and Urquhart (1940) found *C. strictus* inhabiting rather dry sandy pastures but not low-lying humid localities. It occurs in grassy and weedy areas, even in open pine woods.

R. D. Alexander et al. (1972) described the stridulation of *C. strictus* as "soft seedy [*sic*], continuous buzzes audible only a few feet away," lacking ticks, with the buzz changing speed, alternately slowing and speeding up about every 15–30 seconds. The main carrier frequency of the stridulation is about 40 kHz. The tympanic organs are capable of receiving ultrasonic sounds. Weever and Vernon (1959) found that the tympanic nerves of *C. strictus* respond to signals up to 80–120 kHz with maximum sensitivity at 10–40 kHz.

Life history. Univoltine, with wintering in the egg stage. Adults have been collected in Ontario from 24 July to 25 September.

Conocephalus spartinae (Fox)

Figs. 318, 325; Map 67

Xiphidium spartinae H. Fox, 1912*a*:111.
Conocephalus spartinae; J. A. G. Rehn and Hebard 1915*c*:212.

Diagnosis. Male cercus blunt, flattened apically, with tooth basal, small. Ovipositor about length of hind femur on tidal flats and salt-marsh grasses.

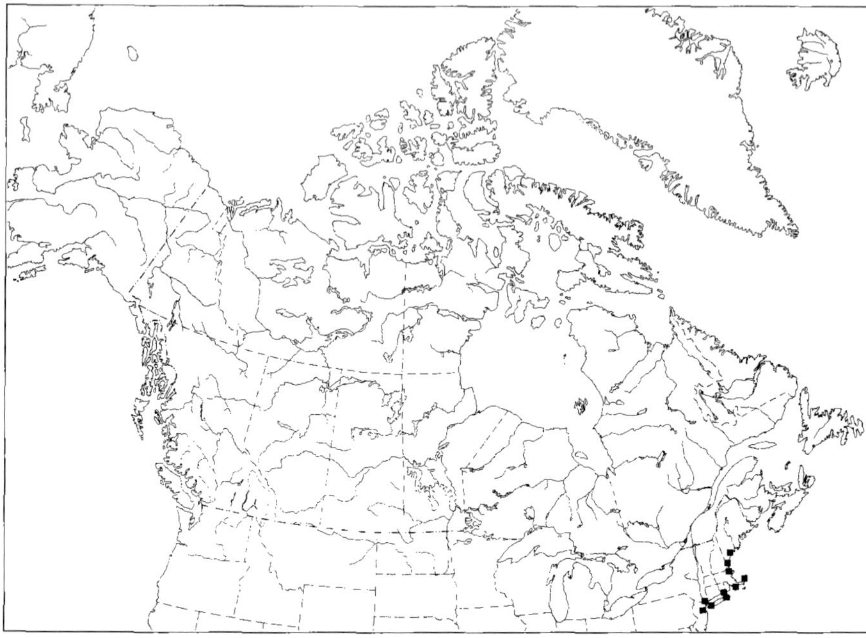

Map 67. Collection localities for *Conocephalus spartinae.*

Description. Body small, slender (pronotal length, male 2.3–3, female 2.5–3 mm). Lateral pronotal lobes with anterior margin rounded to lower margin; posterior angle broadly rounded; humeral sinus absent. Tegmina of male reaching apical abdominal tergum, of female covering about three-quarters of abdomen; tips narrowly rounded (occasional specimens macropterous). Hind femur with 1–4 spines beneath. Male cercus (Fig. 318) slender, swollen in middle, flattened in apical half, incurved, with tooth small, ventral, near base. Ovipositor (Fig. 325) as long as hind femur or slightly shorter (7.1–9.9 mm), broad, slightly curved upward. Color generally grass green to pale brown; tegmina brown, tinged with green near apices; dorsal stripe dark brown; lateral pronotal lobes brown, bordered by yellow.

Range. Maine along the coast to Florida and Texas; Bermudas.

Behavior and habitats. *C. spartinae* is often found on or near tidal flats in coastal areas. Many are found on *Spartina*, a salt-marsh grass.

Life history. Univoltine, with wintering only in the egg stage, so far as is known. In northern localities, adults have been found from 23 August to 9 September in Maine (Morse 1921).

Conocephalus nigropleurum (Bruner)

Figs. 316, 323; Map 68

Xiphidium nigropleurum Bruner, 1891*a*:58.

Conocephalus nigropleurum; Hebard 1925*a*:136; Cantrall 1943:144; R. D. Alexander et al. 1972:45, 52; Vickery et al. 1974:70.

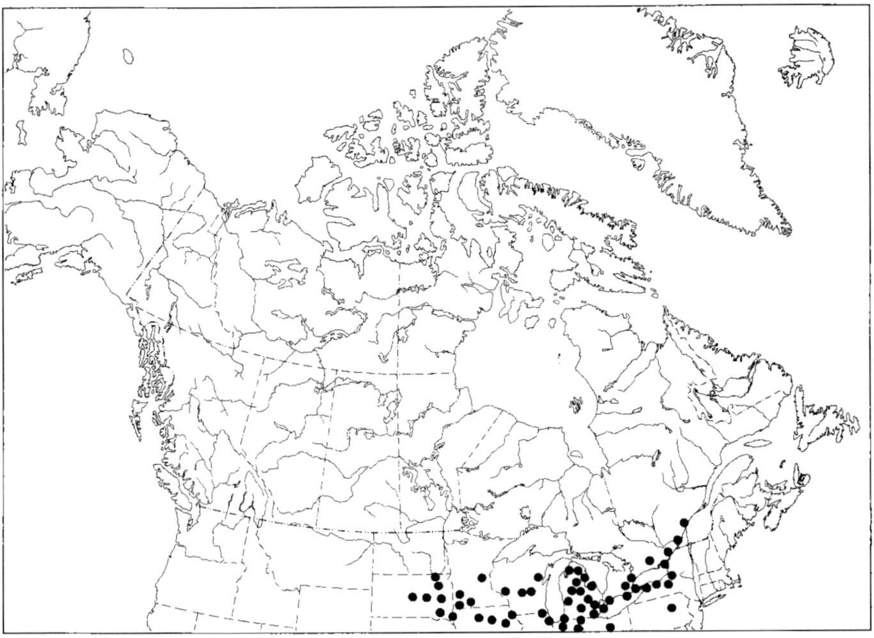

Map 68. Collection localities for *Conocephalus nigropleurum*.

Diagnosis. Sides of head and abdomen black.

Description. Body size medium (pronotal length, male 3.0–3.5, female 3.2–3.6 mm). Lateral pronotal lobes with anterior edge broadly rounded to ventral edge and with humeral sinus obsolete. Tegmina normally short, in male covering four-fifths, in females two-thirds, of abdomen. Male cercus swollen at middle, with slender ventral tooth; apical part flattened (Fig. 316). Ovipositor (Fig. 323) straight, rather broad, slightly longer than hind femur. Color dimorphic; either tegmina, legs, margins of lateral pronotal lobes and metazona bright grass green, or these parts brownish yellow; dorsal brownish stripe clearly defined in green form, faint in brown form; abdomen shining black laterally. Macropterous specimens rare (in such specimens, tegmina surpassing apices of hind femora by about 1.5 mm, and hind wings extending about 3.5 mm beyond tegmina).

Range. South Dakota and Kansas to Québec and Pennsylvania.

Behavior and habitats. The characteristic habitat includes wet, permanent, and semipermanent marshes, and areas near streams, ponds, and lakes. Specimens have been collected in tall rank grasses and sedges in southwestern Québec. Cantrall (1943) reports occurrence in similar situations, but also higher up on bushes and shrubbery. That author recorded 75% of specimens to be of the green, and 25% of the brown phase in southern Michigan; all females were green.

The stridulation lacks the tick phase; the buzz sequence is steady and even, not varying in speed, and may last up to 12 minutes. The intense carrier frequency lies between 28 and 60 kHz. This varies slightly (about 5 kHz) during the prolonged sequence (Pipher and Morris 1974), but this is not noticeable to the human ear. G. K. Morris (1971) described aggressive interactions between males.

The sphecid wasp *Tachytes validus* Cresson has been found to be predaceous on *C. nigropleurum* near Aubrun, NY, using this species extensively to stock its nests (Kurczewski and Ginsberg 1971).

Life history. Univoltine, with overwintering in the egg stage. Like *C. brevipennis, C. nigropleurum* females often oviposit in galls of willow (*Salix* sp.). Pipher and Morris (1974) found that, in Schuyler County, NY, most of the galls containing eggs, contained eggs of *C. nigropleurum* rather than of *C. brevipennis*. Adults have been collected from 1 July to 24 October in Indiana (Blatchley 1920). In Québec, specimens have been taken from 8 to 26 August, but there is no doubt that adults are present in the field much longer than this.

Conocephalus attenuatus (Scudder)

Figs. 317, 324; Map 69

Xiphidium attenuatum Scudder, 1869*a*:305.

Xiphidium scudderi Blatchley, 1892*a*:26.

Conocephalus attenuatus; J. A. G. Rehn and Hebard 1915*c*:207; R. D. Alexander et al. 1972:45, 52; Vickery et al. 1974:69.

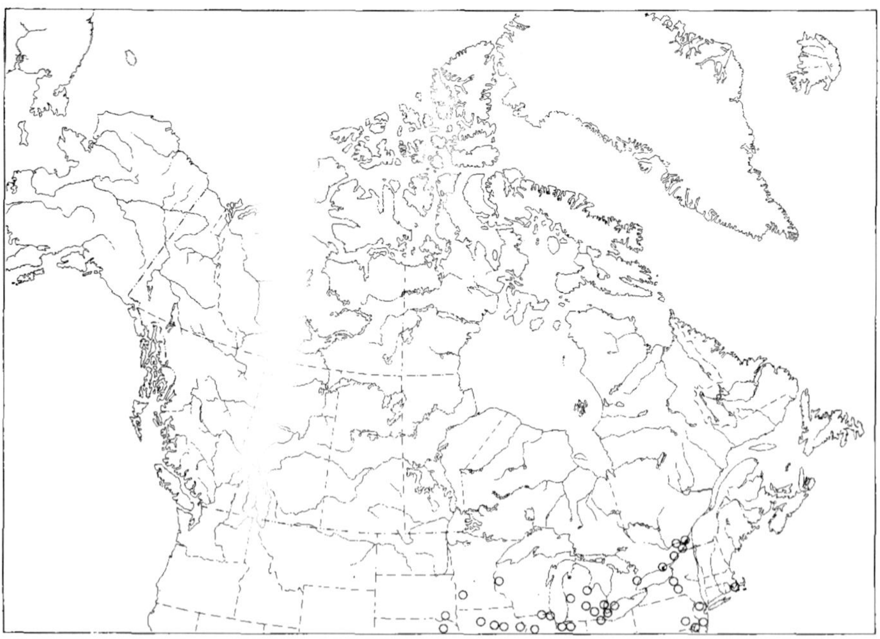

Map 69. Collection localities for *Conocephalus attenuatus*.

Diagnosis. Male cercus broad, with tooth basal, tiny. Ovipositor nearly twice length of body. Hind femur with spines beneath.

Description. Ovipositor long (1.5 times to nearly twice as long as entire body). Body moderately large, slender (pronotal length, male 2.6–3.0, female 2.9–3.2 mm). Lateral pronotal lobes with posteroventral angle sharply rounded and with margins strongly upcurved; humeral sinus obsolete. Tegmina and hind wings commonly dimorphic, with tegmina in brachypterous form in male covering entire abdomen, in female three-quarters of abdomen, in macropterous forms extending twice length of abdomen, about 4 mm shorter than hind wings. Hind femur long, slender, with 1–5 spines on outer lower carina. Male cercus broad, with apical third curved outward, with tooth basal, tiny, slender (Fig. 317). Ovipositor (Fig. 324) long, slender, nearly straight. Color dull yellowish brown, in some specimens tinged with green; dorsal stripe reddish brown; abdomen of male pale brown, of female reddish brown.

Range. South Dakota and Nebraska to Québec and Maryland.

Behavior and habitats. *C. attentuatus* lives among rank grasses, sedges, and rushes growing in shallow water along river banks, lake shores, and marshes (Cantrall 1968). In Québec, most of the specimens at hand were collected by boat, as it was difficult to approach the habitat from land. Eggs are laid between stems and leaves of the tall plants where the insect lives (Blatchley 1920).

The stridulation is very much like that of *C. nigropleurum*, that is, soft "seedy" continuous buzzes continued uniformly for indefinite periods. The song, as well as the habitat, is similar to that of *nigropleurum*, but, in Québec, the two species have not been found together.

Life history. Univoltine, with overwintering in the egg stage. The adult stage is present from 28 July to 26 September in Michigan (Cantrall 1968). All specimens known from Québec were collected in August.

Family Tettigoniidae

This family includes the true tettigoniids, or "great green" grasshoppers of Europe, the false "cigales" and the true "sauterelles" in French, the wart-biters and other "decticids," shield-back katydids, and the "kirigirisu" of Japan. Members of the family are of diverse form. They are, however, distinguishable from other members of the superfamily by means of the following combination of characters: head generally short, with narrower vertex forming short, truncated, or rounded rostrum. Antennae usually longer than body, inserted between the eyes, their insertions ("scrobes") but weakly marginated. Prothoracic spiracle large, elongate, largely concealed by pronotum; prosternum with pair of spines. Anterior tibiae usually with apicodorsal spine; basal auditory organ covered; hind tibiae with apical spines on either side (unless front and middle legs heavily spined below); tarsi with first and second segments laterally grooved. Wings fully developed or reduced (rarely entirely lacking and then only in females); tegmina, when fully developed, often narrowly and elongately leaflike. Ovipositor usually stout, swordlike or daggerlike, straight or somewhat upwardly curved.

Approximately 80 genera and 375 species are known, the great majority falling into the subfamily Tettigoniinae, *sensu lato*. Most of these belong to the tribe Decticini, the members of which generally have a somewhat wider rostrum, are brownish or variegated, rather than more uniform green, in coloration, and, more often than not, are brachypterous. Tettigoniinae species in general are predominantly Palaearctic. A second subfamily, Onconotinae, comprises two flightless species in a single genus from southeastern Europe and southwestern Siberia. Members of the third subfamily, Saginae (recognizable by the raptorial front and middle legs), consisting of about 12, mainly flightless, genera and about 40 species (some with completely wingless females), are known mainly from the eastern Mediterranean and southwestern Asia, although a few are known from other parts of the Palaearctic region and in eastern and southern Africa; one genus each is recorded for Australia and Chile (the last is particularly in need of confirmation); one European species became established in Michigan and is further referred to in the present work.

Members of the Tettigoniidae are mainly ground-living, except for those belonging to the tribe Tettigoniini and the subfamily Onconotinae, which live chiefly on shrubs and bushes. The great majority lay their eggs in the soil, even if they are not ground-dwellers. Some Saginae are parthenogenetic. Most, if not all species are, to a greater or lesser extent, carnivorous, the Saginae being exclusively voracious predators of fairly large insects. Many other Tettigoniidae species do, however, consume some vegetable matter, and a few, such as the Mormon cricket, *Anabrus simplex* Haldeman, and the coulee cricket, *Peranabrus scabricollis* (Thomas), of western North America, are prone to become excessively abundant, when they may form "armies" and become pests of great, if sporadic, importance. However, species of *Tettigonia* are among the favorite singing insects of Europe, where they may be caged for their song. Species of the decticine genus *Gampsocleis*, notably *G. buergersi* (Haan), are similarly a favorite singer in the Orient. This last is called "kirigirisu" in Japan.

Key to subfamilies of Tettigoniidae

1. Front and middle tibiae and femora strongly spined below; tibiae without apicodorsal spines .... **Saginae** (p. 223)
   Tibiae and femora not strongly spined below; tibiae with 1 or 2 apicodorsal spines .... **Tettigoniinae** (p. 224)

Clé des sous-familles de Tettigoniidae

1. Tibias et fémurs antérieurs et médians à face ventrale pourvue de nombreuses épines; tibias dépourvus d'épines apico-dorsales .... **Saginae** (p. 223)
   Tibias et fémurs à face ventrale ne portant pas beaucoup d'épines, tibias armés d'une ou deux épines apico-dorsales .... **Tettigoniinae** (p. 224)

Subfamily Saginae

The characteristics of this subfamily are briefly indicated under the family Tettigoniidae and in the key to subfamilies. A single introduced species concerns us here.

Genus *Saga* Charpentier

Saga pedo (Pallas)

Fig. 276; Map 70

Gryllus (*Tettigonia*) *pedo* Pallas, 1771:467.
Gryllus (*Tettigonia*) *giganteus* Villiers, 1789:451, pl. 3, fig. 7.
Locusta serrata Fabricius, 1793:43.
Saga pedo; Yakobson and Bianki 1903:388.

Saga pedo was reported as an introduction from Europe into Jackson County, MI, in 1970 (Cantrall 1972); Cantrall hypothesized that eggs were introduced on poorly decontaminated plowshares that had been used in Europe in plowing contests. Only six specimens were found from 1970 to 1972, in August and September, and none has been seen since. It is probable that the species did not survive much longer in Michigan and that it is now extinct in North America (I. J. Cantrall, pers. com. 1977).

Saga pedo is parthenogenetic; no male is known. It is not phytophagous, and is well known to be a voracious predator. Cantrall (1972) reported that a captive Michigan specimen fed avidly on grasshoppers.

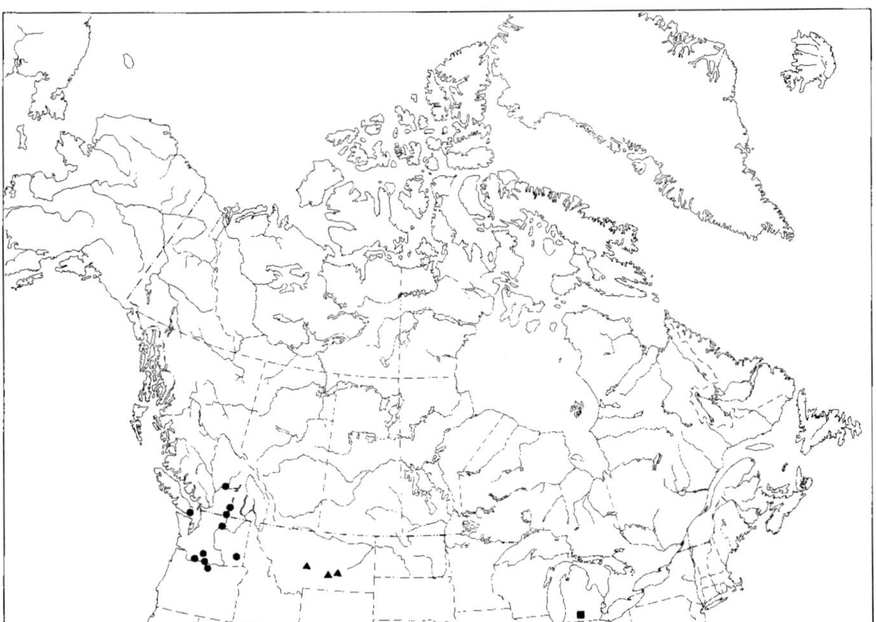

Map 70. Collection localities for *Apote robusta* (●), *Pediodectes stevensonii* (▲), and *Sago pedo* (■).

Subfamily Tettigoniinae

Key to genera of Tettigoniinae, tribe Decticini

1. Prosternum armed with pair of spines 2
   Prosternum without spines 5

2(1). Lateral carinae of pronotum absent or obscure 3
   Lateral carinae of pronotum sharp, distinct 4

3(2). Hind femur short, not or but little surpassing abdomen. Ovipositor decurved (Fig. 245) .... ***Apote*** (p. 225)
   Hind femur long, considerably surpassing abdomen. Ovipositor curved slightly upward .... ***Pediodectes*** (p. 226)

4(2). Hind tibia armed beneath with 4 apical spines (Fig. 248). Ovipositor straight. Found in eastern United States and southern Ontario .... ***Atlanticus*** (p. 226)
   Hind tibia armed beneath with 2 apical spines (Fig. 240). Ovipositor curved upward. Found in Pacific region .... ***Neduba*** (p. 231)

5(1). Pronotum without lateral carinae on anterior half, or these indicated only by color 6
   Pronotum with lateral carinae obvious 7

6(4). Pronotal disc smooth. Fore tibia armed above on both margins. Male cercus apically furcate, with lower branch long, acute ..... ***Anabrus*** (p. 233)
   Pronotal disc rough, scabrous. Fore tibia armed above on outer margin only. Male cercus apically not furcate, with lower branch short .... ***Peranabrus*** (p. 239)

7(5). Lateral carinae of pronotum bowed inward on anterior half, divergent posteriorly (Figs. 264, 266); pronotal disc strongly divergent posteriorly. Median carina scarcely evident 8
   Lateral carinae of pronotum parallel or nearly so; pronotal disc little, if at all, divergent posteriorly. Median carina distinct, continuous .... ***Steiroxys*** (p. 241)

8(6). Pronotum distinctly saddle-shaped (Fig. 264), twice as broad posteriorly as anteriorly, with lateral lobes bent downward at sharp angle. Ovipositor regularly curved at base, tapered gradually (Fig. 271). Found in northern boggy areas .... ***Sphagniana*** (p. 242)
   Pronotum not saddle-shaped (Fig. 266), not twice as broad posteriorly as anteriorly, with lateral lobes not sharply bent downward. Ovipositor bent upward at base, more sharply tapered to apex (Fig. 275). Found in southeastern open grassy fields .... ***Metrioptera*** (p. 244)

Clé des genres de Tettigoniinae, tribu des Decticini

1. Prosternum armé d'une paire d'épines 2
   Prosternum dépourvu d'épines 5

2(1). Carènes latérales du pronotum absentes ou peu visibles 3
   Carènes latérales du pronotum aiguës et distinctes 4

3(2). Fémurs postérieurs courts, ne dépassant pas l'abdomen ou ne le dépassant qu'à peine. Ovipositeur incurvé vers le bas (fig. 245) .... ***Apote*** (p. 225)
   Fémurs postérieurs longs, dépassant de beaucoup l'abdomen. Ovipositeur incurvé légèrement vers le haut .... ***Pediodectes*** (p. 226)

4(2). Tibias postérieurs armés, en face ventrale, de 4 épines apicales (fig. 248). Ovipositeur rectiligne. Espèces de l'est des États-Unis et du sud de l'Ontario .... *Atlanticus* (p. 226)

Tibias postérieurs armés, en face ventrale, de 2 épines apicales (fig. 240). Ovipositeur incurvé vers le haut. Espèces de la côte du Pacifique .... *Neduba* (p. 231)

5(1). Pronotum dépourvu de carènes latérales sur sa portion antérieure, ces carènes étant remplacées par des lignes de couleur différente 6

Pronotum pourvu de carènes latérales distinctes 7

6(4). Disque du pronotum lisse. Tibias antérieurs armés des deux côtés de leur face dorsale. Cerques des mâles fourchus, la branche inférieure étant longue et aiguë .... *Anabrus* (p. 233)

Disque du pronotum rugueux. Tibias antérieurs armés du côté extérieur de la face dorsale seulement. Cerques des mâles non fourchus, à branche inférieure courte .... *Peranabrus* (p. 239)

7(5). Carènes latérales du pronotum inclinées vers l'intérieur sur la portion antérieure, divergeant postérieurement (fig. 264 et 266); disque du pronotum divergeant fortement vers l'arrière. Carène médiane à peine visible 8

Carènes latérales du pronotum parallèles ou presque; disque du pronotum divergeant à peine postérieurement. Carène médiane visible, continue .... *Steiroxys* (p. 241)

8(6). Pronotum en forme de selle (fig. 264), deux fois plus large à l'arrière qu'à l'avant et à lobes latéraux fortement incurvés vers le bas. L'ovipositeur suit une courbure régulière et va s'amenuisant vers l'extrémité (fig. 271). Espèces des régions nordiques marécageuses .... *Sphagniana* (p. 242)

Pronotum non en forme de selle (fig. 266), pas deux fois plus large en arrière qu'en avant et aux lobes latéraux non fortement incurvés vers le bas. L'ovipositeur est incurvé vers le haut à la base et va s'amenuisant rapidement vers l'apex (fig. 275). Espèces des champs herbeux du sud-est .... *Metrioptera* (p. 244)

Genus *Apote* Scudder

Description. General appearance as in Fig. 245. Lateral and median carinae of pronotum present only on posterior part; lateral carinae parallel, blunt; median carina low, sharp. Hind femur not longer than twice pronotal length. Tegmina short, developed in both sexes, convex, overlapping, projecting beyond pronotum by less than length of pronotum, with veins dark and conspicuous. Male subgenital plate triangularly emarginate apically, with a pair of short blunt styles. Female subgenital plate without styles. Male cercus cylindrical with acute subapical tooth on inner face (Fig. 246). Ovipositor more than twice as long as pronotum, rather strongly curved downward (Fig. 247).

Apote robusta Caudell

Figs. 246, 247; Map 70

Apote notabilis var. *robusta* Caudell, 1907*b*:330.
Apote notabilis robusta; Gurney 1939:8, pl. (4), fig. 13.
Apote robusta; Rentz and Birchim 1968:98.

Diagnosis. Pronotum elongate, narrow; disc with V-shaped depression.

Description. As for genus. Pronotal length, male 13.5, female 13 mm. Pronotum elongate, narrow; disc rounded, with transverse sulcus at posterior quarter; marked anterior to sulcus with V-shaped depression; apex of V nearly touching sulcus. Male cercus with point of inner tooth about half the distance from base (Fig. 246). Ovipositor (Fig. 247) at most only slightly longer than hind femur. Color generally brown, marked with ash gray; lateral lobes of pronotum showing general pale posteroventral patch, but without pale ventral margin or this poorly developed; legs yellowish, with infuscations on hind femur; ovipositor yellow with black V at tip.

Range. British Columbia to Oregon.

Behavior and habitats. Virtually nothing is known for this species. It probably inhabits bushes. Rentz and Lightfoot (1976) reported great difficulty in capturing *A. notabilis* on sagebrush in Oregon. Male stridulation was reported for that species as a loud rapid "tsit-tsit-tsit" of about 4 seconds' duration at 2–3 minute intervals. Singing continued both at night and in daylight.

Life history. Unknown. Adults have been collected from June until August.

Genus *Pediodectes* Rehn & Hebard

The taxonomy of the species included in this genus is very confusing. Until a revision is undertaken, it is not possible to define the species with accuracy. One species, *P. stevensonii* (C. Thomas 1870), has been reported by Hebard (1925*a*) from Huntley and Park City, MT (Map 70).

P. stevensonii is said to be characterized as follows (Caudell 1907*b*): size small; pronotum 5–8 mm long, with prozona slightly flattened posteriorly and with posterior margin truncate. Hind femur short, with inner carina possessing few tiny spines. Male cercus with tooth situated much beyond middle, the tooth approximately as long as apical part of cercus, with apex rounded. Another species of the genus is illustrated, by way of example, in Fig. 248.

Genus *Atlanticus* Scudder

Description. Tegmina rudimentary, concealed beneath pronotum. Pronotum large, produced posteriorly, with lateral carinae sharp and distinct (Figs. 226, 249), and with median carina absent or indicated only on posterior part of pronotal disc; prosternum with pair of sharp spines. Hind tibia armed beneath with 4 apical spines (Fig. 250). Male cerci simple, cylindrical, tapered, with small tooth on inner side (Figs. 251–253). Female cerci cylindrical, tapered. Ovipositor more or less straight (Figs. 257, 259), distinctly curved only in *A. monticola* (Fig. 258).

Key to species of *Atlanticus*

1. Subgenital plate notched in both sexes, but not deeply and narrowly cleft, with lobes on either side of notch not sublanceolately produced (Figs. 254, 255). Male cerci rather short, robust, with tooth distinctly beyond middle (Figs. 251, 252) 2

Subgenital plate of both sexes deeply and narrowly cleft, with lobes on either side of cleft of female sublanceolately produced distad (Fig. 256). Male cerci rather elongate, substyliform, acuminate, with tooth a little nearer middle (Fig. 253) .... ***americanus* (Saussure)** (p. 228)

2(1). Length of exposed part of male tegmen more than half length of pronotum. Hind femur less than twice length of pronotum. Notch of female subgenital plate small, shallow, narrowly V-shaped, with lobes on either side of notch short and truncated to broadly rounded (Fig. 254). Ovipositor straight, with apex tapering only on dorsal side and with apex ventral (Fig. 257) .... ***testaceus* (Scudder)** (p. 229)

Length of exposed part of male tegmen about one-quarter length of pronotum. Hind femur twice as long as pronotum. Notch of female subgenital plate deeper, usually more U-shaped, sometimes rather broadly V-shaped, with lobes on either side of notch obtusely pointed (Fig. 255). Ovipositor usually distinctly curved, with apex tapering both dorsally and ventrally and with apex not ventral (Fig. 258) .... ***monticola* Davis** (p. 230)

Clé des espèces d'*Atlanticus*

1. Plaque sous-génitale encochée chez les deux sexes, mais l'encoche n'est ni profonde ni étroite; lobes de part et d'autre de cette encoche non sublancéolés (fig. 254 et 255). Cerques des mâles plutôt courts, robustes et pourvus d'une dent sur leur portion proximale (fig. 251 et 252) 2

Plaque sous-génitale des deux sexes marquée d'une encoche profonde et étroite; lobes de part et d'autre de cette encoche sublancéolés chez les femelles vers les extrémités (fig. 256). Cerques des mâles plutôt longs, substyliformes, pointus et pourvus d'une dent plus près du milieu (fig. 253) .... ***americanus* (Saussure)** (p. 228)

2(1). Longueur de la partie exposée du tegmen du mâle égale à plus de la moitié de la longueur du pronotum. Fémurs postérieurs de longueur égale à moins du double de celle du pronotum. Encoche de la plaque sous-génitale des femelles petite, peu profonde, en forme de V étroit; lobes situés de part et d'autre de cette encoche courts et tronqués ou très arrondis (fig. 254). Ovipositeur rectiligne et ne s'amenuisant que du côté dorsal pour donner un apex ventral (fig. 257) .... ***testaceus* (Scudder)** (p. 229)

Longueur de la partie exposée du tegmen du mâle égale au quart environ de la longueur du pronotum. Fémurs postérieurs deux fois plus longs que le pronotum. Encoche de la plaque sous-génitale des femelles plus profonde, habituellement en forme de U, parfois en forme de V large; lobes de part et d'autre de cette encoche obtus (fig. 255). L'ovipositeur est en général clairement incurvé et va s'amenuisant à la fois dorsalement et ventralement; apex non ventral (fig. 258) .... ***monticola* Davis** (p. 230)

Atlanticus americanus (Saussure)

Map 71

Orchesticus americanus Saussure, 1859:201.
Decticus derrogatus F. Walker, 1869*b*:260.
Atlanticus americanus; J. A. G. Rehn and Hebard 1916*a*:72, pl. VI, figs. 13, 14, 21, pl. VII, figs. 5, 14, 23, pl. VIII, figs. 7, 15; Helfer 1963:294.

Diagnosis. Exposed tegmen of male about one-quarter length of pronotum. Subgenital plate in both sexes deeply cleft rather than notched. Male cerci fairly slender. Ovipositor tapered on dorsal edge only.

Description. Body medium-sized to moderately large (pronotal length of male 9.4–11 mm, of female 10–11.3 mm). Pronotum and tegmina similar to those of *A. monticola*, with disc sometimes subcarinate beyond middle. Male cercus rather long, slender, acuminate apically, with tooth short, sharp, located just beyond middle (Fig. 253). Male subgenital plate deeply and narrowly cleft, of female similar, with lobes on either side of cleft sublanceolately produced apically (Fig. 256). Ovipositor 24–28.3 mm long, as long as or slightly longer than hind femur, more or less straight, with only dorsal edge tapered apically and with point ventral (Fig. 259). Color brownish yellow to dark reddish brown, with lateral lobes of pronotum often partly darker; hind femora and abdomen occasionally with some small dark patches.

Range. Massachusetts, south to Mississippi and Florida.

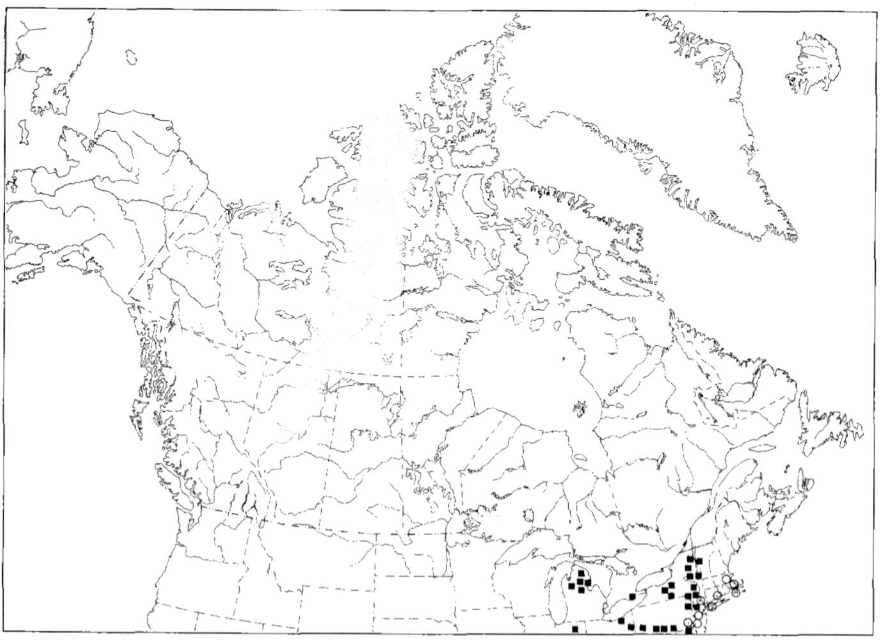

Map 71. Collection localities for *Atlanticus americanus* (○) and *A. monticola* (■).

Behavior and habitats. J. A. G. Rehn and Hebard (1916*a*) state that this species "is a frequenter of the areas of dead leaves and low undergrowth in pine and deciduous forest, occasionally . . . more numerous along the edges of the timber than in the depth of the woods. Its presence is often signaled by the patter on the leaves as it jumps away from the disturbing footsteps. The insects are so thoroughly protected by their coloration that it is often difficult to see them, even when moving, much less when stationary. Their activities are chiefly nocturnal. . . ." Davis (1911) reports several specimens as prey of the burrowing wasp *Sphex ichneumoneus* (Linnaeus) (as *Chlorion ichneumonea* Linnaeus). He also states that the species can be readily taken in pitfall traps baited with molasses (W. T. Davis 1915*a*).

Life history. The species presumably passes the winter in the egg stage, nymphs appearing in spring. J. A. G. Rehn and Hebard (1916*a*) state that the earliest date for a mature adult in Pennsylvania is 15 July, although it matures a little earlier farther south. In New Jersey, adults have been collected as late as 10 October.

Atlanticus testaceus (Scudder)

Figs. 251, 254, 257; Map 72

Engoniaspis testacea Scudder, 1900*a*:97.

Atlanticus testaceus; J. A. G. Rehn and Hebard 1916*a*:48; Cantrall 1943:144; Helfer 1963:293.

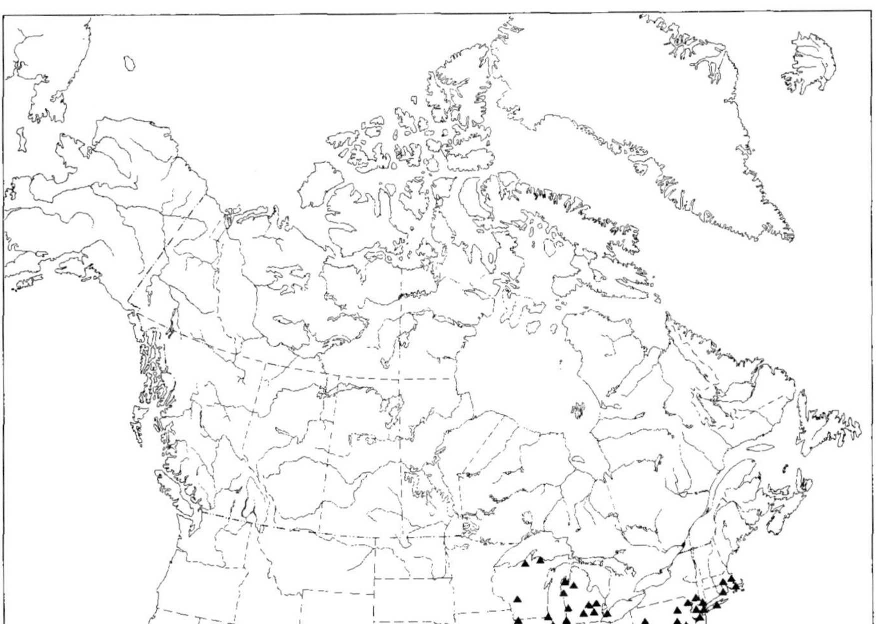

Map 72. Collection localities for *Atlanticus testaceus*.

Diagnosis. Exposed tegmen of male half as long as pronotum. Subgenital plate notched. Ovipositor tapering only on dorsal edge, with apex ventral.

Description. Body size medium (pronotal length, male 9.0–10.8, female 9.2–10.4 mm). Hind femur short (male 14.6–16.5, female 16.2–18.7 mm). Pronotal disc large, about two-thirds as long as wide, with lateral carinae sharp, converging from anterior end on basal fifth then diverging to posterior end. Tegmina of male long, projecting beyond pronotum for a distance equal to or greater than posterior pronotal width, broadly rounded at apices; tegmina of female reaching posterior edge of pronotum but not protruding. Male cercus stout, nearly cylindrical, tapered with short blunt tooth at about two-thirds of its length (Fig. 251). Apex of male subgenital plate broadly and shallowly emarginate, with styles short (but variable) and blunt; female subgenital plate with small shallow V-shaped notch, the lobes on either side subtruncated or broadly rounded apically (Fig. 254). Ovipositor straight, with upper valve tapering apically, appearing almost downturned (Fig. 257). Color generally grayish to blackish brown (usually suffused with grayish dorsally), darker laterally.

Range. Minnesota to Massachusetts, south to Georgia.

Behavior and habitats. E. M. Walker (1905*b*) found the species at Arner, Ont., in "more open parts of a dry upland wood consisting chiefly of oak. Most of them were found on short grass which was growing on the slopes of a ravine in the wood." Gangwere (1966, 1967) described the feeding behavior.

W. T. Davis (1893) reported stridulation to be similar to that of *Orchelimum vulgare*, a continuous z-e-e-e. Allard (1911) reported the song as sh-sh-sh-sh-sh, brief pulses repeated rapidly, followed by irregular periods of silence. An investigation of stridulation in *Atlanticus* is needed.

Life history. Blatchley (1920) recorded *A. testaceus* in Indiana from 6 June until frost. E. M. Walker (1905*b*) found it at Arner on 9 August. Apparently the winter is passed in the egg stage.

Atlanticus monticola Davis

Figs. 226, 249, 250, 252, 255, 258; Map 71

Atlanticus monticola W. T. Davis, 1915*a*:104.

Atlanticus monticola; J. A. G. Rehn and Hebard 1916*a*:67.

Atlanticus davisi J. A. G. Rehn and Hebard, 1916*a*:58, pl. VI, figs. 7–9, pl. VII, figs. 3, 12, 21, pl. VIII, figs. 3–5, 13.

Diagnosis. Exposed tegmen of male about one-quarter length of pronotum. Subgenital plates notched only. Ovipositor tapering on both edges.

Description. Body size medium (pronotal length of male 7.7–9.0 mm, of female 7.8–9.9 mm). Lateral carinae of pronotum converging at anterior third, diverging thereafter, with posterior margin subtruncate, broadly rounded laterally. Male tegmina short, projecting beyond pronotum for a distance equal to about half posterior width of pronotum (Fig. 249); female tegmina vestigial and concealed. Male cercus short, feebly curved, stout at base, with apical third tapered to acute apex and with tooth short, at about three-fifths of length, inwardly curved (Fig. 252). Male subgenital plate with broadly triangular notch; female subgenital plate with narrow U-shaped or broadly V-shaped notch and with lobes on either side of notch obtusely pointed (Fig. 255). Ovipositor 17.6–21.7 mm long, straight or, more typically, slightly curved upward in apical half, with both valves narrowing to point (Fig. 258). Color brownish, flecked or streaked with gray, darker laterally, with hind femur often marked with black.

Range. Michigan to Ontario and Vermont, south to Georgia.

Behavior and habitats. The species occurs among dead leaves and scattered undergrowth of deciduous woods in dry locations.

Life history. Unknown. Probably overwinters in the egg stage, with nymphs hatching in spring, maturing in late spring or early summer, and persisting until frost. Adult specimens have been taken from early June until early October.

Genus *Neduba* F. Walker

Description. Vertex prominent, narrow. Prosternum with 2 long sharp spines; pronotum large, nearly flat dorsally (Fig. 227), with lateral carinae distinct, and with median carina complete and more or less distinct; pronotal disc narrowest at anterior fourth, increasing in breadth anteriorly and to nearly twice anterior breadth posteriorly (Fig. 241); posterior margin semicircular. Tegmina and hind wings developed only in males, these concealed by pronotum. Males with pseudocerci ("infracercal plates" of Caudell 1907*b*), which are simple elongate platelike structures. Ovipositor short, gently curved upward, with apical teeth above and below and with 12, or more, sharp elongate teeth on dorsal valve, fewer on ventral valve.

Neduba steindachneri (Hermann)

Figs. 225, 240–244; Map 73

Arytropteris steindachneri Hermann, 1874:204, pl. VI, figs. 98–102.
Tropizaspis picturata Scudder, 1899*g*:83, 85.
Neduba carinata var. *picturata* Caudell, 1907*b*:299, fig. 8.
Neduba carinata var. *convexa* Caudell, 1907*b*:300, fig. 9.
Neduba steindachneri; Rentz and Birchim 1968:42.

Diagnosis. Pronotum large, nearly flat dorsally, with lateral carinae distinct.

Description. Pronotum elongate (Figs. 225, 241), nearly half length of head and body combined (Fig. 240); lateral carinae converging posteriorly on anterior fifth, then diverging, twice as wide posteriorly as anteriorly (Fig. 241). Tegmina concealed. Prosternal spines usually present. Legs long, with hind femur more than twice as long as pronotum; hind tibia with 2 apical spurs (Fig. 242). Cerci of both sexes simple, rounded, tapering to apex (Fig. 243). Ovipositor short, less than two-thirds as long as hind femur, gently curved upward, armed at apex with 12, or more, sharp teeth on dorsal valve,

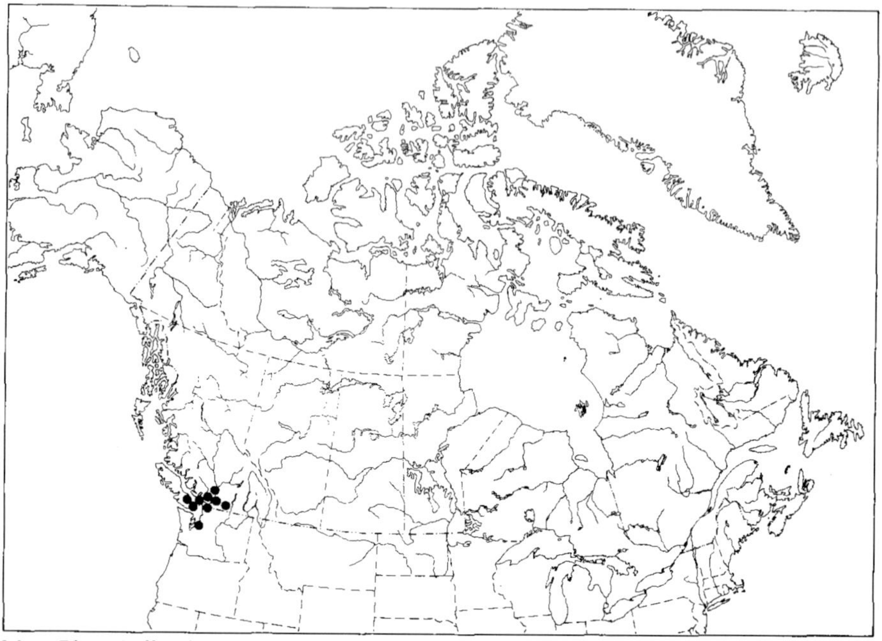

Map 73. Collection localities for *Neduba steindachneri*.

fewer on ventral valve (Fig. 244). Color light brown with darker mottlings or uniformly yellowish; lateral lobes of pronotum usually showing infuscation; fore and mid femora and tibiae usually with broad black bands; hind femur infuscate, mottled on outer face; abdomen nearly always with paired broad dark subdorsal stripes on basal half, covering sides of abdomen toward apex.

Range. British Columbia and Washington.

Behavior and habitats. Buckell (1930) states that *N. steindachneri* (as *N. carinata*) seemed to be confined to oak-*Arbutus* habitats on the south and east coasts of Vancouver Island and on the mainland only inland as far as Boston Bar in the Fraser River canyon. Specimens are difficult to locate during the day. They are well camouflaged and extremely hard to see. Fulton (1928) used the common name "camouflage cricket" for another species of *Neduba* in Oregon.

Stridulation, which occurs only at night, begins about 20:00 hours and continues until about 02:00 hours. The song consists of two sharp metallic clicks, followed by three or four "rather drawn-out sleepy-sounding notes; tick-tick-zeer-zeer-zeer" (Buckell 1930). The song is feeble and is not audible for more than a few metres. The insect jumps to the ground when disturbed, but it is not disturbed by a flashlight as close as 30 cm.

Life history. Unknown. Small nymphs have been found early in the spring, and adults have been collected in August.

Genus *Anabrus* Haldeman

Description. Body rather large, bulky. Head large, prominent. Pronotum produced posteriorly; pronotal disc smooth, evenly rounded, with lateral carinae evident only on posterior half, obscure, or blunt, with transverse sulcus on anterior half inconspicuous and marked near middle with conspicuous V-shaped depression. Abdomen appearing somewhat swollen. Legs short, stout; hind femur with 1–several small spines on each ventral carina; fore tibia with spines dorsally (usually 1 or 2) and ventrally (usually 4 or 5). Male cerci furcate, forming 2 sharp incurved claws (Figs. 227, 229, 231, 232). Female subgenital plate with apical hooklike projections. Ovipositor straight to slightly curved.

Key to species of *Anabrus*

1. Male cercus with inner tooth greatly developed and always at right angles to main shaft of cercus (Fig. 231) .... ***cerciata* Caudell** (p. 234)
   Male cercus with inner tooth less exaggerated, not produced at right angles to main shaft of cercus (Figs. 227, 229, 232) 2

2(1). Hind femur short, less than twice pronotal length. Male cercus with inner tooth stout and with tip of tooth at least half the length of cercus from the base (Fig. 227). Lateral carinae of pronotum feebly indicated, with lateral lobes rounding into disc .... ***simplex* Haldeman** (p. 235)
   Hind femur twice, or more, pronotal length. Male cercus with inner tooth not robust, varying in position. Lateral carinae of pronotum variable, weakly indicated to well-developed 3

3(2). Lateral and median carinae of pronotum weakly indicated. Male cercus with long sweeping curve at apex and with inner tooth usually rather small, located near base (Fig. 229). Found in Oregon to British Columbia .... ***longipes* Caudell** (p. 237)
   Lateral and median carinae well-indicated. Male cercus without sweeping curve at apex and with inner tooth prominent, not close to base (Fig. 232). Found in western Montana, northern Idaho, and probably southern British Columbia ...... ***spokan* Rehn & Hebard** (p. 239)

Clé des espèces d'*Anabrus*

1. Cerques des mâles pourvus d'une dent interne très développée et formant toujours un angle droit avec ces derniers (fig. 231) .... ***cerciata* Caudell** (p. 234)
   Cerques du mâle pourvus d'une dent interne moins grande et ne formant pas un angle droit avec ces derniers (fig. 227, 229 et 232) 2

2(1). Fémurs postérieurs courts et de longueur inférieure au double de celle du pronotum. Cerques du mâle pourvus d'une dent interne forte et, avec le bout de la dent, d'une longueur au moins égale à la moitié de celle du cerque (fig. 227). Carènes latérales du pronotum peu visibles; lobes latéraux formant un disque .... ***simplex* Haldeman** (p. 235)
   Fémurs postérieurs de longueur au moins égale au double de celle du pronotum. Cerques du mâle pourvus d'une dent interne peu robuste, de position variable. Carènes latérales du pronotum variant de peu visibles à très développées 3

3(2). Carènes latérales et médiane du pronotum peu visibles. Cerques du mâle décrivant une longue courbe à l'apex et pourvus d'une dent interne en général plutôt petite, située près de la base (fig. 229). Espèce vivant de l'Oregon jusqu'en Colombie-Britannique .... ***longipes* Caudell** (p. 237)

Carènes latérales et médiane bien développées. Cerques du mâle ne décrivant pas une courbe à l'apex et pourvus d'une dent interne proéminente et plus éloignée de la base (fig. 232). Espèce vivant dans l'ouest du Montana, le nord de l'Idaho et probablement dans le sud de la Colombie-Britannique .... ***spokan* Rehn & Hebard** (p. 239)

Anabrus cerciata Caudell

Fig. 231; Map 74

Anabrus cerciata Caudell, 1907*b*:361, fig. 48.

Anabrus cerciata; Gurney 1939:4, pl. (3), fig. 7, pl. (4), fig. 9; Helfer 1963:300.

Diagnosis. Inner tooth of male cercus long and robust, at right angles to main shaft.

Description. Body large (pronotal length 15–18 mm). Lateral carinae of pronotum rounded; median carina nearly obsolete; posterior margin of disc evenly rounded. Fore tibia with 5 spines on outer carina and 2 or 3 on

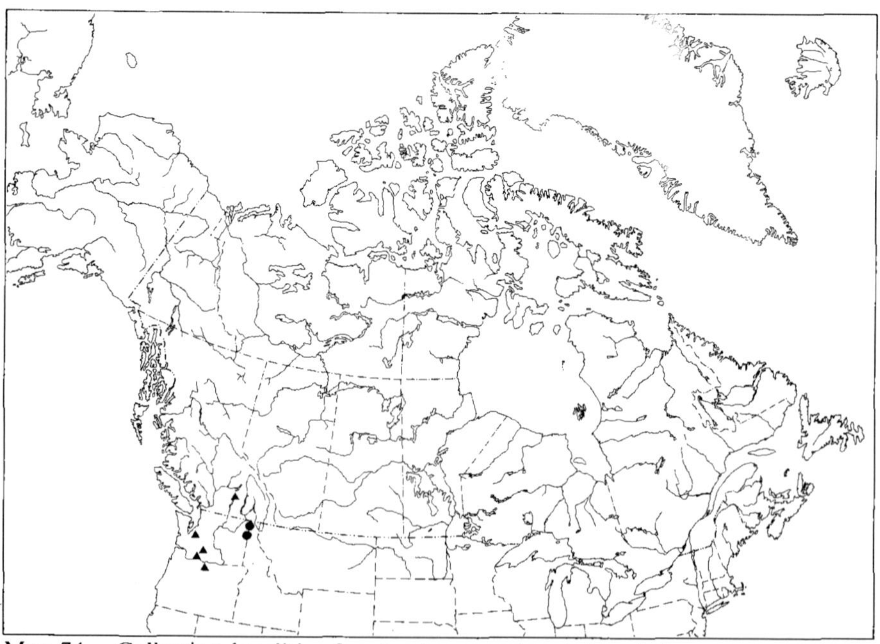

Map 74. Collection localities for *Anabrus cerciata* (▲) and *A. spokan* (●).

inner carina. Inner tooth of male cercus long and robust, set downward at right angles to shaft of cercus (Fig. 231); female cercus long, swollen at base, attenuate, sharply pointed. Ovipositor as long as hind femur, slightly curved upward.

Range. British Columbia to Oregon.

Behavior, habitats, and life history. Not known, presumably similar to *A. longipes*.

Anabrus simplex Haldeman

Figs. 222, 223, 227, 228, 233, 235, 261(1); Map 75

Anabrus simplex Haldeman, 1852:372, pl. 10, fig. 4.

Anabrus purpurascens Uhler, 1864:550.

Acheta nigra Lord, 1866:248, 265.

Anabrus simplex; Glover 1872*b*; pl. IX, fig. 1; C. Thomas 1872:438; Helfer 1963:299.

Anabrus similis Scudder, 1872:249.

Anabrus coloradus C. Thomas, 1872:440.

Anabrus simplex var. *nigra* Caudell, 1907*b*:355.

Anabrus simplex var. *maculatus* Caudell, 1907*b*:356.

Anabrus simplex var. *coloradus* Caudell 1907*b*:356.

Common name. Mormon cricket.

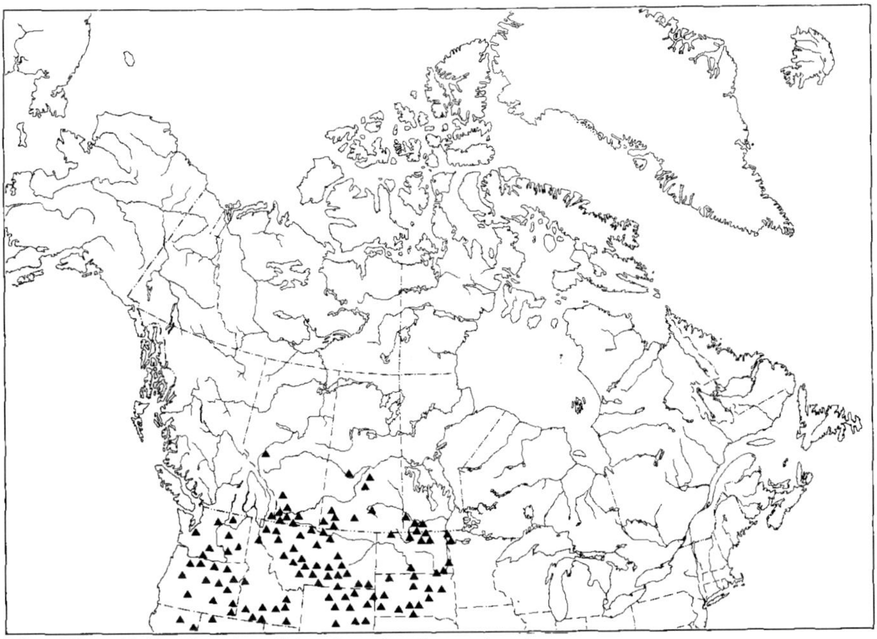

Map 75. Collection localities for *Anabrus simplex*.

Diagnosis. Body bulky. Male cercus with stout inner tooth not at right angles to shaft of cercus. Hind femur short. Tegmina and wings short.

Description. As for genus. Pronotal length, male 7–14.5, female 9–15 mm. Tegmina and hind wings short, those of female broad but concealed by posterior extension of pronotum, those of male overlapping, strongly convex and projecting slightly behind the pronotum. Legs short, with hind femur never quite twice as long as pronotum. Male cerci with branches nearly parallel, both strongly curved inward, with lower branch acute and considerably the longer and with upper branch blunt to nearly pointed (Fig. 227). Ovipositor (Fig. 235) variable in length, usually curved upward, occasionally nearly straight. Color variable, from light yellow to shining black, or grass green; body sometimes uniform in color but more usually mottled. The chromosome system is described by Ueshima and Rentz (1979).

Range. British Columbia to Manitoba, south to California.

Behavior and habitats. General accounts are given by Cowan (1929) and by Wakeland (1959). *A. simplex* breeds mainly in hilly places where vegetation is scarce. It prefers clay soils, and females deposit eggs in surface cracks (Caudell 1907*b*). Adult males stridulate in the morning, ceasing about 10:00 hours. When disturbed, males produce a sharp warning signal that causes nearby individuals to hop away; normally their form of locomotion is by walking. At night all activity ceases and the insects rest beneath or on bushes.

A. simplex is omnivorous. Caudell (1907*b*) reported that it eats sagebrush but prefers moister food, often attacking garden crops, grain, and grasses as well as succulent weed plants. The species also has cannibalistic tendencies, often attacking injured companions or females that have been weakened by oviposition. It will also eat other wounded or dead insects, dead snakes, and any kind of dead flesh, raw or cooked. The insects are also fond of fresh manure. *A. simplex* itself may be part of the diet of other animals, including bears, wolves, hogs, and birds of prey. Some Indian tribes are reported to have utilized them extensively as part of their diet (see Wakeland (1959) for references).

During copulation, the male lies curled under the female or on his back beneath her with genitalia attached. The female does not necessarily remain stationary, but may walk about dragging the male beneath her. Gillette (1905) reported females to be carrying spermatophores ("mass of white jelly-like material," "white sacs" or "blubber") as early in the day as 08:00 hours. This continued until 12:00 hours. Mating and egg-laying continued from early July until 10 September 1903, in Colorado.

Male stridulation is tegminal, although the tegmina are very short, extending but little beyond the posterior edge of the pronotum. During stridulation the pronotum is raised, exposing the wings. The sound produced was described by Gillette (1905) as a "squeaking noise"; Caudell (1907*b*) reported chirping in the morning hours but also during the middle of the day.

Life history. Eggs are laid in soil on dry hills, sometimes at a density of 18 000 to 30 000 per square metre. Each female probably produces 100, or more, eggs. Nymphs begin hatching early in the spring and are able to withstand very cold temperatures. Newly hatched nymphs are pale tan with black eyes, but they soon become entirely blackish, except for a broad pale dorsal stripe with two narrow longitudinal dark lines near the middle of the stripe. Adult coloration is more variable. The posterior margin of the pronotum is pale yellow in all stages (Gillette 1905).

Economic importance. Mormon crickets occasionally occur in outbreak numbers, "vast hordes . . . laying waste large areas of cultivated fields" (Caudell 1907*b*). Millions of them march slowly by walking and may advance 0.75 km/day. When on the march, masses may extend over an area of nearly 2 km^2. In such numbers, they consume everything edible in their path. Early Mormon settlers in Utah apparently suffered considerably from their ravages, hence the common name for the insect. There is a statue and plaque in Salt Lake City commemorating the timely appearance of great flocks of gulls that ate the insects and, legend has it, terminated the first attack on the settlers' crops. Gulls occurring in the area of the Great Salt Lake have, on several occasions, decimated (if not exterminated) armies of *A. simplex* (Glover 1872*a*). In modern times the species has not caused such severe depredations as in the past. Control with chemical insecticides has been effective (Wakeland and Shull 1936; Wakeland 1959). Beirne (1972) reported on an infestation in Saskatchewan in 1962 that "damaged truck crops."

Anabrus longipes Caudell

Figs. 229, 230, 234, 236; Map 76

Anabrus longipes Caudell, 1907*b*:361, 362, fig. 49.
Anabrus longipes; Fletcher and Gibson 1908:131; Helfer 1963:300.

Diagnosis. Male cercus greatly developed in long sweeping curve, with tooth small, near base.

Description. Body small (pronotal length 10–12 mm). Head not broader than pronotum and inserted into its anterior part; vertex prominent, convex. Pronotum smooth, shiny, with disc marked at middle by 2 short posteriorly convergent sulci that sometimes unite to form U-shaped sulcus; hind margin of pronotal disc broadly rounded. Legs long, with hind femur at least twice as long as pronotum, armed below with short stout spines; fore femur with both margins spinose; outer carina with 4 or 5 spines and inner carina with 1–3 spines. Tegmina and hind wings strongly convex, concealed or extending slightly behind pronotum. Male cerci with inner tooth only moderately developed, near base with outer part possessing a long sweeping curve at the apex (Fig. 229). Female subgenital plate broadly concave apically, with side angles sharp, elongate, bent inward (Fig. 234). Ovipositor slightly curved upward, about as long as hind femur (Fig. 236). Color dark brown, with at least posterior half of dorsum of pronotum yellowish; hind femora usually yellowish near apex. N. Criddle (1926*a*) described the nymphal stages.

Range. British Columbia to Montana, south to Oregon.

Behavior and habitats. N. Criddle (1926*a*) reported *Androsace occidentalis* (Primulaceae) as a favored host plant. *A. longipes* also readily eats plants of the Crucifereae. Later instars and adults will feed upon leaves of grain or grasses, but early instars will not accept these as food. Like *A. simplex*,

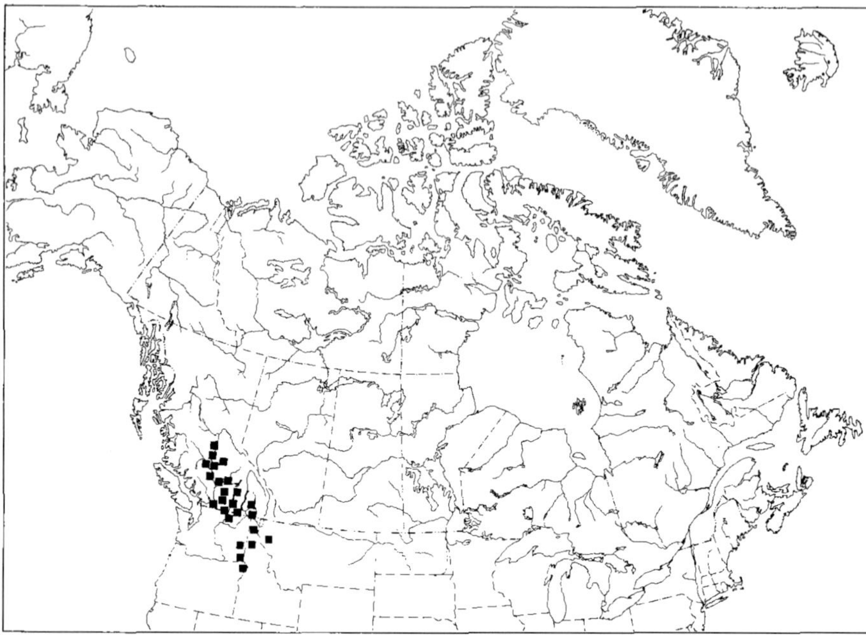

Map 76. Collection localities for *Anabrus longipes*.

A. longipes species are somewhat cannibalistic and will attack members of their own species, especially when these are injured or vulnerable during molting.

Criddle (1926*a*) also gave some account of mating activity. Males attract females by stridulating. When a female comes near, the male approaches and follows, stroking her with his antennae while stridulating continuously. After antennal contact, the female mounts the male, bends her ovipositor down into a groove between the male cerci, and forces genitalic contact. A spermatophore is passed to the female and the pair separates after about 2 minutes. The female carries the spermatophore for several hours and then devours it. Mating takes place three or four times during a pre-oviposition period of several days and is resumed following deposition of each batch of eggs. Females lay 100–200 eggs individually in the soil, each time stabbing the ovipositor several times into the soil.

Life history. N. Criddle (1926*a*) studied reared specimens of *A. longipes* and described each of the seven instars. Maturity is reached in about 43 days. Nymphs are active in cold weather; eggs hatch early in the spring and adults are active from July to September.

Economic importance. Many published reports refer to Mormon crickets in the collective sense. *A. longipes* is probably much less important than *A. simplex*, although an outbreak of the former in British Columbia occurred in 1911 causing damage to onions and range grasses (Beirne 1972).

Range. Washington and Idaho.

Anabrus spokan Rehn & Hebard

Fig. 232; Map 74

Anabrus spokan J. A. G. Rehn and Hebard, 1920:248–251.
Anabrus spokan; Gurney 1939:5, pl. (5), fig. 21.

Diagnosis. Weak median carina present on pronotum.

Description. Disc demarcated by lateral carinae, these rounded but apparent and well-defined posteriorly. Male cercus (Fig. 232) with arm of main shaft beyond internal tooth to point of inward curvature more than twice as long as internal tooth. Color chestnut brown, slightly mottled, dark brown on produced part of lateral pronotal lobes; hind femur with blotches of blackish brown at bases of ventral spines; green phase also known.

Range. Washington and Idaho.

Behavior and habitats. Little is known about this species. The types were taken in a low tangle of raspberry canes and thistles overgrowing charred logs and stumps. All were crawling on the ground and were much less active than *A. simplex*.

Stridulation was faint, only a weak buzzing sound, much like that of *Conocephalus fasciatus* (De Geer) (J. A. G. Rehn and Hebard 1920).

Life history. Unknown. Adults reported by J. A. G. Rehn and Hebard (1920) were taken in late July or early August.

Genus *Peranabrus* Scudder

Description. Pronotum distinctive due to rough pebbly appearance (Figs. 224, 237); lateral carinae well-developed to anterior margin, straight, but diverging posteriorly; median pronotal carinae low, rounded, evident for entire pronotal length. Fore tibia with 4 or 5 spines on ventral margins, unarmed in front. Tegmina of males overlapping, convex, black with yellow margins; tegmina of females not overlapping, reduced to lateral pads.

Peranabrus scabricollis (Thomas)

Figs. 224, 237–239; Map 77

Thamnotrizon scabricollis C. Thomas, 1872:441.
Peranabrus scabricollis; Scudder 1894*b*:181; Helfer 1963:298.

Common name. Coulee cricket.

Diagnosis. Pronotum rough, pebbly.

Description. As for genus. Body medium large (pronotal length, male 7.9–9.5, female 7.5–10.5 mm). Male cerci stout, depressed at tip, broadened so that inner angle bears a short sharp incurved spine (Fig. 238). Ovipositor much longer than hind femur, curved on basal third, with apical two-thirds straight; specimens reported by Buckell (1925) from British Columbia were bright apple green.

Range. British Columbia to Montana, south to Oregon.

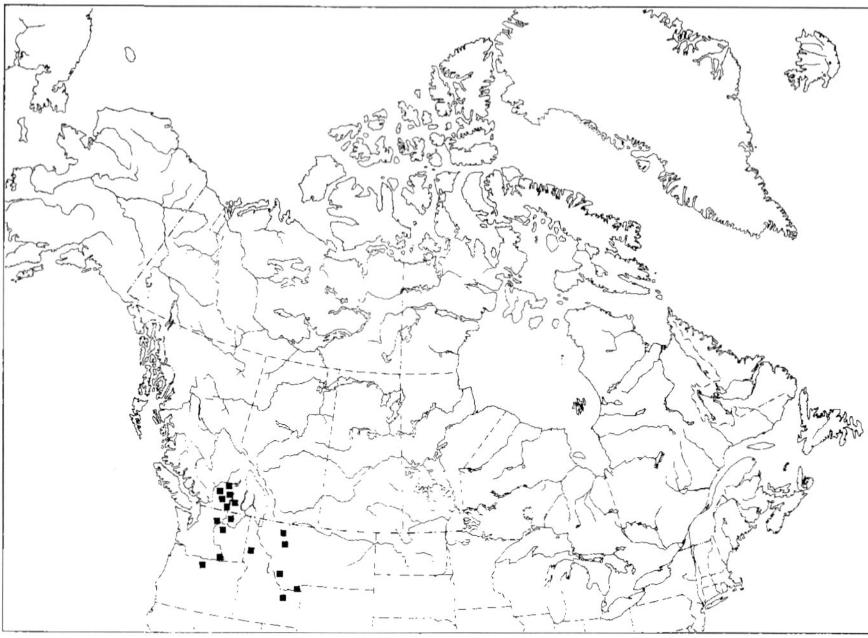

Map 77. Collection localities for *Peranabrus scabricollis*.

Behavior and habitats. Melander and Yothers (1917) reported the occurrence of *P. scabricollis* in desert areas, where it feeds mainly on sagebrush (*Artemisia frigida*, which grows only on poor soil, not *A. tridentata*, which grows on fertile soil). They describe also the migratory habit of this species. Migrations occur when late-instar nymphs, approaching maturity, search for food. This occurs mainly in April, May, and June in Washington State. *P. scabricollis* is omnivorous and cannibalistic, especially in late nymphal instars and as adults. Young nymphs are often found in great numbers under sagebrush, which protects them from frost.

Snodgrass (1905) described mating behavior. The male attracts females by stridulating. The "song" is produced as pulses in slow succession (Snodgrass 1905). When ready to mate, the male pushes his abdomen beneath that of a female and seizes her with his cerci. A bilobed spermatheca is passed to the female within a few minutes. Mating occurs in the morning, seldom at any other time of day. Eggs are laid singly in the bases of grass-stools and near these.

Life history. Melander and Yothers (1917) reported oviposition in late May. Females deposit approximately 40–60 eggs. In Washington, hatching has been observed as early as early March in sheltered southern exposures. Maturity is reached in May and migrations are common throughout May and June. Few specimens survive until mid-July.

Economic importance. *P. scabricollis* can become a serious pest when its migrations lead through cropland. When migrating, it will attack almost any plant in its path. It will feed readily on alfalfa, wheat, and most field and vegetable crops, and on tree foliage, but apparently not on peas, mustard, parsley, lilac, maple, pine, or willow. Beirne (1972) stated that it was suspected of causing harm to grasslands in southern British Columbia. Melander and Yothers (1917) reported that thousands of dollars had been spent on control in Washington between 1910 and 1920.

Genus *Steiroxys* Hermann

Description. General form as indicated in Fig. 260. Head small, not prominent, deeply inserted into pronotum, with vertex broad. Eyes small. Pronotum moderately produced posteriorly, with lateral lobes nearly as deep as long and somewhat sinuate on posterior margins; lateral and median carinae distinct, with lateral carinae diverging slightly posteriorly; disc flat, without noticeable sulci, sometimes with V-shaped sulcus near center. Prosternum without spines. Male tegmina broad, overlapping, projecting beyond pronotum about one-half length of pronotum; female tegmina small, with rounded lateral pads, nearly concealed by pronotum. Male cercus cylindrical, with pointed apex sharp, toothlike, bent inward, with sharp tooth on inner side. Ovipositor slightly curved upward.

The taxonomy of *Steiroxys* is in a confused state. Two species have been recorded for Canada (British Columbia and Alberta), namely, the type-species, *S. trilineatus* (Thomas), and *S. borealis* Scudder. The former is a valid species and the latter probably is, but neither seems to be Canadian. The latter was described from northern California (Scudder 1894*b*:182), and *S. trilineatus* possibly reaches Montana, although this is unlikely.

A revision of the genus, begun by D. C. Rentz and being continued by D. Lightfoot, Department of Entomology, Oregon State University, Corvallis, OR, has shown that "over 30 undescribed species are involved . . . Specimens. . . from Canada represent 2 or 3 species which are undescribed" (D. Lightfoot, pers. com. 1979).

Mr. Lightfoot has supplied the following information: "*Steiroxys* occurs primarily in open grassy areas, both in sage brush-steppe and mountainous areas. They prefer low herbaceous growth and dry, sparse, grasses growing on loose soils. The preferred elevational range appears to be from 2000–7000 feet. *Steiroxys* is not found abundantly over its entire range, but rather as localized populations where individual numbers are often quite high. *Steiroxys* is a diurnal genus and best collected during the daytime. Two general color forms are found, green and gray-brown, the latter being most common. Both color forms are usually found in the same species from a given population." Ueshima and Rentz (1979) describe the chromosome patterns of *S. strepens* Fulton.

Known populations of *Steiroxys* species in Canada are indicated together in Map 78.

Steiroxys strepens Fulton

Fig. 260

Steiroxys strepens Fulton, 1930:627, fig. 2.
Steiroxys strepens; Helfer 1963:298.

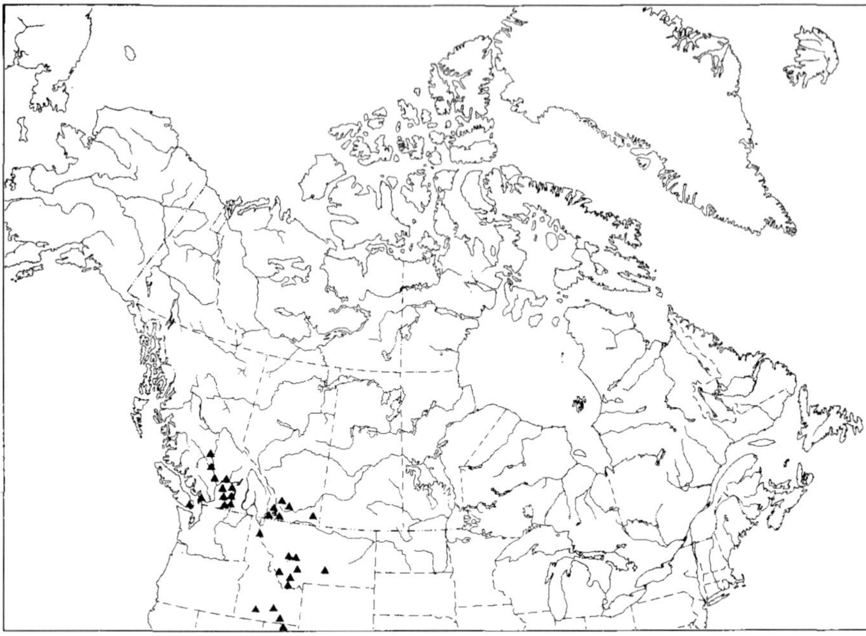

Map 78. Collection localities for *Steiroxys* spp.

This seems to be a distinctive species that is unlikely to be found much beyond the bounds of the state of Oregon. It is noted here as being one species, at least, that can be recognized without much difficulty, for which reason it is used to illustrate the genus.

Steiroxys trilineatus (Thomas)

Fig. 261(2)(?)

Thamnotrizon trilineatus C. Thomas, 1870:76.
Steiroxys trilineata; Hermann 1874:207, pl. V, figs. 64–69.

This species may reach as far north as southern Montana, and Hebard (1928) so records it, as well as from Fort MacLeod, Alta. Hebard (1931*a*) was "satisfied that only a single species, *trilineatus*, occurs on the eastern slopes of the Rocky Mountains."

Genus *Sphagniana* Zeuner

Description. Pronotum broad, enlarged posteriorly, with lateral lobes bent sharply downward (Figs. 263, 264). Legs short. Male supra-anal plate narrowly emarginate, with lobes covering cerci (Fig. 268). Male cercus cone-shaped, with sharp black curved tooth near base (Fig. 269). Female subgenital plate as in Fig. 270. Ovipositor strongly upcurved (Fig. 271).

Sphagniana sphagnorum (F. Walker)

Figs. 262–265, 268–271; Map 79

Decticus sphagnorum F. Walker, 1869*b*:258.
Idionotus brevipes Caudell, 1907*b*:397.
Platycleis fletcheri Caudell, 1907*b*:404.
Metrioptera sphagnorum; Hebard 1931*a*:401; Helfer 1963:291, fig. 458.
Sphagniana sphagnorum; Zeuner 1941:43.
Metrioptera (*Sphagniana*) *sphagnorum*; Kevan et al. 1962:71; Vickery et al. 1974:62.

Diagnosis. Pronotum large, broad, extended posteriorly, with lateral lobes bent sharply downward.

Description. As for genus and in key. Body small to medium-sized (pronotal length, male and female, 5.2–5.5 mm). Tegmina dimorphic, in brachypterous form covering one-quarter of abdomen in female and about one-half in male, in macropterous form (known only in males) extending past the apices of the hind femora. Color pale red brown; lateral pronotal lobes blackish fuscous with margins and lower third pale yellow (Figs. 263, 264); abdomen with row of black spots laterally; hind femur with black lines (Fig. 265); female black on sides of vertex and on hind tarsi.

Range. Western Northwest Territories to western Quebec.

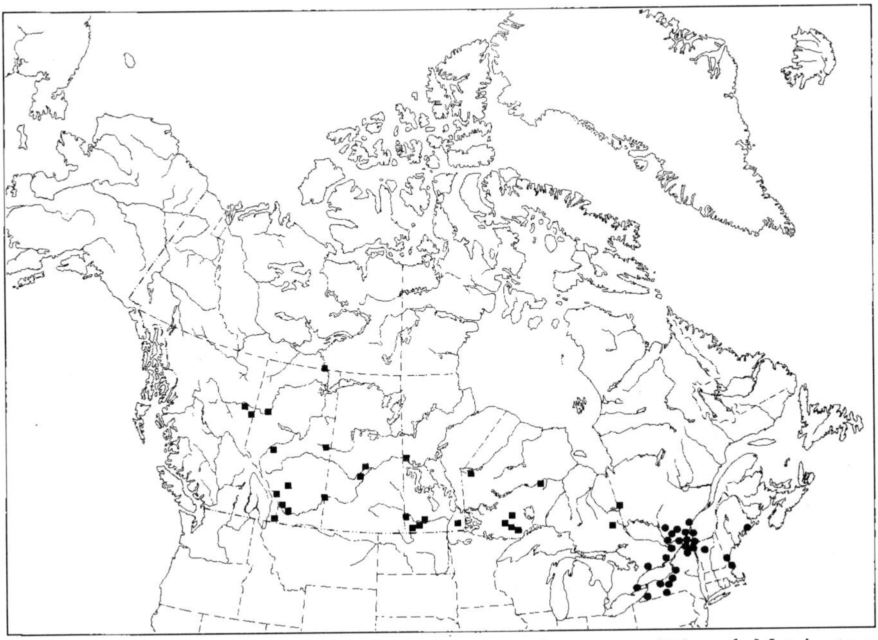

Map 79. Collection localities for *Sphagniana sphagnorum* (■) and *Metrioptera roeselii* (●).

Behavior and habitats. E. M. Walker (1911) gave a good account of *S. sphagnorum* at Fort William, Ont. He found it in open grassy places and along paths and roads in swampy areas where black spruce was the predominant tree. Specimens were located on shrubs near the ground but also high in trees.

All male specimens were traced by E. M. Walker (1911) through stridulation, which he described as "a soft trill of little volume." G. K. Morris (1970) described the stridulation as a prolonged trill alternating regularly between two intensities. G. K. Morris and Pipher (1972) analyzed stridulation and found a complex pulse train pattern, with two different pulse types produced at different regions of the stridulatory file.

Life history. Unknown. Presumably overwinters in egg stage. Adults have been taken from mid-July to September.

Genus *Metrioptera* Wesmael

Description. Body size medium. Tegmina normally covering one-half to two-thirds of abdomen, with apices broad; in macropterous form, extending beyond femoral apices in both sexes. Male subgenital plate broad and long. Male cercus prominent, with tooth on apical third (Fig. 273). Female subgenital plate large, deeply bilobate, not touching ovipositor (Fig. 274). Ovipositor short, strongly curved (Fig. 275). Coloration green or brown, with margins of lateral pronotal lobes bright yellow or yellowish green.

Metrioptera roeselii (Hagenbach)

Figs. 266, 267, 272–27[illegible] Map 79

Locusta roeselii Hagenbach, 1822:39, [illegible]. 24.
Locusta brevipennis Charpentier, 182[illegible] 14.
Locusta diluta Charpentier, 1825:11[illegible] [illegible]acropterous).
Decticus sinuatus Fischer von Waldh[illegible], 1846*b*:170, pl. [illegible]XIX, fig. 6.
Metrioptera roeselii; Chopard 1922:8[illegible] [illegible]elfer 1963:292; Vickery et al. 1974:61.
Metrioptera roeseli; Chopard 1951:1[illegible] 151.
Metrioptera roeseli f. *diluta*; Kevan [illegible]1*b*:605–607 (macropterous).

Diagnosis. Pronotum short, with borders of lateral lobes conspicuously pale.

Description. Body size medium (pronotal length, male 3.7–5.0, female 4.1–5.5 mm), robust in appearance. Legs long and slender (Fig. 272). Tegmina and hind wings dimorphic, in brachypterous form, reduced to short lobes and hing wings vestigial; in macropterous form (Fig. 272), both tegmina and hind wings extending well beyond apex of abdomen. Male cercus with tooth nearer apex than base (Fig. 273). Male subgenital plate with pair of styles. Ovipositor short, upwardly curved, acute at apex (Fig. 275). Color generally brown or green with yellow markings laterally, with borders of lateral pronotal lobes conspicuously pale, yellow to yellow green; mesothorax and metathorax with distinct pale patches.

Beaudry (1973) reported the chromosome complement of *M. roeselii* in Québec as $2n$ male = 15. The X chromosome is longer than the autosomes and is positively heteropycnotic at the beginning of the diplotene stage. At metaphase, 15–18 chiasmata were observed.

Range. Ontario to Maine, south to New York and Massachusetts.

Behavior and habitats. *M. roeselii* prefers microhabitats with high humidity, feeding actively during the cooler times of the day and remaining near plant bases during the heat of midday. It feed upon weeds and grasses. It appears to be especially fond of cultivated and wild timothy grass (*Phleum pratense* L.). It also eats other plants including succulent weeds, and will consume small caterpillars (Kevan et al. 1962). When disturbed, it jumps down in dense grass where it is difficult to capture. It can be collected by hand ("quick grab" method) more successfully than by sweeping vegetation (Kevan et al. 1962). Copulation usually occurs on twigs or in grass, usually lasting from 30–50 minutes. Both sexes may mate more than once and macropterous and brachypterous forms mate indiscriminately. Females consume the spermatophore sac, beginning immediately following copulation, but only a small piece is bitten off at one time (Kevan et al. 1962).

M. roeselii in Québec is heavily attacked by parasites. Kevan et al. (1962) found 80% of adults parasitized by larval mermithid nematodes.

Males stridulate, and females are attracted. The carrier frequency varies between 13.5 and 40 kHz with a peak of 25 kHz (Sales and Pye 1974). The song is somewhat monotonous, often continuous for several minutes, with short interruptions.

Life history. Similar to that recorded in southern England (Ragge 1965). Adults are found from early July to October. Eggs are deposited in softer parts of plants. The female first bites at the stem, then inserts her ovipositor and deposits in a single row. These hatch (in Québec) in late May. There are at least six nymphal instars.

Kevan (1961*b*) reported the presence of the macropterous form "*diluta*." Vickery (1965) reported 83% macropterous forms at Sainte-Anne-de-Bellevue, Qué., and suggested that this represented a dispersal phase. Subsequent observations indicate that macropterism is indeed a dispersal mechanism. Populations in new areas revert to a mainly brachypterous condition until the population becomes dense, then produce generations that are mainly macropterous, these dispersing to new areas. In areas with well established populations, the macropterous form is uncommon (see also Ede 1974).

Economic importance. Although it is capable of causing considerable loss in grass in undisturbed areas, *M. roeselii* appears in the field in late nymphal instars only shortly before hay-cutting operations begin, so that damage to cultivated hay crops is never serious (Vickery 1965).

Suborder Gryllodea

Members of the Gryllodea are the true crickets, the mole crickets, and, as understood here, also the splay-footed crickets, or Schizodactylidea. They may be characterized by the following combination of characters: body moderately large (maximum about 6 cm) to small (although adults rarely less than 1.5 mm without appendages). Color black, brown, yellowish, or gray, but not commonly green. Head usually short, round, with 3 ocelli (not in Gryllotalpoidea, with 2, or Oecanthidae, with none). Antennae nearly always much longer than body (except in Gryllotalpoidea, in which they are very short). Pronotum generally quadrate or even transverse (again with the exception of the Gryllotalpoidea), not usually much longer than wide, with the posterior margin more or less truncated, not usually produced backward. Legs with 3-segmented (rarely 4-segmented) tarsi, sometimes strongly modified for digging or in other ways; anterior tibiae typically with distinct auditory tympanic organs; hind legs only occasionally short and not adapted for leaping (Gryllotalpoidea). Tegmina and hind wings usually present, even if greatly reduced, but sometimes totally lacking; tegmina typically held flat across dorsum with anterior fields bent downward at angle along sides of abdomen, and in males of most (but not all) species, with greater part of dorsal surface, from base outward, forming stridulatory organ, even in brachypterous and strongly micropterous forms; left and right tegmina with more or less similar venation (although stridulation is virtually always with right tegmen over left); females without such stridulatory organs; hind wings at rest typically projecting in spike beyond front pair, often reduced (frequently dimorphic or polymorphic within a species). Abdominal stridulatory mechanisms not developed. Cerci long, flexible, unsegmented (occasionally annulated), beset with long sensory hairs (Hartman et al. 1979). Ninth abdominal sternum (subgenital plate) of males without styli even in immature stages. Ovipositor typically needlelike, sometimes rather swordlike, reduced, or absent (last notably in Gryllotalpoidea), with inner valves always vestigial except in immature stages.

The Gryllodea comprise about 333 world genera and more than 2400 species. The great majority of Gryllodea are tropical, although many occur in temperate regions throughout the world. They do not, however, reach very high latitudes or altitudes. They occur in virtually all terrestrial environments from deserts and savannahs to bogs and swamps, some even being subaquatic (including a few species that skate on the surface of water and others that are intertidal, although not in the region here considered), and from subterranean burrows and caves to treetops; ground-dwelling species are, however, probably the most familiar. Some species are commensals, living in the habitations of other animals (including ants, termites, rodents, and man). However, this is seldom obligatory except in the case of the ant-loving crickets (Myrmecophilidae). Eggs are usually desposited singly, typically in the ground, but many species lay their eggs in, or on, plant tissues. Crickets are characteristically omnivorous, but some are more inclined to a plant or insect diet.

A number of species are crop pests and, under certain circumstances, "population explosions" sporadically occur with devastating results. In such cases, the crickets may invade human habitations in large numbers and cause considerable consternation, as much by their interminable chirping and omnipresence as by any material damage done (the latter can occasionally be severe, not only to foodstuffs but also to furnishings). In addition to such infrequent invasions, one or two species in different parts of the world take up residence in human habitations, where they may be welcomed or abhorred according to local culture or individual taste. There is a great deal of folklore relating to crickets, particularly to the widely distributed house cricket *Acheta domesticus* (Linnaeus), "the cricket on the hearth."

In many parts of the world, and since ancient times, various kinds of crickets have been kept in cages for their songs (some are particularly musical). The cult of fighting crickets is particularly, but not exclusively, Chinese.

Key to superfamilies of Gryllodea

1. Body molelike, heavy, usually covered with velvety pubescence. Head small in relation to thorax, somewhat conical, with only 2 (instead of 3) ocelli. Antennae short. Pronotum forming convex shield, considerably longer than wide, curving downward and ventrad almost to coxae. Legs modified for digging, the front pair with bladelike apophyses on trochanters and femora, and tibial spines modified to form elongate bladelike "dactylar processes"; hind legs short, not modified for jumping. Tegmina (when present) with male stridulatory organ not forming a "mirror." Ovipositor virtually lacking ... **Gryllotalpoidea** (p. 247)
   Characters not as above .... **Grylloidea** (p. 251)

Clé des superfamilles de Gryllodea

1. Corps rappelant celui d'une taupe, fort, en général recouvert d'une pubescence veloutée. Tête petite par rapport au thorax, quelque peu conique, ne comptant que 2 ocelles (au lieu de 3). Antennes courtes. Pronotum formant un bouclier convexe, beaucoup plus long que large, incurvé vers le bas et ventralement presque jusqu'aux coxas. Pattes adaptées pour creuser, les pattes antérieures étant munies d'apophyses en forme de lame sur les trochanters et les fémurs et les épines tibiales étant modifiées pour former de longs appendices en forme de lame; pattes postérieures courtes, non adaptées au saut. Chez les mâles, tegmina (lorsque présents) à organe stridulant ne formant pas un «miroir». Ovipositeur presque inexistant .... **Gryllotalpoidea** (p. 247)
   Combinaison de caractères différente de celle donnée ci-dessus .... **Grylloidea** (p. 251)

Superfamily Gryllotalpoidea

Family Gryllotalpidae

These are the mole crickets, "churr-worms," "eve-churrs," or "changas," characterized as follows: body large, heavy, usually densely covered with short pubescent hairs, tawny or brownish. Head small, somewhat conical, much narrower than pronotum. Antennae short. Ocelli 2 only. Pronotum much longer than wide, forming strongly convex dorsal shield, which curves downward and ventrad to coxae. Legs modified for burrowing, with enlarged trochanter and femur bearing ventral bladelike apophyses, and with tibial and tarsal spines modified to form flattened or elongate–conical bladelike "dactylar" digging processes; hind legs short, not modified for jumping. Wings fully developed, or (the hind pair particularly) somewhat reduced, rarely lacking; tegmina of males with stridulatory organs, lacking "mirror"; main part of tegmen having large triangular cell, and having apical part reduced; female tegminal venation somewhat similar to that of male but not forming stridulatory organ. Ovipositor lacking or reduced to minute vestige.

Mole crickets spend most of their lives tunneling in soil, emerging and dispersing at night. Eggs are laid in the tunnels, sometimes in special chambers, where they may be brooded. Most species feed both on underground parts of plants and on soil invertebrates. A few may become serious crop pests.

Key to genera of Gryllotalpidae

1. Hind tibiae armed dorsally on distal half of inner (posterior) margin with 4 spines (Fig. 329) .... ***Gryllotalpa*** (p. 248)
   Hind tibiae not so armed (Fig. 327) .... ***Neocurtilla*** (p. 249)

Clé des genres de Gryllotalpidae

1. Tibias postérieurs armés dorsalement, sur la portion distale de leur bord intérieur (postérieur), de 4 épines (fig. 329) ... ***Gryllotalpa*** (p. 248)
   Tibias postérieurs non armés de la sorte (fig. 327) .... ***Neocurtilla*** (p. 249)

Genus *Gryllotalpa* Latreille

Description. As in key to genera.

Gryllotalpa gryllotalpa (Linnaeus)

Fig. 329; Map 80

Gryllus Acheta Gryllotalpa Linnaeus, 1758:428 (partim).
gryllotalpa [*sic*] *vulgaris* Latreille, 1804:122.
Gryllotalpa gryllotalpa; Burr 1897:IV, 67, pl. V, fig. 7; Helfer 1963:324.

Diagnosis. As in key to genera.

Description. Body large (35–40 mm long). Trochanter of fore leg with long process (Fig. 329*b*); distal half of hind femur with four strong spines (Fig. 329*a*).

Range. Massachusetts and New Jersey; Europe.

Behavior, habitats, and life history. Much the same as for *Neocurtilla* (q.v.), but with larger brood chamber. *G. gryllotalpa* has a 2-year life cycle; for details, see Ragge (1965). Stridulation, courtship, and mating behavior are noted by him. Information may also be obtained from Harz (1969).

Economic importance. *G. gryllotalpa* may sometimes be a serious horticultural pest in Europe, and is occasionally troublesome in New Jersey

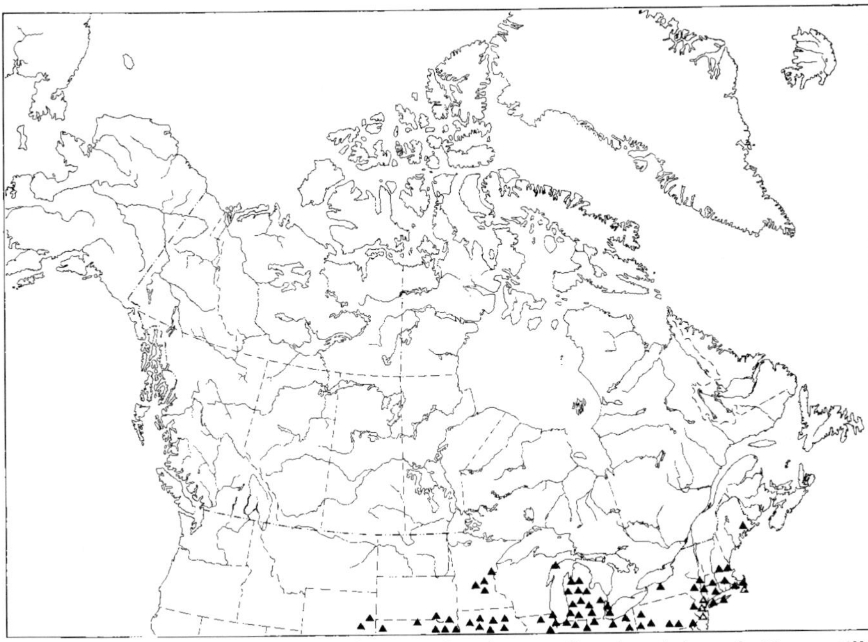

Map 80. Collection localities for *Gryllotalpa gryllotalpa* (△) and *Neocurtilla hexadactyla* (▲).

by shearing through the roots of plants or by otherwise upsetting them as is noted under *Neocurtilla*. The first documented account of important economic loss in North America as a result of attacks by this species is given by Weiss and Dickerson (1918).

Genus *Neocurtilla* Kirby

Description. As for family. Body pubescent, elongate. Pronotum strongly convex. Fore legs short, with tibiae strongly modified for digging; hind femora much shorter than abdomen, not adapted for leaping; hind tibiae not spined above on inner side, with 8 apical spines.

Neocurtilla hexadactyla (Perty)

Figs. 327, 328; Plate III*A*; Map 80

Gryllotalpa hexadactyla Perty, 1832:119, pl. 23.

Gryllotalpa borealis Burmeister, 1838:740.

Gryllotalpa brevipennis Audinet-Serville, 1838:308.

Gryllotalpa longipennis (*nec* Haan, 1842) Scudder, 1863*a*:426.

Gryllotalpa Columbia Scudder, 1869*b*:26 (nomen novum for *longipennis* Scudder, preoccupied).

Gryllotalpa hexadactyla; Lugger 1898:347; Helfer 1963:174.

Neocurtilla hexadactyla; R. D. Alexander and Otte 1967:29; R. D. Alexander et al. 1972:47; Vickery et al. 1974:60.

Common name. Northern mole cricket.

Diagnosis. As for genus.

Description. As for genus (Figs. 327, 328). Process of anterior trochanter short; apical spines of hind tibia arranged in two groups, with 4 long spines on inner face and 4 shorter ones externally. Tegmina usually covering one-half to three-quarters of abdomen; hind wings slightly to much longer than tegmina; stridulatory file with about 25 teeth (Plate III*A*). Cerci nearly 1.5 times as long as pronotum.

Range. Wyoming to Ontario and Maine, possibly south to South America.

Behavior and habitats. Individuals of *N. hexadactyla* are nocturnal, soil dwelling, usually difficult to find in the field, but attracted to ultraviolet light traps (Beck and Skinner 1967). Since they stridulate during the hours of darkness, they may be located by listening for sequences of chirps produced by males after sunset and during early evening in areas where the earth is moist, along streams or ponds, in crop fields, or in plant nursery areas. Individuals tunnel extensively underground, their working sometimes being visible as slight ridges on the soil surface. Eggs are laid in special subterranean broad chambers.

Baumgartner (1910) reported that an anal (defensive?) secretion was produced by this species, but its function against enemies is unknown. Such a secretion had been reported previously by McNeill (1891) who stated, in error and somewhat fancifully, that the "grayish viscid substance" was ejected from the cerci. He stated further that no bad odor was detected and that, if the substance was used to repel enemies, it was probably by covering the eyes or entangling the legs of these enemies!

The calling song is reported by R. D. Alexander (1960) to have a low carrier frequency, 2.0 kHz and [illegible]5 pulses per second. Males remain sedentary in burrows and attract females by stridulating. Some females fly readily and have no difficulty in recognizing males of their own species; the frequency and sound pulse rate are specific. When a female lands near a calling male it enters the male's burrow, and an elaborate premating courtship ritual takes place (R. D. Alexander and Otte 1967).

Life history. R. D. Alexander (1968) reported a 2-year life cycle in North America, with diapause in a nymphal stage, followed the next winter by diapause as an adult. Fulton (1951) had previously suggested a 2-year life cycle in North Carolina: eggs laid in spring or early summer; nymphs maturing the following summer; adults mating in the late summer and fall; males dying and females diapausing during winter, then depositing their eggs the following spring. There appears to be little difference in the life cycle of *G. gryllotalpa* and the present species, at least in the northern part of the range of the latter.

Economic importance. Mole crickets, including the present species, frequently cause significant crop damage. Such damage was reported in Canada as early as the 1890s by Fletcher (1892) who gave an excellent account of the structure and behavior of a captive specimen from Leamington, Ont. *N. hexadactyla* (or *N. borealis*) is more or less omnivorous, but it can cause serious damage to crops where it occurs in dense populations, because it feeds upon the roots and occasionally the stems of a wide variety of plants. It also feeds upon roots of trees but generally prefers the softer roots of herbaceous plants. It also uproots plants or causes the roots to dry out by burrowing in the soil beneath them. Ulagaraj (1975) working with southern mole crickets, *Scapteriscus* species, has shown that the crops of flying females are empty. He suggested that these hungry mole crickets, arriving at a new site, could cause severe damage by feeding on nearby plants. Although this has not been investigated for the present species, the latter may well behave similarly.

Superfamily Grylloidea

The insects in this superfamily are the true crickets. They may be characterized as follows: antennae filiform, longer than body. Tegmina, when present, flat above abdomen, bent abruptly downward laterally. Hind femora stout, with tarsi 3-segmented (occasionally appearing to be 4-segmented). Ovipositor straight or upwardly curved, needlelike, rarely awllike or daggerlike, with tip sometimes slightly enlarged. Stridulating organ of males large, extending across proximal part of each tegmen; hind wings reduced and nonfunctional in many species. Male stridulation by the rubbing of a file on the underside of one tegmen against a scraper on the edge of the other. Tegmina are virtually identical, with the right one almost invariably overlying the left (the relative positions being fixed for a given individual), which is contrary to the arrangement adopted by katydids and other tettigonioids.

The songs of most species are distinctive and are useful in identification. Most species produce eggs in the summer and autumn, and these hatch in the following spring, but one of our species hibernates in the late nymphal stages. Food preferences are variable; some species are omnivorous, whereas others are mainly phytophagous. "Tree crickets" (Oecanthidae) are at least partially predaceous, feeding upon aphids.

The superfamily Grylloidea includes about 325 world genera and more than 2300 known species.

Key to families of Grylloidea

1. Body small (adult 1.5–4 mm long). Eyes present but with few facets. Hind femora stout; tarsi with second segment minute, compressed. Wings lacking. Ovipositor rather short, with splayed daggerlike valves. Commensal with ants .... **Myrmecophilidae** (p. 252)
   Not as above 2

2(1). Body larger (adult 1–1.5 cm long). Ocelli absent. Hind femora long, not stout; tarsi appearing 4-segmented. Wings present. Ovipositor needlelike, strongly denticulate apically. Not commensal .... **Oecanthidae** (p. 254)
   Not as above 3

3(2). Body small (seldom more than 1 cm long). Head generally wider than anterior margin of pronotum. Eyes prominent. Hind legs rather long and slender; hind tibiae with 3 pairs of spines and 5 apical spurs. Hind wings normally projecting well beyond tegmina when at rest. Ovipositor laterally compressed, usually curved upward, with apex acute and usually denticulate, not enlarged .... **Trigonidiidae** (p. 273)

Body small to large (usually 1–2.5 cm long). Head not usually wider than anterior margin of pronotum. Eyes not prominent. Hind legs generally rather stout; hind tibiae with more than three pairs of strong spines. Hind wings, when present, often reduced. Ovipositor needlelike, more or less straight, with apex usually somewhat enlarged but with denticulation weak .... **Gryllidae** (p. 275)

Clé des familles de Grylloidea

1. Corps petit (de 1,5 à 4 mm de longueur chez les adultes). Yeux présents mais petits avec quelques facettes. Fémurs postérieurs forts; deuxième segment des tarses minuscule et comprimé. Ailes absentes. Ovipositeur plutôt court, en forme de poignard. Insectes commensaux des fourmis .... **Myrmecophilidae** (p. 252)

Combinaison de caractères différente de celle décrite ci-dessus 2

2(1). Corps plus grand (de 1 à 1,5 cm de longueur chez les adultes). Les ocelles absents. Fémurs postérieurs longs, pas particulièrement forts; tarses à quatre segments visibles. Ailes présentes. Ovipositeur en forme d'aiguille, fortement denticulé à l'apex. Insectes non commensaux .... **Oecanthidae** (p. 254)

Combinaison de caractères différente de celle décrite ci-dessus 3

3(2). Corps petit (rarement plus de 1 cm de longueur). Tête en général plus large que le bord antérieur du pronotum. Yeux proéminents. Pattes postérieures plutôt longues et minces; tibias postérieurs portant 3 paires d'épines et 5 éperons apicaux. Ailes postérieures dépassant normalement de beaucoup les tegmina en position de repos. Ovipositeur aplati latéralement, en général incurvé vers le haut, à apex aigu et d'habitude denticulé, mais pas agrandi .... **Trigonidiidae** (p. 273)

Corps de taille variable (en général de 1 à 2,5 cm de longueur). Tête d'habitude pas plus large que le bord antérieur du pronotum. Yeux non proéminents. Pattes postérieures généralement plutôt fortes; tibias postérieurs portant plus de 3 paires de fortes épines. Ailes postérieures, lorsque présentes, souvent de taille réduite. Ovipositeur en forme d'aiguille, plus ou moins rectiligne, à apex en général quelque peu agrandi, mais faiblement denticulé .... **Gryllidae** (p. 275)

Family Myrmecophilidae

These minute insects are the ant-loving crickets, or ant crickets. They constitute a small distinct group of Grylloidea that is probably fairly closely related to the subfamily Bothryophylacinae of the family Mogoplistidae. It is possible that these insects do not deserve independent family status, owing their peculiarities to their specialized mode of life. They are, however, easily recognizable from all other crickets by the following combination of characters: appearance rather uniform. Size minute (the smallest of all orthopteroids), with body seldom as much as 2 mm, sometimes less than 1.5 mm, long, broadly oval. Head round, small in relation to pronotum. Eyes small, with few facets. Antennae not much longer than body. Tarsi with second segments minute, compressed; hind femora short and enormously enlarged; hind tibia with few fairly long movable spines in distal parts. Wings absent. Cerci comparatively short, annulated, appearing almost segmented. Ovipositor rather short, with somewhat splayed daggerlike valves.

Five genera and some 45 species have been described within the single included subfamily. They occur throughout most of the world, particularly in temperate and subtropical parts, but they are apparently lacking from the rain forests of the Neotropical and Ethiopian regions. They are always associated either with ants (in the "nests" or hopping along with them) or, occasionally, with termites; they seem unable to live independently for long; some species have a wide range of ant hosts, others are associated with only one species. Some, at least, appear to be habitually, if not exclusively, parthenogenetic. Eggs are relatively large, and (at least in some species) postembryonic development is of long duration (more than a year). The only comprehensive work on North American species remains that of Hebard (1920*a*). The classic study, including morphology, is that of Schimmer (1909).

Genus *Myrmecophilus* Berthold

Description. As for family Myrmecophilidae.

Myrmecophilus oregonensis Bruner

Fig. 369; Map 81

Myrmecophila oregonensis Bruner, 1884*a*:43.
Myrmecophila oregonensis; Scudder 1899*e*:426; Helfer 1963:327.
Myrmecophila formicarum Scudder, 1899*e*:426.

Diagnosis. Body small (1.5–3.7 mm), broadly oval. Associated with ants.

Description. Body size 1.5–3.7 mm (northern specimens averaging larger than those from southern part of range). Hind tibia with 1 external and 3 or 4 internal spines dorsally, and with 3 pairs of distal spurs, the ventral pair often small but equal in length; external dorsal spine of hind tibia as long as or longer than one-half length of tarsal segments; hind tarsal spurs slightly longer than distal tarsal segment.

Range. British Columbia to California.

Behavior and habitats. There is some indication that size variation may be more a response to the size of the host ants than to any other single factor. In this genus, most species associate with certain species of ants. *M. oregonensis* is no exception. Both adults and juveniles are found in nests of various ant species of the subfamilies Formicinae and Myrmecinae, particularly those which live in rotten stumps and logs under stones in damp locations, namely *Camponotus maccooki* Forel, *C. vicinus* Mayr, *C. pennsylvanicus modoc* Wheeler, *Formica fusca* Linnaeus, *F. pilicornis* Emery, *F. neorufibarbis* Emery, *F. subpolita camponoticeps* Wheeler, *F. rufibarbis occidua* Wheeler, *Manica bradleyi* (Wheeler), *Pheidole hyatti hyatti* Emery, *Prenolepsis imparis californica* Wheeler, *Pogonomyrmex californicus californicus*

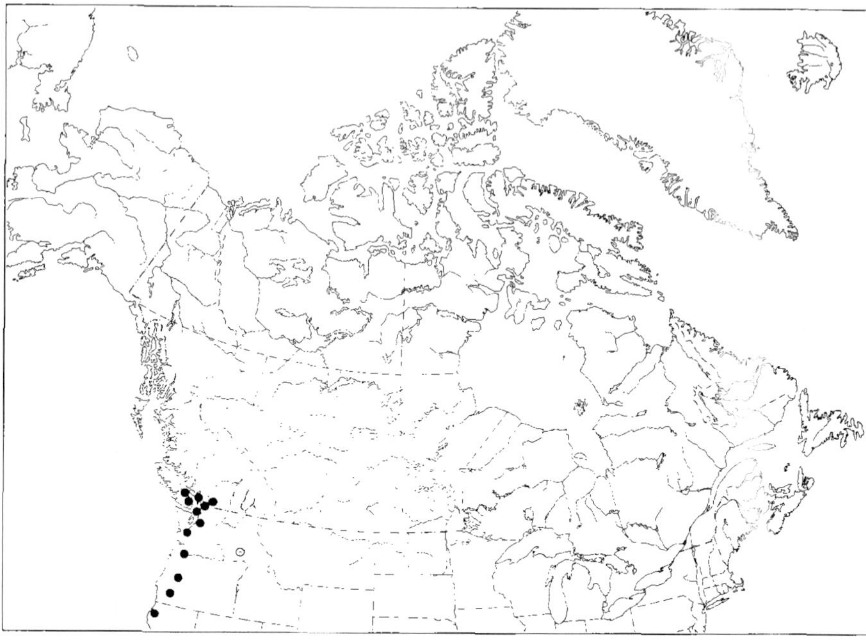

Map 81. Collection localities for *Myrmecophilus oregonensis* (●) and *M. manni* (○).

(Buckley), *Tapinoma sessile* (Say), and *Veramessor andrei andrei* (Mayr). *M. oregonensis* has been taken from nests of *Formica subpolita camponoticeps* Wheeler at Wawawai, WA, the type locality of that ant subspecies, which is also the type locality of *Myrmecophilus manni* Schimmer. *M. manni* is distinguished from *M. oregonensis* by being more pubescent (Schimmer 1911).

M. oregonensis is probably an unwelcome guest in ant nests. It does not cause injury to the hosts but is entirely dependent upon them for food, which appears to consist partly of the secretions that lubricate the bodies of the ants and are found on walls of passageways.

Family Oecanthidae

Members of the family Oecanthidae are collectively the so-called tree crickets. The family is recognizable from other crickets by the following combination of characters: body small (about 1.0–1.5 cm long), delicate, usually greenish white, and of rather uniform appearance. Head horizontal, with mouthparts projecting distinctly forward (prognathous). Ocelli absent. Pronotum distinctly longer than wide; metanotum of male thorax with large glandular sex-attractant pits. Legs slender; metatarsi tending to be subdivided so that tarsi may superficially appear to be 4-segmented; true second tarsal segments small, compressed; hind femora long, little expanded basally, not well-adapted for leaping; hind tibiae serrulate throughout, with or without spines. Wings well-developed; tegmina semitransparent, in females rather narrow, in males greatly widened and occupied almost entirely by large stridulatory apparatus, which has a relatively enormous "mirror" divided by a main vein and 3 or 4 oblique veins. Ovipositor needlelike, strongly denticulate apically. Sex chromosomes rod-shaped.

The family is worldwide in distribution (except for the cooler zones), but it is particularly abundant in the Ethiopian and Neotropical regions. There are, however, only three genera. Although these insects are called tree crickets, many do not occur on trees but rather on bushes, shrubs, and rank herbage. Eggs in all species (so far as known) are deposited in bark, twigs, or stems. Although oecanthids seem to be largely predaceous on small insects such as aphids, one or two species are locally important as crop pests (e.g., of soft fruits) because of the damage caused by oviposition punctures and/or the transmission of plant diseases (Parrott and Fulton 1914; Gloyer and Fulton 1916; L. M. Smith 1930; Elliott and Dhanvantari 1973). Males are regarded in many human societies as melodious, and captive specimens are sometimes kept for their song.

There are poems in several languages that refer to species of the genus *Oecanthus*; for example, the following excerpt from W. B. Carman's *The Pipes of Pan*:

> Pale tree-cricket with his bell
> Ringing ceaselessly and well,
> Sounding silver to the bass
> Of his cousin in the grass.
>
> (Carman 1902)

There is one subfamily of Oecanthidae. Two of its three genera occur in the region considered here. For an account of Canadian and United States species, see T. J. Walker (1962, 1963).

Key to genera of Oecanthidae

1. Hind tibia armed with apical spurs only (Fig. 372). Basal antennal segment (scape) with small prominent tubercle on outer side (Fig. 371); first antennal segment (scape), and second (pedicel) not marked ventrally with black .... ***Neoxabea*** (p. 256)

   Hind tibia armed distally with several long spurs and many short spines (Fig. 373). Basal antennal segment (scape) without prominent tubercle on outer edge (Figs. 375–386); first antennal segment (scape), and sometimes second (pedicel) with or without black ventral markings .... ***Oecanthus*** (p. 257)

Clé des genres d'Oecanthidae

1. Tibias postérieurs armés seulement d'éperons apicaux (fig. 372). Segment basilaire de l'antenne (scape) muni d'un petit tubercule du côté extérieur (fig. 371); côté ventral du premier (scape) et du deuxième (pédicelle) segments de l'antenne non marqué de noir .... ***Neoxabea*** (p. 256)

Tibias postérieurs armés, en position distale, de plusieurs longs éperons et de plusieurs courtes épines (fig. 373). Segment basilaire de l'antenne (scape) sans tubercule proéminent du côté extérieur (fig. 375 à 386); côté ventral du premier (scape) et parfois du deuxième (pédicelle) segments de l'antenne avec ou sans marques de couleur noire .... ***Oecanthus*** (p. 257)

Genus *Neoxabea* W. F. Kirby

Description. Basal antennal segments without ventral black markings; basal segment (scape) with small prominent tubercle on outer side (Fig. 371). Hind tibia with apical spurs only, lacking long and short spines toward distal end (Fig. 372).

Neoxabea bipunctata (De Geer)

Figs. 370-372; Map 82

Gryllus bipunctatus De Geer, 1773:523, pl. 43, fig. 7.

Gryllus (*Acheta*) *binotatus* Gmelin, 1789:2062.

Neoxabea bipunctata; W. F. Kirby 1906:76; T. J. Walker 1962:306; Helfer 1963:337.

Diagnosis. Tegmina of female each with pair of elongate transverse blackish spots.

Map 82. Collection localities for *Neoxabea bipunctata*.

Description. Basal segment of antenna with lateral toothlike projection (Fig. 371). Head depressed between eyes. Pronotum long, with sides nearly parallel but with posterior fifth expanded. Hind wings long, extending well beyond tip of abdomen. General color brownish with slight pinkish cast in fresh specimens; antennae pale yellow; legs pale; tegmina of female each with pair of elongate transverse black spots, one at middle and one near base (Fig. 371).

Range. Minnesota to Massachusetts, south to Mexico and Central America.

Behavior and habitats. *N. bipunctata* occurs on a wide variety of deciduous trees and shrubs, such as apple, wild cherry, ash, elm, oak, sugar maple, willow, hackberry, and sassafras, and on tangled undergrowth and vines of wild grape, from August to the onset of frost in northern regions.

T. J. Walker (1962) recorded the calling song. He called it a broken trill—"a pulse sequence of uniform pulse rate is discontinued and begun at irregular intervals of from one to several seconds." Single sequences may last up to 10 seconds before being interrupted. At a temperature of 24°C, about 105 pulses per second are produced; at 18°C, this is reduced to about 70 pulses per second. This species, like most singing Grylloidea and many Tettigonioidea, exhibits a definite response to temperature in its stridulation.

Life history. As far as is known, there is but a single annual generation. Winter is passed in the egg stage.

Genus *Oecanthus* Audinet-Serville

Description. Basal antennal segment without prominent tubercle on outer edge. Hind tibia armed distally with several long and many short spines in addition to apical spurs (Fig. 373). Posterior fifth of pronotum not expanded. Hind wings little, if any, longer than tegmina.

Key to species of *Oecanthus*

(Adapted from T. J. Walker (1962, 1963))

1. Inner edge of ventral face of basal antennal segment with pale swelling marked with black. Stridulatory file of male with fewer than 30 teeth per millimetre. Calling song a broken trill or series of regular chirps 2

   Inner edge of ventral face of basal antennal segment without pale swelling. Stridulatory file of male usually with more than 30 teeth per millimetre. Calling song a continuous trill 5

2(1). Black mark on basal antennal segment (scape) round or oval; second segment (pedicel) with similar mark. Male tegmina broad, with dorsal field more than two-fifths as wide as long. Stridulatory file with more than 35 teeth. Calling song a regular chirp 3

   Black mark on basal antennal segment (scape) neither round nor oval; second segment (pedicel) with elongate black mark. Male tegmina narrower, with dorsal field usually less than two-fifths as wide as long. Stridulatory file with less than 35 teeth. Calling song a trill . . . . (*niveus* group) . . . 4

3(2). Dark mark on second antennal segment (pedicel) usually reduced (length less than half that of segment) or absent, with center of mark near distal border of segment (Fig. 378). Occurring in British Columbia and Washington to California. Calling song at 21°C about 90 chirps per minute .... *rileyi* **Baker** (p. 261)

Dark mark on second antennal segment (pedicel) usually at least half as long as segment, with center of mark near midpoint of segment (Fig. 377). Widespread, Québec and New England to west coast. Calling song at 21°C at least 120 chirps per minute (in western United States, at least 150 chirps per minute) .... *fultoni* **T. J. Walker** (p. 262)

4(2). Mark on first antennal segment (scape) straight, rarely slightly curved (Fig. 376). Vertex without prominent orange marking. Tegmina of male 12–15 mm long. Stridulatory file with 19.0–23.4 teeth per millimetre .... *exclamationis* **Davis** (p. 264)

Mark on first antennal segment (scape) strongly curved toward inner side or J-shaped (Fig. 375). Vertex yellow or orange in fresh specimens (fades on drying). Tegmina of male 9.8–12.2 mm long. Stridulatory file with 23.7–29.7 teeth per millimetre .... *niveus* **(De Geer)** (p. 265)

5(1). First antennal segment (scape) not marked ventrally with black (sometimes with narrow line along inner edge); frons and basal antennal segments usually tinged with red. Male tegmina broad, with width of dorsal field about half its length. Subgenital plate of female with broad posterior notch that is one-fourth to one-half as broad as widest part of plate 6

First antennal segment (scape) usually marked with more than one black line; frons and basal antennal segments not tinged with red. Male tegmina narrower, with width of dorsal field rarely more than two-fifths its length. Subgenital plate of female with narrow posterior notch that is no wider than one-fifth greatest width of plate 7

6(5). Third antennal segment usually much darker than second segment; first and second segments reddish brown, frequently with distinct dark lines on inner edge (Figs. 379, 380). Stridulatory file with 40–58 teeth. Occurring in western United States and British Columbia .... *californicus* **Saussure** (p. 266)

Third antennal segment seldom darker than second segment; antennae pink, dark red or purple basally, fading rapidly distally, without distinct dark line on first and second segments. Stridulatory file with 36–47 teeth. Occurring in eastern half of North America ... *latipennis* **Riley** (p. 267)

7(5). Pronotum black or with 1–3 black or dark longitudinal stripes; underside and distal parts of legs and antennae black .... *nigricornis* **F. Walker** (in part) (p. 268)

Pronotum neither black nor with dark longitudinal stripes (*O. laricis* may have brown stripes with lateral ones somewhat dusky); underside and distal parts of legs and antennae sometimes dusky or brown, but not black 8

8(7). Head, pronotum, and legs largely brown; tegmina contrasting green or dusky green in fresh specimens (green fades to pale brown in dried specimens); black marks on basal antennal segments usually weak and not extensive (Fig. 386). Found only in (rarely beneath) coniferous trees 9

Head, pronotum, and legs without brown pigment and not contrasting in color with tegmina; black marks on basal antennal segments strong (Figs. 381–385). Seldom found in coniferous trees 10

9(8). Tegmina exceeding 12 mm. Brown areas light and slightly reddish. Found in pines (*Pinus*) and balsam fir (*Abies*) .... *pini* **Beutenmüller** (p. 269)

Tegmina shorter than 12 mm. Brown areas dark, occasionally dusky. Found in tamarack (*Larix*) and hemlock (*Tsuga*) .... *laricis* **T. J. Walker** (p. 271)

10(8). Black marks on second antennal segment confluent, contiguous, or separated by less than one-third width of inside mark (Fig. 385) .... *argentinus* **Saussure** (p. 271)

Black marks on second antennal segment distinctly separated by more than one-third width of inside mark (Figs. 381–384) 11

11(10). Outside marks on first and second antennal segments less heavily pigmented than inner marks; outside mark on first segment often round (Fig. 384). Hind tibia and apex of hind femur not conspicuously darkened. Stridulatory file with 50–67 teeth .. *quadripunctatus* **Beutenmüller** (p. 272)

Outside marks on first and second antennal segments pigmented as heavily as inner marks; outside mark on first segment never round (Figs. 381–383). Hind tibia and apex of posterior femur conspicuously darkened; stridulatory file with 35–51 teeth .... *nigricornis* **F. Walker** (in part) (p. 268)

Clé des espèces d'*Oecanthus*

(adaptée de T.J. Walker, 1962, 1963)

1. Bord intérieur de la face ventrale du segment basilaire de l'antenne portant un renflement pâle marqué de noir. Organe stridulant du mâle portant moins de 30 dents par millimètre. Stridulation émise par saccades ou faite d'une série d'appels réguliers 2

Bord interne de la face ventrale du segment basilaire de l'antenne dépourvu d'un renflement pâle. Organe stridulant du mâle portant d'habitude plus de 30 dents par millimètre. Stridulation continue 5

2(1). Marque noire sur le segment basilaire de l'antenne (scape) ronde ou ovale; deuxième segment (pédicelle) portant une marque semblable. Tegmina des mâles larges; largeur de la région dorsale égale à plus des 2/5 de sa longueur. Organe stridulant comptant plus de 35 dents. Stridulation faite d'une série d'appels réguliers 3

Marque noire du segment basilaire de l'antenne (scape) ni ronde ni ovale; deuxième segment (pédicelle) portant une marque noire allongée. Tegmina des mâles plus étroits; largeur de la région dorsale en général inférieure aux 2/5 de sa longueur. Organe stridulant comptant moins de 35 dents. Stridulation continue ...... (groupe des *niveus*) ... 4

3(2). Marque noire du deuxième segment de l'antenne (pédicelle) habituellement réduite (moins de la moitié de la longueur du segment) ou absente, son centre étant situé près du bord distal du segment (fig. 378). Espèce vivant en Colombie-Britannique et de l'État de Washington jusqu'en Californie. À 21°C la stridulation compte environ 90 appels par minute .... *rileyi* **Baker** (p. 261)

Marque noire du pédicelle de longueur habituellement au moins égale à la moitié de celle de ce segment; centre de cette marque situé près du milieu du segment (fig. 377). Espèce répandue du Québec et de la Nouvelle-Angleterre jusqu'à la côte Ouest. À 21°C, stridulation comptant au moins 120 appels par minute (dans l'ouest des États-Unis, au moins 150 appels par minute) .... *fultoni* **T.J. Walker** (p. 262)

4(2). Marque du scape rectiligne, parfois légèrement incurvée (fig. 376). Vertex sans marque orangée proéminente. Tegmina des mâles de 12 à 15 mm de longueur. Organe stridulant comptant de 19,0 à 23,4 dents par millimètre .... ***exclamationis*** **Davis** (p. 264)

Marque du scape fortement incurvée vers le côté interne ou en forme de J (fig. 375). Vertex jaune ou orangé chez les spécimens frais (la couleur s'atténue en séchant). Tegmina des mâles de 9,8 à 12,2 mm de longueur. Organe stridulant comptant de 23,7 à 29,7 dents par millimètre ... . ***niveus*** **(De Geer)** (p. 265)

5(1). Scape de l'antenne sans marque noire du côté ventral (parfois marqué d'une ligne étroite le long du bord interne); front et segments basilaires de l'antenne en général teintés de rouge. Tegmina des mâles larges; largeur de la région dorsale égale à la moitié environ de sa longueur. Plaque sous-génitale des femelles marquées d'une encoche dont la largeur varie du quart à la moitié de la largeur maximale de la plaque 6

Scape de l'antenne habituellement marqué de plus d'une ligne noire; front et segments basilaires de l'antenne non teintés de rouge. Tegmina des mâles plus étroits; largeur de la région dorsale rarement égale à plus des 2/5 de sa longueur. Plaque sous-génitale des femelles portant une encoche postérieure dont la largeur ne dépasse pas le cinquième de la largeur maximale de la plaque 7

6(5). Troisième segment de l'antenne habituellement beaucoup plus foncé que le deuxième; premier et deuxième segments brun rougeâtre, souvent marqués, du côté interne, de lignes noires distinctes (fig. 379 et 380). Organe stridulant comptant de 40 à 58 dents. Espèce vivant dans l'ouest des États-Unis et en Colombie-Britannique .... ***californicus*** **Saussure** (p. 266)

Troisième segment de l'antenne rarement plus foncé que le deuxième; antennes rose, rouge foncé ou pourpre à la base, la coloration s'atténuant rapidement vers la partie distale; pas de ligne noire distincte sur les premiers et deuxième segments. Organe stridulant comptant de 36 à 47 dents. Espèce vivant dans la partie est de l'Amérique du Nord .... ***latipennis*** **Riley** (p. 267)

7(5). Pronotum noir ou marqué d'une à trois bandes longitudinales noires ou foncées; dessous et région distale des pattes et des antennes noires .... ***nigricornis*** **F. Walker** (en partie) (p. 268)

Pronotum ni noir ni marqué de bandes longitudinales foncées (*O. laricis* peut porter une série de bandes brunes dont les plus latérales sont assez foncées); dessous et portion distale des pattes et des antennes parfois sombres ou bruns, mais jamais noirs 8

8(7). Tête, pronotum et pattes de couleur principalement brune; tegmina vert ou vert foncé, contrastant chez les spécimens frais (avec le séchage, le vert disparaît et les spécimens deviennent brun pâle); marques noires des segments basilaires des antennes habituellement pas très grosses (fig. 386). Insectes n'habitant que dans les peuplements de conifères (rarement plus au sud) 9

Tête, pronotum et pattes sans pigmentation brune, à coloration semblable à celle des tegmina; marques noires des segments basilaires des antennes très développées (fig. 381 à 385). Espèces vivant rarement dans les peuplements de conifères 10

9(8). Tegmina de longueur dépassant 12 mm. Régions à coloration brun pâle et légèrement rougeâtre. Espèce vivant dans les peuplements de pins (*Pinus*) et de sapins baumiers (*Abies*) .... ***pini*** **Beutenmüller** (p. 269)

Tegmina de longueur inférieure à 12 mm. Régions à coloration brun foncé, parfois obscurci. Espèce vivant dans les peuplements de mélèzes (*Larix*) et de pruches (*Tsuga*) .... ***laricis*** **T.J. Walker** (p. 271)

10(8). Marques noires du deuxième segment des antennes confluentes, contiguës ou séparées par une distance inférieure au tiers de la largeur de la marque interne (fig. 385) .... ***argentinus*** **Saussure** (p. 271)

Marques noires du deuxième segment des antennes distinctement séparées par une distance supérieure au tiers de la largeur de la marque interne (fig. 381 à 384) 11

11(10). Marques externes des premier et deuxième segments antennaires moins pigmentées que les marques internes; marque externe du premier segment souvent arrondie (fig. 384). Tibias postérieurs et apex des fémurs postérieurs pas particulièrement foncés. Organe stridulant comptant de 50 à 67 dents .... ***quadripunctatus*** **Beutenmüller** (p. 272)

Marques externes des premier et deuxième segments antennaires aussi foncées que les marques internes; marque externe du premier segment jamais arrondie (fig. 381 à 383). Tibias postérieurs et apex des fémurs postérieurs foncés; organe stridulant comptant de 35 à 51 dents .... ***nigricornis*** **F. Walker** (en partie) (p. 268)

Oecanthus rileyi Baker

Fig. 378; Map 83

Oecanthus Rileyi Baker, 1905:81.
Oecanthus rileyi; T. J. Walker 1962:309.

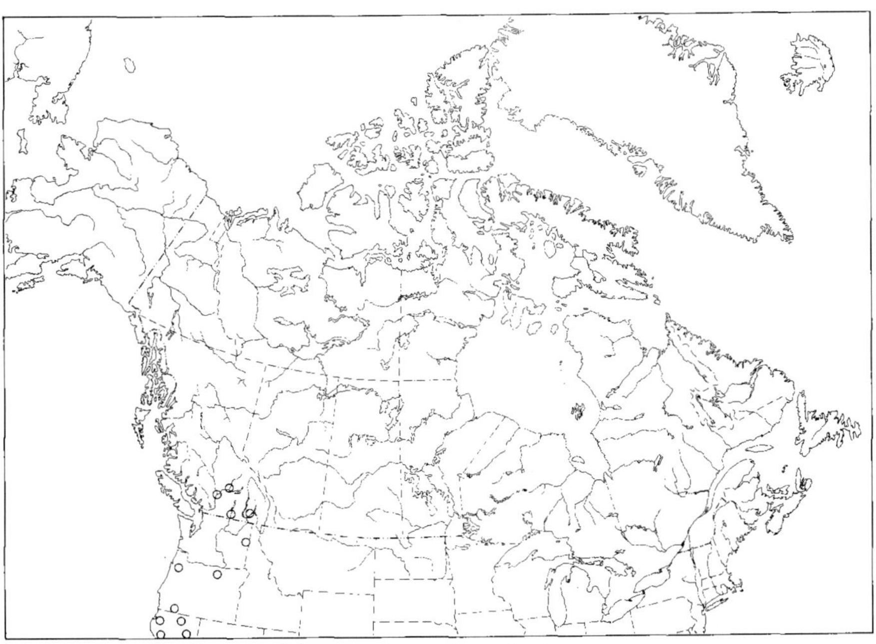

Map 83. Collection localities for *Oecanthus rileyi.*

Diagnosis. Black marks on basal antennal segments small. Cricket western in range.

Description. Black marks on undersides of basal antennal segments usually small (compare Figs. 377, 378), in some specimens reduced to mere dots; mark on basal segments confined to basal half of segment. Stridulatory file 1.59–1.84, mean 1.75 mm long; number of teeth 38–46, mean 42.6; number of teeth per millimetre 22.8–25.8, mean 24.3.

Range. British Columbia to California.

Behavior and habitats. These are practically the same as for *O. fultoni*. The habitat was reported in eastern Oregon by Fulton (1925) as bushes, including loganberry, raspberry, and wild rose, and with cliff-bracken and other plants in burned over areas. He found *O. fultoni* in the same region associated with prune, apple, oak, and ash. Host records apparently help in distinguishing between the two species where they are sympatric, but T. J. Walker (1962) found inconsistencies in California.

Stridulation in *O. rileyi* is slower than in *O. fultoni*, about 90 chirps per minute at 21°C. Chirps usually consist of 11 pulses; 8–13 pulses per chirp have been noted. As later mentioned, *O. fultoni* usually has an eight-pulse chirp.

Life history. *O. rileyi* has only one generation each year.

Economic importance. L. M. Smith (1930) reported this species as causing economic damage to cultivated raspberries in California.

Oecanthus fultoni T. J. Walker

Fig. 377; Plate V*D*; Map 84

Oecanthus fultoni T. J. Walker, 1962:309.
Oecanthus fultoni; Vickery et al. 1974:59.

Common name. Snowy tree cricket.

Diagnosis. Vertex yellow; basal antennal segment with circular black spot.

Description. Body length, male 12.3, female 9.2 mm; ovipositor 4.0 mm. Vertex of head orange yellow; basal antennal segment yellow except for white prominent swelling on inner ventral face, this marked with a black, nearly circular, spot; second segment with ovate to quadrate black spot occupying most of middle of segment (Fig. 377). Length of male stridulatory file 1.49–1.77, mean 1.64 mm; number of teeth 38–45, mean 41.3; number of teeth per millimetre 21.8–25.3, mean 23.0 (T. J. Walker 1962). Plate V*D* shows the shape and spacing of the stridulatory teeth.

Range. British Columbia to Maine, south to California and Texas.

Behavior and habitats. In the northeastern United States, *O. fultoni* is common in vines and shrubbery, sometimes numerous on blackberry (T. J. Walker 1962). In Québec and Ontario, it has been taken on *Rubus odoratus* Linnaeus and other *Rubus* species.

The stridulations of *O. fultoni* and *O. rileyi* are similar, but the chirp rate of the latter species is somewhat slower (T. J. Walker 1962). *O. fultoni*

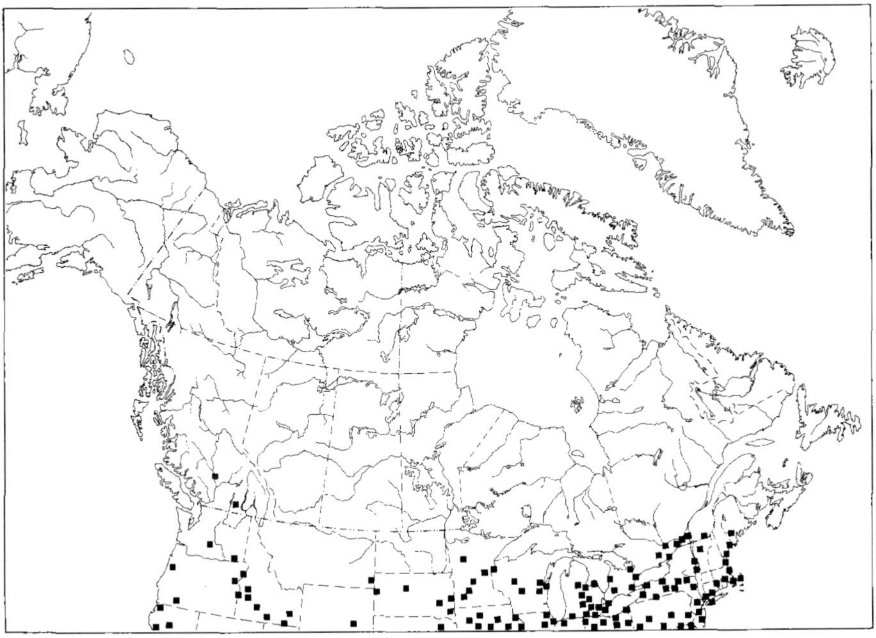

Map 84. Collection localities for *Oecanthus fultoni*.

males produce a regular series of chirps of approximately equal duration with approximately uniform intervals. Each chirp consists of 2–11 pulses, most often 8. T. J. Walker (1969) reported that *O. fultoni* males, singing in proximity to one another, synchronize their songs. Thus, the crickets in a given habitat usually produce a chorus by synchronizing their chirps (but not their pulses).

R. D. Alexander (1975) discussed the evolutionary significance of choruses in acoustical insects. Synchrony of stridulation is "restricted to species in which the normal calling song contains a precise or highly uniform chirp or phrase rate within the range of two to five per second." It is possible that synchrony may tend to improve ability to attract females from greater distances. Aggregation of males might also be brought about by selection of habitats that offer some protection from predators, or by selection of abundance of plants suitable for oviposition by females. Females of chorusing species may be more responsive in the presence of numerous males. Such an aggregation may also be advantageous to individuals that can not produce strong signals in competition. Also, some males do not sing, but can intercept females moving toward the singers. In general, synchronous stridulation is not yet well understood; answers to the questions raised by these possibilities remain to be found. It is known that, in nonchorusing species, where males are singing in close proximity, one male may move away, may try to drive the other male away, or may stay and try to outperform the other male. Sometimes, males near other males sing only during the intervals when the latter are silent.

O. fultoni is sometimes called the "thermometer cricket," because of the ease with which its chirp rate can be correlated with temperature. T. J. Walker (1962) stated that a reasonable approximation of temperature in degrees Fahrenheit can be obtained by counting the number of chirps in 13 seconds and adding 40.[1] At 64°F, *O. fultoni* (in the east) chirps about 24 times in 13 seconds: 24 chirps plus 40° = 64°F. A more complex calculation is necessary to determine degrees Celsius.

Life history. *O. fultoni* has one generation each year throughout its range. At the most northern points of the range, adults appear late in July and some individuals persist until they are killed by frost.

Oecanthus exclamationis Davis

Fig. 376; Map 85

Oecanthus exclamationis W. T. Davis, 1907:173.

Oecanthus exclamationis; Fulton 1915:30; T. J. Walker 1962:309; Helfer 1963:335.

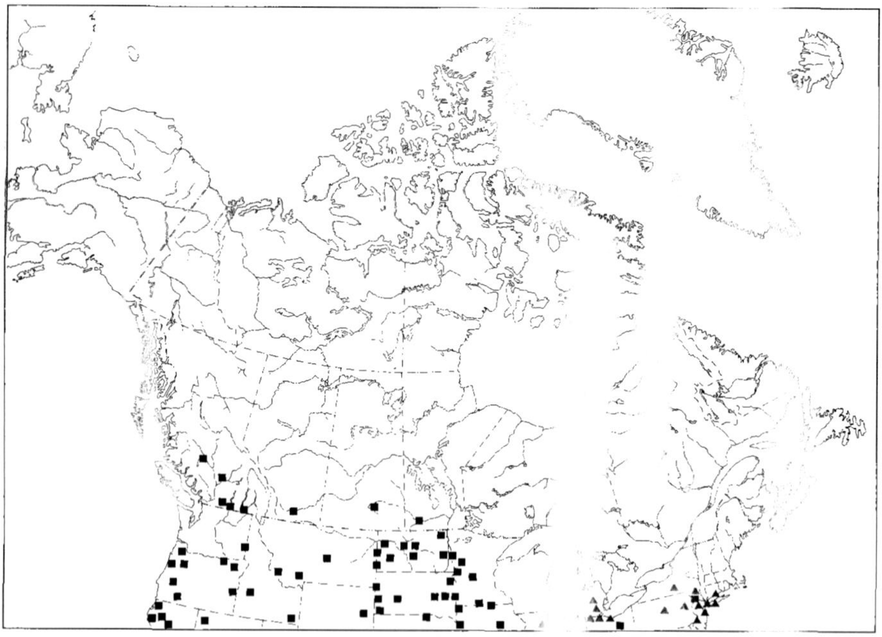

Map 85. Collection localities for *Oecanthus exclamationis* (▲) and *O. argentinus* (■).

[1] This was not Walker's original discovery. It has been attributed to the astronomer Harlow Shapley of Harvard, who used a base line of 14 (not 13) seconds for his "recent" calculations, predicting 75% success to within 1°F. Allard (1930), who reviewed the whole history of the "thermometer cricket" as far back as 1881, refers neither to Shapley nor to this particular formula.

Diagnosis. Black mark on basal antennal segment elongate but not curved (Fig. 376).

Description. Body length 12 mm; ovipositor 5 mm. Antenna with black mark on basal segment; vertex not orange or yellow, occasionally darker than dorsum of prothorax, which is pale greenish white. Stridulatory file 0.86–1.08 mm, mean 0.97 mm long; number of teeth 17–23, mean 21.1.

Range. Illinois to Connecticut, south to Arizona and Florida.

Behavior and habitats. T. J. Walker (1962) states that this species is strictly arboreal, even more so than *O. niveus*, and has been collected from apple, oak, elm, wild cherry, maple, hackberry, and catalpa, usually high in the crowns.

The stridulation is a broken trill, which is difficult to distinguish by ear from that of related species such as *O. niveus*, although the pulse rate is slightly faster.

Life history. Cantrall (1968) reported it to be active in Michigan from 2 August to 17 September.

Oecanthus niveus (De Geer)

Fig. 375; Map 86

Gryllus niveus De Geer, 1773:522, pl. 43, fig. 6.

Oecanthus angustipennis Fitch, 1856:413; Cantrall 1943:168; Helfer 1963:335.

Oecanthus niveus; T. J. Walker and Gurney 1960:10; T. J. Walker 1962:307, 308.

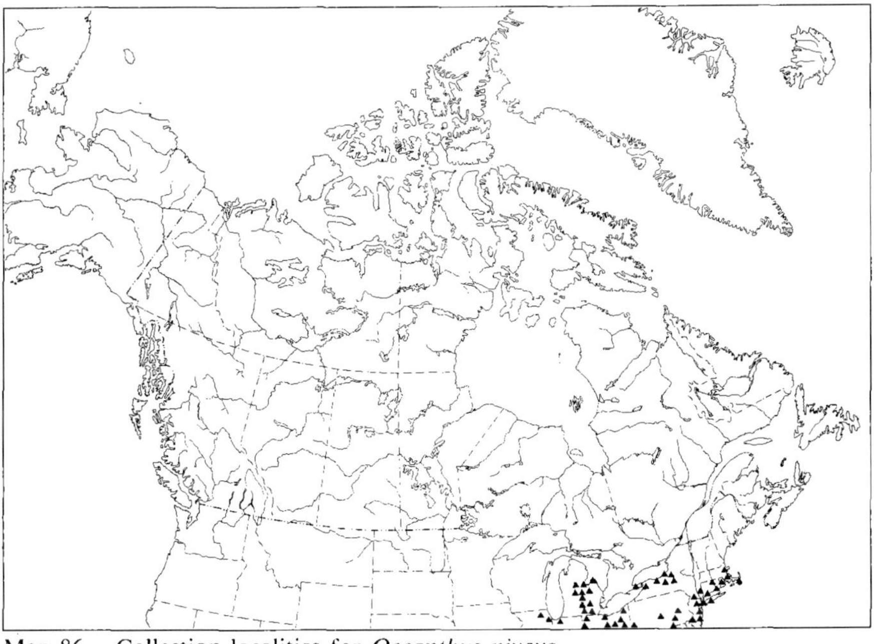

Map 86. Collection localities for *Oecanthus niveus*.

Diagnosis. Black mark on basal antennal segment J-shaped.

Description. Antenna with black marks ventrally on 2 basal segments, that on basal segment J-shaped, distinctly curved toward inner side, that on second segment irregularly ovate (Fig. 375); in fresh specimens, vertex often yellowish to orange, this fading in stored specimens. Male stridulatory file 0.88–1.13 mm, mean 1.00 mm long; number of teeth 22–30, mean 26; number of teeth per millimetre 23.7–29.7, mean 25.8.

Range. Illinois to Ontario and Massachusetts, south to Texas and Florida.

Behavior and habitats. *O. niveus* lives in deciduous trees, often too high for easy capture, but occasionally it is found on low rank vegetation such as blackberry, goldenrod, or ironweed.

Life history. There is a single annual generation in the northern part of the range, the adult stage occurring from early August to October. In Florida, there are two generations each year (T. J. Walker 1962).

Oecanthus californicus Saussure

Figs. 379, 380; Map 87

Oecanthus californicus Saussure, 1874:462.

Oecanthus californicus; Hebard 1923*a*:340; Helfer 1963:335; T. J. Walker 1962:317.

Oecanthus californicus pictipennis Hebard, 1935*b*:78; T. J. Walker 1962:317.

Map 87. Collection localities for *Oecanthus californicus*.

Oecanthus californicus californicus; Hebard 1935*d*:315.

Diagnosis. Third segment of antenna usually much darker than second.

Description. Depression in distal part of terminal segment of maxillary palp usually more than one-half length of segment. Male tegmen less than 14 mm long. Basal two antennal segments often with black line on inner edge of ventral surface (Fig. 379); third segment darker than second segment. Stridulatory file 1.20–1.85 mm, mean 1.44 mm long; number of teeth 40–58, mean 47.5; number of teeth per millimetre 29.5–37.4, mean 33.0.

Range. British Columbia to Wyoming, south to California and Mississippi.

Behavior and habitats. *O. californicus* has been taken from scrub oak, and from low shrubby plants such as rose bushes, in the northern part of its range.

Stridulation is a continuous trill, differing little from that of *O. latipennis*, except that at any given pulse rate the frequency of *O. californicus* songs is higher. The geographic ranges of the two species, however, are distinctly separated.

Oecanthus latipennis Riley

Map 88

Oecanthus latipennis Riley, 1881:61.

Oecanthus latipennis; Beutenmüller 1894*c*:272; Helfer 1963:335; T. J. Walker 1962:318.

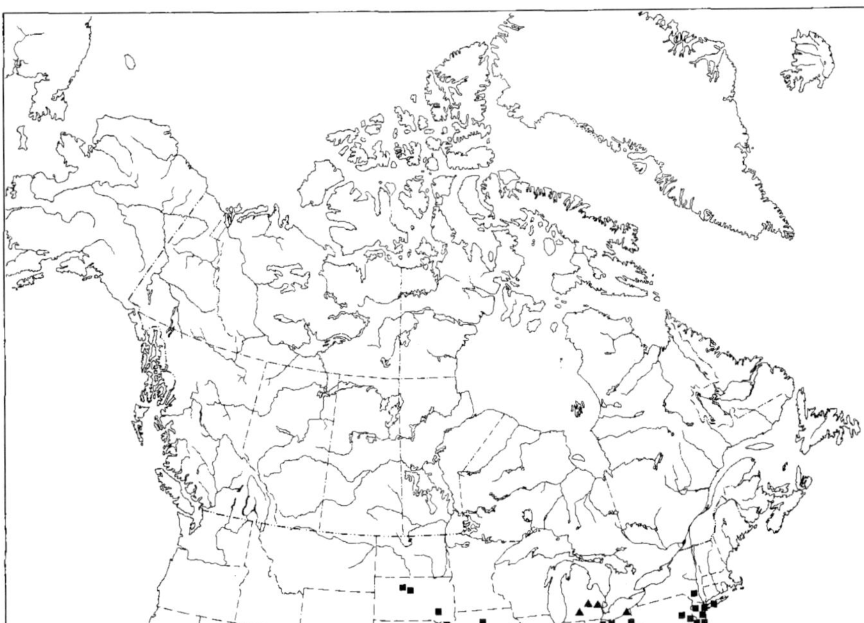

Map 88. Collection localities for *Oecanthus latipennis* (■) and *O. laricis* (▲).

Diagnosis. Basal antennal segments unmarked.

Description. Depression of distal part of terminal segment of maxillary palp less than two-fifths length of segment. Male tegmen more than 14 mm long. Basal antennal segments not marked ventrally with black; third antennal segment usually not darker than second segment. Stridulatory file 1.32–1.68 mm, mean 1.48 mm long; number of teeth 36–47, mean 42.1; number of teeth per millimetre 27.3–30.5, mean 28.4

Range. South Dakota to New York, south to southeastern United States.

Behavior and habitats. The species is found on shrubs and low trees, particularly scrubby oaks in dry open woods. It may be abundant in thickets of vines and among coarse weeds on woodland edges and fencerows, or on plants such as goldenrod, blackberry, and horseweed in abandoned fields.

Stridulation is a continuous trill with a relatively low pulse rate, about 45 pulses per second at 21°C and 60 pulses per second at 27°C.

Life history. There is one generation each year, adults appearing during late August.

Oecanthus nigricornis F. Walker

Figs. 373, 374, 381–383; Plate V*A*; Map 89

Oecanthus nigricornis F. Walker, 1869*a*:93.

Oecanthus nigricornis; Beutenmüller 1894*b*:250; Helfer 1963:336; T. J. Walker 1963:772; R. D. Alexander et al. 1972:47.

Oecanthus nigricornis nigricornis; Hebard 1928:305; Cantrall 1943:169.

Common name. Blackhorned tree cricket.

Diagnosis. In the present region, the only species of *Oecanthus* with black legs.

Description. Ventral side of 2 basal antennal segments with paired black marks (Figs. 382, 383); in very dark specimens, markings on basal segment sometimes obscured by general black color (Fig. 381); eastern specimens normally conspicuously blackened on legs, head, and pronotum, but some specimens, particularly from western part of range, lacking dark color, although distal end of hind femur and hind tibia almost always somewhat darkened. Male stridulatory file (Plate V*A*) with 37–51 teeth, mean 45.8 (*O. forbesi*, 35–45 teeth, mean 38.6); length of file 1.06–1.39 mm, mean 1.28 mm (*O. forbesi*, 1.03–1.35 mm, mean 1.16 mm); teeth per millimetre 32.8–39.8, mean 35.9 (*O. forbesi*, 29.4–35.5, mean 33.15) (T. J. Walker 1963).

Range. British Columbia to Maine, south to West Virginia.

Behavior and habitats. *O. nigricornis* is found on coarse-stemmed weeds, brambles, and small trees such as willow and sumac, usually in fertile areas and near water. In Québec, the species is commonly associated with goldenrod (*Solidago* spp.) and blackberry (*Rubus*). Oviposition occurs in the stems of coarse herbaceous and woody plants. Elliott and Dhanvantari (1973) reported that eggs are inserted into the pith of plants, closely spaced in rows, resulting in splits in the bark and elongate areas of dead sapwood. Eggs are usually deposited on the upper sides of horizontal or inclined twigs; they are

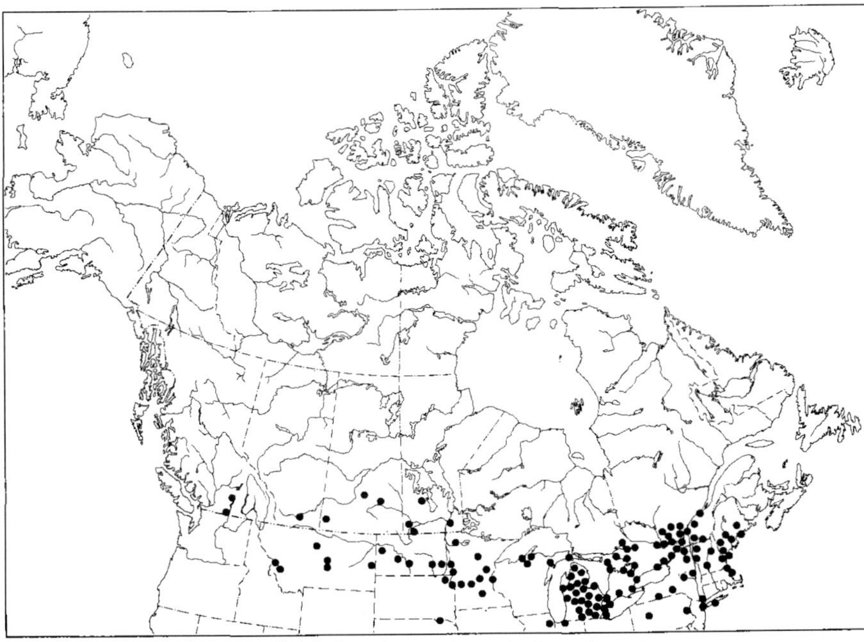

Map 89. Collection localities for *Oecanthus nigricornis*.

deposited in any position on vertical stems or twigs. Bell (1980) describes some abnormal sexual behavior.

The stridulation of *O. nigricornis* has a carrier frequency of 3.7–3.9 kHz.

Life history. *O. nigricornis* has one generation each year. Eggs are laid in September and the winter is passed in the egg stage. Nymphs appear in late June and July and adults through most of August and September, some persisting into October if not killed by frost before then. Bell (1979) discusses laboratory rearing.

Economic importance. Damage to cultivated plants has been reported for many years, particularly in Ontario (Fletcher 1896). Such plants as tobacco, raspberry, blackberry, gooseberry, and, to some extent, twigs of apple, plum, and peach (Elliott and Dhanvantari 1973) have been reported as being damaged by oviposition. Many of these reports list *O. niveus* or *O. angustipennis* as causing the injury, but all of them from Ontario undoubtedly refer to *O. nigricornis*. Oviposition punctures weaken the stems and twigs, so that they eventually break at the injured spots. Fungus diseases are also allowed to penetrate the tissues (Parrott and Fulton 1914) (Gloyer and Fulton 1916).

Oecanthus pini Beutenmüller

Fig. 386; Map 90

Oecanthus pini Beutenmüller, 1894*a*:56.

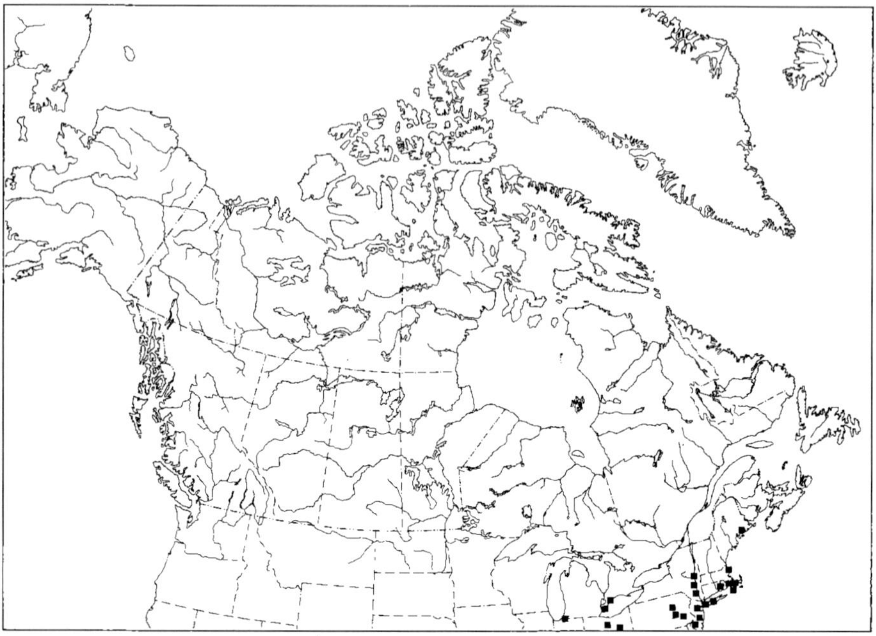

Map 90. Collection localities for *Oecanthus pini.*

Oecanthus pini; Beutenmüller 1894*c*:271, fig. 6; Helfer 1963:337; T. J. Walker 1963:774; R. D. Alexander et al. 1972:35.

Diagnosis. Tegmina green; legs and pronotum brown. Found on pine trees.

Description. Tegmina longer than 12 mm. Basal segments of antennae without distinct black markings ventrally, these obscured by brown pigment (Fig. 386); head, pronotum, and legs brown, with tegmina contrasting green (fading in preserved specimens to light brown). Stridulatory file length 1.48–1.74 mm, mean 1.59 mm; number of teeth 45–56, mean 49.4; number of teeth per millimetre 28.2–34.8, mean 31.1.

Range. Michigan to Maine, south to Texas and Florida.

Behavior and habitats. *O. pini* is usually found on pine (*Pinus*), although specimens have been taken on balsam fir (*Abies*) in New York State. It usually remains in the crowns of trees and for this reason is difficult to collect. It is never found on low herbaceous plants and never in company with other species of the genus.

The song of *O. pini* is a trill with pulse rate approximately 44 pulses per second at 25°C.

Life history. T. J. Walker (1963) reported that *O. pini* appears to be univoltine throughout its range. Adults should appear early in August in the north and persist until killed by frost.

Economic importance. There is no indication that *O. pini* is a pest species, as no damage to the hosts has been reported.

Oecanthus laricis T. J. Walker

Map 88

Oecanthus laricis T. J. Walker, 1963:773.
Oecanthus pini; Cantrall 1943:171.
Oecanthus laricis; T. J. Walker 1966:271.

Diagnosis. Tegmina dusky green; head and pronotum brown. Found on larch (*Larix*) trees.

Description. Tegmina shorter than 12 mm. Head, pronotum, and legs dark brown, sometimes slightly fuscous; tegmina dusky green (in fresh specimens). Stridulatory file length 1.19–1.33 mm, mean 1.27 mm; number of teeth 43–49, mean 46.4; number of teeth per millimetre 35.3–37.6, mean 36.4.

Range. Ohio and Michigan.

Behavior and habitats. *O. laricis* is found on tamarack (or larch) (*Larix*) and occasionally on hemlock (*Tsuga*) trees, sometimes high up in the branches.

The song is a trill with approximately 39 pulses per second at 27°C, similar to the song of *O. quadripunctatus*.

Life history. According to T. J. Walker (1963), *O. laricis* is univoltine. Adults have been found only during the period 24 July to 4 September.

Oecanthus argentinus Saussure

Fig. 385; Plate V*C*; Map 85

Oecanthus argentinus Saussure, 1874:460.
Oecanthus argentinus; Caudell 1903*a*:166; T. J. Walker 1963:773.
Oecanthus rehnii Baker, 1905:82.
Oecanthus nigricornis argentinus; Fulton 1926:13; Helfer 1963:336.

Diagnosis. Black markings on 2 basal antennal segments broad, confluent or nearly so.

Description. Ventral surface of 2 basal antennal segments with black markings broad, on second segment usually confluent, rarely separated, on first segment confluent or separated by no more than one-third width of inner mark (Fig. 385). Stridulatory file (Plate V*C*) with length 1.09–1.45 mm, mean 1.26 mm; number of teeth 42–53, mean 47.6; number of teeth per millimetre 32.8–43.7, mean 37.9.

Range. British Columbia to Manitoba and Connecticut, south to Mexico; Central and South America.

Behavior and habitats. *O. argentinus* is known to occur mainly on low herbaceous plants, although males are sometimes heard singing in low trees. Oviposition is in fine plant stems, much less woody than those in which *O. nigricornis* oviposits. T. J. Walker (1963) reports collecting *O. argentinus* on the following plants: ragweed, wild carrot, daisy fleabane, goldenrod, Joe-Pye weed, alfalfa, and others. *O. quadripunctatus* is found on many of these same plant species.

The song of *O. argentinus* is a trill, resembling those of *O. nigricornis* and *O. quadripunctatus*, but the pulse rate is somewhat different, approximately 50 pulses per second at 25°C. This is similar to *O. forbesi*, greater than *O. quadripunctatus*, and less than *O. nigricornis*.

Life history. T. J. Walker (1963) records this species as having two generations each year in Ohio, adults appearing in July and again in mid-September to October. It is apparent that this species will be able to move northward only to a limited extent; the prevalence of frosty periods in late September in Ontario, would not allow late-maturing individuals to reproduce, unless only a single annual generation were produced in northern areas. There is, however, evidence of this in the northwestern United States, as adults have been found there only later than late July. There is probably only a single generation in Western Canada.

Oecanthus quadripunctatus Beutenmüller

Fig. 384; Plate V*B*; Map 91

Oecanthus 4-punctatus Beutenmüller, 1894*b*:250.

Oecanthus quadripunctatus; E. M. Walker 1904*a*:225; T. J. Walker 1963:773; Vickery et al. 1974:58.

Oecanthus nigricornis quadripunctatus; E. M. Walker in Faull, ed. 1913:302; Cantrall 1943:170; Helfer 1963:337.

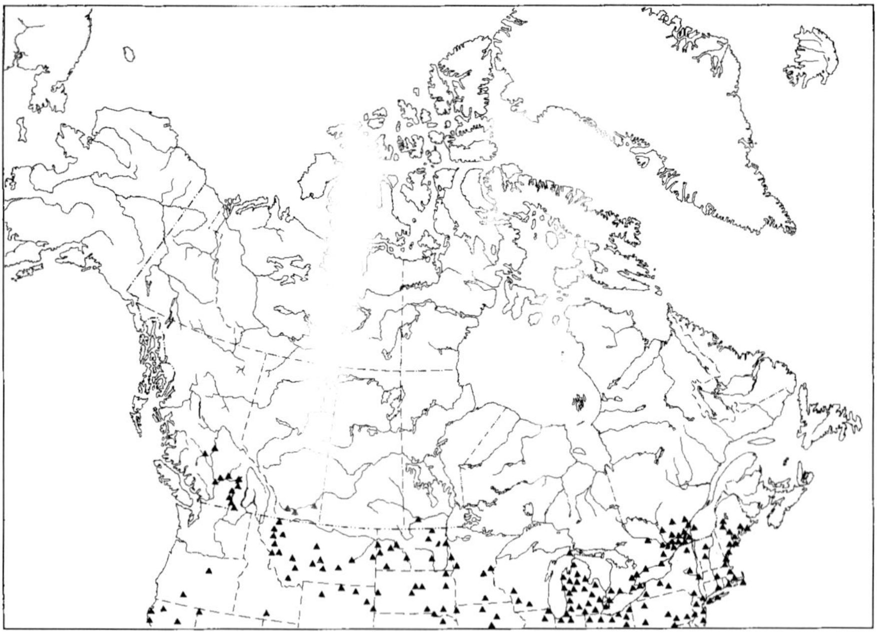

Map 91. Collection localities for *Oecanthus quadripunctatus*.

Common name. Fourspotted tree cricket.

Diagnosis. Inner marks on 2 basal antennal segments less distinct than outer marks. Tibiae and tarsi lacking black markings.

Description. First and second basal antennal segments with blackish markings, with outer markings not as heavily pigmented as inner markings; outer mark on basal segment often round (Fig. 384). Pronotum pale. Male stridulatory file (Plate V*B*) with length 1.16–1.78 mm, mean 1.47 mm; number of teeth 50–67, mean 56.7; number of teeth per millimetre 34.8–44.8, mean 38.7.

Range. British Columbia to Québec, south to California and Florida.

Behavior and habitats. *O. quadripunctatus* usually inhabits low-growing weeds and other herbaceous plants, and has been found on alfalfa at several localities in Québec and Ontario. It is not known if eggs are deposited in alfalfa. Oviposition is known in small twigs of *Rubus* and *Rosa* species.

The song is a long continued trill, somewhat lower in pitch than that of *nigricornis* or *forbesi*, with approximately 40 pulses per second at 25°C. T. J. Walker (1963) showed that the differential in pulse rates of *quadripunctatus* and *nigricornis* is consistent geographically.

Life history. *O. quadripunctatus* has a single annual generation in northern areas; adults are present from early to mid-August until frost. Farther south, it has two generations each year (T. J. Walker 1963).

Economic importance. Elliott and Dhanvantari (1973) reported this species in southern Ontario as causing damage to twigs of peach trees by oviposition.

Family Trigonidiidae

These insects are known as sword-tailed crickets and, sometimes, together with the Eneopteridae, are called bush crickets, although this is a term that may be confused with bush-crickets for Tettigonioidea. They constitute a family of small Grylloidea distinguishable from other members of the superfamily by the following combination of characters: size small (body seldom more than 1 cm long). Head generally wider than anterior margin of pronotum. Eyes prominent. Legs moderately long, slender; tarsi with second segment depressed, heart-shaped in dorsal view; hind tibia long, without denticles or serrulation, armed with 3 pairs of long movable dorsal spines and 5 apical spurs (only 2 internal ones). Wings usually fully developed, with hind wings typically much longer than tegmina, although hind wings often short; tegmina, in male, usually with well-developed stridulatory organs; "mirror," when present, large, undivided, with 1 oblique vein. Ovipositor laterally compressed, usually curved upward, with apex acute and usually denticulate, not enlarged.

There are some 25 known world genera and about 275 species (with doubtless many more awaiting discovery). In these, the pronotum is transverse to slightly longer than wide (Fig. 365), and the body is neither very hairy nor metallic.

Trigonidiidae species are found mainly in humid habitats, living on rank herbage, shrubs, bushes, and trees, often near water (including mangrove swamps, where these occur). Some exotic species (with greatly enlarged hind tibial spines) skate on the surface of the water. So far as is known all deposit their eggs in plant tissues. Only one genus is found in Canada and the adjacent United States.

Genus *Anaxipha* Saussure

Description. Small dusky to brown crickets of general form indicated in Fig. 365. Head vertical, with front oblique, forming protuberance between antennae. Ocelli small, arranged in triangle. Apical segment of maxillary palps longer than preapical segment, dilated, truncate at apex. Dorsal field of male tegmina broad, of female tegmina narrow and closely appressed to sides of abdomen. Legs long, pubescent; hind tibia with spines long; hind tarsus with long inner apical spine on first segment (metatarsus), and with brushlike adhesive pad on ventral surface of second segment (Fig. 366).

Anaxipha exigua (Say)

Figs. 365, 366; Plate III*E* Map 92

Acheta exigua Say, 1825*a*:309.

Anaxipha exigua; W. T. Davis 1888:1148; Cantrall 1943:172; R. D. Alexander et al. 1972:48.

Diagnosis. Body yellowish brown to clay yellow; hind femur with dark stripe on lower half of outer face.

Description. Size small, male (Fig. 365) 5–6, female 6–7 mm. Antennae long, about five times body length. Wings macropterous or brachypterous, although only brachypterous forms are known from northern areas; tegmina of males slightly surpassing, those of females not quite reaching tip of abdomen. Ovipositor about one-half length of hind femur. Male stridulatory file as in Plate III*E*. Coloration dull yellowish brown to clay yellow; hind femur in both sexes usually with dark stripe on lower half of outer face; head and pronotum in fresh specimens dark reddish brown; face usually with 3 stripes, and dots and dashes, of reddish brown; tegmina and legs pale brown.

Range. South Dakota to southern Ontario, south to Texas and Florida.

Behavior and habitats. *A. exigua* is found on shrubs on margins of lakes, rivers, and marshes, and among *Sphagnum* mosses in dense larch (*Larix*) swamps (Blatchley 1920). In coastal areas in Connecticut and Virginia, the species is found on bushes and grasses in tidal marshes.

Fulton (1951, 1956) reported on acoustical behavior. Adults sing day and night, except during the hottest part of the day. Fulton also reported on three different types of song and indicated that any one individual produced only one kind of song. Furthermore, the stridulatory peg counts of individuals producing each type of song differed significantly from one another. T. J. Walker (1964) indicated geographical differences as well, and

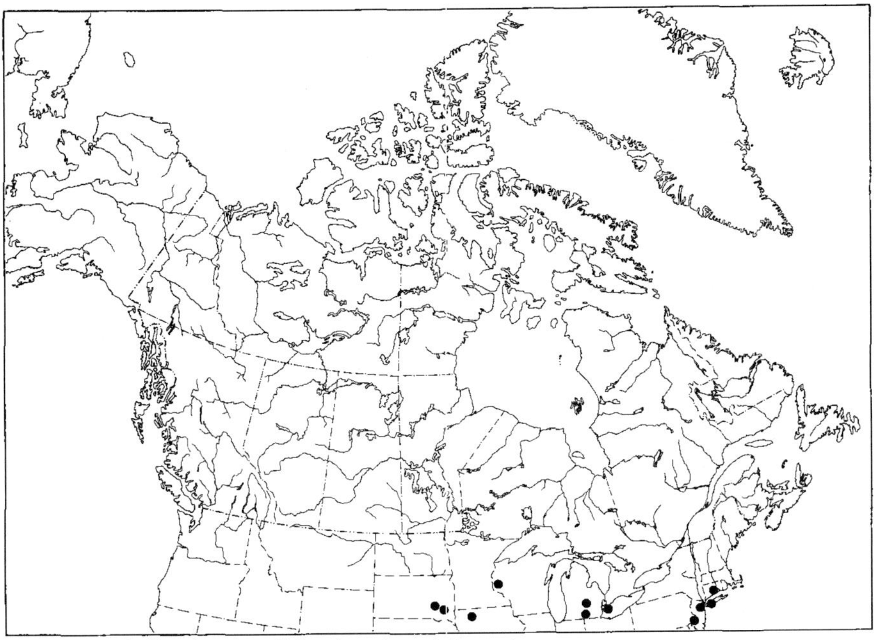

Map 92. Collection localities for *Anaxipha exigua*.

concluded that each of the three types belongs to a distinct species, but he did not name them. The northern form, which he called a "triller," has an average of 121 teeth in the stridulatory file and produces a trill of 39 pulses per second at 24°C. It is not certain that the name *A. exigua* truly applies to this taxon, but Walker states that the "triller" occurs farther north and west than the other two forms. It is probable, therefore, that the name *exigua* is properly applicable to the species here treated, the type locality of the species so named being in Kansas. Critical examination of material from that state will, however, be necessary before this can be stated with certainty.

Life history. In all parts of its range, adults of *A. exigua, sensu lato*, are found late in the season, in August and throughout September, or until first frost in northerly regions. There is only one generation each year. Overwintering is in the egg stage.

Family Gryllidae

Members of the family Gryllidae, *sensu stricto*, include the field, house, ground, and short-tailed crickets. They constitute the principal family of the Grylloidea, distinguishable from other members of the superfamily by the following combination of characters: size variable, from rather small to medium (4.5–24 mm). Coloration generally blackish, brownish, or tawny. Head generally globular. Antennae inserted above middle of face. Hind tibiae strongly spined, without sawlike denticles between spines; tarsi with second segment compressed. Wings usually moderately developed, often reduced; male tegmina rather broad, with well-developed basal stridulatory apparatus. Ovipositor needlelike, with apex slightly enlarged, without obvious denticles.

Nearly 800 world species and a little over 100 genera are known.

Most Gryllidae species are ground-inhabiting, often (except Nemobiinae) living in burrows for at least part of their lives, and some species (particularly of Brachytrupinae) making extensive galleries. Some of these last-named species may sometimes be serious crop pests, destroying plant roots and shoots. Some Gryllinae species may also be occasional crop pests or may be generally obnoxious in other ways.

Key to subfamilies of Gryllidae

1. Body small (less than 15 mm), rather bristly. Hind tibiae with articulated spines in addition to apical spurs. Males with well-developed sex glands at bases of tegmina, or with specialized "bleeding" spines on hind tibiae. Ovipositor with apex of dorsal valve crenulate or denticulate (Figs. 358–364) .... **Nemobiinae** (p. 276)

   Body larger, not noticeably bristly. Hind tibiae with fixed spines. Males without or with less well-developed sex glands at bases of tegmina, lacking "bleeding" spines on hind tibiae. Ovipositor with apex of dorsal valve smooth (Fig. 334) .... **Gryllinae** (p. 289)

Clé des sous-familles de Gryllidae

1. Corps petit (moins de 15 mm) et plutôt poilu. Tibias postérieurs armés d'épines articulées en plus des éperons apicaux. Mâles pourvus de glandes sexuelles bien développées à la base des tegmina ou à tibias postérieurs armés d'épines spécialisées « saignantes ». Valve dorsale de l'ovipositeur crénelée ou denticulée (fig. 358 à 364) .... **Nemobiinae** (p. 276)

   Corps plus gros et pas très poilu. Tibias postérieurs armés d'épines fixes. Chez les mâles, glandes sexuelles de la base des tegmina réduites ou absentes, et tibias postérieurs dépourvus d'épines spécialisées. Valve dorsale de l'ovipositeur à apex uni (fig. 334) ..... **Gryllinae** (p. 289)

Subfamily Nemobiinae

The subfamily Nemobiinae includes the ground crickets and wood crickets, of which there are about 24 genera and over 200 species, even more widely distributed than the Gryllinae. They are generally rather small, distinguishable particularly by the long movable spines on the hind tibiae. For general remarks on the Nemobiinae, see Vickery and Johnstone (1973).

Key to genera of Nemobiinae

1. Distoventral spurs of posterior tibiae equal or nearly equal in length (Fig. 349); dorsal spines of posterior tibiae without serrulations beneath. Ovipositor with dentations on ventral valve (Fig. 364) .... *Eunemobius* (p. 277)

Distoventral spurs of posterior tibiae distinctly unequal in length (Figs. 347, 348); dorsal spines of posterior tibiae concave beneath with margins minutely serrulate. Ovipositor with underside of ventral valve smooth 2

2(1). Length nearly always greater than 6 mm. Face and/or occiput nearly always with distinct markings. Ovipositor straight (Figs. 358–362), nearly as long as to longer than posterior femur .... ***Allonemobius*** (p. 279)

Length, rarely more than 6 mm, never more than 7 mm. Face and occiput nearly always immaculate. Ovipositor gently curved (Fig. 363), not more than two-thirds as long as posterior femur ...... ***Neonemobius*** (p. 288)

Clé des genres de Nemobiinae

1. Éperons disto-ventraux des tibias postérieurs de longueur égale ou presque égale (fig. 349); épines dorsales des tibias postérieurs non serrifères. Valve ventrale de l'ovipositeur denticulée (fig. 364) .... ***Eunemobius*** (p. 277)

Éperons disto-ventraux des tibias postérieurs de longueur clairement inégale (fig. 347 et 348); épines dorsales des tibias postérieurs à dessous concave et à bords finement serrifères. Valve ventrale de l'ovipositeur à dessous uni 2

2(1). Longueur presque toujours supérieure à 6 mm. Face ou occiput portant presque toujours des marques distinctes. Ovipositeur rectiligne (fig. 358 à 362) et presque aussi long ou plus long que les fémurs postérieurs .... ***Allonemobius*** (p. 279)

Longueur rarement supérieure à 6 mm, jamais supérieure à 7 mm. Face et occiput presque toujours immaculés. Ovipositeur légèrement incurvé (Fig. 363) et de longueur ne dépassant pas les deux tiers de celle des fémurs postérieurs .... ***Neonemobius*** (p. 288)

Genus *Eunemobius* Hebard

Description. Body size, males 5.2–9.8, females 5.8–9.9 mm. Distoventral spurs of hind tibiae nearly equal in length (Fig. 349); other spines of hind tibiae in both sexes concave beneath with margins of sulcations smooth, not serrulate; internal dorsal and median spurs with prominent and regular fringes of hairs; distointernal spur of male somewhat swollen at base and shaft bent (Fig. 349); proximointernal spine specialized. Ovipositor short (Figs. 342, 343, 346), nearly always less than two-thirds length of hind femur, gently curved, with apex of dorsal margin armed with heavy teeth; ventral margin with minute widely spaced serrulations (Fig. 364). Ventral segments of abdomen in male widened, making abdomen appear unusually broad.

Eunemobius carolinus (Scudder)

Nemobius carolinus Scudder, 1877*a*:36.

For synonymy and so forth, see under the nominate subspecies below.

Eunemobius carolinus carolinus (Scudder)

Figs. 339, 342, 343, 346, 349, 357, 364; Plate IV*F*; Map 93

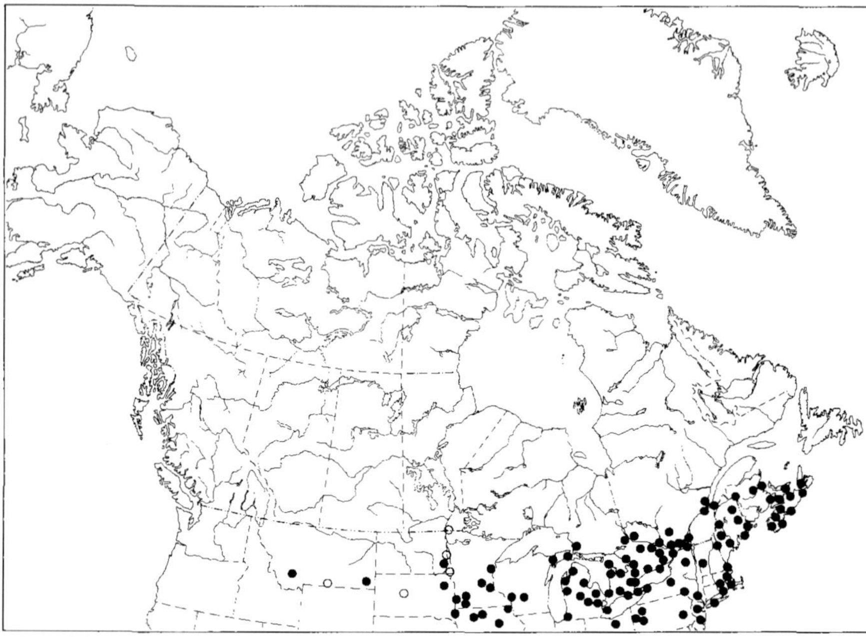

Map 93. Collection localities for *Eunomobius carolinus carolinus* (●) and *E. carolinus brevicaudus* (○).

Nemobius carolinus Scudder, 1877*a*:36.
Nemobius (*Anaxipha*) *septentrionalis* Provancher, 1877:292, 299.
Nemobius affinis Beutenmüller, 1894*b*:249.
Nemobius angusticollis E. M. Walker, 1904*a*:186.
Nemobius carolinus; E. M. Walker in Faull, ed. 1913:302.
Nemobius (*Eunemobius*) *carolinus carolinus*; Hebard 1913:473.
Nemobius carolinus carolinus; Hebard 1925*a*:149; Cantrall 1943:166.
Nemobius macdunnoughi Urquhart, 1938:101.
Eunemobius carolinus carolinus; Vickery and Johnstone 1970:1746; 1973:642; Vickery et al. 1974:56.
Neonemobius [*sic*] *carolinus*; R. D. Alexander et al. 1972:48.

Diagnosis. Body small, shining. Ovipositor (3–3.2 mm), with coarse teeth dorsally and serrulations ventrally.

Description. Body medium (males 7.5–8.5, females 6.5–8.5 mm long). Head prominent, as wide as pronotum. Mouthparts large, strongly developed, with mandibles large and strong, and with laciniae of maxillae long, curved, acute, often protruding below labrum (Fig. 357). Fastigium in lateral aspect angulate. Pronotum one-third wider than long, slightly narrower anteriorly than posteriorly. Tegmina glossy, covering three-quarters of abdomen in males (Fig. 339) and about two-thirds in females (Figs. 342, 346) (macropterous specimens, particularly females (Fig. 343), in which hind wings are much longer than the abdomen, are not rare). Tibial spurs (Fig. 349) and ovipositor (Fig. 364) as described for genus (the coarse teeth on the dorsal ovipositor valve and serrulations on the ventral valve make females of this species easy to identify). Male stridulatory vein (Plate IV*E*) with 58–69 teeth, mean 62. Color generally dark brownish, shining, sometimes with yellowish tinge on head, pronotum, and femora; maxillary palpi whitish or pale yellow, at least on apical segments; dorsal surface of abdomen black or dark brown, in females with small pale spots on exposed part, paler on ventral surface. Chromosome number, $2n$ male = 7, with submetacentric X chromosome and 3 pairs of metacentric autosomes.

Range. Utah to Nova Scotia, south to California and Florida.

Behavior and habitats. *E. carolinus carolinus* is found in many types of damp habitats, such as dense grassy areas, near lakes and streams, under stones and debris, around foundations of houses, and in *Sphagnum* moss (but not in open *Sphagnum* bogs).

Male stridulation takes the form of a wavering continuous trill made up of two types of pulse sequence, smooth and even, together with periodic de-emphasized faster pulses; during even sequences there are 83 wing strokes per second at a frequency of 5.3–6.6 kHz, median at 6.0 kHz, at 27°C and 80% RH.

Life history. In Canada, specimens occur in the adult stage from early August until September or early October. Cantrall (1968) reported adults in Michigan from 19 July until 6 November. Winter is passed in the egg stage in the soil.

Eunemobius carolinus brevicaudus (Bruner)

Map 93

Nemobius brevicaudus Bruner, 1904*a*:57.
Nemobius carolinus brevicaudus; Hebard 1925*a*:150.

Diagnosis and description. Similar to *E. carolinus carolinus* but larger and paler, with shorter ovipositor (2.3–2.5 instead of 3–3.2 mm). This form has not been sufficiently studied to reveal its status.

Range. Montana to Minnesota, south to Colorado and Kansas.

Behavior, habitats, and life history. Insufficient is known to make comment here, other than to suggest a general similarity to *E. carolinus carolinus* but with an adaptation to drier environments.

Genus *Allonemobius* Hebard

Description. Length in males 6.5–11.0, in females 7.5–12.0 mm. Hind tibiae with distoventral spurs unequal in length, and with inner spurs much longer (Fig. 347); dorsal spines concave beneath, with sulcated margins minutely serrulate in both sexes; males with specialized proximointernal glandular spines. Tegmina variable in length, with relatively dull surface. Ovipositor nearly straight, as long as or longer than hind femur, with margins of dorsal valves more or less obliquely subtruncate and armed with teeth or denticles; margins of ventral valves smooth (Figs. 358–362). Male genitalic characters as described and illustrated by Vickery and Johnstone (1973). Chromosomes as described by Lim (1971).

Key to species and subspecies of *Allonemobius*

1. Tegmina brachypterous; hind wings absent. Ovipositor not exceeding length of hind femur .... ***maculatus* (Blatchley)** (p. 281)
   Tegmina sometimes macropterous; hind wings usually present. Ovipositor longer than hind femur 1

2(1). Body grayish to grayish black (dark brown in western specimens; face below antennae uniform dark piceous in eastern specimens); lateral margins of dorsal field of male tegmen usually narrowly whitish to pale yellow. Ovipositor broad, with tip of dorsal valve bent downward (Fig. 360) .... ***griseus griseus* (E. M. Walker)** (p. 282)
   Body generally brownish to black, usually with dorsal longitudinal stripes on vertex of head; face below antennae not contrasting with area above 2

3(2). Vertex of head with distinct dorsal longitudinal dark and lighter stripes. Head wide (often more than 2.8 mm), well-rounded. Pronotum barrel-shaped, with head and anterior edge of pronotum as broad as posterior edge of pronotum. Stridulatory vein of right tegmen of male with 100–150 teeth; part of vein mesad of ulnar vein 0.65–0.80 mm in length. Stridulation a steady repetition of short buzzing chirps .... ***fasciatus* (De Geer)** (p. 284)
   Vertex of head with indistinct dorsal stripes. Head narrow, retracted, not greater than 2.8 mm in width. Pronotum narrower in front so that head and anterior edge of pronotum are narrower than posterior edge of pronotum. Stridulatory file of right tegmen of male with more than 160 teeth; part of vein mesad of ulnar vein 0.85–1.10 mm in length. Stridulation a slow trill with occasional breaks or a slow continuous tinkling trill 4

4(3). General color pale, reddish, especially on top of head. Song a slow, continuous tinkling trill. Insect found in woodlands and leaf litter .... ***tinnulus* (Fulton)** (p. 285)
   General color dark, red brown to black. Song a slow trill with occasional breaks, not tinkling. Insect found in grasslands .... ***allardi* (Alexander & Thomas)** (p. 286)

Clé des espèces et des sous-espèces d'*Allonemobius*

1. Tegmina brachyptères; ailes postérieures absentes. Longueur de l'ovipositeur inférieure à celle des fémurs postérieurs .... ***maculatus* (Blatchley)** (p. 281)
   Tegmina parfois macroptères; ailes postérieures habituellement présentes. Ovipositeur plus long que les fémurs postérieurs 2

2(1). Corps variant de grisâtre à noir grisâtre (brun foncé chez les espèces de l'ouest; portion de la face située sous les antennes de teinte très noire et uniforme chez les espèces de l'est); bords latéraux de la région dorsale des tegmina des mâles variant habituellement d'à peine blanchâtre à jaune pâle. Ovipositeur large; valve dorsale à extrémité incurvée vers le bas (fig. 360) .... ***griseus griseus* (E.M. Walker)** (p. 282)

Corps en général brunâtre à noir; vertex marqué habituellement d'une bande longitudinale dorsale; portion de la face située sous les antennes de couleur semblable au reste 3

3(2). Vertex marqué de bandes dorsales longitudinales noires et plus pâles. Tête large (souvent plus de 2,8 mm), arrondie. Pronotum en forme de baril, la tête et le bord antérieur du pronotum aussi larges que le bord postérieur de ce dernier. Nervure stridulante du tegmen de droite des mâles portant de 100 à 150 dents; partie de cette nervure située près de la portion médiane de la nervure ulnaire mesurant de 0,65 à 0,80 mm de longueur. Chant fait d'une répétition ininterrompue de courts appels bourdonnants .... ***fasciatus* (De Geer)** (p. 284)

Vertex marqué de bandes dorsales floues. Tête étroite, retirée, de largeur ne dépassant pas 2,8 mm. Pronotum plus étroit en avant; tête et bord antérieur du pronotum plus étroits que le bord postérieur de ce dernier. Nervure stridulante du tegmen droit du mâle comptant plus de 160 dents; portion de cette nervure située près de la portion médiane de la nervure ulnaire mesurant 0,85 à 1,10 mm de longueur. Le chant est une stridulation lente, interrompue à l'occasion, ou rappelle un tintement de clochettes ininterrompu 4

4(3). Couleur généralement pâle, rougeâtre, surtout sur le dessus de la tête. Le chant rappelle un tintement ininterrompu de clochettes. Insecte vivant en forêt, sur le sol .... ***tinnulus* (Fulton)** (p. 285)

Couleur généralement foncée, de brun rougeâtre à noir. Le chant est une stridulation lente, interrompue à l'occasion, mais ne rappelle pas un tintement de clochettes. Insecte vivant dans les prairies .... ***allardi* (Alexander & Thomas)** (p. 286)

Allonemobius maculatus (Blatchley)

Figs. 355, 362; Plate IV*D*; Map 94

Nemobius maculatus Blatchley, 1900:52.

Allonemobius maculatus; Vickery and Johnstone 1970:1746; 1973:632; R. D. Alexander et al. 1972:48.

Diagnosis. Tegmina apically nearly truncate. Body pale brown, mottled.

Description. Body medium (males 6.5–8.4, females 7.6–9.1 mm long). Head as broad as pronotum. Tegmina of both sexes nearly truncate, those of male covering two-thirds, those of female covering only one-third of abdomen; hind wings lacking. Ovipositor (Fig. 362) straight, robust, with apex slender, tapering, acute; teeth of dorsal valve minutely needlelike, evenly spaced, with tips directed slightly posteriorly. Male with 114 stridulatory pegs (Plate IV*D*). Coloration generally pale brown; entire face from labrum to vertex mottled (Fig. 355), or buff with few markings (never with striped pattern); head with pale yellow occipital stripe extending to vertex, and with yellowish ring around each compound eye; pronotum brown with numerous darker brown spots and a broad piceous lateral stripe; hind femur mottled brown and buff; abdomen with distinctive ebony brown markings on buff to light brown base.

Range. South Dakota to southern Ontario, south to Georgia.

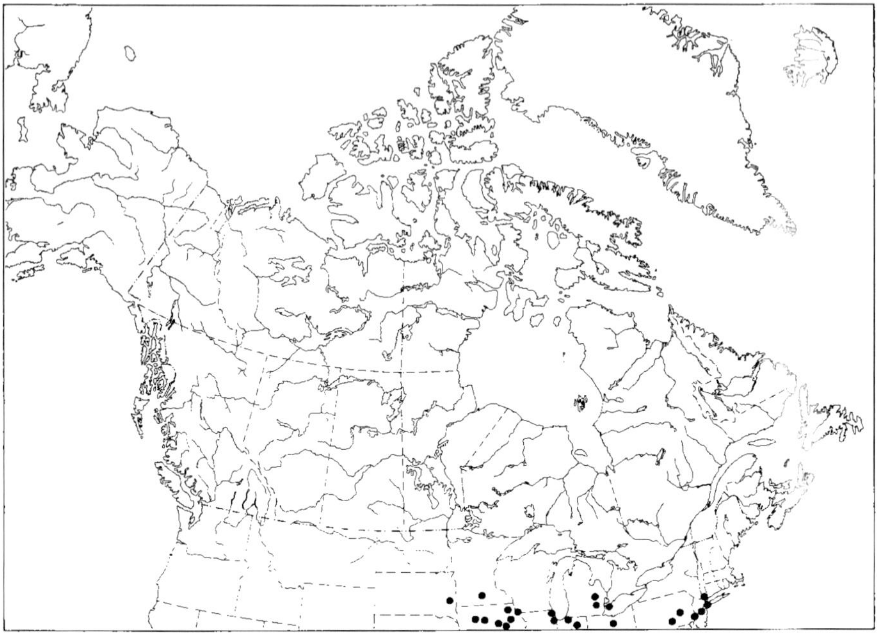

Map 94. Collection localities for *Allonemobius maculatus.*

Behavior and habitats. The species frequents low dry open woodlands and thickets, and has been found among dead leaves and beneath logs (Blatchley 1920).

Stridulation is described by R. D. Alexander (1957*a*) as "a jerky trill in which the catches occur regularly at about 3–8 per second (depending on temperature)."

Life history. As for other species of *Allonemobius*. Cantrall (1968) reported it as being adult in Michigan from 9 August to 23 September.

Allonemobius griseus griseus (E. M. Walker)

Figs. 336, 354, 360, 361; Plate IV*C*; Map 95

Nemobius griseus E. M. Walker, 1904*a*:182.

Nemobius fasciatus abortivus Caudell, 1904*c*:248; R. D. Alexander et al. 1972:48.

Nemobius griseus griseus; Hebard 1932*a*:52; Cantrall 1943:164.

Pteronemobius griseus abortivus; Chopard 1967:175.

Allonemobius griseus griseus; Vickery and Johnstone 1970:1746; 1973:629; Vickery et al. 1974:54.

Diagnosis. Body small, slender. Head as broad as pronotum. Color grayish, with lower half of face usually shining black.

Description. Size small, slender (males 6.8–8.0, females 7.6–8.6 mm long). Head as broad as anterior margin of pronotum. Pronotum slightly

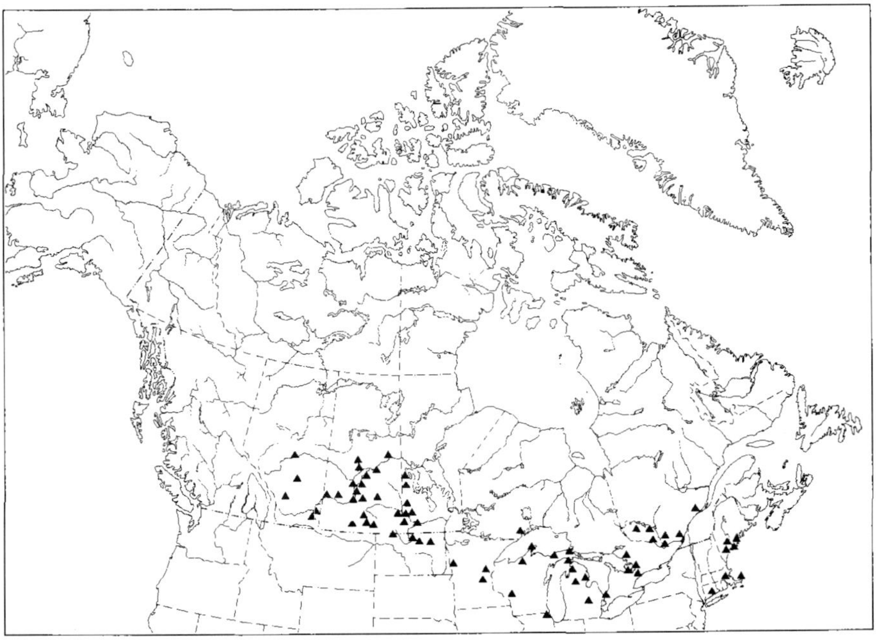

Map 95. Collection localities for *Allonemobius griseus griseus*.

broader posteriorly. Insect usually micropterous and virtually without hind wings; tegmina of female dull, dusty, covering about one-half of abdomen; male tegmina covering about three-quarters of abdomen (occasional macropterous individuals have hind wings considerably exceeding tegmina in length and extending well beyond apices of hind femora). Ovipositor long (7.2–8.9 mm), robust, usually straight, with apex short, slightly swollen, and with upper edges of apices of dorsal valves slightly convex, often distinctly sloping to subacute tip; teeth fine, conical, acute, becoming more widely spaced toward apices (Figs. 360, 361); robust in western populations, with upper edges of apices of dorsal valves more convex (Fig. 361). Stridulatory file of male (Plate IV*C*) with 93–116 teeth, mean 104. Coloration generally rather dull grayish black, with lower half of face usually shining black, contrasting with part above antennal sockets (Fig. 354).

Range. Alberta to Québec, south to Indiana and Massachusetts.

Behavior and habitats. Unlike most other Nemobiinae found in the present region, *A. griseus griseus* shows a decided preference for dry sandy areas rather than for damp localities. It sometimes becomes numerous in the prairies (MacNay 1961*a*), and it is frequently found together with *A. allardi* in sandy habitats where grass clumps occur sparsely on otherwise uncovered soil.

Stridulation takes the form of a trill, unlike that of most other Nemobiinae in that it consists of paired pulses, the first shorter than the second, probably representing two cycles of wing movement. The calling song has 34 double-wing strokes per second at a mean frequency of 7.5 kHz at 27°C.

The courtship song is a much slower trill with only 4 double-wing strokes per second, about 7.0 kHz at 27°C.

Life history. Univoltine, with overwintering in the egg stage. Cantrall (1968) reported adults from 26 July to 7 October in Michigan.

Allonemobius fasciatus (De Geer)

Figs. 347, 350, 352, 358; Plate IV*A*; Map 96

Gryllus fasciatus De Geer, 1773:522.

Nemobius vittatus Scudder, 1863*a*:430.

Nemobius socius Scudder, 1877*a*:37.

Nemobius fasciatus; Provancher 1876:61; 1877:299; Cantrall 1943:159; Helfer 1963:328.

Nemobius fasciatus vittatus; Piers 1896:210.

Nemobius fasciatus fasciatus; Morse 1919*b*:29; Cantrall 1943:160.

Nemobius fasciatus socius; Urquhart 1941*a*:116.

Allonemobius fasciatus; Vickery and Johnstone 1970:1746; 1973:626; R. D. Alexander et al. 1972:49; Vickery et al. 1974:52.

Diagnosis. Occiput of head swollen; head with 5 brown to fuscous longitudinal stripes.

Description. Size medium (males 7.0–10.0, females 8.0–11.0 mm long). Male head in frontal view broadly ovate (Fig. 352); female head generally more oblong-oval; occiput somewhat tumid; head in dorsal aspect broad,

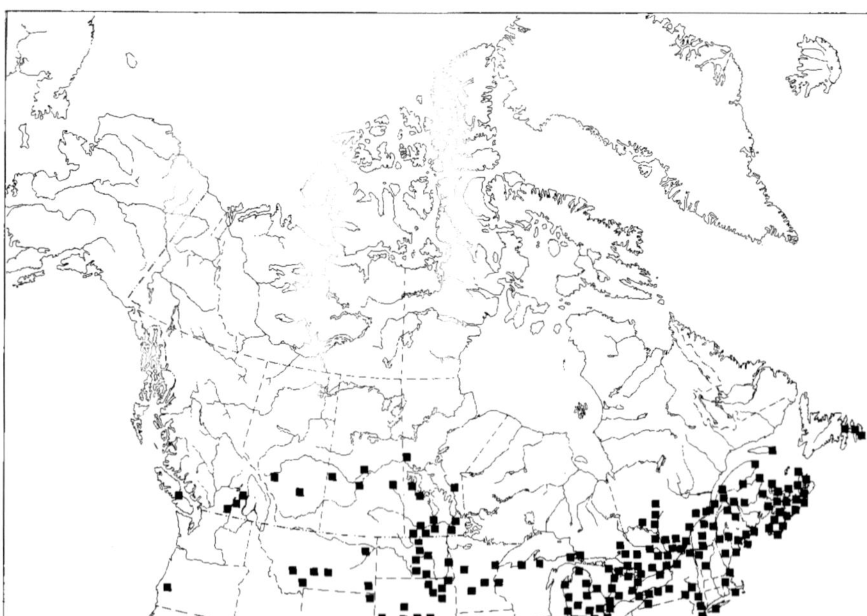

Map 96. Collection localities for *Allonemobius fasciatus*.

rounded; pronotum barrel-shaped so that, in brachypterous specimens, the head and anterior margin of the pronotum are as broad as the posterior edge of the pronotum (posterior margin of pronotum broader in macropterous form). In brachypterous form (much the more common), tegmina of females (cf. Fig. 340) covering about half, and those of males (cf. Fig. 337) about two-thirds of abdomen; macropterous females with wings extending to tip of ovipositor; hind wings of males correspondingly long. Ovipositor of females 5.9–10.0 mm long (longer than 7.5 mm only in specimens with head width behind the eyes greater than 2.6 mm), nearly straight, often with slight upward curve before apical third; apex of dorsal valve not swollen but slender and tapering, with denticles or teeth evenly spaced and relatively short (Fig. 358). Stridulatory file of males (Plate IV*A*) with 106–128 teeth, mean 114. Color generally brownish; head with 5 distinct dark brown to fuscous stripes dorsally from vertex to occiput (Fig. 350); face (Fig. 352) without distinct color pattern; body nearly black in some northern localities.

Range. British Columbia to Newfoundland, south to New Mexico and Georgia.

Behavior and habitats. *A. fasciatus* and *A. allardi* are frequently found together in moist areas or in microhabitats that tend to remain humid. *A. fasciatus* appears to be confined to such habitats and is even found in *Sphagnum* bogs with *Neonemobius palustris*, whereas *A. allardi* is frequently found in much drier places. *A. fasciatus* is omnivorous, feeding equally readily upon vegetable and animal matter, although it appears that individuals require both in their diets. Nielsson and Bass (1967) give a general account of the biology.

In its stridulation, the calling song of *A. fasciatus* consists of short buzzing chirps, 3 per second, with 10 wing strokes per chirp, about a mean of 7.4 kHz at 27°C and 80% RH.

Life history. Like all Nemobiinae in Canada, *A. fasciatus* overwinters as a diapausing egg. It hatches in June, begins to mature in mid-July to early August, and is present until late September, usually disappearing before the first frost. Macropterous individuals occasionally are taken at night, at light.

Economic importance. *A. fasciatus* has been recorded as destroying clover seedlings (Nielsson and Bass 1967).

Allonemobius tinnulus (Fulton)

Map 97

Nemobius fasciatus tinnulus Fulton, 1931:210–216.
Nemobius fasciatus tinnulus; Cantrall 1943:164.
Allonemobius tinnulus; Vickery and Johnstone 1970:1746.

Diagnosis. Head and dorsum of pronotum orange brown, tending to reddish on top of head; head not striped.

Description. Body length, male 9.15, female 9.60 mm. Male tegmen rectangular, with ratio of width (2.9 mm) to length (5.4 mm) large; hind wings reduced (flightless species). Ovipositor equal in length to hind femur, about 7.5 mm. Male stridulatory file 1.5 mm long, with 187–239 teeth. Color as indicated in diagnosis; male tegmina infuscate basally.

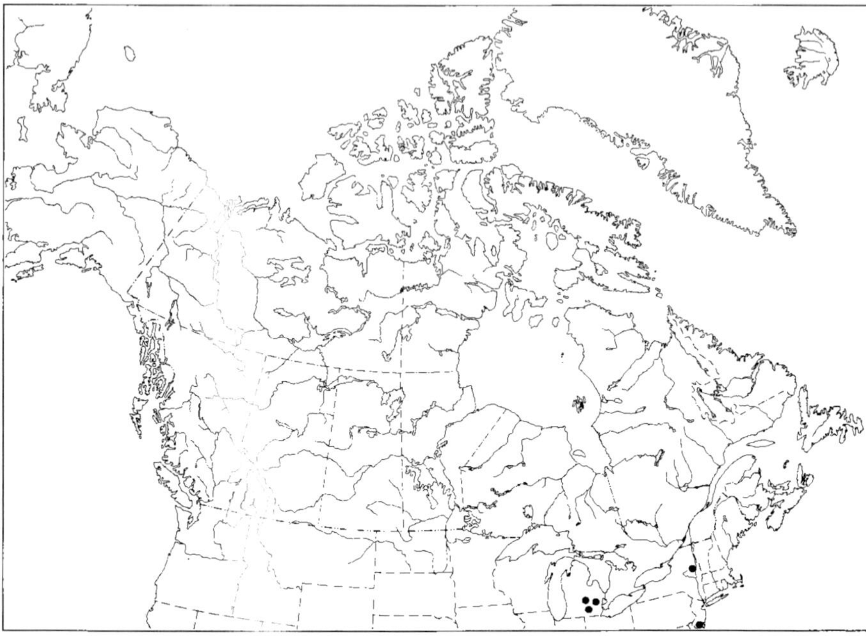

Map 97. Collection localities for *Allonemobius tinnulus*.

Range. Iowa to New York, south to Alabama and North Carolina.

Behavior and habitats. *A. tinnulus* is confined to dry woodlands and woodland borders in ground litter, dry leaves, or pine needles in sunny or lightly shaded oak–pine, oak–hickory, or oak–hickory–chestnut woods. Mating behavior is described by Fulton (1931) and by R. D. Alexander and Thomas (1959).

Male stridulation consists of a slow uninterrupted "tink-tink-tink," 2–3 per second at 10°C, 9–10 per second at 32°C (R. D. Alexander and Thomas 1959).

Life history. Similar to that of other species of the genus, with overwintering in the egg stage in the soil.

Allonemobius allardi (Alexander & Thomas)

Figs. 337, 340, 344, 351, 353, 359; Plate IV*B*; Map 98

Nemobius allardi R. D. Alexander and Thomas, 1959:592.

Allonemobius allardi; Vickery and Johnstone 1970:1746; 1973:628; R. D. Alexander et al. 1972:49; Vickery et al. 1974:53–54.

Diagnosis. Occiput of head narrow, retracted into pronotum. Dorsal stripes on head indistinct.

Description. Size medium (males 7.2–11.1, females 7.6–12.0 mm long). Face as in Fig. 353. Head from dorsal aspect narrow, retracted. Pronotum

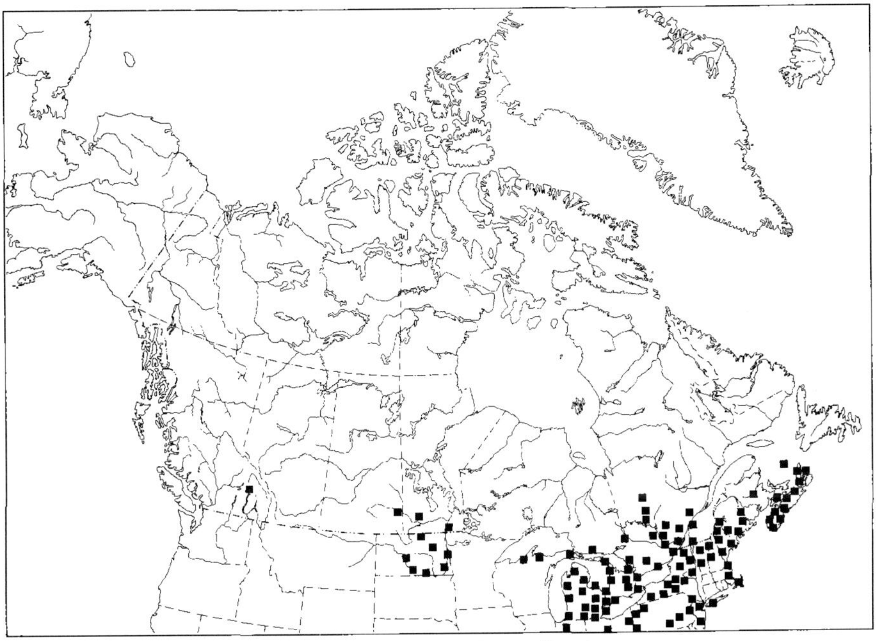

Map 98. Collection localities for *Allonemobius allardi*.

narrow anteriorly so that, in either brachypterous or macropterous specimens, the head and anterior edge of the pronotum are narrower than the posterior edge of the pronotum (Figs. 337, 340). Wings mainly brachypterous but occasionally macropterous; ovipositor (Fig. 359) long in proportion to head width behind eyes (Fig. 340); apex of dorsal valve somewhat variable in shape, generally tapering and sublanceolate, with small evenly spaced teeth (Fig. 359), grading to slightly convex, oblique. Stridulatory file of male (Plate IV*B*) with 163–228 teeth, mean 194. General color fuscous to black, without distinctive markings; frontal aspect of head dark (Fig. 353), and dorsal head stripes not distinct.

Range. British Columbia to Nova Scotia, south to southern United States.

Behavior and habitats. *A. allardi* frequents a wider range of microhabitats than does *A. fasciatus*, ranging from damp stream banks to dry sandy areas. Like that species, it also is an omnivorous feeder.

The stridulation of males is a trill, the calling song having 18 wing strokes per second, about a mean of 8.3 kHz at 27°C and 80% RH; the courtship song is much slower, with 7 wing strokes per second at about 8.0 kHz.

Life history. In Eastern Canada, *A. allardi* reaches the adult stage 7–10 days earlier than does *A. fasciatus*, that is, in early July to the end of that month. It can usually be heard stridulating until the first severe frost. Eggs and egg diapause were studied by Rakshpal (1962*b*, 1964*b*).

Genus *Neonemobius* Hebard

Description. Size range smaller than in *Allonemobius* (males 5.2–8.5, females 5.9–9.4 mm long). Distoventral spurs of hind tibiae (Fig. 348) distinctly unequal. Ovipositor shorter than hind femur and usually distinctly curved upward (Fig. 363), its apex with margins of dorsal valves armed with distinct teeth or with minute serrulations (*N. palustris*); margins of ventral valves unarmed. Concealed male genitalia as described by Vickery and Johnstone (1973).

Neonemobius palustris (Blatchley)

Figs. 338, 341, 345, 348, 356, 363; Plate IV*E*; Map 99

Nemobius palustris Blatchley, 1900:53.

Nemobius palustris aurantius J. A. G. Rehn and Hebard, 1911*a*:597.

Nemobius cubensis palustris; Urquhart 1941*d*:80–81; Cantrall 1943:166.

Neonemobius palustris; Vickery and Johnstone 1970:1746; 1973:637; R. D. Alexander et al. 1972:48; Vickery et al. 1974:55.

Diagnosis. Body small. Head unicolorous. Ovipositor curved, shorter than hind femur.

Description. Body small (males 5.2–6.5, females 5.9–6.8 mm long). Mouthparts small, narrower than head (Fig. 356). Ovipositor (Fig. 363) shorter than hind femur, curved upward, with apex possessing very fine

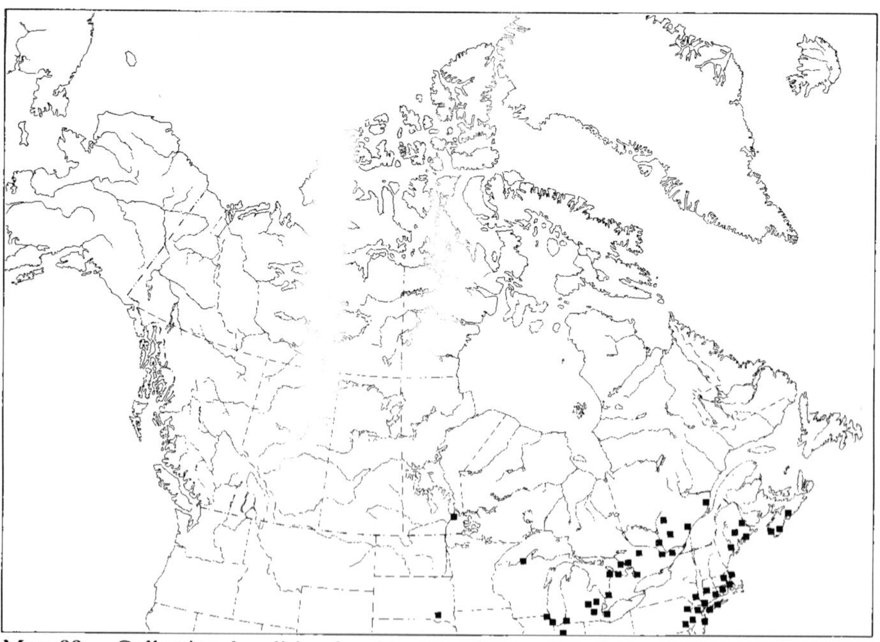

Map 99. Collection localities for *Neonemobius palustris*.

denticulations on upper valve. Male stridulatory file as in Plate IV*E*, with 101–143 teeth, mean 119. Color of head nearly uniform; general color variable, from pale to brightly colored, golden buff to cinnamon, brownish or black. Chromosome number, $2n$ male = 19, karyotype with V-shaped X chromosome and 3 large pairs, 4 medium pairs, and 2 small pairs of acrocentric autosomes; male testes usually rhomboid but elongate-oval in some individuals.

Range. Manitoba to Nova Scotia, south to Florida.

Behavior and habitats. *N. palustris* is confined to *Sphagnum* bogs. It feeds upon *Sphagnum* mosses and is peculiar in that it deposits its eggs in the foliage of this genus of plants (Vickery 1969*a*:23, fig. 1). It occurs in scattered colonies, each isolated from the others. This isolation, because it is due to habitat type, is undoubtedly of long standing. Genetic divergence is certainly taking place in the colonies, which would explain the wide color variation. The karyotype previously mentioned applies to a colony at Lac Carré, Qué.; other colonies may differ somewhat.

Stridulation involves a calling song consisting of a continuous high-pitched trill with 44 wing strokes per second at a frequency of 8.0–8.8 kHz at 27°C and 80% RH.

Life history. Similar to that of other Nemobiinae; univoltine, with over-wintering in the egg stage. Cantrall (1968) reported adults from 12 July to 4 October in Michigan.

Subfamily Gryllinae

Members of this subfamily include insects such as the field crickets and house crickets, of which there are some 70 genera and about 500 species distributed throughout most of the world, except in the coolest regions. They differ from the Nemobiinae as previously indicated and may be characterized as follows: robust crickets of small to fairly large size. Head relatively large. Eyes prominent; 3 ocelli arranged in triangle. Pronotum broader than long. Hind tibiae armed above on each side with 5–8 stout fixed spines and with 3 pairs of apical spurs. Ovipositor as long as or longer than hind femora, with dorsal valves having apices smooth, not denticulate or crenulate (Fig. 334).

Randell (1964) gives the most recent general classification of the Gryllinae of the world. Only three representatives of the subfamily are established in the present region, two species of field cricket and the house cricket.

Key to genera of Gryllinae

1. Field specimen. General color, particularly on head, typically dark, usually blackish .... *Gryllus* (p. 290)
   Synanthropic specimen. General color, including head, paler .... *Acheta* (p. 294)

Clé des genres de Gryllinae

1. Spécimen capturé en nature. Couleur générale, notamment celle de la tête, typiquement foncée et habituellement noirâtre .... ***Gryllus*** (p. 290)
   Spécimen synanthrope. Couleur générale, y compris la tête, plus pâle .... ***Acheta*** (p. 294)

Genus *Gryllus* Linnaeus

Description. Body size medium to fairly large, generally rather robust. Color black or blackish brown (Figs. 331, 333). Bases of antennae widely spaced, with interval between them more than twice width of antennal scape. Tegmina and hind wings fully developed to rather abbreviate, often dimorphic.

Key to species of *Gryllus*

1. Adults from mid-May to late July; nymphal stages from mid-August onward, overwintering in late instar .... ***veletis*** **(Alexander & Bigelow)** (p. 290)
   Adults from early August until frost; nymphal stages from late May to August, overwintering as eggs .... ***pennsylvanicus*** **Burmeister** (p. 292)

Clé des espèces de *Gryllus*

1. Stade adulte de la mi-mai à la fin de juillet; nymphes apparaissant à partir de la mi-août; hiverne à un stade nymphal avancé .... ***veletis*** **(Alexander & Bigelow)** (p. 290)
   Stade adulte du début d'août jusqu'aux premiers gels; nymphes trouvées de la fin mai au mois d'août; passe l'hiver à l'état d'œufs .... ***pennsylvanicus*** **Burmeister** (p. 292)

Gryllus veletis (Alexander & Bigelow)

Fig. 333; Plate IIIC; Map 100

Acheta veletis R. D. Alexander and Bigelow, 1960:335.
Acheta veletis; Bigelow 1960*b*:[illegible]73–988.
Gryllus veletis; Bigelow 1962:39[illegible]-406; R. D. Alexander et al. 1972:49; Vickery et al. 1974:51.

Diagnosis. Large black species, adult in spring. *G. veletis* (Fig. 333) is virtually indistinguishable from *G. pennsylvanicus* in size, structure, and appearance, and in male stridulation.

Description. As for *G. pennsylvanicus*, which follows.

Range. British Columbia to Maine, south to Mexico.

Behavior and habitats. *G. veletis* is never as numerous as *G. pennsylvanicus* and tends to be a much more solitary, sedentary, aggressive, burrow-inhabiting species. It is often found in debris or rock piles, and is much more difficult to capture than is the other species.

R. D. Alexander and Bigelow (1960) suggested a mode of speciation which they called allochronic (in this case, seasonal), based upon *G. veletis*

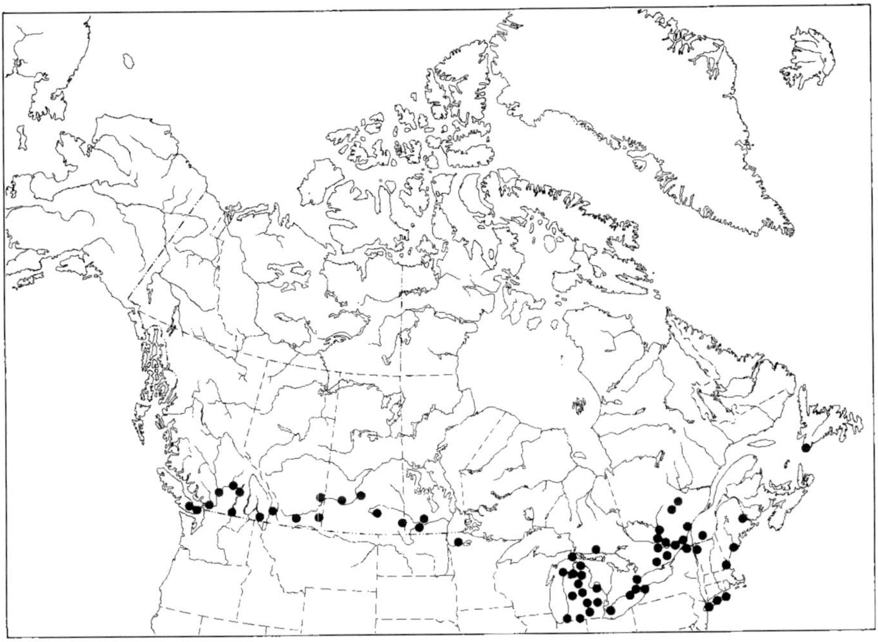

Map 100. Collection localities for *Gryllus veletis*.

and *G. pennsylvanicus*. Mayr (1963) criticized this suggestion, and rebuttals followed by R. D. Alexander (1963, 1968). Zoologists now recognize the allochronic speciation model as a reasonable hypothesis (Bush 1976). The primary impediment to acceptance of the theory is lack of experimental evidence that environmental factors can produce genetic changes resulting in asynchrony in the seasonal reproductive cycle of crickets. R. G. Harrison (1977) studied enzyme variation in sympatric and allopatric populations of the two species and found that sympatric populations of both species had similar mean frequencies of particular electrophoretic variants, which also exhibited a common pattern of geographic variation. He expressed the view that "parallel selection" provides the only satisfactory explanation for this. Another, earlier, impediment to acceptance of the theory of allochronic speciation, at least in the present situation, was that Randell (1964), on morphological grounds, placed *G. pennsylvanicus* and *G. veletis* in different groups of the genus. If the theory of allochronic speciation applied, this would not be done. If Randell is correct, then the theory is difficult to support!

Life history. Mature in spring and early summer; eggs give rise to nymphs, which reach the penultimate or final instar in autumn; these hibernate in burrows, then emerge early in the following spring, when they soon molt to the adult stage.

Economic importance. There seems to be no authentic report of injury to crops or property by *G. veletis*. Indeed, such is not to be anticipated, as the species is a more or less solitary cricket, unlike the more gregarious *G. pennsylvanicus*.

Gryllus pennsylvanicus Burmeister

Figs. 331, 332; Plate III*B*; Map 101

Gryllus pennsylvanicus Burmeister, 1838:734.
Gryllus luctuosus Audinet-Serville, 1838:335.
Gryllus abbreviatus Audinet-Serville, 1838:336.
Acheta nigra T. W. Harris, 1841:123.
Gryllus angustus Scudder, 1863*a*:427.
Gryllus neglectus Scudder, 1863*a*:428.
Gryllus pennsylvanicus; Lugger 1898:354; R. D. Alexander et al. 1972:49; Vickery et al. 1974:49.
Gryllus arenaceus Blatchley, 1903:434.
Gryllus assimilis pennsylvanicus; Blatchley 1920:701 (partim).
Gryllus assimilis luctuosus; Blatchley 1920:701.
Gryllulus assimilis; Cantrall 1943:150 (partim).
Acheta assimilis; Gurney 1950*b*:412 (partim); Helfer 1963:331 (partim).

Diagnosis. Large black species, adult in fall.

Description. General appearance as in Fig. 331. Body large (length, male 15–21 mm; female 14.5–24 mm), broad, with width of abdomen about one-third body length. Head nearly as broad as pronotum. Pronotal width 1.5 times pronotal length, with front margin slightly concave, and with hind margin slightly convex. Tegmina of female usually covering three-fifths to

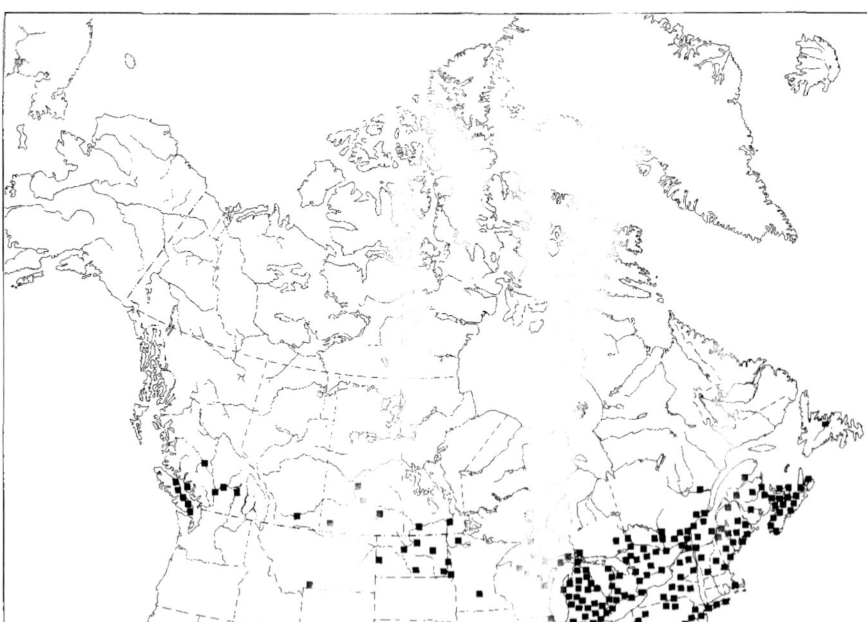

Map 101. Collection localities for *Gryllus pennsylvanicus*.

four-fifths of abdomen, never longer than abdomen, of male usually nearly covering abdomen, rarely slightly longer; inner edges of tegmina overlapping for entire length in males, somewhat separated posteriorly in females. Color black with fine gray pubescence on pronotum and femora; head shining black; tegmina shining black with pale line at edge of dorsal field; ovipositor brownish; femora with brownish or reddish brown areas on inner and outer faces; specimens in some localities piceous brown.

Range. British Columbia to Nova Scotia, south to Mexico; adventive in Newfoundland and Labrador.

Behavior and habitats. This species matures in late July and August, producing eggs that overwinter in the soil and hatch in May and June. It prefers grassy areas and seldom is found in forests. It is probable that the southern edge of the northern forests marks the general northern limit of the range, although it does occur in some grassy areas in more or less isolated pockets north of the forest margin. It is not known whether these are relict populations or modern intrusions that were transported or encouraged by human activity.

The stridulation of *G. pennsylvanicus* is a slow "chirp, chirp, chirp," which may be heard on sunny days and also on warm summer evenings. R. D. Alexander (1957*b*) listed the pulse rate, at 29°C, as having a mean of 25 pulses per second and 150–240 chirps per minute. The frequency is approximately 5 kHz. Although the data are probably based upon both *G. pennsylvanicus* and *G. veletis* (which was not described until 3 years later) the conclusions are valid, because there is little if any detectable difference in the stridulation of the two species. Over most of the range of these, seasonal isolation prevents any attempt at interbreeding. If such matings were to occur, no hybrids would result, since the species are intersterile (Bigelow 1958, 1960*a*, 1960*b*; R. D. Alexander and Bigelow 1960; Vickery, unpublished, 1980).

G. pennsylvanicus is reported to be an intermediate host for a nematode, *Physaloptera maxillaris* Molin, which infests a species of skunk, *Mephitis mephitis* Schreber. The parasite does not cause noticeable differences in the appearance or longevity of the crickets. Mermithid nematodes, probably *Agamermis decaudata* Cobb, Steiner & Christie and/or *Mermis nigrescens* Dujardin, frequently infest *G. pennsylvanicus*, with fatal results. On account of their dark color, there is little doubt that the "eels" described long ago from Montréal as emerging from crickets (J. Hale 1829) refer, not to a nematode, but to a gordiid ("horsehair") worm [Nematomorpha].

Life history. The species matures in late July and August and produces eggs that pass the winter in the soil and hatch the following May or June.

Economic importance. Beirne (1972) has outlined the economic importance of *G. pennsylvanicus*, stating that it "may be blamed for more damage than it really does." It has been reported at various times as damaging ripening tomatoes (Goble 1966), especially cracked fruit; clipping flax bolls, but apparently eating seeds only in bolls injured by other insects; damaging clover and alfalfa by leaf feeding and by clipping heads and seed pods. Occasional damage to wheat and oats and to beans, corn, and other vegetable crops has been reported. It does feed upon eggs of grasshoppers and pupae of Lepidoptera or Diptera in the soil, and it is known, on occasion, to rob spiders of their prey (Corbett *in* Vickery et al. 1974). It is likely that the species is more beneficial than harmful, as it does not seem to multiply to the extent of producing really severe "plagues," as occasionally occurs elsewhere with other species of field crickets. Nevertheless, Kelleher (1979) notes that, in 1978, "Field crickets . . . invaded many homes, particularly in southern and western areas of Saskatchewan." This is not a rare occurrence in the West. Perhaps its economically most important trait in Canada was in the Prairie Provinces, where it made a practice of chewing through the twine used to bind wheat sheaves, causing them to fall apart. Synthetic twine and changing agricultural practices have now largely eliminated the problem.

This cricket has featured in North American poetry. The earliest example that has come to our notice is by Andrew John Ramsay (*alias* J. R.) of Montréal. The following lines are from stanza 4 of his *November—A Dirge*:

> The little cricket singing
> Sounds lonely in the crisp yellow leaves,
> Like by-gone tones of tenderness up-bringing
> A thought that grieves:
> A bell upon a ruined turret ringing
> On Sabbath eves.
>
> (Ramsay 1859)

In more cheerful vein, the Nova Scotia poet, Richard Huntington, in *Sunrise on the Tusket* (I, stanza 3), written in the 1870s, says:

> No zephyr's wing the leaf hath stirred,
> No sound to break the calm is heard
> Save cricket's chirp or trill of bird.
>
> (Huntington 1874)

Toward the end of the nineteenth century, four Canadian poets, in particular, Charles George Douglas Roberts, Archibald Lampman, Duncan Campbell Scott, and William Bliss Carman, contributed a great deal of "nature" verse, often referring to crickets, meaning, for the most part *Gryllus pennsylvanicus*.

Genus *Acheta* Fabricius

Description. Domiciliary or at least synanthropic. Color yellowish brown to stramineous, with head possessing distinct dark brown to blackish transverse bars (Fig. 334). Bases of antennae widely separated, with basal segments (scapes) projecting slightly beyond front of head. Tegmina and hind wings well-developed.

Acheta domesticus (Linnaeus)

Fig. 334; Plate III*D*; Map 102

Gryllus (*Acheta*) *domesticus* Linnaeus, 1758:428.

Acheta domesticus; Stephens 1829*b*:303; Helfer 1963:331; Vickery et al. 1974:47.

Gryllus transversalis F. Walker, 1871:6.

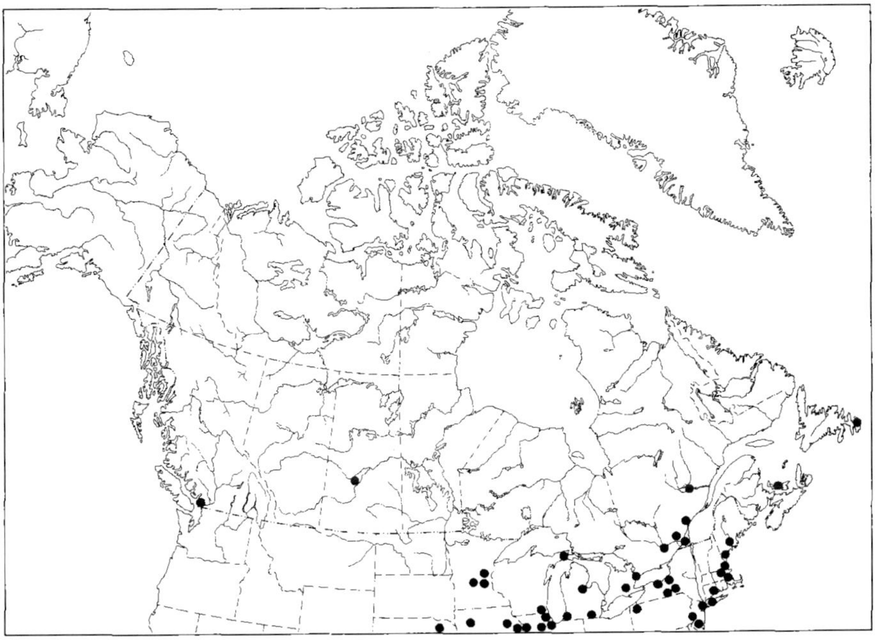

Map 102. Collection localities for *Acheta domesticus*.

Common name. House cricket.

Diagnosis. Body pale yellowish brown; head with four reddish brown transverse bars.

Description. Body size medium (length, males 14.4–19.5, mean 15.9; females 14.0–17.7, mean 15.7 mm), rather slender. Tegmina nearly reaching end of abdomen; hind wings either short and covered by tegmina or extending considerably beyond them. Posterior femur short, slender. Male stridulatory file as in Plate III*D*. Ovipositor slightly longer than posterior femur. General color pale yellowish brown; head with 4 dark reddish brown transverse bars, one on occiput, one dorsally between eyes, one between antennal bases, and one across labrum; pronotum with several irregular dark reddish brown spots dorsally and with narrow stripe of same color laterally; tegmina sometimes with reddish spot on anterior third; ovipositor pale brown, darker at tip.

Range. British Columbia to Newfoundland, south to Mexico; Europe, Asia, North Africa.

Behavior and habitats. In the most northerly part of its range, *A. domesticus* is essentially domiciliary. Vickery et al. (1974) reported well-established colonies living out-of-doors near buildings in Québec, but none of these was known to survive the winter. A single specimen was taken by a survey team from the Biosystematics Research Institute, Agriculture Canada, Ottawa, on Thwartway Island, St. Lawrence Islands National Park

(9 September 1976), where there are no buildings. It is possible that this specimen was a member of a colony that survived winter there, but more likely it was transported to the island in equipment by the survey team and then captured. Survival out-of-doors in these northern latitudes is thought to be possible only in very favorable locations such as close to foundations of heated buildings or in city refuse dumps where spontaneous combustion keeps them warm. E. A. Back (1936) and Caesar and Dustan (1939) record *A. domesticus* from such situations in the northern United States and in Ontario (Oshawa), respectively. A detailed study of the association between *A. domesticus* and rubbish dumps in England was made by Bate (1969). Wild colonies are known to survive throughout the year in California (Weissman and Rentz 1977), and it seems that such feral colonies also persist in many parts of the southern United States, particularly in the neighborhood of "cricket farms," where they are extensively reared for such purposes as fish bait and pet food.

Life history. Ghouri and McFarlane (1958*a*) ("Canadian race") and Bate (1971) summarized the life cycle under laboratory conditions. These conditions are probably optimum or nearly so; feral colonies would have longer cycles, dependent upon the outside conditions where they occur (Bate 1969), since temperature changes cause significant alteration of the life cycle (Bate 1972).

Economic importance. *A. domesticus* may cause economic damage (Caesar 1941). It is known to injure fabrics and especially to chew upon wet clothing; indeed the first reports for Canada (Kalm 1756*b*, 1761) record their destructiveness to cloth. Most often the insects are treated as merely a nuisance inside dwellings, their nocturnal chirping disturbing the peace of mind of the occupants. However, this is entirely dependent upon the likes and dislikes of individual householders. Even in North America, a few people keep crickets (including *A. domesticus*) in cages and enjoy the cheery chirping. Sometimes, however, the insects invade premises in large numbers from city rubbish dumps (E. A. Back 1936; Caesar 1941). A massive invasion of houses at Oshawa, Ont., is reported by Caesar and Dustan (1939). Rearing house crickets ("gray crickets") for fish bait and pet food is a large industry in some parts of the United States, and the species is used extensively in experimental laboratory studies of insect physiology.

Order Orthoptera—locusts and grasshoppers

To this order belong the true, or short-horned, grasshoppers, the true locusts, and their relatives. These are still generally regarded as a suborder (Caelifera) of Orthoptera, *sensu lato*, but, although they share some features with the Grylloptera, the differences between the two groups are numerous and many of them fundamental. True orthopteran (i.e., caeliferan) characters include the following: antennae typically comparatively thick, not whiskerlike, rarely, if ever, with more than 30 segments, usually cylindrical and simple (although often flattened or triangular in section, particularly basally), occasionally serrated, clubbed, or expanded (rarely pectinate or lamellate) apically, seldom as long as body. Mandibles of chewing type, somewhat asymmetrical, with rather heavy molar teeth, never greatly enlarged.

Thoracic pleura not largely concealed by lateral pronotal lobes, with visible parts extensive; prothoracic spiracles and associated tracheal system not enlarged or having auditory function. Tarsi typically with 3 segments, sometimes less, never 4; front tibiae never with tympanic ("auditory") organs; hind femur often with longitudinal stridulatory ridge (with or without single row of denticles) on inner face. Tegmina rarely resembling broad-leaved (dicotyledonous) foliage but commonly resembling grass blades and stems; if modified for stridulation, such modification does not involve tegminal base; stridulation (when it occurs) achieved by striking the tegmen with ridge on inside of hind femur, never by rubbing tegmina together; hind wing with "fan" confined to the anal region (i.e., apical margin normally discontinuously rounded, unequally biarcuate, the division coming behind the cubital veins). Base of abdomen generally with lateral tympanal ("auditory") organs; abdominal stridulatory ridges occasionally present. Cerci neither long nor flexible. Ovipositor of female short, usually stout, with inner valves reduced so that it is composed of 4 main valves, which are independent and not "tongued and grooved" together to form a single unit, rarely reduced or absent. Male subgenital plate (ninth abdominal sternum) devoid of styles; phallic structures including an intromittant aedeagus (rarely reduced). Salivary glands and proventriculus ("gizzard") not well-developed, former lacking reservoirs, latter with, at most, minute teeth; gastric caeca typically comprising 3 pairs; intestine straight. Malpighian tubules entering gut separately but closely together. Male ejaculatory duct without prostate glands. Accessory glands of female reproductive system tubular, arising from anterior end of each lateral oviduct.

Orthoptera, *sensu stricto*, also differ from Grylloptera in certain important features of their embryology, and in the method of water-uptake by the eggs (unlike Grylloptera, these have a distinct hydropyle at the posterior pole). The number of chromosomes, although variable, is, compared with the Grylloptera, fairly constant—at 12 or 10 pairs (diploid female). It is sometimes lower than this (as in Tetrigodea and most Tridactylodea, in which it is seven pairs), but (so far as known) there are never more than 13 pairs.

There are an estimated 11 000 described species, involving at least 2000 genera, distributed among 4 suborders.

Members of the Orthoptera, *sensu stricto*, are largely, but not exclusively, diurnal, being particularly active in bright sunshine. Males of many species (particularly in north temperate regions) stridulate (females seldom do so, at least not audibly to the human ear), either typically as indicated previously or by various other mechanisms (Kevan 1955). It is probable that the majority of species do not stridulate (except perhaps in Europe), despite a popular impression to the contrary.

Although the majority are tropical, they are better represented in temperate regions and extend farther into higher altitudes and latitudes than do the Grylloptera. Even though most are generally known collectively as "grasshoppers," this is often a misnomer, for, although Orthoptera are prominent in grasslands, a great many species prefer other vegetation to live in and feed upon. Arbusticolous and thamnicolous species are numerous except in temperate countries, and arboreal species form an appreciable minority in some tropical regions. There are even a few tree dwellers in temperate countries. Very few species descend into soil crevices or bury themselves shallowly under sand or dust. None lives in caves. Subaquatic forms frequently swim (their hind tibiae may sometimes be adapted for this), and a few species may actually spend most of their lives in water; however, this does not apply to the region here covered. Practically all species are exclusively phytophagous, although some animal matter may sometimes be ingested, and some cannibalism may occur at high population densities. None is commensal with other animals. Aposematic (warning) coloration and/or the discharge of irritant and/or otherwise unpleasant secretions are fairly common (not, however, in the region here considered). The "drooling" of a dark, regurgitated fluid (not a secretion), widespread among orthopteroids, is an almost universal defensive reaction in the largest family, Acrididae. When Orthoptera, *sensu stricto*, copulate, the males (unlike those of most Grylloptera) cling to the backs of the females. The spermatophores are greatly elongated and fully taken into the female genital tract; the sacs are not devoured. Eggs are normally laid in "pods" held together by a foamy secretion and buried in the ground. They are rarely laid in living plant material, but a few species lay their eggs in rotten wood.

Gregarious behavior and sustained migration, both on the ground (mainly immature stages) and in flight, are common and may be spectacular, for instance locust swarms; individual migration is also frequent. Migratory flights not infrequently continue at night if the upper air remains warm. True locusts belong to Orthoptera, *sensu stricto*, and to a single family (Acrididae), but a number of other species, representing several families, may be injurious to crops to differing degrees, with or without gregarious behavior in the immature stages. Species of this order are not considered indoor pests—except those individuals that enter houses in large numbers in the course of migration, as graphically described by the prophet Joel (Joel 2:9—see Kevan 1978*b*). A few species are poisonous if eaten. Only one species has been seriously used (and another considered) in an attempt at biological control (of water weeds (*Salvinia*) and water hyacinth, *Eichhornia*, respectively).

There is a considerable amount of folklore associated with grasshoppers generally, in respect of both injurious and innocuous species, and of "musicians" in particular. That relating to the last, however, frequently has its origins outside the order Orthoptera, *sensu stricto*, and may stem from tettigonioid and/or cicada lore.

For a general scientific account of Orthoptera, *sensu stricto*, (mainly Acridodea), the reader is referred to the comprehensive works of Uvarov (1966*a*, 1977). A compilation of natural enemies is given by Rees (1973).

Key to suborders of Orthoptera

1. Size small (can be less than 5 mm long). Legs modified for digging; tarsi with less than 3 segments. Ovipositor often lacking .... **Tridactylodea** (p. 299)

   Size variable, but mostly moderate. Legs not or little modified for digging; tarsi typically 3-segmented, sometimes (Tetrigodea) with fore and middle tarsi 2-segmented. Ovipositor present 2

2(1). Size generally rather small (maximum about 2 cm, usually much less). Pronotum greatly extended backward, almost always nearly to or beyond end of abdomen, typically drawn out into point, rarely blunt. Fore and middle tarsi 2-segmented; hind tarsi 3-segmented (second segment very small), without arolia between claws. Tegmina, if present, reduced to small oval lateral lobes even when hind wings well-developed (these, at rest, largely covered by extension of pronotum). Ovipositor valves slender, usually strongly denticulate .... **Tetrigodea** (p. 304)

Not as above .... **Acridodea** (p. 321)

Clé des sous-ordres d'Orthoptera

1. Petite taille (la longueur peut être inférieure à 5 mm). Pattes adaptées pour creuser; tarses comptant moins de trois segments. Ovipositeur souvent absent .... **Tridactylodea** (p. 299)

Taille variable, le plus souvent moyenne. Pattes peu ou pas modifiées pour creuser; tarses à trois segments, les tarses antérieurs et médians n'en comptant parfois que deux (Tetrigodea). Ovipositeur présent ... 2

2(1). Taille généralement assez petite (maximum d'environ 2 cm, mais habituellement beaucoup moins). Pronotum très développé vers l'arrière, rejoignant presque ou dépassant presque toujours l'extrémité de l'abdomen, se terminant typiquement en pointe, rarement obtus. Tarses antérieurs et médians à deux segments; tarses postérieurs à trois segments (le deuxième étant très petit) et sans arolium entre les griffes. Tegmina, lorsque présents, limités à de petits lobes latéraux même lorsque les ailes postérieures sont bien développées (ces dernières, en position de repos, sont en grande partie recouvertes par le pronotum). Valves de l'ovipositeur minces et en général fortement denticulées .... **Tetrigodea** (p. 304)

Combinaison de caractères différente de celle donnée ci-dessus .... **Acridodea** (p. 321)

Suborder Tridactylodea

Members of this suborder (now comprising 3 families, 9 genera, and just over 140 species) were at one time thought by most authors to be related to the mole crickets (now order Grylloptera, superfamily Gryllotalpoidea). Various workers since E. M. Walker (1919*a*), in Canada, and Crampton (1919), in the United States, and in agreement with an earlier diagram published by Handlirsch (1908), however, have regarded at least the principal included group (now superfamily Tridactyloidea) as closer to Acridodea, and this is now generally accepted.

Superfamily Tridactyloidea

These are small insects varying in size from less than 4 mm (among the smallest of all orthopteroids except for the Myrmecophilidae) to 1.5 cm in length: color variegated black and whitish, grayish, or yellowish. Head rounded. Eyes well-developed. Antennae 9- to 12-segmented, often somewhat beaded ("moniliform"). Ocelli 3. Pronotum rounded, with posterior margin convex; prosternum moderately broad; mesosterna and metasterna wide. Front legs with tibiae expanded apically, typically spined and modified for digging; tarsi (like those of less specialized middle tibiae) 2-segmented; hind legs well-adapted for jumping, their femora stout, compressed; hind tibiae normally terminating in elongate paired movable flattened spurs (or calcaria), often preceded by series of paired narrow leaflike labellae; hind tarsal segments reduced to 1, sometimes vestigial and appearing to be absent. Wings always present in adults; anterior pair (tegmina) normally much shorter than hind pair (these sometimes abbreviated), leathery and lobelike or flaplike, sometimes possessing, in males, an apical row of teeth, these interacting with others on subcostal vein at base of hind wing to form a stridulatory mechanism; hind wing with narrow anterior part (remigium) and wide posterior part (vannus); vannus with single series of crossveins linking longitudinal veins; fourth abdominal tergum sometimes with vertical stridulatory ridges that interact with teeth on tegmina. Cerci with 2 segments. Terminalia of abdomen also bearing pair of cercuslike "paraproctal" plates. Ovipositor absent.

There are about 135 species in 7 genera, divided between 2 families: Tridactylidae, which is more or less worldwide in occurrence, and Ripipterygidae, which is Neotropical. Most species are apparently associated with damp or wet mud or sand beside water. Most, if not all, tend to be gregarious and burrow, some, at least on occasion, deeply, where they make "nests" or, in temperate countries, hibernate. Many species can move on the surface of water and some are also active swimmers. They are mainly nocturnal and apparently feed upon small particles of vegetable matter, probably including algae, retrieved from mud and sand.

Of the two included families, Tridactylidae, which was given separate ordinal status by Dirsh (1973, 1975), is represented in southern Canada.

Family Tridactylidae

Members of this family are the true pygmy hoppers. They are characterized by their generally minute size (seldom reaching 1 cm in length, and not reaching 1 cm in length in the region here considered) and by the general features indicated for the superfamily above. They are less slender, of shinier appearance, and have a more variegated black-and-white or gray coloration than the Ripipterygidae. They have 11-segmented, moniliform antennae. They also differ from the Ripipterygidae by the following characters: smaller eyes, more forwardly directed mouthparts, posteriorly rounded but scarcely produced posterior margin of the pronotum, shorter legs, which are more strongly modified for digging, males possessing a tegmino-alary (Figs. 822, 824), and sometimes an abdomino-tegminal, stridulatory mechanism, 2-segmented cerci, lack of an ovipositor in females, and reduced phallic structures.

There are 3 American and 2 Old World genera and probably more than 70 species, 38 of them known from the Americas. Most are tropical or subtropical, but some penetrate temperate latitudes (including the southern Palaearctic and southernmost Canada). Urquhart (1937, 1940*b*) gives an account of their biology (for one of the Canadian species). Food seems to consist of minute particles (e.g., algae) ingested with quantities of mud or silt. The family has no economic significance. Two species, now placed in different genera, occur in Canada and the neighboring United States.

Key to genera of Tridactylidae

1. Prosternum with median conical process (Fig. 818). Tarsus of hind leg large, nearly as long as subapical spurs (calcaria) of hind tibia; apical spurs of hind tibia about one-third as long as hind tibia (Fig. 819) .... ***Neotridactylus*** (p. 301)
   Prosternum without median conical process. Tarsus of hind leg vestigial, scarcely visible, much shorter than subapical spurs of hind tibia; apical spurs of hind tibia almost half as long as hind tibia (Fig. 820) .... ***Ellipes*** (p. 303)

Clé des genres de Tridactylidae

1. Prosternum prolongé en un cône médian (fig. 818). Tarses des pattes postérieures grands, presque aussi longs que les éperons subapicaux des tibias postérieurs (calcars); longueur des calcars égale au tiers environ de celle des tibias postérieurs (fig. 819) .... ***Neotridactylus*** (p. 301)
   Prosternum dépourvu de cône médian. Tarses des pattes postérieures rudimentaires, à peine visibles, beaucoup plus courts que les calcars des tibias postérieurs; calcars des tibias postérieurs atteignant presque la moitié de la longueur de ces derniers (fig. 820) ... ***Ellipes*** (p. 303)

Genus *Neotridactylus* Günther

Description. Small insects, 5.5–9.5 mm long. Pronotal disc constricted in anterior quarter, with faint transverse groove near anterior margin; prosternum with median conical process (Fig. 818). Apices of tegmina broadly rounded. Hind tarsus almost as long as apical spurs (calcaria) of hind tibia (Fig. 819). Color mainly black or dark brown.

Neotridactylus apicialis (Say)

Figs. 817–819; Map 103

Tridactylus apicialis Say, 1825*a*:310.
Tridactylus tibialis Guérin-Méneville, 1844:336, 337.
Xya mixa Haldeman, 1853:364.
Tridactylus illinoiensis C. Thomas, 1862:104.
Tridactylus terminalis Scudder, 1863*a*:425.
Tridactylus fissipes Saussure, 1874:352.
Tridactylus incertus Saussure, 1896:206.
Tridactylus apicalis; E. M. Walker 1904*a*:143; Helfer 1963:325.
Neotridactylus apicialis; K. K. Günther 1975:308, fig. 4; 309, fig. 6; 331, 332, fig. 33; 333, figs. 34, 35; 334, fig. 36.

Diagnosis. Prosternum with median conical process. Hind tarsus nearly as large as spurs of hind tibia.

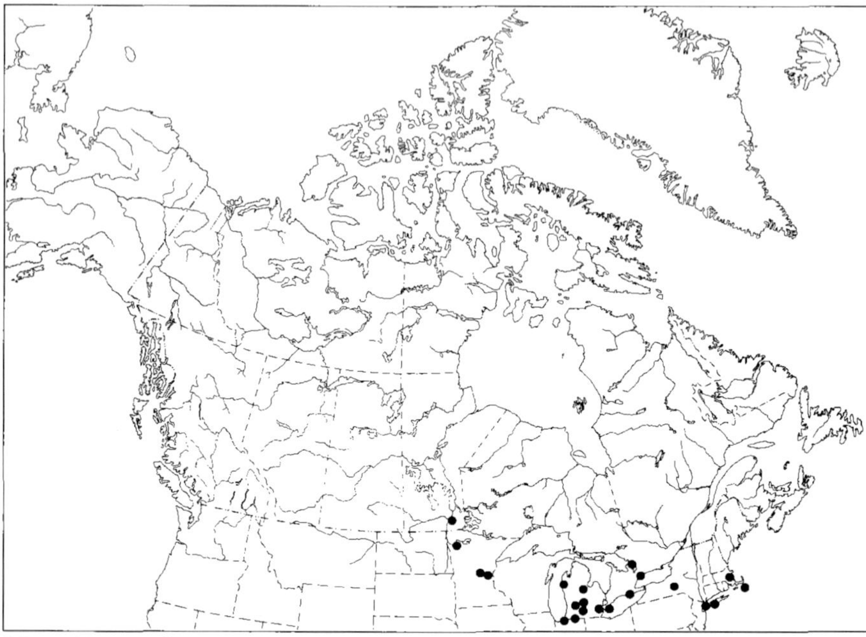

Map 103. Collection localities for *Neotridactylus apicialis*.

Description. As for genus *Neotridactylus* (Fig. 817). Head and pronotum with pale markings; tegmina with yellow dorsal area and yellow spot on apical half; legs dull yellow; hind femur with upper half either brown or with 3 dark cross-bars.

Range. Manitoba to Massachusetts, south to Mexico; South America.

Behavior and habitats. *N. apicialis* occurs mainly on damp sand bars or in sandy or silty areas with sparse vegetation near water, including the edges of ditches. Small pits are excavated in the sand or silt and the insects often remain in these for considerable periods with only their heads and pronota protruding. The pits lead into tunnels 2–3 cm below the surface. At the ends of these, females excavate brood chambers in which they deposit batches of 8–18 eggs. There is some evidence of maternal care of eggs and young and of differential development rates. If disturbed, the insects retreat into their burrows. When not in these, they leap vigorously, as much as 1.5 m high and 3 m horizontally. To avoid capture they may leap into water and then swim back to shore or to an exposed bar, whence they will leap again. The biology of *N. apicialis* is discussed by Urquhart (1937, 1940*b*), who examined the gut contents and found these to be composed largely of fine "sand" particles, from which he deduced that they fed by ingesting quantities of such matter, digesting the minute fragments of organic matter and possibly microscopic algae that they contained.

Life history. Urquhart (1937) found adults throughout the year near Toronto; they were commonest in late summer. In winter, the insects hibernate, mainly as nymphs, some 4.5–6.0 cm below the surface. Disturbed hibernating nymphs were very active, but adults taken at the same time "only showed signs of life when taken indoors." Eggs seem to be laid mainly in early summer and midsummer (Urquhart 1940*b*). There is probably only one rather protracted generation each year. In Michigan, Cantrall (1968) records adults as occurring most commonly from early May to early October.

Genus *Ellipes* Scudder

Description. Size small (often less than 5 mm long). Pronotum short, not constricted anteriorly; prosternum without median conical process. Hind tarsus vestigial, always shorter than smaller subapical spurs (calcaria) of hind tibia; larger apical calcaria of hind tibia long, finely fringed dorsally with hairs (Fig. 820).

Ellipes minutus minutus (Scudder)

Figs. 820–824; Map 104

Tridactylus minutus Scudder, 1863*a*:425.
Tridactylus (*Heteropus*) *histrio* Saussure, 1896:207, 208.
? *Tridactylus* (*Heteropus*) *histrionicus* Saussure, 1896:207.
Ellipes minuta; Scudder 1902:309, 310.

Diagnosis. Hind tarsus vestigial, shorter than smaller spur of hind tibia. Fore and middle tibiae with pale cross-bars.

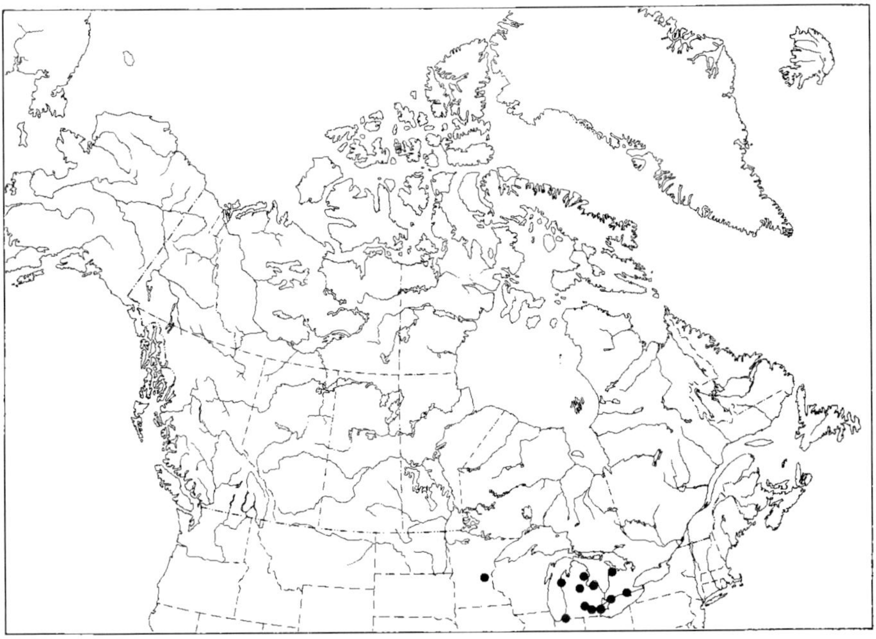

Map 104. Collection localities for *Ellipes minutus minutus*.

Description. As for genus *Ellipes* (Fig. 821). Body length 4–5 mm. Black or dark brown; head and pronotum with paler markings; margins of lateral pronotal lobes pale; femora and middle tibiae with pale cross-bars. Tegmina short, covering only half of abdomen. Hind wings usually longer than tegmina (rarely up to twice as long).

Range. Minnesota to southern Ontario, south to Mexico; South America.

Behavior and habitats. Unknown.

Life history. Cantrall (1968) recorded adults in Michigan from 8 May to 8 September. In southern Ontario, the period of major activity seems to be about the same.

Suborder Tetrigodea

Members of this suborder are known as groundhoppers or pygmy grasshoppers; in North America, they are known as grouse locusts or grouse grasshoppers. They may be distinguished by the following combination of characters: size small, seldom exceeding 2 cm in length, usually much shorter. Coloration usually grayish or brownish, sometimes blackish or variegated, but never green. Head generally small. Eyes prominent, often protruding. Antennae filiform. Mouthparts enveloped by anterior margin of prosternum. Pronotum characteristic, greatly extended backward (Figs. 796–798) to apex of abdomen and typically drawn out into attenuated point, sometimes blunt, with lateral lobes small. Legs usually rather short, with anterior and middle tarsi 2-segmented; posterior tarsi 3-segmented (the middle segment very small); all tarsi lacking arolia; hind femora stout. Wings typically present, sometimes absent; anterior pair, when present, represented merely by abbreviate oval lateral lobes with reduced venation; hind pair, when present, largely concealed when at rest by backward extension of pronotum, frequently extending considerably beyond this, often shorter, some species being polymorphic for this character (short hind wings normally associated with a shorter pronotal process, "brachypronotal"; longer hind wings with a longer pronotal process, "macropronotal"); specialized stridulatory mechanism absent (although there have been reports of sound production in some species). Auditory tympana lacking. Male subgenital plate large and divided by transverse groove, the apical part subconical and movable; phallic structures without epiphallus. Ovipositor similar to that of Acridodea, but valves proportionately longer, more slender, usually strongly denticulate. Chromosome number, generally $2n$ male = 13.

There are about 185 genera and nearly 1000 species distributed throughout the world, a few even at high latitudes. They are, however, predominantly tropical, especially in the Indomalayan–Papuan region. Most species belong to the family Tetrigidae, *sensu stricto*. The great majority live on the ground, particularly in damp muddy situations. Some, particularly among the Oriental tetrigine tribe Scelimenini, are semiaquatic or aquatic, swimming vigorously (even under water) by means of their expanded hind tibiae. Most species are cryptically colored; some closely mimic dead vegetation or lichens. Tetrigodea species feed on small fragments of vegetation and probably mainly on microscopic algae. So far as known, eggs are for the most part laid in soil or mud; they are peculiar in that they have a terminal filament, directed upward, when laid. The economic importance of the group is negligible although there have been reports of some damage being done to rice crops in the Far East.

The North American species of Tetrigodea are treated by J. A. G. Rehn and Grant (1961). A list of Nearctic genera and species is given by Steinmann (1971).

Key to families of Tetrigodea

1. Antennae with less than 15 segments. Anterior femora keeled above. Pronotum (except in immature individuals) usually not distinctly tectiform, with keel not usually pronounced and not projecting forward over head beyond middle of eyes in dorsal view .... **Tetrigidae** (p. 305)
   Antennae with at least 16 and normally 20–22 segments. Anterior femora grooved above. Pronotum more or less tectiform, with keel pronounced and projecting forward over head beyond middle of eyes in dorsal view .... **Batrachideidae** (p. 318)

Clé des familles de Tetrigodea

1. Antennes comptant moins de 15 segments. Fémurs antérieurs carénés dorsalement. Pronotum (sauf chez les spécimens immatures) habituellement non tectiforme, à carène d'habitude non prononcée et ne se prolongeant pas au-dessus de la tête au-delà du milieu des yeux lorsque vu du dessus .... **Tetrigidae** (p. 305)
   Antennes comptant au moins 16 et habituellement 20 à 22 segments. Fémurs antérieurs pourvus d'un sillon dorsal. Pronotum plus ou moins tectiforme, caréné, se prolongeant au-dessus de la tête, au-delà du milieu des yeux lorsque vu du dessus .... **Batrachideidae** (p. 318)

Family Tetrigidae

The Tetrigidae species constitute the great majority of the members of the superfamily. They have the general characters already indicated under Tetrigodea. They are of variable form but may be distinguished from the Batrachideidae as noted in the key to families.

There are about 170 genera and more than 850 species.

The general remarks under suborder Tetrigodea are fully applicable to the Tetrigidae. Only the widely distributed Tetriginae species occur in Canada and adjacent territories, all of them in the relatively unspecialized tribe Tetrigini.

Key to genera of Tetrigidae

1. Antenna 12- or 13-segmented. Pronotum with median carina arched in lateral view; upper sinus of lateral lobe about half as deep as lower sinus (Fig. 797) .... *Nomotettix* (p. 306)

Antenna 14-segmented. Pronotum (except in immature forms) with median carina low, rather flat in lateral view; upper sinus of lateral lobe almost as deep as lower sinus (Fig. 798) 2

2(1). Fastigium of vertex produced in front of eyes (Figs. 806–809), as seen in dorsal view much wider than width of an eye (Figs. 800–803) .... *Tetrix* (p. 308)

Fastigium of vertex not produced in front of eyes (Fig. 810), as seen in dorsal view not wider than width of an eye (Fig. 804) .... *Paratettix* (p. 316)

Clé des genres de Tetrigidae

1. Antennes à 12 ou 13 segments. Pronotum à carène médiane arquée en vue latérale; sinus supérieur des lobes latéraux de profondeur approximativement égale à la moitié de celle du sinus inférieur (fig. 797) .... *Nomotettix* (p. 306)

Antennes à 14 segments. Pronotum (sauf chez les formes immatures) à carène médiane basse et plutôt aplatie en vue latérale; sinus supérieur des lobes latéraux presque aussi profond que le sinus inférieur (fig. 798) 2

2(1). Fastigium du vertex développé en avant des yeux (fig. 806 à 809); vu du dessus, beaucoup plus large que l'œil (fig. 800 à 803) .... *Tetrix* (p. 308)

Fastigium du vertex non développé en avant des yeux (fig. 810); vu du dessus, pas plus large que l'œil (fig. 804) .... **Paratettix** (p. 316)

Genus *Nomotettix* Morse

Description. Antennae short, filiform, with 12 or 13 segments. Fastigium of vertex produced in front of eyes, rounded to slightly angulate, wider than an eye; occiput usually with pair of minute tubercles behind middle of eyes. Pronotum with median carina strongly tectiform or subcristate, arched longitudinally, with anterior margin produced over posterior part of head; tegminal (upper) sinus of lateral lobe about half as deep as lower sinus; posterior process usually short, reaching to about apices of hind femora (rarely much longer in macropronotal form). Tegmina elongate-ovate. Hind wings not usually reaching beyond apex of pronotal process, rarely slightly longer.

Key to species and subspecies of *Nomotettix*

1. Fastigium projecting in front of eyes for a distance of less than half width of an eye (Figs. 799, 805). Pronotum distinctly arcuate in lateral view, with highest point above middle of lateral lobes (Fig. 797). Found in Eastern Canada and United States .... ***cristatus cristatus*** **(Scudder)** (p. 307)

Fastigium projecting in front of eyes for a distance equal to width of an eye. Pronotum relatively flatter in dorsal view, never strongly and convexly arcuate. Found in southeastern Minnesota .... ***parvus*** **Morse** (p. 308)

Clé des espèces et des sous-espèces de *Nomotettix*

1. Fastigium développé en avant des yeux sur une distance inférieure à la moitié de la largeur d'un œil (fig. 799 et 805). Pronotum clairement arqué en vue latérale, son point le plus élevé situé au-dessus du milieu des lobes latéraux (fig. 797). Espèce vivant dans l'est du Canada et des États-Unis .... ***cristatus cristatus* (Scudder)** (p. 307)

   Fastigium développé en avant des yeux sur une distance égale à la largeur d'un œil. Pronotum relativement plus aplati en vue dorsale; jamais fortement convexe. Espèce vivant dans le sud-est du Minnesota ... . ***parvus* Morse** (p. 308)

Nomotettix cristatus cristatus (Scudder)

Figs. 796, 797, 799, 805; Map 105

B(atrachidea) cristata Scudder, 1863*a*:478.
B(atrachidea) carinata Scudder, 1863*a*:479.
Nomotettix sinuifrons Hancock, 1899*b*:279; Helfer 1963:87 (partim).
N(omotettix) cristatus carinatus; Hancock 1906:62.
Nomotettix borealis E. M. Walker, 1909:173.
Nomotettix cristatus cristatus; J. A. G. Rehn and Hebard 1916*b*:128; J. A. G. Rehn and Grant 1961:36; Vickery et al. 1974:81.
Nomotettix cristatus sinuifrons; Hebard 1925*a*:44.

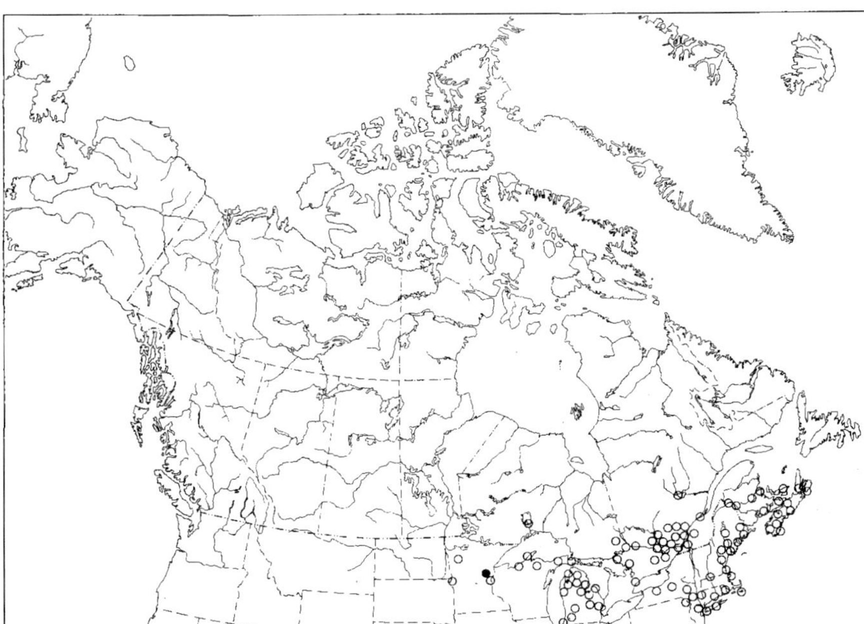

Map 105. Collection localities for *Nomotettix cristatus cristatus* (○) and *N. parvus* (●).

Diagnosis. Median carina of pronotum strongly tectiform, arched; tegminal sinus of lateral pronotal lobe about half as deep as lower sinus. Fastigium projecting ahead of eyes by less than half length of eye.

Description. As for genus (Fig. 796). Pronotal length, males 7.6–8.4, females 8.7–10.3 mm (macropronotal and brachypronotal forms known). Color usually dull gray to sooty black, with velvety black patches on lateral areas of pronotal disc.

Range. Minnesota to Nova Scotia, south to northern Georgia.

Behavior and habitats. *N. cristatus cristatus* tolerates drier locations than most tetrigids and consequently is found in a wide variety of habitats. It occurs in old fields, at the edges of woodland clearings, and along roadsides, in pastures, and even on newly burned barren lands. It prefers areas near deciduous woods, with *Vaccinium* and other undergrowth plants. It is somewhat lethargic and can usually be collected by hand without difficulty.

Life history. Univoltine, with overwintering as adults and, in most areas, also as late-instar nymphs. Adults are common in April and persist to July; adults of the succeeding generation appear in numbers in September.

Nomotettix parvus Morse

Map 105

Nomotettix parvus Morse, 1895:14.
Nomotettix acuminatus Hancock, 1899*a*:8; 1902:56, pl. II, figs. 2, 2*a*.
Nomotettix parvus; Hancock 1902:55, pl. II, figs. 4, 4*a*; Helfer 1963:67.

Diagnosis. Fastigium projecting ahead of eyes by length of an eye.

Description. Pronotal length, males 5.5–6.3, females 6.5–7.6 mm; only brachypronotal form known.

Range. Minnesota, Iowa, Illinois, and Kansas.

Behavior and habitats. *N. parvus* seems to be a prairie species, as it is known only from actual prairie or from relict prairie areas. It is relatively uncommon and little has been recorded concerning it.

Life history. Presumably similar to *N. cristatus*. Adults have been found from April to September.

Genus *Tetrix* Latreille

Description. Body robust to slender. Dorsal surface of head and pronotum finely granulate to rugose. Antennae filiform, 14-segmented. Eyes not distinctly globose and protruding. Fastigium of vertex carinate, produced in front of eyes to a variable degree (Figs. 806–809), anteriorly arcuate, angulate, or transversely truncate in dorsal aspect, much wider than width of an eye (Figs. 800–803); facial outline in lateral aspect usually sinuate, moderately convex between antennal bases and concave between paired ocelli (Figs. 806–809). Pronotum of variable length (macropronotal or brachypronotal), but often long, nearly flat to strongly tectate dorsally, occasionally subcristate (immature forms), usually with paired triangular impressions behind humeral area; median carina usually evident; lateral lobe with hind margin bisinuate; tegminal (upper) sinus almost as deep as lower sinus. Tegmina ovate. Hind wings fully developed, usually distinctly exceeding apex of pronotal process.

Key to species and subspecies of *Tetrix*

1. Anterior margin of fastigium rectangulate to acute-angulate (dorsal aspect); median carina not projecting (Figs. 800, 801) 2
   Anterior margin of fastigium truncate to convex (dorsal aspect); median carina distinctly projecting beyond sides (Figs. 802, 803) 3

2(1). Body slender. Angle of fastigium with vertex nearly rectangulate; facial outline weakly emarginate at region of paired ocelli (Fig. 806). Middle femur slender, about 3.8–4.7 times as long as broad .... ***subulata* (Linnaeus)** (p. 310)
   Body more robust. Angle of fastigium with vertex acute-angulate; facial outline decidedly emarginate at region of paired ocelli (Fig. 807). Middle femur broad, about 3.0–3.4 times as long as broad .... ***brunnerii* (Bolívar)** (p. 311)

3(1). Anterior margin of fastigium truncate in outline (dorsal aspect, Fig. 802). Pronotum relatively flat in cross-section, with median carina slightly elevated anteriorly, weakly concave in outline posteriorly .... ***arenosa angusta* (Hancock)** (p. 313)
   Anterior margin of fastigium arcuate to subarcuate in outline (dorsal aspect, Fig. 803). Pronotum distinctly tectate in cross-section, somewhat arcuate in outline (*ornata, sensu lato*) 4

4(3). Anterior margin of fastigium, particularly in males, projecting laterally anterior to eyes (dorsal aspect); fastigio-frontal angle projecting (lateral aspect). Found in southern British Columbia, western Montana .... ***ornata occidua* Rehn & Grant** (p. 314)
   Anterior margin of fastigium projecting laterally very little anterior to eyes (dorsal aspect, Fig. 803); fastigio-frontal angle less produced (lateral aspect). Found over much of North America .... ***ornata ornata* (Say)** (p. 314)

Clé des espèces et des sous-espèces de *Tetrix*

1. Bord antérieur du fastigium de forme variant de rectangulaire à aiguë ou angulaire (vu du dessus); carène médiane non protubérante (fig. 800 et 801) 2
   Bord antérieur du fastigium de forme variant de tronquée à convexe (vu du dessus); carène médiane clairement protubérante (fig. 802 et 803) 3

2(1). Corps mince. Angle ménagé entre le fastigium et le vertex presque droit; profil de la face très légèrement émarginé au niveau de la paire d'ocelles (fig. 806). Fémurs médians minces, de 3,8 à 4,7 fois plus longs que larges environ .... ***subulata* (Linnaeus)** (p. 310)
   Corps plus robuste. Angle ménagé entre le fastigium et le vertex aigu ou angulaire; profil de la face clairement émarginé au niveau de la paire d'ocelles (fig. 807). Fémurs médians larges, environ de 3,0 à 3,4 fois plus longs que larges .... ***brunnerii* (Bolívar)** (p. 311)

3(1). Bord antérieur du fastigium tronqué (vu du dessus, fig. 802). Pronotum de section relativement plane, à carène médiane légèrement plus élevée vers l'avant et légèrement concave vers l'arrière .... **_arenosa angusta_ (Hancock)** (p. 313)

Bord antérieur du fastigium arqué ou subarqué (vu du dessus, fig. 803). Section du pronotum clairement tectiforme; pronotum quelque peu arqué (*ornata*, *sensu lato*) 4

4(3). Bord antérieur du fastigium, en particulier chez les mâles, développé latéralement en avant des yeux (vu du dessus); angle fastigio-frontal porté en avant (vu de côté). Espèce vivant dans le sud de la Colombie-Britannique et l'ouest du Montana .... **_ornata occidua_ Rehn & Grant** (p. 314)

Bord antérieur du fastigium peu développé latéralement au-delà des yeux (vu du dessus, fig. 803); angle fastigio-frontal moins porté vers l'avant (vu de côté). Espèce vivant presque partout en Amérique du Nord .... **_ornata ornata_ (Say)** (p. 314)

Tetrix subulata (Linnaeus)

Figs. 800, 806, 811; Map 106

Gryllus subulatus Linnaeus, 1761:236.
Acrydium granulatum W. Kirby, 1837:251.
Acrydium granulatum; A. White 1851:360; Cantrall 1943:77.
Tettix incurvatus Hancock, 1895:761.
Tettix luggeri Hancock in Lugger, 1898:199.
Tettix morsei Hancock, 1899*b*:280.
Tettix granulatus variegatus (var.) Hancock, 1902:38, 66.
Acrydium granulatum granulatum; Morse 1919*b*:38.
Acrydium granulatum incurvatum; Morse 1919*b*:38.
Acrydium subulatum; Hebard 1936*a*:27.
Acrydium subulatum granulatum; Procter 1946:40.
Tetrix subulata; J. A. G. Rehn and Grant 1955*a*:145; 1961:52; Brooks 1958:62; Helfer 1963:89; Vickery et al. 1974:83.

Diagnosis. Fastigium obtuse-angulate (dorsal aspect). Middle femur slender, not more than 3.8 times as long as broad.

Description. Body slender (Fig. 811). Pronotal length, males 10.0–10.5, females 11.9–12.5 mm. Fastigio-frontal angle nearly rectangulate in profile, weakly emarginate at paired ocelli (Fig. 806); lateral carinae of frontal costa evenly divergent or subparallel, not concave at antennal fossae; fastigium definitely obtuse-angulate in dorsal aspect (Fig. 800). Pronotum long (macropronotal, with about 6% of individuals brachypronotal; proportion varying in different areas, apparently greater in northern and western locations). Middle femur slender, 3.8–4.7 times as long as broad. Color variable, grayish to reddish brown, sometimes darker, often with median longitudinal whitish band on pronotum.

Range. Alaska to Newfoundland, south to Mexico; Europe and Asia.

Behavior and habitats. *T. subulata* prefers moist locations on sandy soils and occasionally is found in drier, cultivated or forested areas. Cantrall (1943) indicated that individuals migrate to new territory when water levels

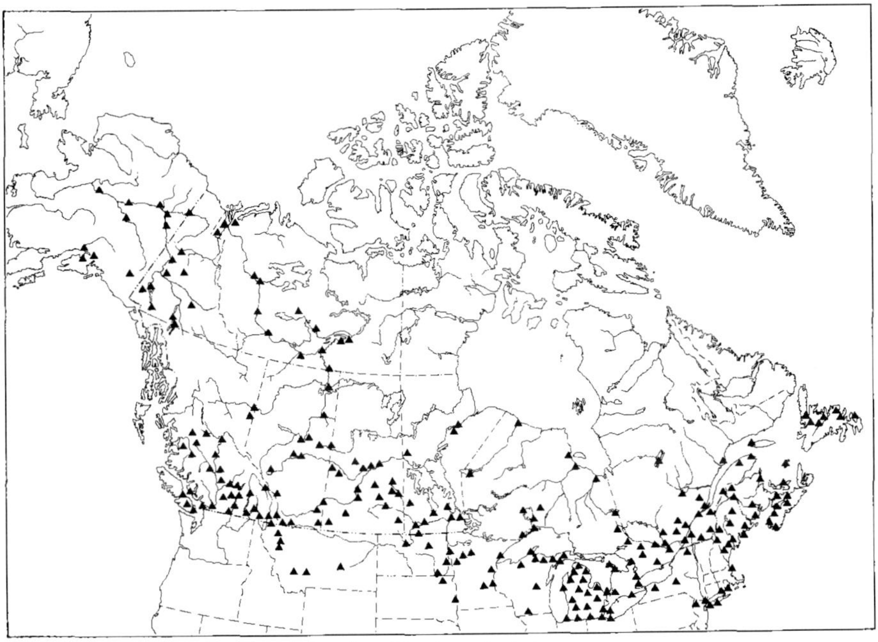

Map 106. Collection localities for *Tetrix subulata*.

fall, so that it might be found 100 m distant from standing water. Flight is rapid and the insect flies readily when disturbed. Following flight, a specimen may drop hidden into vegetation or it may make several leaps in different directions in attempting to avoid capture. It is more frequently attracted to light than are other North American species of *Tetrix*. Details of biology, postembryonic development, and spermatophore formation and transfer for a British population are given by Farrow (1963, 1964*a*, 1964*b*).

Life history. Adults have been taken from 5 April to 4 October in Nova Scotia (Vickery 1961). The earliest record in Québec is 22 March. There is one generation each year, which overwinters in the adult stage in the northern part of the range. Farther south, some late-instar nymphs have been found in the spring, so that they must have overwintered in that stage. Adult population levels become low in midsummer as the winter survivors die, and peak again in late August and September as adults of the succeeding generation mature. In the far north, owing to the abbreviated summer season, adults appear later in spring, in June or even July, and the next generation seeks winter quarters during August; nymphal growth is accelerated (J. A. G. Rehn and Grant 1955*b*).

Tetrix brunnerii (Bolívar)

Figs. 801, 807; Map 107

T(ettix) brunnerii Bolívar, 1887:259, 266.
Tettix tentatus Morse, 1899:200.

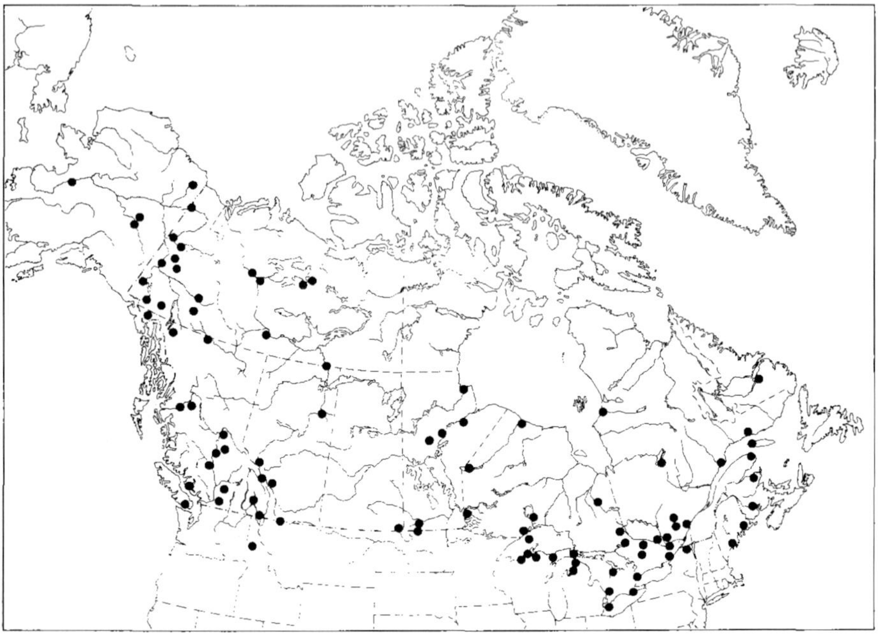

Map 107. Collection localities for *Tetrix brunnerii*.

Tettix brunneri f. *caudata* Hancock, 1906:59.

Tetrix brunneri; E. M. Walker 1909:174; Helfer 1963:89.

Acrydium acadicum brunneri; Hebard 1925*a*:47.

? *Clinotettix ussuriensis* Beĭ-Bienko, 1933:327, fig. 9, 328, fig. 10, 329.

Tetrix brunnerii; J. A. G. Rehn and Grant 1961:46; Vickery and Kevan 1967:42; Vickery et al. 1974:83.

Diagnosis. Fastigium angulate in dorsal aspect. Middle femur robust, 3.0–3.4 times longer than broad.

Description. Pronotal length, males 8.3–11.8, females 9.5–13.5 mm. Fastigio-frontal angle angular, never rectangulate; frontal profile definitely emarginate at region of median ocellus (Fig. 807); fastigium angulate in dorsal aspect (Fig. 801); lateral carinae of frontal costa slightly concave at antennal fossae. Pronotum with low but distinct hump on median carina (about 80% of known specimens brachypronotal, this proportion decreasing in eastern localities). Middle femur robust, 3–3.4 times as long as broad. Color varying from blackish to buffy gray to yellowish, often with pale dorsal longitudinal stripe and usually with paired triangular or irregular dark blotches on humeral area of pronotum.

Range. Alaska to Labrador, south to northern United States; possibly Asia.

Behavior and habitats. *T. brunnerii* is found in meadow or bog habitats, often in forested regions. Buckell (1922) reported it "among fallen leaves and mosses under birch and willow around an upland spring'' and ''in spruce swamps and along streams'' in British Columbia.

Life history. Similar to that of *T. subulata*, with overwintering in the adult stage. Adults appear in spring, disappear in midsummer, and are succeeded by adults of the next generation in late summer and autumn.

Tetrix arenosa angusta (Hancock)

Figs. 798, 802, 808; Map 108

T(ettix) angustus Hancock, 1896:238.
T(ettix) inflatus Hancock, 1896:238.
T(ettix) obscurus Hancock, 1896:239.
T(ettix) gibbosus Hancock, 1896:239.
T(ettix) fluctuosus Hancock, 1896:240.
T(ettix) decoratus Hancock, 1896:240.
Tettix a(renosus) var. *costatus*; Hancock 1902:38, 68.
Acrydium arenosum angustum; Gooderham 1918:35; Cantrall 1943:81.
Acrydium arenosum obscurum; Blatchley 1920:169.
Tetrix arenosus angustus; Froeschner 1954:199.
Tetrix arenosa angusta; J. A. G. Rehn and Grant 1956*c*:121; 1961:65; Vickery et al. 1974:85.

Diagnosis. Anterior margin of vertex nearly truncate; pronotum with prehumeral hump, then flat to concave posteriorly.

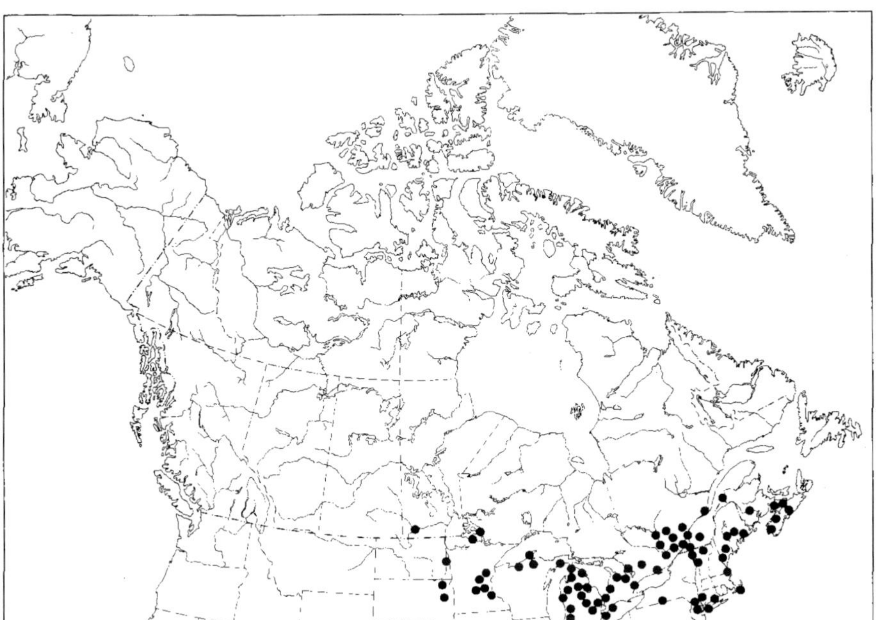

Map 108. Collection localities for *Tetrix arenosa angusta*.

Description. Body small to average (pronotal length, male 8.4–10.7, female 10.1–12.3 mm). Anterior margin of vertex nearly truncate in outline from dorsal aspect (Fig. 802); frontal profile sinuous, deeply emarginate at region of paired ocelli and protruding arcuately between antennal fossae (Fig. 808). Pronotum with pronounced prehumeral hump; carina posterior to this not raised, and dorsum flat to slightly concave (Fig. 798). Color generally light brown, somewhat darker on pronotal disc, occasionally with other pronotal markings.

Range. Manitoba to Nova Scotia, south to the Gulf Coast states and northern Georgia.

Behavior and habitats. Although *T. arenosa angusta* appears to prefer moist situations, such as stream banks, it has been found in numbers in fallen leaves in dry open woodland areas (Vickery et al. 1974). Other authors (see J. A. G. Rehn and Grant 1956*c*) have also reported finding this insect in a wide variety of habitat types. It is apparently tolerant and adaptable in this respect. Moisture is not a limiting factor.

Life history. Univoltine throughout the northern part of range, with overwintering as adults (J. A. G. Rehn and Grant 1956*c*). Adults appear in April in the north, and may be found nearly every month until September, although they are scarce during midsummer.

Tetrix ornata occidua Rehn & Grant

Map 109

Tetrix ornata occidua J. A. G. Rehn and Grant, 1956*c*:45.
Tetrix ornata occidua; J. A. G. Rehn and Grant 1961:60.

Diagnosis. Anterior margin of fastigium projecting laterally in front of eyes.

Description. Anterior margin of fastigium, in dorsal aspect, projecting laterally for distance appreciably in front of eyes, especially in males. Festigio-frontal angle, in lateral aspect, produced. Frontal profile retreating. Middle femur of male not inflated.

Range. Interior British Columbia, Washington, and Idaho.

Behavior and habitats. *T. ornata occidua* is similar in these aspects to *T. ornata ornata*. The specimen reported by Vickery and Nagy (1973) from Naramata Mountain was found, along with *T. subulata* (Linnaeus), in a localized damp depression of an otherwise dry area in a large disturbed clearing, in ponderosa pine forest at 1300 m elevation.

Life history. Presumably the same as for *T. ornata ornata*, i.e., univoltine, with overwintering as adults. Adult specimens have been collected from May to August.

Tetrix ornata ornata (Say)

Figs. 803, 809, 812; Map 109

Acrydium ornatum Say, 1824*a*:(27), pl. 5, left fig.

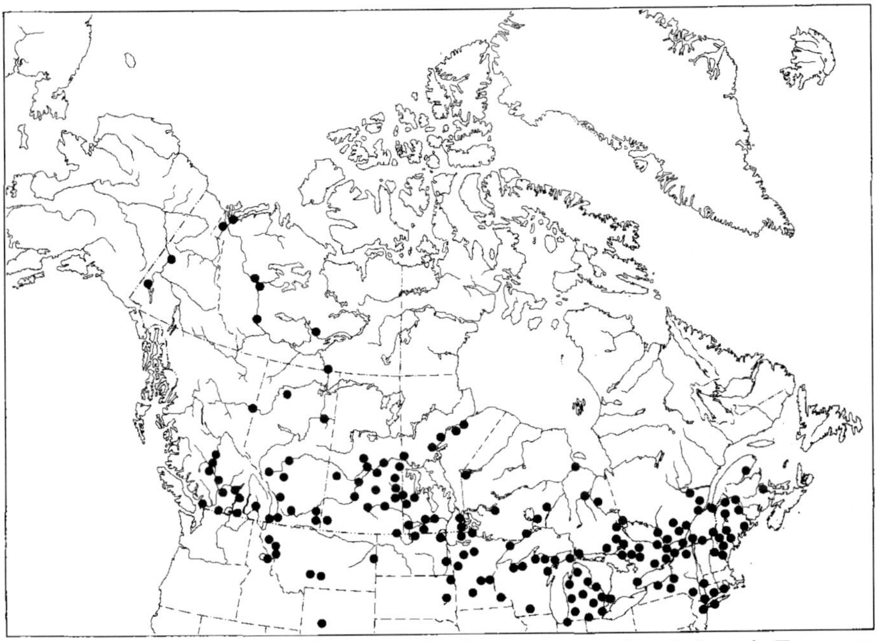

Map 109. Collection localities for *Tetrix ornata occidua* (○) and *T. ornata ornata* (●).

Tetrix ornata; T. W. Harris 1835:577; Brooks 1958:62; Helfer 1963:80 (partim); Vickery et al. 1974:84.

Tetrix dorsalis T. W. Harris, 1841:151.

Tetrix quadrimaculata T. W. Harris, 1841:151.

Tetrix bilineata T. W. Harris, 1841:151.

Tettix triangularis Scudder, 1863*a*:475.

Tettigidea acadica Scudder, 1875*b*:345.

Tettix hancocki Morse, 1899:200.

Tettix hancocki abbreviatus; Hancock 1902:38, 83.

Tettix ornatus triangularis; E. M. Walker 1902*b*:253.

Acrydium ornatum ornatum; Morse 1919*b*:38.

Acrydium ornatum hancocki; Morse 1919*b*:39.

Acrydium ornatum; Blatchley 1920:165; Cantrall 1943:79; Brooks 1958:62.

Acrydium acadicum acadicum; Hebard 1928:222; Cantrall 1943:78; Brooks 1958:62.

Tetrix ornata ornata; J. A. G. Rehn and Grant 1956*c*:123; 1961:55.

Diagnosis. Vertex arcuate to subarcuate in outline (dorsal aspect). Pronotum distinctly tectate in cross-section and arcuate in outline.

Description. (*a*) Typical form *ornata* (Fig. 812): Pronotal length, males 7.6–8.2, females 8.7–9.8 mm. Eyes large, prominent in both dorsal and lateral aspects (Figs. 803, 809). Frontal costa with margins narrowly separated, regularly diverging from fork between antennal fossae; costa with median sulcation narrow, incised, not V-shaped in section. Middle femur of male narrow, not inflated. (*b*) Form *hancocki*: Pronotal length, males 8.1–10.0, females 9.9–10.4 mm. Eyes smaller, not prominent. Frontal costa with margins more widely separated than in typical *ornata*, sharply diverging from fork to ocelli, then subparallel to point between antennal fossae; costa with median sulcation broad, V-shaped in section. Middle femur of male appreciably inflated.

Range. Yukon Territory to New Brunswick, south to Arizona and South Carolina.

Behavior and habitats. *T. ornata ornata*, in all its forms, is found in a wide variety of habitats. In general, grassy places seem to be preferred, especially if they are somewhat damp. Nevertheless, the insects also occur in rather dry areas, such as along the edges of open woods or on sandy hills.

Life history. Univoltine throughout the range, with overwintering as adults. Adults are active from early spring (usually late March and April in most of the northern areas) to the fall, with peaks of abundance in spring and fall. Immature individuals are most common during the period late spring to midsummer.

Genus *Paratettix* Bolívar

Description. Body slender. Antennae filiform, 14-segmented. Eyes prominent, globose. Fastigium short, not extending anteriorly in front of eyes (Fig. 810), narrow, not wider than width of an eye, with anterior margin truncate; disc concave on both sides of median carina (Fig. 804); frontal costa feebly sinuate between eyes and prominent between antennal bases (Fig. 810). Pronotum usually rather long (either macropronotal or brachypronotal), flattened, convex in humeral area; anterior margin truncate, enclosing back of head; median carina low; lateral lobe much as in *Tetrix*. Tegmina oval to elongate-oval, punctate. Hind wings fully developed. Hind tibia with apical third gradually enlarged, spined. Chromosome number, $2n$ male = 13 (where known).

Paratettix cucullatus (Burmeister)

Figs. 804, 810, 813; Map 110

T(etrix) cucullata Burmeister, 1838:658.
T(etrix) oxycephala Burmeister, 1838:659.
Paratettix cucullatus; E. M. Walker 1898*b*:123; J. A. G. Rehn and Grant 1961:82.
Paratettix texanus Hancock, 1902:39, 109, 113.
Paratettix texanus nanus Hancock, 1902:39, 109, 113.
Paratettix cucullatus cucullatus; Cantrall 1943:82.
Paratettix cuculatus [*sic*]; Helfer 1963:85.

Diagnosis. Fastigium short, not extending past eyes, narrow, not wider than width of an eye. Eyes prominent, globose.

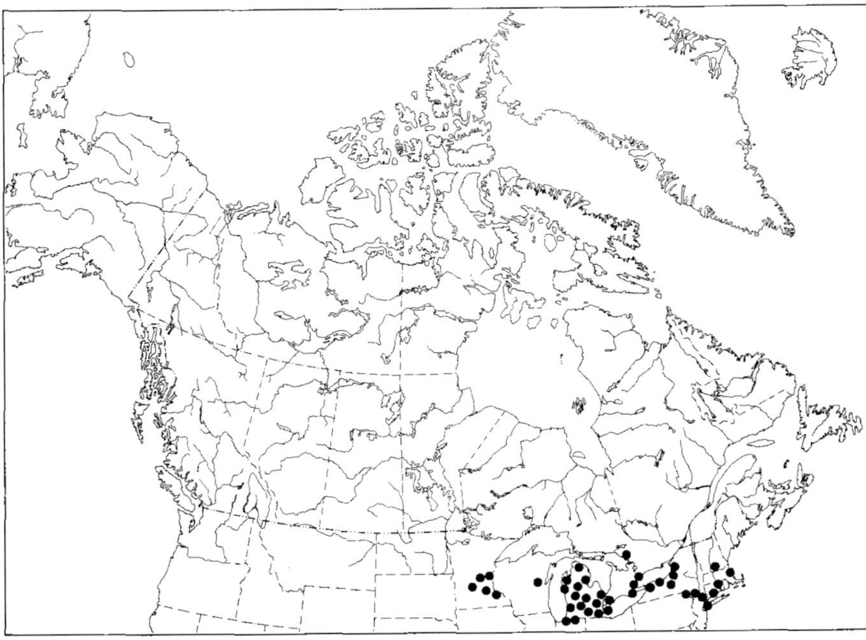

Map 110. Collection localities for *Paratettix cucullatus*.

Description. Body rather robust (pronotal length, male 9.2–10.7, female 11.1–12.7 mm) (Fig. 813). Fastigium of vertex slightly narrower than width of an eye (Fig. 804); median carinula faint, not projecting anteriorly (Figs. 804, 810). Pronotum smoothly and evenly granulate, with dorsum nearly flat; median carina indistinct anteriorly, low posteriorly. Tegmina elongate, narrow, with apices somewhat pointed. Hind wings extending about 2 mm beyond apex of pronotum. Middle femur sinuate on dorsal margin, straight to undulate beneath. Hind femur stout, finely granulate with numerous indistinct oblique ridges on outer face. Color variable, usually grayish brown or blackish with small black markings, sometimes with dark patches above tegmina; dorsum of head and pronotum often lighter than rest of body.

Range. Minnesota to southern Ontario and New Hampshire; Washington to New Mexico.

Behavior and habitats. *P. cucullatus* is found near margins of streams, ponds, and lakes. It usually remains in moist sandy areas or where there is enough standing water to produce growths of algae, which make up at least a large part of its food supply. Occasional specimens are found in grasses or wet leaves in swampy meadows. They often rest on stones projecting above water level in ponds or streams. Nabours (1947) found that females mate repeatedly and that the sperm of the last male tended to father a predominant proportion of the offspring.

Life history. Univoltine in northern areas, with overwintering as adults, which usually become active in April. These are more common in May, after which adult populations decline in midsummer, become more numerous again in August, and move to winter quarters in September before killing frosts occur.

Family Batrachideidae

Members of this family may be called "frog groundhoppers." They are usually regarded as a subfamily of Tetrigidae, but they differ significantly by having longer, always filiform antennae with more numerous (usually 20–22) segments, and the anterior femora grooved, not keeled, above. The pronotum, also, is always more or less tectiform, with a pronounced keel, and projects forward over the head.

There are 16 described genera with about 90 known species. Except for one genus and three species, all are American and mainly tropical (subfamily Batrachideinae). Little is known of the biology of this family.

Genus *Tettigidea* Scudder

Description. Body robust, with males distinctly more slender than females. Head large in proportion to body. Antennae with 20–22 segments. Eyes large, protuberant, subtriangulate in lateral aspect (Fig. 814). Pronotum either macropronotal or brachypronotal, moderately tectate, arcuate anteriorly (Fig. 816). Tegmina and hind wings typically well-developed, with tegmina elongate, rather narrow. Hind femora robust. Chromosome number, $2n$ male = 13 (where known).

Key to species of *Tettigidea*

1. Anterior margin of pronotal disc with distinct acute median point or process .... ***armata* Morse** (p. 318)
   Anterior margin of pronotal disc arcuate or obtuse-angulate (Fig. 814), never with acute median process .... ***lateralis* (Say)** (p. 319)

Clé des espèces de *Tettigidea*

1. Bord antérieur du disque du pronotum muni d'une pointe médiane aiguë distincte .... ***armata* Morse** (p. 318)
   Bord antérieur du disque du pronotum arqué ou obtus et angulaire (fig. 814), ne formant jamais de pointe aiguë médiane .... ***lateralis* (Say)** (p. 319)

Tettigidea armata Morse

Map 111

Tettigidea armata Morse, 1895:107.

Tettigidea armata depressa Morse, 1895:107.

Tettigidea armata; Hancock 1902:40, 142; Cantrall 1943:88; J. A. G. Rehn and Grant 1961:109; Helfer 1963:91.

Tettigidea davisi Morse, 1908:25.

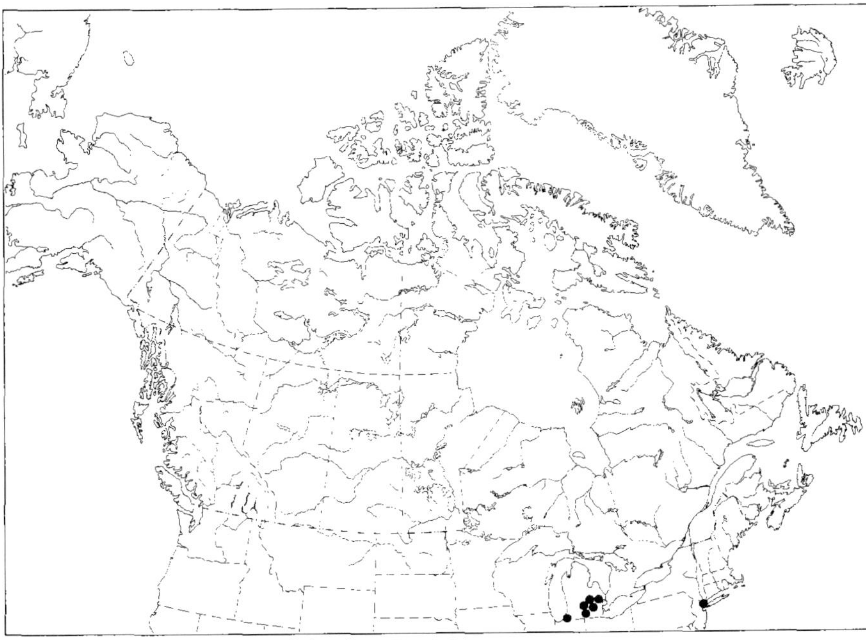

Map 111. Collection localities for *Tettigidea armata*.

Diagnosis. Anterior margin of pronotum produced as sharp process.

Description. As for genus *Tettigidea*. Body small (pronotal length, in northern region, males 9.3–10.0, females 11.4–14.7 mm). Anterior margin of pronotum produced medially between eyes into acute process; pronotum granulate and with branching longitudinal rugae on disc; both brachypronotal (60%) and macropronotal forms (40%) occur in northern regions. Tegmina and hind wings well-developed. Color blackish or grayish to light brown or fuscous dorsally, darker laterally, with hind femur sometimes irregularly barred.

Range. Michigan to New York, south to Texas and Florida.

Behavior and habitats. *T. armata* shows a distinct preference for wet mud, generally in shade, often in timbered swamps (J. A. G. Rehn and Grant 1958).

Life history. Bivoltine in the southern United States and possibly in the north as well. Not enough evidence is available to clarify this point. Cantrall (1968) reports *T. armata* as "adult the year around."

Tettigidea lateralis (Say)

Figs. 814–816; Map 112

Acrydium laterale Say, 1824*a*:(28), pl. 5, upper and right figs.
T(etrix) polymorpha Burmeister, 1838:659.
Tetrix parvipennis T. W. Harris, 1841:152.

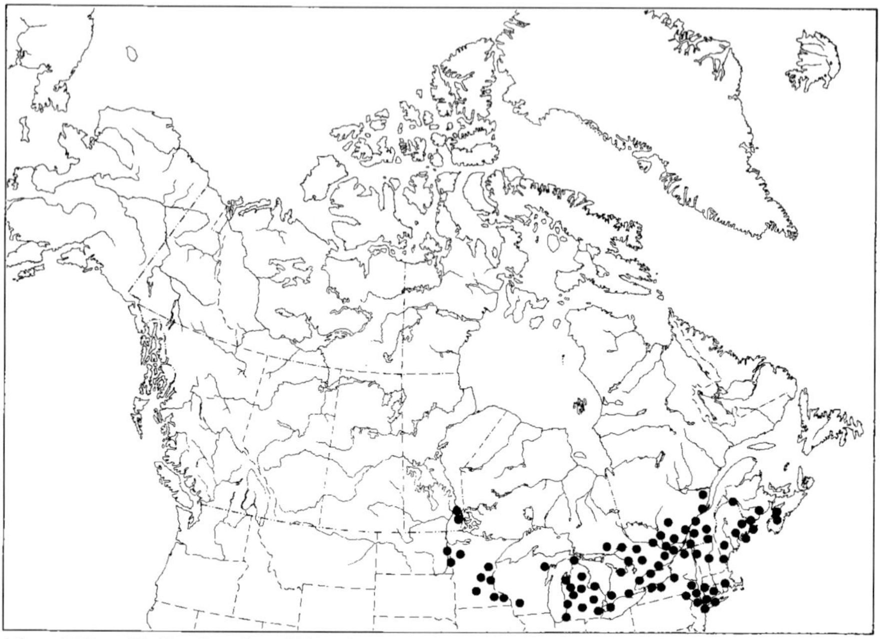

Map 112. Collection localities for *Tettigidea lateralis*.

Tettigidea lateralis; Scudder 1863*a*:477; Helfer 1963:92, 93; Vickery et al. 1974:86.

Tettigidea parvipennis Morse, 1895:108, 109.

Tettigidea parvipennis pennata Morse, 1895:109.

Tettigidea medialis Hancock, 1902:40, 140, 152, 165.

Tettigidea lateralis parvipennis; Morse 1919*b*:39; Cantrall 1943:83; Brooks 1958:62.

Tettigidea lateralis lateralis; J. A. G. Rehn and Grant 1958:25; 1961:103.

Diagnosis. Anterior margin of pronotum angulate, not produced.

Description. As for genus *Tettigidea*. Body robust (pronotal length, in northern region, males 7.2–11.3, females 9.5–14.2 mm). Median carina of head higher than eyes in lateral aspect (Fig. 815). Anterior margin of pronotum angulate, not forming median acute process between eyes (Fig. 814); pronotum granulate with distinct longitudinal branching rugae; brachypronotal form predominating (53–75%) in northern regions. Tegmina and hind wings well-developed; former rather elongate, with apices obliquely rounded. Color varying from blackish or grayish to brown, tan, or creamy dorsally, often much darker laterally; hind femur often with pale spot. Chromosome number, $2n$ male = 13 (Fontana and Vickery 1973, 1975).

Range. Manitoba to Nova Scotia, south to Arizona and Florida.

Behavior and habitats. *T. lateralis, sensu lato*, occupies a wide variety of habitats in both dry and moist areas. It has often been collected in dry open woods as well as in damp places near bodies of water. J. A. G.

Rehn and Grant (1961) also record its occurrence on dry boulder-strewn slopes and sand dunes. It is apparent that soil moisture is not a limiting factor in survival of this species.

Life history. Univoltine, with overwintering as adults in the north. In southern areas *T. lateralis* is bivoltine, adults remaining active throughout the year. Cantrall (1968) reported adults taken in all but the coldest months in Michigan. In Québec, specimens have been taken from 2 April to 12 September but mainly in May and June (Larochelle 1978).

Suborder Acridodea

The suborder Acridodea includes true grasshoppers, locusts, bush-hoppers, and related Orthoptera. It is distinguished from other suborders by the characters given in the key to families of Acridodea.

Key to families of Acridodea

1. Hind femur with dorsal and ventral proximal lobes subequal in length; hind tibia with apical fixed dorsal spine before movable spur on each side (i.e., both inner and outer series of tibial spines reach distal end of tibia). Male stridulatory apparatus, if present, consisting of unique arrangement of denticles on arched transverse veinlets of narrowed anterior area of fan of hind wing, which rubs against sharp vein on underside of tegmen .... **Romaleidae** (p. 322)
   Hind femur with dorsal proximal lobe projecting forward beyond ventral proximal lobe; hind tibia usually lacking external apical dorsal spine before movable spurs. Male stridulatory apparatus, if present, not as above .... **Acrididae** (p. 324)

Clé des familles d'Acridodea

1. Fémurs postérieurs à lobes proximaux dorsal et ventral de longueur pratiquement égale; tibias postérieurs armés, à l'extrémité apicale, d'une épine dorsale fixe avant l'éperon mobile, de chaque côté (les séries internes et externes d'épines des tibias rejoignent l'extrémité distale). Organe stridulant des mâles, lorsque présent, fait d'un seul groupe de denticules sur les petites nervures transverses arquées de la région antérieure rétrécie des ailes postérieures, qui frottent contre l'arête aiguë d'une nervure du dessous du tegmen .... **Romaleidae** (p. 322)
   Fémurs postérieurs à lobe proximal dorsal faisant saillie au-delà du lobe ventral proximal; tibias postérieurs habituellement dépourvus d'une épine dorsale apicale externe avant les éperons mobiles. Organe stridulant du mâle, lorsque présent, différent de celui décrit ci-dessus ... . **Acrididae** (p. 324)

Family Romaleidae

Members of this family are variously known by such names as lubber grasshoppers, giant "locusts," and American tree "locusts." Until recently they were regarded as a subfamily of Acrididae. They may be characterized in general by the following combination of characters: size moderate to large. Body form variable, usually rather robust. Coloration (especially of hind wings) frequently striking. Integument smooth or rugose, usually without prominent tubercles or spinelike adornments. Head variable, generally short and rounded, but sometimes subconical or elongate-conical, with frons usually concave in profile and with fastigium strongly produced or not, its apex typically somewhat angular, although lacking a longitudinal furrow and either dorsal or lateral impressions or "foveolae." Antennae filiform, sometimes a little flattened or somewhat ensiform. Pronotum variously shaped, often dorsally tectiform, strongly keeled, or crested; prosternum with median conical process. Hind femora varying from short and stout to long and slender, but with lower basal lobe usually projecting as far forward as upper one; hind tibia with external apical spine generally present. Wings abbreviated, occasionally lacking; tegmina usually with dense reticulation; hind wings often brightly colored, characteristically possessing (even when reduced) a unique stridulatory mechanism formed by numerous fine denticles on arched transverse veinlets of narrowed anterior area of fan; associated longitudinal veins may also be serrated, with denticles of hind wing rubbed against sharp vein on underside of tegmen to produce sound. Base of abdomen with pair of tympanic organs, except in apterous species. Cerci conical in both sexes, or occasionally lobed in male. Phallic structures basically similar to those of Acrididae, *sensu stricto*, but valves of cingulum reduced or absent, and flexure between basal and apical aedeagal valves stronger and thicker. Ejaculatory sac not opening directly into genital chamber. Spermatophore sac ventral. Epiphallus bridgelike, robust, with fairly large associated lateral "oval" sclerites.

This family is now regarded as comprising nearly 80 genera and at least 200 described species in 3 subfamilies. The great majority are South and Central American, but a few extend well into North America (one species just reaching our region) and there are some scattered Old World forms.

Romaleids occur in many different habitats from desert to tropical rain forest. Some live on the ground, closely imitating their background, others on shrubs, bushes, or trees. Some show aposematic (warning) coloration. In addition to stridulating with their wings, a few species may, when alarmed, make hissing sounds by forcing air and watery bubbles from the thoracic spiracles. In view of the wide diversity of habitat, the food plants of Romaleidae are clearly diverse, but there do not appear to be many graminivorous species. Occasionally, some species are unusually abundant and become serious, if sporadic, crop pests. This has even happened with the largest of all "grasshoppers" (species of *Tropidacris* and *Titanacris*), which have been known to defoliate plantation trees, earning themselves the appellation "giant locusts," although they do not actually form proper swarms. The nymphal stages of some species of *Chromacris*, however, are often dark colored and occur in dense aggregations, which disperse later.

A general, but incomplete, account of the Romaleidae is given by J. A. G. Rehn and Grant (1959). J. A. G. Rehn and Grant (1961) give a full account of those species occurring in North America north of Mexico, including *Brachystola magna*, which belongs to the subfamily Romaleinae.

Genus *Brachystola* Scudder

Description. Body robust. Occiput of head broad, smooth, convex; fastigium short, broad, slightly sulcate. Pronotum with median and lateral carinae about equally distinct; prosternum lacking tubercle. Tegmina and hind wings rudimentary, with tegmina lobate, meeting in males, but distinctly separated in females. Legs robust.

Brachystola magna (Girard)

Fig. 390; Map 113

Brachypeplus magnus Girard, 1853:260, pl. 15, figs. 1-4.
Brachystola magna; Scudder 1876*a*:267; J. A. G. Rehn and Grant 1961:251; Helfer 1963:99.
Brachystola ponderosa Bruner, 1906*a*:193, 194.
Brachystola intermedia Bruner, 1906*a*:193, 194.

Common name. Lubber grasshopper.

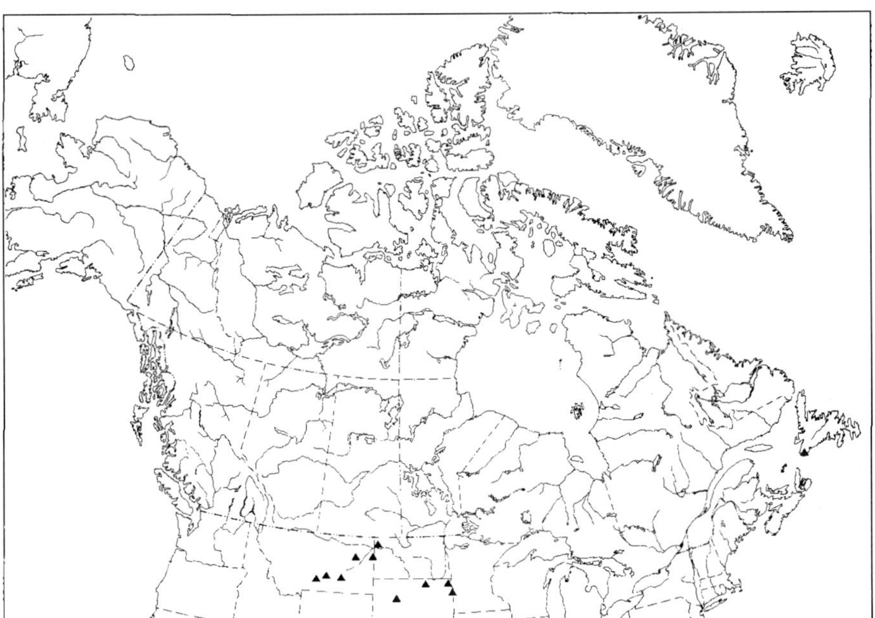

Map 113. Collection localities for *Brachystola magna*.

Diagnosis. Body and legs robust. Lateral carinae of pronotum dark green, often contrasting with rest of body.

Description. As for genus *Brachystola*. Dimorphic for color with both green and brown phases, with green phase apparently predominating; pronotum with darker green markings; lateral carinae dark green, bordered mesally by pinkish line; tegmina tawny with distinct dark brown to black spots; venter pale.

Range. Montana to Minnesota, south to Mexico; adventive in Newfoundland (R. F. Morris 1981).

Behavior and habitats. J. A. G. Rehn and Hebard (1906) report members of this species as being awkward insects that "seemed to have almost no control over their movements." The color apparently serves as good protection. They are usually found in prairie regions, where they are rare in the north. In the south they may be abundant at times. A general account of the habitats occupied throughout the range of the species is given by J. A. G. Rehn and Grant (1961). The eggs are described by Tuck and Smith (1940) and by Onsager and Mulkern (1963).

Life history. J. A. G. Rehn and Grant (1961), referring to *B. magna* throughout its range in the United States, say that adults and immatures are found in Texas from June to November, the latter mainly from June to August; adults alone have been found as early as March and as late as December, although these dates are unusual. "The date of appearance of adults is clearly correlated with climate and latitude." No adult is known from northern regions before the 2nd week of September. Overwintering is in the egg stage. Late-maturing adults in the north are probably killed by frost before they can mate, but clearly a few manage to survive long enough to reproduce their kind, even within a short distance of the Canadian border.

Economic importance. *B. magna*, in the south, occasionally becomes extremely abundant, damaging rangeland and attacking various crops (Parker et al. 1932). In northern latitudes, however, the species is a rarity and not likely to be at all injurious, although, even in Iowa, damage to gardens was recorded long ago by Bessey (1877). Its main economic significance is probably as an article of commerce, large numbers being sold to educational institutions in the United States and Canada.

Family Acrididae

This is the largest family of Orthoptera, and, indeed, of all the orthopteroids. It includes all the true locusts and grasshoppers and formerly was considered to encompass virtually the whole of the superfamily Acridoidea. It has recently been split again by Dirsh (1975) into three families. There are, however, insufficient differences between these to merit such treatment (Kevan 1976*a*). As here understood, the Acrididae may be distinguished from other Acridoidea by the following combination of characters: size small to large, varying from short and stout to long, slender, and stemlike, from depressed to cylindrical or compressed, not usually strongly rugose or tuberculate. Coloration often partly or almost wholly green or stramineous. Head short to long, round to elongate-conical, with or without pronounced fastigium, the apex of which is without an apical furrow or dorsal areola (although it may be "dished" behind a callous anterior margin), sometimes with subapical lateral or dorsolateral foveolae, which are not found in other Acridoidea; frontal profile variable but most often moderately receding. Antennae generally filiform. Pronotum not usually wider than long, varying from subcylindrical to saddle-shaped or tectiform, occasionally crested; prosternum with or without median tubercle or process (which is typically conical but which may be long and backwardly curved or of other form). Hind femora short and stout to slender and greatly elongate, with upper anterior basal lobe almost always at least as long as ventral one and usually projecting forward beyond it; hind tibia with external apical spine before the movable spur generally absent or vestigal. Tegmina and hind wings fully developed, brachypterous, micropterous, or totally lacking; tegmina often with intercalate vein (which may be denticulate) used in stridulation by being struck by a longitudinal ridge (which may be denticulate if the vein is not so) on inner face of the hind femur; either tegmen or hind wing often with an expanded "speculum" or resonating area (other types of stridulatory mechanism—but not of the precise forms found in other families—may be developed, or, in perhaps the majority, no such mechanism is apparent). Base of abdomen usually with pair of lateral tympanal organs. Abdominal terminalia variable, usually unspecialized, but cerci often specialized in male; membranous, with strongly sclerotized cingulum, its "arches" well-differentiated; aedeagal valves divided into apical and basal parts, either separate or joined by weak flexure; ejaculatory sac not opening directly into genital chamber; spermatophore sac ventral; epiphallus of variable shape, normally bridgelike with pair of anterior and pair of posterior processes associated lateral "oval" sclerites present or absent.

Nymphs of a few eastern species have been described by Vickery et al. (1981).

Acrididae species occur throughout the world, wherever Orthoptera members are found. Although many groups have restricted geographic distributions and the majority of species are tropical, the family is well represented in temperate climates and reaches "frigid" latitudes and altitudes. There are probably nearly 1100 described genera and about 7000 known species. These are distributed unevenly amongst numerous subfamilies, which, on a world basis, can scarcely total less than 30 (despite the impression gained from general textbooks that there are but a few). Even without those subfamilies, formerly included, which are now assigned to other families, and without a critical consideration of certain Australian elements, 40 subfamilies of Acrididae have been recognized by various authors (most of them by Dirsh 1975). Thirty-three are recognized by us.

With such a vast and diverse assemblage of species, it would be profitless to attempt a generalized account of acridid biology, beyond what has already been indicated under the Orthoptera, *sensu stricto*, and the reader is referred to the texts of Uvarov (1966*a*, 1977), which deal primarily with Acrididae. Members of the family occur in all manner of habitats (including sand burrows, treetops, or even, as in certain species of the South American leptismine genera *Stenopola* and *Cornops*, in water). The term "grasshopper" is customarily retained, although it is appropriate only to a minority of species. It should perhaps be stressed that the majority of economically important, saltatorial orthopteroids, and all true locusts belong to this family, in particular to the subfamilies Calopteninae, Cyrtacanthacridinae, *sensu stricto*, Melanoplinae, Locustinae (= Oedipodinae), and Gomphocerinae, the first of which is unrepresented in the Americas and the last two of which do not include American true locusts.

Key to subfamilies of Acrididae

1. Prosternum with median tubercle or spine. Hind femur without stridulatory ridge on internal face 2

Prosternum usually lacking median tubercle or spine. Hind femur usually with stridulatory ridge (at least in males) 3

2(1). Lateral lobes of mesosternum approximately rectangular, longer than wide, with straight inner margins. Insect large, fully winged. Male subgenital plate notched apically .... **Cyrtacanthacridinae** (p. 327)

Lateral lobes of mesosternum at least as wide as long and with rounded or obtuse angles and curved inner margins. Insect small to medium, with wings often reduced. Male subgenital plate not notched apically .... **Melanoplinae** (p. 333)

3(1). Body compressed, rather elongate. Antennae ensiform. Head strongly conical but not greatly elongate, lacking lateral foveolae below fastigium. Prosternal tubercle absent. Male hind wings with resonant speculum in anterior cubital area .... **Hyalopteryginae** (p. 437)

Body of various forms, but, if sometimes in general agreement with the above, and antennae ensiform, then with lateral foveolae below fastigium. Prosternal tubercle occasionally present. Male hind wings without speculum 4

4(3). Frontal profile usually nearly vertical (angle measured from occiput to fastigium to clypeus approximately a right angle), sometimes more oblique. Tegmen with intercalary vein present in median (discal) cell, stronger and serrate in male. Hind femur on inner face (of male) with stridulatory ridge lacking pegs, teeth, serration, or any distinct indication thereof. Tegmina and hind wings almost always fully developed; hind wings commonly colored or dark, or with dark bands or apices, or a combination of these, less commonly hyaline .... **Locustinae (= Oedipodinae)** (p. 439)

Frontal profile almost always at least slightly oblique (angle measured from occiput to fastigium to clypeus less than a right angle). Tegmen without intercalary vein in median (discal) cell. Hind femur on inner face (at least of male) with stridulatory ridge bearing articulated pegs or teeth in males, or setalike indications of same in females. Tegmina and hind wings frequently reduced; hind wings generally hyaline, only occasionally slightly colored, infumated, or banded with dark pigmentation. .... **Gomphocerinae** (p. 530)

Clé des sous-familles d'Acrididae

1. Prosternum muni d'une épine ou d'un tubercule médian. Fémurs postérieurs dépourvus d'une crête stridulante sur la face interne 2
   Prosternum habituellement dépourvu d'une épine ou d'un tubercule médian. Fémurs postérieurs en général munis d'une arête stridulante (au moins chez les mâles) 3

2(1). Lobes latéraux du mésosternum à peu près rectangulaires, plus longs que larges et à bords internes rectilignes. Insectes de grande taille et ailés. Plaque sous-génitale des mâles munie d'une encoche apicale .... **Cyrtacanthacridinae** (p. 327)
   Lobes latéraux du mésosternum au moins aussi larges que longs, à angles arrondis ou obtus et à bords internes incurvés. Insectes de taille petite à moyenne et à ailes souvent réduites. Plaque sous-génitale des mâles dépourvue d'une encoche apicale .... **Melanoplinae** (p. 333)

3(1). Corps aplati et plutôt allongé. Antennes ensiformes. Tête très conique mais pas très allongée, dépourvue de fovéoles latérales sous le fastigium. Prosternum sans tubercule. Ailes postérieures des mâles munies d'une plaque sonore (speculum) dans la région cubitale antérieure .... **Hyalopteryginae** (p. 437)
   Corps de formes diverses, mais s'il est aplati et si les antennes sont ensiformes, des fovéoles sont présentes sous le fastigium. Tubercule parfois présent sur le prosternum. Ailes postérieures des mâles sans speculum 4

4(3). Profil du front habituellement presque vertical (angle mesuré de l'occiput au fastigium et au clypéus approximativement droit) et parfois plus oblique. Tegmen muni, dans la région médiane, d'une nervure intercalaire plus forte et denticulée chez les mâles. Fémurs postérieurs munis, sur leur face interne (chez les mâles), d'une crête stridulante dépourvue de toute forme de denticulation. Tegmina des ailes postérieures presque toujours entièrement développés; ailes postérieures habituellement colorées ou foncées ou marquées de bandes ou d'apex foncés ou d'une combinaison des deux; moins communément hyalines .... **Locustinae** (= **Oedipodinae**) (p. 439)
   Profil du front presque toujours au moins légèrement oblique (angle mesuré de l'occiput au fastigium et au clypéus inférieur à 90°). Région médiane du tegmen dépourvue de nervures intercalaires. Fémurs postérieurs portant, sur leur face interne (au moins chez les mâles), une crête stridulante munie de chevillettes ou de dents articulées chez les mâles, le tout remplacé chez les femelles par une rangée de soies. Tegmina et ailes postérieures fréquemment réduites; ailes postérieures généralement hyalines, ou légèrement colorées ou enfumées ou marquées de bandes pigmentées à l'occasion .... **Gomphocerinae** (p. 530)

Subfamily Cyrtacanthacridinae

The subfamily name Cyrtacanthacr(id)inae has for long been used, particularly among North American authors, to embrace virtually all acridid grasshoppers possessing a prosternal spine or tubercle. In recent times, the limits of the Cyrtacanthacridinae have been severely restricted, so that the name is now applied only to a relatively small independent subfamily.

Members of this subfamily have been called "bird-locusts" on account of their generally large size and strong flight. Their diagnostic features are noticed in the foregoing key to subfamilies. The largest included genus is *Schistocerca*, which is predominantly American, although nearly all of the 35 genera are from the Old World tropics. There are probably about 120 species, which include the notorious desert locust of the Old World, *Schistocerca gregaria gregaria* (Forskål), at least two other species of the same genus, the South American locust, *S. cancellata* (Audinet-Serville), and the Central American locust, *S. vicaria* (Walker), the African red locust, *Nomadacris septemfasciata* (Audinet-Serville), and a number of other important pest species such as the so-called Bombay locust, *Patanga succincta* (Pallas). The occurrence of the subfamily in the region here covered, however, is infrequent and, especially in Canada, largely restricted to immigrants.

Genus *Schistocerca* Stål

Description. Body medium to large (males 26.0–48.7, females 35.0–68.7 mm; pronotal length, males 6.0–10.6, females 7.8–13.7 mm), with surface finely rugose and pitted. Pronotum slightly tectiform to weakly saddle-shaped, with median carina obtuse, elevated, or linear, sometimes indistinct on prozona; metazona longer than prozona, its posterior margin acute-angled to rounded. Prosternal tubercle cylindrical, spinelike, straight or curved backward. Tegmina and hind wings fully developed, at least as long as and usually surpassing abdomen. Hind femur moderately slender. Male cercus broad, compressed, with apex truncate, angular, or slightly emarginate. Male subgenital plate notched or emarginate at apex. Female subgenital plate truncate, obtuse-angulate, or widely rounded apically, with egg-guide long, tongue-shaped; ovipositor valves short, robust to rather long and slender, with curved tips; lower valve with lateral projection on outer margin.

Key to species of *Schistocerca*

1. Antennae slightly shorter (females) to slightly longer (males) than combined length of head and pronotum .... ***americana* (Drury)** (p. 329)
   Antennae at least as long as (females) to much longer (males) than combined length of head and pronotum 2

2(1). Fastigium narrow, strongly projecting anteriorly (Fig. 395). Pronotum subparallel in prozona, divergent, flaring abruptly in metazona. Male cercus large, subquadrate, with apical margin more or less distinctly truncate-emarginate (Fig. 400). Ovipositor slender, elongate, with scoop of dorsal valve evenly curved (Fig. 410) .... ***alutacea* (Harris)** (p. 331)
   Fastigium broader, less projecting anteriorly (Figs. 393, 397). Pronotum with sides more parallel, less expanded posteriorly, not flaring. Male cercus smaller, tapering or nearly subquadrate, with apical margin moderately to deeply notched (Figs. 405–408). Ovipositor short, with dorsal valve deeply scooped (Fig. 412) .... ***emarginata* (Scudder)** (p. 332)

Clé des espèces de *Schistocerca*

1. Antennes légèrement plus courtes (femelles) ou légèrement plus longues (mâles) que la longueur combinée de la tête et du pronotum .... ***americana* (Drury)** (p. 329)

 Antennes au moins aussi longues (femelles) ou beaucoup plus longues (mâles) que la longueur combinée de la tête et du pronotum 2

2(1). Fastigium étroit et saillant (fig. 395). Pronotum à côtés presque parallèles dans la zone antérieure, puis divergeant de façon très marquée dans la zone médiane. Cerques des mâles grands, subquadrangulaires, à extrémité apicale plus ou moins tronquée ou concave (fig. 400). Ovipositeur étroit et long; dessus de la valve dorsale décrivant une courbe régulière (fig. 410) .... ***alutacea* (Harris)** (p. 331)

 Fastigium plus large et moins saillant (fig. 393 et 397). Côtés du pronotum plus parallèles, moins divergents postérieurement. Cerques des mâles plus petits, à côtés convergents ou presque subquadrangulaires et portant, à l'apex, une encoche plus ou moins profonde (fig. 405 à 408). Ovipositeur court; dessus de la valve dorsale fortement incurvé (fig. 412) .... ***emarginata* (Scudder)** (p. 332)

Schistocerca americana (Drury)

Figs. 392*a*, 392*b*, 394, 398, 399, 409; Map 114

Americanus . . . *Libell(ula*—error for *Gryllus*) Drury, 1773:Index to Vol. 1:(ii), and *Grylli Locustae* (*americana*).

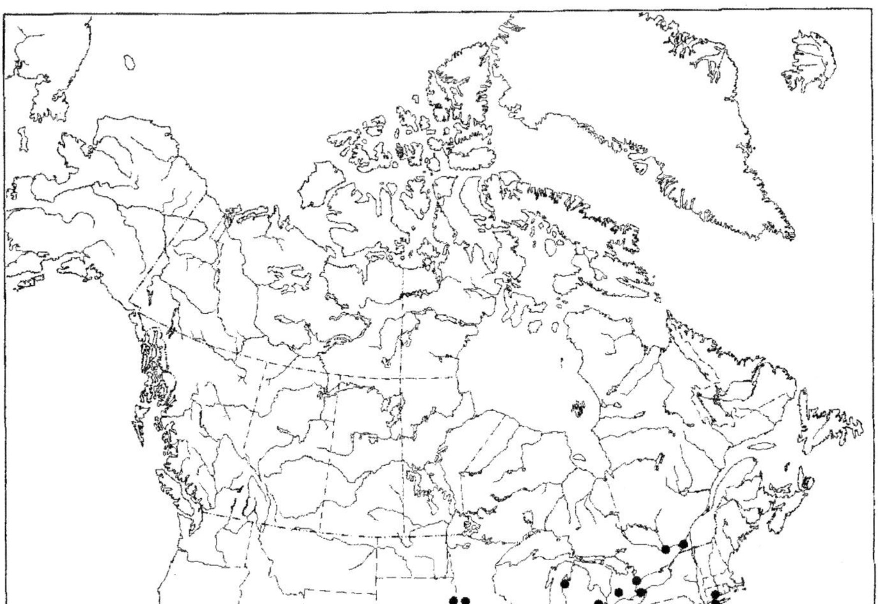

Map 114. Collection localities for *Schistocerca americana*.

Schistocerca americana; Scudder 1899*d*:447; Helfer 1963:186; Vickery and Kevan 1964*b*:1555; Vickery et al. 1974:88.

Schistocerca americana americana; Froeschner 1954:246; 319; Dirsh 1974:45.

Common name. American grasshopper.

Diagnosis. Antennae about same length as head and pronotum. Size large.

Description. Body large (males 27–47, females 39–94 mm long). Antennae slightly shorter to slightly longer than combined length of head and pronotum. Male cercus narrowed toward apex (Figs. 398, 399). Ovipositor slender, narrow, with dorsal valve regularly curved (Fig. 409). Color dark brown, yellow brown to reddish; head and pronotum with median dorsal longitudinal buffy stripe; lateral pronotal lobe with large brown patch interrupted by narrow longitudinal whitish stripe; tegmina covered with well-defined brown spots; hind femur with upper carina black, this sometimes interrupted to form smaller stripes or dots; hind tibia red to buff.

Although Dirsh (1974) included Palaearctic, Nearctic, and Neotropical taxa as subspecies of *S. americana*, this has been shown by Jago et al. (1979) to be incorrect. Interbreeding experiments clearly indicated separate specific status for *S. americana*.

Range. South Dakota to Québec and New England (as migrants); southeastern United States to South America.

Behavior and habitats. *S. americana*, in the northern part of its range, is usually found in damp localities in tall grasses and sedges, near marshes or lakes (Blatchley 1920). Farther south, it occurs mainly in dry upland localities, roadsides, old fields, and weedy areas. It is a strong flier and often migrates for long distances, carried by prevailing winds. Vickery and Kevan (1967) provided records of migrants from a number of localities in southern Ontario, and Cantrall (1968) listed localities in Michigan. Migrants were reported from Massachusetts by Morse (1919*b*, 1920). Adventive specimens, brought in on imported produce, were reported from Sainte-Anne-de-Bellevue, Qué. (Vickery and Kevan 1964*b*; Vickery et al. 1974), and Ottawa, Ont. (Vickery and Kevan 1967).

Life history. In its breeding area in the southern United States, *S. americana* appears to have two generations each year. In Florida, adults are found throughout the year, but nymphs are scarce in the winter (Blatchley 1920).

Economic importance. *S. americana* is virtually of no importance in the north, since only a few migrants occur. Should a large flight of migrant *S. americana* arrive at a locality, it could cause serious crop damage. Considerable damage has been reported to citrus trees in Florida, where young trees have been defoliated and terminal buds and twigs on larger trees have been destroyed. The species feeds upon many crop plants as well as weeds, but it avoids certain others. Potatoes are not attacked, nor are solanaceous weeds or knotweed (*Polygonum*).

Schistocerca alutacea (Harris)

Figs. 395, 400, 410; Map 115

Acrydium alutaceum T. W. Harris, 1841:139.

Schistocerca alutacea; Lugger 1898:262; Cantrall 1943:108; Helfer 1963:191.

Schistocerca alutacea alutacea; Dirsh 1974:194.

Diagnosis. Male antennae long, considerably longer than combined length of head and pronotum. Both sexes with bright yellow dorsal stripe.

Description. Antennae of male considerably longer than combined length of head and pronotum. Fastigium narrow, strongly projecting anteriorly (Fig. 395). Pronotum with prozonal margins subparallel and with metazonal margins divergent posteriorly, flaring. Male cercus large, subquadrate, with apical margin truncate-emarginate (Fig. 400). Ovipositor slender, elongate, with scoop of dorsal valve evenly curved (Fig. 410). Color dark brown to olivaceous on head and thorax; tegmina dark brown, sometimes purplish brown, usually unspotted in males; in females, unspotted or with poorly defined spots or blotches, these sometimes joined; dorsum of both sexes with strongly contrasting bright yellow longitudinal stripe, this often tinged with green, sometimes orange yellow.

Range. Minnesota to Massachusetts, south to Oklahoma and Florida.

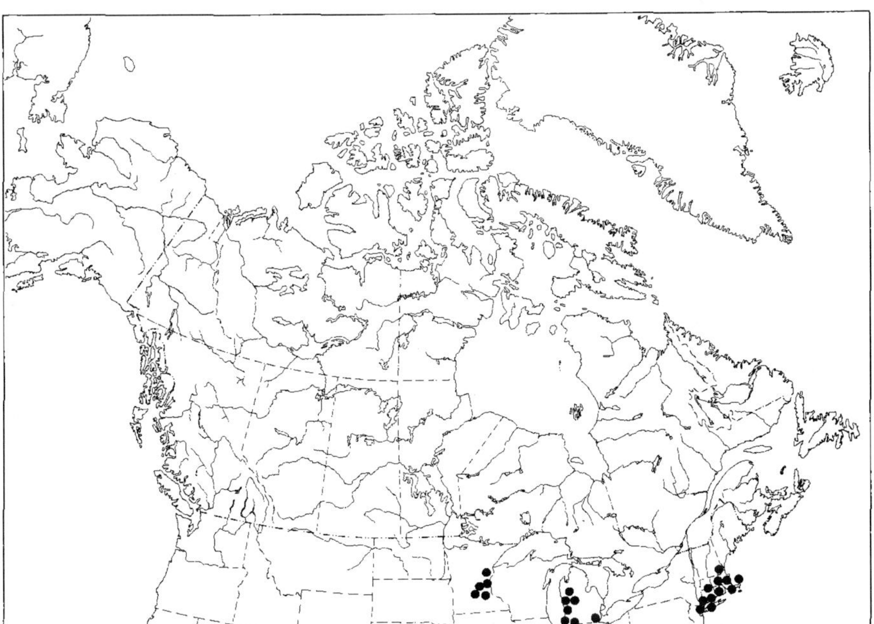

Map 115. Collection localities for *Schistocerca alutacea*.

Behavior and habitats. Hubbell (1960) listed "marshes, bogs, shrubby swamps, thickets of bushes and weeds in wet or moist environments, marginal thickets of mesic forest" as normal habitats for this species. He also stated that it generally occurred on sandy soils. Cantrall (1943) had reported it in dry locations in grass-herbaceous habitats in southern Michigan. *S. alutacea* is a strong flier and may fly for more than 100 m when flushed, often into trees where it is difficult to capture. It does not seem to migrate for such great distances as does *S. americana*.

Life history. Adults are present in southern Michigan from 30 June to 7 October, with peak abundance during August (Cantrall 1968). There is one generation each year.

Economic importance. Morse (1920) reported destruction of growing cranberries by this species in New England, the grasshoppers biting open the berries and consuming the seeds.

Schistocerca emarginata (Scudder)

Figs. 393, 397, 405–408, 412; Map 116

Acridium emarginatum Scudder, 1872:250.
Schistocerca emarginata; Lugger 1898:263.
Schistocerca lineata Scudder, 1899*d*:465; Brooks 1958:31; Helfer 1963:190.
Schistocerca alutacea; Lugger 1898:262; Cantrall 1943:108.
Schistocerca scudderi Bruner, 1906*b*:676.
Schistocerca alutacea lineata; Dirsh 1974:204.

Diagnosis. Dorsal stripe present or absent; pronotum usually with yellow spot at junction of median carina and principal sulcus.

Description. General appearance as in Fig. 393. Head broad. Eyes prominent (Fig. 397). Fore and middle femora of males inflated; hind femur stout. Male cercus tapering or nearly subquadrate, with apical margin moderately to deeply notched (Figs. 405–408), variable. Ovipositor (Fig. 412) short, with dorsal valve deeply scooped. Coloration extremely variable; base color yellowish brown, reddish brown, dark brown, olivaceous brown, olive green, or even yellowish green; dorsal longitudinal stripe either present or absent (usually present, broader in western populations), (when present, often bordered by blackish on pronotum and tegmina, especially in western populations; when absent, pronotum usually with small yellow spot at junction of median carina and principal sulcus); tegmina immaculate to distinctly spotted; hind femur unmarked above or with carinae black or with dark cross-bars; general coloration duller in eastern populations and brighter in western populations. For descriptions of the egg and nymphal stages, see N. Criddle (1932), Tuck and Smith (1940), and Onsager and Mulkern (1963).

Range. Alberta to Ontario, south to New Mexico and North Carolina.

Behavior and habitats. *S. emarginata* has been found in a variety of habitats: dry grassland to open dry forested areas, weedy thickets, and tall-grass prairie. It appears to prefer dry areas and often occurs in areas with sandy soils, such as grassland, oak scrub, dunes, and old fields. It may, how-

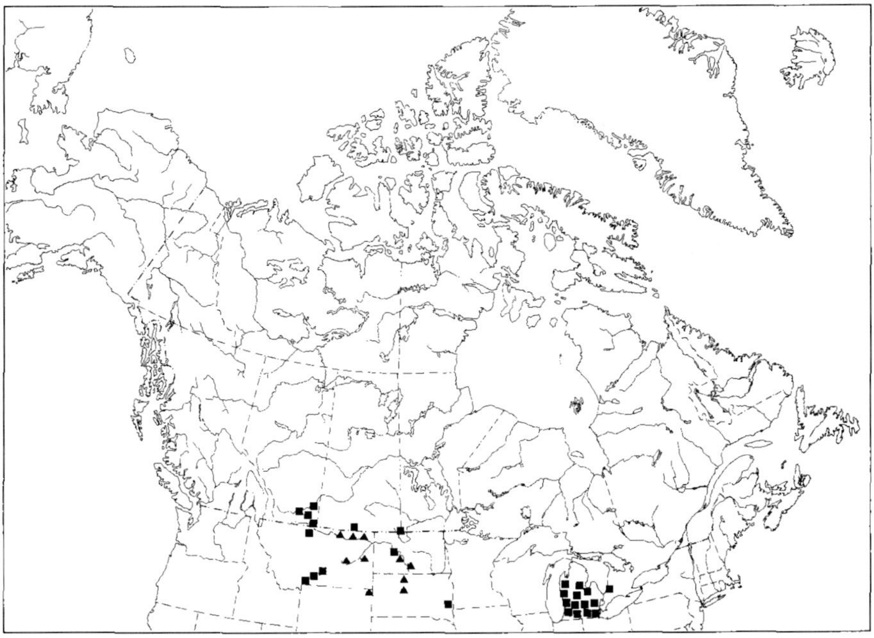

Map 116. Collection localities for *Schistocerca emarginata* (■) and *Dactylotum bicolor pictum* (▲).

ever, also occur in damp areas, such as along stream margins and in irrigated locations. Wild liquorice (*Glycyrrhiza lepidota*) is a favorite food plant. *S. emarginata* also eats vetch, wild peas, beans, sweet clover, alfalfa, dandelion, and wild rose. It does not appear to feed extensively upon grasses (N. Criddle 1932; Gangwere 1961).

Reproductive behavior was discussed by Otte (1970). He described "courtship" and also aggressive behavior of males. Males locate females by sighting their movements, then stalking them until close enough to mount. Females sometimes struggle and may dislodge males.

Life history. N. Criddle (1932) reared this species to maturity in Manitoba, from eggs collected in Alberta, and gave excellent descriptions of egg and nymphal stages. The species is univoltine, wintering in the egg stage. Egg masses are laid in the soil, each pod containing 35-64 eggs. Each female produces about 200 eggs. Nymphs hatch in spring and maturity is reached in late June or early July. Cantrall (1968) reported adults in Michigan from 13 July to 15 October, with a peak in mid-August.

Subfamily Melanoplinae

Members of Melanoplinae constitute one of the most characteristic and also one of the largest subfamilies of New World Acrididae. Until recently they tended to be lumped with the Catantopinae by European authors, or they were treated simply as a tribe of Cyrtacanthacridinae by American writers. Amedegnato (1974) and Cohn and Cantrall (1974) established independence at the subfamily level. They are the principal nonstridulating "grasshoppers" of North America. There are over 150 known genera and a very large, but undetermined, number of species. They are divisible into two tribes: Podismini (about 33 genera and 90 species; Holarctic) and Melanoplini (including Dactylotina; about 120 genera and a large number of species; the great majority North and South American). Fontana and Vickery (1976) and Vickery (1978) show how members of the two tribes can be separated.

Key to genera of Melanoplinae

1. Light green, occasionally pale gray, species with variable markings .. 2
   Brownish, brownish gray, or blackish species 4

2(1). Median carina present as a ridge for entire length of pronotum (Fig. 416), cut only by principal sulcus. Prosternal tubercle short, acute. Tegmina short (rarely long in females) .... ***Hypochlora*** (p. 337)
   Median carina of pronotum absent, or present as a ridge only behind principal sulcus. Prosternal tubercle larger or rounded at apex. Tegmina nearly reaching end of abdomen 3

3(2). Body rather short and squat. Hind femora stout; anterior and middle femora strongly enlarged, and anterior and middle tibiae curved in male. Median carina of pronotum present as ridge behind principal transverse sulcus. Color greenish to greenish brown, with variable degrees of darker greenish brown markings; posterior femur with conspicuous bands on outer face (Fig. 417) .... ***Aeoloplides*** (p. 338)
   Body neither short nor squat. Hind femora not stout; anterior and middle femora and tibiae of male not noticeably enlarged or curved. Median carina of pronotum only feebly developed, marked only by whitish line (Fig. 415). Color light green with pinkish suffusions on legs and thorax; posterior femur not banded .... ***Hesperotettix*** (p. 340)

4(1). Body brightly colored, mottled with black and pale yellow, the yellow with orange or pink markings (Fig. 413) .... ***Dactylotum*** (p. 343)
   Not colored as above 5

5(4). Apterous 6
   Tegmina present, variable in length 8

6(5). Eastern (not west of Ontario or eastern Minnesota) .... ***Booneacris*** (p. 345)
   Western (southwestern Alberta, western Montana, southern British Columbia, and northern Washington) 7

7(6). Body robust, toadlike. Width of head across genae markedly greater than width across eyes. Posterior margin of pronotum with central emargination. Furculae of male ninth abdominal tergum not developed .... ***Bradynotes*** (p. 349)
   Body less robust. Width of head across genae not much greater than width across eyes. Posterior margin of pronotum not centrally emarginate. Furculae of male ninth abdominal tergum distinctly developed .... ***Buckellacris*** (p. 352)

8(5). Tegmina separated at their bases by distance greater than their length .... ***Asemoplus*** (p. 356)
   Tegmina separated at their bases by not more than half their length 9

9(8). Head large, appearing swollen in relation to thorax (Figs. 418, 424). Face noticeably slanting, especially below. Sides of pronotum appearing to converge posteriorly .... ***Phoetaliotes*** (part) (p. 358)
Head smaller, not appearing swollen in relation to thorax (Fig. 426). Face less slanting. Sides of pronotum diverging posteriorly 10
10(9). Tegmina short, not extending beyond fourth abdominal segment ... 11
Tegmina longer 14
11(10). Eastern species (New York, Michigan, and Ontario) 12
Western species (southern Montana) 13
12(11). Male cercus short, sublamellate, twisted apically (Fig. 620). Female subgenital plate with distal margin produced obtusely rectangulate ... . ***Dendrotettix*** (part) (p. 359)
Male cercus simple, tapering to acute apex (Fig. 621). Female subgenital plate with distal margin produced acutely (Fig. 622) .... ***Appalachia*** (p. 362)
13(11). Body slightly depressed. Vertex of head broad, blunt. Eyes small, much shorter than genae. Pronotum with disc rounding into lateral lobes and with posterior margin broadly obtuse-angulate (Fig. 631) .... ***Argiacris*** (p. 363)
Body feebly depressed. Vertex of head narrower, with fastigium rounding into frontal costa. Pronotum with lateral lobes nearly vertical, with posterior margin truncate or rounded .. ***Melanoplus*** (part) (p. 364)
14(10). Tegmina covering only about two-thirds of abdomen .... ***Paroxya*** (p. 434)
Tegmina longer, nearly reaching or exceeding apex of abdomen 15
15(14). Head large, appearing swollen in relation to thorax (Figs. 418, 425). Face noticeably slanting, especially below. Pronotum slender, with sides appearing convergent posteriorly. Body slender, with bluish to dull greenish posterior tibiae .... ***Phoetaliotes*** (part) (p. 358)
Head smaller, not appearing swollen in relation to thorax. Face vertical or nearly so. Pronotum thicker, with sides diverging posteriorly .. 16
16(15). Interspace between mesosternal lobes much wider than long, as broad as or broader than lobes. Tegmina long, exceeding apex of abdomen. Color yellowish or greenish yellow with darker markings; head usually with median dorsal black stripe; tegmina dull brownish yellow; posterior tibiae greenish yellow with brighter yellow basal annulus .... ***Dendrotettix*** (part) (p. 359)
Interspace between mesosternal lobes distinctly longer than broad in males, usually quadrate in females, always much narrower than lobes. Tegmina variable in shape and size. Color dark; head without median dorsal black stripe; tegmina transparent to dark, immaculate, speckled, or mottled, usually shining; posterior tibiae variable but without yellow basal annulus .... ***Melanoplus*** (part) (p. 364)

Clé des genres de Melanoplinae

1. Insectes vert pâle, parfois gris pâle, portant divers types de marques 2
Insectes brunâtres, gris brunâtre ou noirâtres 4
2(1). Pronotum muni, sur toute sa longueur, d'une crête médiane (fig. 416), coupé seulement par le sulcus principal. Tubercule du prosternum court et aigu. Tegmina courts (dans de rares exceptions, longs chez les femelles) .... ***Hypochlora*** (p. 337)

Crête médiane du pronotum absente ou visible seulement derrière le sulcus principal. Tubercule du prosternum plus gros ou arrondi à l'apex. Tegmina rejoignant presque l'extrémité de l'abdomen 3

3(2). Corps plutôt court et trapu. Fémurs postérieurs forts; fémurs antérieurs et médians très gros et tibias antérieurs et médians incurvés chez les mâles. Crête médiane présente sur le pronotum, en arrière du sulcus principal. Coloration variant de verdâtre à brun verdâtre, avec diverses marques brun verdâtre plus foncées; fémurs postérieurs marqués de bandes sur la face externe (fig. 417) .... ***Aeoloplides*** (p. 338)

Corps ni court ni trapu. Fémurs postérieurs pas particulièrement forts; fémurs antérieurs et médians et tibias chez les mâles ni particulièrement grands ni incurvés. Crête médiane du pronotum très peu développées, marquée seulement par une ligne blanchâtre (fig. 415). Coloration vert clair parsemée de marques rosâtres sur les pattes et le thorax; fémurs postérieurs non marqués de bandes .... ***Hesperotettix*** (p. 340)

4(1). Corps aux couleurs voyantes, marqué de bandes noires et jaune pâle, le jaune étant parsemé d'orangé et de rose (fig. 413) .... ***Dactylotum*** (p. 343)

Coloration différente de celle décrite ci-dessus 5

5(4). Espèces aptères 6

Tegmina présents, de longueur variable 8

6(5). Espèces de l'est (pas dans l'ouest de l'Ontario ni dans l'est du Minnesota) .... ***Booneacris*** (p. 345)

Espèces de l'ouest (sud-ouest de l'Alberta, ouest du Montana, sud de la Colombie-Britannique et nord de l'État de Washington) 7

7(6). Corps robuste, de forme rappelant celle d'un crapaud. Tête beaucoup plus large au niveau des gênes qu'au niveau des yeux. Bord postérieur du pronotum concave. Furcula du neuvième tergite de l'abdomen des mâles non développée .... ***Bradynotes*** (p. 349)

Corps moins robuste. Tête pas beaucoup plus large au niveau des gênes qu'au niveau des yeux. Bord postérieur du pronotum sans concavité. Furcula du neuvième tergite de l'abdomen des mâles clairement développée .... ***Buckellacris*** (p. 352)

8(5). Tegmina séparés à leur base par une distance supérieure à leur longueur .... ***Asemoplus*** (p. 356)

Tegmina séparés à leur base par une distance au plus égale à la moitié de leur longueur 9

9(8). Tête large, d'apparence renflée par rapport au thorax (fig. 418 et 424). Face clairement oblique, surtout en bas. Côtés du pronotum semblant converger vers l'arrière .... ***Phoetaliotes*** (en partie) (p. 358)

Tête plus petite, pas renflée par rapport au thorax (fig. 426). Face moins oblique. Côtés du pronotum divergeant vers l'arrière 10

10(9). Tegmina courts, ne dépassant pas le quatrième segment abdominal 11

Tegmina plus longs 14

11(10). Espèces de l'est (New York, Michigan et Ontario) 12

Espèces de l'ouest (sud du Montana) 13

12(11). Cerques des mâles courts, sublamellés, tordus à l'extrémité (fig. 620). Plaque sous-génitale des femelles à bord distal formant un rectangle aux coins obtus .... ***Dendrotettix*** (en partie) (p. 359)

Cerques des mâles simples, plus ou moins aigus à l'apex (fig. 621). Plaque sous-génitale des femelles à bord distal effilé (fig. 622) .... ***Appalachia*** (p. 362)

13(11). Corps légèrement aplati. Vertex large et émoussé. Yeux petits, beaucoup plus courts que les gênes. Pronotum à disque formant deux lobes latéraux et à bord postérieur formant un angle obtus (fig. 631) ... . ***Argiacris*** (p. 363)

Corps très légèrement aplati. Vertex plus étroit, à fastigium arrondi pour former une costa frontale. Pronotum à lobes latéraux presque verticaux et à bord postérieur tronqué ou arrondi .... ***Melanoplus*** (en partie) (p. 364)

14(10). Tegmina ne recouvrant que les deux tiers environ de l'abdomen .... ***Paroxya*** (p. 434)

Tegmina plus longs, rejoignant presque ou dépassant l'apex de l'abdomen 15

15(14). Tête grosse, d'apparence renflée par rapport au thorax (fig. 418 et 425). Face clairement oblique, surtout en-dessous. Pronotum étroit, aux côtés semblant converger vers l'arrière. Corps mince, à tibias postérieurs bleuâtres ou verdâtre mat .... ***Phoetaliotes*** (en partie) (p. 358)

Tête plus petite, ne semblant pas renflée par rapport au thorax. Face verticale ou presque. Pronotum plus épais dont les côtés divergent vers l'arrière 16

16(15). Espace séparant les lobes mésosternaux beaucoup plus large que long, aussi large ou plus large que les lobes. Tegmina longs, dépassant l'apex de l'abdomen. Coloration jaunâtre ou jaune verdâtre avec des marques vert plus foncé; tête habituellement marquée d'une bande dorsale médiane noire; tegmina d'un jaune brunâtre mat; tibias postérieurs jaune verdâtre avec un anneau basal jaune plus clair .... ***Dendrotettix*** (en partie) (p. 359)

Espace séparant les lobes mésosternaux clairement plus long que large chez les mâles, habituellement aussi long que large chez les femelles, toujours beaucoup plus étroit que les lobes. Tegmina de forme et de taille variables. Coloration foncée; tête sans bande médiane dorsale noire; tegmina transparents à foncés, immaculés, mouchetés, en général luisants; tibias postérieurs variables, mais dépourvus d'un anneau basal jaune .... ***Melanoplus*** (en partie) (p. 364)

Genus *Hypochlora* Brunner von Wattenwyl

Description. Body small (15–20 mm), rather slender (Fig. 424). Tegmina short (rarely long in females only), pointed. Median carina entire, sharply ridged. Prosternal tubercle short, acute.

Hypochlora alba (Dodge)

Figs. 416, 424; Map 117

Pezotettix alba Dodge, 1876:10.

Hypochlora alba; Brunner von Wattenwyl 1893:145; Brooks 1958:19; Helfer 1963:206.

Diagnosis. Greenish yellow to grayish. Found on *Artemisia ludoviciana*.

Description. As for genus *Hypochlora*. Male subgenital plate with tubercle at apex. Male cercus slender, except at base. Furculae of ninth abdominal tergum of male short, slender. Color greenish yellow to grayish.

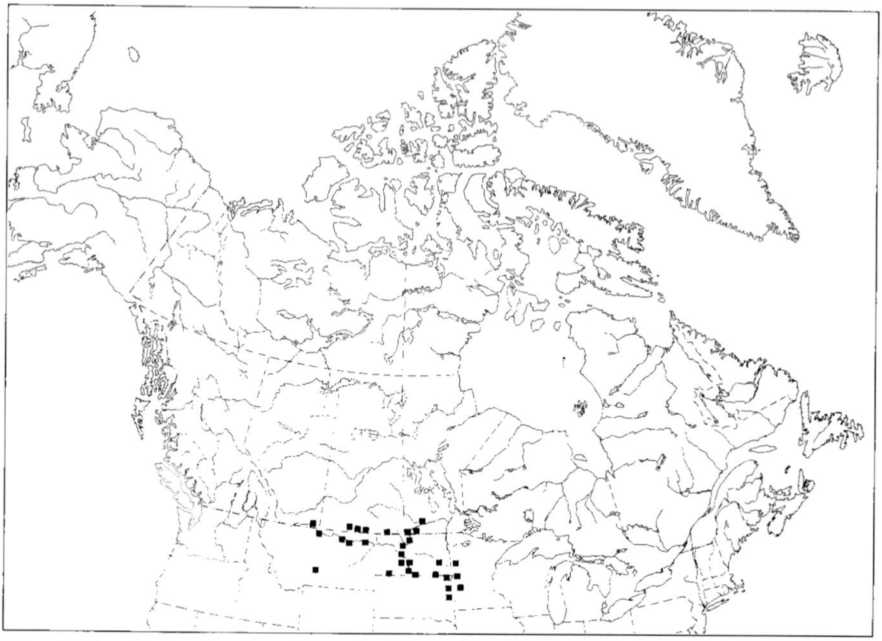

Map 117. Collection localities for *Hypochlora alba*.

Range. Alberta to Manitoba, south to Colorado and Texas.

Behavior and habitats. *H. alba* feeds almost exclusively on *Artemisia ludoviciana*, with some limited feeding on other *Artemisia* species and on alfalfa (Mulkern et al. 1969).

Life history. Univoltine, with overwintering in the egg stage. Adults have been found in August and September.

Genus *Aeoloplides* Caudell

Description. Body rather deep through metathorax so that head and pronotum appear disproportionately small in lateral view (Fig. 423). Median carina of pronotum present as ridge behind principal transverse sulcus. Prosternal tubercle rounded at apex. Hind femur short, stout. Males with anterior and middle femora strongly enlarged and anterior and middle tibiae curved. Male cercus broad at base, with apex narrowly rounded or pointed. Color generally brownish green, usually with dark striped pattern dorsally; outer face of hind femur conspicuously banded (Fig. 417).

Aeoloplides turnbulli turnbulli (Thomas)

Figs. 417, 420, 423; Map 118

C[aloptenus] turnbulli C. Thomas, 1872:452.

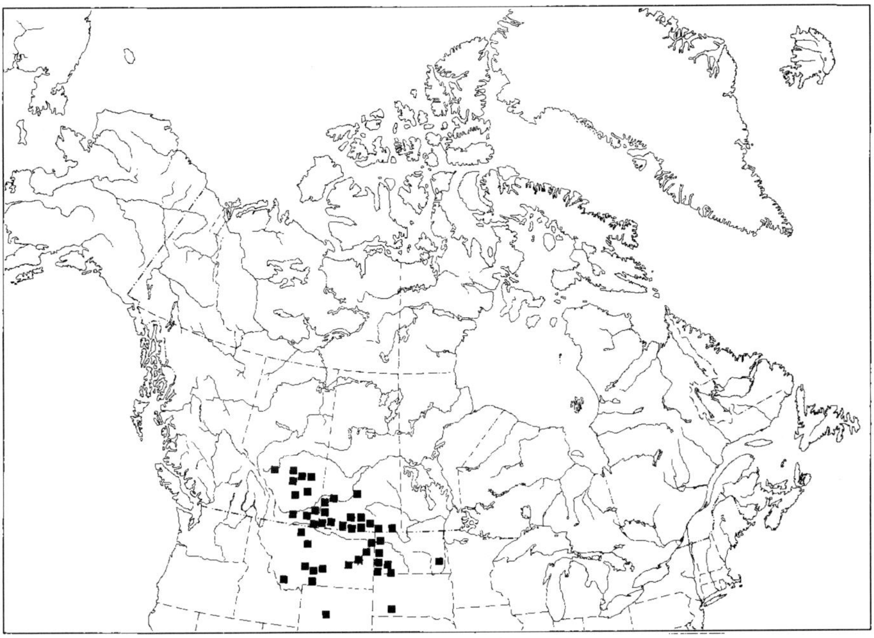

Map 118. Collection localities for *Aeoloplides turnbulli turnbulli*.

Pezotettix plagosus Scudder, 1876*c*:504.
Aeoloplus plagosus; Scudder 1897*a*:205.
Aeoloplus turnbulli turnbulli; Hebard 1925*a*:97; Brooks 1958:17.
Aeoloplides turnbulli turnbulli; Wallace 1955:477.

Diagnosis. Tegmina short, rounded. Body brown with greenish markings.

Description. General appearance as in Fig. 417, small (males 11–19, females 16–21 mm long). Tegmina not reaching apex of abdomen, gradually converging to small rounded apices. Male cercus long, tapering (Fig. 420). Color largely brown to blackish brown (Wallace 1955); hind femur banded (Fig. 417).

Brooks (1958) described "major" and "minor" forms of *A. turnbulli turnbulli*: minor form with body length 14–17 mm, tegmina short, color greenish yellow with green or olive green markings; major form with body length 16–22 mm, tegmina nearly reaching apices of hind femora, color brownish with brown or brownish green markings. He found many intermediates between these forms.

Range. Alberta to North Dakota, south to New Mexico.

Behavior and habitats. *A. turnbulli turnbulli* is largely confined to alkaline areas and localities where host plants grow. Mulkern et al. (1969) found that it feeds mainly on a few closely related plants of the family Chenopodiaceae. In the absence of preferred plants, others might be eaten. A distinct preference was exhibited for *Kochia scoparia, Atriplex* species, and *Sarcobatus vermiculatus* (greasewood). The species does not migrate and develops as local populations, each with its distinguishing characteristics. The "major" form discussed by Brooks (1958) tends toward *A. turnbulli bruneri*, which does not occur north of South Dakota or Wyoming (Wallace 1955), although Hebard (1936*a*) had recorded *bruneri* from southwestern North Dakota.

Life history. Univoltine, with overwintering in the egg stage. Adults have been collected from early July until the end of September.

Genus *Hesperotettix* Scudder

Description. Size small to medium (15–34 mm) with sexes nearly equal in size, resembling *Aeoloplides*, but pronotum with median carina only feebly developed and lateral carinae absent; prozona 1.5 times as long as metazona (Fig. 415). Fore and middle tibiae not short or noticeably curved. Tegmina variable in length but usually not exceeding apex of abdomen. Hind femur slender, long, exceeding apex of abdomen. Furculae of ninth abdominal tergum of male present, distinct; male subgenital plate with small subapical tubercle. Ovipositor strongly exserted. Chromosome number, $2n$ male = 22 or 23.

Key to species and subspecies of *Hesperotettix*

1. Prozona of pronotum smooth; metazona punctate; pronotal disc slightly tectiform. Tegmina twice as long as pronotum 2

   Prozona and metazona of pronotum finely rugose; pronotal disc distinctly tectiform. Tegmina elongate, 3–5 times as long as broad .... **speciosus (Scudder)** (p. 341)

2(1). Posterior margin of metazona obtuse-angulate. Prosternal spine long, slender, subcylindrical, with tip rounded. Tegmina and hind wings long, usually reaching or surpassing tip of abdomen. Found in Minnesota and Manitoba, west to Montana and British Columbia .... ***viridis pratensis*** **Scudder** (p. 342)

   Posterior margin of metazona broadly rounded. Prosternal spine short, conical. Tegmina and hind wings shorter, covering about two-thirds of abdomen. Found in Massachusetts .... ***viridis brevipennis*** **(Thomas)** (p. 342)

Clé des espèces et des sous-espèces d'*Hesperotettix*

1. Zone antérieure du pronotum lisse; zone médiane ponctuée; disque du pronotum légèrement tectiforme. Tegmina deux fois plus longs que le pronotum 2

   Zones antérieure et médiane du pronotum marquées d'une fine rugosité; disque du pronotum clairement tectiforme. Tegmina longs, de 3 à 5 fois plus longs que larges .... **speciosus (Scudder)** (p. 341)

2(1). Bord postérieur de la zone médiane angulaire et obtus. Épine du prosternum longue, mince, subcylindrique et à extrémité arrondie. Ailes postérieures et tegmina longs, rejoignant habituellement ou dépassant l'extrémité de l'abdomen. Espèce vivant au Minnesota et au Manitoba et vers l'ouest, jusqu'au Montana et en Colombie-Britannique .... *viridis pratensis* **Scudder** (p. 342)

Bord postérieur de la zone médiane arrondi. Épine du prosternum courte et conique. Ailes postérieures et tegmina plus courts, recouvrant environ les deux tiers de l'abdomen. Espèce vivant au Massachusetts .. .. *viridis brevipennis* **(Thomas)** (p. 342)

Hesperotettix speciosus (Scudder)

Map 119

Pezotettix speciosa Scudder, 1872:250.
Acridium frontalis C. Thomas, 1872:448.
Hesperotettix speciosus; Scudder 1897*a*:205.

Diagnosis. Pronotum tectiform; median carina purplish.

Description. Dorsum of head rugulose, with vertex impressed. Prozona and metazona finely rugose; pronotal disc distinctly tectiform, obtuse-angulate posteriorly. Tegmina elongate, covering two-thirds to three-quarters of abdomen. Body green, with median carina of pronotum purplish; femora usually marked with purple. Chromosome number, $2n$ male = 22.

Range. Montana to Minnesota, south to New Mexico and Texas.

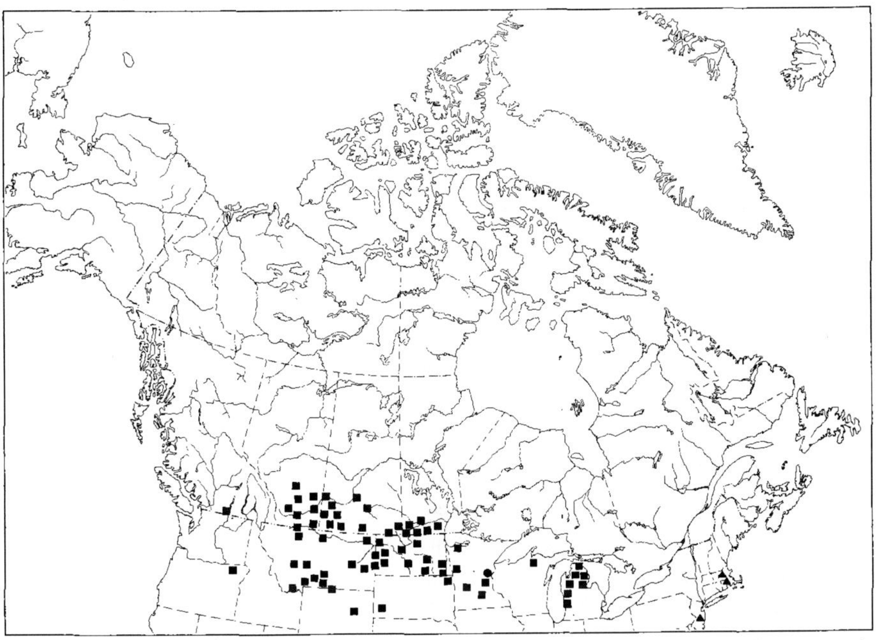

Map 119. Collection localities for *Hesperotettix speciosus* (●), *H. viridis pratensis* (■), and *H. viridis brevipennis* (▲).

Behavior and habitats. The usual habitat for this species is among sparse weeds on dry soils (Froeschner 1954). Blatchley (1920) stated that *H. speciosus* feeds on Compositae. It does not feed upon crop plants and has no economic significance. Mulkern et al. (1969) reported that, at North Platte, Nebraska, its numbers appeared to fluctuate in direct proportion to the abundance of *Helianthus annuus* and *Iva xanthifolia*.

Hesperotettix viridis pratensis Scudder

Figs. 414, 415, 419; Map 119

Hesperotettix pratensis Scudder, 1897*d*:64.
Hesperotettix viridis; Somes 1914:70.
Hesperotettix brevipennis pratensis; Hebard 1925*a*:98.
Hesperotettix viridis pratensis; Hebard 1931*a*:392; Brooks 1958:18; Helfer 1963:210.

Diagnosis. Pronotum green with yellow median stripe; tegmina green with pinkish or purplish stripes at base.

Description. Pronotum with prozona smooth; metazona punctate, with posterior margin obtuse–angulate (Fig. 415). Tegmina twice as long as pronotum, usually reaching or surpassing tip of abdomen. Male cercus triangular, with apex rounded (Fig. 419). Pronotum green with pale yellow median longitudinal stripe, sometimes with pink suffusion on either side of stripe; tegmina green with short narrow poorly defined pinkish or purplish stripes near base; hind femur green with pink pre-apical band. McClung (1905, 1914, 1917) gives the chromosome number as $2n$ male = 22.

Range. British Columbia to Michigan, south to Oklahoma.

Behavior and habitats. *H. viridis pratensis* is primarily a forb feeder. Mulkern et al. (1969) found that it fed predominantly on *Ambrosia, Solidago,* and *Artemisia* species, and only rarely on grasses. Brooks (1958) reported feeding on *Solidago, Helianthus, Gutierrezia, Grindelia,* and *Aster*. Hubbell (1922*a*) reported it as fairly common, but not abundant, in dry grassy fields and pastures, and in sparse vegetation on barren clay and sandy soils in North Dakota. An account of the biology is given by Connin (1964). Eggs are described by Tuck and Smith (19[illegible]) and by Onsager and Mulkern (1963).

Life history. Univoltine, with overwintering in the egg stage. Eggs are deposited in soil. Cantrall (1968) reported adults from 11 July to 16 September in Michigan.

Hesperotettix viridis brevipennis (Thomas)

Map 119

Ommatolampis brevipennis C. Thomas, 1874:67.
Hesperotettix brevipennis; Scudder 1897*a*:205.
Hesperotettix brevipennis brevipennis; Morse 1919*b*:36.
(*Hesperotettix viridis*) *brevipennis*; Hebard 1931*a*:393.

Diagnosis. Body bright pea green, with median carina of pronotum and tegminal stripe purplish red.

Description. Body small (15–24 mm). Prosternal spine short, conical; posterior margin of metazona broadly rounded. Tegmina and wings short, covering about two-thirds of abdomen. Color mainly bright pea green; median carina of pronotum and narrow stripe on tegmina purplish red; femur and apical third of hind tibia purplish red; antennae reddish; spot on vertex, stripe beneath eye, and broad area on upper third of lateral pronotal lobes fuscous. McClung (1917) gives the chromosome number as $2n$ male = 23.

Range. Kansas to Massachusetts, south to Alabama and Georgia.

Behavior and habitats. *H. viridis brevipennis* is usually found in association with wild grasses (*Andropogon scoparius* Michx.) on gravelly soils.

Life history. Not recorded but undoubtedly univoltine, with overwintering in the egg stage. Morse (1920) reported adults from 10 July to 30 August in Massachusetts.

Genus *Dactylotum* Charpentier

Description. Size moderate (18–36 mm), with males distinctly smaller than females. Head with weakly developed fastigium; interocular space broad. Pronotum rounded dorsally in transverse section, without longitudinal carinae but with transverse sulci distinct and rather deep; posterior margin of disc convexly arcuate. Tegmina and hind wings strongly abbreviated; tegmina coarsely reticulate, not meeting dorsally over base of abdomen. Male with furculae weakly developed; supra-anal plate small, simple. Cerci slender, rodlike. Coloration strikingly contrasting darker (blackish, purplish, bluish, or dark olive) and lighter (crimson, red, orange, yellow, or whitish). Chromosome number, $2n$ male = 17.

Dactylotum bicolor pictum (Thomas)

Fig. 413; Map 116

Pezotettix picta C. Thomas, 1870:78.

Pezotettix flavoannulatus La Munyon, 1877:1 [sep.].

Dactylotum pictum; Scudder 1900*a*:67.

Dactylotum bicolor pictum; H. R. Roberts 1947:220, pl. 6, fig. 25; Helfer 1963:198; Mulkern in Mulkern et al. 1969:16.

Diagnosis. Tegmina lobelike, ovate-lanceolate. Coloration brightly contrasting blackish and yellow to red.

Description. As for genus *Dactylotum* (Fig. 413). Body length of males 20–24, of females 29–35 mm. Tegmina ovate-lanceolate, short (males 4–5, females 5–6 mm), widely separated dorsally (in subspecies *D. bicolor variegatum* (Scudder), the tegmina are broader, rounder, and less widely separated dorsally). Coloration strikingly marked blackish and yellow, the latter usually ornamented with orange and/or pink (in subspecies *D. bicolor variegatum* the coloration is even more brilliant, the blackish being replaced by blue, and the red, white, and yellow markings highly contrasting, with white in addition); markings on abdominal segments giving abdomen annulated appearance. (In dried or fluid-preserved material, the reds in particular lose much of their brilliance, but there is never any difficulty in recognizing the species and subspecies by pattern.) Chromosome number, $2n$ male = 17 (Helwig 1942), although McClung (1914) indicated it as being 23 for a member of the genus.

Range. Montana and North Dakota, south to New Mexico and Texas.

Behavior and habitats. Gillette (1904) notes that, in Colorado, *D. pictum* is fairly common on the plains in the eastern part of the state, where it occurs in dry exposed areas extending a short distance into the eastern foothills of the Rocky Mountains. Farther north it becomes less common. Gillette (1904) also quotes from other sources that indicate *Aster multiflora* to be the principal food plant, but that it also feeds commonly upon American laurel (*Kalmia glauca*) and possibly on *Senecio douglasi*. Gillette himself occasionally found it on alfalfa. There seem to be differences in food preferences between regions. Isely (1938*b*), in Texas, found it to be "a selective feeder. In field checking it was only found as an occasional specimen unless definitely associated with its primary host-plant *Baccharis texana* (T. & G.) Gray." He also lists *Aster* as a host plant, but most other plants investigated in cage tests were apparently unattractive or rejected. However, Mulkern (*in* Mulkern et al. 1969) says that "a variety of forbs, including *Sphaeralcea coccinea, Petalostemum purpureum* and *Psoralea argophylla*, plus a few grasses were identified in the crops of specimens from sparse hillsides in western North Dakota." The favored host plant genus, *Baccharis*, does not extend its range nearly so far north (H. R. Roberts 1947). It is also said by D. Van Horn of Colorado Springs (pers. com. 1979) to be attracted to pink-flowered Malvaceae. The subspecies *D. bicolor variegatum* may be found in even more arid places than is *D. bicolor pictum*. Ball et al. (1942) say that "*D. variegatum*" is "omnivorous" occasionally abundant in alfalfa fields in Arizona, "but more common on plants like *Baccharis sarothroides, Grindellia*, and *Haplopappus*, usually on rocky or gravelly soil sparsely covered with grass and other plants." Keasey (1974) says that *D. bicolor variegatum* is only found in desert grasslands, where it feeds on desert broom and low-growing sunflowers. It is one of our few aposematic (warning) species of orthopteroids, and *D. bicolor pictum* has been demonstrated to be distasteful to birds (Isely 1938*a*). Despite its striking coloration, however, *D. bicolor* is usually difficult to detect against its natural backgrounds because of the "fragmented" nature of the pattern. The eggs are described by Tuck and Smith (1940) and by Onsager and Mulkern (1963).

Life history. Univoltine, with overwintering in the egg stage in the ground. In the southern part of its range, nymphs appear about June and adults from about the end of July until October (Gillette 1904), although Keasey (1974) says that adults of *D. bicolor variegatum* disappear soon after they are first seen. In the northern part of its range, *D. bicolor pictum* adults are probably similarly short-lived. Adults in Montana and North Dakota have been found in early August.

Economic importance. Although both Gillette (1904) and Ball et al. (1942) reported that *D. bicolor* was frequently found in alfalfa fields, there is no suggestion that the species could be regarded as a pest, although the potential is clearly there. Mulkern et al. (1969) say that it feeds "mainly on low value forbs and [is] of little economic importance to grasslands." In the region covered by the present work, it is unlikely to have any economic significance.

Genus *Booneacris* Rehn & Randell

Description. Of the general form indicated in Fig. 639, apterous. Size medium to small, subcylindrical. Head slightly broader than prozona; fastigium narrow, feebly declivent; face nearly vertical; frontal costa narrow; eyes large, prominent. Pronotum with prozona long, smooth; median carina of pronotum low, nearly obsolete on prozona; lateral carinae obsolete. Prosternal spine short, conical. Interspace between mesosternal lobes broader than long, as broad (female) or nearly as broad (male) as a lobe (Fig. 640).

Key to species and subspecies of *Booneacris*

1. Antennae of male longer than hind femur by nearly one-quarter. Anterodorsal margin of pronotum truncate. Hind femur distinctly bifasciate fuscous above and on outer face. Male cercus slender, strongly narrowed at middle (Fig. 644). Ovipositor with dorsal valves apically deep in profile (Fig. 648). .... ***variegata* (Scudder)** (p. 346)

   Antennae of male no longer than hind femur. Anterodorsal margin of pronotum slightly but obviously rounded, feebly emarginate at middle. Hind femur nearly uniform green or with faint banding above. Male cercus somewhat to distinctly stouter, slightly narrowed at middle (Figs. 641–643). Ovipositor with dorsal valves apically shallow in profile (Figs. 645–647) .... ***glacialis*, *sensu lato*** ... 2

2(1). Antennae of male distinctly shorter than hind femur, of female only about three-quarters as long. Hind femur usually without darker banding above. Male cercus stout, scarcely narrower at middle than at apex (Fig. 641). Ovipositor valves rather short (Fig. 645). Found in eastern Québec and in montane areas of New Brunswick, New England, and New York .... ***glacialis glacialis* (Scudder)** (p. 347)

   Antennae of male about as long as hind femur, of female about five-sixths as long. Hind femur usually transversely marked. Male cercus long, attenuated, slightly narrower at middle than at base or apex (Fig. 643). Ovipositor valves moderately long (Fig. 647). Found in western Québec, Ontario, Michigan, Wisconsin, and Minnesota .... ***glacialis canadensis* (E. M. Walker)** (p. 349)

Clé des espèces et des sous-espèces de *Booneacris*

1. Antennes des mâles d'une longueur atteignant presque une fois et quart la longueur des fémurs. Bord antérodorsal du pronotum tronqué. Fémurs postérieurs en deux couleurs distinctes; brun foncé mat en haut et sur la face externe. Cerques des mâles fins, fortement étranglés au milieu (fig. 644). Valves dorsales de l'ovipositeur hautes à l'apex (vues de côté) (fig. 648) .... ***variegata* (Scudder)** (p. 346)

Antennes des mâles pas plus longues que les fémurs postérieurs. Bord antérodorsal du pronotum légèrement quoique clairement arrondi, légèrement concave au milieu. Fémurs postérieurs d'un vert pratiquement uniforme ou marqués d'une bande peu visible au-dessus. Cerques des mâles plus ou moins forts, légèrement rétrécis au milieu (fig. 641 à 643). Valves dorsales de l'ovipositeur pas très hautes à l'apex (vues de côté) (fig. 645 à 647) .... ***glacialis***, *sensu lato* ... 2

2(1). Antennes des mâles clairement plus courtes que les fémurs postérieurs; antennes des femelles n'atteignant que les trois quarts environ de la longueur des fémurs. Fémurs postérieurs habituellement sans bande foncée au-dessus. Cerques des mâles forts, à peine plus étroits au milieu qu'à l'apex (fig. 641). Valves de l'ovipositeur plutôt courtes (fig. 645). Espèce vivant dans l'est du Québec et dans les régions montagneuses du Nouveau-Brunswick, de la Nouvelle-Angleterre et de l'État de New York .... ***glacialis glacialis*** **(Scudder)** (p. 347)

Antennes des mâles à peu près aussi longues que les fémurs postérieurs; antennes des femelles d'une longueur atteignant les 5/6 environ de celle des fémurs postérieurs. Fémurs postérieurs habituellement marqués de bandes transversales. Cerques des mâles longs, minces et légèrement étranglés au milieu (fig. 643). Valves de l'ovipositeur modérément longues (fig. 647). Espèce vivant dans l'ouest du Québec, en Ontario, au Michigan, au Wisconsin et au Minnesota .... ***glacialis canadensis*** **(E.M. Walker)** (p. 349)

Booneacris variegata (Scudder)

Figs. 644, 648; Map 120

Podisma variegata Scudder, 1897*d*:97, 101, pl. VII, fig. 4.

Podisma glacialis variegata; E. M. Walker 1903*a*:300, pl. 6, figs. 32–52, 57–60, 64, 65.

Boonacris variegata; J. A. G. Rehn and Randell 1962:162, pl. V, figs. 9, 10, 17, 23, 35, 36, 48, 54, 60, 70, 71 and pl. VI, figs. 9, 10; pl. VIII, figs. 21, 22.

Diagnosis. Brownish. Male cercus attenuate, narrowest at middle, expanded apically.

Description. Apterous. Body size medium to small (length, male 15–21, female 20–28 mm). Eyes prominent, round-ovate. Male furculae short; male cercus attenuate, clavate, narrowest at middle, slightly expanded apically, distinctly bowed inward apically (Fig. 644); male subgenital plate with dorsolateral margins straight. Dorsal ovipositor valves apically rather deep in profile (Fig. 648). Color generally chocolate brown with olivaceous or greenish tinge; males with broad pale median pronotal stripe extending to apex of abdomen, strongly contrasting with brown lateral color; females without contrasting dorsal stripe; face with cheeks (genae) and lower halves of lateral pronotal lobes pale, cream-colored to yellowish; ventral face of hind femur yellowish to carmine red; outer face of hind femur usually with two distinct pale to creamy cross-bars and pale subapical annulus.

Range. Ontario and Michigan, south to Tennessee and Virginia.

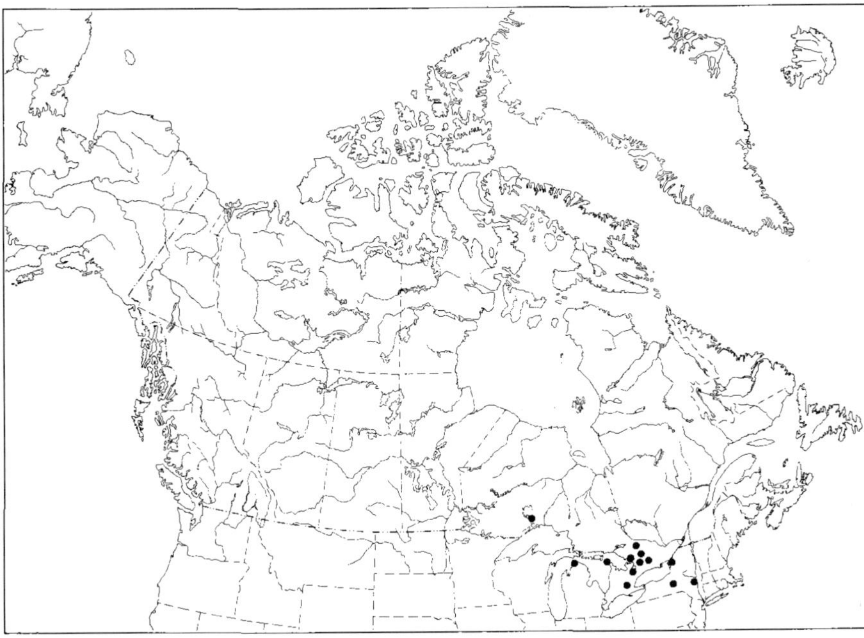

Map 120. Collection localities for *Booneacris variegata*.

Behavior and habitats. E. M. Walker (1903*a*) found *B. variegata* in Ontario on bushes such as raspberry, but also on arborvitae, sometimes 2–3 m aboveground. Other authors (J. A. G. Rehn and Randell 1962) mentioned records from hemlock, huckleberry, aspen, wild cherry, and golden rod. *B. variegata* is a secretive sluggish species, difficult to find because it hides and, when in the open, tends to remain immobile.

Life history. Univoltine, with overwintering in the egg stage. Adults are active from mid-August to late September. In the southern part of the range, adults have been found both earlier and later than in the north.

Booneacris glacialis glacialis (Scudder)

Figs. 641, 645; Map 121

Pezotettix glacialis Scudder, 1863*b*:630, pl. 14, figs. 9, 10.

Booneacris glacialis glacialis; J. A. G. Rehn and Randell 1962:122, 135, 137, 139; Vickery et al. 1974:104.

Booneacris glacialis; Vickery et al. 1981:43.

Zubovskya glacialis; Helfer 1963:192 (partim).

Diagnosis. Apterous. Dark green above. Hind femur coral red beneath and on inner face. Male cercus stout, feebly tapering on apicale two-thirds.

Description. Apterous. Body size medium (length, male 16.5–20, female 20–27.5 mm). Pronotum smooth, without lateral carinae, with low

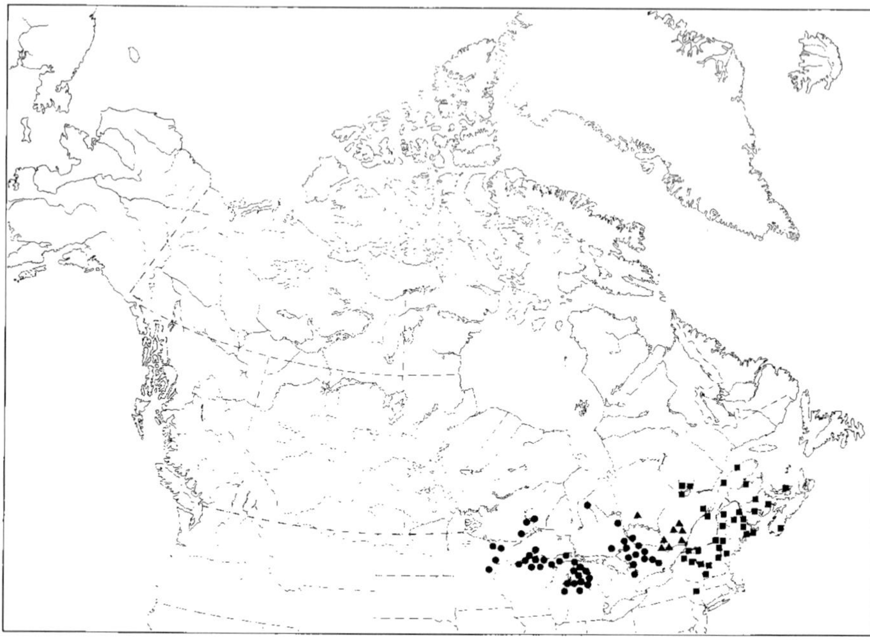

Map 121. Collection localities for *Booneacris glacialis glacialis* (■), *B. glacialis canadensis* (●), and intermediates (▲).

median carina (in male, cut by 3 sulci, in female by 1); prozona nearly twice length of metazona. Abdomen of male subcylindrical, upwardly curved apically. Male supra-anal plate long, triangular, with sides sinuous and with apex acute; furculae slender, subparallel; male cercus stout, erect, feebly tapering beyond broad basal third, with apex oblique (Fig. 641). Abdomen of female compressed, carinate dorsally; ovipositor valves rather short (Fig. 645). Color dark greenish above, greenish yellow beneath; lateral stripe from eye to abdomen shining black, fainter on metazona and abdomen; lower and inner faces of hind femur coral red; hind tibia bluish green with black-tipped spines.

Range. Québec, south to southern New York.

Behavior and habitats. A large number of specimens typical of this subspecies were taken on *Rubus* sp. near the summit of Mont Albert, Gaspésie Park, Qué., in 1971. E. M. Walker (1915*b*) recorded the subspecies from a "black spruce swamp" on Prince Edward Island. It has also been reported "resting on rocks," and Scudder's type series was found on branches of dwarf birch (*Betula nana*) in New Hampshire (Scudder 1863*b*). It has frequently been found in association with *Vaccinium* and other ericaceous plants. Morse (1919*b*) reported *B. glacialis glacialis* as "a strictly boreal species locally common from sea-level in eastern Maine to subalpine thickets on the highest mountains of New England. It is found in shrubby thickets in cold bogs, moist woodlands, and at timber-line on mountains." The same author (Morse

1920) again noted that it was a "thicket-lover," characteristic of the subalpine regions of the higher mountains of New England, but equally common at low levels in similar conditions. He called it a sluggish insect that often escapes notice because of its inconspicuous green color. It tends to hide on dull days but emerges when the sun is shining.

Life history. Univoltine, with overwintering in the egg stage. Adults have been found from 23 July (E. M. Walker 1915*b*) in Québec to 9 October—collected by one of us (V. R. V.)—at Barnum Pond, northern New York state (J. A. G. Rehn and Randell 1962).

Booneacris glacialis canadensis (E. M. Walker)

Figs. 639, 640, 643, 647; Map 121

Podisma glacialis; Scudder 1896*c*:63.

P(odisma) glacialis canadensis E. M. Walker, 1903*a*:300, pl. 6, figs. 10–31, 50, 54–56, 62, 63.

Podisma glacialis canadensis; E. M. Walker 1906*b*:66.

Booneacris glacialis canadensis; J. A. G. Rehn and Randell 1962:122, 145; Vickery et al. 1974:106.

Diagnosis. Hind femora often trifasciate with buff. Male cercus slender, tapered.

Description. Male cercus attenuate, slender, evenly tapered, with apex weakly spatulate (Fig. 643). Female cercus long, apically attenuate; dorsal ovipositor valves slender, acute (Fig. 647).

Range. Minnesota to western Québec.

Behavior and habitats. E. M. Walker (1903*a*) collected the type specimens from "open woods on bushes, chiefly the common beaked hazel (*Corylus rostrata*) and the red raspberry." Specimens at hand (both atypical specimens from Québec and typical specimens from the type locality) were collected mainly on *Rubus* species. *B. glacialis canadensis* has been found in dry open woods and also in bogs on *Vaccinium* species and other ericaceous plants. Males are agile and leap downward when disturbed. In boggy areas they may dig into layers of *Sphagnum* moss when pursued. Females are less agile but their green colors make detection difficult.

Life history. Adults are known from 9 July to 18 September in Michigan. In Canada, specimens have been collected in August and September.

Genus *Bradynotes* Scudder

Description. Apterous, robust, especially females (Fig. 632). Fastigio-facial profile broadly rounded. Eyes broad, ovate, flattened anteriorly in lateral aspect, more prominent in males than in females. Pronotum broad, with lateral margins diverging posteriorly, and with lateral carinae present but usually nearly obsolete on metazona, cut by three transverse sulci; prozona twice as long as metazona, with posterior margin truncate to shallowly concave. Prosternal spine represented by low tubercle. Interspace between mesosternal lobes strongly transverse in females, subquadrate in males. Fore and middle femora short, inflated in males; hind femur moderately robust. Tympanum on first abdominal tergum of both sexes well-developed, prominent.

Key to subspecies of *Bradynotes obesa*

1. Hind femur greenish on outer face .... *obesa* **(Thomas)** (p. 350)
   Hind femur either entirely fuscous or with faint oblique bars on outer face .... *caurus* **Scudder** (p. 351)

Clé des sous-espèces de *Bradynotes obesa*

1. Fémurs postérieurs à face externe verdâtre ... *obesa* **(Thomas)** (p. 350)
   Fémurs postérieurs soit entièrement gris brunâtre, soit marqués de bandes obliques peu visibles sur la face extérieure *caurus* **Scudder** (p. 351)

Bradynotes obesa obesa (Thomas)

Fig. 632; Map 122

P[*ezotettix*] *obesa* [*sic*] C. Thomas, 1872:454, pl. II, figs. 13, 14.
Bradynotes obesa; Scudder 1880:76.
Bradynotes obesa obesa; J. A. G. Rehn 1964:146, figs. 1, 12, 18, 24, 30, 35, 36, 38, pl. 4, figs. 1, 2, pl. 6, figs. 13, 14.

Diagnosis. Apterous, robust. Eyes prominent. Hind femur greenish on outer face, trifasciate dorsally and on inner face. Note: All specimens from Canada are intermediate between this and the following subspecies.

Description. As for genus *Bradynotes* (Fig. 632). Body robust. Head of both sexes in frontal aspect inflated, subglobose. Median carina of pronotum obvious only on metazona. Male supra-anal plate elongate, triangular, slightly concave laterally, with apex narrowly rounded (cf. Fig. 633). Male cercus long, tapering from broad base to slender apical part, with margins concave and with apex rounded (cf. Fig. 634); male subgenital plate short, broad at base, with dorsal margins straight in lateral aspect, and with apex angulate. Color dull brownish above, buffy beneath, sometimes with buffy dorsal median abdominal stripe; hind femur greenish on outer face, trifasciate dorsally and on inner face; ventral sulcus orange red with narrow median longitudinal black bar; hind tibia blackish basally to coral red apically.

Range. Southwestern Montana, northwestern Wyoming, and eastern Idaho.

Behavior and habitats. *B. obesa obesa* is usually found on mountainsides at altitudes from 1300 to 2400 m (J. A. G. Rehn 1964). Little is known about habitats except for notes by J. A. G. Rehn and Hebard (1906) on specimens taken on "gravelly tops of foothills where vegetation was almost

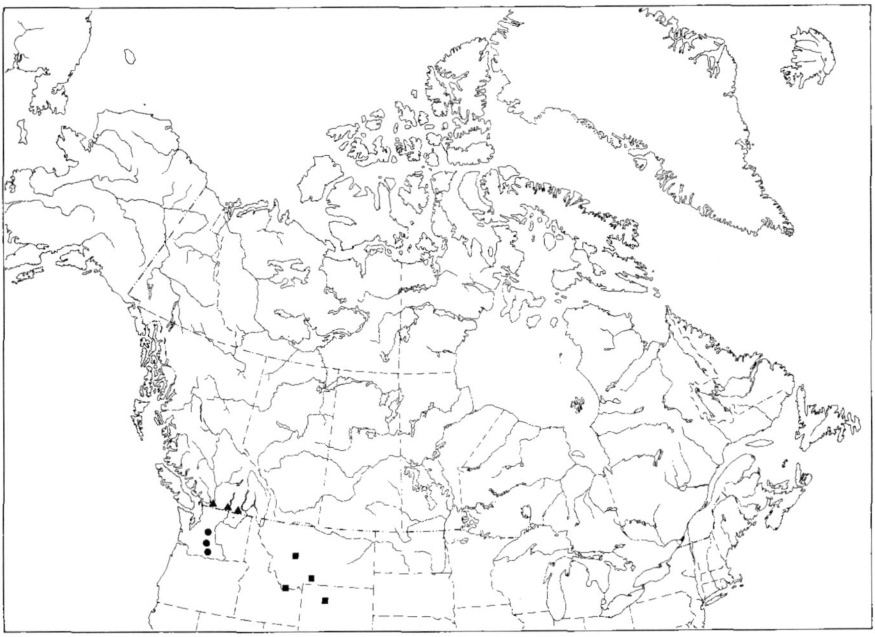

Map 122. Collection localities for *Bradynotes obesa obesa* (■), *B. obesa caurus* (●), and intermediates (▲).

absent" at Mammoth Hot Springs, Yellowstone National Park, WY, and by Hebard (1928) of specimens from Garrison, MT, on "rolling hills above bluffs, covered with grass but rather bare in spots."

Life history. Virtually unknown. Adults have been found from 17 June to 16 September in Montana (J. A. G. Rehn 1964).

Bradynotes obesa caurus Scudder

Map 122

Bradynotes caurus Scudder, 1897*d*:81, 83, pl. VI, fig. 6.

Bradynotes expleta Scudder, 1897*d*:81, 84, pl. VI, fig. 7.

Bradynotes obesa caurus; J. A. G. Rehn 1964:165, figs. 4, 5, 14, 20, 26, 39, 40, pl. 4, figs. 5, 6, pl. 6, figs. 17, 18.

Diagnosis. As for *B. obesa obesa*, but hind femur either entirely fuscous or with faint oblique bars on outer face.

Description. Slender. Head narrow. Pronotum long in proportion to width. Male subgenital plate moderately produced apically. Color generally dark; hind femur with outer face either solid fuscous or crossed diagonally by poorly defined bars; hind tibia usually entirely pinkish red.

Range. Yakima, Kittitas, and Klickitat counties, WA.

Behavior, habitats, and life history. See *B. obesa obesa*.

Genus *Buckellacris* Rehn & Rehn

Description. Size small (15–22 mm). Apterous. Pronotum with sides nearly parallel, diverging slightly posteriorly, rounding broadly into lateral lobes, with lateral carinae indicated, if at all, only by color pattern; transverse sulci deep in male, less so in females; prozona twice as long as metazona. Fore and middle femora moderately inflated; hind femur slender, with apex reaching or slightly exceeding apex of abdomen. Male supra-anal plate slightly longer than broad, with apex obtuse-angulate; furculae ranging from minute rounded projections to well-developed triangulate lobes; male cercus styliform, tapering; male subgenital plate scooplike. Ovipositor valves short, stout; female cercus short, stout, conical; female subgenital plate with apical margin produced.

Key to species and subspecies of *Buckellacris*

(Adapted from J. A. G. Rehn and J. W. H. Rehn 1945)

1. Body robust. Hind legs stout, robust. Fastigio-frontal angle broadly rounded in lateral aspect (Fig. 626). Pronotum with transverse sulci shallow, poorly marked (Fig. 626). Coloration usually dull, poorly contrasting. Male cercus narrow at base, broadening, then tapering to rounded-acute apex, with upper and lower edges not distinctly concave (Fig. 630); male furculae large, forming triangular widely separated projections (Fig. 628). Females with interspace between eyes at vertex about equal to transverse width of a single eye from dorsal aspect .... *chilcotinae chilcotinae* **(Hebard)** (p. 353)

   Body more slender. Hind legs slender, elongate. Fastigio-frontal angle more pronounced, less broadly rounded in lateral aspect (Fig. 625). Pronotum with transverse sulci strongly marked and well-impressed (Fig. 625). Coloration with distinct contrast. Male cercus broad at base, tapering to apex, with one or both edges distinctly concave (Fig. 629). Females with interspace between eyes at vertex distinctly narrower than transverse width of a single eye from dorsal aspect 3

2(1). Male furculae minute; supra-anal plate with continuity of margins appearing broken due to pronounced median shoulder (cf. Fig. 628). Prosternal spine in both sexes sharp at apex, acute, produced. Fastigio-frontal angle in both sexes obtuse-angulate in lateral aspect. Found in British Columbia and northern Washington .... ***hispida* (Bruner)** (p. 354)

   Male furculae forming well-developed trigonal plates (Fig. 627); supra-anal plate with margins regularly continuous (Fig. 627). Prosternal spine in both sexes blunt at apex, much less acute. Fastigio-frontal angle broadly rounded in lateral aspect. Found in British Columbia, Washington, Idaho, Montana, and Alberta .... ***nuda nuda* (E. M. Walker)** (p. 355)

Clé des espèces et des sous-espèces de *Buckellacris*

(adaptée de J.A.G. Rehn et J.W.H. Rehn, 1945)

1. Corps robuste. Pattes postérieures grosses et robustes. Angle fastigio-frontal très arrondi en vue latérale (fig. 626). Pronotum à sulcus transverses peu profonds et peu visibles (fig. 626). Coloration habituellement mate

et peu contrastante. Cerques des mâles étroits à la base, devenant plus gros, puis allant s'amenuisant jusqu'à donner un apex aigu mais arrondi, ni le bord supérieur ni le bord inférieur ne présentant de concavité distincte (fig. 630); furcula des mâles grande, formant deux prolongements triangulaires largement séparés l'un de l'autre (fig. 628). Chez les femelles, espace séparant les yeux au niveau du vertex environ égal à la largeur transversale d'un de ces yeux vu du dessus .... ***chilcotinae chilcotinae* (Hebard)** (p. 353)

Corps plus mince. Pattes postérieures plus minces et plus longues. Angle fastigio-frontal plus prononcé, moins arrondi en vue latérale (fig. 625). Pronotum à sulcus transverses très visibles et profonds (fig. 625). Coloration contrastante. Cerques des mâles larges à la base et allant s'amenuisant vers l'apex, un des bords ou les deux étant clairement concaves (fig. 629). Chez les femelles, l'espace séparant les yeux au niveau du vertex est clairement plus étroit que la largeur transversale d'un des yeux vu du dessus 3

2(1). Furcula des mâles minuscule; plaque supra-anale à bords brisés par un renflement médian prononcé (fig. 628). Épine du prosternum aiguë et proéminente chez les deux sexes. Angle fastigio-frontal angulaire et obtus chez les deux sexes, en vue latérale. Espèce vivant en Colombie-Britannique et au nord de l'État de Washington .... ***hispida* (Bruner)** (p. 354)

Furcula des mâles formant des plaques de forme triangulaire bien développées (fig. 627); plaque supra-anale à bords continus (fig. 627). Épine du prosternum émoussée à l'apex et beaucoup moins aiguë chez les deux sexes. Angle fastigio-frontal très arrondi en vue latérale. Espèce vivant en Colombie-Britannique, dans l'État de Washington, en Idaho, au Montana et en Alberta .... ***nuda nuda* (E.M. Walker)** (p. 355)

Buckellacris chilcotinae chilcotinae (Hebard)

Figs. 624, 626, 628, 630; Map 123

Bradynotes chilcotinae Hebard, 1922*a*:58, pl. 3, figs. 6–8.

Bradynotes chilcotinae; Buckell 1922:29.

Buckellacris chilcotinae chilcotinae; J. A. G. Rehn and J. W. H. Rehn 1945:15.

Diagnosis. Body dull, with color poorly contrasting. Sulci of pronotum shallow, poorly marked. Female broad between eyes.

Description. Body size large (length 16.8–25.0 mm; mean 19.9 mm). Apterous (Fig. 624). Fastigio-frontal angle gradually and broadly rounded in lateral aspect (Fig. 626). Females with interspace between eyes about equal to width of a single eye from dorsal aspect. Transverse sulci of pronotum shallow, poorly marked (Fig. 626). Male furculae consisting of fairly large triangular widely separated projections (Fig. 628); supra-anal plate broad, as long as basal width, with lateral margins weakly convergent and weakly and irregularly convex to apex (Fig. 628); male cercus short, narrow at base, broadening, then tapering to slender rounded–acute apex (Fig. 630); male subgenital plate conical, with lateral margins weakly convex to blunt apex.

Range. British Columbia and Washington.

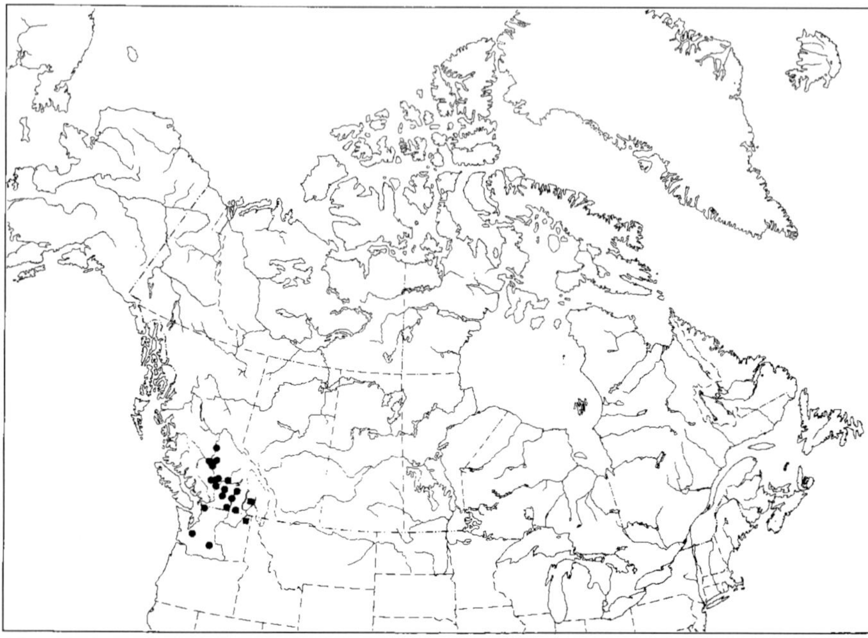

Map 123. Collection localities for *Buckellacris chilcotinae chilcotinae* (●) and *B. hispida* (■).

Behavior and habitats. The species is adaptable, being found in meadows as well as in montane fir forests and in sagebrush areas. Buckell *in* J. A. G. Rehn and J. W. H. Rehn (1945) stated that "it is a very common species on almost any warm bushy slope in the dry interior of the Province [British Columbia]." Buckell (1922) reported both adults and nymphs feeding upon balsamroot (*Balsamorhiza sagittata* (Pursh) Nutt.). Males are active and agile, but females are heavy and do not move much. Nagy *in* Vickery and Nagy (1973) found it in bushy mixed vegetation, mainly snowbrush and juniper, on the south side of Mount Acland, near Summerland, B.C.

Ants and ground spiders prey upon young nymphs. Treherne and Buckell (1924) found dipterous parasitic larvae in adult specimens.

Life history. Adults have been taken from 30 May to 29 August, and are most numerous from mid-June to late July. Oviposition occurs in July. Nymphs hatch early in May of the following year.

Economic importance. Treherne and Buckell (1924) reported damage to vegetation under poplar trees on the fringes of open rangeland. No other reports of damage are known, so that *B. chilcotinae chilcotinae* must be regarded as of negligible economic importance.

Buckellacris hispida (Bruner)

Map 123

Pezotettix hispidus Bruner, 1885*b*:12.

Buckellacris hispida; J. A. G. Rehn and J. W. H. Rehn 1945:28, figs. 9–12, 17, pl. I, figs. 24, 27, pl. II, figs. 30, 33.

Diagnosis. Like previous species but with male furculae reduced. Fastigio-frontal angle obtuse.

Description. Fastigio-frontal angle in both sexes obtuse-angulate in lateral aspect. Prosternal spine in both sexes produced to acute apex. Male supra-anal plate with continuity of margins appearing broken due to pronounced median shoulder (cf. Fig. 628); male furculae reduced to small widely separated lobes; aedeagal valves distinctive (see J. A. G. Rehn and J. W. H. Rehn 1945:15, fig. 17).

Range. British Columbia and Washington.

Behavior and habitats. Little is known about this species. E. M. Walker (1898*c*) reported taking it (as *Asemoplus nudus*) "on the grassy path of snowslide" at Sandon, B.C.

Life history. Unknown. The known adult specimens, 3 males, 3 females, were collected from 17 August to 16 September.

Buckellacris nuda nuda (E. M. Walker)

Figs. 625, 627, 629; Map 124

Asemoplus nudus E. M. Walker, 1898*c*:197 (partim), pl. 6, figs. A–D.

Asemoplus nudus; Caudell 1907*a*:134.

Asemoplus somesi Hebard, 1919*b*:271.

Buckellacris nuda nuda; J. A. G. Rehn and J. W. H. Rehn 1945:35; Brooks 1958:18.

Diagnosis. Prosternal spine blunt. Fastigio-frontal angle broadly rounded. Male furculae trigonal, lobelike.

Description. Both sexes apterous. Fastigio-frontal angle pronounced, rather abruptly rounded in lateral aspect (Fig. 625). Females with interspace between eyes narrower than width of eye in dorsal aspect. Transverse sulci of pronotum strongly marked and well-impressed (Fig. 625). Prosternal spine in both sexes blunt, not produced to acute apex. Male furculae in form of well-developed trigonal plates (Fig. 627); male supra-anal plate with continuity of margins appearing regularly continuous (not appearing broken due to median shoulder) (Fig. 627); male cercus broad at base, tapering to acute apex, with one or both margins distinctly concave (Fig. 629); aedeagal valves distinctive (see J. A. G. Rehn and J. W. H. Rehn 1945:15, fig. 19). Color pattern of males with distinct dorsal paired longitudinal pale lines laterally, extending from eyes to genitalia in males, obvious only anteriorly in females; hind femur usually with definite dark bars dorsally in males, these less apparent to nearly obsolete in females; hind tibia yellowish.

Range. Interior British Columbia, Washington, Alberta, Idaho, and Montana.

Behavior and habitats. *B. nuda nuda* usually occurs in coarse herbage and alpine grasses among rocks in mountains. The species has been recorded

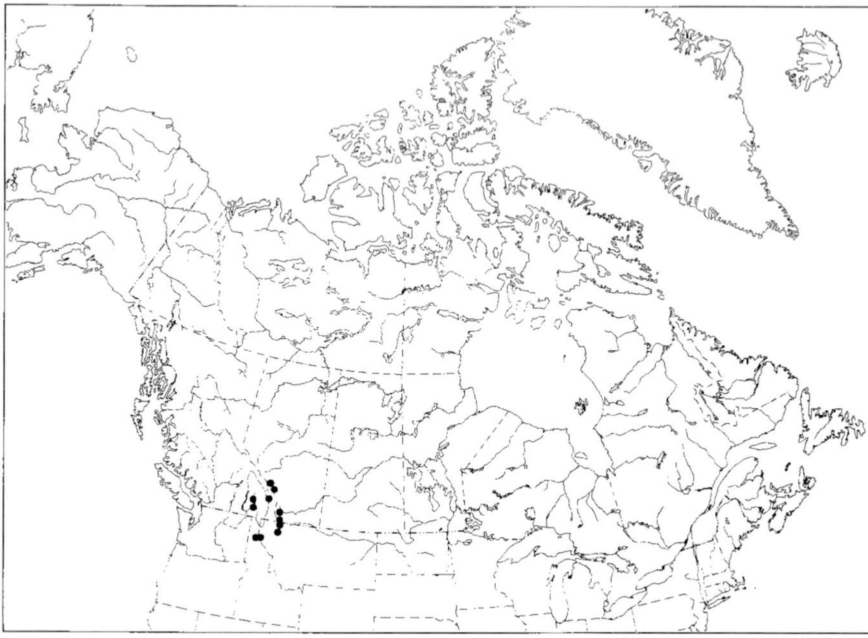

Map 124. Collection localities for *Buckellacris nuda nuda*.

at elevations from 800 to 2750 m, sometimes well above timberline. It occurs in discrete and disjunct populations, and there is a tendency for these to exhibit variations in characteristics that are relatively constant within populations.

Life history. Adults have been taken from 5 July to 19 September. Last-instar nymphs have been reported as late as 9 August in northwestern Montana. It appears that the life cycle is univoltine, with overwintering in the egg stage. The early appearance of adults together with prolonged presence of nymphs, however, indicate that there may, at least sometimes, be a 2-year life cycle.

Genus *Asemoplus* Scudder

Description. Body slender. Pronotum short, flaring on metazona, with lateral carinae absent (Fig. 635); median carina feeble, on metazona only, with anterior margin truncate and with posterior margin shallowly emarginate. Tegmina linear, separated, lateral in position, shorter than pronotum.

Asemoplus montanus (Bruner)

Figs. 635–638; Map 125

Bradynotes montanus Bruner, 1885*b*:16.

Asemoplus montanus; Scudder 1897*a*:206; Brooks 1958:17; Helfer 1963:199.

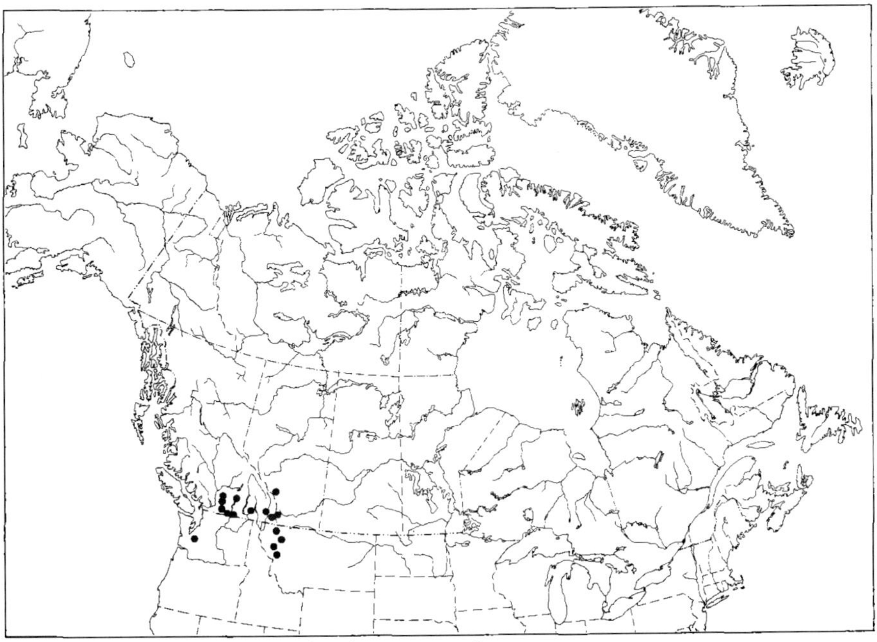

Map 125. Collection localities for *Asemoplus montanus*.

Diagnosis. Pronotum short, flaring posteriorly. Tegmina linear, lateral in position, shorter than pronotum.

Description. Body small (length 15–19 mm), robust (Fig. 638). Vertex and frontal costa broad; vertex weakly convex. Mesosternal lobes transverse, at least as wide as long, with inner margins rounded. Tegmina present as short linear scales separated at bases by distance greater than their length (Fig. 635). Male furculae absent (Fig. 636); male cercus conical, tapering, with acute apex directed backward and downward (Fig. 637); male subgenital plate pointed to rounded apically. Female cercus short, triangular. Color dark reddish brown with darker and paler markings; both sexes with single median longitudinal pale dorsal stripe; lateral lobes of pronotum with oblique pale streak, this sometimes interrupted; hind femur pale on outer face, darker above; hind tibia yellow with black-tipped spines.

Range. Southwestern Alberta, southeastern British Columbia, Washington, and northern Montana.

Behavior and habitats. E. M. Walker (1906*a*) collected *A. montanus* in "a small low wood of poplar, Douglas fir, etc., surrounded by dry barren hills" near Vernon, B.C. Bruner (1885*b*) originally collected it near Helena, Montana, "among the trailing junipers on north mountain slopes at moderate elevations."

Life history. Adults have been collected from 16 June to 30 August, but they probably persist for a few weeks later than this. The life cycle is probably univoltine.

Genus *Phoetaliotes* Scudder

Description. Body elongate, slender, weakly compressed. Head large, prominent, long, 1.5 times as deep as prozona is long (Figs. 418, 425); vertex prominent. Pronotum enlarged anteriorly, with posterior margin broadly obtuse-angled and with median carina sharp, entire. Tegmina dimorphic, usually brachypterous, little longer than pronotum, broadly lanceolate, often well-separated, sometimes touching, or else long, extending beyond apices of hind femora in either sex.

Phoetaliotes nebrascensis (Thomas)

Figs. 418, 421, 425; Map 126

Pezotettix nebrascensis C. Thomas, 1872:455.
Pegotettix [*sic*] *autumnalis* G. M. Dodge, 1876:10.
Caloptenus volucris G. M. Dodge, 1877:112.
Phoetaliotes nebrascensis nebrascensis; Scudder 1897*d*:377.
Phoetaliotes nebrascensis volucris; Scudder 1897*d*:378.
Melanoplus phoetaliotiformis Scudder, 1898*c*:179.
Melanoplus harrisii Morse, 1909:12.
Phoetaliotes nebrascensis; Somes 1914:95; Cantrall 1943:122; Brooks 1958:31; Helfer 1963:196.

Diagnosis. Head large. Pronotum narrowing posteriorly.

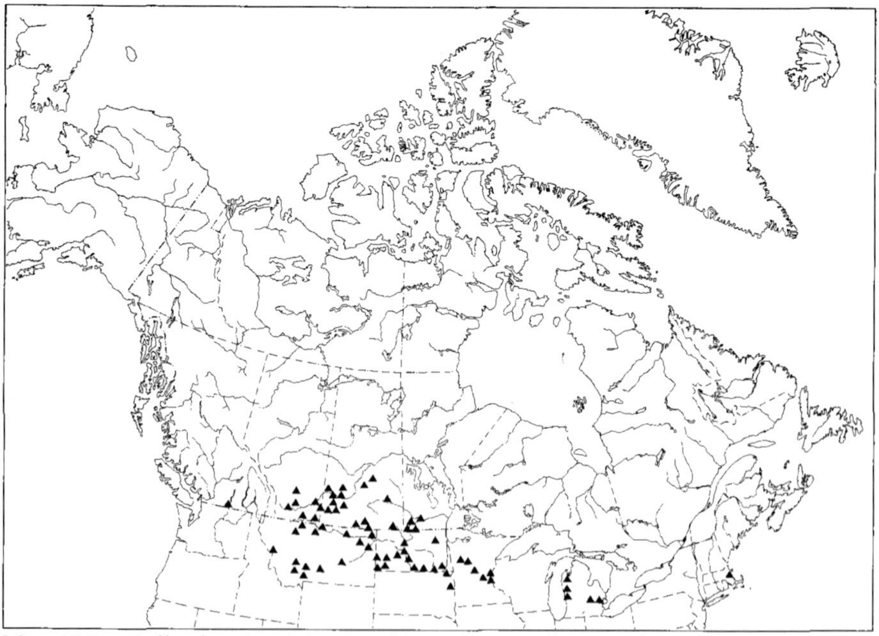

Map 126. Collection localities for *Phoetaliotes nebrascensis*.

Description. Head exceptionally large as compared with thorax, broad and deep, with face noticeably slanted (Figs. 418, 425). Pronotum slender, with lateral margins appearing to converge posteriorly. Tegmina dimorphic, usually short, extending only to second or third abdominal segment, pointed; occasional specimens with tegmina long, extending well beyond apices of hind femora. Median carina of pronotum entire. Furculae of ninth abdominal tergum of male small; male subgenital plate pointed or subtubercular at apex; male cercus symmetrical, slender apically (Fig. 421). Color generally brownish, with dorsum of head and pronotum pale; hind tibiae blue to purple.

Range. British Columbia to Massachusetts, south to Arizona and Texas.

Behavior and habitats. *P. nebrascensis* has been found in a wide variety of habitats on various types of soils. Presence of food plants seems to be the main factor determining where it occurs. It has a wide range of acceptable food plants, both grasses and forbs. Mulkern et al. (1969) found it feeding upon 17 species of grasses and 27 species of forbs in Kansas, Nebraska, and North Dakota. N. Criddle (1935) classed it as a general grass feeder in Manitoba, and Anderson and Wright (1952) found it to feed mainly on grasses in Montana. Males are active jumpers, but females are sluggish and inactive. Only the macropterous form has flight ability; therefore increase in range is very slow.

Life history. Univoltine, with overwintering in the egg stage.

Economic importance. Mulkern et al. (1969) state that the species has the potential to be destructive "to grassland and some cultivated crops."

Genus *Dendrotettix* Packard

Description. Body medium (length, males 24–27, females 28–32 mm), robust. Head large, broad. Eyes small, widely separated. Fastigium wide, distinctly concave between eyes. Face distinctly, but not strongly, oblique. Pronotum rugose, short, with anterior third flaring to receive head; metazona much shorter but broader than prozona, strongly rugoso-punctate, with posterior margin truncate to broadly rounded. Tegmina and wings variable in development, brachypterous or macropterous. Hind femur slender, not reaching apex of abdomen (or extending only slightly beyond it); hind tibia strongly spined.

Dendrotettix quercus Packard

Figs. 619, 620; Map 127

Dendrotettix quercus Packard, 1890:214.
Dendrotettix quercus; Bruner 1893*b*:28; Helfer 1963:207.

Diagnosis. Tegmina usually no longer than pronotum (but can also be macropterous). Pronotum flaring on anterior third to receive large head.

Description. As for genus. Tegmina usually strongly brachypterous, no longer than pronotum, with apices rounded, occasionally nearly as long as abdomen or fully macropterous (Fig. 619) in either sex, extending beyond

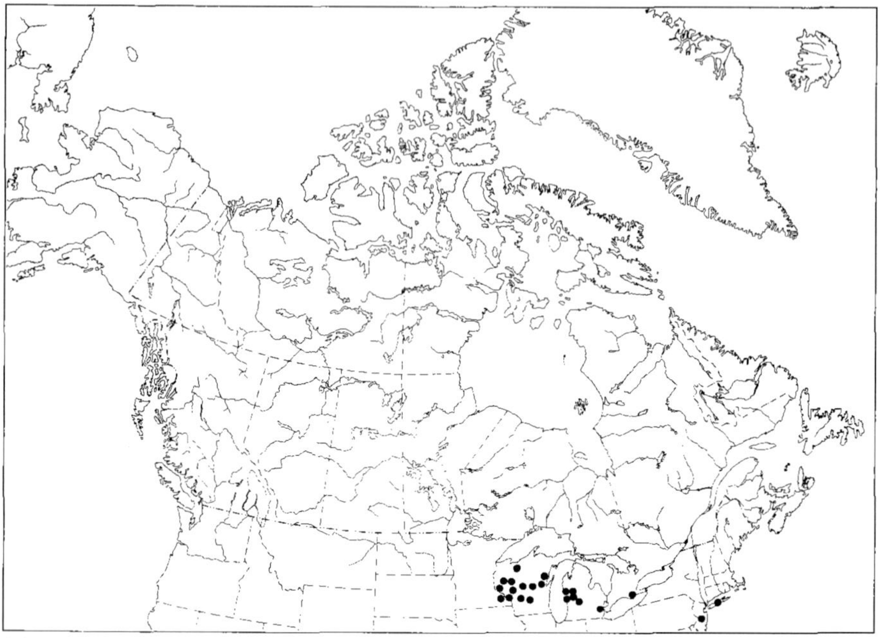

Map 127. Collection localities for *Dendrotettix quercus*.

abdominal apex, broad in apical half. Male supra-anal plate long, triangular; furculae in form of small triangular projections. Male cercus subquadrate, with apical half twisted, depressed, and with apices truncate (Fig. 620); male subgenital plate scoop-shaped with bluntly rounded apex. Ovipositor valves short, broad, not compressed. Color yellow to greenish yellow with darker markings; occiput usually with black median stripe; postocular black stripe extending across lateral pronotal lobes, covering upper half; tegmina brownish yellow; abdomen laterally and subgenital plate black; hind femur dull greenish yellow with two broad fuscous cross-bars; hind tibia greenish yellow with bright yellow basal annulus.

Cytology. Fontana and Vickery (1976) reported on the cytogenetic system of this species. The chromosome complement is $2n$ male = 23 and $2n$ female = 24, with *XO*, *XX* sex determination. The X chromosome is the third largest. Autosomes include two large, six medium, and three small pairs. A mean of 14.47 cell chiasma frequency was found in a Michigan population. Two males were found with supernumerary chromosomes, one with a single *B*, the other with two *B*'s. In both specimens these were small and highly stable. They were found to divide reductionally at first meiotic division and to move randomly to either pole with respect to the X chromosome at Anaphase I.

Range. Nebraska to southern Ontario, south to Texas and Tennessee.

Behavior and habitats. *D. quercus* lives among and feeds upon oaks. Cantrall (1943) stated that, in Michigan, it seemed to be "partial to *Quercus ellipsoidalis*, the northern pin oak." It is also found on other species of

Quercus (Cantrall 1968; Valek and Coppel 1972*a*, 1972*b*; Fontana and Vickery 1976)—the "red" oaks being preferred to the "white" oaks (Valek and Coppel 1972*a*). It is also found on other vegetation (J. A. G. Rehn and J. W. H. Rehn 1938) including hazel (*Corylus americana* Marsh.) (Valek and Coppel 1972*a*). Red maple (*Acer rubrum* L.) does not seem to be favored (Valek and Coppel 1972*a*). Adults are active and will run or leap readily when disturbed. Macropterous forms make relatively long flights when flushed, but Valek and Coppel (1972*a*) state that the species appears to have no migratory tendency. According to Bruner (1893*a*), egg pods, in Texas, are laid in the ground under host trees or are scattered on the surface among decaying leaves. Valek and Coppel (1972*a*, 1972*b*), who described the eggs and nymphal stages, confirmed the general findings regarding oviposition sites in Wisconsin. They note that these are normally located at the edges of oak woods. *D. quercus* has long been known to increase its numbers irregularly and suddenly to "outbreak" proportions, as discussed by J. A. G. Rehn and J. W. H. Rehn (1938). Its numbers decline with equal suddenness, sometimes assisted by heavy rains and frosts to which late final-instar nymphs seem particularly susceptible.

Several species of parasites have been reported to attack *D. quercus*. During an outbreak in New Jersey, J. A. G. Rehn and J. W. H. Rehn (1938), reported a heavy infestation by maggots of *Sarcophaga atlanis* (Aldrich) (Diptera: Sarcophagidae). Valek and Coppel (1972*a*) report two other species of sarcophagid parasites and an 18% infestation of adult *Dendrotettix* by a species of *Agamermis* (Nematoda: Mermithidae). This caused a considerable reduction in fecundity and longevity of the parasitized grasshoppers.

Life history. Eggs hatch, in Wisconsin, in May or early June (Valek and Coppel 1972*a*) and the young nymphs, after the first molt, climb into the host trees (or young saplings) and begin feeding. J. A. G. Rehn and J. W. H. Rehn (1938) expressed surprise that mating, but no oviposition, occurred in August in New Jersey, but, according to Valek and Coppel (1972*a*, 1972*b*), eggs are laid in Wisconsin in mid-July. These authors give a fairly detailed account of the life history of *D. quercus*. They observe that the species has a biennial life cycle, the eggs usually passing two winters before hatching. There is, however, a lack of synchrony, so that some nymphs and adults are present every season, but they are much more abundant every other year. They report adults in Wisconsin from mid-July until September, with a few lingering until early October. In Michigan, Cantrall (1968) recorded adults from 10 July to 8 September. The specimens found in 1940 by Urquhart (1942) at Turkey Point, Ont., were collected on 25 August.

Economic importance. When *D. quercus* populations become large and reach "outbreak" proportions, the insects can defoliate young oak trees. Various authors have referred to the damage (e.g., Bruner 1893*a*; Packard 1890; J. A. G. Rehn and J. W. H. Rehn 1938; Cantrall 1968; Fontana and Vickery 1976). Nevertheless, because the outbreaks seem to decline as suddenly as they arise and do not apparently persist for more than one season, at least in the northern part of the species' range, it is improbable that lasting harm is done, except to very young seedlings and saplings. A comprehensive account of an outbreak in New Jersey is given by J. A. G. Rehn and J. W. H. Rehn (1938).

Genus *Appalachia* Rehn & Rehn

Description. Body surface, particularly legs, covered with numerous erect short hairs. Pronotum in lateral aspect arcuate with shallow median emargination; median carina evident throughout but more marked on metazona; transverse sulcus continuous and deep in male, less so in female. Tegmina lobate, lateral in position; hind wings almost lacking, represented by short vestiges. Abdomen with fine median longitudinal carina. Fore femur inflated in both sexes; hind femur slender.

Appalachia arcana Hubbell & Cantrall

Figs. 621-623; Map 128

Appalachia arcana Hubbell and Cantrall, 1938:1–22, pl. I.

Appalachia arcana; J. A. G. Rehn and J. W. H. Rehn 1939:95; Helfer 1963:205.

Diagnosis. Found in Michigan only. Tegmina lobate, lateral. Abdomen with conspicuous broad pale stripe dorsally.

Description. As for genus *Appalachia* (Fig. 623). Body size small to medium (length, male 17.2–19.1, female 24.4–29.5 mm), robust. Male supra-anal plate with lateral margins elevated, usually arcuate; cercus straight, conical, acute (Fig. 621); subgenital plate weakly compressed, bluntly conical.

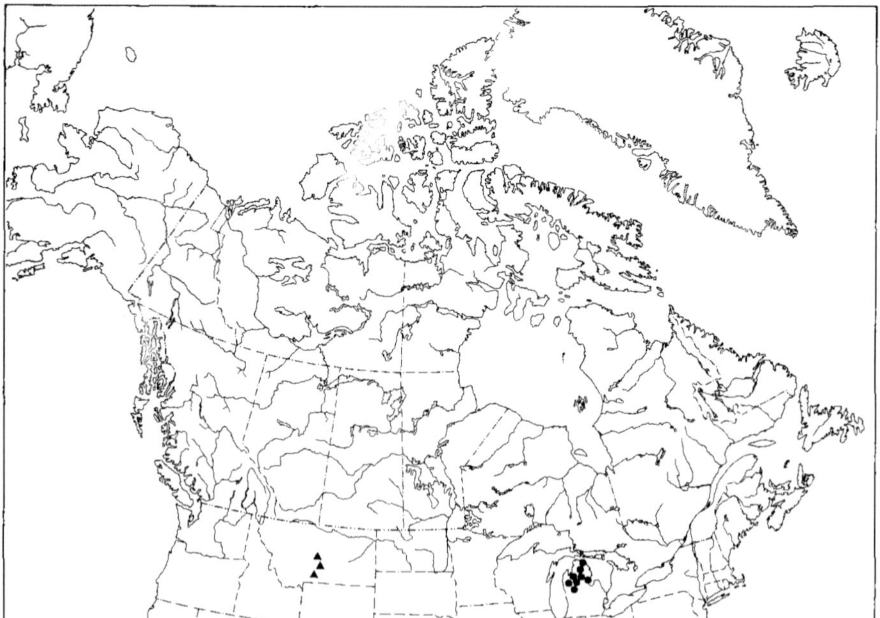

Map 128. Collection localities for *Appalachia arcana* (●) and *Argiacris rehni* (▲).

Ovipositor valves slender (Fig. 622), with dorsal valves having sharp angle, and with apices weakly upcurved; female subgenital plate tapering to acute point (Fig. 622). Color brownish gray dorsally; laterally with strongly contrasting black stripes from head nearly to abdominal apex; abdomen with conspicuous broad median pale stripe; legs greenish suffused with brown; hind femur red beneath and on inner face; hind tibia yellowish green on ventral face, bluish green apically, yellow green on sides and dorsal face, and with pale annulus basally.

Range. Lower peninsula of Michigan.

Behavior and habitats. This grasshopper is secretive and difficult to find except when abundant; also, it tends to remain immobile and its cryptic coloration makes it difficult to see. It is found in bogs where leatherleaf (*Chamaedaphne calyculata*) and Labrador-tea (*Ledum groenlandicum*) occur in heavy stands. Jack pine (*Pinus banksiana*) and tamarack (*Larix* sp.) occur in and near such bogs, and specimens may be found on trunks and foliage of these trees. It has also been taken on eastern bracken (*Pteridium aquilinum* var. *latiusculum*), sweet fern (*Myrica asplenifolia*), and in open groves of aspen and pines. It probably feeds upon a variety of plants. Males often climb shrubbery and sun themselves near the tips of branches. When they move, their movements are jerky and "nervous." When disturbed the insects may become immobile. If they attempt to escape, they leap two or three times in rapid succession in zigzag fashion, moving downward, and may burrow into moss or debris, thus escaping capture. Females generally remain at lower levels and are usually hidden from view. It is often necessary to trample bushes in order to find them (Hubbell and Cantrall 1938). Mating behavior was observed and described by Hubbell and Cantrall (1938). Females oviposit on twigs, not in the soil, leaving frothy material that hardens into brown globose masses 8–12 mm in diameter.

Life history. Univoltine, with overwintering in the egg stage. Adults have been taken from 1 July to 18 September (Cantrall 1968).

Genus *Argiacris* Hebard

Description. Body stout, slightly depressed. Interocular space wide, with fastigium of vertex blunt. Eyes small, with diameter much shorter than cheeks. Pronotum with disc rounding rather broadly into lateral lobes, with posterior margin obtuse-angulate. Tegmina short, broad, meeting dorsally at base in male, slightly separated in female. Male genitalia with small furculae. Ovipositor valves fully exposed.

Argiacris rehni Hebard

Fig. 631; Map 128

Argiacris rehni Hebard, 1918:167, pl. VIII, fig. 18.
Argiacris rehni; Helfer 1963:212.

Diagnosis. Fairly large, robust, strongly brachypterous. Pronotal disc obtuse-angulate behind. Coloration obscure except for brilliant pink ventral interior faces of hind femora and pink hind tibiae.

Description. Body rather large, robust (male 16.9–22.3, female 27.9–29.4 mm long) (Fig. 631). Interocular space one-half (male) or two-thirds (female) as wide as greatest diameter of an eye; eye three-quarters (male) or three-fifths (female) as long as cheek. Pronotal spine short, heavy (particularly in female), acutely conical, with apex slightly rounded in female. Space between mesosternal lobes longer than least width in male, about one-third wider than long in female. Tegmina broadly lanceolate (less so in female), barely (male) or not (female) surpassing base of hind femur, considerably shorter than pronotum. Furculae of male represented by 2 minute rectangular projections with angles rounded, separated by space slightly wider than a projection; supra-anal plate simple, triangularly shield-shaped, with apex rather acute; male cercus simple, about 2.5 times as long as its basal width, tapering, slightly curved dorsad, slightly narrowed about middle, with apex broadly rounded; male subgenital plate with acute conical dorsomedian projection. Color generally brownish, paler on face and ventrally, darker dorsally; hind femur with 2 broad transverse blackish bands running obliquely and rather obscurely across dorsal half of outer surface; ventral surface bright red; hind tibia similarly bright red.

Range. Southern Montana.

Behavior and habitats. Hebard (1918) stated that the type series was collected "on the ridge of a slope of a bare hog-back showing numerous cherty exposures. The ground there showed rather scant vegetation with tufts of a peculiar woolly plant all about." It would seem that the species occurs in very localized colonies.

Life history. Nothing is known.

Genus *Melanoplus* Stål

Description. Body moderately robust. Head not prominent, about as long as prozona of pronotum; face nearly vertical; interocular space little if any wider than frontal costa; fastigium declivent, merging into frontal costa, which is usually sulcate below median ocellus. Lateral carinae of pronotum obsolete. Tegmina present, variable in shape and size, often reduced. Furculae of ninth abdominal tergum of male present, variable in degree of development. Male cerci simple, conical to elaborately expanded. Ovipositor usually strongly exserted.

Key to species and subspecies of *Melanoplus*

1. Tegmina somewhat reduced, not reaching tip of abdomen (Fig. 427), or abbreviate 2
   Tegmina exceeding tip of abdomen (Figs. 428–430) 23

2(1). Hind tibiae green. Eastern species 3
   Hind tibiae not green, or if so, not eastern species 5

3(2). Hind femur without dark markings; hind tibia green, but darker at base. Male cercus extending downward at apex ....
   .... ***gracilis* (Bruner)** (p. 378)

Hind femur heavily marked with brown; hind tibia not darkened at base. Male cercus tapered, with rounded apex (Figs. 478, 479) (***viridipes***, *sensu lato*) 4

4(2). Male cercus triangular, tapered, with rounded apex (Fig. 479). Found in Vermont (and probably southern Québec), Massachusetts, and Pennsylvania[1] .... ***viridipes eurycercus*** **Hebard** (p. 379)

Male cercus longer, less tapered, slightly convex subapically above, slightly concave below (Fig. 478). Found in southern Ontario,[1] Michigan, Illinois, Indiana, and southward[1] .... ***viridipes viridipes*** **Scudder** (p. 380)

5(2). Hind tibiae pink 6

Hind tibiae red, buff, blue, or slightly greenish 7

6(5). Tegmina in form of ovate pads, not exceeding second abdominal segment. Male cercus triangular, short (cf. Fig. 479) .... ***scudderi scudderi*** **(Uhler)** (p. 380)

Tegmina longer, reaching at least fourth abdominal segment. Male cercus spatulate (cf. Fig. 482) .... ***walshii*** **Scudder** (p. 381)

7(5). Hind tibiae buff to blue .... 8[2]

Hind tibiae red 11

8(7). Male cercus not broadened distally, narrowing to broad rounded apex, not or scarcely spatulate (cf. Figs. 495, 505) (***indigens***, *sensu lato*) .. 9

Male cercus somewhat broadened distally, with distal part distinctly but not strongly spatulate (cf. Figs. 480, 481) (***oregonensis***, *sensu lato*) 10

9(8). Ventral surface of hind femur bright orange to deep red. Male cercus short, not spatulate (cf. Fig. 495, but a little longer). Found in northern Montana (Missoula) .... ***indigens missoulae*** **Hebard** (p. 383)

Ventral surface of hind femur dull yellowish buff, tinged with orange. Male cercus longer, slightly spatulate (cf. Fig. 505). Found in Rocky Mountain Divide and southern Montana .... ***indigens indigens*** **Scudder** (p. 383)

10(8). Male supra-anal plate distinctly longer than broad, with apex broadly rounded; furculae large (Fig. 444). Male cerci long in relation to width (Fig. 480). Found in southern Montana, southern Idaho, and southernmost Alberta .... ***oregonensis oregonensis*** **(Thomas)** (p. 384)

Male supra-anal plate triangularly shield-shaped, only slightly longer than broad; furculae smaller (Fig. 445). Male cerci shorter (Fig. 481). Found in northwestern Montana, northern Idaho, southwestern Alberta, and southeastern British Columbia .... ***oregonensis triangularis*** **Hebard** (p. 385)

11(7). Tegmina represented by short pads, not extending past fourth abdominal segment (rarely longer in *M. dawsoni*, but if so, abdominal terga banded) 12

Tegmina longer, not exceeding abdomen (abdominal terga not banded, except in *dawsoni*) 16

12(11). Abdomen banded, each segment black anteriorly and yellow posteriorly. Prosternal tubercle blunt. Male cercus curved to broadly rounded apex (Fig. 490) .... ***dawsoni*** **(Scudder)** (p. 385)

Abdomen not banded. Prosternal tubercle pointed 13

13(12). Eastern species (Maine to eastern Manitoba) 14

Western species (southwestern Alberta and Montana to Washington and British Columbia) (***montanus***-group) 15

14(13). Lateral carinae of pronotum nearly obsolete. Tegmina (particularly of females) with tips broadly and evenly rounded. Male cercus elongate, broad at base, tapering from middle to apex (Fig. 482). Male furculae moderately long, subcylindrical (Fig. 446) .... ***mancus*** **(Smith)** (p. 386)

Lateral carinae of pronotum distinct. Tegmina (particularly of females) with tips narrowly rounded or subangulate. Male cercus short, strongly tapered, concave on upper apical fourth (Fig. 483). Male furculae short, flat, oblong (Fig. 447) .... ***islandicus*** **Blatchley** (p. 388)

15(13). Male cercus somewhat subspatulate (Fig. 484). Aedeagal valves not bent outward at apical third .... ***montanus*** **(Thomas)** (p. 389)

Male cercus not subspatulate, with apex truncate (Fig. 485). Aedeagal valves suddenly divergent at dorsal third, tapered to acute apices .... ***washingtonius*** **(Bruner)** (p. 390)

16(11). Tegmina extending to third or fourth abdominal segment, tapered to rounded apices .... ***dodgei*** **(Thomas)** (p. 391) .... ***huroni*** **Blatchley** (p. 392)

Tegmina longer, extending beyond sixth abdominal segment 17

17(16). Abdominal segments banded .... ***dawsoni*** **(Scudder)** (p. 385)

Abdominal segments not banded 18

18(17). Large dark olive green species with narrow yellow lines from eyes to tips of tegmina; hind femur with narrow black longitudinal stripes on outer face and on upper inner edge of flange .. ***bivittatus*** **(Say)** (p. 393)

Dorsal stripes absent, or restricted to head and thorax only; hind femur usually crossbanded at least dorsally 19

19(18). Hind femur banded on outer face 21

Hind femur not banded on outer face 20

20(19). Large species. Found in Magdalen Islands, Qué. .... ***madeleineae*** **Vickery & Kevan** (p. 395)

Smaller species. Widely distributed .... ***borealis borealis*** **(Fieber)** (p. 397) .... ***borealis palaceus*** **Fulton** (p. 398)

21(19). Hind femur dark with 2 pale oblique streaks (Fig. 440). Found on Mont Albert, Gaspé Peninsula, Qué. ...... ***gaspesiensis*** **Vickery** (p. 399)

Hind femur with definite crossbands. Range broad 22

22(21). Male cercus straplike, recurved at apex (Fig. 505). Hind femur mainly dark with paler complete crossbands. Robust species. Range broad .... ***fasciatus*** **(F. Walker)** (p. 399)

Male cercus broad at base, narrowing to blunt apex (Fig. 506). Hind femur mainly pale with darker crossbands. Range in extreme north (northern Alaska, northern Yukon Territory, and northern Mackenzie District) .... ***frigidus frigidus*** **(Boheman)**[3] (p. 401)

23(1). Males (without ovipositor) 24

Females (with ovipositor) 47

24(23). Cercus with apex divided 25

Cercus with apex not divided 28

25(24). Dorsal arm of cercus broad, large, with rounded apex, longer than ventral arm 26

Dorsal arm of cercus thumblike, much shorter than ventral arm ... 27

26(25). Ventral arm of cercus thumblike (Fig. 507). Hind femur with longitudinal dark stripe; hind tibia dark red with basal black band or spot .... ***keeleri luridus*** **(Dodge)** (p. 402)

Ventral arm of cercus triangular, projecting (Fig. 492). Hind femur without black stripe; hind tibia red to yellowish or greenish blue .... ***confusus*** **Scudder** (p. 403)

27(25). Ventral arm of cercus nearly straight (Fig. 500). Apex of subgenital plate notched (Fig. 536) .... ***infantilis*** **Scudder** (p. 405)

Ventral arm of cercus strongly curved (Fig. 501). Apex of subgenital plate not notched, instead truncate and dished behind (Fig. 537) .... ***alpinus*** **Scudder** (p. 406)

28(24). Cercus spatulate and broader distally than at base 29

Cercus as broad as or broader at base than distally 31

29(28). Large species (male 28–34, female 34–44 mm). Cercus with rounded ventral projection (Fig. 489). .... ***differentialis*** **(Thomas)** (p. 407)

Medium-sized species (male 18–25, female 26–30 mm). Cercus with apex greatly expanded (Fig. 487). Hind tibia dull reddish to gray, with black spines (***punctulatus***, *sensu lato*) 30

30(29). Eyes prominent (anterior view). Pronotum distinctly flared outward anteroventrally (dorsal view). Male cercus with dorsal lobe nearly as deep as depth of remainder of cercus and with dorsal margin broadly and more evenly rounded. Found in Michigan,[4] Minnesota, and Wisconsin .... ***punctulatus griseus*** **(Thomas)** (p. 409)

Eyes less prominent (anterior view). Pronotum less flared outward anteroventrally (dorsal view). Male cercus with dorsal lobe distinctly shallower than depth of remainder of cercus and with dorsal margin bluntly angular (Fig. 487). Found in New England, New York, Québec, Ontario, and Michigan[4] .... ***punctulatus punctulatus*** **(Scudder)** (p. 410)

31(28). Cercus broad, straplike, with sides nearly parallel (Fig. 491). Hind femur with lower flange absent from basal third (Fig. 438). Tegmina conspicuously mottled .... ***gladstoni*** **Scudder** (p. 411)

Cercus tapered, subquadrate, or spatulate but distinctly broader at base. Hind femur with lower flange complete. Tegmina not noticeably mottled 32

32(31). Cercus symmetrical, spatulate, or slightly lobed below, narrowest at middle 33

Cercus asymmetrical, broadest at base, tapered, curved dorsally, lobelike or subquadrate 38

33(32). Apex of subgenital plate notched .... ***angustipennis angustipennis*** **(Dodge)** (p. 412)

Apex of subgenital plate broadly rounded or acute 34

34(33). Apex of subgenital plate acute (Fig. 551). Cercus small (Fig. 514) .... ***cinereus cinereus*** **Scudder** (p. 414)

Apex of subgenital plate broadly rounded (Figs. 456, 457). Cercus larger (Fig. 510) 35

35(32). Aedeagus with anterior and posterior processes projecting to same level (Figs. 554, 555)[5] (***packardii***, *sensu lato*) 36

Aedeagus with anterior process much shorter and projecting less than posterior process (Figs. 552, 553)[5] 37

36(35). Length 27–32 mm. Hind femur with bands at top of upper ridge and on inner face; hind tibia blue, purple, or pink; general color bluish gray. Found in grasslands and parklands .... ***packardii packardii*** **Scudder** (p. 415)

Length 23–31 mm. Hind femur with bands extending on both sides of upper ridge; hind tibia dark red; general color dark brownish black. Found in Jack pine forests (Saskatchewan, Alberta, and British Columbia) .... ***packardii brooksi*** **Vickery** (p. 417)

37(35). Hind tibiae dark red; pronotal stripes indistinct; tegmina with row of spots basally; hind femora usually with conspicuous oblique crossbands on upper half of outer face .... ***stonei*** **Rehn** (p. 417)

Hind tibiae blue or purple; pronotal stripes either conspicuous or absent; tegmina usually without spots, or spotted only on central line toward base; outer face of hind femora without crossbands below upper flange, but usually with longitudinal discoloration .... *foedus foedus* **Scudder** (p. 418)

38(32). Cercus strongly tapered in apical half (Figs. 493, 512, 513) 39

Cercus not tapered in apical half, often broad, earlike, or quadrate 41

39(38). Subgenital plate bulbous, with apex concave (Fig. 456) .... *femurrubrum femurrubrum* **(De Geer)** (p. 420)

Subgenital plate normal, with apex rounded, pointed, cleft, or trilobate .. 40

40(39). Hind femur with crossbands above. Tegmina not finely speckled. Color blended black and brown. Aedeagus as in Figs. 556, 557 .... *flavidus* **Scudder** (p. 421)

Hind femur without crossbands. Tegmina finely speckled. Color gray to bluish gray. Aedeagus as in Figs. 558, 559 .... *bowditchi canus* **Hebard** (p. 423)

41(38). Cercus earlike or quadrate (Figs. 503, 504) 42

Cercus broad, not earlike (Figs. 494, 498, 499, 502) 43

42(41). Cercus large, earlike (Fig. 503), converging above supra-anal plate. Tegmina extending to or slightly beyond apices of hind femora .... *occidentalis occidentalis* **(Thomas)** (p. 424)

Cercus large, broad, with apex quadrate (Fig. 504). Tegmina variable, slightly surpassing apex of abdomen to much exceeding apices of hind femora .... *rugglesi* **Gurney** (p. 425)

43(41). Cercus large, rectangular with rounded apex bent dorsally (Fig. 502). Apex of subgenital plate truncate or rounded-acute .... *kennicottii kennicottii* **Scudder** (p. 427)

Cercus smaller, flat or nearly so (Figs. 494, 498, 499). Apex of subgenital plate notched or trilobate 44

44(43). Apex of subgenital plate trilobate, with outer lobes projecting and obliquely truncate (Fig. 530) .... *gordonae* **Vickery** (p. 427)

Apex of subgenital plate notched (Figs. 534, 535) 45

45(44). Subgenital plate long, produced upward posteriorly (Fig. 535). Furculae large, long, nearly parallel (Fig. 463) .... *bruneri* **Scudder** (p. 428)

Subgenital plate shorter, broader than long (Fig. 534). Furculae smaller, distinctly diverging 46

46(45). Tegmina relative to hind femora short (Fig. 430). Hind tibia red, pink, buff, or blue. Cercus nearly equally broad throughout, rounded at end, with length not greater than twice middle breadth (Fig. 498) .... *sanguinipes sanguinipes* **(Fabricius)** (p. 430)

Tegmina relative to hind femora longer (Fig. 610). Hind tibia dark red. Cercus shorter, broader, more pointed ... *spretus* **(Walsh)** (p. 432)

47(23). Dorsal angle of dorsal ovipositor valve right-angled or nearly so (90–105°) (Figs. 573, 589–592) 48

Dorsal angle or curve of dorsal ovipositor valve greater than 105° (Figs. 560–572, 574–588, 595) 56

48(47). Notches on either side of eighth sternum conspicuous, deep, with angle about 90° (Figs. 589–591) 49

Notch of eighth sternum shallower, with angle wider (125–150°) ... 53

49(48). Hind tibiae red. Forest species 50

Hind tibiae blue, purple, or pink. Grassland and parkland species .. 51

50(49). Pronotal stripes distinct; hind femur yellow below, banded above, these extending below upper ridge on both inner and outer faces. Found in

forested (Jack pine) areas of Saskatchewan and Alberta ....
.... *packardii brooksi* **Vickery** (p. 417)

Pronotal stripes indistinct or absent; hind femur conspicuously banded on upper half. Found in forested areas of Manitoba to Québec ......
.... *stonei* **Rehn** (p. 417)

51(49). Hind femur reddish or orange beneath; pronotal stripes indistinct or absent
.... *foedus foedus* **Scudder** (p. 418)

Hind femur yellow beneath; pronotal stripes distinct 52

52(51). Hind femur banded above, but not on outer face; general coloration bluish gray, grading to yellowish in southern areas ....
.... *packardii packardii* **Scudder** (p. 415)

Hind femur banded above, but not below upper flanges, usually with longitudinal discoloration on outer face; general coloration brownish yellow or greenish gray .... *foedus foedus* **Scudder** (p. 418)

53(48). Cercus convex on both edges (cf. Fig. 579*a*) ....
.... *confusus* **Scudder** (p. 403)

Cercus not convex on both sides 54

54(53). Cercus slightly concave above, convex below (cf. Fig. 576*a*) ....
.... *angustipennis angustipennis* **(Dodge)** (p. 412)

Cercus with both edges conspicuously concave (cf. Fig. 574*a*), sometimes rather flask-shaped 55

55(54). Hind femur with faint crossbands above; tegmina without spots or fine speckling; general coloration blended brown and black ....
.... *flavidus* **Scudder** (p. 421)

Hind femur without crossbands; tegmina finely speckled ....
.... *bowditchi canus* **Hebard** (p. 423)

56(47). Hind femur reddish below 57

Hind femur yellow or pink below 63

57(56). Hind femur banded 59

Hind femur not banded on outer face 58

58(57). Size small (males 16–20, females 20–24 mm) ....
.... *borealis borealis* **(Fieber)** (p. 397)

Size larger (males 19–21, females 20–30 mm) ....
.... *borealis palaceus* **Fulton** (p. 398)

59(57). Banding on hind femur confined to upper half; hind tibia blue 60

Banding on hind femur complete; hind tibia usually red, sometimes grayish 61

60(59). Hind femur bright red below and on inner face ....
.... *occidentalis occidentalis* **(Thomas)** (p. 424)

Hind femur no more than dull pinkish below and on inner face ......
.... *rugglesi* **Gurney** (p. 425)

61(59). Lower ovipositor valve with lateral tooth (Fig. 586). Tegmina nearly uniform in color. Woodland areas but not arboreal ....
.... *fasciatus* **(F. Walker)** (p. 399)

Lower ovipositor valve without lateral tooth (Fig. 568). Tegmina with round fuscous spots. Associated with pine, arboreal (*punctulatus*, *sensu lato*)
.... 62

62(61). Eyes prominent in anterior view. Pronotum flaring outward anteroventrally in dorsal aspect. Found in Michigan,[6] Minnesota, and Wisconsin .
.... *punctulatus griseus* **(Thomas)** (p. 409)

Eyes less prominent in anterior view. Pronotum less flaring outward anteroventrally in dorsal aspect. Found in New England, New York, western Québec, Ontario, and Michigan[6] ....

. . . . *punctulatus punctulatus* **(Scudder)** (p. 410)

63(56). Hind femur flattened below near base (Fig. 438) . . . . *gladstoni* **Scudder** (p. 411)

Hind femur not flattened below 64

64(63). Large species (34-44 mm). Hind femur usually with distinct "herringbone" pattern (Fig. 427) . . . *differentialis differentialis* **(Thomas)** (p. 407)

Smaller species (less than 34 mm). Hind femur not as above 65

65(64). Hind femur with bands at least on upper half, or with longitudinal discoloration 67

Hind femur not banded on outer face 66

66(65). Cercus concave on both edges. Distribution general . . . . *femurrubrum femurrubrum* **(De Geer)** (p. 420)

Cercus convex on both edges. Found in Alaska . . . . *gordonae* **Vickery** (p. 427)

67(65). Cercus convex on both edges 70

Cercus concave or straight on both edges 68

68(67). Small species (not more than 25 mm). Lateral black markings at least on first two abdominal segments 69

Larger species (up to 29 mm). Dull-colored, without contrasting markings on abdominal segments . . . . *cinereus cinereus* **Scudder** (p. 414)

69(68). Small species (16-19 mm). Abdominal terga with prominent black markings laterally. Found in grasslands . . . . *infantilis* **Scudder** (p. 405)

Larger (18-25 mm). Abdomen with black markings on first two or three terga only. Confined to foothills . . . . *alpinus* **Scudder** (p. 406)

70(67). Tegmina not exceeding apex of hind femur 71

Tegmina exceeding apex of hind femur 72

71(70). External face of hind femur banded on upper half only; hind tibia blue or yellow . . . . *kennicottii kennicottii* **Scudder** (p. 427)

External face of hind femur with longitudinal blackish stripe; hind tibia red . . . . *keeleri luridus* **(Dodge)** (p. 366)

72(70). Hind femur entirely yellowish below. Upper edge of cercus straight . . . . *bruneri* **Scudder** (p. 428)

Hind femur with pink or reddish longitudinal stripe on outer surface of lower flange. Both edges of cercus convex 73

73(72). Tegmina tapered toward apex and with few spots, or narrow throughout their length and speckled. Species widespread . . . . *sanguinipes sanguinipes* **(Fabricius)** (p. 430)

Tegmina broad to apex even when folded, conspicuously speckled. Species western (apparently extinct) . . . . *spretus* **(Walsh)** (p. 432)

---

[1] Intermediates are found in Michigan and Wisconsin; the only specimens found to date in Canada (Turkey Point, Ont.) are also intermediates.

[2] There is no reliable character for separating females of the subspecies of *M. oregonensis* and *M. indigens*. Females should be associated with males from the same locality.

[3] This species is now placed in *Bohemanella* Ramme.

[4] Michigan specimens are intermediate between the two subspecies.

[5] There is, unfortunately, no easy method of distinguishing between the species involved here; examining the concealed genitalia is necessary.

[6] Michigan specimens are intermediate between the two subspecies.

Clé des espèces et des sous-espèces de *Melanoplus*

1. Tegmina quelque peu réduits, ne rejoignant pas l'extrémité de l'abdomen (fig. 427) ou écourtés 2
Tegmina dépassant l'extrémité de l'abdomen (fig. 428 à 430) 23

2(1). Tibias postérieurs verts. Espèces de l'est 3
Tibias postérieurs non verts, ou dans le cas contraire, espèces ne vivant pas dans l'est 5

3(2). Fémurs postérieurs sans marques foncées; tibias postérieurs verts, mais plus foncés à la base. Cerques des mâles incurvés vers le bas à l'apex .. .. *gracilis* **(Bruner)** (p. 378)
Fémurs postérieurs fortement marqués de brun; tibias postérieurs non foncés à la base. Cerques des mâles allant s'amenuisant pour donner un apex arrondi (fig. 478 et 479) (***viridipes***, *sensu lato*) 4

4(2). Cerques des mâles triangulaires, allant s'amenuisant pour donner un apex arrondi (fig. 479). Espèce vivant au Vermont (et probablement dans le sud du Québec), au Massachusetts et en Pennsylvanie[1] .... *viridipes eurycercus* **Hebard** (p. 379)
Cerques des mâles plus longs, aux côtés formant un angle moins prononcé, légèrement convexes au-dessus vers l'apex et légèrement concaves en-dessous (fig. 478). Espèce vivant dans le sud de l'Ontario[1], au Michigan, en Illinois, en Indiana et plus au sud[2] .... *viridipes viridipes* **Scudder** (p. 380)

5(2). Tibias postérieurs roses 6
Tibias postérieurs rouges, jaune clair, bleus ou légèrement verdâtres 7

6(5). Tegmina de forme ovale, ne dépassant pas le deuxième segment abdominal. Cerques des mâles triangulaires et courts (fig. 479) .... *scudderi scudderi* **(Uhler)** (p. 380)
Tegmina plus longs, rejoignant au moins le quatrième segment abdominal. Cerques des mâles spatulés (fig. 482) .... *walshii* **Scudder** (p. 381)

7(5). Tibias postérieurs jaune clair à bleus .... 8[2]
Tibias postérieurs rouges 11

8(7). Cerques des mâles non élargis distalement, allant s'amenuisant pour donner un apex arrondi, à peine spatulé ou pas du tout (fig. 495 et 505) (***indigens***, *sensu lato*) 9
Cerques des mâles quelque peu élargis distalement, à partie distale distincte, mais pas très spatulée (fig. 480 et 481) (***oregonensis***, *sensu lato*) 10

9(8). Surface ventral des fémurs postérieurs variant d'orange brillant à rouge foncé. Cerques des mâles courts et non spatulés (cf. fig. 495, mais un peu plus long que dans la figure). Espèce vivant dans le nord du Montana (Missoula) .... *indigens missoulae* **Hebard** (p. 383)
Surface ventral des fémurs postérieurs d'un jaune clair mat et teinté d'orange. Cerques des mâles plus longs et légèrement spatulés (fig. 505). Espèce vivant à la ligne de partage des eaux des Rocheuses et dans le sud du Montana .... *indigens indigens* **Scudder** (p. 383)

10(8). Plaque supra-anale des mâles clairement plus longue que large et à apex très arrondi; furcula grande (fig. 444). Cerques des mâles longs par rapport à leur largeur (fig. 480). Espèce vivant dans le sud du Montana, le sud de l'Idaho et l'extrême sud de l'Alberta .... *oregonensis oregonensis* **(Thomas)** (p. 384)
Plaque supra-anale des mâles de forme triangulaire et à peine plus longue que large; furcula plus petite (fig. 445). Cerques des mâles plus courts (fig. 481). Espèce vivant au nord-ouest du Montana, au nord de l'Idaho,

au sud-ouest de l'Alberta et au sud-est de la Colombie-Britannique .... ***oregonensis triangularis*** **Hebard** (p. 385)

11(7). Tegmina courts, ne dépassant pas le quatrième segment abdominal (rarement plus longs chez le *M. dawsoni*, mais dans ce cas, tergites de l'abdomen marqués de bandes) 12

Tegmina plus longs, mais ne dépassant pas l'extrémité de l'abdomen (pas de bandes sur les tergites de l'abdomen sauf chez le *M. dawsoni*) 16

12(11). Abdomen marqué de bandes transversales, chaque segment à portion antérieure noire et à portion postérieure jaune. Tubercule du prosternum émoussé. Cerques des mâles incurvés et à apex très arrondi (fig. 490) .... ***dawsoni*** **(Scudder)** (p. 385)

Abdomen non rayé transversalement. Tubercule du prosternum pointu ... 13

13(12). Espèces vivant dans l'est (du Maine à l'est du Manitoba) 14

Espèces vivant dans l'ouest (du sud-ouest de l'Alberta et du Montana à l'État de Washington et à la Colombie-Britannique) (groupe ***montanus***) ... 15

14(13). Crêtes latérales du pronotum presque invisibles. Tegmina (surtout chez les femelles) à extrémités larges et arrondies également. Cerques des mâles longs, larges à la base et allant s'amenuisant du milieu à l'extrémité (fig. 482). Furcula des mâles modérément longue et subcylindrique (fig. 446) .... ***mancus*** **(Smith)** (p. 386)

Crêtes latérales du pronotum distinctes. Tegmina (surtout chez les femelles) à extrémités étroites et arrondies ou subangulaires. Cerques des mâles courts, à côtés très convergents, concaves sur le dessus, au niveau du quart apical (fig. 483). Furcula des mâles courte, plate, oblongue (fig. 447) .... ***islandicus*** **Blatchley** (p. 388)

15(13). Cerques des mâles quelque peu subspatulés (fig. 484). Valves de l'édéage non recourbées vers l'extérieur au tiers apical .... ***montanus*** **(Thomas)** (p. 389)

Cerques des mâles non subspatulés et à apex tronqué (fig. 485). Valves de l'édéage soudainement divergentes au tiers dorsal et à apex plus ou moins aigu .... ***washingtonius*** **(Bruner)** (p. 390)

16(11). Tegmina rejoignant le troisième ou le quatrième segment abdominal et allant s'amenuisant pour se terminer par des apex arrondis. .... ***hodgei*** **(Thomas)** (p. 391) ***huroni*** **Blatchley** (p. 392)

Tegmina plus longs, dépassant le sixième segment abdominal 17

17(16). Segments abdominaux rayés .... ***dawsoni*** **(Scudder)** (p. 385)

Segments abdominaux non rayés 18

18(17). Grosse espèce d'un vert olive foncé marqué de fines lignes jaunes des yeux à l'extrémité des tegmina; fémurs postérieurs marqués d'étroites bandes noires longitudinales sur la face externe et sur le côté interne du rebord supérieur .... ***bivittatus*** **(Say)** (p. 393)

Bandes dorsales absentes ou limitées à la tête et au thorax; fémurs postérieurs habituellement rayés transversalement au moins sur le dessus 19

19(18). Fémurs postérieurs rayés sur le côté extérieur 21

Fémurs postérieurs non rayés sur le côté extérieur 20

20(19). Espèce de grande taille, vivant aux Îles-de-la-Madeleine (Québec) .... ***madeleineae*** **Vickery & Kevan** (p. 395)

Espèce plus petite et très répandue .... ***borealis borealis*** **(Fieber)** (p. 397) .... ***borealis palaceus*** **Fulton** (p. 398)

21(19). Fémurs postérieurs foncés et marqués de deux rayures obliques pâles (fig. 440). Espèce vivant au mont Albert, en Gaspésie (Québec) ... . ***gaspesiensis*** **Vickery** (p. 399)

Fémurs postérieurs clairement marqués de bandes transversales. Espèce très répandue 22

22(21). Cerques des mâles aux côtés presque parallèles, incurvés à l'apex (fig. 505). Fémurs postérieurs principalement foncés et marqués de bandes transversales pâles complètes. Espèce robuste et très répandue .... *fasciatus* **(F. Walker)** (p. 399)

Cerques des mâles larges à la base et allant s'amenuisant pour donner un apex émoussé (fig. 506). Fémurs postérieurs principalement pâles et à bandes transversales plus foncées. Espèce du Grand Nord (nord de l'Alaska, nord des Territoires du Yukon et nord du district de Mackenzie) .... *frigidus frigidus* **(Boheman)**[3] (p. 401)

23(1). Mâles (sans ovipositeur) 24

Femelles (avec ovipositeur) 47

24(23). Cerques à apex divisé 25

Cerques à apex non divisé 28

25(24). Bras dorsal des cerques large, grand, à apex arrondi, plus long que le bras ventral 26

Bras dorsal des cerques en forme de pouce et beaucoup plus court que le bras ventral 27

26(25). Bras ventral des cerques en forme de pouce (fig. 507). Fémurs postérieurs à bandes foncées longitudinales; tibias postérieurs rouge foncé à bandes ou à taches noires à la base .... *keeleri luridus* **(Dodge)** (p. 402)

Bras ventral des cerques de forme triangulaire (fig. 492). Fémurs postérieurs sans bandes noires; tibias postérieurs rouge à bleu jaunâtre ou verdâtre .... *confusus* **Scudder** (p. 403)

27(25). Bras ventral des cerques presque rectiligne (fig. 500). Apex de la plaque sous-génitale encoché (fig. 536) .... *infantilis* **Scudder** (p. 405)

Bras ventral des cerques fortement incurvé (fig. 501). Apex de la plaque sous-génitale non encoché, mais tronqué et concave postérieurement (fig. 537) .... *alpinus* **Scudder** (p. 406)

28(24). Cerques spatulés et plus larges à l'extrémité qu'à la base 29

Cerques aussi larges ou plus larges à la base qu'à l'extrémité 31

29(28). Espèce de grande taille (mâles de 28 à 34 mm, femelles de 34 à 44 mm). Cerques munis d'une protubérance ventrale arrondie (fig. 489) ... . *differentialis* **(Thomas)** (p. 407)

Espèce de taille moyenne (mâles de 18 à 25 mm, femelles de 26 à 30 mm). Cerques à apex très développé (fig. 487). Tibias postérieurs de couleur mate, rougeâtre à grise, armés d'épines noires (***punctulatus***, *sensu lato*) 30

30(29). Yeux proéminents (vus du devant). Pronotum à bords divergents antéroventralement (vu du dessus). Cerques des mâles à lobe dorsal presque aussi profond que le reste du cerque et à bord dorsal décrivant une large courbe uniforme. Espèce vivant au Michigan, au Minnesota et au Wisconsin .... *punctulatus griseus* **(Thomas)** (p. 409)

Yeux moins proéminents (vus du devant). Pronotum à bords moins divergents antéroventralement (vu du dessus). Cerques des mâles à lobe dorsal clairement moins profond que le reste du cerque et à bord dorsal formant un angle aigu (fig. 487). Espèce vivant en Nouvelle-Angleterre, dans l'État de New York, au Québec, en Ontario et au Michigan[4] .... *punctulatus punctulatus* **(Scudder)** (p. 410)

31(28). Cerques larges, à bords presque parallèles (fig. 491). Fémurs postérieurs à rebord inférieur absent du tiers proximal (fig. 438). Tegmina clairement mouchetés .... *gladstoni* **Scudder** (p. 411)

Cerques allant s'amenuisant vers l'extrémité, subquadrangulaires ou spatulés, mais clairement plus larges à la base. Fémurs postérieurs à rebord inférieur complet. Tegmina pratiquement non mouchetés 32

32(31). Cerques symétriques, spatulés ou légèrement lobés en-dessous, plus étroits au milieu 33

Cerques asymétriques, plus larges à la base, allant s'amenuisant vers l'extrémité, à bord dorsal décrivant une courbe, en forme de lobe ou subquadrangulaires 38

33(32). Apex de la plaque sous-génitale encoché .... *angustipennis angustipennis* **(Dodge)** (p. 412)

Apex de la plaque sous-génitale décrivant une large courbe ou un angle aigu 34

34(33). Apex de la plaque sous-génitale aigu (fig. 551). Cerques petits (fig. 514) .... *cinereus cinereus* **Scudder** (p. 414)

Apex de la plaque sous-génitale décrivant une large courbe (fig. 456 et 457). Cerques plus grands (fig. 510) 35

35(32). Lobes antérieur et postérieur de l'édéage à extrémités situées sur le même plan (fig. 554 et 555)[5] (*packardii*, *sensu lato*) 36

Édéage à lobe antérieur beaucoup plus court que le lobe postérieur et ne rejoignant pas ce dernier (fig. 552 et 553)[5] 37

36(35). Longueur de 27 à 32 mm. Fémurs postérieurs marqués de bandes sur le dessus du rebord supérieur et sur la face interne; tibias postérieurs bleus, pourpres ou roses; coloration générale gris bleuâtre. Sous-espèce vivant dans les prairies et les prairies-parcs .... *packardii packardii* **Scudder** (p. 415)

Longueur de 23 à 31 mm. Fémurs postérieurs marqués de bandes s'étendant de part et d'autre du rebord supérieur; tibias postérieurs rouge foncé; coloration générale noir brunâtre foncé. Sous-espèce vivant dans les peuplements de pin gris (Saskatchewan, Alberta et Colombie-Britannique) .... *packardii brooksi* **Vickery** (p. 417)

37(35). Tibias postérieurs rouge foncé; bandes du pronotum indistinctes; tegmina marqués d'une rangée de taches basales; fémurs postérieurs habituellement marqués de bandes obliques claires sur la moitié supérieure de la face externe .... *stonei* **Rehn** (p. 417)

Tibias postérieurs bleus ou pourpres; bandes du pronotum très claires ou absentes; tegmina ne portant habituellement pas de tache ou n'en portant que sur la ligne médiane, vers la base; face externe des fémurs postérieurs sans bande transversale sous le rebord postérieur, mais marquée en général d'une décoloration longitudinale .... *foedus foedus* **Scudder** (p. 418)

38(32). Cerques à bords très convergents dans la moitié apicale (fig. 493, 512 et 513) 39

Cerques à bords non convergents dans la moitié apicale, souvent larges, en forme d'oreille ou quadrangulaires 41

39(38). Plaque sous-génitale bulbeuse, à apex concave (fig. 456) .... *femurrubrum femurrubrum* **(De Geer)** (p. 420)

Plaque sous-génitale normale, à apex arrondi, pointu, bilobé ou trilobé ... 40

40(39). Fémurs postérieurs marqués de bandes transversales sur le dessus. Tegmina non finement mouchetés. Coloration faite d'un mélange de noir et de brun. Édéage comme dans les figures 556 et 557 .... *flavidus* **Scudder** (p. 421)

Fémurs postérieurs dépourvus de bandes transversales. Tegmina finement mouchetés. Coloration grise ou grise bleuâtre. Édéage comme dans les figures 558 et 559 .... *bowditchi canus* **Hebard** (p. 423)

41(38). Cerques en forme d'oreille ou quadrangulaires (fig. 503 et 504) 42
Cerques larges, non en forme d'oreille (fig. 494, 498, 499 et 502) 43

42(41). Cerques grands, en forme d'oreille (fig. 503) et à côtés convergents au-dessus de la plaque supra-anale. Tegmina rejoignant ou dépassant légèrement les apex des fémurs postérieurs .... *occidentalis occidentalis* **(Thomas)** (p. 424)
Cerques grands, larges, à apex quadrangulaire (fig. 504). Tegmina de longueur variable, pouvant dépasser légèrement l'extrémité de l'abdomen et aller jusqu'à dépasser de beaucoup l'extrémité des fémurs .... *rugglesi* **Gurney** (p. 425)

43(41). Cerques grands, rectangulaires, ayant un apex arrondi et incurvé vers le haut (fig. 502). Apex de la plaque sous-génitale tronqué ou à angle aigu et arrondi .... *kennicottii kennicottii* **Scudder** (p. 427)
Cerques plus petits, aplatis ou presque (fig. 494, 498 et 499). Apex de la plaque sous-génitale bilobé ou trilobé 44

44(43). Apex de la plaque sous-génitale trilobé, les lobes externes étant saillants et tronqués obliquement (fig. 530) .... *gordonae* **Vickery** (p. 427)
Apex de la plaque sous-génitale encoché (fig. 534 et 535) 45

45(44). Plaque sous-génitale longue, incurvée postérieurement vers le haut (fig. 535). Furcula grande, longue et à bords presque parallèles (fig. 463) .... *bruneri* **Scudder** (p. 428)
Plaque sous-génitale plus courte, plus large que longue (fig. 534). Furcula plus petite et à côtés divergents 46

46(45). Tegmina courts comparativement aux fémurs postérieurs (fig. 430). Tibias postérieurs rouges, roses, jaune clair ou bleus. Cerques de largeur presque égale sur toute leur longueur, arrondis à l'extrémité et de longueur ne dépassant pas le double de leur diamètre (fig. 498) ... . *sanguinipes sanguinipes* **(Fabricius)** (p. 430)
Tegmina plus longs comparativement aux fémurs postérieurs (fig. 610). Tibias postérieurs rouge foncé. Cerques plus courts, plus larges et moins pointus (apparemment éteints) .... *spretus* **(Walsh)** (p. 432)

47(23). Angle dorsal de la valve dorsale de l'ovipositeur à angle droit ou presque (de 90 à 105°) (fig. 573, 589 à 592) 48
Angle dorsal ou courbe de la valve dorsale de l'ovipositeur plus grand que 105° (fig. 560 à 572, 574 à 588 et 595) 56

48(47). Encoches, situées de part et d'autre du huitième sternite, très visibles, profondes et formant un angle d'environ 90° (fig. 589 et 591) 49
Encoches du huitième sternite moins profondes, formant un angle plus grand (de 125 à 150°) 53

49(48). Tibias postérieurs rouges. Espèces vivant en forêt 50
Tibias postérieurs bleus, pourpres ou roses. Espèces vivant dans les prairies et les prairies-parcs 51

50(49). Pronotum marqué de bandes distinctes; fémurs postérieurs jaunes en dessous, rayés au-dessus, ces rayures se prolongeant sous le rebord supérieur du côté interne et du côté externe. Insecte vivant dans les forêts (pin gris) de la Saskatchewan et de l'Alberta .... *packardii brooksi* **Vickery** (p. 417)
Bandes du pronotum peu visibles ou absentes; fémurs postérieurs clairement rayés sur la moitié supérieure. Insecte vivant dans les forêts, du Manitoba au Québec .... *stonei* **Rehn** (p. 417)

51(49). Fémurs postérieurs rougeâtres ou orangés en dessous; bandes du pronotum peu visibles ou absentes .... *foedus foedus* **Scudder** (p. 418)
Fémurs postérieurs jaunes en dessous; bandes du pronotum visibles 52

52(51). Fémurs postérieurs rayés sur le dessus, mais non du côté extérieur; coloration générale d'un gris bleuâtre, tournant au jaunâtre dans les régions du sud .... *packardii packardii* **Scudder** (p. 415)
Fémurs postérieurs rayés sur le dessus, mais non en dessous du rebord supérieur, habituellement marqués d'une décoloration longitudinale du côté extérieur; coloration générale d'un jaune brunâtre ou d'un gris verdâtre .... *foedus foedus* **Scudder** (p. 418)
53(48). Cerques convexes des deux côtés (fig. 579*a*) .... *confusus* **Scudder** (p. 403)
Cerques non convexes des deux côtés 54
54(53). Cerques légèrement concaves au-dessus, convexes en dessous (fig. 576*a*) .... *angustipennis angustipennis* **(Dodge)** (p. 412)
Cerques aux deux côtés clairement concaves (fig. 574*a*), parfois en forme de flacon 55
55(54). Fémurs postérieurs marqués au-dessus de bandes transversales floues; tegmina sans tache ni moucheture fine; coloration générale d'un brun mêlé de noir .... *flavidus* **Scudder** (p. 421)
Fémurs postérieurs sans bandes transversales; tegmina finement mouchetés .... *bowditchi canus* **Hebard** (p. 423)
56(47). Fémurs postérieurs à dessous rougeâtre 57
Fémurs postérieurs à dessous jaune ou rose 63
57(56). Fémurs postérieurs rayés 59
Fémurs postérieurs non rayés sur la face externe 58
58(57). Insecte de petite taille (mâles de 16 à 20 mm, femelles de 20 à 24 mm) .... *borealis borealis* **(Fieber)** (p. 397)
Insecte de plus grande taille (mâles de 19 à 21 mm, femelles de 20 à 30 mm) .... *borealis palaceus* **Fulton** (p. 398)
59(57). Rayures des fémurs postérieurs limitées à la portion supérieure; tibias postérieurs bleus 60
Fémurs postérieurs complètement rayés; tibias postérieurs habituellement rouges, parfois grisâtres 61
60(59). Fémurs postérieurs rouge clair en dessous et sur la face interne .... *occidentalis occidentalis* **(Thomas)** (p. 424)
Fémur postérieurs pas plus foncé que rosâtre mat en dessous et sur la face interne .... *rugglesi* **Gurney** (p. 425)
61(59). Valve inférieure de l'ovipositeur armée d'une dent latérale (fig. 586). Tegmina de couleur presque uniforme. Insecte vivant dans les régions boisées, mais non arboricole .... *fasciatus* **(F. Walker)** (p. 399)
Valve inférieure de l'ovipositeur dépourvue de dents latérales (fig. 568). Tegmina marqués de taches gris brunâtre. Insecte arboricole, vivant en association avec le pin .... **(*punctulatus*, *sensu lato*)** 62
62(61). Yeux proéminents vus du devant. Pronotum à côtés divergents antéroventralement, en vue dorsale. Espèce vivant au Michigan[6], au Minnesota et au Wisconsin .... *punctulatus griseus* **(Thomas)** (p. 409)
Yeux moins proéminents vus du devant. Côtés du pronotum moins divergents antéroventralement, en vue dorsale. Espèce vivant en Nouvelle-Angleterre, dans l'État de New York, dans l'ouest du Québec, en Ontario et au Michigan[6] .... *punctulatus punctulatus* **(Scudder)** (p. 410)
63(56). Fémurs postérieurs aplatis en dessous, près de la base (fig. 438) .... *gladstoni* **Scudder** (p. 411)
Fémurs postérieurs non aplatis en dessous 64
64(63). Espèce de grande taille (de 34 à 44 mm). Fémurs postérieurs habituellement ornés d'un dessin rappelant des arêtes de poisson (fig. 427) .... *differentialis differentialis* **(Thomas)** (p. 407)

Espèces de plus petite taille (moins de 34 mm). Fémurs postérieurs différents de la description donnée ci-dessus 65

65(64). Fémurs postérieurs marqués de bandes au moins sur la portion supérieure ou marqués d'une décoloration longitudinale 67

Fémurs postérieurs non rayés sur leur surface extérieure 66

66(65). Cerques concaves des deux côtés. Espèce très répandue .... *femurrubrum fermurrubrum* **(De Geer)** (p. 420)

Cerques convexes des deux côtés. Insecte vivant en Alaska .... *gordonae* **Vickery** (p. 427)

67(65). Cerques convexes des deux côtés 70

Cerques concaves ou rectilignes des deux côtés 68

68(67). Espèces de petite taille (pas plus de 25 mm). Marques latérales noires au moins sur les deux premiers segments de l'abdomen 69

Espèce de taille plus grande (jusqu'à 29 mm). De coloration mate, sans marque contrastante sur les segments de l'abdomen .... *cinereus cinereus* **Scudder** (p. 414)

69(68). Espèce de petite taille (de 16 à 19 mm). Tergites de l'abdomen marqués de taches noires évidentes sur les côtés. Insecte vivant dans les Prairies .... *infantilis* **Scudder** (p. 405)

Espèce plus grande (de 18 à 25 mm). Abdomen ne portant des marques noires que sur les deux ou trois premiers tergites. Espèce confinée aux contreforts des Rocheuses .... *alpinus* **Scudder** (p. 406)

70(67). Tegmina ne dépassant pas l'apex des fémurs postérieurs 71

Tegmina dépassant l'apex des fémurs postérieurs 72

71(70). Face externe des fémurs postérieurs rayée sur la portion supérieure seulement; tibias postérieurs bleus ou jaunes .... *kennicottii kennicottii* **Scudder** (p. 427)

Face externe des fémurs postérieurs marquée d'une bande noirâtre longitudinale; tibias postérieurs rouges .... *keeleri luridus* **(Dodge)** (p. 366)

72(70). Fémurs postérieurs entièrement jaunâtres en dessous. Bord supérieur des cerques rectiligne .... *bruneri* **Scudder** (p. 428)

Fémurs postérieurs à bande longitudinale rose ou rougeâtre sur la surface extérieure du rebord inférieur. Les deux côtés des cerques convexes ...73

73(72). Tegmina allant s'amenuisant vers l'apex et marqués de quelques taches, ou étroits sur toute leur longueur et mouchetés. Espèce très répandue .... *sanguinipes sanguinipes* **(Fabricius)** (p. 430)

Tegmina larges à l'apex, même en position repliée, clairement mouchetés. Espèce vivant dans l'ouest (apparemment éteinte) .... *spretus* **(Walsh)** (p. 432)

---

[1] On trouve des formes intermédiaires au Michigan et au Wisconsin; les seuls spécimens trouvés jusqu'à maintenant au Canada (Turkey Point, Ont.) appartiennent à cette catégorie.

[2] Il n'existe pas de caractère fiable pour distinguer les femelles des sous-espèces de *M. oregonensis* et *M. indigens*. L'identification doit porter sur les mâles recueillis au même endroit.

[3] Cette espèce est maintenant dans le genre *Bohemanella* Ramme.

[4] Les spécimens provenant du Michigan représentent une forme intermédiaire entre les deux sous-espèces.

[5] Il n'y a malheureusement pas de moyen facile de distinguer l'une de l'autre les espèces dont il est question ici. Il faut en effet pour cela examiner les génitalia dissimulés.

[6] Les spécimens provenant du Michigan constituent une forme intermédiaire entre les deux sous-espèces.

Melanoplus gracilis (Bruner)

Map 129

Pezotettix gracilis Bruner, 1876:124.
Pezotettix minutipennis C. Thomas, 1876*b*:66.
Melanoplus gracilis; Scudder 1897*d*:327; Helfer 1963:247.

Diagnosis. Tegmina narrow, oblong, widely separated. Hind femur bright green; hind tibia green with black spines.

Description. Body slender, small (length, male 14–16, female 18.5–20 mm; pronotal length, male 4.5, female 5.0–5.5 mm), subcylindrical, compressed. Frontal costa prominent, slightly broader than interocular space, sulcate beneath ocellus. Pronotum nearly cylindrical, slightly expanded on metazona, with median carina low, distinct; prozona twice as long as metazona; posterior margin of metazona with shallow median notch. Tegmina narrowly oblong, widely separated, with apices pointed in male but narrowly rounded in female. Hind femur slender, long, exceeding abdomen in both sexes. Male cercus long, narrow, with middle third half as wide as base, and with apical third expanded and feebly curved downward. Furculae of male minute, oblong, lobelike. General color dull ashy brown dorsally, greenish yellow beneath; laterally with shining blackish bar from eye to metazona; hind femur bright green, with lower margin yellowish and with geniculae black; hind tibia green with black spines.

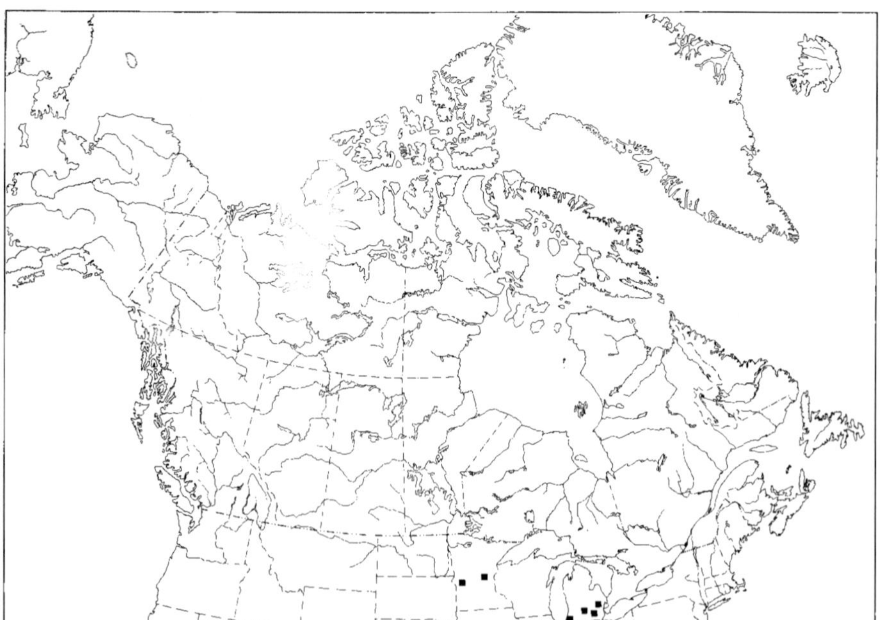

Map 129. Collection localities for *Melanoplus gracilis*.

Range. Minnesota to Michigan, south to Mississippi and Tennessee.

Behavior and habitats. *M. gracilis* is found in old fields, grasses along roadsides, blue-grass pastures, tall weeds and grasses along edges of marshes, and open wooded areas. Blatchley (1920) reported that it was found frequently on ironweed (*Vernonia* spp.) and also occurred on jewelweed (*Impatiens* spp.) and giant ragweed *Ambrosia trifida*. It does not seem to feed upon crop plants to any extent.

Life history. Univoltine in the north, overwintering in the egg stage. In the southern part of the range, there may be two generations each year. Adults occur in Michigan from 13 July to 9 September (Cantrall 1968), but they are present from early June to November in the south.

Melanoplus viridipes eurycercus Hebard

Figs. 479, 516, 560; Map 130

Melanoplus viridipes eurycercus Hebard, 1920*b*:392.

Diagnosis. Male cercus short, triangular. Hind femur distinctly banded.

Description. Male cercus heavy, short (about twice as long as broad), triangular, strongly tapered throughout from base to bluntly rounded apex (Fig. 479). Bands on hind femora distinct and often fused ventrally.

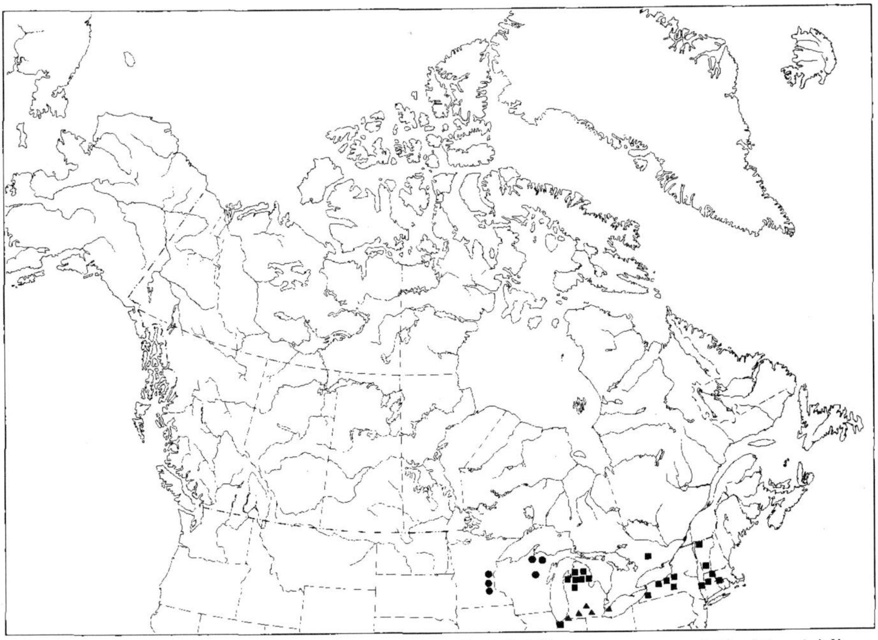

Map 130. Collection localities for *Melanoplus viridipes eurycercus* (■), *M. viridipes viridipes* (●), and intermediates (▲).

Range. Michigan to Vermont, south to North Carolina.

Behavior, habitats, and life history. As for *M. viridipes viridipes.*

Melanoplus viridipes viridipes Scudder

Figs. 443, 478, 515; Map 130

Melanoplus viridipes Scudder, 1897*d*:255, pl. XVII, fig. 4.
Melanoplus juvencus Scudder, 1897*d*:266.
Melanoplus viridipes viridipes; Hebard 1920*b*:393.

Diagnosis. Tegmina padlike. Fore and middle coxae and all tibiae greenish. Male cercus long.

Description. Body small (length, male 17.5–18.5, female 21.5–25.5 mm; pronotal length, male 3.8–4.2, female 4.8–5.2 mm). Male cercus elongate, 2.3–3 times as long as broad, conspicuously tapering, somewhat concave in apical half (Fig. 478). Male subgenital plate (Fig. 515) with small median tubercle. Hind femur pale, with narrow fuscous bands, these never fused ventrally. Tegmina reduced to small pads, in females about as long as pronotum, in males slightly longer. Color generally brownish above, yellowish beneath; black stripe extending from eye across pronotum and abdomen; fore and middle femora and all tibiae greenish; hind femur conspicuously fuscous.

Range. Minnesota, Wisconsin, Illinois, Iowa, and Indiana.

Behavior and habitats. *M. viridipes viridipes* is usually found in low flat open woods and clearings, sometimes in fairly populous colonies but usually scarce. Males jump well, but females are much more sluggish.

Life history. Univoltine, with overwintering in the egg stage. The species matures rapidly in spring and adults are present early in June, usually disappearing by mid-August.

Melanoplus scudderi scudderi (Uhler)

Map 131

Pezotettix scudderi Uhler, 1864:555.
Pezotettix unicolor C. Thomas, 1873:51.
Melanoplus scudderi; Scudder 1897*d*:212; Helfer 1963:249 (partim).
Melanoplus scudderi scudderi; J. A. G. Rehn and Hebard 1916*b*:230.

Diagnosis. Tegmina as long as pronotum. Hind tibia bright red.

Description. Body small (length, male 16.2–18.5, female 22.0–24.0 mm). Pronotal disc broadly concave, with metazona densely punctate; anterior margin of pronotum truncate to faintly medially nötched, with posterior margin obtuse-angulate. Tegmina about as long as pronotum, elongate-ovate in males, with edges slightly overlapping, in females broadly ovate and separated. Male cercus broad at base, with ventral margin convex, and with upper margin concave, tapering to rounded apex (cf. Fig. 479). Male subgenital plate small, conical. Ovipositor exserted, with valves long, acute.

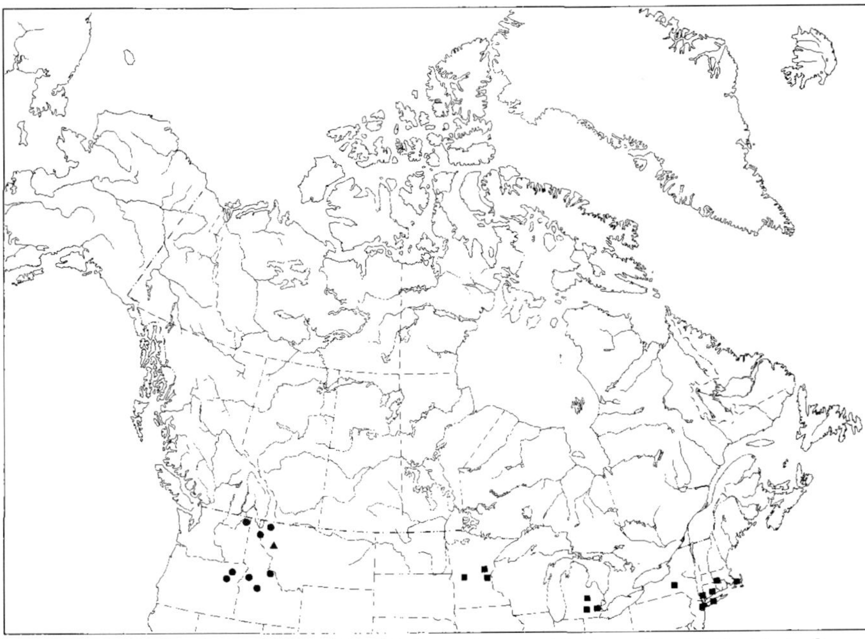

Map 131. Collection localities for *Melanoplus scudderi scudderi* (■), *M. indigens missoulae* (▲), and *M. indigens indigens* (●).

Coloration dull, light to dark brown above, paler on abdomen, yellowish beneath; lateral longitudinal dark stripe found on male prozona only, lacking in female; hind femur without spots or showing traces of bands; hind tibia bright red, occasionally paler basally but without dark or pale ring at base.

Range. Minnesota to Massachusetts, south to Texas and Florida.

Behavior and habitats. *M. scudderi scudderi* is found in open places along edges of woodlands, in bushy pastures, and on rocky slopes of hills. Blatchley (1920) reports this species as a dry-area species that often sits on logs or stones in sunny places. Mulkern et al. (1969) reported it as mainly a forb feeder. Gangwere (1961) reported its feeding upon *Aster, Achillea*, and *Solidago* in Michigan.

Life history. Univoltine, with overwintering in the egg stage. Adults are found late in the season. Cantrall (1968) recorded these from 22 August to 6 November in Michigan.

Melanoplus walshii Scudder

Map 132

Melanoplus walshii Scudder, 1897*d*:235, pl. XV, fig. 10.
Pezotettix occidentalis (*nec* C. Thomas, 1872), Bruner 1876:124.
Melanoplus amplectens Scudder, 1897*d*:280.

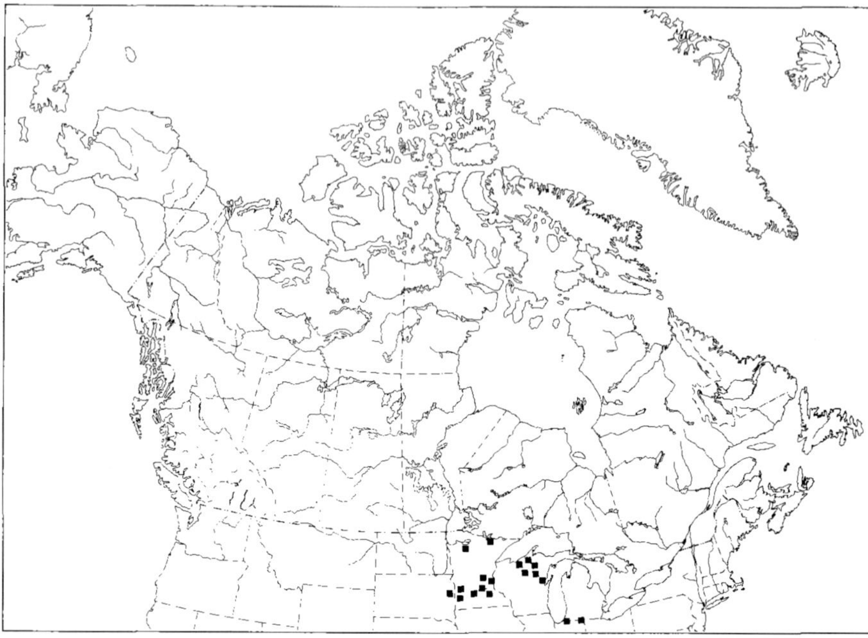

Map 132. Collection localities for *Melanoplus walshii*.

Melanoplus blatchleyi Scudder, 1897*d*:322.
Melanoplus walshii; Blatchley 1920:403.

Diagnosis. Tegmina short, sublanceolate. Male fore and middle femora swollen. Hind tibia red, with pale basal ring.

Description. Body size medium (length, male 19–25, female 23–29 mm), robust. Pronotal disc flat, with lateral margins feebly diverging posteriorly; posterior margin broadly obtuse-angulate. Tegmina sublanceolate, acute apically, longer than pronotum by about one-quarter, with inner edges overlapping slightly. Fore and middle femora swollen in male; hind femur extending slightly beyond abdomen in male, to ovipositor in female. Abdomen of male strongly recurved apically. Male cercus slightly spatulate, curved abruptly upward, concave apically, with apex broadly rounded (cf. Fig. 482). General color dark grayish brown dorsally, yellowish beneath; postocular dark stripe narrow, often indistinct or obsolete in females; lower half of lateral lobes grayish brown with dark flecks; tegmina pale brown dorsally; hind femur reddish to yellowish brown, with two crossbars on upper and outer faces; hind tibia red with pale ring near base; spines black.

Range. Minnesota and South Dakota to Michigan, south to Georgia.

Behavior and habitats. Cantrall (1968) reports *M. walshii* as having a spotty distribution in Michigan, among "thickets of herbs, low shrubs, seedlings, vines and brambles in or near deciduous woodland." Blatchley (1920) reported it as more arboreal than many other species, often being found in shrubbery well above the ground.

Life history. Cantrall (1968) recorded adults from 16 July to 4 September in Michigan. The species is undoubtedly univoltine, overwintering in the egg stage.

Melanoplus indigens missoulae Hebard

Map 131

Melanoplus indigens missoulae Hebard, 1936*b*:175.

Diagnosis. Hind femur red or orange beneath.

Description. Hind femur rich orange to deep red beneath. Furculae of male less than one-third as long as supra-anal plate. Male cercus scarcely spatulate (cf. Fig. 495). Aedeagus with projecting parts of dorsal lobes small, short, rounded (see Hebard 1936*b*, pl. XIII, figs. 1, 2).

Range. Known only from Missoula, Montana.

Behavior and habitats. Not known.

Life history. Adults were collected in August.

Melanoplus indigens indigens Scudder

Map 131

Melanoplus indigens Scudder, 1897*d*:211, pl. XIV, fig. 4.
Melanoplus indigens indigens; Hebard 1936*b*:177.
Melanoplus indigens; Helfer 1963:256 (partim).

Diagnosis. Tegmina elongate-ovate, shorter than pronotum. Male cercus broad, spatulate. Note: For accurate determination of this species it is necessary to examine the concealed male genitalia, as this is the only way by which the different taxa of the *indigens*-group (*M. oregonensis, M. marshalli, M. indigens*, and their subspecies) can be identified with certainty (see Hebard 1936*b* for figures). Females are more difficult to determine and should be associated with males with which they are collected.

Description. Body size medium (length, male 18.0–19.3, female 22.0–23.7 mm; pronotal length, male 4.7–4.9, female 5.0–5.6 mm). Pronotal disc with weak carina on metazona, absent on prozona in males, rarely weakly indicated in females. Tegmina elongate-ovate, shorter than pronotum, separated, with apices broadly to sharply rounded. Male cercus broad, straight, slightly spatulate apically and broader than at middle, with ventral margin rounding more sharply to apex than dorsal margin (cf. Fig. 510). Male subgenital plate produced to broad blunt rounded apex; aedeagus with dorsal lobes broad, erect, with lateral surfaces flattened, and with dorsal outline rounded-angulate in lateral aspect; apices not meeting above parameres.

Range. Washington, Oregon, Idaho, and Montana.

Behavior and habitats. *M. indigens indigens* is found in areas with grass, wild strawberry, and sagebrush, and in thickets of wild rose. It occurs only in small colonies, usually at altitudes above 1800 m.

Life history. Not studied, but probably similar to that of other alpine species of *Melanoplus*. Adults have been found in August and September.

Melanoplus oregonensis oregonensis (Thomas)

Figs. 444, 480, 517, 561; Map 133

Pezotettix oregonensis C. Thomas, 1875:888.

Melanoplus oregonensis oregonensis; Hebard 1928:268, pl. 25, fig. 4; Brooks 1958:29, fig. 7.

Diagnosis. Tegmina short, Hind femur yellow beneath, not banded; hind tibia buff to blue.

Description. Body small (length 18–23 mm). Tegmina short, reaching only to second abdominal segment. Male supra-anal plate (Fig. 444) distinctly longer than broad, with apex broadly rounded, and with furculae long. Male cercus nearly equally broad throughout, with apex broadly rounded, slightly spatulate (Fig. 480). Antennal crescent of female (cf. Fig. 431) complete. Ovipositor (Fig. 561) with dorsal angle 135°; lateral notch of female eighth abdominal sternum 110°. Color dark brown to grayish brown; hind femur yellow beneath, not conspicuously banded on outer face; hind tibia buff to blue.

Range. British Columbia, Alberta, Idaho, and Montana.

Behavior and habitats. *M. oregonensis oregonensis* usually occurs above 1500 m in "tall grass and in weedy places" (Hebard 1928). The authors have collected it at an altitude of 2560 m in southwestern Montana.

Life history. Adults have been found from mid-June to mid-August. The insects are probably univoltine, with overwintering in the egg stage.

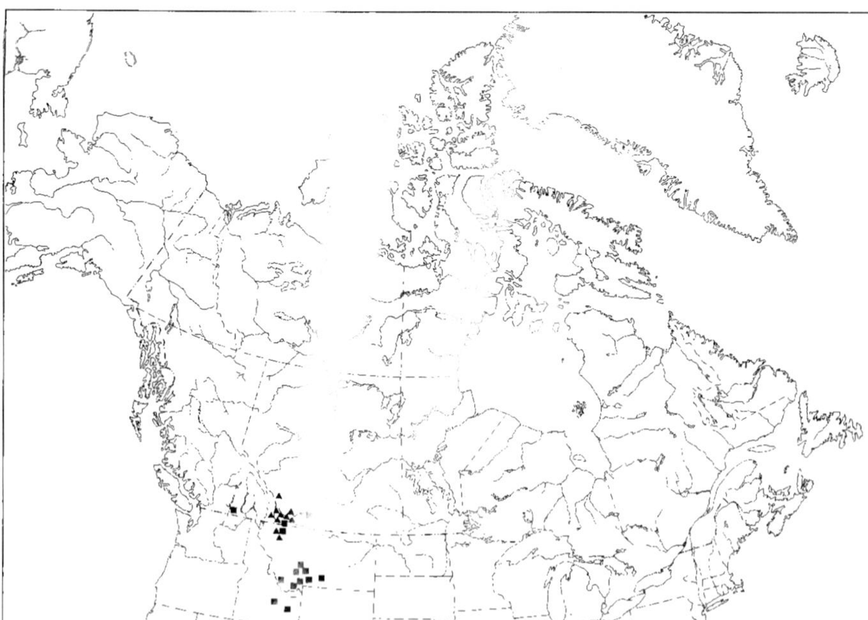

Map 133. Collection localities for *Melanoplus oregonensis oregonensis* (■) and *M. oregonensis triangularis* (▲).

Melanoplus oregonensis triangularis Hebard

Figs. 445, 481, 562; Map 133

Melanoplus oregonensis triangularis Hebard, 1928:269.
Melanoplus oregonensis triangularis; Hebard 1936*b*:174; Brooks 1958:29.

Diagnosis. Male supra-anal plate triangular, shield-shaped. Females are indistinguishable from those of *M. oregonensis oregonensis.*

Description. Male supra-anal plate (Fig. 445) triangularly shield-shaped. Male cercus short, heavy, distinctly spatulate (Fig. 481).

Range. British Columbia, Alberta, and Montana.

Behavior and habitats. *M. oregonensis triangularis* occurs only at high elevations on mountain slopes.

Life history. Probably univoltine; adults have been collected in late August.

Melanoplus dawsoni (Scudder)

Figs. 453, 490, 526, 571; Map 134

Pezotettix dawsoni Scudder, 1875*b*:343.
Pezotettix tellustris Scudder, 1876*c*:502.
Pezotettix abditum Dodge, 1877:113.

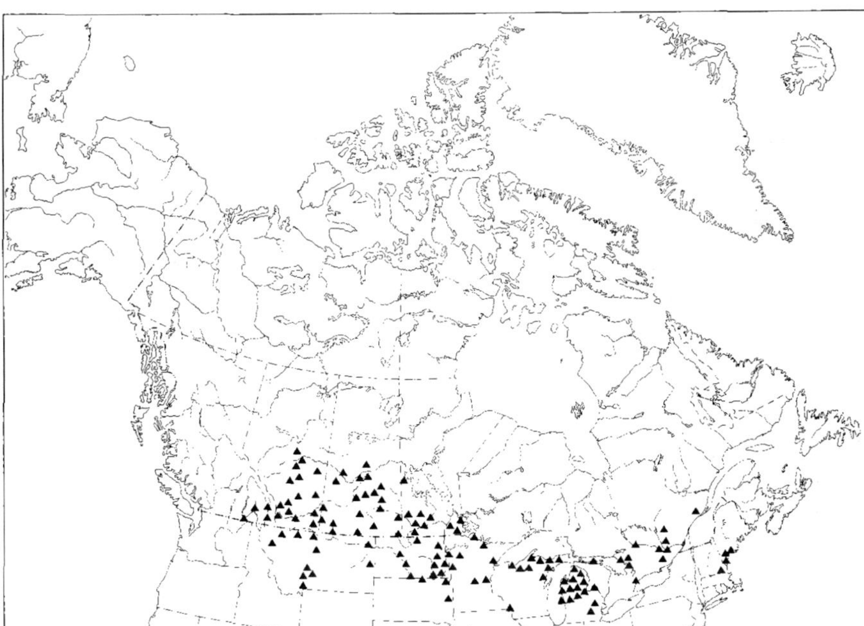

Map 134. Collection localities for *Melanoplus dawsoni*.

Melanoplus dawsoni; Scudder 1896*c*:64; Brooks 1958:24; Helfer 1963:221; Vickery et al. 1974:96.

Melanoplus dawsoni completus Scudder, 1897*d*:227.

Melanoplus acutus Scudder, 1898*c*:171.

Diagnosis. Easily recognized by abdominal color pattern: anterior half of each tergum black, and posterior half yellow.

Description. Body medium-small (length, male 14–17, female 19–22 mm). Antennal crescent (cf. Fig. 431) divided. Pronotum short, with disc convex; prozona slightly longer than broad; posterior margin of metazona obtuse-angulate. Tegmina usually brachypterous, ovate-lanceolate, slightly longer than pronotum (occasionally macropterous, with tegmina extending beyond apices of hind femora). Male supra-anal plate triangular, somewhat flared toward apex (Fig. 453). Male cercus short, with apical third less than half basal width, feebly concave, with apex rounded (Fig. 490). Male subgenital plate small, longer than broad, tapering to compressed, with truncate to slightly notched apex (Fig. 526). Ovipositor narrow, with angle of dorsal valve about 150° (Fig. 571); eighth abdominal sternum of female with shallow lateral notch (Fig. 571). Color generally brown to grayish brown, bright yellow beneath; abdominal terga conspicuously banded; hind femur greenish ventrally, banded dorsally; hind tibia red. Nymphs are described by N. Criddle (1926*b*) and by Handford (1946).

Range. British Columbia to Québec, south to Utah and Pennsylvania.

Behavior and habitats. Shrubby bushy areas along woodland margins, clearings, and grassy pastures in the eastern part of the range (Vickery et al. 1974) are favored. Brooks (1958) stated that it preferred "grassy pastures where short shrubs are growing" in Western Canada.

Life history. Adults are found in the field from midsummer to late summer, 10 July to 18 October in Michigan (Cantrall 1968). There is one generation each year, with overwintering in the egg stage in soil. In the laboratory the elapsed time between hatching of the eggs and molting to the adult stage was about 50 days.

Melanoplus mancus (Smith)

Figs. 446, 482, 518, 563; Map 135

Pezotettix manca S. I. Smith, 1868:149.

Melanoplus mancus; Scudder 1896*c*:66; Helfer 1963:254; Vickery et al. 1974:92.

Diagnosis. Tegmina ovate, separated. Hind femur reddish brown, with basal narrow black streak.

Description. Body small, short (length, male 14–17.5, female 18–25 mm). Interocular space as wide as frontal costa between antennae; fastigium strongly declivent. Posterior margin of pronotum subtruncate (sometimes emarginate) to broadly rounded. Tegmina ovate, two-thirds length of pronotum, separated by half their width and broadly, evenly rounded in female, separated by less than one-quarter their length in male. Male cercus long, with apical two-thirds narrow; sides nearly parallel, then rounded

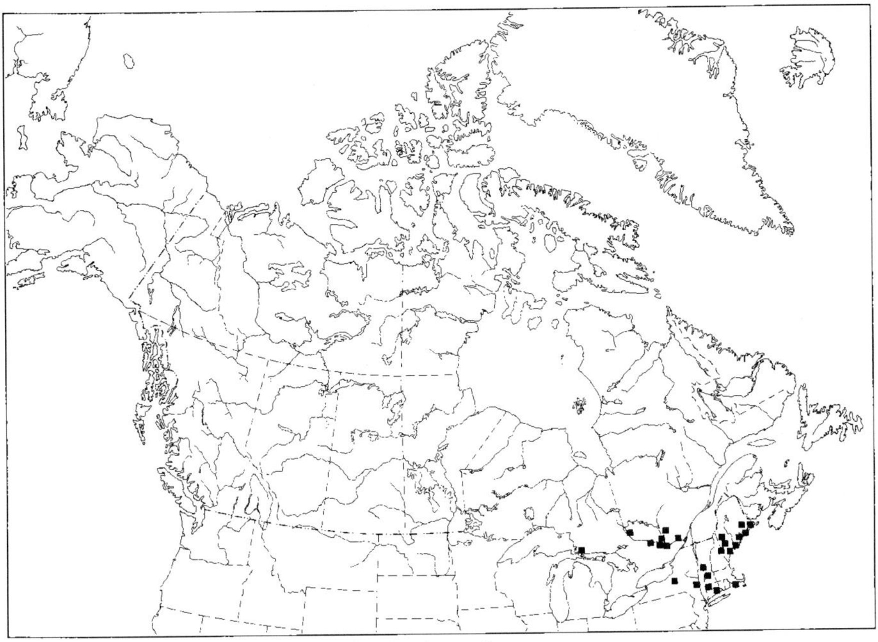

Map 135. Collection localities for *Melanoplus mancus*.

apically, slightly concave on outer apical face (Fig. 482). Male subgenital plate (Fig. 446) narrowed to prolonged conical apex; male furculae in form of cylindrical parallel projections about twice as long as width of 10th abdominal tergum (Fig. 446). Ovipositor valves (Fig. 563) narrow, with angle of dorsal valve obtuse (about 155°); lateral notch of eighth sternum of female shallow (about 150°); female cercus elongate–triangular with sides concave. Color dark brown above, yellowish beneath; abdomen reddish brown with broad paler stripe dorsally; hind femur reddish brown, often with proximal narrow black streak; hind tibia red with black spines.

Range. Ontario, Québec, and Maine, south to Virginia.

Behavior and habitats. *M. mancus* is rare almost throughout its range, occurring as small scattered colonies in open areas. Morse (1920) recorded it from exposed rocky mountain slopes usually in association with *Vaccinium* [*angustifolium* and *myrtilloides*]. Specimens are often, but not always, found at higher elevations in the eastern mountains, 600–1000 m above sea level. They may feed on *Vaccinium* or on plants that are commonly found with it. Morse (1920) also found them to be agile; they frequently leap suddenly for some distance and often make several leaps in succession. We have also noted its agility, particularly of males.

Life history. Adults have been taken from late July to mid-September, indicating a univoltine life cycle, with overwintering in the egg stage. In the southern part of the range (Virginia), adults have been found as early as 10 July (Hebard 1937*a*), but the adult stage is attained later in the north.

Melanoplus islandicus Blatchley

Figs. 447, 483, 519, 564; Map 136

Melanoplus islandicus Blatchley, 1898*b*:196.
Melanoplus abortivus E. M. Walker, 1898*a*:90–92.
Melanoplus islandicus; Scudder 1898*c*:177; Cantrall 1943:110; Brooks 1958:28; Vickery et al. 1974:93.
Melanoplus mancus islandicus; Hebard 1932*a*:34.

Diagnosis. Hind femur lacking conspicuous crossbands; hind tibia red. Male furculae short. Male cercus broad at base, strongly tapering.

Description. Body small (length, male 14.5–17, female 19–24 mm); males slender and females moderately robust. Male cercus broad at base, strongly tapering (Fig. 483). Male subgenital plate (Fig. 447) conical, with sides compressed and with apex acute; male furculae short, flat, oblong (Fig. 447). Ovipositor (Fig. 564) slender, with dorsal valve at angle of about 156°; eighth sternum of female with shallow lateral notch (135°); female cercus triangular, with sides nearly straight. Color generally dark brown above and greenish yellow beneath; abdomen with broad paler dorsal line; hind femur not conspicuously crossbanded on outer face; hind tibia red, somewhat darkened proximally. Nymphs are described by Handford (1946).

Range. Manitoba to Québec, south to Virginia.

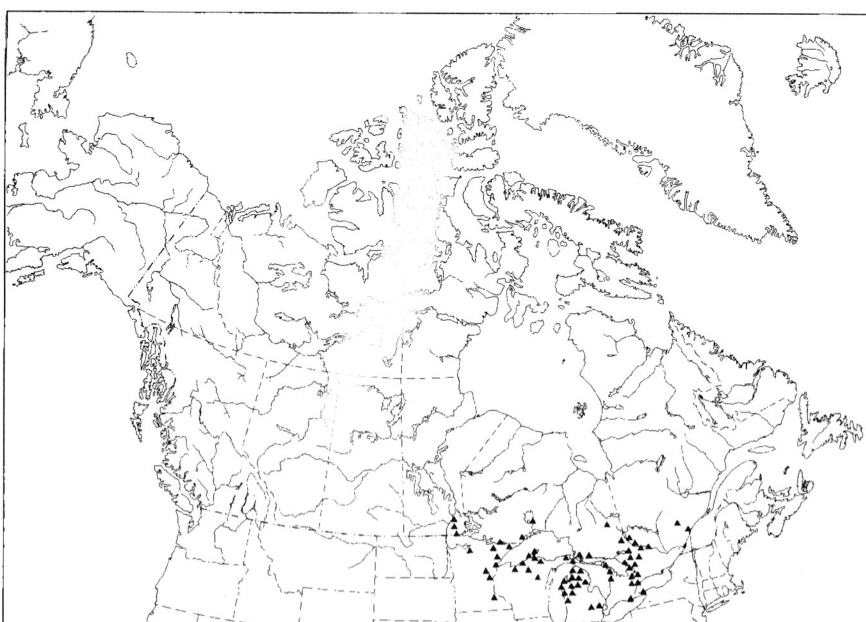

Map 136. Collection localities for *Melanoplus islandicus*.

Behavior and habitats. *M. islandicus* is primarily a forest-dwelling species, often being found in shady woods near clearings and pathways. The series recorded by Blatchley (1898*b*) was found near a clearing and pathway through dense spruce, birch, and hemlock woods. E. M. Walker (1898*a*) found the species in similar locations at De Grassi Point, Lake Simcoe, Ont. Colonies are usually small in numbers and rather scattered. The species may be much more common than collections indicate.

Life history. Univoltine, with overwintering in the egg stage. Cantrall (1968) recorded adults from 8 July to 8 September in Michigan.

Melanoplus montanus (Thomas)

Figs. 484, 520, 565; Map 137

Pl[*atyphyma*] *montana* C. Thomas, 1873:155.

Melanoplus montanus; Scudder 1896*c*:66; Brooks 1958:29; Helfer 1963:261.

Diagnosis. Tegmina small, oval, padlike. Hind femur conspicuously banded; hind tibia red.

Description. Body small (17–25 mm long). Antennal crescent (cf. Fig. 432) divided. Tegmina in form of small oval pads extending only to second abdominal tergum. Male cercus narrow, straplike, tapered, with round apex (Fig. 484). Male subgenital plate rounded-acute apically. Furculae of

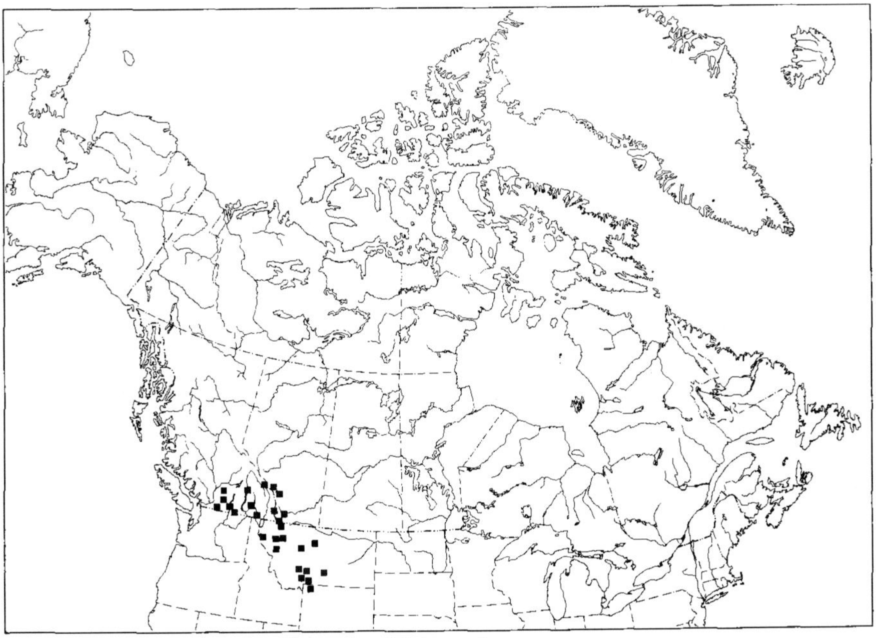

Map 137. Collection localities for *Melanoplus montanus*.

male small. Female cercus triangular, with sides slightly convex. Upper ovipositor valve robust, with dorsal angle about 136° (Fig. 565); lateral notch of eighth abdominal sternum of female shallow, about 140° (Fig. 565). Color generally dark brown to blackish; hind femur conspicuously banded, mottled beneath; hind tibia red with black spot near base.

Range. British Columbia and Alberta, south to Wyoming.

Behavior and habitats. *M. montanus* is confined to slopes of mountains, and is often found on bare western exposures with only occasional bushes. Occurrence is discontinuous. The species is regarded as rare. Specimens are difficult to capture, because they usually move rapidly down slopes by powerful leaps until they find a tangle in which they hide (Hebard 1928).

Life history. Hebard (1928) recorded adults from 29 July to 27 August. In Alberta, specimens have been found from 19 July to 30 August. There is probably one generation each year, with overwintering in the egg stage.

Melanoplus washingtonius (Bruner)

Figs. 448, 485, 521, 566; Map 138

Pezotettix washingtonius Bruner, 1885*b*:14.
Melanoplus validus; E. M. Walker 1910:337.
Melanoplus washingtonius; Buckell 1924:12.

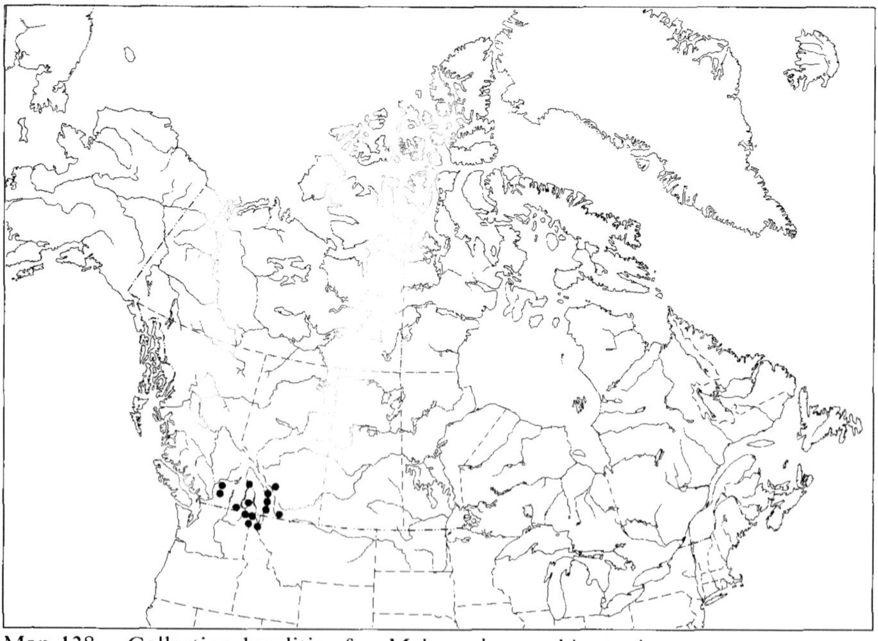

Map 138. Collection localities for *Melanoplus washingtonius*.

Diagnosis. Pronotum with lateral markings nearly parallel in males, diverging posteriorly in females.

Description. Body size medium (length, male 18.5–20, female 22–24 mm). Eyes large, prominent. Pronotum with lateral markings nearly parallel in males, diverging posteriorly in females, with median carina distinct on metazona, obsolete on prozona in females, and only at anterior margin in males; lateral carinae sharp, with anterior margin of pronotum truncate, and with posterior margin obtusely rounded. Tegmina about two-fifths length of pronotum, acute apically. Male cercus (Fig. 485) long, three times as long as broad, directed inward posteriorly, with apex rounded. Color brownish fuscous above, yellowish beneath; hind femur yellowish brown with 2 fuscous crossbars; hind tibia pale reddish. Note: For accurate determination of this species it is necessary to examine the concealed male genitalia (see Hebard 1935*a*:pl. XXVI, fig. 9).

Range. British Columbia, Alberta, Washington, and Idaho.

Behavior and habitats. Not reported. Found on mountain slopes.

Life history. Probably univoltine, with overwintering in the egg stage. Adults have been found in July and August.

Melanoplus dodgei (Thomas)

Map 139

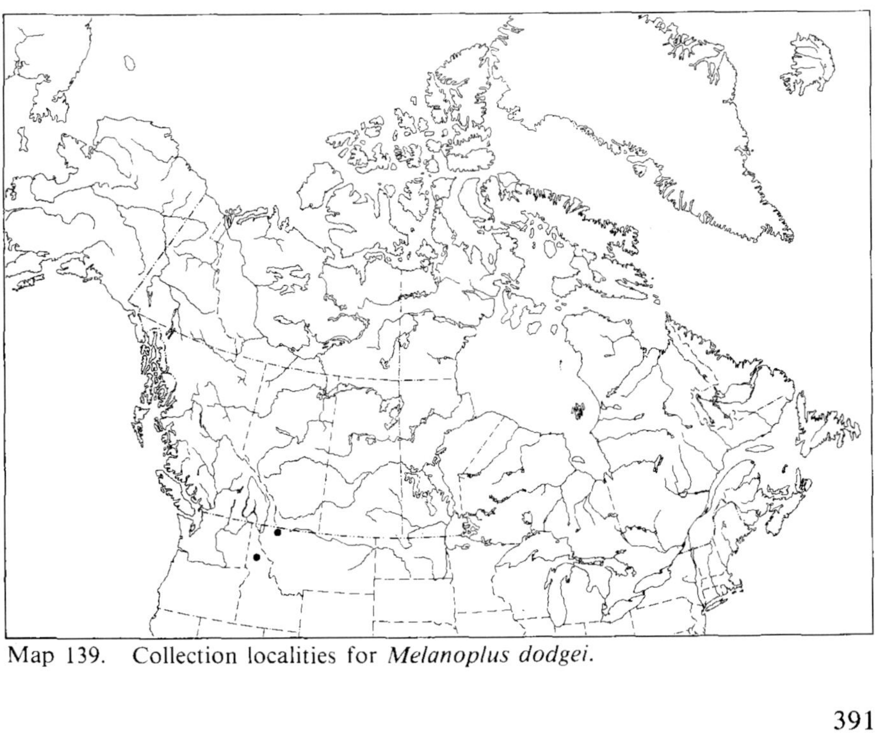

Map 139. Collection localities for *Melanoplus dodgei*.

Caloptenus dodgei C. Thomas, 1871*b*:168.

Pezotettix aspirans Scudder, 1879*b*:85.

Melanoplus dodgei dodgei; Hebard 1929*b*:376.

Melanoplus dodgei; Newton and Gurney 1956:1092, map 66; Helfer 1963:246.

Diagnosis. Hind femur yellowish beneath, black apically and basally. Differing from *M. huroni* in range.

Description. Hind femur usually yellowish beneath, conspicuously marked with black basally and apically, with 2 intermediate moderately broad transverse dark bands; hind tibia pale red, marked with fuscous basally. Male supra-anal plate, cercus, and subgenital plate, and female ovipositor as in Figs. 449, 486, 522, and 567.

Range. Idaho and Montana, south to Colorado.

Behavior and habitats. *M. dodgei* is alpine, found on mountain slopes and valleys at altitudes of 1800–4000 m (D. Van Horn 1965).

Life history. Probably univoltine, like that of others in this group of species. It is possible that more than one year may be required to complete development, as was reported for some other high-altitude species of *Melanoplus* by Kreasky (1960).

Melanoplus huroni Blatchley

Figs. 449, 486, 522, 567; Map 140

Melanoplus huroni Blatchley, 1898*b*:195.

Melanoplus huroni; Scudder 1898*c*:182; Brooks 1958:27; Helfer 1963:247.

Melanoplus dodgei huroni; Blatchley 1920:407.

Diagnosis. Tegmina extending to third or fourth abdominal segment, conspicuously spotted dorsally. Note: Accurate differentiation of *M. huroni* and *M. dodgei* (Thomas) requires examination of the concealed male genitalia (Hebard 1935*a*). No illustration of the genitalia of either of these two species has yet been published. At present, geographical information is probably the most reliable method of distinguishing them.

Description. Body size medium (length, male 19–21, female 28–32 mm), robust. Pronotum with prozona slightly longer than metazona; metazona coarsely and densely punctate, with posterior margin obtuse-angulate; lateral carinae evident but rounded in cross section. Tegmina 1.3–1.5 times as long as pronotum, extending only to third or fourth abdominal segment (rarely macropterous, with tegmina extending to or beyond apices of hind femora), conspicuously spotted dorsally. Male cercus broad at base, slender apically (Fig. 486). Male subgenital plate (Fig. 522) with pointed apical tubercle. Male furculae (Fig. 449) small, blunt. Ovipositor (Fig. 567) with upper valve heavy but narrow, with dorsal angle about 150°; eighth abdominal sternum of female without lateral notches. Color generally dark gray to blackish; hind femur red ventrally and on inner face, with narrow dark crossbands on outer face; hind tibia dark red.

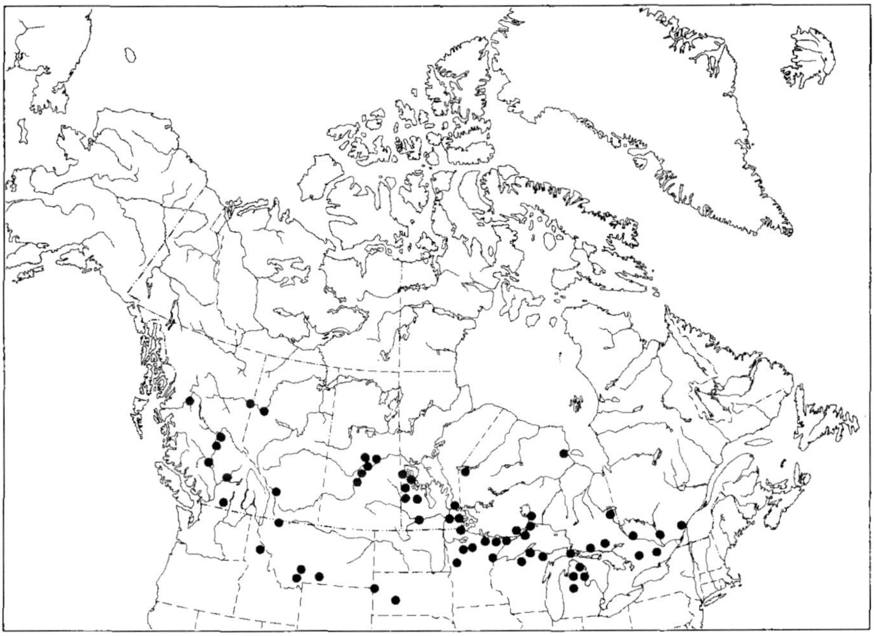

Map 140. Collection localities for *Melanoplus huroni*.

Range. British Columbia to Québec, south to Nebraska and Michigan.

Behavior and habitats. *M. huroni* occurs in undergrowth of woodlands, often near clearings and open paths where sunlight penetrates, in coniferous, deciduous, and mixed forests. Brooks (1958) recorded it from parklands near Saskatoon, Sask., far removed from evergreen trees, as well as from along roadsides in spruce and Jack pine forests. He also stated that it was known from a "few localities in northern Alberta," but apparently it has not been found in southern Alberta. These grasshoppers are clumsy and inactive, often remaining motionless even when approached. They are well camouflaged when on either open ground or forest litter.

Life history. Univoltine, with overwintering in the egg stage. In Michigan, adults are known from 14 July to 17 August (Cantrall 1968).

Melanoplus bivittatus (Say)

Figs. 4, 433, 434, 451, 488, 524, 569; Map 141

Gryllus bivittatus Say, 1825*a*:308.

Caloptenus femoratus Burmeister, 1838:638.

Acrydium flavo-vittatum T. W. Harris, 1841:140.

Melanoplus bivittatus; Scudder 1874:376; Cantrall 1943:111; Brooks 1958:21; Helfer 1963:220; Vickery et al. 1974:95.

Melanoplus bivittatus femoratus; Morse 1894*b*:106.

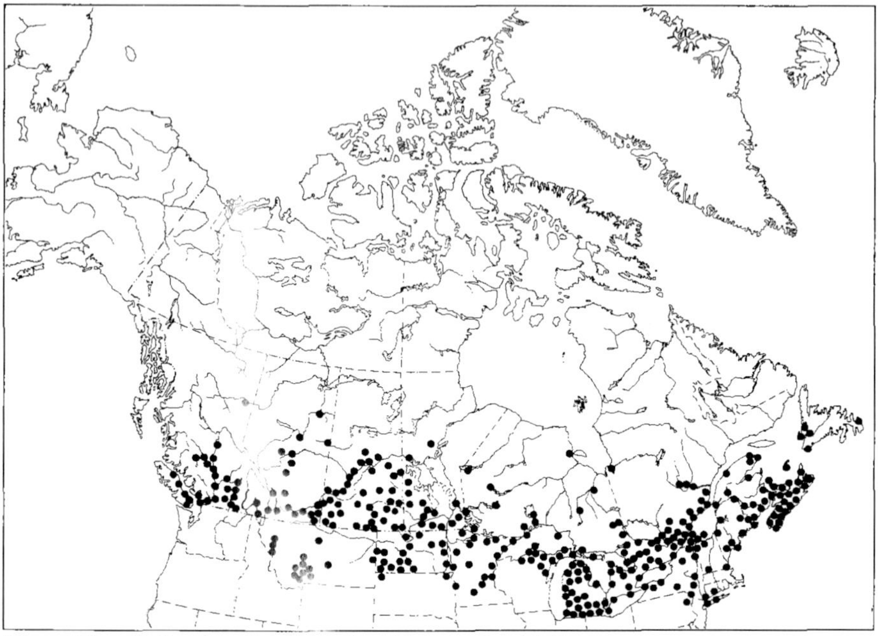

Map 141. Collection localities for *Melanoplus bivittatus*.

Common name. Twostriped grasshopper.

Diagnosis. Body apple green with pale lateral lines. Male cercus boot-spaced.

Description. Body large (length, male 26.5–28, female 31–35 mm), robust. Tegmina of male reaching or surpassing apices of hind femora, of female usually not quite reaching apex of abdomen. Male furculae short, swollen, triangular (Fig. 451). Male cercus with apical half expanded, boot-shaped (Fig. 488). Angle of dorsal ovipositor valve obtuse, about 156° (Fig. 569); lateral notch of female eighth abdominal sternum conspicuous, about 90° (Fig. 569). Color generally bright apple green above, yellowish beneath, occasionally pale greenish yellow (Fig. 4), with distinct pale testaceous to yellowish lines extending from upper region of eyes along lateral pronotal margins and along anal area of tegmina; hind femur greenish black on upper half of outer face, darker at base; hind tibia either poppy red with black spines or greenish buff to bluish, usually darker basally. Nymphs are described by N. Criddle (1924, 1926*b*) and by Handford (1946).

Range. British Columbia to Newfoundland, south to California and Georgia.

Behavior and habitats. *M. bivittatus* prefers locations that are relatively moist and is seldom found abundantly in dry areas. It is to some extent omnivorous but feeds mainly upon plants. It is said to be mainly a forb feeder (Brooks 1958), but it also eats grasses in areas where forbs occur. It prefers forbs, including many broad-leaved weeds, or cultivated plants such as rhubarb, and is seldom found where only grasses are available. Females are rather clumsy and heavy-bodied, and, having short wings and tegmina, they do not fly to any extent. They may, however, become airborne in strong winds and be carried for long distances. Males fly well when disturbed but, unless assisted by wind, do not fly far. As reported by Vickery et al. (1974) this species was well represented in the "invasions" of the city of Montréal in 1948, 1960, and 1971. Although males were much more numerous, females were also represented and the flights had come distances of 8–25 km. General biology is given by N. Criddle (1924), and eggs are described by N. Criddle (1918, 1924, 1935) and by Onsager and Mulkern (1963) and others.

Life history. Over most of the North American continent, adults are found from late June until September or October, when they are killed by frost. In most parts of Canada, and presumably also in most lowland areas in the northern United States, there is one generation each year, with overwintering in the egg stage. In British Columbia, at altitudes above 900 m, the life cycle takes 2 years, whereas below that level, the species is univoltine, and it is probable that a 2-year cycle also occurs in other mountainous areas (Putnam and Handford 1958*b*). Higher population levels appear in alternate years where the 2-year life cycle occurs, with few individuals seen in the intervening years (Beirne 1972).

Many natural enemies attack *M. bivittatus*. Adults and nymphs are affected by at least 12 species of parasitic insects, two nematode species, parasitic mites, and also bacterial pathogens (Beirne 1972). Birds may destroy a large proportion of a population when the grasshopper density is high. Dry conditions reduce egg-hatching. The eggs appear to be susceptible to desiccation, and this probably prevents permanent establishment in dry locations.

Economic importance. Although this species is classed as highly important economically, the damage caused by its feeding is usually confined to broad-leaved crops. Occasionally it will cause damage to grain crops, but, if other vegetation is available, the crop damage is minimal and usually peripheral. Alfalfa seed crops, tobacco, corn, and strawberry may be severely damaged by *M. bivittatus*, especially when grown under irrigation.

Melanoplus madeleineae Vickery & Kevan

Figs. 460, 497, 533, 578; Map 142

Melanoplus madeleineae Vickery and Kevan, 1978:188–192, figs. 1–5, 7.
Melanoplus madeleineae; Kevan and Vickery 1978:202.

Diagnosis. Hind femur crimson beneath. Found on Magdalen Islands only.

Description. Body medium-large (length, male 21.0–25.5, female 24.0–29.0 mm). Lower edge of hind femur of striking crimson color, this extending above lower carinae on both inner and outer faces; tibial spines robust, black. Male furculae and supra-anal plate as in Fig. 460. Male cercus as in Fig. 497. Male subgenital plate as in Fig. 533; apices of aedeagal valves

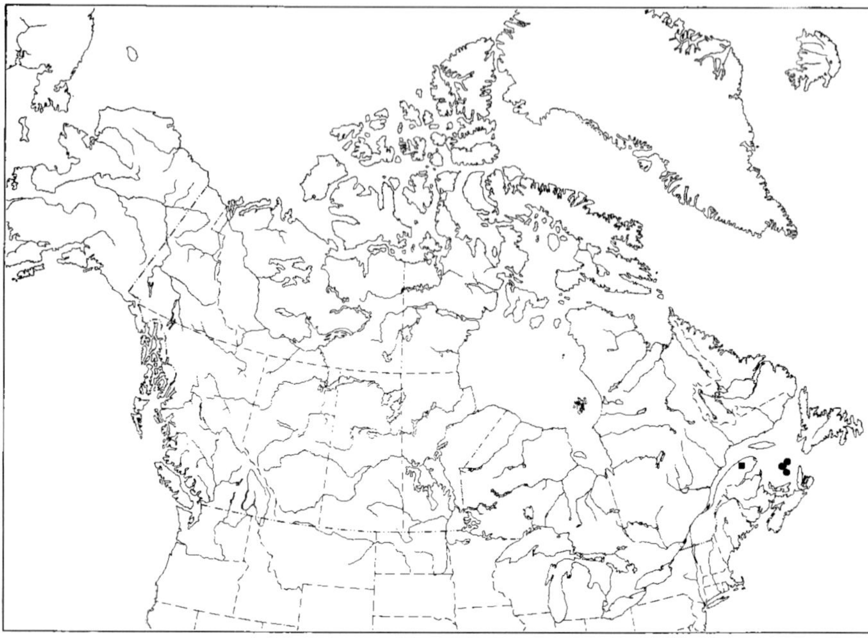

Map 142. Collection localities for *Melanoplus madeleineae* (●) and *M. gaspesiensis* (■).

large, broad, with apices of dorsal valves broadly rounded (Vickery and Kevan 1978:190, fig. 7); epiphallus large, heavily sclerotized (Vickery and Kevan op. cit., fig. 5). Ovipositor with distinct denticulation on dorsal valves (Fig. 578); notch of eighth abdominal sternum of female about 139° (Fig. 578).

Range. Magdalen Islands, Québec.

Behavior and habitats. Specimens were collected mainly in damp areas where grasses and sedges grow. *M. madeleineae* is not a strong flier, and apparently is not as agile as *M. borealis borealis*.

Life history. Adults have been taken from 26 July to 11 September. They probably persist until near the end of September. The life cycle is almost certainly univoltine, as no immatures were found late in the season. Like others of the group, *M. madeleineae* presumably overwinters in the egg stage in the soil.

Comments. Prest (*in* Howden 1969) indicated that the Magdalen Islands had formed an unglaciated nunatak during the entire Wisconsin glaciation period. The presence of *M. madeleineae* on these islands, and nowhere else, supports this contention, as does evidence published later by Prest et al. (1976). The species is undoubtedly derived from *M. borealis* (which is not found [now] in these islands) and has undergone significant changes. The male genitalia have come to differ from those of *M. borealis* to the extent that breeding between that species and *M. madeleineae* is probably not possible, even if the insects were given the opportunity.

Melanoplus borealis borealis (Fieber)

Figs. 436, 437, 439, 458, 495, 531, 576, 576*a*, 596–599; Map 143

Caloptenus borealis Fieber, 1853:120.
Pezotettix septentrionalis Saussure, 1861:159.
Caloptenus extremus F. Walker, 1870:681.
Pezotettix junius G. M. Dodge, 1876:9.
Caloptenus parvus Provancher, 1876:110.
Melanoplus borealis; Caulfield 1888:71.
Melanoplus monticola Scudder, 1897*c*:24, 34.
Melanoplus extremus scandens Scudder, 1897*d*:289.
Melanoplus borealis monticola; Buckell 1920:31.
Melanoplus borealis junius; Hubbell 1922*a*:40.
Melanoplus borealis borealis; Buckell 1929:13; Brooks 1958:22; Vickery et al. 1974:98.

Diagnosis. Tegmina slightly shorter than abdomen. Outer face of hind femur dull yellow to brown, not barred.

Description. Body medium-small (length, male 16–20, female 20–24 mm). Tegmina not reaching apices of posterior femora, generally reaching apex of abdomen in males, covering about three-quarters of abdomen in females (some macropterous specimens occur, and in these the tegmina are much longer than the abdomen). Male furculae large, parallel,

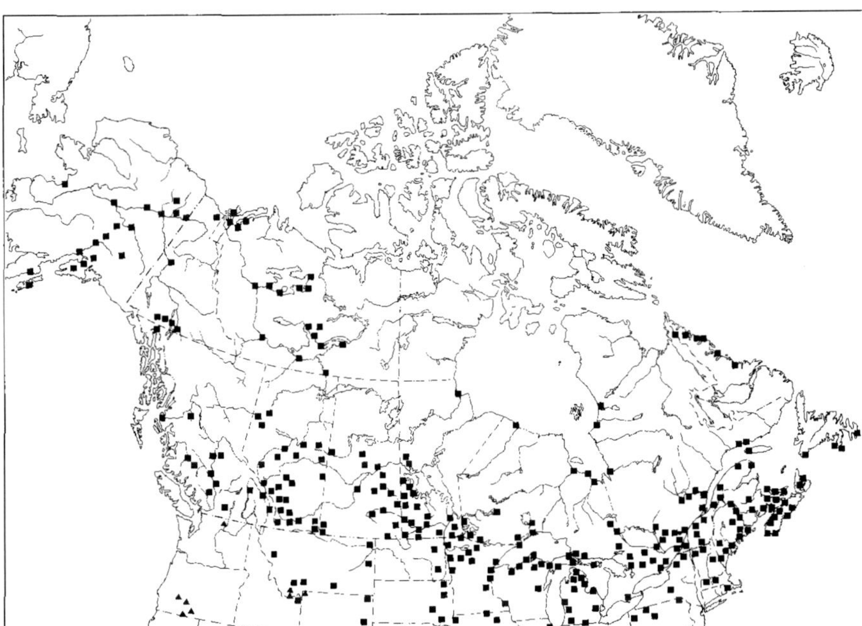

Map 143. Collection localities for *Melanoplus borealis borealis* (■) and *M. borealis palaceus* (▲).

tapering, about half as long as supra-anal plate (Fig. 458). Male cercus short, broad, with apical half distinctly more than half as broad as base, gently curved, rounded at apex (Fig. 495). Male subgenital plate rounded, not bulbous, not notched apically (Fig. 530). Ovipositor with angle of dorsal valve about 145° (Fig. 576); female cercus triangular, short, with upper edge concave (Fig. 576*a*). General color dark green with yellowish tinge; hind femur with outer face dull yellow, tinged with brown, never barred with black, with lower face light orange; hind tibia reddish to yellowish, with black spines. Nymphs are described by Handford (1946).

Range. Alaska to Newfoundland, south to Colorado and Massachusetts.

Behavior and habitats. *M. borealis borealis* is indeed boreal, occurring north of the tree line and also at elevations above the timberline. It is often found in cool damp situations, bogs, moist open pastures, and stream margins, and near ponds. Cantrall (1943) reported its "secretive" nature and the great agility with which it avoided capture. In southern Québec it is often found in swampy or marshy areas, near lakes or streams, usually where grass growth is luxuriant but also where it is sparse. W. J. Brown collected it on "gently rolling, treeless, Arctic tundra that consisted of damp cold sphagnum which thawed only from 11 to 19 inches during summer" (Gurney and Brooks 1959). It feeds primarily on grasses but will also feed on forbs (Mulkern et al. 1969). Dr. S. C. Palmer (*in* J. A. G. Rehn 1931) notes that on the inhospitable northern coast of Labrador, specimens were found on "warmer open gently sloping hillsides" (Nain) and in "grassy meadows as they neared the sea" (Cape Mugford), but not elsewhere. Eggs are described by Onsager and Mulkern (1963).

Life history. Adults appear in Québec about mid-June and persist to mid-August. Nymphs are present during the last week of April. For Michigan, Cantrall (1968) recorded adults from 18 June to 27 September, but none has been seen so late in Québec. I[t ha]s been suggested that a 2-year life cycle occurs in elevated parts of Briti[sh Co]lumbia (Gurney and Brooks 1959). This has not been verified. That a 2[-yea]r life cycle occurs throughout the northern part of the range of the su[bspe]cies is not an unreasonable supposition.

Melanoplu[s bor]ealis palaceus Fulton

[M]ap 143

Melanoplus borealis pala[ceus] Fulton, 1930:619, fig. 1:*F*, *G*, *H*.

Diagnosis. Tegmina of [male] extending to apices of hind femora, those of female slightly shorter.

Description. Body lengt[h, m]ale 19–21, female 26–30 mm; pronotal length, male 4.2–4.8, female [ ]-5.9 mm. Usually fully alate, with male tegmina reaching apices of hi[nd] femora, those of female slightly shorter, rarely surpassing hind knees. [M]ale abdomen strongly turned upward and recurved in apical third. Mal[e c]ercus variable, with apical margin sinuate in some specimens. Male sub[ge]nital plate produced dorsally.

Range. Washington to [M]ontana, south to California and Utah.

Behavior and habitats. The type specimens were collected at the edge of a large meadow, near pine woods. Most specimens have been found in swampy areas, some of these at relatively high altitudes (up to 3500 m). Specimens from Montana were taken in "wiregrass" and "bluegrass" swales (Gurney and Brooks 1959).

Life history. Adult specimens have been taken from 23 June to 31 August. The life cycle probably is univoltine at low altitudes, but may require 2 years at high elevations.

Melanoplus gaspesiensis Vickery

Figs. 427, 440, 459, 496, 532, 577; Map 142

Melanoplus gaspesiensis Vickery, 1970:7–13, figs. 1–9, 12, 14, 15.
Melanoplus gaspesiensis; Johnstone 1973:41; Vickery et al. 1974:99.

Diagnosis. Vertex higher than prothorax. Hind femur dark, with 2 narrow oblique pale bands. Found on Mont Albert, Gaspé Peninsula, only.

Description. General aspect as in Fig. 427. Large, robust (body length, male 19.5–24, female 24–28.5 mm; pronotal length, male 4.1–4.8, female 4.8–5.6 mm). Vertex of head elevated above prothorax. Metazonal area of pronotum deeper than broad. Tegmina short, in males nearly reaching apex of abdomen, in females covering two-thirds to three-quarters of abdomen. Male furculae short (Fig. 459). Male cercus broad at base, tapered to rounded apex (Fig. 496). Male subgenital plate reflexed narrowly at apex (Fig. 532); aedeagal valves as illustrated by Vickery (1970:11, figs. 19, 20). Ovipositor valves thick, with angle of dorsal valve about 155° (Fig. 577). Female cercus triangular, concave on dorsal side. Color generally brownish to fuscous, brown on underside; hind femur with 2 narrow oblique pale bands on black (Fig. 440); about 10% of individuals exhibiting bright yellowish white areas on genae, on ventral half of lateral pronotal lobe, and on basal half of dorsum of hind femur.

Range. Confined to higher parts of Mont Albert, Gaspé Peninsula, Qué.

Behavior and habitats. *M. gaspesiensis* is found in grassy areas near small ponds. The preferred food may be grasses but could also be some ericaceous plants, such as *Vaccinium*. This has not been investigated.

Comments. This species is undoubtedly derived from *Melanoplus borealis* but must be of long standing. It probably evolved from a population segment of that species which became isolated for a long period of time. The area has again been occupied by *M. borealis borealis* and, although both this and *M. gaspesiensis* now occur together on the mountain, apparently no interbreeding occurs. No specimen that could be called intermediate has been found, and laboratory rearing has produced no hybrid individuals.

Melanoplus fasciatus (F. Walker)

Figs. 441, 469, 505, 541, 586, 611–617; Map 144

Caloptenus fasciatus F. Walker, 1870:680.

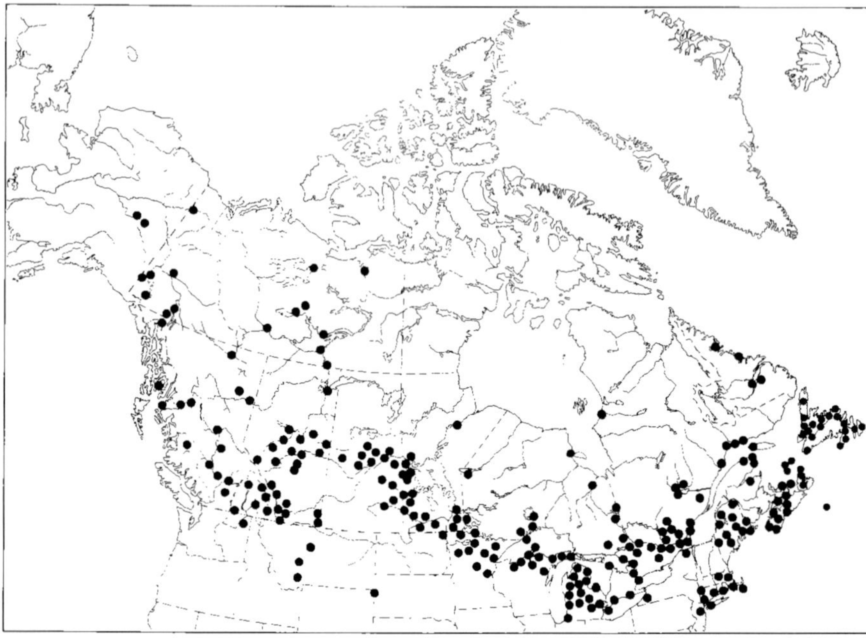

Map 144. Collection localities for *Melanoplus fasciatus*.

Melanoplus rectus Scudder, 1878*a*:284.
Melanoplus curtus Scudder, 1879*a*:70.
Melanoplus fasciatus; Caulfield 1888:71; Cantrall 1943:119; Brooks 1958:24; Vickery et al. 1974:100.
Melanoplus fasciatus curtus; Scudder 1897*d*:268, 270.
Melanoplus fasciatus volaticus Scudder, 1897*d*:268, 270.

Diagnosis. Tegmina not reaching apex of abdomen. Hind femur red beneath, banded. Male cercus broad, downturned at apex.

Description. Body medium-small (length, male 17–20, female 20–25.5 mm), robust. Tegmina covering about three-quarters of abdomen in males and about two-thirds in females (macropterous individuals occur, in which both tegmina and wings are longer than the abdomen in both sexes). Male furculae small, blunt, widely separated (Fig. 469). Male cercus broad, straplike, with broadly rounded slightly downturned apex (Fig. 505). Male subgenital plate (Fig. 541) longer than broad, with apical margin rounded, feebly elevated at middle. Ovipositor with angle of dorsal valve about 142° (Fig. 586). Female cercus long, triangular, with concave sides. Color generally dark brown above and clay yellow below; hind femur usually red below and on inner surface, dull yellowish brown with conspicuous broad black bars on outer face (Fig. 441); hind tibiae typically dull red, lighter near base, with spines black. An unusual, variegated (neotenic) color morph has been described by Johnstone (1973)—see Fig. 617.

Range. Alaska to Newfoundland, south to Washington, Colorado, and New Jersey.

Behavior and habitats. *M. fasciatus* occurs in association with heath plants. Vickery (1961) reported the presence of *Vaccinium* species wherever *M. fasciatus* was found in Nova Scotia. Brooks (1958) reported finding it under spruce or pine or on ground-cedar or bearberry. Cantrall (1943) stated that this species preferred blueberry or huckleberry shrubs in open areas of woodlands. It is a sedentary species, preferring to hide or remain immobile. At most, it makes only one or two short leaps when disturbed. At night, it climbs to the tops of shrubs and remains there until warmed by the sun the following day.

Life history. Univoltine, at least in the southern parts of its range. The life cycle may require 2 years, or more, in the north and at high elevations. Adults were recorded from 24 June to 1 September in Michigan (Cantrall 1968), from 9 August to 27 October in Nova Scotia (Vickery 1961), and as early as 16 July in northern British Columbia (R. M. White and Rock 1945).

Melanoplus frigidus frigidus (Boheman)[1]

Figs. 470, 506, 542, 587, 618; Map 145

Gryllus frigidus Boheman, 1846:80.
P[ezotettix] (Melanoplus) frigidus; Stål 1873*b*:79.
Podisma baicalensis Uvarov, 1914:171.

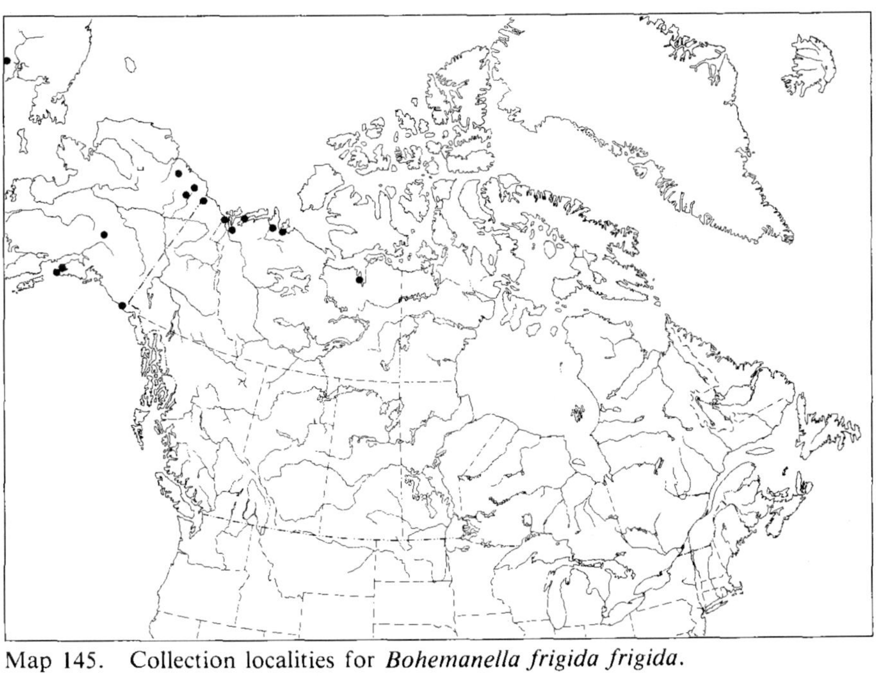

Map 145. Collection localities for *Bohemanella frigida frigida*.

[1] This species is now placed in the genus *Bohemanella* Ramme, in the tribe Podismini (Vickery 1984).

Melanoplus frigidus; Hebard 1919*b*:259.
Melanoplus frigidus frigidus; Mishchenko 1951:233, figs. 510–512.

Diagnosis. Brachypterous. Hind femur banded; hind tibia red with black basal annulus; arctic and subarctic only.

Description. Body small–medium (length, male 18–21, female 22–27 mm). Tegmina short, reaching only to middle of hind femora (Fig. 618), rarely to near apices. Supra-anal plate triangular (Fig. 470). Male cercus slightly curved upward apically (Fig. 506). Male subgenital plate conical, with apex distinctly produced (Fig. 542). Male aedeagal valves, see Gurney (1941) and Vickery (1967*a*:258, fig. 8). Female abdominal terminalia as in Fig. 587. Color blackish brown or greenish brown; hind femur banded; hind tibia red with black basal annulus, with black spines. Some data on nymphal stages are given by Beĭ-Bienko (1928), Rubtsov (1932), and Ander (1950).

Range. Alaska to Northwest Territories; Scandinavia, northern Asia.

Behavior and habitats. The behavior and habitats are practically unknown in North America, except that the species is found on the tundra.

Life history. Adults have been taken in northern North America in July and August, as early as 5 July in Yukon Territory. Ander (1950) outlines seasonal occurrence in Lapland (where conditions closely approach those in which the species occurs in North America). Eggs are laid there in the autumn, hatch the following June or July, and mature in about 4 weeks. The reported time from hatching to maturity seems to be very short and has been questioned by Uvarov (1977), but it may be accounted for by the extremely long northern day-length in summer, which would prolong the diurnal period of feeding activity.

Melanoplus keeleri luridus (Dodge)

Figs. 442, 471, 507, 543, 588; Map 146

Caloptenus luridus G. M. Dodge, 1876:11.
Melanoplus collinus Scudder, 1878*a*:285.
Melanoplus luridus; Scudder 1896*c*:66.
Melanoplus keeleri; Scudder 1897*d*:341.
Melanoplus luridus keeleri; J. A. G. Rehn and Hebard 1916*b*:242.
Melanoplus keeleri luridus; Blatchley 1920:437; Cantrall 1943:121; Brooks 1958:24; Helfer 1963:228; Vickery et al. 1974:103.

Diagnosis. Male cercus forked; dorsal arm broad, with rounded apex, ventral arm slender, acute. Color brownish fuscous to rusty brown.

Description. Body size medium (length, male 17–20, female 19–27 mm). Tegmina reaching or slightly surpassing apices of hind femora, narrow and gently tapering. Male supra-anal plate and furculae as in Fig. 471. Male cercus distinctly forked, with dorsal arm large, broad, with rounded apex, ventral arm slender and pointed (Fig. 507). Male subgenital plate truncate at apex. Male furculae in form of minute triangular lobes (Fig. 543). Ovipositor with angle of dorsal valve about 137° (Fig. 588). Female cercus short, triangular, with dorsal edge concave. Color generally brownish fuscous

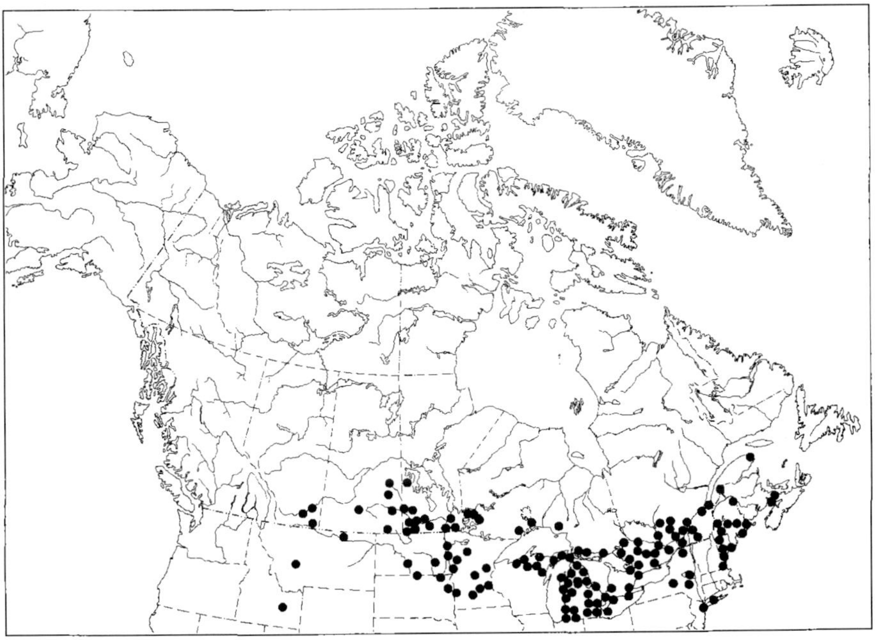

Map 146. Collection localities for *Melanoplus keeleri luridus*.

and rusty brown; hind femur (Fig. 442) with outer face yellowish brown; dorsal area blackish fuscous broken by 3 yellowish spots that suffuse the whole inner face, lower part yellow to bright orange; hind tibia coral red with narrow fuscous basal annulus, with spines black apically.

Range. Alberta to Nova Scotia, south to Utah and Colorado.

Behavior and habitats. This species is found mainly on grassy wastelands, often near forests, usually on sandy or gravelly soils. It is much more sluggish than most species of *Melanoplus*. E. M. Walker (1899) reported it as plentiful in late summer on sandy or gravelly uplands and open rocky unsettled Laurentian areas of Ontario. Mulkern et al. (1969) found that it feeds on many species of forbs and to a lesser extent upon grasses. It has been reported as feeding upon alfalfa.

Life history. Adults appear rather late in the season. They have been reported in Canada from 9 July to 27 October. Cantrall (1968) recorded adults from 10 July to 7 November in Michigan. There is one generation each year, with overwintering in the egg stage. In the authors' laboratory, nymphs progressed from hatching to maturity in 48 days at 26.5 ± 1.5°C, with a 15-hour photophase.

Melanoplus confusus Scudder

Figs. 455, 492, 528, 573; Map 147

Melanoplus confusus Scudder, 1897*d*:339, pl. XXII, fig. 10.
Caloptenus minor Scudder, 1875*d*:478 (homonym).

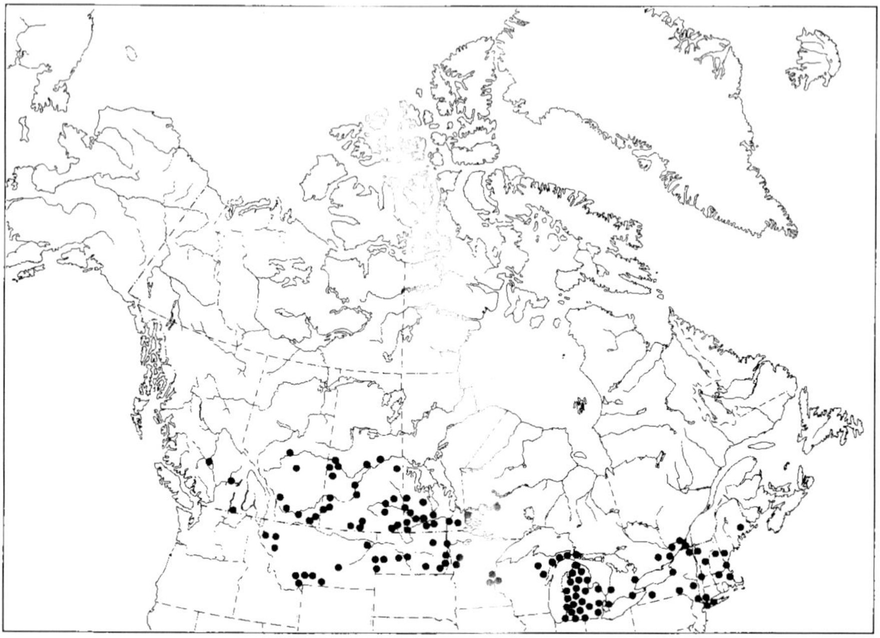

Map 147. Collection localities for *Melanoplus confusus*.

Melanoplus mutatus Caudell, 1915:30.

Melanoplus confusus; Morse 1919*b*:37; Cantrall 1943:117; Brooks 1958:24; Helfer 1963:225; Vickery et al. 1974:96.

Diagnosis. Hind femur yellow beneath, not banded. Male cercus with ventral tooth, and with broad dorsal arm.

Description. Antennal crescent (cf. Fig. 43[illegible]) broad, entire. Pronotum short, with lateral margins diverging on metazona; posterior margin broadly obtuse-angulate. Tegmina long, of male extending well beyond, and of female reaching apices of hind femora. Male supra-anal plate triangular; furculae moderately long, pointed (Fig. [illegible]). Male cercus with broad prominent triangular ventral tooth and with broadly rounded dorsal arm (Fig. 492). Male subgenital plate short, with apex low, rounded-truncate (Fig. 528). Dorsal valve of ovipositor with angle of [illegible]0° (Fig. 573); lateral notch of female eighth abdominal sternum moderate, about 130° (Fig. 573). Female cercus convex above and below (cf. Fig. 57[illegible]). Color grayish brown; hind femur yellow beneath, lacking conspicuous banding on outer face; hind tibia red to yellowish or greenish blue, usually pinkish apically. Note: Freshly captured nymphs may have eyes heavily spotted with white and have two irregular whitish bars across middle.

Range. British Columbia to Québec, south to Oklahoma and Virginia.

Behavior and habitats. Brooks (1958) called this species a mixed feeder, preferring forbs but also feeding readily on sedges. It is found in dry sandy wastelands, on hillsides, or in sandy clearings in forests. Morse (1920) found it common in well-drained grassy areas, particularly in bushy old pastures in New England.

Life history. Like most species of the genus, *M. confusus* has a univoltine life cycle, with overwintering in the egg stage. Eggs are usually deposited in sandy soil. It matures rapidly in the spring. Cantrall (1968) recorded adults from 30 May to 6 September in Michigan. Adults are most common in June and early July.

Melanoplus infantilis Scudder

Figs. 464, 500, 536, 581; Map 148

Melanoplus infantilis Scudder, 1879*a*:65.

Melanoplus infantilis; Caulfield 1888:71; Brooks 1958:17; Helfer 1963:227.

Diagnosis. Abdominal terga dark laterally at bases. Male cercus with small thumblike dorsal projection and straight tapered ventral arm. Hind tibia blue.

Description. Body small (length 14–19 mm). Antennal crescent (cf. Fig. 431) divided. Width of fastigium slightly greater than width between eyes. Tegmina of male extending well beyond, of female reaching apices of hind femora. Male supra-anal plate triangular with deep median groove; furculae small (Fig. 464). Male cercus with small thumblike dorsal projection

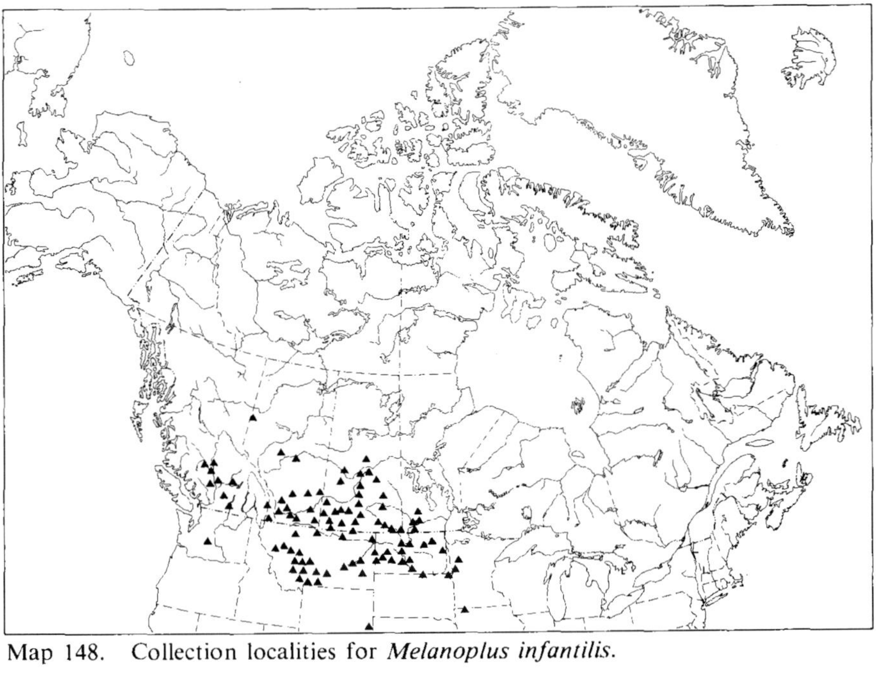

Map 148. Collection localities for *Melanoplus infantilis*.

and straight tapered ventral arm (Fig. 500). Male subgenital plate slightly produced to notched apex (Fig. 536). Ovipositor with dorsal valve angle about 130° (Fig. 581); eighth abdominal sternum of female with moderately deep lateral notch, about 138° (Fig. 581). Female cercus slender with concave margins. General color pale grayish; abdominal terga dark laterally at bases; hind femur yellow beneath, obliquely crossbanded on outer face; hind tibia pale blue.

Range. British Columbia to Manitoba, south to New Mexico.

Behavior and habitats. Brooks (1958) calls this species a mixed feeder, preferring grasses. It is found throughout the prairie region and extends into forests and parklands in dry grassy areas. Mulkern et al. (1969) found it to feed almost entirely on grasses. Buckell (1922) reported it as common in 1920 and 1921 throughout the Chilcotin District of British Columbia on open rangeland. *M. infantilis* is usually associated with short-grass prairie on treeless sandy or gravelly soils. It exhibits flight migration and often flies into cities and towns.

Life history. Univoltine, with overwintering in the egg stage.

Economic importance. Hubbell (1922*a*) stated that it is often unnoticed among the larger species of grasshoppers, but "is probably responsible for a considerable amount of damage to crops in the central and western parts of the state [North Dakota]." Hardman and Smoliak (1980) indicate potential damage to rangeland in Alberta.

Melanoplus alpinus Scudder

Figs. 465, 501, 537, 582; Map 149

Melanoplus alpinus Scudder, 1897*d*:333, pl. XXII, fig. 7.

Melanoplus pilatus Scudder, 1898*c*:192.

Melanoplus alpinus; Scudder 1901*a*:172; Brooks 1958:19; Helfer 1963:227.

Diagnosis. Tegmina long. Male cercus with dorsal thumblike projection, and with long slender curved ventral arm.

Description. Body small (length 18–25 mm). Antennal crescent (cf. Fig. 432) divided. Tegmina long, extending beyond apices of hind femora. Male supra-anal plate long, triangular, with median longitudinal groove; furculae small (Fig. 465). Male cercus medium–large, long, with apical part divided into dorsal thumblike projection and long slender strongly curved ventral projection (Fig. 501). Male subgenital plate truncate at apex, concave behind (Fig. 537). Ovipositor with dorsal valve angle about 120° (Fig. 582); lateral notch of eighth abdominal sternum of female moderately prominent, about 123° (Fig. 582). Female cercus long, slender, concave on both margins. General color gray with prominent black markings laterally; hind femur yellow beneath without conspicuous crossbands; hind tibia blue, rarely pink.

Range. British Columbia and Alberta, south to northern California and Colorado.

Behavior and habitats. *M. alpinus* is a grass feeder, found in foothills, grasslands, and montane parklands.

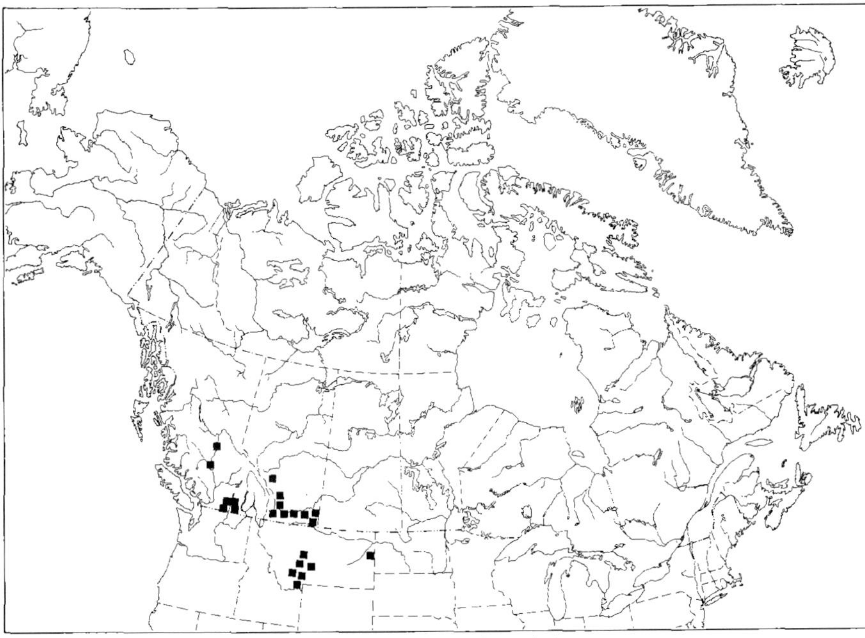

Map 149. Collection localities for *Melanoplus alpinus*.

Life history. Not reported but probably requiring at least 2 years to complete the life cycle. Kreasky (1960) reared this species in the Big Horn Mountains in Montana and suggested that a 3-year developmental period was required, eggs hatching after two winters in the soil, and adults maturing, mating, and depositing eggs during the same season.

Melanoplus differentialis differentialis (Thomas)

Figs. 428, 452, 489, 525, 570; Map 150

Acridium differentiale C. Thomas, 1865:450.

Cyrtacanthacris diversifera F. Walker, 1870:611.

Melanoplus differentialis; Scudder 1897*d*:349; Helfer 1963:219 (partim).

Melanoplus differentialis differentialis; H. R. Roberts 1942:154, pl. X, fig. 2, pl. XI, figs. 8–11.

Common name. Differential grasshopper.

Diagnosis. Brownish green; hind femur yellowish with distinct "herringbone" pattern (Fig. 428).

Description. Body robust, large (body length, male 28–34, female 34–44 mm). Pronotal disc nearly flat, with median carina visible but not distinct on anterior half of prozona, distinct and sharp on metazona; prozona nearly quadrate, slightly longer than metazona; metazona finely rugose. Tegmina long, extending beyond apices of hind femora in both sexes. Male cercus boot-shaped, with apical part as long as basal part, and with "toe"

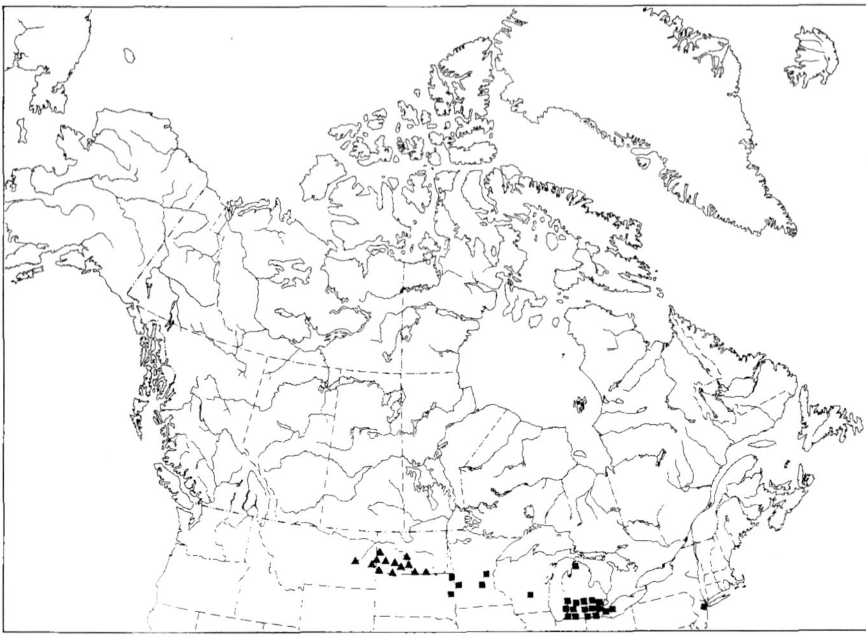

Map 150. Collection localities for *Melanoplus differentialis differentialis* (■) and *M. differentialis nigricans* (▲).

of boot relatively narrow, rounding into "heel" (Fig. 489). Male subgenital plate short, with middle of apical margin thickened and produced slightly upward (Fig. 525). Male aedeagus with dorsal processes produced, slender, with cavity obsolete and with channel shallow (see H. R. Roberts 1942:154, pl. X, figs. 2*c*, 2*d*; or Froeschner 1954:349, fig. 59). Ovipositor dorsal valve short, stout, strongly upcurved apically, with outer margins usually crenulate (Fig. 570). Color usually nearly uniform brownish green to olive brown dorsally, dull or bright yellow beneath; transverse sulci of pronotum black; tegmina olive brown, unspotted; basal abdominal segments black laterally; hind femur yellowish with distinct black "herringbone" pattern; hind tibia yellow with narrow black basal ring. Handford (1946) described the nymphs.

Range. South Dakota to Michigan and New Jersey, south to Mississippi and West Virginia.

Behavior and habitats. *M. differentialis differentialis* is classed as a forb feeder. Blatchley (1920) reports it as especially fond of giant ragweed, *Ambrosia trifida* L. and also smartweeds (*Polygonum* spp.) in or near water. Specimens become progressively darker in color as the season advances. At night, individuals climb tall weeds or shrubs and shelter in the leaves until warmed by the sun the following morning. Eggs are laid in loose or compacted soil, old roads, deserted fields, weedy areas, and well-grazed areas. Females deposit 1–3 egg masses, each containing about 175 eggs.

Blatchley (1920) reported extensive feeding upon this grasshopper by the loggerhead shrike, *Lanius ludovicianus* L. The bird impales the insects on thorns or on the barbs of barbed wire fences.

Life history. Univoltine, with overwintering in the egg stage. Adults occur in Michigan from 25 July to 25 September (Cantrall 1968).

Economic importance. This subspecies has the potential to cause severe crop losses. Corn, alfalfa, fruit trees, grape vines, poplar trees, and ornamental plants have suffered moderate to severe damage from its feeding. Cultivation of soils has caused it to become much more numerous, presumably because it can oviposit more readily in soil that has been loosened.

Melanoplus differentialis nigricans Cockerell

Map 150

[*Melanoplus differentialis*] var. *nigricans* Cockerell, 1917:247.
Melanoplus differentialis; Hebard 1925*a*:101.
Melanoplus differentialis nigricans; H. R. Roberts 1942:154, pl. X, fig. 3, pl. XI, figs. 12–15; Helfer 1963:220.

Diagnosis. Pale brown to buff (or black). Male cercus broad.

Description. Essentially as for *M. differentialis differentialis.* Male cercus broad at base and "toe"; aedeagus with dorsal valves short, broad, with cavity deep and channel conspicuous (see H. R. Roberts 1942:154, pl. X, figs. 3*c*, 3*d*; Froeschner 1954:349, fig. 60). Color as for the nominate subspecies, but usually paler with yellow tending more to buff; a melanic form is known to appear occasionally in many populations; nymphs as well as adults entirely black above, yellowish beneath; antennae reddish, fore and middle legs paler; hind femur yellowish with shining black basal, premedian, and apical areas; hind tibia dark with yellow basal ring. This form has been shown to be recessive and non-sexlinked, not appearing in F_1 generations, but in a 1:3 ratio in F_2 generations (King 1933, 1942).

Range. Montana and North Dakota, south to Mexico.

Behavior and habitats. *M. differentialis nigricans* is similar to if not identical in behavior and habitat with the nominate subspecies.

Life history. As for *M. differentialis differentialis.*

Economic importance. As for *M. differentialis differentialis.*

Melanoplus punctulatus griseus (Thomas)

Map 151

C[*aloptenus*] *griseus* C. Thomas, 1872:454.
Caloptenus helluo Scudder, 1875*d*:476.
Melanoplus punctulatus griseus; J. A. G. Rehn 1946*a*:257.

Diagnosis. Pronotum flaring anteriorly. Male cercus boot-shaped, with dorsal margin broadly rounded.

Description. Male cercus boot-shaped but with dorsal margin broadly rounded (see J. A. G. Rehn 1946*a*:247, fig. 9). Eyes prominent in anterior aspect. Pronotum flaring anteriorly to receive head. Male aedeagus with dorsal lobes inwardly curved apically, rounded, acute (see J. A. G. Rehn 1946*a*:247, fig. 10). Females with eyes prominent in dorsal aspect, usually determinable by association with males from same population.

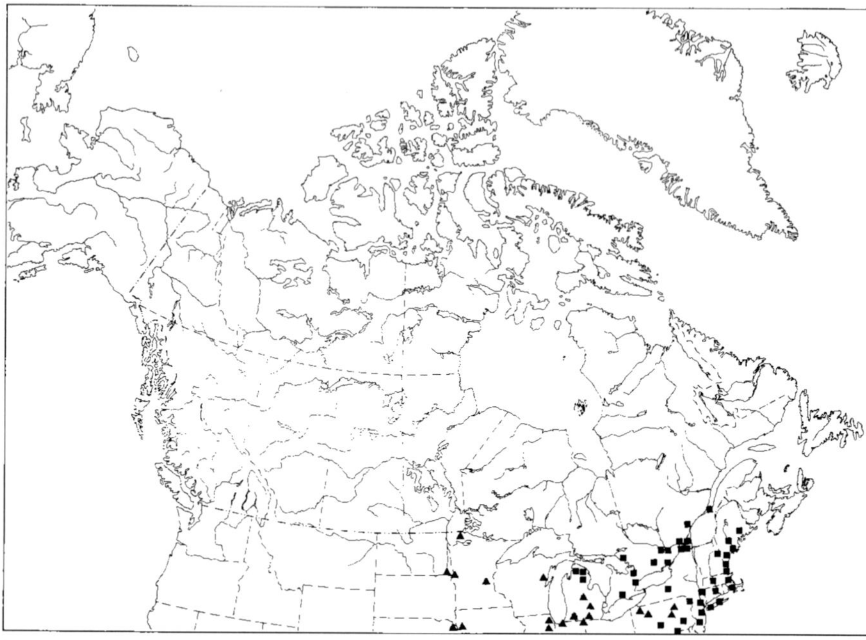

Map 151. Collection localities for *Melanoplus punctulatus griseus* (▲) and *M. punctulatus punctulatus* (■).

Range. North Dakota to Pennsylvania, south to Texas.

Behavior and habitats. *M. punctulatus griseus* is often found on tamarack (*Larix*), although in Indiana, Minnesota, and Nebraska it has been recorded mainly from willow, oak, beech, and maple.

Life history. As for *M. punctulatus punctulatus*. Cantrall (1968) recorded adults from 19 July to 14 September in Michigan.

Melanoplus punctulatus punctulatus (Scudder)

Figs. 3, 450, 487, 523, 568; Map 151

Caloptenus punctulatus Scudder, 1863*a*:465.

Melanoplus punctulatus; Scudder 1874:376; Helfer 1963:236 (partim).

Melanoplus punctulatus punctulatus; J. A. G. Rehn and Hebard 1916*b*:246; Cantrall 1943:113; Vickery et al. 1974:94.

Diagnosis. Occiput swollen. Male cercus large, boot-shaped. Ventral ovipositor valve lacking lateral tooth. Arboreal on pines.

Description. Body size medium (length, male 19–24, female 27–29 mm), with less size differential between sexes than for most species of the genus. Occiput swollen, elevated above pronotum; frontal costa convex, prominent, with sides parallel, slightly sulcate below ocellus; eyes large, prominent in male. Pronotal disc scarcely widening posteriorly, with posterior margin broadly obtuse-angulate. Tegmina extending beyond posterior femora in both sexes, tapered and rounded at apices. Male furculae virtually absent (Fig. 450). Male cercus large, broad, somewhat boot-shaped (Fig. 487). Male subgenital plate (Fig. 523) abruptly elevated at apex as conical tubercle. Aedeagus with dorsal lobes obliquely truncate apically, not strongly incurved (see J. A. G. Rehn 1946*a*: 247, fig. 3). Ovipositor with ventral valve almost straight, lacking lateral tooth, with upper valve possessing obtuse dorsal angle, about 170° (Fig. 568); lateral notch of eighth abdominal sternum ventral in position and acute-angled, about 60° (Fig. 568); female cercus subquadrate, rounded–truncate, tuberculate at tip. Color generally dark grayish brown, mottled with fuscous above, lighter beneath (Fig. 3); pronotal disc with sides of metazona and tegmina sprinkled with numerous rounded to quadrate fuscous spots; posterior femur barred with reddish brown and bright to dull yellowish on upper and outer faces; hind tibiae dull reddish to grayish, with black spines. The chromosome number was reported by Beaudry (1973) to be $2n$ male = 23.

Range. Michigan to Québec, south to northern Georgia.

Behavior and habitats. *M. punctulatus punctulatus* is not common anywhere. It occurs in isolated scattered colonies in forested areas in association with pine and other coniferous trees. Specimens are seldom found on the ground but usually rest upon the trunks of trees, a metre or more above the ground. Occasionally they are seen on the foliage, and several specimens have been seen in Québec on mature fruit of staghorn sumac. A colony in the Morgan Arboretum, Macdonald College, Sainte-Anne-de-Bellevue, Qué., is confined to white pine (*Pinus strobus*). E. M. Walker (1901) found females ovipositing in dead stumps and logs in southern Ontario. The female does not excavate holes in the wood but inserts her eggs in cracks, or in holes bored by beetles, either in sound or in partly decayed wood. *M. punctulatus punctulatus* is sluggish. When disturbed, individuals may leap to the ground, make one or two more short leaps, and then remain motionless, at which point they become difficult to observe because of their cryptic coloration.

Life history. Univoltine, with overwintering in the egg stage.

Melanoplus gladstoni Scudder

Figs. 438, 454, 491, 527, 572; Map 152

Melanoplus gladstoni Scudder, 1897*d*:229, pl. I, fig. 6, pl. XV, fig. 6.
Melanoplus corpulentis Scudder, 1897*d*:313.
Melanoplus conspersus Scudder, 1897*d*:315.
Melanoplus compactus Scudder, 1897*d*:316.
Melanoplus gladstoni; Lugger 1898:284; Helfer 1963:221.
Melanoplus bicoloratus Scudder in Scudder and Cockerell, 1902:48.

Diagnosis. Hind femur flattened ventrally near base (Fig. 438). This character distinguishes this species from all others in the genus.

Description. Antennal crescent (cf. Fig. 432) divided. Pronotal disc nearly flat, with lateral carinae rounded; median carina distinct on metazona, less distinct, blunt, on prozona. Tegmina long, extending at least to

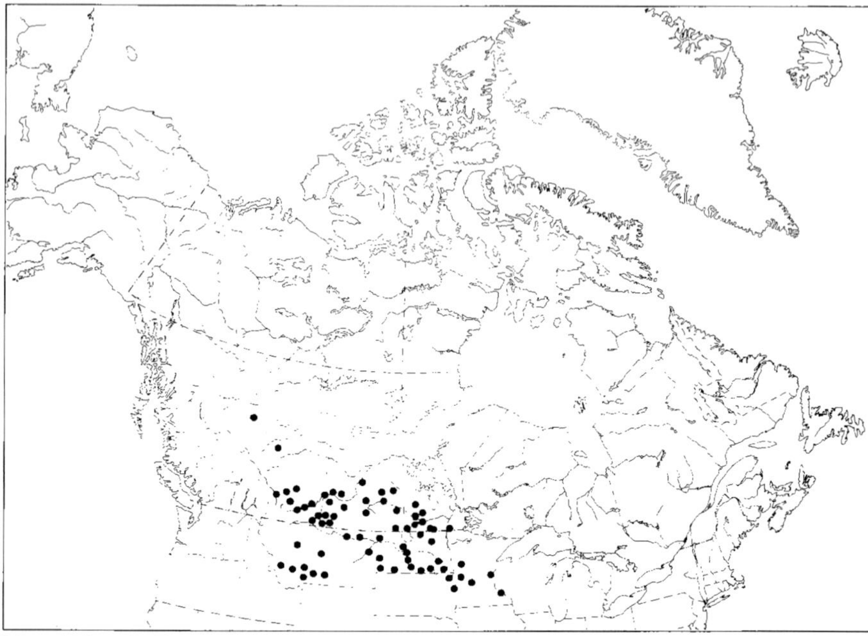

Map 152. Collection localities for *Melanoplus gladstoni*.

apices of hind femora, usually beyond. Hind femur flattened ventrally near base, with lower flange or carina obsolete on basal third (Fig. 438). Male supra-anal plate broad, with conspicuous but shallow depressions (Fig. 454); furculae small (Fig. 454). Male cercus broad, straplike, with sides nearly parallel (Fig. 491). Male subgenital plate broad, scarcely produced apically (Fig. 527). Ovipositor with dorsal valve angle about 136° (Fig. 572). Eighth abdominal sternum of female with conspicuous lateral notch, about 122° (Fig. 572). Color generally mottled brownish gray; pronotum pale, or with pale lines at lateral carinae; hind femur yellow below, banded on upper half of outer face; hind tibia usually red.

Range. Alberta to Manitoba, south to Arizona and New Mexico.

Behavior and habitats. *M. gladstoni* inhabits dry uplands and grassy areas, usually on sandy to gravelly soils. It is never associated with trees or forested areas. It feeds extensively upon grasses and also on forbs. It will feed almost entirely on alfalfa if this plant is available.

Life history. Univoltine, with overwintering in the egg stage. This is a late-maturing species. Adults are found from mid-August to late October.

Melanoplus angustipennis angustipennis (Dodge)

Figs. 474, 511, 548, 592; Map 153

Caloptenus angustipennis G. M. Dodge, 1877:111.
Caloptenus angustipennis; Bruner 1877:145.
Melanoplus comptus Scudder, 1897*d*:302.

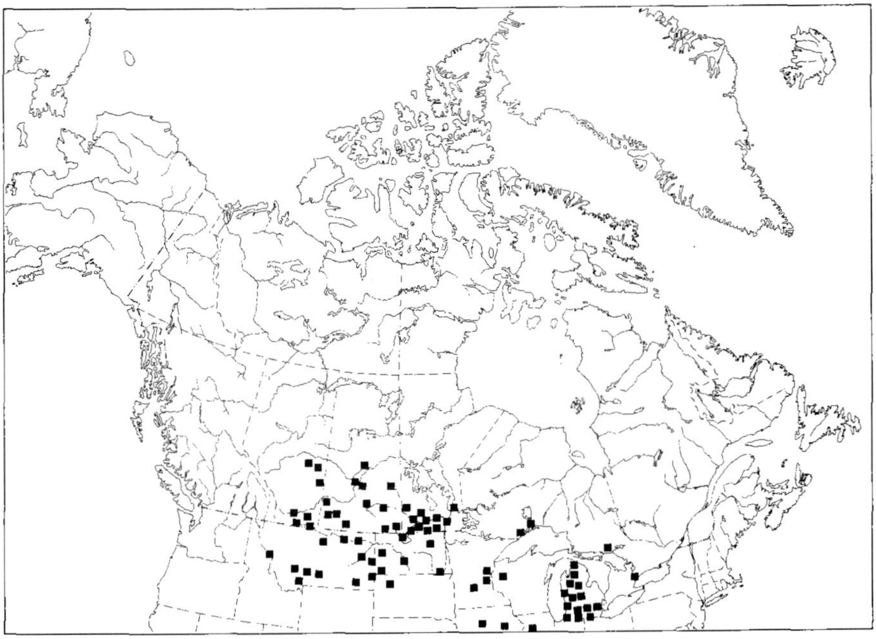

Map 153. Collection localities for *Melanoplus angustipennis angustipennis.*

Melanoplus angustipennis; Scudder 1897*d*:305; Cantrall 1943:122; Helfer 1963:231.

Melanoplus angustipennis coccineipes; E. M. Walker in Fletcher and Gibson 1909:114.

Melanoplus angustipennis angustipennis; Froeschner 1954:262; Brooks 1958:20.

Diagnosis. Body brownish gray; hind femur yellow with indistinct banding. Male cercus small, spatulate.

Description. Body size medium (length, male 19–22, female 21–22 mm), slender. Tegmina slender, tapering, reaching or extending slightly beyond apices of hind femora. Antennal crescent (cf. Fig. 432) divided or narrow. Pronotum with sides parallel in male, slightly diverging posteriorly in female, with metazona closely punctate. Male supra-anal plate narrow, pointed in apical third; furculae moderately large, with apices divergent (Fig. 473); male cercus small, symmetrical, nearly spatulate, narrowest at middle, with apex rounded (Fig. 511); apex of male subgenital plate usually notched (Fig. 548). Dorsal angle of ovipositor about 90° (Fig. 592); eighth abdominal sternum of female with prominent lateral notch, about 127° (Fig. 592); female cercus slightly concave above, convex below (cf. Fig. 576*a*). Color generally brownish gray; hind femur pale yellow, with banding on dorsal part of outer face indicated but obscure; hind tibia red to blue.

M. angustipennis angustipennis bears a superficial resemblance to *M. sanguinipes sanguinipes*, but it is easy to separate by the characters given in the key to species and in the relevant descriptions.

Range. Alberta to Ontario, south to Colorado.

Behavior and habitats. Cantrall (1943) found this species in mixed grass–herbaceous habitats but never in heavy stands of upland vegetation. Hubbell (1922*b*) found it to be common in sandy fields and dune areas. Although it had been reported as a mixed feeder preferring grasses by N. Criddle (1935), Mulkern et al. (1969) found that it feeds mainly upon forbs such as species of *Ambrosia, Artemisia, Helianthus*, and *Solidago*. Many adults were found to feed upon floral parts of *Helianthus petiolaris*. Maw (1976) collected this species by sweeping *Cirsium arvense, Centaurea diffusa*, and *Verbascum thapsus*.

R. W. Smith (1958) reported on parasites, which included several species of Diptera (Sarcophagidae, Tachinidae, and Muscidae).

Life history. Univoltine, with overwintering in the egg stage. Adults have been found from 13 June to 28 September in Michigan (Cantrall 1968).

Economic importance. Minor; of little economic significance in grasslands (Mulkern et al. 1969).

Melanoplus cinereus cinereus Scudder

Figs. 477, 514, 551, 595; Map 154

Melanoplus cinereus Scudder, 1878*b*:288.
Melanoplus cinereus; Scudder 1881:(24), pl. XVII, figs. 1, 4, 5.
Melanoplus cinereus cinereus; Hebard 1928:289.

Diagnosis. Pale gray with head and pronotum rusty; tegmina speckled. Male cercus narrow, tapered, twisted.

Description. Body medium–small (length 23–26 mm). Tegmina long, usually considerably exceeding hind femora. Supra-anal plate elongate-triangular; furculae broad, long, tapered (Fig. 477). Male cercus narrow, tapering, twisted, bent inward at middle and upward at tip, broadened apically, with apex rounded-truncate (Fig. 514). Subgenital plate longer than broad, somewhat flared and feebly elevated at apex (Fig. 551). Ovipositor as in Fig. 595. Color pale gray, usually with reddish black on head and pronotum, often rusty dorsally; tegmina finely speckled with reddish brown; hind femur rusty with darker bars on outer face; hind tibia pale blue with spines pale at bases and black at tips.

Range. British Columbia, south to California, east to Idaho and Montana.

Behavior and habitats. Buckell (1922) stated that this insect "makes its home among the Sage-brush and Rabbit-brush in the warmest sections of British Columbia." It remains on sagebrush during the day and is difficult to capture. At night, it climbs to the tops of the plants and remains on the uppermost twigs until warmed by the sun the following day. Hebard (1928) reported it as usually being associated with sagebrush.

Life history. Probably univoltine, with overwintering in the egg stage. Adults have been found from late July through August, but they probably persist through most of September also.

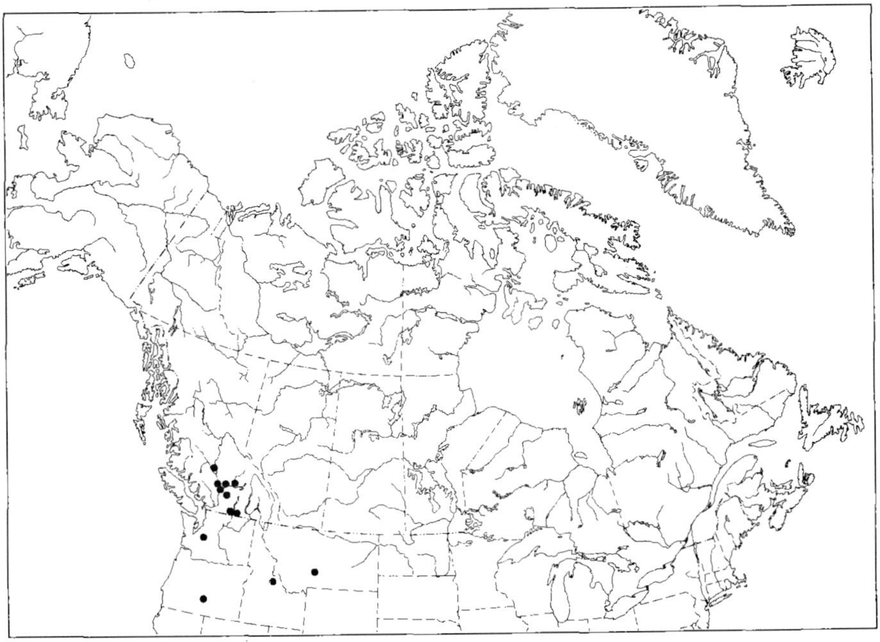

Map 154. Collection localities for *Melanoplus cinereus cinereus*.

Melanoplus packardii packardii Scudder

Figs. 473, 510, 546, 554, 555, 591; Map 155

Caloptenus fasciatus Scudder, 1875*d*:477 (homonym).
Melanoplus packardii Scudder, 1878*b*:287 (replacing *Caloptenus fasciatus*); Bruner 1885*b*:18; Brooks 1958:30; Helfer 1963:240.
Melanoplus packardi packardi; Hebard 1928:286.
Melanoplus packardii packardii; Brooks 1958:30.

Common name. Packard grasshopper.

Diagnosis. Blue gray to yellowish brown, usually with distinct lateral pronotal stripes; hind femur with bands dorsally only, these not distinct. Grasslands inhabitant. The aedeagal valves distinguish males of this species from those of *M. foedus*; females of these species not distinguishable.

Description. Body size medium to medium-large (length 27–32 mm). Tegmina extending slightly beyond apices of hind femora, those of male narrow. Male supra-anal plate with apical third depressed, and with apex acute (Fig. 473). Male cercus subspatulate, narrowest at middle of apex, broadly rounded, with outer face concave on apical third (Fig. 510). Male subgenital plate broadly rounded, with small median tubercle projecting posteriorly (Fig. 546); aedeagal valves about equal in length (Figs. 554, 555). Ovipositor with angle of dorsal valve about 105° (Fig. 591); lateral notch of eighth abdominal sternum of female about 90° (Fig. 591). Color generally bluish gray to yellowish brown (usually more yellowish in southern part

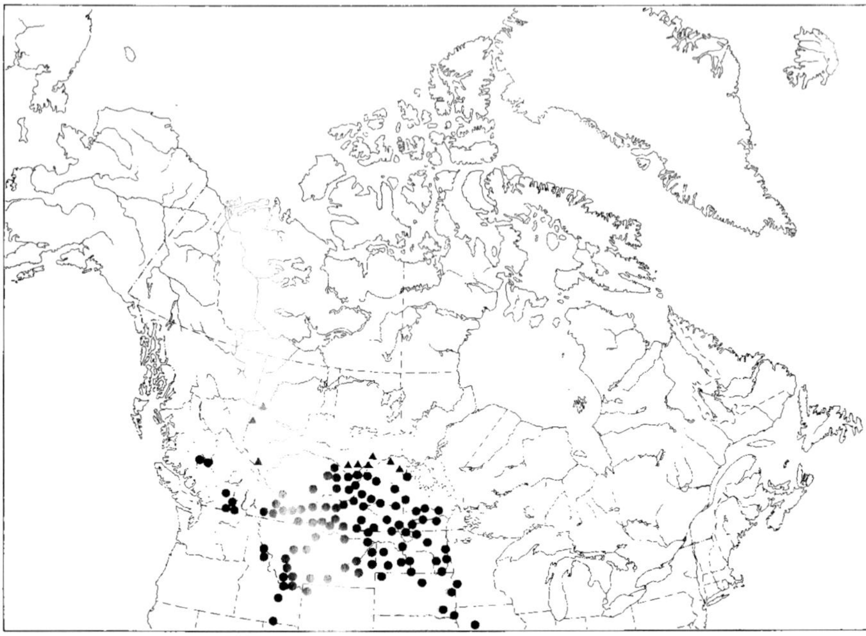

Map 155. Collection localities for *Melanoplus packardii packardii* (●) and *M. packardii brooksi* (▲).

of range); usually with distinct lateral pronotal stripes; hind femur banded only at top of inner ridge and on inner face, with upper edge of outer face often possessing narrow longitudinal fuscous stripe; hind tibia usually yellowish, but varying from blue to purple, pink, or reddish.

Range. British Columbia to Manitoba, south to California and Texas.

Behavior and habitats. *M. packardii packardii* is found almost exclusively in grasslands, especially on light, sandy, gravelly, or drift soils (Brooks 1958). Mulkern et al. (1969) found it to be primarily a forb feeder, but also that "grasses were ingested in proportion to their abundance." In Nebraska, preference was shown for legumes. The insect is known to feed on alfalfa.

Life history. Univoltine, with overwintering in the egg stage. Adults are active in the field during July and August and probably are present from late June to mid-September.

Economic importance. N. Criddle (1935) stated that *M. packardii packardii* had the potential to become a serious pest. Mitchener (1954*a*, 1954*b*) reported that it was "destructive locally," but not on a widespread basis like other pest species. Mulkern et al. (1969) stated that "it is frequently a pest of cultivated crops (in North Dakota) but is probably of little importance in grasslands." Beirne (1972) described it as "being on occasion the second most important species of grasshopper in Alberta and the third most important in Saskatchewan." He also reported considerable damage to fall rye and winter wheat. Hardman and Smoliak (1980) indicate potential damage to rangeland in Alberta.

Melanoplus packardii brooksi Vickery

Fig. 547; Map 155

Melanoplus packardii brooksi Vickery, 1979:699.

Diagnosis. Hind femur heavily banded dorsally. Inhabitant of forests and parklands in Saskatchewan and Alberta.

Description. Body small (length 23–31 mm). Male subgenital plate as in Fig. 547. Pronotal stripes not distinct; dark banding on outer face of hind femur distinct, extending on to upper half of outer face, giving femur distinctly banded appearance; hind tibia dark red.

Range. Alberta, Saskatchewan, and British Columbia.

Behavior and habitats. Occurs in Jack pine forests and parklands.

Life history. Presumably similar to that of *M. packardii packardii*. Adults have been collected from 27 July to 3 September.

Melanoplus stonei Rehn

Figs. 472, 509, 545, 590; Map 156

Melanoplus stonei J. A. G. Rehn, 1904*a*:85.
Melanoplus stonei; Blatchley 1920:432; Vickery et al. 1974:103.

Diagnosis. Tegmina with row of dark spots on basal half. Northern forest species.

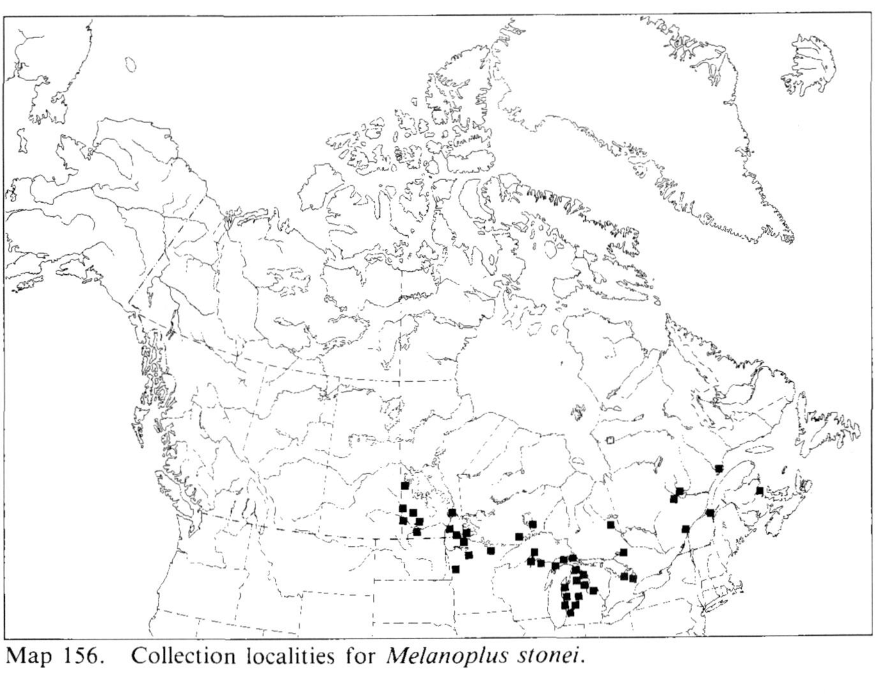

Map 156. Collection localities for *Melanoplus stonei*.

Description. Body size medium (length, male 22–23.5, female 26–28 mm). Occiput swollen, elevated above pronotum. Frontal costa broad, sulcate below ocellus. Tegmina slightly exceeding posterior femora. Male furculae widely separated, divergent, flat basally, scarcely tapering apically (Fig. 472). Male cercus subspatulate, with ventral margin straight and with dorsal margin broadly concave, incurved, and concave apically on outer surface (Fig. 509). Male subgenital plate scoop-shaped, quadrate, with apical margin not elevated, with small apical tubercle (Fig. 545); aedeagus much as in *M. foedus* (cf. Figs. 552, 553). Ovipositor broad, with angle of dorsal valve about 105° (Fig. 590); lateral notch of eighth abdominal sternum of female about 90° (Fig. 590). Color dark fuscous brown above, dull reddish to clay yellow beneath; tegmina with row of dark spots on basal half; hind femur dull yellow with 2 conspicuous broad black bands, sometimes fused to form dark streak on upper half of outer face; hind tibia dark red, sometimes with dark basal annulus.

Range. Manitoba to New Brunswick, south to Michigan.

Behavior and habitats. Cantrall (1968) reported *M. stonei* to be associated with Jack pine barrens in northern Michigan. Vickery et al. (1974) regarded it as a forest species found mainly in or near pine trees on well-drained sandy soils.

Life history. Undoubtedly univoltine, with overwintering in the egg stage. In Michigan, Cantrall (1968) recorded adults from 20 June to 13 September. Other available records from elsewhere fall between these dates.

Melanoplus foedus foedus Scudder

Figs. 508, 544, 552, 553, 589; Map 157

Melanoplus foedus Scudder, 1879*a*:69.

Melanoplus foedus; Scudder 1897*d*:311, pl. XX, fig. 9; Brooks 1958:26; Helfer 1963:240.

Melanoplus fluviatilis Bruner, 1897:136.

Melanoplus foedus foedus; Hebard 1932*a*:39; Brooks 1958:26.

Melanoplus foedus fluviatilis; Hebard 1932*a*:39; Brooks 1958:26.

Diagnosis. Resembles *M. packardii*. Requires examination of male genitalia for positive identification (Figs. 552, 553) (cf. Figs. 554, 555).

Description. Body size medium (length 22–32 mm). Male slender, female more robust. Tegmina of male narrow, tapering from base to rounded apex, of female broader with rounded apex, in both sexes surpassing apices of hind femora. Male supra-anal plate depressed on apical third, narrowed gradually apically; furculae short, widely divergent, well-separated. Male cercus symmetrical, spatulate, or apically slightly lobed beneath, narrowest at center (Fig. 508). Male subgenital plate extended, well-rounded at apex (Fig. 544); aedeagus with anterior process short, much less projected than posterior process (Figs. 552, 553). Dorsal ovipositor valve with dorsal angle about 90° (Fig. 589); lateral notch of eighth abdominal sternum of female prominent, deep, about 90° (Fig. 589). Color variable, pale to dark greenish

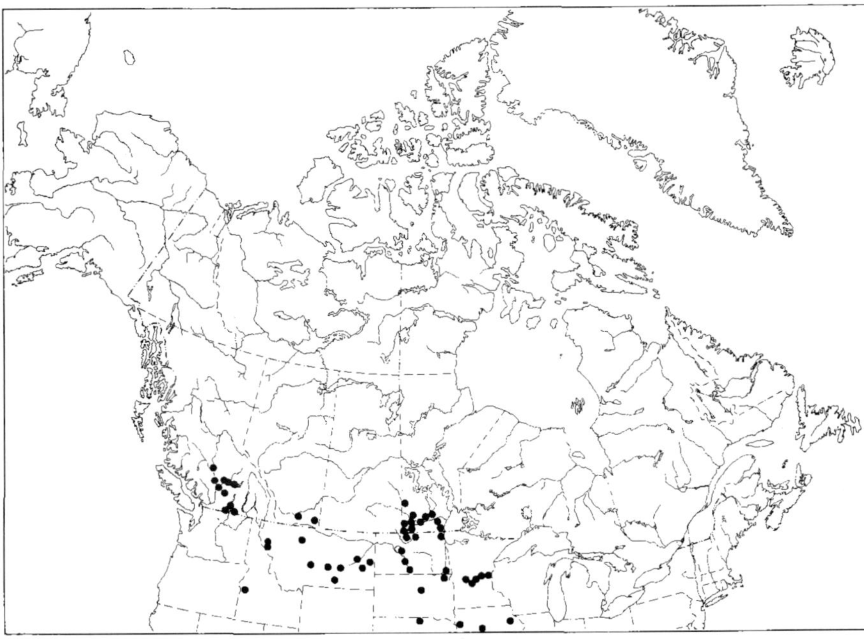

Map 157. Collection localities for *Melanoplus foedus foedus*.

brown or greenish gray; lateral pronotal stripes present or absent; tegmina without spots or with spots only on center near base; hind femur greenish to yellow beneath or with a reddish streak on outer face of ventral flange, and with outer face not usually banded, but with longitudinal discoloration on upper half; hind tibia blue to purple. Handford (1946) described the nymphs.

Range. British Columbia to Manitoba, south to Colorado and Oklahoma.

Behavior and habitats. *M. foedus foedus* usually inhabits sandy areas or lighter soils, particularly along rivers, streams, and lakes. N. Criddle (1935) found it on "soft" soils in southern Manitoba, such as mounds made by pocket-gophers, or in cultivated grain or stubble fields. Mulkern et al. (1969) reported it as mainly forbivorous in North Dakota, although this species also used small quantities of several types of grasses as food. The eggs are described by Onsager and Mulkern (1963).

Life history. Univoltine, with overwintering in the egg stage. Adults have been reported from July to September.

Economic importance. *M. foedus foedus* is not generally regarded as being of much economic significance. J. E. Mitchell and Pfadt (1974) found that it consumed only about half a gram of food grasses per gram of insect per day in short-grass prairie. Nevertheless, like many other species of the genus, it destroys more than it eats by cutting and dropping grasses.

Melanoplus femurrubrum femurrubrum (De Geer)

Figs. 429, 431, 435, 456, 493, 529, 574, 574*a*: Map 158

Acrydium (*femur rubrum*) De Geer, 1773:498, pl. 42, fig. 5.
Melanoplus femur-rubrum; Scudder 1874:375.
Caloptenus devorator Scudder, 1875*d*:474.
Caloptenus sanguinolentus Provancher, 1876:109.
Melanoplus interior Scudder, 1879*a*:71.
Melanoplus femur-rubrum femur-rubrum; Cantrall 1943:117; Brooks 1958:17.

Melanoplus femurrubrum femurrubrum; Vickery et al. 1974:97; Meloche et al. 1980:86.

Common name. Redlegged grasshopper.

Diagnosis. Hind femur black above, pale beneath on outer face; hind tibia bright red. Male cercus long, tapered on upper side.

Description. Body medium-large (length, male 16–23, female 18–28 mm). Tegmina longer than abdomen. Prosternal tubercle nearly cylindrical, with tip bluntly rounded. Male cercus at least three times as long as middle breadth, with apical half less than half as broad as base (Fig. 493). Male subgenital plate rounded, bulbous, not notched at dorsal apex (Fig. 529). Ovipositor with angle of dorsal valve about 133° (Fig. 574); notch of eighth abdominal sternum shallow (Fig. 574). Female cercus triangular, with both sides concave (Fig. 574*a*). Color brownish above, grayish white

Map 158. Collection localities for *Melanoplus femurrubrum femurrubrum*.

to yellow beneath; outer face of hind femur black above, with lower edge yellowish; hind tibia bright red with black spines.

Range. Western Northwest Territories to Nova Scotia, south to California and Texas.

Behavior and habitats. *M. femurrubrum femurrubrum* is one of the most numerous grasshoppers in North America. It has wide ecological diversity and great adaptability. It is mainly a forb feeder and will feed extensively upon alfalfa if available. Mulkern et al. (1969) recorded *Poa pratensis* as the main grass utilized as food, although other grasses were fed upon to a lesser extent. It feeds mainly during morning hours. Although it is found on nearly all soil types, it is most abundant in damper areas.

M. femurrubrum femurrubrum is attacked by a range of natural enemies, including parasitic flies, nematodes, and a mite. Numerous specimens parasitized by an externally feeding mite have been seen in southwestern Québec, and occasional counts of nematode parasitism in the same area have been as high as 35%. The nematode species in this instance was *Mermis nigrescens* Dujardin.

Life history. Univoltine, with overwintering in the egg stage. Cantrall (1968) recorded adults from 12 July to 18 November in Michigan. In Québec, specimens of this species have often been active in late October and early November, following several severe frosts.

Economic importance. Occasional outbreaks have caused moderate crop damage. Beirne (1972) reported 20–25% crop damage to oats in Eastern Canada. Other crops, such as bean, potato, turnip, corn, buckwheat, rye, and grass have been damaged by this species, usually at field margins. Clover and alfalfa are damaged by high population levels of this insect. Meloche et al. (1980) report damage to tobacco crops in Québec. It is probable, however, that some reports of damage by *M. femurrubrum femurrubrum* should be referred to other species. It is much less important in grasslands than the other major "pest" species.

Melanoplus flavidus Scudder

Figs. 475, 512, 549, 556, 557, 593; Map 159

Melanoplus flavidus Scudder, 1879*a*:74.

Melanoplus flavidus; Scudder 1897*d*:158; Brooks 1958:25; Helfer 1963:235.

Melanoplus incisus Scudder, 1898*c*:163.

Melanoplus flavidus flavidus; Hebard 1925*a*:118.

Diagnosis. Male cercus broad at base, with slender rounded apex. Dorsal angle of ovipositor 90°. Dark species, with hind femur yellow beneath. Male genitalia distinguish it from *M. bowditchi canus* Hebard.

Description. Body size medium (length, male 21–26, female 25–32 mm), slender. Antennal crescent (cf. Fig. 432) divided. Pronotal disc slightly convex, with sides parallel, rounded into lateral lobes; metazona nearly flat, with sides feebly divergent posteriorly; posterior margin obtuse-angulate, with median carina distinct on metazona, indistinct on prozona.

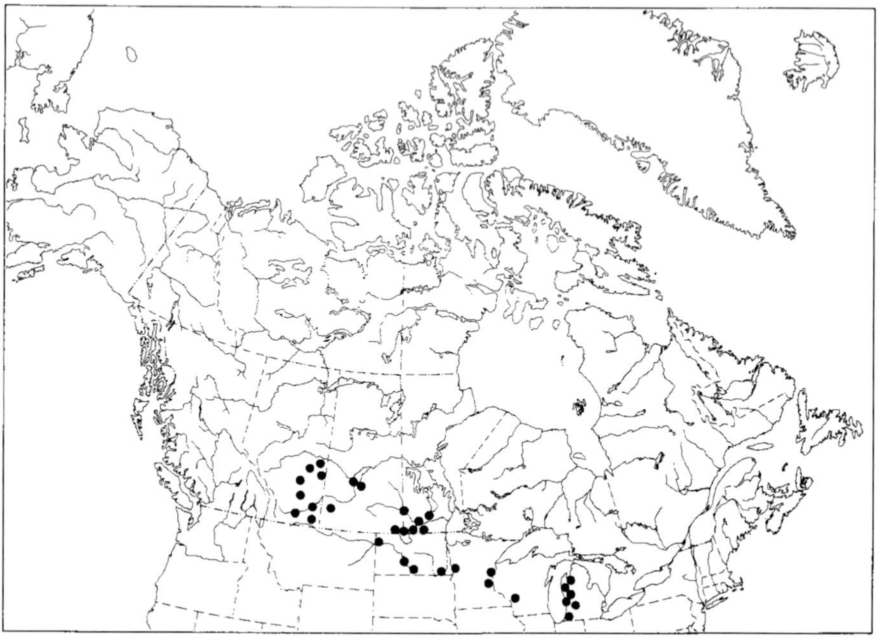

Map 159. Collection localities for *Melanoplus flavidus*.

Tegmina narrow, surpassing apex of abdomen in males, reaching ovipositor base in females. Male supra-anal plate broad, concave with apex acute; furculae broad, long, flat, with apices truncate (Fig. 475). Male cercus broad at base, slender distally to rounded apex (Fig. 512). Male subgenital plate obtusely pointed apically (Fig. 549); aedeagus as in Figs. 556, 557. Dorsal angle of ovipositor valve about 90° (Fig. 593); eighth abdominal sternum of female with shallow lateral notch, about 148° (Fig. 593); female cercus long, slender, with sides concave. Color generally brown and black, with lateral pronotal stripe black, broad, continuous; tegmina not speckled; hind femur yellow beneath, with crossbands present dorsally on outer face; hind tibia blue or purple. In parts of the United States, the general coloration is much paler; Froeschner (1954) reported it as being "olive to yellowish-brown."

Range. Alberta to Michigan, south to Colorado and Kansas.

Behavior and habitats. *M. flavidus* usually occurs in open sandy areas with sparse vegetation in grasslands or parklands. Mulkern et al. (1969) found it to be entirely a forb feeder, stating also that the mandibles were distinctly of the forbivorous type. Like *M. angustipennis angustipennis*, it was found to feed upon *Helianthus* and to eat the floral parts of *H. petiolaris*. N. Criddle (1935) and R. M. White and Rock (1945) had previously reported it as preferring grasses in Manitoba and Alberta, respectively. It is probable that it feeds on grasses only when forbs are scarce.

Life history. Univoltine, with overwintering in the egg stage. Adults were reported from 25 July to 13 September in southwestern Michigan by Cantrall (1968).

Melanoplus bowditchi canus Hebard

Figs. 476, 513, 550, 558, 559, 594; Map 160

Melanoplus bowditchi canus Hebard, 1925*a*:120, figs. 1, 2.
Melanoplus bowditchi canus; Hebard 1928:288; Brooks 1958:23.
Melanoplus bowditchi; Mills and Pepper 1938:21.

Diagnosis. Tegmina speckled. Male genitalia distinguish it from *M. flavidus* Scudder.

Description. Body small (length, male 19–23.5, female 22.5–25.5 mm), slender. Tegmina extending well beyond apices of hind femora. Antennal crescent (cf. Fig. 432) divided. Median carina of pronotum prominent on metazona, only faintly indicated on prozona. Male supra-anal plate broad, short, concave; furculae large, flat, with apices truncate (Fig. 476). Male cercus broad at base, slender on apical half, with apex rounded (Fig. 513). Male subgenital plate with apex obtusely pointed (Fig. 550); aedeagus as in Figs. 558, 559. Angle of dorsal ovipositor valve about 90° (Fig. 594); lateral notch of eighth abdominal sternum of female about 148° (Fig. 594); female cercus long, slender, with sides concave. Color generally gray and bluish gray; lateral pronotal stripe not continuous; tegmina finely speckled with darker spots; hind femur pale beneath, not banded on outer face; hind tibia blue.

Range. Alberta and Saskatchewan, south to western Colorado.

Behavior and habitats. *M. bowditchi canus* occurs in "soft" soils, particularly at the bases of steeply eroded banks. Brooks (1958) reported that

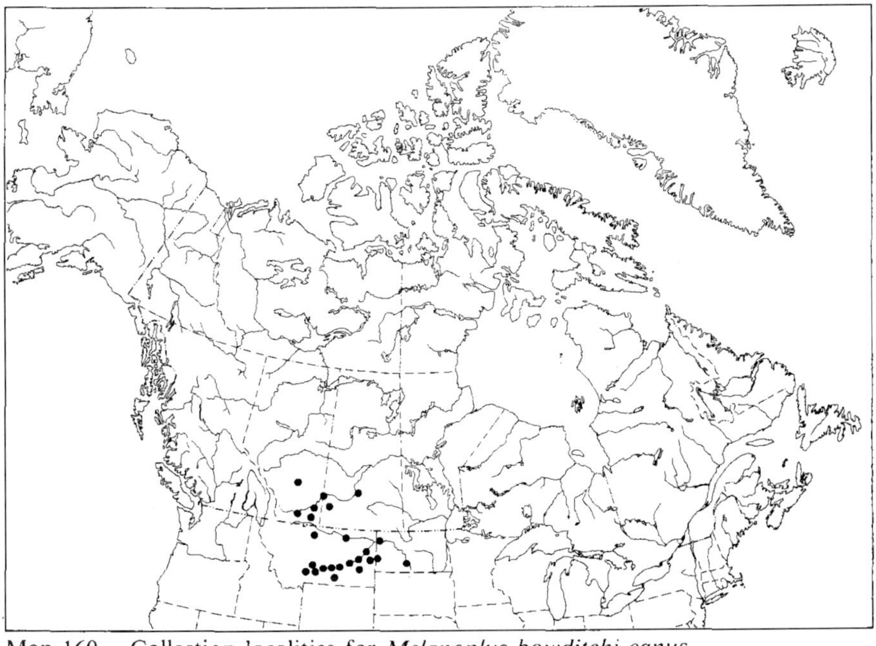

Map 160. Collection localities for *Melanoplus bowditchi canus*.

it was seldom found more than a few metres away from this type of habitat, even though the host plant, sagebrush (*Artemisia cana* Pursh) was generally distributed in the area. Mulkern et al. (1969) found it to feed almost entirely on *Artemisia filifolia* in Nebraska, although *A. cana* was the main food plant in western North Dakota, with minor feeding on *A. frigida*.

Life history. Univoltine, with overwintering in the egg stage. Adults are known to be active from 20 July to 5 September.

Melanoplus occidentalis occidentalis (Thomas)

Figs. 467, 503, 539, 584; Map 161

Caloptenus occidentalis C. Thomas, 1872:453.
Melanoplus variolosus Scudder, 1879*a*:67.
Melanoplus flabellifer Scudder, 1879*a*:69.
Melanoplus cuneatus Scudder, 1897*d*:147.
Melanoplus occidentalis; Scudder 1896*c*:66; Helfer 1963:239.
Melanoplus occidentalis occidentalis; Hebard 1925*a*:109; Brooks 1958:29.

Diagnosis. Hind femur red beneath and on inner face; hind tibia bright blue. Male cercus earlike.

Description. Body medium-small (length 18–22 mm). Antennal crescent (cf. Fig. 432) divided. Tegmina extending to or slightly beyond apices of hind femora. Male supra-anal plate subquadrate; furculae short, blunt

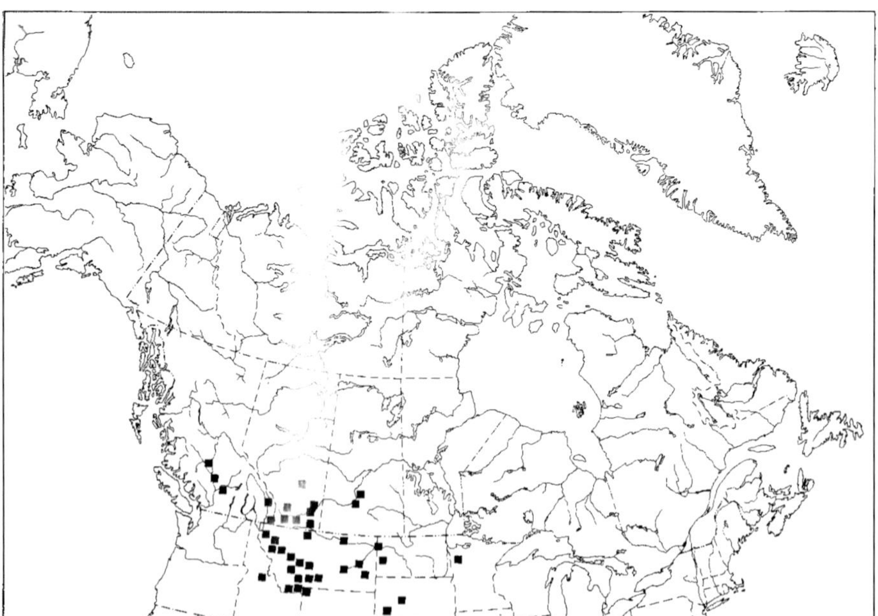

Map 161. Collection localities for *Melanoplus occidentalis occidentalis*.

(Fig. 467). Male cercus large, earlike, bent inward above supra-anal plate (Fig. 503). Male subgenital plate produced, rounded-acute apically (Fig. 539). Angle of dorsal ovipositor valve about 132° (Fig. 584); eighth abdominal sternum of female not notched (Fig. 584); female cercus short, broad, flat. Color generally dark gray; hind femur bright red beneath and on inner face, conspicuously banded on outer face; hind tibia bright blue.

Range. British Columbia to Saskatchewan and Minnesota, south to Arizona and Texas.

Behavior and habitats. *M. occidentalis occidentalis* is said to be a grass feeder, preferring *Agropyron smithii* Rydb. (N. Criddle 1935). Buckell (1922) reported it (as *M. flabellifer*) in considerable numbers in a sagebrush flat in the Chilcotin River Valley of British Columbia. Although it is common on short-grass plains, Mulkern et al. (1969) found it to feed primarily on forbs. It appears to be variable in this respect, able to survive on whatever food is available. As well as having a plains distribution, it has also been taken in open grassy areas in the mountains above the timberline at 3890 m in the southern part of its range (Hebard 1928).

Life history. Adults were recorded between 26 June and 5 October in Montana by Hebard (1928). The life cycle is probably completed in 1 year.

Economic importance. This species has not been reported as being of any economic significance, but, like many others, it could compete with livestock for grazing or browsing under drought conditions.

Melanoplus rugglesi Gurney

Figs. 468, 504, 540, 585; Map 162

Melanoplus rugglesi Gurney, 1949:267–272.
Melanoplus rugglesi; Gurney 1953*a*:306; Helfer 1963:238.

Common name. Nevada sage grasshopper.

Diagnosis. Tegmina long. Hind femur banded; hind tibia purplish. Closely resembling *M. occidentalis occidentalis* (Thomas).

Description. *Migratory phase*: tegmina long, extending beyond apex of abdomen. Pronotum with anterior margin broadly and weakly emarginate, flaring on metazona, with posterior margin obtuse-angulate. Male cercus broad, with apex quadrate (Fig. 504). Aedeagus with fleshy cuplike rim enclosing three pairs of vertical structures (see Gurney and Buxton 1965: pl. 3, fig. 14). Color generally reddish brown with contrasting blackish and pale areas; hind tibia purplish; hind femur banded.

Solitary phase: somewhat smaller. Tegmina shorter, of male usually extending to apex of abdomen, of female usually shorter than abdomen, in some British Columbia specimens not reaching apices of hind femora. Color usually more grayish, with yellow and reddish tinges on pronotum and on hind femur much less developed.

Range. British Columbia to California, inland to Montana and Utah.

Behavior and habitats. *M. rugglesi* is mainly a shrub feeder but will also feed upon forbs and grasses. Sagebrush (*Artemisia*) appears to be the

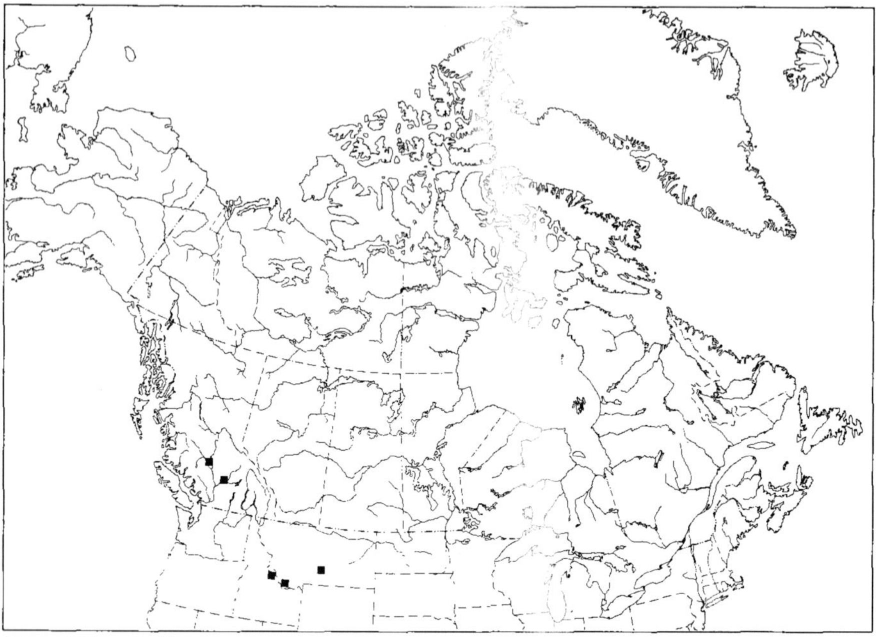

Map 162. Collection localities for *Melanoplus rugglesi*.

usual food plant. Cowan (1971) reported that grasses were fed upon much less than either shrubs or forbs and the amount of grass eaten was negligible. The migratory swarming phase sometimes travels as much as 120 km in a season. Migration flights appear to be somewhat aimless, except that temperature, sunlight, and especially wind influence the direction in which swarms will travel. Moderate to strong winds cause them to come to the ground. Gallaway (1948) recorded that swarms move only when the temperature is above 27°C and when the wind is light. Flight direction is said usually to be upwind but occasionally across the wind. The true flight direction in relation to air currents, however, is difficult to judge from the ground, just as it is with other migratory locusts.

Life history. *M. rugglesi* is an early season species. It overwinters in the egg stage, hatching early in May. Adults occur from early June and usually disappear by early September.

Economic importance. This species, when it is abundant in the migratory swarming phase, can be destructive, and Gallaway (1948) reviewed this aspect in the late 1940s (as *M. occidentalis* (Thomas)). Some early reports of the Rocky Mountain locust, *M. spretus* (Walsh), in Nevada probably referred instead to *M. rugglesi* (Gurney and Brooks 1959). As it is largely a sagebrush feeder, however, it is not usually a serious pest of rangeland, except under drought conditions when livestock is sometimes reduced to feeding on this plant. In the region here covered, it is not usually of much economic significance.

Melanoplus kennicottii kennicottii Scudder

Figs. 466, 502, 538, 583; Map 163

Melanoplus kennicottii Scudder, 1878*b*:287, 289, 290.
Melanoplus kennicottii; Scudder 1896*c*:63.
Melanoplus kennicotti kennicotti; Hebard 1928:277; Brooks 1958:28.

Diagnosis. Tegmina speckled dorsally. Male cercus large, rectangular, depressed dorsally at apex.

Description. Body small (length 17–22 mm). Antennal crescent (cf. Fig. 432) divided. Tegmina extending to apices of hind femora, speckled dorsally. Male supra-anal plate triangular; furculae short, acute (Fig. 466). Male cercus large, rectangular, with rounded apex, depressed dorsally at apex (Fig. 502). Male subgenital plate produced, apically truncate to acutely rounded (Fig. 538). Angle of dorsal ovipositor valve about 120° (Fig. 583); eighth abdominal sternum of female with shallow lateral notch, about 128° (Fig. 583); female cercus triangular. Color generally gray, with greenish tinge; posterior part of pronotum often yellow; hind femur yellow beneath, with crossbands obvious only on upper half of outer face; hind tibia buff or blue.

Range. Alaska to Saskatchewan, south to Montana and South Dakota.

Behavior and habitats. *M. kennicottii kennicottii* prefers forbs but will eat grasses and sedges (Brooks 1958), although N. Criddle (1935) listed it as a grass-feeding species. Hebard (1925*a*) called it a "scarce boreal species." Some specimens have been collected at high altitudes on grassy slopes.

Life history. Adults have been collected during July and August. The length of the life cycle is unknown.

Melanoplus gordonae Vickery

Figs. 457, 494, 530, 575; Map 163

Melanoplus gordonae Vickery, 1969*b*:268, figs. 9–20.
Melanoplus gordonae; Vickery 1969*b*:271.

Diagnosis. Male subgenital plate trilobate apically.

Description. Frontal costa of head broad and of nearly uniform breadth, with slight enlargement at median ocellus pronounced and abrupt. Median carina of pronotum elevated on metazona. Male furculae long, broad, subparallel (Fig. 457). Male cercus broad throughout, not distinctly narrowed apically, with apical margin shallowly emarginate (Fig. 494). Male subgenital plate distinctly trilobate apically, with lateral lobes obliquely truncate (Fig. 530); epiphallus large, heavily sclerotized, with bridge and lophi broad and heavy; aedeagal valve narrow apically in dorsal aspect, obtusely rounded in lateral aspect (see Vickery 1969*b*:269, figs. 17, 18). Dorsal ovipositor valves with angle about 130° (Fig. 575). Female cercus triangular, with margins strongly convex.

Range. Known only from the type locality, near Fairbanks, Alaska.

Behavior and habitats. Unknown.

Life history. Unknown.

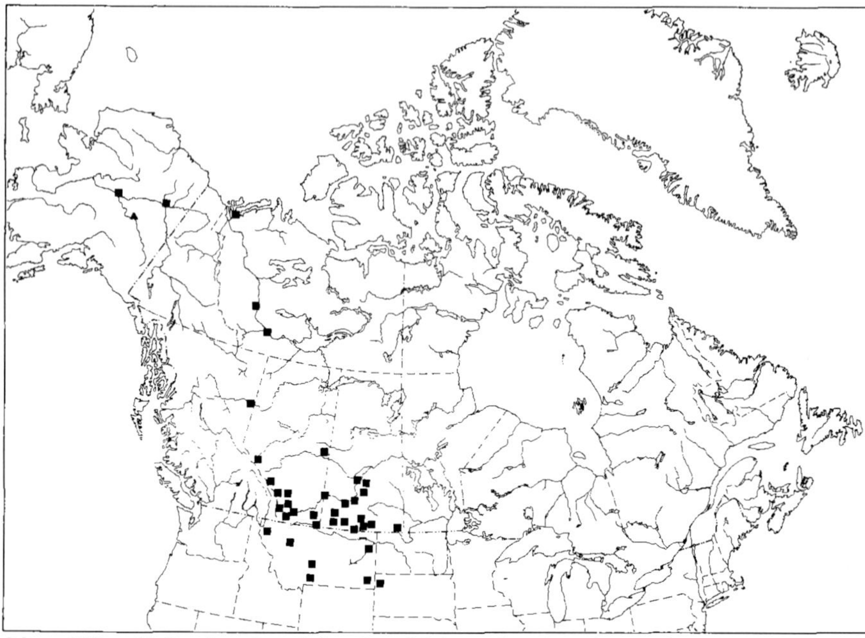

Map 163. Collection localities for *Melanoplus kennicottii kennicottii* (■) and *M. gordonae* (▲).

Melanoplus bruneri Scudder

Figs. 463, 499, 535, 580; Map 164

Melanoplus bruneri Scudder, 1896*c*:64.

Melanoplus bruneri; Scudder 1897*c*:19, 32; Brooks 1958:24; Helfer 1963:216; Vickery et al. 1974:102.

Melanoplus excelsus Scudder, 1897*c*:19, 32.

Melanoplus alaskanus Scudder, 1897*c*:20, 32.

Melanoplus bruneri alaskanus; E. M. Walker 1909:206.

Diagnosis. Male subgenital plate apically notched, produced upward posteriorly. Upper margin of female cercus straight.

Some male specimens of *M. bruneri* closely resemble *M. sanguinipes sanguinipes* and can be identified with certainty only by comparison of male aedeagal valves. Females are usually more robust and have shorter tegmina than *M. sanguinipes sanguinipes* females taken at the same location.

Description. Body size medium (length, male 20–25, female 22–29 mm; pronotal length, male 3.6–5.1, female 3.6–6.0 mm), robust. Tegmina reaching or exceeding apices of hind femora. Male supra-anal plate long, rectangular, with pointed apex. Furculae large, broad (Fig. 463). Male cercus moderately large, flat, curved upward apically (Fig. 499). Male subgenital plate long, produced upward posteriorly and distinctly depressed medially, appearing bilobed (Fig. 535); aedeagus short, with dorsal valve large, striated,

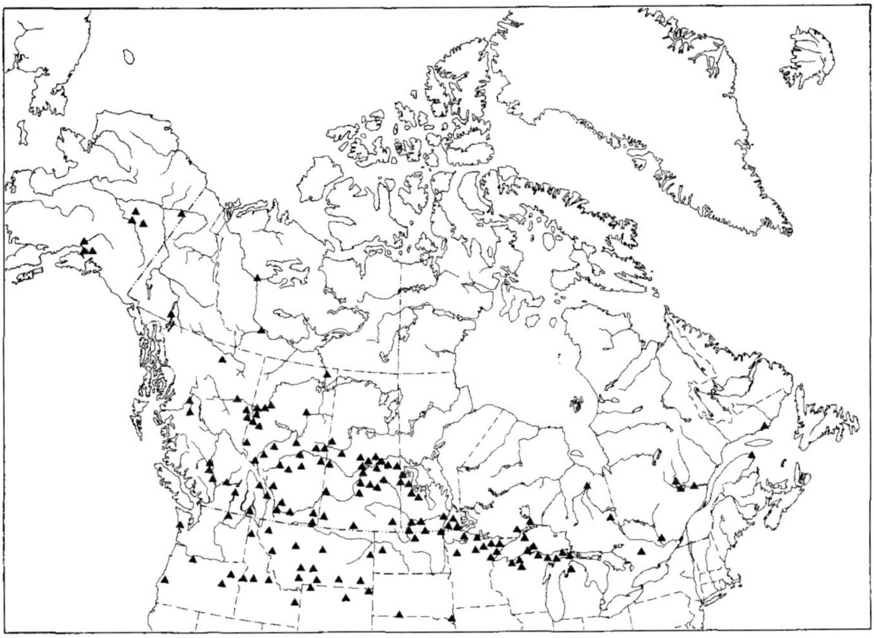

Map 164. Collection localities for *Melanoplus bruneri*.

moderately sclerotized, appearing parchmentlike, with triangular projection of sclerotized dorsal valve apparent in dorsal aspect (see Gurney and Brooks 1959: 17, figs. 2*a*, 2*b*, 19, figs. 3*k*, 3*l*). Ovipositor with dorsal valve angle about 130° (Fig. 580); female cercus triangulate, with upper margin nearly straight. Color pale brown to blackish; hind femur usually distinctly banded (less apparent in pale specimens); hind tibia usually pink to red, occasionally pale greenish yellow.

Range. Alaska to Québec, south to Arizona.

Behavior and habitats. *M. bruneri* is found in forested areas, grassy mountain slopes, and rangelands where bushes and shrubs are common. Brooks (1958) reported it as a mixed feeder preferring grasses. Treherne and Buckell (1924) reported it as a range pest following outbreaks in 1920 and 1921, when range grasses were heavily damaged. Kreasky (1960) investigated the plant species that were eaten by *M. bruneri* on rangelands on the eastern slope of the Crazy Mountains in Montana. Although 13 dicotyledenous and 11 monocotyledenous plant species were present, only 8 of these were eaten, and not all of them at the same time. Lupin (*Lupinus sericeus*) was eaten until depleted, after which *M. bruneri* fed mainly on timothy grass (*Phleum pratense*) and yarrow (*Achillea lunulosa*).

Life history. Cantrall (1968) reported adults from 5 July to 3 October in the Upper Peninsula of Michigan. In southern areas there is probably a single generation each year, with overwintering in the egg stage. After outbreaks in 1952 and 1954 (but not in 1953) in parts of British Columbia,

Gurney and Brooks (1959) suggested the life cycle there may require 2 years. This may also be true for all northern populations. Kreasky (1960) suggested that *M. bruneri* possibly has a 2- or 3-year life cycle at higher altitudes in Montana.

Economic importance. Outbreaks causing appreciable damage to rangeland grasses occurred in British Columbia in 1920 and 1921 (Treherne and Buckell 1924) and in 1952 and 1954 (Gurney and Brooks 1959). The latter authors also recorded severe damage on Mount Moriah, Nevada, from 1936 to 1938. The population of this species during this last infestation was decimated by cold and a severe hail storm. Beirne (1972) recorded reports of damage to seed crops of timothy grass, alsike clover, and alfalfa in north central British Columbia and to oats in Alberta. Outbreaks are not likely to occur in open rangeland areas, but future problems could well develop where grassland is reasonably well covered with shrubby growth.

Melanoplus sanguinipes sanguinipes (Fabricius)

Figs. 430, 432, 461, 498, 534, 579, 579*a*, 600–609; Map 165

Gryllus sanguinipes Fabricius, 1798:195.
Caloptenus bilituratus F. Walker, 1870:679.
Caloptenus scriptus F. Walker, 1870:680.
Caloptenus arcticus F. Walker, 1870:681.
Caloptenus atlanis Riley, 1875:169.
Melanoplus bilituratus; Scudder 1896*c*:64; Brooks 1958:20.
Melanoplus mexicanus atlanis; Morse 1920:499.
Melanoplus mexicanus bilituratus; Buckell 1930:41.
Melanoplus bilituratus atlanis; Brooks 1958:20.
Melanoplus bilituratus bilituratus; Brooks 1958:20.
Melanoplus sanguinipes sanguinipes; Gurney 1962:189; Vickery et al. 1974:101.
Melanoplus sanguinipes; Anderson 1964:736.

Common name. Migratory grasshopper.

Diagnosis. Tegmina long. Apex of male subgenital plate notched. Female cercus triangular, with margins convex. Hind femur with 2 oblique dark crossbars.

Description. Body size medium (length, male 17–21, female 16–27 mm). Dorsal carinae of vertex moderately prominent. Prosternal tubercle tapering, with apex acute. Tegmina long, much exceeding apex of abdomen in both sexes (Fig. 430). Male furculae distinctly diverging at apices (Fig. 461). Male cercus nearly equally broad throughout, rounded apically, with length not greater than twice width at middle (Fig. 498). Apex of male subgenital plate not strongly elevated, but with distinct median notch (Fig. 534). Angle of dorsal ovipositor valve about 117° (Fig. 579); female cercus short, triangular, with margins convex (Fig. 579*a*). Color grayish yellow to dark grayish brown or blackish brown dorsally, yellowish beneath; tegmina spotted or not; hind femur dirty yellowish brown with 2 oblique blackish crossbars on upper and outer faces; hind tibia variable, dark red to pink or yellowish, buff glaucous, or blue.

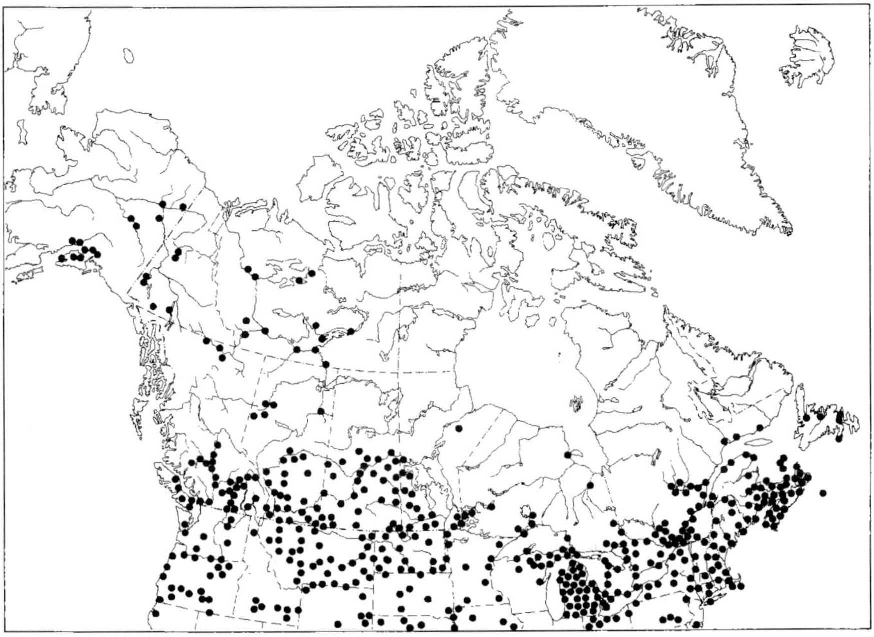

Map 165. Collection localities for *Melanoplus sanguinipes sanguinipes.*

Range. Alaska to Newfoundland, south to northern California and New Jersey.

Behavior and habitats. King and Slifer (1955) report breeding data. Habitats are variable, from sea level to mountainous elevations of up to 2700 m, and from southern plains to northern tundra (in more favorable locations). *M. sanguinipes sanguinipes* inhabits grassy fields, pastures, roadsides, arid lands, and boggy situations. Occasionally it occurs in open areas in forests. Mulkern et al. (1969), reporting on food plants, stated that nymphs appeared to prefer forbs, but that adults feed more extensively upon grasses. A wide range of food plants is accepted as food, but the insect is also to some extent omnivorous (N. Criddle 1935). This, and its propensity to migrate over considerable distances, are probably at least part of the reason for its ability to colonize such a wide range of habitats. It must also be very tolerant of conditions of temperature and humidity.

Egg predators, such as field crickets (*Gryllus* spp., usually *G. pennsylvanicus*) may destroy up to 70% of the eggs. Both nymphs and adults may be parasitized by at least 12 species of Diptera, two kinds of nematodes, and by external parasitic erythraeid mites. These mites are bright red and are noticeable, even at a distance, when infestations are heavy. Nevertheless, even heavy mite infestations do not appear to have a drastic effect upon either individuals or populations.

Life history. According to Parker (1930), females deposit an average of 8.8 (up to 15) egg pods. Pickford (*in* Gurney and Brooks 1959) obtained an average of 20 eggs per pod from 43 pods produced by one female. The life cycle is univoltine over most of the range, with overwintering in the egg stage in soil. There is an indication that the life cycle takes 2 years in elevated parts of British Columbia. This may also be true for most northern and high altitude localities.

Economic importance. Beirne (1972) has summarized the salient points regarding economic damage caused by this insect in Canada. It is now usually ranked as the most important pest species of grasshopper in Western Canada, where it often attacks cereal crops. Damage is usually more severe to a depth of about 20 m from field margins. Up to 70% of open range grasses in British Columbia have been destroyed during outbreaks. In Eastern Canada, it is less important, ranking only fourth in economic importance among grasshopper pest species. Beirne (1972) refers to egg pod densities of up to 230 per square metre. The potential for increase is so great that *M. sanguinipes sanguinipes* may become a pest of major proportions in any season and almost anywhere when conditions are favorable.

Melanoplus spretus (Walsh)

Figs. 462, 610; Plate I*A*; Map 166

Caloptenus spretus Walsh, 1866:1.

Melanoplus spretus; Scudder 1878*a*:287; Brooks 1958:30; Helfer 1963:218.

Caloptenus spretus caeruleipes Cockerell, 1889:127.

Common name: Rocky Mountain grasshopper.

Diagnosis. Pronotum constricted anterior to principal sulcus, shallowly V-shaped in lateral aspect. Tegmina long. Treated as extinct.

Comments. Faure (1933) tried breeding experiments with *M. sanguinipes sanguinipes*, rearing it under crowded and under solitary conditions. Specimens reared under crowded conditions developed color patterns like those of *M. spretus*. Several authors, including Hebard (1931*a*, 1932*a*, 1936*a*), Froeschner (1954), and Cantrall and Young (1954) regarded *spretus* as a "migratory phase" of *M. mexicanus mexicanus* (= *M. sanguinipes sanguinipes*). Nevertheless, the male genitalia of the two differ sufficiently so that they are now treated as distinct species (cf. Gurney and Brooks 1959:17, fig. 2*n*, and 19, fig. 3*a*, with 17, figs. 2*e*, 2*f*, and 19, figs. 3*e*, 3*m*). Even if one could construe the difference as "phase" variation, this is not reconcilable with the fact that *M. sanguinipes sanguinipes* not only migrates but occurs naturally at high population densities without any indication of transforming into *M. spretus*.

Description. Body medium–large (length 30–36 mm; pronotal length, male 4.2–4.5, female 4.8–5.4 mm), rather robust. Fastigium projecting well ahead of eyes. Dorsal carinae of vertex inconspicuous. Eyes not bulging. Tegmina long (Fig. 610, Plate I*A*), extending beyond apex of hind femur (4.5–10 mm in male, 5–10 mm in female). Pronotum constricted anterior to principal transverse sulcus; in lateral aspect, shallowly V-shaped, lowest at principal sulcus. Male supra-anal plate as in Fig. 462. Male cercus short,

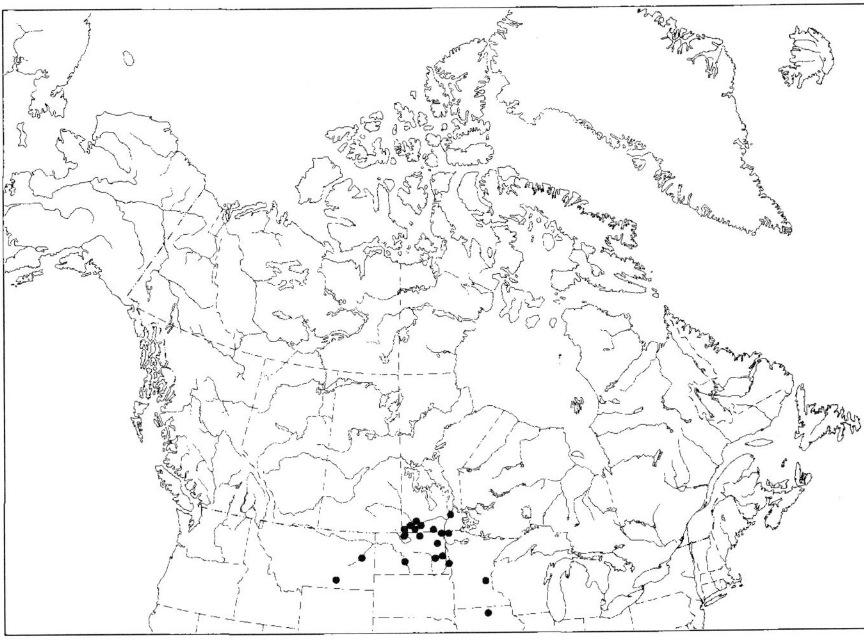

Map 166. Collection localities for *Melanoplus spretus*.

broad, somewhat pointed. Male subgenital plate short in proportion to width, notched apically; dorsal valve of aedeagus (see Gurney and Brooks 1959:17, figs. 2*n*, 3*a*), concave posteriorly, with pointed apex, and with valve twisted laterally and posteriorly around edge of large membranous posterior valve. Dorsal valve of ovipositor with well-developed scoop. Color generally brownish yellow to grayish, with contrasting black markings on head and lateral pronotal lobes; hind femur with 4 dark bands (including band at apex); hind tibia dark red.

Range. Now extinct, but formerly from Alberta to Manitoba, south to Arizona and Texas.

Behavior and habitats. *M. spretus* was migratory, in the past often occurring in vast numbers. It was apparently able to breed continuously only in favorable locations, probably mainly in Montana and Wyoming, but to some extent in associated regions of all neighboring provinces and states, along river banks, sunny upland slopes, and subalpine grassy areas. During flights, swarms flew from 32 to 240 km/day at altitudes up to 3350 m. Swarms from permanent breeding sites were said to have traveled primarily west and southwest, although swarms were reported many times in southern Manitoba. Temporary breeding areas were reported outside the main breeding regions and swarms from these sites often migrated to the north or northwest. D. Van Horn (1965) discusses anatomical and ecological variation.

Life history. In the northern part of the range, only one generation each year occurred, but in Texas a partial second generation was reported.

Hatching occurred from mid-May to early June in Montana. Nymphs behaved gregariously and tended to "march" during warm hours of the day, stopping to feed in the evening and early morning. Marching swarms of nymphs ("hopper-bands"), as much as 1.6 km across the front, were reported by Riley (1877*a*, 1877*b*). Adults were reported from 1 July to 6 October in the north.

Economic importance. For an historical résumé of the ravages of this species in Manitoba from 1800 ("1799") until it apparently became extinct in 1903, we refer the reader again to Mitchener's (1954*a*, 1954*b*, 1956) accounts.

The reason for the disappearance of *M. spretus* remains a mystery. N. Criddle (1917) was of the opinion that it still existed in restricted breeding grounds. Gurney (1958) and Gurney and Brooks (1959) wondered if it might still exist in the "solitary phase." Such a phase would probably be smaller and with shorter wings than those which caused so much damage in the past. Undoubtedly it would look much like *M. sanguinipes sanguinipes* and could be confused with that species. As already noted, several authors in the past (Hebard 1931*a*, 1932*a*, 1936*a*; Froeschner 1954, and Cantrall and Young 1954) regarded *spretus* as the migratory phase of *M. sanguinipes sanguinipes*. The male aedeagi of the two are different, however, and examination of this structure may be the only means of distinguishing between them, should a population of *M. spretus* be found.

Specimens of *M. spretus* are still to be found in small numbers frozen in the ice of "Grasshopper Glacier," Park County, Montana (Gurney 1953*a*; Gurney and Brooks 1959).

Genus *Paroxya* Scudder

Description. Body size medium (length, male 16–32, female 22–44 mm), nearly cylindrical. Head moderately large, with face distinctly oblique; eyes large, prominent, 1.5 times as long as cheek area beneath; vertex strongly narrowed between eyes; lateral foveolae present as small oval or rounded pits. Pronotal disc flat, twice as long as broad, with lateral margins parallel and with posterior margin obtuse-angulate to broadly rounded. Tegmina and wings variable in length but always longer than pronotum. Furculae on ninth abdominal tergum of male well-developed; male subgenital plate short. Ovipositor valves strongly exerted, with dorsal valves crenate; lower valves smooth.

Key to species of *Paroxya*

1. Tegmina as long as abdomen (slightly shorter in female). Male furculae slender, subcylindrical. Male cercus with apices broadly rounded. Found in Atlantic United States ..... ***P. clavuliger* (Audinet-Serville)** (p. 435)

   Tegmina shorter, at most about two-thirds as long as abdomen (Fig. 425). Male furculae stout, somewhat depressed. Male cercus with apices subtruncate or feebly emarginate (Fig. 426). Found in inland regions .... ***P. hoosieri* (Blatchley)** (p. 436)

Clé des espèces de *Paroxya*

1\. Tegmina aussi longs que l'abdomen (légèrement plus courts chez les femelles). Furcula des mâles étroite et subcylindrique. Cerques des mâles à apex très arrondi. Espèce vivant sur la côte est des États-Unis .... ***P. clavuliger* (Audinet-Serville)** (p. 435)

Tegmina plus courts, de longueur atteignant au plus les deux tiers de la longueur de l'abdomen (fig. 426). Furcula des mâles forte, quelque peu aplatie. Cerques des mâles à apex subtronqué ou légèrement concave (fig. 422). Insecte vivant dans les régions intérieures .... ***P. hoosieri* (Blatchley)** (p. 436)

Paroxya clavuliger (Audinet-Serville)

Map 167

Acridium clavuliger Audinet-Serville, 1838:676, pl. 14, fig. 11.

Caloptenus floridanus C. Thomas, 1874:68.

Paroxya clavuliger; J. A. G. Rehn and Hebard 1916*b*:251; Helfer 1963:213, 214.

Diagnosis. Very like *P. hoosieri* but tegmina normally extending to near end of abdomen or slightly beyond.

Description. Body medium-large (length, male 20–32, female 28–44 mm). Tegmina covering most, if not all, of abdomen (usually slightly shorter in females). Male furculae relatively slender, subcylindrical, with inner

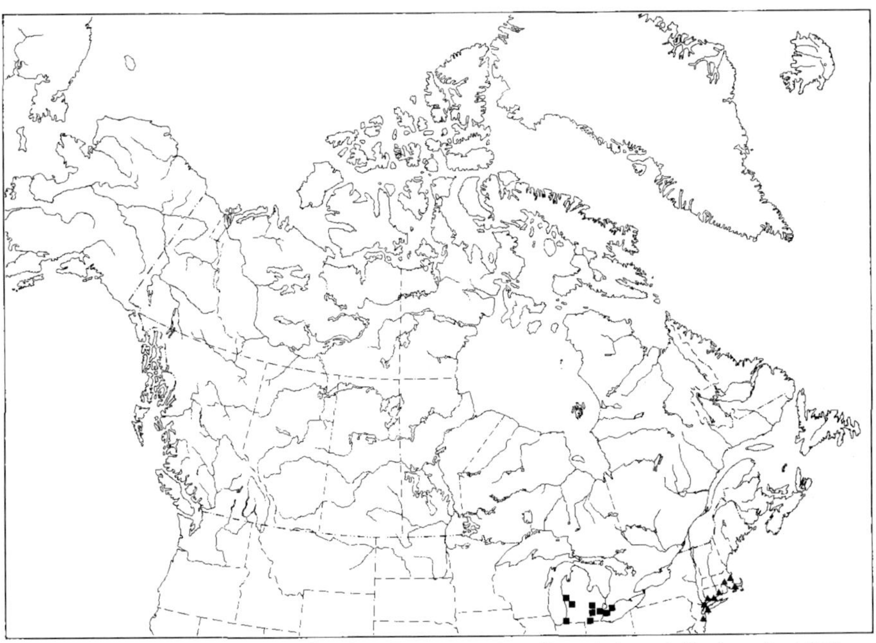

Map 167. Collection localities for *Paroxya clavuliger* (▲) and *P. hoosieri* (■).

margins not touching at bases. Male cercus with apices broadly rounded. Male subgenital plate without flared margins.

Range. Massachusetts to Florida.

Behavior and habitats. *P. clavuliger* is similar to *P. hoosieri*, but it includes saltwater marshes and tidewater ditches as well as freshwater marshes and bogs in its habitats. Morse (1920), as well as discussing the habitats in New England, also notes the reluctance of the species to fly and a similar escape manoeuvre to that described later by Cantrall (1943) for *P. hoosieri*. When *P. clavuliger* does take flight, it leaps with alacrity, but it does not fly far.

Life history. As for *P. hoosieri*. In New England, *P. clavuliger* is adult from late July until mid-October; immatures were plentiful in Massachusetts in late July (Morse 1920).

Economic importance. Long ago Scudder (1897*d*) recorded that S. I. Smith (1868) had reported the species as being injurious to cranberries in New Jersey.

Paroxya hoosieri (Blatchley)

Figs. 422, 426; Map 167

Pezotettix hoosieri Blatchley, 1892*b*:31.
Paroxya hoosieri; Cantrall 1943:123.
Paroxya clavuliger hoosieri; Blatchley 1920:355.

Diagnosis. Tegmina considerably shorter than abdomen, oblong. Body yellowish green. Antennae reddish brown.

Description. As for genus *Paroxya* (Fig. 426). Body length, male 18–22, female 25–34 mm. Tegmina oblongate, short, covering one-half or two-thirds of abdomen. Male furculae oblong, close together, somewhat depressed, feebly convex, with inner margins abruptly narrowed and forming blunt tooth. Male cercus long, expanded apically with shallow apical emargination (Fig. 422). Male subgenital plate with distolateral margins somewhat flared upward. Color generally yellowish green; male antennae pale reddish brown, darker apically, with apical sixth of each segment yellowish; face green; mouthparts yellow; vertex, pronotum, and tegmina olive brown; underside yellow; abdomen of female with conspicuous black lateral stripe; femora green with genicular area black; hind tibia greenish with black spines.

Range. Indiana, Michigan, Ohio, and southern Ontario.

Behavior and habitats. *P. hoosieri* is seldom found far from water. It inhabits grasses and sedges around edges of small lakes and ponds (Blatchley 1920). Cantrall (1968) reported similar habitats, as well as moist marshes and swales associated with heavier soils. Cantrall (1943) found it to be "surprisingly inactive for so graceful a grasshopper." The insects seldom hop when approached but sidle around stems or drop into the vegetation below, where they remain quiet. Otte (1970) was not able to observe mating behavior, but he noted that bursts of femur shaking were aggressive signals in encounters between males.

Life history. There is one generation each year, with overwintering in the egg stage. Cantrall (1968) recorded adults as being active in Michigan between 30 June and 17 October.

Subfamily Hyalopteryginae

This subfamily, which is exclusively American and mainly tropical, was formerly placed in the Acridinae, *sensu stricto*, largely on the basis of the characteristic "speculum" of the male hind wing (Fig. 721). In the true Acridinae (now restricted to three genera and about 25 Old World species), however, the modified part of the wing is the medial and not the anterior cubital area, so that the evolution of this form of resonator was presumably independent in the two groups. Some exotic species of *Orphulella* (Gomphocerinae, Orphulellini) have a similar "speculum" (Otte 1979*a*), so that convergent evolution has probably occurred several times; see also Dirsh (1975). There are about 14 described genera and probably 50 known species of Hyalopteryginae.

The single species of *Metaleptea*, the one genus that occurs in Canada, has one of the widest American ranges known for orthopteroids; it occurs, in one form or another, southward as far as southern Chile. Because of its Pan-American range, *Metaleptea brevicornis* (Johansson) was chosen by the Pan-American Acridological Society (Sociedad Panamericana de Acridologia) as the emblem of the society (Fig. 719), whose publication, bearing the emblem, is called "Metaleptea."

Genus *Metaleptea* Brunner von Wattenwyl

Description. Vertex of head horizontal, parabolic, projecting in front of eyes by a distance equal to length of eyes (Fig. 722); disc with sides flattened, and with median carina fine. Antennae as long as head and pronotum combined, tapered, flattened at base. Pronotal disc flat, with all 3 carinae distinct, parallel; median carina cut by sulcus much behind middle (Fig. 722). Tegmina and hind wings fully developed. Hind femur slender. Male subgenital plate conical, with apex acute.

Metaleptea brevicornis brevicornis (Johansson)

Figs. 719–722; Map 168

Gryllus brevicornis Johansson, 1763:398.
Acrydium ensicornu De Geer, 1773:499, pl. 42, fig. 7.
Truxalis notochlorus Palisot de Beauvois, 1807:80, pl. III, fig. 3.
Truxalis viridulus Palisot de Beauvois, 1807:81, pl. III, fig. 4.
Pyrgomorpha punctipennis C. Thomas, 1873:68.
Metaleptea brevicornis; Gurney 1940*a*:92; Helfer 1963:152; R. D. Alexander et al. 1972:52.
Metaleptea brevicornis brevicornis; J. A. G. Rehn 1944*b*:224.

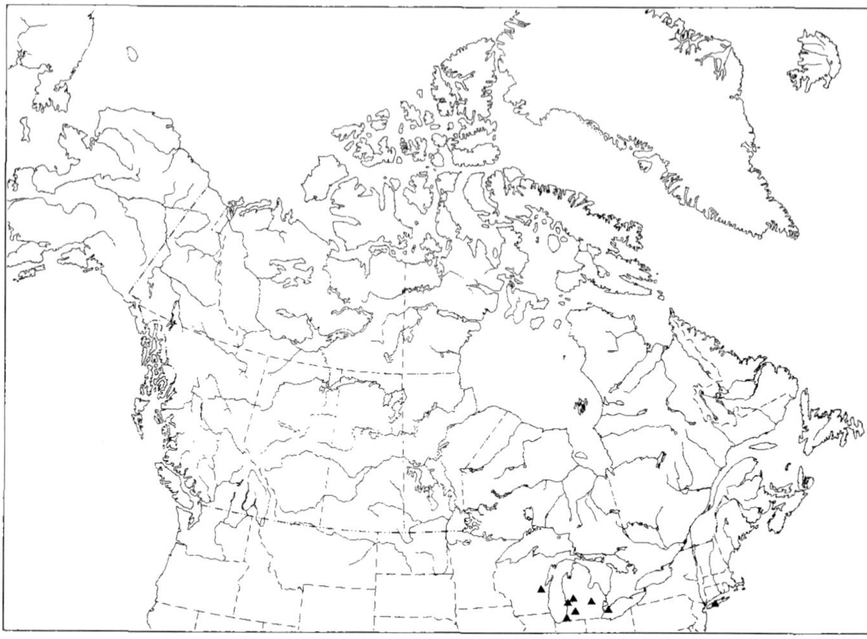

Map 168. Collection localities for *Metaleptea brevicornis brevicornis*.

Diagnosis. Vertex of head horizontal, projecting in front of eyes by a distance equal to length of an eye.

Description. As for genus *Metaleptea*. Size medium (length, male 19–25, female 32–35 mm). Male usually with dorsum, face, and fore and middle tibiae bright green, elsewhere pale brown; hind wings transparent with greenish veins; females dimorphic, pale green with tegmina dotted with brown, or entirely rusty brown with darker line laterally on head and pronotum.

Range. Wisconsin to southern Ontario, south to Mexico.

Behavior and habitats. *M. brevicornis brevicornis* inhabits sedges and tall grasses in swampy areas, especially along margins of streams, ponds, and lakes, and in salt marshes on the Atlantic coast. Although distribution is discontinuous, it may be numerous in the locations where it occurs. It is a shy insect. It does not leap when disturbed, but it takes flight, which is zigzag and noiseless, alighting on a stem some distance away. It feeds on the grasses and sedges in its confined habitat.

As is characteristic of the subfamily, stridulation is achieved in males by means of rubbing the tegmina on (or with) the specialized ridgelike distal parts of the anterior longitudinal veins of the hind wing, the speculum of the anterior cubital area acting as a resonator. Details of the specific stridulatory pattern for *M. brevicornis* are lacking.

Life history. Univoltine, with overwintering in the egg stage. Adults are known from 9 August to 15 September in Michigan (Cantrall 1968). E. M. Walker (1902*b*, 1902*c*, 1902*d*) reported collecting this species at Point Pelee, Ont., on 8 August.

Subfamily Locustinae

This subfamily includes the so-called banded-winged grasshoppers, although the characteristic dark bands on the hind wings are lacking in many genera, including *Locusta* Linnaeus itself, and, conversely, a few excluded genera also have banded wings. The subfamily is a large one, occurring in most parts of the world, even in frigid zones. There are about 185 known genera and numerous species. The most infamous-included member of the subfamily is *Locusta migratoria* Linnaeus, various subspecies of which appear from time to time in devastating swarms over much of the Old World. There are also other species that can cause catastrophic crop damage, as for example the Australian plague locust, *Chortoicetes terminifera* (F. Walker), as well as numerous other less spectacular, although none the less serious, pests, such as the North American *Camnula pellucida* (Scudder).

Many species are well known for their association with dry or even arid conditions, and these generally live on the surface of the ground between, rather than on, vegetation. A high proportion of the species exhibit "flash coloration" in flight, drawing attention to themselves by their brightly colored hind wings (red, yellow, or blue) and then suddenly "disappearing" when they settle. This assists in defeating predators, but it may have been developed primarily as a recognition signal. The visual stimulus appears to be utilized by some species, along with sound production, in courtship behavior. In addition to more "orthodox" methods of stridulation by rubbing a ridge on the inner face of the hind femur against a serrated, intercalary vein on the tegmen, and movements of the hind legs, known as femur-tipping, males of many (and females of some) species make a loud clattering noise when they fly. This is known as "crepitation." It may be imitated by rapidly opening and closing an ordinary folding fan, and may be intensified by striking the edge of the fan on a nearby object at the end of each stroke (as if the hind wings were striking the tegmen). In some genera the long veins of the hind wings are extra strong, and as a result, the crepitation is unusually loud; other modifications of the hind wing venation may increase the volume of sound even further.[1] Crepitation by members of this subfamily has even attracted the attention of a few poets, including federal public servant Charles Mair. In his *August*, he refers to the "cracker," almost certainly *Trimerotropis verruculata* (W. Kirby):

> And watch the dusky crackers snap their wings
> Whilst gangs of blue-flies fetch a buzzing tease.
>
> (Mair 1868)

[1] Crepitation of this kind is not confined to the present subfamily; some members of other subfamilies also employ similar mechanisms.

We here recognize two tribes of Locustinae; the predominantly Old World Epacromiini (with about 10 genera—one shared with North America, and some 45 species) and the worldwide Locustini (or Oedipodini), including the bulk of the subfamily. The position of the Epacromiini and, in particular, of the one genus that occurs in North America, *Stethophyma*, is somewhat anomalous, and some authors view the presence of a serrated intercalary stridulatory vein on the tegmen as a result of convergent evolution rather than of a common origin with the Locustini. Such authors place *Stethophyma*, at least, with the Gomphocerinae, although it "sits" with that subfamily no more easily than it does here. Indeed the auxiliary movements of the hind femur of *Stethophyma* are distinctly more "locustine" than "gomphocerine."

Key to tribes of Locustinae

1. Frontal profile more or less vertical (angle measured from occiput to fastigium to clypeus approximately 90°), sometimes slightly oblique. Pronotal disc with lateral carinae poorly or not developed. Crossveins connecting with intercalary vein of male tegmen not rough or "pebbly" and playing no part in sound production. Body commonly robust, rugose. Color of most species brownish or grayish, procryptic against soil, sand, and gravel .... **Locustini (Oedipodini)** (p. 441)

   Frontal profile somewhat oblique (angle measured from occiput to fastigium to clypeus distinctly less than 90°). Pronotal disc with well-defined lateral carinae. Crossveins connecting with intercalary vein of male tegmen rough or "pebbly" and playing a role in sound production. Body rather slender, smooth. Color not as above, frequently at least partly green, stramineous, or brownish, blending with vegetation .... **Epacromiini** (p. 525)

Clé des tribus de Locustinae

1. Profil du front plus ou moins vertical (angle mesuré de l'occiput au fastigium et du fastigium au clypeus égal à environ 90°), parfois légèrement oblique. Disque du pronotum à crêtes latérales à peine visibles ou inexistantes. Nervures transversales reliées à la nervure intercalaire des tegmina des mâles ni rudes ni granuleuses et ne jouant aucun rôle dans la production de son. Corps habituellement robuste et rugueux. Coloration de la plupart des espèces brunâtre ou grisâtre, se fondant avec la couleur du sol, du sable et du gravier ..... **Locustini (Oedipodini)** (p. 441)

   Profil du front quelque peu oblique (angle mesuré de l'occiput au fastigium et du fastigium au clypeus clairement inférieur à 90°). Disque du pronotum à crêtes latérales bien définies. Nervures transversales reliées à la nervure intercalaire des tegmina des mâles rudes ou granuleuses et jouant un rôle dans la production du son. Corps plutôt mince et lisse. Coloration différente de celle décrite ci-dessus, souvent verte, au moins en partie, jaune paille ou brunâtre, se fondant dans la végétation .... **Epacromiini** (p. 525)

Tribe Locustini

Prosternal spine absent. Intercalary vein of median area of tegmen strong, serrated. Face vertical, seldom oblique. Hind femur lacking stridulatory pegs on inner face. Hind wings often colored. Median carina of pronotum usually cut by one or two sulci.

Key to genera of Locustini (= Oedipodini)

1. Median carina of pronotum entire, not cut by sulcus (Figs. 649-651) .. .. ***Arphia*** (in part) (p. 444)
   Median carina of pronotum cut by one or two sulci 2

2(1). Median carina of pronotum cut by one sulcus 3
   Median carina of pronotum cut by two sulci, the front one often poorly impressed 13

3(2). Incision of principal sulcus deep, the carina anterior to the sulcus high, sharp, entire 4
   Incision of principal sulcus shallow, or, if deeper, the region anterior to sulcus rough, pitted, wrinkled, or tuberculate 6

4(3). Hind tibia yellow or buff, without bands. Hind wing black with pale yellow border, or transparent with darker suffusion in outer anal area ... . ***Dissosteira*** (p. 451)
   Hind tibia orange or red. Hind wing yellow or white with black crossband. . 5

5(4). Tegmina narrow, with straight, subobsolete intercalary veins. Pronotal crest low, straight, not deeply cut by principal transverse sulcus (small species, males less than 20 mm, females less than 25 mm in length) .... ***Scirtetica*** (p. 454)
   Tegmina wider, with more obvious less straight intercalary veins. Pronotal crest higher, deeply cut by principal sulcus (larger species, males greater than 20 mm, females greater than 25 mm in length) .... ***Spharagemon*** (p. 456)

6(3). Pronotum roughened, especially in front, with posterior half possessing small tubercles 7
   Pronotum smooth or slightly wrinkled in front, with posterior half lacking tubercles 11

7(6). Median carina of pronotum high, sharp, nearly entire (Figs. 652, 653) .... ***Arphia*** (in part) (p. 444)
   Median carina of pronotum low, especially in front, the incision of the sulcus conspicuous 8

8(7). Posterior margin of lateral lobe of pronotum concave (Fig. 663). Depression of vertex divided in front by ridge (Fig. 667) .... ***Cratypedes*** (p. 462)
   Posterior margin of lateral lobe of pronotum nearly vertical (Figs. 659, 661, 662). Depression of vertex not divided in front by ridge (Fig. 666) (vertex not prominent, but flattened and rugose) 9

9(8). Median carina of pronotum distinct throughout; tubercles of pronotum low, pebblelike 10
   Median carina of pronotum not distinct in front of sulcus; tubercles of pronotum large .... ***Xanthippus*** (p. 463)

10(9). Prozona and metazona of pronotal disc subequal in length .... ***Hippiscus*** (p. 470)
   Prozona considerably shorter than metazona ... ***Pardalophora*** (p. 471)

11(6). Median carina of pronotum smoothly elevated, crestlike, with incision of sulcus superficial (Fig. 655) .... *Chortophaga* (p. 475)
Median carina of pronotum low, with incision of sulcus usually conspicuous 12
12(11). Hind tibia blue or fuscous brown, with pale basal annulus. Hind wing generally clear, clouded on apical third .... *Encoptolophus* (p. 477)
Hind tibia yellowish to brown. Hind wing not clouded .... *Camnula* (p. 480)
13(2). Posterolateral lobe of pronotum rounded or slightly angulate; median carina obsolete on central part 14
Posterolateral lobe of pronotum distinctly angulate; median carina distinct 15
14(13). Hind wings clear, darkened on veins and toward base or with inner half darkened, and with crossband absent; anal veins thickened (Fig. 713) .... *Aerochoreutes* (p. 482)
Hind wings at least faintly yellow, red, or bluish, with black crossband .. 16
15(13). Hind wings clear, or faintly yellow basally, with evidence of darker band .... *Trachyrhachys* (p. 485)
Hind wings yellow or red with black crossband ...... *Metator* (p. 488)
16(14). Posterolateral lobe of pronotum produced, toothlike (Fig. cf. 686) 17
Posterolateral lobe of pronotum not produced (Fig. cf. 685) 18
17(16). Median carina elevated on metazona .... *Conozoa* (p. 491)
Median carina low on metazona ...... *Trimerotropis* (in part) (p. 492)
18(16). Metazona rugose tuberculate with protuberances near median carina .. .. *Derotmema* (p. 516)
Metazona smooth or with scattered granulations 19
19(18). Median carina of pronotum cut near middle by posterior sulcus .... *Mestobregma* (p. 517)
Median carina of pronotum cut considerably ahead of middle by posterior sulcus 20
20(19). Body robust. Antennae unusually long. Inner face of hind femur bluish black .... *Hadrotettix* (p. 519)
Body slender. Antennae not unusually long. Inner face of hind femur not bluish 21
21(20). Disc of hind wing pink to red. Small species (15–25 mm) .... *Psinidia* (p. 520)
Disc of hind wing pale blue to bright yellow but not red. Larger species (18–35 mm) 22
22(21). Hind wing with anal veins thickened, and posterior (vannal) lobe distinctly produced .... *Circotettix* (p. 522)
Hind wing with anal veins not thickened, and posterior (vannal) lobe not distinctly produced .... *Trimerotropis* (in part) (p. 492)

Clé des genres de Locustini (= Oedipodini)

1. Crête médiane du pronotum entière, non coupée par un sulcus (fig. 649 à 651) .... *Arphia* (en partie) (p. 444)
Crête médiane du pronotum coupée par un ou deux sulcus 2
2(1). Crête médiane du pronotum coupée par un sulcus 3
Crête médiane du pronotum coupée par deux sulcus, le sulcus antérieur étant souvent peu visible 13

3(2). Sulcus principal profond, la portion de la crête antérieure à ce dernier élevée, acérée et entière 4

Sulcus principal peu profond ou dans le cas contraire, région antérieure à ce sulcus rude, parsemée de trous, ridée ou tuberculée 6

4(3). Tibias postérieurs jaunes ou jaune brunâtre, sans bande. Ailes postérieures noires à bord jaune pâle ou transparentes et à région anale extérieure plus foncée .... ***Dissosteira*** (p. 451)

Tibias postérieurs orangés ou rouges. Ailes postérieures jaunes ou blanches, à bande transversale noire 5

5(4). Tegmina étroits, à nervures intercalaires rectilignes presque invisibles. Crête du pronotum basse, rectiligne, non coupée profondément par le sulcus transversal principal (petites espèces, mâles de moins de 20 mm, femelles de moins de 25 mm) .... ***Scirtetica*** (p. 454)

Tegmina plus larges, à nervures intercalaires plus évidentes et moins rectilignes. Crête du pronotum plus élevée, profondément coupée par le sulcus principal (espèces plus grandes, mâles de plus de 20 mm, femelles de plus de 25 mm) .... ***Spharagemon*** (p. 456)

6(3). Pronotum rude, surtout en avant, recouvert de petits tubercules sur sa moitié postérieure 7

Pronotum lisse ou légèrement ridé en avant, sans tubercule sur sa portion postérieure 11

7(6). Crête médiane du pronotum haute, aiguë, presque entière (fig. 652) .... ***Arphia*** (en partie) (p. 444)

Crête médiane du pronotum basse, surtout en avant, clairement séparée par le sulcus 8

8(7). Bords postérieurs des lobes latéraux du pronotum concaves (fig. 663). Dépression du vertex divisée en avant par une crête (fig. 667) .... ***Cratypedes*** (p. 462)

Bords postérieurs des lobes latéraux du pronotum presque verticaux (fig. 659, 661 et 662). Dépression du vertex non divisée en avant par une crête (fig. 666) (vertex non proéminent, mais aplati et rugueux) 9

9(8). Crête médiane du pronotum distincte sur toute sa longueur; tubercules du pronotum petits, granuleux 10

Crête médiane du pronotum non visible en avant du sulcus; tubercules du pronotum gros .... ***Xanthippus*** (p. 463)

10(9). Zones antérieure et postérieure du disque du pronotum de longueur presque égale .... ***Hippiscus*** (p. 470)

Zone antérieure beaucoup plus courte que la zone postérieure .... ***Pardalophora*** (p. 471)

11(6). Crête médiane du pronotum unie, à peine incisée par le sulcus (fig. 655) .... ***Chortophaga*** (p. 475)

Crête médiane du pronotum basse, en général clairement incisée par le sulcus 12

12(11). Tibias postérieurs bleus ou brun grisâtre, avec un annulus basal pâle. Ailes postérieures généralement claires et enfumées sur le tiers apical ... . ***Encoptolophus*** (p. 477)

Tibias postérieurs jaunâtres à bruns. Ailes postérieures non enfumées .... ***Camnula*** (p. 480)

13(2). Lobes postérolatéraux du pronotum arrondis ou légèrement angulaires; crête médiane absente sur la partie centrale 14

Lobes postérolatéraux du pronotum clairement angulaires; crête médiane distincte 15

14(13). Ailes postérieures claires, à nervures et à base foncées ou à moitié interne foncée, sans bande transversale; nervures anales épaissies (fig. 713) . . . . *Aerochoreutes* (p. 482)

Ailes postérieures au moins légèrement jaunes, rouges ou bleuâtres, à bande transversale noire 16

15(13). Ailes postérieures claires ou légèrement jaunies à la base, à bande légèrement plus foncée . . . . *Trachyrhachys* (p. 485)

Ailes postérieures jaunes ou rouges avec une bande transversale noire . . . . *Metator* (p. 488)

16(14). Lobes postérolatéraux du pronotum développés, en forme de dent (fig. cf. 686) 17

Lobes postérolatéraux du pronotum non développés (fig. cf. 685) 18

17(16). Crête médiane élevée sur la zone postérieure . . . . *Conozoa* (p. 491)

Crête médiane basse sur la zone postérieure . . . . *Trimerotropis* (en partie) (p. 492)

18(16). Zone postérieure rugueuse, tuberculée, portant des protubérances près de la crête médiane . . . . *Derotmema* (p. 516)

Zone postérieure lisse ou légèrement granulée 19

19(18). Crête médiane du pronotum coupée près du milieu par le sulcus postérieur . . . . *Mestobregma* (p. 517)

Crête médiane du pronotum coupée clairement en avant du centre par le sulcus postérieur 20

20(19). Corps robuste. Antennes de longueur inhabituelle. Face interne des fémurs postérieurs d'un noir bleuté . . . . *Hadrotettix* (p. 519)

Corps mince. Antennes de longueur normale. Face interne des fémurs postérieurs non bleutée 21

21(20). Disque des ailes postérieures de coloration rose à rouge. Petites espèces (de 15 à 25 mm) . . . . *Psinidia* (p. 520)

Disque des ailes postérieures bleu pâle à jaune clair, mais pas rouge. Espèces plus grandes (de 18 à 35 mm) 22

22(21). Ailes postérieures à nervures anales épaissies et à vannus développés . . . . *Circotettix* (p. 522)

Ailes postérieures à nervures anales non épaissies et à vannus pas particulièrement développés . . . . *Trimerotropis* (en partie) (p. 492)

Genus *Arphia* Stål

Description. Body robust, medium-large (length, male 17–27, female 26–34 mm), compressed. Face vertical; vertex with distinct concavity, often divided by transverse curved impression behind middle; foveolae present, shallow, large, triangular to rhomboidal. Anterior margin of pronotum bluntly angulate, produced; median carina compressed, high, weakly cut or not by principal sulcus; lateral carina rounded, distinct only on metazona. Tegmina long, leathery, densely veined, with distinct cells only in apical third. Hind wings brightly colored, red or yellow at base. Hind femur stout, depressed, dilated basally.

Key to species and subspecies of *Arphia*

1. Frontal costa of face nearly parallel-sided to vertex, not noticeably narrowed above, more than half as wide at vertex as below median ocellus 2
   Frontal costa of face strongly narrowed above to vertex, less than half as wide at vertex as below median ocellus 3

2(1). Median carina of pronotum strongly elevated, cristate, longitudinally arched (Fig. 651). Base of hind wing deep yellow to orange red, with spur of dark crossband extending about one-third distance to base .... ***xanthoptera* (Burmeister)** (p. 445)
   Median carina of pronotum elevated but not cristate, at most only faintly arched longitudinally (Fig. 652). Base of hind wing pink to deep red, with spur of dark crossband extending more than halfway to base .... ***pseudonietana pseudonietana* (Thomas)** (p. 447)

3(1). Depressed area of vertex of head longer than broad; vertex meeting frontal fastigium at indistinct obtuse angle. Base of hind wing pale yellow to pink or orange .... ***conspersa* Scudder** (p. 448)
   Depressed area of vertex of head as broad as long; vertex meeting frontal fastigium at distinct angle. Base of hind wing deep yellow .... ***sulphurea* (Fabricius)** (p. 450)

Clé des espèces et des sous-espèces d'Arphia

1. Costa frontale de la face presque parallèle au vertex, pas particulièrement rétrécie au-dessus; largeur à la hauteur du vertex égale à plus de la moitié de la largeur sous l'ocelle médian 2
   Costa frontale de la face fortement rétrécie au-dessus du vertex; largeur à la hauteur du vertex égale à moins de la moitié de la largeur sous l'ocelle médian 3

2(1). Crête médiane du pronotum très élevée, crêtée et arquée dans le sens longitudinal (fig. 651). Base des ailes postérieures jaune foncé à rouge orangé; éperon de la bande transversale foncée et s'étendant sur le tiers environ de la distance à la base .... ***xanthoptera* (Burmeister)** (p. 445)
   Crête médiane du pronotum élevée, mais non crêtée, au plus légèrement arquée dans le sens longitudinal (fig. 652). Base des ailes postérieures de coloration rose à rouge foncé; éperon de la bande transversale foncée s'étendant sur plus de la moitié de la distance à la base .... ***pseudonietana pseudonietana* (Thomas)** (p. 447)

3(1). Région concave du vertex plus longue que large; jonction du vertex et du fastigium frontal formant un angle obtus indistinct. Base des ailes postérieures jaune pâle à rose ou orangé .... ***conspersa* Scudder** (p. 448)
   Région concave du vertex aussi large que longue; jonction du vertex et du fastigium frontal formant un angle distinct. Base des ailes postérieures jaune foncé .... ***sulphurea* (Fabricius)** (p. 450)

Arphia xanthoptera (Burmeister)

Fig. 651; Map 169

Oedipoda xanthoptera Burmeister, 1838:643.
Oedipoda carinata Scudder, 1869*a*:306.

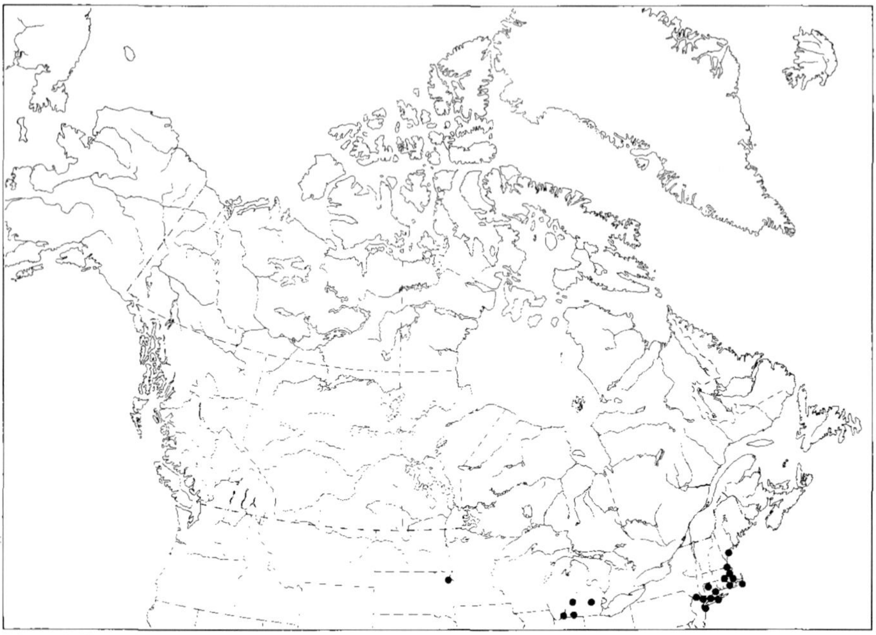

Map 169. Collection localities for *Arphia xanthoptera*.

Tomonotus carinatus; C. Thomas 1873:106.
Arphia xanthoptera; Lugger 1898:228; Helfer 1963:111.

Diagnosis. Vertex with concavity wide, subtruncate in front. Median carina of pronotum high, cristate, arched.

Description. Body size medium (length, male 21–27, female 28–34 mm), robust. Vertex with dorsal concavity wide, subtruncate anteriorly, with sides continuous with those of frontal costa. Foveolae small, triangular. Pronotum arched (Fig. 651), with posterior margin distinctly acute. Color pale to dark reddish brown; head and pronotum pale; tegmina pale (but less so than head and pronotum), in female sprinkled with many dark brown spots; wings deep yellow to orange red on basal two-thirds, with outer third having fuscous band and spur extending one-third distance to base; hind femur dull gray brown, with 2 or 3 oblique dark bars dorsally; genicular area dark with pale ring; hind tibia dusky with pale basal ring.

Range. South Dakota to Maine, south to Texas and Florida.

Behavior and habitats. Like *A. sulphurea*, this species is a dry-land inhabitant, occurring in old fields, wastelands, and pastures that are overgrown with weeds. It is also found in scrub and coarse herbs near woodlands. Food plants are probably the same or similar to those mentioned for *A. sulphurea*.

Otte (1970) reported a crepitation rate of about half that of *A. sulphurea*.

Life history. Univoltine, with overwintering in the egg stage. Cantrall (1968) reported adults in Michigan from 31 August to 23 September. In contrast to *A. sulphurea*, this is a late-season species. The two species do not occur together as adults; *A. sulphurea* disappears before *A. xanthoptera* reaches maturity.

Arphia pseudonietana pseudonietana (Thomas)

Figs. 652, 654; Map 170

Tomonotus pseudo-nietanus C. Thomas, 1870:82.
Oedipoda tenebrosa Scudder, 1872:251.
Arphia sanguinaria Stål, 1873*b*:119.
Tomonotus nietanus C. Thomas, 1880*a*:108.
Arphia ovaticeps Saussure, 1888:165.
Tomonotus theresiae Brunner von Wattenwyl, 1895*a*:277.
Arphia pseudonietana; Caudell 1903*d*:786; Helfer 1963:111; R. D. Alexander et al. 1972:54.
Arphia calida Bruner, 1905:121, 127.
Arphia pseudonietana pseudonietana; Hebard 1935*d*:286; Cantrall 1943:97; Brooks 1958:36.

Diagnosis. Body dark, speckled. Disc of hind wing bright red.

Description. Body large (length, male 19–24, female 27–33 mm). Frontal costa not noticeably narrowed above, more than half as wide at vertex as below median ocellus. Median carina of pronotum elevated, not cristate, only faintly arched longitudinally (Fig. 652). Color brown to nearly black,

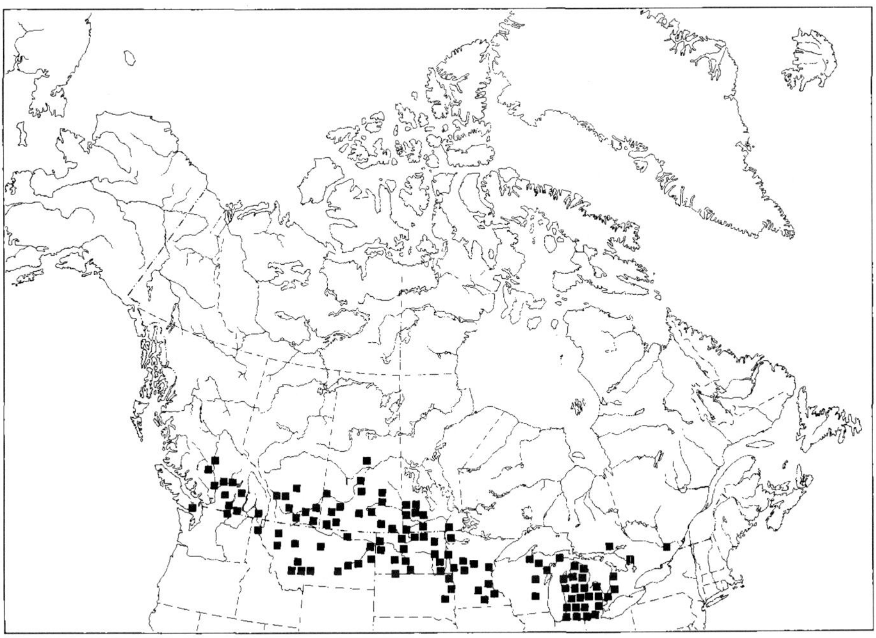

Map 170. Collection localities for *Arphia pseudonietana pseudonietana*.

heavily speckled, usually with contrasting white markings on head and pronotum; pronotum dark brown to black or entirely white dorsally; tegmina pale gray, closely sprinkled with black; disc of hind wing brilliant red, with crossband black and with long sharp spur, apically clear with dark flecks or entirely dark; hind femur brown on outer face with traces of 3 pale bands, with inner face and lower sulcus black as far as preapical pale band; hind tibia dark purplish brown with pale subbasal ring (Fig. 654). Cantrall (1968) reported a yellow-winged form of *A. pseudonietana pseudonietana* in the Upper Peninsula of Michigan.

Range. British Columbia to Ontario, south to New Mexico and Texas.

Behavior and habitats. The preferred habitat is a dry sparsely weeded area, especially on sandy or gravelly hills (Froeschner 1954). N. Criddle (1935) reported the species to be more confined to prairie uplands than *A. conspersa.* It is a grass feeder. Cantrall (1943) stated that the behavior of *A. pseudonietana pseudonietana* is similar to that of *A. conspersa.*

Cantrall (1943) also gave an excellent description of flight and crepitation. Otte (1970) reported courtship and aggressive signals. The crepitation rate is 21.2–23.5 pulses per second at 30°C. Males also stridulate on the ground, thus attracting females.

Life history. Univoltine, with overwintering in the egg stage. Cantrall (1968) recorded adults in Michigan from 10 July to 28 September, later in the season than *A. conspersa.*

Arphia conspersa Scudder

Fig. 653; Map 171

Arphia conspersa Scudder, 1875*f*:514.
Arphia frigida Scudder, 1875*b*:344.
Arphia teporata Scudder, 1876*c*:508.
Arphia arcta Scudder, 1876*a*:263.
Arphia infernalis Saussure, 1884:66, 70.
Arphia canora J. A. G. Rehn, 1904*d*:564.
Arphia conspersa; Scudder 1901*a*:31; Brooks 1958:35; Helfer 1963:112.

Diagnosis. Vertex with depressed area longer than broad. Base of hind wing yellow, pink, or orange.

Description. Body size medium (length, male 19–21, female 22–28 mm). Frontal costa strongly narrowed above to vertex, less than half as wide at vertex as below median ocellus; depressed area of vertex longer than broad. Color light brown to brownish black, usually with pale median dorsal line along tegmina when these at rest; tegmina usually dark brown, not speckled, but occasionally light brown, faintly speckled with black, with apical third clear to smoky, and with posterior edge often pale yellow; base of hind wing variable in color, bright red, pale pink, reddish orange, or yellow, with dark crossband narrow and with spur extending nearly to wing base; hind femur light brown on outer face with faint preapical pale band, or brownish black with 3 prominent white bands; inner face of hind femur and lower sulcus black as far as preapical pale band, or dark area broken

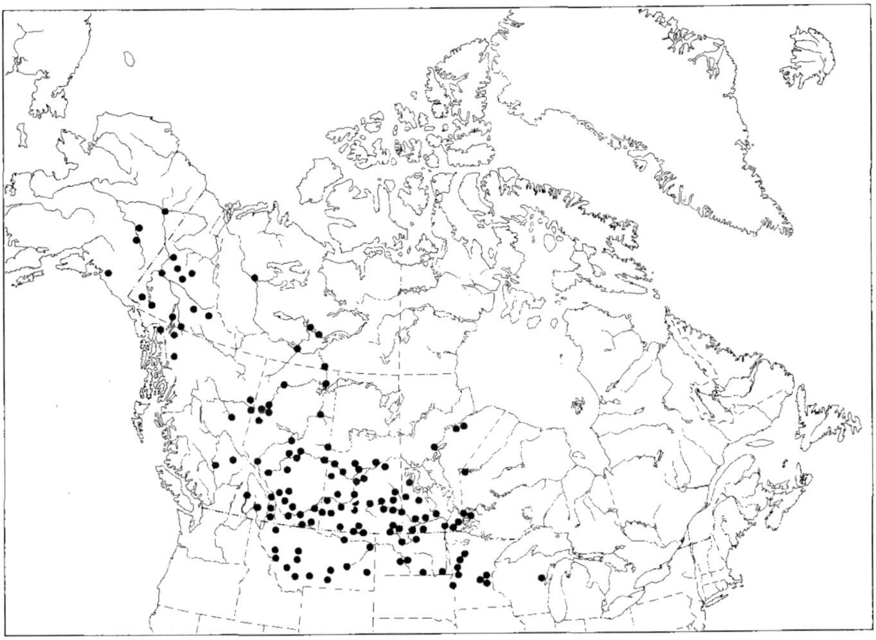

Map 171. Collection localities for *Arphia conspersa*.

in middle; hind tibia usually pale greenish brown with dark apex and ring at basal third (Fig. 653).

Note: Hebard (1937*b*) discussed five "phases" of this species with geographical correlation but declined to recognize them as subspecies. His "phase 1," characterized by small size, moderately robust form, and dark coloration, is the only one occurring in the region covered by this work. A detailed analysis of wing color variation and 14 anatomical characters on a geographical basis was recently done in the southwestern United States by Schennum and Willey (1979). They conclude that "moderate to high levels of interpopulation differentiation appear to be more common than expected and indicate that local selective pressures in the mountains are not offset by gene flow between populations."

Range. Alaska to Ontario, south to Mexico.

Behavior and habitats. Brooks (1958) reported *A. conspersa* as an inhabitant of forest clearings and grasslands, generally on sandy or gravelly soils. He also stated that it is a mixed feeder, preferring grasses and sedges such as *Stipa, Agropyron*, and *Carex* species. N. Criddle (1935) listed it as a grass-feeding species.

Otte (1970) observed that both males and females crepitate in flight, but he recorded crepitation only from individuals that had been disturbed. During courtship, he observed "tipping" and shaking of the femora and "ordinary" and "vibratory" stridulation. Male stridulation was seen to attract females, which signaled by femur-tipping prior to mounting by a male. Crepitation rate by disturbed individuals was 33.3 pulses per second at 28°C.

Life history. Winter is passed in late nymphal instars; adults are present from May to July, with occasional specimens surviving to September. Brooks (1958) stated that the "life cycle may be completed in one or two years." Willey and Willey (1971) found a 2-year life cycle, with both egg and nymphal overwintering stages, in the mountains of Colorado. The usual life cycle, however, is 1 year, with overwintering occurring as nymphs.

Arphia sulphurea (Fabricius)

Figs. 649, 650; Map 172

Gryllus sulphureus Fabricius, 1781:369.

Arphia sulphurea; Caulfield 1888:70; Cantrall 1943:96; Helfer 1963:111; R. D. Alexander et al. 1972:37; Vickery et al. 1974:107.

Diagnosis. Vertex with fastigial cavity deep and with anterior half ascending. Male tegmen often with pale band on posterior margin.

Description. Body small (length, male 17–22, female 26–30 mm), that of male slender. Vertex with anterior half of deep concavity strongly ascending, with median carina distinct but low. Pronotum with anterior margin rounded to obtuse-angulate, with disc rugose. Color pale brown to dark brown; tegmen of male often with yellowish band along posterior margin; hind wings sulfur yellowish on basal two-thirds, with outer third having a fuscous curved band and dark spur extending almost to base; hind femur

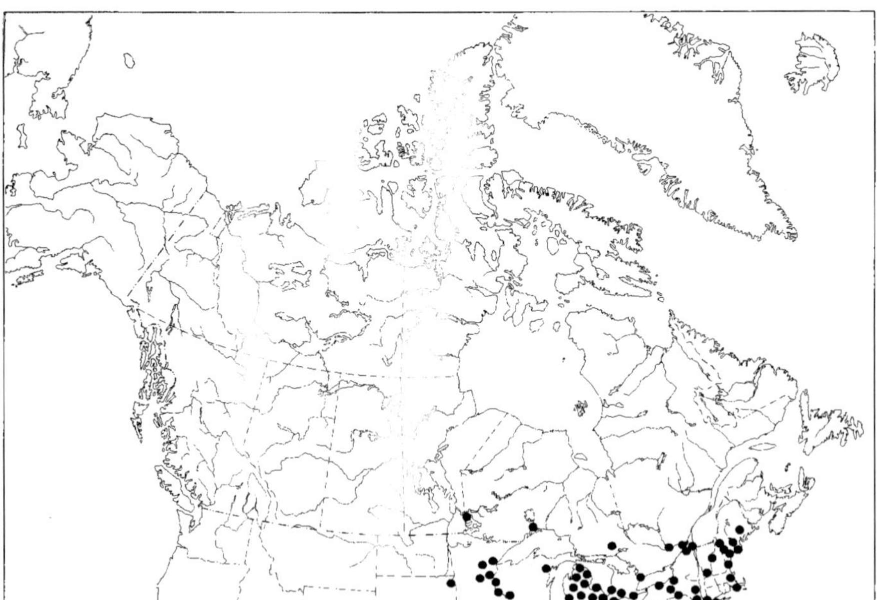

Map 172. Collection localities for *Arphia sulphurea*.

variable, dark brown with pale distal ring or with alternating light and dark bands; hind tibia dark with pale basal ring.

Range. South Dakota to Ontario, south to Texas and Florida.

Behavior and habitats. *A. sulphurea* usually occurs in pastures, along roadsides, in open woodlands, or occasionally in meadows. Gangwere (1965) reported it to be most common in Michigan in "the field-wood ecotone where there is sparse vegetation, some bare ground, and leaf litter." He found that *A. sulphurea* preferred grasses as food, but that it could also survive on forbs. *Poa pratensis* was preferred where it was available.

Otte (1970) reported upon sound production and mating behavior. Males crepitate by snapping their wings while in flight. A series of such flights attract responsive females, which then alight close by. On the ground males produce short crepitations and "tipping" movements of the hind femora, then approach the females. Crepitation rate of males in flight is about 33 pulses per second at 24°C. Willey (1975) analyzed the chirp stridulation (on substrate) by ultra-high-speed photography and also produced excellent scanning electron photomicrographs of the elements of the stridulatory apparatus (intercalary vein of the tegmen and the scraper of the femoral ridge) (Willey 1975:329, figs. 3–6).

Life history. Univoltine, with nymphal stages occurring in fall, winter, and early spring. Adults first appear in May and are most common in June, although some persist to late July (Gangwere 1965).

Genus *Dissosteira* Scudder

Description. Body medium to fairly large, slender, strongly compressed. Pronotum with disc of prozona tectiform; median carina strongly elevated and sharp, arcuate posterior to incision of sulcus, with metazonal disc flat, and with posterior margin obtuse-angulate; tegmina broad, considerably surpassing apex of abdomen; hind femur slender, shorter than abdomen in both sexes.

Key to species of *Dissosteira*

(Adapted from Hewitt and Barr (1967))

1. Hind wings with disc brown to black, and with margin yellow to white. Tegmina unicolorous or with many small indistinct maculations. Widespread .... ***carolina* (Linnaeus)** (p. 452)
   Hind wings with disc clear to pale yellow, with partly obliterated fuscus border near margin. Tegmina usually with large maculations. Found in British Columbia and southward .... ***spurcata* Saussure** (p. 454)

Clé des espèces de *Dissosteira*

(adaptée de Hewitt et Barr, 1967)

1. Ailes postérieures à disque variant de brun à noir et à bord variant de jaune à blanc. Tegmina unicolores ou recouverts de nombreuses petites taches floues. Espèce très répandue .... ***carolina* (Linnaeus)** (p. 452)

Ailes postérieures à disque clair à jaune pâle et à bord brun grisâtre partiellement effacé. Tegmina habituellement marqués de larges taches. Insecte vivant en Colombie-Britannique et plus au sud .... *spurcata* **Saussure** (p. 454)

Dissosteira carolina (Linnaeus)

Figs. 5, 664, 669, 670; Map 173

Gryllus Locusta carolina Linnaeus, 1758:433.

Dissosteira carolina; Caulfield 1888:70; Cantrall 1943:103; Brooks 1958:39; Helfer 1963:115; R. D. Alexander et al. 1972:37; Vickery et al. 1974:113.

Common name. Carolina grasshopper.

Diagnosis. Median carina of pronotum high and sharp, deeply notched. Hind wing dark with pale margin.

Description. Body medium-large (length, male 26–30, female 34–40 mm, occasionally less or greater than these measurements), not robust. Disc of pronotum with anterior margin nearly truncate, with hind margin obtuse-angulate; median carina high, cut before middle by deep narrow notch, with front lobe almost level, and with hind lobe arched (Fig. 664). Tegmina extending about one-third their length beyond apex of abdomen.

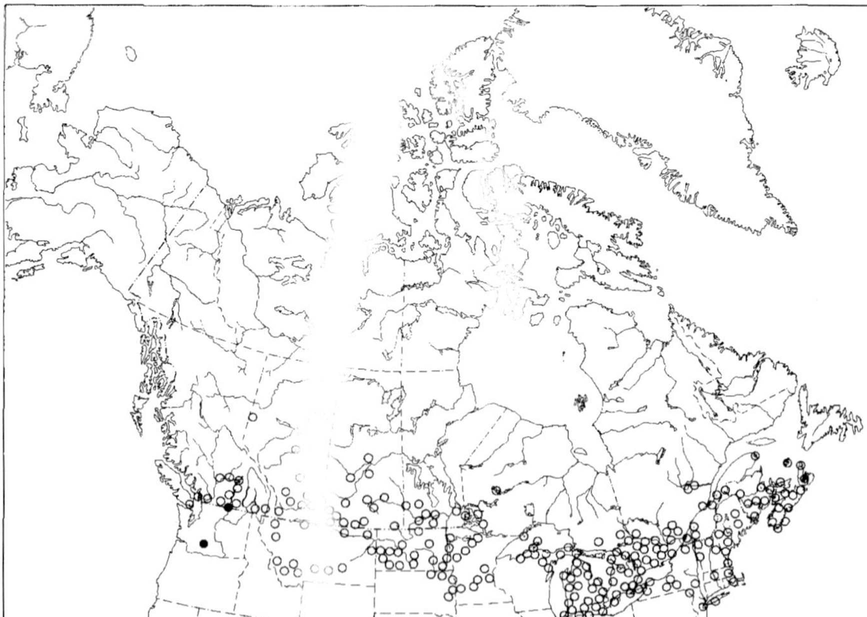

Map 173. Collection localities for *Dissosteira carolina* (○) and *D. spurcata* (●).

Color variable, depending upon that of substrate on which the specimen developed and matured, from light grayish yellow through bright reddish brown to nearly black, with underside lighter (Fig. 5); tegmina brown, clouded, usually spotted with darker brown; hind wing dark brown or black with pale yellow outer border, and with apex smoky with several darker spots; hind femur with inner face light, crossed by 3 blackish bands, the first nearly covering basal half; outer face of femur usually reddish brown with faint apical annulus; hind tibia yellowish to brownish yellow.

Range. British Columbia to Nova Scotia, south to Mexico; Bermudas.

Behavior and habitats. *D. carolina* prefers sandy and gravelly areas where bare spots occur. It frequently rests on roads, along railway cuttings, or in gravel pits. It is most common near inhabited areas and appears to gain access to the more remote of these by following the tracks provided by human colonization (Vickery et al. 1974). It often occurs in the streets of towns and even relatively large cities. Wherever it occurs, it is easily recognizable by the flashing of its yellow-rimmed black wings. It flies readily and can hover more than a metre above the ground. It is a mixed-feeder, utilizing both grasses and forbs as food (N. Criddle 1935; Mulkern et al. 1969).

Cantrall (1943) reported that flight in *D. carolina* is a "slightly undulating, wavering fluttering from side to side, and very similar to that of a butterfly." In flight, individuals produce a slight whirring sound and also a clear crepitation "*chick-a - - - - - chick-a - - - - - chick-a*." Copulation may last up to 2.5 hours. Cantrall observed one pair that remained in the mating position overnight—more than 14 hours. Both Otte (1970) and Kerr (1974) studied pair formation, courtship, and the acoustics of this species. Kerr observed no fewer than seven different visual signals, seven different visual-acoustical signals, and three different behavior patterns. Otte found the most common behavior pattern to be as follows: males move toward other moving individuals by flying or walking; the male begins to court the female when he has approached within 15 cm, the male always tipping the femur during his approach, either silently or striking it on the tegmen; stridulation is more effective in inducing females to copulate than is silent femur-tipping. Both types of femur-tipping are also used as aggressive signals between males. Nonreceptive females use the femora to kick away males attempting to mount and copulate.

Life history. In the north, *D. carolina* has a univoltine life cycle, with overwintering in the egg stage in the soil. N. Criddle (1918) described the egg pod, which is about 0.5 cm long, of which more than one-third made up the "neck" of the pod; one pod contained 42 eggs. He timed ovipositing females and found a range of 48–81 minutes required for the female to excavate a hole, deposit the eggs, and fill in the hole, using both hind legs to cover the egg chamber.

Economic importance. Although *D. carolina* may become numerous, it is not generally regarded as an economic pest. C. J. S. Fox and Stirrett (1953), however, indicate that it may be important for tobacco crops in Ontario. Beirne (1972) stated that it has been recorded as causing minor damage to cereals, alfalfa, and grasses in Canada. A fungus disease usually decimates populations that reach high densities.

Dissosteira spurcata Saussure

Map 173

Dissosteira spurcata Saussure, 1884:137.
Oedipoda obliterata C. Thomas, 1880*b*:221 (name preoccupied).
Dissosteira spurcata; Scudder 1901*a*:102; Helfer 1963:116.

Diagnosis. Pronotum with dark areas margined by white. Disc of hind wing clear to faintly yellow.

Description. Body large (length 32–50 mm), slender. Color usually brownish; pronotum usually with contrasting light and dark areas, with dark dorsal areas on prozona and metazona margined with white, or pronotum entirely brownish. Tegmina with irregular dark spot near middle and smaller dark spots near apex, giving a banded appearance, or entirely light brown to brown. Base of hind wing clear to faintly yellow, with crossband light brown, short, and having spur short or absent; hind femur with 2 dark bands on both outer and inner faces; hind tibia pale yellow or tan, sometimes white.

Range. British Columbia, south to California and Utah.

Behavior and habitats. "Dry desert areas, waste places and hilly country. It appears to be a mixed feeder but is never numerous enough to be of economic importance. The flight is much like that of *D. carolina*" (Hewitt and Barr 1967).

Life history. Probably univoltine, with overwintering as eggs in the soil. Adults are known throughout July until the end of September.

Genus *Scirtetica* Saussure

Description. Resembling *Spharagemon*, but distinguished by the narrow tegmina with straight, subobsolete intercalary veins. Pronotum with crest low, straight, not deeply cut by principal sulcus; prozona strongly constricted, with anterior margin truncate. Hind femur slender, slightly surpassing apex of abdomen in male, slightly shorter in female.

Scirtetica marmorata marmorata (Harris)

Fig. 678; Map 174

Locusta marmorata T. W. Harris, 1841:145.
Scirtetica marmorata; Morse 1897:37, 89, pl. 2, figs. 27, 27*a*, 27*b*; Helfer 1963:118; R. D. Alexander et al. 1972:37.
Scirtetica marmorata marmorata; Hebard 1932*a*:30.

Diagnosis. Tegmina narrow. Pronotal carina low, straight; metazona with pale "X" mark.

Description. Body small–medium (length, male 15–19, female 22–25 mm). Eyes prominent. Posterior margin of pronotum obtuse to nearly truncate. Color variable, pale ash gray, mottled with brown or reddish brown; pronotum grayish yellow with black bar behind eye and another irregular

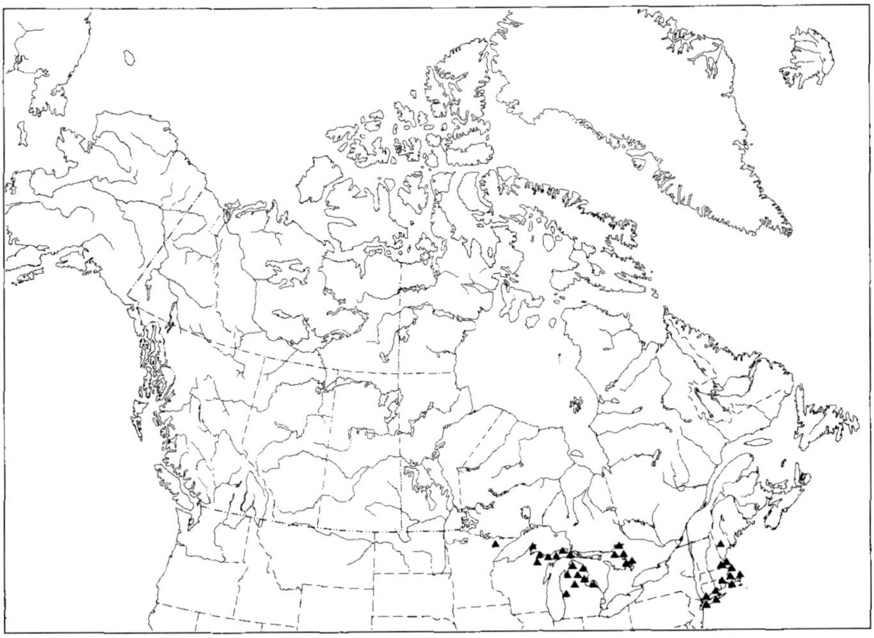

Map 174. Collection localities for *Scirtetica marmorata marmorata*.

bar midway between upper bar and lower margin; middle of disc of metazona and anterior margin dusky, making a pale "X" mark; tegmina brown to blackish with 2 or 3 large subquadrate pale spots on apical two-thirds of costal area; hind wing disc sulfur yellow, with crossband broad; spur short; apex of wing dusky beyond pale subapical band.

Range. Minnesota to Ontario, south to North Carolina.

Behavior and habitats. *S. marmorata marmorata* is often found near pine woods or in open growth among *Vaccinium*, other heath plants, and bracken fern. The species may be associated with lichens and mosses in open locations on sandy soils with sparse vegetation (Cantrall 1968).

Otte (1970) reported loud and long crepitating flights by *S. marmorata* (subsp. *picta*), as did Blatchley (1920). Males fly in a zigzag fashion for up to half a minute. During courtship on the ground, *S. marmorata* males perform unique and complicated movements. The behavior of *S. marmorata marmorata* probably is similar to that of *S. marmorata picta*. In addition to "femur-tipping" and "ordinary" stridulation, Otte (1970) observed that males employ a type of femur-jerking that appears to be peculiar to the species. First there are two silent movements, followed by alternate striking of the tegmina by the femora, resulting in three or four bursts of vibratory stridulation that are always preceded by a two-pulse burst of ordinary stridulation. Femur-tipping, femur-shaking, and striking the substrate with the hind tibiae are used as aggressive signals between males. Females reject males by jerking the hind femora and by kicking.

Life history. Univoltine, with overwintering as eggs in soil. Adults occur in Michigan between 12 July and 12 September (Cantrall 1968). Most specimens taken in Canada have been found in August.

Genus *Spharagemon* Scudder

Description. Medium-sized species. Body slender. Head somewhat swollen. Pronotum with median carina high, sharp, particularly in front of deep incision of the principal sulcus; carina posterior to incision more or less arched; lateral carinae distinct only on metazona. Hind wing yellow with darker crossband and short blunt spur, and with apical third clear to hyaline; hind femur short and stout; hind tibia with at least apical half red or orange.

Key to species and subspecies of *Spharagemon*

1. Hind tibia red, with broad brown band next to pale basal band (Fig. 676) .... ***bolli bolli*** **Scudder** (p. 457)
   Hind tibia orange or red, without a dark band (cf. Fig. 677) 2

2(1). Median carina of pronotum low, tectiform; incision of sulcus shallow, not oblique (Fig. 672) .... ***equale*** **(Say)** (p. 458)
   Median carina of pronotum high, sharp; incision of sulcus deep and oblique (Figs. 674, 675) 3

3(2). Median carina of pronotum cristate, straight, or feebly arched (Fig. 674); posterior margin of pronotum acute-angled. Hind tibia without pale basal annulus .... ***collare*** **(Scudder)** (p. 459)
   Median carina of pronotum slightly sinuate (Fig. 675); posterior margin of pronotum obtuse or right-angled. Hind tibia often with pale basal annulus .... ***saxatile*** **Morse** (p. 461)

Clé des espèces et des sous-espèces de *Spharagemon*

1. Tibias postérieurs rouges, marqués d'une large bande brune accolée à une bande pâle basale (fig. 676) .... ***bolli bolli*** **Scudder** (p. 457)
   Tibias postérieurs orangés ou rouges, sans bande foncée (fig. 677) ...2

2(1). Crête médiane du pronotum basse, tectiforme; incision du sulcus peu profonde et non oblique (fig. 672) .... ***equale*** **(Say)** (p. 458)
   Crête médiane du pronotum haute et aiguisée; incision du sulcus profonde et oblique, (fig. 674 et 675) 3

3(2). Crête médiane du pronotum prononcée, droite ou faiblement arquée (fig. 674); bord postérieur du pronotum formant un angle aigu. Tibias postérieurs sans annulus basal pâle ..... ***collare*** **(Scudder)** (p. 459)
   Crête médiane du pronotum légèrement sinueuse (fig. 675); bord postérieur du pronotum obtus ou à angle droit. Tibias postérieurs souvent pourvus d'un annulus basal pâle .... ***saxatile*** **Morse** (p. 461)

Spharagemon bolli bolli Scudder

Figs. 673, 676; Map 175

Spharagemon bolli Scudder, 1875*c*:469.

Spharagemon bolli; Blatchley 1892*b*:30; Cantrall 1943:107; Helfer 1963:114; R. D. Alexander et al. 1972:37.

Spharagemon bolli bolli; Hebard 1937*b*:365; Brooks 1958:42; Vickery et al. 1974:115.

Diagnosis. Hind tibia red, with broad brown band next to pale basal band.

Description. Body size medium (length, male 20–28, female 27–36 mm). Head with vertex almost flat, and with sides low and converging anteriorly and meeting frontal costa; foveolae triangular, distinct. Pronotum with median carina equally compressed throughout, with incision of sulcus not oblique (Fig. 673), and with hind margin of pronotum obtuse to right-angled. Color rusty brown to dark gray; tegmina reddish brown, without conspicuous markings; disc of hind wing yellow with broad black cross-band and short spur; apical third of disc clear; hind femur pale with 2 dark bands on outer face; hind tibia red apically, with broad brown band (Fig. 676).

Range. Manitoba to Québec, south to southeastern United States.

Behavior and habitats. *S. bolli bolli* is usually found in sandy areas or in grasslands on sandy soil, roadsides, or old fields. Brooks (1958) recorded it "on sandy soil in oak bluffs" in Manitoba. N. Criddle (1935) found it

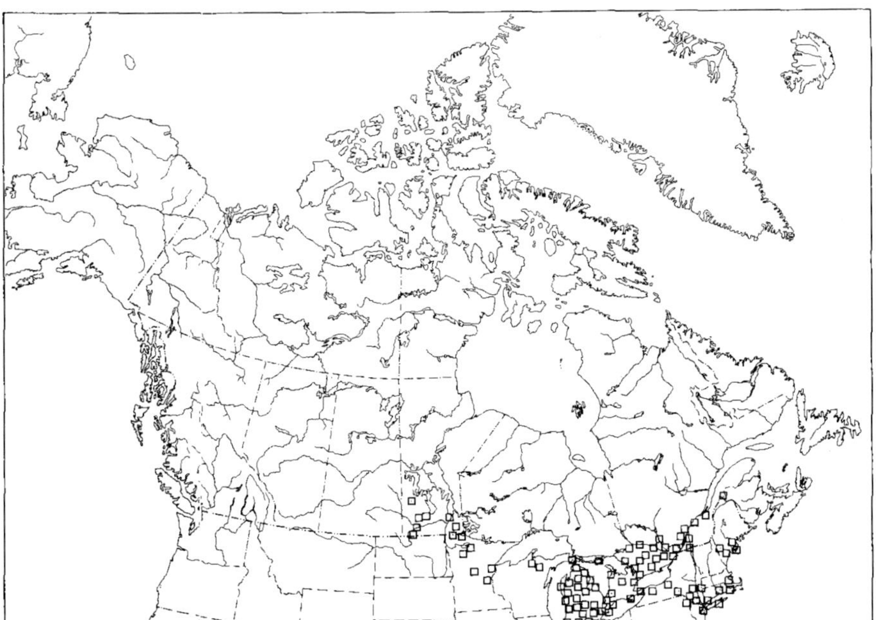

Map 175. Collection localities for *Spharagemon bolli bolli*.

in "open glades or around the margins of shrubby areas." Collections in Québec were from similar habitats. Criddle stated that it is not a grass eater but prefers broad-leaved plants, such as dandelion and vetches.

Cantrall (1943) reported crepitation in flight. Otte (1970) stated that males make frequent solitary crepitation flights. During courtship, *S. bolli bolli* uses "ordinary" stridulation and incorporates femur-shaking in its repertoire. During encounters between males, femur-tipping, femur-shaking, and striking the substrate (drumming) are aggressive signals. Nonreceptive females repel courting males by femur-tipping.

Life history. Univoltine, with overwintering in the egg stage in the ground. Adults are known to occur from 1 July to 18 October in Michigan (Cantrall 1968). In Québec, these appear about 10 July; they do not usually survive the frost in September.

Economic importance. Beirne (1972) reported moderate damage to upland pastures. It is a minor pest even when numerous.

Spharagemon equale (Say)

Fig. 672; Map 176

Gryllus equalis Say, 1825*a*:307.

Spharagemon aequale; Scudder 1875*c*:467, 468.

Spharagemon equale; Hebard 1925*a*:82; Brooks 1958:42; Helfer 1963:114.

Diagnosis. Median carina of pronotum low; incision of sulcus conspicuous but not oblique. Western species.

Description. Body medium-large (length 27–46 mm), robust. Median carina of pronotum low throughout, with incision of principal sulcus conspicuous, not oblique (Fig. 672). Color of head contrasting brown or gray, usually with area below eyes and antennae pale; pronotum gray or brownish, with metazona often white with dark spots dorsally; tegmina reddish brown to light gray, with 3 darker bands; disc of hind wing pale yellow to nearly white, with crossband broad, having short spur; hind femur pale with 2 darker bands dorsally on outer face; inner face yellow to orange, with 2 incomplete bands; hind tibia pale orange, paler at base on outer face.

Range. British Columbia to Saskatchewan, south to Texas.

Behavior and habitats. *S. equale* prefers open sandy areas and short sparse grasslands. It is a mixed feeder preferring forbs belonging to the Cruciferae. It has been reported from riverine flats and sheltered valley slopes. It is seldom found at elevations above 750 m.

Otte (1970) did not observe crepitation in this species. He did, however, observe vibratory stridulation in bursts of 0.05–0.15 second duration, always produced by a single femur and with a nearly constant interval between bursts. In aggressive encounters, femur-tipping and femur-shaking, as well as striking the substrate, were observed.

Life history. Univoltine, with overwintering in the egg stage. Hewitt and Barr (1967) reported adults to be most abundant in August and September in southern Idaho.

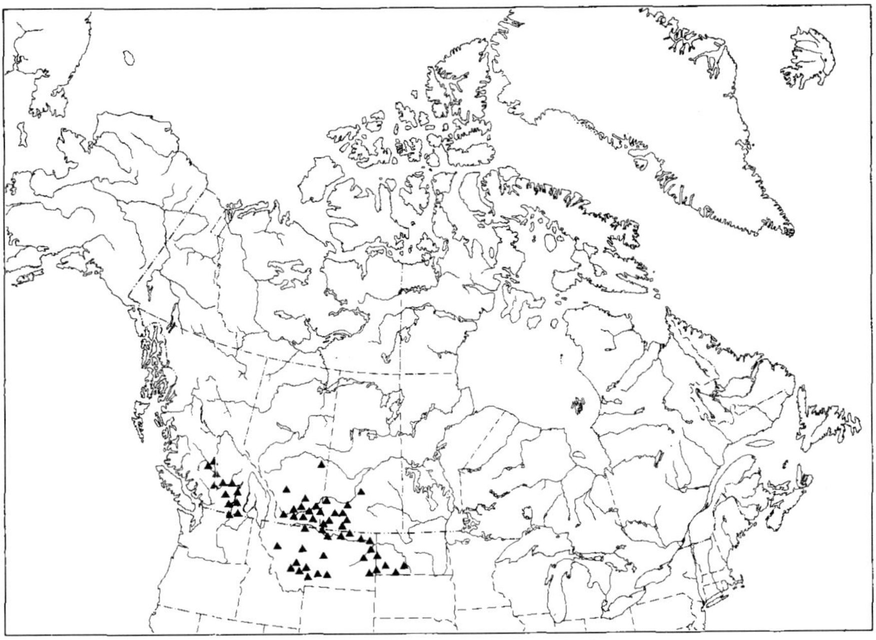

Map 176. Collection localities for *Spharagemon equale*.

Spharagemon collare (Scudder)

Figs. 671, 674, 677; Map 177

Oedipoda collaris Scudder, 1872:250.

Oedipoda wyomingiana C. Thomas, 1872:462.

Spharagemon collare; Lugger 1898:250; Brooks 1958:42; Helfer 1963:113; R. D. Alexander et al. 1972:37.

[*Spharegemon aequale*] *scudderi* Morse, 1894*c*:225.

Spharagemon collare scudderi Lugger, 1898:252.

Spharagemon collare wyomingianum E. M. Walker, 1901:20; Vickery et al. 1974:116.

Spharagemon collare collare; Cantrall 1943:107.

Diagnosis. Median carina of pronotum high, with notch deep and oblique. Hind tibia red or orange, yellowish at base, not banded.

Description. Body size average (length, male 19–25, female 24–30 mm). Pronotum with median carina high, having deep oblique notch, this often closed above by overlapping of ends of carina (Fig. 674), with posterior margin acute-angled. Color generally grayish, sprinkled with darker spots; tegmina gray, speckled, indistinctly barred; hind wing disc yellow, with cross-band dark and broad, and with spur short; hind femur pale with 3 dark cross-bands; hind tibia red to orange, somewhat yellowish near base, with no pale or dark ring (Fig. 677).

Range. British Columbia to Maine, south to Mexico and Florida.

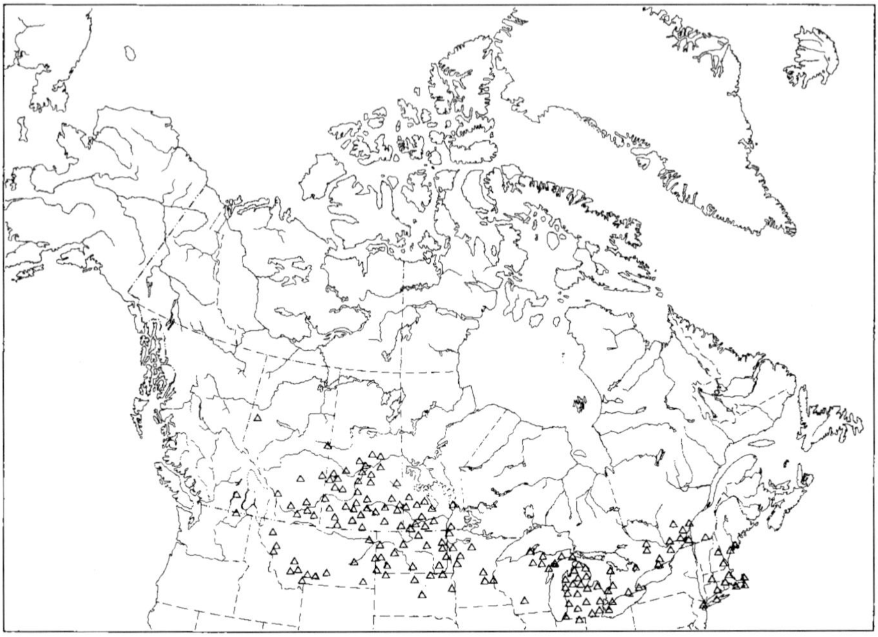

Map 177. Collection localities for *Spharagemon collare*.

Behavior and habitats. Like other species of *Spharagemon*, this species is usually associated with sandy areas and grassland on sandy soil. Mulkern et al. (1969) reported it from "sparsely vegetated sandy areas where it appeared to feed on most vegetation availac'e with no distinct preference." N. Criddle (1935) found that it ingested mostly grasses in Manitoba.

Otte (1970) reported that *S. collare* crepitated in flight only infrequently, possibly mainly when disturbed. During courtship, males produced stridulation on both upstroke and downstroke of the femora. Femur-tipping was not used during courtship but appeared to have a purely aggressive function. Drumming by striking the tibiae on the substrate is also an aggressive act. Nonreceptive females repel courting males by raising or shaking the femora and occasionally by femur-tipping and striking the substrate with the hind tibiae.

Life history. Univoltine, with overwintering as eggs in soil. N. Criddle (1918) described oviposition of *S. collare*. The egg pods that he examined contained 11, 2, and 12 eggs, but it is probable that the average number is higher. Adults are active in Michigan from 10 June to 1 November (Cantrall 1968). Hewitt and Barr (1967) called the insect a late-summer species, most abundant in August and September in Idaho. In Québec, adults have been found from 11 August to 30 October, but they probably reach adulthood as early as the end of July.

Spharagemon saxatile Morse

Fig. 675; Map 178

Spharagemon saxatile Morse, 1894*c*:229, 234, figs. 4, 5.
Spharagemon saxatile; Morse 1919*b*:35.
Spharagemon saxatile saxatile; Otte 1970:44.

Diagnosis. Median carina of pronotum high, arched, with notch suboblique, often nearly closed above. Found in New England and southward.

Description. Body size medium (length, male 20–24, female 28–34 mm). Vertex of head elongate-oval, with disc concave; median carina evident but faint; foveolae small, shallow, open beneath. Pronotum with high carina, distinctly arched on metazona; notch suboblique, with tip of carina of prozona projected back over notch (Fig. 675); prozona constricted, rugose; metazona broad, flat, with posterior margin obtuse to right-angled. Tegmina long, surpassing apex of abdomen. Color grayish brown, paler beneath; pronotum gray, mottled with black, usually with X-shaped ashy gray blotch; tegmina brownish basally, usually with 3 fuscous crossbars; hind wing sulfur yellow at base, with crossband broad, and with spur extending halfway to base; hind femur ashy gray with fuscous bars, with inner face yellow with 4 black bands (basal two partly united); hind tibia coral red, usually with pale basal ring, and with spines black-tipped.

Range. Massachusetts to Maryland.

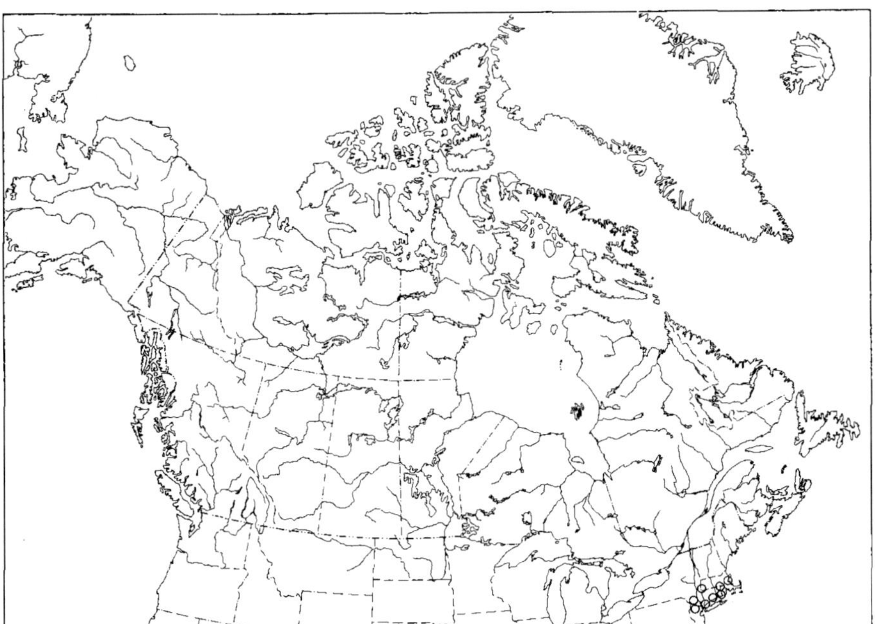

Map 178. Collection localities for *Spharagemon saxatile*.

Behavior and habitats. *S. saxatile* is almost always associated with rough land and rocky outcroppings. It is often found with *Trimerotropis verruculata* (W. Kirby) where the ranges of the two overlap.

Otte (1970) did not observe crepitation in flight, such as was reported by Morse (1920). He found "ordinary" as well as vibratory stridulation during courtship. Femur-jerking was used as an aggressive signal.

Life history. Adults have been reported from mid-July to late October (Morse 1920). The life cycle is univoltine, with overwintering in the egg stage in soil.

Genus *Cratypedes* Thomas

Description. Similar to *Xanthippus* and *Pardalophora*. Depression of vertex divided in front by ridge. Metazona 1.5–1.75 times as long as prozona; lateral pronotal lobes produced lateroposteriorly, wider at bottom than at middle; pronotum of irregular form, rough, with median carina cut by 2 sulci (Fig. 663).

Cratypedes neglectus (Thomas)

Figs. 663, 667, 668; Map 179

Oedipoda neglecta Thomas, 1870:81.

Hippiscus (*X*[*anthippus*]) *obscurus* Scudder, 1892:359.

Cratypedes neglectus; W. F. Kirby 1910:205; Brooks 1958:38; Helfer 1963:122.

Diagnosis. Depression of vertex divided in front by ridge. Lateral pronotal lobes produced posterolaterally, rounded.

Description. As for genus (Fig. 668), robust. Body length 28–41 mm. Color dark gray to brownish pink; pronotum variable, entirely dark, dark gray with light gray lateral lines, dark with pale apical border, or pale; tegmina of males dark with 2 pale gray spots before middle, spotted toward apex, often with dorsal yellow lines that converge posteriorly; tegmina of females paler; hind wing disc yellow, with crossband dark and broad, and with spur reaching three-quarters of distance to base of wing; hind femur with dorsal and outer faces banded, with inner ventral face tinged with orange; hind tibia reddish, orange, or occasionally pale yellow.

Range. British Columbia to Manitoba, south to California and New Mexico.

Behavior and habitats. Brooks (1958) reported *C. neglectus* as an inhabitant of short mixed-grass prairie in dry sandy locations. He called it primarily a grass feeder. Hewitt and Barr (1967) found it in dry areas where sagebrush occurred but primarily in mountainous areas in meadows and grassy hillsides in Idaho. General biology is outlined by Scoggan and Brusven (1972). Flights are usually straight, about 3–9 m in distance. During flight, crepitation is noticeable, "a loud clacking sound" according to Hewitt and Barr (1967).

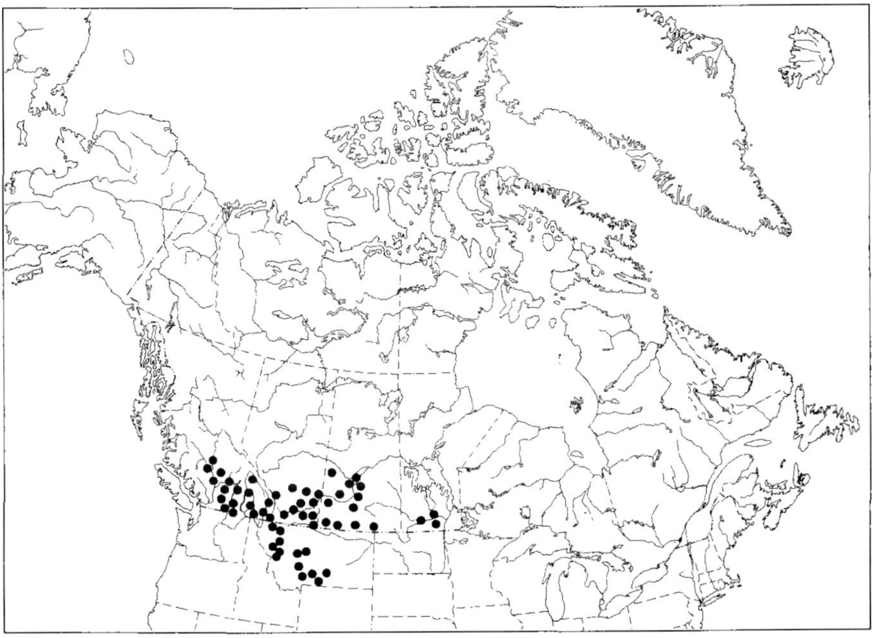

Map 179. Collection localities for *Cratypedes neglectus*.

Life history. Unlike the life cycle of species of *Hippiscus* and *Xanthippus*, that of *Cratypedes neglectus* is univoltine. The species overwinters in the egg stage and is adult from June to late August.

Genus *Xanthippus* Saussure

Description. Size medium–large, robust. Vertex of head with distinct median carina in depression, the latter open anteriorly; frontal costa constricted above antennal bases, strongly sulcate around and below ocellus. Pronotum with disc rugose or tuberculate, or both; median carina low, distinctly cut by 2 sulci ahead of middle; carina between sulci variable in degree of development, distinct, depressed, deformed, or obliterate; posterior margin of pronotum acute-angled, with lateral carinae well-defined only on metazona; lateral lobes with anterior and posterior margins nearly parallel. Tegmina and wings surpassing apex of abdomen.

Key to species and subspecies of *Xanthippus*

(Adapted from Brooks (1958))

1. Hind tibia yellow. Found in southern British Columbia and adjacent United States, not Northwest Territories ..... ***vitellinus*** **Saussure** (p. 465)

   Hind tibia pink, orange, or reddish, at least on apical half or on inner face (if completely yellow, then from Northwest Territories and not from British Columbia) 2

2(1). Hind tibia yellow on basal half, yellowish to pink apically. Tegmina speckled .... *corallipes brooksi* **Vickery** (p. 466)

Hind tibia reddish or orange, at least on inner face. Tegmina yellowish, with larger regular markings 3

3(2). Hind femur with inner face bluish black on flange of basal half and often above; hind tibia with inner face entirely red, and with outer face partly to entirely yellow .... *corallipes buckelli* **Hebard** (p. 467)

Hind femur with inner face entirely red; hind tibia red to orange, usually yellowish on basal third of outer face 4

4(3). Hind tibia deep red, either entirely or on inner face. Depression of vertex with or without conspicuous cross-ridges. Hind wing generally without spots at apex. Found in grasslands and eroded areas .... *corallipes latefasciatus* **(Scudder)** (p. 467)

Hind tibia orange to reddish orange. Depression of vertex without conspicuous cross-ridges. Hind wing spotted at apex. Found in forested and montane areas .... *montanus* **(Thomas)** (p. 468)

Clé des espèces et des sous-espèces de *Xanthippus*

(adaptée de Brooks, 1958)

1. Tibias postérieurs jaunes. Insecte vivant dans le sud de la Colombie-Britannique et les États adjacents des États-Unis, mais non dans les Territoires du Nord-Ouest .... *vitellinus* **Saussure** (p. 465)

Tibias postérieurs roses, orangés ou rougeâtres au moins sur leur portion apicale ou sur leur face interne (si les tibias sont complètement jaunes, l'insecte provient des Territoires du Nord-Ouest et non de Colombie-Britannique) 2

2(1). Tibias postérieurs jaunes sur leur moitié basale, jaunâtres à roses sur la moitié apicale. Tegmina mouchetés ... *corallipes brooksi* **Vickery** (p. 466)

Tibias postérieurs rougeâtres ou orangés, au moins sur leur face interne. Tegmina jaunâtres, portant de grandes marques régulières 3

3(2). Fémurs postérieurs à face interne noir bleuâtre sur le rebord de la portion basale et souvent au-dessus; tibias postérieurs à face interne entièrement rouge et à face externe partiellement à entièrement jaune ... . *corallipes buckelli* **Hebard** (p. 467)

Fémurs postérieurs à face interne entièrement rouge; tibias postérieurs rouges à orangés, habituellement jaunâtres sur le tiers basal de la face externe 4

4(3). Tibias postérieurs rouge foncé, entièrement ou sur la face interne seulement. Dépression du vertex avec ou sans crête transversale. Ailes postérieures généralement dépourvues de taches à l'apex. Insecte vivant dans les prairies et les zones érodées .... *corallipes latefasciatus* **(Scudder)** (p. 467)

Tibias postérieurs orangés à orangé rougeâtre. Dépression du vertex sans crête transversale. Ailes postérieures marquées de taches à l'apex. Insecte vivant dans les régions forestières et montagneuses .... *montanus* **(Thomas)** (p. 468)

Xanthippus vitellinus Saussure

Map 180

Hippiscus (*Xanthippus*) *vitellinus* Saussure, 1884:90, 94.
Xanthippus vitellinus; Buckell 1922:23; Brooks 1958:48.

Diagnosis. Hind tibia yellow. Found in British Columbia, Washington, and Oregon.

Description. Median carina of pronotum usually distinct throughout prozona and nearly as distinct before the principal sulcus. Tegmina mottled with rather irregular small and large dark blotches, not tending to form bands; hind wing pinkish, with dark brownish crossband.

Range. British Columbia and Idaho, south to Nevada.

Behavior and habitats. Not known.

Life history. Presumed to be similar to that of other members of the genus but not studied.

Xanthippus corallipes Haldeman, *sensu lato*

Oedipoda corallipes Haldeman, 1852:371, pl. 10, fig. 2.

Hind tibia pink, orange, or reddish (yellow only in far northern Canada). Hind wing not spotted apically. Found in grasslands and eroded areas in southern Canada and in the United States; in northern Canada, found on eroded slopes. The species needs revising.

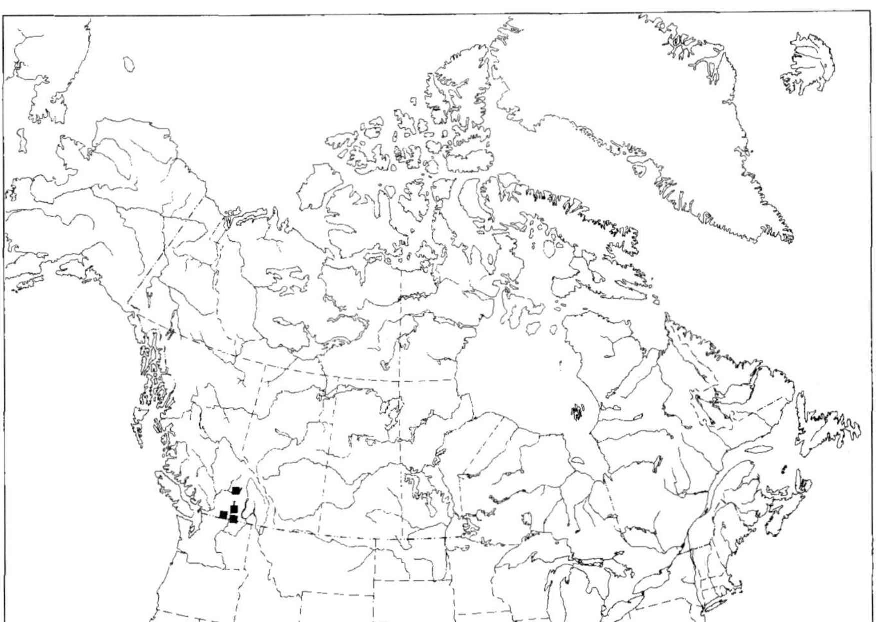

Map 180. Collection localities for *Xanthippus vitellinus*.

Xanthippus corallipes brooksi Vickery

Map 181

Xanthippus corallipes brooksi Vickery, 1967*a*:263, figs. 21, 24, 25, 28, 32.

Xanthippus corallipes; Brooks 1958:48.

Diagnosis. Found in Northwest Territories. Hind tibia yellowish at base to pinkish apically. Hind wing disc pale yellow.

Description. Body small, slender (length, male 22.7, female 36.8 mm). Head rugose; vertex with median carina obsolete between eyes; area beside and anterior to carina deeply impressed between lateral carinae, with depression dividing into 2 lateral lines at antennal bases; lateral foveolae deep. Prozona rugose; lateral carinae distinct only on metazona; median carina distinct, cut by 2 sulci on anterior half. Tegmina and wings surpassing abdomen. Color dull blackish brown with dirty gray markings; head and pronotum dull dusty brown with paler M-shaped line on head; tegmina with buff line between dorsal and lateral fields, with dorsal area dark, and with lateral area speckled with small irregular dark spots that show no evidence of banding; hind wing pale yellow, with black crossband that does not reach anal angle; hind femur with 3 oblique dark bands on outer face; inner face, including that of ventral flange, red as far as pale preapical ring; hind tibia yellowish at base to pink on distal half.

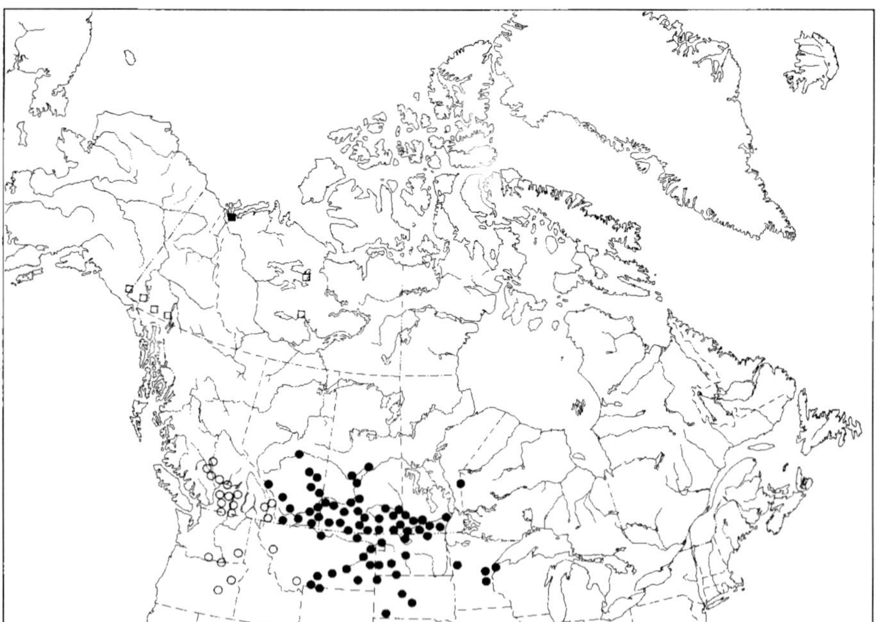

Map 181. Collection localities for *Xanthippus corallipes brooksi* (■), *X. corallipes buckelli* (○), *X. corallipes latefasciatus* (●), and atypical *X. corallipes brooksi* (□).

Note: Vickery (1967*a*) pointed out slight differences between the male genitalia of *X. corallipes brooksi, X. corallipes buckelli*, and *X. corallipes latefasciatus*.

Range. Known in typical form only from the type locality.

Behavior and habitats. *X. corallipes brooksi* is found in grassy areas in northern tundra. No other information is available.

Life history. Presumably a 2-year cycle occurs.

Xanthippus corallipes buckelli Hebard

Fig. 665; Map 181

Xanthippus corallipes buckelli Hebard, 1928:241, pl. 24, fig. 1, pl. 25, fig. 1.

Xanthippus corallipes buckelli; Strohecker 1952:684; Brooks 1958:49; Helfer 1963:123.

Diagnosis. Depression of vertex distinct, without conspicuous cross-ridges. Median carina distinct only on posterior half. Found from British Columbia to western Montana.

Description. Body large (length 29–47 mm), robust (Fig. 665). Depression of vertex distinct, without conspicuous cross-ridges, with median carina distinct only on posterior half. Pronotum moderately tuberculate, with median carina weak posteriorly. Color generally yellowish with 2 pale dorsal stripes; head usually paler than pronotum; pronotum with pale lines near lateral carinae; tegmina brown with irregular dark blotches, these smaller and less contrasting apically, with dorsal pale lines converging posteriorly; hind wing with disc red or pink, with crossband dark, broad; spur extending nearly to wing base, usually with 3–6 dark cells apically; hind femur vaguely obliquely banded, with inner face having basal half of lower flange and area above bluish black, occasionally reddish; hind tibia reddish on inner face, yellowish on outer face.

Range. British Columbia and Montana, south to Nevada.

Behavior and habitats. Hewitt and Barr (1967) state that *X. corallipes buckelli* inhabits "sage-brush-grass and salt-desert shrub zones" in Idaho. They also state that the only flight noise is a "low whirring sound made by the females." Apparently males make no sound while in flight, as was stated by Otte (1970) for *X. corallipes pantherinus* (Scudder).

Life history. *X. corallipes buckelli* has a 2-year cycle, spending the first winter as eggs in the soil and the second as partly grown nymphs. Adults were reported to be active in early spring and to persist until August (Hewitt and Barr 1967). The two females reported from Fairview, B.C., by Buckell (1920) were taken in May. Hebard (1928) recorded adults from 2 May to 8 August, but adults are most numerous in May and June.

Xanthippus corallipes latefasciatus (Scudder)

Figs. 662, 666; Map 181

Hippiscus (*X*[*anthippus*]) *latefasciatus* Scudder, 1892:273, 359.

Xanthippus corallipes latefasciatus; Hebard 1925*a*:78; Brooks 1958:49.
Xanthippus corallipes; Mills and Pepper 1938:14; Helfer 1963:123.

Diagnosis. Hind tibia red. Pronotum tuberculate. Depression of vertex rough, with median carina distinct only on posterior half.

Description. Body large (length 29–41 mm), robust. Depression of vertex rough, with median carina distinct only on posterior half, sometimes with cross-ridges (Fig. 666). Pronotum tuberculate, with median carina prominent (Fig. 662). Color generally brown, mottled; head gray and brown, darker dorsally; pronotum grayish to brownish; tegmina usually with conspicuous maculations, and with pale dorsal stripes converging posteriorly, sometimes entirely gray; hind wings yellow at base, with dark crossband narrow to broad, and with spur reaching nearly to wing base; apex of wing without dark spots; hind femur with 3 dark bands that are prominent dorsally, somewhat broken on outer face, and with inner face usually deep red, occasionally yellow or orange yellow; hind tibia deep red on inner face, usually yellowish on basal third of outer face.

Brooks (1958) distinguished "major" and "minor" forms of this subspecies. The major form, which occurs only in eroded areas, is paler and has the pronotum less tuberculate, the spur of the hind wing crossband shorter, and the band itself narrow and often broken. The minor form, which is found in grasslands, has a strongly tuberculate pronotum, the spur of the hind wing crossband long, and the band itself broad and unbroken. These are probably environmental phenotypes, or ecophenes.

Range. Alberta to Manitoba, south to Colorado.

Behavior and habitats. This insect is generally found in grasslands. Very little is known regarding its food plants. Brooks (1958) called it primarily a grass feeder, but it will eat some forbs. Mulkern et al. (1969) found a grass, *Stipa comata*, in the crop of one individual.

Otte (1970) described behavior of *X. corallipes pantherinus* (Scudder) in Texas. He did not observe flight displays during pair formation, but both "ordinary" and vibratory stridulations were recorded, each type produced by movement of one femur. The two types of sound were often produced sequentially. Both femur-tipping and silent femur-shaking were used by males as aggressive signals.

Life history. The life cycle of this and other subspecies of *X. corallipes* lasts for 2 years. Adults occur in June and July. The eggs deposited by females do not hatch until the following spring. The progeny spend the next winter as partly grown nymphs and reach maturity in the following spring and summer.

Xanthippus montanus (Thomas)

Map 182

Oedipoda montana Thomas, 1872:462.
Xanthippus montanus; Hebard 1925*a*:78.
Xanthippus corallipes montanus; Brooks 1958:49.

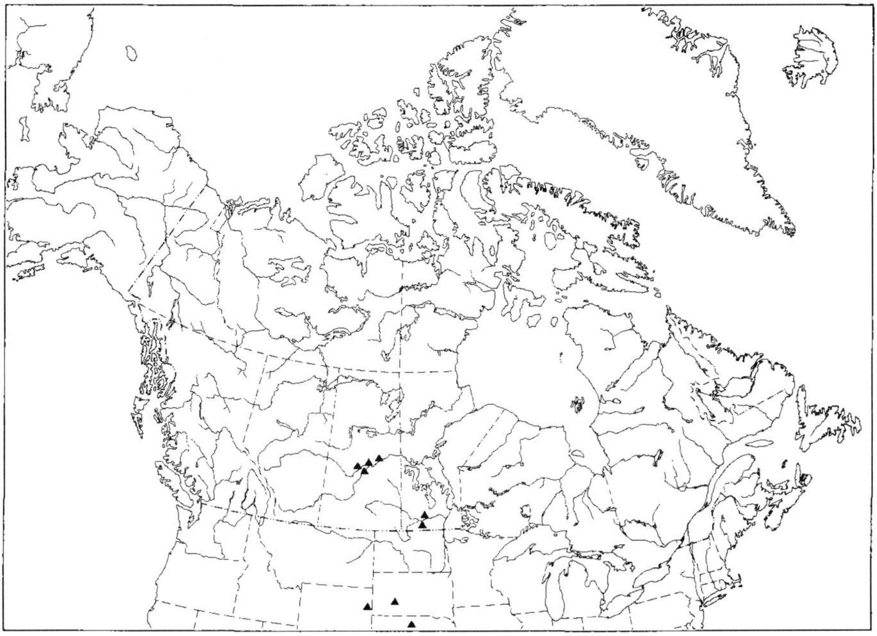

Map 182. Collection localities for *Xanthippus montanus*.

Diagnosis. Depression of vertex deep, narrow, with irregular protuberances; margins sharp.

Description. Body size medium (length 28–50 mm). Head moderately large, rough, with depression of vertex deep, narrow, and with irregular protuberances; margins sharply defined. Pronotum distinctly sulcate anteriorly, strongly tuberculate. Color mottled gray to brown; tegmina with longitudinal blotches, paler at apex, these distinct in northern specimens but less well-defined in those from southern areas; hind wing disc red to pink, with crossband and dark spur reaching wing base; hind femur indistinctly banded on outer face, coral red on inner face; hind tibia variable, yellowish to pink on outer face, usually orange to reddish orange on inner face.

Range. Saskatchewan and Manitoba, south to Arizona and Oklahoma.

Behavior and habitats. *X. montanus* is found in sandy areas, including spruce and pine forests in Saskatchewan (Brooks 1958). Hewitt and Barr (1967) recorded it from "flat dry areas where *Atriplex* spp. and *Artemisia* spp. are found." They called it a mixed feeder. There is some discrepancy between reports of flight behavior. Brooks (1958) stated that it "flies only a few yards and tends to move erratically." Hewitt and Barr (1967) said that it flies "some distance" with a low whirring sound. There is a possibility that the *X. corallipes montanus* reported by Brooks (1958) does not belong to the same taxon as that reported from Idaho by Hewitt and Barr (1967). We have not been able to investigate this.

Life history. *X. montanus*, like other members of the genus, has a 2-year life cycle. Adults have been recorded from April until the end of July in Idaho (Hewitt and Barr 1967), although they appear to be most common from late May to early July.

Genus *Hippiscus* Saussure

Description. Similar to *Pardalophora*, but vertex of head with cross-ridges, and disc of vertex divided into four parts (Fig. 661). Pronotal disc with prozona and metazona about equal in length.

Hippiscus ocelote Saussure

Figs. 660, 661; Map 183

Oedipoda (*Hippiscus*) *ocelote* Saussure, 1861:398.
Oe[dipoda] rugosa Scudder, 1863*a*:469.
Hippiscus compactus Scudder, 1892:288.
Hippiscus variegatus Scudder, 1892:301.
Hippiscus suturalis Scudder, 1892:301.
Hippiscus citrinus Scudder, 1901*d*:88.
Hippiscus immaculatus Morse, 1906:119.
Hippiscus ocelote; Hebard 1945:86.

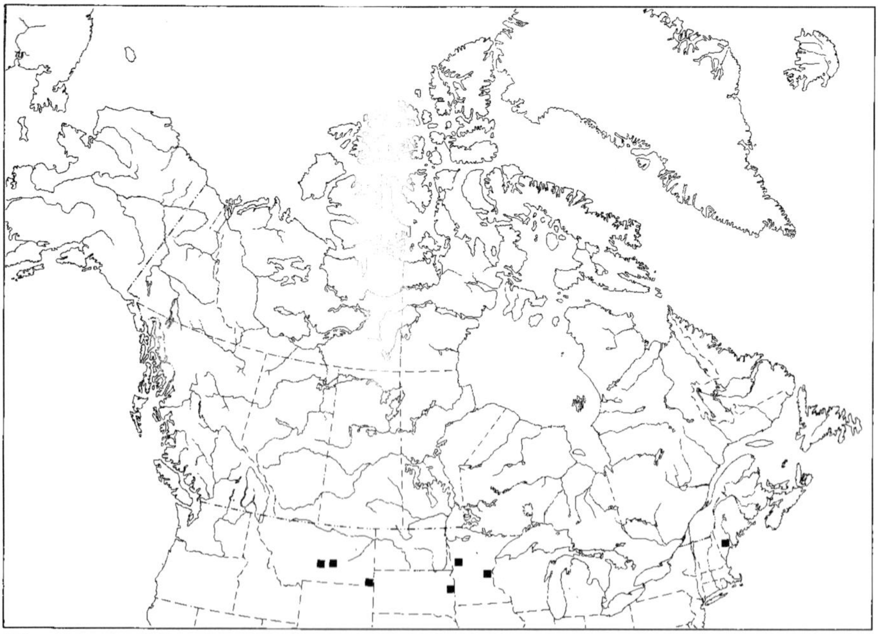

Map 183. Collection localities for *Hippiscus ocelote*.

Diagnosis. Vertex short, divided into four areas by low carinae. Prozona rugose; metazona tuberculate.

Description. Body large (length, male 28–36, female 37–42 mm) (Fig. 660). Head rounded, with vertex short, divided into four parts by low ridges, smoothly rounded to face, and with occiput convex. Pronotal disc with prozona and metazona about equal in length; prozona strongly rugose; metazona coarsely tuberculate. Color brown to clay brown; tegmina heavily and irregularly spotted with fuscous; hind wing yellowish white, yellow, or pinkish to bright vermillion red, with fuscous crossband narrow, and with spur extending about two-thirds of distance to base; hind femur with 3 oblique dark bands on outer face; hind tibia yellow with paler basal ring.

Range. Montana to Maine, south to Mexico and Florida.

Behavior and habitats. *H. ocelote* is found only in dry upland areas, such as open woodland pastures and along roadsides and rail fences. Mulkern et al. (1969) listed *H. ocelote* (as *H. rugosus*) as a grass feeder, adults preferring tall grasses such as *Agropyron smithii, Andropogon scoparious*, and *Bouteloua curtipendula*. Young nymphs fed mainly on *B. gracilis* in Montana.

Blatchley (1920) and Otte (1970) remarked on the lack of flight displays by males of this species. These may or may not produce "ordinary" stridulation on the substrate during courtship. Males tap the heads of females and may produce a distinct sideways rocking of the head prior to copulation. Aggressive signals between males include femur-tipping and femur-shaking.

Life history. The species becomes mature in late July or August and adults persist until late October. It is not known if the life cycle requies 2 years or is completed in one season.

Economic importance. Seldom numerous enough to cause significant damage, but occasional injury to grazing has been recorded.

Genus *Pardalophora* Saussure

Description. Body large, robust. Head large, subglobose with foveolae shallow, triangular, and with disc of vertex lacking transverse ridges; frontal costa constricted above antennal bases. Pronotum roughened, with metazona distinctly longer than prozona. Color usually ashy brown to gray; tegmina with large dark spots; hind wing red at base with black crossband.

Key to species of *Pardalophora*

1. Basal half of median area of tegmen with large continuous blotch. Northern species .... ***apiculata* (Harris)** (p. 472)
   Basal half of median area of tegmen with several well-separated dark blotches. Found in southern Manitoba .. ***haldemanii* (Scudder)** (p. 473)

Clé des espèces de *Pardalophora*

1. Moitié basale de la zone médiane du tegmen marquée d'une grande tache. Espèce nordique .... ***apiculata* (Harris)** (p. 472)
   Moitié basale de la zone médiane du tegmen marquée de plusieurs taches foncées distinctes. Espèce vivant dans le sud du Manitoba .... ***haldemanii* (Scudder)** (p. 473)

Pardalophora apiculata (Harris)

Fig. 659; Plate I*A*; Map 184

Locusta apiculata T. W. Harris, 1835:576.
Oedipoda obliterata Burmeister, 1838:643.
Locusta corallina T. W. Harris, 1841:142.
Pardalophora apiculata; Morse 1919*b*:34; Cantrall 1943:100; Brooks 1958:41; Helfer 1963:119; R. D. Alexander et al. 1972:37; Vickery et al. 1974:112.

Diagnosis. Large species. Pronotal disc flat, granulate to tuberculate. Hind wing disc red. Adult in spring.

Description. Body large (length, male 30–36, female 40–46 mm), robust. Females much larger than males. Pronotum with disc flat, granulate, bearing a few small tubercles, with median carina low, distinct, incised well forward of middle by principal sulcus; lateral carinae distinct only on metazona. Hind femur broad, with both upper and lower carinae prominent. Tegmina long, broad, apically rounded. Color generally ashy brown to gray; hind wing red at base, bordered by black band, with long anterior spur, infumate at apex; inner face of hind femur and hind tibia bright red, the former becoming darkened on basal half.

Range. Western Northwest Territories to New Brunswick, south to Colorado and North Carolina.

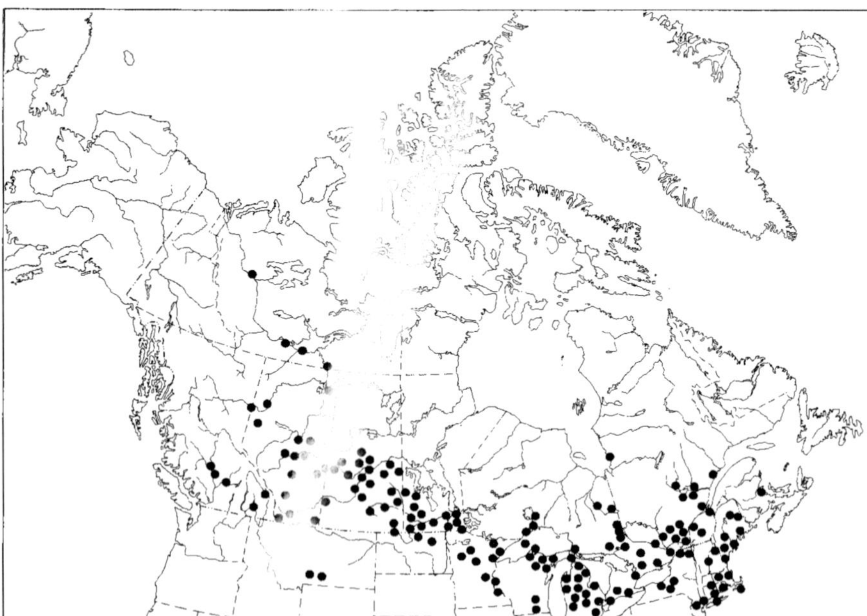

Map 184. Collection localities for *Pardalophora apiculata*.

Behavior and habitats. *P. apiculata* usually occurs in small colonies and is rarely numerous. In Québec it has been found in sandy scrubby fields in the Laurentian foothills. N. Criddle (1935) recorded it in or around openings in woodlands. He said it was rarely found on open prairie. Although it will eat grass it prefers broad-leaved plants, including dandelion (*Taraxacum* sp.). Adults are strong fliers and occasionally fly beyond their usual habitats (Cantrall 1943).

Blatchley (1920) described crepitation of *P. apiculata* as "a low rattling note" while the insect is in flight. Otte (1970) reported both "ordinary and vibratory stridulation" by males on the ground. Ordinary stridulation is produced by 1–4 strokes of both femora and is usually produced by males approaching stationary females, by males following retreating females, and also by isolated males, whether or not they had recently encountered a female. Vibratory stridulation consists of a series of rapid strokes by a single femur, producing 4–18 pulses, usually produced by stationary males just before they approach females. Otte also stated that crepitation in flight does not occur. The statement of Blatchley (1920) to the contrary is in error. The brilliantly colored hind wings render the male conspicuous to females and the visual component is apparently the important factor in short flights prior to pair formation.

Life history. Cantrall (1943) reported hibernation of fourth-instar nymphs under rosettes of panic grass (*Panicum* spp.) in Michigan. Pickford (1953) found that *P. apiculata* has a 2-year life cycle in Saskatchewan, passing the first winter as eggs in the soil, and the second winter as partially grown nymphs. Recent studies in Québec (Crozier 1977) have shown that the life cycle requires 2 years in that province also. This is probably true throughout the range of the species. Adults appear in May; they have usually disappeared before mid-July. Eggs hatch about the 1st week of the following July. Winter is passed (in Québec) as fourth- or fifth-instar nymphs.

Pardalophora haldemanii (Scudder)

Map 185

Oedipoda Haldemanii Scudder, 1872:251.
Hippiscus nanus Saussure, 1884:83, 85.
Pardalophora saussurei Scudder, 1892:268, 302.
Hippiscus (*X*[*anthippus*]) *tigrinus* Scudder, 1892:334.
Pardalophora haldemanii; Hebard 1925*a*:77; Cantrall 1943:101; Helfer 1963:119; R. D. Alexander et al. 1972:37.

Diagnosis. Tubercles on metazona forming ridges parallel to posterior margin.

Description. Body large (length, male 27–35, female 42–46 mm). Pronotum with metazonal tubercles variable in number but tending to form ridges parallel to posterior margin. Color ashy to brownish gray, paler below; pronotal and tegminal maculations small, numerous, and ulnar area with series of separated spots; hind wings yellow or reddish basally, with spur of crossband nearly reaching base; hind femur with upper and outer faces

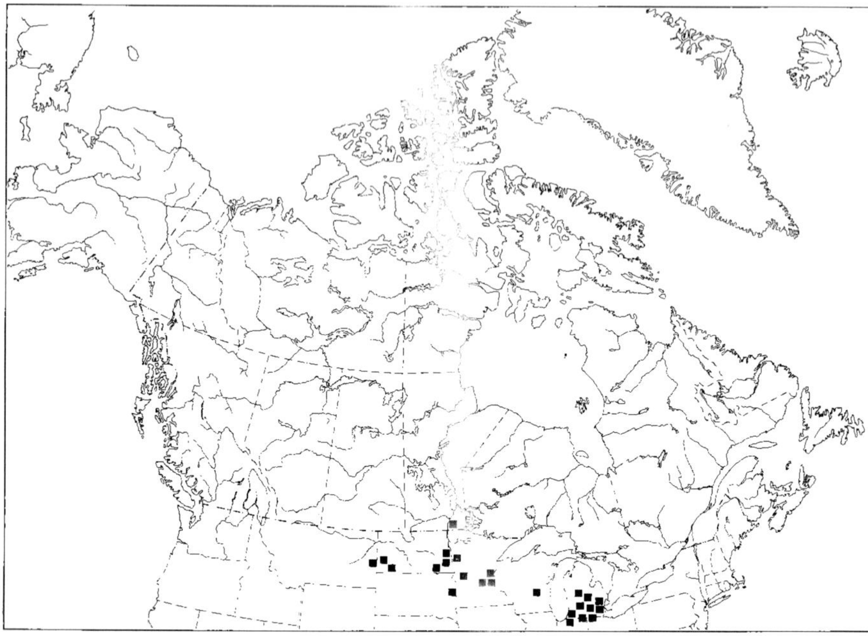

Map 185. Collection localities for *Pardalophora haldemanii.*

vaguely trifasciate, and with inner and lower faces unicolorous; hind tibia yellow, orange, or reddish.

Range. Montana to Michigan, south to Arizona and Texas.

Behavior and habitats. Mulkern et al. (1969) recorded *P. haldemanii* as a grass feeder, showing little preference for any particular kinds. The species is not usually numerous although it may, at times, for no known reason, become very numerous. Parker and Connin (1964), for a night in July 1932, reported "a flight of great magnitude throughout southern South Dakota and northeastern Nebraska." Both the abundance and night flight in large numbers were remarkable. Cantrall (1968) noted that adults are attracted to street lights on warm nights in July in Michigan.

Otte (1970) reported only femur-tipping and shaking by males during courtship. He did not observe stridulation in 30 courtship sequences. During femur-shaking, the apical ends of the tibiae were struck vigorously against the substrate, producing an audible drumming sound. Otte did not observe such behavior in other species of *Pardalophora.*

Life history. Not studied. Cantrall (1968) reported adults in Michigan from 27 May to 30 August. Nymphal overwintering appears to be the rule, and it is possible, but not certain, that the life cycle extends over 2 years, as in *P. apiculata.*

P. haldemanii may ingest appreciable quantities of food; it is not usually regarded as being particularly injurious. Even on the occasions of mass flights, the species has not been recorded as damaging crops. The only claim to fame as an economically important insect seems to have been on the occasion of the mass night flight previously mentioned (Parker and Connin 1964), when it invaded a number of towns and caused considerable inconvenience to the human inhabitants.

Genus *Chortophaga* Saussure

Description. Size medium. Antennae flattened. Body slender, compressed. Face slightly slanted. Pronotum with disc tectiform; median carina forming high sharp crest, only faintly cut by principal sulcus. Tegmina narrow, with apical halves membranous.

Chortophaga viridifasciata (De Geer)

Figs. 655, 656; Map 186

Acrydium viridi-fasciatum De Geer, 1773:498.
Gryllus virginiana Fabricius, 1775:291.
Locusta (Tragocephala) infuscata T. W. Harris, 1841:147.
Tomonotus zimmermanni Saussure, 1861:320.
Tragocephala infuscata viridifasciata; Caulfield 1888:70.
Chortophaga viridifasciata; Cantrall 1943:98; Brooks 1958:37; Helfer 1963:110.
Chortophaga viridifasciata viridifasciata Vickery et al. 1974:108.

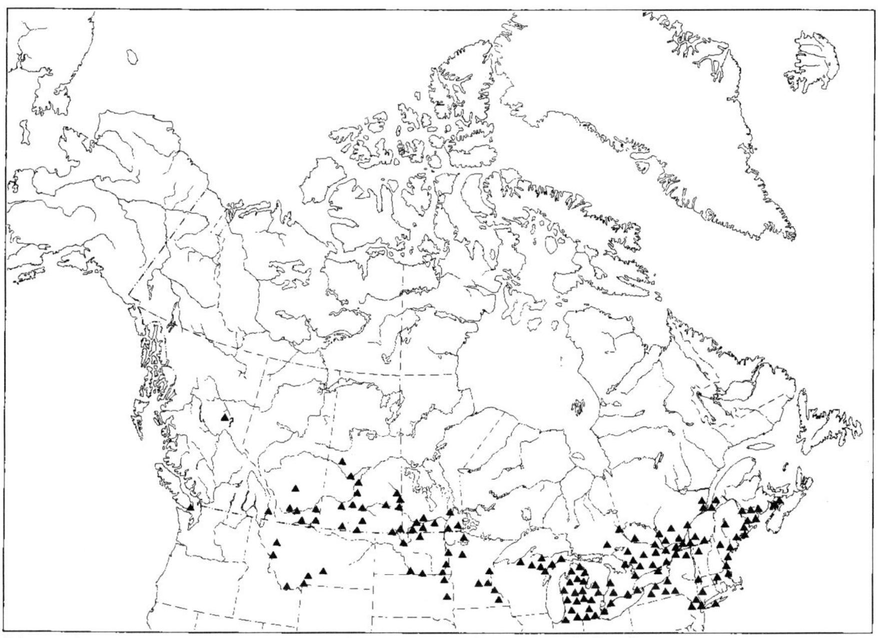

Map 186. Collection localities for *Chortophaga viridifasciata*.

Common name. Greenstriped grasshopper.

Diagnosis. Pronotal disc tectiform. Body brown or green. Early spring species.

Description. Body size medium (length, male 17–24, female 24–30 mm) (Fig. 656). Vertex of head horizontal, triangular; apex truncate; foveolae shallow, elongate-triangular. Pronotum with disc tectiform; anterior margin angulate, projected; posterior margin acute, with median carina not prominent, straight, faintly notched (Fig. 655). Tegmina long, narrow, longer than abdomen with apical part membranous. Color dimorphic, either brown or green; green form with head, pronotum, tegminal bases, and outer faces of hind femora grass green, and with apical part of tegmina and narrow costal stripe pale brown; abdomen reddish brown; brown form with apical halves of tegmina darker; hind tibia brown with whitish basal annulus.

In eastern populations, the brown form predominates in males, whereas most females are green. About 18% of females are brown, but green males are uncommon. In Texas, Otte and Williams (1972) found green forms to predominate in green habitats (i.e., grassy areas), whereas brown morphs were more abundant in dry brown habitats. No such correlation has been observed in northern areas. Rearing in brown or green cages did not produce significant changes in color. Laboratory-reared nymphs tend to turn gray or brown irrespective of background, food, or humidity.

Range. British Columbia to Nova Scotia, south to New Mexico and Virginia.

Behavior and habitats. Vickery et al. (1974) recorded this species from grassy areas in Eastern Canada. Mulkern et al. (1969) found it to feed mainly on grasses.

Visual and acoustical displays have been described by Otte (1970) and by Steinberg and Willey (1974). Males crepitate loudly during flight. Since the wings are transparent, the audible component seems to be more important than the visual one. Most flights are very short, lasting only 1–2 seconds. Females are attracted to crepitating males and approach them by flying or walking. Females also produce crepitations while in flight. During courtship on the substrate, several behavioral sequences may be involved. Males may flick the femora upward, then produce a stridulatory chirp. If a female approaches, the male mounts her and copulation follows. The entire sequence may last only a few seconds or may be prolonged for up to 45 minutes. Femur-tipping is more often used as an aggressive signal between males rather than to attract females. Various male/female behaviors were analyzed by Steinberg and Conant (1974).

Life history. Somes (1914) called *C. viridifasciata* "the harbinger of spring," as it is one of the first common grasshopper species to appear in the adult stage. It overwinters as partly grown nymphs, adults appearing by late May (sometimes as early as April in southern New England). They usually disappear by the end of June. Small nymphs are common in grassy areas during late summer and early autumn. Older nymphs are found late in the season and, in unusually mild years, they may emerge from "hibernation" before all the snow has melted. Indeed, deceived by extraordinarily mild winter conditions, large numbers of late-instar nymphs of this species (judging by a published photograph) were found hopping about on the scanty snow during the 1st week of January 1980, near Barrys Bay, Ont. (Stamplecoski 1980). Throughout the region here treated, the life cycle is univoltine, but, in Oklahoma and Texas, it is bivoltine or even trivoltine.

Economic importance. Recorded as a minor pest of tobacco in Québec (C. J. S. Fox and Stirrett 1953).

Genus *Encoptolophus* Scudder

Description. Body rather short, robust. Head slightly swollen; vertex broadly triangular, its disc with deep concavity and having median carina on posterior half; foveolae distinct, elongate-triangular; frontal costa narrow, strongly sulcate in male, sulcate only above the median ocellus in female. Pronotum with anterior margin truncate and with posterior margin right-angled; median pronotal carina prominent, slightly higher on prozona, cut at center by principal sulcus; lateral carinae confined to metazona. Tegmina broad, short, with tips broadly rounded, extending beyond abdomen in both sexes.

Key to species of *Encoptolophus*

1. Head swollen, distinctly broader than anterior margin of pronotum. Pronotal disc tectate, usually rugose or tuberculate, with lateral carinae strongly interrupted and dislocated ahead of principal sulcus. Found in New England and Québec, west to western Ontario and Minnesota .... ***sordidus* (Burmeister)** (p. 478)

   Head little broader than anterior margin of pronotum. Pronotal disc nearly flat, smooth, with lateral carinae only slightly dislocated ahead of principal sulcus. Found in Manitoba, Saskatchewan, Alberta, North Dakota, and Montana .... ***costalis* (Scudder)** (p. 479)

Clé des espèces d'*Encoptolophus*

1. Tête renflée, clairement plus large que le bord antérieur du pronotum. Disque du pronotum tectiforme, habituellement rugueux ou tuberculé, à crêtes latérales clairement coupées et morcelées en avant du sulcus principal. Espèce vivant en Nouvelle-Angleterre et au Québec, et jusque dans l'ouest de l'Ontario et du Minnesota .... ***sordidus* (Burmeister)** (p. 478)

   Tête légèrement plus large que le bord antérieur du pronotum. Disque du pronotum presque plat, lisse, à crêtes latérales seulement légèrement morcelées en avant du sulcus principal. Insecte vivant au Manitoba, en Saskatchewan, en Alberta, au Dakota-nord et au Montana .... ***costalis* (Scudder)** (p. 479)

Encoptolophus sordidus (Burmeister)

Fig. 657; Map 187

O[edipoda] sordida Burmeister, 1838:643.

Locusta nebulosa T. W. Harris, 1841:146.

Oedipoda sordida; Provancher 1876:114.

Encoptolophus sordidus; Caulfield 1888:66, 70; Helfer 1963:108; R. D. Alexander et al. 1972:37.

Encoptolophus sordidus sordidus; Hebard 1937*c*:274; Cantrall 1943:99; Vickery et al. 1974:109.

Diagnosis. Body dull brown. Pronotum with X-shaped pale mark on disc. Late summer species. Eastern.

Description. Body size medium-large (length, male 21.0–28.5, female 28.0–32.5 mm), robust (Fig. 657). Head large. Pronotum with disc flat, with median carina sharp but low, cut by principal sulcus slightly ahead of middle; lateral carinae continuous; posterior margin obtuse-angulate. Tegmina exceeding apices of hind femora, broad, obliquely truncate at apices. Color dull brown, mottled with lighter and darker markings; pronotum with X-shaped pale mark on disc; tegmina with 2 transverse pale bars alternating with dark patches; wings transparent, yellowish at base, with outer third dark; hind femur indistinctly banded with dull yellow and dark brown on outer face; hind tibia fuscous brown with pale subbasal pale ring.

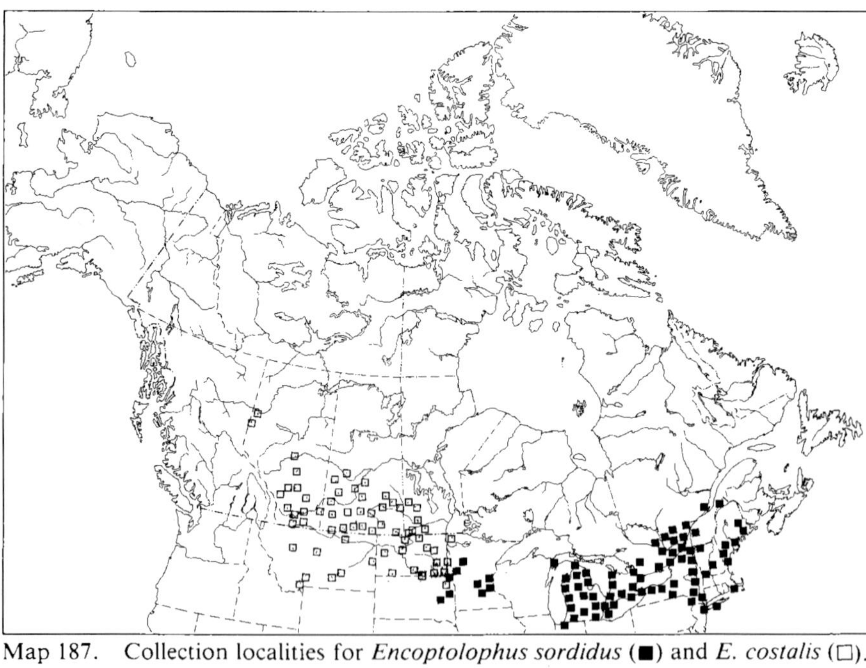

Map 187. Collection localities for *Encoptolophus sordidus* (■) and *E. costalis* (□).

Range. South Dakota to Maine, south to Tennessee and North Carolina.

Behavior and habitats. *E. sordidus* is common in dry grassy areas in Québec during the fall (Vickery et al. 1974). Blatchley (1920) reported it from dry upland meadows and pastures, roadsides, and sunny spots in woodland pastures. Hebard (1925*a*) stated that it, "unlike *E. costalis*, prefers waste weedy areas rather than prairie or plains conditions" in South Dakota.

Otte (1970) observed two types of flight by males, crepitating flights and short low soundless flights. Crepitations attract females to the vicinity of the male. These occur at the rate of 48 per second. During courtship, males produce three kinds of acoustical signals. The first, called by Otte (1970) a "stuttering" signal, was the only courtship sound produced in some cases, immediately followed by mounting of the female. Other males, or the same males at other times, produced all three types of sounds, stuttering, ticks (produced by alternating action of the hind femora striking substrate), and trills produced by stationary males by holding and vibrating one or both hind femora against the tegmina. In courtship, all three signals may be used; in such cases they are always in the sequence: "tick," "trill," "stutter." A female may respond in any of four different ways. She may remain motionless and be mounted by the male; she may exhibit femur-tipping at a slow rate until the male approaches, then remain motionless; she may expose her abdomen by lowering the hind femur closer to the male; or she may tip up the near femur, in which case she is not receptive and the male turns away.

Judd (1955*b*) found specimens of *E. sordidus* impaled on thorns during the fall in western Ontario. This was probably done by shrikes (*Lanius* spp.). It is not known whether this is a significant factor in population regulation, but it probably is not.

Life history. Univoltine in the present region, with overwintering in the egg stage. Eggs hatch in late spring and early summer. Some nymphs molt to adults about mid-July, but the latter are much more common during late August and September. In Michigan adults have been recorded from 8 July to 10 November (Cantrall 1968). The life cycle in southern areas has not been studied, but it may be bivoltine or even trivoltine.

Economic importance. C. J. S. Fox and Stirrett (1953) indicate that this species, in late summer, is occasionally injurious to tobacco in southern Ontario.

Encoptolophus costalis (Scudder)

Map 187

Oe[dipoda] costalis Scudder, 1863*a*:473.
Encoptolophus parvus Scudder, 1875*e*:479.
Encoptolophus costalis; Scudder 1901*a*:105.
Encoptolophus coloradensis Bruner, 1904*a*:58.
Encoptolophus sordidus costalis; Hebard 1936*a*:35; Brooks 1958:40; Helfer 1963:109.

Diagnosis. Body grayish to greenish white. Late summer species. Western. Concealed genitalia of male similar or identical to those of *E. sordidus* (see Barnum 1959).

Description. Color varying from greenish white to gray. Body length 20–27 mm.

Range. Alberta to Manitoba, south to New Mexico and Texas.

Behavior and habitats. N. Criddle (1935) reported *E. costalis* as a grass feeder in Manitoba. This was also stated to be the case by Mulkern et al. (1969), but these authors found that it would also feed upon forbs, its feeding being greatly influenced by the food available in the habitat. Alfalfa was readily accepted and heavy feeding upon this crop was reported. The species is found in open grasslands, dry hillsides, and rangelands, and in forb crops seeded in such areas.

Otte (1970) reported behavioral characteristics of *E. costalis*. Solitary males crepitate in flight with a crepitation rate of 62.5 pulses per second (*E. sordidus* rate is 48 pulses per second). *E. costalis* also produces single pulses of "ordinary" stridulation (*E. sordidus* seldom produces such pulses). In other respects, courtship and mating behavior were similar in the two species.

Life history. Not reported but undoubtedly similar to that of *E. sordidus*.

Economic importance. C. G. Bailey and Riegert (1971) found that feeding by *E. costalis* removed 10% of the green vegetation in a grassland ecosystem. Nevertheless, only 2% of this was ingested by the grasshoppers; the remainder was dropped to the ground (C. G. Bailey and Riegert 1973). This species can cause injury at high population levels but seldom causes more than minor damage to grasses and alfalfa.

Genus *Camnula* Stål

Description. Size small. Head compressed, with apex rounded; foveolae shallow; frontal costa flat, not prominent, narrow near vertex. Pronotum with disc flat; median carina low, of equal height throughout, faintly incised by principal sulcus. Tegmina narrow, extending beyond tip of abdomen.

Camnula pellucida (Scudder)

Fig. 658; Map 188

Oedipoda pellucida Scudder, 1863*a*:472.

Camnula pellucida: Caulfield 1886:212; Cantrall 1943:100; Brooks 1958:36; Helfer 1963:108; R. D. Alexander et al. 1972:37; Vickery et al. 1974:110.

Common name. Clearwinged grasshopper.

Diagnosis. Pronotal disc flat. Tegmina brownish, with pale stripe laterally in resting specimen.

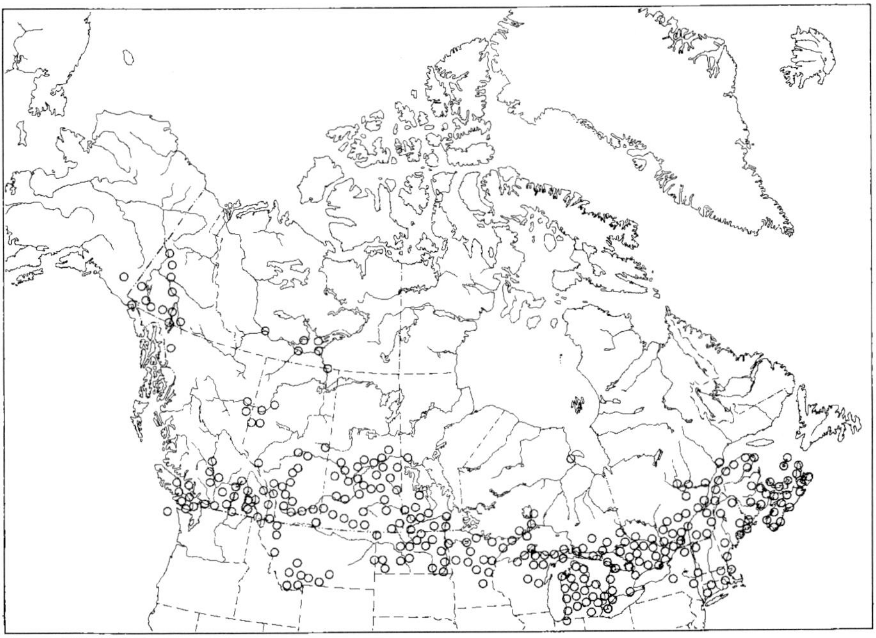

Map 188. Collection localities for *Camnula pellucida*.

Description. Body small (length, male 17–21, female 19.5–28.5 mm) (Fig. 658). Head compressed. Pronotum with disc flat, smooth, wider behind, truncate anteriorly; median carina low, of equal height throughout, faintly cut by one sulcus ahead of middle; lateral lobes deeper than long. Tegmina narrow, reaching beyond apex of abdomen. Color generally light brown; head with dark triangular spot behind eye; lateral lobe of pronotum with dark vertical spot on front half; tegmina smoky brown, with light and dark patches on sides, darker on dorsal surface with yellowish stripe along humeral angle; hind wings transparent with dark veins; hind femur yellowish brown, darker at apex, faintly marked with dark bars; tibia yellowish brown; abdomen yellowish beneath, with sides darker. A gynandromorphic individual was recorded by Paul (1941) in Saskatchewan.

Range. Alaska to Nova Scotia, south to Mexico.

Behavior and habitats. The habitats are somewhat variable, but the species is common on undisturbed soils such as roadsides and grassy areas (one of its unofficial common names is "roadside grasshopper"). When it occurs in crop fields, it has usually migrated from the undisturbed locations and usually returns to such places and oviposits. Putnam (1963*a*) estimated that nymphs from an egg site of about 800 m^2 dispersed over an area of about 75 ha before reaching maturity.

Neither sex crepitates in flight, but resting males produce "ordinary" stridulation using one or both femora. Courting males attract nearby females by femur-tipping. Femur-tipping and shaking are also used as aggressive signals between males.

Life history. Univoltine, with overwintering as eggs in the soil. When *C. pellucida* reaches maturity in crop fields, the adults eventually fly to a nearby breeding site, where they mate and oviposit. Riegert (1967*a*, 1967*b*) and Pickford (1974) described the reproductive biology of this species. Eggs are laid from midsummer to early fall. The date of hatching the following year is strongly influenced by the date of oviposition. Eggs laid in the fall undergo considerably less embryonic development before winter than those laid earlier. If they are deposited so late (late September or October) that practically no embryonic development occurs before winter, they do not usually survive until spring. Eggs laid in midsummer hatch first, those laid later hatch later so that hatching in spring occurs over a period of more than a month. Gage et al. (1976) produced mathematical models to predict the seasonal occurrence of *C. pellucida* (as well as of two other grasshopper pest species) in Saskatchewan. The mathematical model for *C. pellucida* is based upon fall embryonic development and rate of egg and nymphal development during the following spring. Heat units accumulated above 10°C in fall and spring play a significant role in development and survival.

Economic importance. In Canada, *C. pellucida* is one of the three most important species causing damage to crops. It is mainly a grass feeder and consequently causes severe damage to cereal crops, particularly wheat (Pickford 1963; Riegert et al. 1965), and to barley and forage crops (Misra and Putnam 1966); it also attacks alfalfa, turnip, and tobacco. As well as being of major economic significance in the West, *C. pellucida* ranks as the third most important pest grasshopper in Ontario (Vickery and Kevan 1967) and in Québec (Vickery et al. 1974).

Genus *Aerochoreutes* Rehn

Description. Body large, robust. Metazona of pronotum twice as long as prozona. Hind wings transparent except for darkened basal area, with hind margins lobed and with all radial veins enlarged and thickened (Fig. 713). Color gray brown to reddish, occasionally purplish.

Key to subspecies of *Aerochoreutes carlinianus*

1. Tegmina long (at least twice as long as hind femora), slender, rounded at apices. Eyes large, prominent in dorsal aspect (Fig. 715). Found in British Columbia, Washington, and Idaho .... ***carlinianus strepitus*** **Rehn** (p. 483)

   Tegmina shorter (barely twice as long as hind femora), broader, with rather obliquely truncate apices. Eyes smaller, not prominent in dorsal aspect (Fig. 714). Found from Manitoba to Alberta and Montana .... ***carlinianus carlinianus*** **(Thomas)** (p. 484)

Clé des sous-espèces d'*Aerochoreutes carlinianus*

1. Tegmina longs (au moins deux fois la longueur des fémurs postérieurs), étroits, arrondis à l'apex. Yeux grands, proéminents vus du dessus (fig. 715). Insecte vivant en Colombie-Britannique, dans l'État de Washington et en Idaho ...... ***carlinianus strepitus*** **Rehn** (p. 483)

Tegmina plus courts (à peine deux fois plus longs que les fémurs postérieurs), plus larges, à apex tronqué obliquement. Yeux plus petits, non proéminents vus du dessus (fig. 714). Insecte vivant du Manitoba à l'Alberta ainsi qu'au Montana ... ***carlinianus carlinianus*** **(Thomas)** (p. 484)

Aerochoreutes carlinianus strepitus Rehn

Fig. 715; Map 189

Aerochoreutes carlinianus strepitus J. A. G. Rehn, 1921:174.

Aerochoreutes carlinianus strepitus; Buckell 1922:25; Brooks 1958:35; Helfer 1963:145.

Diagnosis. Prozona short. Hind wing transparent, with radial veins thickened and darkened. Eyes large. Tegmina with slender apices. Found in British Columbia southward.

Description. Body robust (length 35–43 mm). Hind wings extending beyond apex of abdomen by an amount equal to length of head and pronotum combined; tegmina at least twice as long as hind femora and rather narrowed and rounded at apices. Eyes prominent. Fastigium narrow. Color brownish to purplish, mottled; tegmina banded, often indistinct; hind wing disc variable, clear to entirely darkened; hind femur on outer face with darker ventral markings, on inner face with basal half and preapical band dark; hind tibia yellow, paler at base.

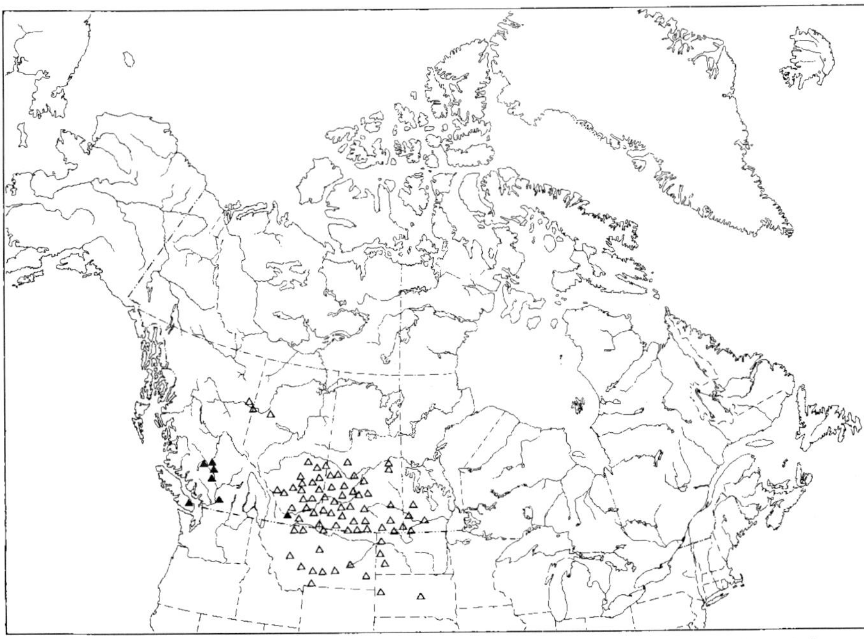

Map 189. Collection localities for *Aerochoreutes carlinianus strepitus* (▲) and *A. carlinianus carlinianus* (△).

Range. British Columbia and Alberta, south to Colorado.

Behavior and habitat. This subspecies is similar to *A. carlinianus carlinianus.* Hewitt and Barr (1967) described similar hovering flight and crepitation for the present subspecies. Buckell (1922) described the crepitation at take-off from the ground as "extremely similar to the buzzing produced by the rattle-snakes which frequent the warmer parts of southern British Columbia."

Life history. Presumably the same as for *A. carlinianus carlinianus.* Adults are known to be active during August and September.

Economic importance. Beirne (1972) notes *A. carlinianus* [sensu lato] among grasshoppers injurious in Canada (citing references), but says that "it is normally found in numbers only in the interior of British Columbia. . . in severely overgrazed cattle ranges. . . ."

Aerochoreutes carlinianus carlinianus (Thomas)

Figs. 712–714; Map 189

Oedipoda carliniana C. Thomas, 1870:81.

Aerochoreutes carlinianus carlinianus; J. A. G. Rehn 1921:173; Brooks 1958:34.

Diagnosis. Prozona half length of metazona. Hind wing largely transparent, with radial veins thickened and darkened. Eyes small. Tegmina with obliquely truncate apices.

Description. Body robust (length 32–44 mm). Tegmina and hind wings surpassing apex of abdomen by an amount less than length of head and pronotum together, barely twice as long as hind femora; tegmina rather broad, somewhat obliquely truncate at apices. Color light grayish brown through pink or reddish to dark purple; tegmina with or without evident bands; hind wing transparent, darkened at base and along main veins, often with entire basal half black.

Brooks (1958) distinguished between two forms, *major* and *minor*, occurring from Alberta to Manitoba. These appear to be fairly stable ecologically induced phenotypes ("ecophenes") rather than subspecies (Vickery 1979). They are distinguished as follows: *major form* in sandy alkaline-eroded areas near the United States border, with tegmen evidently banded, hind wing with apex produced, and with inner half usually black, contrasting with apical half (size larger, 38–45 mm); *minor form* on dark mud outcrops and in summerfallow fields west of Assiniboine Valley, with tegmen more speckled than banded, with hind wing not produced, and with inner half dark but dark color also extending along main veins (size somewhat smaller, 32–40 mm). Chromosome number, $2n$ male = 21.

Range. Alberta to Manitoba, south to Colorado.

Behavior and habitat. *A. carlinianus carlinianus* is a grass feeder, preferring *Agropyron smithii* Rydh. (N. Criddle 1935). Habitat types are as aforementioned, differing for *major* and *minor* forms (Brooks 1958).

Brooks (1958) remarked that *A. carlinianus carlinianus*, in flight, remains hovering for many minutes poised 7.5–15 m above the earth. Crepitation was said to be continuous in hot dry weather. Further, he distinguished between the type of crepitation produced by the two forms, that of the *minor* form being decidedly more even than that of the *major* form, so that the two may be distinguished by their crepitation alone. Helfer (1963) also notes that the form on dark mud outcrops and in summerfallow fields (Brooks's *minor* form) is "noisier in flight." Otte (1970) reported crepitation as being composed of long bursts (0.75–1.25 seconds) interspersed with short bursts (0.07 second). Males hover over one spot during crepitation, moving upward and downward, but not forward.

Life history. Univoltine, with overwintering as eggs in soil. Adults are active during the latter half of July, August, and most of September. Tinkham (1939) recorded *A. carlinianus carlinianus* in Alberta on 17 July and on 15 September.

Economic importance. The subspecies is a potential threat to rangeland (Hardman and Smoliak 1980).

Genus *Trachyrhachys* Scudder

Description. Similar to *Metator*. Size medium. Foveolae large. Pronotum distinctly constricted, with median carina strongly elevated; prozona and metazona about equal in length; metazona tuberculate (Fig. 684); lateral lobes produced posteriorly, angulate. Tegmina definitely banded. Hind wing disc clear to slightly yellowish, with or without crossband.

Key to subspecies of *Trachyrhachys kiowa*

1. Hind wing hyaline or slightly yellowish on basal third, with crossband completely absent. Found in Manitoba and Minnesota, west to Montana and British Columbia .... ***kiowa kiowa*** **(Thomas)** (p. 486)
   Hind wing yellow on basal third, with crossband present. Found in eastern states and provinces 2

2(1). Crossband of hind wing dark, broad, complete. Found in Pennsylvania, Ohio, Michigan, and Illinois ..... ***kiowa fuscifrons*** **(Stål)** (p. 487)
   Crossband of hind wing not fully developed, represented by faint central clouding or by narrow broken band. Found in Minnesota and Wisconsin .... ***kiowa thomasi*** **(Caudell)** (p. 487)

Clé des sous-espèces de *Trachyrhachys kiowa*

1. Ailes postérieures hyalines ou légèrement jaunies sur le tiers basal, sans trace de bande transversale. Insecte vivant au Manitoba et au Minnesota, vers l'ouest jusqu'au Montana et à la Colombie-Britannique .... ***kiowa kiowa*** **(Thomas)** (p. 486)
   Ailes postérieures jaunes sur le tiers basal et marquées d'une bande transversale. Insectes vivant dans les États et les provinces de l'Est 2

2(1). Bande transversale de l'aile postérieure noire, large et complète. Insecte vivant en Pennsylvanie, en Ohio, au Michigan et en Illinois .... ***kiowa fuscifrons*** **(Stål)** (p. 487)

Bande transversale de l'aile postérieure non entièrement développée, prenant la forme d'une tache centrale diffuse ou d'une bande étroite et morcelée. Insecte vivant au Minnesota et au Wisconsin .... *kiowa thomasi* **(Caudell)** (p. 487)

Trachyrhachys kiowa kiowa (Thomas)

Figs. 684, 687; Map 190

Oedipoda kiowa C. Thomas, 1872:461.
Trachyrhachys kiowa; Caudell 1908:75; Helfer 1963:129 (partim).
Trachyrhachys kiowa kiowa; Hebard 1931*a*:388; Brooks 1958:44.

Diagnosis. Hind wing transparent to slightly yellowish at base, without dark markings.

Description. As for genus. Body small (length 20–30 mm). Color gray to brightly contrasting green; head and pronotum dark gray to brown without contrasting pattern, or greenish yellow with broad black bar on side of head, with narrow transverse black bar on vertex, and with black central line on anterior part of pronotum; tegmina dark gray to nearly black with 2 incomplete white bands, or light grayish green with 3 darker bands (at rest tegmina often show 2 pale dorsal stripes); hind wing transparent, without markings, occasionally slightly yellow at base; hind femur brownish on outer face with darker or white markings; inner face of femur black on basal half,

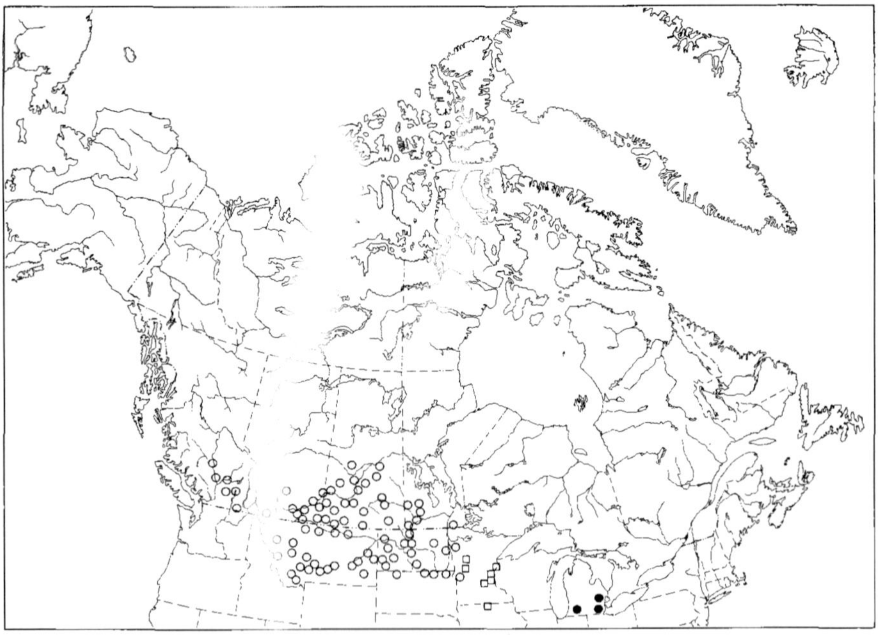

Map 190. Collection localities for *Trachyrachys kiowa kiowa* (○), *T. kiowa fuscifrons* (●), and *T. kiowa thomasi* (■).

usually yellow with black band on apical half; hind tibia blue to black, occasionally buffy, paler at base.

Range. British Columbia to Manitoba, south to Arizona.

Behavior and habitats. *T. kiowa kiowa* is primarily a grassland grasshopper, although it is often found in areas where vegetation is scarce. N. Criddle (1935) listed it as a grass feeder in Manitoba. Mulkern et al. (1969) recorded it as preferring *Bouteloua gracilis* but also feeding on other grasses such as *Agropyron smithii* and *Poa pratensis*.

Otte (1970) obtained only inconclusive results in observations of courtship, but he did detect femur-shaking and bursts of vibratory stridulation. He observed femur-tipping and striking of the substrate with the hind tibia employed as aggressive signals.

Life history. Adults appear during the latter half of July and persist well into September. *T. kiowa kiowa* has a univoltine life cycle, with overwintering in the egg stage.

Trachyrhachys kiowa fuscifrons (Stål)

Map 190

Psinidia fuscifrons Stål, 1873*b*:134.
Trachyrhachys kiowa fuscifrons; Froeschner 1954:241; Brooks 1958:44.
Trachyrhachys kiowa; Newton and Gurney 1957:316 (partim); R. D. Alexander et al. 1972:54.

Diagnosis. Hind wing disc yellow, with complete broad dark crossband.

Description. As for *T. kiowa kiowa*. Hind wing disc yellow and crossband dark, broad, complete.

Range. Michigan to Pennsylvania, south to Texas and Georgia.

Behavior and habitats. Cantrall (1968) reported *T. kiowa fuscifrons* from open sandy areas with sparse vegetation in Michigan. Its behavior is probably similar to that of *T. kiowa kiowa*.

Life history. Univoltine, with overwintering as eggs in soil. Adults were recorded in Michigan from 21 July to 26 September (Cantrall 1968).

Trachyrhachys kiowa thomasi (Caudell)

Map 190

Mestobregma thomasi Caudell, 1904*g*:125.
Trachyrhachys kiowa thomasi; Hebard 1931*b*:161; Brooks 1958:44.

Diagnosis. Hind wing disc yellow, with fuscous median crossband.

Description. As for *T. kiowa kiowa*. Hind wing with yellow disc, and with fuscous median crossband, this not fully developed but indicated by faint central clouding or narrow broken band extending across wing.

Range. Minnesota to Ohio, south to Texas and Georgia.

Behavior and habitat. Presumably similar to those of *T. kiowa kiowa*.

Life history. As for *T. kiowa kiowa*.

Genus *Metator* McNeill

Description. Head large, smooth, with disc of vertex quadrate, having high lateral and frontal carinae. Antenna longer than hind femora. Pronotum slightly constricted at middle, with disc relatively smooth; median carina moderately high, cut by two sulci; anterior lobe slightly higher than posterior lobe (Figs. 681, 682); lateral lobes produced, acutely angulate postero-ventrally.

Key to species of *Metator*

(Adapted from Hewitt and Barr 1967)

1. Tegmina with dark maculations over entire surface, and with longitudinal converging yellow stripes; inner face of hind femur blue; hind tibia bluish. Found in Manitoba to Alberta and Montana .... ***pardalinus* (Saussure)** (p. 488)

   Tegmina with dark maculations confined to costal area, and without yellow stripes; inner face of hind femur brown and yellow; hind tibia brownish. Found in British Columbia, Idaho, Washington, and southward .. .. ***nevadensis* (Bruner)** (p. 490)

Clé des espèces de *Metator*

(adaptée de Hewitt et Barr, 1967)

1. Tegmina entièrement maculés de noir et à bandes jaunes longitudinales convergentes; face interne des fémurs postérieurs bleue; tibias postérieurs bleuâtres. Insecte vivant du Manitoba à l'Alberta ainsi qu'au Montana .... ***pardalinus* (Saussure)** (p. 488)

   Tegmina maculés seulement dans la région costale et sans bandes jaunes; face interne des fémurs postérieurs brune et jaune; tibias postérieurs brunâtres. Insecte vivant en Colombie-Britannique, en Idaho, dans l'État de Washington et plus au sud .... ***nevadensis* (Bruner)** (p. 490)

Metator pardalinus (Saussure)

Fig. 682; Map 191

Ps[*inidia*] *pardalina* Saussure, 1884:162.
Ps[*inidia*] *maculosa* Saussure, 1884:162.
Metator pardalinus; J. A. G. Rehn and Hebard 1906:381; Brooks 1958:41; Helfer 1963:127.

Diagnosis. Lateral lobes of pronotum acute-angulate posteroventrally; hind femur gray on outer face with oblique fascia dorsally, and with inner face blue.

Description. As for genus. Body large (length 27–42 mm). Color yellowish gray, mottled; pronotum dark with diamond pattern on metazona; abdomen brownish, often bluish laterally; tegmina grayish with dark blotches

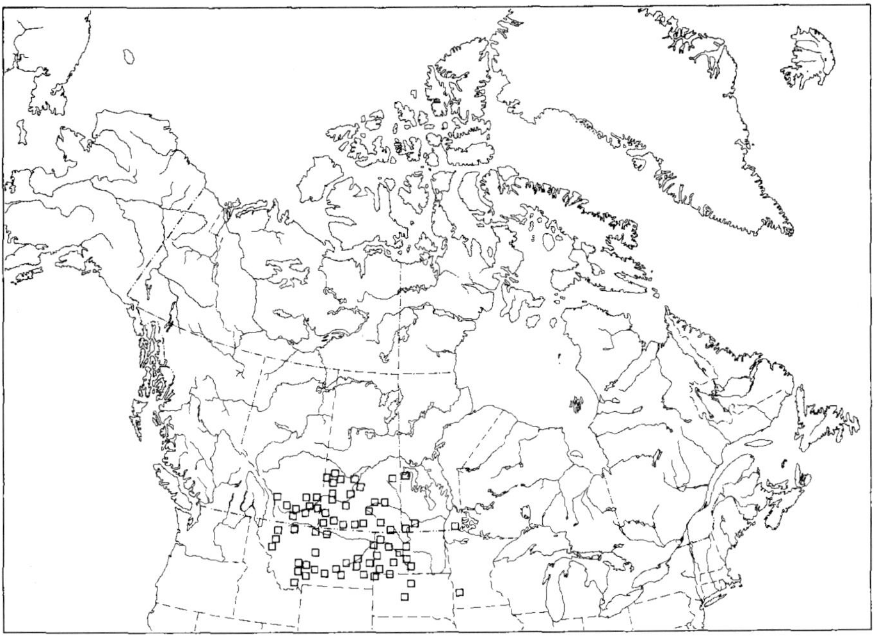

Map 191. Collection localities for *Metator pardalinus*.

over entire surface, not banded, with yellowish line dividing dorsal and lateral area; hind wing with disc yellow or red, with crossband dark, broad, and with spur extending more than halfway to base; apex of wing clear or with few opaque cells; hind femur gray to yellowish with dark oblique lines dorsally on outer face, with inner face blue; hind tibia dark blue to pale blue (paler on large females), with pale basal ring.

Range. Alberta to Manitoba, south to Arizona and Texas.

Behavior and habitats. *M. pardalinus* occurs throughout the short-grass prairie and in Saskatchewan north into parklands. It is a grass and sedge feeder. It shows a preference for *Agropyron smithii* Ryd., but it also eats *Andropogon, Stipa*, and *Carex* species (N. Criddle 1935; Brooks 1958). Anderson and Wright (1952) reported that older nymphs and adults are gregarious. Adults appear to fly only infrequently, but they are in nearly constant motion, crawling on the ground. They are easily disturbed. Feeding behavior was noted to be somewhat different from that of other range species. An individual crawls up grasses, then cuts off a green leaf and returns to the substrate to eat it, feeding on an end of the leaf rather than at the side as most species do. Adults as well as nymphs also ingest dry materials found on the ground, including large amounts of fine sand and clay.

No report of courtship or of aggressive behavior is available.

Life history. Univoltine, with overwintering as eggs in the soil. Hewitt and Barr (1967) reported adults from July to September in Idaho. Anderson and Wright (1952) recorded the first adult in Montana in 1950 on 6 July.

Metator nevadensis (Bruner)

Figs. 681, 683; Map 192

[*Trachyrhachis*] *nevadensis* Bruner, 1905:175.
Metator nevadensis; Buckell 1922:24; Helfer 1963:127.

Diagnosis. Lateral pronotal lobes acute-angulate posteroventrally. Inner face of hind femur yellow, with two brownish bands.

Description. As for genus. Body size medium (length 24–36 mm). Color grayish brown, appearing streaked dorsally; head with oblique dark stripes through eyes; pronotum with dorsal longitudinal dark band; tegmina with dark maculations on costal area, these becoming smaller and fainter toward apex; hind wing with disc either red or yellow, with crossband dark, broad, and with spur broad and long, reaching nearly to base; apex of wing smoky, with darkened veins; hind femur gray on outer face with 2 oblique dark lines dorsally; inner face of femur yellow, with 2 brownish bands; hind tibia buff to blue, usually darker at base and apex.

Range. British Columbia, south to California and Arizona.

Behavior and habitats. *M. nevadensis* usually occurs in dry areas where sagebrush and grasses occur. It is a grass feeder, preferring species of *Agropyron*. Flight is silent and sluggish, extending only about 3–6 m (Hewitt and Barr 1967).

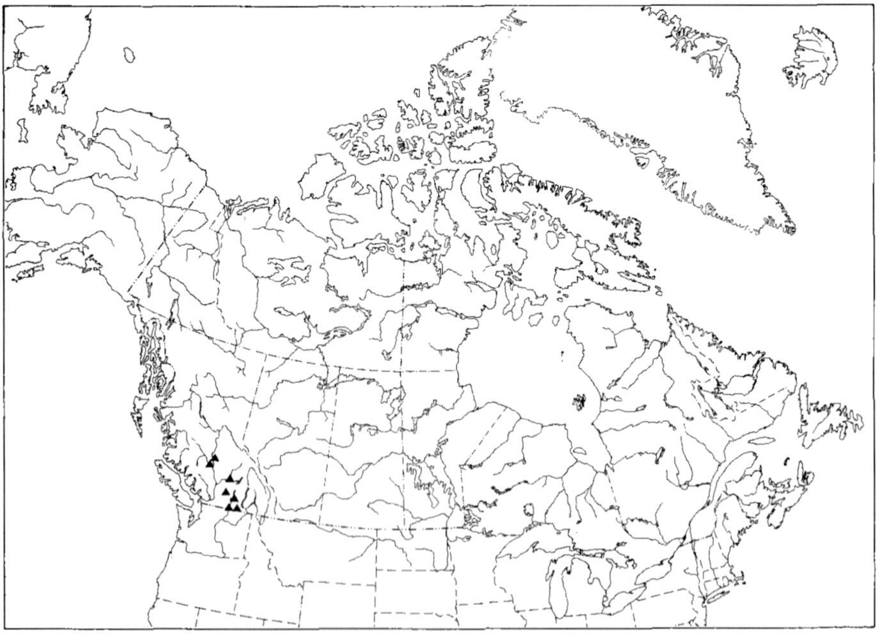

Map 192. Collection localities for *Metator nevadensis*.

Life history. Hewitt and Barr (1967) reported adults in Idaho during July and August. The life cycle is probably univoltine, with overwintering in the egg stage.

Economic importance. In British Columbia, Buckell (1922) reported damage to grass on winter cattle ranges by this species together with *Spharagemon equale* (Say) (see also Beirne 1972).

Genus *Conozoa* Saussure

Description. Body slender. Pronotum with median carina feebly elevated, notched by 2 sulci, with dorsal surface irregularly roughened, not distinctly tuberculate; metazona 1–1.75 times as long as prozona; posterior angles of lateral lobes produced in form of broad tooth (Fig. 686). Color gray or brown; tegmina with 2 prominent dark areas on costal margin. The genus needs revising.

Conozoa wallula (Scudder)

Figs. 686, 689; Map 193

Psinidia wallula Scudder, 1881:27.
Conozoa behrensi Saussure, 1884:165.
Conozoa wallula; Saussure 1884:166; Helfer 1963:130.

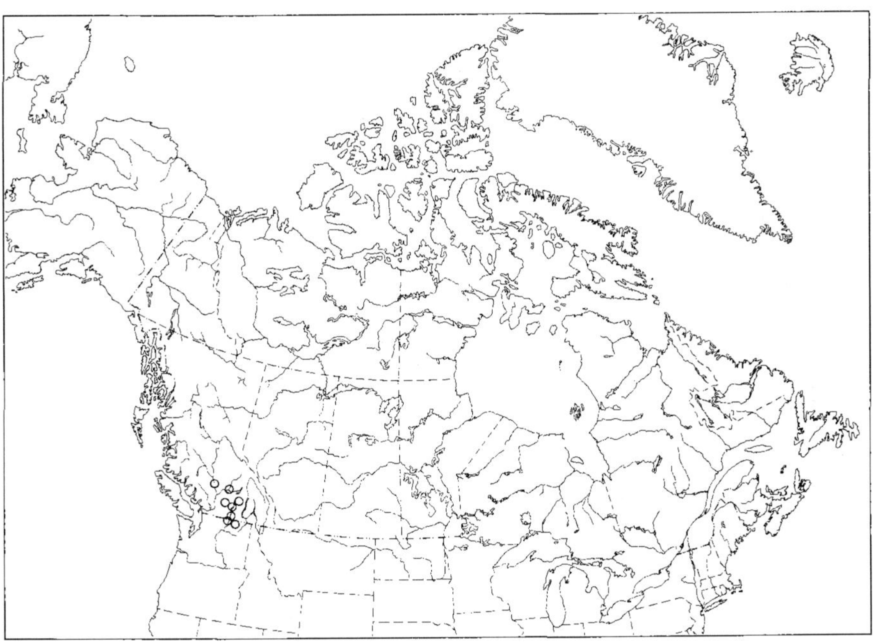

Map 193. Collection localities for *Conozoa wallula*.

Diagnosis. Posterolateral angles of lateral pronotal lobes acute-angulate. Hind tibia orange to yellow, with pale basal band.

Description. Body slender, medium-sized (length 19–35 mm). Head with lateral carinae well-developed. Posterior angles of lateral pronotal lobes produced, acute-angulate to somewhat rounded (Fig. 686). Color variable; yellowish brown or gray, unicolorous dorsally, or with metazona much paler than prozona; tegmina with 2 dark bands on costal margin; hind wing disc yellow, with crossband distinct, dark, reaching costal margin, with spur short, reaching only halfway to wing base, and with apex clear; hind femur usually gray on outer face, occasionally with faint bands, these always present dorsally, and with inner face having 2 contrasting dark bands; hind tibia orange to yellow with pale band near base.

Range. Interior British Columbia to Wyoming, south to California.

Behavior and habitats. Buckell (1920) found *C. wallula* to be common on dry, mostly flat, sandy spots near Okanagan Lake and other parts of the Okanagan Valley in British Columbia, in fairly large aggregations. They were inconspicuous on the substrate, but were easily disturbed. Flights were short. Hewitt and Barr (1967) recorded *C. wallula* as an inhabitant of the sage grass zone and of dry grassy areas, as well as along the edges of rivers and lakes in Idaho. They stated that it is a mixed feeder on grasses and forbs, but that it appears to prefer the former.

Both sexes crepitate in flight, producing a "moderately loud crackling sound" (Hewitt and Barr 1967). Otte (1970) observed femur-tipping and femur-shaking (with striking of the substrate with the hind tibiae) by males during courtship, although some males exhibited only femur-tipping prior to mounting females. The only aggressive signal seen was femur-tipping.

Buckell (1920) reported that in the Okanagan Valley of British Columbia, many individuals of *C. wallula* were heavily infested with red trombidiid mites—named as *Trombidium locustarum* Walsh [= *Eutrombidium trigonum* (Hermann) = *E. rostratum* (Scopoli)]—some carrying so many of the parasites that they were unable to fly or even to close their tegmina.[1] Larvae of tachinid and sarcophagid flies were also observed to be parasitic upon this species.

Life history. Hewitt and Barr (1967) reported adults from July until the end of September in Idaho. In British Columbia, Buckell (1920) observed adults in July and August and thought that they persisted until killed by frost. The life cycle is probably univoltine, with overwintering as eggs in the soil.

Genus *Trimerotropis* Stål

Description. Body elongate, rather slender. Disc of vertex distinctly impressed, usually longer than broad. Pronotum with median carina prominent, cut by two sulci before middle; metazona nearly twice as long as

[1] The same phenomenon was observed by us in late July 1979, on White Sulphur Mountain, central Montana. A detailed account of the biology of this mite is given by H. C. Severin (1944).

prozona, not rugose or tuberculate (or only faintly so on metazona) (Figs. 691, 692). Tegmina and hind wings surpassing abdomen. Cytology: see M. J. D. White (1951*a*, 1951*b*, 1951*c*, 1954, 1973) and Weissman and Rentz (1980).

Key to species and subspecies of *Trimerotropis*

1. Hind tibia red or orange (except specimens from Great Lakes region) 2
   Hind tibia yellow or bluish (orange in specimens from Great Lakes region) 5

2(1). Posteroventral angle of lateral pronotal lobe produced as rounded point (Fig. 691) .... ***agrestis* McNeill** (p. 496)
   Posteroventral angle of lateral pronotal lobe broadly rounded, not produced or pointed 3

3(2). Bands of tegmina definite, conspicuous; spur of hind wing band extending less than halfway to wing base (Figs. 705, 706) 4
   Bands of tegmina usually indefinite; spur of hind wing band extending more than halfway to wing base (Fig. 704) .... ***campestris* McNeill** (p. 497)

4(3). Inner face of hind femur reddish yellow with distinct black band on apical half continuous with band on outer face .... ***pistrinaria* Saussure** (p. 498)
   Inner face of hind femur black or with preapical pale band, or black area broken in center .... ***latifasciata* Scudder** (p. 500)

5(1). Hind tibia brown to bluish brown with pale basal band 6
   Hind tibia not mainly brown 7

6(5). Hind wing with black crossband extending to apex (Fig. 709). Found in western mountains and west coast .... ***suffusa* Scudder** (p. 501)
   Hind wing with clear area outside black crossband, with tip darkened (Fig. 708). Found from east coast to Rocky Mountains .... ***verruculata* (Kirby)** (p. 502)

7(5). Hind tibia yellow, buff, or reddish 9
   Hind tibia blue 8

8(7). Hind tibia blue, with pale basal band; head without dark transverse bands. Found in British Columbia, east to Wyoming, Colorado, and southward .... ***fontana* Thomas** (p. 503)
   Hind tibia entirely pale blue; head with transverse dark bands above and beneath antennae. Found in Montana, western Dakotas, northwestern Nebraska, and southward .... ***cincta* (Thomas)** (p. 505)

9(7). Hind wing clear with faint bluish tinge; crossband faint or absent (Fig. 698) .... ***sparsa* (Thomas)** (p. 505)
   Hind wing disc yellow with black crossband, with latter usually distinct ... 10

10(9). Black crossband of hind wing indistinct, represented by infuscated veins (Fig. 695) .... ***gracilis* (Thomas)** (p. 507)
   Black crossband of hind wing distinct 11

11(10). Area between ventral carinae of hind femur black as far as apical pale band 12
   Area between ventral carinae of hind femur pale with 2 dark spots or with dark bar at apical third 14

12(11). Tegmina variable, not strongly banded, usually without bands (but not reddish) (Fig. 697) .... ***huroniana* E. M. Walker** (p. 507)
   Tegmina evidently banded (sometimes entirely reddish) 13

13(12). Body robust. Crossband of hind wing broad, with disc pale yellow; tegminal banding broken, not conspicuous (Fig. 701). Found in Alberta and eastward .... ***pallidipennis salina*** **McNeill** (p. 509)

Body less robust. Crossband of hind wing not excessively broad, with disc yellow; tegminal banding complete, conspicuous (Fig. 700). Found in British Columbia and southward .... ***pallidipennis pallidipennis*** **(Burmeister)** (p. 510)

14(11). Tegmina not conspicuously banded 15

Tegmina conspicuously banded 17

15(14). Tegmina not banded, with dark flecks along veins (Fig. 696). Western species .... ***sordida*** **E. M. Walker** (p. 511)

Tegmina plain to faintly barred with fuscous spots on basal third (Fig. 702). Eastern species 16

16(15). Found in sand dunes on Atlantic coast .... ***maritima maritima*** **(Harris)** (p. 512)

Found in sandy areas in Great Lakes region .... ***maritima interior*** **E. M. Walker** (p. 513)

17(14). Spur of hind wing crossband extending at least halfway to wing base. Found in interior British Columbia and southward .... ***longicornis*** **E. M. Walker** (p. 514)

Spur of hind wing crossband not extending halfway to wing base. Found in Montana and Wyoming .... ***diversellus*** **Hebard** (p. 515)

Clé des espèces et des sous-espèces de *Trimerotropis*

1. Tibias postérieurs rouges ou orangés (sauf pour les spécimens provenant de la région des Grands Lacs) 2

Tibias postérieurs jaunes ou bleuâtres (orangés chez les spécimens provenant de la région des Grands Lacs) 5

2(1). Angle postéroventral des lobes latéraux du pronotum formant une pointe arrondie (fig. 691) .... ***agrestis*** **McNeill** (p. 496)

Angle postéroventral des lobes latéraux du pronotum très arrondi, ne formant pas de pointe 3

3(2). Bandes des tegmina claires et nettes; éperon de la bande de l'aile postérieur s'étendant sur moins de la moitié de la distance à la base (fig. 705 et 706) 4

Bandes des tegmina habituellement floues; éperon de la bande de l'aile postérieure s'étendant sur plus de la moitié de la distance à la base (fig. 704) .... ***campestris*** **McNeill** (p. 497)

4(3). Face interne des fémurs postérieurs jaune rougeâtre; bande noire distincte sur la moitié apicale, rejoignant la bande de la face externe ...... ***pistrinaria*** **Saussure** (p. 498)

Face interne des fémurs postérieurs noire ou marquée d'une bande pré-apicale pâle, ou à région noire brisée au centre .... ***latifasciata*** **Scudder** (p. 500)

5(1). Tibias postérieurs bruns à brun bleuâtre, à bande basale pâle 6

Tibias postérieurs à coloration brune non dominante 7

6(5). Ailes postérieures à bande transversale noire rejoignant l'apex (fig. 709). Espèce vivant dans les Rocheuses et sur la côte Ouest .... ***suffusa*** **Scudder** (p. 501)

Ailes postérieures à région claire en dehors de la bande transversale noire, à extrémité assombrie (fig. 708). Espèce vivant de la côte Est aux Rocheuses .... ***verruculata*** **(Kirby)** (p. 502)

7(5). Tibias postérieurs jaunes, chamois ou rougeâtres 9
Tibias postérieurs bleus 8
8(7). Tibias postérieurs bleus, à bande basale pâle; tête sans bande transversale noire. Espèce vivant en Colombie-Britannique, vers l'est jusqu'au Wyoming et au Colorado et plus au sud .... *fontana* **Thomas** (p. 503)
Tibias postérieurs entièrement bleu pâle; tête marquée de bandes transversales foncées au-dessus des antennes et entre ces dernières. Insecte vivant au Montana, dans l'ouest des Dakotas, au nord-ouest du Nebraska et plus au sud .... *cincta* **(Thomas)** (p. 505)
9(7). Ailes postérieures transparentes, légèrement teintées de bleu; bande transversale faible ou absente (fig. 698) ...... *sparsa* **(Thomas)** (p. 505)
Disques des ailes postérieures jaunes à bande transversale noire, cette dernière étant habituellement claire 10
10(9). Bande transversale noire des ailes postérieures floue, marquée seulement par les nervures enfumées (fig. 695) .... *gracilis* **(Thomas)** (p. 507)
Bande transversale noire des ailes postérieures distincte 11
11(10). Zone située entre les crêtes ventrales des fémurs postérieurs noire jusqu'à la bande apicale pâle 12
Zone située entre les crêtes ventrales des fémurs postérieurs pâle et marquée de deux taches foncées ou marquée d'une bande noire au tiers apical 14
12(11). Tegmina variables, à bandes peu visibles ou le plus souvent inexistantes (jamais rougeâtres) (fig. 697) .... *huroniana* **E.M. Walker** (p. 507)
Tegmina à bandes claires (parfois entièrement rougeâtres) 13
13(12). Corps robuste. Bande transversale de l'aile postérieure large, à disque jaune pâle; bandes des tegmina brisées et pas très claires (fig. 701). Espèce vivant en Alberta et plus à l'est .... *pallidipennis salina* **McNeill** (p. 509)
Corps moins robuste. Bande transversale de l'aile postérieure pas spécialement large et ayant un disque jaune; bandes des tegmina complètes et claires (fig. 700). Espèce vivant en Colombie-Britannique et plus au sud .... *pallidipennis pallidipennis* **(Burmeister)** (p. 510)
14(11). Tegmina non marqués de bandes claires 15
Tegmina marqués de bandes claires 17
15(14). Tegmina dépourvus de bandes, marqués de taches noires le long des nervures (fig. 696). Espèce vivant dans l'Ouest .... *sordida* **E.M. Walker** (p. 511)
Tegmina unis ou légèrement marqués de taches brunâtres sur le tiers basal (fig. 702). Espèce vivant dans l'Est 16
16(15). Espèce vivant dans les dunes de sable de la côte de l'Atlantique .... *maritima maritima* **(Harris)** (p. 512)
Espèce vivant dans les zones sablonneuses de la région des Grands Lacs .... *maritima interior* **E.M. Walker** (p. 513)
17(14). Éperon de la bande transversale de l'aile postérieure s'étendant sur la moitié au moins de la distance à la base. Espèce vivant dans les régions intérieures de la Colombie-Britannique et plus au sud .... *longicornis* **E.M. Walker** (p. 514)
Éperon de la bande transversale de l'aile postérieure ne s'étendant pas sur la moitié de la distance à la base. Espèce vivant au Montana et au Wyoming .... *diversellus* **Hebard** (p. 515)

Trimerotropis agrestis McNeill

Figs. 691, 703; Map 194

Trimerotropis agrestis McNeill, 1900*a*:32.

Trimerotropis agrestis; McNeill 1901:433; Brooks 1958:44; Helfer 1963:137.

Diagnosis. Posteroventral angle of lateral pronotal lobes produced as rounded points. Tegmina speckled.

Description. Body size medium (length 23–35 mm). Pronotum with median carina prominent anterior to principal sulcus, moderately prominent on metazona to posterior margin, with posteroventral angle of lateral lobe produced as blunt point (Fig. 691). Color sandy, reddish to gray brown, speckled with black; tegmina sandy with numerous scattered dark spots, but without indication of banding (Fig. 703); hind wing disc yellow, with cross-band dark, broad; spur broad, short, extending about halfway to wing base (Fig. 703); hind femur sandy on outer face, with small dark spots and faint indications of 3 incomplete darker bands; inner face of femur reddish with dark spot at middle, or reddish with basal third entirely dark, with narrow black band on apical third; hind tibia red, becoming yellowish basally on outer face.

Range. Alberta to Manitoba, south to New Mexico.

Behavior and habitats. Hebard (1928, 1936*a*) reported *T. agrestis* as being found only on sand. Brooks (1958) also reported it only from "bare

Map 194. Collection localities for *Trimerotropis agrestis*.

yellow sand." Tinkham (1959) reports that this species inhabits (pink) sand dunes and that it excavates protective depressions in the sand. It is a mixed feeder of no economic importance.

Flight is short and rather slow; both sexes crepitate just before landing. Otte (1970) reported that courting males produced four strokes of "ordinary" stridulation from a distance of about 15 cm from the females, which they then mounted. Femur-tipping sometimes, but not always, preceded stridulation. Aggressive movements included femur-tipping and femur-shaking, together with striking the substrate with the hind tibiae.

Life history. Univoltine, with overwintering in the egg stage. Adults have been collected from 20 July to 9 September, but it is probable that they are to be found later than this.

Trimerotropis campestris McNeill

Fig. 704; Map 195

Trimerotropis campestris McNeill, 1900*a*:31.

Trimerotropis campestris; McNeill 1901:432; Brooks 1958:45; Helfer 1963:139.

Diagnosis. Tegminal banding indistinct. Spur of hind wing band long. Hind tibia red. This species, in general appearance, resembles *Spharagemon equale* (Say), *S. collare* Scudder, and *Trimerotropis latifasciata* Scudder, q.v.

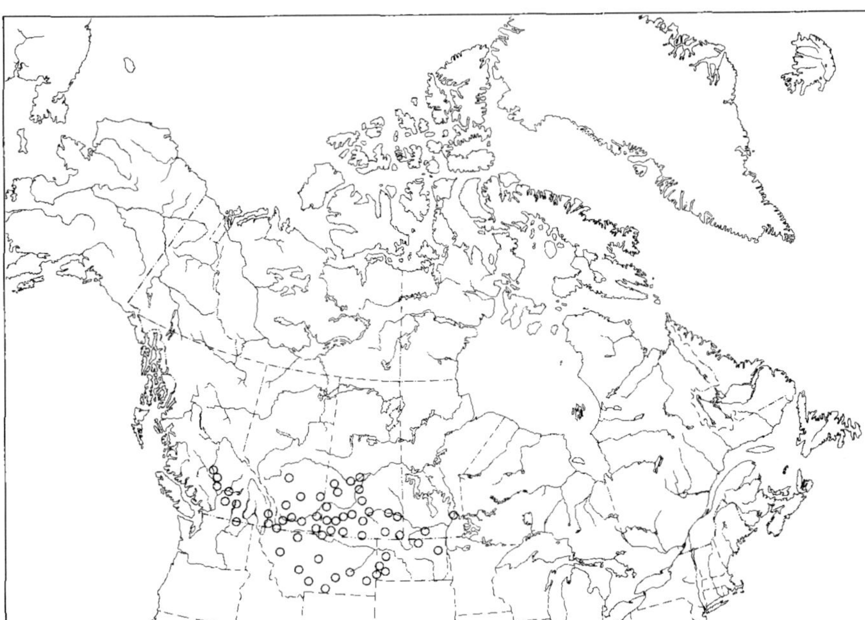

Map 195. Collection localities for *Trimerotropis campestris*.

(E. M. Walker 1902*a*; Hewitt and Barr 1967). The resemblance to certain species of *Spharagemon* is most marked in specimens in which the anterior incision of the median carina of the pronotum is nearly obsolete. Some *Spharagemon* specimens have a slight depression in the carina anterior to the incision of the principal sulcus. In *Spharagemon* the tegminal bands are usually distinct, but in *T. campestris* they are always irregular. *T. latifasciata* is distinguishable from it by the broader crossband of the hind wing, with the spur short and blunt, and by many of the veins on the outer edge of the crossband being whitish.

Description. Body medium–large (length 26–40 mm). Lateral carinae of vertex of head prominent; median carina linear. Pronotum with median carina prominent on prozona, linear to feebly elevated on metazona, with anterior incision shallow, sometimes obsolete, and with incision of primary sulcus deep. Color usually dark gray, occasionally reddish, mottled, and with black spots; head and pronotum often pale yellow without spots, or ventral part of head and posterior half of pronotum whitish with black spots; tegmina brown, with darker spots forming basal and central irregular bands; apex of tegmina with scattered spots (Fig. 704); hind wing disc yellow, with crossband broad, black, sometimes broken at spur; spur broad, extending slightly more than halfway to wing base, with apex clear (Fig. 704); hind femur grayish white (occasionally pinkish) with 3 dark crossbands; inner face of hind femur reddish orange, with median dark spots and narrow dark band at apical third; hind tibia orange to red, yellowish basally on outer face.

Range. British Columbia to Manitoba, south to Arizona and New Mexico.

Behavior and habitats. Brooks (1958) stated that this species is common on open grasslands and he also found it locally on dry hillsides and gravelly spots in parklands. In Idaho it seems to occur only on hillsides and gravelly habitats in the mountains where grasses and sagebrush occur. It feeds mainly on species of *Astragalus* (Leguminosae) rather than on grasses (Hewitt and Barr 1967), although Brooks (1958) indicated that it is a mixed feeder, preferring grasses. Otte (1970) found it at elevations above 3050 m.

When they were disturbed, both sexes crepitated in flight; undisturbed males also crepitated (Otte 1970). Crepitation consisted of single bursts of sound. During courtship males perform three types of movements: bursts of vibratory stridulation by the action of a single femur; femur-tipping, sometimes preceding stridulation; and trills of "ordinary" stridulation, usually produced when females made repelling movements or moved away. Femur-tipping and femur-shaking were noted as aggressive signals (Otte 1970).

Life history. As in other species of the genus, the life cycle is univoltine, with overwintering in the egg stage in soil. There is a possibility that the cycle may be extended over 2 years in high montane areas, but we have no data on this. Adults are found from early July until late September.

Trimerotropis pistrinaria Saussure

Fig. 705; Map 196

Tr[*imerotropis*] *pistrinaria* Saussure, 1884:173.

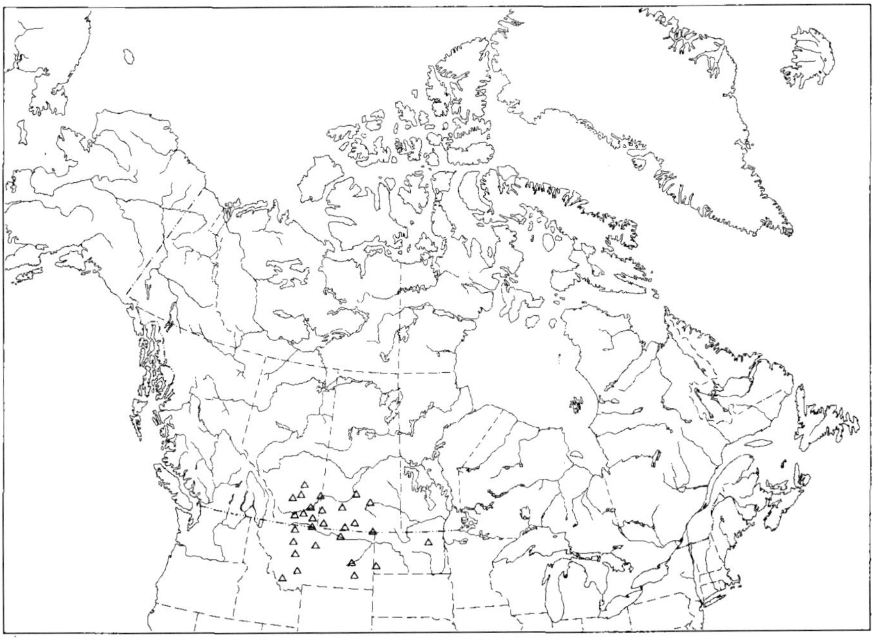

Map 196. Collection localities for *Trimerotropis pistrinaria.*

Trimerotropis bruneri McNeill, 1900*a*:31; 1901:423.

Trimerotropis pistrinaria; McNeill 1900*a*:32; Brooks 1958:47; Helfer 1963:138.

Diagnosis. Tegmina conspicuously banded; hind wing crossband broad, black; inner face of hind femur reddish yellow, with black band on apical half.

T. pistrinaria somewhat resembles *Hadrotettix trifasciatus* (Say), but it is easily distinguished from that species by the reddish inner face of the hind femur, which, in *H. trifasciatus*, is deep blue to nearly black. The two species often occur together, but they may also be distinguished by their flight (see following text).

Description. Body medium-large (length, male 21–30, female 26–36 mm). Median carina of pronotum low, linear, with impression of anterior sulcus shallow to obsolete, and with metazona twice length of prozona, acute-angled at posterior margin. Hind wings broad. Color brownish gray; tegmina prominently banded, with dark bands widely spaced, pale gray to whitish, with basal area dark gray to brown and with median pale band not constricted at middle; apical third of tegmen grayish or with few dark spots (Fig. 705); hind wing disc pale yellow, with crossband black, broad, and with spur reaching less than halfway to wing base; apex of wing clear (Fig. 705); hind femur pale on outer face with moderately prominent black band, and with inner face entirely reddish yellow with prominent black band at apical third, this continuous with that on outer face; hind tibia red, yellowish basally on outer face.

Range. Alberta and Saskatchewan, south to Mexico.

Behavior and habitats. Brooks (1958) reported *T. pistrinaria* from gravelly spots and hillsides. It is a mixed feeder, preferring forbs such as *Astragalus* species. Brooks called it a moderately strong and erratic "circle-back" flier, with a brief, but loud, crepitation. *Hadrotettix trifasciatus*, with which it might be confused (see preceding text), is a weak flier, moving for only short distances at a time.

Life history. As in other species of *Trimerotropis*, *T. pistrinaria* is univoltine, with overwintering in the egg stage in soil. Adults have been reported from 26 June to 14 October in South Dakota (Hebard 1925*a*), from 30 July to 14 October in Montana (Hebard 1928), and from 25 July to 30 September in Alberta (Hebard 1931*a*).

Trimerotropis latifasciata Scudder

Fig. 706; Map 197

Trimerotropis latifasciata Scudder, 1881:26.

Trimerotropis latifasciata; Tinkham 1947:139–141; Brooks 1958:46; Helfer 1963:136.

Diagnosis. Tegmina conspicuously banded; hind wing crossband broad, black; inner face of hind femur black or with pale preapical band.

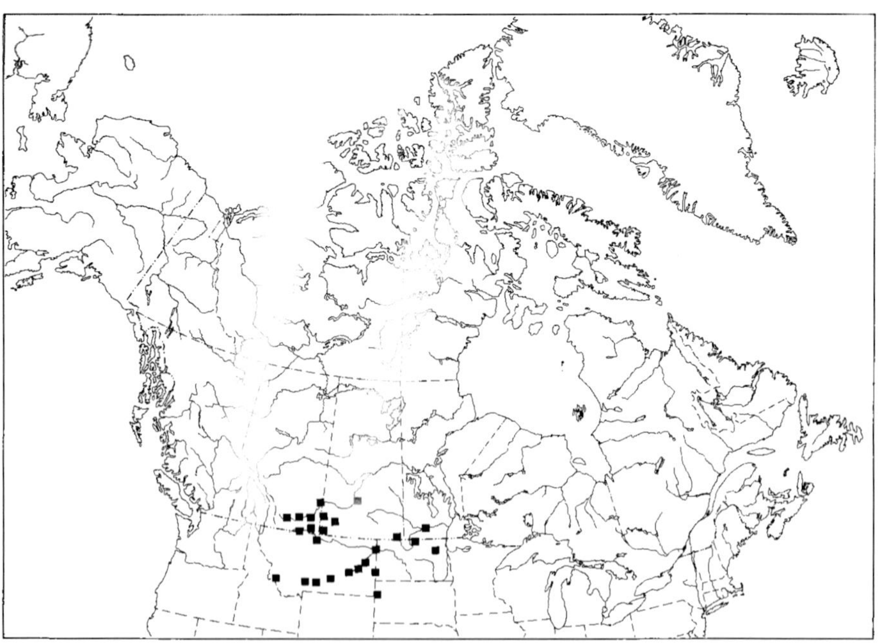

Map 197. Collection localities for *Trimerotropis latifasciata*.

Description. Body large (length 29–46 mm). Vertex of head with disc deeply impressed; lateral carinae prominent. Pronotum with median carina usually low, linear, occasionally moderately prominent throughout, with incisions of sulci shallow. Color grayish to reddish brown; abdomen yellowish; tegmina gray or black, prominently banded, with dark bands conspicuous and with pale area between bands constricted at middle; apical third of tegmina gray with scattered dark spots; hind wing disc pale yellow to nearly white, with crossband dark, broad; spur blunt, short, extending less than halfway to wing base; hind femur with outer face pale gray, with faint band and 4 spots; inner face of femur entirely black as far as apical pale band, or black broken in middle by reddish area; hind tibia red to bright orange, yellowish basally on outer face.

Range. Alberta to Manitoba, south to Montana and South Dakota.

Behavior and habitats. *T. latifasciata* usually is found on bare sandy alkaline flats where the chief vegetation is sagebrush (*Artemisia*) and greasewood (*Sarcobatus*). It is a mixed feeder, preferring bunchgrass (*Agropyron smithii*) (Brooks 1958). It is a strong erratic flier, difficult to capture. N. Criddle (1935) remarked that it is a "rather quiet species, its noise when flying being soft." No information on courtship is available.

Life history. Univoltine, with overwintering in the egg stage. Hebard (1928) reported adults from 12 July to 8 September in Montana. Hewitt and Barr (1967) recorded this species from 7 July to 8 September in southern Idaho.

Trimerotropis suffusa Scudder

Fig. 709; Map 198

Trimerotropis suffusa Scudder, 1876*a*:265.
Trimerotropis obscura Scudder, 1876*c*:514.
Trimerotropis nubila McNeill, 1901:442.
Trimerotropis suffusus; Hebard 1928:261.
Trimerotropis suffusa; M. J. D. White 1949:550; Brooks 1958:48; Helfer 1963:135.
Trimerotropis verruculatus suffusa; Vickery and Nagy 1973:30.

Diagnosis. Body dark; tegmina not banded; hind wing crossband extending to apex of wing.

Description. Body size medium (length 31–40 mm). Median carina elevated on prozona; incision of sulci moderately deep. Color black or brownish, speckled; pronotum with contrasting whitish and dark markings; tegmina uniformly dark brown to black with darker flecking (Fig. 709); hind wing disc bright yellow, with crossband broad, usually unbroken, somewhat darker than apex, dark smoky black to somewhat paler from crossband to apex but always somewhat darkened next to crossband and darker at apex (Fig. 709); hind femur brownish on outer face, with banding diffuse to nearly absent (although usually apparent dorsally); inner face of femur banded; hind tibia pale blue to greenish blue, mottled with brown, with prebasal band paler, darkened at extremities.

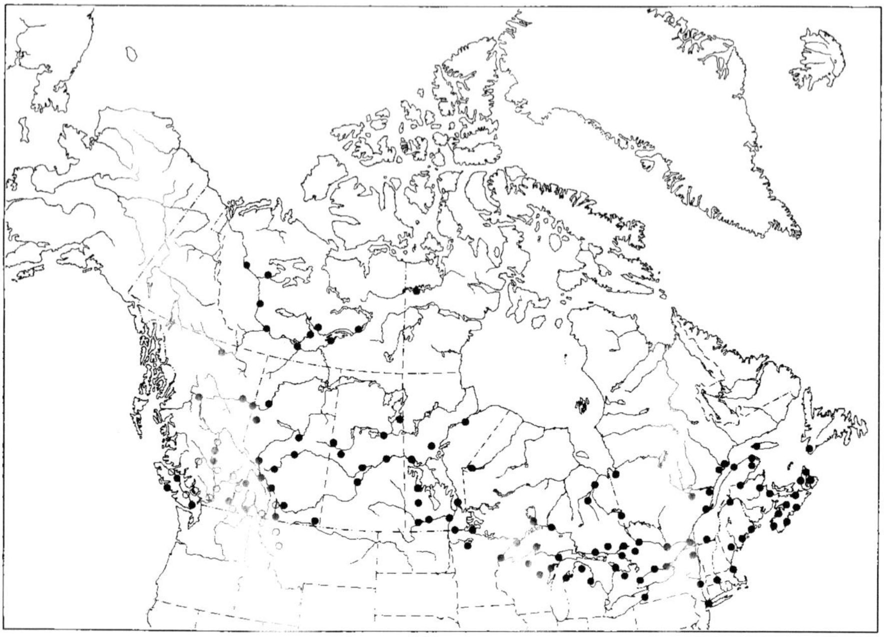

Map 198. Collection localities for *Trimerotropis verruculata* (●) and *T. suffusa* (○). (Note: Inaccurate in British Columbia.)

Range. British Columbia and southwestern Alberta, south to California and Colorado.

Behavior and habitats. Same as for *T. verruculata.*

Life history. Same as for *T. verruculata.* Hewitt and Barr (1967) reported adults from July to October in Idaho.

Trimerotropis verruculata (Kirby)

Figs. 662, 708; Map 198

Locusta verruculata W. Kirby, 1837:250.
Locusta latipennis T. W. Harris, 1841:144.
Oedipoda verruculata; Provancher 1876:113.
Trimerotropis verruculata; Caulfield 1888:70; R. D. Alexander et al. 1972:54.
Trimerotropis verruculatus; Vickery 1967*a*:268.

Diagnosis. Body dark; tegmina not banded; hind wing with clear area outside black crossband.

Description. Body size medium (length, male 21–25, female 25–28 mm). Pronotum flattened dorsally, with median carina elevated on prozona; incision of anterior sulcus shallow; posterior margin of metazona right-angled to acute-angled. Color variable, usually blackish above, brown beneath; tegmina gray to brownish gray, blotched with black, not banded (Fig. 708); hind wing nearly transparent basally, with pale yellow tinge, and with crossband faint, narrow at costal margin, widening at anal margin; apical quarter of wing hyaline, and apex fuscous (Fig. 708); fore and middle tibiae annulated; hind femur dull yellow to grayish on outer face, crossed by oblique blackish bands; hind tibia dull yellow tinged with fuscous, darker at extremities.

Range. Western Northwest Territories to Newfoundland, south to Minnesota and Massachusetts.

Behavior and habitats. *T. verruculata* prefers exposed rocky or gravelly places. Brooks (1958) reported it from "exposed gravel, rock, sand, or clay banks along river courses and roadsides." We have taken it on rocky ledges, gravel outcroppings, and gravel pits in Nova Scotia, Québec, and Ontario, sometimes fairly deep in forested areas. It feeds almost exclusively on forbs, and only rarely on grasses (Brooks 1958).

Otte (1970) reported dramatic male flight displays, lasting up to a minute, the insect hovering over a spot, then darting and hovering over another while crepitating nearly continuously. The crepitations are "snapping sounds" produced in pairs with a relatively constant interval between them, although the interval between pairs of snaps varies somewhat, this being related to altitude. At higher elevations the intervals between snap pairs is shorter than at lower altitudes. Females also crepitate, producing a similar but less intense sound. This was observed long ago by E. M. Walker (1898*b*). According to Otte (1970), courting males produce "ordinary" stridulation in bursts when they are near females on the ground. Femur-tipping and femur-shaking are not employed in courtship but are used as aggressive signals in male-to-male encounters, the femur-shaking always accompanied by a striking of the substrate with the hind tibiae.

Life history. There is one generation each year, with overwintering as eggs in soil. Adults are active in Québec from 14 July to 9 September (Larochelle 1978). Cantrall (1968) recorded adults in Michigan from 8 July to 22 September. In most other areas, adult specimens have been reported to be active during this period.

Trimerotropis fontana Thomas

Fig. 707; Map 199

Trimerotropis fontana C. Thomas, 1876*a* [June]:255, pl. 36, fig. 5.
Trimerotropis juliana Scudder, 1876*c* [November]:294.
Trimerotropis caeruleipes Scudder, 1881:27.
Trimerotropis ferruginea McNeill, 1901:420.
Trimerotropis fontana; Hebard 1928:259; Helfer 1963:143.

Diagnosis. Hind wing disc pale yellowish to blue green; hind tibia blue with pale basal band.

Description. Body size medium (length 25–37 mm). Depression of vertex with faint median carina; lateral carinae prominent. Median carina

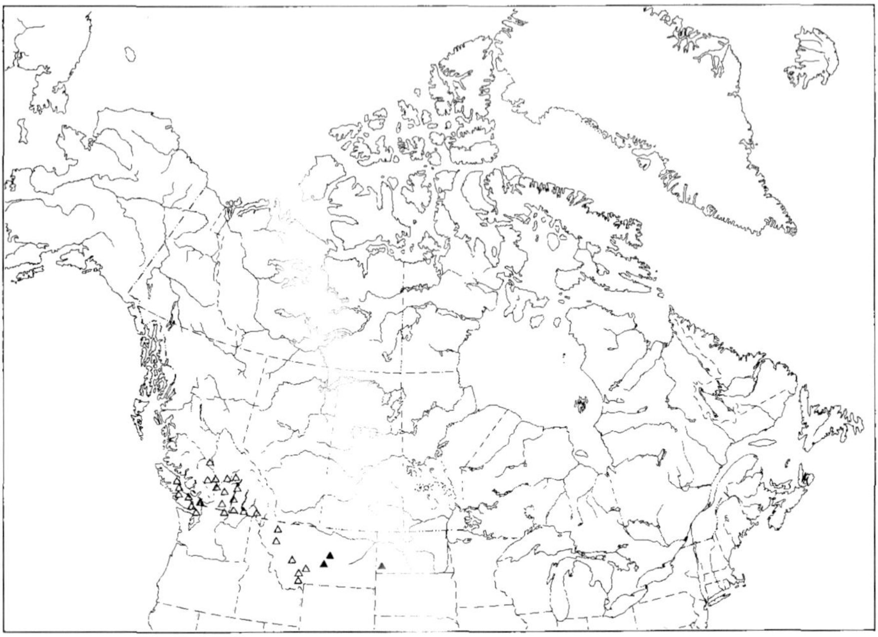

Map 199. Collection localities for *Trimerotropis fontana* (△) and *T. cincta* (▲).

of pronotum distinct, with anterior incision shallow, and with posterior incision deep. Tegmina narrow. Color brownish, with head paler between antennal bases; pronotum irregularly marked with light and dark areas, or metazona completely pale; tegmina variable, essentially light brown, gray, or pale reddish, with bands brown, incomplete, made up of aggregates of spots, and with outer band often obscure or obsolete; apical third of tegmen with scattered spots (Fig. 707); hind wing disc pale, translucent, lemon yellow to bluish green or nearly colorless, with crossband narrow, irregular, often indistinct or absent, and with spur reaching halfway to wing base; apical third of wing clear or with few darker cells near apex (Fig. 707); hind femur brownish on outer face, dark apically with preapical dark band; inner face of femur dark on basal half and preapical band; hind tibia blue with distinct pale prebasal ring, with base black.

Range. British Columbia, south to California and Colorado.

Behavior and habitats. Buckell (1922) reported *T. fontana* as being common in dusty cultivated land such as clear-cultivated orchards. Hewitt and Barr (1967) found it in a variety of habitats from wooded areas to sagebrush–grass areas. They remarked that it was commonly found on rocky or bare hilltops. It is a mixed feeder.

Otte (1970) recorded crepitation only by individuals that had been disturbed. Femur-shaking, producing a low-pitched hum, was the only courtship signal produced by males. Neither wings nor substrate were struck during this activity. The only aggressive signal was femur-tipping. He noted that the function of femur-shaking had changed entirely to courtship in this species, contrary to its function in most species of *Trimerotropis*.

Life history. Univoltine, with overwintering in the egg stage. Hewitt and Barr (1967) recorded adults from 15 June to 21 October in Idaho, but the great majority were taken from late July to late September. Buckell (1920) found adults in the Okanagan Valley, B.C., from 20 July until the end of September, stating that heavy frosts in the latter month caused no visible decrease in numbers.

Trimerotropis cincta (Thomas)

Map 199

Oe[dipoda] cincta C. Thomas, 1870:80.
Trimerotropis cincta; McNeill 1901:414 (partim); Helfer 1963:134.

Diagnosis. Face with transverse dark bands above and beneath antennae.

Description. Face with transverse dark bands above and below antennae. Hind tibia pale blue with broad whitish preapical ring, and with apex darkened.

Range. Oregon to North Dakota, south to Arizona and New Mexico.

Behavior and habitats. Otte (1970) collected *T. cincta* in exposed gravelly areas in pine woods on mountain slopes in Colorado. It is primarily a montane species.

Otte (1970) observed crepitation, in solitary males, to consist of one to several bursts of irregular length. Disturbed females did not crepitate. Males employed both silent shaking of the hind femora and "ordinary" stridulation when courting females. Shaking was always evident but was not always followed by stridulation, which usually was a single pulse of sound. Occasionally, femur-tipping preceded femur-shaking in courtship and was the usual behavior in aggressive encounters.

Life history. Univoltine, with overwintering as eggs in soil. Adults were recorded from 20 July to 7 September in South Dakota and Montana (Hebard 1925*a*, 1927).

Trimerotropis sparsa (Thomas)

Fig. 698; Map 200

Oedipoda sparsa C. Thomas, 1875:883; pl. 45, fig. 6.
Circotettix sparsus; Hebard 1927:6.
Trimerotropis sparsa; Hebard 1928:252; Brooks 1958:47; Helfer 1963:144.

Diagnosis. Hind wing clear with faint bluish tinge; crossband faint to absent.

Description. Body size medium (length 26–37 mm). Disc of vertex deeply impressed, with lateral carinae high. Median carina of pronotum distinctly bilobed on prozona, low and linear on metazona. Color generally

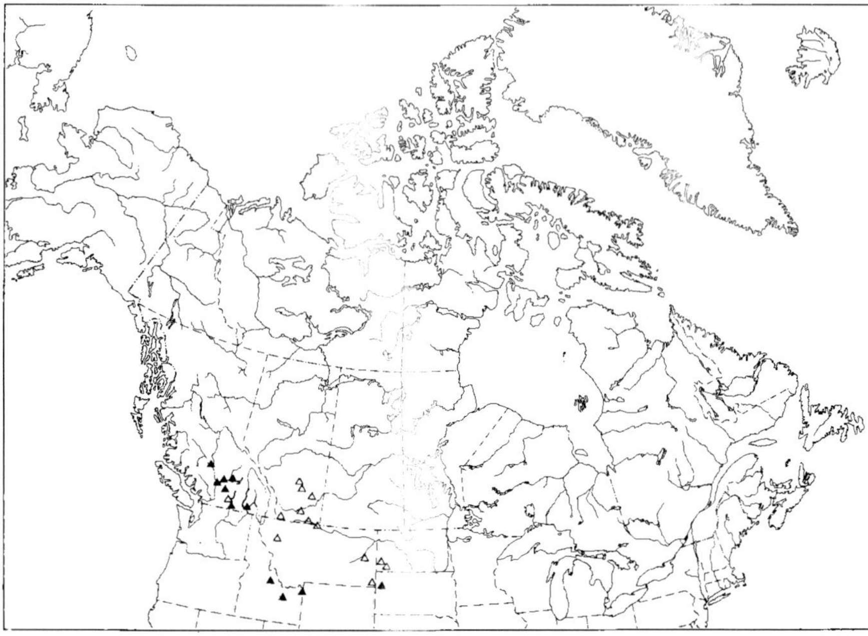

Map 200. Collection localities for *Trimerotropis sparsa* (△) and *T. gracilis* (▲).

gray with darker markings; head grayish, sometimes marked with orange on vertex and beneath eyes; pronotum often with orange markings; tegmina gray with scattered darker spots, these forming poorly defined bands, with apex little spotted (Fig. 698); hind wing disc clear, occasionally with slight bluish tinge on lower part of basal area, with anal veins darkened, and with apex clear or with veins darkened; crossband absent (Fig. 698); hind femur pale, without bands on outer face, with diffuse bands dorsally only; inner face of femur black on basal half, pale with black band on apical half; hind tibia greenish buff to yellow or yellowish brown, mottled with brown, darker basally and apically.

Range. British Columbia to North Dakota, south to Arizona.

Behavior and habitats. *T. sparsa* inhabits bare eroded areas, alkaline flats, and hillsides. Hewitt and Barr (1967) reported it from eroded areas in arid salt-desert and sagebrush-grass zones in Idaho. Vickery and Nagy (1973) recorded it from eroded and disturbed slopes of a sandy gravelly plateau. The vegetation consisted of grasses, *Agropyron* and *Bromus* species, Oregon grape (*Berberis aquifolium* Pursh), antelope bush (*Purshia tridentata* (Pursh) DC.), but with rabbitbrush (*Chrysothamnus nauseosus* (Pall.) Britt.) and sagebrush (*Artemisia tridentata* Nutt.) predominating.

Brooks (1958) and Hewitt and Barr (1967) reported flight as short and erratic, with loud crepitation produced just before alighting.

Life history. Univoltine, with overwintering as eggs in soil. Adults are active in August and September (Hewitt and Barr 1967; Vickery and Nagy 1973).

Trimerotropis gracilis (Thomas)

Fig. 695; Map 200

O[edipoda] gracilis C. Thomas, 1872:461.
Trimerotropis gracilis; C. Thomas 1873:110, 121; Helfer 1963:144.
Trimerotropis gracilis gracilis; Hebard 1936*a*:39.

Diagnosis. Body grayish, appearing streaked; crossband of hind wing indistinct, with disc pale yellow to greenish.

Description. Body size medium (length 26–36 mm). Disc of vertex of head strongly depressed with prominent lateral carinae. Pronotum with median carina bilobed, strongly to feebly elevated on prozona, well-developed on metazona. Color gray, usually with streaky appearance; head pale gray, mottled with brown or orange; pronotum grayish brown with small dark spots; tegmina pale brown with darker veins and darker spots on basal half; hind wing disc pale yellow to faintly greenish or nearly clear, without crossband (in some specimens indicated by darkened veins (Fig. 695); apex of wing clear or with few smoky cells; hind femur grayish on outer face, mottled, with 2 dark spots dorsally, and with inner face dark as far as pale preapical band; hind tibia yellowish, mottled with brown.

Range. British Columbia to South Dakota, south to California and Utah.

Behavior and habitats. Hewitt and Barr (1967) recorded *T. gracilis* as an inhabitant of dry flat country in association with sagebrush, but also that it is often common near roads. They classed it as a mixed feeder. The flight is said by them to be swift and erratic, usually accompanied by a "low clacking" crepitation.

Life history. Univoltine, with overwintering as eggs in soil. Adults are common from July until the end of September.

Economic importance. Although *T. gracilis* is not generally regarded as of economic significance, Hewitt and Barr (1967) stated that, since it is a common range grasshopper, it may have pest potential.

Trimerotropis huroniana E. M. Walker

Figs. 693, 697; Map 201

Trimerotropis huroniana E. M. Walker, 1902*a*:6, pl. 1, figs. 8–13.
Trimerotropis huroniana; E. M. Walker 1902*b*:256; Helfer 1963:132; Alexander et al. 1972:54.

Diagnosis. Inner face of hind femur black, with preapical pale band and with pale dorsomedian patch. *T. huroniana* superficially resembles *T. maritima interior* E. M. Walker, but it is easily distinguished by the color pattern of the inner face of the hind femur. In *T. maritima interior*, this is pale with a darker genicular area and black median and subbasal crossbands (Fig. 694).

Description. Body small (length, male 19–21, female 26–28 mm). Vertex of head distinctly longer than broad in male, only slightly longer in

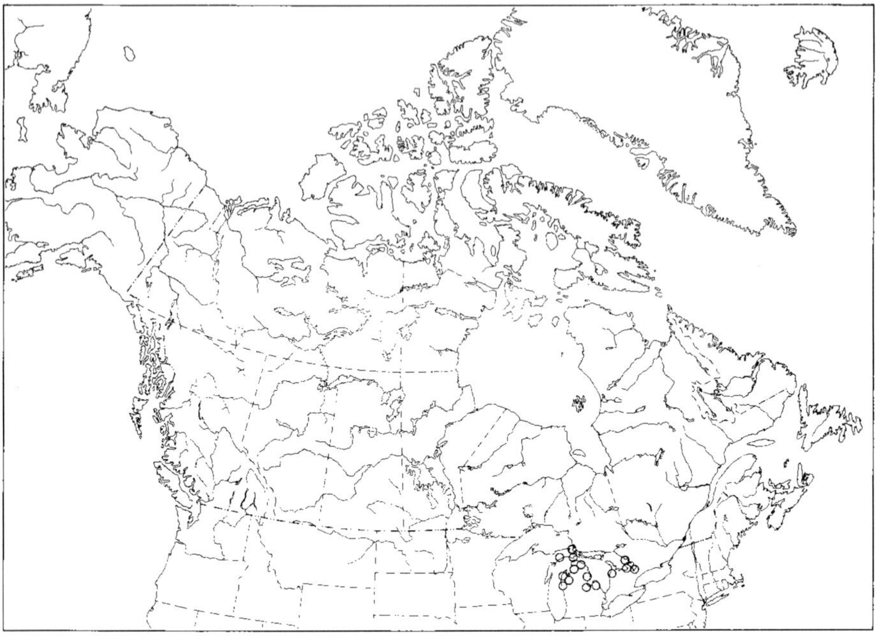

Map 201. Collection localities for *Trimerotropis huroniana.*

female, with median carina low but distinct, and with foveolae small, triangular, but distinct. Pronotum with metazona longer than prozona in ratio of 3:2, with median carina higher on prozona, and with metazona broad with surface finely and densely rugose and punctate; posterior margin of male pronotum right-angled, of female obtuse-angled. Color ashy gray with brownish and white markings; pronotum with pale lateral marginal stripe on both prozona and metazona; sides of prozona with 2 white spots; tegmina ashy gray or brownish with evidence of fuscous crossbars, these often faint or obsolete (when distinct, the median one more pronounced), with apex hyaline, lacking spots; hind wing with disc faded lemon yellow; crossband fuscous, narrow, with spur extending halfway to base, and with apex hyaline without spots (Fig. 697); hind femur pale gray on outer face with pale yellow preapical band and indications of 3 oblique darker crossbands; inner face of femur black with pale preapical band and median area continuously black dorsally along lower sulcus as far as preapical band (Fig. 693); hind tibia pale dull yellow, mottled with brown basally and shading to brownish apically.

Range. Ontario and Michigan.

Behavior and habitats. *T. huroniana* inhabits sandy beaches with scattered grasses along lake shores. E. M. Walker (1902*a*) remarked that it is a very alert species.

The flight sound of this species is "peculiar, being a very rapid but not loud crepitation interrupted about thrice in a second, so that at a little distance it seems to be composed of separate notes. Usually three, sometimes four of these, are produced at a time" (E. M. Walker 1902*a*). Otte (1970) recorded "ordinary" stridulation as bursts made up of 2–8 pulses. In courtship, males often produce a series of stridulatory trills as they approach females. Each burst is produced by a single femur. This type of courtship behavior appears to be typical for the species. In male-to-male encounters, both femur-tipping and femur-shaking are employed as aggressive signals.

Life history. Univoltine, with overwintering as eggs in sandy soil. The type series was collected 20–29 August. Cantrall (1968) reported *T. huroniana* in the adult stage from 6 July to 11 September in Michigan.

Trimerotropis pallidipennis salina McNeill

Fig. 701; Map 202

Trimerotropis salina McNeill, 1900*a*:33.

Trimerotropis salina; McNeill 1901:436.

Trimerotropis pallidipennis salina; Hebard 1928:254; Brooks 1958:46; Helfer 1963:140.

Diagnosis. Tegminal band broken, inconspicuous; hind wing crossband broad, dark.

Description. Body large (28–42 mm). Median carina of pronotum prominent on prozona, linear on metazona. Color grayish brown to grayish black; tegmina pale gray with large dark basal blotch, with irregular narrow median dark band, and with scattered spots on apical third; hind wing disc

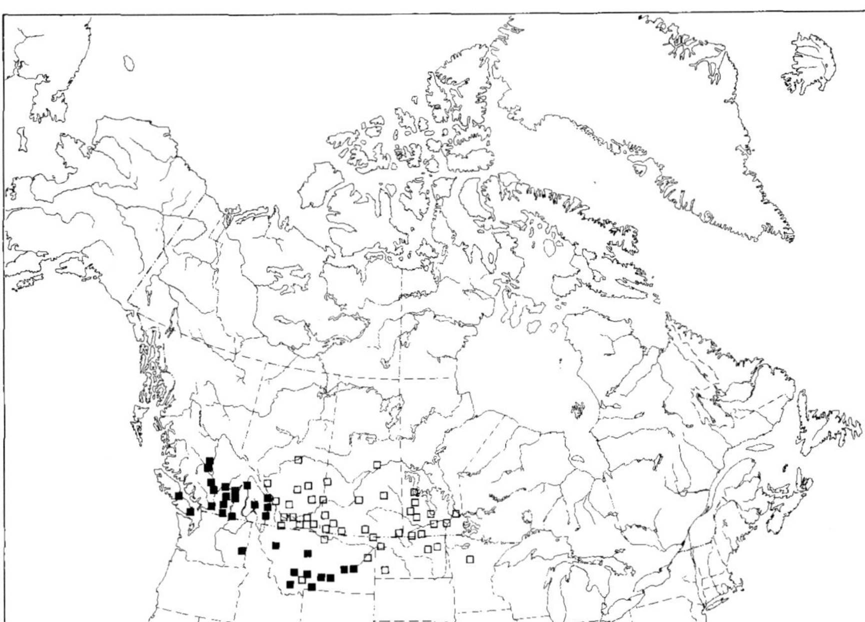

Map 202. Collection localities for *Trimerotropis pallidipennis salina* (□) and *T. pallidipennis pallidipennis* (■).

pale yellow, with crossband broad, black; spur broad, short, extending less than halfway to wing base (Fig. 701); hind femur pale on outer face with darker blotches at base near middle and before a preapical pale band, but dark bands rarely distinct; inner face of femur black on basal half or more, this usually broken by pale spot; hind tibia yellow to buff, with some brown mottling; entire body occasionally pink (Brooks 1958).

Range. Alberta to Manitoba, south to Colorado and Oklahoma.

Behavior and habitats. This subspecies favors bare alkaline flats and margins of alkaline sloughs. *T. pallidipennis salina* is a mixed feeder, preferring to feed on grasses. Flights are short, straight, and close to the ground. Crepitation is soft, sounding like the fluttering of a small bird (Brooks 1958).

Life history. Presumably univoltine, with overwintering in the egg stage in soil. Specimens from Alberta reported by Hebard (1931*a*) were taken from 20 July to 29 August, but there is little doubt that adults are active earlier and later than these dates.

Trimerotropis pallidipennis pallidipennis (Burmeister)

Figs. 690, 700; Map 202

Oedipoda pallidipennis Burmeister, 1838:641.
Trimerotropis vinculata Scudder, 1876*d*:270.
Trimerotropis pallidipennis; McNeill 1901:437; Helfer 1963:140.
Trimerotropis pallidipennis pallidipennis; Buckell 1930:36.

Diagnosis. Tegmina banded; hind wing crossband narrow, with disc yellow; spur of crossband short.

Description. Body large (length 28–42 mm). Disc of vertex of head shallowly impressed, with lateral carinae not prominent. Median carina of pronotum low, linear but distinct on prozona, obsolete on metazona. Color brownish; head brownish gray, mottled, paler beneath eyes; pronotum brownish with darker mottling; tegmina gray or brown, banded but variable, usually with irregular solid basal and median black bands, but sometimes brownish and broken, with apical third sprinkled with brown spots (in reddish specimens, banding nearly obsolete); hind wing disc bright lemon yellow, with crossband narrow, dark, extending along anal margin; spur blunt, short (Fig. 700); hind femur brownish, mottled on outer face, occasionally with faint bands, with 2 broad yellow areas beyond middle on inner face; hind tibia yellow or whitish, occasionally with brown suffusion (especially noticeable in dried specimens), usually paler on inner face and near base on outer face.

Range. British Columbia and Montana, south to Mexico; South America.

Behavior and habitats. Sites with sparse vegetation, such as sandy or rock areas, are preferred, but the insect occurs also in dry locations along fields and roadsides and even in towns and cities. It is a mixed feeder (Hewitt and Barr 1967). Vickery and Nagy (1973) reported it from a disturbed weedy place in Penticton, B.C., and also from dry short-grass pastures.

Otte (1970) recorded crepitation of males in flight, consisting of series of bursts, each composed of 3–6 wing snaps. Females do not crepitate. During courtship, males stridulate, producing a trilling sound. This is produced by 2–14 strokes by a single femur. Femur-tipping and shaking were observed during courtship. These signals are also employed as aggressive signals, but femur-shaking is accompanied by striking the substrate with the hind tibiae in male-to-male encounters.

Life history. Despite the species' wide distribution, virtually nothing is known of the life cycle. Most adults have been found in Canadian localities from July to late September. Hewitt and Barr (1967) reported adults in Idaho from May to late October, but the earliest specimens listed by them were taken 13 June. It is probable that the life cycle is univoltine, with an overwintering egg stage.

Trimerotropis sordida E. M. Walker

Fig. 696; Map 203

Trimerotropis sordida E. M. Walker, 1902*a*:9, pl. 1, figs. 14–18.

Trimerotropis sordida; E. M. Walker 1906*a*:57.

Trimerotropis gracilis sordida; Hebard 1928:250; Brooks 1958:45 (partim).

Diagnosis. Body grayish; tegmina not banded; hind wing disc yellow with black crossband; spur of wing long.

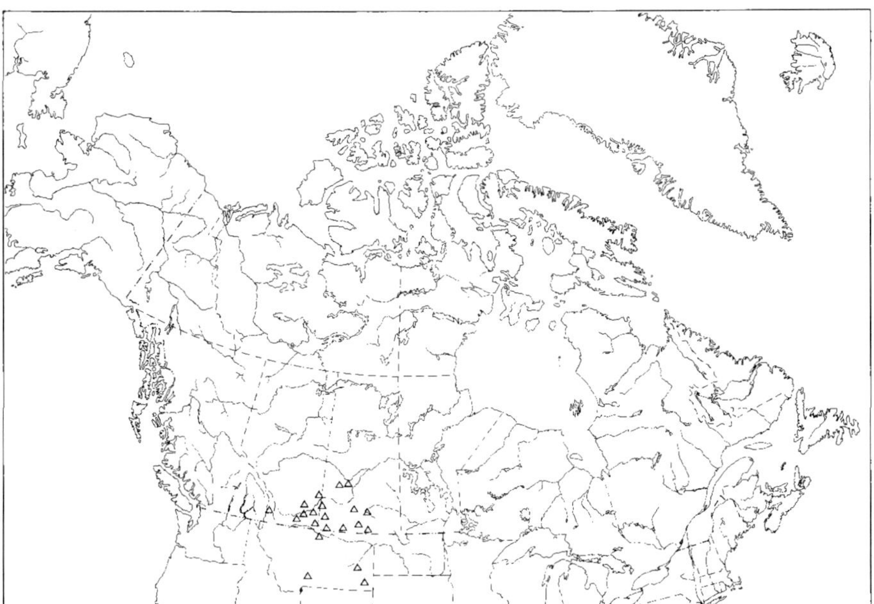

Map 203. Collection localities for *Trimerotropis sordida*.

Description. Body small (length 20–28 mm). Pronotum with median carina prominent throughout, with incisions of sulci deep. Color dull grayish brown without prominent markings; tegmina gray with scattered black spots concentrated toward base; hind wing disc yellow, with crossband moderately broad, black, and with spur extending about two-thirds distance to base (Fig. 696); hind femur grayish brown without banding on outer face, with inner face black as far as preapical pale band, or with black area interrupted by pale spot beyond middle; hind tibia pale yellow to buff, somewhat mottled with brown.

Range. Southwestern Saskatchewan, Alberta, Montana, and Idaho.

Behavior and habitats. Rather bare sandy–alkaline flats seem to be favored by *T. sordida*. It is a mixed feeder, preferring grasses.

Flight is short and erratic, both sexes crepitating and producing a crackling or ticking sound (Brooks 1958). Otte (1970) recorded a very slow crepitation rate for this species; the wings "snap" only every third or fourth beat. Crepitation flights are frequent in both sexes and both will crepitate even when disturbed. "Ordinary" stridulation and femur-tipping are used in courtship, but some males mount females without preliminaries. Femur-tipping and femur-shaking are common in male-to-male encounters (Otte 1970).

Trimerotropis maritima maritima (Harris)

Map 204

Locusta maritima T. W. Harris, 1841:143.

Trimerotropis maritima; McNeill 1900*a*:28 (partim); Helfer 1963:137 (partim).

T[*rimerotropis*] *m*[*aritima*] *maritima*; Vickery and Kevan 1967:61.

Diagnosis. Tegmina not conspicuously banded; hind femur pale on inner face at knee. Inhabiting sand on Atlantic coast.

Description. Body medium-large (length, male 19–25, female 26–35 mm). Pronotum with median carina low on both prozona and metazona, with metazona about twice length of prozona; disc flat with few scattered small rounded tubercles, and with hind margin obtusely angulate. Tegmina long, narrow, tapered toward apex. Hind wings long, pointed at apices. Color light to dark gray or pale reddish brown, mottled with fuscous or brown; often with white markings on sides of head, lower parts of thorax, front legs, and abdomen; tegmina plain gray buff to brownish yellow, sometimes faintly barred with fuscous; disc of hind wing pale yellow, with crossband black, narrow, never more than one-sixth of wing length, and with apex clear; hind femur gray brown with fuscous markings dorsally, the spot nearest the base triangular; inner face of genicular area pale; hind tibia white, yellow, or greenish yellow, with black-tipped spines.

Range. Maine, along Atlantic coast to Florida.

Behavior and habitats. Morse (1920) described the habitat of *T. maritima maritima* as drifting seaside sand regions and adjacent dunes where spotty areas of beach grass (*Ammophila arenaria*) occur. It blends well

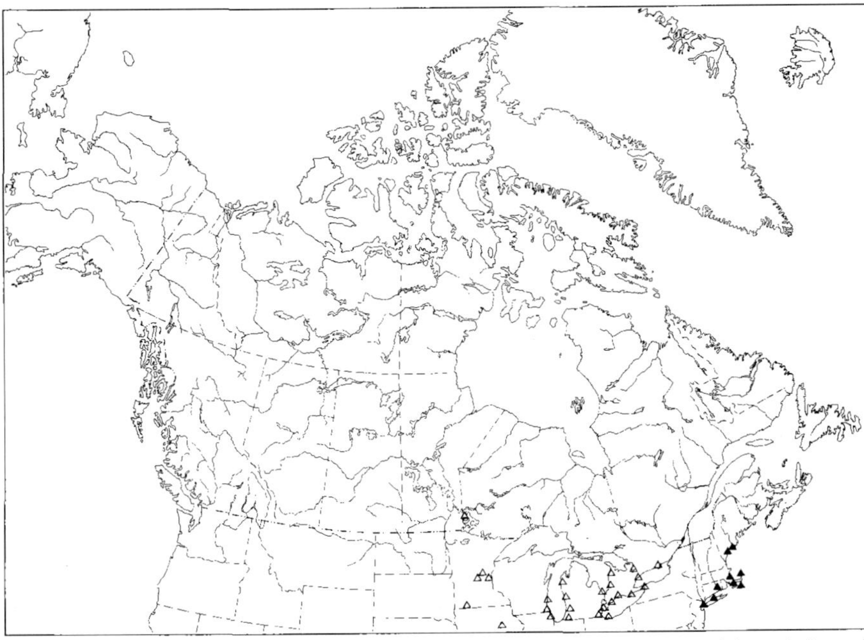

Map 204. Collection localities for *Trimerotropis maritima maritima* (▲) and *T. maritima interior* (△).

with the substrate and is difficult to see when at rest. It is easily disturbed, and flight is prolonged for considerable distances.

The crepitation sounds like a subdued rattling. Otte (1970) recorded the main courting sequence of *T. maritima maritima* from Long Island, New York, as mainly femur-shaking followed by stridulation. This is in contrast to that of *T. maritima interior*, where, in more than half (17) of 28 observations, shaking alone was observed. Aggressive behavior by males involved femur-shaking and "femur-tipping" to a vertical position.

Life history. Univoltine, with overwintering in the egg stage. Morse (1920) reported adults in New England from late July to November, but that the insect was most common in August and September.

Trimerotropis maritima interior E. M. Walker

Figs. 694, 702; Map 204

Trimerotropis maritima interior E. M. Walker, 1898*b*:262.

Trimerotropis maritima; Lugger 1898:258, figs. 101, 102.

Trimerotropis maritima interior; E. M. Walker 1902*a*:2; R. D. Alexander et al. 1972:54.

Diagnosis. Hind wing crossband broad. Found in sandy areas near Great Lakes.

Description. Body size medium (length, male 19–22, female 26–30 mm). Tegmina with distinct traces of 3 bands; hind wing crossband broad, often equal to one-quarter of wing length (Fig. 702).

Range. Manitoba to Ontario, south to Iowa and Michigan.

Behavior and habitats. This subspecies is found on sandy beaches on Lake St. Clair, Lake Erie, Lake Michigan, and other lakes in the area. *T. maritima interior* is easily disturbed, flies for long distances, and is often difficult to capture. It digs and buries itself in sand as dusk approaches and remains buried until the following morning when the sun warms the sand. The method of digging is similar to that described for other locustine grasshoppers such as *Acrotylus junodi* Schulthess in East Africa (Knipper and Kevan 1954), and *A. hirtus* Dirsh in South Africa (Callan 1956). In North America, Tinkham (1959) reported that *Coniana snowi* Caudell, and *Trimerotropis agrestis* McNeill in the southwestern United States, excavate depressions in sand, but he did not observe true self-burial.

Otte (1970) found courtship behavior of males to involve femur-shaking alone in more than half of his observations, rather than shaking followed by stridulation as was the case with *T. maritima maritima*. Other activities appear to be similar in the two subspecies.

Life history. Univoltine, with overwintering as eggs in the soil. Adults are active from mid-July to mid-September. Cantrall (1968) reported earliest and latest dates in Michigan as 9 July and 11 September.

Trimerotropis longicornis E. M. Walker

Map 205

Trimerotropis longicornis E. M. Walker, 1902*a*:4, pl. 1, figs. 1–4.
Trimerotropis longicornis; E. M. Walker 1906*a*:57.

Diagnosis. Tegminal banding distinct; hind wing band spur extending halfway to wing base; hind tibia yellowish to greenish, with pale basal ring. Although this species is similar to *T. campestris* McNeill (McNeill 1900*a*), it differs by its pale yellowish to yellow green color with brown mottling of the hind tibia, whereas the hind tibia of *T. campestris* throughout its range is uniformly bright orange tinged with a pale prebasal ring. The two species occur together in British Columbia.

Description. Body size medium (male, 21 mm). Tegmina distinctly banded, with basal area solid brownish black with well-defined margin followed by a solid median dark band and a narrower somewhat broken outer band; apical third transparent with few scattered brown spots; hind femur grayish brown on outer face, with pale preapical band and crossed by 2 rather obscure darker bands; inner face pale, crossed by 2 black bands; hind tibia pale yellow tending to greenish, with paler basal ring, and with spines black.

Range. Southern British Columbia and, presumably, Washington.

Behavior and habitats. Not known.

Life history. Not known. Specimens have been found in August and September.

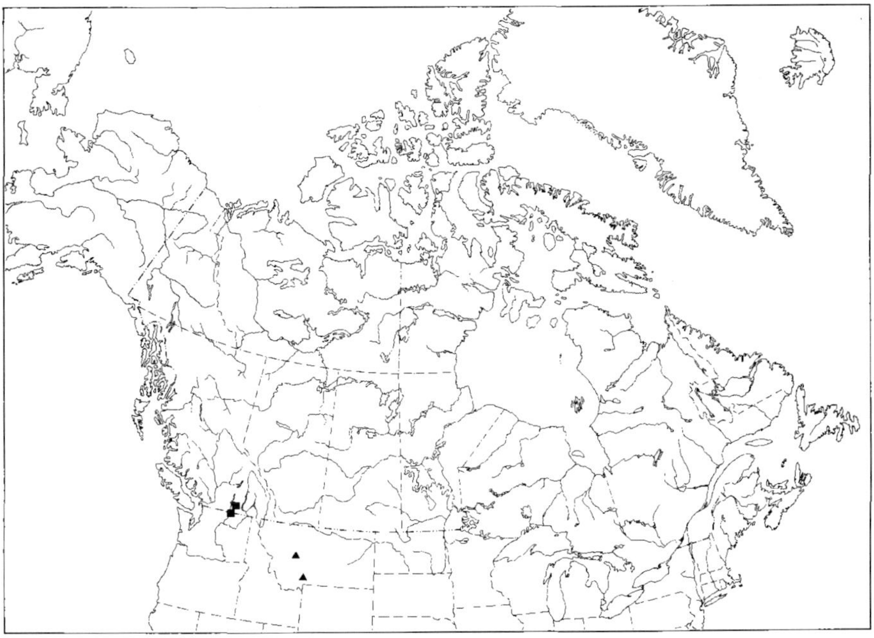

Map 205. Collection localities for *Trimerotropis diversellus* (▲) and *T. longicornis* (■).

Trimerotropis diversellus Hebard

Fig. 699; Map 205

Trimerotropis diversellus Hebard, 1928:255, pl. 25, figs. 2, 3.
Trimerotropis diversellus; Mills and Pepper 1938:16.

Diagnosis. Tegmina banded; hind wing sulfur yellow, with crossband broad, brown; spur of wing short.

Description. Body small (length, male 19.5–20, female 27–33 mm), robust. Vertex with weak median longitudinal carina; lateral foveolae rounded-trapezoidal, distinct; pronotum with median carina bluntly bilobed on prozona, with anterior lobe definitely higher and longer, and with carina low on metazona. Color generally dark grayish brown, occasionally reddish; face grayish white; pronotum gray brown with white fleck on each lateral lobe; tegmina light brown, definitely banded, with dark bands separated by broader bands of grayish white; hind wing disc sulfur yellow to pale greenish yellow; crossband brown, broad in anal region, narrowing at costal margin, with spur short, broad; hind femur brown on outer face, blackish ventrally with pale subapical ring, sometimes with median pale area reaching only to median line; hind tibia buff, tinged with brown, always with gray basal ring.

Range. Montana and Wyoming.

Behavior and habitats. Hebard (1928) found *T. diversellus* to be common around geysers and on geyser deposits in Yellowstone National Park.

He also found the species to be common on bare areas near a railroad at Dailey's, Park County, Montana.

Life history. Probably univoltine, with overwintering in the egg stage. Adults have been collected from 25 July to 29 August (Hebard 1928), but they are probably present later in the season than these dates indicate.

Genus *Derotmema* Scudder

Description. Eyes prominent, bulging. Depression of vertex deep, extended anteriorly, with lateral and median ridges prominent. Dorsal surface of pronotum rough, with longitudinal tubercles and ridges (Fig. 685). Tegmina long, narrow, tapered.

Derotmema haydenii haydenii (Thomas)

Figs. 685, 688; Map 206

Oe[dipoda] haydenii C. Thomas, 1872:460.

Derotmema cupidineum Scudder, 1876*c*:513.

Derotmema haydenii haydenii; J. A. G. Rehn 1919*b*:230; Brooks 1958:39; Helfer 1963:125.

Map 206. Collection localities for *Derotmema haydenii haydenii*.

Diagnosis. Eyes bulging. Dorsal surface of pronotum rough, tuberculate, ridged.

Description. As for genus. Body small (length 20–30 mm). Color gray to brownish gray, speckled with black; tegmina gray or brown, with black on costal and anal areas; hind wing disc yellow or red, with crossband broad, and with spur extending more than halfway to wing base; hind femur gray, mottled on outer face, with 2 dark bands dorsally, brownish and discolored on inner face; hind tibia mottled brownish yellow.

Range. Alberta to Nebraska, south to New Mexico.

Behavior and habitats. Brooks (1958) found *D. haydenii haydenii* only "around margins of wet spots in sandy-alkaline situations, often accompanied by *Trimerotropis pallidipennis salina*." Anderson and Wright (1952) found it in "old prairie dog towns" in Montana. Nymphs were observed feeding upon common purslane (*Portulaca oleracea*) and to a lesser extent on prostrate knotweed (*Polygonum aviculare*). Adults fed upon grass, short dry stubble, and new green growth of western wheat grass (*Agropyron smithii*) only when other plants had become very dry.

Otte (1970) observed femur-tipping and stridulation by males during courtship before they attempted to mount females. During courtship the downstroke of the femur during femur-tipping is much faster than the upstroke. He did not observe flight or crepitation by solitary individuals. Femur-tipping is also employed by males as an aggressive signal, but in such situations is more rapid than in courtship, and the upstroke is much faster than the downstroke. Femur-shaking occurs only when two males touch each other and is accompanied by striking of the substrate with the hind tibiae.

Life history. Anderson and Wright (1952) recorded adults on 31 July in Montana. Tinkham (1939) collected adults in Alberta on 15 September. The life cycle is univoltine, with overwintering in the egg stage in the soil.

Genus *Mestobregma* Scudder

Description. Head large, with vertex tumid, but occiput not strongly elevated as in *Psinidia*, broad between eyes; eyes small, moderately prominent. Pronotum constricted in middle, with metazona little longer than prozona, with oblique rugae and small tubercles; median carina cut twice by sulci, slightly higher anterior to sulcus; lateral carinae prominent on metazona. Tegmina and hind wings surpassing apex of abdomen. Hind femora as long as abdomen.

This genus resembles *Trachyrhachys*, but, in the latter, the posterolateral part of the lateral lobe of the pronotum is angulate and the median carina of the pronotum is distinct between the sulci, whereas in *Mestobregma* the posterolateral angle is rounded and the median carina is obsolete on the central part of the pronotal disc; in addition, in *Mestogregma* the sides of the pronotum are marked with black.

Mestobregma plattei plattei (Thomas)

Fig. 680; Map 207

Oe[dipoda] plattei C. Thomas, 1873:123.
Mestobregma plattei; Scudder 1876*a*:264; Helfer 1963:128.
Mestobregma plattei plattei; J. A. G. Rehn 1919*b*:238.

Diagnosis. Median carina of pronotum cut by two sulci; posteroventral angles of lateral pronotal lobes angulate, rounded.

Description. Vertex broad, open in front. Pronotum with median carina moderately high, cut by two sulci, more prominent anterior to first sulcus; middle lobe short; posterior sulcus at about middle of pronotum; metazona smooth or with scattered granulations (not lateral ridges or elevations); posteroventral angles of lateral lobes weakly angulate, rounded. Color brownish; head with oblique brownish black stripe at lower part of eye; pronotum with oblique brownish black stripe laterally; tegmen with broad dark band on costal margin; hind wing yellow to bright pink with black crossband; hind femur black on inner face and with pale ring near apex.

Range. Montana and South Dakota, south to Colorado.

Behavior and habitats. J. A. G. Rehn and Hebard (1906) reported *M. plattei plattei* as an inhabitant of clifflike outcroppings where it occurred around stunted pines and bushes. Hebard (1928) found it on bare boulder-covered slopes among junipers, and in grasses at the bases of slopes.

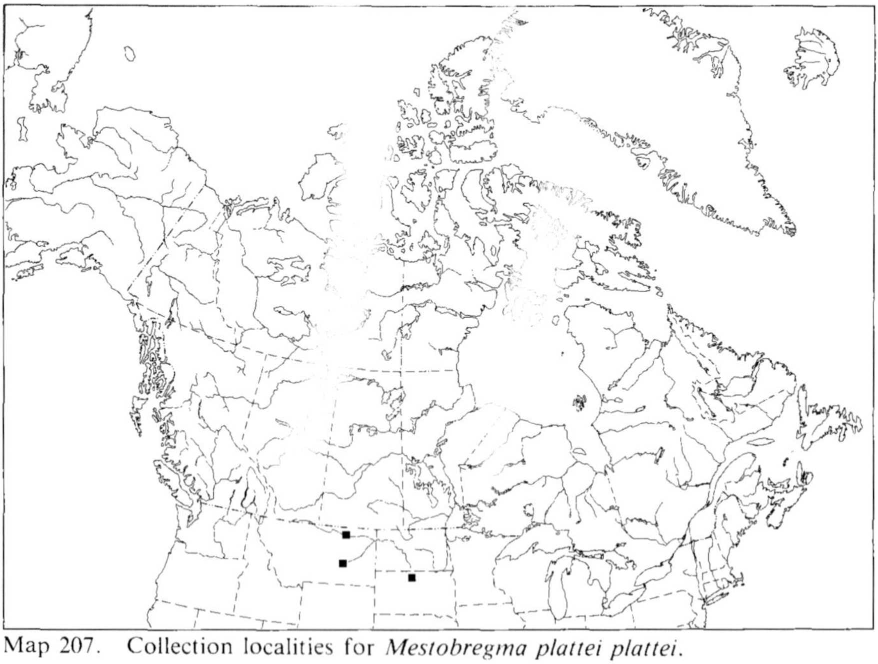

Map 207. Collection localities for *Mestobregma plattei plattei*.

J. A. G. Rehn and Hebard (1906) reported flight as being low and swift. Otte (1970) observed a population, probably of the nominate subspecies, in Colorado. No stridulation on the ground was noted, and no crepitation or flight by solitary individuals occurred. Femur-tipping was the sole component of courtship by males.

Life history. Not known, but probably univoltine, with overwintering in the egg stage. Hebard (1925*a*) reported adults from 25 July to 11 September in South Dakota, and the same author (Hebard 1928) recorded dates in late July and August in Montana.

Genus *Hadrotettix* Scudder

Description. Body large, robust, with conspicuous dark bands on tegmina. Pronotum finely rugose, with median carina nearly obsolete, cut by two sulci, the anterior one shallow, the posterior one deep. Hind wing crossband beyond center of wing, extending along anal margin. Hind femur with both dorsal and ventral margins strongly keeled.

Hadrotettix trifasciatus (Say)

Figs. 716, 717; Map 208

Gryllus trifasciatus Say, 1825*b*:114, pl. 34, bottom figure.

Hadrotettix trifasciatus; J. A. G. Rehn and Hebard 1906:397; Brooks 1958:40; Helfer 1963:123.

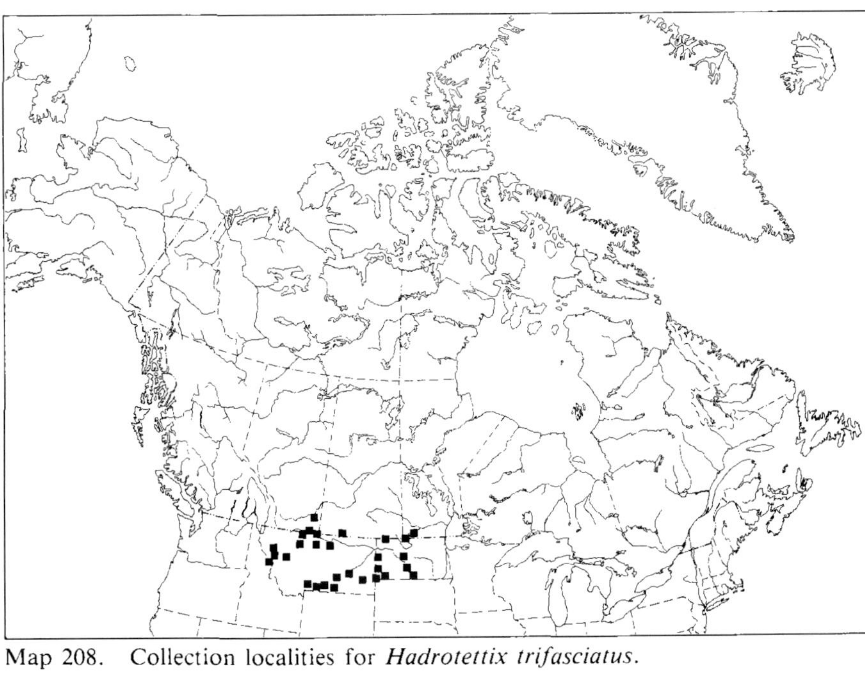

Map 208. Collection localities for *Hadrotettix trifasciatus*.

Diagnosis. Tegmina conspicuously banded; hind wing disc yellow, with crossband on apical half dark, broad, nearly reaching anal margin.

Description. As for genus (Fig. 717). Head large, with antennae long and heavy. Prozona of pronotum dorsally rounded in lateral view (Fig. 716). Color reddish brown to bluish black; vertex of head and dorsum of pronotum usually dark; tegmina reddish brown, dark gray to black with clearly defined complete crossbars; apices of tegmina clear; hind wing disc yellow, with crossband on apical half broad, complete, nearly reaching anal margin, with spur absent and apex clear; hind femur pale on outer face, with prominent dark oblique band on apical side of middle; inner face of femur deep blue to blue black on basal two-thirds, with subapical band pale and with genicular area dark; hind tibia red to orange, with outer face somewhat yellowish to brownish at base.

Range. Alberta to Manitoba, south to Mexico.

Behavior and habitats. Brooks (1958) reported *H. trifasciatus* from gravelly hillsides. It appears to be an exclusive forb feeder, eating various broad-leaved plants (N. Criddle 1935; Anderson and Wright 1952; Brooks 1958), although Hewitt and Barr (1967) stated that it "prefers sparse grass on rocky soil." They also mentioned that flight is weak, extending only for short distances.

Otte (1970) recorded "ordinary" stridulation during courtship, sometimes preceded by femur-tipping. He did not observe crepitation. Male aggressive signals included femur-tipping, femur-shaking, and striking of the substrate with the hind tibiae. Nonreceptive females repelled advancing males by hopping away, holding their hind tibiae horizontally over the back, femur-tipping, and kicking with the hind tibiae.

Life history. Univoltine, with overwintering in eggs in soil. Adults are active from mid-July to mid-September or even later.

Genus *Psinidia* Stål

Description. Body small, slender, compressed. Antennae swollen, depressed. Head with occiput distinctly elevated; disc of vertex short, strongly concave, with frontal costa sulcate throughout, the upper half narrow, feebly expanded below; pronotum strongly constricted in middle, with hind margin acute-angled; median carina sharp, straight, cut before middle by two sulci. Tegmina narrow, long, with many of the cells of anterior part of middle third, two to four times as long as wide.

Psinidia fenestralis fenestralis (Audinet-Serville)

Fig. 679; Map 209

Oedipoda fenestralis Audinet-Serville, 1838:726.
Locusta eucerata T. W. Harris, 1841:145.
Psinidia fenestralis; Stål 1873*b*:133; Helfer 1963:124.

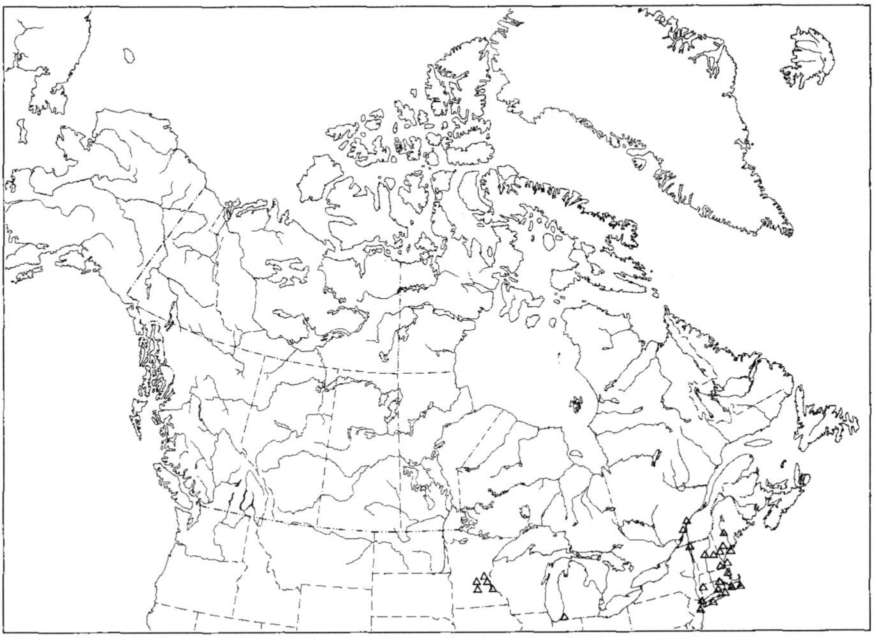

Map 209. Collection localities for *Psinidia fenestralis fenestralis*.

Diagnosis. Occiput of head elevated; head with narrow yellow stripe from eye to pronotal disc. Hind wing disc orange, red, or yellow, with broad black band nearly reaching anal margin; apex clear.

Description. As for genus (Fig. 679). Body small (length, male 15–18, female 19–25 mm). Color varying from pale yellow through reddish brown to blackish; head with narrow yellowish stripe extending from eye to disc of pronotum; pronotum with lateral fuscous bar; tegmina dull yellow to reddish; lower half of lateral area with many alternating light and dark spots, with upper half unspotted; hind wing with basal third orange to red or yellow, with middle third having broad curved black band that nearly reaches anal angle; apical third of wing pellucid; hind femur reddish brown with poorly defined darker bars; hind tibia greenish yellow with distal and basal dark rings and a broad fuscous ring about the middle, with spines black-tipped.

Range. Minnesota to Maine, south to Texas and Florida. Québec localities shown in Map 109 refer to early collections.

Behavior and habitats. *P. fenestralis fenestralis* is an inhabitant of sandy situations, particularly in Atlantic coastal areas, but also in isolated colonies in similar places well removed from the sea, near inland lakes and rivers, as well as in sandy areas far from water. Otte (1970) reported finding individuals of this species always on sand and usually in spots with the least cover. No report regarding food plants has been published, but the species must utilize economically unimportant plants that grow in the very sandy locations where it occurs.

Otte (1970) also reported lack of flight display and crepitation by this species. During courtship, males stridulate on the ground as they approach females and as they attempt to mount them. Stridulation is produced by single strokes of the femora. Femur-tipping is also used in courtship and this, femur-shaking, and striking the substrate with the hind tibiae are employed as aggressive signals between males. Females do not exhibit femur-shaking, but unreceptive females repel males by femur-tipping and kicking.

Life history. Univoltine, with overwintering in the egg stage. Cantrall (1968) recorded adults from 31 August to 9 September in Berrien County, Michigan. Males were recorded from Québec in August (Vickery et al. 1974).

Genus *Circotettix* Scudder

Description. Body moderately large. Metazona twice as long as prozona. Hind wing with outer radial veins enlarged, thickened but not darkened, with wing margin having noticeable lobe; wing usually transparent yellowish to slightly greenish, with crossband narrow, broken.

Key to subspecies of *Circotettix*

(Adapted from Hewitt and Barr (1967))

1. Hind wing with disc yellow; crossband present but fragmented; posterior tibia yellowish. Found in Manitoba westward to British Columbia and Idaho .... ***rabula rabula* Rehn & Hebard** (p. 522)
   Hind wing with disc greenish to bluish; crossband absent; posterior tibia blue or green. Found in British Columbia southward .... ***undulatus undulatus* (Thomas)** (p. 524)

Clé des sous-espèces du *Circotettix*

(adaptée de Hewitt et Barr, 1967)

1. Ailes postérieures à disque jaune; bande transversale présente mais fragmentée; tibias postérieurs jaunâtres. Insecte vivant au Manitoba et plus à l'ouest, jusqu'en Colombie-Britannique et en Idaho .... ***rabula rabula* Rehn & Hebard** (p. 522)
   Ailes postérieures à disque verdâtre à bleuâtre; bande transversale absente; tibias postérieurs bleus ou verts. Insecte vivant en Colombie-Britannique et plus au sud .... ***undulatus undulatus* (Thomas)** (p. 524)

Circotettix rabula rabula Rehn & Hebard

Figs. 710, 711; Map 210

Circotettix rabula J. A. G. Rehn and Hebard, 1906:393, 396, figs. 13, 14.
Circotettix rabula; Buckell 1922:26; Helfer 1963:146 (partim).
Circotettix rabula rabula; Hebard 1928:262; Brooks 1958:37.

Diagnosis. Anal veins of hind wing enlarged. Wing disc yellow with fragmented dark crossband.

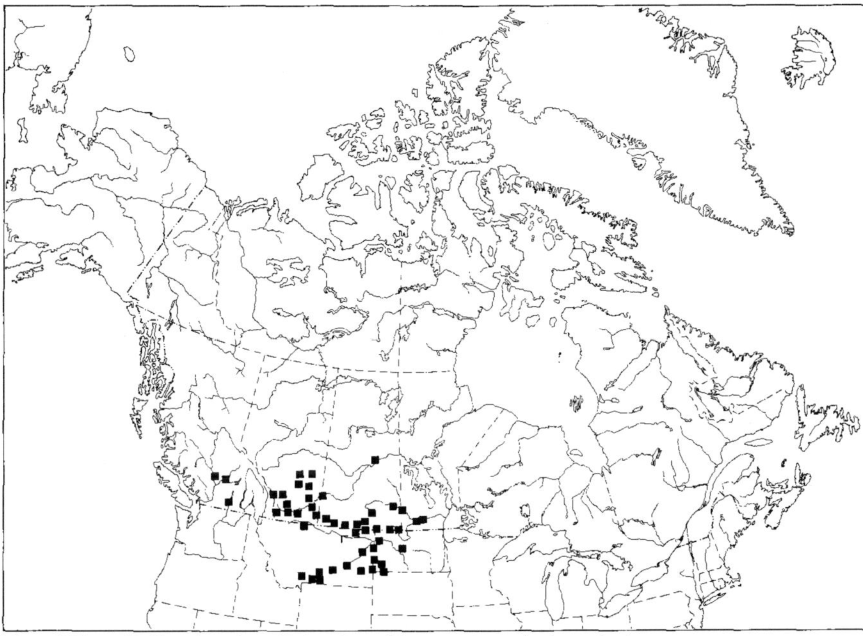

Map 210. Collection localities for *Circotettix rabula rabula*.

Description. Body large (length 34–39 mm). Anal veins of hind wing enlarged only medially; secondary anal veins of first anal pleat nearly equal, closed apically (Fig. 711). Color brown, varying to reddish or light gray with dark speckling (in southern Saskatchewan often pink or purplish); tegmina darker basally, usually with dark spots; hind wing disc (Fig. 711) yellow to faint yellow, with crossband usually broken, in some populations nearly complete (Qu'Appelle Valley, Saskatchewan, and Manitoba); apex of wing clear; hind femur pale brown on outer face, or darker brown with dark longitudinal stripe; inner face of femur dark or banded; hind tibia straw yellow. Chromosome number, $2n$ male = 21.

Range. Interior British Columbia to Manitoba, south to Arizona.

Behavior and habitats. *C. rabula rabula* prefers rough terrain, steep bare south-facing hillsides, and eroded areas (Brooks 1958). Hewitt and Barr (1967) confirm this, but state that it is usually found near heavy vegetation. It is a mixed feeder, preferring forbs; it often feeds on weeds.

Flight is quick and erratic, but *C. rabula rabula* may also hover for as long as 5 or 6 minutes at about 3–5 m above the ground. Crepitation is very loud on both short and hovering flights. Brooks (1958) stated that the sharp loud crackling noise may be heard at a distance of 400 m, or more. Otte (1970) remarked that *C. rabula rabula* produced the loudest crepitation of any of the many species he had studied. Crepitation is not continuous but is broken into bursts of variable duration. During courtship, males stridulate on the ground, using only one femur at a time. Femur-tipping is not associated with courtship but is used extensively as an aggressive signal in encounters with other males. Femur-shaking and striking the substrate with the hind tibiae were also observed, but they were less frequent than femur-tipping. Unreceptive females repelled males by femur-raising and kicking (Otte 1970).

Life history. Univoltine, with overwintering in the egg stage in soil. Adults have been reported in July and August in Idaho (Hewitt and Barr 1967). Tinkham (1939) collected *C. rabula rabula* in July and September in Alberta.

Circotettix undulatus undulatus (Thomas)

Map 211

Oe[*dipoda*] *undulata* C. Thomas, 1872:460.
Circotettix lobatus Saussure, 1888:65, pl. 2, fig. 5.
Circotettix lapidicolus Bruner, 1890:75.
Circotettix undulatus; E. M. Walker in Fletcher and Gibson 1909:113; Brooks 1958:37; Helfer 1963:147.

Diagnosis. Anal veins of hind wing enlarged; wing disc greenish to bluish, without dark crossband. Found in British Columbia southward.

Description. As for genus. Body large (length 35–43 mm). Anal veins of hind wing enlarged throughout, with secondary anal veins of first anal pleat indistinctly closed at apex, not equal; anterior vein more prominent.

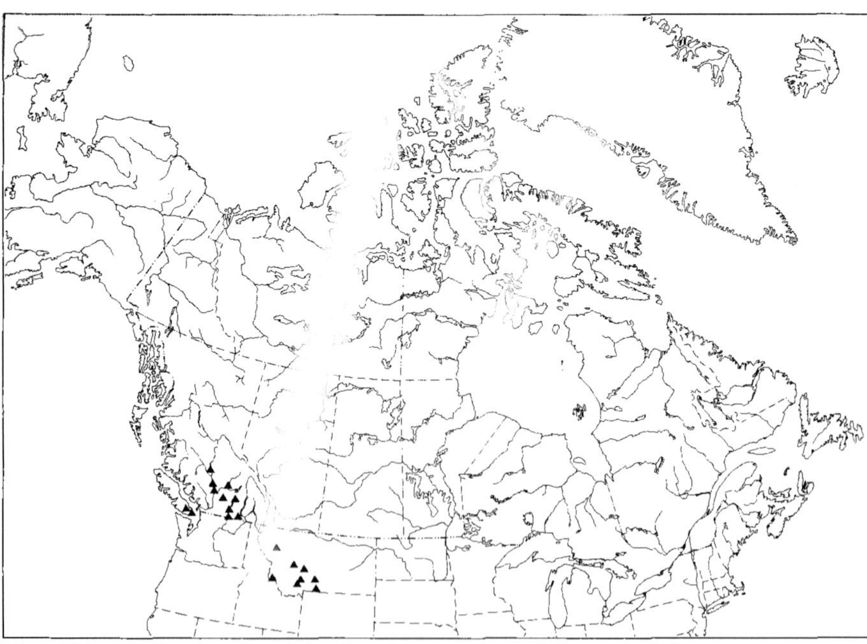

Map 211. Collection localities for *Circotettix undulatus undulatus*.

Color light to dark brown; head paler than thorax and tegmina; tegmina darker at base with dark spots; hind wing disc bluish to greenish, occasionally with yellowish tinge; crossband absent, sometimes faintly indicated at margins; apex of wing clear; hind femur brown with faint darker bands on outer face; inner face of femur with 2 dark bands; hind tibia greenish or blue with pale subbasal ring.

Range. Interior British Columbia and Montana, south to Utah.

Behavior and habitats. Buckell (1920) stated that *C. undulatus* is a common species around rock slides, cliffs, and rocky ravines in the interior dry belt of British Columbia. Hewitt and Barr (1967) reported it from roadsides, irregular terrain, and rocky areas. It is a mixed feeder but prefers forbs and is not considered to be a pest (actual or potential).

Scudder (1898*e*) said of this species that "it is very noisy in flight and may be heard fifty rods [ca. 240 m] away." Buckell (1920) described crepitation as "a loud crackling and snapping sound." Hewitt and Barr (1967) reported flight to be rapid, a series of looping arches. Buckell (1920) described stridulation during courtship. Males "dance up and down in the air, producing five or six sharp clicks followed by a shrill rattling sound, very similar to the noise made by a rattlesnake."

Life history. Univoltine, with adults appearing in mid-July, August, and September.

Tribe Epacromiini

The single North American genus placed in this tribe, namely *Stethophyma* Fischer, is distinctly more "gomphocerinae" in general appearance than any of the Locustini (Oedipodini), so that there has always been, on the part of North American authors, a general reluctance to place it in the Locustinae (= Oedipodinae) rather than in the Gomphocerinae. This has also been largely true in the Old World, where the genus also occurs, but where there are other epacromiine genera, which makes acceptance of the position easier. Among the most recent authors to relocate *Stethophyma* in the Gomphocerinae is Otte (1979*b*), who tentatively invokes "loss" of stridulatory teeth on the inner face of the hind femur to account for their absence. This, he observed, is the situation with some species of the genera *Orphulella* (Otte 1979*a*) and *Achurum*, but the cases are not similar: the relevant species of these genera have not retained their stridulatory powers nor compensated for missing teeth on the femoral ridge by developing denticulations on the tegminal veins (or, indeed, any comparable stridulatory mechanism). Otte also says that "*Stethophyma* is very similar to gomphocerines in behavior (Otte 1970)," but his reference is not documented.

Genus *Stethophyma* Fischer

Description. Frons somewhat slanting (resembling the Gomphocerinae in this respect). Lateral foveolae small, sometimes almost obsolete; frontal costa wide, sulcate, divergent below median ocellus. Pronotum short, with median carina sharp but low, only superficially cut by principal sulcus; lateral carinae entire. Tegmina long, with open venation. Male subgenital plate acutely produced, twice as long as wide. Ovipositor valves strongly exserted, with upper pair possessing minute marginal teeth. Stridulatory mechanism tegmino-femoral, with roughened femoral ridge (Plate VI*A*, *B*).

Key to species of *Stethophyma*

1. Pronotum with lateral carinae nearly parallel, less than 1.5 times as wide posteriorly as anteriorly, with prozona and metazona nearly equal in length .... ***celatum*** **Otte** (p. 526)

Pronotum with lateral carinae distinctly diverging on metazona, more than 1.5 times as wide posteriorly as anteriorly, with prozona distinctly shorter than metazona 2

2(1). Tegmen with conspicuous pale submarginal longitudinal stripe on basal half to three-quarters. Lateral carinae of pronotum cut by 3 sulci ahead of middle .... ***lineatum*** **(Scudder)** (p. 527)

Tegmen without longitudinal pale stripe. Lateral carinae of pronotum cut only by first and third sulci .... ***gracile*** **(Scudder)** (p. 529)

Clé des espèces de Stethophyma

1. Pronotum à crêtes latérales presque parallèles, séparées postérieurement par une distance égale à moins d'une fois et demie la distance qui les sépare antérieurement; zone antérieure et zone postérieure de longueur presque égale .... ***celatum*** **Otte** (p. 526)

Pronotum à crêtes latérales clairement divergentes au niveau de la zone postérieure, la distance les séparant postérieurement égale à plus d'une fois et demie celle qui les sépare antérieurement; zone antérieure clairement plus courte que la zone postérieure 2

2(1). Tegmina marqués d'une bande longitudinale submarginale pâle sur une portion variant de la moitié aux trois quarts de leur longueur à partir de la base. Crêtes latérales du pronotum coupées par trois sulcus sur la moitié antérieure .... ***lineatum*** **(Scudder)** (p. 527)

Tegmina sans bande pâle longitudinale. Crêtes latérales du pronotum coupées seulement par le premier et le troisième sulcus .... ***gracile*** **(Scudder)** (p. 529)

Stethophyma celatum Otte

Map 212

Stethophyma celata [*sic*] Otte, 1979*b*:241.
Stethophyma celata [*sic*]; Otte 1979*c*:158.

Diagnosis. Top third of lateral area of tegmen dark brown to black.

Description. Body large (length, male 26–29, female 39–47 mm). Top third of lateral area of tegmen dark brown to black; ventral margin of hind femur yellow, rather than red, in western part of range; spines of hind tibia yellow with black tips.

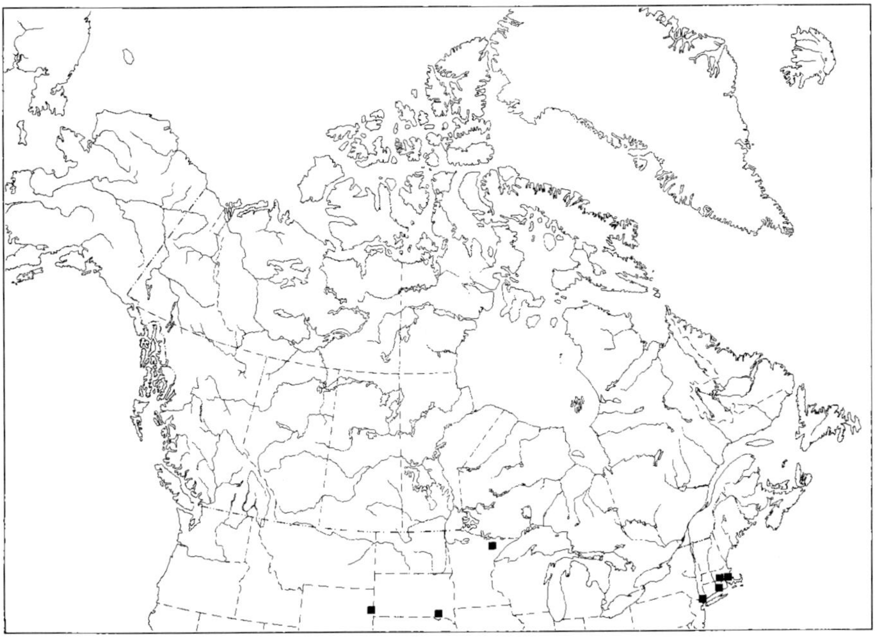

Map 212. Collection localities for *Stethophyma celatum*.

Range. Wyoming to Minnesota; Massachusetts, south to South Carolina.

Behavior and habitats. *S. celatum* is found in tall grasses, moist swales, marshes, and tamarack (*Larix*) swamps. Morse (1920) reported it (as *Mecostethus platypterus*) to be "a less shy and active species than *lineatus*"; females are (like other species) sluggish and seldom fly. He also stated that this is a rare species, found only after persistent searching.

Life history. Presumably univoltine, with overwintering in the egg stage. Morse (1920) reported adults as being taken on 4 and 25 August, but Blatchley (1920) recorded adults from 22 June to 25 August.

Stethophyma lineatum (Scudder)

Fig. 718; Plate VI*A*; Map 213

Arcyptera lineata Scudder, 1863*a*:462.

Stethophyma lineatum; Hebard 1925*a*:65; Brooks 1958:43; Helfer 1963:160; R. D. Alexander et al. 1972:53; Vickery et al. 1974:119.

Diagnosis. Body with pale yellow line extending from eye to scapular area of tegmen, generally olive green. Pronotum with lateral carinae cut by 3 sulci.

Description. Body large (length, male 26, female 36 mm). Foveolae shallow but visible from above. Pronotum with 3 distinct carinae; median

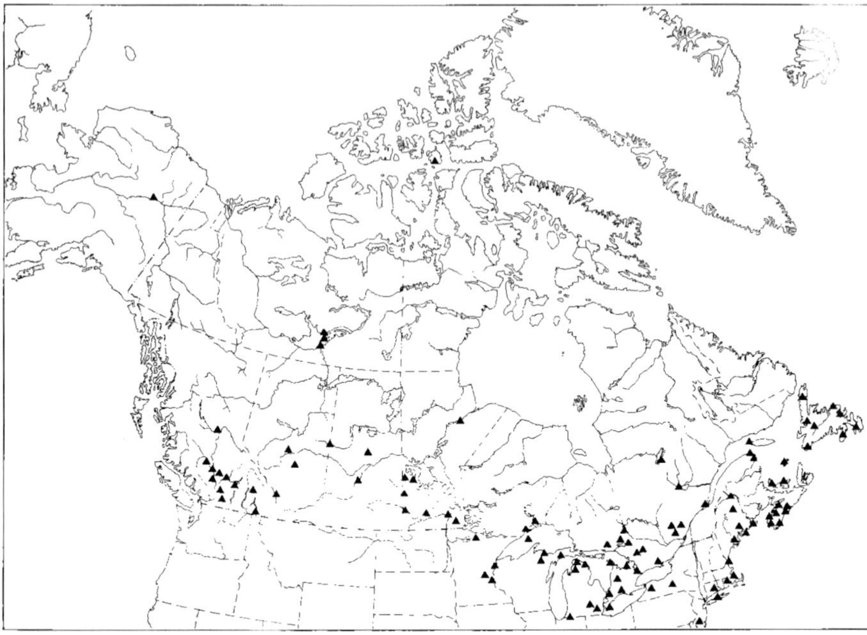

Map 213. Collection localities for *Stethophyma lineatum*.

carina rather high, sharp, cut anterior to middle by principal sulcus; lateral carinae distinctly divergent posteriorly. Tegmina and hind wings long, surpassing apices of hind femora. Color generally olive greenish to purplish brown with yellow markings, green beneath; pale yellow line present from eye across pronotum and continued on to scapular area of tegmen; hind femur olive on outer face, blackish near apex, with 3 whitish spots in middle; upper edge of femur brownish and lower edge maroon with 2 black marks separated by yellow near apex; hind tibia buffy, darker near base and apex, with spines entirely black. Chromosome number, $2n$ male = 23.

Range. Alaska to Newfoundland, south to Washington and New Jersey; adventive in Arctic archipelago.

Behavior and habitats. *S. lineatum* prefers areas such as wet bogs and meadows, and tamarack (*Larix*) swamps. Cool moist situations appear to be preferred, but the species is also found in warmer bogs and in meadows, often in tall grass. Blatchley (1920) commented that "males seem to far outnumber the females," but this is a reflection of the ease with which males are disturbed, in contrast to the sedentary behavior of females. Males are quick to fly, and in flight are swift and noiseless, the insects often alighting more than 30 m from their point of departure. Females may not move at all even when a searcher nearly steps on them. Nevertheless, they fly as well as the males once they are airborne, but, being bulky, they are unable to take off with the same rapidity as can males. We have collected females by observing males from a distance and then searching the tall grass near where one or more have alighted. *S. lineatum* is a sedge feeder and hence not a potential pest.

There is little published information on sound-production by this species, but, in general terms it appears, when it occurs, to be similar to that given for the Palaearctic *S. grossum* (Linnaeus).

Life history. Univoltine, so far as is known. Adults are active from late July until near the end of October (Vickery 1961). Cantrall (1968) listed early and late dates as 30 June and 1 October in Michigan.

Stethophyma gracile (Scudder)

Plate VI*B*; Map 214

Arcyptera gracilis Scudder, 1862:286.

Arcyptera platyptera Scudder, 1863*a*:463.

Stethophyma gracile; Hebard 1925*a*:64; Cantrall 1943:95; Brooks 1958:43; Helfer 1963:160; R. D. Alexander et al. 1972:53; Vickery et al. 1974:120.

Diagnosis. Body olive green, without pale stripes. Pronotum with lateral carinae cut by first and third sulci only.

Description. Body size medium (length, male 19–23, female 26.5–31 mm). Pronotum with lateral carinae cut by first and third sulci only.

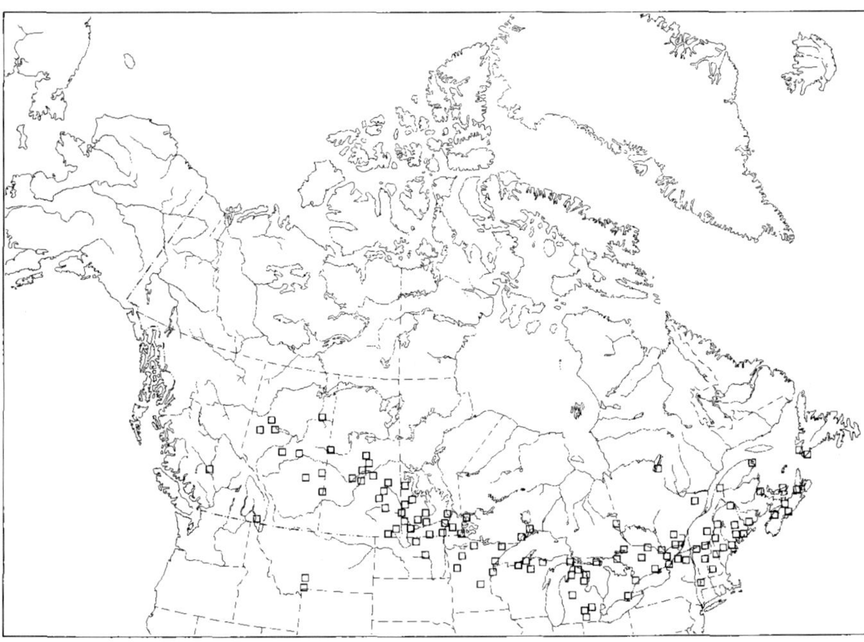

Map 214. Collection localities for *Stethophyma gracile*.

Tegmen lacking pale yellow line; ventral margin of hind femur bright red; hind tibia greenish yellow with darker band on apical third, with spines entirely black.

Range. British Columbia to Newfoundland, south to Colorado and New England.

Behavior and habitats. *S. gracile* feeds upon sedges. It is seldom found far from meadows, swamps, and ponds. Cantrall (1943) refers to its shy behavior. He described flight as rapid, for distances up to 12 m, considerably shorter than flights of *S. lineatum*. When disturbed, the insects may hide by sidling around stems, or, more often, they drop and burrow into heavy vegetation and thus avoid capture. As with *S. lineatum*, females seldom fly but remain low in vegetation and, if alarmed, may remain motionless for long periods of time.

The stridulation of *S. gracile* is loud and clear, more noticeable than that of any other acridid that may be present in the vicinity. The rasping phrases resemble zeeeek! – – – zeeeek! – – – zeeeek! – – – zeeeek!, the first softer than the following phrases and preceding the others by a half-second pause (Cantrall 1943). Females have not been heard to stridulate.

Life history. As for *S. lineatum*. Adults have been reported in Michigan from 18 July until 12 October (Cantrall 1968).

Subfamily Gomphocerinae

Members of this subfamily, the great majority of which are of comparatively small size, might justifiably be called the true grasshoppers, for, although some feed on forbs and other forms of vegetation, they are largely graminivorous and are prominent in the large grassland regions of the Northern Hemisphere. Further, when the word "grasshopper" is used in English (in other than scientific literature), it is usually associated with this rather than with any other subfamily. A common (although not universal) feature in the males is an expanded anterior region of the tegmen with enlarged oblique cells. These cells assist in amplifying the sound that is produced by striking the sharp median and radial veins of the tegmen with the row of pegs on the inner face of the hind femur. All species do not stridulate in this way, however, for some species of *Orphulella*, for example, are virtually mute, the row of stridulatory pegs being reduced. Females may sometimes stridulate but not commonly, usually only in response to a male and (with exceptions) not loudly enough to be readily audible to the human ear.

The songs of grasshoppers of this subfamily have inspired many literary men and women, particularly among poets, from time immemorial and in many parts of the world. Brief excerpts from the works of two nineteenth century Canadian poets, Charles Mair and Charles George Douglas Roberts, must suffice by way of example:

From: *Alice*

For I long for the wild bee's hum,
And the grasshopper's chirp in the grass.

(Mair 1868)

From: *The Mowing*

Or brazen grasshopper with triumphing note
From the long swath proclaimed the fate that smote
The clover and timothy—tops and meadowsweet.
(C. G. D. Roberts 1893)

The subfamily Gomphocerinae is a large one, including at least 100 genera from all parts of the World, except Australasia. The number of species described is very great, but there is no accurate estimate of the total. Jago (1971) made a preliminary attempt to determine which of the world's genera of Acrididae should or should not be assigned to the Gomphocerinae as understood by him, and, despite its few minor inaccuracies, this is a valuable piece of work. Dirsh (1975) split the Gomphocerinae, as previously recognized, into several subfamilies, two of which are of present concern. We do not, however, believe that the differences between them are sufficiently fundamental or constant to warrant their recognition above tribal level. Otte (1979*a*) has recently defined the new tribe Orphulellini.

Key to tribes of Gomphocerinae

1. Males with fore and middle femora somewhat enlarged. Fastigium broader than long, with lateral foveolae (if any) not visible from above (Fig. 745). Antennae filiform or slightly ensiform, not clavate. Lateral pronotal carinae strong, usually complete, rather straight or gradually curved inward at middle .... **Orphulellini** (p. 532)
   Males with fore and middle femora not enlarged. Body lacking above combination of characters 2

2(1). Fastigium without distinguishable foveolae (except very shallow in *Opeia*). Antennae ensiform or distinctly flattened, at least in part (only slightly flattened basally in *Chloealtis*), never clavate. Prosternum usually with small tubercle or swelling .... **Chrysochraontini** (p. 539)
   Fastigium with distinct lateral foveolae, visible from above (Fig. 744) or not. Antennae filiform (clavate apically in *Aeropedellus*), usually slightly flattened near base only. Prosternum lacking tubercle (except in some *Aeropedellus* species that have a small tubercle, but which can be distinguished by their clavate antennae) ... **Gomphocerini** (p. 552)

Clé des tribus de Gomphocerinae

1. Chez les mâles, fémurs antérieurs et médians quelque peu développés. Fastigium plus large que long, à fovéoles latérales (lorsque présentes) non visibles du dessus (fig. 745). Antennes filiformes ou légèrement ensiformes, mais non claviformes. Crêtes latérales du pronotum développées, habituellement complètes, plutôt rectilignes ou graduellement incurvées vers l'intérieur au milieu .... **Orphulellini** (p. 532)
   Chez les mâles, fémurs antérieurs et médians normaux. Corps ne présentant pas la combinaison de caractères décrite ci-dessus 2

2(1). Fastigium dépourvu de fovéoles visibles (sauf chez les *Opeia*, où l'on observe des fovéoles très peu profondes). Antennes ensiformes ou clairement aplaties, au moins en partie (aplatissement très léger de la base chez les *Chloealtis*), jamais claviformes. Prosternum habituellement pourvu de petits tubercules ou de petits renflements .... **Chrysochraontini** (p. 539)

Fastigium à fovéoles latérales distinctes, parfois mais pas toujours visibles du dessus (fig. 744). Antennes filiformes (claviformes à l'extrémité apicale chez les *Aeropedellus*), habituellement légèrement aplaties près de la base. Prosternum dépourvu de tubercule (sauf chez certaines espèces d'*Aeropedellus* qui possèdent un petit tubercule, mais qui se caractérisent aussi par leurs antennes claviformes) .... **Gomphocerini** (p. 552)

Tribe Orphulellini

Key to genera of Orphulellini

1. Fastigium of vertex virtually without lateral foveolae. Disc of pronotum parallel-sided or slightly divergent posteriorly, lacking posterolateral marks .... *Dichromorpha* (p. 532)

Fastigium of vertex with lateral foveolae visible from sides only. Disc of pronotum more or less constricted at center, usually with posterolateral marks .... *Orphulella* (p. 534)

Clé des genres d'Orphulellini

1. Fastigium du vertex pratiquement dépourvu de fovéoles latérales. Disque du pronotum à bords parallèles ou légèrement divergents postérieurement, dépourvu de marques postérolatérales .... *Dichromorpha* (p. 532)

Fastigium du vertex muni de fovéoles latérales visibles du côté seulement. Disque du pronotum plus ou moins rétréci au centre, presentant habituellement des marques postérolatérales ... *Orphulella* (p. 534)

Genus *Dichromorpha* Morse

Description. Antennae ensiform, slightly flattened at base. Vertex of head short, broader than long, with margins strongly elevated; depression semicircular; median carinula and lateral foveolae virtually absent. Pronotum with disc flat; all 3 carinae distinct, parallel, straight, cut behind middle by principal sulcus. Tegmina and hind wings short, expanded on costal margin, usually not extending beyond middle of abdomen (although a macropterous form occurs in which tegmina and wings are twice as long as in brachypterous form; this mainly in females in southern New England; macropterous males rare). Fore and middle femora of male somewhat enlarged. Internal apical spurs of hind tibia about equal.

Dichromorpha viridis (Scudder)

Fig. 790; Map 215

Chloealtis viridis Scudder, 1863*a*:455.
Chloealtis punctulata Scudder, 1863*a*:455 (macropterous).

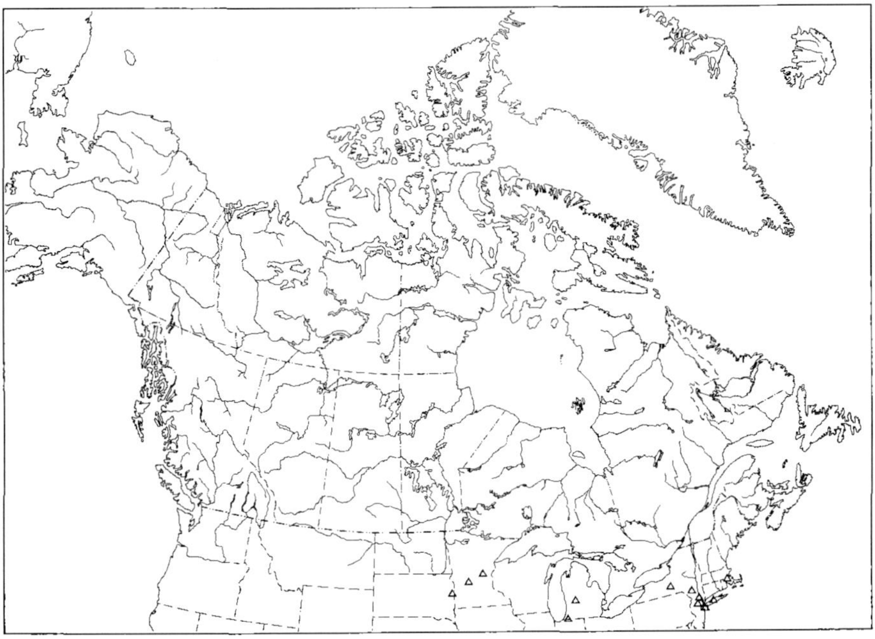

Map 215. Collection localities for *Dichromorpha viridis*.

Dichromorpha viridis; Morse 1896*a*:383; Helfer 1963:178; R. D. Alexander et al. 1972:52; Otte 1979*a*:76.

Diagnosis. Pronotal disc flat; all 3 carinae prominent, parallel. Tegmina short. Color uniform grass green or brown dorsally.

Description. As for genus *Dichromorpha* (Fig. 790). Color dimorphic; uniform grass green or brown dorsally, or green dorsally and brown laterally on head, pronotum, and lateral areas of tegmina (at rest); grayish, yellowish green, or brown ventrally.

Range. South Dakota to southern New England, south to Mexico.

Behavior and habitats. This species is common in rich pastures and fields on heavy soils, particularly in moist areas such as meadows and along margins of brooks and ponds. It prefers deep luxuriant grasses (Morse 1920). Blatchley (1920) noted that females of the brown color phase are more common in the fall, when they blend with fallen leaves. *D. viridis* is a sluggish insect, remaining motionless or hopping away, rather than using its wings, even in the macropterous phase.

Otte (1970) reported that males make no courtship signals. Males approach females stealthily and mount them before the females become aware of their presence. Stridulation takes the form of raising of the femora; kicking motions are used in aggressive encounters. The male of a copulating pair was observed to regurgitate a drop of brown fluid when the pair was molested by another male. The head of the copulating male was pointed toward the offending male until he departed, after which the fluid droplet was withdrawn. This is the only known instance of regurgitation of this fluid in interspecific interaction, although most grasshoppers regurgitate brown fluid when handled.

Life history. The species is univoltine, with overwintering in the egg stage. Adults are active from June to mid-September. Cantrall (1968) reported adults from June to 9 September in Michigan; Froeschner (1954) recorded them from 24 June to 16 September in Iowa; Morse (1920) reported adults in southern New England from the latter half of July until 23 October.

Genus *Orphulella* Giglio-Tos

Description. Antennae slender, filiform. Frons slanted; fastigium of vertex with foveolae narrow, shallow, not visible from above (Fig. 746); depression of vertex without median carinula. Pronotum with strong median carina, cut by one sulcus; lateral carinae evident throughout, slightly to definitely constricted; posterior margin obtuse; prosternum without tubercle. Tegmina slender, extending to apices of hind femora. Males with fore and middle femora somewhat enlarged; male hind femur with stridulatory pegs weak. Inner apical spurs of hind tibia nearly equal. All the species in our region are univoltine, with overwintering as eggs in the soil.

Key to species and subspecies of *Orphulella*

1. Lateral carinae of pronotum evenly curved, not constricted (Fig. 793); posterior margin of pronotum not much broader than anterior margin. Found in Manitoba and North Dakota eastward .... ***speciosa* (Scudder)** (p. 535)
   Lateral carinae of pronotum somewhat constricted before sulcus (Fig. 794); pronotum broader at posterior than at anterior margin. Found transcontinentally (*pelidna, sensu lato*) 2

2(1). Tegmina usually extending beyond apices of hind femora. Lateral carinae of pronotum rather strongly incurved before sulcus (Fig. 794). Found from Atlantic to foothills of Rocky Mountains .... ***pelidna pelidna* (Burmeister)** (p. 537)
   Tegmina shorter, extending only to apices of hind femora. Lateral carinae of pronotum not strongly incurved before sulcus (Fig. 795). Found in British Columbia, Washington, Idaho, and western Montana .... ***pelidna deserta* Scudder** (p. 538)

Clé des espèces et des sous-espèces d'*Orphulella*

1. Crêtes latérales du pronotum décrivant une courbe régulière, sans constriction (fig. 793); bord postérieur du pronotum pas beaucoup plus large que le bord antérieur. Insecte vivant au Manitoba et au Dakota-nord ainsi que plus à l'est .... ***speciosa* (Scudder)** (p. 535)
   Crêtes latérales du pronotum présentant une constriction en avant du sulcus (fig. 794); pronotum plus large au bord postérieur qu'au bord antérieur. Distribution transcontinentale (*pelidna, sensu lato*) 2

2(1). Tegmina dépassant habituellement l'apex des fémurs postérieurs. Crêtes latérales du pronotum fortement incurvées avant le sulcus (fig. 794). Espèce vivant de la côte de l'Atlantique jusqu'aux contreforts des Rocheuses .... ***pelidna pelidna* (Burmeister)** (p. 537)

Tegmina plus courts, rejoignant l'apex des fémurs postérieurs. Crêtes latérales du pronotum non fortement incurvées avant le sulcus (fig. 795). Espèce vivant en Colombie-Britannique, dans l'État de Washington, en Idaho et dans l'ouest du Montana .... ***pelidna desereta* Scudder** (p. 538)

Orphulella speciosa (Scudder)

Figs. 745, 791-793; Plate VII*E*; Map 216

Stenobothrus speciosus Scudder, 1863*a*:458.
Stenobothrus aequalis Scudder, 1863*a*:459.
Stenobothrus bilineatus Scudder, 1863*a*:460.
Stenobothrus gracilis Scudder, 1872:250.
Orphula decora McNeill, 1896:240.
Orphulella obliquata Scudder, 1899*c*:181.
Orphulella picturata Scudder, 1899*c*:182.
Orphulella speciosa; Scudder 1899*c*:183; Cantrall 1943:91; Brooks 1958:59; Helfer 1963:177; R. D. Alexander et al. 1972:53; Otte 1979*a*:59.
Orphulina speciosa; Jago 1971:233; Vickery et al. 1974:121.

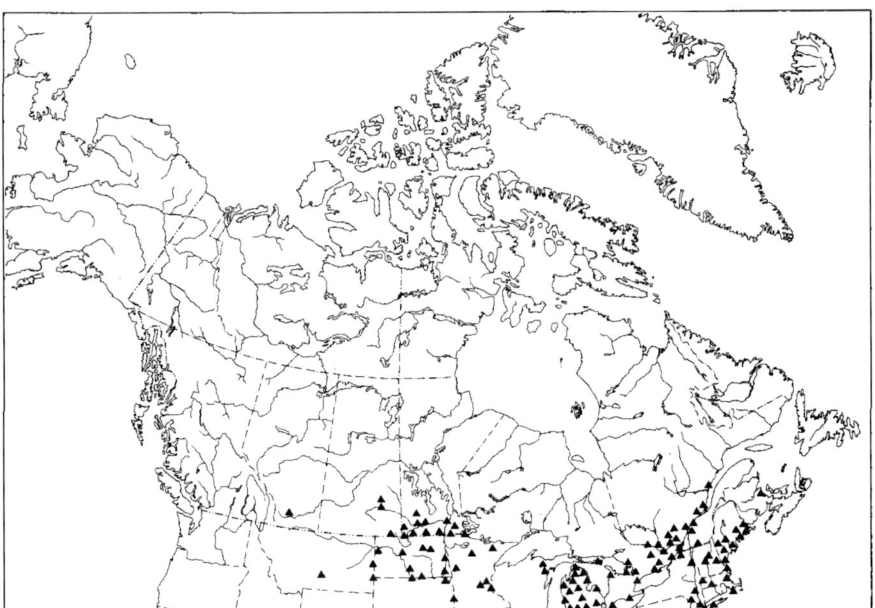

Map 216. Collection localities for *Orphulella speciosa*.

Diagnosis. Lateral foveolae narrow, not visible from above. Lateral carinae of pronotum evenly curved inward, not constricted; anterior and posterior pronotal widths about equal.

Description. Body small (length, male 13–16, female 16–21 mm) (Fig. 791). Frons slanted; vertex bluntly rounded, obtuse in female, right-angled in male; median depression narrow, extending to near apex; lateral foveolae narrow, distinct, not visible from above (Fig. 745). Pronotum with lateral carinae ridgelike, evenly curved, about equally separated anteriorly and posteriorly, cut by one sulcus (Fig. 793). Tegmina usually extending to near apices of hind femora. Male stridulatory pegs on inner face of hind femur as in Plate VII*E*. Color variable, with four distinct phases: (*a*) head, pronotal disc, and tegmina green; (*b*) head, pronotal disc, and tegmina rose red; (*c*) head and pronotum brown, with tegmina rose red; (*d*) head, pronotum, and tegmina brown; green and brown phases (*a* and *d*) predominate, males either brown or green, females predominantly green (proportion varies from area to area); males usually with pale oblique line on lower posterior part of lateral pronotal lobe, this also evident in some brown females; dorsum of pronotum usually with 2 dark posterolateral triangles.

Range. Alberta to New Brunswick, south to Texas and Virginia.

Behavior and habitats. *O. speciosa* is found generally in grasslands or pastures on dry, or semidry, sandy and loamy soils. It is a grass feeder but is not economically important. When disturbed, it usually conceals itself among grasses or leaps away. It does not fly readily, but, when flushed, males can fly rapidly for a distance of 2–3 m, as noted by Cantrall (1943). This author reported the grass *Poa pratensis* as the preferred food plant. The eggs are described by Onsager and Mulkern (1963).

Otte (1970) observed "crepitating" during several very short flights by male *O. speciosa*, but only on one occasion. In southwestern Québec, Hunt (1978) noted the frequent occurrence of short audible flights by males. The flight noises, he suggested, "seemed to be incidental to flight generally, and not comparable with oedipodine [locustine] crepitation." There was no observable response by females to the sounds. Nevertheless, there is a distinct "speculum" similar to that of *Metaleptea* in certain exotic species of *Orphulella* (Otte 1979*a*), so that flight sound may have some significance in *O. speciosa* also. It was stated by Otte (1970, 1979*a*) that males do not signal when they are approaching females but stalk them until they are close enough, then pounce on them and attempt to copulate. Available data from southwestern Québec (Hunt 1978), however, indicate that, although the above behavior pattern is the most usual, courting by males may occur. They sometimes stridulate, making a "tick" sound, repeated 3–10 times in sequence (Hunt, op. cit., provides audiospectrographs), following antennal contact with a female and prior to mounting (unless repulsed by the female). The stridulation is, nevertheless, so faint that it is virtually undetectable by the human ear, and it has apparently been overlooked by previous observers, although Otte (1970) reported "ticking" sounds as well as aggressive kicking in male-to-male encounters. In courtship, the very low audibility of the stridulation would presumably render this virtually useless unless two individuals were very close together. Femur-raising and kicking were methods used by nonreceptive females to repel males (Otte 1970; Hunt 1978). Hunt (1978) notes that successful copulation lasts from 30 to 90 minutes.

Life history. Adults appear in July and persist until September. Cantrall (1968) recorded dates of adult activity as 9 July to 25 September in Michigan. Records in southwestern Québec indicate adult activity from 24 July to 24 September.

Orphulella pelidna pelidna (Burmeister)

Figs. 754, 755, 794; Map 217

Gomphocerus pelidnus Burmeister, 1838:650.
Stenobothrus maculipennis Scudder, 1863*a*:458.
Stenobothrus propinquans Scudder, 1863*a*:461.
Orphulella pratorum Scudder, 1899*c*:184.
Orphulella pelidna; E. M. Walker 1902*b*:254; Helfer 1963:177 (partim); R. D. Alexander et al. 1972:53; Otte 1979*a*:60 (partim).
Orphulella pelidna pelidna; Gurney 1940*a*:117; Brooks 1958:59.

Diagnosis. Lateral carinae of pronotum somewhat constricted anterior to sulcus; pronotum broader at posterior margin. Tegmina exceeding apices of hind femora. Found from Atlantic to Rocky Mountain foothills.

Description. Body small (length 18–25 mm). Depression of vertex not extending close to apex of fastigium, with margin broad and with apex acute.

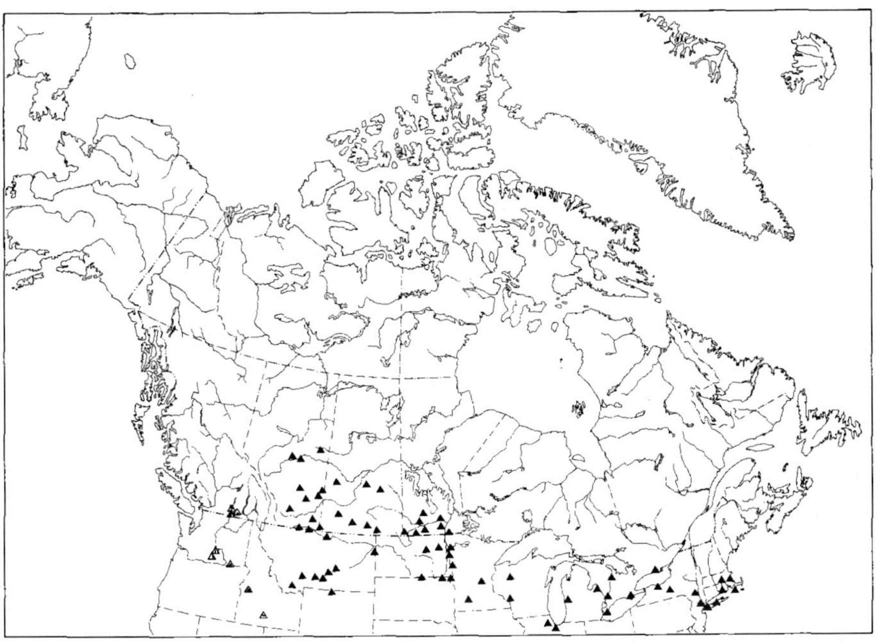

Map 217. Collection localities for *Orphulella pelidna pelidna* (▲) and *O. pelidna desereta* (△).

Pronotum with lateral carinae cut by at least 2, often 3, sulci (Fig. 794). Tegmina usually extending beyond apices of hind femora. Coloration variable, mainly green, brown, or gray; some individuals with prominent longitudinal dark stripes; lateral pronotal lobe lacking pale line, that of male usually dark except for pale ventral band.

Range. Alberta to Ontario and southern New England, south to Mexico.

Behavior and habitats. *O. pelidna pelidna* is a grass feeder. Brooks (1958) called it an inhabitant of dry sandy land on hillsides and in pastures. Hebard (1925*a*) stated that it is abundant in grasses of open or wooded country in the southeastern United States. Later (Hebard 1928), he reported it in the Great Plains only in swales or river valleys. Gurney (1940*a*) recorded its occurrence from sea level to 1700 m elevation. The form *olivacea* (Morse) occurs in maritime situations, especially in salt marshes. Onsager and Mulkern (1963) described the eggs.

Otte (1970) recorded flight crepitation but did not detect its function. Disturbed flying males usually produce a sound. Courtship and aggressive behavior are similar to those of *O. speciosa*. Gurney (1940*a*) reported a case of mismating of *O. pelidna pelidna*: in 1930, T. H. Hubbell had found a female of this species in Florida in attempted copulation with a male *Dichromorpha viridis* (Scudder).

Life history. Adults are known from mid-July to early September. Cantrall (1968) listed early and late dates as 14 July and 2 September in Michigan.

Orphulella pelidna desereta Scudder

Fig. 795; Map 217

Orphulella desereta Scudder, 1899*c*:178, 184.
Orphulella salina Scudder, 1899*c*:179, 185.
Orphulella pelidna; Buckell 1920:55.
Orphulella pelidna desereta; Gurney 1940*a*:125; Brooks 1958:59; Helfer 1963:177.

Diagnosis. As for *O. pelidna pelidna*, but tegmina extending only to apices of hind femora. Found in British Columbia, Washington, Idaho, and western Montana.

Description. Tegmina short, usually reaching only as far as apices of hind femora or not exceeding them by more than 1 mm. Lateral carinae of pronotum not markedly incurved, parallel, or nearly so, on prozona (Fig. 795). Male genitalic characters as indicated by Gurney (1940*a*:pl. XI, fig. 38 for *O. pelidna pelidna*; pl. XI, fig. 44 for *O. pelidna desereta*), with anterior lip on aedeagus greatly reduced.

Range. Interior British Columbia, south to northern California and New Mexico.

Behavior and habitats. Buckell (1920) recorded *O. pelidna desereta* from the edges of ponds and along the banks of the Okanagan River. He found it only where the grass was green, never on dry rangelands. Vickery and Nagy (1973) reported it in mixed vegetation in a residential garden area near Okanagan Lake and also in disturbed clearings in ponderosa pine park-forests and coniferous forest up to 1300 m elevation. Both areas were relatively damp. *O. pelidna desereta* is probably a grass feeder; the dominant grass in the areas referred to is *Hordeum* sp.

Life history. Similar to that of *O. pelidna pelidna*. Adults were taken in British Columbia from early August to September.

Tribe Chrysochraontini

Key to genera of Chrysochraontini

1. Hind wing banded. Frons ascending to pointed cone. Median carina of pronotum high, crestlike behind principal sulcus (Figs. 723, 724); posterior margin of pronotum acute-angled .... ***Acrolophitus*** (p. 540)
   Hind wing not banded. Frons not ascending to pointed cone. Median carina of pronotum low; posterior margin of pronotum truncate, rounded, or obtuse, not acutely angled 2

2(1). Antennae ensiform, wider at base than beyond middle (Figs. 730, 733, 734, 736, 744) 3
   Antennae subfiliform, slightly flattened basally or apically, not distinctly wider at base than beyond middle (Figs. 726, 738, 743) 5

3(2). Tegmina and hind wings extending beyond apex of abdomen. Lateral part of body with strong dark band on upper half; hind tibia reddish. Males at least 24 mm, females at least 30 mm, in body length .... ***Mermiria*** (p. 541)
   Not as above 4

4(3). Apex of abdomen extending well beyond apices of hind femora .... ***Pseudopomala*** (p. 543)
   Apex of abdomen not extending beyond apices of hind femora .... ***Opeia*** (p. 545)

5(2). Head as long as pronotum in male and nearly as long in female (Figs. 725, 727). Antennae of male slightly expanded apically (Fig. 726). Tegmina and hind wings usually surpassing apex of abdomen .... ***Syrbula*** (p. 546)
   Head not as long as pronotum (Figs. 737, 738, 743). Antennae of male not expanded apically. Tegmina and hind wings usually not surpassing apex of abdomen .... ***Chloealtis*** (p. 548)

Clé des genres de Chrysochraontini

1. Ailes postérieures rayées. Front remontant pour former un cône pointu. Crête médiane du pronotum haute, développée en arrière du sulcus principal (fig. 723 et 724); bord postérieur du pronotum formant un angle aigu .... ***Acrolophitus*** (p. 540)
   Ailes postérieures non rayées. Front ne formant pas un cône pointu. Crête médiane du pronotum basse; bord postérieur du pronotum tronqué, arrondi ou obtus, mais ne formant pas d'angle aigu 2

2(1). Antennes ensiformes, plus larges à la base qu'au-delà du milieu (fig. 730, 733, 734, 736 et 744) 3

Antennes subfiliformes, légèrement aplaties à la base ou à l'apex, pas clairement plus larges à la base qu'au-delà du milieu (fig. 726, 738 et 743) 5

3(2). Tegmina et ailes postérieures s'étendant au-delà de l'apex de l'abdomen. Côtés du corps marqués d'une forte bande foncée sur la portion supérieure; tibias postérieurs rougeâtres. Longueur totale d'au moins 24 mm chez les mâles et d'au moins 30 mm chez les femelles .... ***Mermiria*** (p. 541)

Combinaison de caractères différente de celle décrite ci-dessus 4

4(3). Apex de l'abdomen dépassant clairement l'extrémité des fémurs postérieurs .... ***Pseudopomala*** (p. 543)

Apex de l'abdomen ne dépassant pas les extrémités des fémurs postérieurs .... ***Opeia*** (p. 545)

5(2). Tête aussi longue que le pronotum chez les mâles et presque aussi longue chez les femelles (fig. 725 et 727). Antennes légèrement renflées à l'apex chez les mâles (fig. 726). Tegmina et ailes postérieures dépassant habituellement l'apex de l'abdomen .... ***Syrbula*** (p. 546)

Tête pas aussi longue que le pronotum (fig. 737, 738 et 743). Antennes non renflées à l'apex chez les mâles. Tegmina et ailes postérieures ne dépassant pas habituellement l'apex de l'abdomen .... ***Chloealtis*** (p. 548)

Genus *Acrolophitus* Thomas

Description. Antennae long, slender, with basal segments slightly flattened and oval in cross-section. Frons moderately slanted, terminating in conspicuous pointed cone; vertex of head without depression, with lateral foveolae completely absent. Pronotum with high semicircular crest on posterior two-thirds; metazona long, pointed, with lateral carinae obsolete and with median carina cut by 3 sulci (Fig. 724). Tegmina long, opaque, closely reticulate dorsally at base. Hind wings with broad black transverse band.

Acrolophitus hirtipes hirtipes (Say)

Figs. 723, 724; Map 218

Gryllus hirtipes Say, 1825*b*:[113], pl. 34, upper 2 figs.

Acrolophitus hirtipes; C. Thomas 1871*a*:266, 278; Brooks 1958:52; Helfer 1963:105.

A[*crolophitus*] *uniformis* Bruner, 1904*b*:47, 48.

Acrolophitus hirtipes hirtipes; Jago 1969:252.

Diagnosis. Pronotum with high semicircular crest on posterior two-thirds. Body hairy. Color generally green, with antennae red and hind wings banded.

Description. As for genus (Fig. 724). Body large (length 30–40 mm), hairy. Pronotal crest high (Figs. 723, 724). Color rather uniform green; antennae red; hind wings banded yellow and black.

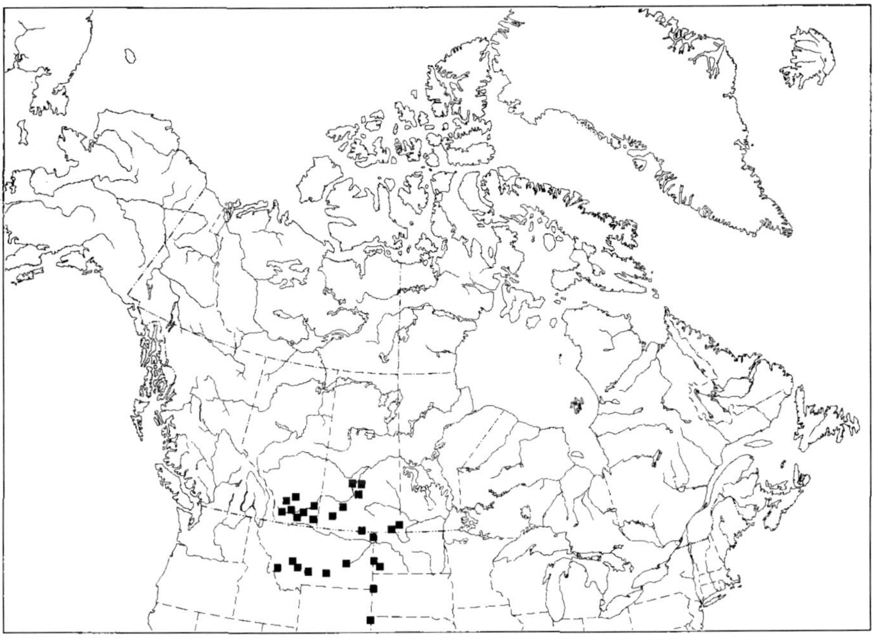

Map 218. Collection localities for *Acrolophitus hirtipes hirtipes.*

Range. Alberta and Saskatchewan, south to New Mexico and Texas.

Behavior and habitats. Brooks (1958) reported *A. hirtipes hirtipes* from sandy areas in river valleys. It is a forb feeder, preferring species of Boraginaceae (*Lithospermum* spp., *Lappula* spp., *Cryptantha* spp., and *Phasuba* spp.) (N. Criddle 1935; Brooks 1958). Anderson and Wright (1952) observed the species feeding upon leaves of western bluebur (*Lappula redowskii*). Adults are easily startled. Early in the season they appear to be poor fliers and only jump or crawl. Later, however, older adults fly readily (Anderson and Wright 1952). Onsager and Mulkern (1963) describe the eggs.

Otte (1970) observed stridulation by males when they were courting females, but more often males mounted females without prior production of sound. Stridulation appears to be employed mainly in pair formation prior to courtship and consists of 12–18 paired pulses.

Life history. N. Criddle (1935) found that eggs of *A. hirtipes* did not hatch in the laboratory until the 2nd year after deposition. That this is the case in the field is not certain. There seems to be no particular reason why such a delay should occur with this species. Adults have been recorded in the field from 7 July to 4 September in Montana (Hebard 1928).

Genus *Mermiria* Stål

Description. Antennae strongly flattened basally (Fig. 732), triangular in cross-section. Face strongly slanted; lateral foveolae shallow, not visible from above; depression of vertex without median carinula. Pronotum narrow, with median carina moderately strong, incised by one sulcus; lateral carinae absent; posterior margin of metazona rounded; prosternum with small tubercle. Tegmina extending to apex of abdomen. Inner apical spurs of hind tibia nearly equal in size. Hind wings clear.

Mermiria bivittata maculipennis Bruner

Figs. 731–733; Map 219

Mermiria maculipennis Bruner, 1890:54.

Mermiria maculipennis mcclungi J. A. G. Rehn, 1919*a*:65, 111, pl. VII, figs. 15–26; Brooks 1958:57.

Mermiria maculipennis; Mills and Pepper 1938:8; Helfer 1963:156.

Mermiria bivittata maculipennis; Jago 1969:272; map fig. 93.

Diagnosis. Long, slender species. Frons strongly slanted. Head and pronotum with broad dark lateral stripes.

Description. As for genus (Fig. 733). Body slender, long (length, male 26–32, female 36–45 mm). Color pale brown to yellowish gray; head and pronotum with broad lateral purplish brown to black stripes (Fig. 731); tegmina pale brown with opaque yellowish subcostal stripe on basal third; hind tibia reddish, with spines tipped with black.

Range. Alberta to Wisconsin, south to Mexico.

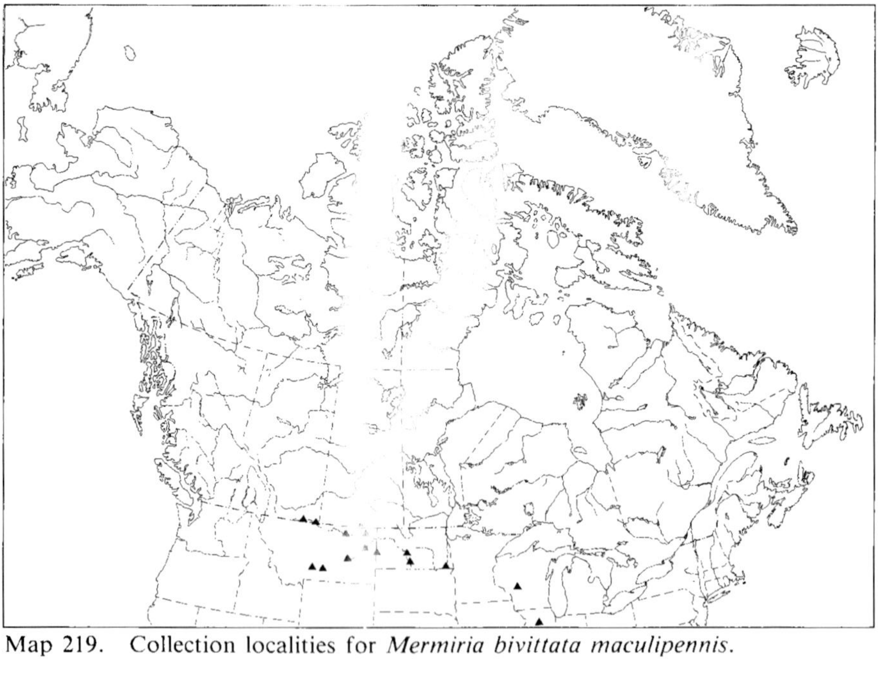

Map 219. Collection localities for *Mermiria bivittata maculipennis*.

Behavior and habitats. Brooks (1958) rated *M. bivittata maculipennis* as a grass feeder, preferring taller grasses such as side-oats grama, *Bouteloua curtipendula* (Michx.) Torr., and sand reed-grass, *Calamovilfa longifolia* (Hook.) Scribn. He found it only in eroded areas south of Manyberries, Alta., and near Coronach, Sask. The food plants were confirmed by Mulkern et al. (1969), who found that it also fed readily on *Agropyron* species. Anderson and Wright (1952) mentioned that both nymphs and adults remained high on vegetation during periods of cold temperatures or reduced light, when most other grasshoppers remained inactive on the ground. Adult males are quite active, moving constantly from plant to plant, eating very little, but females tend to remain in one place, moving only sluggishly when disturbed and eating far greater quantities of food than males. Tuck and Smith (1940) and Onsager and Mulkern (1963) describe the eggs.

Otte (1970) observed stridulation by courting males. This took the form of bursts of 2–5 strokes of the femora, as they approached females. Both single and double pulses were produced. Males employed three types of aggressive actions toward other males; aggressive stridulation, with regularly spaced pulses of short duration produced by a single hind femur striking the tegmen; the hind femora being held horizontally and shaken slightly; and raising of the hind femora to a vertical position.

Life history. Univoltine, with overwintering as eggs. In Montana, eggs hatch about 1 June and adults begin to appear about 25 July (Anderson and Wright 1952). R. M. White and Rock (1945) recorded adults in Alberta and Montana from 25 July to 1 September.

Genus *Pseudopomala* Morse

Description. Antennae conspicuously flattened basally, triangular in cross-section (Fig. 736). Face strongly slanting; lateral foveolae absent; depression of vertex with strong median carinula. Pronotum with median carina strong, cut only superficially behind middle by principal sulcus; lateral carinae nearly parallel, strong throughout; posterior margin of disc truncate; prosternum with low tubercle. Tegmina short, with veins appearing crowded.

Pseudopomala brachyptera (Scudder)

Figs. 734–736; Map 220

O[pomala] brachyptera Scudder, 1863*a*:454.

Opomala aptera Scudder, 1869*a*:305.

Pseudopomala brachyptera; Morse 1896*a*:325, 343, pl. 7, figs. 6, 6*a*, 6*b*; Cantrall 1943:89; Brooks 1958:60; Helfer 1963:158; R. D. Alexander et al. 1972:52.

Diagnosis. Face strongly slanting. Vertex elongate, projecting ahead of eyes by twice eye length.

Description. Body size medium (length, male 23.5–27, female 27.5–35 mm), slender (Figs. 734, 735). Face strongly slanted, with vertex elongate, projecting ahead of eyes by nearly twice eye length, bluntly rounded.

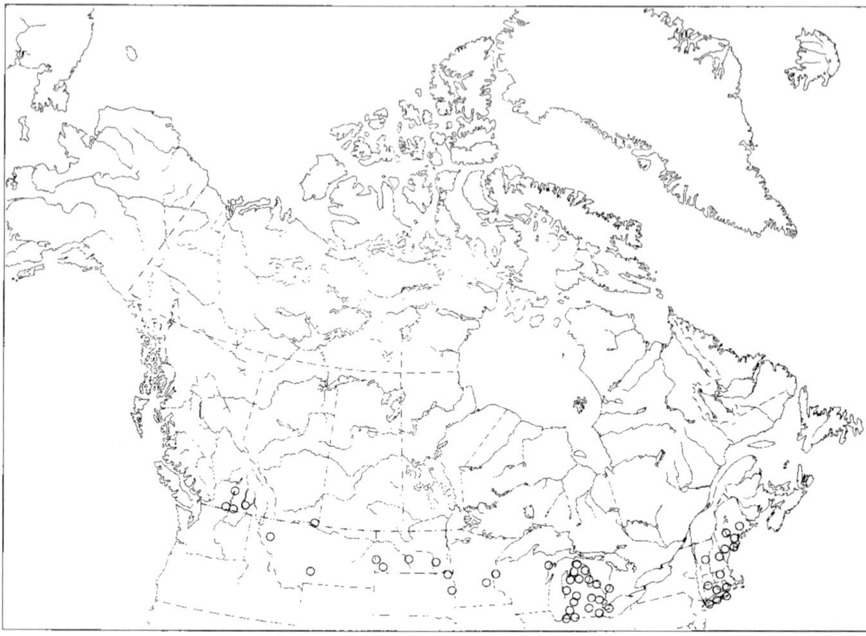

Map 220. Collection localities for *Pseudopomala brachyptera*.

Antennae tapering to apex (Fig. 736). Disc of pronotum nearly flat. Tegmina of male narrowing to rounded apices, usually covering about one-half of abdomen, of female usually covering about one-third of abdomen, tapering to acute apices (macropterous individuals occur in both sexes). Legs long, slender. Male subgenital plate long, conical. Ovipositor barely exserted. Color light brown or pale gray (males usually brown, females gray), usually with paler or darker lines on head, pronotal carinae, and tegmina of females. Nymphs are described by Scoggan and Brusven (1972).

Range. Interior British Columbia to Maine, south to Utah.

Behavior and habitats. Buckell (1922) reported *P. brachyptera* from rank growths of bunch-grass and dry rough pastures in British Columbia. Elsewhere it appears to prefer rather moist habitats. Anderson and Wright (1952) found it in moist bluestem (*Agropyron smithii*) meadows along creeks. A similar habitat was reported in Ontario by E. M. Walker and Urquhart (1940), and in Michigan by Cantrall (1943). Morse (1920) reported it from coarse grass in uncultivated lands, especially on bunch-grass (*Andropogon scoparius*) on poor soils. It jumps well when disturbed, but often sidles around grass stems out of sight. Pickford and Taylor (1963) reported oviposition in plant stems, and Onsager and Mulkern (1963) describe the eggs. Scoggan and Brusven (1972) outline the bionomics.

Otte (1970) studied behavior in this species. Males stridulate by moving the femora against the tegmina, without vibration, but with successive bursts increasing in intensity. As a male approaches a female, the song contains longer and slower pulses. Short bursts of stridulation are also used by males as an aggressive signal.

Life history. Adults appear early in the season, 25 May in Connecticut (Morse 1920). Cantrall (1968) gave dates of adult activity from 29 June to 7 September in Michigan. There is one generation each year. Nymphs and adults have been found in the fall, so it is probable that winter is spent in late nymphal stages in the north and as nymphs and adults farther south.

Genus *Opeia* McNeill

Description. Antennae conspicuously flattened basally, triangular in cross-section (Fig. 730). Face strongly slanting; lateral foveolae shallow, not visible from above; depression of vertex with median carinula. Pronotum with lateral carinae strong, calloused, nearly parallel; median carina strong, cut by principal sulcus behind middle; metazona truncate posteriorly; prosternum with tubercle. Tegmina short, not reaching apex of abdomen. Hind wings clear.

Opeia obscura (Thomas)

Figs. 728–730; Plate VI*C*; Map 221

Ox[*ycoryphus*] *obscurus* C. Thomas, 1872:466.
Opeia obscura; Lugger 1898:208; Brooks 1958:58; Helfer 1963:181.

Map 221. Collection localities for *Opeia obscura*.

Diagnosis. Body small, grayish yellow to greenish. Antennae conspicuously flattened, triangular in cross-section. Depression of vertex with median carinula.

Description. As for genus (Fig. 728). Body small (length 13–25 mm). Male stridulatory pegs on hind femur as in Plate VI*C*. Color grayish yellow to greenish, occasionally slightly brownish with diffuse darker markings on head and pronotum; antennae yellow, reddish, or fuscous; tegmina with median line of blackish spots; hind femur pale with dark dorsal streak; hind tibiae pale blue green to brownish yellow.

Range. Alberta to Manitoba, south to Utah and Nebraska.

Behavior and habitats. Froeschner (1954) recorded *O. obscura* as a short-grass prairie inhabitant. Brooks (1958) found it in eroded parts of mixed prairie. N. Criddle (1935), Anderson and Wright (1952), and Mulkern et al. (1969) all reported it as a grass feeder. Blue grama (*Bouteloua gracilis*) appears to be preferred by both nymphs and adults. Although it feeds on grasses, *O. obscura* has not been recorded as an economic pest. The eggs are described by N. Criddle (1935) and by Onsager and Mulkern (1963).

Life history. Univoltine, with overwintering in the egg stage. Hebard (1928) recorded *O. obscura* from 30 July to 8 September in Montana, and Hebard (1932*a*) gave 17 August and 27 September as early and late dates in Minnesota. Tinkham (1939) found it on 15 September in Alberta, and R. M. White and Rock (1945) extended the period to 25 September in that province.

Genus *Syrbula* Stål

Description. Head long, in male as long as pronotum (Fig. 725), in female nearly so; vertex triangulate, with lateral carinae and median carinula prominent and with lateral foveolae absent. Antennae subfiliform, flattened near base and expanded apically in male (Fig. 726). Pronotum with median carina distinct and with lateral carinae distinct and sinuate, all cut by 3 sulci. Tegmina and hind wings well developed, at least as long as, and usually exceeding, abdomen in both sexes. Hind femur long, slender. Hind tibia armed on outer carina with 1[illegible]–21 small spines. Male subgenital plate conical, acute, not longer than preceding sternum.

Syrbula admirabilis (Uhler)

Figs. 725–727; Map 222

Stenobothrus admirabilis Uhler, 1864:553.

Syrbula admirabilis; McNeill 1896:221; Helfer 1963:173; R. D. Alexander et al. 1972:53.

Diagnosis. Head as long (male) or nearly as long (female) as pronotum. Tegmina and hind wings long. Antennal segments flattened. Sexes dimorphic for color.

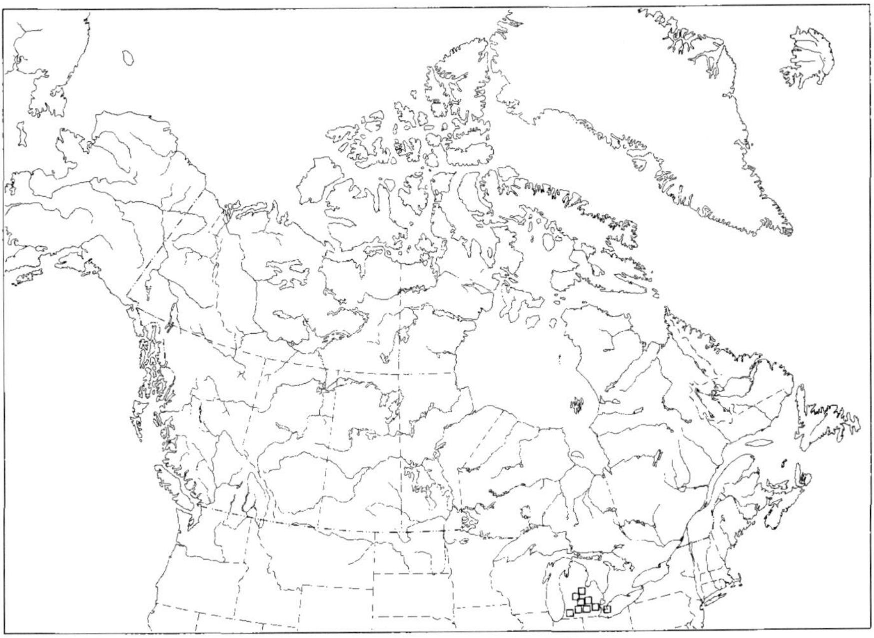

Map 222. Collection localities for *Syrbula admirabilis*.

Description. Body medium-large (length, male 22–27, female 35–40 mm). Antennae slender in basal half with segments flattened, apically expanded in male (Fig. 726). Color differing between sexes: Males olive brown and yellowish; antennae pale yellow basally, with apical third yellow on one side and fuscous on the other; head with yellowish lines laterally; pronotum olive brown with lateral yellow markings as in Fig. 725, and with posterior part possessing reticulate pattern; tegmina brownish; hind femur yellowish with 2 or 3 dusky bars dorsally and with genicular area black; hind tibia pale, with spines tipped with black. Females usually more greenish with reddish brown and fuscous markings; tegmina brownish with bright green stripes in dorsal and lateral areas, the dorsal stripe notched by serrations of a fuscous stripe between 2 green stripes; hind femur with upper carinae whitish, with outer face green above and lower face reddish brown.

Range. Nebraska to southern Ontario, south to Texas and Florida.

Behavior and habitats. High open uplands with sparse vegetation and poor soil appear to be the preferred habitat for this species. Cantrall (1968) found it to be "fairly common in old, fallow fields" in Michigan.

Otte (1970) recorded the behavior of this species. Stridulation of isolated males was heard at several metres distance. This consisted of 15–63 pairs of pulses, each pair seemingly produced by a single femoral stroke with both femora moving synchronously. The apices of the femora are held just above tegminal level throughout the sequence. Courtship in *S. admirabilis* was "more complicated than that of any other species examined in [t] his study."

The four phases which were observed are not given here, and the reader is referred to the original paper by Otte (1970). Aggressive signals include femur-jerking with a soft tegminal strike and occasionally, also, striking of the substrate with the hind tibiae. Otte and Loftus-Hills (1979) discuss "nonsynchronous" chorusing by males, which appears to be very complex.

Life history. Univoltine, with overwintering in the egg stage. Adults were reported from 16 August to 15 September in Michigan by Cantrall (1968), but the species is certainly active in the adult stage earlier and later than these dates. The adult specimens (mentioned above) from Point Pelee, Ont., 10 to 12 August, were taken with many late-instar nymphs, indicating that *S. admirabilis* matures at about that time of year in Canada. Froeschner (1954) recorded adults in Iowa from 3 July to 3 October.

Genus *Chloealtis* Harris

Description. Antennae slender, slightly flattened near base. Face slanting; lateral foveolae absent; depression of vertex with median carinula faint. Pronotum with median carina strong, cut by one sulcus; lateral carinae evident, nearly parallel to somewhat constricted; posterior margin rounded to truncate; prosternum without tubercle or with small protuberance. Inner apical spurs of hind tibia nearly equal. Tegmina short, usually not extending beyond apex of abdomen. Hind wings clear.

Key to species of *Chloealtis*

1. Tegmina of male broadly rounded at apices (Fig. 739), with close longitudinal venation except at costal margin; tegmina of female not broadened, extending to fourth abdominal segment (Fig. 740). Head with area above antennae and patch on side of pronotum usually black .... ***conspersa* Harris** (p. 549)

   Tegmina of male with widely spaced longitudinal venation (Fig. 741); tegmina of female with costal area broad and apices sharply pointed (Fig. 742). Head with area above antennae and on side of pronotum not usually black (only in a few strongly "speckled" specimens) .... ***abdominalis* (Thomas)** (p. 550)

Clé des espèces de *Chloealtis*

1. Tegmina des mâles très arrondis à l'apex (fig. 739), à nervures longitudinales rapprochées, sauf près du bord antérieur; tegmina des femelles non élargis, rejoignant le quatrième segment abdominal (fig. 740). Région de la tête située au-dessus des antennes et tache sur les côtés du pronotum habituellement de couleur noire .... ***conspersa* Harris** (p. 549)

   Tegmina des mâles à nervures longitudinales espacées (fig. 741); tegmina des femelles à région costale large et à apex très pointu (fig. 742). Région de la tête située au-dessus des antennes et côtés du pronotum non habituellement noirs (sauf chez certains spécimens particulièrement mouchetés) .... ***abdominalis* (Thomas)** (p. 550)

Chloealtis conspersa Harris

Figs. 738–740; Plate VI*D*; Map 223

Locusta (*Chloealtis*) *conspersa* T. W. Harris, 1841:149.

Chloealtis conspersa; Scudder 1862:286; Cantrall 1943:92; Brooks 1958:55; Helfer 1963:178; Jago 1969:292; R. D. Alexander et al. 1972:53; Vickery et al. 1974:123.

Stenobothrus melanopleurus Scudder, 1863*a*:456.

Diagnosis. Antennae nearly twice as long as head and pronotum together. Tegmina shorter than abdomen, broadly rounded apically, with longitudinal venation close. Head with area above antennae black.

Description. Body small (length, male 15–20, female 20–28 mm) (Figs. 737, 738). Antennae long, nearly twice as long as combined length of head and pronotum. Tegmina in male (Fig. 739) extending nearly to apex of abdomen, in female (Fig. 740) short, not reaching beyond fourth abdominal tergum, broadly rounded apically, with costal region not broadened; longitudinal venation close (rarely macropterous). Male stridulatory pegs on hind femur as in Plate VI*D*. Color dimorphic between sexes: Male dark brown; area above antennae, pronotum, and base of abdomen black laterally; tegmina brown, not spotted. Female pale brown to gray without black markings laterally (except above antennae); tegmina brown, conspicuously speckled; hind tibiae and underside of hind femur dark red in both sexes. Moens (1978) discusses the chromosomes.

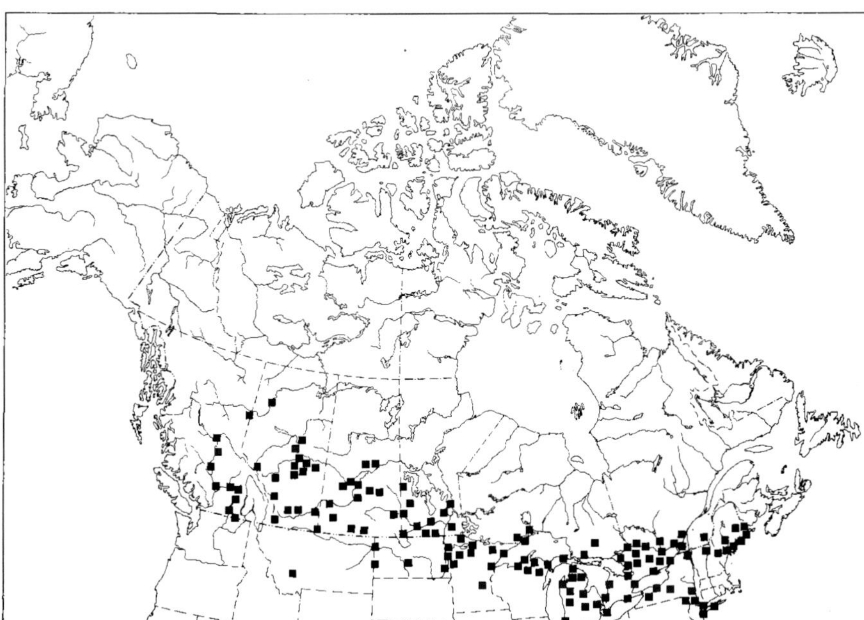

Map 223. Collection localities for *Chloealtis conspersa*.

Range. Interior British Columbia to southern Québec, south to Colorado and Tennessee.

Behavior and habitats. In Québec, *C. conspersa* occurs mainly in hilly areas in the Eastern Townships and along the Ottawa River valley. It is most often found in thickets and along edges of open woods, usually in dry locations (Vickery et al. 1974). This species oviposits in rotting wood such as fallen logs or old stumps; at times it attempts to bore holes in much harder wood (Blatchley 1920; Vickery et al. 1974). Cantrall (1943) observed feeding by this species on Canada blue grass (*Poa compressa*). He also reported it to be very alert and active by day but sluggish at night. Gangwere (1961) found the preferred food grasses to be *Dactylis glomerata* L., *Poa pratensis* L., and *Phleum pratense* L., although these could scarcely be reckoned as being original host plants, as they were all introduced into North America from Europe. Although *C. conspersa* is primarily a grass feeder, it is not of economic importance owing to its limited habitat preference (Mulkern et al. 1969). Blatchley (1920) and Onsager and Mulkern (1963) give accounts of the eggs.

Stridulation is heard only during the warmer hours of the day and consists of a series of "zeeek" or "eeeek" sounds, made only by the downstrokes of the hind femora (Cantrall 1943). Otte (1970) reported up to eight song sequences at one spot, followed by walking and then by repeated stridulating. Females appear to remain motionless until a male comes close. During courtship, stridulation is very soft, produced by alternate or synchronous femoral strokes. The song alters as the male comes closer to a female, from four to 14 regularly spaced pulses being produced just prior to attempted mounting of the female. In aggressive encounters, a third type of stridulation occurs, in which the number of strokes and the stroke rate are greatly reduced. There is no transitional phase between types of stridulation. Some males raise the femora as an aggressive signal.

Life history. Univoltine, with overwintering in the egg stage in rotten wood. Adults were reported by Cantrall (1968), from 20 June to 11 October in Michigan, and adults have been found in Québec over much the same period.

Chloealtis abdominalis (Thomas)

Figs. 741–743; Plate VI*E*; Map 224

Chrysochraon abdominalis C. Thomas, 1873:74.

Chloealtis abdominalis; McNeill 1896:229; R. D. Alexander et al. 1972:53; Vickery et al. 1974:123.

Neopodismopsis abdominalis; Beĭ-Bienko 1932:56; Brooks 1958:58.

Diagnosis. Tegmina short, pointed, with longitudinal venation widely spaced. Head with area above antennae not black.

Description. Body small-medium (length, male 18–19, female 23–28 mm) (Fig. 743). Antennae short, less than twice length of head and pronotum combined. Tegmina short, with apices pointed, especially in female;

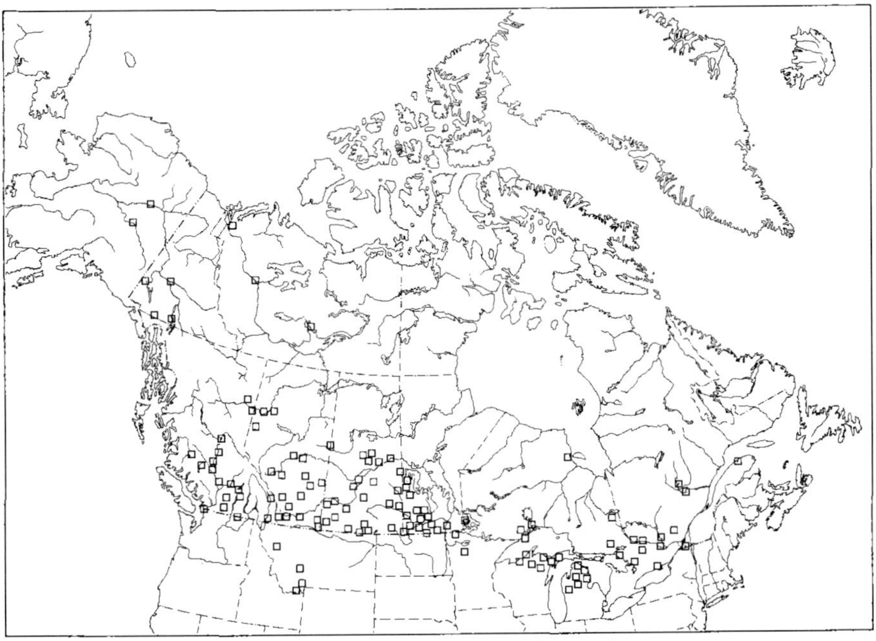

Map 224. Collection localities for *Chloealtis abdominalis*.

costal area broad; longitudinal venation widely spaced (Figs. 741, 742) (occasional macropterous females are found). Male stridulatory pegs as in Plate VI*E*. Color generally grayish brown, with area above antennae not black; sides of pronotum darker above; male tegmen nearly transparent with ill-defined median spots; anal area opaque, yellowish; hind femur red beneath; hind tibia brown, darker toward apex. N. Criddle (1926*b*, 1930*a*) described the nymphal stages. Moens (1978) discusses the chromosomes. An isolated montane colony of *C. abdominalis* from Mt. Albert, Gaspésian Park, Qué., is strikingly maculated dark brown to black, with irregular ebony black patches beneath vertex and around antennae (but not solid black above antennae); upper two-thirds of lateral pronotal lobes shining brown; dorsum of head and pronotum usually ashy black, with irregular dark blotches on anterior third of pronotum (Fig. 743).

Range. Alaska to Québec, south to Arizona and New Mexico.

Behavior and habitats. *C. abdominalis* is known mainly from forested areas and parklands. Brooks (1958) reported it as being confined to depressions and valleys in grassland areas. Buckell (1922) found it to be numerous in woods, especially along trails and in slash areas following logging operations. Vickery and Nagy (1973) recorded the species from ponderosa pine park–forest on dry slopes with short-grass pasture and from undisturbed clearings in coniferous forest in British Columbia. This is a grass and sedge feeder. *Bouteloua, Stipa, Agropyron, Koeleria*, and *Carex* species are reported as hosts by Brooks (1958). The eggs are described by N. Criddle (1930*a*), and by Onsager and Mulkern (1963). A general account of the biology is given by N. Criddle (1930*a*), who notes the propensity of the species to oviposit in cattle manure.

Stridulation and courtship were reported by Otte (1970) to be similar to those of *C. conspersa*.

Life history. The species is univoltine, with overwintering in the egg stage. Adults are found from early July until the end of September. Cantrall (1968) gave early and late dates as 9 July and 18 October for Michigan.

Tribe Gomphocerini

Key to genera of Gomphocerini

1. Lateral foveolae visible from above (Fig. 744) 2
   Lateral foveolae of vertex not visible from above (cf. Fig. 745) 8

2(1). Antennae filiform to slightly flattened basally, not apically clavate .. 3
   Antennae apically clavate (Fig. 766), less strongly so in females .... ***Aeropedellus*** (p. 553)

3(2). Inner apical spurs of hind tibia slightly unequal in size (Fig. 756) ... 4
   Inner apical spurs of hind tibia conspicuously unequal in size (Figs. 760–761) 5

4(3). Lateral foveolae narrow, four times as long as broad (Fig. 746) .... ***Chorthippus*** (p. 556)
   Lateral foveolae broader, no more than three times as long as broad (Fig. 747) .... ***Bruneria*** (p. 558)

5(3). Lateral foveolae triangular (Figs. 750, 751) .... ***Aulocara*** (p. 561)
   Lateral foveolae rectangular (Figs. 749, 752, 753) 6

6(5). Lateral foveolae poorly developed. Lateral carinae of pronotum parallel on prozona, diverging posteriorly on metazona. Depression of vertex often with faint median carinula. Males black ...... ***Boopedon*** (p. 565)
   Lateral foveolae conspicuous (Figs. 749, 752, 753). Lateral carinae of pronotum strongly constricted. Depression of vertex without median carinula. Males not black 7

7(6). Hind tibiae red or orange. Tegmina dull, opaque, not extending to apices of hind femora. Mature in summer .... ***Ageneotettix*** (p. 566)
   Hind tibiae usually greenish brown, sometimes yellowish brown to pale blue. Tegmina extending beyond apices of hind femora. Mature in spring .... ***Psoloessa*** (p. 568)

8(1). Inner apical spurs of hind tibia conspicuously unequal (Fig. 758) .... ***Eritettix*** (p. 571)
   Inner apical spurs nearly equal in size (Fig. 757) 9

9(8). Depression of vertex without median carinula .... ***Cordillacris*** (p. 572)
   Depression of vertex with median carinula, which may be well-developed or faint 10

10(9). Lateral margins of pronotal disc rounded; lateral carinae nearly obsolete. Hind tibia blue to purplish with black-tipped spines .... ***Amphitornus*** (p. 576)
   Lateral margins of pronotal disc fairly sharply defined; lateral carinae weak on metazona but evident throughout. Hind tibia yellow to orange, with black-tipped spines .... ***Phlibostroma*** (p. 578)

Clé des genres de Gomphocerini

1. Fovéoles latérales visibles du dessus (fig. 744) 2
   Fovéoles latérales du vertex non visibles du dessus (fig. 745) 8

2(1). Antennes filiformes, légèrement aplaties à la base, non claviformes à l'apex 3
   Antennes claviformes à l'apex (fig. 766), moins clairement cependant chez les femelles .... *Aeropedellus* (p. 553)

3(2). Éperons apicaux internes des tibias postérieurs de taille légèrement inégale (fig. 756) 4
   Éperons apicaux internes des tibias postérieurs de taille clairement inégale (fig. 760 à 761) 5

4(3). Fovéoles latérales étroites, quatre fois plus longues que larges (fig. 746) .... *Chorthippus* (p. 556)
   Fovéoles latérales plus larges, pas plus de trois fois plus longues que larges (fig. 747) .... *Bruneria* (p. 558)

5(3). Fovéoles latérales triangulaires (fig. 750 et 751) ..... *Aulocara* (p. 561)
   Fovéoles latérales rectangulaires (fig. 749, 752 et 753) 6

6(5). Fovéoles latérales peu développées. Crêtes latérales du pronotum parallèles sur la zone antérieure, divergeant postérieurement sur la zone postérieure. Dépression du vertex portant souvent une faible crête médiane. Mâles noirs .... *Boopedon* (p. 565)
   Fovéoles latérales très visibles (fig. 749, 752 et 753). Crêtes latérales du pronotum marquées d'une forte constriction. Dépression du vertex sans crête médiane. Mâles de coloration autre que noire 7

7(6). Tibias postérieurs rouges ou orangés. Tegmina mats, opaques, ne rejoignant pas les extrémités des fémurs postérieurs. Maturité en été .... *Ageneotettix* (p. 566)
   Tibias postérieurs habituellement brun verdâtre, parfois brun jaunâtre à bleu pâle. Tegmina dépassant les extrémités des fémurs postérieurs. Maturité au printemps .... *Psoloessa* (p. 568)

8(1). Éperons apicaux internes des tibias postérieurs clairement inégaux (fig. 758) .... *Eritettix* (p. 571)
   Éperons apicaux internes de longueur presque égale (fig. 757) 9

9(8). Dépression du vertex dépourvue d'une crête médiane .... *Cordillacris* (p. 572)
   Dépression du vertex pourvue d'une crête médiane développée ou non ... 10

10(9). Bords latéraux du disque du pronotum arrondis; crêtes latérales presque invisibles. Tibias postérieurs bleus à pourpres, armés d'épines à extrémité noire .... *Amphitornus* (p. 576)
   Bords latéraux du disque du pronotum assez bien définis; crêtes latérales peu visibles sur la zone médiane mais évidentes ailleurs. Tibias postérieurs de coloration variant de jaune à orangé et armés d'épines à extrémité noire .... *Phlibostroma* (p. 578)

Genus *Aeropedellus* Hebard

Description. Apical antennal segments broadened, flattened, and darkened, especially in male. Frons slanting; lateral foveolae of vertex rectangular, narrow, conspicuous, visible from above; depression of vertex flat, without a median carinula. Pronotum with median carina evident, entire except where cut by one sulcus; lateral carinae evident, strongly constricted in prozona, cut by one sulcus (Fig. 768); posterior margin of pronotum obtuse (Fig. 767); prosternum with small tubercle. Fore tibiae often inflated apically; inner apical spurs of hind tibia slightly unequal. Tegmina short, with venation open. Hind wing clear.

Key to species of *Aeropedellus*

1. Body slender. Males with short heavily clubbed antennae (Fig. 768). Pronotal disc elongate, less constricted (Fig. 767). Ranging from Minnesota, North Dakota, west to British Columbia and north to Great Slave Lake, Northwest Territories .... ***clavatus*** **(Thomas)** (p. 554)
   Body more robust. Males with longer less heavily clubbed antennae (Fig. 769). Pronotal disc more constricted. Ranging from northern Northwest Territories to Alaska .... ***arcticus*** **Hebard** (p. 556)

Clé des espèces d'*Aeropedellus*

1. Corps mince. Chez les mâles, antennes courtes et à extrémité en forme de massue (fig. 768). Disque du pronotum allongé, moins incourbé (fig. 767). Espèce vivant du Minnesota et du Dakota-nord jusqu'en Colombie-Britannique vers l'ouest et jusqu'au Grand Lac des Esclaves (Territoires du Nord-Ouest) vers le nord .... ***clavatus*** **(Thomas)** (p. 554)
   Corps plus robuste. Chez les mâles, antennes plus longues et moins claviformes (fig. 769). Disque du pronotum plus incourbé. Insectes vivant du nord des Territoires du Nord-Ouest jusqu'en Alaska .... ***arcticus*** **Hebard** (p. 556)

Aeropedellus clavatus (Thomas)

Figs. 748, 766–768; Plate VI*G*; Map 225

Gomphocerus clavatus C. Thomas, 1873:96.
Gomphocerus carpenterii C. Thomas, 1874:65.
Gomphocerus clepsydra Scudder, 1875*b*:344.
Aeropedellus clavatus clavatus; Hebard 1935*e*:204.
Aeropedellus clavatus; Hebard 1936*a*:33; Brooks 1958:52; Helfer 1963:159 (partim).

Diagnosis. Male antennae distinctly clavate; female antennae less so. Lateral foveolae rectangular, narrow, conspicuous, visible from above. Lateral carinae of pronotum strongly constricted in prozona.

Description. Body small (length 12–21 mm, usually 17–19 mm). Antennae of male conspicuously flattened and broadened apically (Fig. 768), of females much less so. Tegmina of males extending to apex of abdomen, of females extending only as far as base of fourth abdominal tergum. Stridulatory pegs on inner face of male hind femur as in Plate VI*G*. Male and female concealed genitalic characters as illustrated by Vickery (1967*a*:272, figs. 35, 39). Color differing somewhat between sexes: Males usually dull blackish

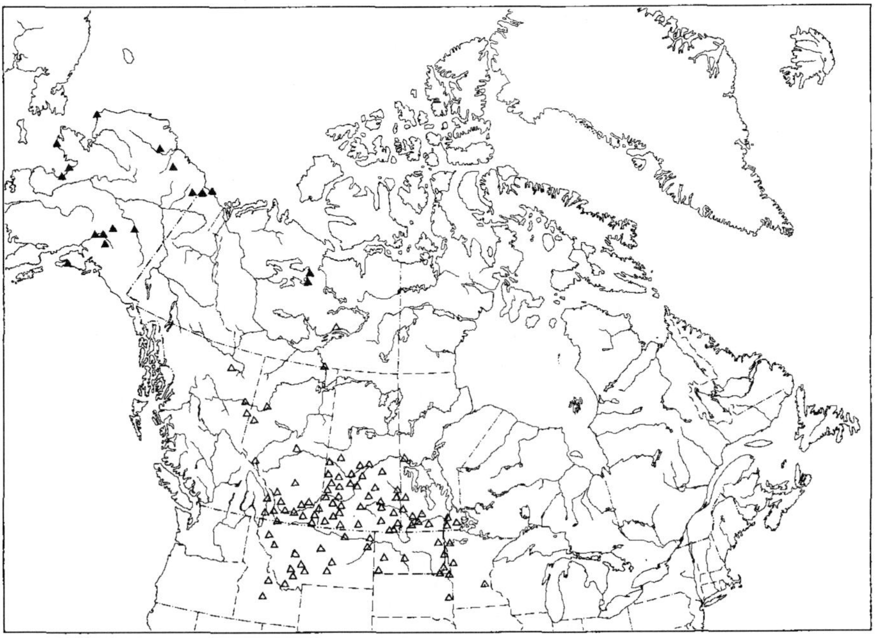

Map 225. Collection localities for *Aeropedellus clavatus* (△) and *A. arcticus* (▲).

green, with apices of antennae black. Females variable, with head and pronotum often conspicuously patterned in black, creamy white, and green. N. Criddle (1926*b*) describes the nymphs.

Range. Western Northwest Territories to Manitoba, south to Arizona and New Mexico.

Behavior and habitats. *A. clavatus* is an inhabitant of dry, somewhat sandy areas, often at high altitudes (in Colorado, to almost 4000 m). N. Criddle (1935) listed it as a grass feeder. Mulkern et al. (1969) confirmed this and found *Poa pratensis* to be the main food plant, although it fed upon other grasses and sedges to a limited extent. Anderson and Wright (1952) reported that *A. clavatus* fed on several species of grasses and on threadleaf sedge (*Carex filifolia*). These authors also reported that males appeared to be easily startled and seemed to be constantly in flight or crawling rapidly on the ground. Females were much slower and remained motionless on the ground for long periods of time. Tuck and Smith (1940) and Onsager and Mulkern (1963) describe the eggs. G. Alexander and Hilliard (1964) studied the biology at high altitudes in Colorado; eggs were laid at the bases of sedges and grasses.

Otte (1970) did not report on behavior of this species, but J. A. G. Rehn and Hebard (1906) described the sound of the stridulation as "sik-sik-sik-sik," similar to but louder than that of *Chorthippus curtipennis curtipennis*. Somes (1914) disagreed that stridulation of these two species was similar, reporting that of *A. clavatus* as a buzzing trill, prefaced by a series of 2–4 clicks. The stridulation is produced by holding a closed hind femur and tibia at a high angle and vibrating them rapidly through a small arc. Bursts were of about 2–4 seconds in duration, separated by intervals of 3–6 seconds.

Life history. Univoltine, with overwintering in the egg stage. Adults have been reported from 24 June to 24 September in the Prairie Provinces of Canada (E. M. Walker 1910). At high altitudes, the life history is abbreviated, there being only four nymphal instars and sexual maturity being attained in 6 weeks (G. Alexander and Hilliard 1964).

Economic importance. Since *A. clavatus* feeds primarily upon high-value grasses, it has considerable pest potential. "Outbreak" populations have sometimes caused extensive damage, such as the destruction of 120 ha of wheat in Saskatchewan in 1936 (Paul and Berg 1948; Beirne 1972). MacNay (1961*a*) reports only a minor outbreak in 1960, but some economic damage since 1957. The report by N. Criddle (*in* Fletcher 1902*a*) of "*Gomphocerus* species" being among destructive grasshoppers in southern Manitoba is probably the earliest such record for Canada.

Aeropedellus arcticus Hebard

Fig. 769; Map 225

Aeropedellus variegatus arcticus Hebard, 1935*e*:207.

Aeropedellus clavatus; Hebard 1935*e*:204, 205; Brooks 1958:52 (partim); Helfer 1963:159 (partim).

Aeropedellus arcticus; Vickery 1967*a*:270.

Diagnosis. Similar to *A. clavatus*, but with antennae less strongly clavate, and with concealed genitalia different (see Vickery 1967*a*). Found in northern Northwest Territories, Yukon, and Alaska.

Description. Much as in *A. clavatus*. Antennae of male enlarged and flattened apically, but not strongly clubbed. Tegmen length, of male 9.5–12.0 mm, of female 8.0–9.6 mm. Coloration much as in *A. clavatus*.

Range. Alaska, Yukon Territory, and northern Northwest Territories.

Behavior, habitats, and life history. Not known.

Genus *Chorthippus* Fieber

Description. Antennae long, slender, slightly flattened near base. Face slanting; lateral foveolae narrow, rectangular, conspicuous, visible from above; depression of vertex flat, lacking median carinula. Pronotum with median carina cut by one sulcus; lateral carinae complete, strong, evenly incurved; posterior margin of pronotum rounded; prosternum without tubercle. Inner apical spurs of hind tibia somewhat unequal. Tegmina variable in length. Hind wing clear.

Chorthippus curtipennis curtipennis (Harris)

Figs. 744, 746, 762, 763; Plate VI*F*; Map 226

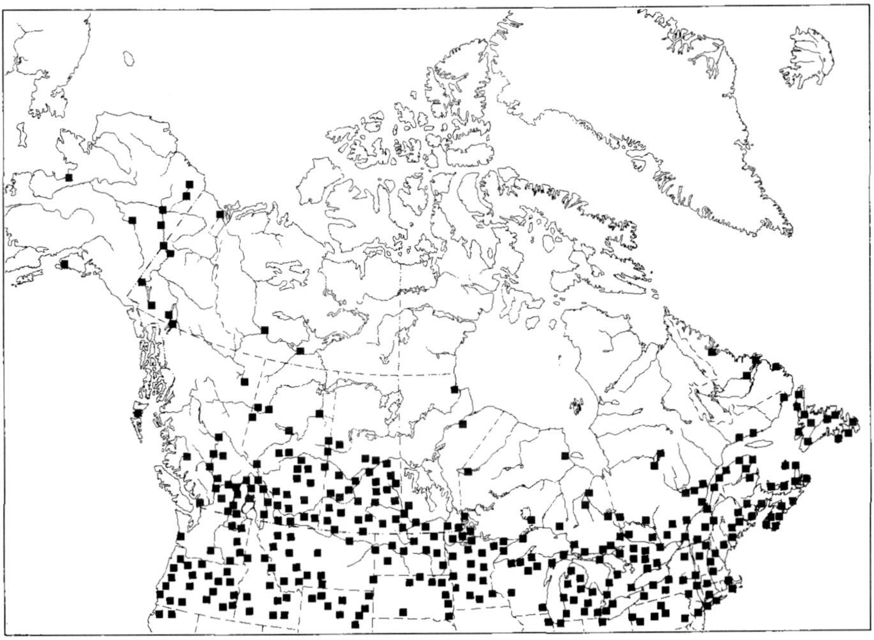

Map 226. Collection localities for *Chorthippus curtipennis curtipennis.*

Locusta (*Chloealtis*) *curtipennis* T. W. Harris, 1841:149.

Stenobothrus longipennis (macropterous) Scudder, 1862:286.

Stenobothrus curtipennis longipennis; C. Thomas 1873:91.

Stenobothrus oregonensis Scudder, 1899*f*:50.

Chorthippus curtipennis; J. A. G. Rehn and Hebard 1910*b*:6; R. D. Alexander et al. 1972:53, fig. 54.

Chorthippus curtipennis curtipennis; Buckell 1929:11; Vickery et al. 1974:125.

Chorthippus longicornis; Hebard 1935*d*:283; Cantrall 1943:52, 55, 62, 94; Brooks 1958:55; Helfer 1963:168.

Diagnosis. Lateral foveolae distinct, narrow, rectangular, visible from above.

Description. Body small (length, male 12.5–16.3, female 16–22 mm; variable on geographic basis, smaller in northern latitudes and higher elevations) (Figs. 762, 763). Antennae long, with 22 segments, slightly dorsoventrally flattened. Frons slanted, with lateral foveolae distinct, visible from above (Figs. 744, 746). Pronotum with median carina distinct, evenly elevated, cut behind middle by principal sulcus; lateral carinae incurved on anterior third, diverging posteriorly, cut by 3 sulci. Tegmina and hind wings variable in length, in brachypterous specimens extending to, or nearly to, apex of abdomen in males, to fifth abdominal segment in females, in macropterous specimens exceeding apices of hind femora in both sexes. Male stridulatory pegs as in Plate VI*F*. Color commonly entirely brown with black markings near pronotal carinae and on abdomen (60% of specimens), or brown dorsally and green laterally (2.5%), or purple dorsally, green laterally (10%), or green dorsally, brown laterally (very rare), or blackish brown dorsally and on upper parts of head and lateral pronotal lobes, lower parts yellowish white (uncommon). N. Criddle (1926*b*) and Scoggan and Brusven (1972) describe the nymphal stages. Moens (1978) discusses the chromosomes.

Range. Alaska to Newfoundland, south to Mexico.

Behavior and habitats. *C. curtipennis curtipennis* is common in thick grass, in damp places, in dry meadows, on roadsides, and even on the Arctic tundra. The species is primarily a grass feeder. Mulkern et al. (1969) recorded it as feeding mainly on *Poa pratensis* in North Dakota, and on *Andropogon gerardi* and *A. scoparius* in Nebraska. N. Criddle (1935), Tuck and Smith (1940), and Onsager and Mulkern (1963) consider the eggs. Scoggan and Brusven (1972) outline the bionomics.

Otte (1970) obtained recordings of male stridulation in Michigan under various circumstances (op. cit.:81, fig. 11), including when solitary, when touched by another male, when responding to stridulation by another male, in initial and advanced courtship, and when mounting females. Similar records have been made in Québec (Hunt 1978). In general, the song of *C. curtipennis curtipennis* is similar to that of its nearest relative, *C. montanus* (Charpentier 1825), in the duration of its chirp (strict term ''echeme''), but more like that of *C. parallelus* (Zetterstedt) in syllable repetition rate, so that it has more syllables per unit time (about 30 per echeme) than in either of these Palaearctic species (Reynolds 1980). Otte (1970) found that the first part of the downstroke of the femora against the tegmina was smooth, but toward the end of the stroke, it seemed to ''vibrate or bounce'' against the femora, producing a distinct modulated effect at the end of each pulse. Females stridulate in answer to males. When mounted on females, the males always produce a whirring sound by holding one leg with the hind femur against the tegmen and with tibia extended, the sound being produced by vibrating the femur. Males also produce aggressive stridulation of much shorter pulses than when calling.

Life history. Over the southern part of the range, and at low elevations, there is one generation each year, with overwintering in the egg stage. Adults are active from mid-June to mid-October. Cantrall (1968) gave 20 June to 18 October as early and late dates in Michigan. At higher latitudes and elevations, the life cycle probably takes 2 years, or more. Kreasky (1960) reported that eggs hatched in the 3rd year after deposition at an altitude of about 2590 m in the Big Horn Mountains of Wyoming.

Economic importance. *C. curtipennis curtipennis* has the potential to cause economic damage to grasslands when population levels are high, but it has not been reported as an important pest.

Genus *Bruneria* McNeill

Description. Antennae slender, slightly flattened near base. Frons slightly slanted; lateral foveolae (Fig. 747) rectangular, conspicuous, visible from above; depression of vertex shallow, its median carinula nearly obsolete. Pronotum with median carina evident, cut by one sulcus (Fig. 765); lateral carinae continuous, incurved near middle; posterior margin of pronotum rounded; prosternum without tubercle. Inner apical spurs of hind tibia somewhat unequal. Tegmina shining, variable in length. Hind wings clear.

Key to species of *Bruneria*

1. Median carina of pronotum cut behind middle. Tegmina of males about half, of females less than half, as long as abdomen; lateral pronotal lobes narrow below .... ***shastana* (Scudder)** (p. 559)
   Median carina of pronotum cut before middle. Tegmina longer than above; lateral pronotal lobes scarcely narrower below than above 2

2(1). Head large in proportion to body. Tegmina slightly surpassing abdomen. Found in Manitoba and North Dakota, west to Montana and British Columbia, but not north of Jasper and Edmonton, Alta., or of Prince Alberta and Lloydminster, Sask. ..... ***brunnea* (Thomas)** (p. 560)
   Head small in proportion to body. Tegmina shorter, barely reaching apex of abdomen. Found in Yukon Territory .... ***yukonensis* Vickery** (p. 561)

Clé des espèces de *Bruneria*

1. Crête médiane du pronotum coupée en arrière de son point médian. Tegmina des mâles de longueur égale à la moitié environ (chez les femelles, moins de la moitié) de la longueur de l'abdomen; lobes latéraux du pronotum étroits vers le bas .... ***shastana* (Scudder)** (p. 559)
   Crête médiane du pronotum coupée avant son point médian. Tegmina plus longs que ci-dessus; lobes latéraux du pronotum à peine plus étroits en bas qu'au-dessus 2

2(1). Tête grosse par rapport au corps. Tegmina dépassant légèrement l'abdomen. Insecte vivant au Manitoba et au Dakota-Nord, vers l'ouest jusqu'au Montana et jusqu'en Colombie-Britannique mais pas plus au nord que Jasper et Edmonton (Alb.) ou que Prince-Albert et Lloydminster (Sask.) .... ***brunnea* (Thomas)** (p. 560)
   Tête petite par rapport au corps. Tegmina plus courts, rejoignant à peine l'apex de l'abdomen. Insecte vivant au Yukon .... ***yukonensis* Vickery** (p. 561)

Bruneria shastana (Scudder)

Map 227

Gomphocerus shastanus Scudder, 1881:25, pl. XVII, figs. 15, 18.
Stenobothrus sordidus McNeill, 1897:263.
Brunneria [*sic*] *shastana*; McNeill 1897:265.
B[*runeria*] *shastana*; Helfer 1963:169.

Diagnosis. Similar to *B. brunnea*, but with lateral pronotal lobes narrowing ventrally.

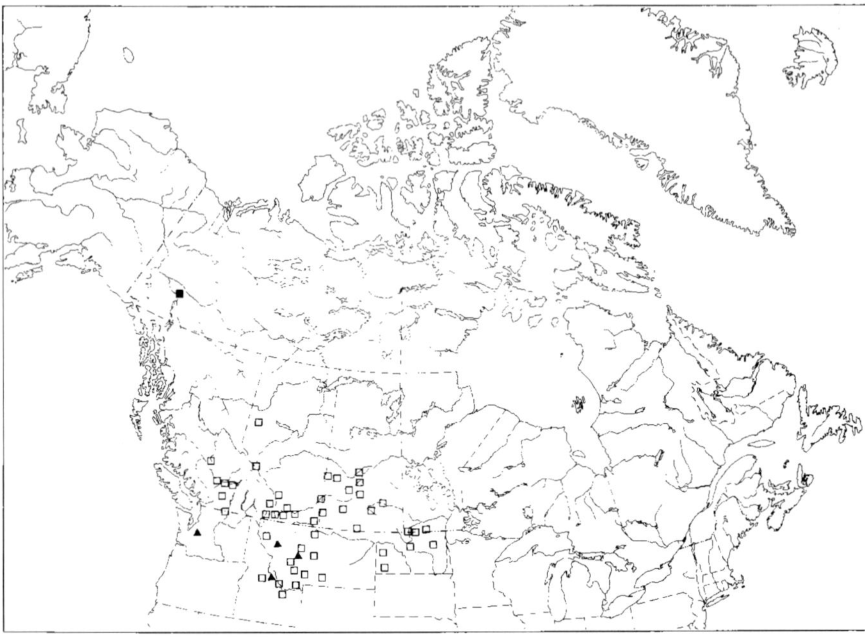

Map 227. Collection localities for *Bruneria shastana* (▲), *B. brunnea* (□), and *B. yukonensis* (■).

Description. Similar to *B. brunnea*. Pronotum with median carina cut behind middle by principal sulcus; lateral lobes narrowing ventrally. Tegmina short, not exceeding half length of abdomen. The color pattern lacks the median longitudinal pale stripe on head and pronotum. Scoggan and Brusven (1972) describe the nymphs.

Range. Washington, Idaho, and western Montana.

Behavior and habits. Individuals extend from the high shadescale community to mountain meadows, frequently with loose rocky soils or talus and perennial bunchgrasses. Varied vegetation is inhabited, but grasses of the genera *Poa, Agropyron, Festuca*, and *Sporobolus* are preferred habitats. The species has been observed to feed on the first three of these. Nymphs are strong jumpers, seeking open areas when disturbed. They may number as many as 4 to 5 per square metre.

Economic importance. The species is seldom of economic importance.

Bruneria brunnea (Thomas)

Figs. 747, 756, 764, 765; Map 227

Stenobothrus brunneus C. Thomas, 1871*a*:280.
Bruneria brunnea; Hebard 1926:56; Brooks 1958:54; Helfer 1963:169.

Diagnosis. Lateral foveolae rectangular, conspicuous, visible from above, but no more than three times as long as broad. Pronotum short. Head and pronotum usually black with yellow median line.

Description. Body small (length 18–25 mm). Pronotum short, little longer than head, slightly constricted at middle (Figs. 764, 765). Tegmina long, slightly exceeding apex of abdomen. Male epiphallus and female subgenital plate as figured by Vickery (1969*b*:267, figs. 6, 8). Color variable; head and pronotum usually black dorsally with yellow median longitudinal line, sometimes entirely whitish or entirely brown without contrasting markings; tegmina shining, conspicuously spotted; hind femur with 3 oblique brown bands on outer face and dorsal area, and with similar bands on upper half of inner face; hind tibia pale with dark area at basal third.

Range. Interior British Columbia to Manitoba, south to Colorado.

Behavior and habitats. *B. brunnea* is a grassland species that feeds upon several species of grasses and sedges (*Koeleria* spp., *Carex* spp., *Bouteloua* spp., *Stipa* spp., and *Agropyron* spp.) (Brooks 1958). N. Criddle (1935) listed it as an inhabitant of hilly country, among gravelly glacial drift. Feeding behavior has not been adequately studied. Onsager and Mulkern (1963) describe the eggs.

Life history. Univoltine, with overwintering as eggs. Adults are known from June to mid-September.

Economic importance. At high population levels, *B. brunnea* can cause economic damage to rangeland grasses, such as was reported in British Columbia in 1920 (Buckell 1922; Beirne 1972).

Bruneria yukonensis Vickery

Map 227

Bruneria yukonensis Vickery, 1969*b*:265, 266, figs. 1–4, 267, figs. 5, 7.

Diagnosis. Like *B. brunnea* but with head relatively smaller, and concealed genitalia different. Known only from Yukon Territory.

Description. Frontal costa narrow at vertex, sulcate below median ocellus, with lateral foveolae narrow, decreasing in width toward vertex. Pronotum with median carina faintly cristate on prozona, elevated but not arcuate on metazona. Tegmina nearly reaching but not exceeding apex of abdomen. Color and pattern as for *B. brunnea*.

Range. Yukon Territory.

Behavior and habitats. Virtually unknown.

Life history. Not known. Type series collected 23 August.

Genus *Aulocara* Scudder

Description. Antennae slender, slightly flattened toward base. Head large, tumid, with frons slanting; lateral foveolae triangular, visible from above; depression of vertex without median carinula. Pronotum with median carina continuous, cut by one sulcus; lateral carinae weak or absent; posterior margin obtuse to rounded; prosternum without tubercle. Inner apical spurs of hind tibia unequal in size. Tegmina opaque. Hind wings clear.

Key to species of *Aulocara*

(Adapted from Brooks (1958))

1. Tegmen extending to apex of hind femur. Inner spurs of hind tibia only slightly unequal (Fig. 759). Color greenish or brownish gray. Found in grasslands .... ***elliotti* (Thomas)** (p. 562)
   Tegmen extending only to basal third of hind femur. Inner spurs of hind tibia conspicuously unequal (Fig. 760). Color yellowish brown. Found in eroded areas .... ***femoratum* Scudder** (p. 564)

Clé des espèces d'*Aulocara*

(adaptée de Brooks, 1958)

1. Tegmina rejoignant l'apex des fémurs postérieurs. Éperons internes des tibias postérieurs seulement légèrement inégaux (fig. 759). Coloration verdâtre à gris brunâtre. Insecte vivant dans les prairies .... ***elliotti* (Thomas)** (p. 562)
   Tegmina ne rejoignant que le tiers de la base des fémurs postérieurs. Éperons internes des tibias postérieurs clairement inégaux (fig. 760). Coloration brun jaunâtre. Insecte vivant dans les régions érodées .... ***femoratum* Scudder** (p. 564)

Aulocara elliotti (Thomas)

Figs. 750, 759, 778, Pl. VII*A*; Map 228

St[*auronotus*] *elliotti* C. Thomas, 1870:82.
Aulocara caeruleipes Scudder, 1876*a*:266.
Aulocara elliotti; Bruner 1885*b*:10; Brooks 1958:54; Helfer 1963:165.

Common name. Bigheaded grasshopper.

Diagnosis. Lateral foveolae conspicuous, triangular, visible from above. Tegmina at rest showing pale median stripe. Inner apical spurs of hind tibia slightly unequal. Hind tibia usually blue.

Description. Body small-medium (length, male 15–20, female 21–27 mm) (Fig. 778). Head large in proportion to body, with lateral foveolae conspicuous (Fig. 750). Pronotum with lateral carinae lacking or nearly so, with margins cut by 3 sulci, and with median carina weak in front of single sulcus which cuts it. Tegmina extending beyond apex of abdomen. Inner apical spurs of hind tibia slightly unequal (Fig. 759). Male stridulatory pegs on inner face of hind femur as in Plate VII*A*. Color dark grayish brown, usually with pale median dorsal line evident when tegmina at rest; tegmina opaque with diffuse brown spots; hind femur with inconspicuous bars on outer face; inner face and lower flange often blue; hind tibia usually blue, occasionally buff to yellowish. Nymphs are described by Scoggan and Brusven (1972).

Range. Interior British Columbia to Manitoba, south to Texas.

Behavior and habitats. *A. elliotti* is found in grasslands, especially on dry grassy hillsides. Buckell (1920) reported it to be a "powerful jumper"

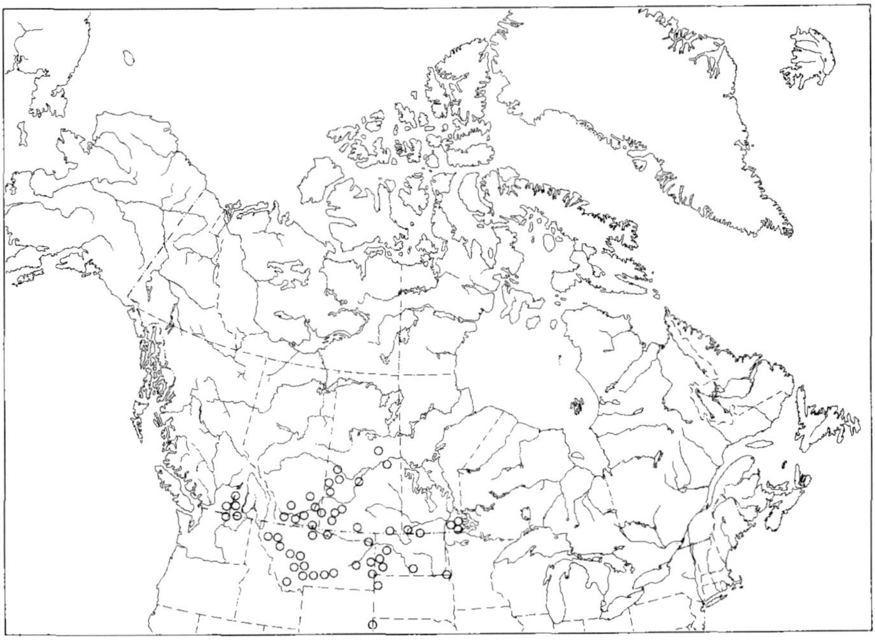

Map 228. Collection localities for *Aulocara elliotti*.

that does not fly much. It is primarily a grass feeder, seeming to prefer bluestem (*Agropyron smithii*), although it also feeds readily on grama grass (*Bouteloua gracilis*), *Sporobolus cryptandrus, Stipa comata*, and on the sedge *Carex filifolia* (Mulkern et al. 1969). No feeding on forbs or shrubs has been observed. Eggs are described by Onsager and Mulkern (1963). The general bionomics are outlined by Scoggan and Brusven (1972); see also S. N. Van Horn (1965, 1968).

Otte (1970) reported on acoustic and courtship behavior of this species. Males run rapidly after females, stopping occasionally to stridulate. Rejected males are kicked away by nonreceptive females. Stridulation is also used aggressively when one male approaches and touches another.

Life history. There is one generation each year, with overwintering in the egg stage in soil. There is an average of eight eggs per pod. Adults were recorded in Montana from 26 June to 5 October (Hebard 1928), and in North Dakota from 15 July to 7 September (Hebard 1936*a*). Specimens from Canadian localities are known from mid-July to the end of September.

Economic importance. Beirne (1972) referred to reports of damage to range grasses in Alberta and British Columbia. Crop losses have also been reported on rangelands in the United States (Hewitt 1978). *A. elliotti* has the potential to cause serious losses whenever it occurs at high population densities.

Aulocara femoratum Scudder

Figs. 751, 760; Map 229

Aulocara femoratum Scudder, 1899*f*:55.

Aulocara femoratum; Bruner 1905:111; Jago 1971:262.

Drepanoptera femoratum; J. A. G. Rehn 1927:226; Brooks 1958:56; Helfer 1963:165.

Diagnosis. Inner apical spurs of hind tibia conspicuously unequal. Tegmina not reaching apex of abdomen. Hind tibiae blue to purple, black at base.

Description. Body small (length 18–22 mm). Depression of vertex deep; lateral foveolae shallow (Fig. 751). Pronotum [illegible] lateral carinae weak or absent ahead of sulcus, apparent behind it; median carina evident throughout, with hind margin rounded. Tegmina short, [illegible] reaching apex of abdomen. Inner apical spurs of hind tibia conspicuously unequal (Fig. 760). Color brownish yellow, with pronotum dark laterally a[illegible] of sulcus; hind femur pale with conspicuous black bars on both inner [illegible] outer faces; hind tibia blue to purple, black at base.

Range. Alberta and Saskatchewan, south [illegible] Mexico.

Behavior and habitats. Brooks (1958) rep[illegible] *A. femoratum* from only a few localities in deeply eroded valleys of t[illegible]k, Lost, and Frenchman rivers. Anderson and Wright (1952) foun[illegible] chiefly on "clay-soil, bluestem flats." They reported the favored f[illegible]lant to be bluestem

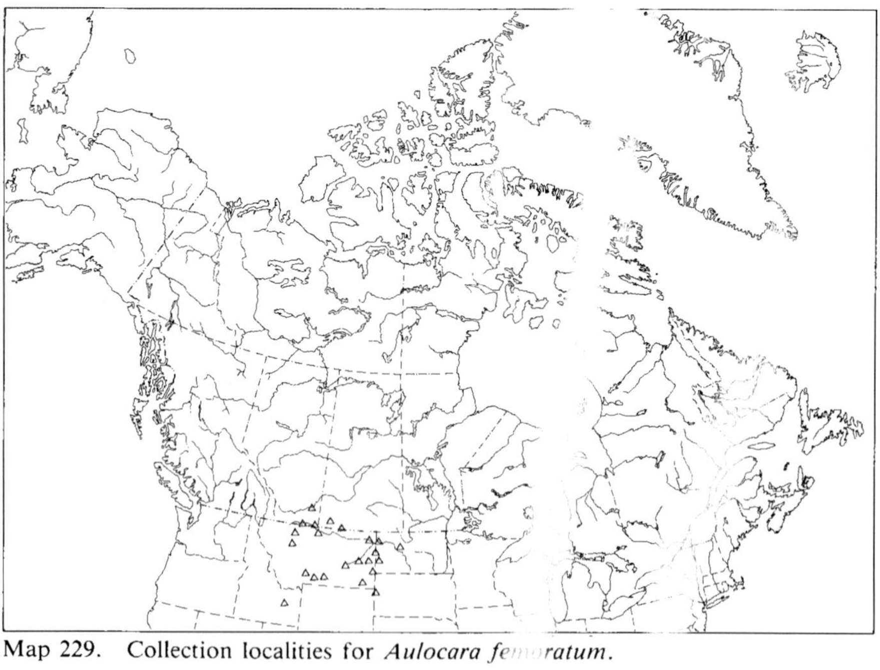

Map 229. Collection localities for *Aulocara femoratum*.

(*Agropyron smithii*), although other grass species were eaten by both nymphs and adults. Older nymphs and especially adults are gregarious, although males are on the move almost continuously. When disturbed, individuals either fly for short distances or jump or crawl on the ground (Anderson and Wright 1952). Otte (1970) observed that males move across bare ground by a series of short hops, each of about 5 cm distance. Onsager and Mulkern (1963) describe the eggs.

In respect of male courtship behavior, Otte (1970) noted femur-tipping at a slow rate, rapid femur-tipping when the male came near a female, and ticking sounds and bursts of vibratory stridulation. These behavioral functions were used in courtship and to some extent as aggressive signals.

Life history. There is one generation each year, with overwintering as eggs in the soil. There are 6–19 eggs in a pod (Ball et al. 1942). Nymphs and adults are found from June onward (in the south), nymphs occurring until August and adults until frost.

Economic importance. *A. femoratum* has the potential to cause economic damage to grasses when population densities are high. Ball et al. (1942) listed this species as one of the three most important rangeland grasshoppers in Arizona. It is unlikely to be of more than minor concern in Canada.

Genus *Boopedon* Thomas

Description. Vertex of head convex, sometimes flat, usually with faintly indicated median carinula anteriorly; lateral foveolae of vertex present, poorly developed, but visible from above. Pronotum with median carina distinct throughout, cut by one sulcus; lateral carinae nearly obsolete, diverging on metazona; lateral edges of disc cut by 2 or 3 sulci. Inner apical spurs of hind tibia unequal, one twice as long as others. Tegmina broadest at middle, usually short, not longer than combined length of head and pronotum, in males rarely extending beyond apex of abdomen, but not beyond apices of hind femora. Pronotal disc never with triangular black markings.

Boopedon nubilum (Say)

Fig. 777; Map 230

Gryllus nubilus Say, 1825*a*:308.
Boopedon nigrum C. Thomas, 1870:83.
Boopedon flavo-fasciatum C. Thomas, 1870:84.
Boopedon nubilum; C. Thomas 1871*a*:265.

Diagnosis. Males shiny black; females pale brown. Inner apical spurs of hind tibia unequal, one twice as long as others.

Description. As for genus (Fig. 777). Body moderate (length 25–39 mm). Coloration dimorphic: Males shiny black; face dark brown to black; hind femur entirely black or with pale preapical ring; hind tibia sometimes entirely black, or black and red, occasionally black and red with pale area. Females pale brown to straw-colored, sometimes with green on head,

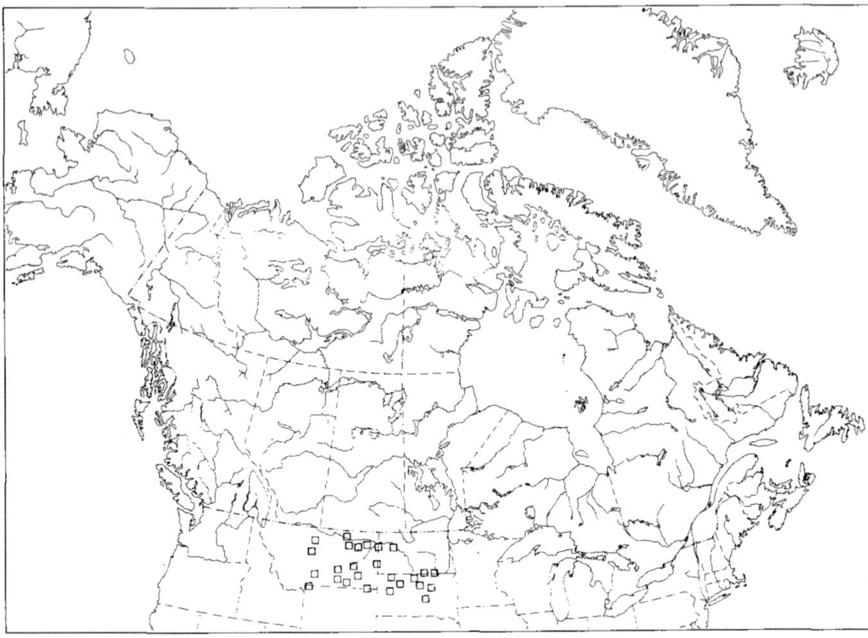

Map 230. Collection localities for *Boopedon nubilum*.

pronotum, and hind femora; sulci of lateral pronotal lobes usually black; genicular area of hind femora and tibiae black.

Range. Montana and North Dakota, south to Mexico.

Behavior and habitats. *B. nubilum* is mainly confined to short-grass prairie. Males are active and conspicuous, but the females, which are larger, well camouflaged, and not usually active, are not seen unless sought. Even "long-winged" males cannot fly readily, but all use their wings to some extent to assist leaping. Females usually dive into grass and hide when disturbed (Hebard 1928). The eggs are described by Onsager and Mulkern (1963).

Life history. Presumably univoltine, with overwintering as eggs. Adults are known from July to September in North Dakota and Montana (Hebard 1928, 1932*c*, 1936*a*).

Genus *Ageneotettix* McNeill

Description. Antennae slender, slightly flattened toward base. Frons slanting; lateral foveolae of vertex rectangular, conspicuous, visible from above; depression of vertex without median carinula. Pronotum strongly constricted, with median carina of pronotum strong throughout, cut by one sulcus, with lateral carinae evident behind sulcus but weak anteriorly, with lateral margins cut by 3 sulci, and with posterior margin rounded; prosternum without tubercle. Inner apical spurs of hind tibia conspicuously unequal (Fig. 761). Tegmina opaque, short, not reaching apices of hind femora. Hind wings clear, transparent.

Ageneotettix deorum (Scudder)

Figs. 752, 753, 761, 779; Plate VII*B*; Map 231

Chrysochraon deorum Scudder, 1876*a*:262.

Ageneotettix occidentalis Bruner, 1904*a*:58.

Ageneotettix deorum; J. A. G. Rehn and Hebard 1906:371; Helfer 1963:167; R. D. Alexander et al. 1972:53.

Ageneotettix deorum deorum; Hebard 1935*d*:283; Brooks 1958:53.

Diagnosis. Lateral foveolae rectangular, conspicuous, visible from above. Inner apical spurs of hind tibia conspicuously unequal. Body with pale dorsal stripe from head to apices of tegmina; hind tibia red or orange with prebasal pale and basal black rings.

Description. As for genus (Fig. 779). Body small (length, male 10–16, female 15–22 mm). Male stridulatory pegs on inner face of hind femur as in Plate VII*B*. Color mottled brown above, yellowish beneath, often with pale median dorsal stripe extending from head to apices of tegmina; antennae often whitish; lateral carinae of pronotum usually marked by pale line, margined on both sides with black; tegmina opaque, whitish with diffuse brown spotting; hind femur with 3 more or less distinct dark crossbars dorsally and on outer face, with apices polished black; hind tibia red or orange with pale prebasal ring, black basally. Nymphs are described by N. Criddle (1926*b*) and by Scoggan and Brusven (1972).

Range. Interior British Columbia to Michigan, south to New Mexico and Texas.

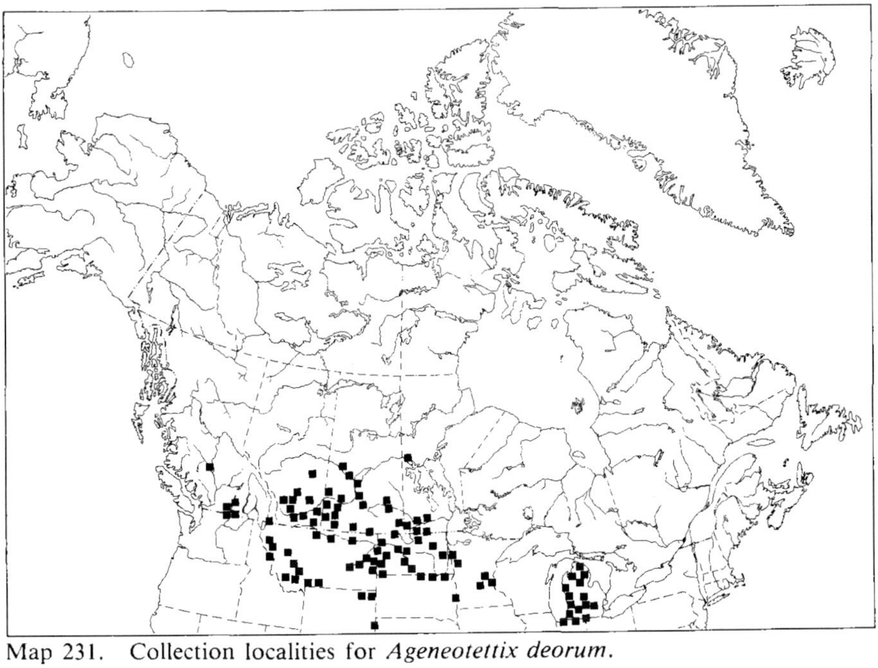

Map 231. Collection localities for *Ageneotettix deorum*.

Behavior and habitats. *A. deorum* is primarily a grass feeder, but it also feeds extensively on dry plant debris. *Bouteloua gracilis* is one of the main food grasses, but the species also feeds readily on many other grasses and sedges, on a few forbs, and even on cattle droppings. It usually inhabits areas with open sandy soils with relatively poor vegetation. Some account of the biology of the species is given by N. Criddle (1926*b*) and by Scoggan and Brusven (1972). The eggs are described by Tuck and Smith (1940) and by Onsager and Mulkern (1963).

Buckell (1922) remarked that males "make a shrill squeaky stridulation while in pursuit of the females." Otte (1970) stated that pair formation occurs only when males approach moving females. Femur-tipping by courting males is common; the downstroke is always faster than the upstroke. It is also used as an aggressive signal but more slowly and less precisely than in courtship. Males also raise and lower their antennae repeatedly and rapidly as they move about. Females that are not responsive repel males by kicking with the hind femora, by raising the femora to a vertical position, or by stridulating. The stridulation of females is similar to that of males, but softer.

Life history. There is only one generation each year, with overwintering in the egg stage. Over most of the range, adults are found from mid-July to late September. They have been reported as being active from 12 July to 10 September in Michigan (Cantrall 1968), and from 11 July to 25 October in Montana (Hebard 1928).

Genus *Psoloessa* Scudder

Description. Antenna slender, slightly flattened near base. Frons slanted, but not conspicuously so; lateral foveolae of vertex (Fig. 749) rectangular, conspicuous, visible from above; depression of vertex deep, without median carinula. Pronotum with lateral carinae strongly constricted before middle, nearly obsolete on metazona; median carina strong, cut by one sulcus; posterior margin obtuse–angulate; prosternum without tubercle. Inner apical spurs of hind tibia not equal in size. Tegmina long. Hind wings clear.

Key to subspecies of *Psoloessa delicatula*

1. Head, in dorsal aspect, much produced anterior to eyes (Fig. 775). Pronotum more elongate, with lateral carinae somewhat less incurved (Fig. 775). Found in extreme southern British Columbia and western Columbia Plains of Washington .... ***delicatula buckelli*** **Rehn** (p. 569)

   Head, in dorsal aspect, much less produced anterior to eyes (Fig. 776). Pronotum less elongate, with lateral carinae strongly incurved. Found in Manitoba to Alberta and Montana .... ***delicatula delicatula*** **(Scudder)** (p. 570)

Clé des sous-espèces de *Psoloessa delicatula*

1. Tête, vue du dessus, très développée en avant des yeux. Pronotum plus allongé, à crêtes latérales un peu moins incurvées. Espèce vivant à l'extrême sud de la Colombie-Britannique et dans l'ouest des plaines Columbia, dans l'État de Washington .... ***delicatula buckelli*** **Rehn** (p. 569)

   Tête, vue du dessus, beaucoup moins développée en avant des yeux (fig. 776). Pronotum moins long, à crêtes latérales fortement incurvées. Insecte vivant du Manitoba à l'Alberta et au Montana .... ***delicatula delicatula*** **(Scudder)** (p. 570)

Psoloessa delicatula buckelli Rehn

Figs. 749, 775, 776; Map 232

Psoloessa delicatula buckelli J. A. G. Rehn, 1937*b*:326.

Psoloessa delicatula buckelli; J. A. G. Rehn 1942:213; Brooks 1958:61; Helfer 1963:170.

Diagnosis. Head more produced ahead of eyes than in *P. delicatula delicatula*. Body slender. Found in British Columbia and Washington.

Description. Body long, narrow, attenuate, deep. Frons slanting. Fastigium of vertex produced ahead of eyes (Fig. 775), in male, with margins rectangulate in dorsal aspect. Pronotal disc short; lateral carinae constricted (Fig. 775). Hind femur slender, elongate.

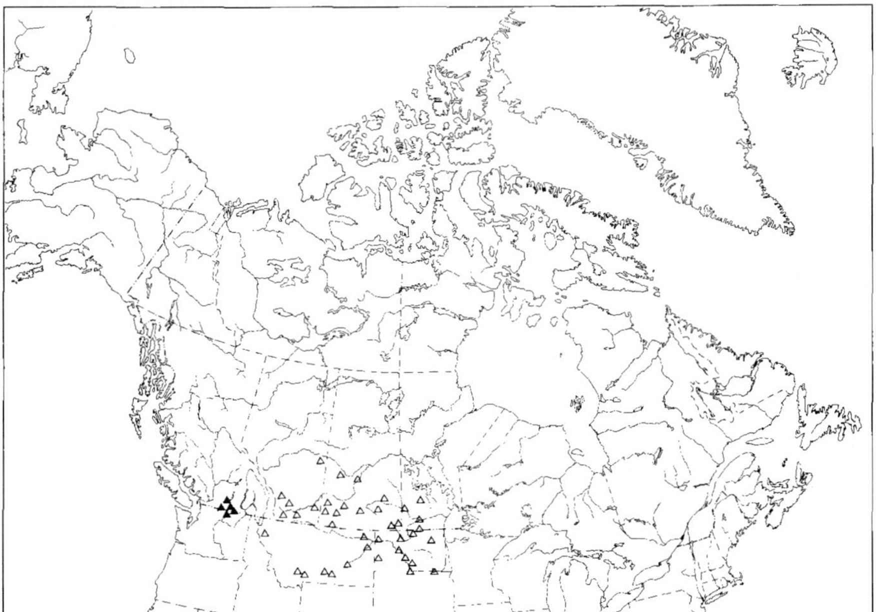

Map 232. Collection localities for *Psoloessa delicatula buckelli* (▲) and *P. delicatula delicatula* (△).

Range. Interior British Columbia to interior California.

Behavior and habitats. This species inhabits stony ground and sagebrush areas as well as open bunch-grass plains (Buckell 1920). Presumably the behavior and food plants are the same as, or similar to, those of *P. delicatula delicatula*.

Life history. As for *P. delicatula delicatula* but maturing slightly earlier in spring. Buckell (1920) reported adults on 26 April in a sheltered spot where the snow had melted early. They were common by 4 May and were ovipositing from 19 May to mid-June. Numbers decreased rapidly after mid-June, and the insects had disappeared by the end of that month.

Psoloessa delicatula delicatula (Scudder)

Map 232

Scyllina delicatula Scudder, 1876*a*:263.
Stirapleura decussata Scudder, 1876*c*:510.
Psoloessa delicatula delicatula; Hebard 1928:231; Brooks 1958:61.

Diagnosis. Inner apical spurs of hind tibia not equal in size. Body dark gray to green with conspicuous black markings. Lateral carinae of pronotum strongly constricted.

Description. As for genus (Fig. 776). Body small (length 16–18 mm). Color dark gray to green with conspicuous black markings behind eye, on sides of pronotum, and next to lateral carinae on metazona; tegmina grayish brown, shining, with 4–6 rectangular black spots on median line; hind femur with incomplete black bands dorsally; hind tibia yellow, brown, or pale blue. N. Criddle (1926*b*) and Scoggan and Brusven (1972) describe the nymphs.

Range. Alberta to Manitoba, south to Colorado.

Behavior and habitats. N. Criddle (1935) described this as one of the dominant grasshopper species in sandy uplands. Brooks (1958) stated that it feeds upon grasses and sedges (*Stipa, Agropyron, Festuca*, and *Carex* spp.) and also on *Phlox hoodi* Richards. Onsager and Mulkern (1963) describe the eggs. Scoggan and Brusven (1972) outline the bionomics.

Otte (1970) recorded two courtship acoustic signals by males; the first of rather short duration produced when other nearby individuals moved; the second comprising two to four widely but uniformly spaced pulses produced when the male was approaching a female. No femur-tipping was observed during courtship. Aggressive signals, in the form of ticking sounds, are produced by both sexes; by males when touched by other males, or in response to their songs, and by unreceptive females when repelling males which had attempted to mount them.

Life history. Pickford (1953) indicated a 2-year life cycle in Saskatchewan as follows: eggs laid in June and early July, overwintering in this stage, hatching in August (the following year) and developing to late-instar nymphs before the second winter, eventually reaching adulthood early in the following season. Adults were reported in Montana from 12 May to 18 June (Hebard 1928), and from 6 May to 4 July in Alberta (Hebard 1931*a*).

Genus *Eritettix* Bruner

Description. Antennae flattened, especially at base and apex (Fig. 773), oval in cross-section. Lateral foveolae shallow, not visible from above; depression of vertex with median carinula. Pronotum with median carina low, cut by one sulcus, with lateral carinae evident, slightly constricted at middle to nearly parallel; posterior margin obtuse-angulate (Fig. 774). Inner apical spurs of hind tibia conspicuously unequal (Fig. 758). Tegmina extending nearly to apex of abdomen. Hind wings clear.

Eritettix simplex tricarinatus (Thomas)

Figs. 758, 772–774; Map 233

Gomphocerus simplex Scudder, 1869*a*:305 (partim).
St[*enobothrus*] *tricarinatus* C. Thomas, 1873:84.
Eritettix simplex; McNeill 1896:219 (partim); Helfer 1963:182 (partim).
Eritettix simplex tricarinatus; Hebard 1936*a*:29; Brooks 1958:57.

Diagnosis. Body grayish. Male head and pronotum with pair of narrow dark lines dorsally; female gray with contrasting dark and white markings. Lateral carinae of pronotum only slightly constricted.

Description. As for genus. Dimorphic. Male as in Fig. 772: body small (length 15–17 mm). Color gray to grayish brown; dorsum of head and pronotum with pair of narrow dark lines; lateral lobes of pronotum, mesothorax,

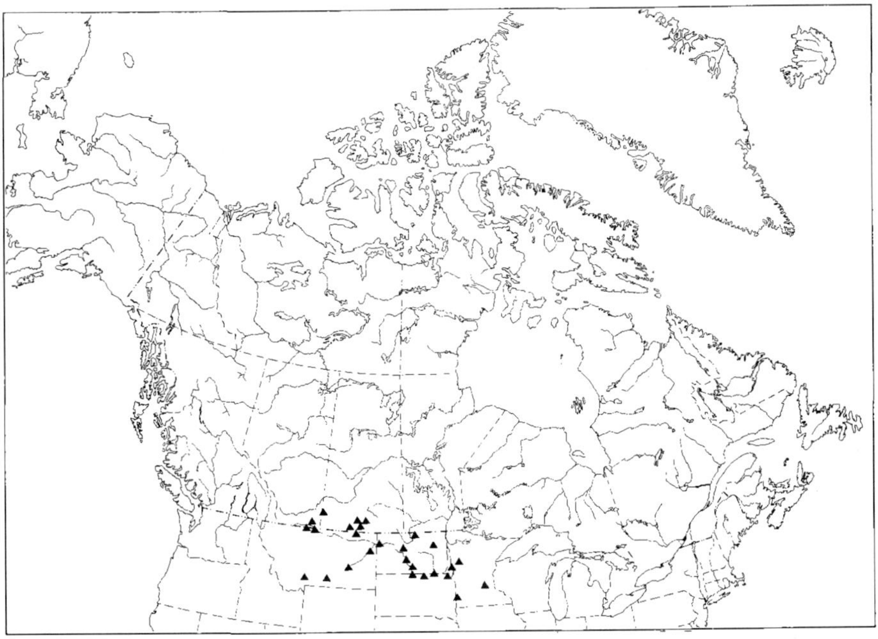

Map 233. Collection localities for *Eritettix simplex tricarinatus*.

and metathorax dark; tegmina uniform dark gray with pale streaks on costal margin and near apex. Female: small, but larger and more robust than male (length 20–24 mm). Color gray with contrasting dark and creamy white markings, or bright green with creamy white and dark lines; tegmina brownish to green with row of dark markings; hind femur dark or green on at least upper half of outer face, pale below; hind tibia buff, mottled with brown, with spines black-tipped.

Range. Alberta to Minnesota, south to Colorado.

Behavior and habitats. *E. simplex tricarinatus* is a feeder on grasses, principally *Bouteloua gracilis*, according to Mulkern et al. (1969). These authors state that it sometimes becomes very numerous. Brooks (1958), however, states that, although it is fairly common on grassy hillsides in eastern Alberta and southwestern Saskatchewan, it is rarely collected there. The eggs are described by Onsager and Mulkern (1963).

Otte (1970) studied acoustic and courtship behavior of *E. simplex*, presumably *E. simplex tricarinatus*. Males on grasses stridulate up to three times in succession, then move about 30 cm and stridulate again. During a courtship observation, a male was seen to follow a female up a grass stem until he was within 3 cm of her; then he rocked from side to side, stopped, stridulated, and rushed upward toward the female. Bursts of stridulation involve the use of one hind leg at a time, but legs are switched during each burst. Stridulation by one male causes nearby males to stridulate also. Irregular bursts of stridulation, wild jumping, and jerking of the hind femora were all observed as being aggressive acts.

Life history. There appears to be one generation each year, but the cycle differs from that of most acridids in that winter is passed as partly grown nymphs. Such immature individuals have been reported in April and May and again in September. Adults appear early in spring and most of them disappear before summer is over. Hebard (1925*a*) recorded adults from South Dakota from 11 May to 24 August.

Economic importance. Mulkern et al. (1969) state that, as this insect feeds upon high-value grasses and may become very abundant, it has the potential to become a pest in the Great Plains region of the United States. Brooks's (1958) information, however, leads one to suppose that this is not so in Canada.

Genus *Cordillacris* Rehn

Description. Antennae slender, somewhat flattened near base. Frons slanting; lateral foveolae of vertex shallow, large, open below, not visible from above; depresssion of vertex without median carinula. Pronotum with median carina nearly absent anterior to sulcus, low posterior to sulcus; lateral carinae obsolete, indicated only by color lines, considerably constricted at middle; posterior margin rounded; prosternum without tubercle. Inner apical spurs of hind tibia nearly equal. Tegmina long, opaque, exceeding apex of abdomen but not apices of hind femora. Color of lateral pronotal lobes usually dark on upper half, pale gray to white on lower half; hind femora barred, but not strongly so, usually with upper half dark and lower half pale.

Key to species of *Cordillacris*

1. Tegmen with dorsal and dorsolateral areas brown, with pale markings of costal margin encroaching on the brown as a series of broad regular "teeth" (Fig. 789). Found in southeastern Alberta and Montana .. .. *crenulata* **(Bruner)** (p. 573)

Tegmen gray to brown, with series of elongate spots (Fig. 788). Found in Manitoba and North Dakota to Alberta and Montana .... *occipitalis* **(Thomas)** (p. 574)

Clé des espèces de *Cordillacris*

1. Tegmen à régions dorsale et dorsolatérale brunes, la zone plus pâle du bord antérieur s'y prolongeant en une série de "dents" pâles, larges et régulières (fig. 789). Espèce vivant au sud-est de l'Alberta et au Montana .... *crenulata* **(Bruner)** (p. 573)

Tegmen de couleur variant de gris à brun, marqué d'une série de taches allongées (fig. 788). Espèce vivant au Manitoba et au Dakota-nord, jusqu'en Alberta et au Montana .... *occipitalis* **(Thomas)** (p. 574)

Cordillacris crenulata (Bruner)

Figs. 784, 786, 787, 789; Plate VII*D*; Map 234

Orchrilidia crenulata Bruner, 1890:51.
Cordillacris crenulata; Bruner 1904*b*:70; Jago 1969:281.
Cordillacris crenulata crenulata; Hebard 1935*d*:281; Brooks 1958:56.

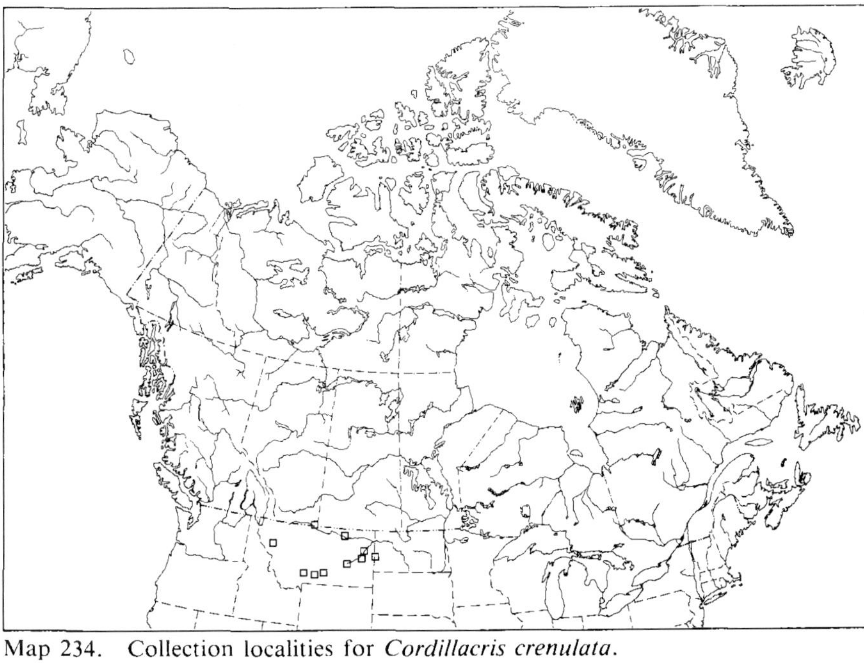

Map 234. Collection localities for *Cordillacris crenulata*.

Diagnosis. Tegmina opaque, whitish, with 4 or 5 broad dark crenulations extending from anal to median area.

Description. Body small (length 13–18 mm) (Fig. 784). Lateral carinae obsolete. Male stridulatory pegs on inner face of hind femur as in Plate VII*D*. Tegmina whitish, opaque; median area with 4 or 5 broad dark crenulate maculations that project toothlike from the darkened anal area (giving the species its name); costal region without spots (Fig. 789). Lateral carinae of pronotum marked as yellow lines, conspicuously constricted at middle in both sexes (Fig. 786). Hind femur pale ventrally; hind tibia buff.

Range. Alberta to North Dakota, south to Arizona and Texas.

Behavior and habitats. Hebard (1925*a*) reported that *C. crenulata* preferred short-plains grasses, particularly in gravelly areas. Anderson and Wright (1952) found the species to be most numerous on blue grama grass (*Bouteloua gracilis*), which appears to be the favored food plant. It was also found feeding on thread-leaf sedge (*Carex filifolia*). Otte (1970) observed that individuals, when flushed, tended to land on bare ground.

According to Otte (1970), courtship involves both femur-tipping and stridulation. Femur-tipping is similar to that of *C. occipitalis*, but stridulation, produced by both hind femora operating simultaneously, has a much slower pulse rate. Quick short dipping movements of the antennae accompany femur-tipping and may serve as an additional signaling procedure. Aggressive behavior is similar to that of *C. occipitalis*, with femur-tipping as the primary mechanism in preservation of male territoriality.

Life history. There is one generation each year, with overwintering as eggs in soil. Adults are active in Montana from 8 July to 8 September (Hebard 1928). R. M. White and Rock (1945) recorded the adult at Onefour, Alta., on 15 July.

Cordillacris occipitalis (Thomas)

Figs. 783, 785, 788; Map 235

Stenobothrus occipitalis C. Thomas, 1873:81.

Orchrilidia cinerea Bruner, 1890:52.

Cordillacris occipitalis occipitalis; Hebard 1925*a*:55.

Cordillacris occipitalis cinerea; Hebard 1925*a*:56.

Cordillacris occipitalis; Mills and Pepper 1938:9; Helfer 1963:171; Jago 1969:275, 276, 278.

Diagnosis. Lateral carinae of pronotum obsolete; pronotum constricted at middle. Antennae white; tegmina opaque, whitish. Inner apical spurs of hind tibia nearly equal.

Description. As for genus (Fig. 783). Body small (length 15–22 mm). Lateral margins of pronotum only moderately constricted, this more pronounced in males than in females (Fig. 785). Antennae usually white; tegmina opaque, whitish, usually with dark spots in costal region (male) and a row of discrete dark spots medially (Fig. 785). Nymphs are described by Scoggan and Brusven (1972).

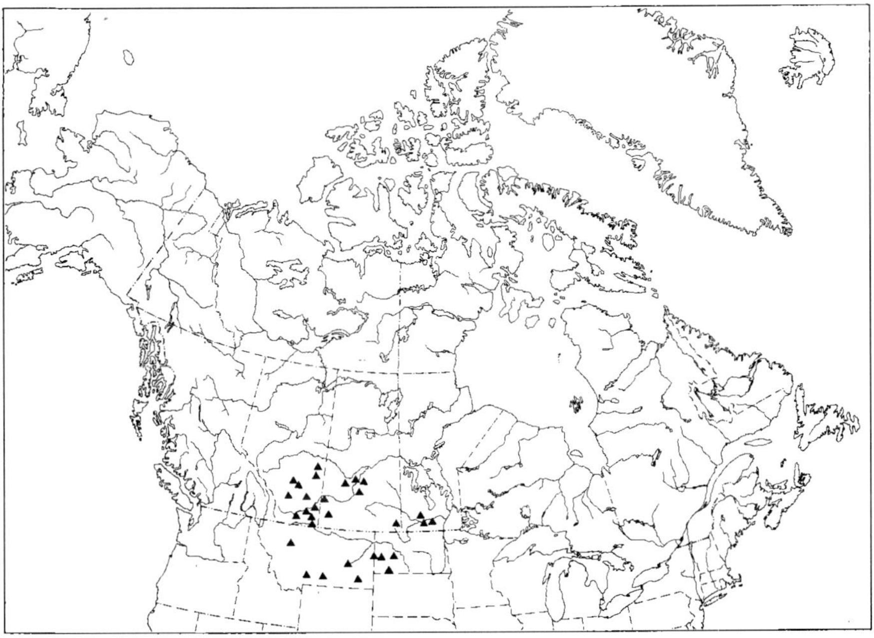

Map 235. Collection localities for *Cordillacris occipitalis*.

Range. Alberta to Manitoba, south to Arizona and Texas.

Behavior and habitats. N. Criddle (1935) called *C. occipitalis* a "sand-loving insect, usually found on dry gravelly hills or on the margins of drifting sand." Brooks (1958) listed grass species of the genera *Sporobolus, Agropyron*, and *Stipa* as food plants. This was confirmed by the study of Mulkern et al. (1969). Eggs are described by Onsager and Mulkern (1963). The bionomics of the species are outlined by Scoggan and Brusven (1972).

Otte (1970) found pair formation to be brought about by stridulation and also by rapid femur-tipping by males. In addition, the antennae are raised and lowered to the substrate. The white antennae undoubtedly perform a signaling function in pair formation, especially when the two sexes are close together. Femur-tipping does not always accompany stridulation and sometimes precedes, and at other times follows, stridulation. Courtship usually consists of femur-tipping by males to 25–30° from horizontal, in groups of two to five (at 2–3 strokes per second). Males assume a position at right-angles to the females prior to mounting and copulation. Femur-tipping and femur-raising are used as aggressive signals in male-to-male encounters. Nonreceptive females kick, raise their hind femora to a vertical position, or hop away from unacceptable ardent males.

Life history. There is one generation each year, with overwintering in the egg stage in soil. In Canada, adults have been recorded from 7 July to 15 September (E. M. Walker *in* Fletcher 1906; Tinkham 1939).

Economic importance. Mulkern et al. (1969) stated that this species could be of economic importance, but there is little evidence that it is so in Canada.

Genus *Amphitornus* McNeill

Description. Antennae slender, slightly flattened toward base. Frons slanting; lateral foveolae shallow, not visible from above; depression of vertex with median carinula. Pronotum with median carina low, cut by one sulcus; lateral carinae nearly obsolete; lateral margins of disc rounded; posterior margin convex; prosternum without tubercle. Inner spurs of hind tibia nearly equal (Fig. 757). Tegmina exceeding apex of abdomen. Hind wings clear.

Key to subspecies of *Amphitornus coloradus*

1. Median carina of pronotum cut at middle by principal sulcus. Body dull-colored, with markings not highly contrasting. Found in Manitoba to Alberta and Montana ..... ***coloradus coloradus* (Thomas)** (p. 576)
   Median carina of pronotum cut distinctly behind middle by principal sulcus. Body brightly colored, high contrasting. Found in British Columbia and southward .... ***coloradus ornatus* McNeill** (p. 578)

Clé des sous-espèces d'*Amphitornus coloradus*

1. Crête médiane du pronotum coupée au milieu par le sulcus principal. Corps de coloration mate, portant des marques peu contrastantes. Insecte vivant du Manitoba jusqu'en Alberta, ainsi qu'au Montana .... ***coloradus coloradus* (Thomas)** (p. 576)
   Crête médiane du pronotum clairement coupée en arrière du point médian par le sulcus principal. Corps à coloration brillante, très contrastante. Insecte vivant en Colombie-Britannique et plus au sud .... ***coloradus ornatus* McNeill** (p. 578)

Amphitornus coloradus coloradus (Thomas)

Fig. 757; Map 236

S[*tenobothrus*] *bicolor* C. Thomas, 1872:465 (name preoccupied).
St[*enobothrus*] *coloradus* C. Thomas, 1873:82 (replacement name).
Stenobothrus coloradus var. *unicolor* C. Thomas, 1873:83.
Amphitornus coloradus; E. M. Walker 1910:273 (partim); Helfer 1963:175 (partim).
Amphitornus coloradus coloradus; Froeschner 1954:208; Brooks 1958:53.

Diagnosis. Lateral foveolae shallow; depression of vertex with median carinula. Body brownish yellow, with paired broad dark lines dorsally from vertex to near apices of tegmina.

Description. As for genus (Figs. 770, 771). Body small (length, male 17–19, female 20–24 mm). Color brownish yellow, with pair of broad dark lines dorsally from vertex to, or nearly to, apices of tegmina; tegmen with yellowish to whitish line on basal third; hind femur pale with conspicuous black bands on upper half of outer face; hind tibia blue to purplish with

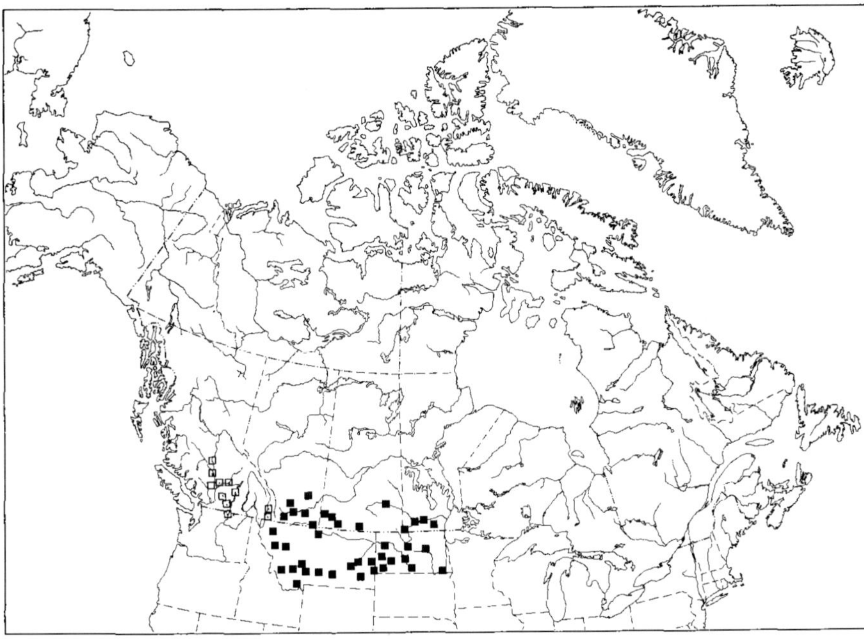

Map 236. Collection localities for *Amphitornus coloradus coloradus* (■) and *A. coloradus ornatus* (□).

black-tipped spines. Nymphs are described by N. Criddle (1926*b*) and by Scoggan and Brusven (1972).

Range. Alberta to Manitoba, south to Utah and Colorado.

Behavior and habitats. The usual habitat is dry grassy areas. Brooks (1958) found *A. coloradus coloradus* to be confined to "small, grassy spots, mostly along river valleys" in eastern Saskatchewan parklands, but common in grasslands elsewhere. Both nymphs and adults feed on grasses of the genera *Bouteloua, Stipa, Poa*, and *Agropyron* (Mulkern et al. 1969). In Montana, *A. coloradus coloradus* showed a definite preference for *Stipa comata*, according to Anderson and Wright (1952). These authors also mentioned that males were quite active, constantly moving from plant to plant, whereas females tended to remain feeding in one place for long periods of time. N. Criddle (1935) listed *Artemisia* and other forbs as alternative food plants, but the species is primarily a grass feeder. N. Criddle (1935), Tuck and Smith (1940), and Onsager and Mulkern (1963) describe the eggs.

Hubbell (1922*a*) described stridulation as a rather faint "zzzzz-zzzzz-zzzzz" at a rate of about three "notes" in 2 seconds.

Life history. There is one generation each year, with overwintering as eggs in soil. Hebard (1928) reported adults in Montana from 26 June to 19 September. All known specimens from Canadian localities were collected within these limits.

Amphitornus coloradus ornatus McNeill

Figs. 770, 771; Plate VI*H*; Map 236

Amphitornus ornatus McNeill, 1896:225.

Amphitornus nanus J. A. G. Rehn and Hebard, 1908:376.

Amphitornus coloradus ornatus; Hebard 1937*b*:357, 360; Helfer 1963:176.

Diagnosis. Median carina of pronotum cut behind middle of sulcus. Coloration contrasting.

Description. Similar to *A. coloradus coloradus*. Male stridulatory pegs on inner face of hind femur as in Plate VI*H*.

Range. British Columbia, south to Nevada and California.

Behavior and habitats. As for *A. coloradus coloradus*. Buckell (1920) found *A. coloradus ornatus* common on "dry bunch-grass ranges" in British Columbia.

Life history. As for *A. coloradus coloradus*. In British Columbia, adults occur from 27 June to the end of August (Buckell 1920).

Genus *Phlibostroma* Scudder

Description. Antennae slender, slightly flattened near base. Frons slanting, but not strongly so; lateral foveolae of vertex large, shallow, not visible from above; vertex of head with shallow depression and faint median carinula. Pronotum with median carina entire, strong, cut by one sulcus; lateral carinae weak on posterior half, strongly constricted ahead of middle; posterior margin obtuse; prosternum without tubercle. Inner apical spurs of hind tibia slightly unequal. Tegmina short, not reaching apex of abdomen. Hind wings clear.

Phlibostroma quadrimaculatum (Thomas)

Figs. 780–782; Plate VII*C*; Map 237

St[*enobothrus*] *quadri-maculatus* C. Thomas, 1871*a*:280.

Phlibostroma pictum Scudder, 1875*f*:517.

Phlibostroma quadrimaculatum; Bruner 1904*b*:72; Brooks 1958:60; Helfer 1963:181.

Stethophyma [error] *quadrimaculata* [*sic*!] Jago, 1969:297.

Diagnosis. Lateral foveolae large, shallow, not visible from above. Lateral carinae of pronotum strongly constricted ahead of middle. Tegmina with 4 or 5 irregular median dark blotches.

Description. As for genus (Fig. 780). Body small (length 14–26 mm). Pronotum constricted as indicated in Fig. 781. Male stridulatory pegs on inner face of hind femur as in Plate VII*C*. Color yellowish brown to greenish; head with black stripe behind each eye; pronotum with narrow black stripes or blotches; tegmina with 4 or 5 irregular dark blotches on median area (Fig. 782); hind femur faintly barred (bars often obsolete in female); hind tibia yellow to orange or pinkish, with spines black at tips.

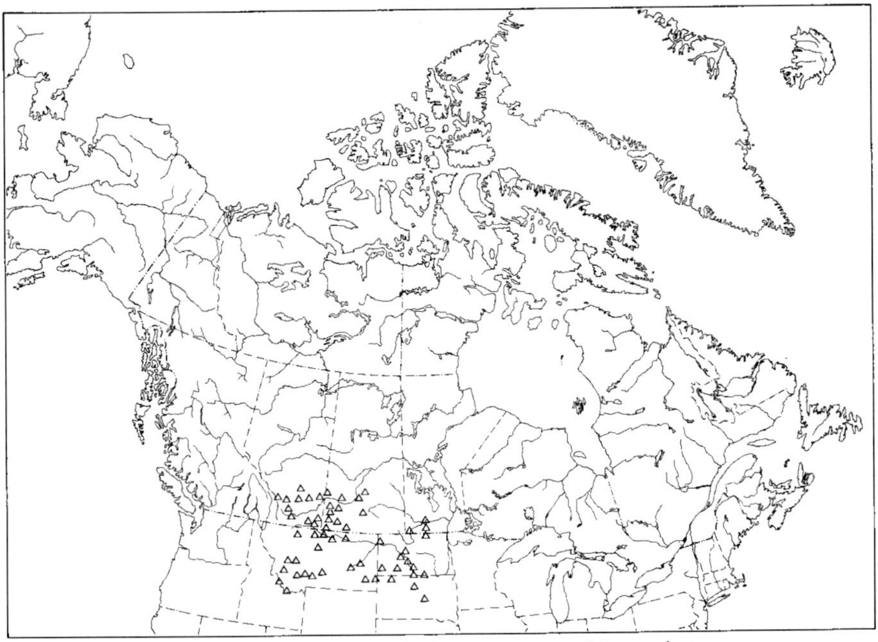

Map 237. Collection localities for *Phlibostroma quadrimaculatum*.

Range. Alberta to Manitoba, south to Mexico.

Behavior and habitats. Brooks (1958) reported this species as being fairly common on dry mixed prairies. Anderson and Wright (1952) stated that, in Montana, it is confined to localized areas in which blue grama (*Bouteloua gracilis*) forms the dominant vegetation. Similar circumstances were recorded by Mulkern et al. (1969) who note that a few other species of grasses were fed upon, but that blue grama made up most of the food supply. The eggs are described by Tuck and Smith (1940) and by Onsager and Mulkern (1963).

Otte (1970) reported femur-tipping by males as they approached females, as being the primary courtship signal in *P. quadrimaculatum*. This was rapid (duration about 0.3 second) and often was interspaced with bursts of stridulation. Males reacted to contact by other males with bursts of vibratory stridulation. Nonreceptive females rejected males by producing vibratory stridulation that was similar to but fainter than that of males. Jerking and shaking the femora were also used by females when repelling males.

Life history. Univoltine, with overwintering in the egg stage. Adults have been reported from 19 July to 8 September in North Dakota (Hebard 1936*a*), 20 July to 25 October in Montana (Hebard 1928), and 27 July to 15 September in Alberta (E. M. Walker 1910; Hebard 1931*a*; Tinkham 1939).

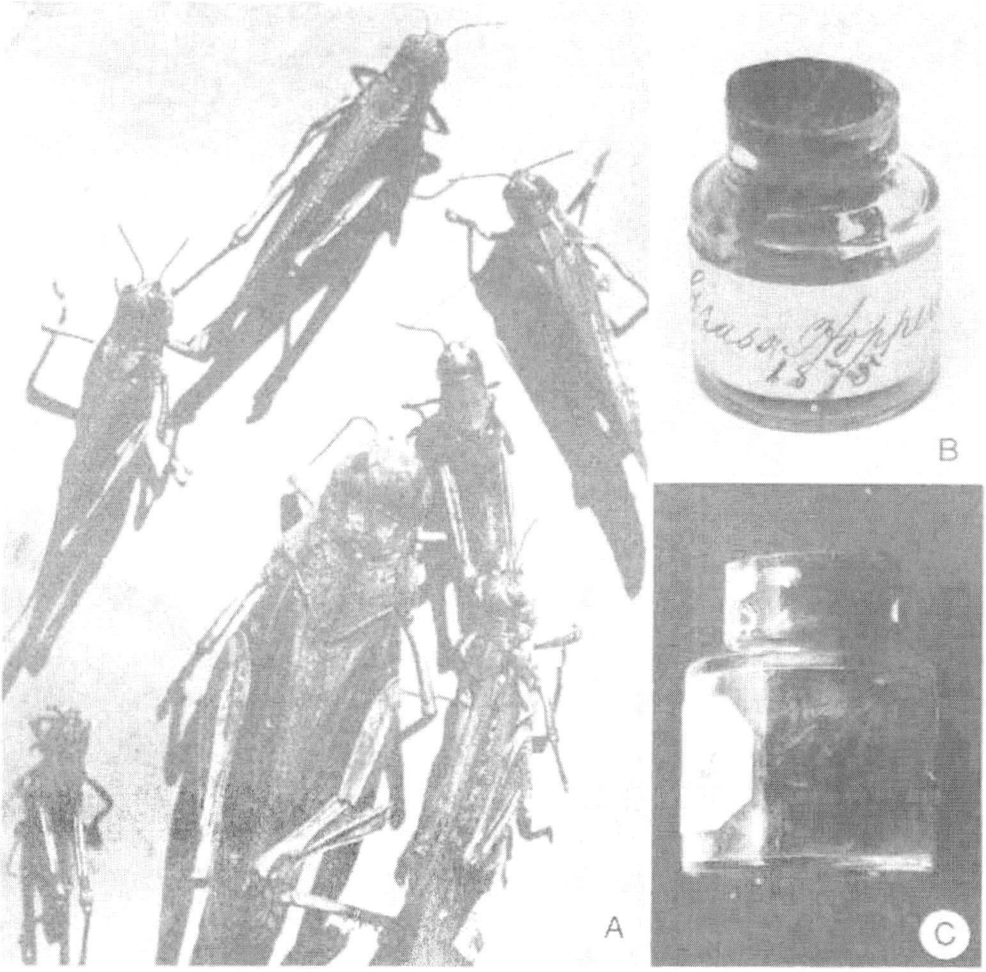

Plate I. Historical relics. *A*, Rocky Mountain grasshopper, *Melanoplus spretus*, and a female of *Pardalophora, apiculata* (large specimen) photographed during 1874 outbreak at Red River, Man. (S. Duffin). *B,C*, Bottle containing Rocky Mountain grasshoppers and other orthopteroids, buried in cornerstone of old city hall, Winnipeg (with above photograph) in 1875 and recovered in 1962. *A*, courtesy of Manitoba Provincial Archives, Winnipeg. *B,C*, courtesy of R. L. Randell.

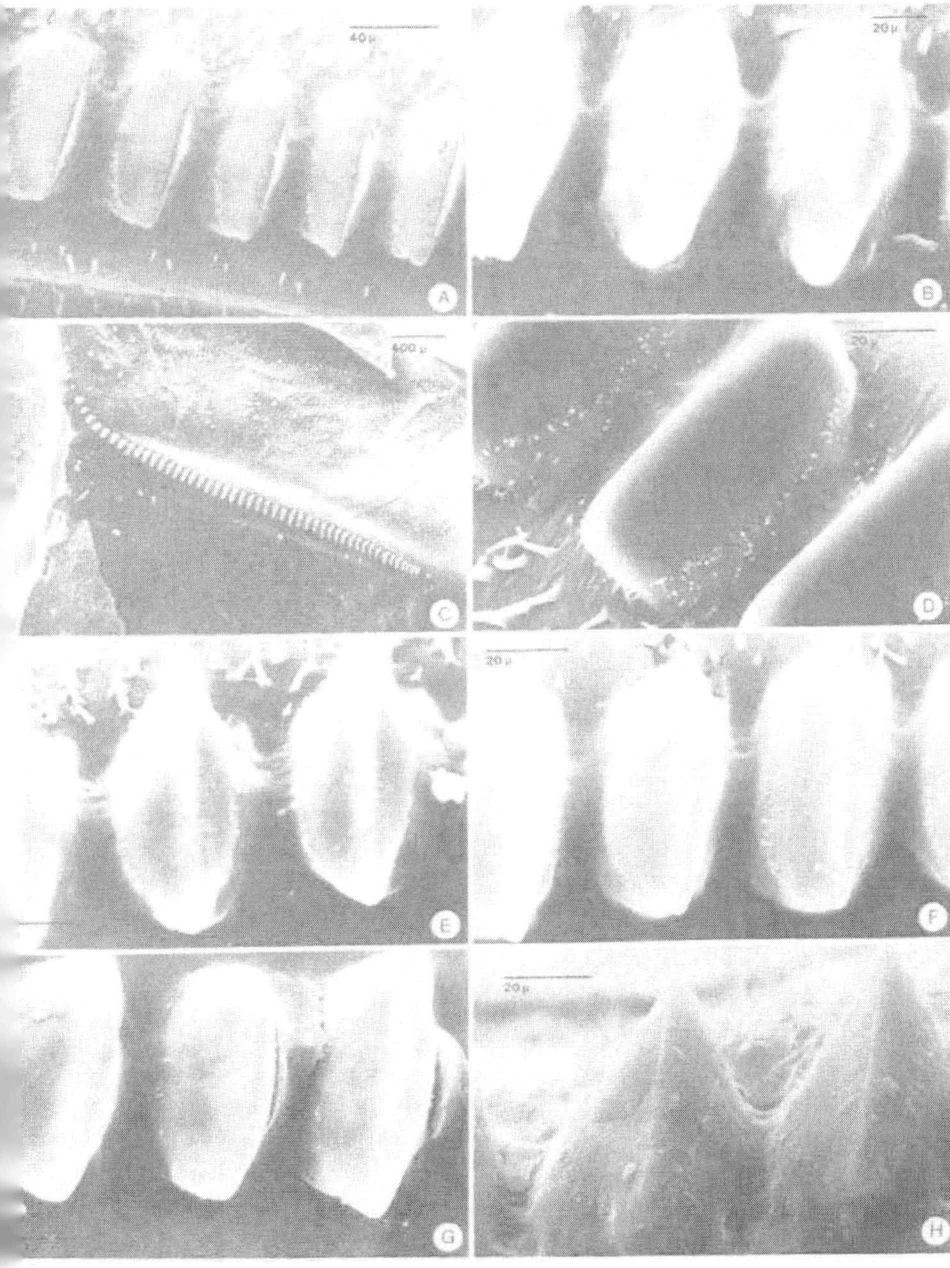

Plate II. Stridulatory files on tegmina of *Orchelimum* species (Tettigonioidea: Conocephalidae). *A*, *O. vulgare*; *B*, *O. gladiator*; *C*, the same, less greatly enlarged; *D*, *O. nigripes*; *E*, *O. concinnum*; *F*, *O. delicatum*; *G*, *O. campestre*; *H*, *O. volantum* (lateral).

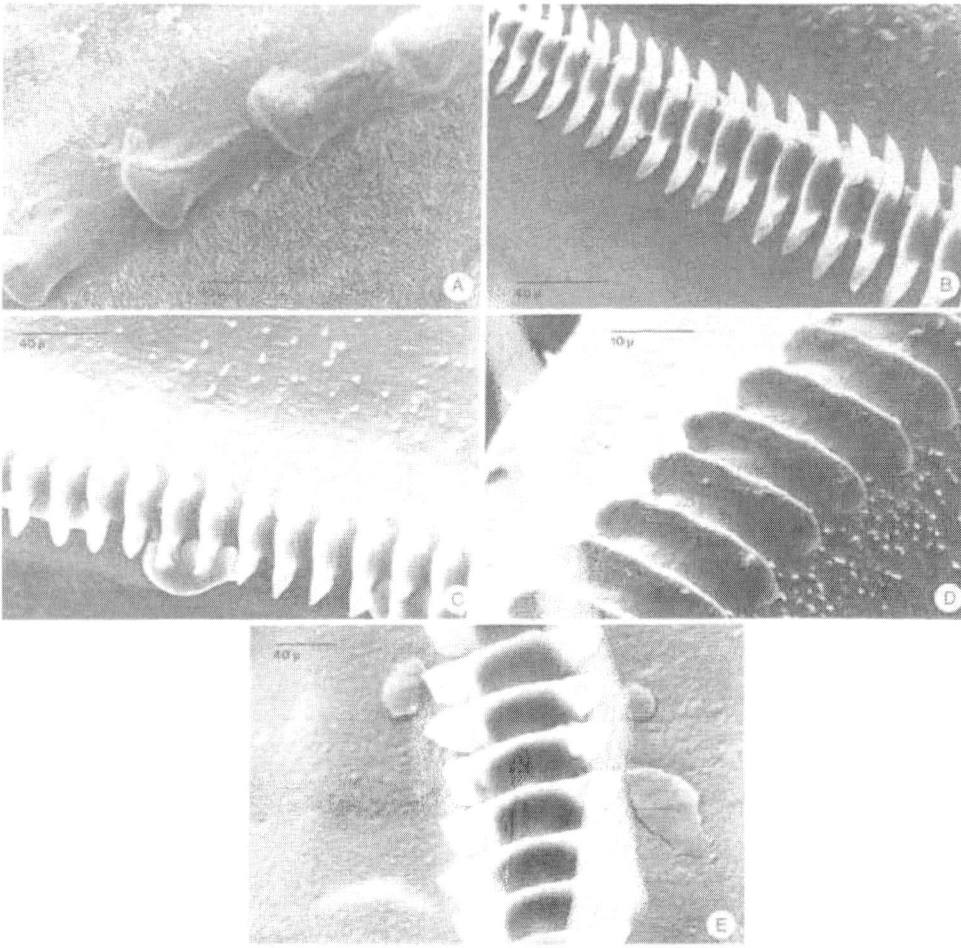

Plate III. Stridulatory files on tegr a of us Gryllodea. *A*, *Neocurtilla hexadactyla*; *B*, *Gryllus pennsylvanic* *C*, *letis*; *D*, *Acheta domesticus*; *E*, *Anaxipha exigua*.

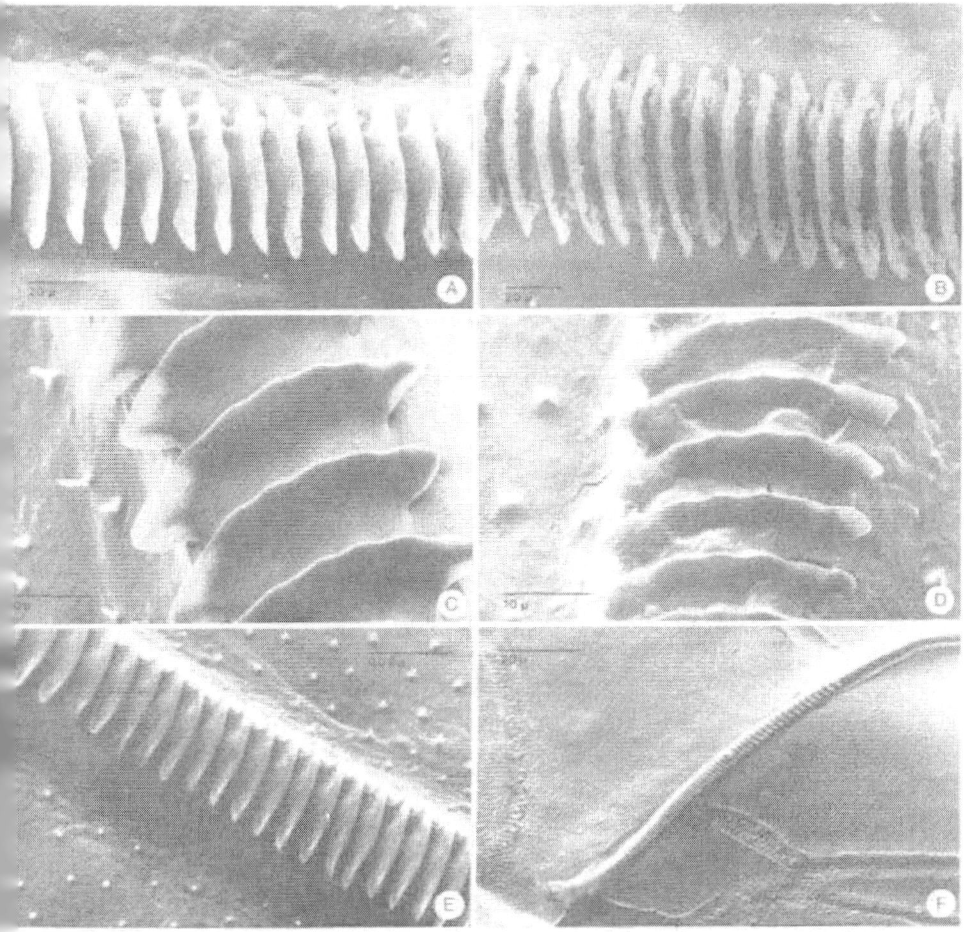

Plate IV. Stridulatory files on tegmina of Nemobiinae (Grylloidea: Gryllidae). *A*, *Allonemobius fasciatus*; *B*, *A. allardi*; *C*, *A. griseus griseus*; *D*, *A. maculatus*; *E*, *Neonemobius palustris*; *F*, *Eunemobius carolinus carolinus*.

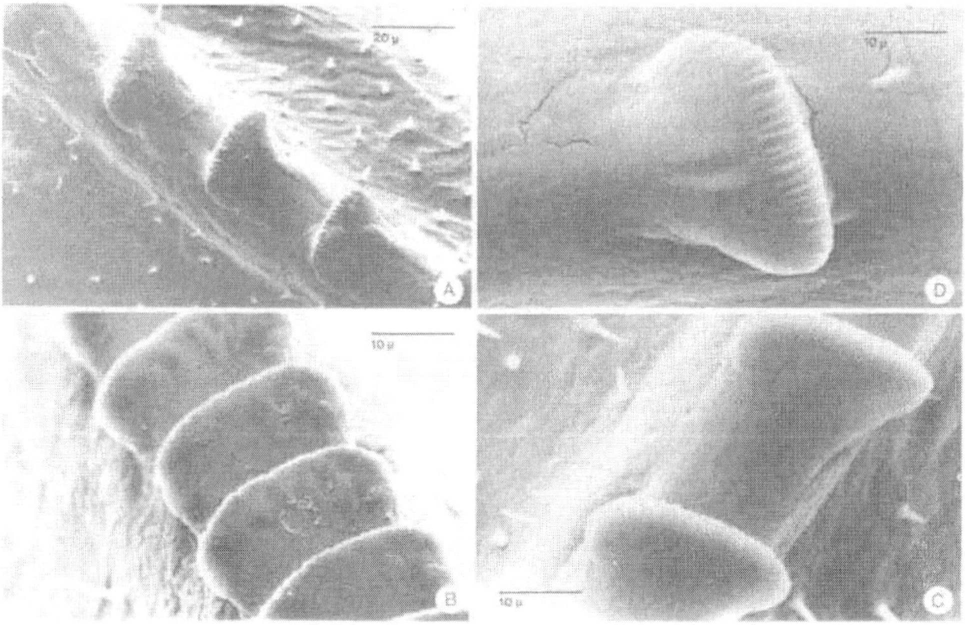

Plate V. Stridulatory files and pegs on tegmina of *Oecanthus* species (Grylloidea: Oecanthidae). *A*, *O. nigricornis*; *B*, *O. quadripunctatus*; *C*, *O. argentinus*; *D*, *O. fultoni*.

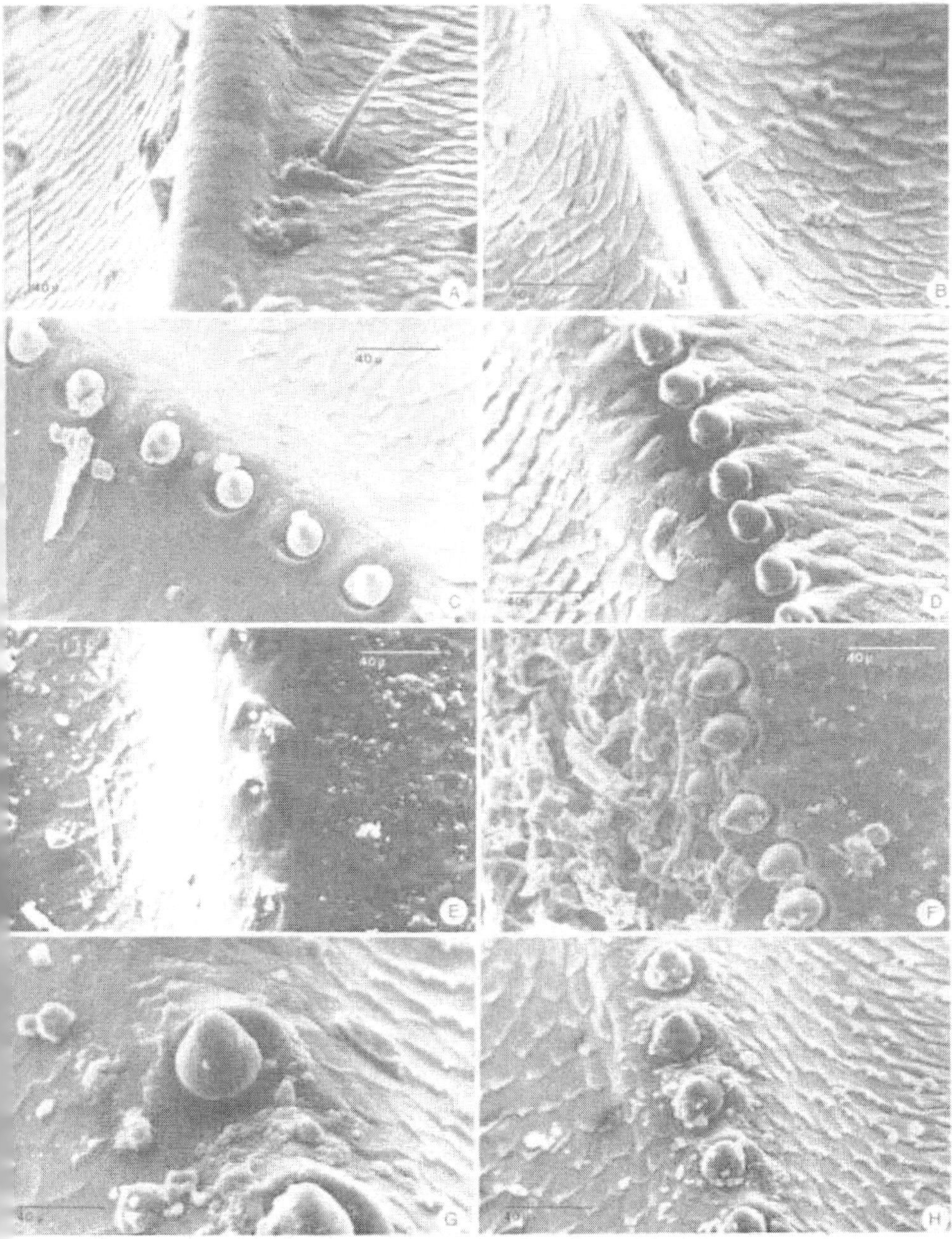

Plate VI. Stridulatory ridges on hind femora of Acrididae, without (*A*, *B*, Locustinae) or with (*C–H*, Gomophocerinae) stridulatory pegs. *A*, *Stethophyma lineatum*; *S. gracile*; *C*, *Opeia obscura*; *D*, *Chloealtis conspersa*; *E*, *C. abdominalis*; *F*, *Chorthippus curtipennis curtipennis*; *G*, *Aeropedellus clavatus*; *H*, *Amphitornus coloradus ornatus*.

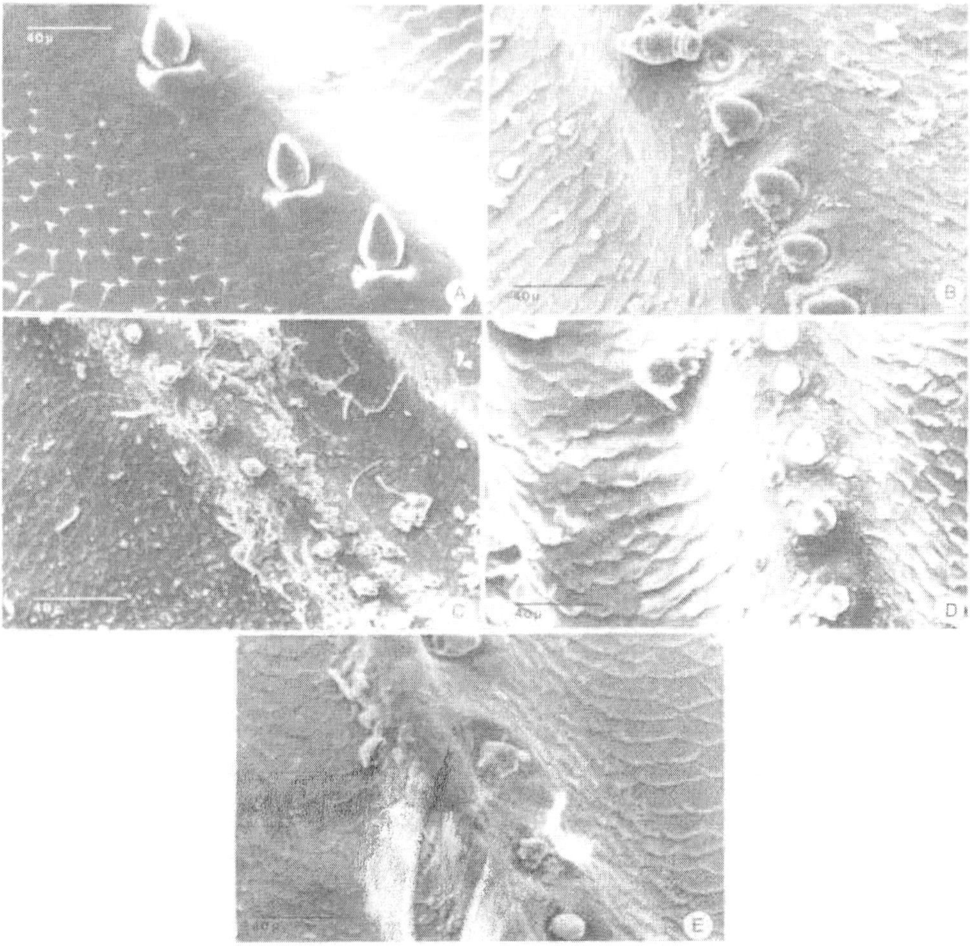

Plate VII. Stridulatory ridges (with pegs) on hind femora of Acrididae (Gomphocerinae). *A*, *Aulocara elliotti*; *B*, *Ageneotettix deorum*; *C*, *Phlibostroma quadrimaculatum*; *D*, *Cordillacris crenulata*; *E*, *Orphulella speciosa*.

Figs. 1–5. 1, A katydid, *Amblycorypha oblongifolia*, from Greenfield Park, Québ. (S. M. Ulagaraj). 2*a*, European mantid, *Mantis religiosa religiosa*, female of green phase, from Greenfield Park, Québ. (S. M. Ulagaraj). 2*b*, European mantid, *M. religiosa religiosa*, female of brown phase, from Osoyoos, B.C. (S. G. Cannings). 3, A grasshopper, *Dendrotettix quercus*, adult male amid lichens on trunk of elm tree, 6 km north of Baldwin, Lake Co., MI. (L.M. Crozier). 4, Twostriped grasshopper, *Melanoplus bivittatus*, male, from Pincourt, Québ. (V. R. Vickery). 5, Carolina grasshopper, *Dissosteira carolina*, resting on rock of same color, from Québec (V. R. Vickery).

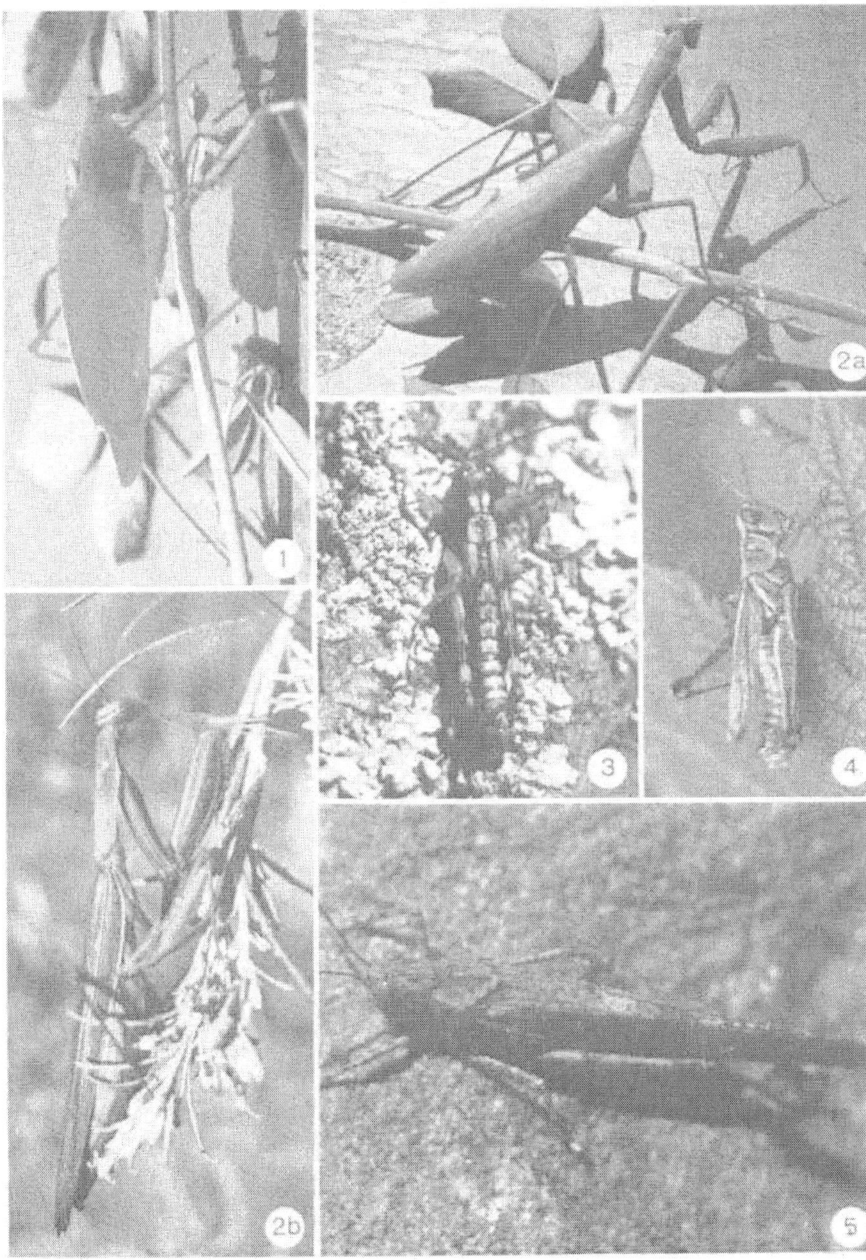
1
2a
3
4
2b
5

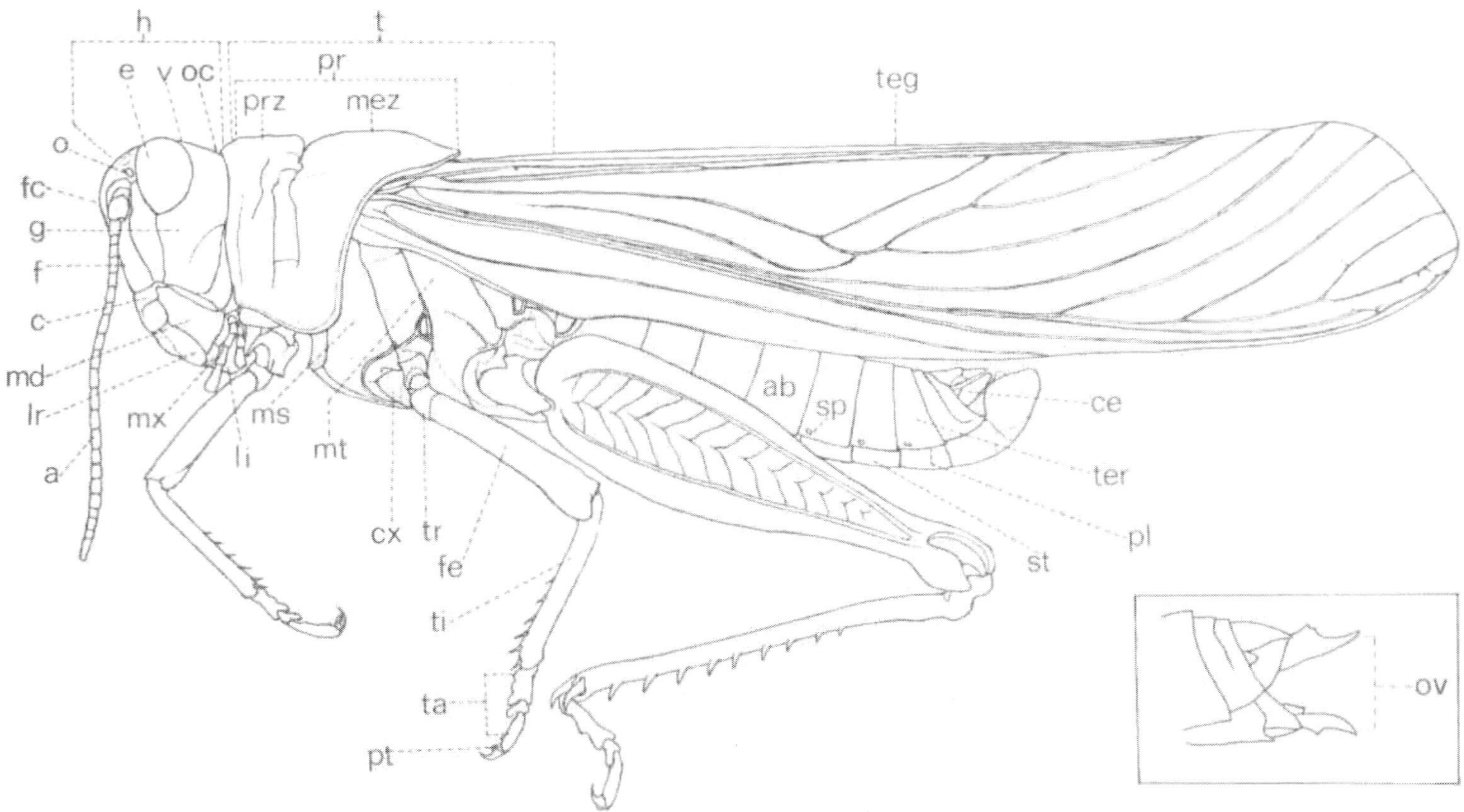

Fig. 6. Generalized diagrams of an orthopteroid (grasshopper) illustrating anatomical features. Modified from Vickery et al. (1974). For explanation of lettering, see text, pp. 30, 31.

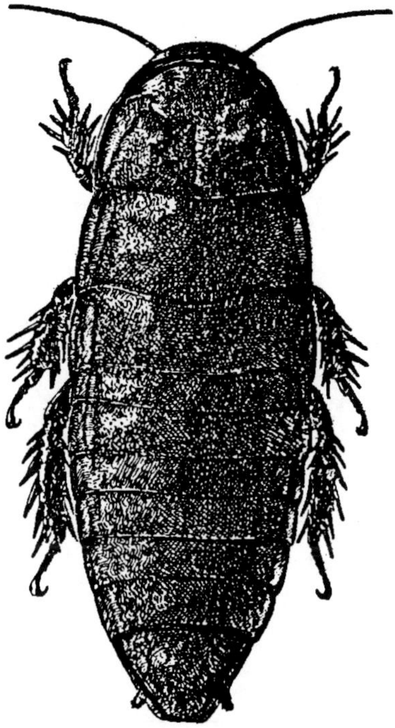

Fig. 7. *Cryptocercus punctulatus* male. After Chopard (1949), courtesy of Editions Lechevalier, Paris.

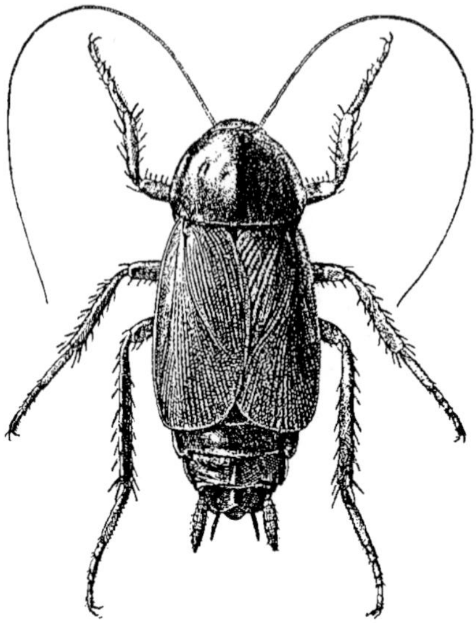

8

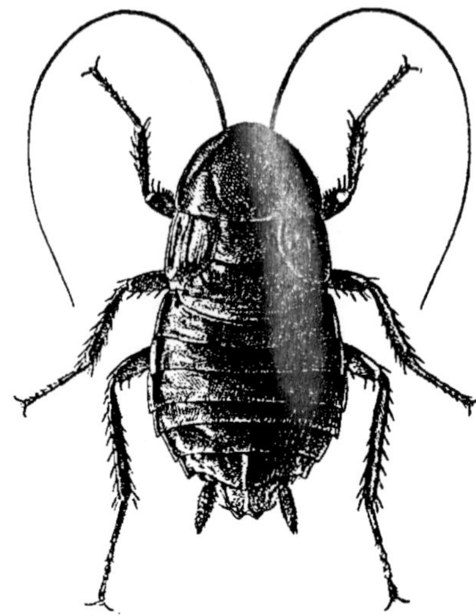

9

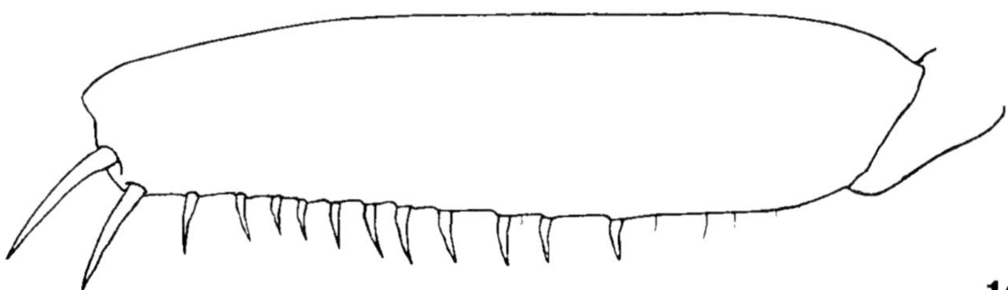

10

Figs. 8–10. *Blatta orientalis*. 8, male; 9, female; 10, anterior femur. 8, 9, after Laing (1946); 10, after Vickery et al. (1974).

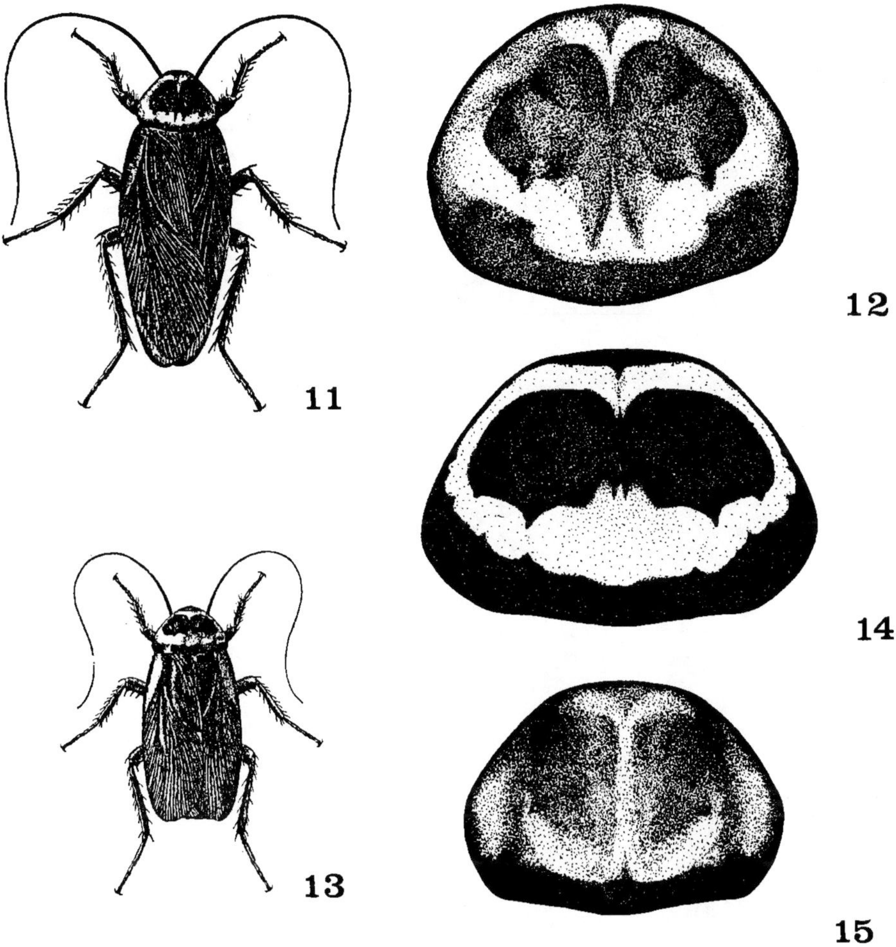

Figs. 11–15. *Periplaneta* spp. 11, 12, *P. americana*; 11, male; 12, pronotum. 13, 14, *P. australasiae*; 13, female; 14, pronotum. 15, *P. brunnea*, pronotum. 11, 13, after Laing (1930); 12, 14, after Vickery et al. (1974).

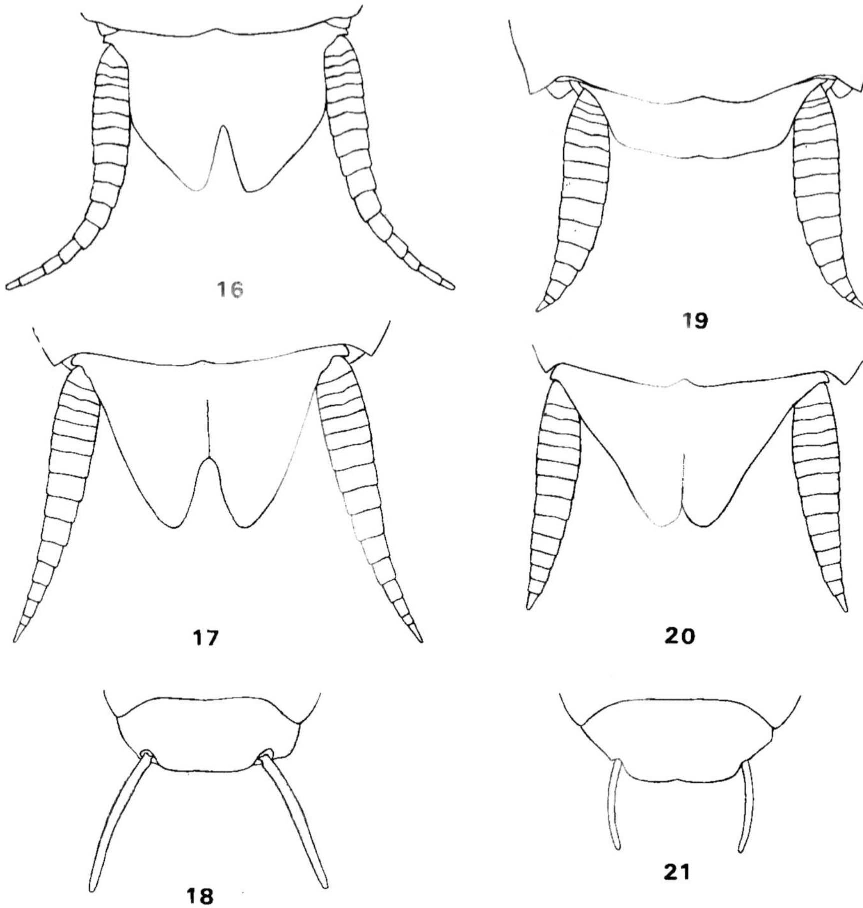

Figs. 16–21. *Periplaneta* spp. 16–18, *P. americana*; 19–21, *P. brunnea*. 16, 19, supra-anal plate and cerci of male, dorsal; 17, 20, the same, female; 18, 21, subgenital plate and styles of male, ventral. 16, 17, 19, 20, after Vickery et al. (1974).

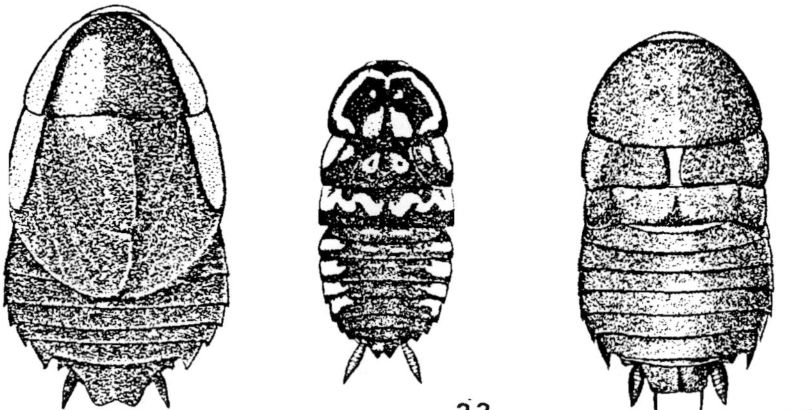

22 23 24

Figs. 22–24. *Blattidae*. 22, *Pelmatosilpha larifuga*; 23, *Stylopyga rhombifolia*; 24, *Eurycotis floridana*. 22, after Ragge (1965), courtesy of the Trustees of the British Museum (Natural History), London; 23, 24, after Hebard (1917*c*).

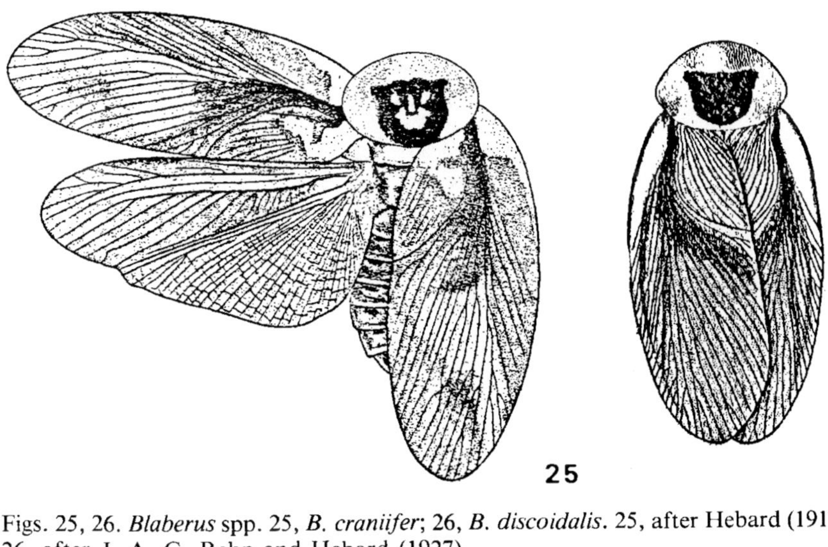

25 26

Figs. 25, 26. *Blaberus* spp. 25, *B. craniifer*; 26, *B. discoidalis*. 25, after Hebard (1917*c*); 26, after J. A. G. Rehn and Hebard (1927).

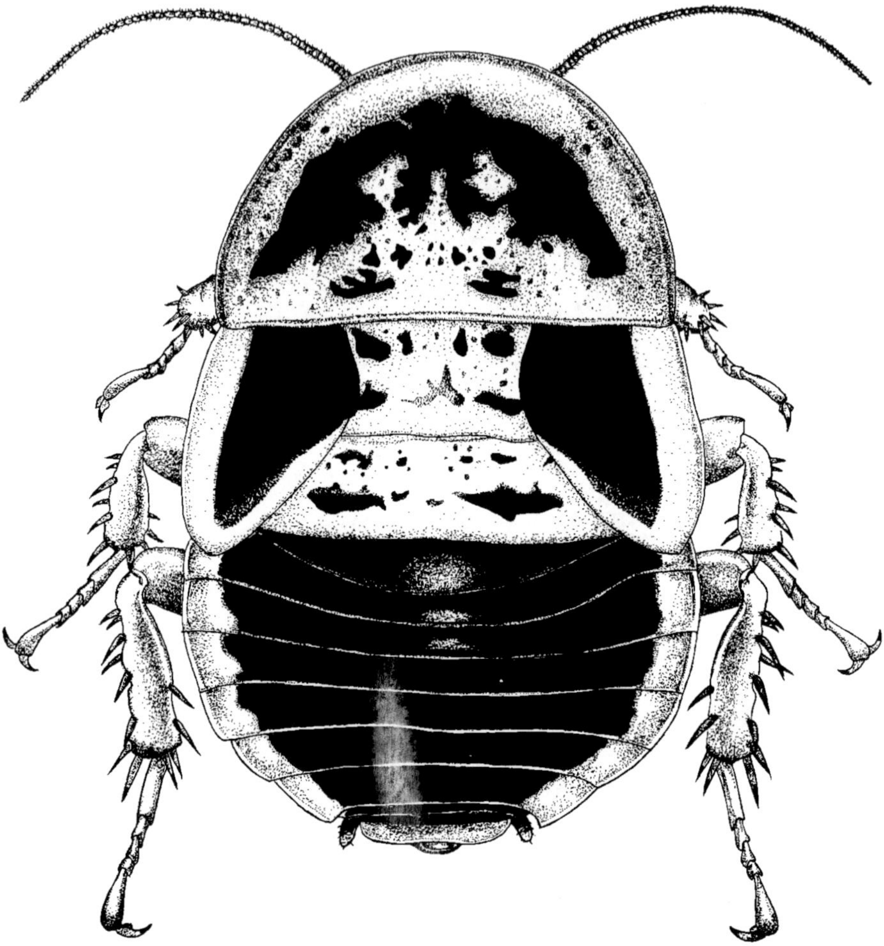

Fig. 27. *Sibylloblatta panesthoides*.

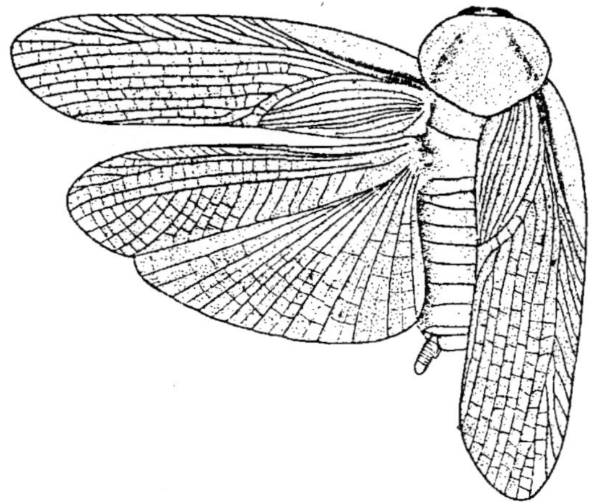

28

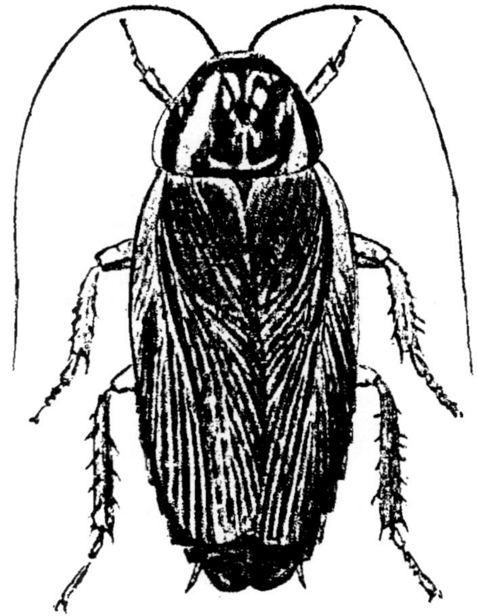

29

Figs. 28, 29. Nauphoetidae. 28, *Panchlora nivea*; 29, *Nauphoeta cinerea*. 28, after Hebard (1917*c*); 29, adapted from Ragge (1965).

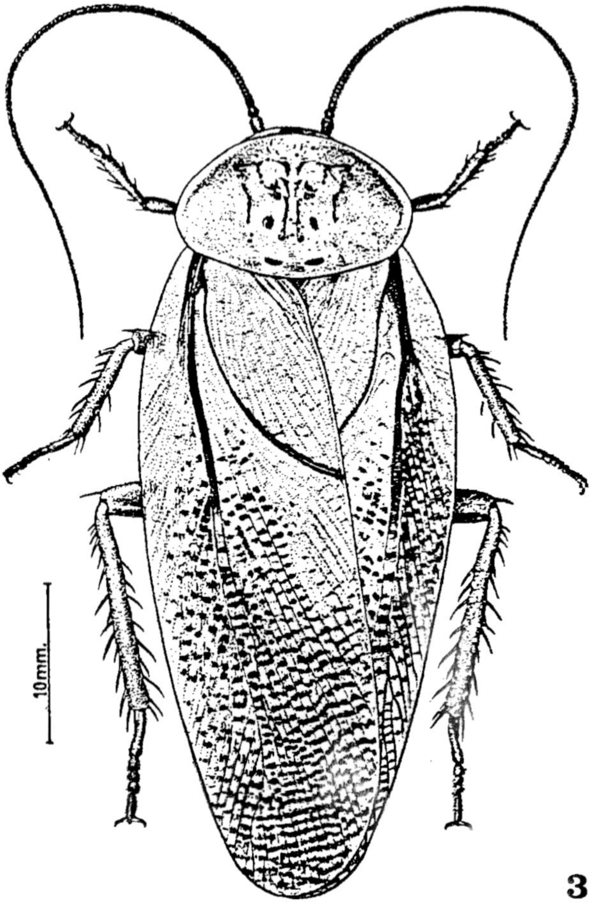

30

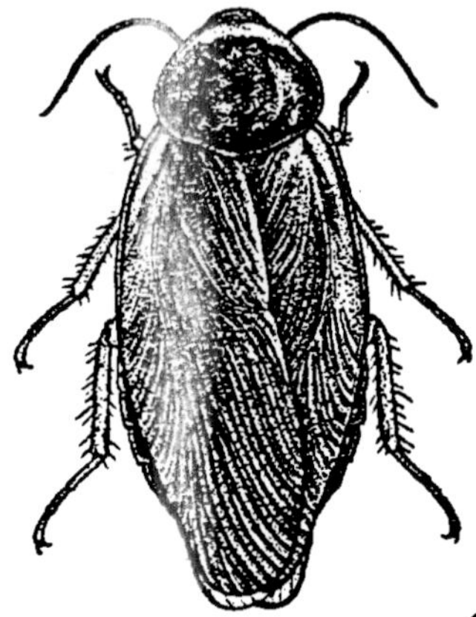

31

Figs. 30, 31. Nauphoetidae. 30, *Rhyparobia maderae*; 31, *Pycnoscelus surinamensis*. 30, after Ragge (1973); 31, after Shiraki (1932).

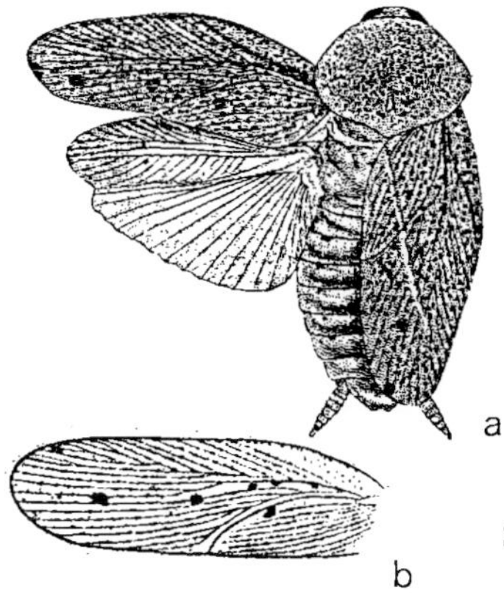

32

33

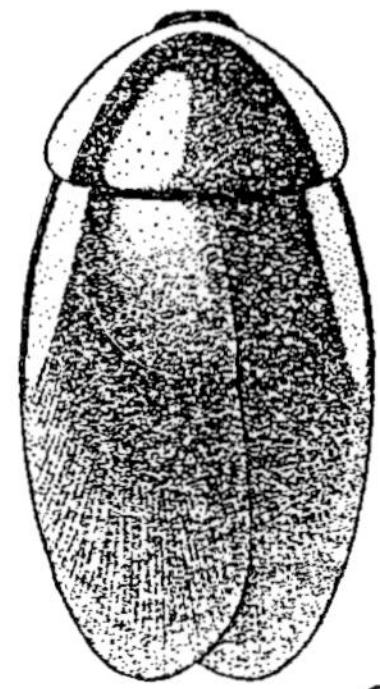

34

Figs. 32–34. Epilampridae and Nyctiboridae. 32, *Epilampra maya*: *a*, *E. maya brachyptera*, from Mexico, male; *b*, tegmen of typical *E. maya maya*; 33, *Nauclidas nigra*; 34, *Nyctibora laevigata*; 32*a*, after Hebard (1921); 32*b*, after Hebard (1920*c*); 32, 33, after Ragge (1965), courtesy of the Trustees of the British Museum (Natural History), London.

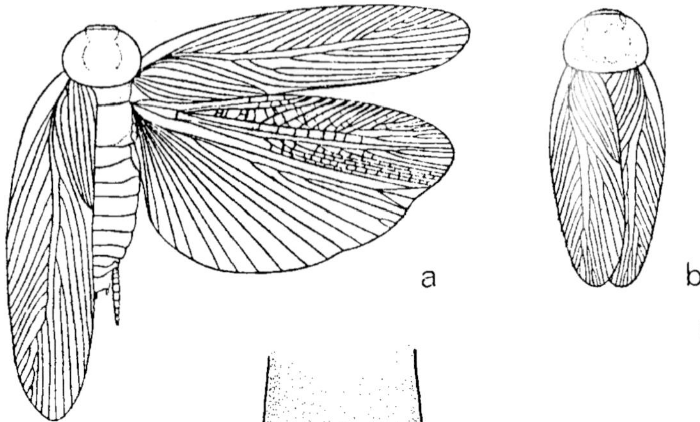

35

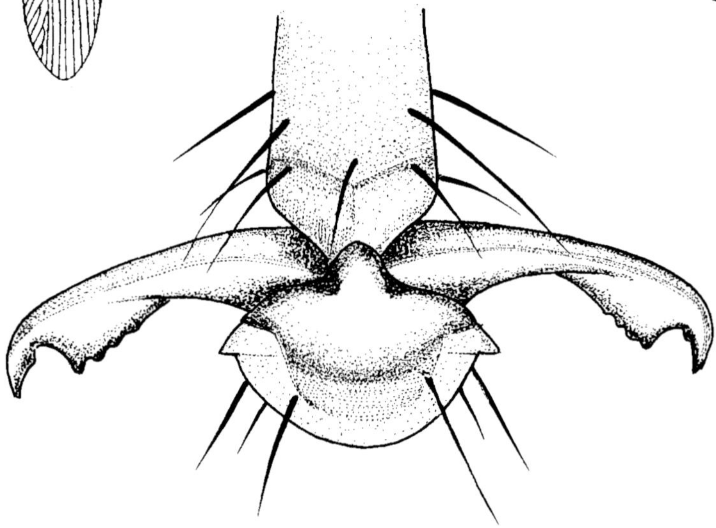

36

Figs. 35, 36. *Neoblattella* spp. 35, general appearance, as represented by the Colombian *N. carrikeri*: *a*, male; *b*, female; 36, specialized tarsal claws, as represented by the West Indian *N. carcina*. 35, after Hebard (1919*a*).

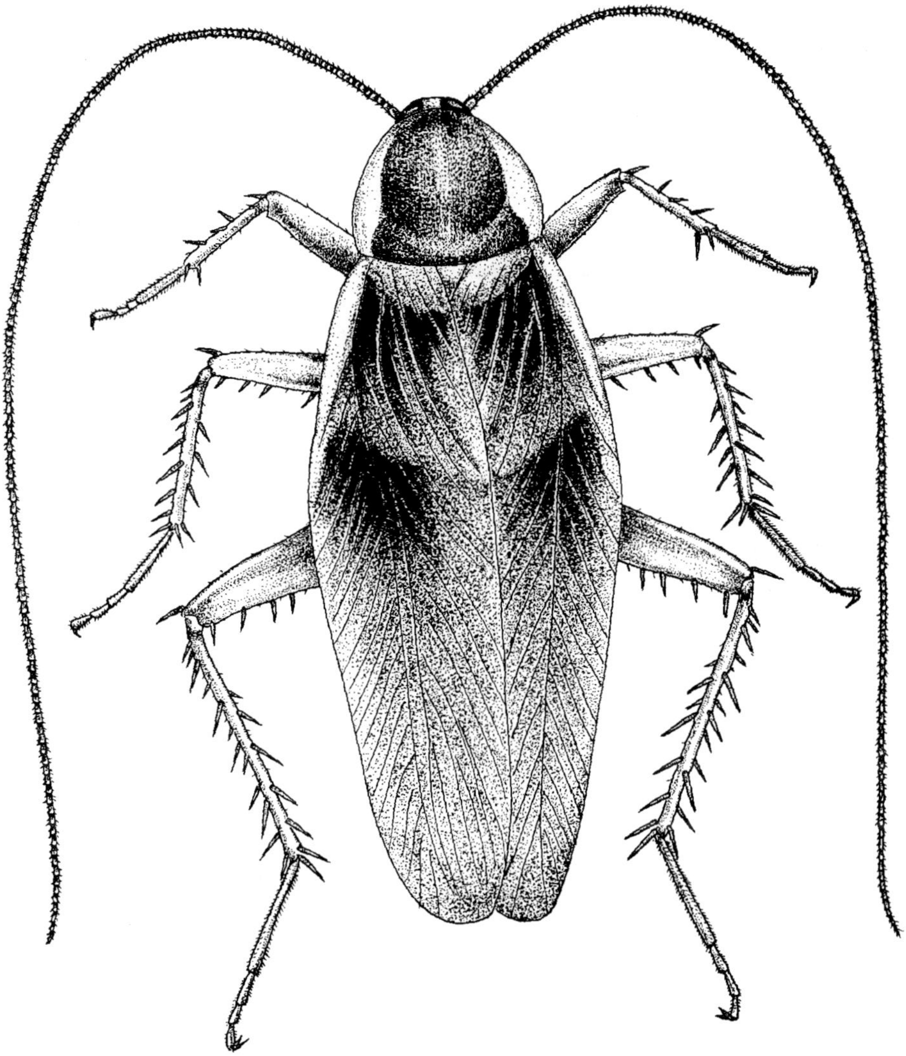

Fig. 37. *Supella longipalpa*, male.

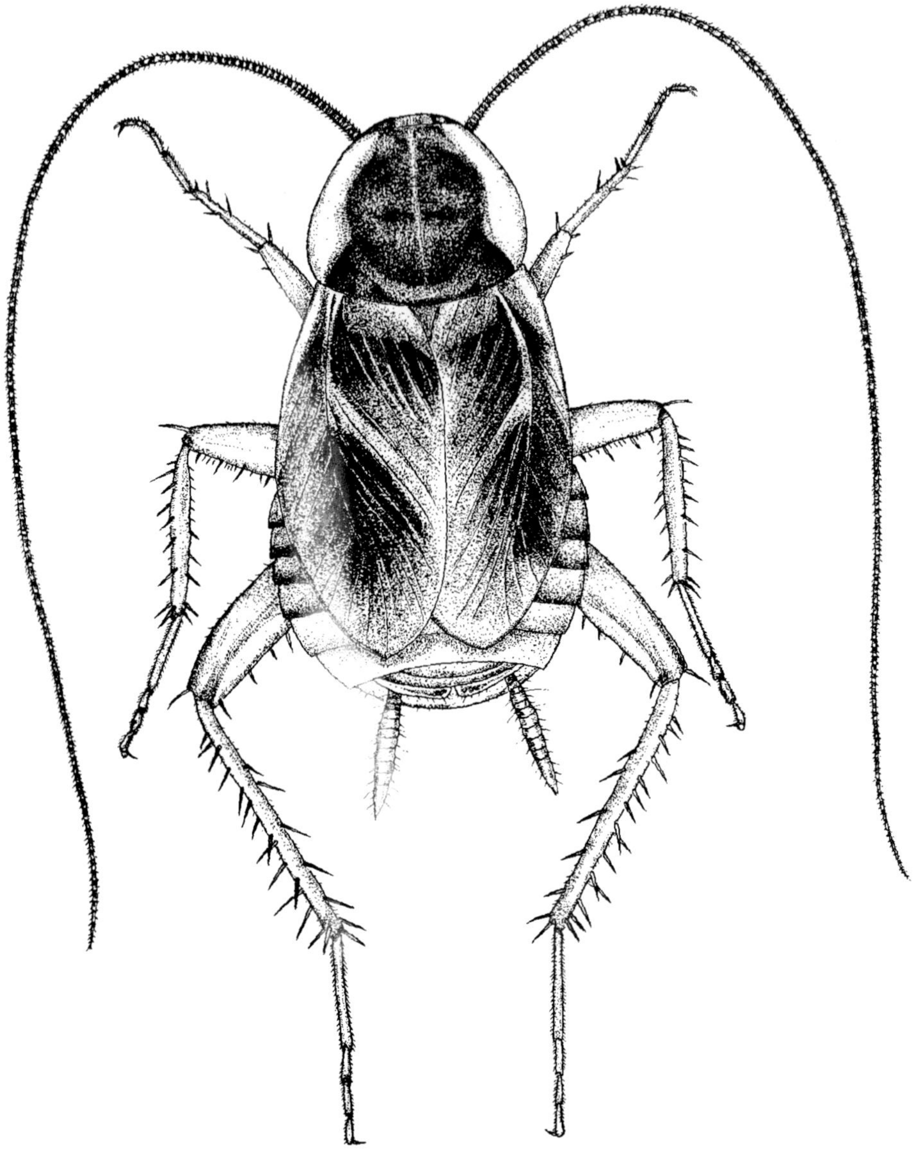

Fig. 38. *Supella longipalpa*, female.

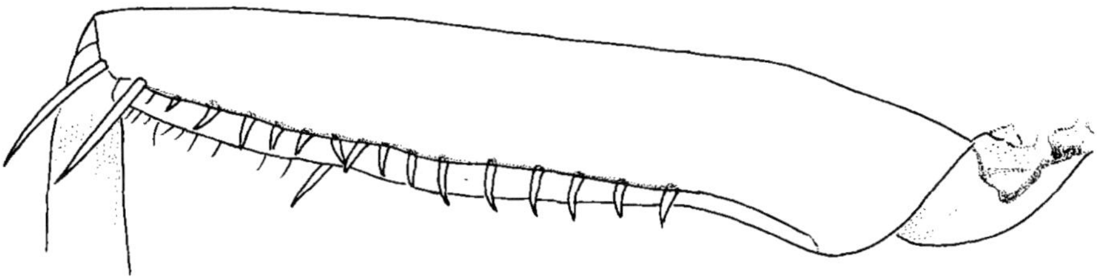

Fig. 39. *Supella longipalpa*, anterior femur.

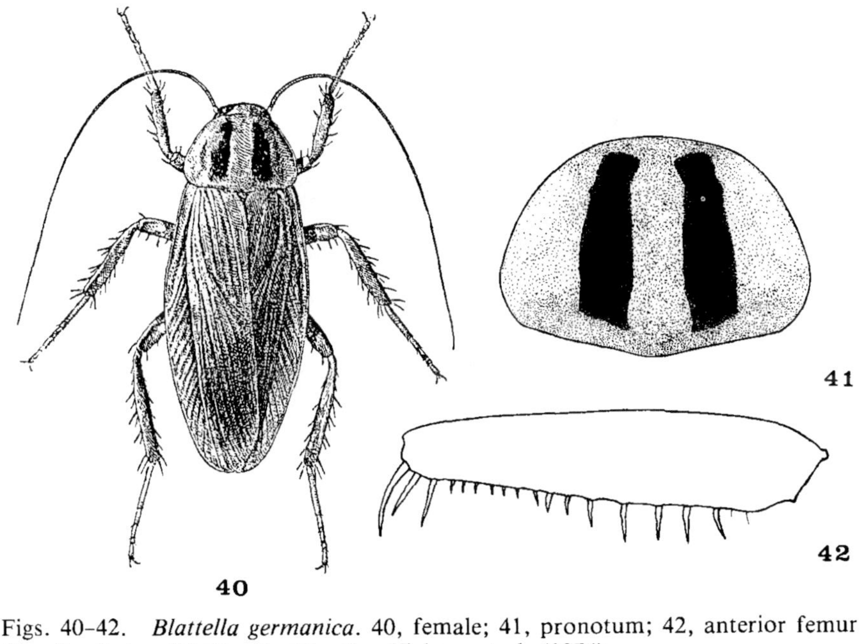

40

41

42

Figs. 40–42. *Blattella germanica*. 40, female; 41, pronotum; 42, anterior femur. 40, after Laing (1946); 41, 42, after Vickery et al. (1974).

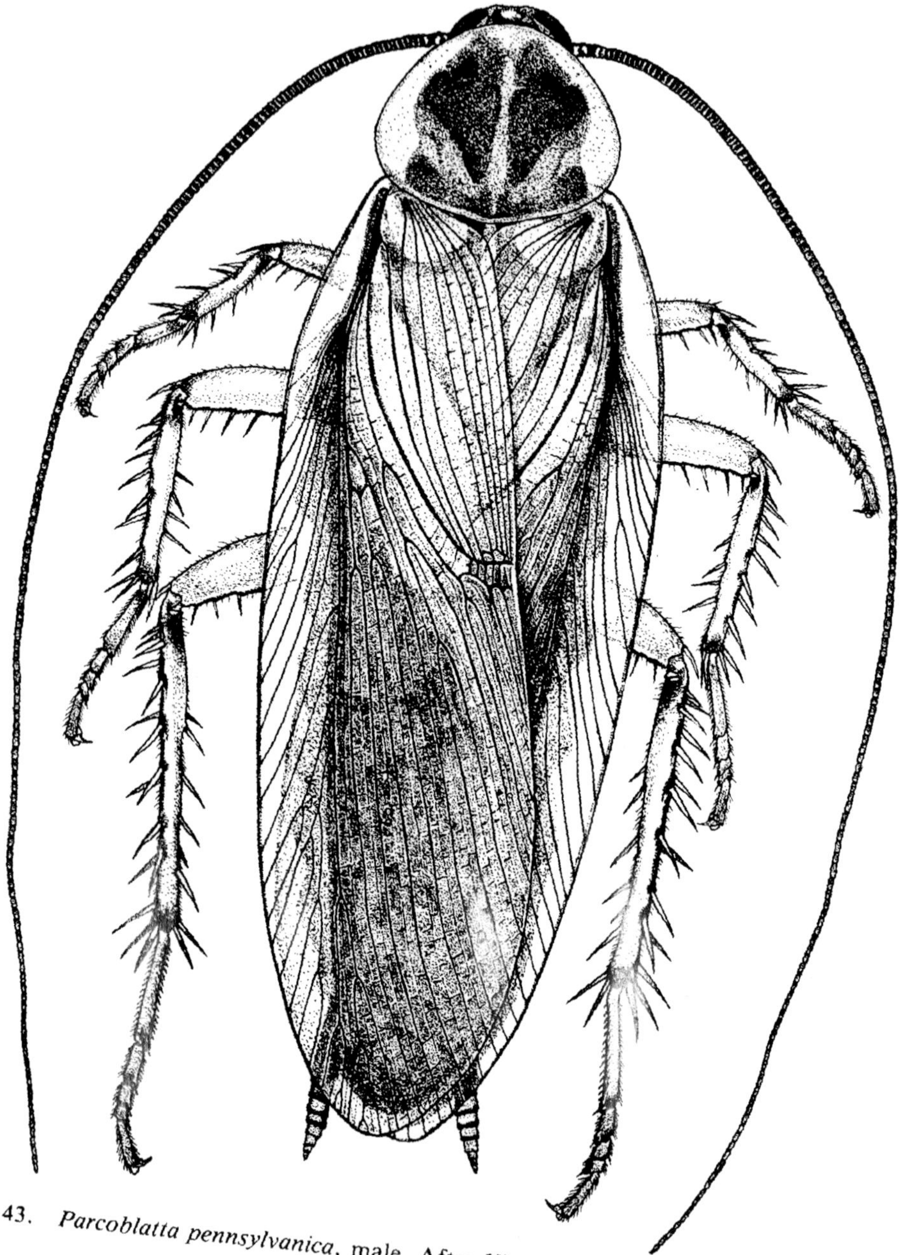

Fig. 43. *Parcoblatta pennsylvanica*, male. After Vickery et al. (1974).

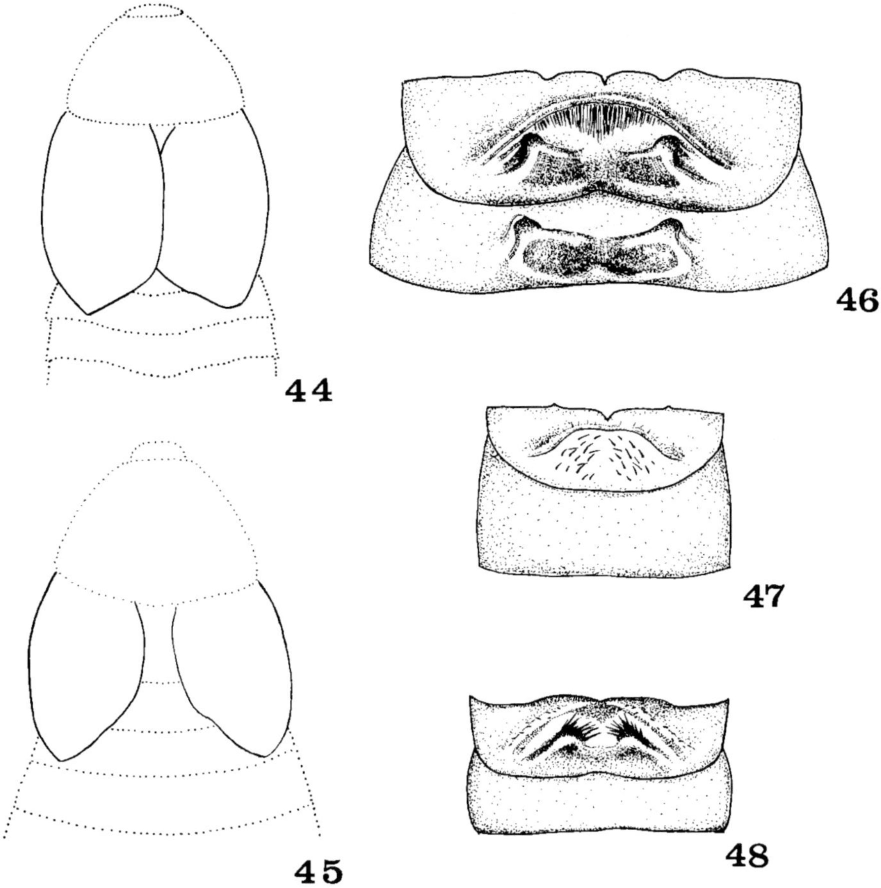

Figs. 44–48. *Parcoblatta* spp. 44, female tegmina of *P. virginica*; 45, the same, *P. uhleriana*; 46, median segment and true first abdominal tergum of *P. pennsylvanica*; 47, the same *P. virginica*; 48, the same, *P. uhleriana*.

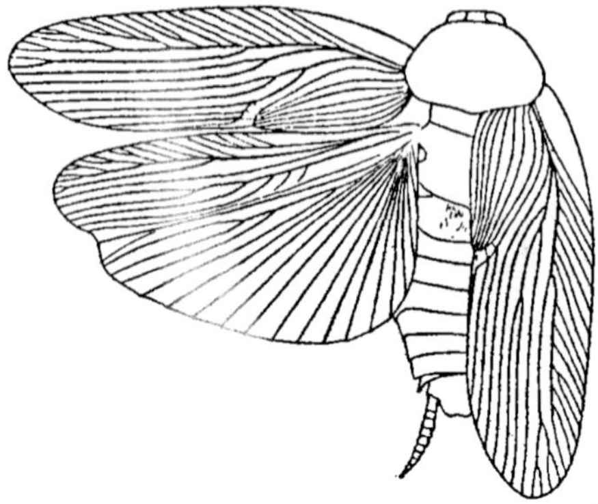

49

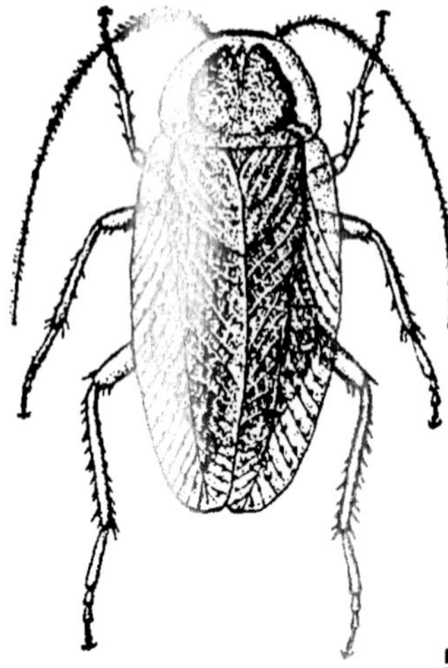

50

Figs. 49, 50. Blattellidae and Ectobiidae. 49, *Symploce pallens*; 50, *Ectobius pallidus*. 49, after Hebard (1916*a*) [Hebard's *S. capitata* was actually a closely related species, now named *S. hebardi* Princis, but the figure is sufficiently close to true *S. pallens*]; 50, adapted from Bazyluk (1956).

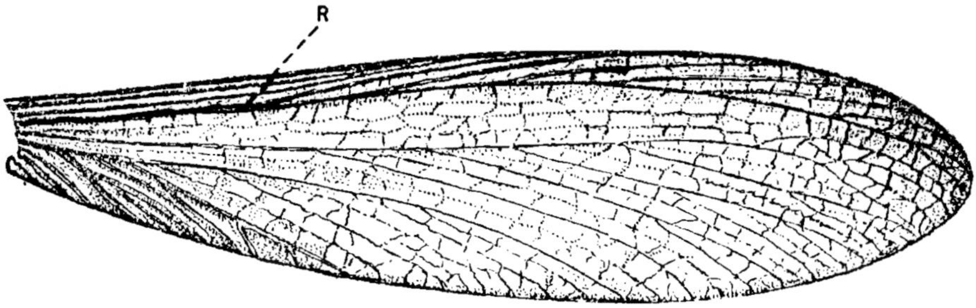

51

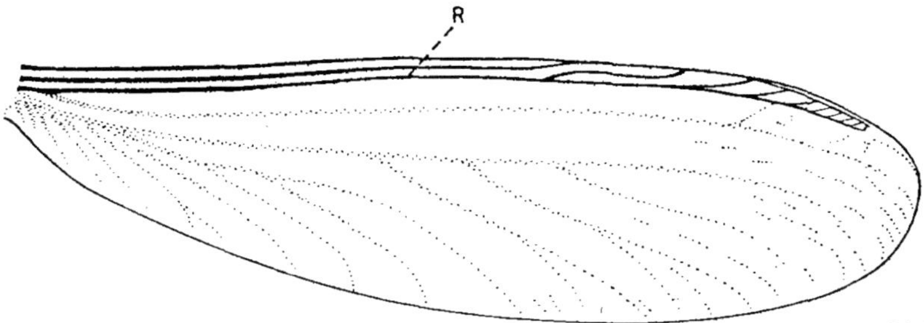

52

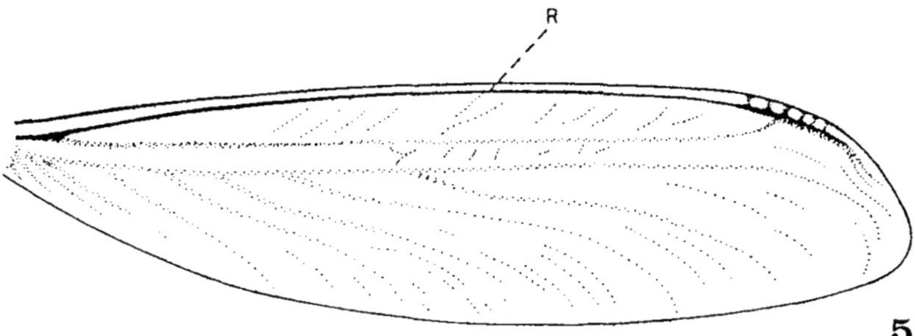

53

Figs. 51–53. Venation of fore wings of Termitoidea. 51, Termopsidae; 52, Kalotermitidae; 53, Rhinotermitidae. *R* = radial vein. 52, 53, adapted from Borror and DeLong (1971).

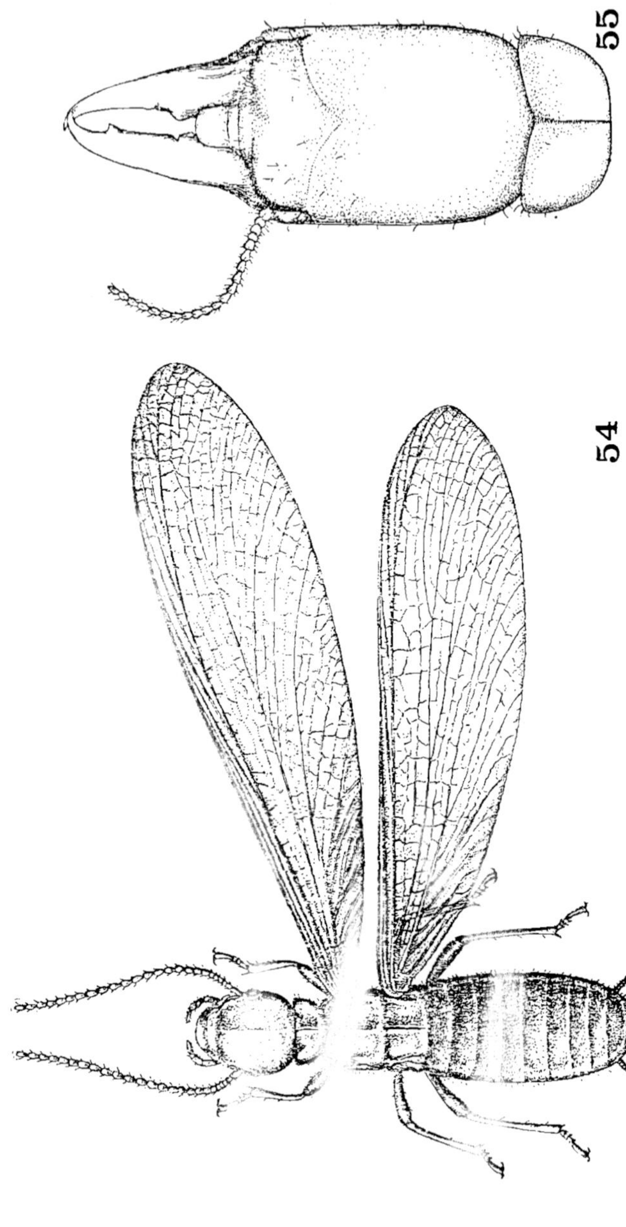

Figs. 54, 55. *Zootermopsis nevadensis.* 54, alate sexual adult; 55, soldier, head and pronotum, dorsal. After Banks and Snyder (1920).

Figs. 56–59. *Zootermopsis angusticollis*, figures not to same scale. 56, gravid primary queen (dealated); 57, reproductive form nymph; 58, secondary sexual male (note rudimentary wing pads); 59, soldier. 56–58, after Banks and Snyder (1920); 59, adapted from W. V. Harris (1961).

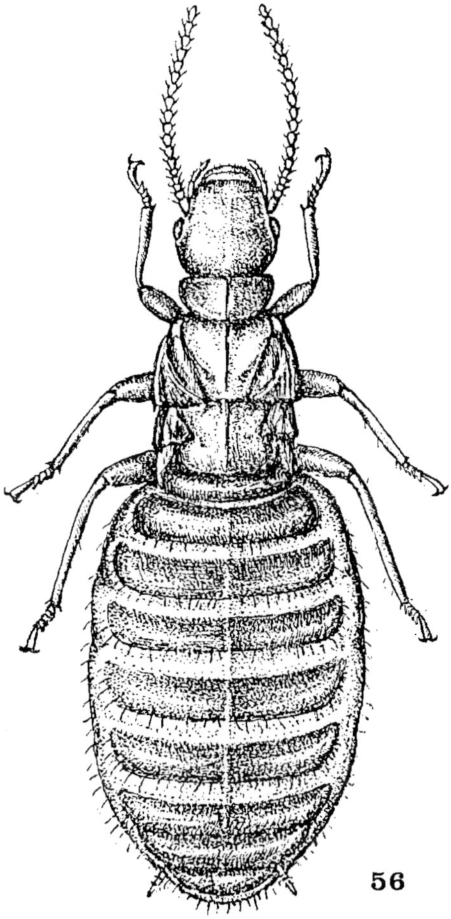

56

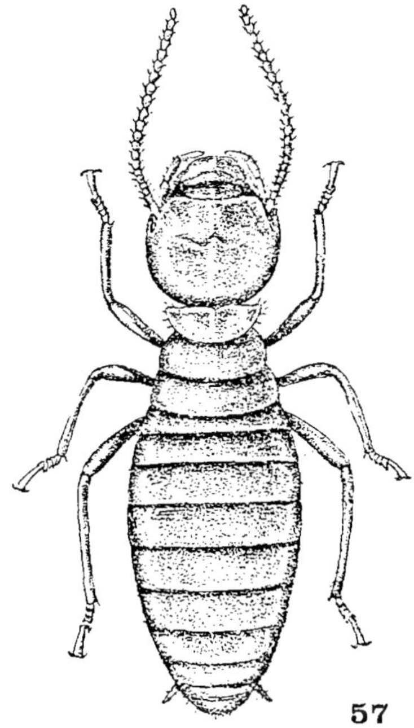

57

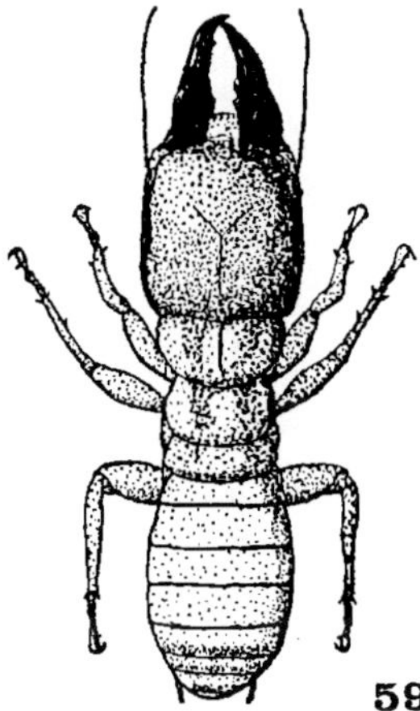

59

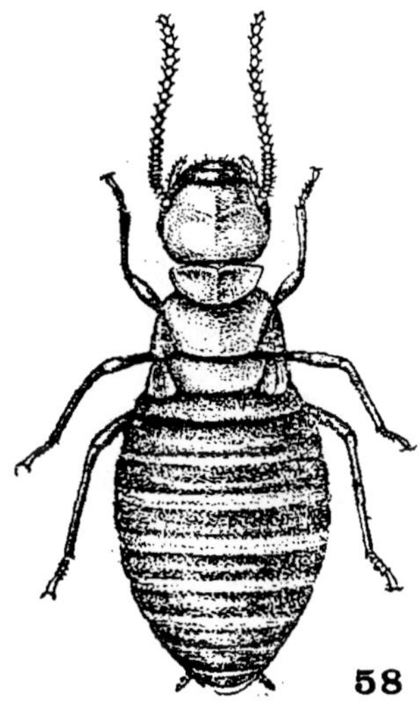

58

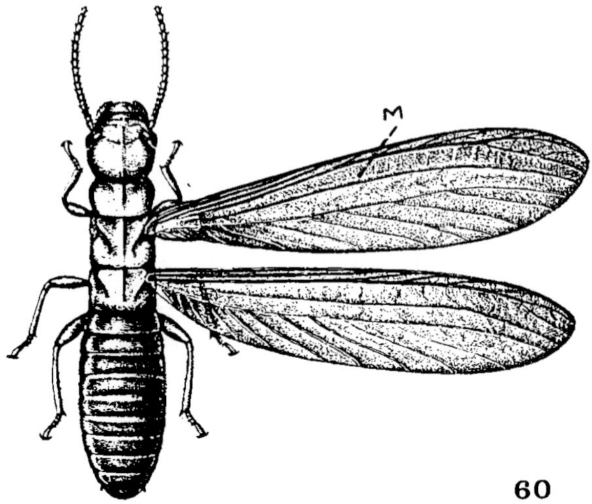

60

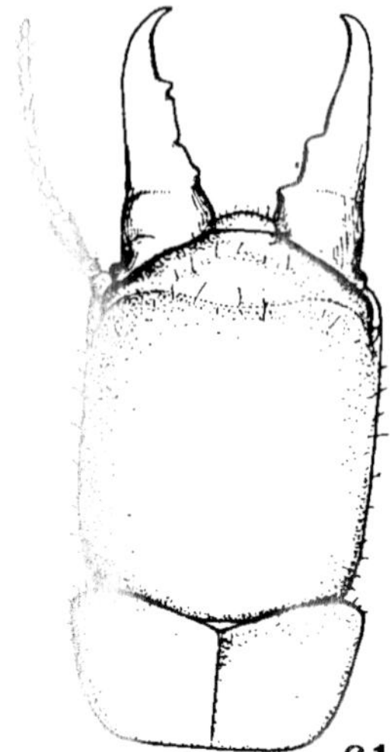

61

Figs. 60, 61. *Incisitermes minor*. 60, alate adult: *M* = median vein; 61, soldier, head and pronotum, dorsal. After Banks and Snyder (1920).

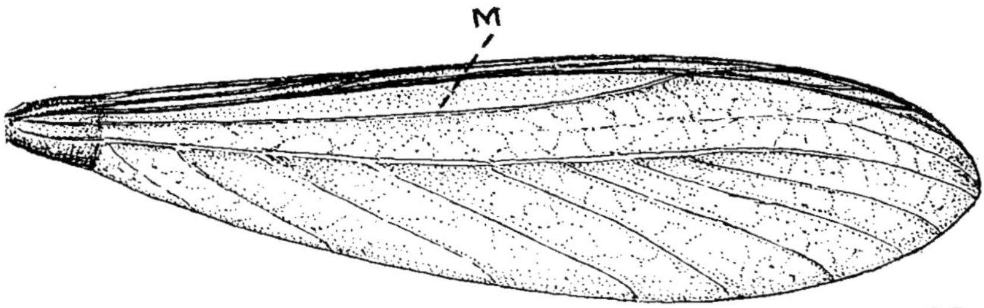

62

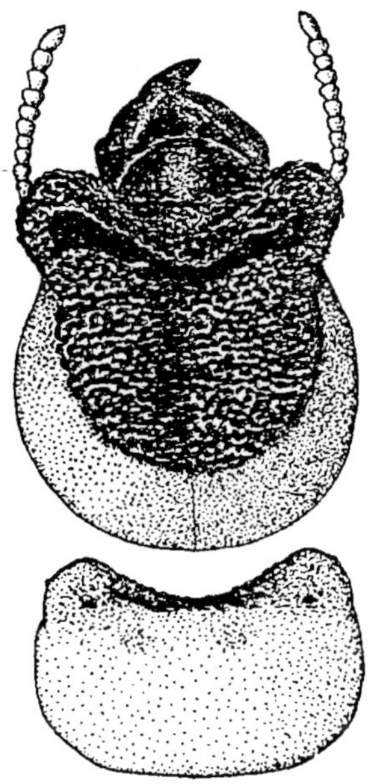

63

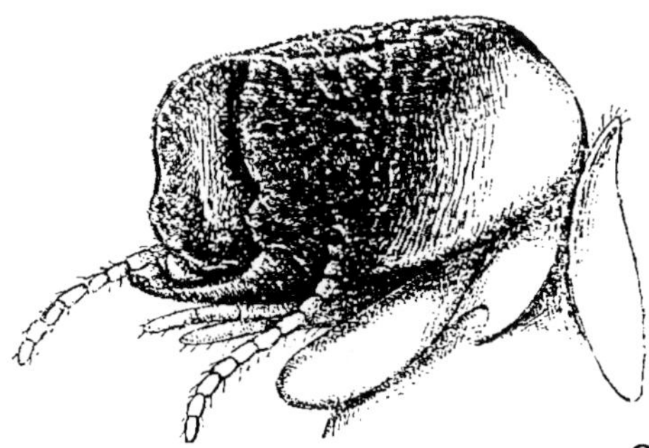

64

Figs. 62–64. *Cryptotermes* sp. 62, typical fore wing venation: *M* = median vein; 63, *C. brevis* soldier, head and pronotum, dorsal; 64, the same, head from left. 63, after W. V. Harris (1971); 64, after Banks and Snyder (1920).

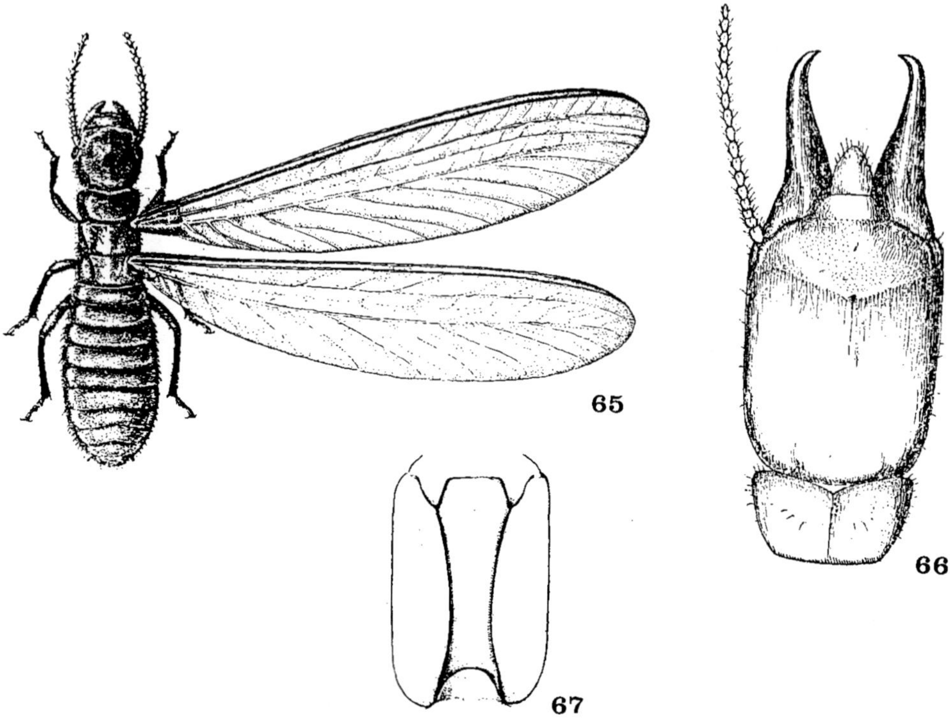

Figs. 65–67. *Reticulitermes tibialis*. 65, alate adult; 66, soldier, head and pronotum, dorsal; 67, soldier, head ventral, gular region. After Banks and Snyder (1920).

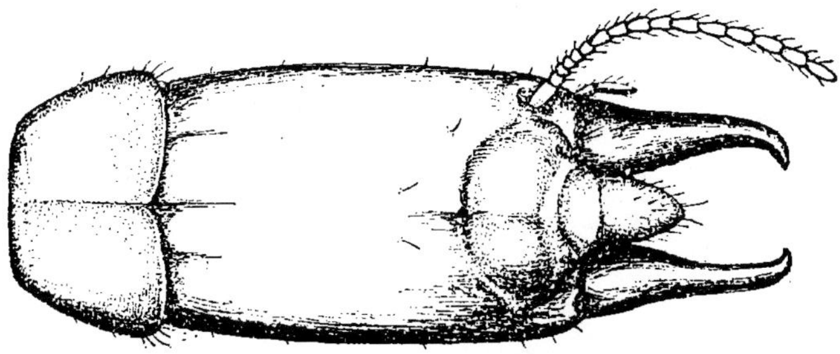

Fig. 68. *Reticulitermes hesperus*, soldier, head and pronotum, dorsal. After Banks and Snyder (1920).

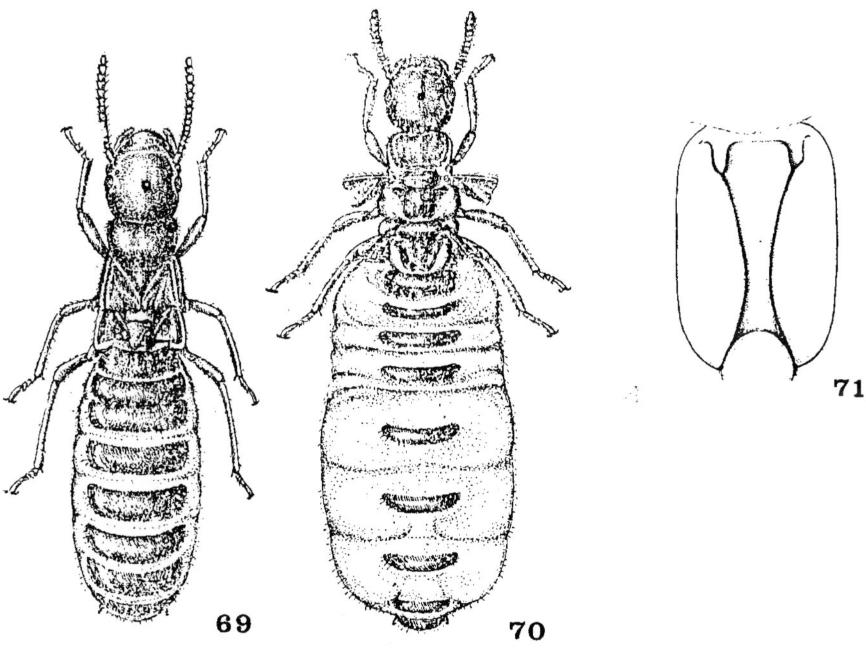

Figs. 69–71. *Reticulitermes flavipes*, not to same scale. 69, mature, dealated primary male (king); 70, gravid, dealated primary female (queen); 71, soldier, gular region of ventral side of head. After Banks and Snyder (1920).

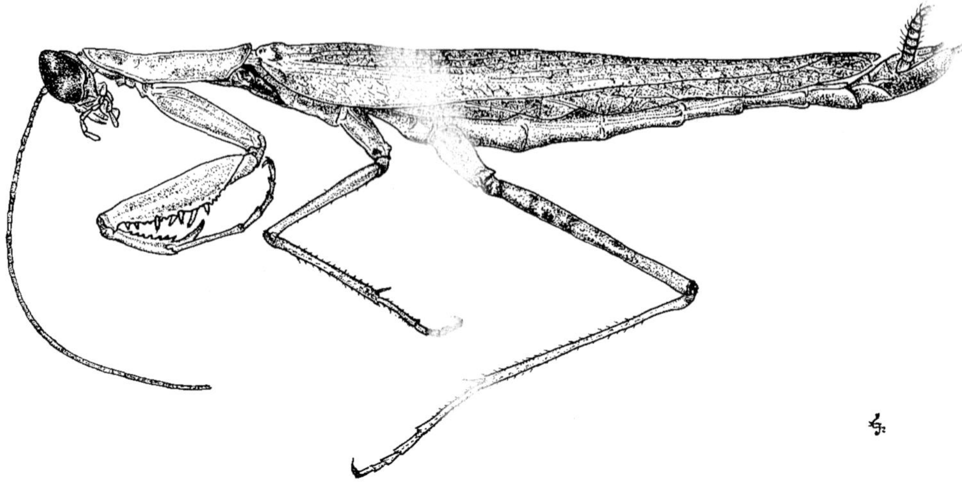

Fig. 72. *Litaneutria minor*, male.

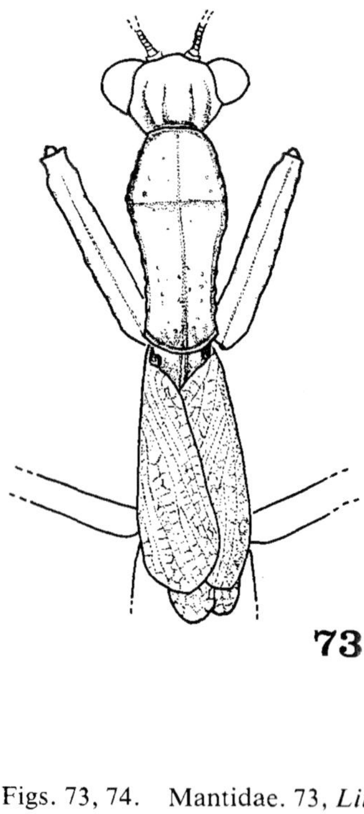

73

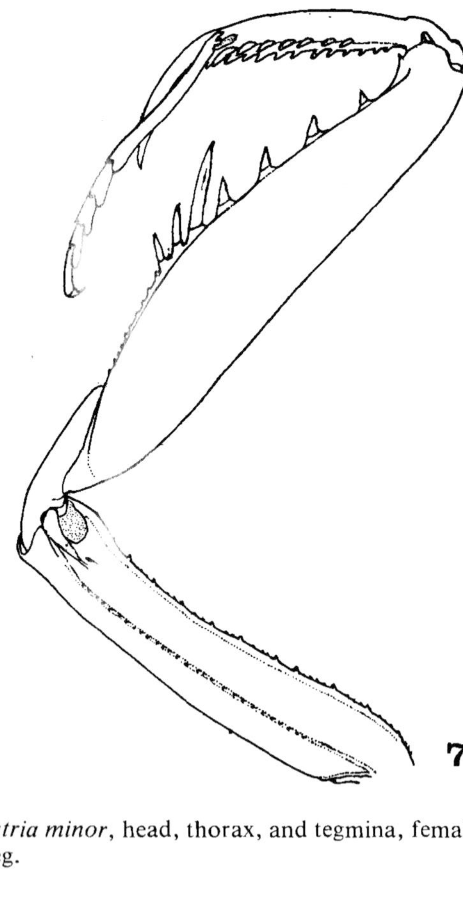

74

Figs. 73, 74. Mantidae. 73, *Litaneutria minor*, head, thorax, and tegmina, female; 74, *Mantis religiosa religiosa*, foreleg.

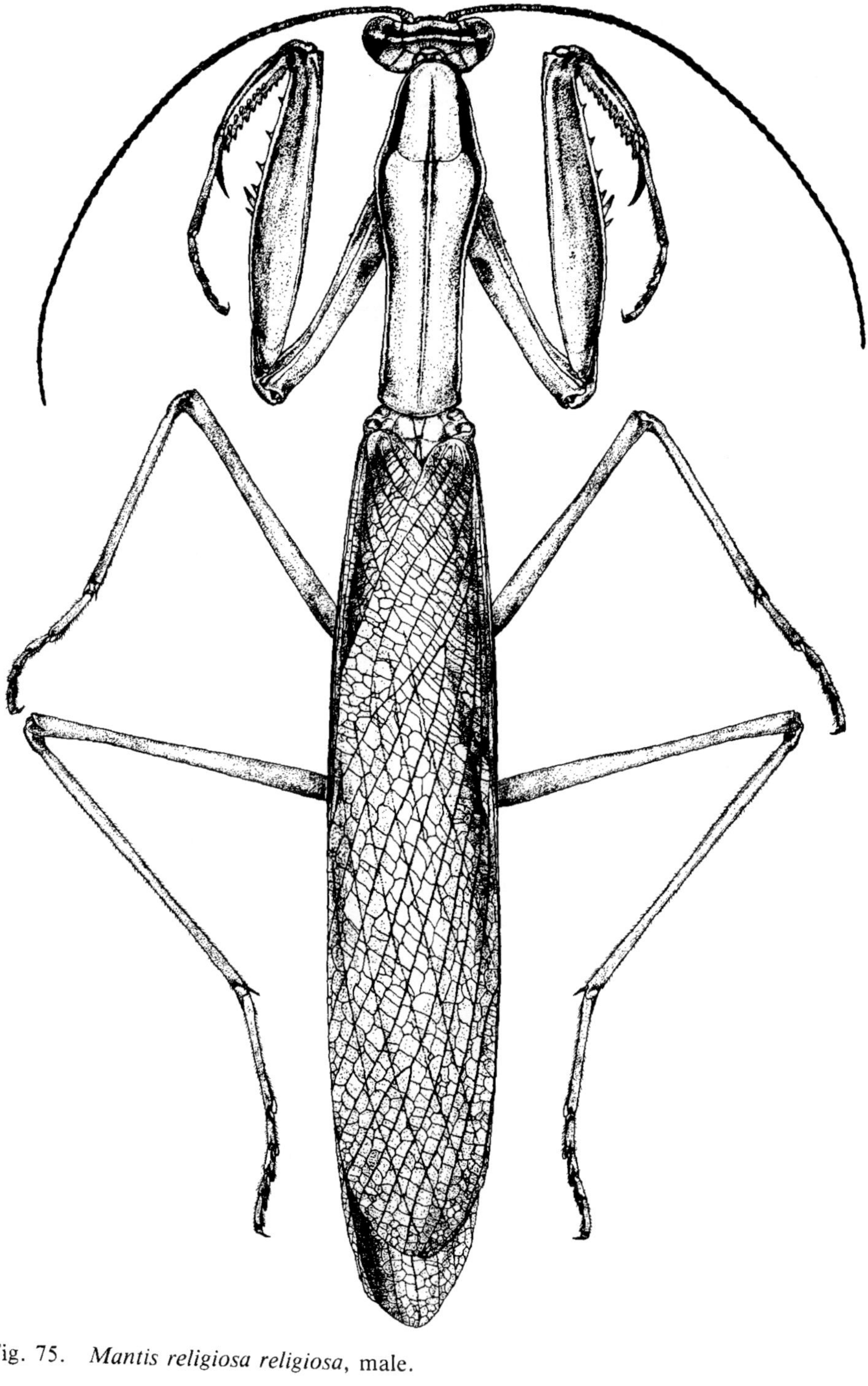

Fig. 75. *Mantis religiosa religiosa*, male.

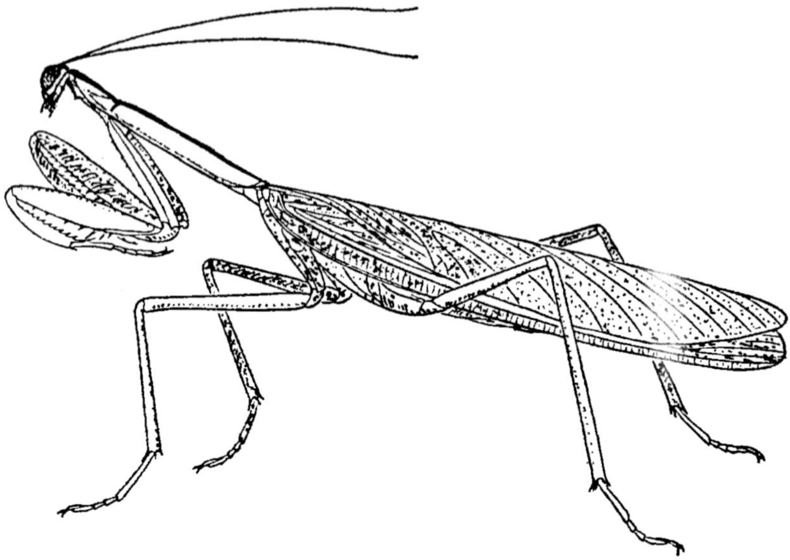

Fig. 76. *Tenodera aridifolia sinensis*, male. Based on Essig (1942).

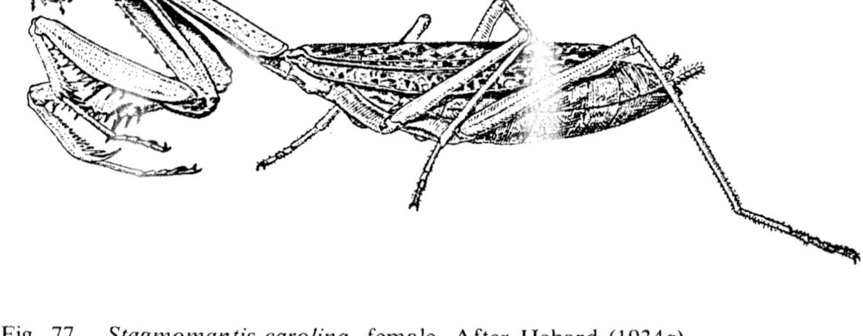

Fig. 77. *Stagmomantis carolina*, female. After Hebard (1934*a*).

79

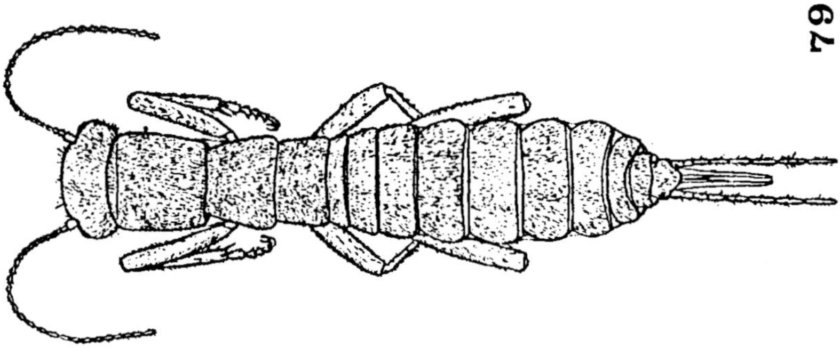

78

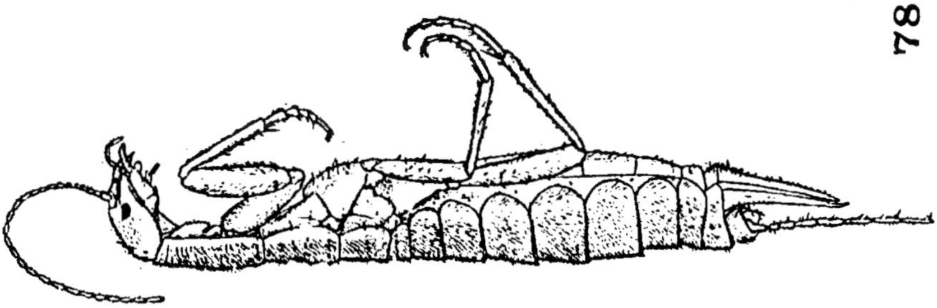

Figs. 78, 79. The original illustrations of *Grylloblatta campodeiformis campodeiformis*, female holotype. 78, lateral; 79, dorsal. After E. M. Walker (1914*a*) (lettering deleted).

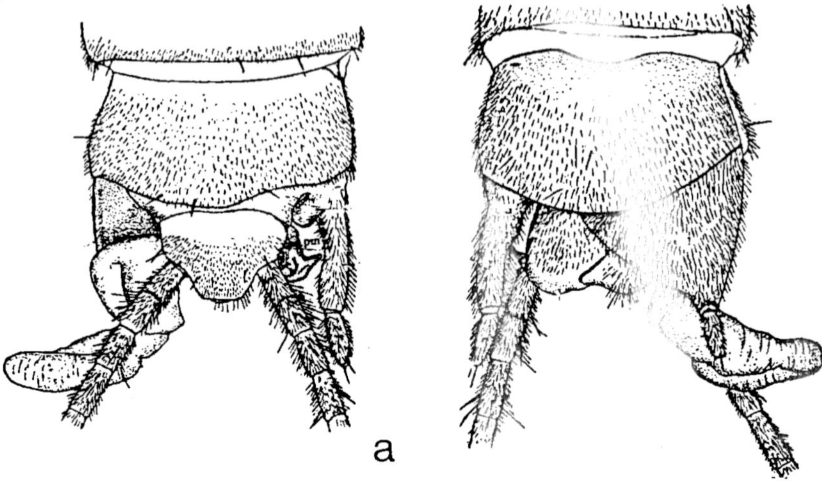

80

81

Figs. 80, 81. *Grylloblatta* male terminalia. 80, *G. campodeiformis campodeiformis*, entire: *a*, dorsal; *b*, ventral; 81, *G. scudderi*, supra-anal plate, dorsal. 80, after E. M. Walker (1919*b*) (lettering deleted); 81, after Kamp (1979).

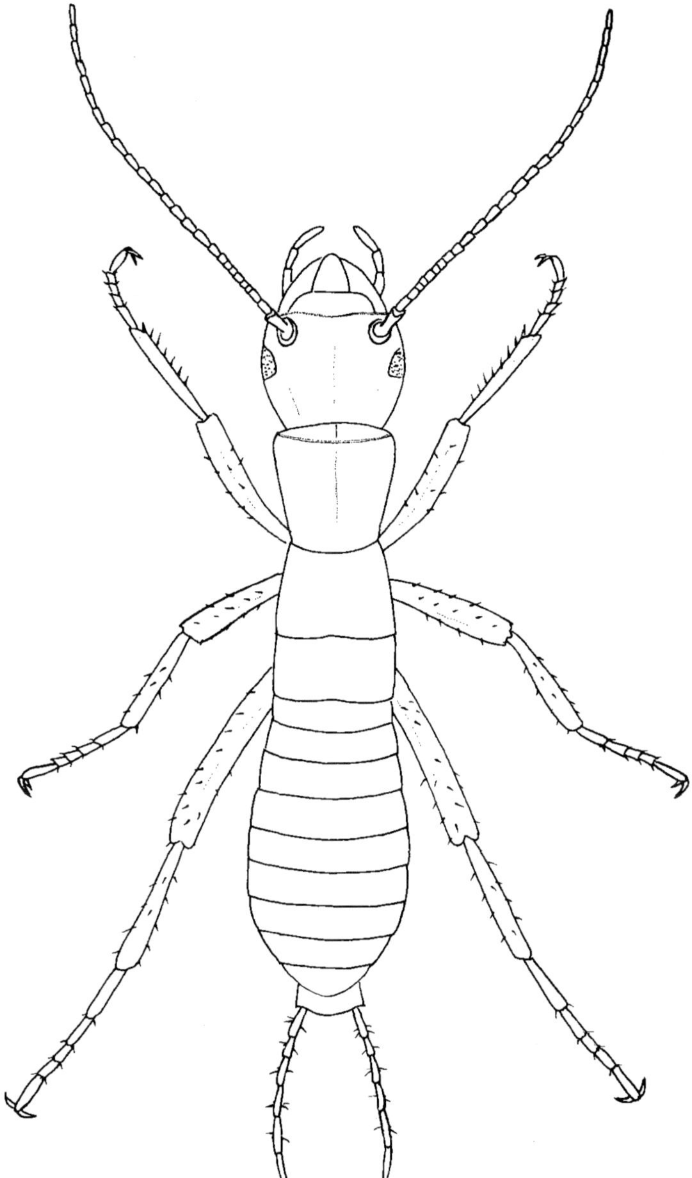

Fig. 82. *Grylloblatta campodeiformis campodeiformis*, juvenile.

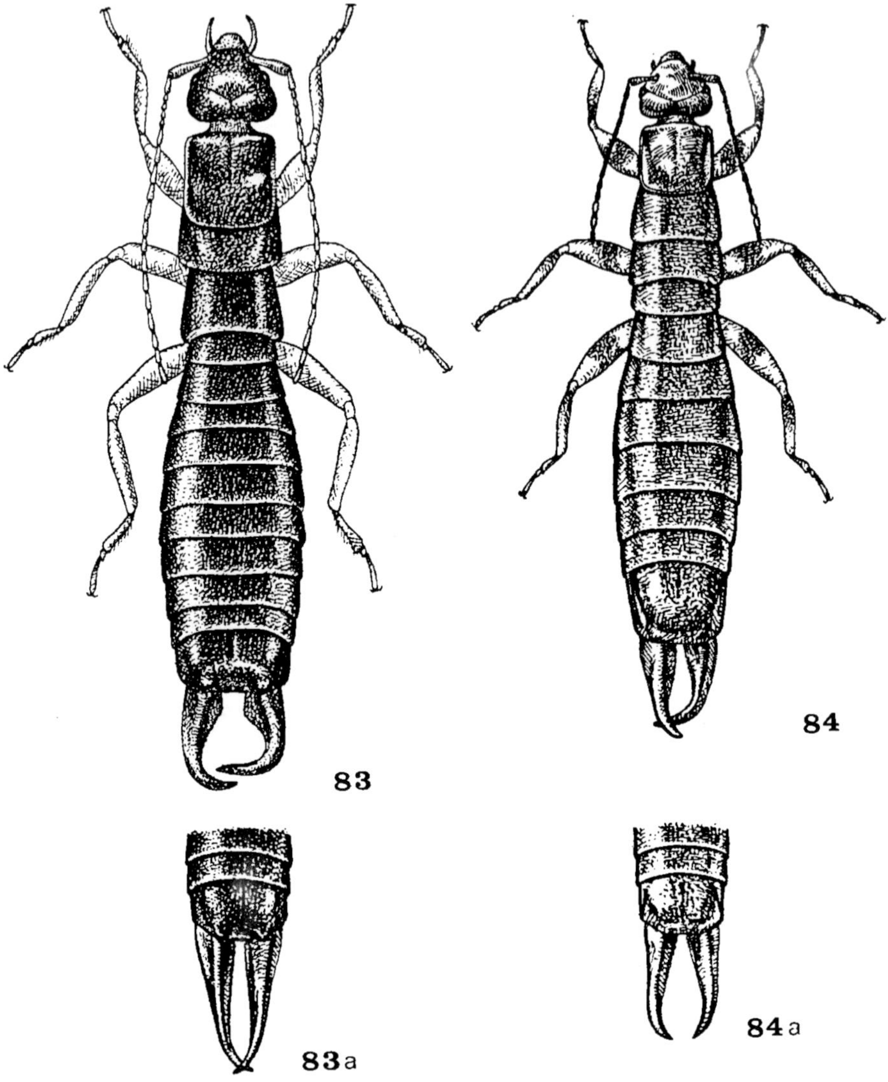

Figs. 83, 84. Anisolabididae. 83, *Anisolabis maritima*, male: *a*, end of female abdomen; 84, *Euborellia annulipes*: *a*, end of female abdomen. After Langston and Powell (1975), courtesy of University of California Press.

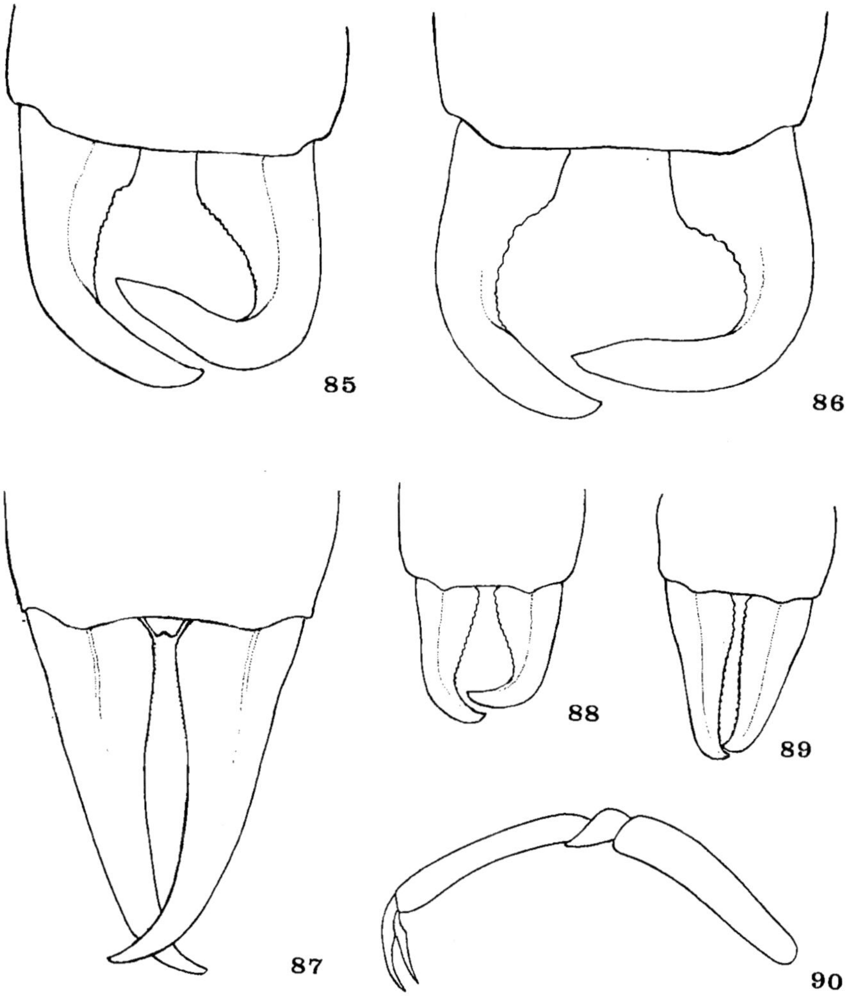

Figs. 85–90. Anisolabididae. 85, 86, male forceps of *Anisolabis maritima*, variations; 87, female forceps of the same; 88, 89, male forceps of *Euborellia annulipes*, variations; 90, tarsus of the same.

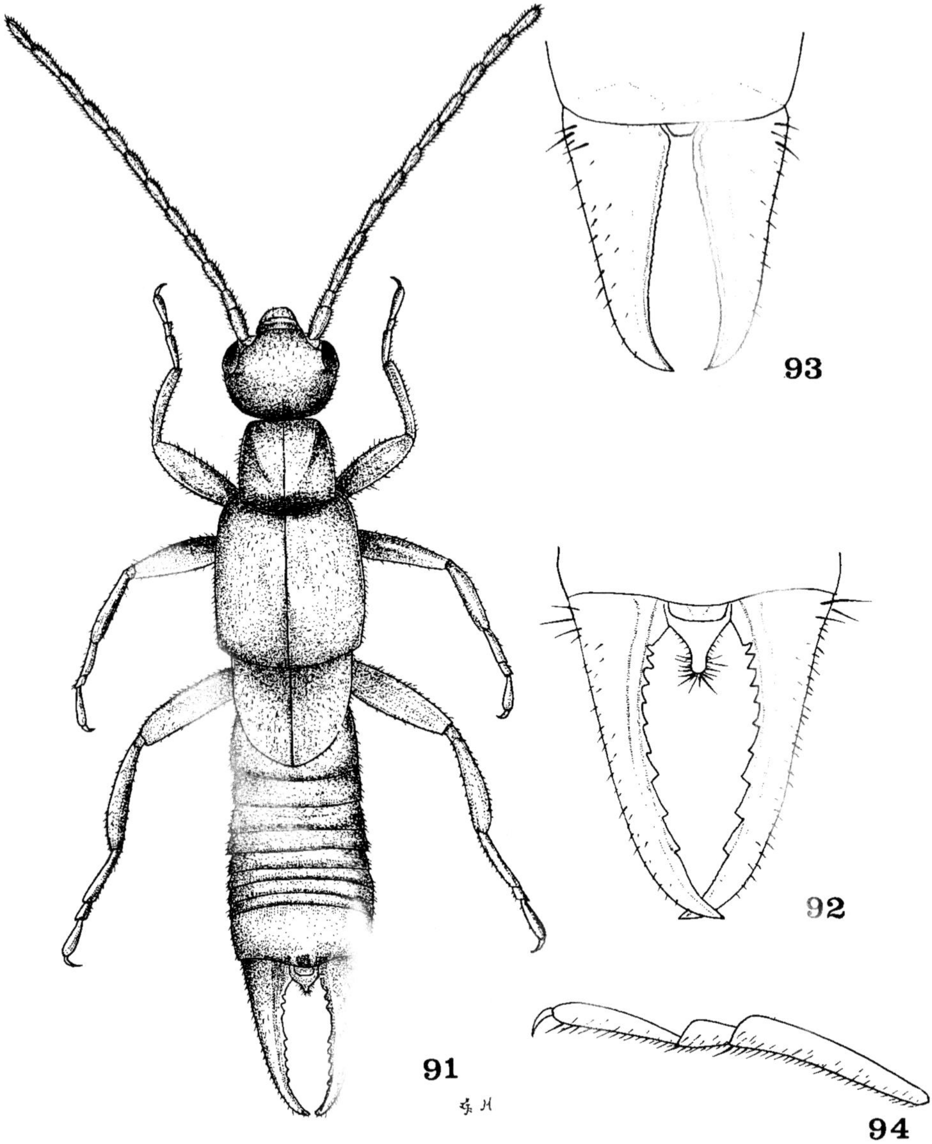

Figs. 91–94. *Labia minor*. 91, male; 92, male forceps; 93, female forceps; 94, tarsus.

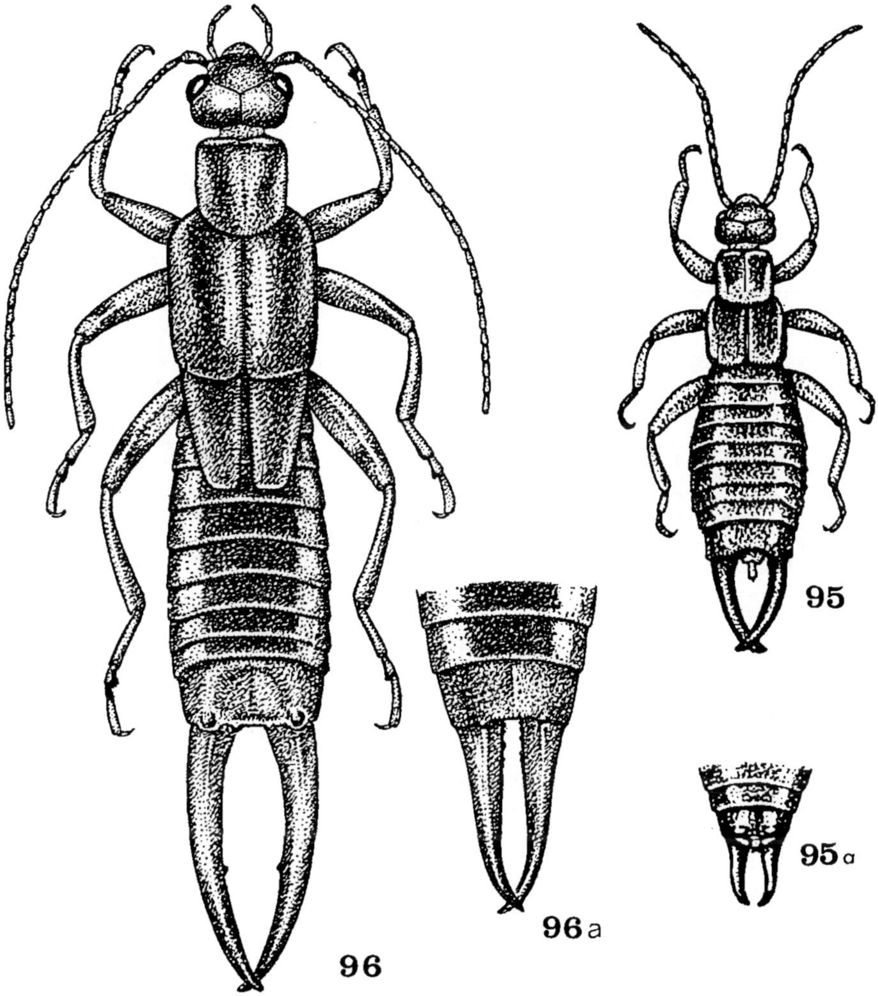

Figs. 95, 96. *Marava arachidis* and *Labidura riparia*. 95, *L. riparia*, male: *a*, end of female abdomen; 96, *M. arachidis*, male; *a*, end of female abdomen. After Langston and Powell (1975), courtesy of University of California Press.

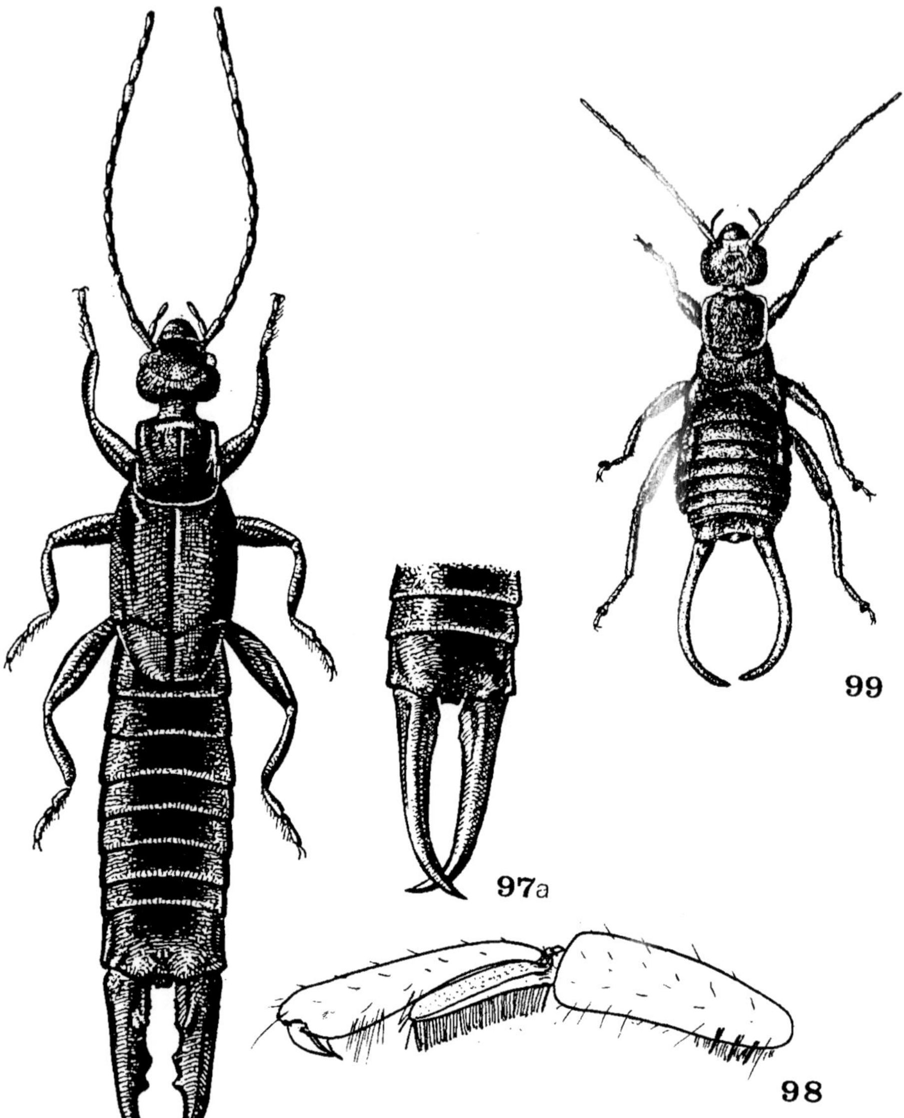

97 97a 98 99

Figs. 97–99. Structures of *Chelisoches* sp. and *Chelidurella* sp. 97, 98, *Chelisoches morio*. 97, male: *a*, end of female abdomen; 98, tarsus; 99, *Chelidurella acanthopygia*, male. 97, after Langston and Powell (1975), courtesy of University of California Press; 99, after Holst (1970), courtesy of Editor, Danmarks Fauna.

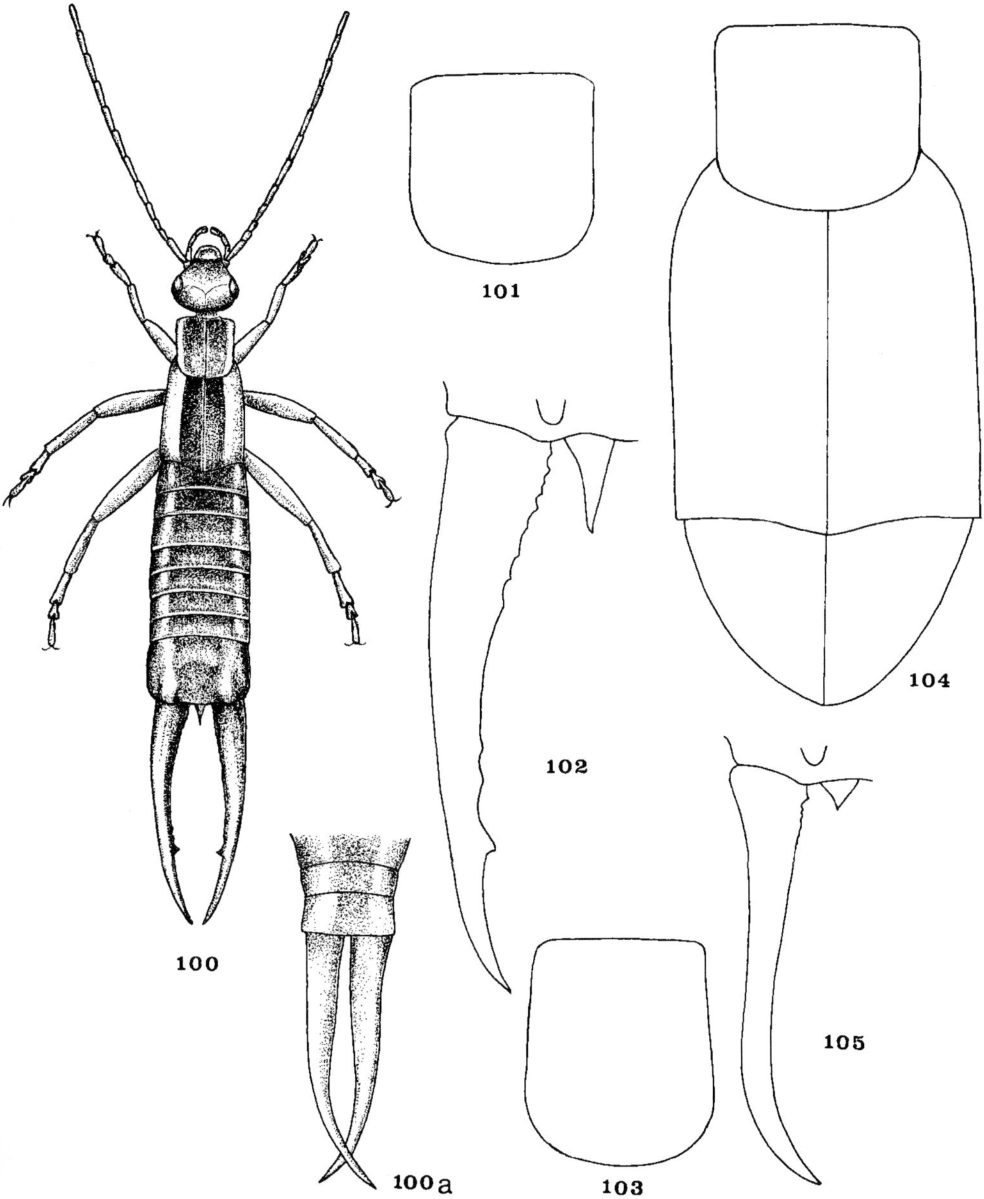

Figs. 100–105. *Doru* spp. 100, *D. aculeatum*, male: *a*, end of female abdomen; 101, the same, outline of pronotum; 102, left forceps of male; 103, *D. davisi*, outline of pronotum; 104, *D. taeniatum*, outline of pronotum, tegmina and folded hind wings; 105, the same, left forceps of male.

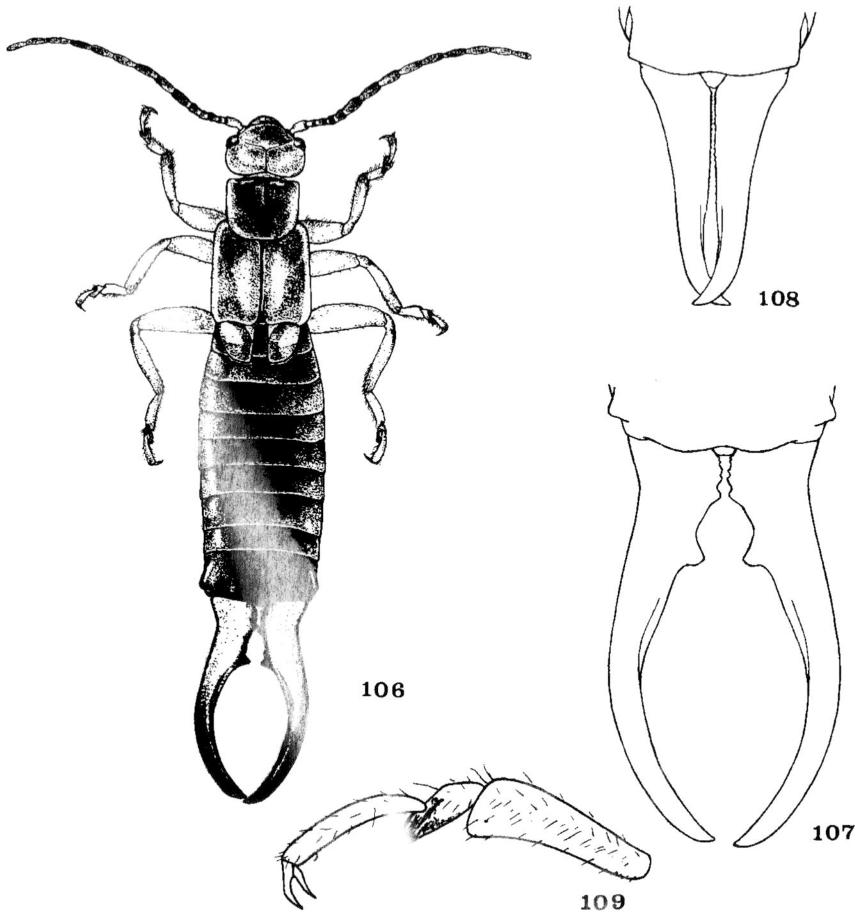

Figs. 106–109. *Forficula auricularia*. 106, male; 107, male forceps; 108, female forceps; 109, tarsus. 106–108, after Vickery et al. (1974).

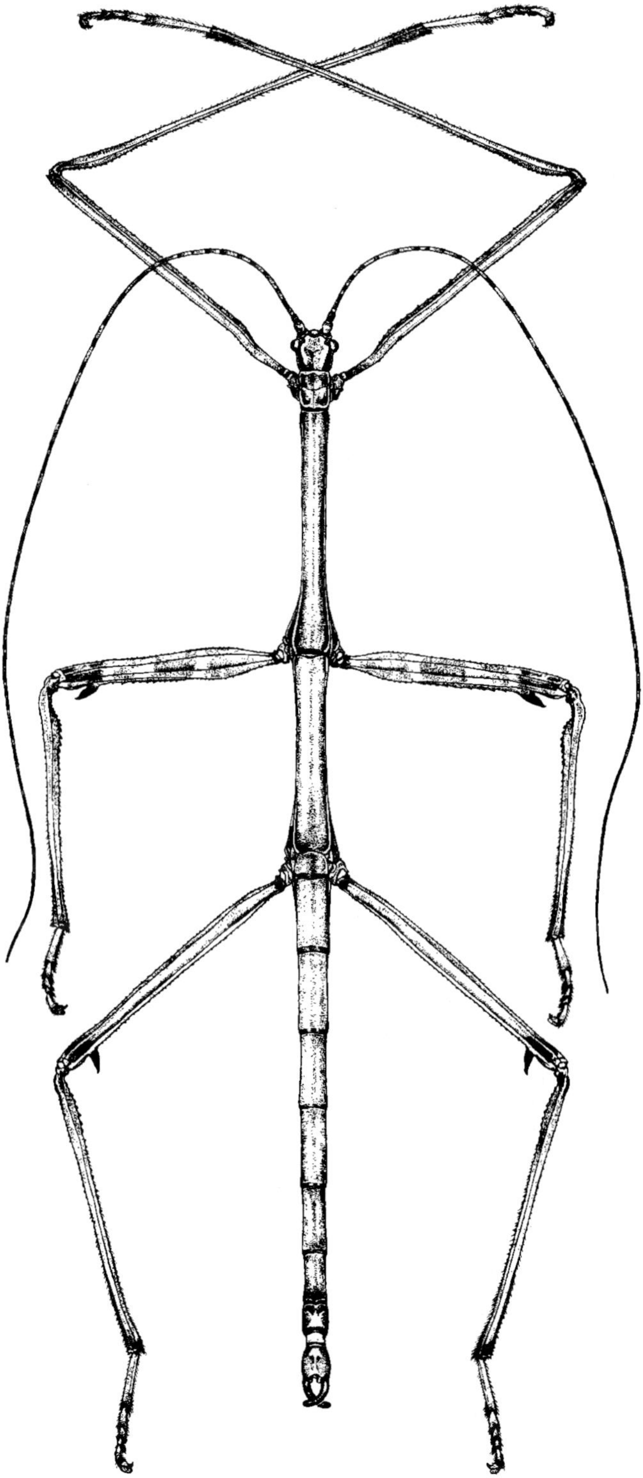

Fig. 110. *Diapheromera femorata*, male. After Vickery et al. (1974).

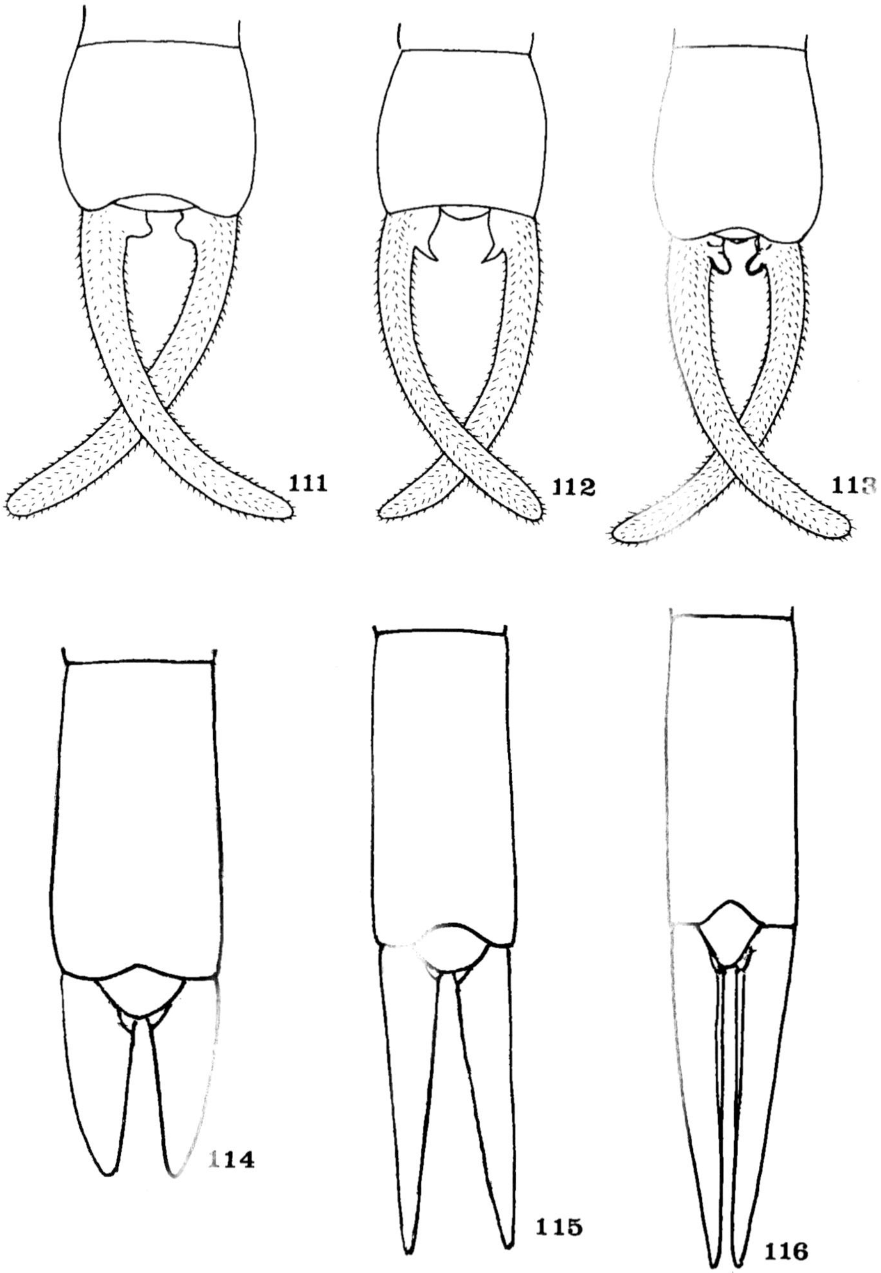

Figs. 111–116. *Diapheromera* spp., abdominal terminalia showing cerci. 111–113, male; 114–116, female; 111, 114, *D. femorata*; 112, 115, *D. velii velii*; 113, 116, *D. blatchleyi blatchleyi*.

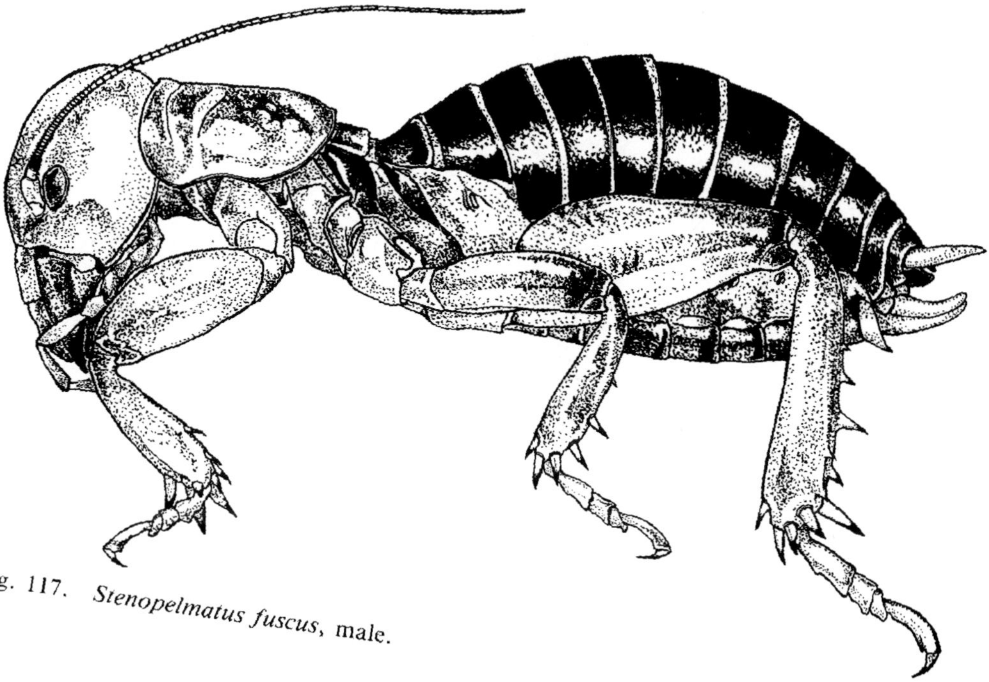

Fig. 117. *Stenopelmatus fuscus*, male.

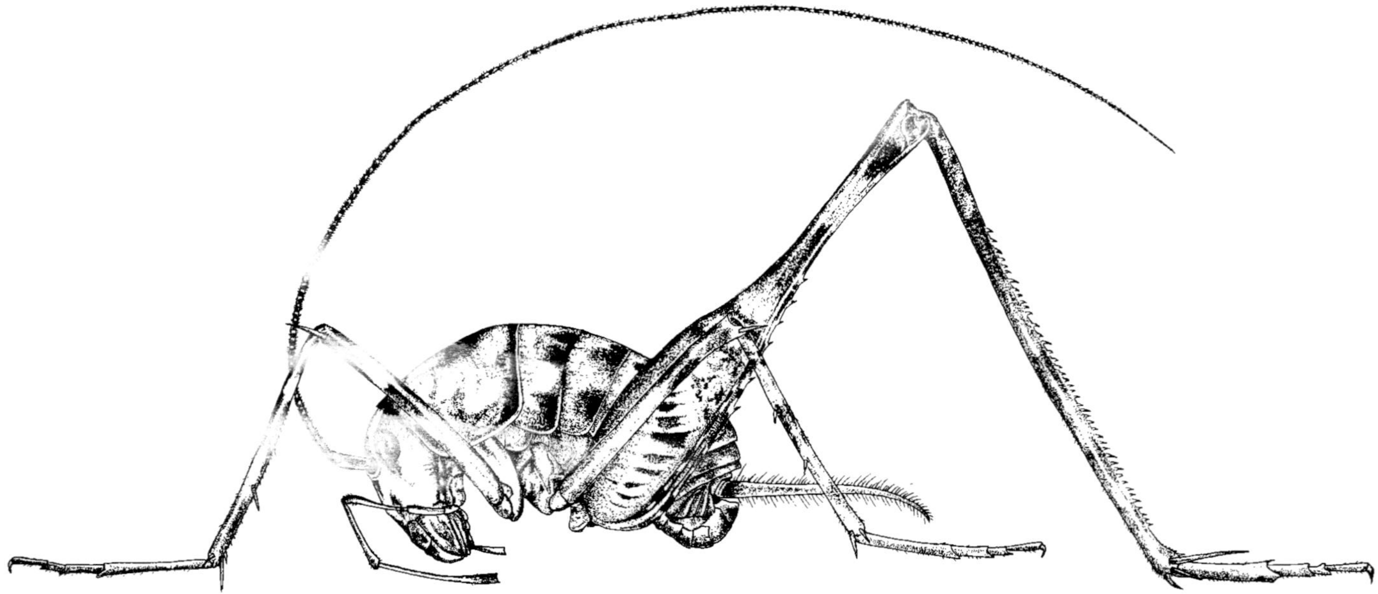

Fig. 118. *Tachycines asynamorus*, male.

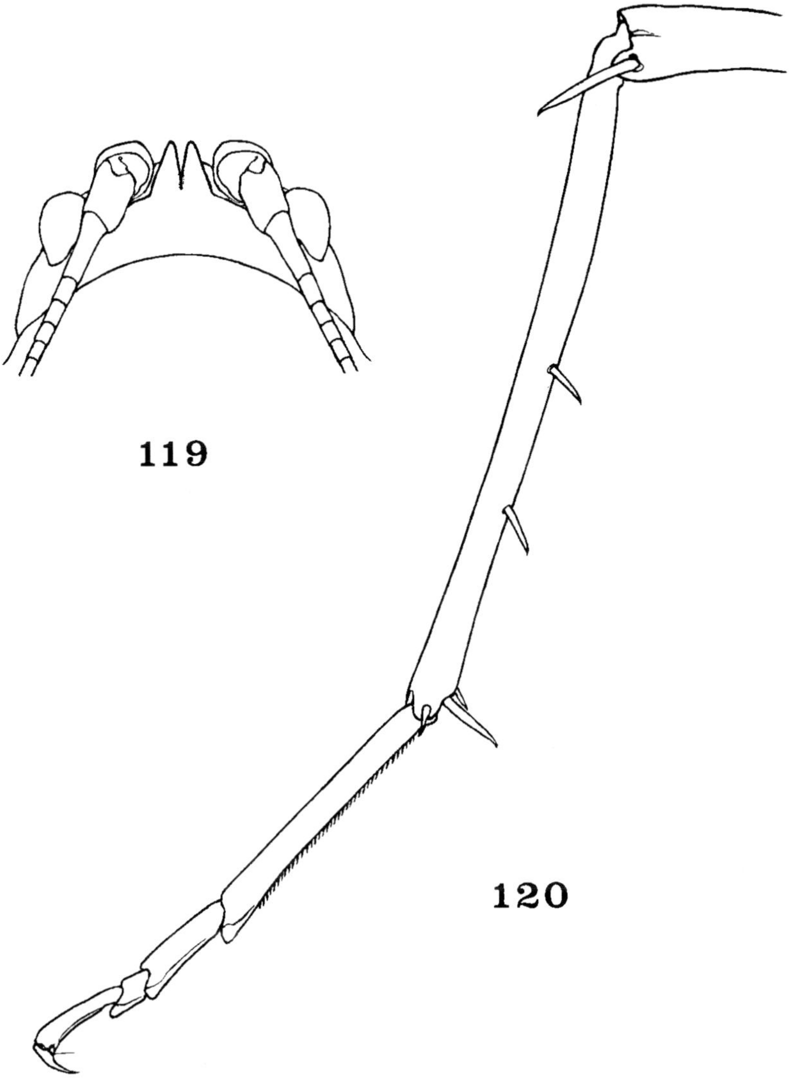

119

120

Figs. 119, 120. *Tachycines asynamorus*. 119, head, dorsal; 120, apex of anterior femur, tibia and tarsus, female: note apical spur on femur, apical spine between apical spurs of hind tibia, and ventroproximal seta associated with tarsal claw.

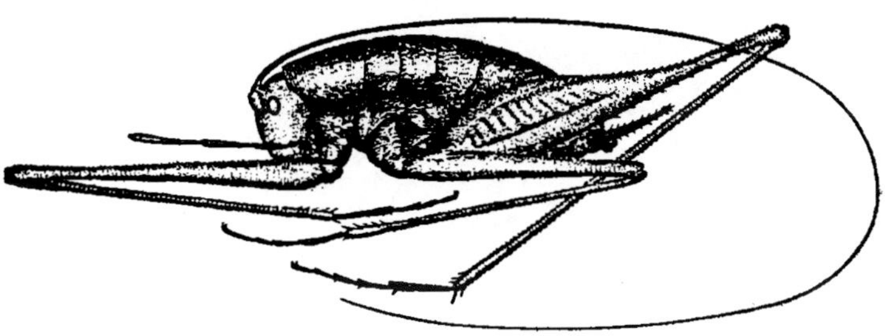

Fig. 121. *Tropidischia xanthostoma*, male. After Helfer (1963), courtesy of J. R. Helfer.

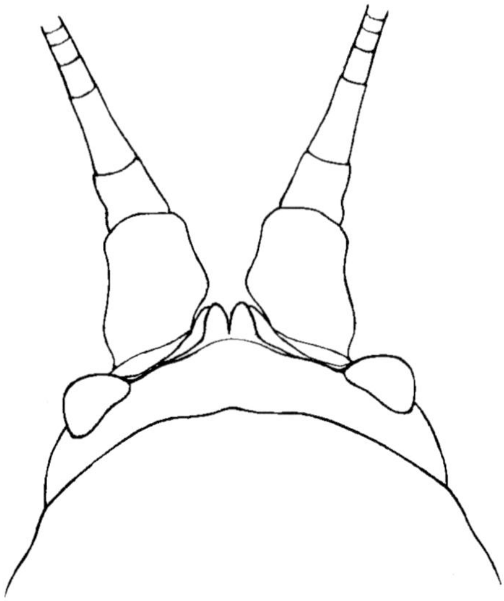

122

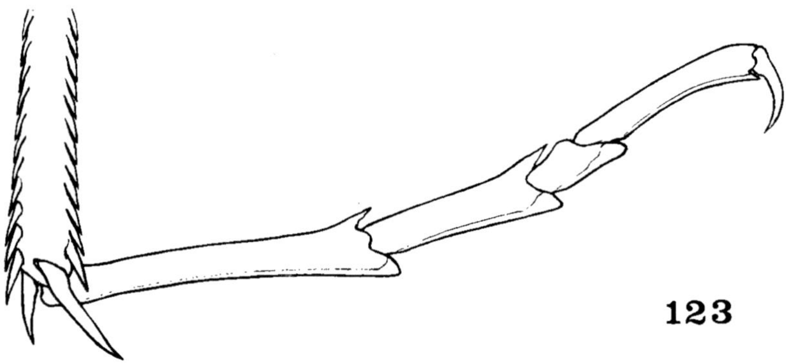

123

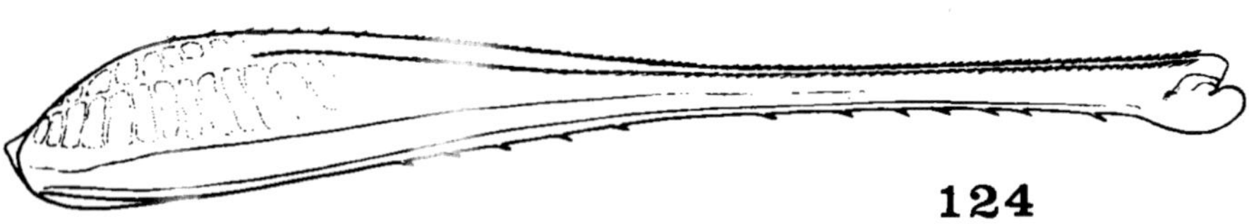

124

Figs. 122–124. *Tropidischia xanthostoma*. 122, head, dorsal, female; 123, apex of hind tibia and tarsus, outer face; 124, hind femur, male, outer face.

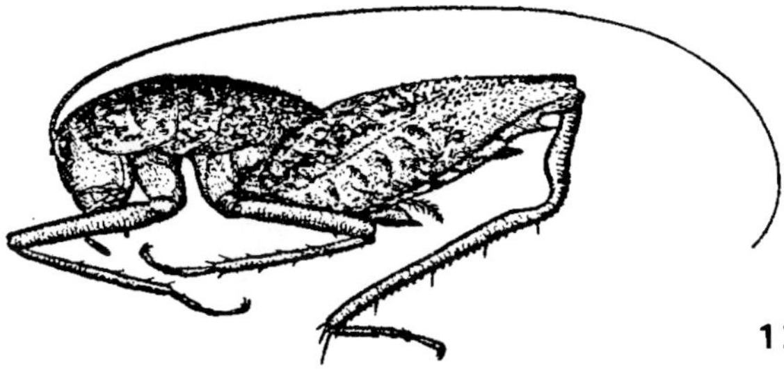

125

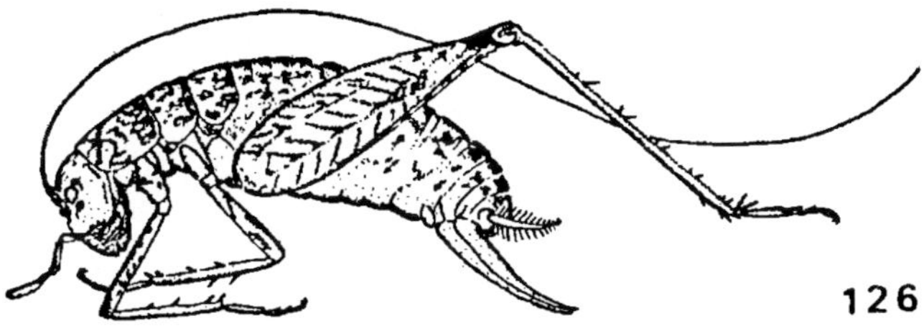

126

Figs. 125, 126. *Pristoceuthophilus pacificus*. 125, male; 126, female. After Helfer (1963), courtesy of J. R. Helfer.

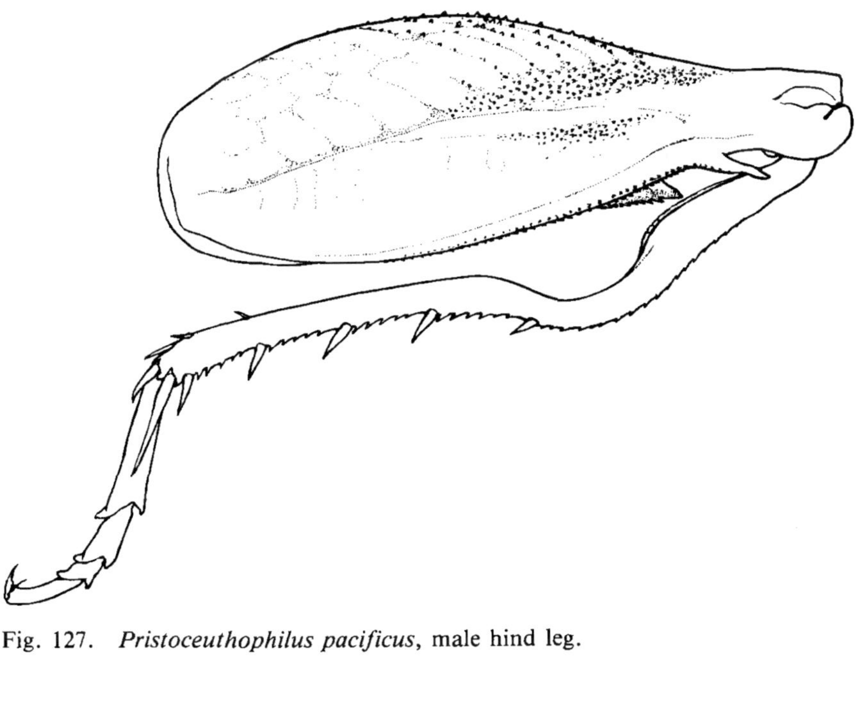

Fig. 127. *Pristoceuthophilus pacificus*, male hind leg.

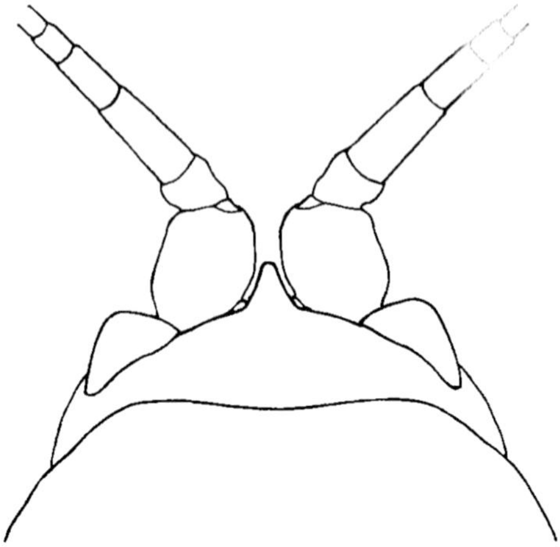

128

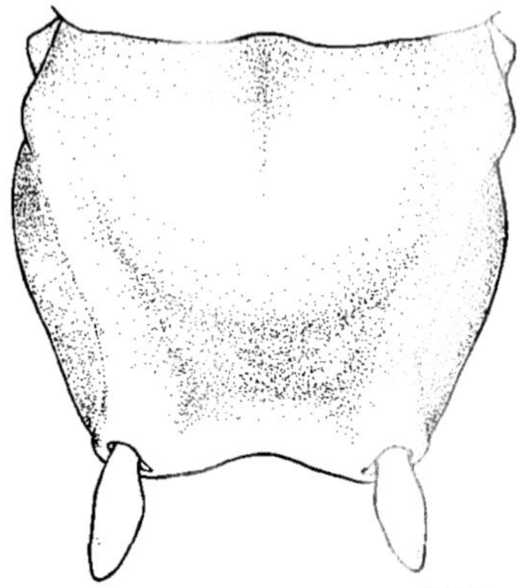

129

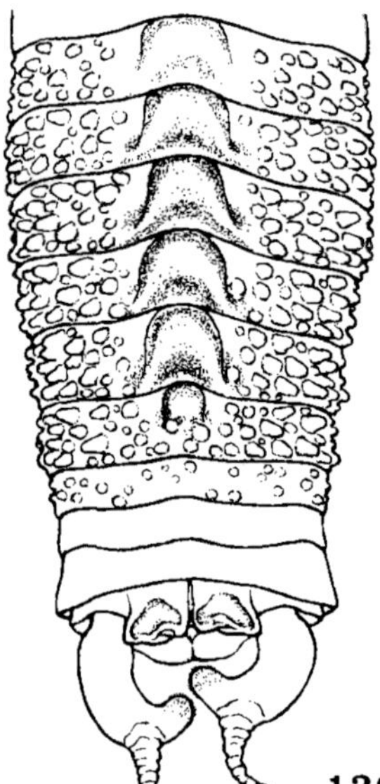

130

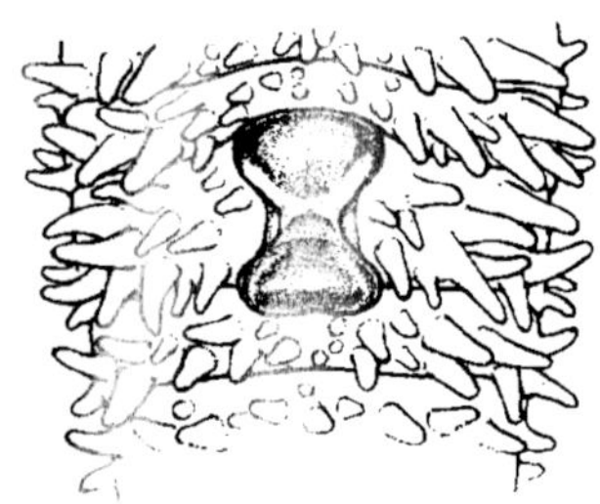

131

Figs. 128–131. *Pristoceuthophilus* spp. 128–130, *P. celatus*; 128, head, dorsal; 129, male subgenital plate; 130, male abdomen, dorsal; 131, *P. gaigei*, terga III–V (with parts of II and VI) of male abdomen.

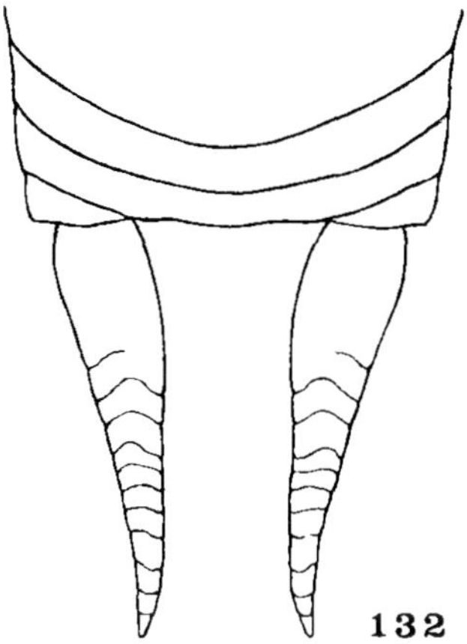

132

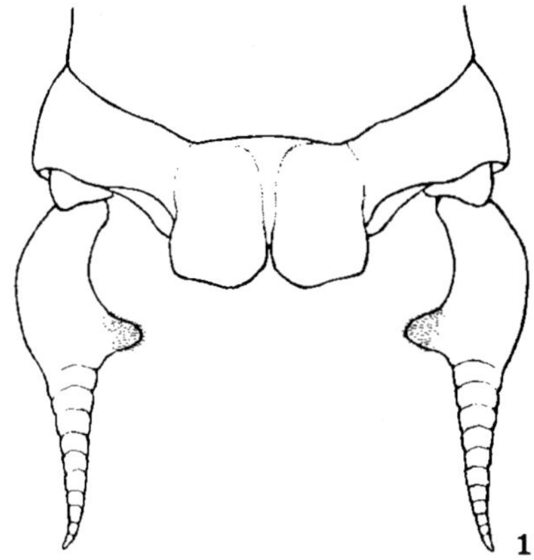

133

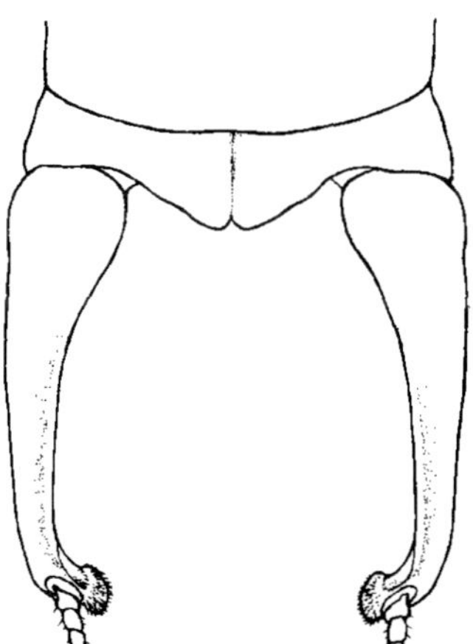

134

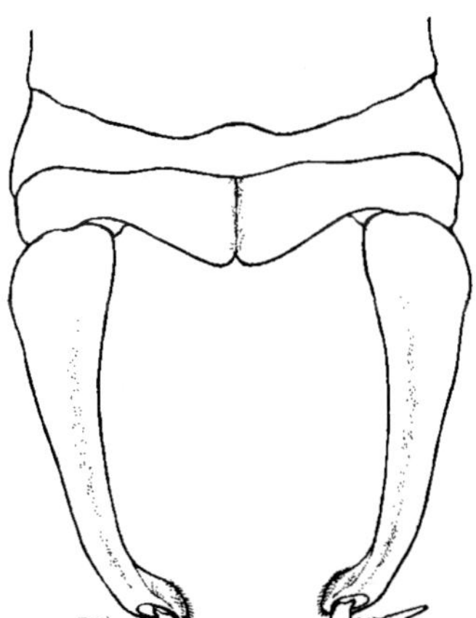

135

Figs. 132–135. *Pristoceuthophilus* spp., male abdominal terminalia showing cerci. 132, *P. pacificus*; 133, *P. celatus*; 134, *P. gaigei*; 135, *P. cercalis*.

136

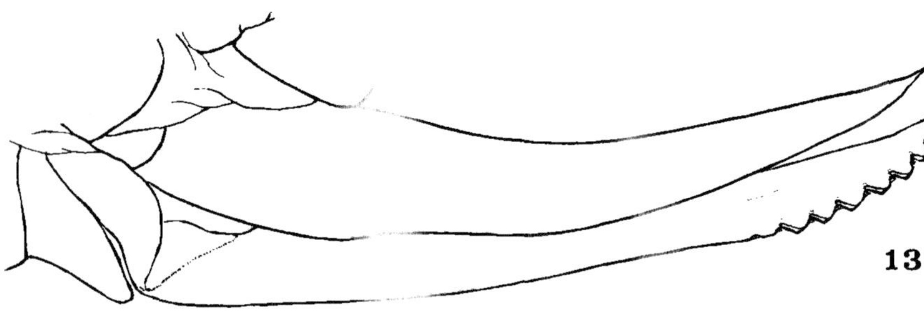

137

138

Figs. 136–138. *Pristoceuthophilus* spp., ovipositors. 136, *P. pacificus*; 137, *P. celatus*; 138, *P. cercalis*.

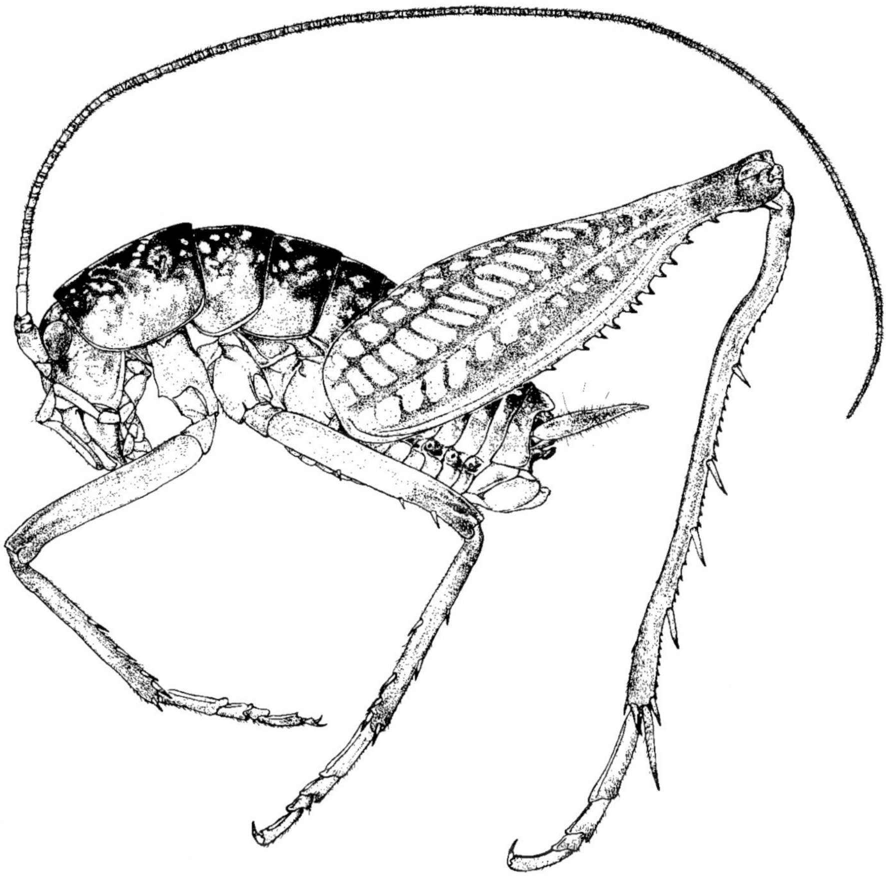

Fig. 139. *Ceuthophilus maculatus*, male.

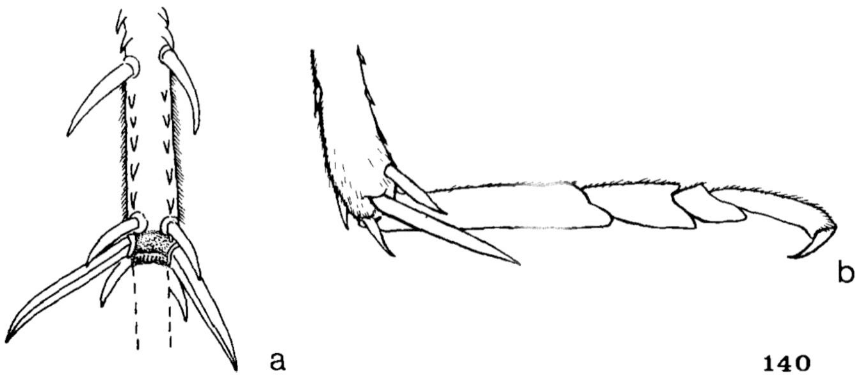

140

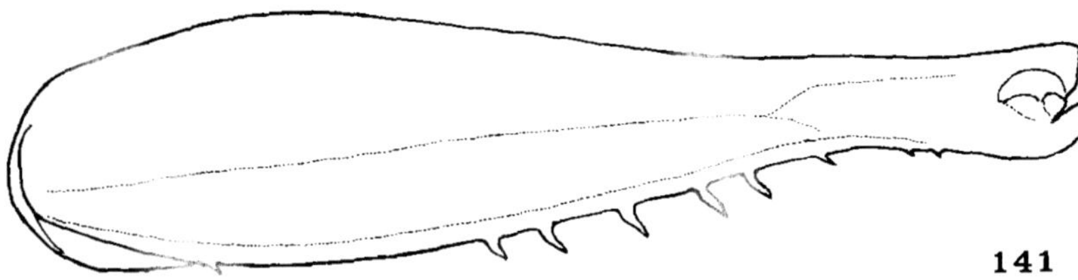

141

142

143

Figs. 140–143. *Ceuthophilus* spp., details of legs. 140, *C. maculatus*: *a*, apex of hind tibia; *b*, apex of hind tibia and tarsus; 141, *C. gracilipes gracilipes*, male hind femur; 142, *C. meridionalis*, apex of hind tibia and tarsus; 143, *C. fusiformis*, apex of hind tibia and tarsus.

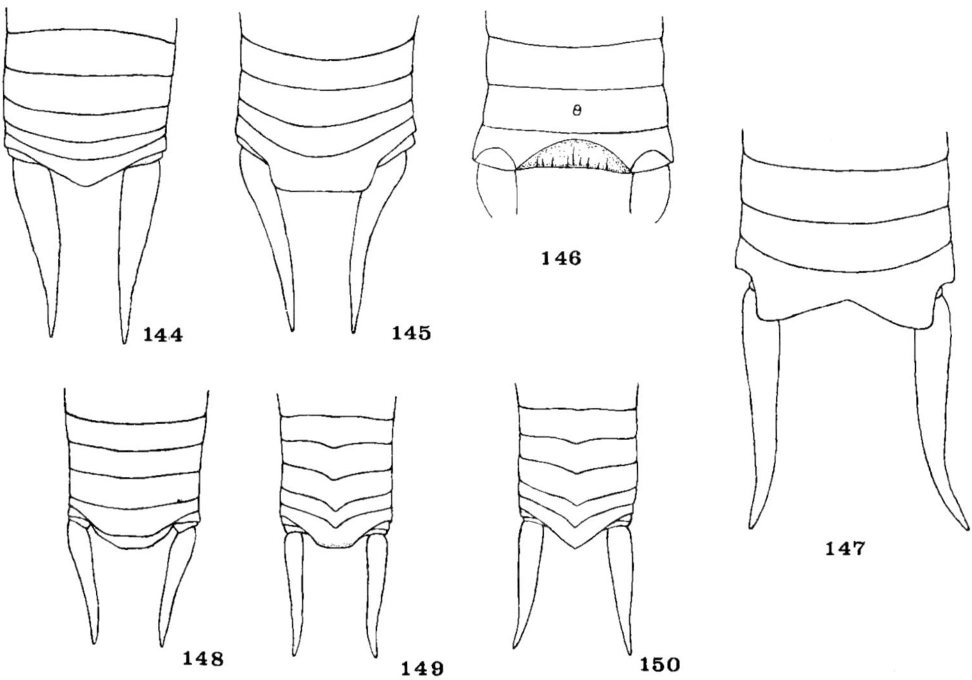

Figs. 144–150. *Ceuthophilus* spp., male abdominal terminalia. 144, *C. brevipes*; 145, *C. pallidipes*; 146, *C. latens*; 147, *C. maculatus*; 148, *C. guttulosus guttulosus*; 149, *C. guttulosus*, intermediate between *C. guttulosus guttulosus* and *C. guttulosus thomasi*; 150, *C. guttulosus thomasi*. Except for Fig. 146, after Vickery et al. (1974).

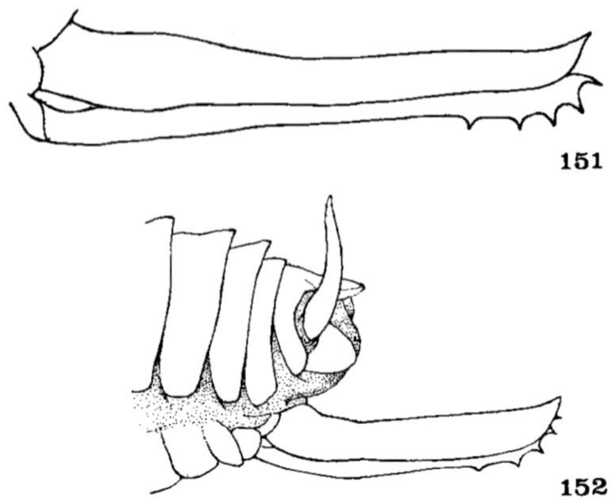

151

152

Figs. 151, 152. *Ceuthophilus* spp., ovipositors. 151, *C. brevipes*; 152, *C. guttulosus guttulosus*. After Vickery et al. (1974).

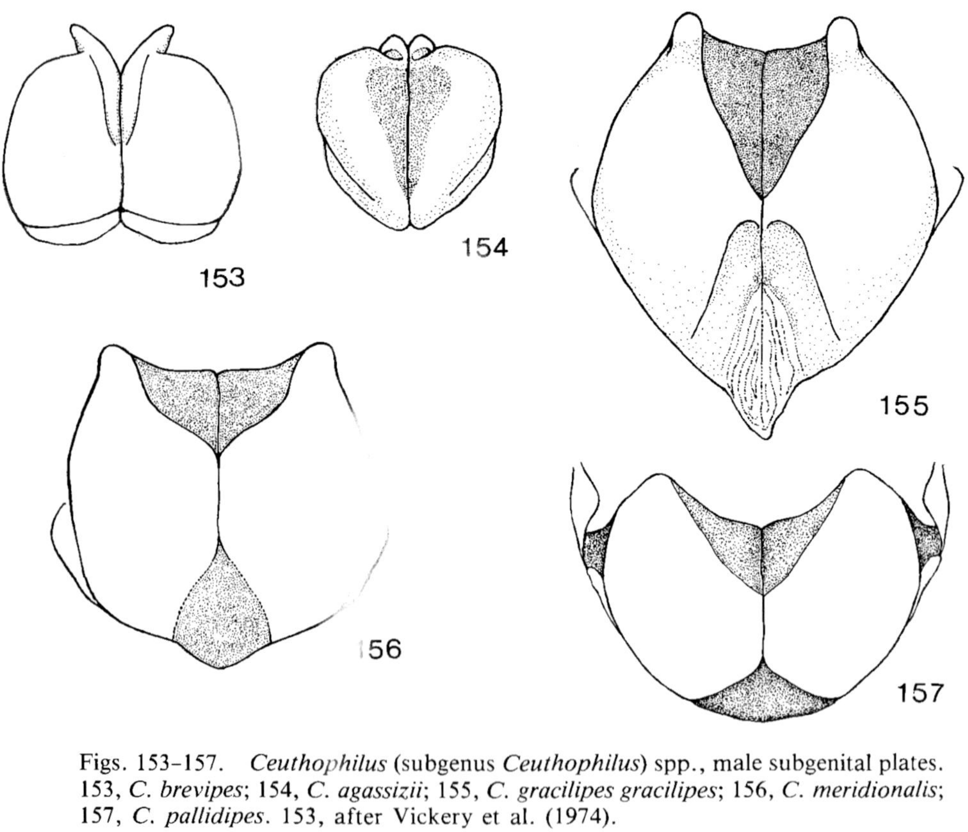

153

154

155

156

157

Figs. 153–157. *Ceuthophilus* (subgenus *Ceuthophilus*) spp., male subgenital plates. 153, *C. brevipes*; 154, *C. agassizii*; 155, *C. gracilipes gracilipes*; 156, *C. meridionalis*; 157, *C. pallidipes*. 153, after Vickery et al. (1974).

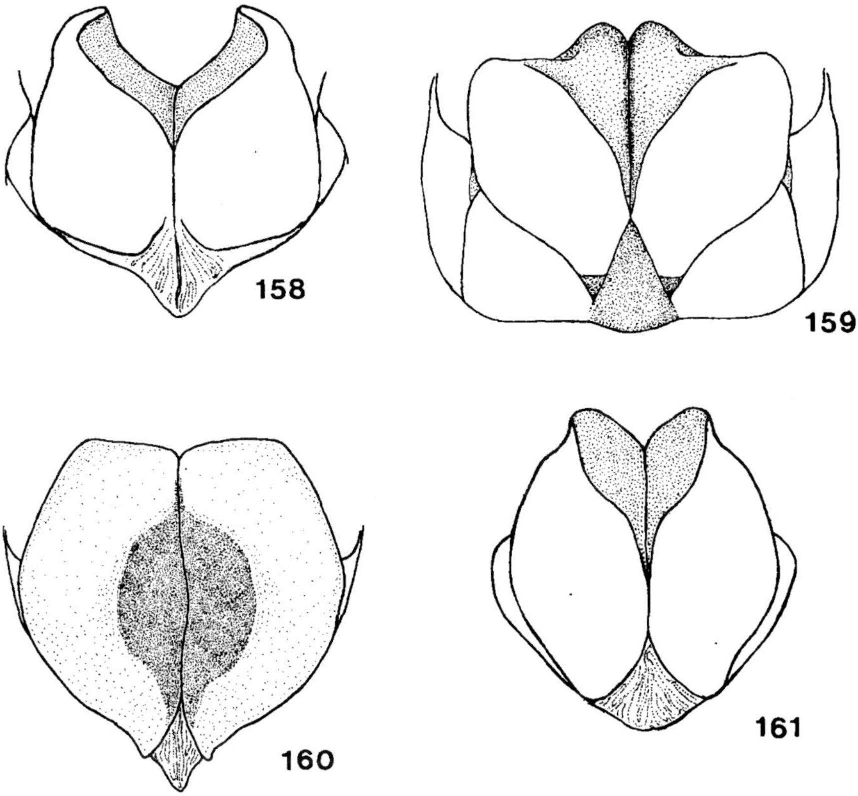

Figs. 158–161. *Ceuthophilus* (subgenus *Ceuthophilus*) spp., male subgenital plates. 158, *C. latens*; 159, *C. maculatus*; 160, *C. pallidus*; 161, *C. uhleri*.

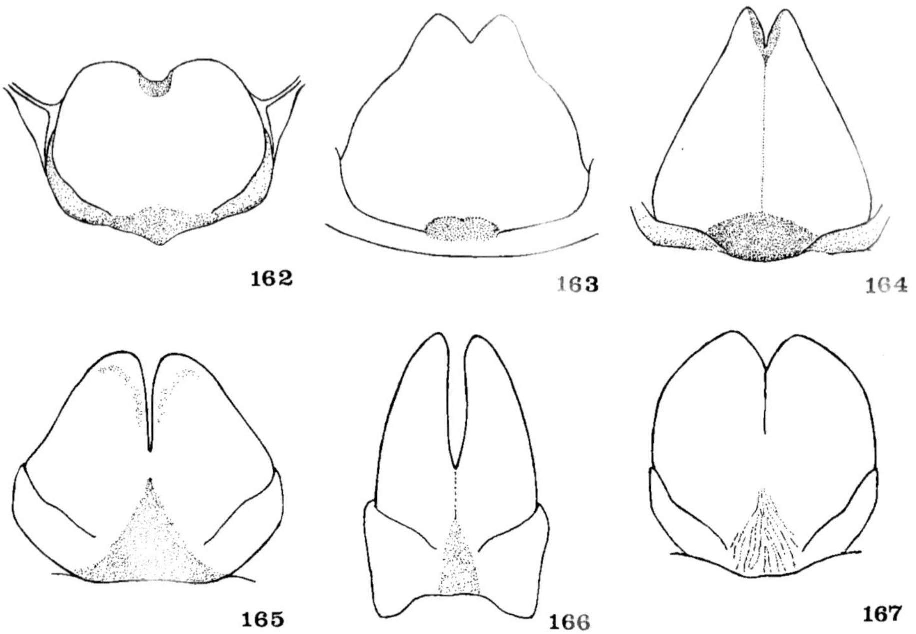

Figs. 162–167. *Ceuthophilus* (subgenus *Geotettix*) spp., male subgenital plates. 162, *C. guttulosus guttulosus*; 163, *C. guttulosus*, intermediate between *C. guttulosus guttulosus* and *C. guttulosus thomasi*; 164, *C. guttulosus thomasi*; 165, *C. fusiformis*; 166, *C. silvestris*; 167, *C. alpinus*. 162–164, after Vickery et al. (1974).

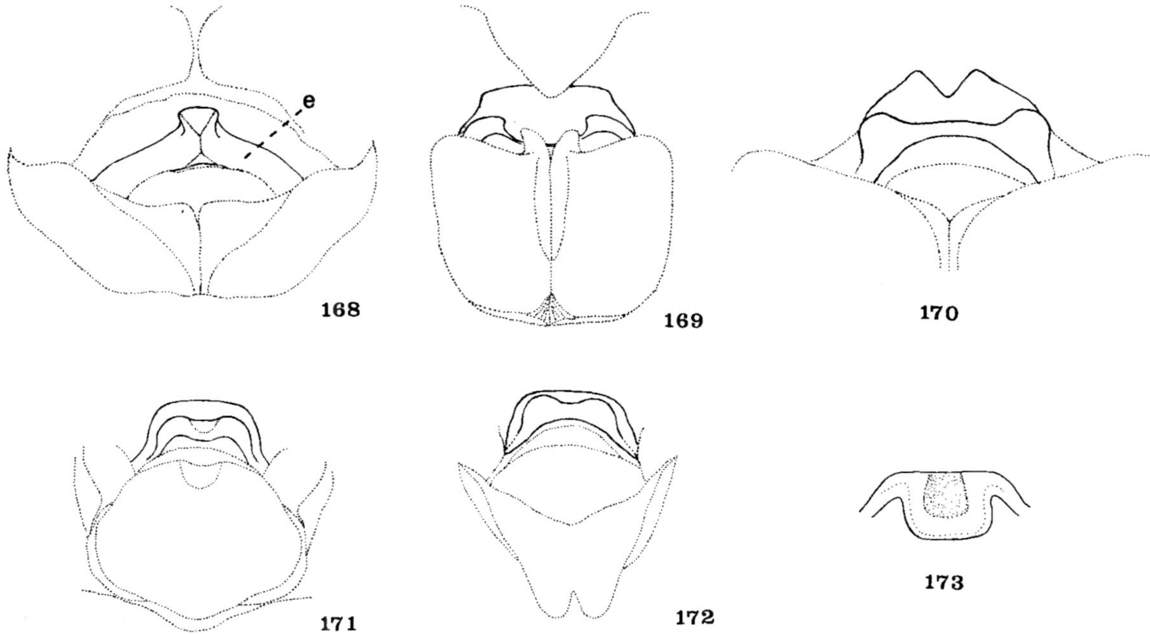

Figs. 168–173. *Ceuthophilus* spp., epiphalli (*in situ*, except Fig. 173). 168, *C. maculatus*, *e* = epiphallus; 169, *C. brevipes*; 170, *C. pallidipes*; 171, *C. guttulosus guttulosus*; 172, *C. guttulosus thomasi*; 173, *C. alpinus*. 168, 169, 171, 172, after Vickery et al. (1974).

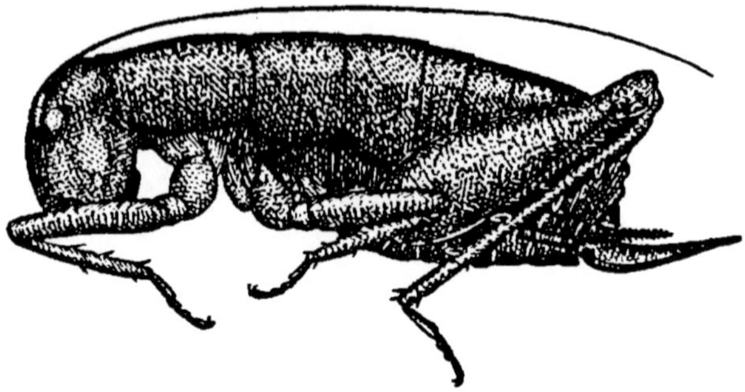

Fig. 174. *Udeopsylla robusta*, female. After Helfer (1963), courtesy of J. R. Helfer.

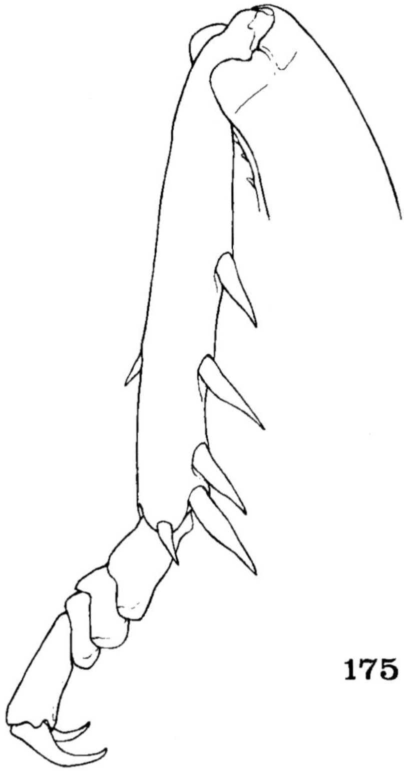

175

176

Figs. 175, 176. *Udeopsylla robusta*. 175, anterior tibia and tarsus of female; 176, ovipositor.

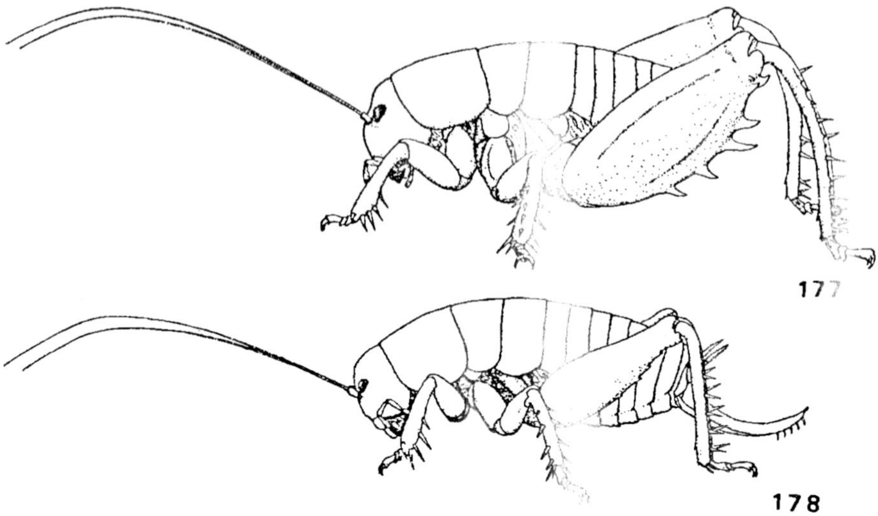

177

178

Figs. 177, 178. *Daihinia brevipes*. 177, male; 178, female. After Whitehead and Miner (1944).

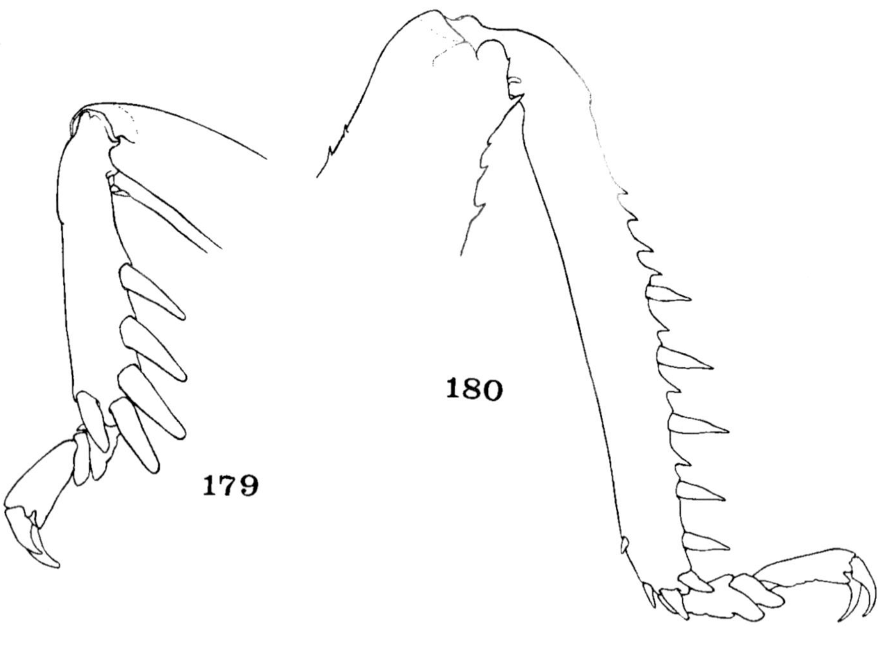

179

180

Figs. 179, 180. *Daihinia brevipes*, female. 179, anterior, tibia, and tarsus; 180, posterior, tibia, and tarsus.

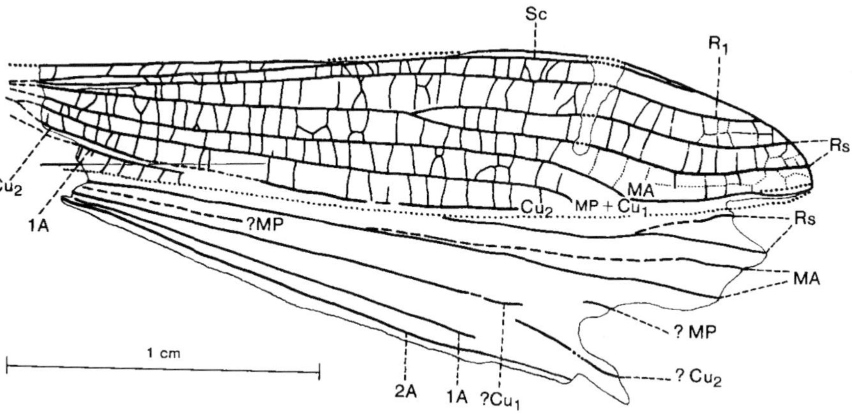

Fig. 181. *Phasmomimella paskapoensis*, (Phasmomimidae) from the Paskapoo (Palaeocene) deposits (fine argillaceous limestone), Red Deer River, Alta. After Kevan and Wighton (1981). Notation of veins after Ragge (1955). 1A, 2A, first and second anals; Cu_1, Cu_2, first and second branches of cubitus; MA, anterior median; MP, posterior median; R_1, first branch of radius; Rs, radial sector (rest of radius); Sc, subcosta. Note: according to the notation of Sharov (1968, 1971) Cu_1 and Cu_2 are CuA and CuP (anterior and posterior cubitus) in fore wing and MP = MP & CuA_1, Cu_1 = CuA_2, and Cu_2 = CuP in hind wing.

Fig. 182. *Phasmomimella paskapoensis*, Phasmomimidae, as in Fig. 181, specimen *in situ*. After Kevan and Wighton (1981).

Fig. 183. *Cyphoderris monstrosa*, male. A "living fossil."

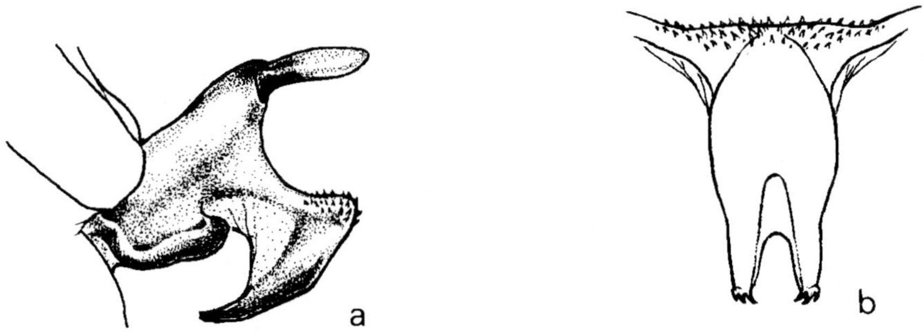

184

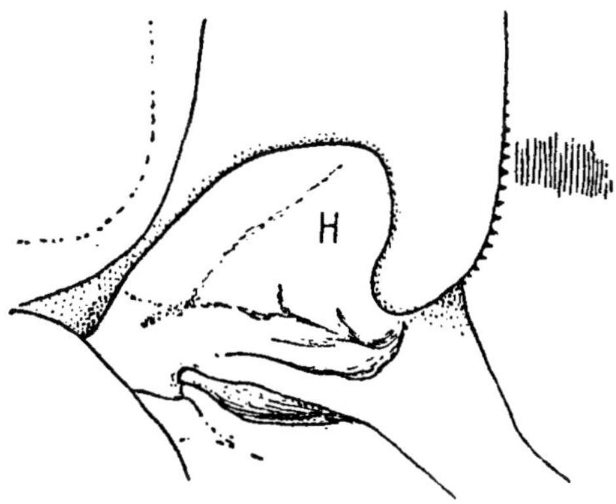

185

Figs. 184, 185. *Cyphoderris monstrosa*. 184*a*, male subgenital plate and left stylus, from left; 184*b*, male subgenital plate, posteroventrad aspect; 185, female metathorax and first abdominal tergum from left (adapted from Ander 1938). *H* = hind wing vestige. Note denticles on posterior margin of metanotum and ridges (Ander's organ) at base of first abdominal tergum; they were originally thought to operate together as a stridulatory mechanism, but this is still unconfirmed (Morris and Gwynne 1979).

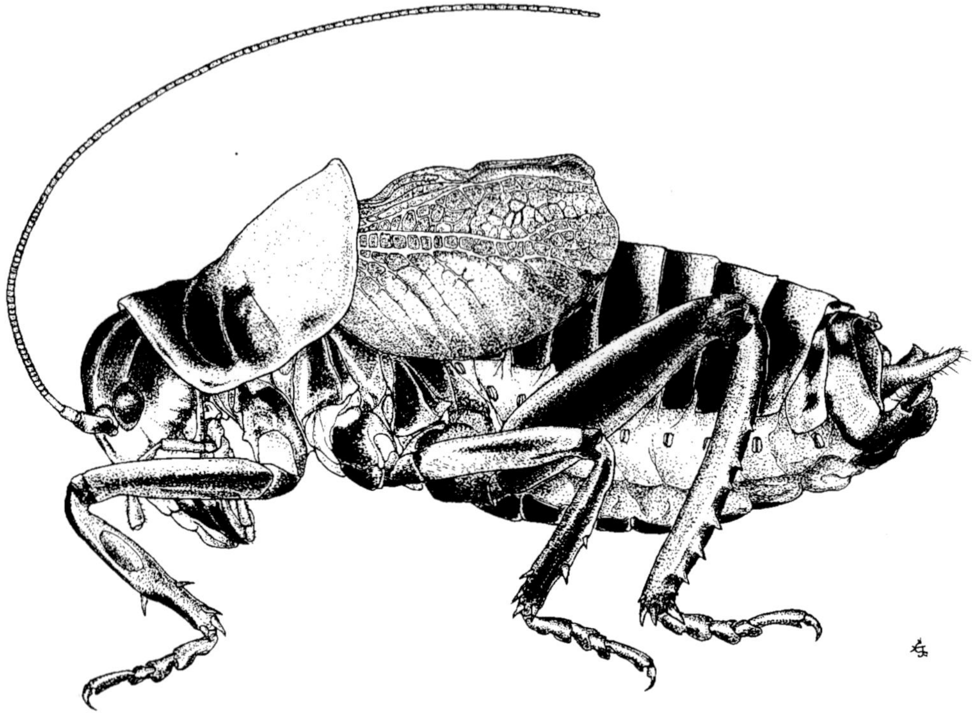

Fig. 186. *Cyphoderris buckelli*, male.

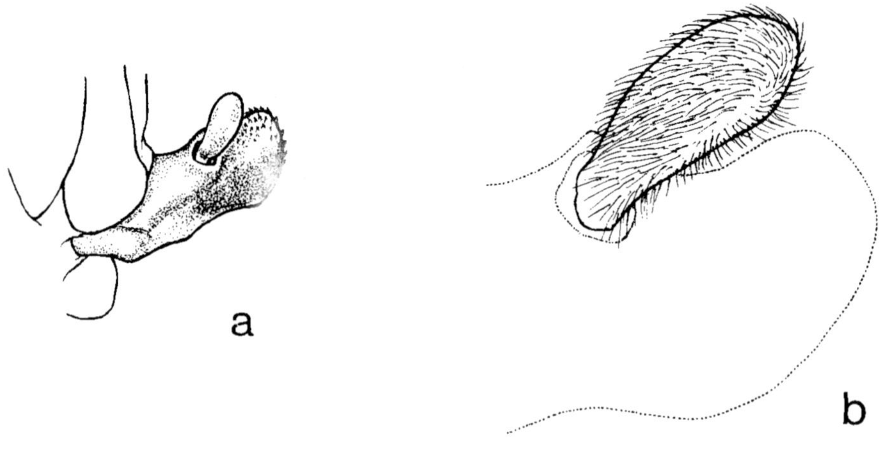

Fig. 187. *Cyphoderris buckelli*. *a*, male subgenital plate and left stylus, from left; *b*, male left stylus from left, enlarged, showing arrangement of setae.

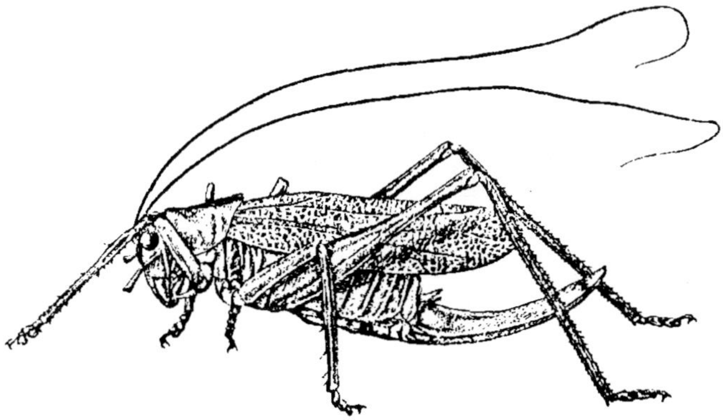

Fig. 188. *Meconema thalassinum*, female. After Holst (1970).

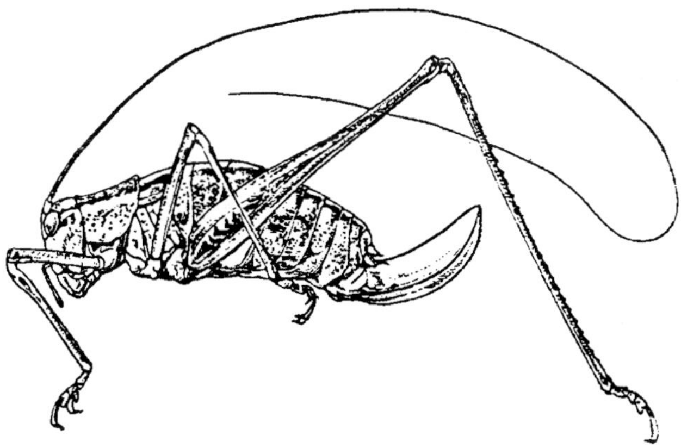

Fig. 189. *Leptophyes punctatissima*, female. After Chopard (1951), courtesy of Editions Lechevalier, Paris.

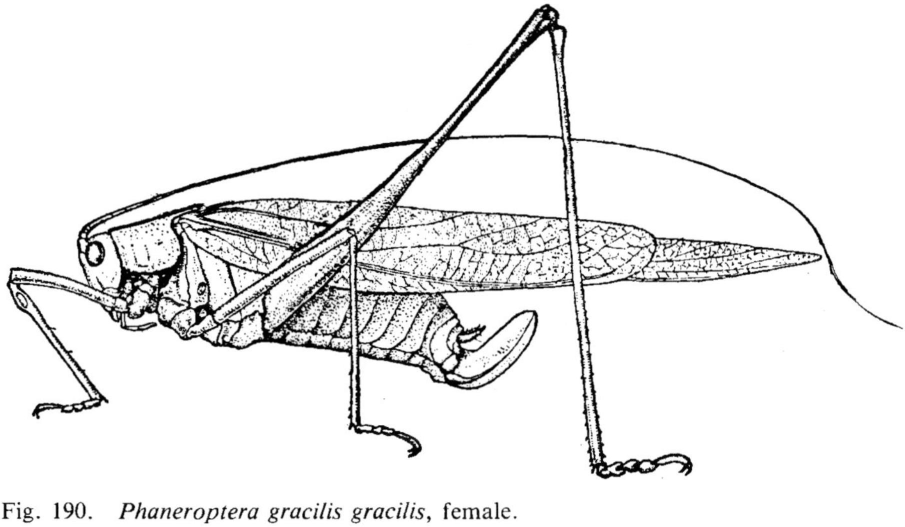

Fig. 190. *Phaneroptera gracilis gracilis*, female.

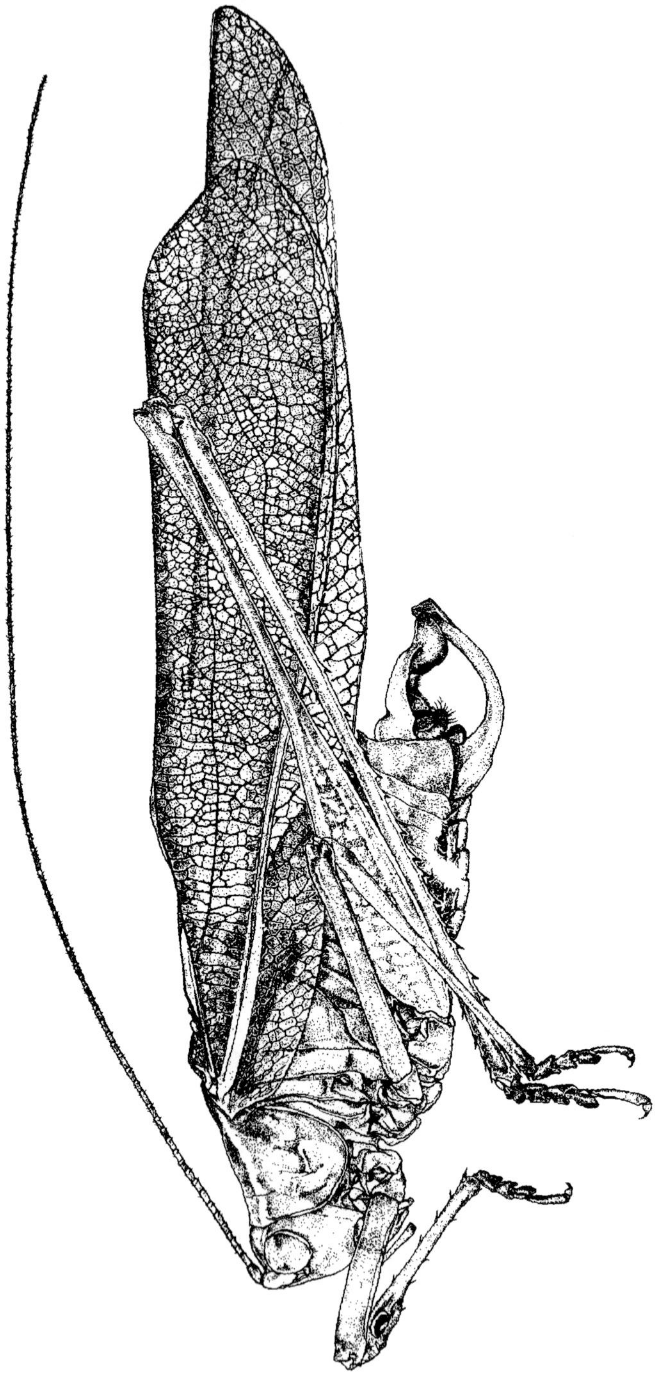

Fig. 191. *Scudderia furcata furcata*, male.

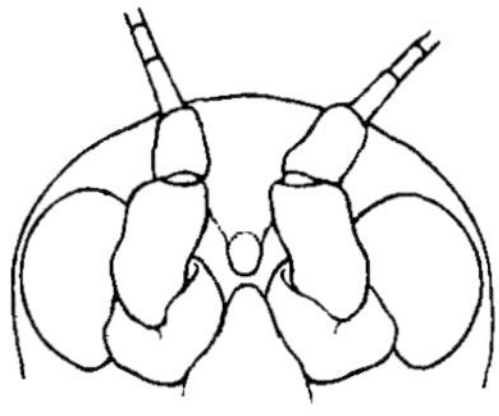
192

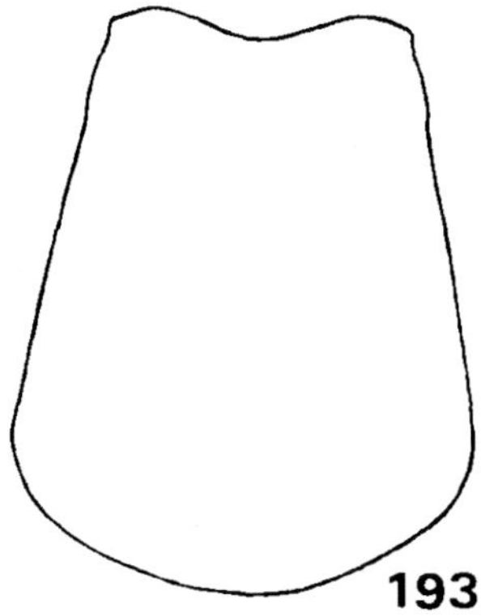
193

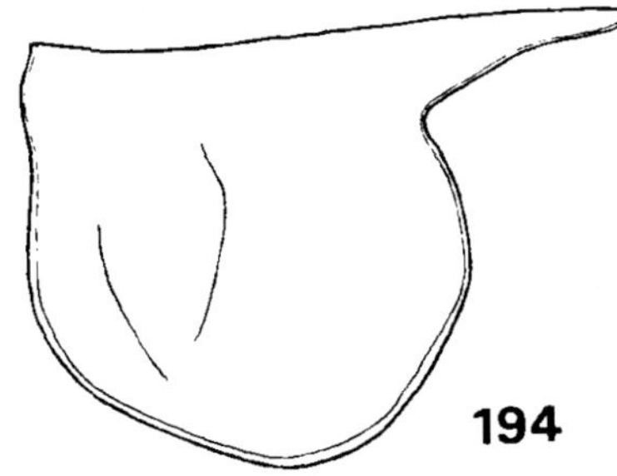
194

Figs. 192–194. *Scudderia pistillata*. 192, vertex of head, anterior; 193, pronotum, dorsal; 194, pronotum, lateral. 192, after Vickery et al. (1974).

195

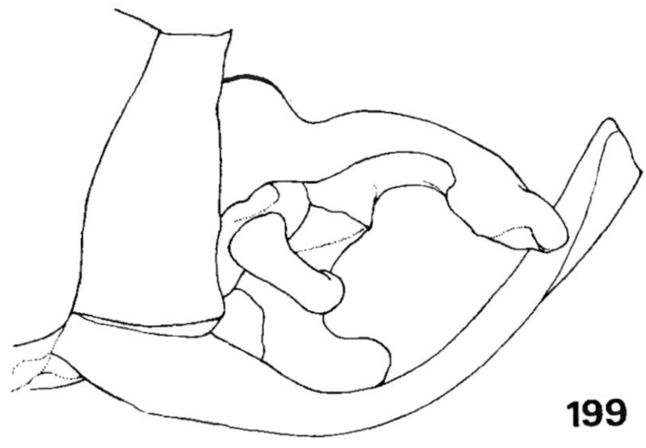
199

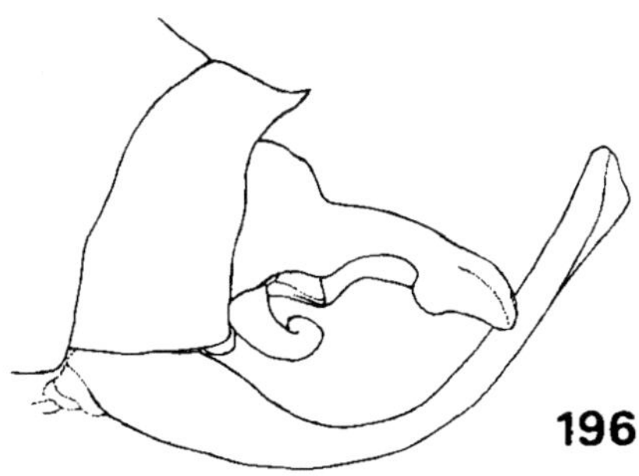
196

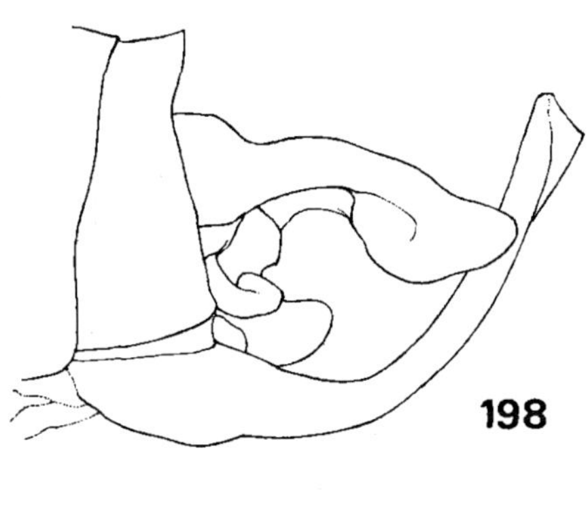
198

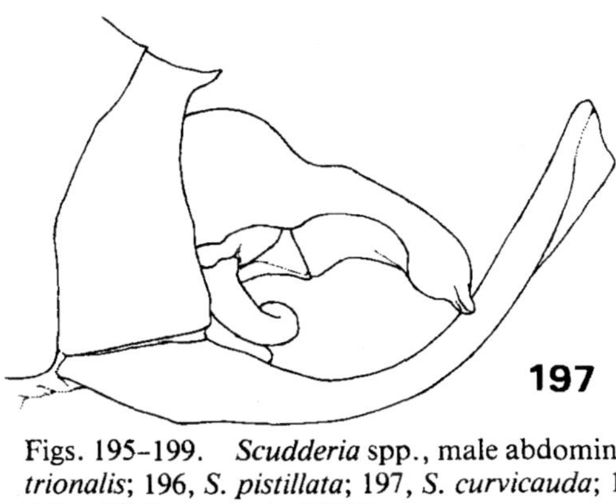
197

Figs. 195–199. *Scudderia* spp., male abdominal terminalia, from left. 195, *S. septentrionalis*; 196, *S. pistillata*; 197, *S. curvicauda*; 198, *S. furcata furcata*; 199, *S. texensis*. After Vickery et al. (1974).

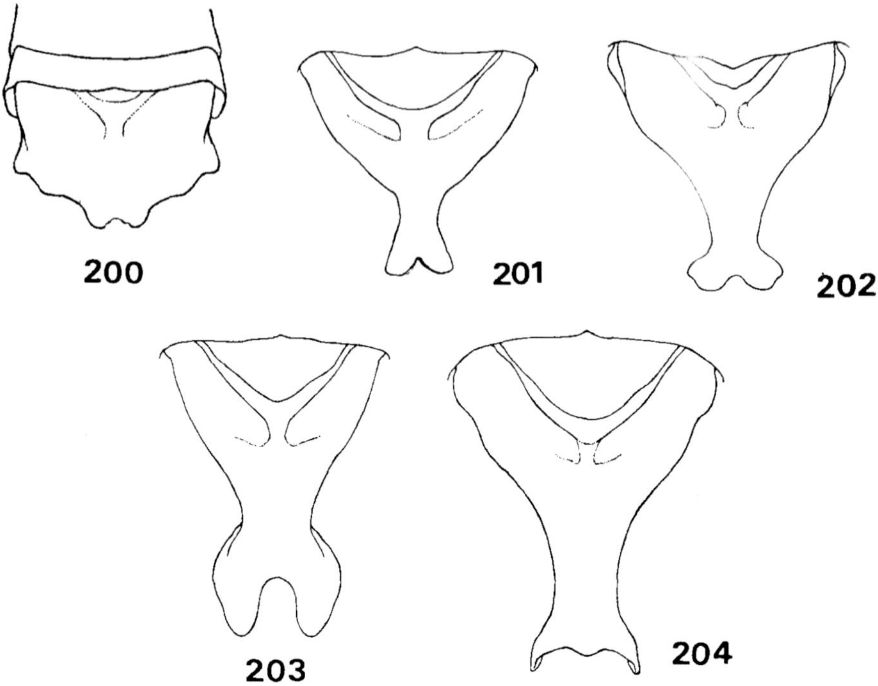

Figs. 200–204. *Scudderia* spp., male supra-anal plates, dorsal. 200, *S. septentrionalis*; 201, *S. pistillata*; 202, *S. curvicauda*; 203, *S. furcata furcata*; 204, *S. texensis*.

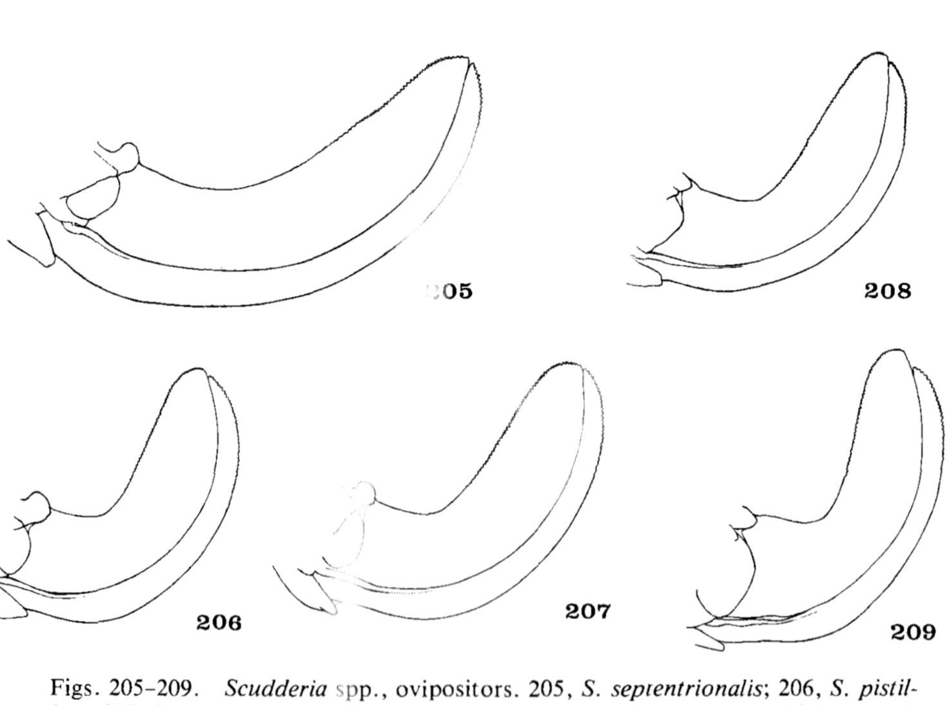

Figs. 205–209. *Scudderia* spp., ovipositors. 205, *S. septentrionalis*; 206, *S. pistillata*; 207, *S. curvicauda*; 208, *S. furcata furcata*; 209, *S. texensis*. After Vickery et al. (1974).

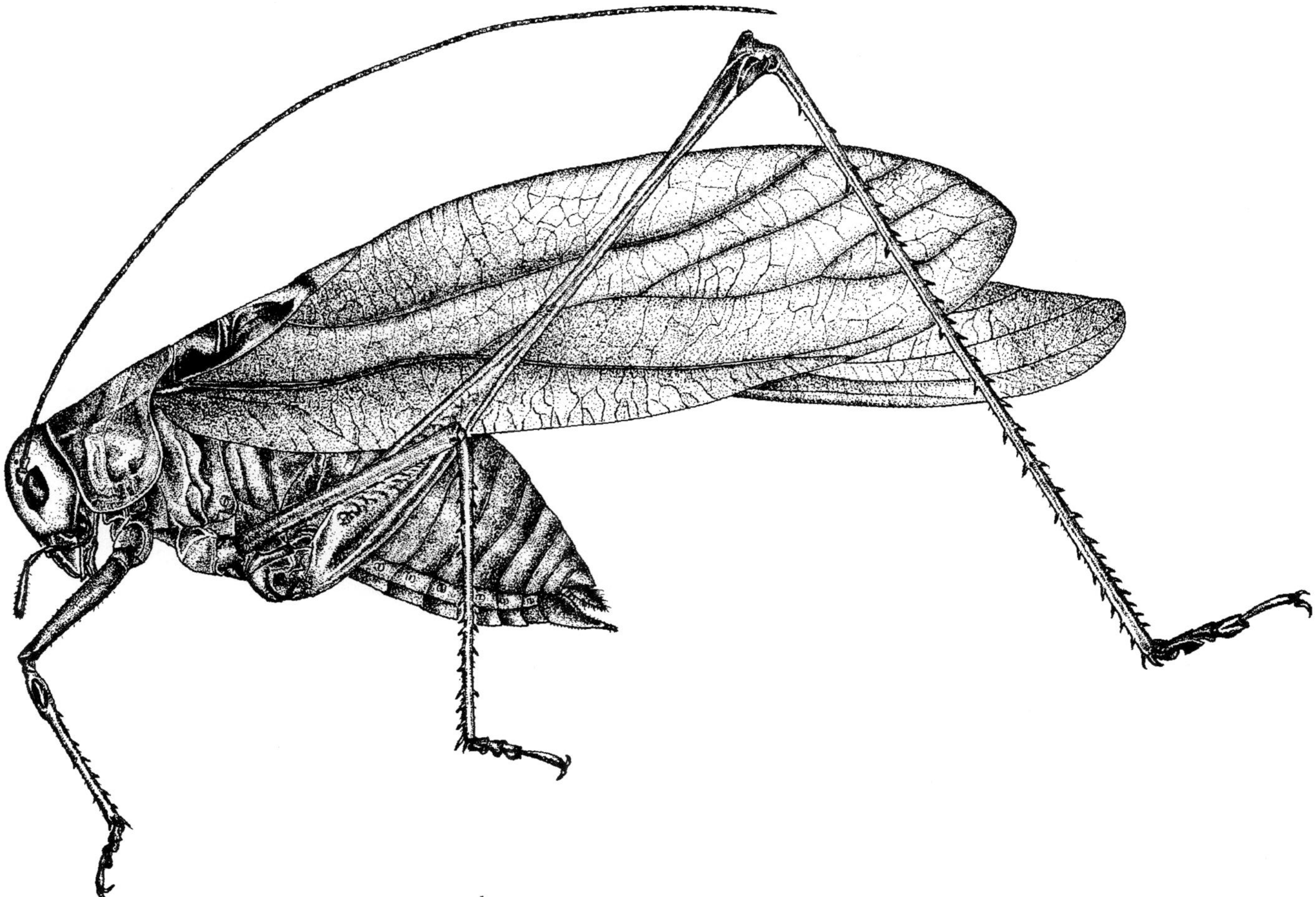

Fig. 210. *Amblycorypha oblongifolia*, male.

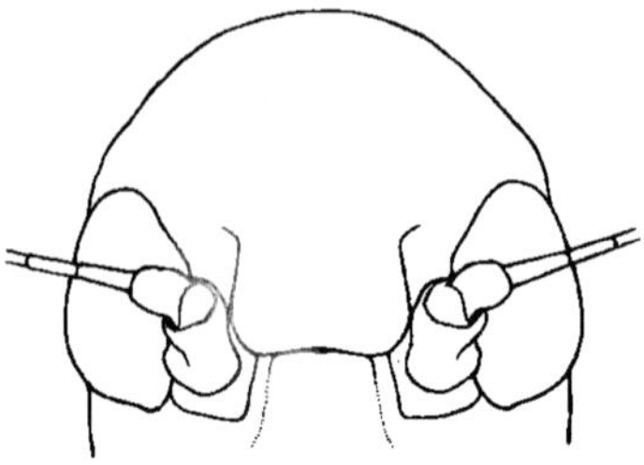

211

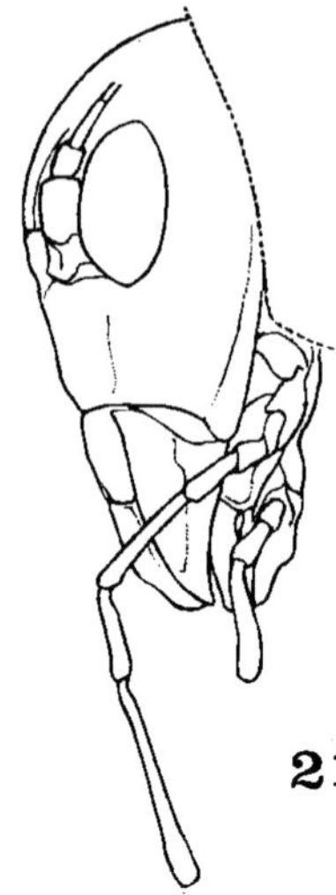

212

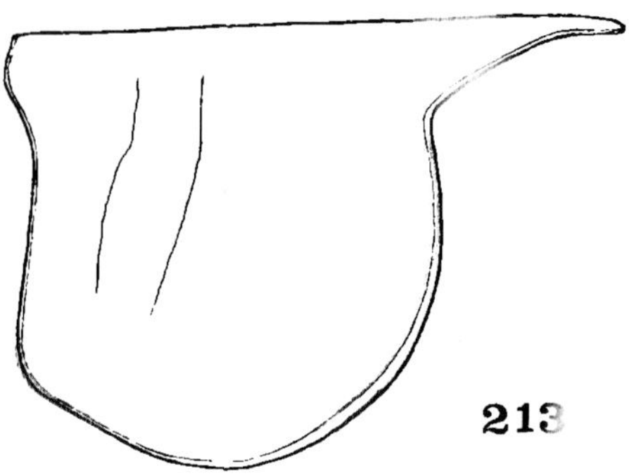

213

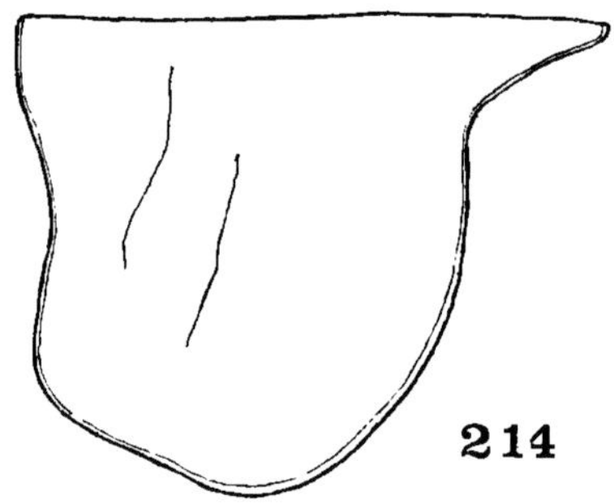

214

Figs. 211–214. *Amblycorphya* spp. 211, *A. oblongifolia*, vertex of head, anterior; 212, the same, head, lateral; 213, the same, pronotum, lateral; 214, *A. rotundifolia rotundifolia*, pronotum, lateral. 211, after Vickery et al. (1974).

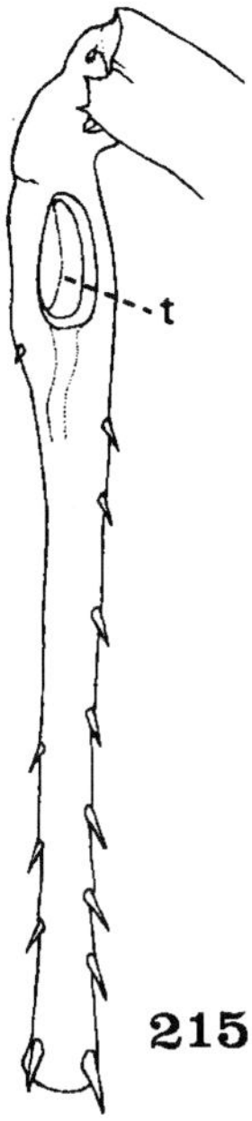

215

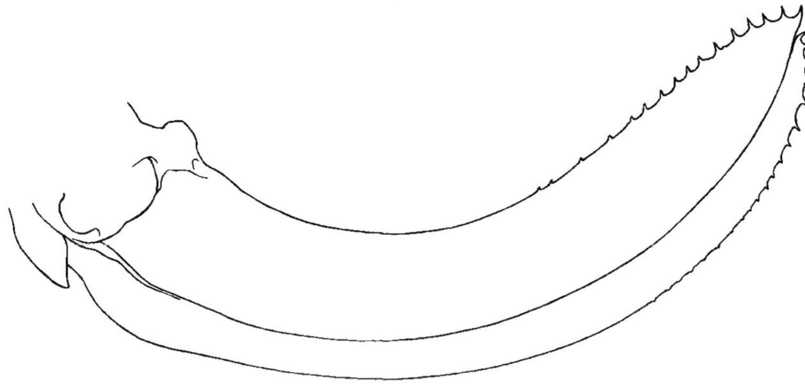

216

Figs. 215, 216. *Amblycorypha oblongifolia*. 215, anterior tibia showing tympanic (auditory) organ (*t*); 216, ovipositor. 215, after Vickery et al. (1974).

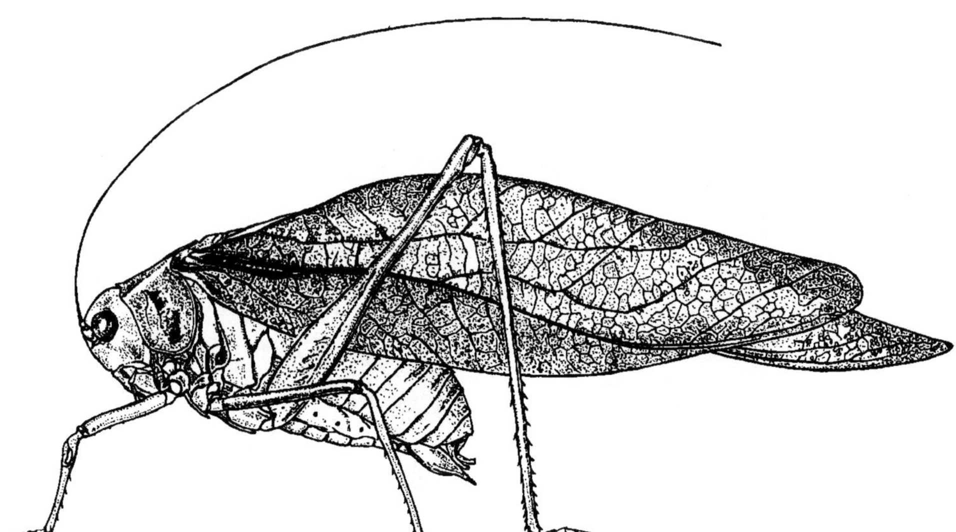

Fig. 217. *Microcentrum rhombifolium*, male.

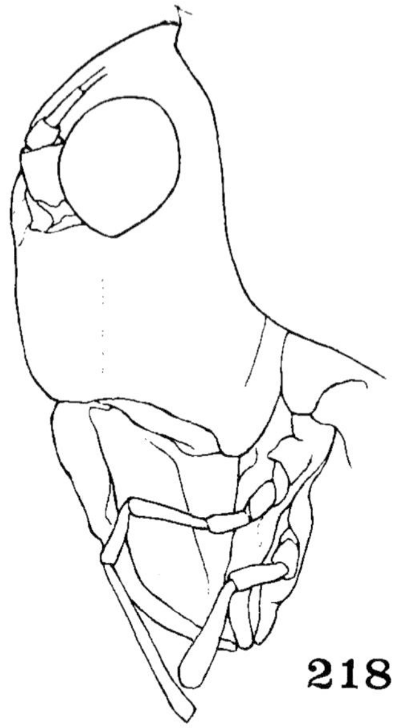

218

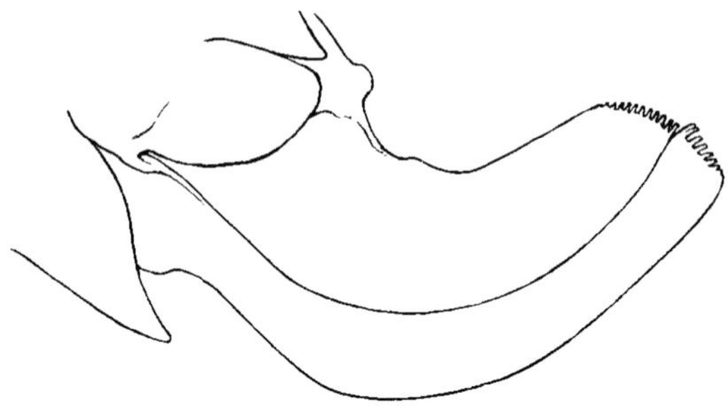

219

Figs. 218, 219. *Microcentrum rhombifolium*. 218, head and anterior margin of pronotum, female, lateral; 219, ovipositor.

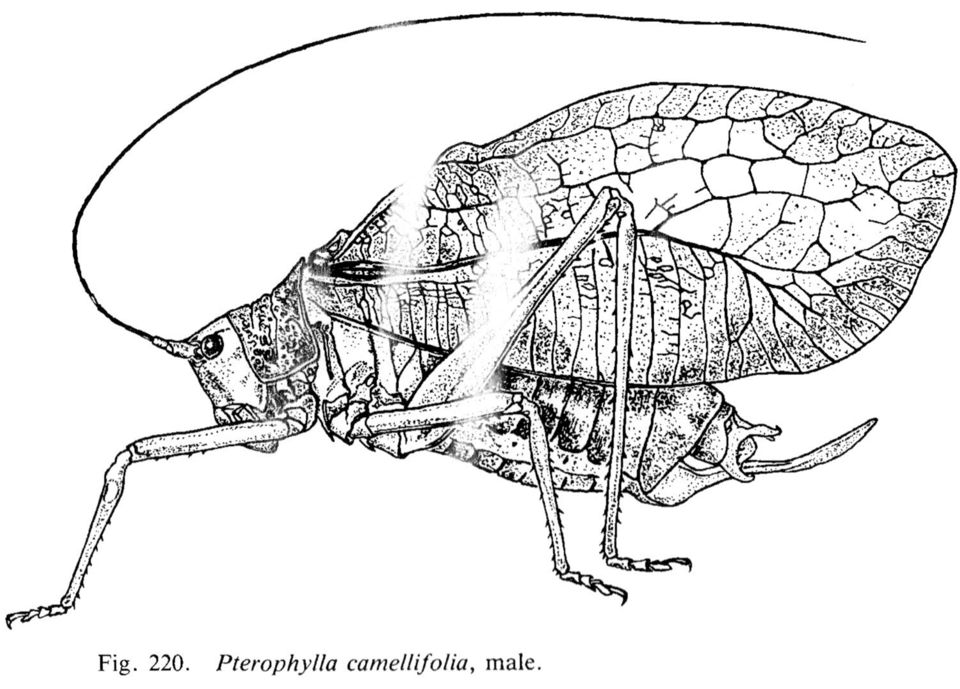

Fig. 220. *Pterophylla camellifolia*, male.

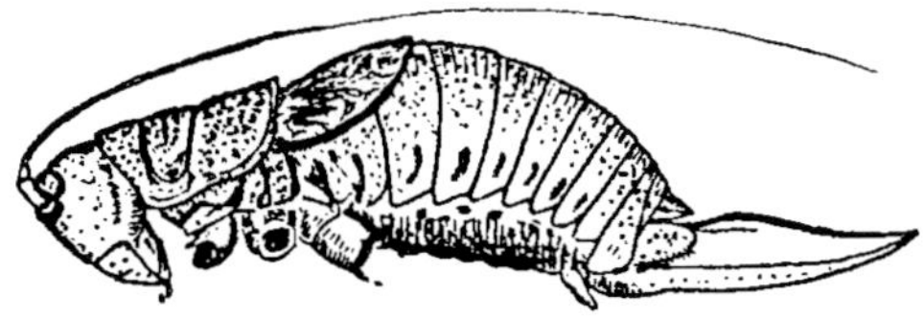

Fig. 221. *Nesoecia nigrispina*, female. Adapted from Beier (1960).

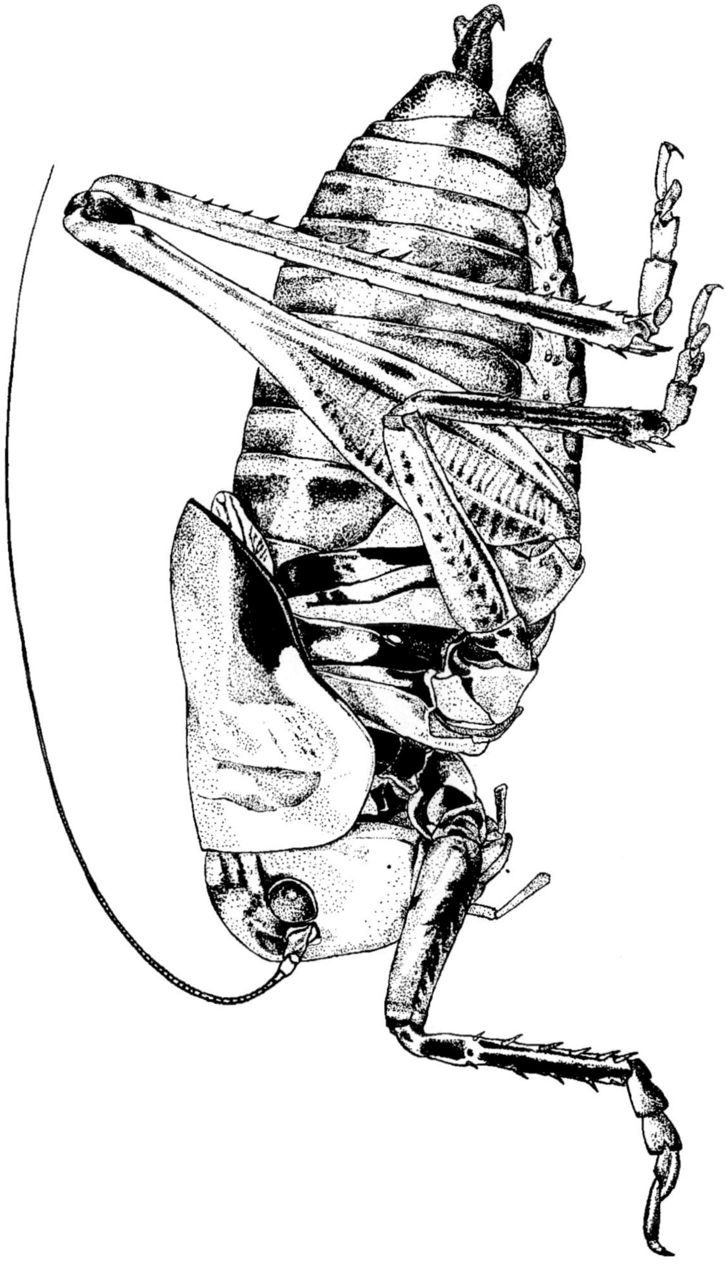

Fig. 222. *Anabrus simplex* Haldeman, male.

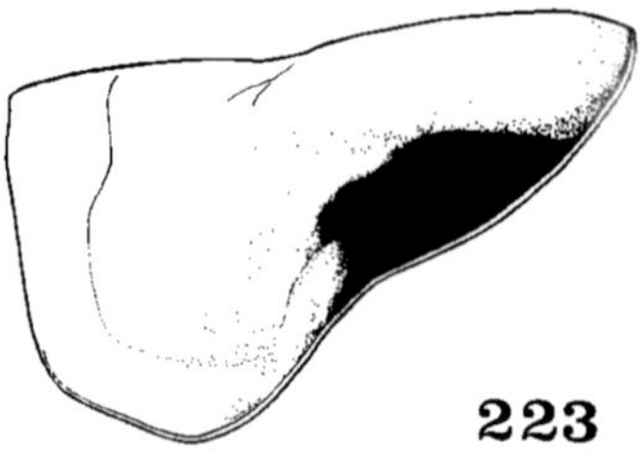

223

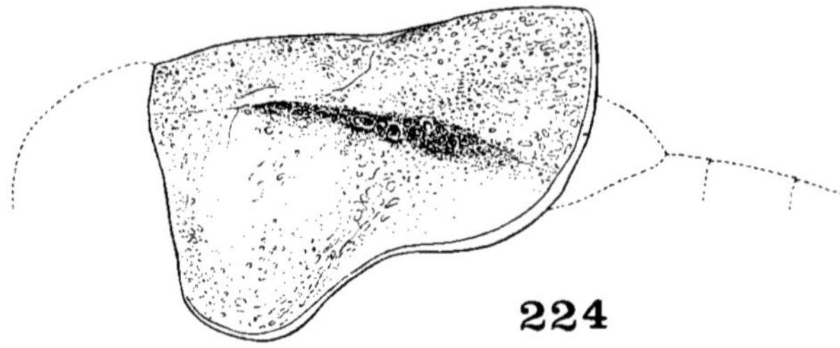

224

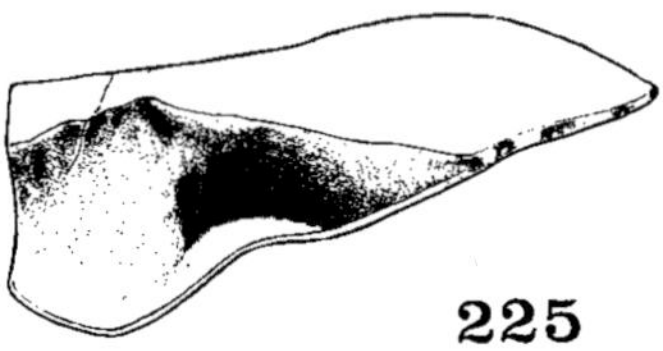

225

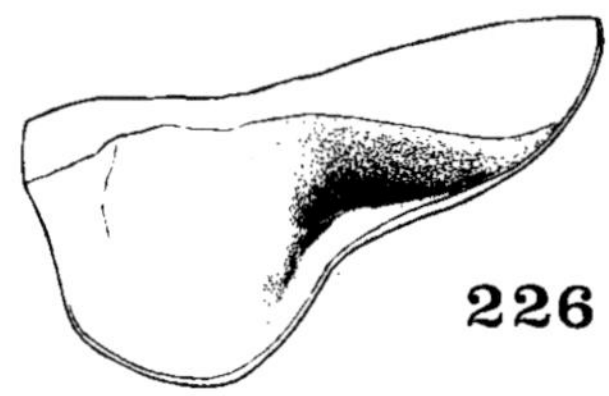

226

Figs. 223–226. Decticini, pronota. 223, *Anabrus simplex*; 224, *Peranabrus scabricollis*; 225, *Neduba steindachneri*; 226, *Atlanticus monticola*.

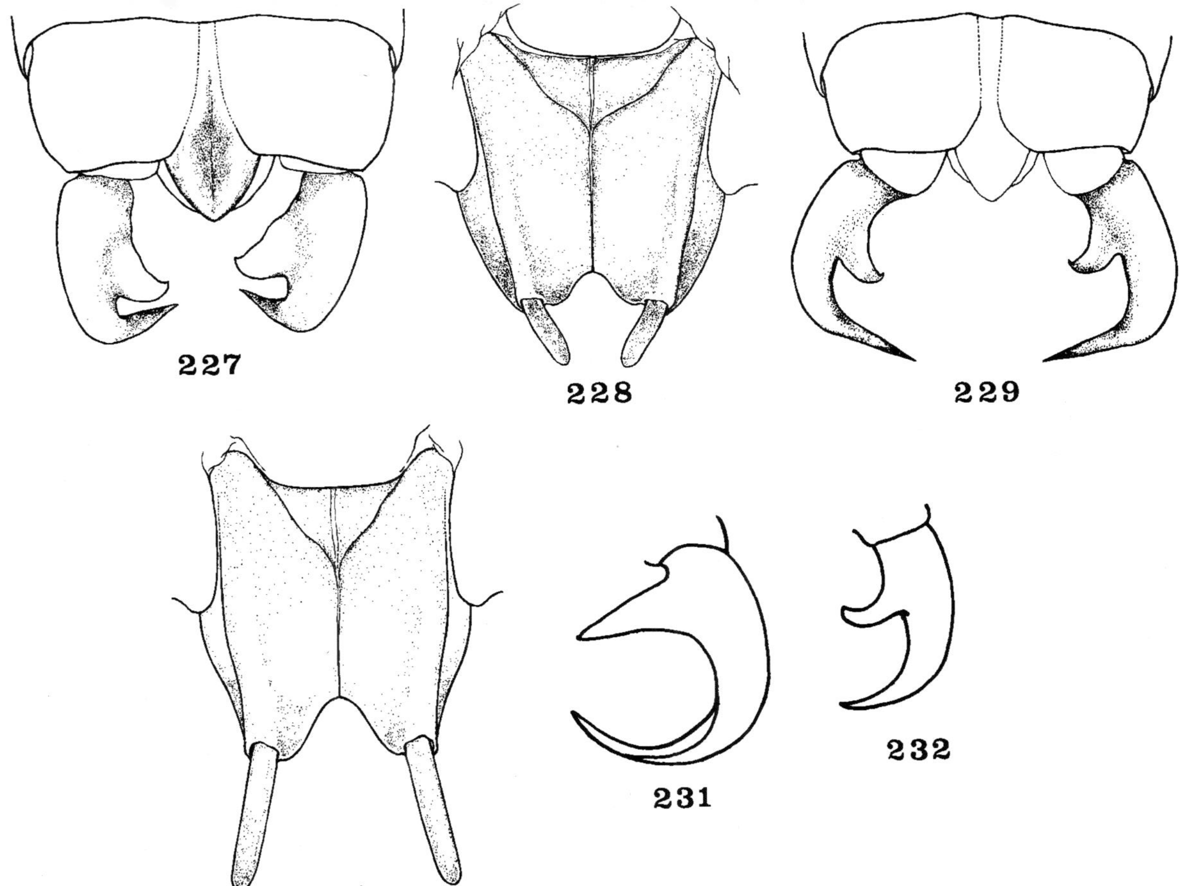

227 228 229 230 231 232

Figs. 227–232. *Anabrus* spp. 227, *A. simplex*, male abdominal terminalia; 228, *A. simplex*, male subgenital plate, ventral; 229, *A. longipes*, male abdominal terminalia, dorsal; 230, *A. longipes*, male subgenital plate, ventral; 231, right male cercus of *A. cerciata*; 232, the same, *A. spokan*.

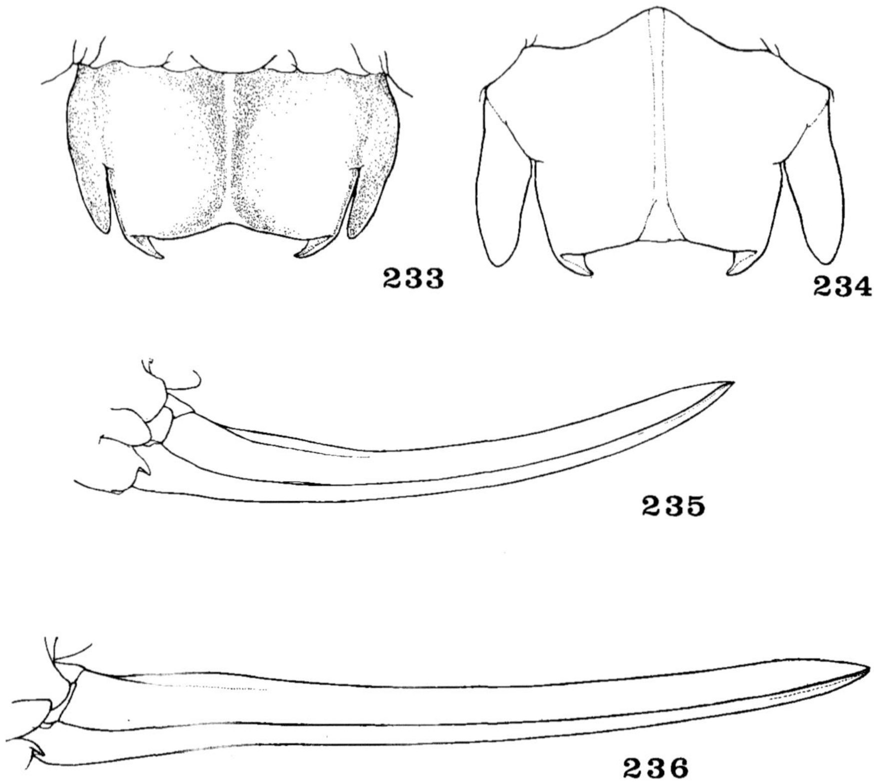

Figs. 233–236. *Anabrus* spp. 233, *A. simplex*, female subgenital plate, ventral; 234, *A. longipes*, female subgenital plate, ventral; 235, *A. simplex*, ovipositor; 236, *A. longipes*, ovipositor.

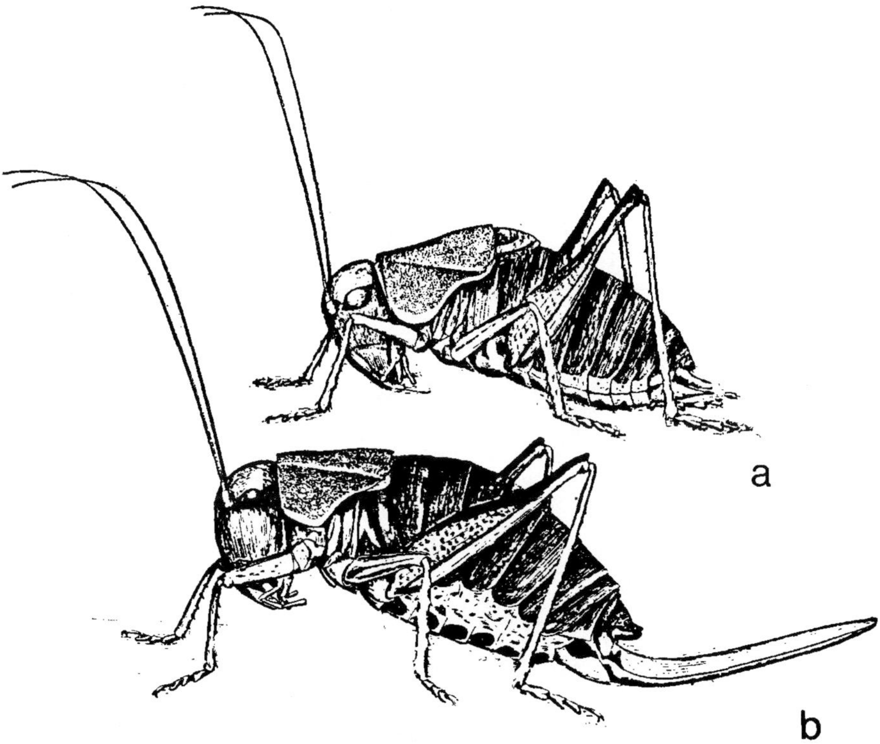

Fig. 237. *Peranabrus scabricollis*. *a*, male; *b*, female. Adapted from Snodgrass (1905).

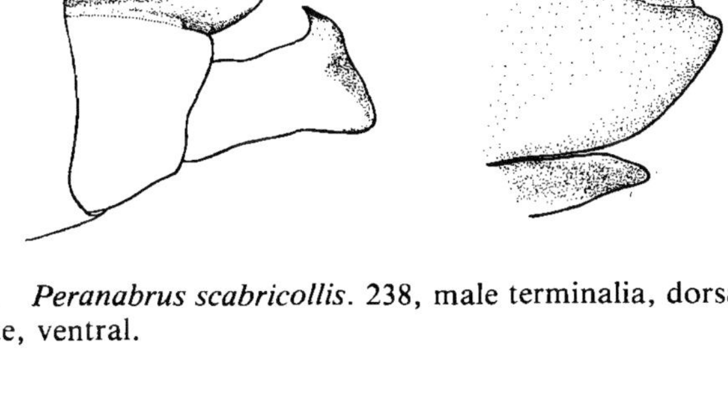

Figs. 238, 239. *Peranabrus scabricollis*. 238, male terminalia, dorsal; 239, female subgenital plate, ventral.

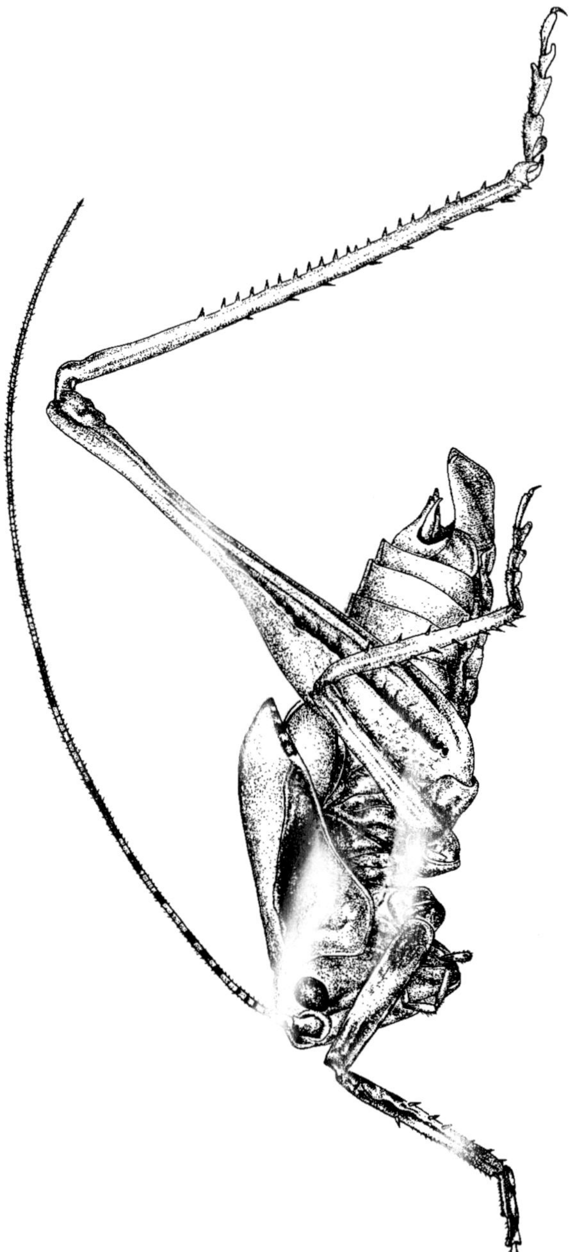

Fig. 240. *Neduba (Neduba) steindachneri*, male.

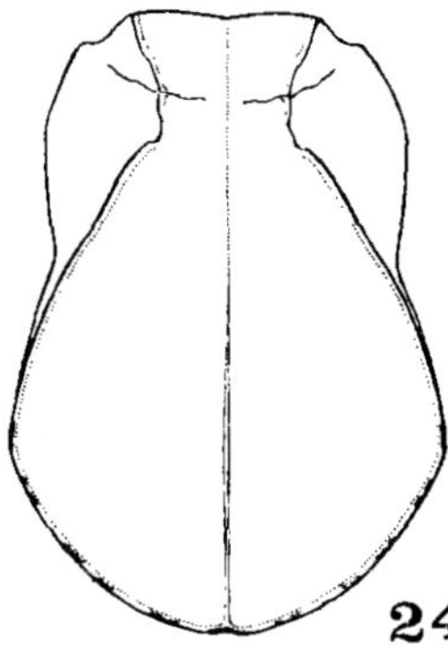

241

242

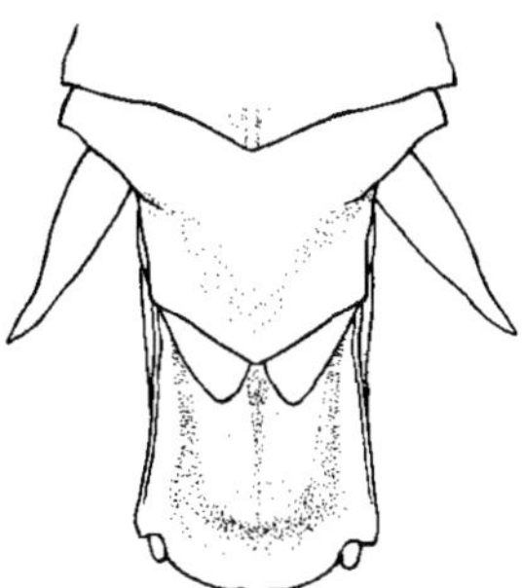

243

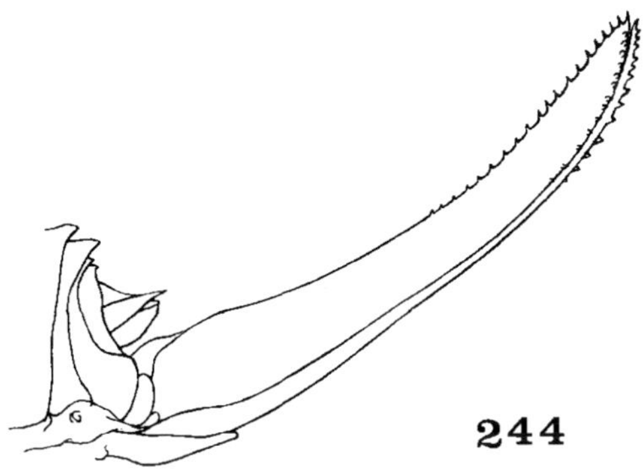

244

Figs. 241–244. *Neduba* (*Neduba*) *steindachneri*. 241, male pronotum, dorsal; 242, apex of hind tibia; 243, male abdominal terminalia, dorsal; 244, ovipositor.

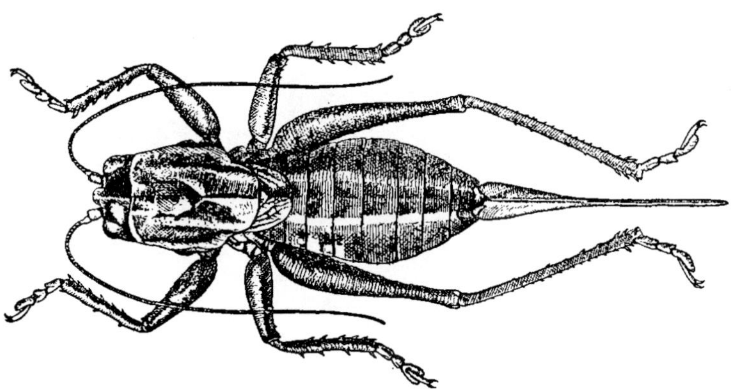

Fig. 245. *Apote notabilis*, female, after Gurney (1939).

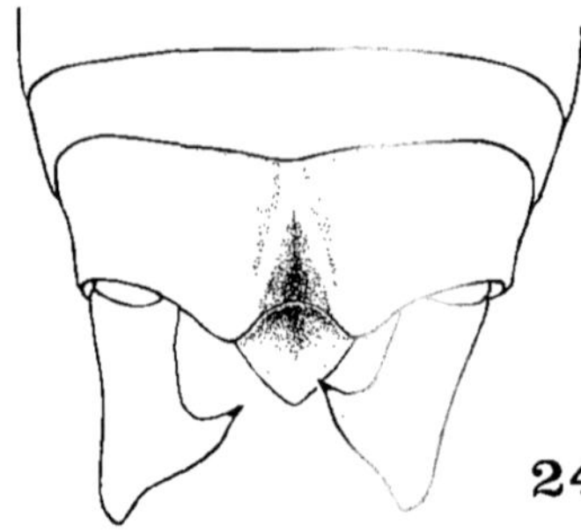

246

247

Figs. 246, 247. *Apote robusta*. 246, male abdominal terminalia, dorsal; 247, ovipositor.

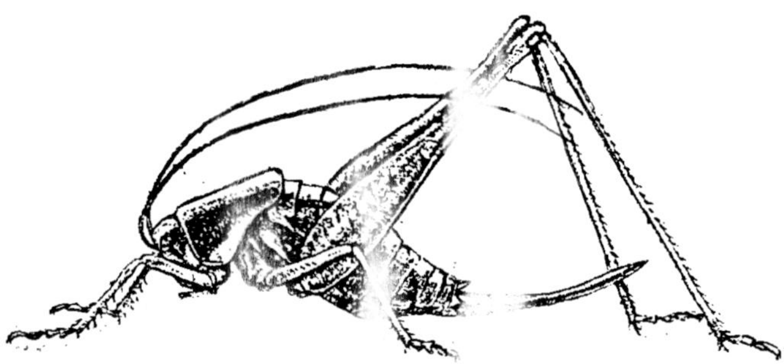

Fig. 248. *Pediodectes* sp., general appearance of female (*P. nigromarginatus*, originally described from Texas), adapted from Caudell (1907*b*).

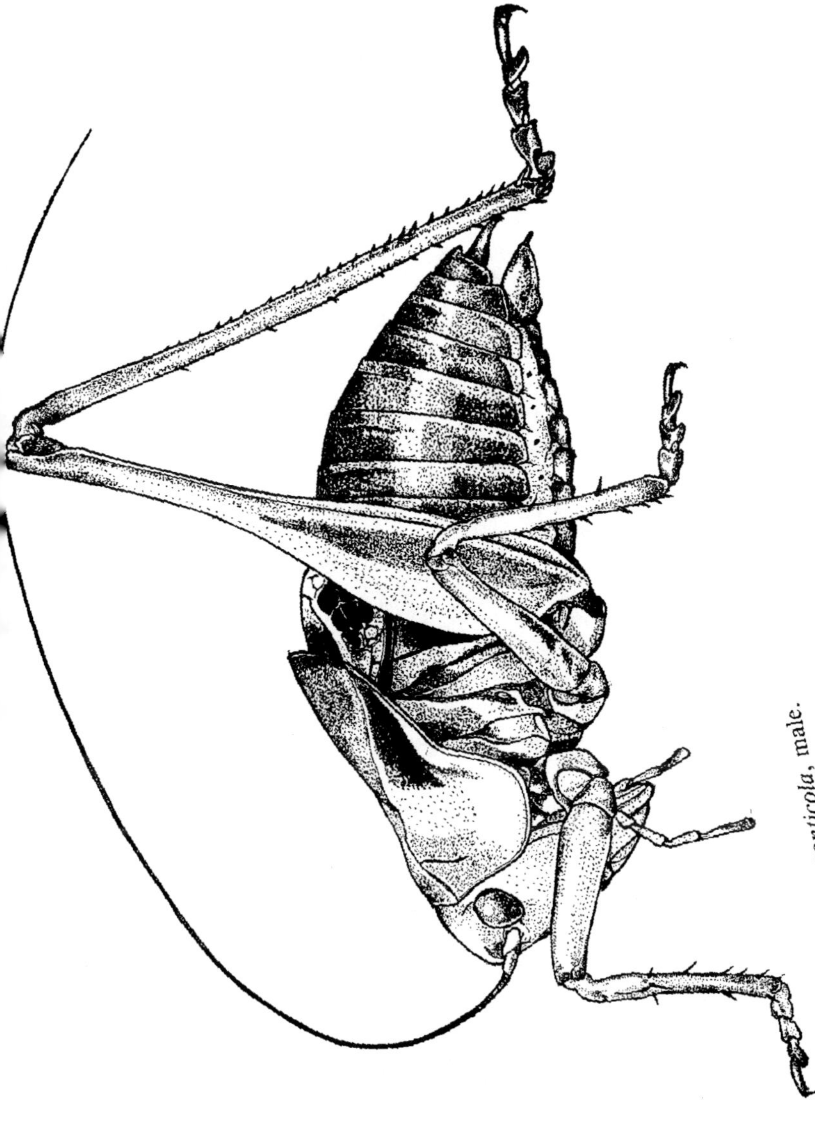

Fig. 249. *Atlanticus monticola*, male.

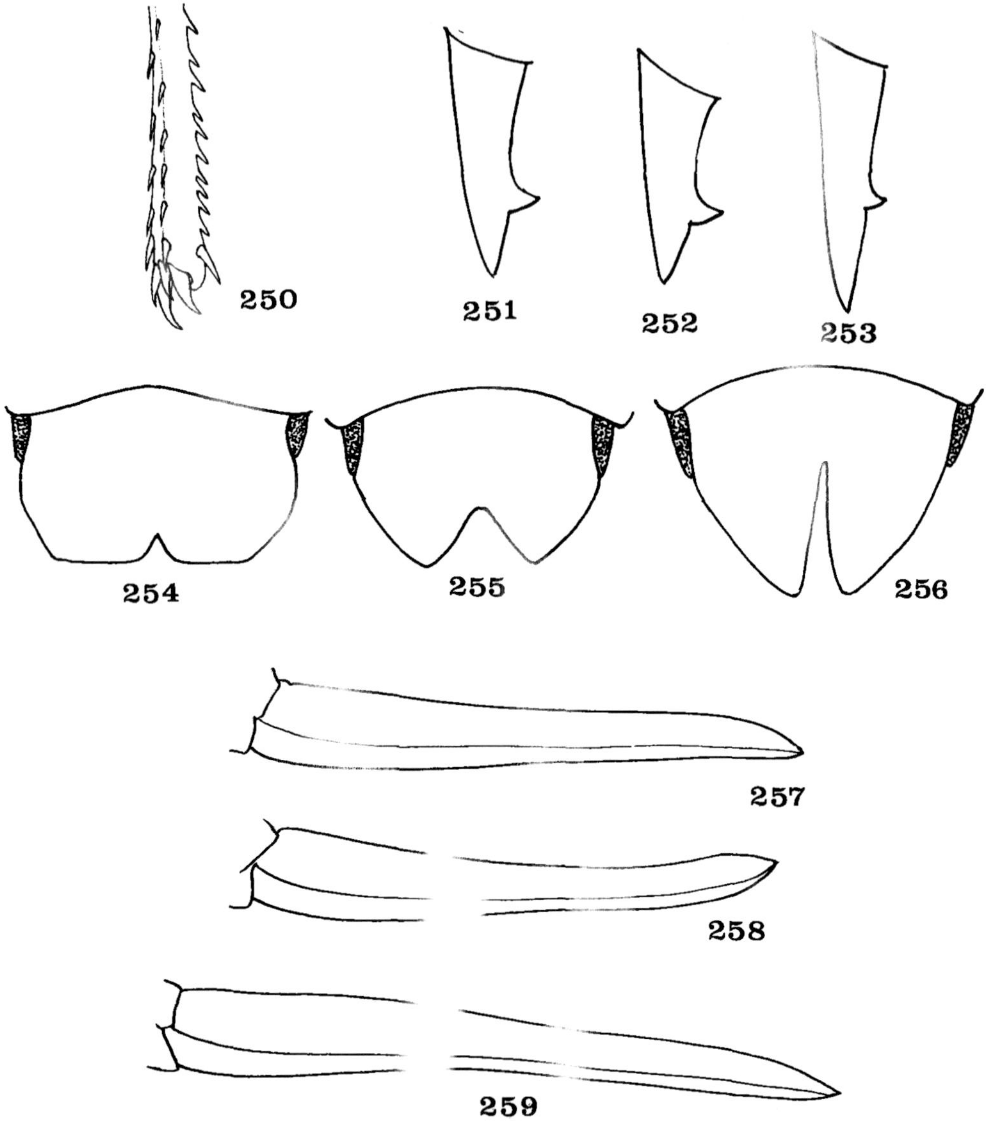

Figs. 250–259. *Atlanticus* spp. 250, 252, 255, 258, *A. monticola*; 251, 254, 257, *A. testaceus*; 253, 256, 259, *A. americanus*; 250, apex of hind tibia; 251–253, male cerci, left; 254–256, female subgenital plates, ventral; 257–259, ovipositors, from left.

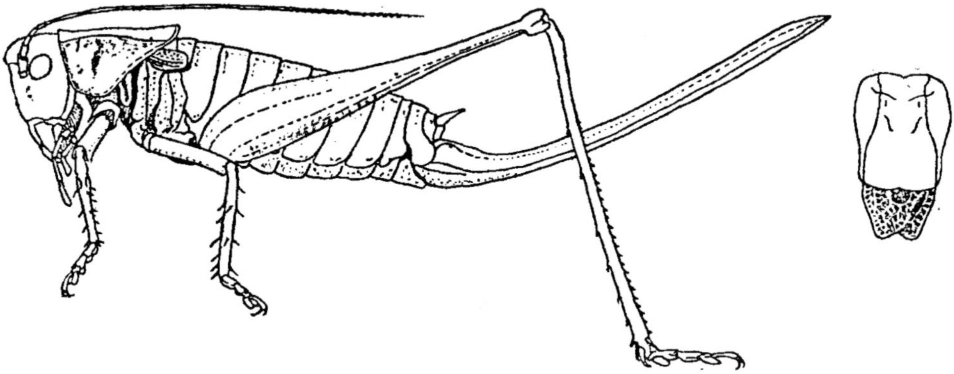

Fig. 260. *Steiroxys strepens*, female; right, pronotum and tegmina, male. Adapted from Fulton (1930).

Fig. 261. Original (unpublished) drawings of (1) *Anabrus simplex* and (2) *Steiroxys trilineatus* made by Otto Hermann, at the time he proposed the genus *Steiroxys*.

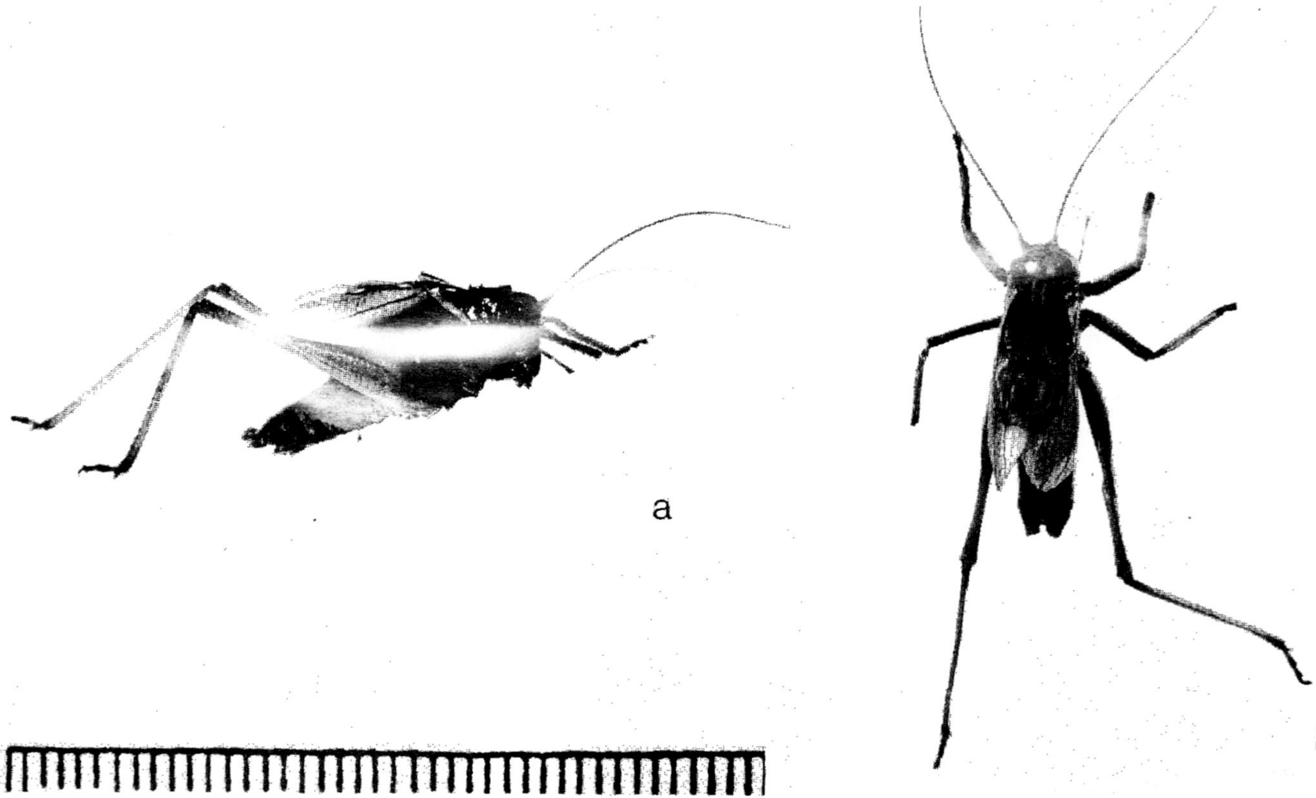

Fig. 262. *Sphagniana sphagnorum*, male holotype. *a*, lateral; *b*, dorsal. Scale in millimetres. Photos courtesy of British Museum (Natural History), London.

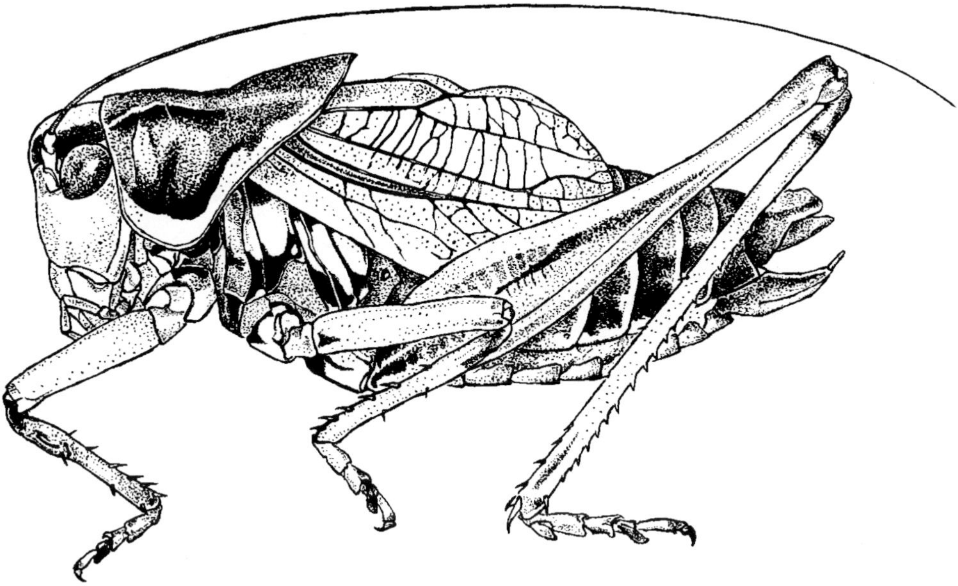

Fig. 263. *Sphagniana sphagnorum*, male.

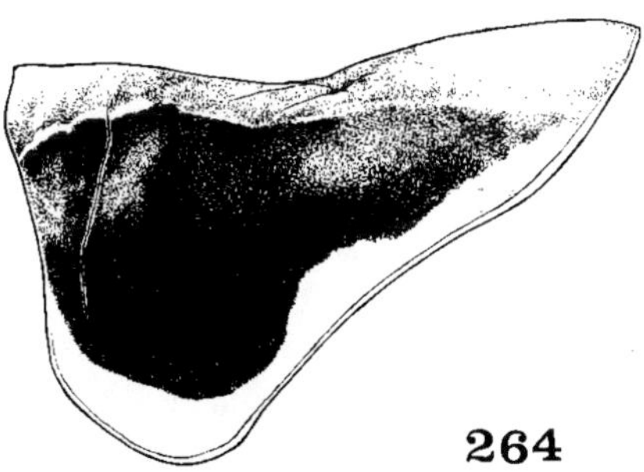

264

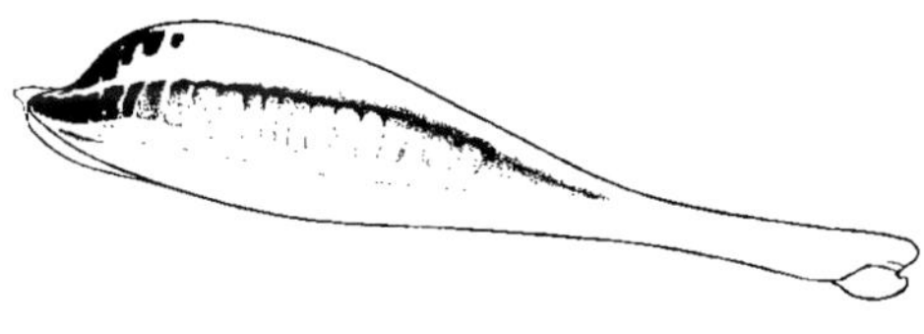

265

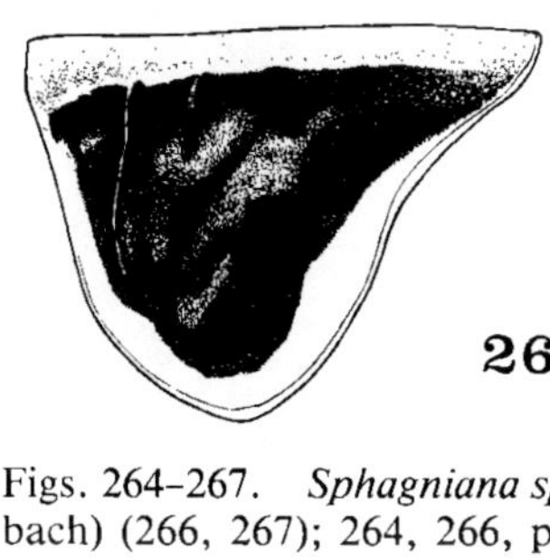

266

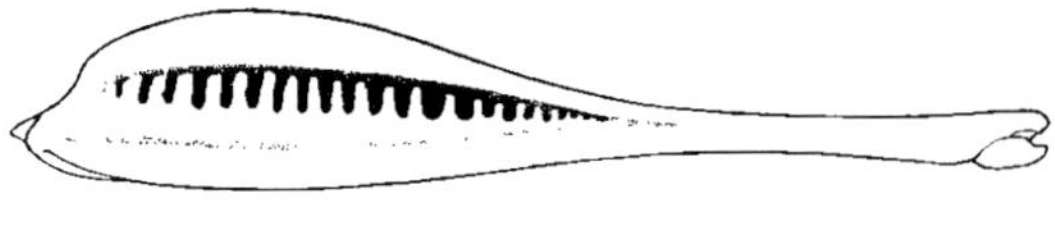

267

Figs. 264–267. *Sphagniana sphagnorum* (264, 265) and *Metrioptera roeselii* (Hagenbach) (266, 267); 264, 266, pronota, lateral; 265, 267, hind femora.

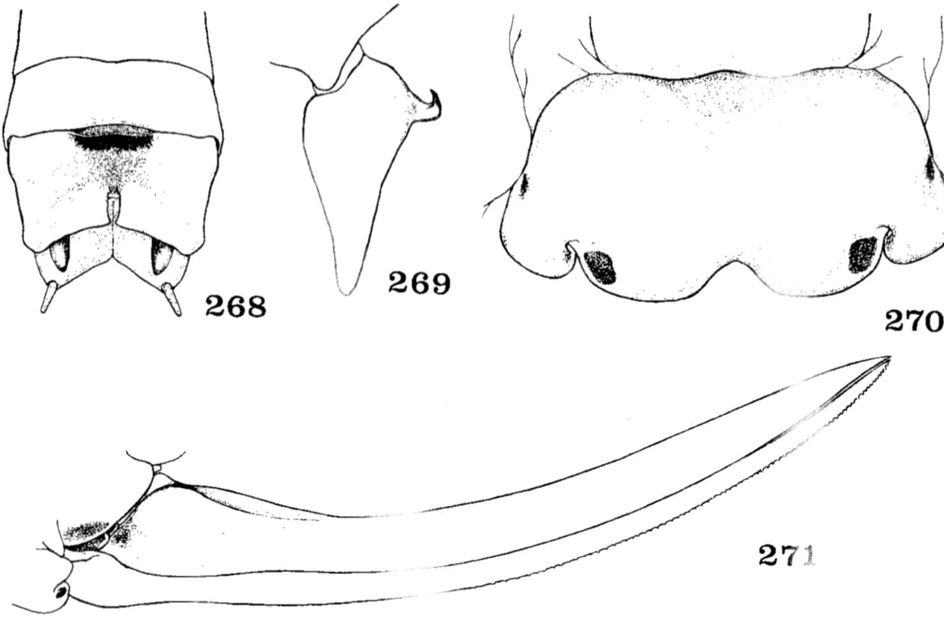

Figs. 268–271. *Sphagniana sphagnorum*. 268, male abdominal terminalia; 269, male left cercus extruded, dorsal; 270, female subgenital plate, ventral; 271, ovipositor.

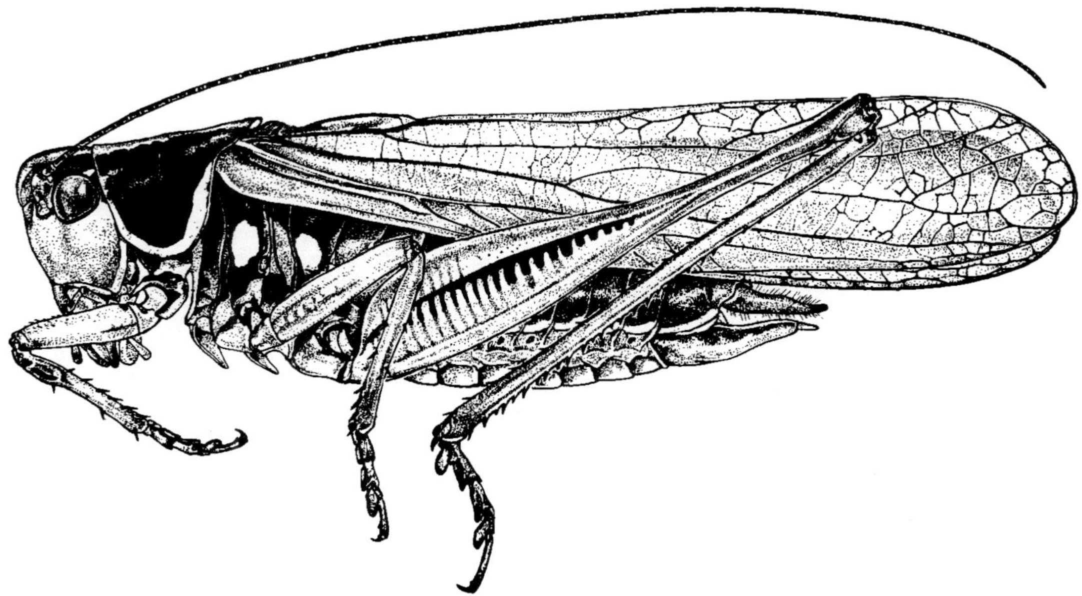

Fig. 272. *Metrioptera roeselii*, macropterous male. After Vickery et al. (1974).

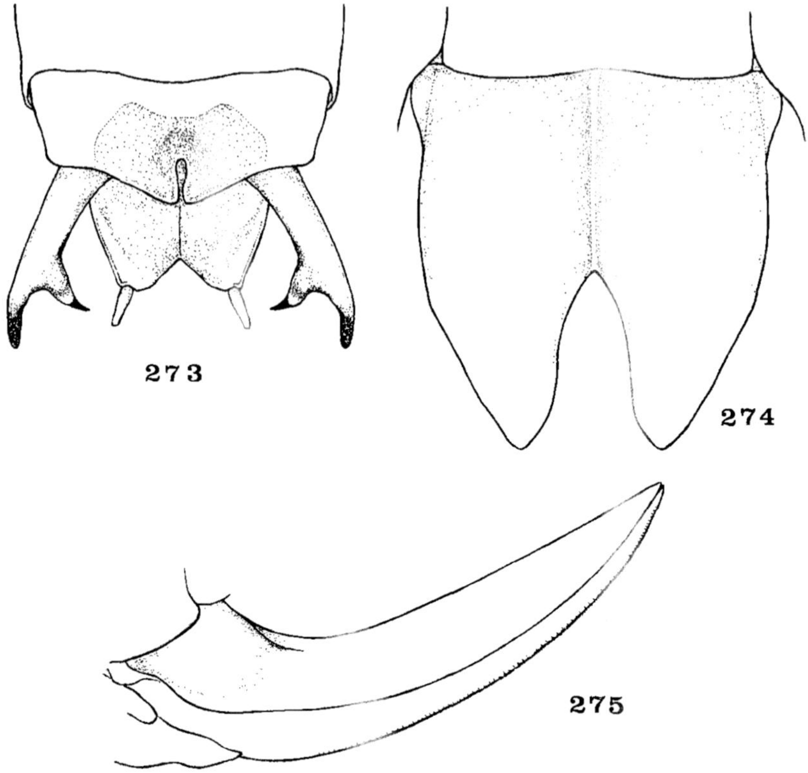

Figs. 273–275. *Metrioptera roeselii*. 273, male abdominal terminalia, dorsal; 274, female subgenital plate, ventral; 275, ovipositor.

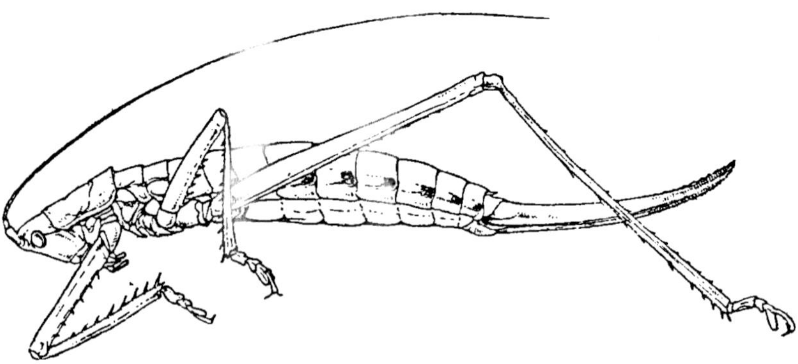

Fig. 276. *Saga pedo*, female. After Chopard (1951), courtesy of Editions Lechevalier, Paris.

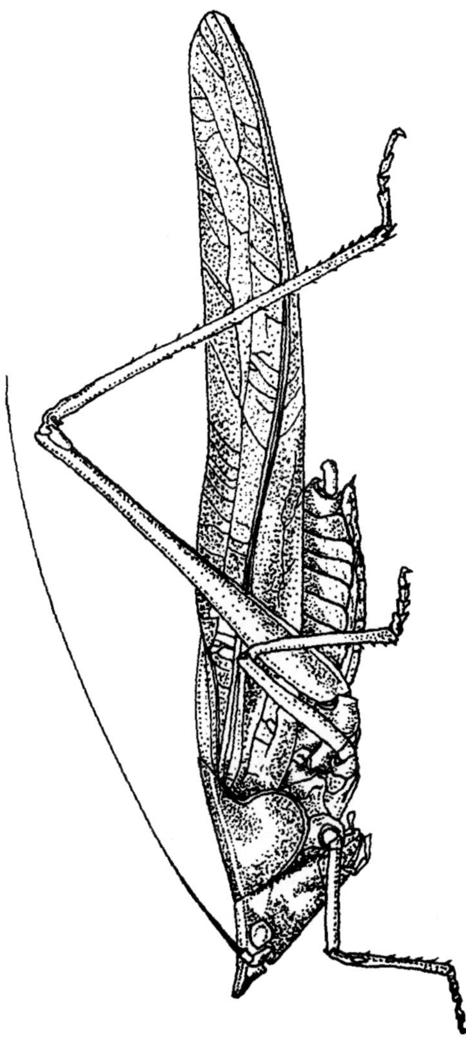

Fig. 277. *Neoconocephalus ensiger*, male.

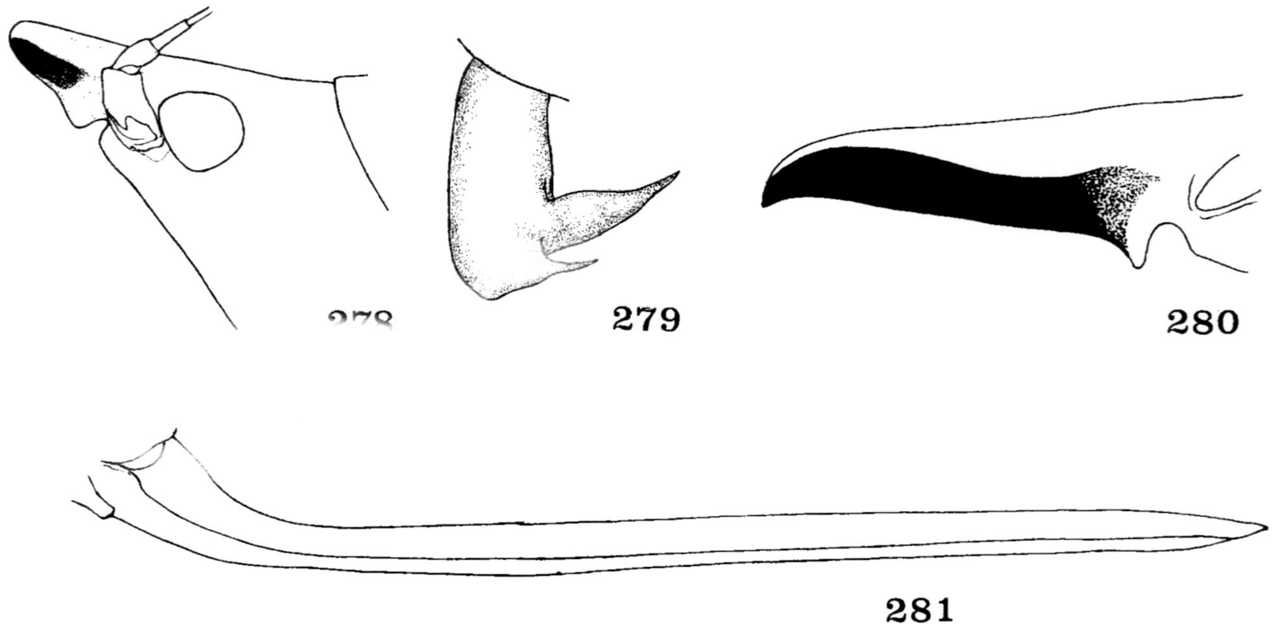

Figs. 278–281. *Neoconocephalus* spp. *Neoconocephalus ensiger*. 278, head, lateral; 279, left male cercus; 280, *N. exiliscanorus*, fastigium of vertex, from left; 281, *N. ensiger*, ovipositor. 281, after Vickery et al. (1974).

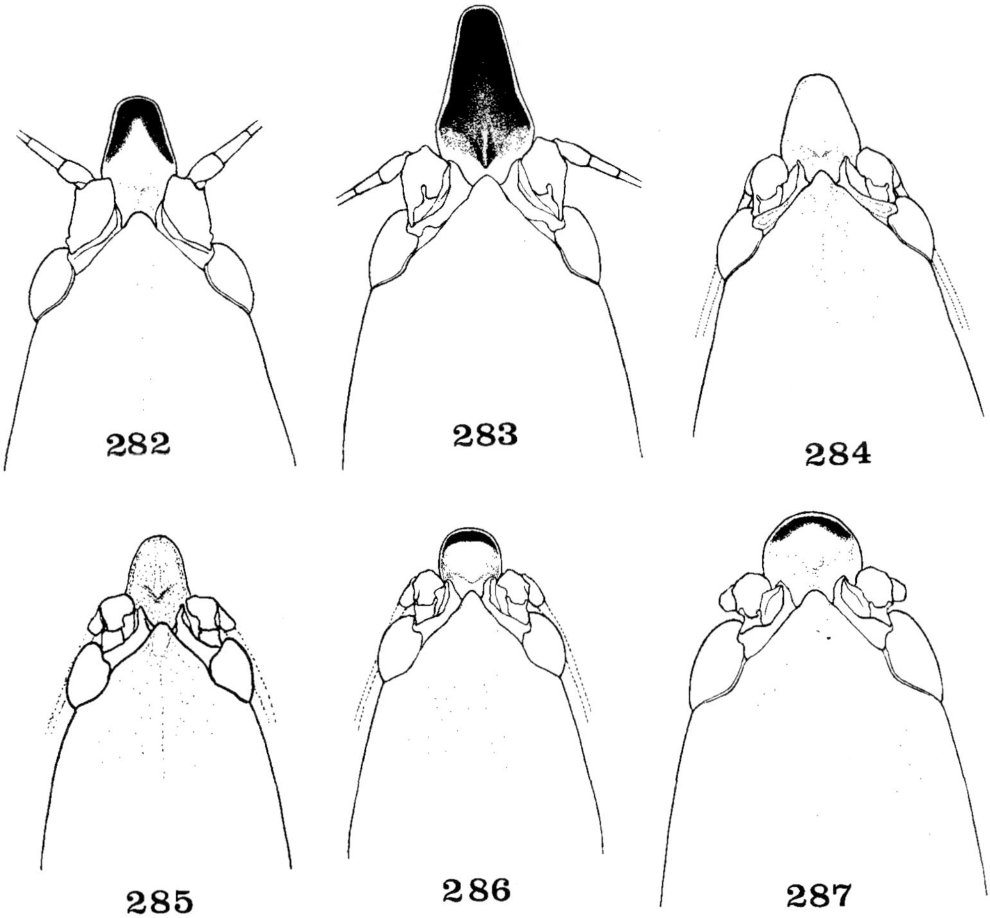

Figs. 282–287. *Neoconocephalus* spp., frontes and fastigia, from below. 282, *N. ensiger*; 283, *N. lyristes*; 284, *N. robustus*; 285, *N. palustris*; 286, *N. retusus*; 287, *N. triops*.

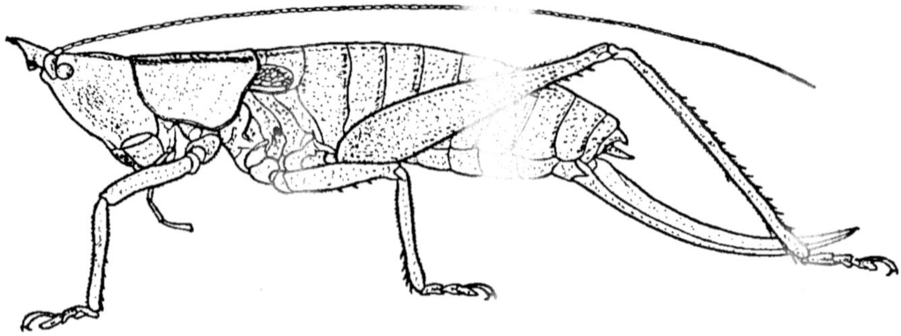

Fig. 288. *Belocephalus sabalis*, female (adventive). Adapted from H. Fox *in* Blatchley (1920). (Similar to *B. subapterus*.)

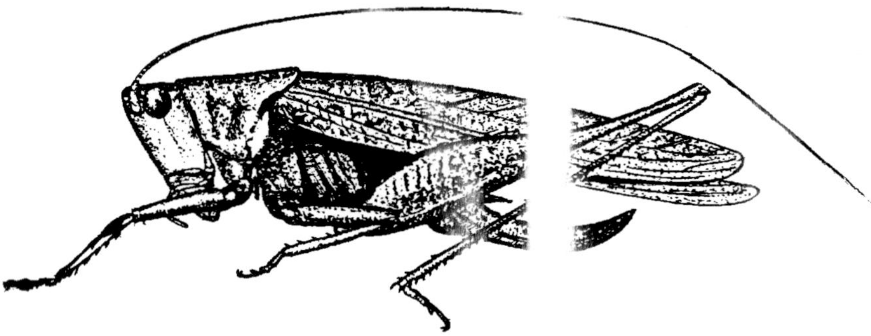

Fig. 289. *Orchelimum vulgare*, female.

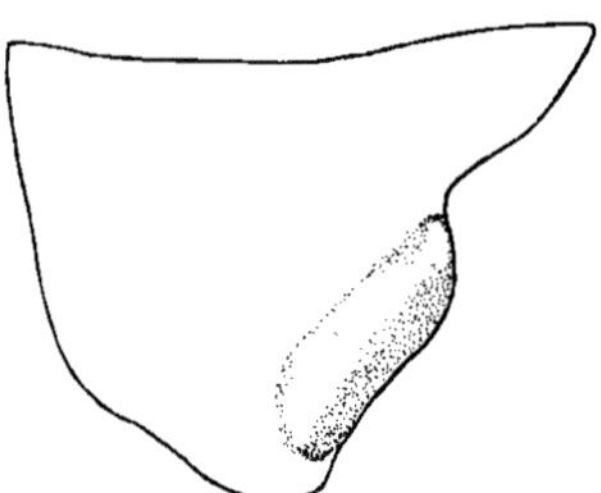

290

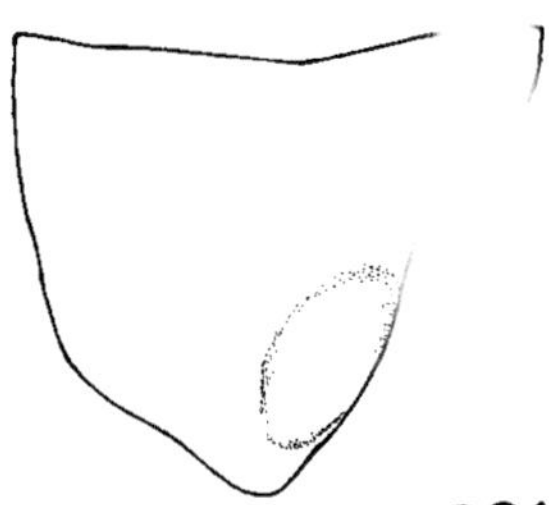

291

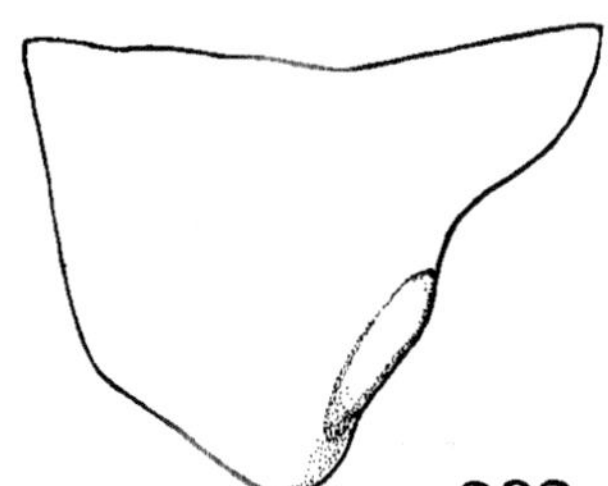

292

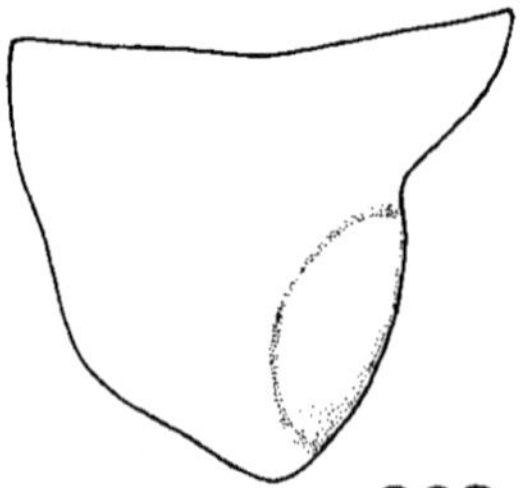

293

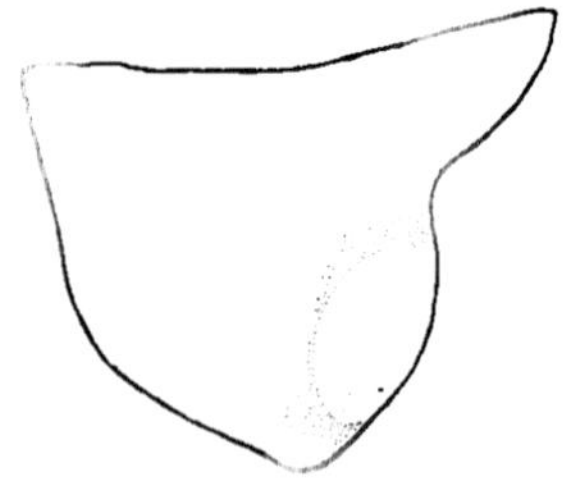

294

Figs. 290–294. *Orchelimum* spp., pronota, from left. 290, *O. vulgare*; 291, *O. gladiator*; 292, *O. nigripes*; 293, *O. concinnum*; 294, *O. volantum*. 290, 291, after Vickery et al. (1974).

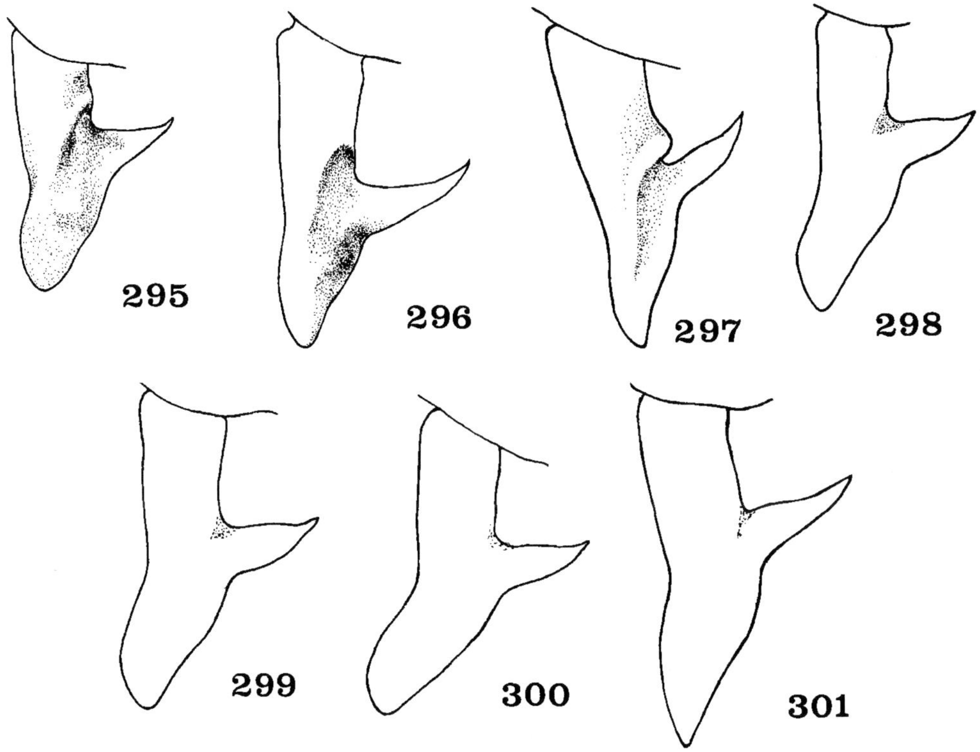

Figs. 295–301. *Orchelimum* spp., left male cerci. 295, *O. vulgare*; 296, *O. gladiator*; 297, *O. nigripes*; 298, *O. concinnum*; 299, *O. delicatum*; 300, *O. campestre*; 301, *O. volantum*. 295, 296, after Vickery et al. (1974).

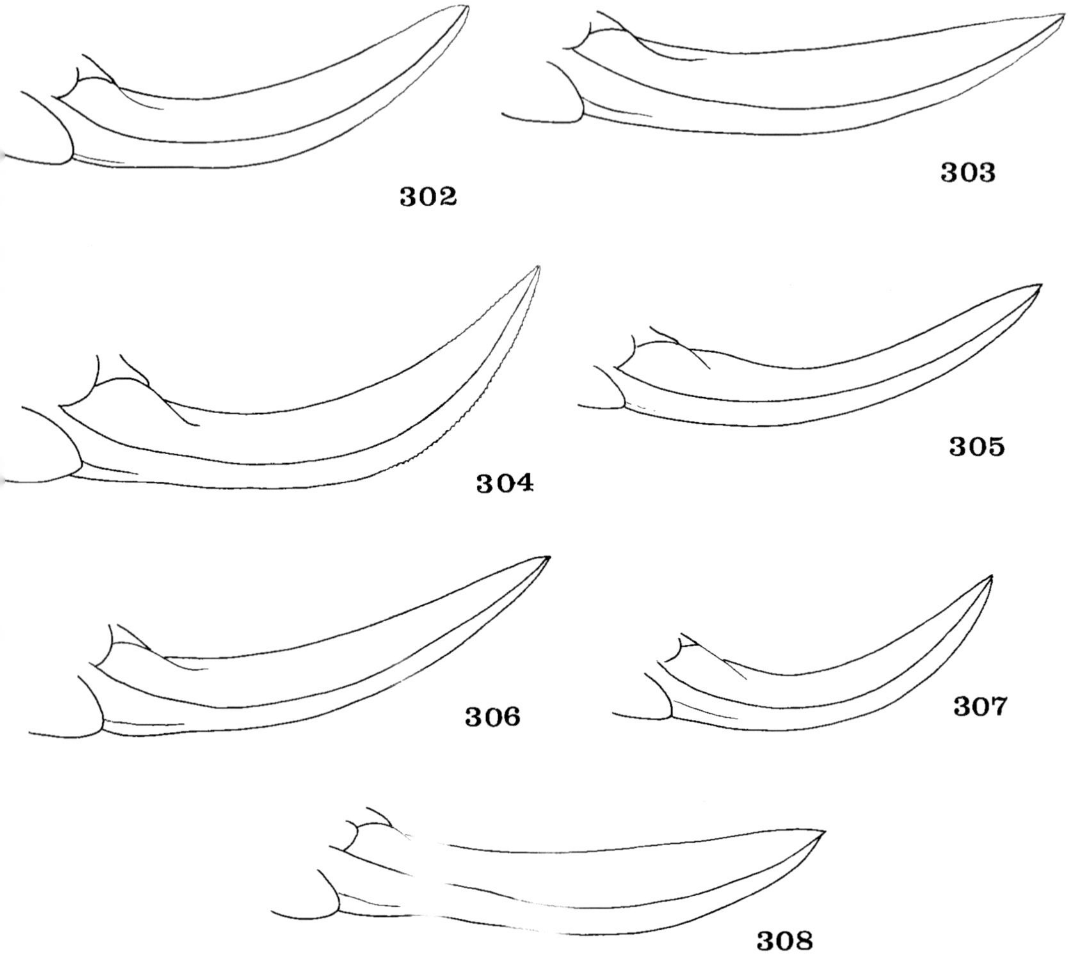

Figs. 302–308. *Orchelimum* spp., ovipositors. 302, *O. vulgare*; 303, *O. gladiator*; 304, *O. nigripes*; 305, *O. concinnum*; 306, *O. delicatum*; 307, *O. campestre*; 308, *O. volantum*.

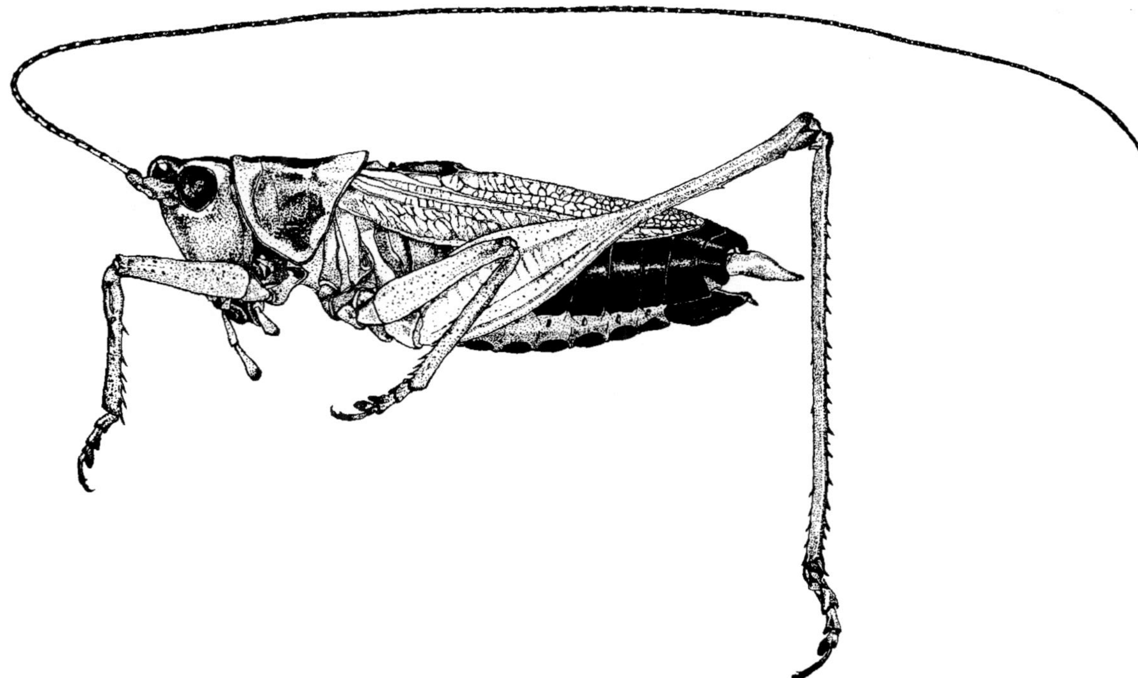

Fig. 309. *Conocephalus nigropleurum*, male.

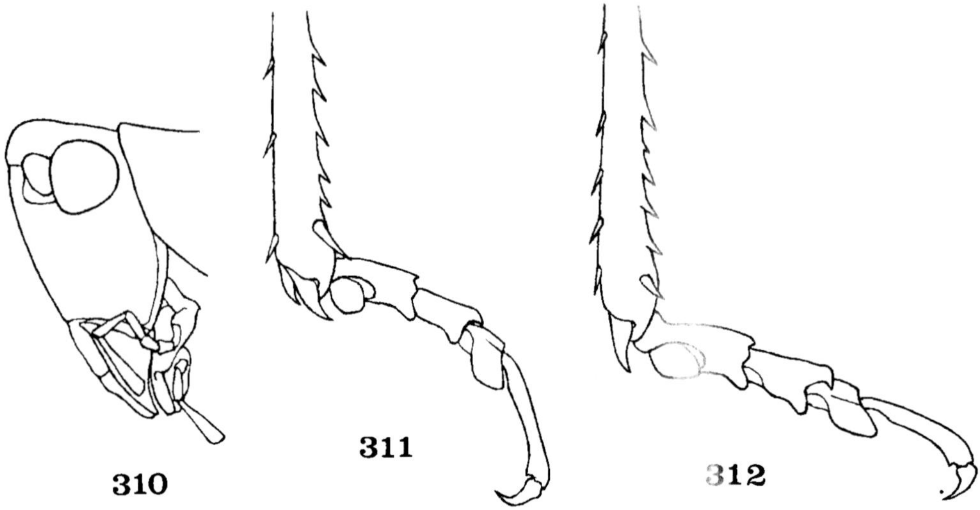

Figs. 310–312. *Conocephalus* spp. 310, *C. fasciatus*, head, lateral; 311, *C. fasciatus*, apex of hind tibia and tarsus; 312, *C. saltans*, the same.

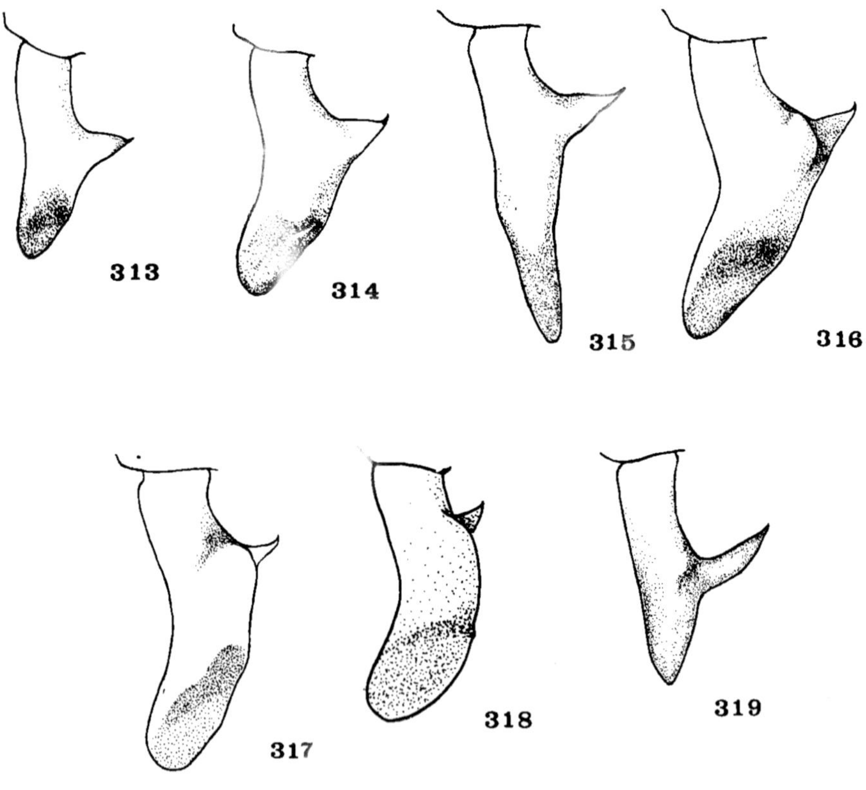

Figs. 313–319. *Conocephalus* spp., left male cerci. 313, *C. fasciatus*; 314, *C. brevipennis*; 315, *C. strictus*; 316, *C. nigropleurum*; 317, *C. attenuatus*; 318, *C. spartinae*; 319, *C. saltans*. 313, 314, 316, 317, after Vickery et al. (1974).

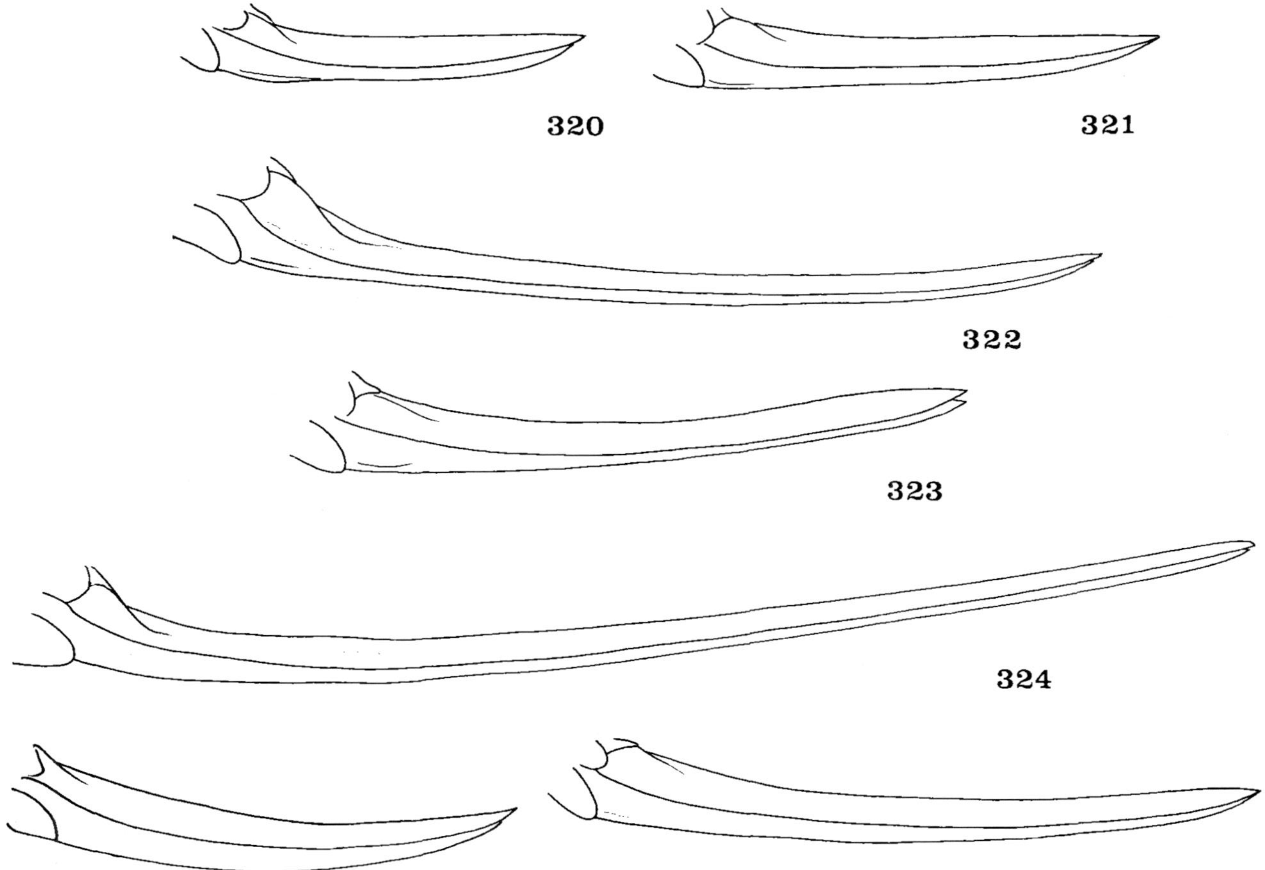

Figs. 320–326. *Conocephalus* spp., ovipositors, from left. 320, *C. fasciatus*; 321, *C. brevipennis*; 322, *C. strictus*; 323, *C. nigropleurum*; 324, *C. attenuatus*; 325, *C. spartinae*; 326, *C. saltans*.

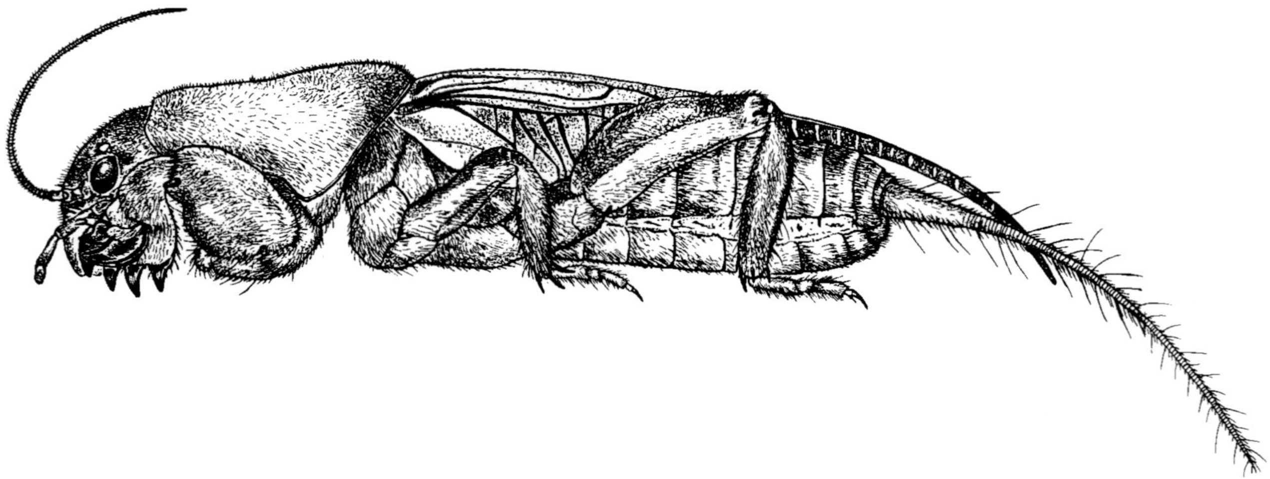

Fig. 327. *Neocurtilla hexadactyla* [o V. *borealis*], female.

Fig. 328. Catesby's (1731: pl. 8) illustration of *"Grillotalpa"* being pursued by *"Caprimulgus: The goat-sucker of Carolina"* (= *Chordeiles minor*, the nighthawk). This figure was referred to by Linnaeus (1758) in his description of *Gryllotalpa gryllotalpa* (Linnaeus), and the insect that formed the basis of the illustration is the holotype of *Gryllotalpa borealis* Burmeister, 1838 (? = *Neocurtilla hexadactyla* (Perty)), although the author of the species never saw the actual specimen.

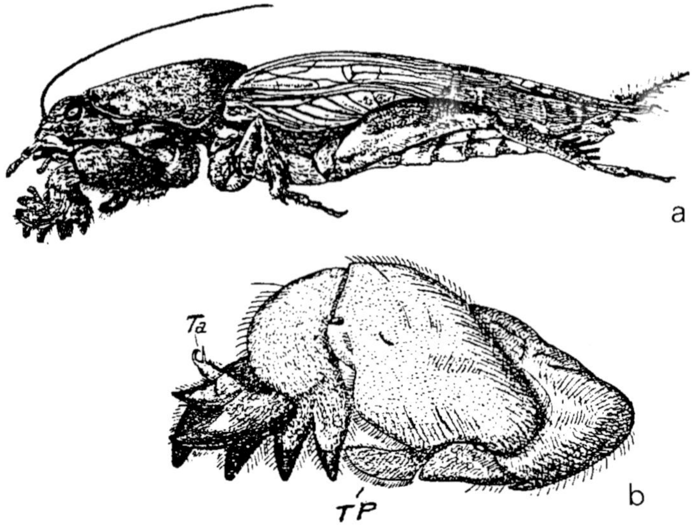

Fig. 329. *Gryllotalpa gryllotalpa*. *a*, male, after Chopard (1951), courtesy of Editions Lechevalier, Paris; *b*, left foreleg, adapted from Beier (1955). *Ta* = tarsus, *TP* = process of trochanter.

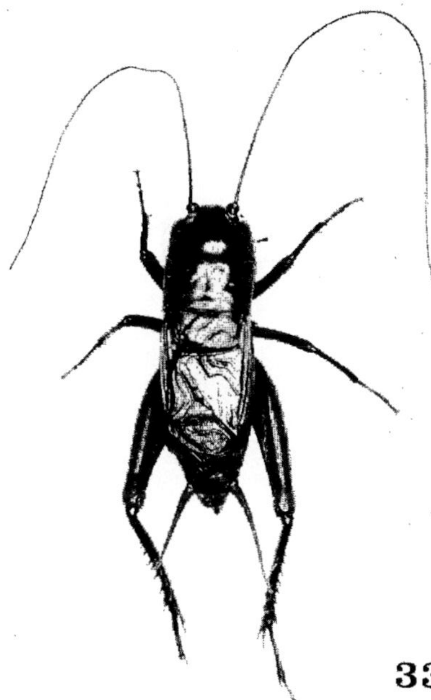

330

331

Figs. 330, 331. *Gryllus pennsylvanicus*. 300, male, dorsal; 331, female, lateral.

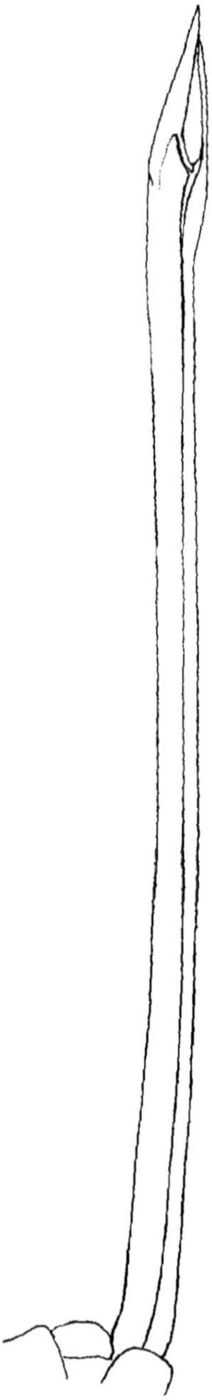

Fig. 332. *Gryllus pennsylvanicus*, ovipositor. After Vickery et al. (1974).

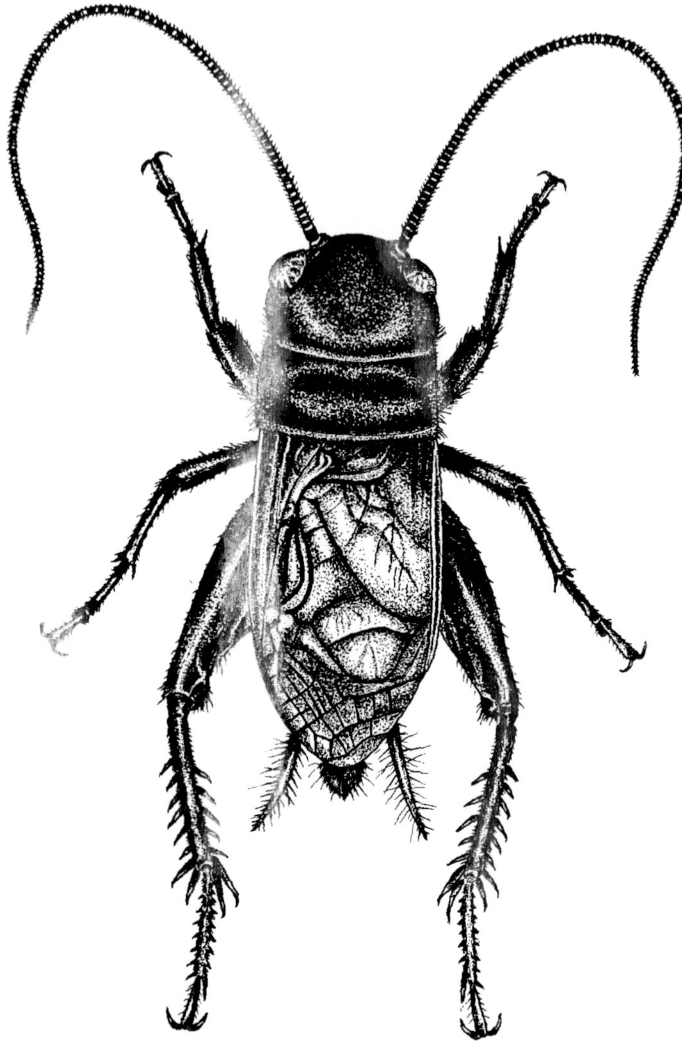

Fig. 333. *Gryllus veletis*, male.

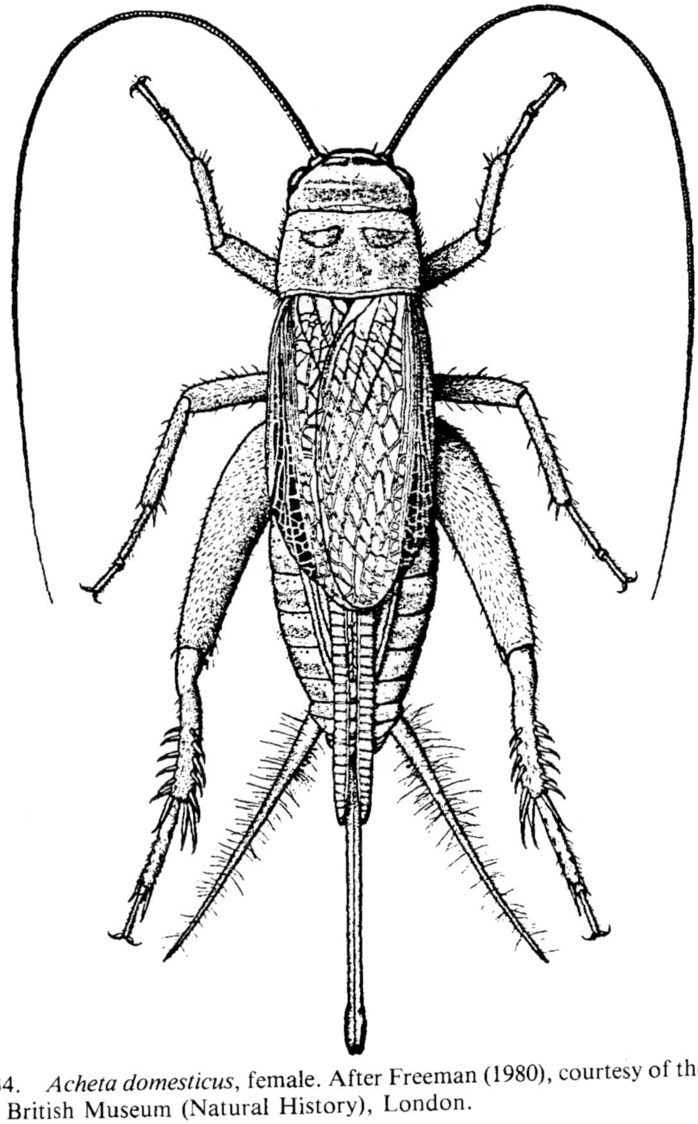

Fig. 334. *Acheta domesticus*, female. After Freeman (1980), courtesy of the Trustees of the British Museum (Natural History), London.

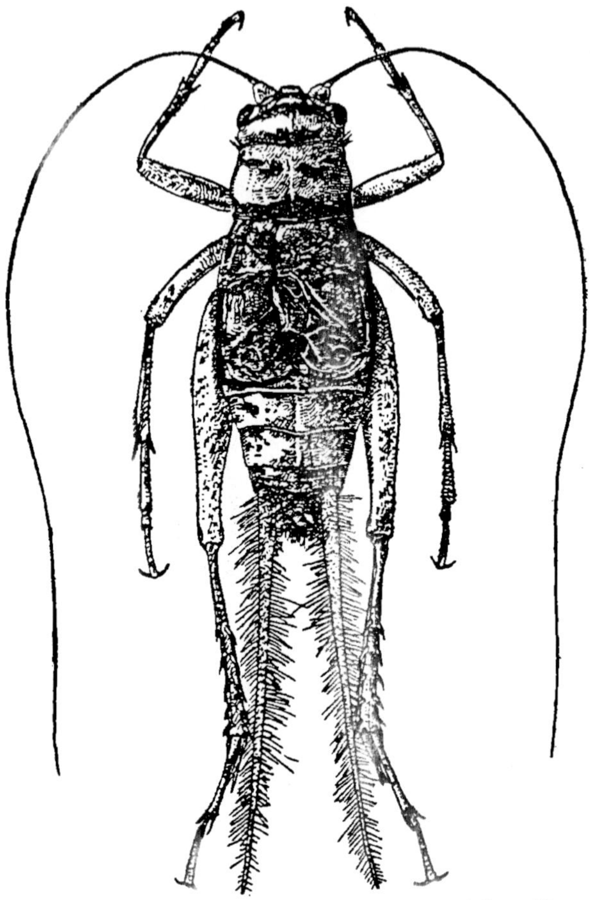

Fig. 335. *Gryllodes supplica[illegible]* form *sigillatus*, male. After Chopard (1938).

Fig. 336. *Allonemobius griseus griseus*, male. After Vickery and Johnstone (1973).

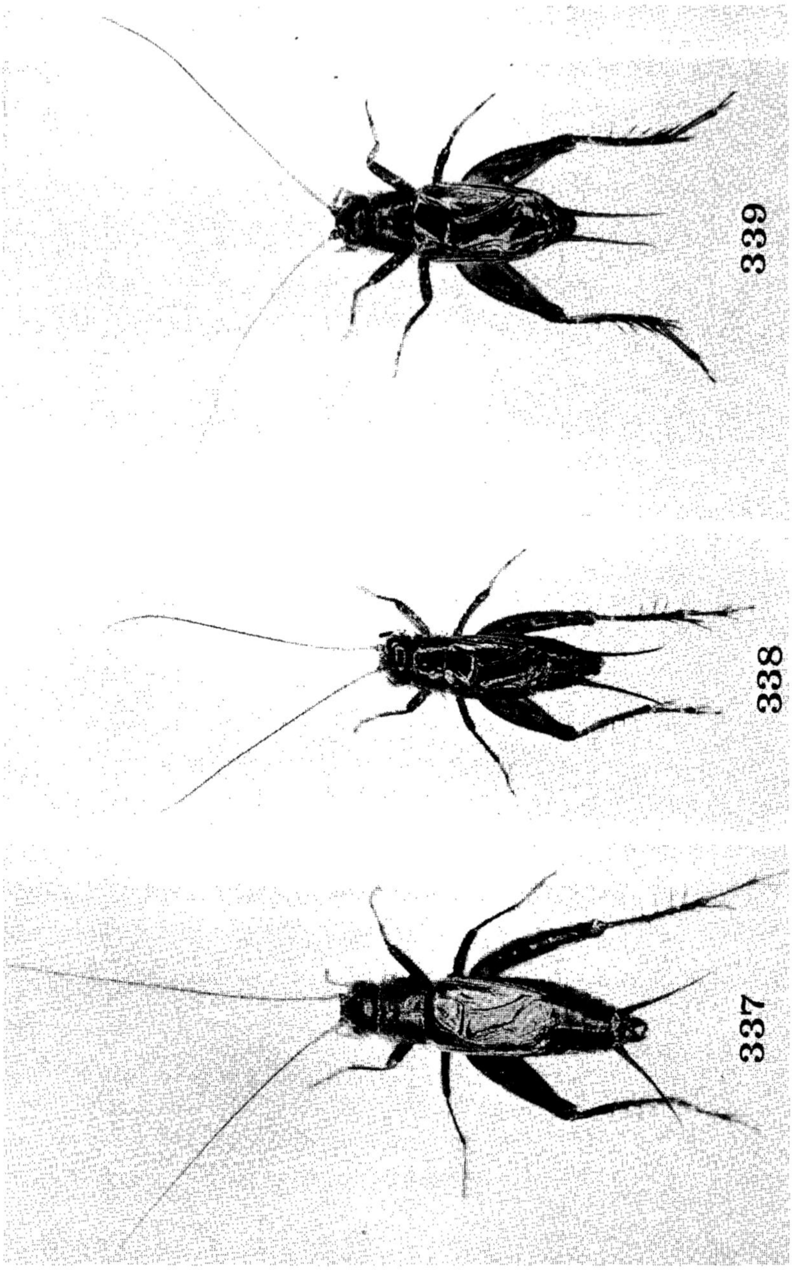

337 338 339

Figs. 337–339. Nemobiinae, males. 337, *Allonemobius allardi*; 338, *Neonemobius palustris*; 339, *Eunemobius carolinus carolinus*. 338, 339, after Vickery and Johnstone (1973).

Figs. 340–343. Nemobiinae, females, dorsal. 340, *Allonemobius allardi*; 341, *Neonemobius palustris*; 342, *Eunemobius carolinus carolinus*, brachypterous; 343, the same, macropterous. After Vickery and Johnstone (1973).

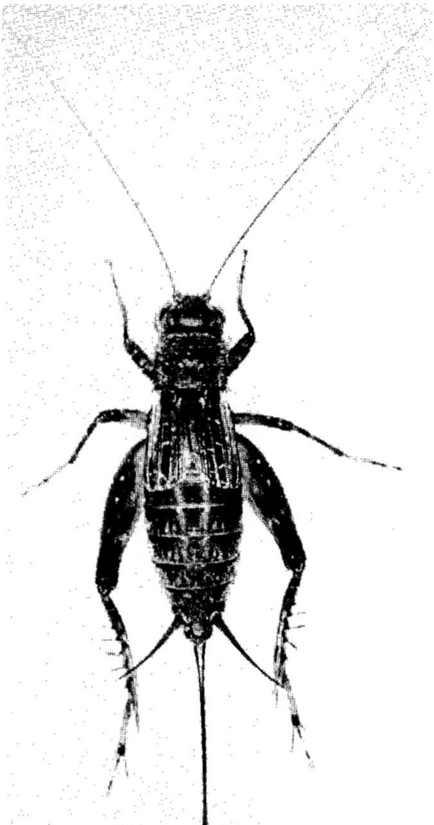

340

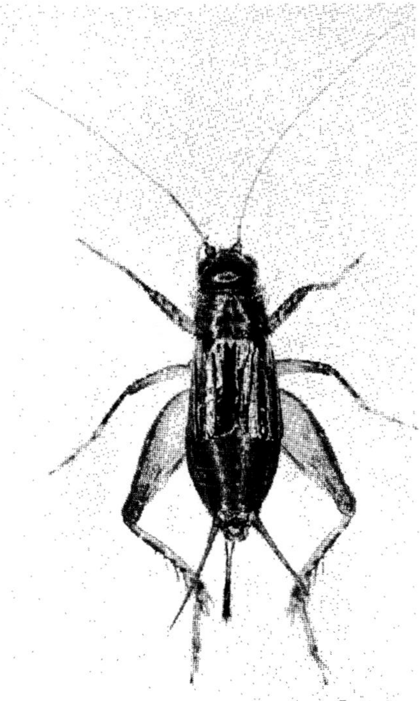

342

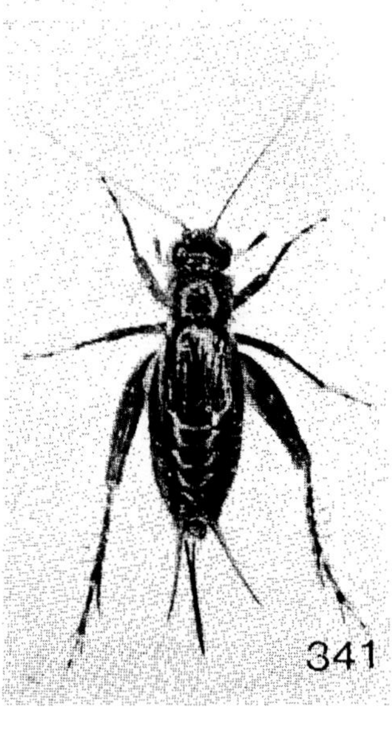

341

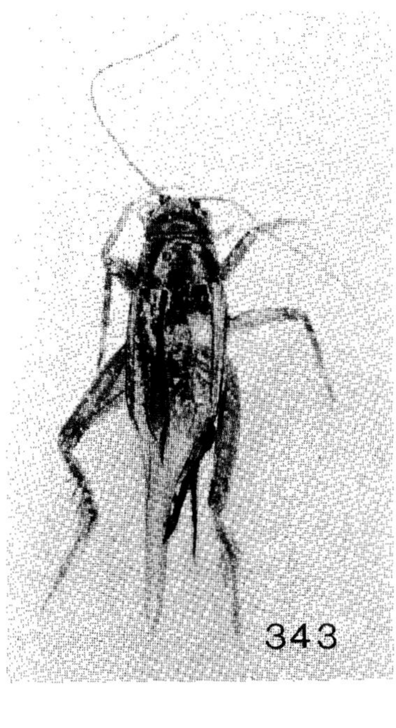

343

344

345

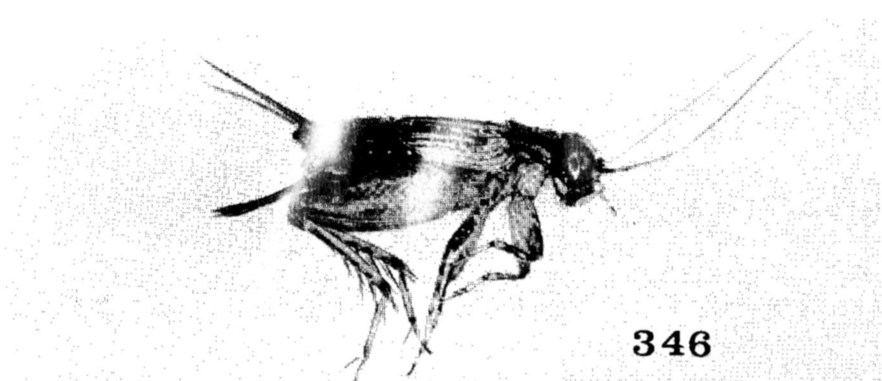

346

Figs. 344–346. Nemobiinae, females, lateral, showing ovipositors. 344, *Allonemobius allardi*; 345, *Neonemobius palustris*; 346, *Eunemobius carolinus carolinus*. After Vickery and Johnstone (1973).

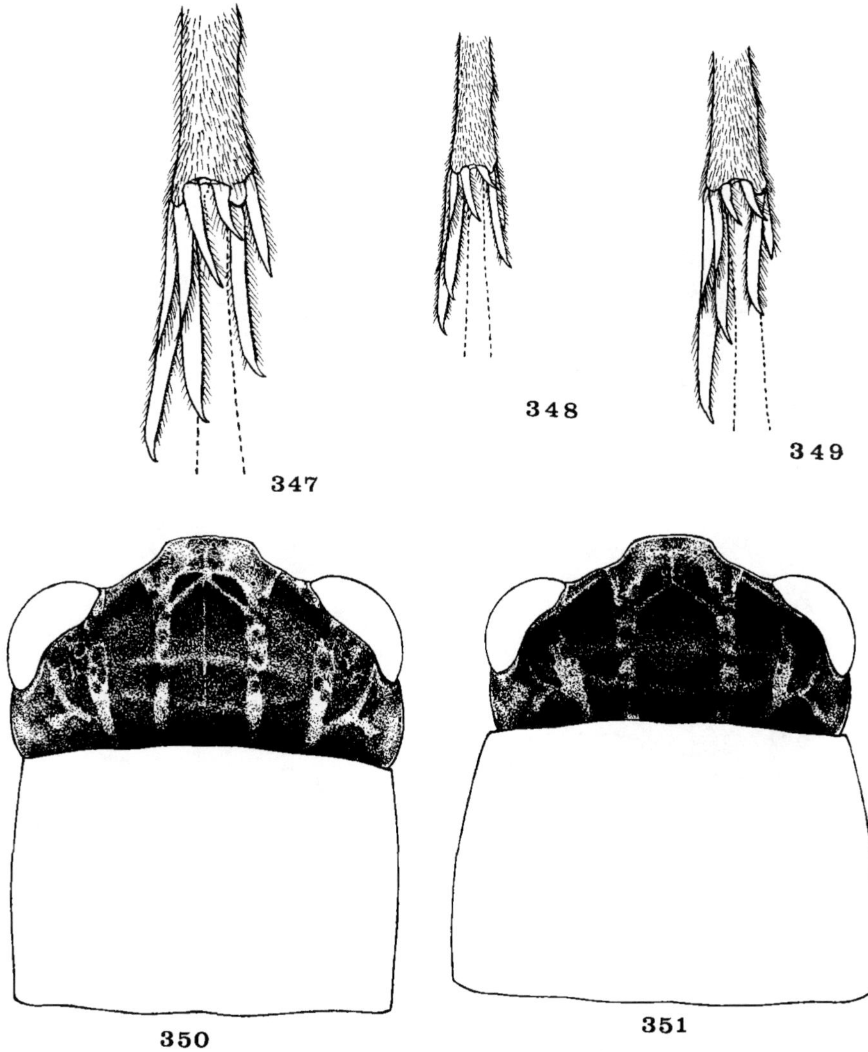

Figs. 347–351. Nemobiinae. 347–349, hind tibial spurs, dorsal; 350, 351, heads, dorsal. 347, *Allonemobius fasciatus*; 348, *Neonemobius palustris*; 349, *Eunemobius carolinus carolinus*; 350, *Allonemobius fasciatus*; 351, *A. allardi*. After Vickery and Johnstone (1973).

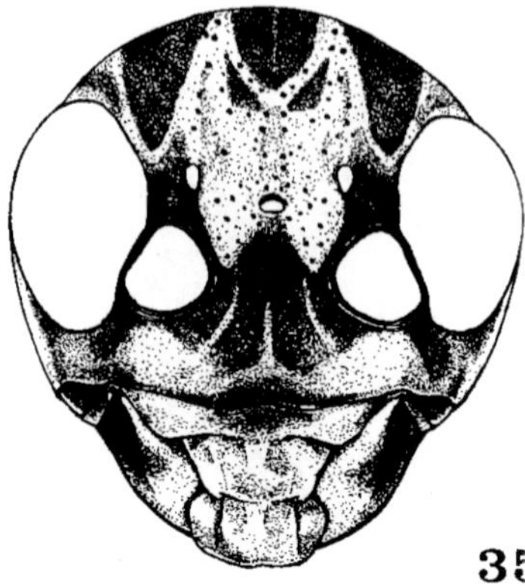

352

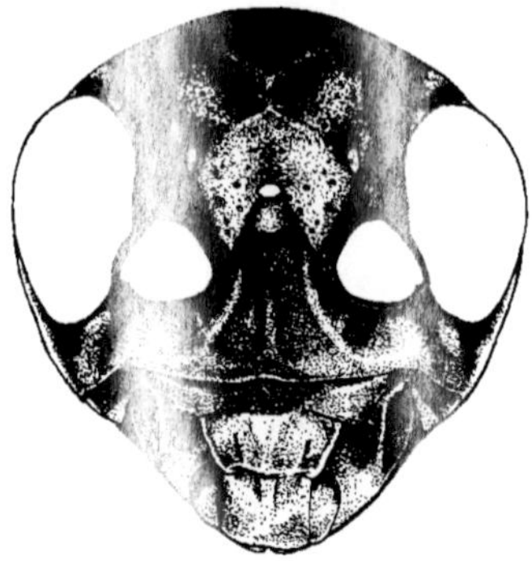

353

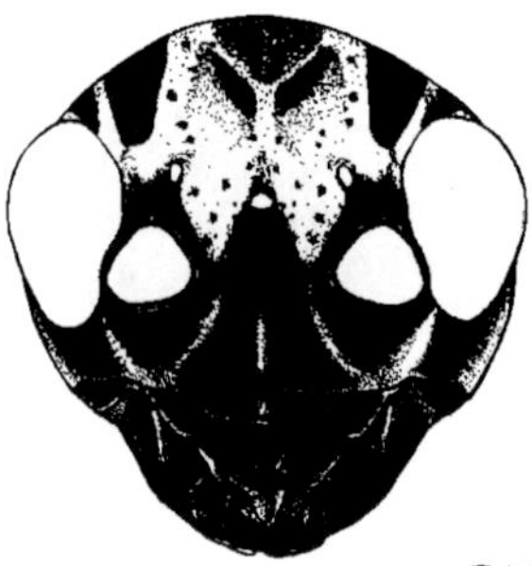

354

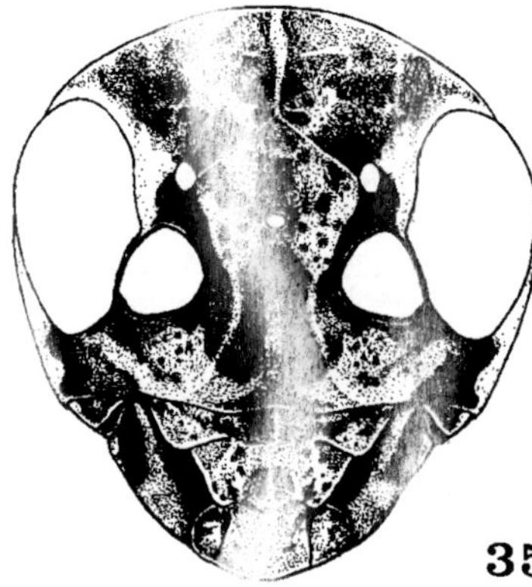

355

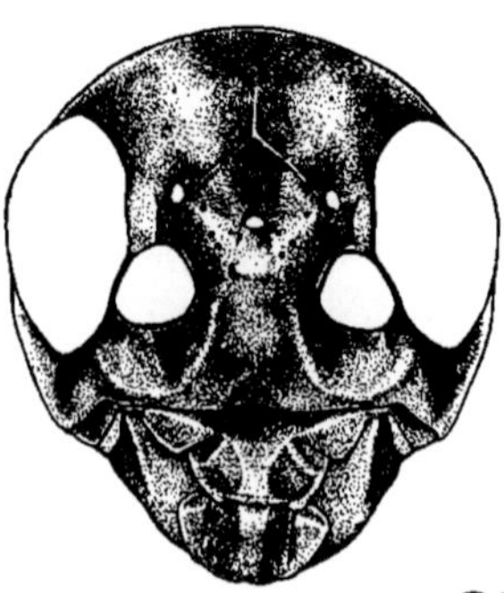

356

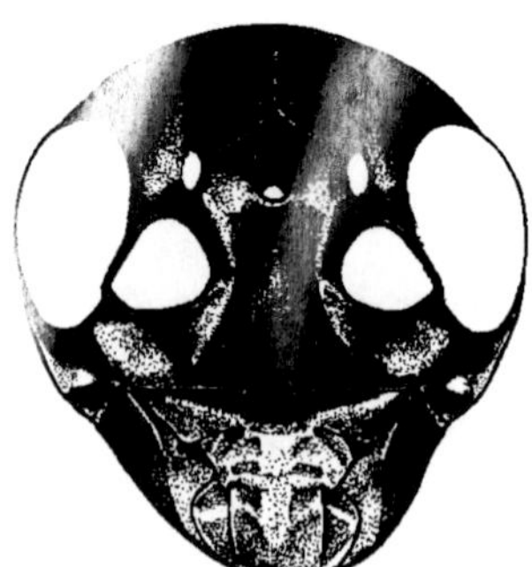

357

Figs. 352–357. Nemobiinae, male heads, frontal. 352, *Allonemobius fasciatus*; 353, *A. allardi*; 354, *A. griseus griseus*; 355, *A. maculatus*; 356, *Neonemobius palustris*; 357, *Eunemobius carolinus carolinus*. After Vickery and Johnstone (1973).

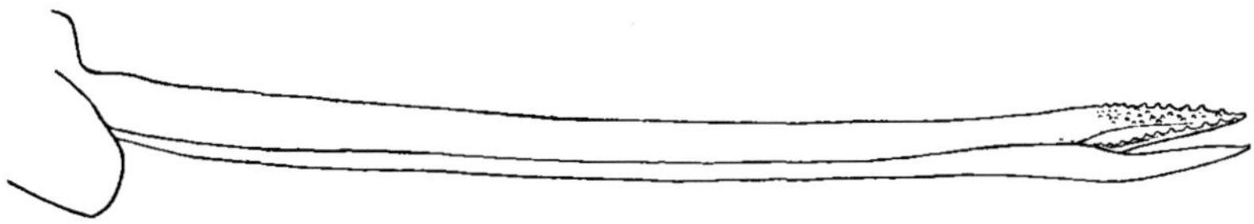

358

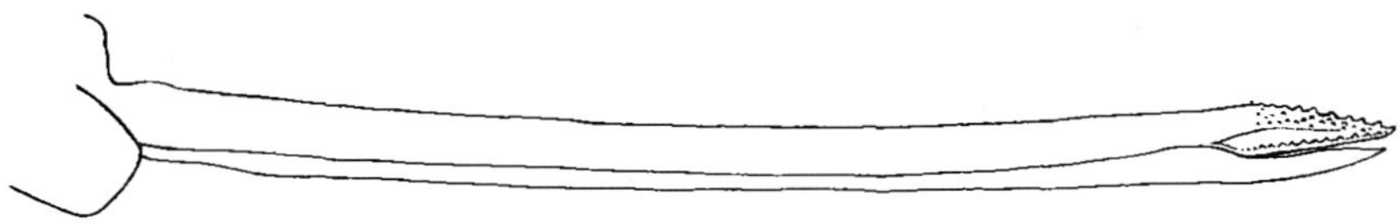

359

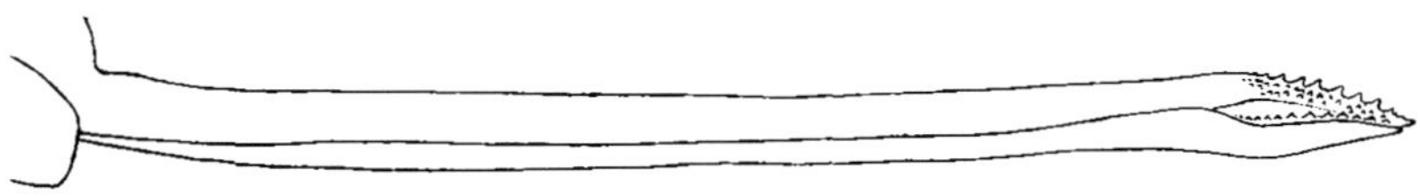

360

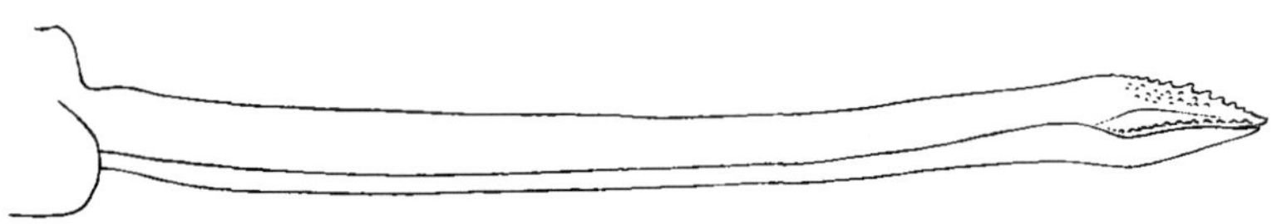

361

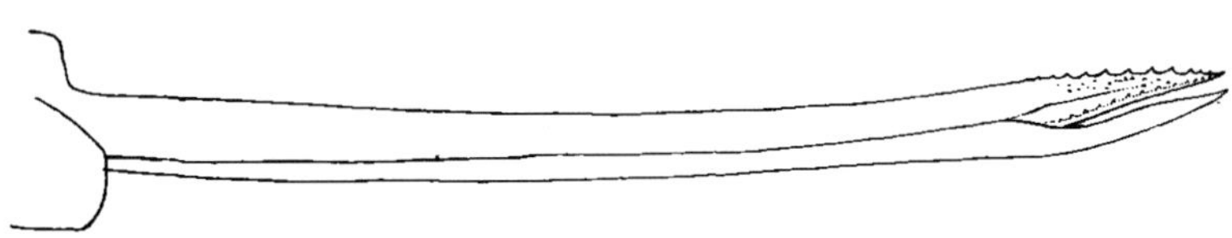

362

363

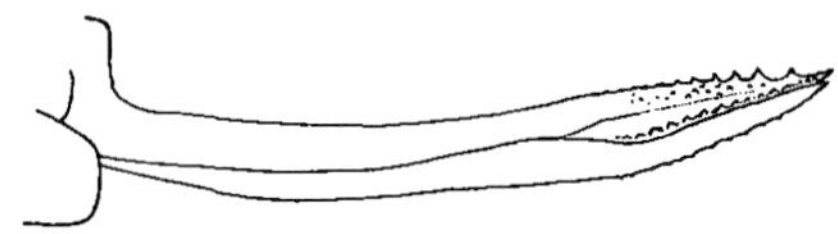

364

Figs. 358–364. Nemobiinae, ovipositors, from left. 358, *Allonemobius fasciatus*; 359, *A. allardi*; 360, *A. griseus griseus*, eastern (Québec) population; 361, the same, western (Saskatchewan) population; 362, *A. maculatus*; 363, *Neonemobius palustris*; 364, *Eunemobius carolinus carolinus*. After Vickery and Johnstone (1973).

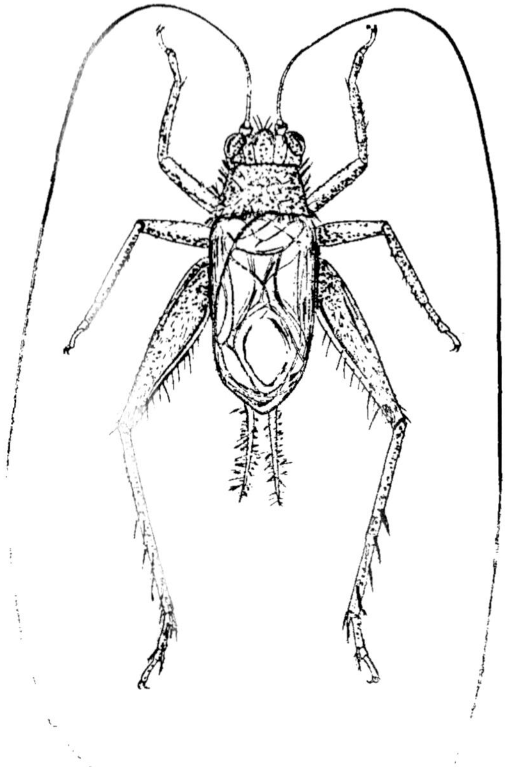

365

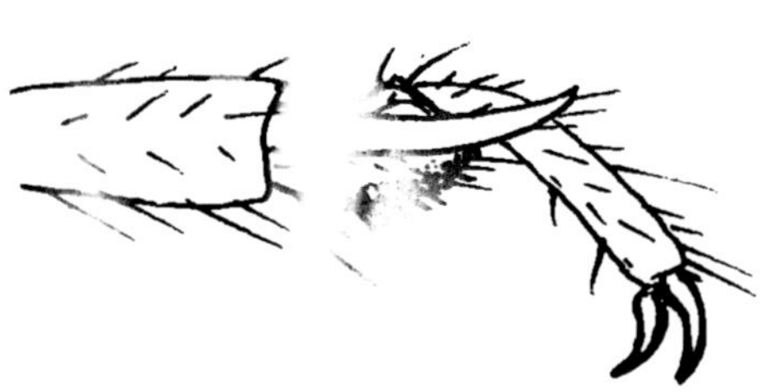

366

Figs. 365, 366. *Anaxipha exigua*. 365, Brachypterous male, dorsal; 366, hind tarsus, showing brushlike pad on second segment.

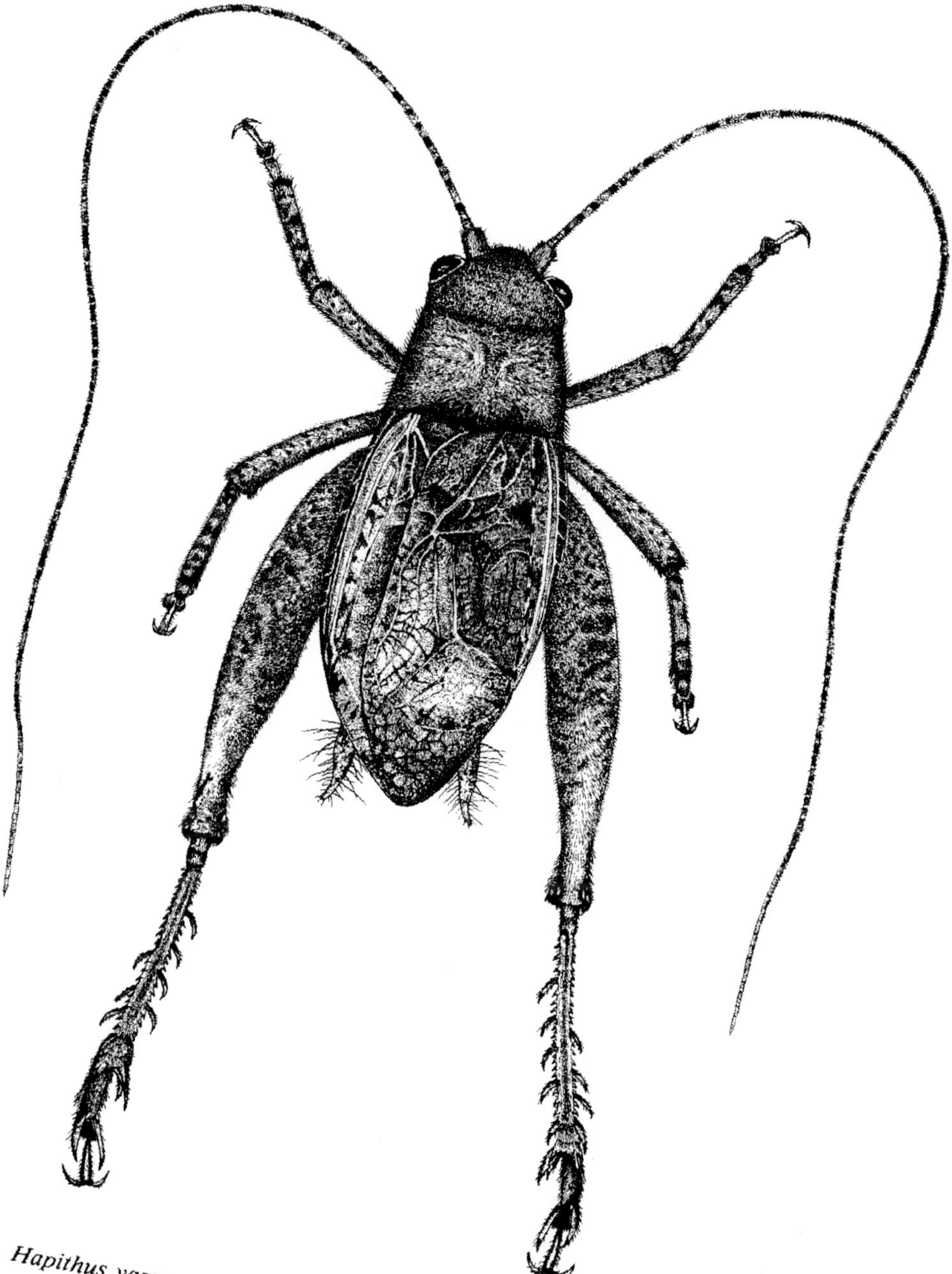

ig. 367. *Hapithus vagus*, male.

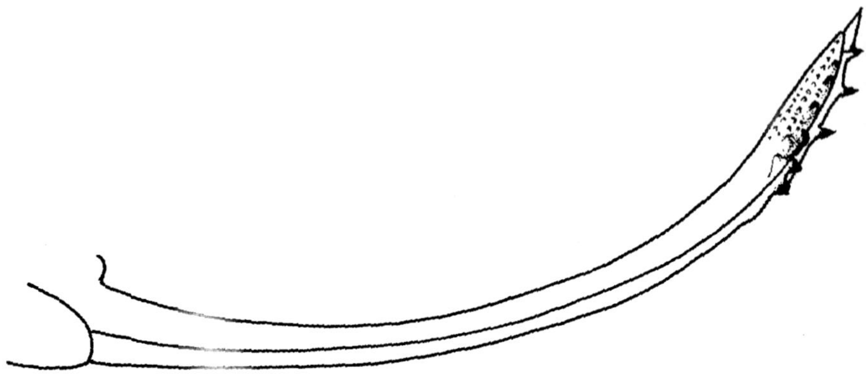

Fig. 368. *Hapithus vagus*, ovipositor, lateral.

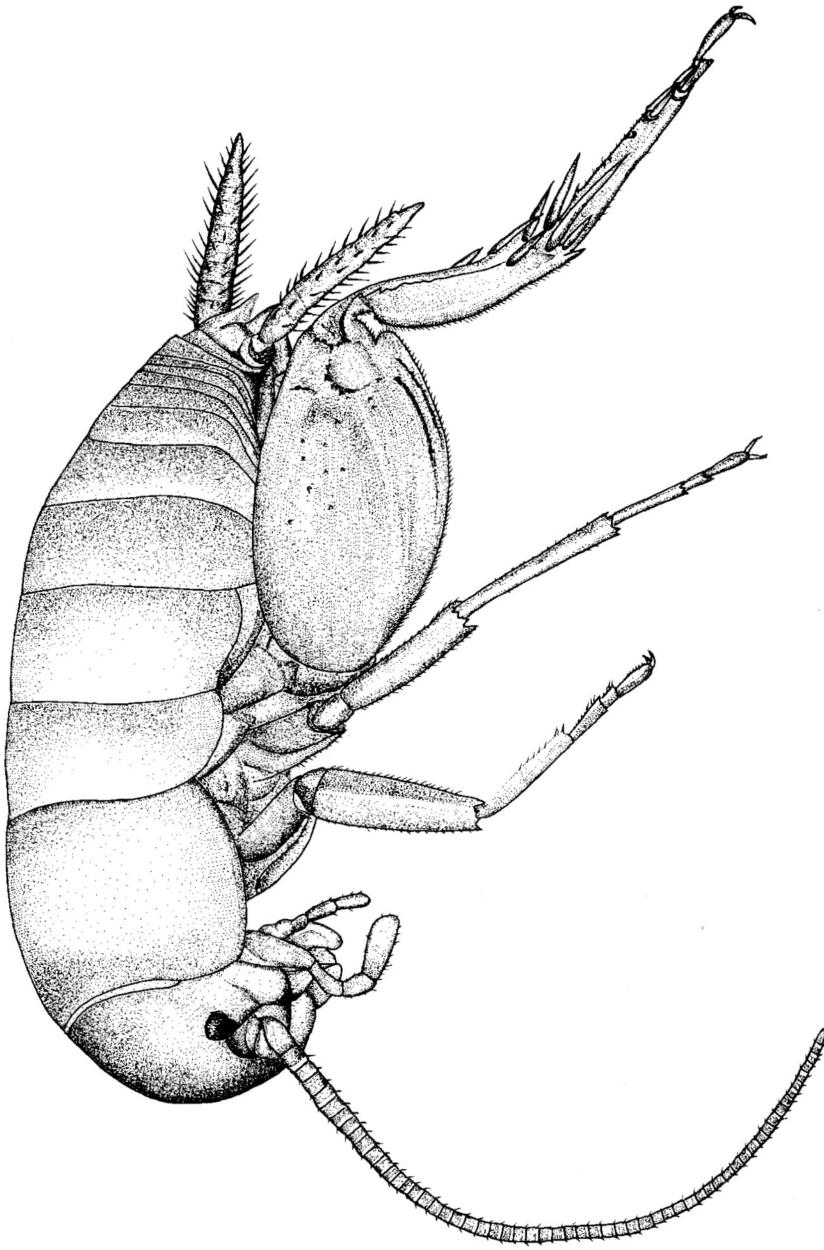

Fig. 369. *Myrmecophilus oregonensis*, male.

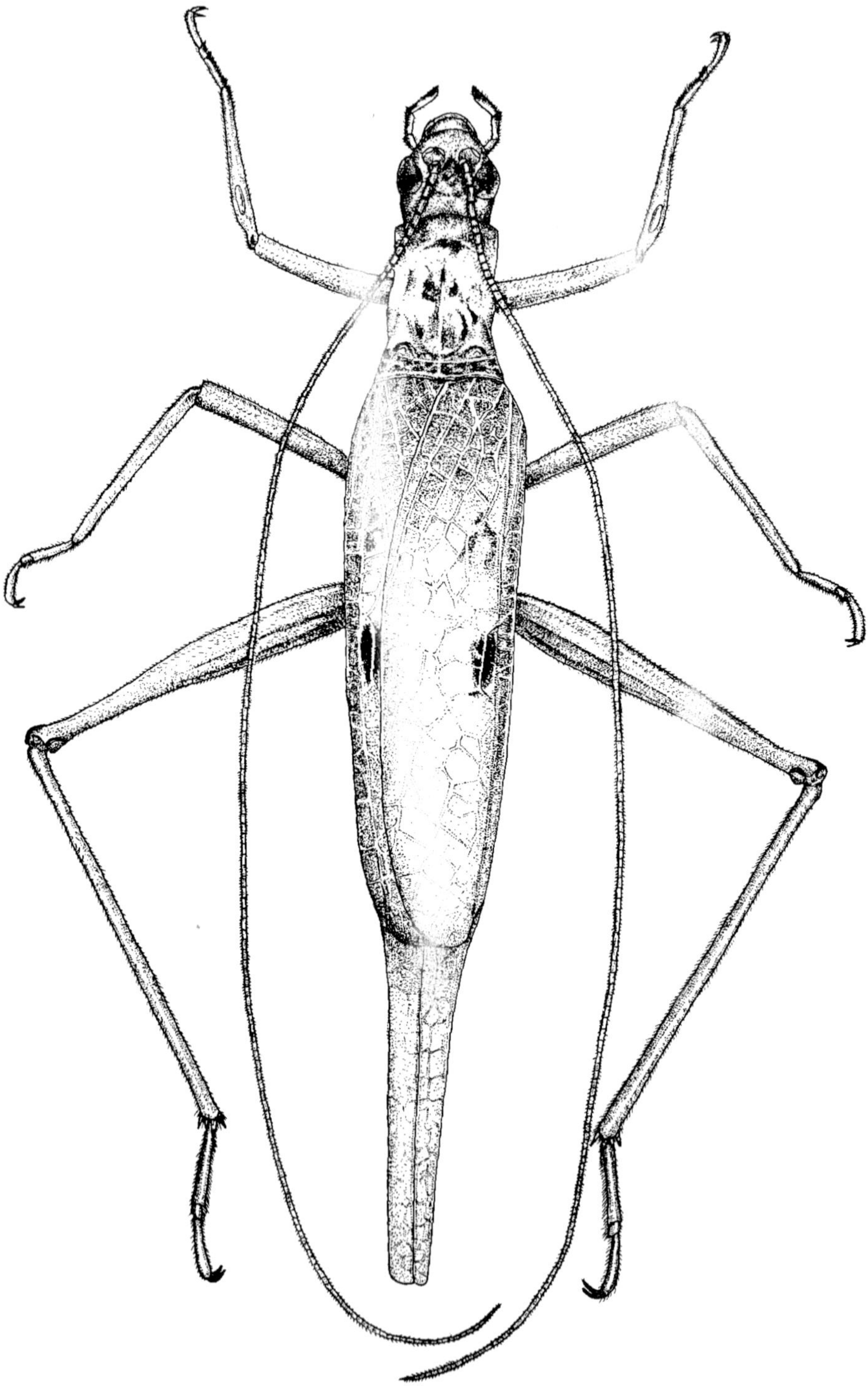

Fig. 370. *Neoxabea bipunctata*, female.

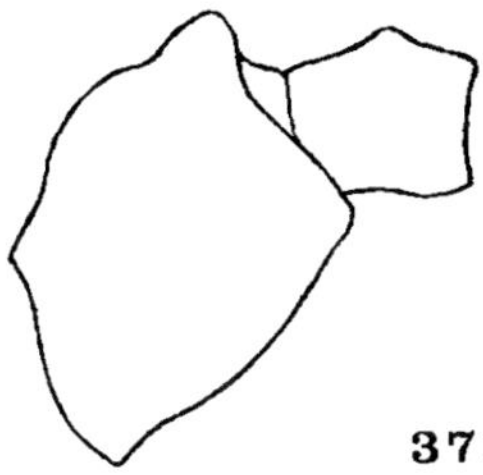

371

372

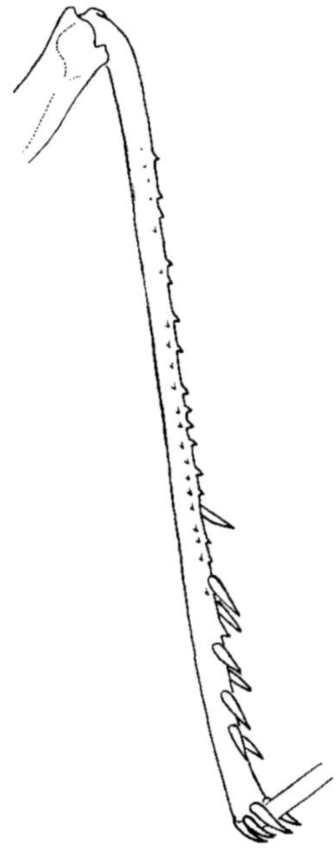

373

Figs. 371–373. Oecanthidae. 371, *Neoxabea bipunctata*, basal antennal segments, dorsal; 372, *N. bipunctata*, hind tibia; 373, *Oecanthus nigricornis*, hind tibia.

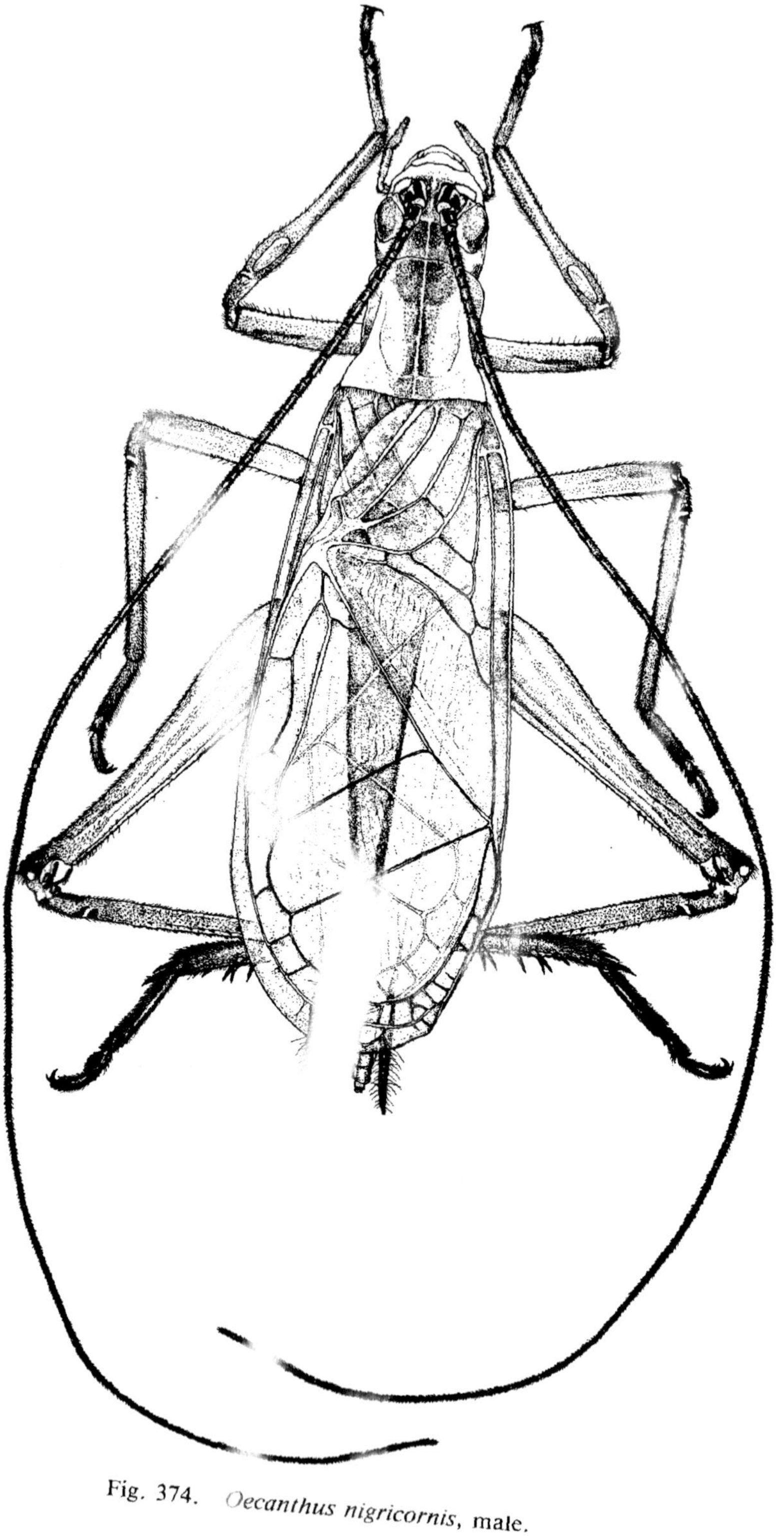

Fig. 374. *Oecanthus nigricornis*, male.

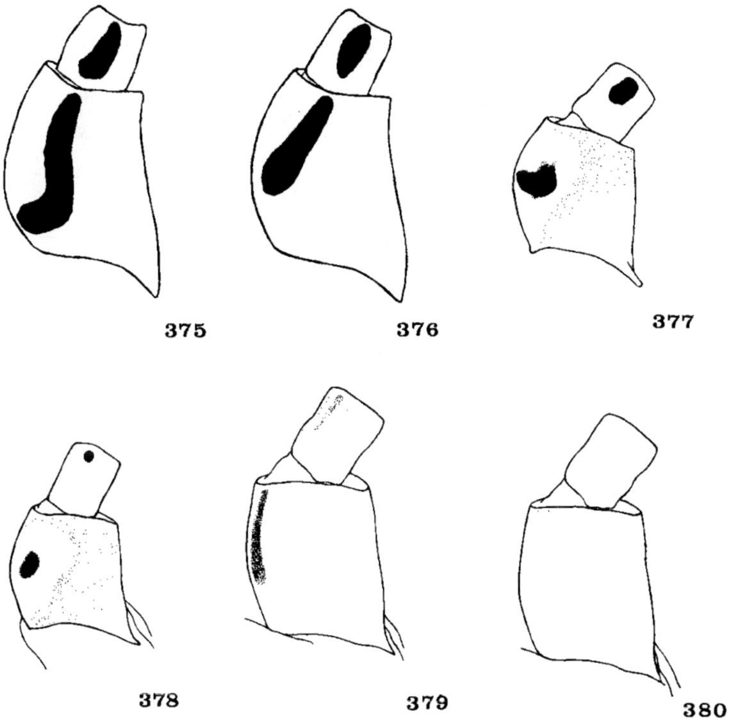

Figs. 375–380. *Oecanthus* spp., basal antennal segments, ventral. 375, *O. niveus*; 376, *O. exclamationis*; 377, *O. fultoni*; 378, *O. rileyi*; 379, 380, *O. californicus*. 377, after Vickery et al. (1974).

381

382

383

384

385

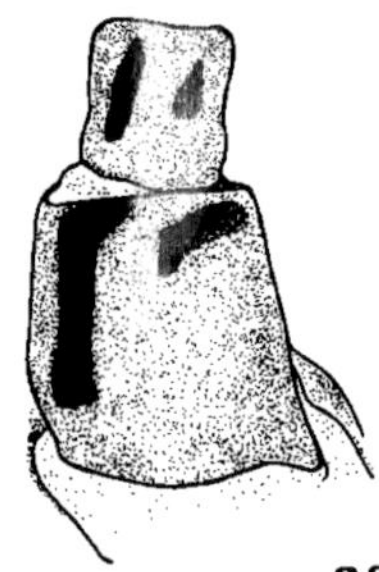

386

Figs. 381–386. *Oecanthus* spp., basal antennal segments, ventral. 381–383, *O. nigricornis*; 384, *O. quadripunctatus*; 385, *O. argentinus*; 386, *O. pini*. 381, 383, 384, after Vickery et al. (1974).

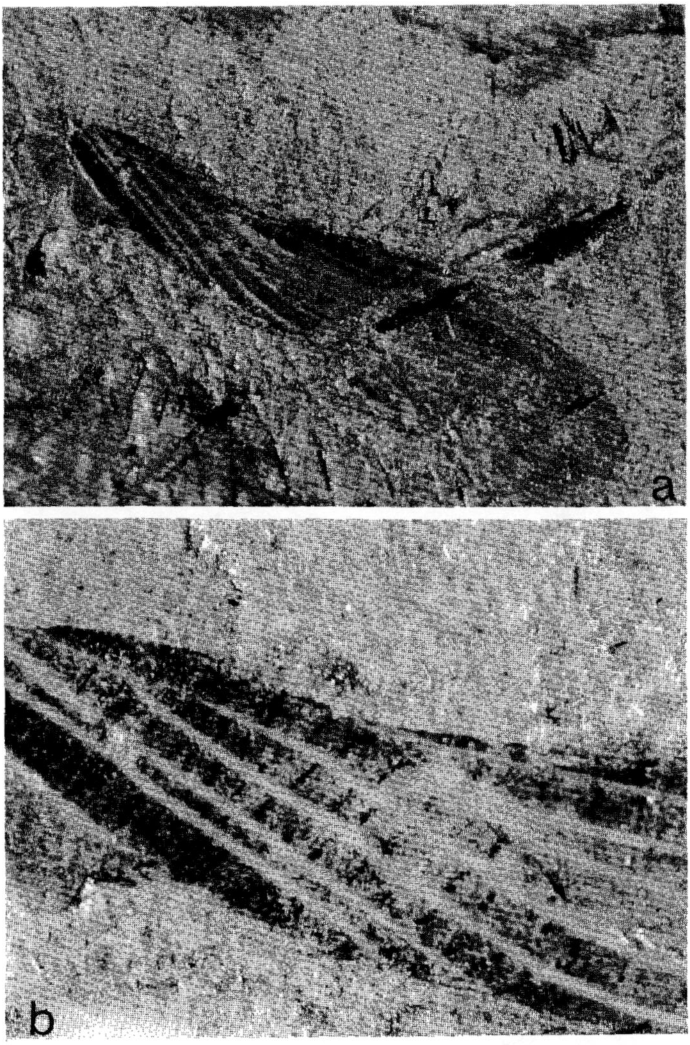

Fig. 387. Tegmen of *Promastacoides albertae* Kevan & Wighton (Promastacidae) from the Paskapoo (Palaeocene) deposits (fine argillaceous limestone), Red Deer River, Alta. *a*, entire tegmen; *b*, enlargement of principal venation. After Kevan and Wighton (1981).

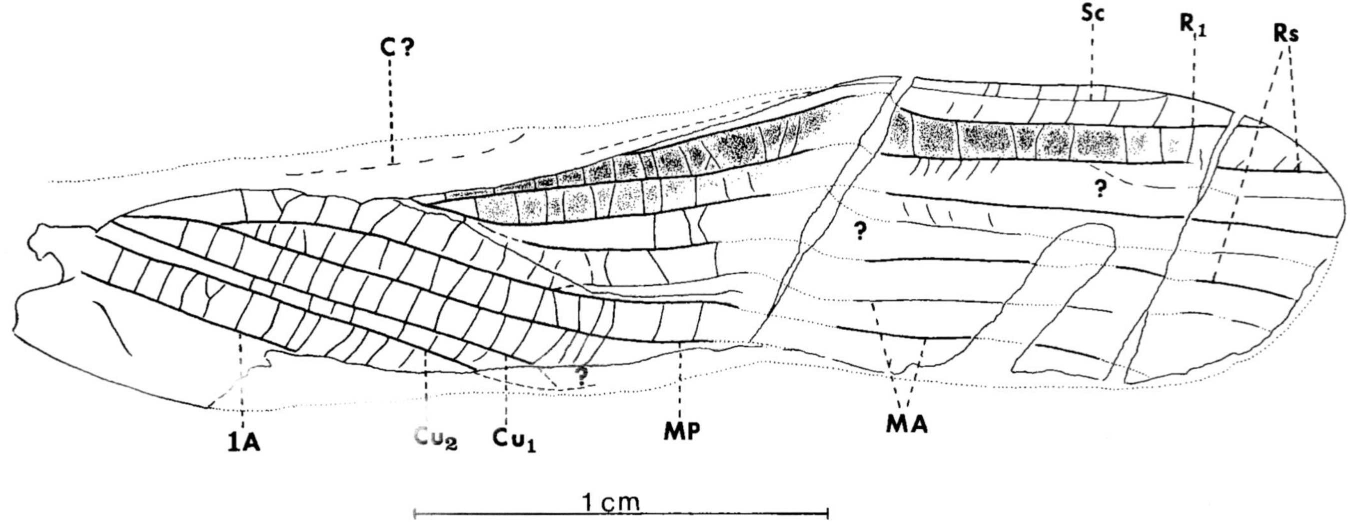

Fig. 388. Tegmen of *Promastacoides albertae*, (Promastacidae) from Alberta (see Fig. 387). After Kevan and Wighton (1981). Notation after Ragge (1955). Note: according to notation of Sharov (1968, 1971), Cu_1 and Cu_2 are CuA and CuP. (See caption for Fig. 181 for definition of symbols.)

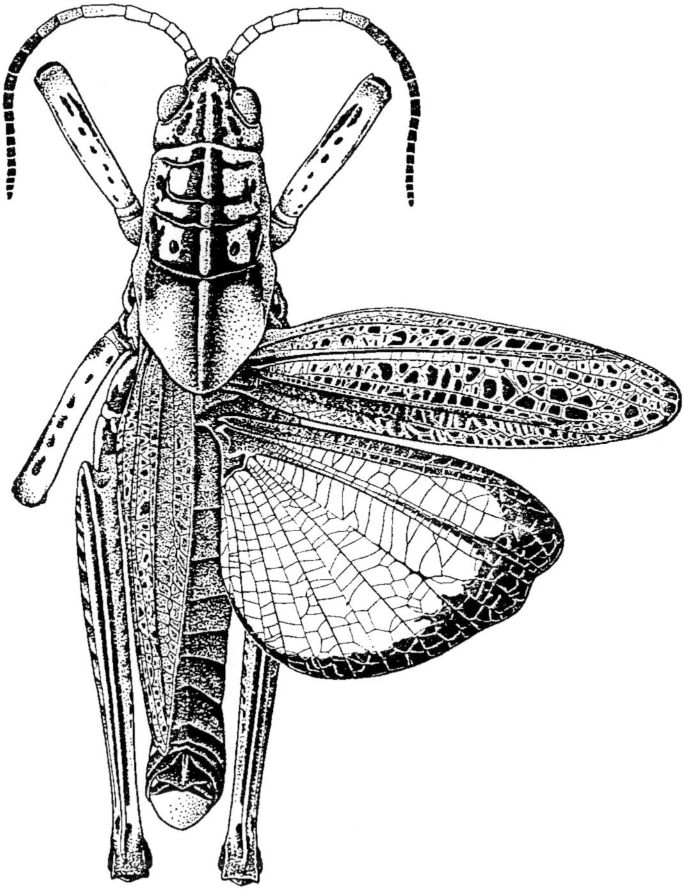

Fig. 389. *Romalea guttata*, male. After Dirsh (1961).

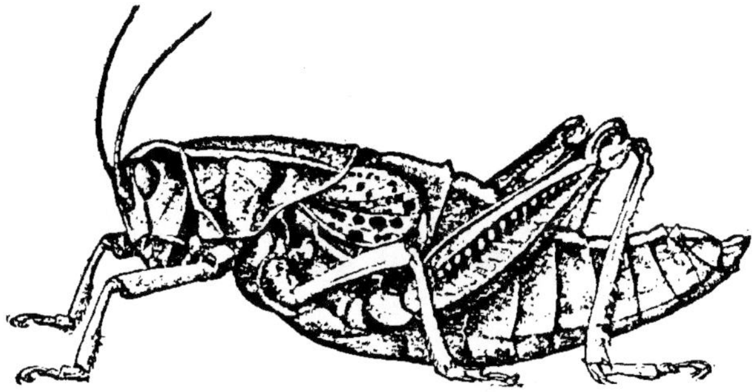

Fig. 390. *Brachystola magna*, female.

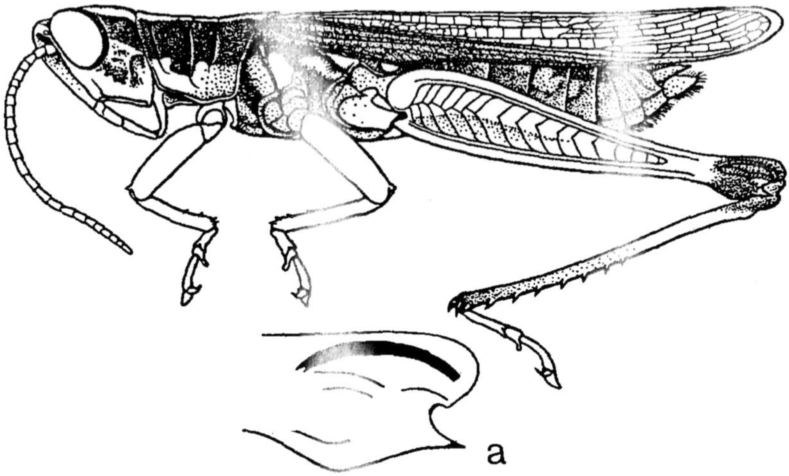

Fig. 391. *Oxya hyla*, male: *a*, hind knee. After Dirsh (1961).

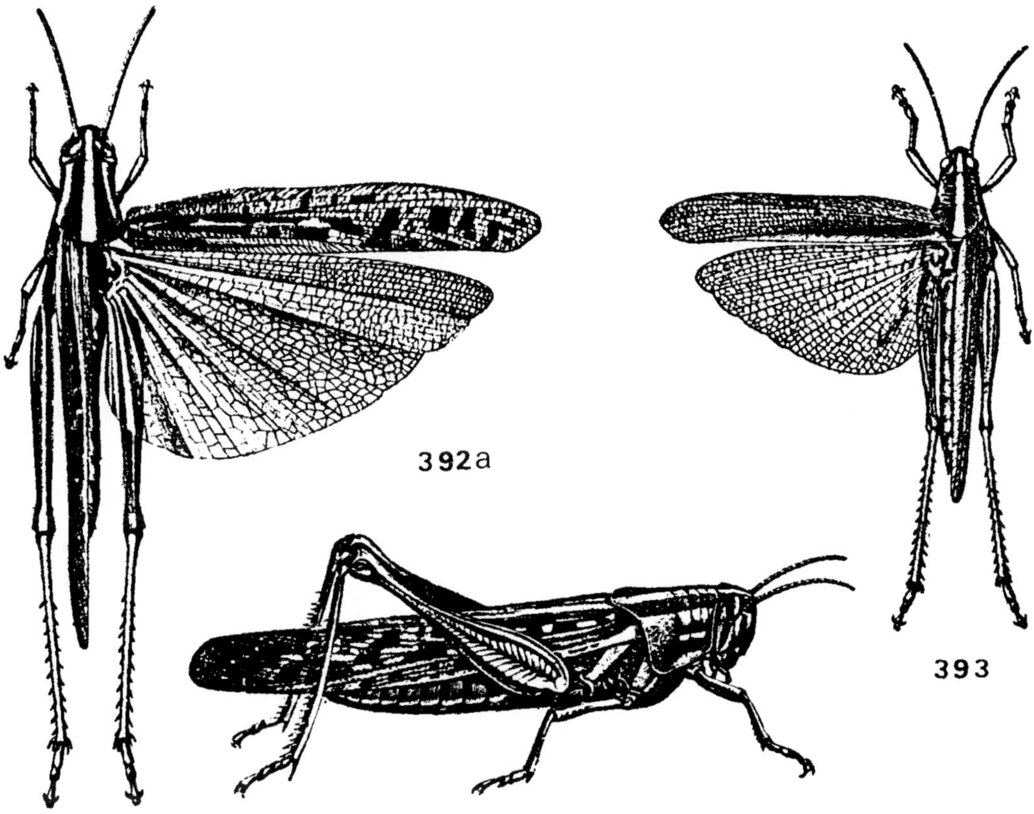

Figs. 392, 393. *Schistocerca* spp. 392*a*, *S. americana*, female; 392*b*, the same; 393, *S. emarginata*, female. 392*a*, 393, after Lugger (1898); 392*b*, adapted from Riley (1877*a*).

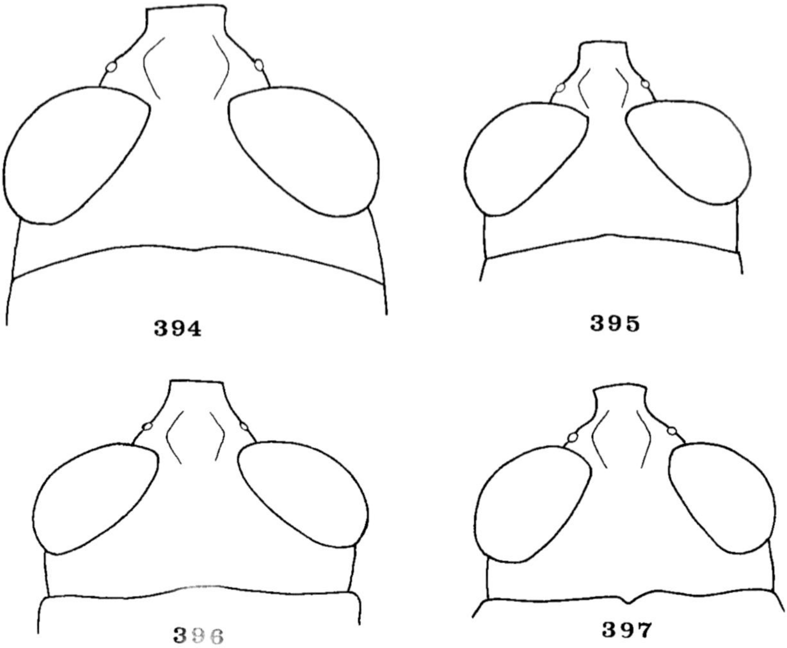

Figs. 394–397. *Schistocerca* spp., male heads, dorsal (all to same scale). 394, *S. americana*; 395, *S. alutacea*; 396, *S. rubiginosa*; 397, *S. emarginata*. 395–397 redrawn from Hubbell (1960).

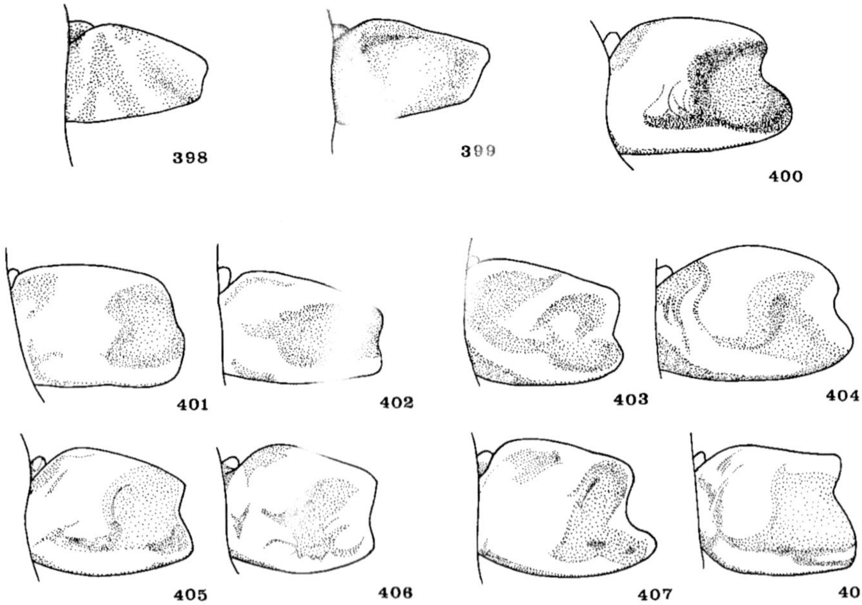

Figs. 398–408. *Schistocerca* spp., male cerci, from left (all to same scale). 398, *S. americana*, Georgia; 399, the same, Florida; 400, *S. alutacea*; 401–404, *S. rubiginosa*; 405–408, *S. emarginata*. 400–408, redrawn from Hubbell (1960).

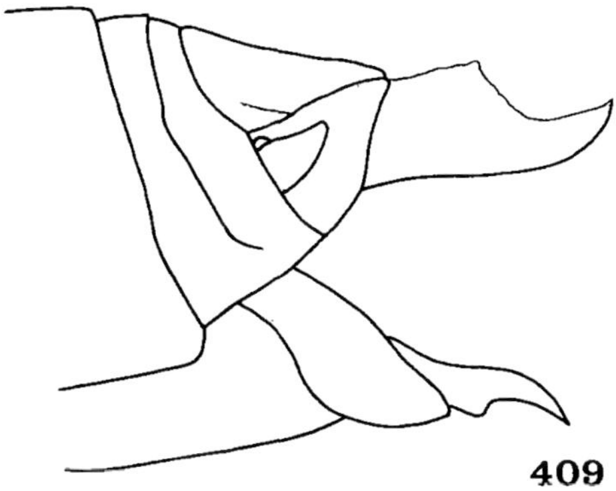

409

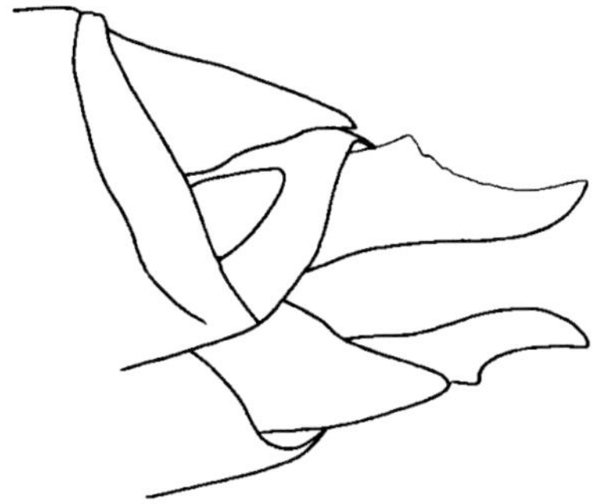

410

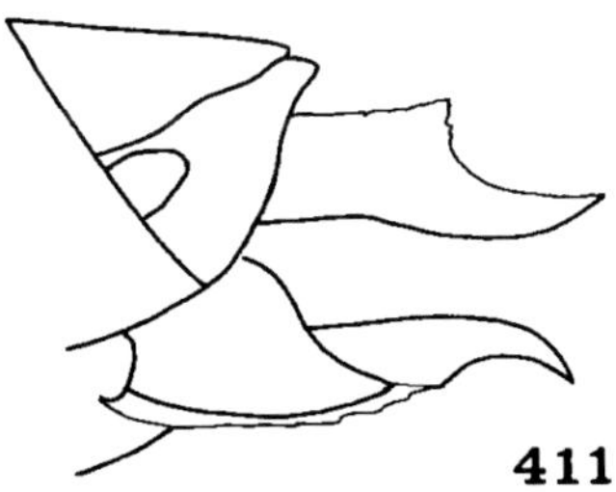

411

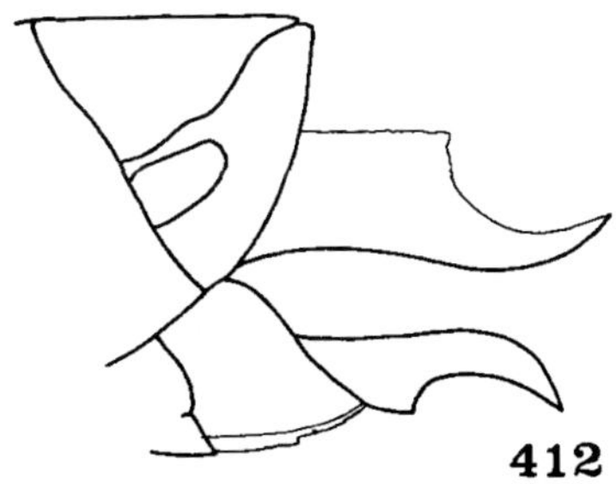

412

Figs. 409–412. *Schistocerca* spp., ovipositors, from left (all to same scale). 409, *S. americana*, Florida; 410, *S. alutacea*; 411, *S. rubiginosa*; 412, *S. emarginata*. 411, 412, redrawn from Hubbell (1960).

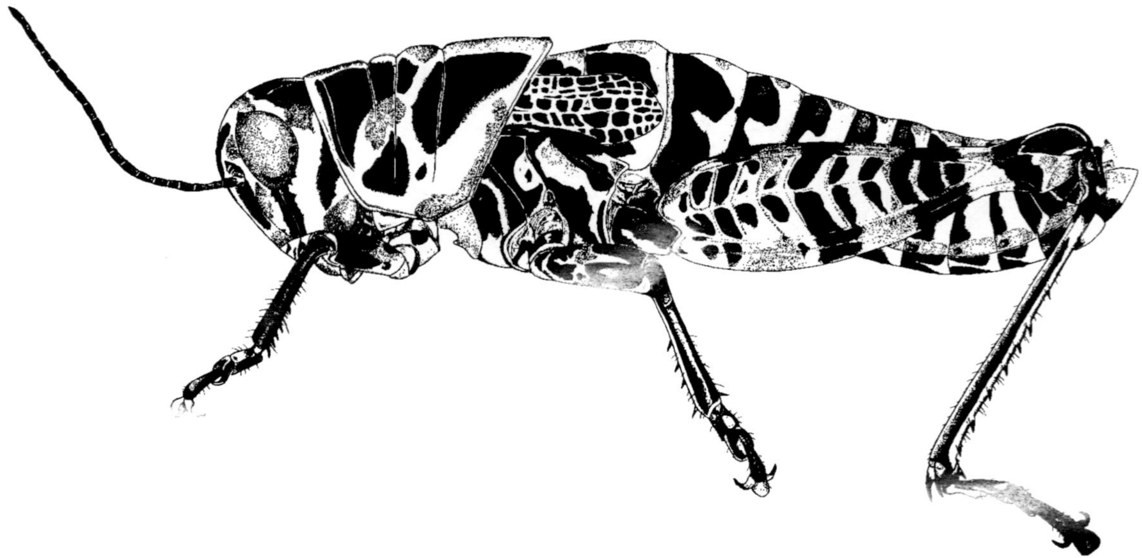

Fig. 413. *Dactylotum bicolor pictum*, male.

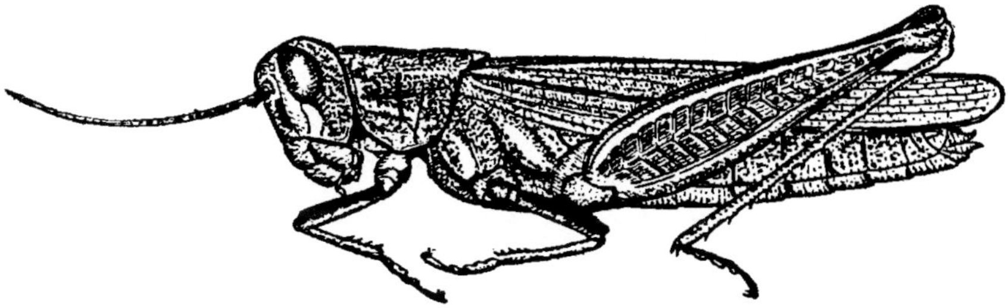

Fig. 414. *Hesperotettix viridis pratensis*, female. Adapted from Helfer (1963), courtesy of J.R. Helfer.

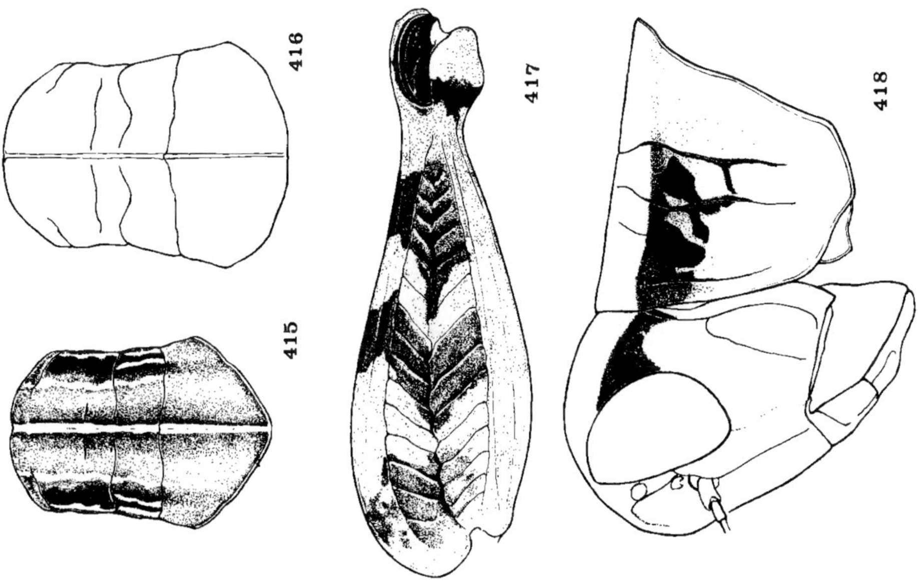

Figs. 415–418. Melanoplinae, Melanoplini. 415, *Hesperotettix viridis pratensis*, pronotum, dorsal; 416, *Hypochlora alba*, pronotum, dorsal; 417, *Aeoloplides turnbulli turnbulli*, hind femur (male); 418, *Phoetaliotes nebrascensis*, head and pronotum, lateral.

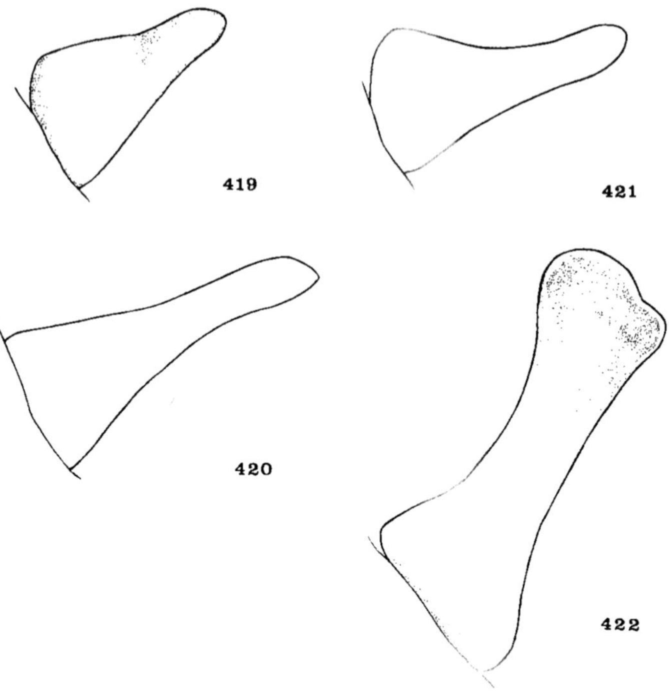

Figs. 419–422. Melanoplinae, Melanoplini, male left cerci. 419, *Hesperotettix viridis pratensis*; 420, *Aeoloplides turnbulli turnbulli*; 421, *Phoetaliotes nebrascensis*; 422, *Paroxya hoosieri*.

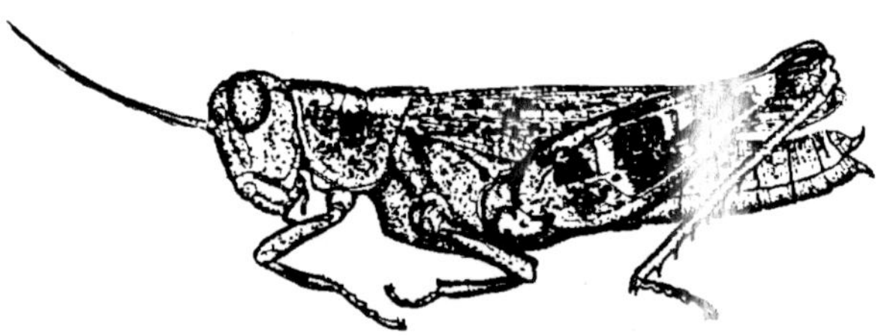

Fig. 423. *Aeoloplides turnbulli turnbulli*, female.

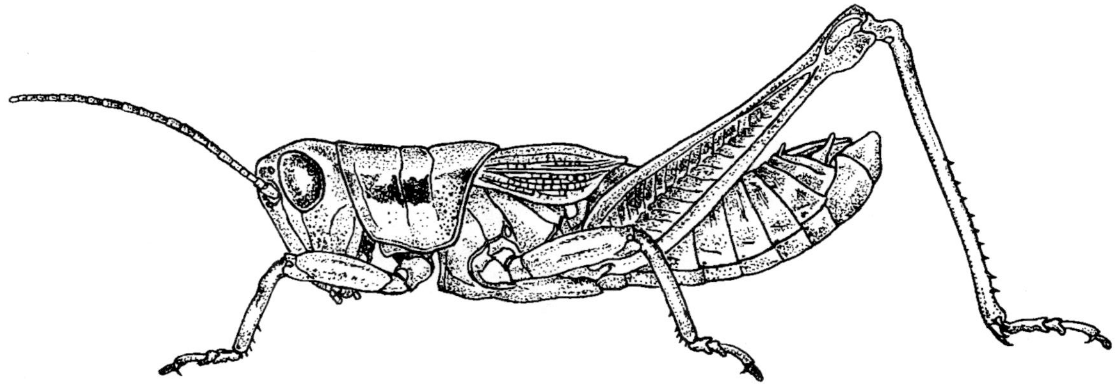

Fig. 424. *Hypochlora alba*, male.

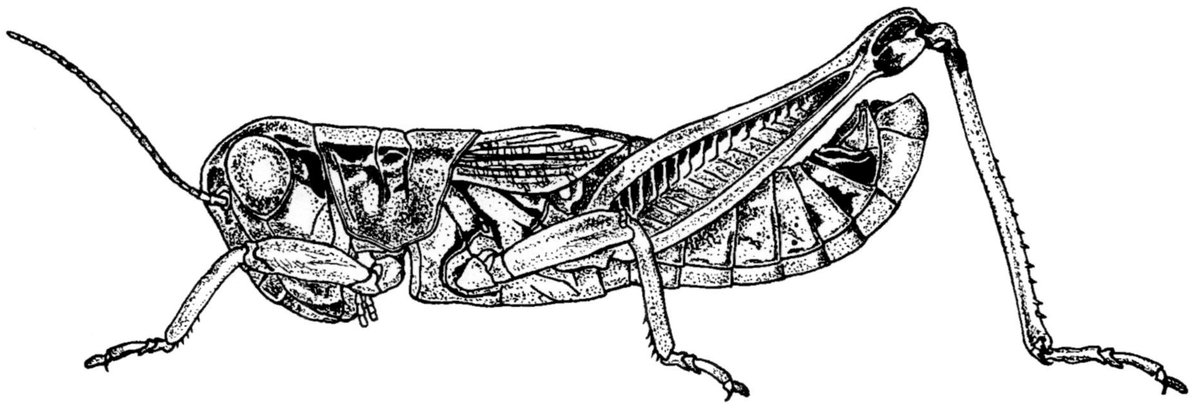

Fig. 425. *Phoetaliotes nebrascensis*, male.

Fig. 426. *Paroxya hoosieri*, male.

Fig. 427. *Melanoplus gaspesiensis*. After Vickery et al. (1974).

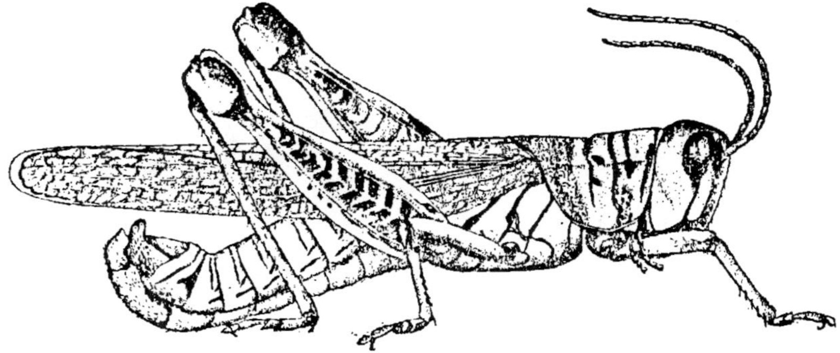

428

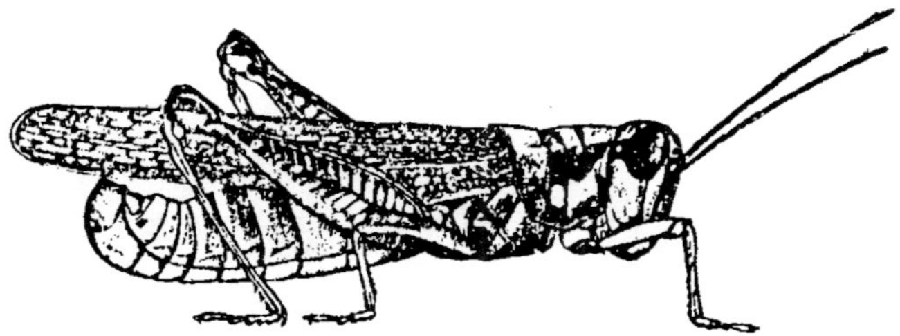

429

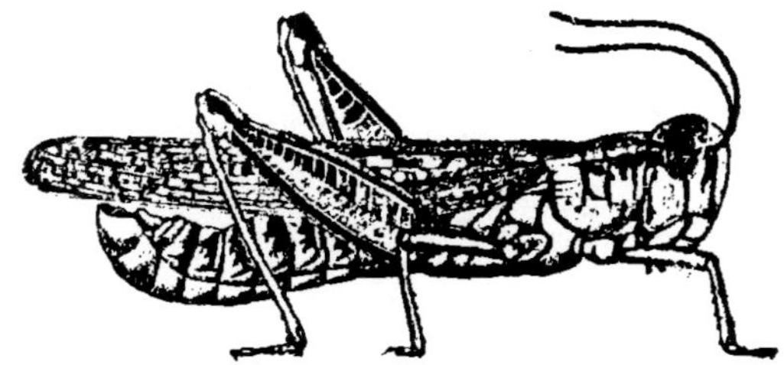

430

Figs. 428–430. *Melanoplus* spp. 428, *M. differentialis differentialis*, male; 429, *M. femurrubrum femurrubrum*, male; 430, *M. sanguinipes sanguinipes*. After Parker et al. (1932).

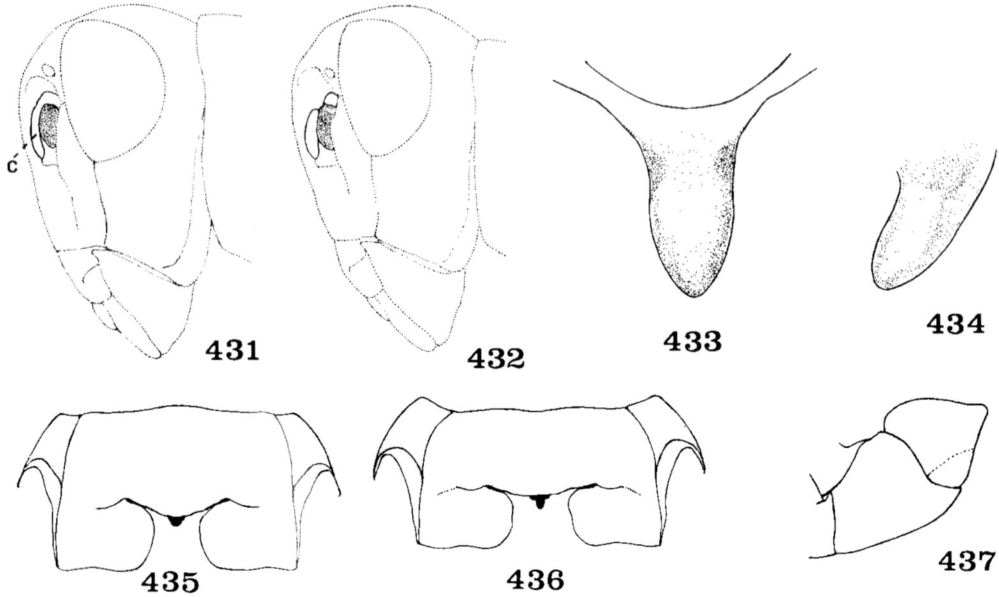

Figs. 431–437. *Melanoplus* spp. 431, *M. femurrubrum femurrubrum*, head, lateral, showing complete antennal crescent (*c*); 432, *M. sanguinipes sanguinipes*, showing divided antennal crescent; 433, *M. bivittatus*, prosternal spine, anterior view; 434, the same, view from right; 435, *M. femurrubrum femurrubrum*, mesosternum; 436, *M. borealis borealis*, mesosternum; 437, *M. borealis borealis*, male abdominal terminalia, from left. 431, 432, 434–436, after Vickery et al. (1974); 437, after Vickery (1970).

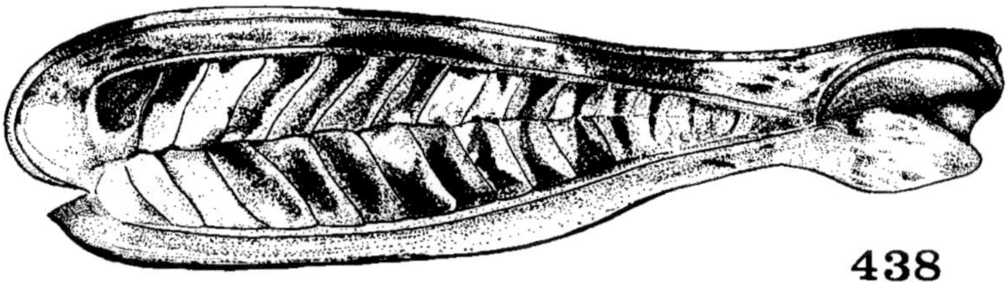

438

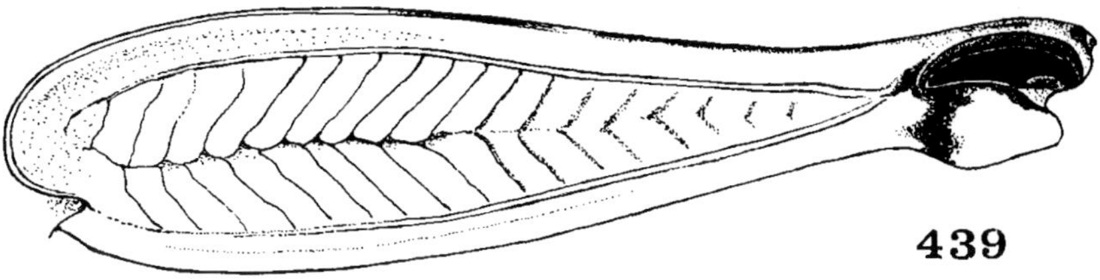

439

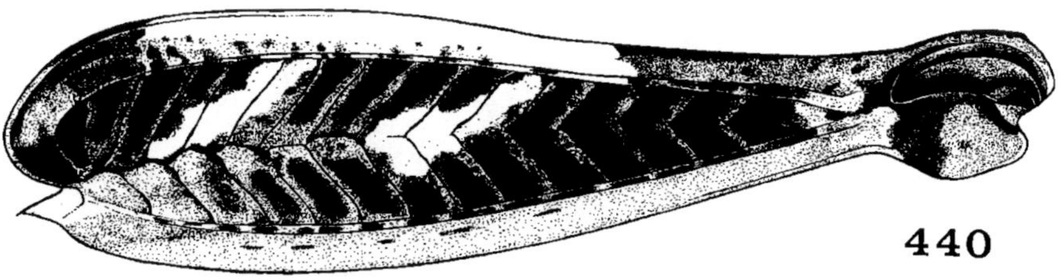

440

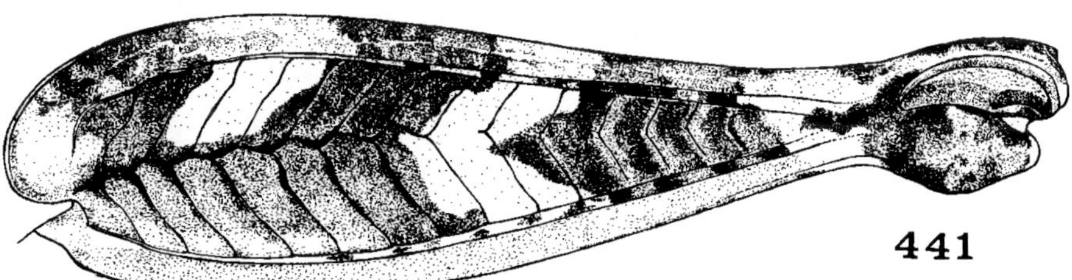

441

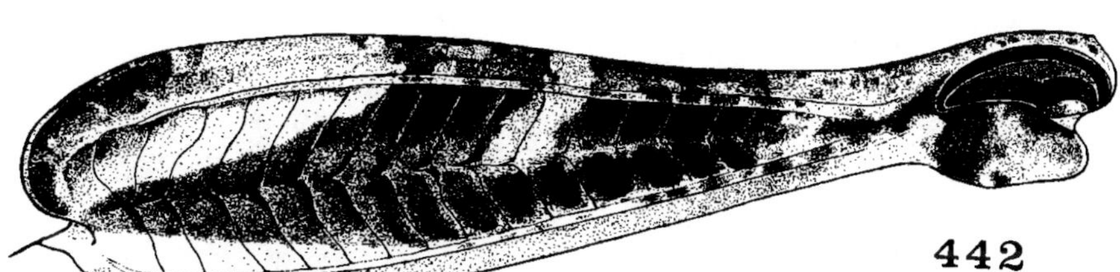

442

Figs. 438–442. *Melanoplus* spp., hind femora. 438, *M. gladstoni*; 439, *M. borealis borealis*; 440, *M. gaspesiensis*; 441, *M. fasciatus*; 442, *M. keeleri luridus*. 439, after Vickery (1970); 440–442, after Vickery et al. (1974).

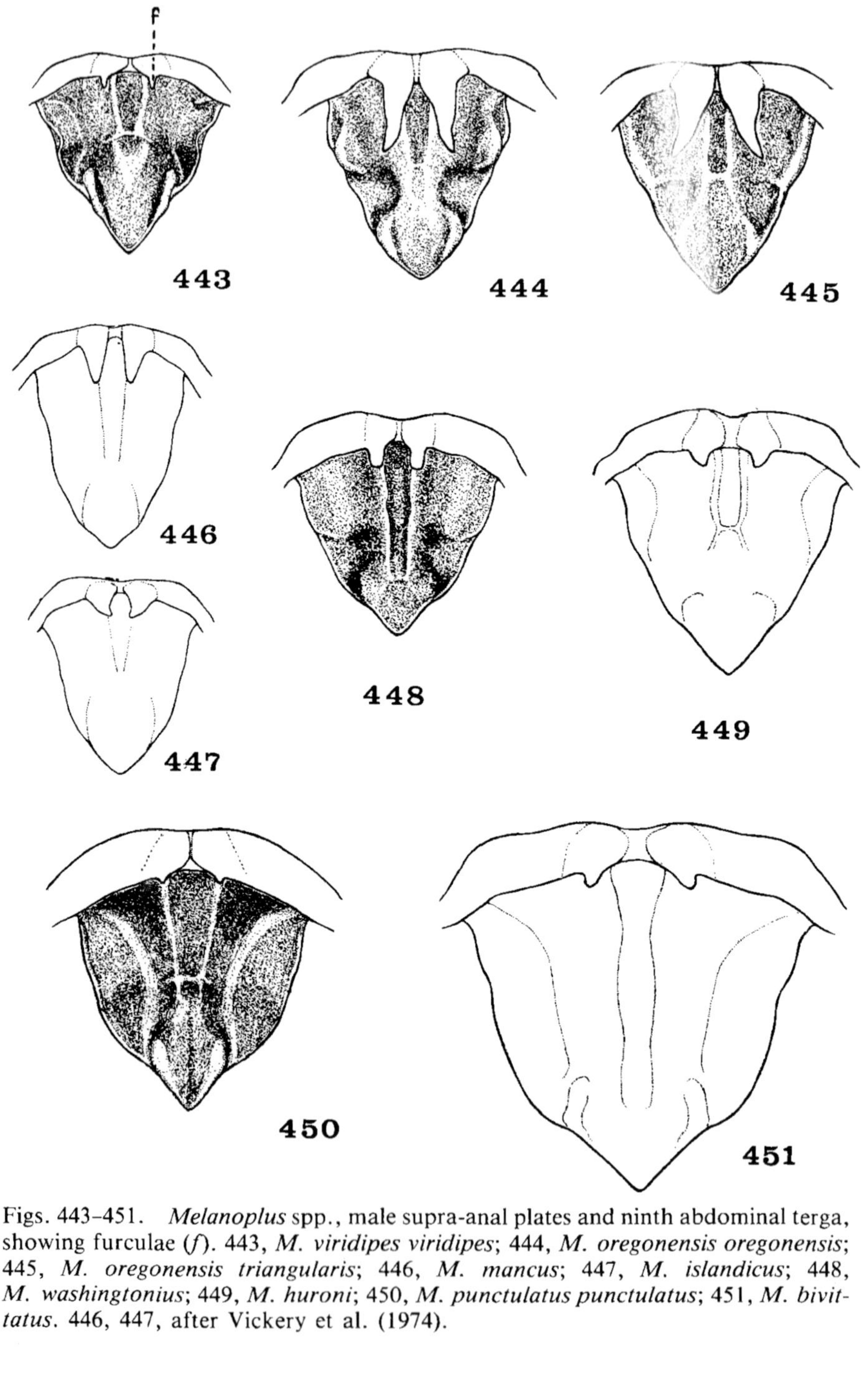

Figs. 443–451. *Melanoplus* spp., male supra-anal plates and ninth abdominal terga, showing furculae (*f*). 443, *M. viridipes viridipes*; 444, *M. oregonensis oregonensis*; 445, *M. oregonensis triangularis*; 446, *M. mancus*; 447, *M. islandicus*; 448, *M. washingtonius*; 449, *M. huroni*; 450, *M. punctulatus punctulatus*; 451, *M. bivittatus*. 446, 447, after Vickery et al. (1974).

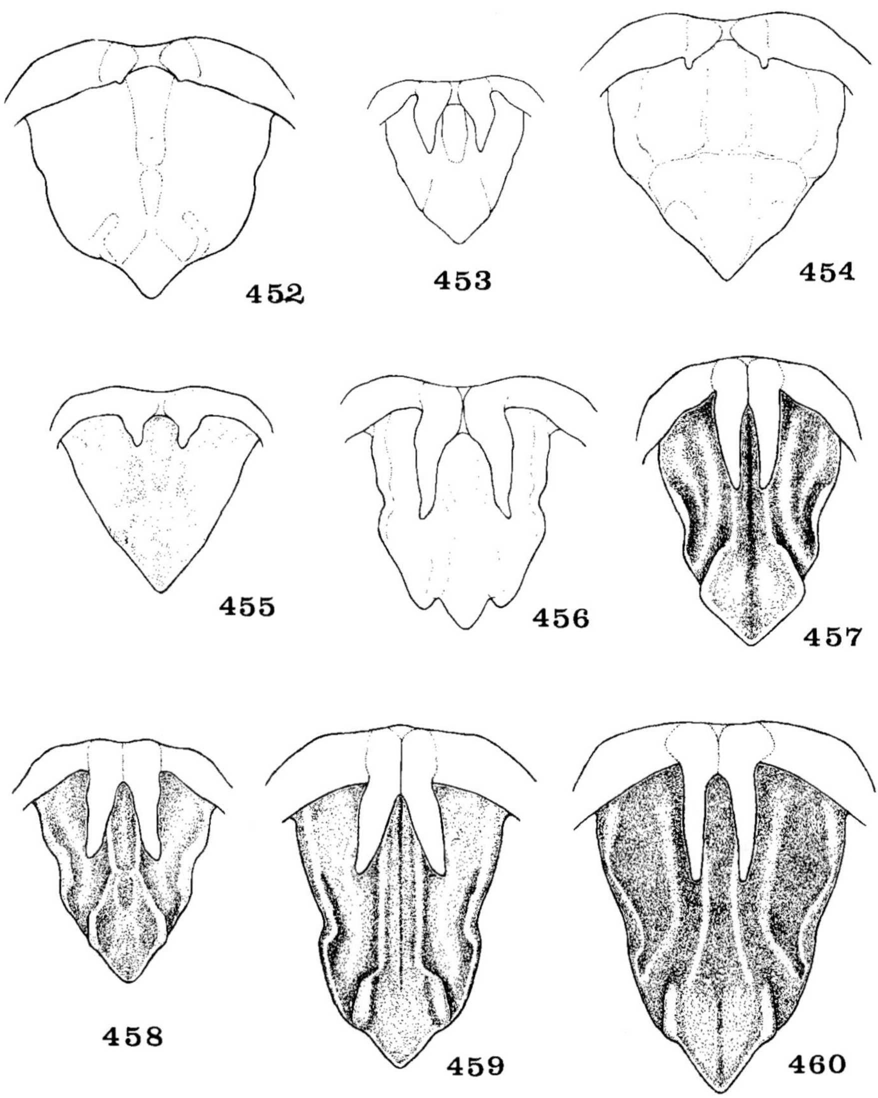

Figs. 452–460. *Melanoplus* spp., male supra-anal plates and ninth abdominal terga, showing furculae. 452, *M. differentialis differentialis*; 453, *M. dawsoni*; 454, *M. gladstoni*; 455, *M. confusus*; 456, *M. femurrubrum femurrubrum*; 457, *M. gordonae*; 458, *M. borealis borealis*; 459, *M. gaspesiensis*; 460, *M. madeleineae*. 453, after Vickery et al. (1974).

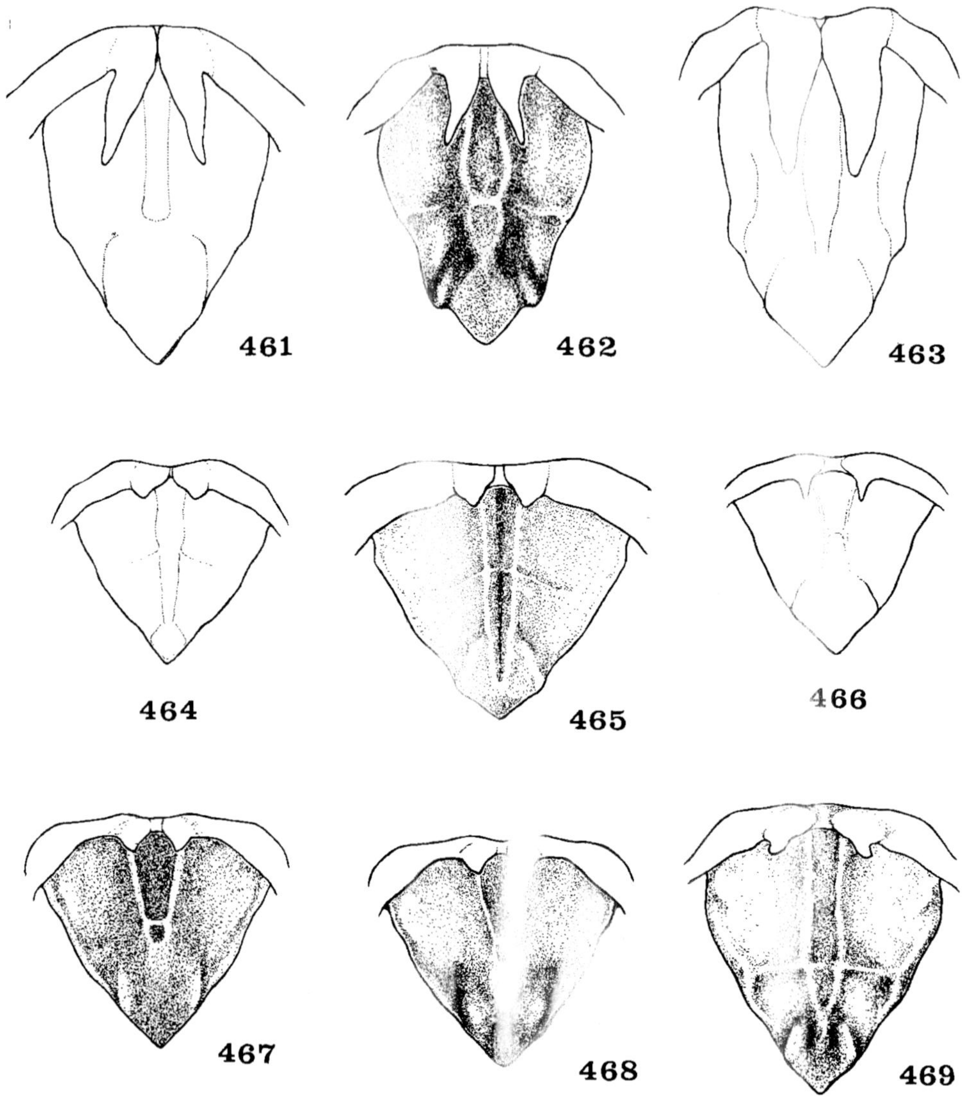

Figs. 461–469. *Melanoplus* spp., male supra-anal plates and ninth abdominal terga, showing furculae. 461, *M. sanguinipes sanguinipes*; 462, *M. spretus*; 463, *M. bruneri*; 464, *M. infantilis*; 465, *M. alpinus*; 466, *M. kennicottii kennicottii*; 467, *M. occidentalis occidentalis*; 468, *M. rugglesi*; 469, *M. fasciatus*. 461, 463, after Vickery et al. (1974).

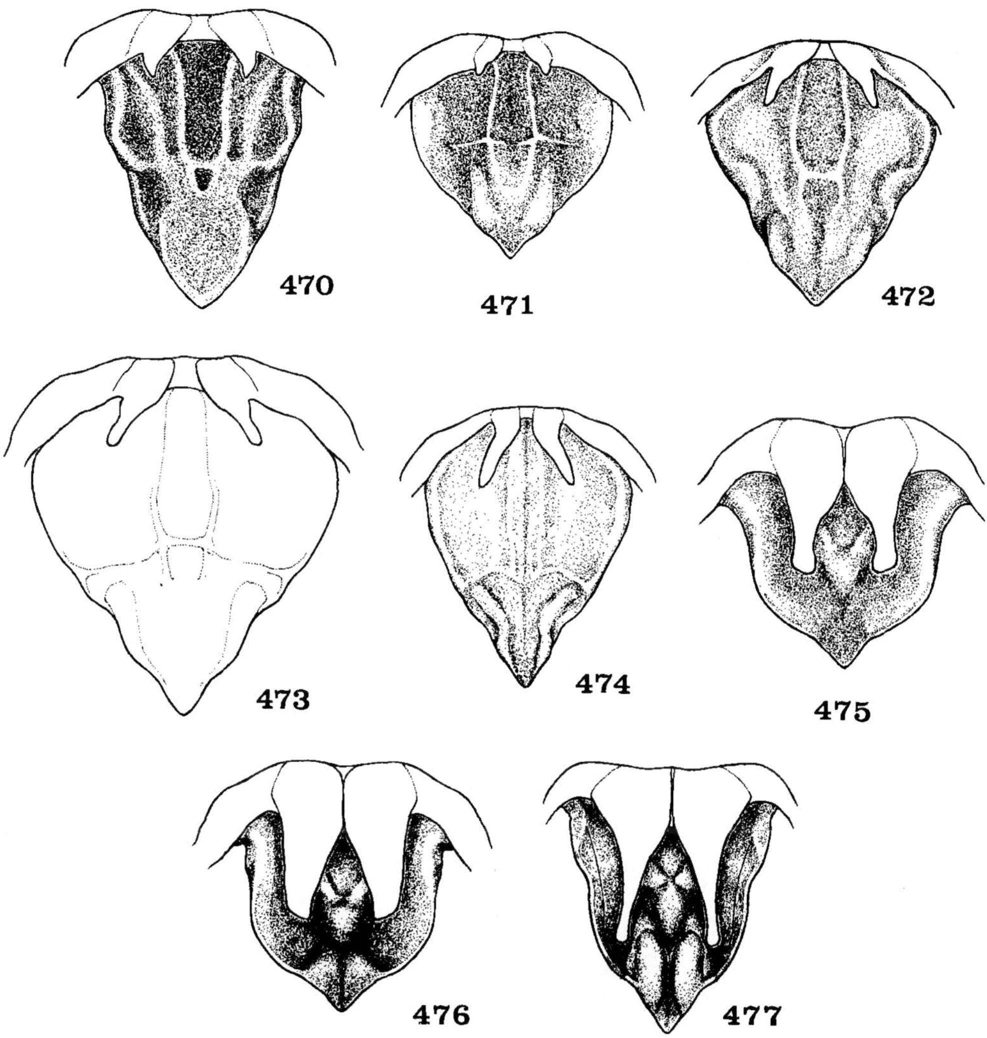

Figs. 470–477. *Bohemanella* and *Melanoplus* spp., male supra-anal plates and ninth abdominal terga, showing furculae. 470, *B. frigida frigida*; 471, *M. keeleri luridus*; 472, *M. stonei*; 473, *M. packardii packardii*; 474, *M. angustipennis angustipennis*; 475, *M. flavidus*; 476, *M. bowditchi canus*; 477, *M. cinereus cinereus*.

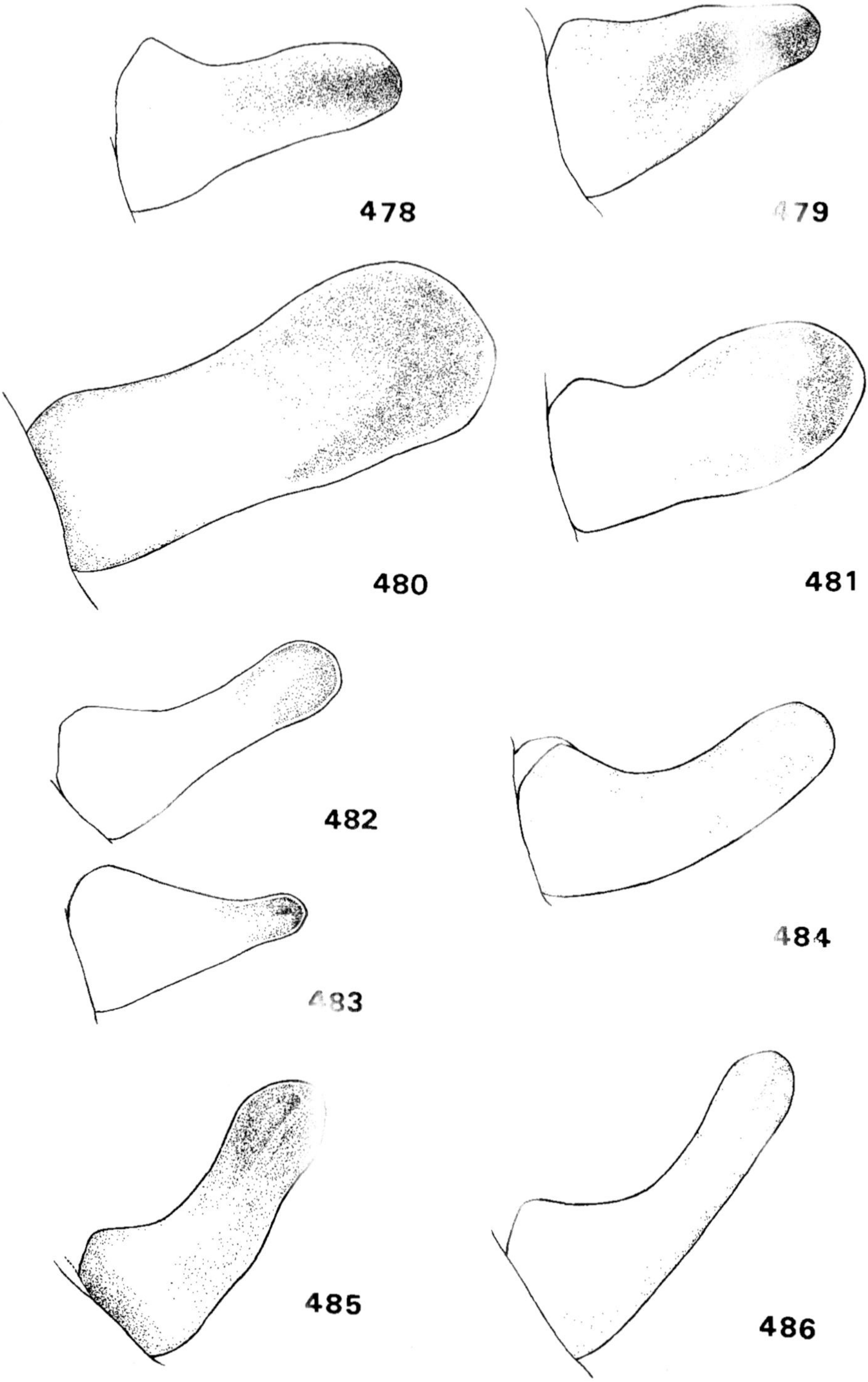

Figs. 478–486. *Melanoplus* spp., male cerci, from left. 478, *M. viridipes viridipes*; 479, *M. viridipes eurycercus*; 480, *M. oregonensis oregonensis*; 481, *M. oregonensis triangularis*; 482, *M. mancus*; 483, *M. islandicus*; 484, *M. montanus*; 485, *M. washingtonius*; 486, *M. huroni*. 482, 483, after Vickery et al. (1974).

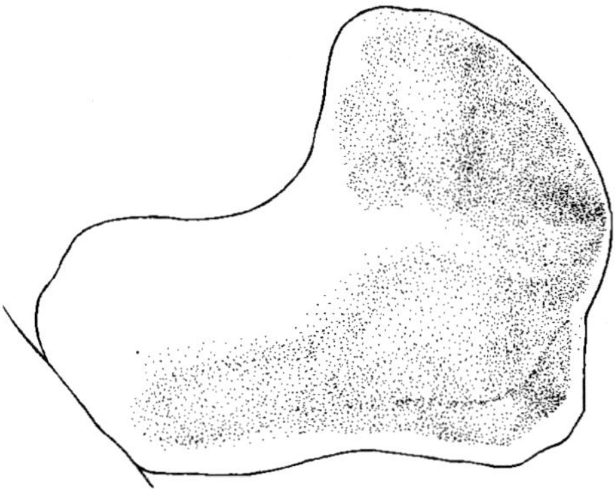

487

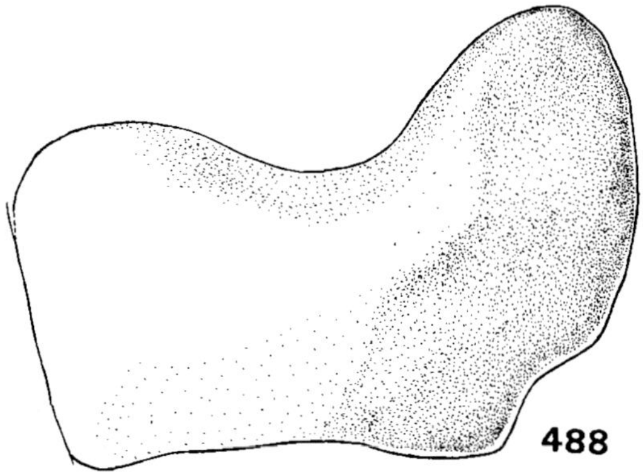

488

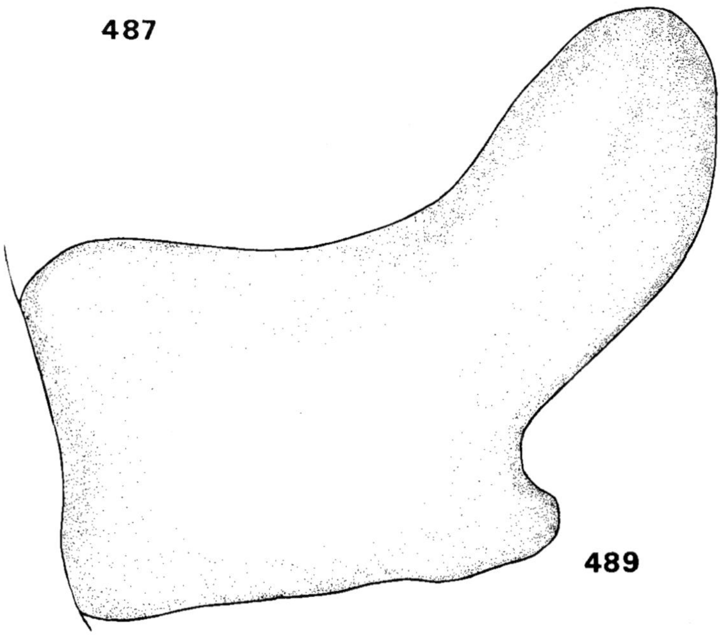

489

Figs. 487–489. *Melanoplus* spp., male cerci, from left. 487, *M. punctulatus punctulatus*; 488, *M. bivittatus*; 489, *M. differentialis differentialis*. 487, 488, after Vickery et al. (1974).

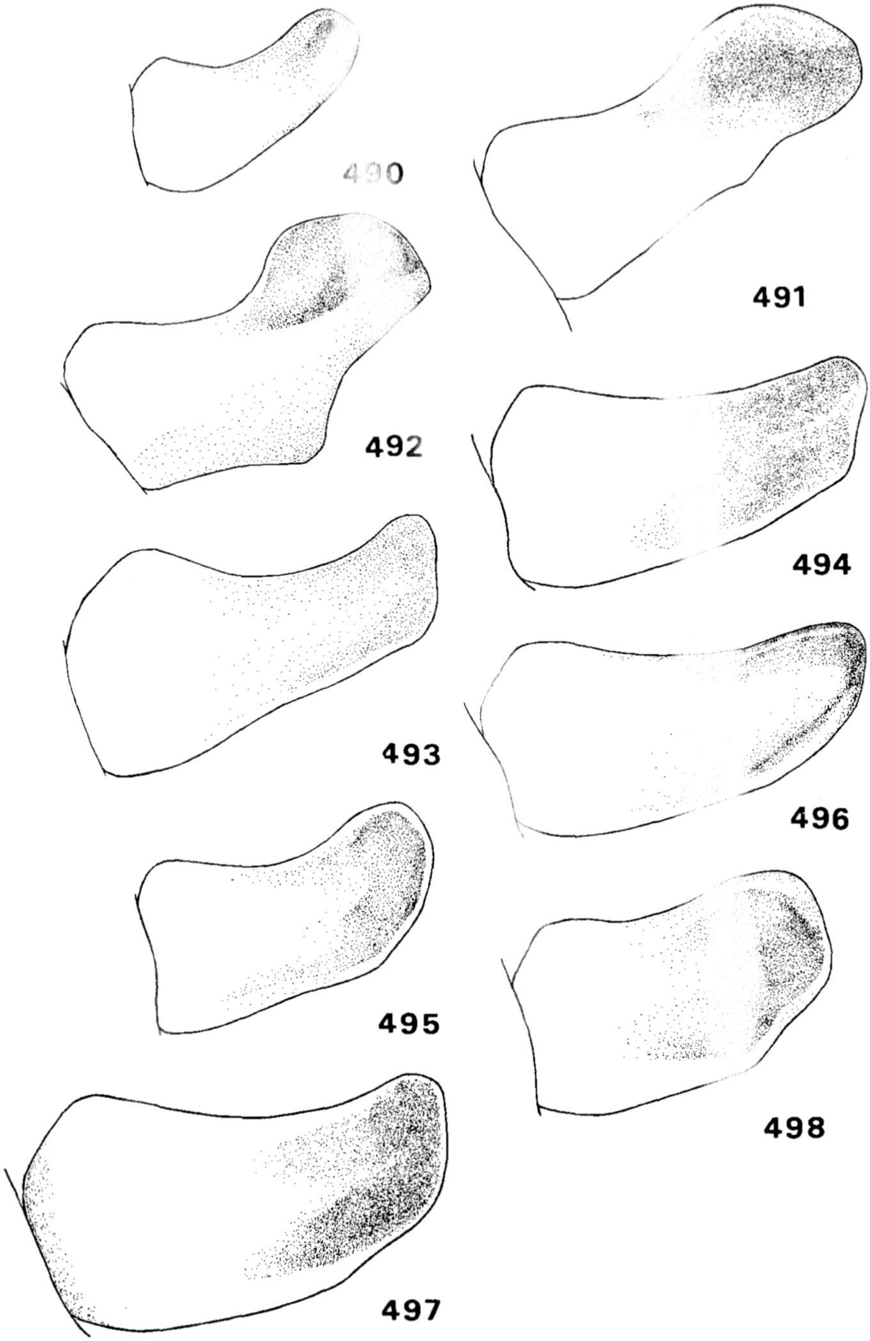

Figs. 490–498. *Melanoplus* spp., male cerci, from left. 490, *M. dawsoni*; 491, *M. gladstoni*; 492, *M. confusus*; 493, *M. femurrubrum femurrubrum*; 494, *M. gordonae*; 495, *M. borealis borealis*; 496, *M. gaspesiensis*; 497, *M. madeleineae*; 498, *M. sanguinipes sanguinipes*. 490, 492, 493, 495, 496, 498, after Vickery et al. (1974).

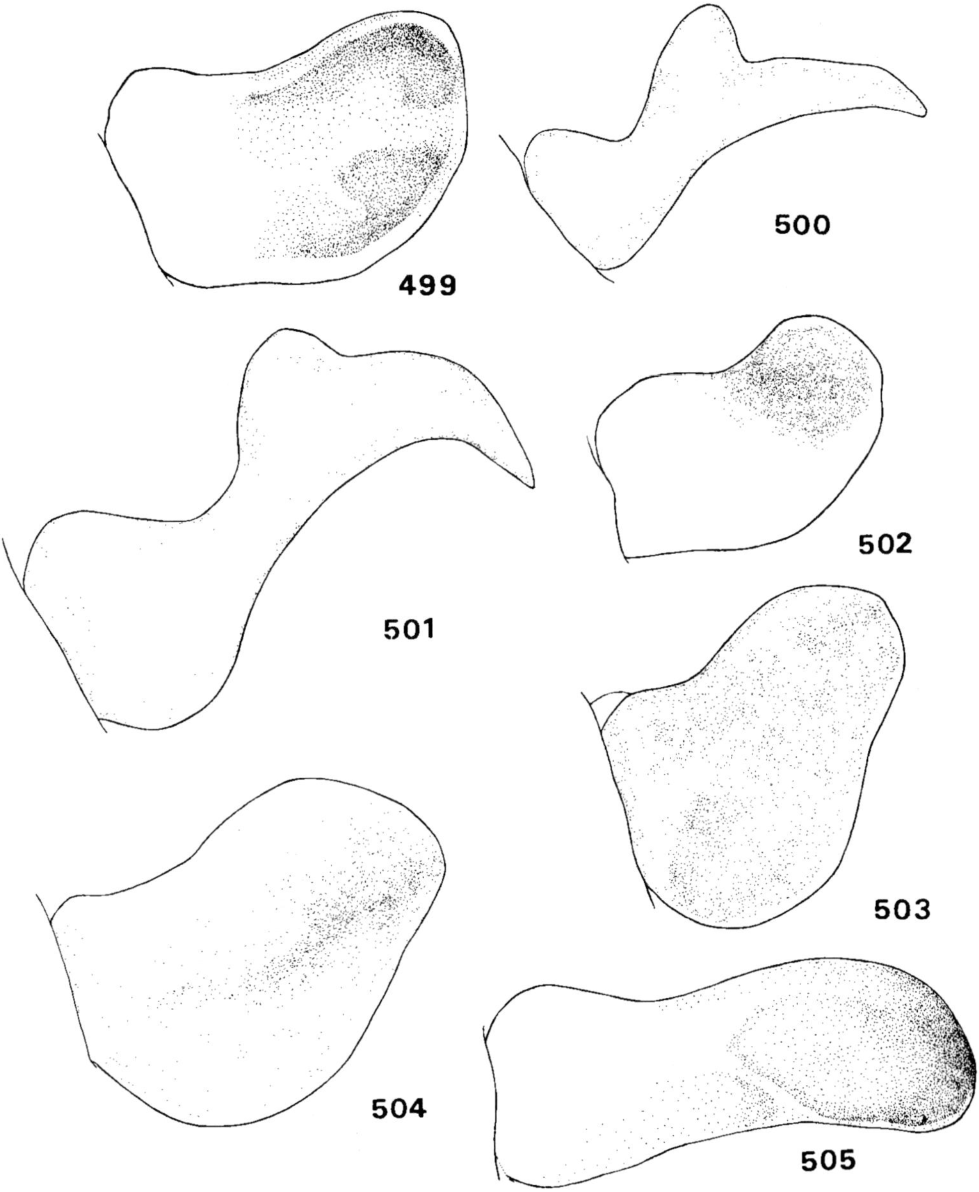

Figs. 499–505. *Melanoplus* spp., male cerci, from left. 499, *M. bruneri*; 500, *M. infantilis*; 501, *M. alpinus*; 502, *M. kennicottii kennicottii*; 503, *M. occidentalis occidentalis*; 504, *M. rugglesi*; 505, *M. fasciatus*. 499, 505, after Vickery et al. (1974).

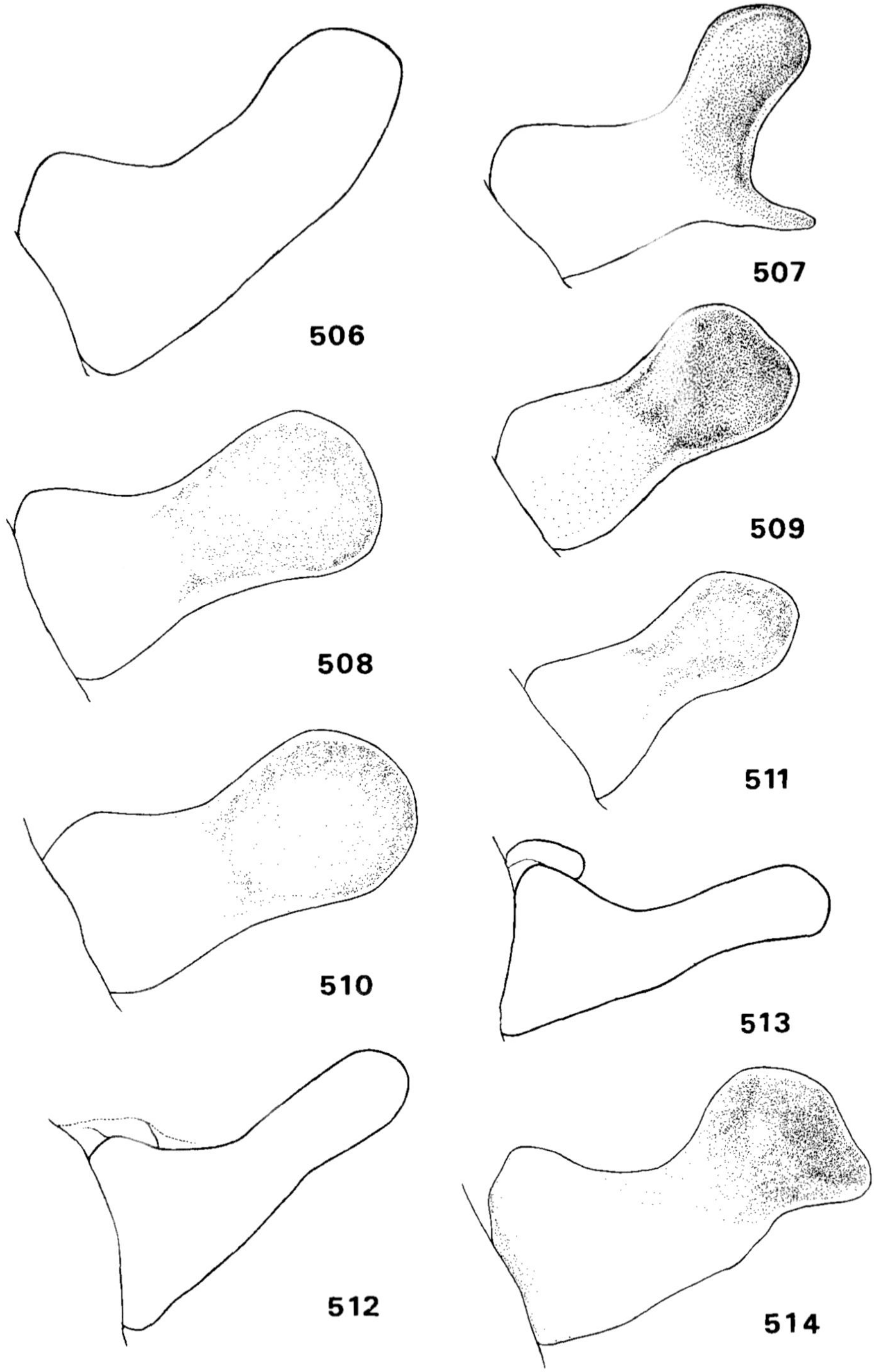

Figs. 506–514. *Bohemanella* and *Melanoplus* spp., male cerci, from left. 506, *B. frigida frigida*; 507, *M. keeleri luridus*; 508, *M. foedus foedus*; 509, *M. stonei*; 510, *M. packardii packardii*; 511, *M. angustipennis angustipennis*; 512, *M. flavidus*; 513, *M. bowditchi canus*; 514, *M. cinereus cinereus*. 507, 509, after Vickery et al. (1974).

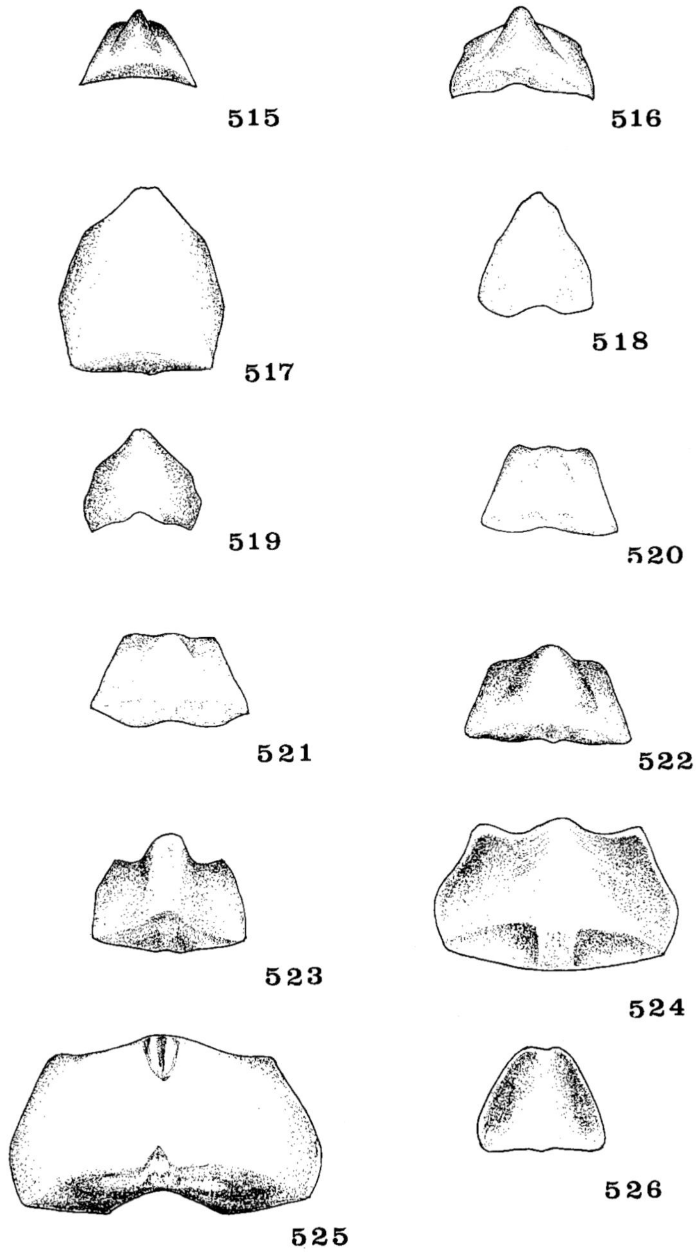

Figs. 515–526. *Melanoplus* spp., male subgenital plates, from rear. 515, *M. viridipes viridipes*; 516, *M. viridipes eurycercus*; 517, *M. oregonensis oregonensis*; 518, *M. mancus*; 519, *M. islandicus*; 520, *M. montanus*; 521, *M. washingtonius*; 522, *M. huroni*; 523, *M. punctulatus punctulatus*; 524, *M. bivittatus*; 525, *M. differentialis differentialis*; 526, *M. dawsoni*.

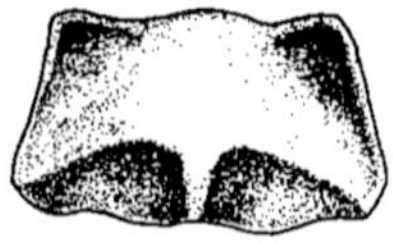

527

528

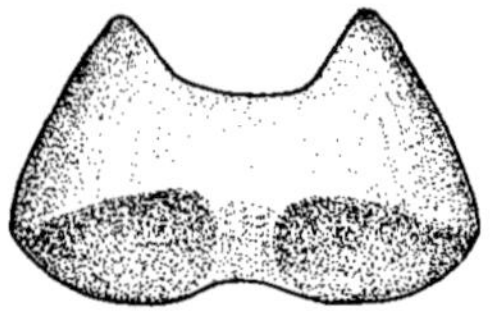

529

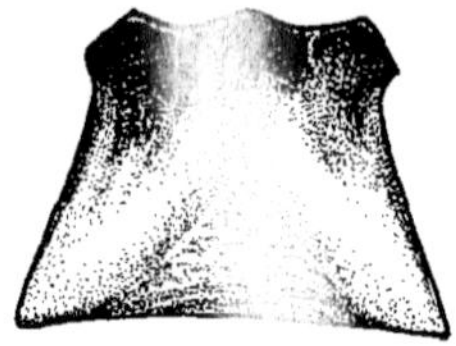

530

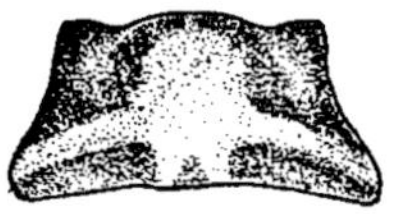

531

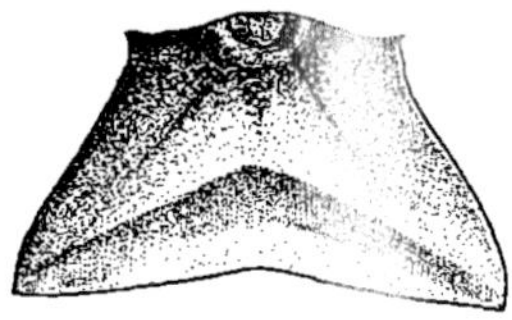

532

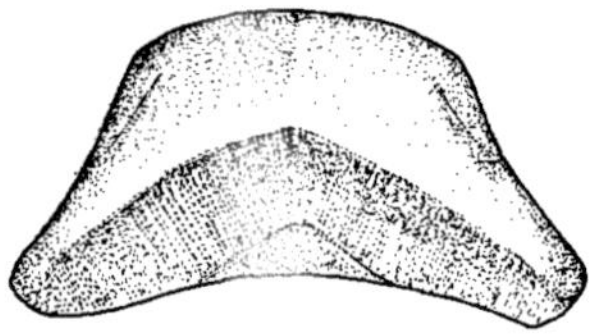

533

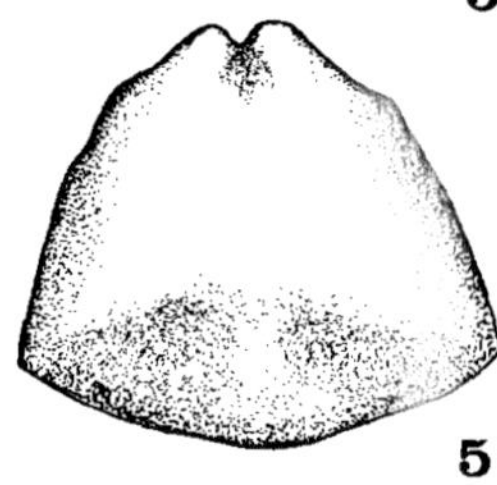

534

535

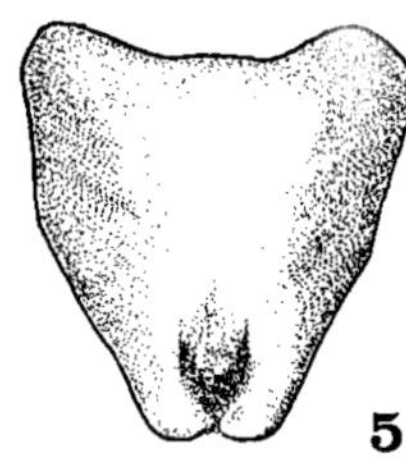

536

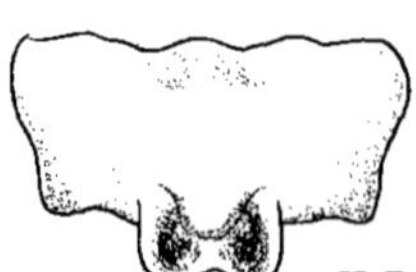

537

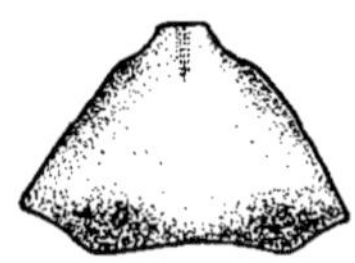

538

Figs. 527–538. *Melanoplus* spp., male subgenital plates, from rear. 527, *M. gladstoni*; 528, *M. confusus*; 529, *M. femurrubrum femurrubrum*; 530, *M. gordonae*; 531, *M. borealis borealis*; 532, *M. gaspesiensis*; 533, *M. madeleineae*; 534, *M. sanguinipes sanguinipes*; 535, *M. bruneri*; 536, *M. infantilis*; 537, *M. alpinus*; 538, *M. kennicottii kennicottii*.

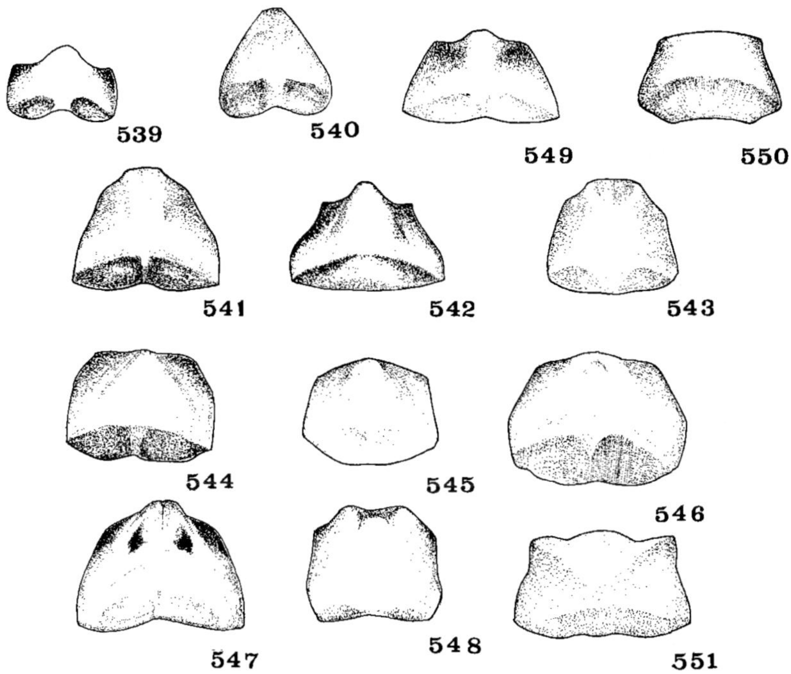

Figs. 539–551. *Melanoplus* and *Bohemanella* spp., male subgenital plates, from rear. 539, *M. occidentalis occidentalis*; 540, *M. rugglesi*; 541, *M. fasciatus*; 542, *B. frigida frigida*; 543, *M. keeleri luridus*; 544, *M. foedus foedus*; 545, *M. stonei*; 546, *M. packardii packardii*; 547, *M. packardii brooksi*; 548, *M. angustipennis angustipennis*; 549, *M. flavidus*; 550, *M. bowditchi canus*; 551, *M. cinereus cinereus*.

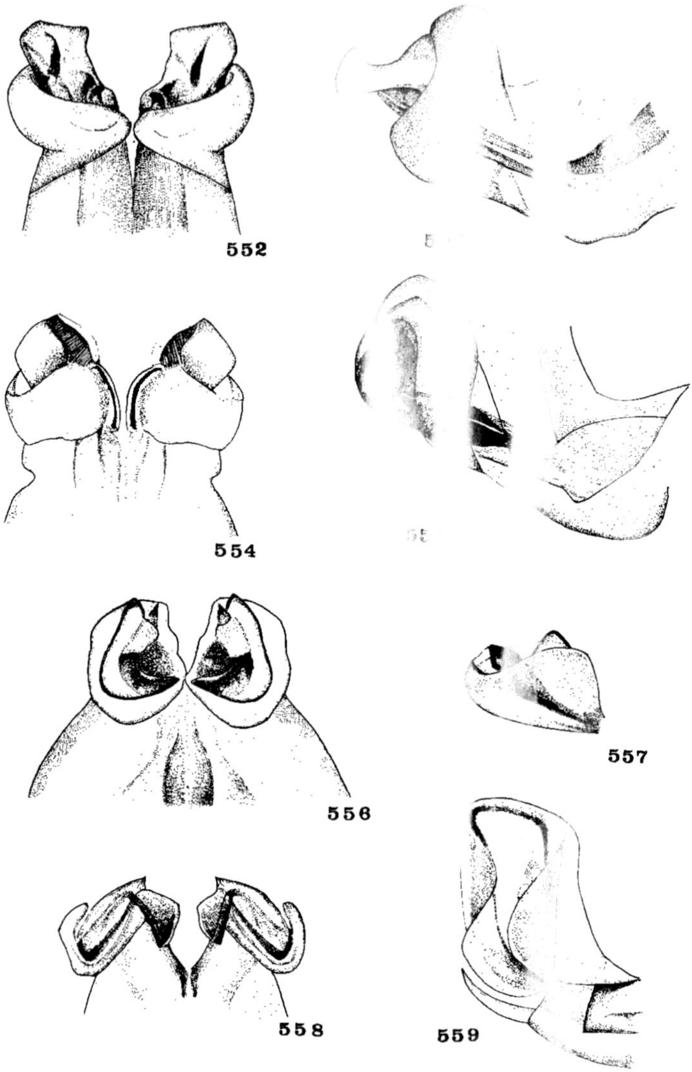

Figs. 552–559. *Melanoplus* spp., aedeagi. [illegible] 553, *M. fo[illegible]dus foedus*; 554, 555, *M. packardii packardii*; 556, 557, *M. flavid*[illegible] 558, 559, *M.* [illegible]*owditchi canus*. 552, 554, 556, 558, dorsal; 553, 555, 557 (apex[illegible] [illegible]9, from right.

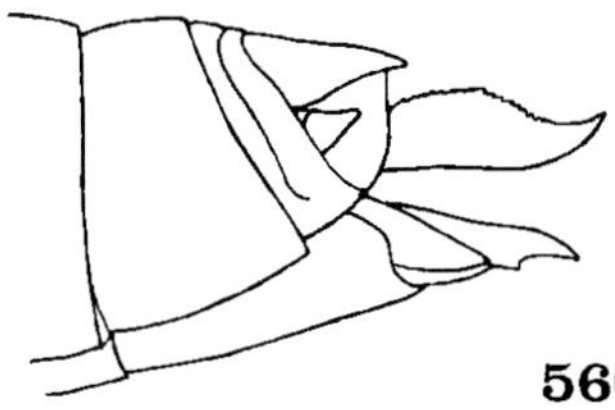

560

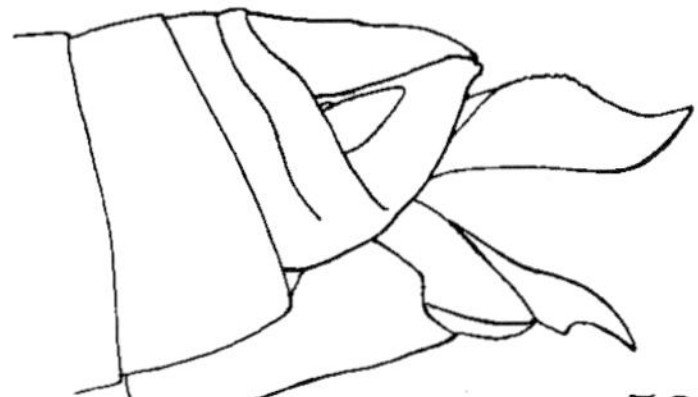

561

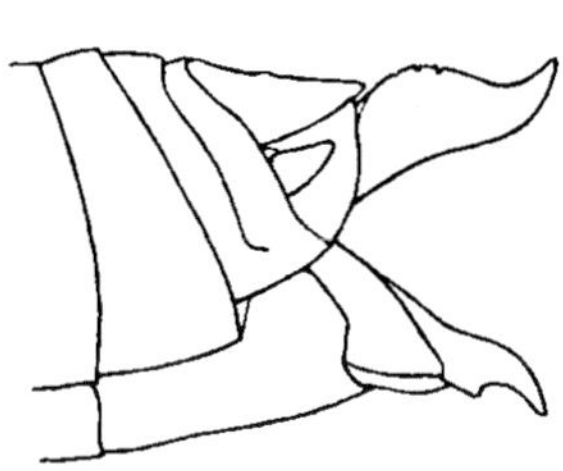

562

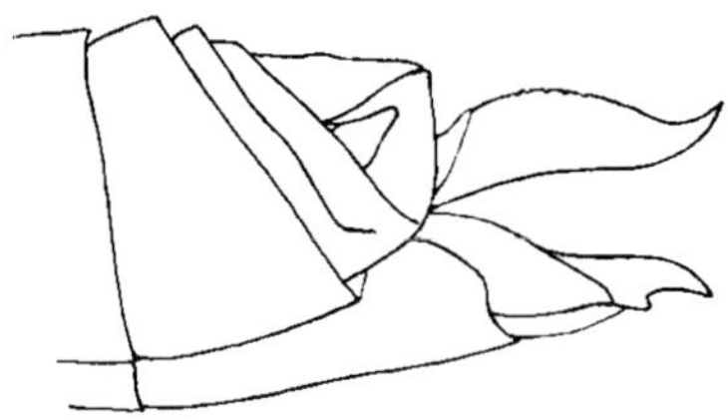

563

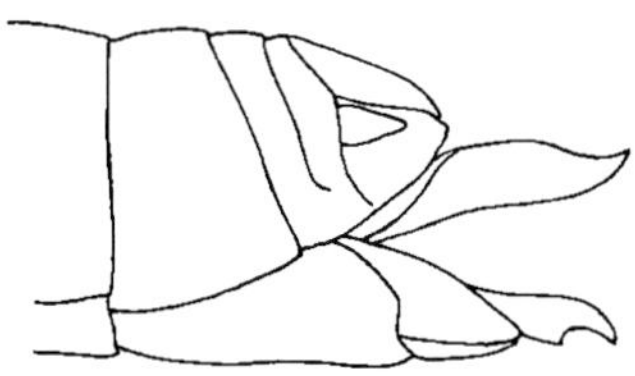

564

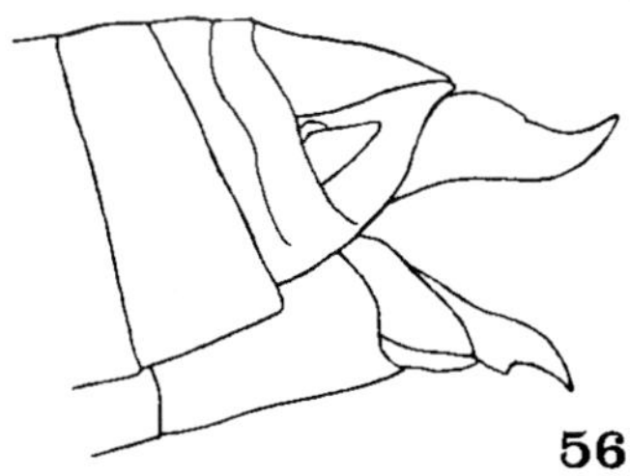

565

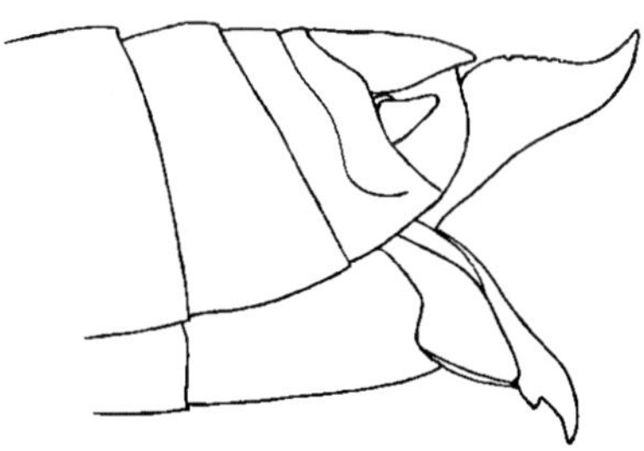

566

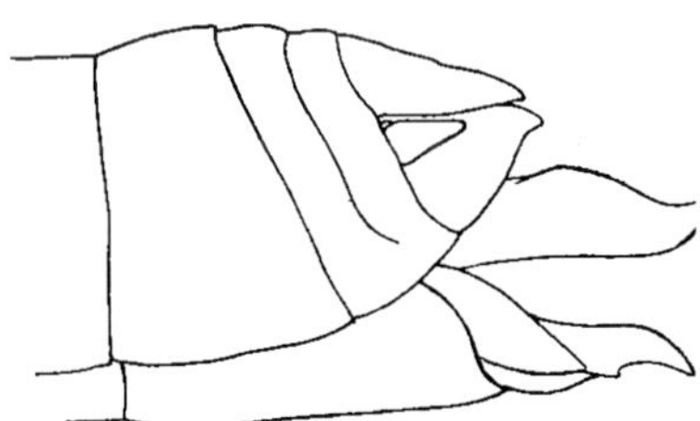

567

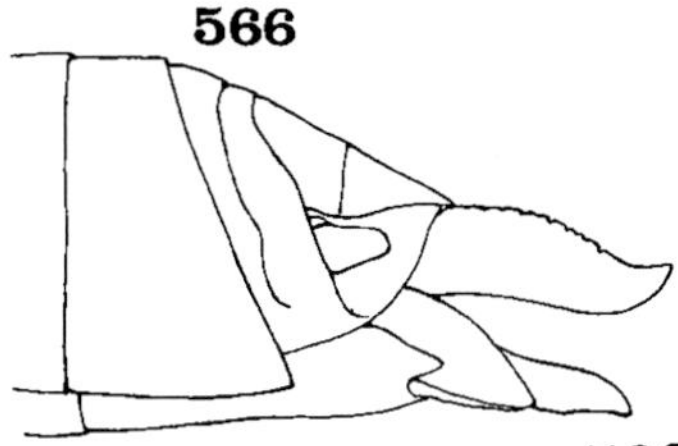

568

Figs. 560–568. *Melanoplus* spp., female abdominal terminalia, showing ovipositors and cerci. 560, *M. viridipes eurycercus*; 561, *M. oregonensis oregonensis*; 562, *M. oregonensis triangularis*; 563, *M. mancus*; 564, *M. islandicus*; 565, *M. montanus*; 566, *M. washingtonius*; 567, *M. huroni*; 568, *M. punctulatus punctulatus*. Fig. 564 after Vickery et al. (1974).

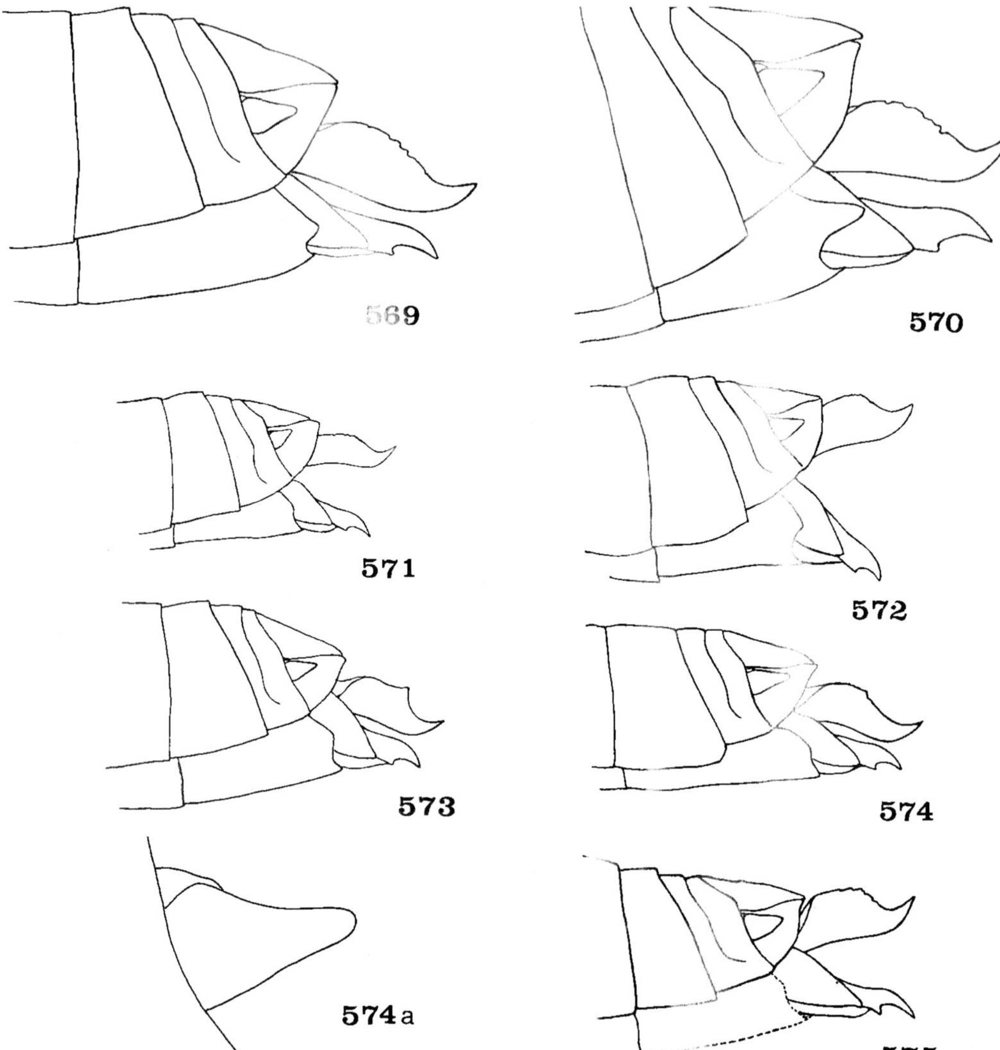

Figs. 569–575. *Melanoplus* spp., female abdominal terminalia, showing ovipositors and cerci. 569, *M. bivittatus*; 570, *M. differentialis differentialis*; 571, *M. dawsoni*; 572, *M. gladstoni*; 573, *M. confusus*; 574, *M. femurrubrum femurrubrum*; 574*a*, the same, left cercus enlarged; 575, *M. gordonae*. 574*a*, after Vickery et al. (1974).

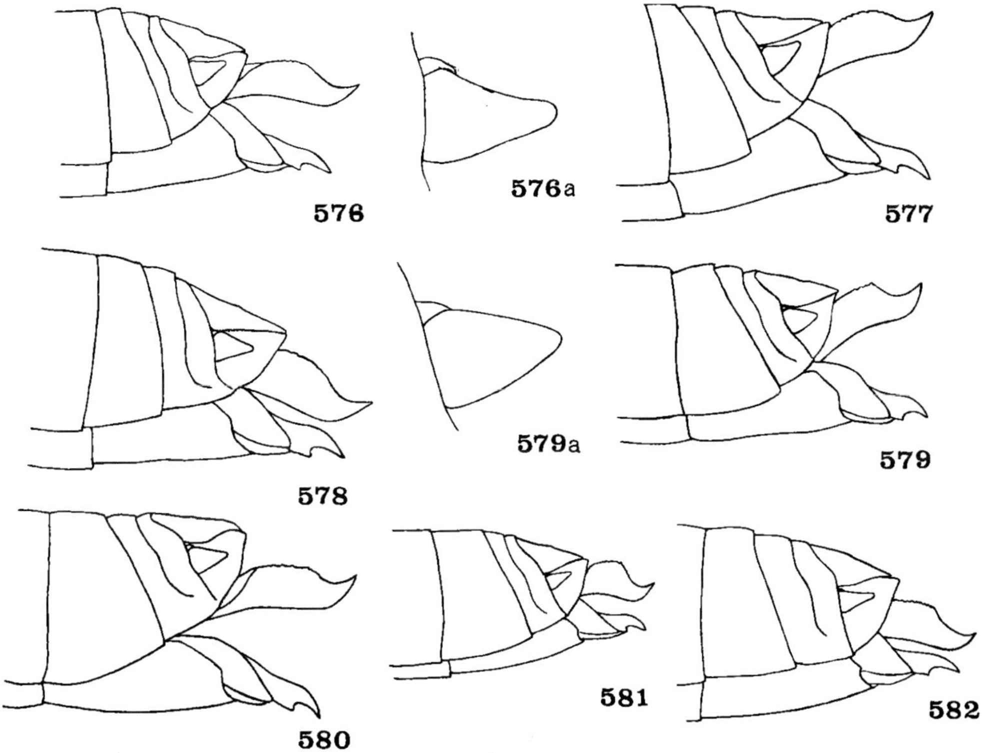

Figs. 576–582. *Melanoplus* spp., female abdominal terminalia, showing ovipositors and cerci. 576, *M. borealis borealis*; 576*a*, the same, left cercus enlarged; 577, *M. gaspesiensis*; 578, *M. madeleineae*; 579, *M. sanguinipes sanguinipes*; 579*a*, the same, left cercus enlarged; 580, *M. bruneri*; 581, *M. infantilis*; 582, *M. alpinus*. 576*a*, 579*a*, after Vickery et al. (1974).

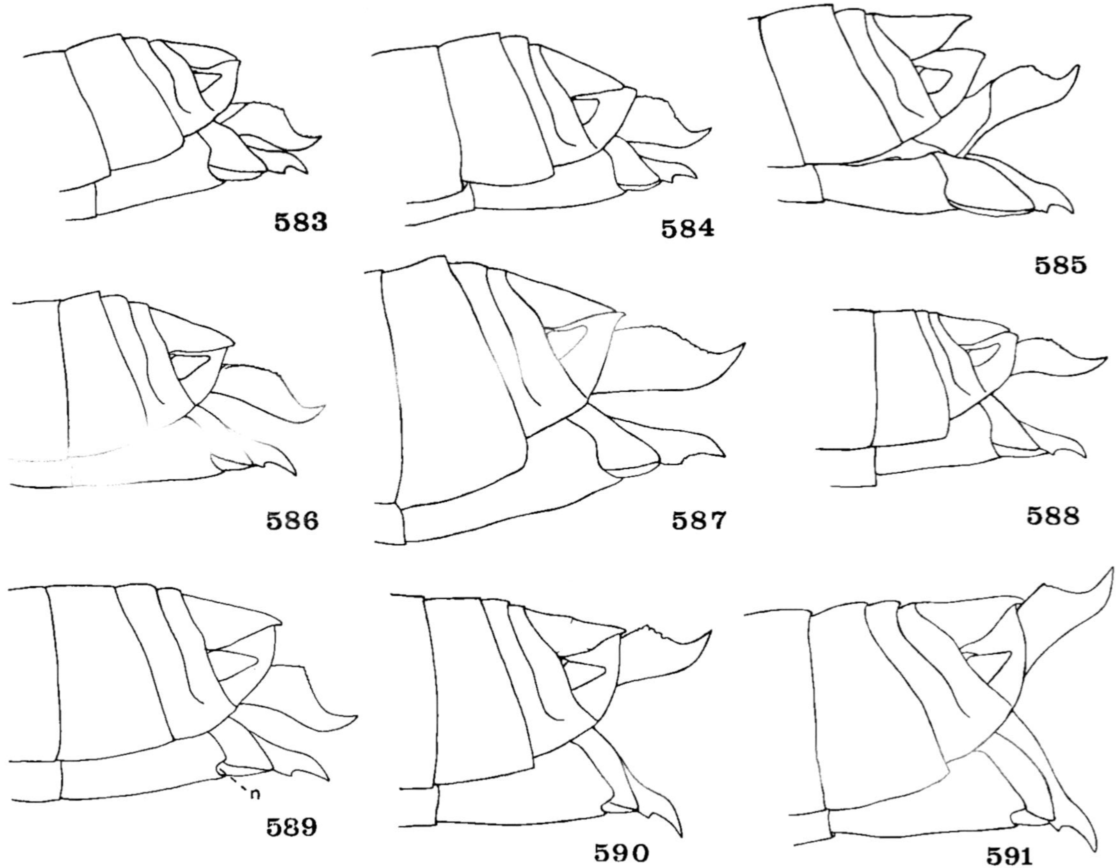

Figs. 583–591. *Melanoplus* and *Bohemanella* spp., female abdominal terminalia, showing ovipositors and cerci. 583, *M. kennicottii kennicottii*; 584, *M. occidentalis occidentalis*; 585, *M. rugglesi*; 586, *M. fasciatus*; 587, *B. frigida frigida*; 588, *M. kennicotti luridus*; 589, *M. foedus foedus* (*n* = deep lateral notch of eighth abdominal sternum, cf. also Figs. 590, 591); 590, *M. stonei*; 591, *M. packardii packardii*.

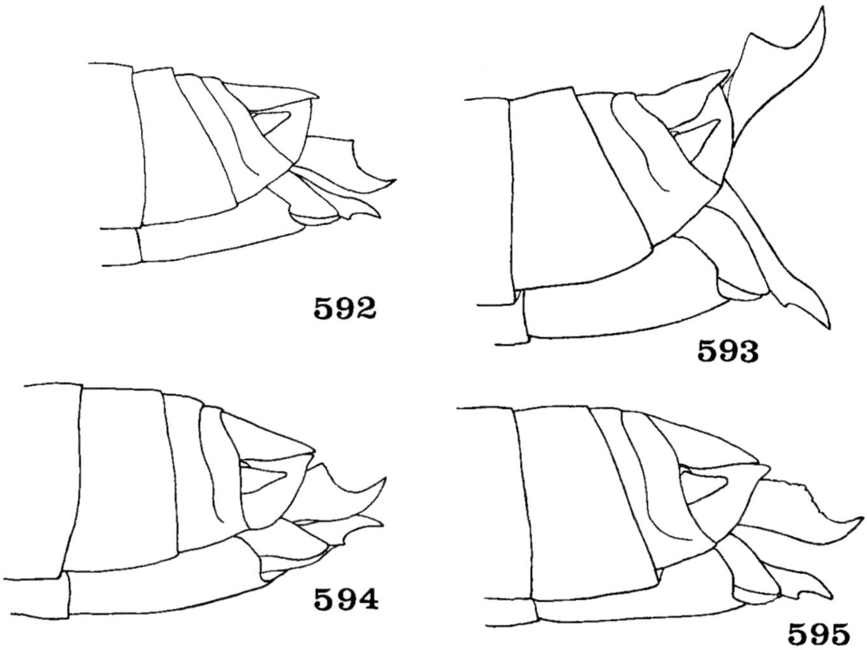

Figs. 592–595. *Melanoplus* spp., female abdominal terminalia, showing ovipositors and cerci. 592, *M. angustipennis angustipennis*; 593, *M. flavidus*; 594, *M. bowditchi canus*; 595, *M. cinereus cinereus*.

Figs. 596–599. *Melanoplus borealis borealis* (female syntypes of *Caloptenus extremus*). 596, 597, lectotype, dorsal and lateral; 598, 599, paralectotype, dorsal and lateral. Photos courtesy of British Museum (Natural History), London.

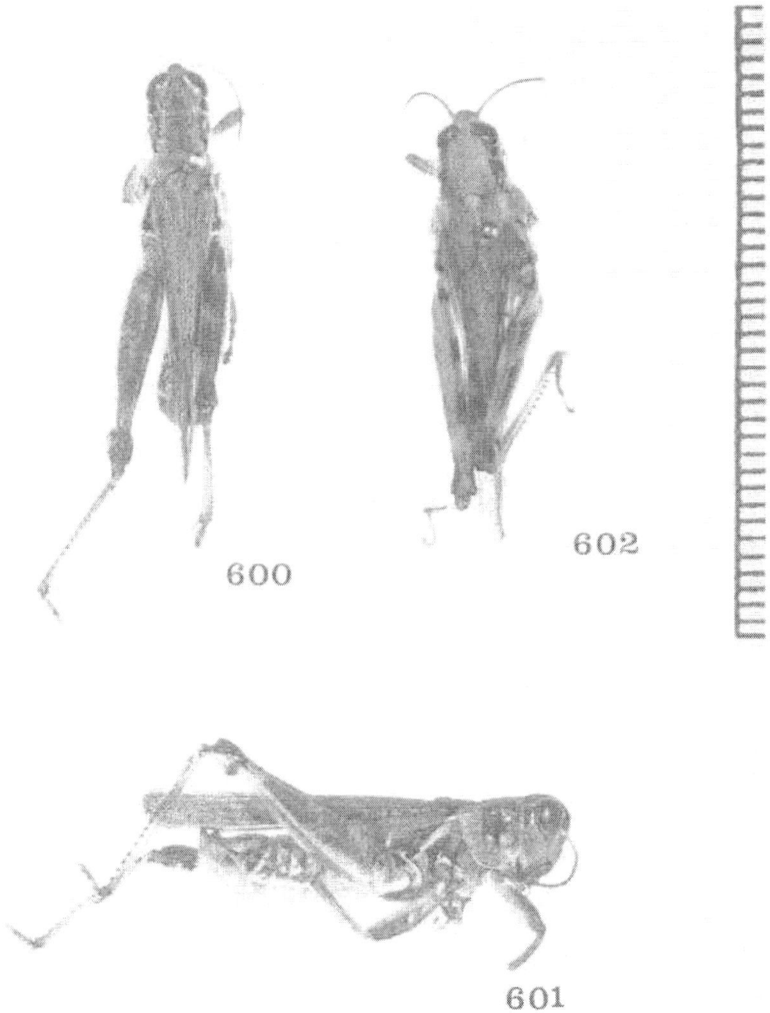

Figs. 600–602. *Melanoplus sanguinipes* (syntypes of *Caloptenus bilituratus*). 600, 601, male lectotype, dorsal and lateral; 602, female paralectotype, dorsal. Photos courtesy of British Museum (Natural History), London.

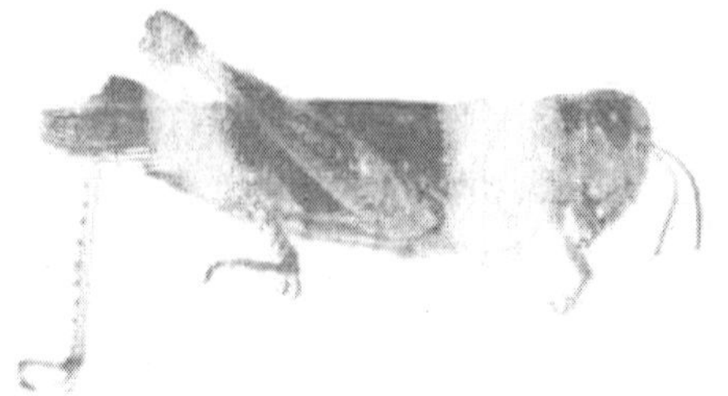

603

604

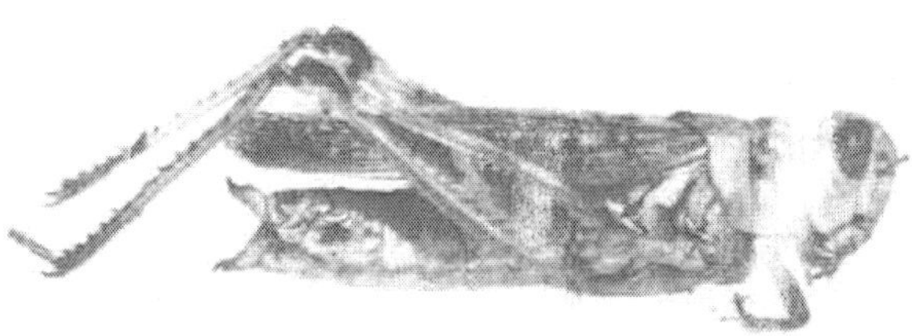

605

Figs. 603–605. *Melanoplus sanguinipes*. 603, *Caloptenus bilituratus*, female paralectotype, lateral; 604, *Caloptenus scriptus*, female lectotype, lateral; 605, the same, female paralectotype. Photos courtesy of British Museum (Natural History), London.

606

607

608

609

Figs. 606–609. *Melanoplus sanguinipes*. 606, *Caloptenus scriptus*, female lectotype, dorsal; 607, the same, female paralectotype; 608, 609, *Caloptenus arcticus*, female holotype, dorsal and lateral. Photos courtesy of British Museum (Natural History), London.

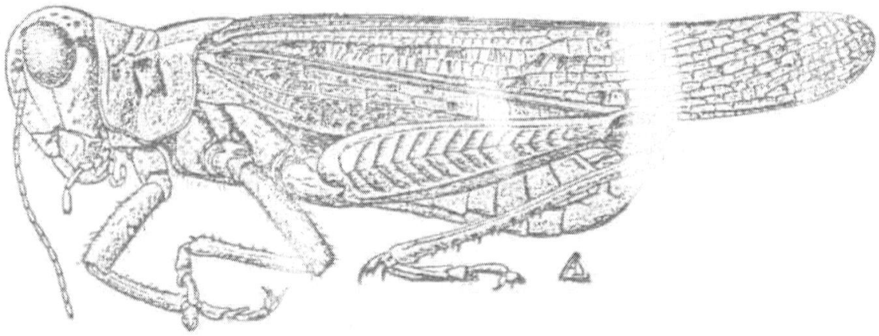

Fig. 610. *Melanoplus spretus*, Rocky Mountain grasshopper, male. After Gurney and Brooks (1959).

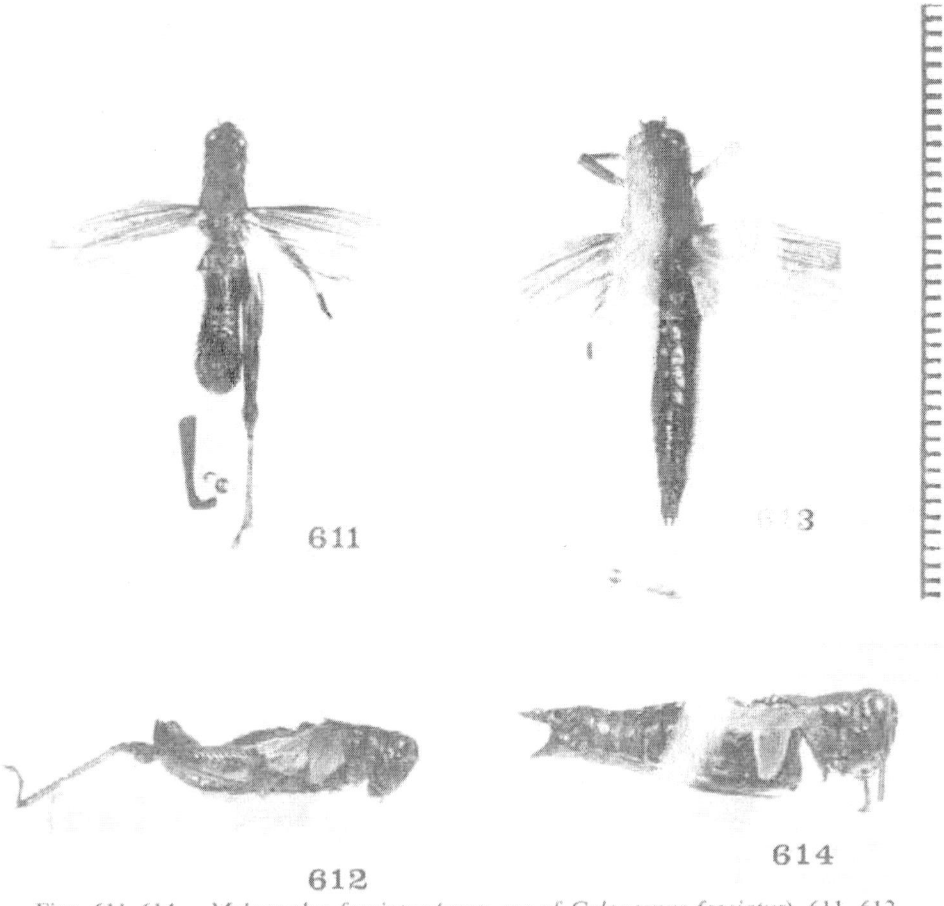

Figs. 611–614. *Melanoplus fasciatus* (syntypes of *Caloptenus fasciatus*). 611, 612, male lectotype, dorsal and lateral; 613, 614, female paralectotype, dorsal and lateral. Photos courtesy of British Museum (Natural History), London.

615

616

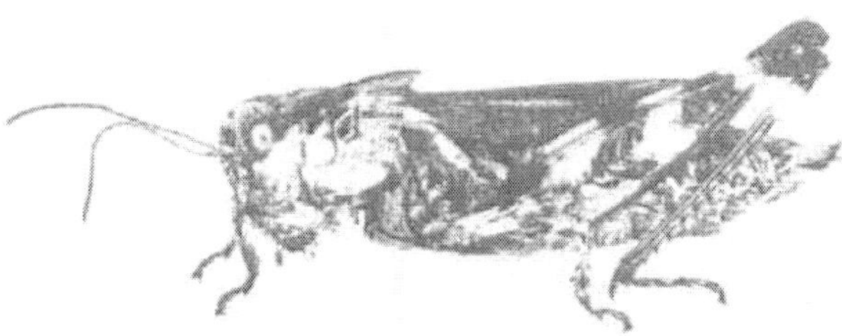

617

Figs. 615–617. *Melanoplus fasciatus*. 615, male of normal color form; 616, female of the same; 617, female of unusual noetenic color form. These are the same specimens as illustrated by Johnstone (1973).

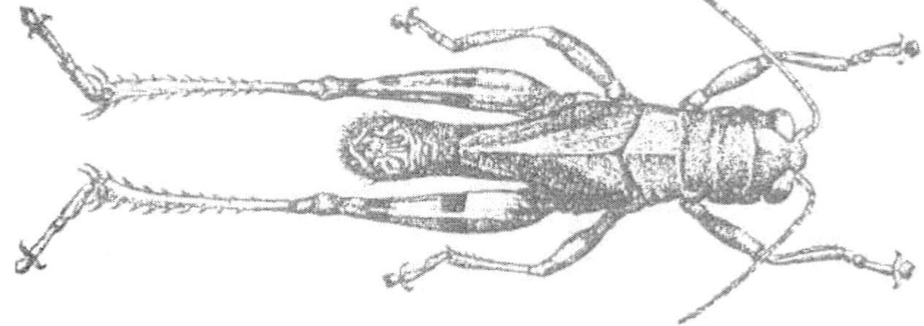

Fig. 618. *Bohemanella frigida frigida*. After Mishchenko (1951).

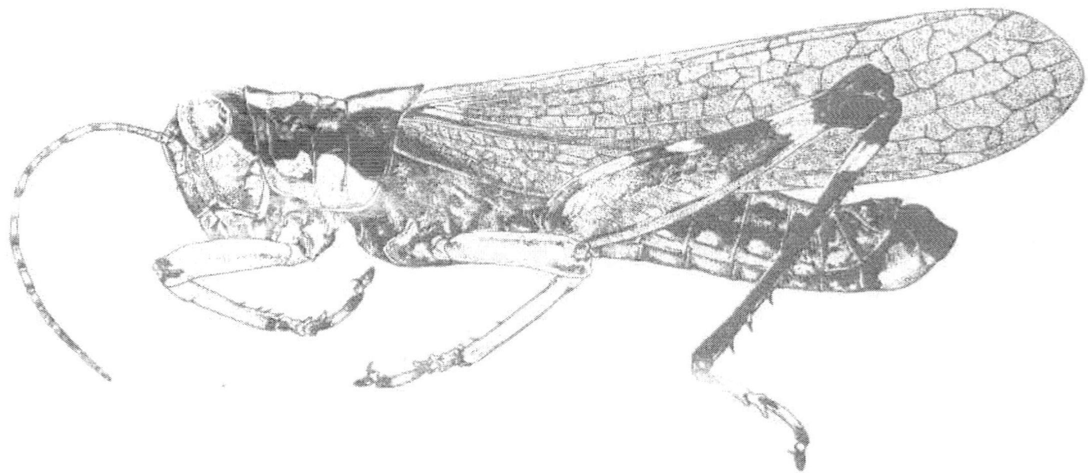

Fig. 619. *Dendrotettix quercus*, macropterous male.

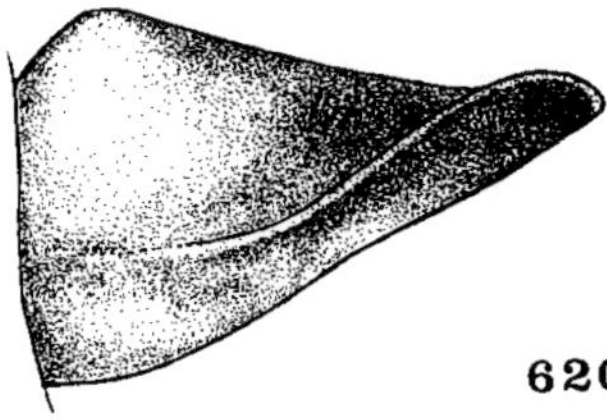

620

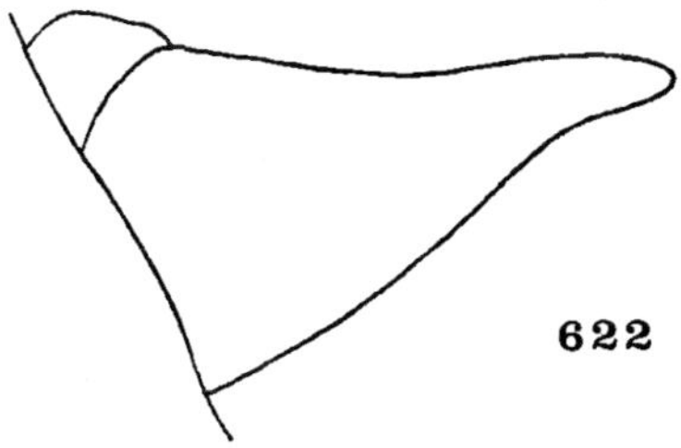

622

621

Figs. 620–622. *Dendrotettix* and *Appalachia* spp. 620, *D. quercus*, left male cercus, lateral; 621, *A. arcana*, left male cercus, lateral; 622, *A. arcana*, female abdominal terminalia, ventral.

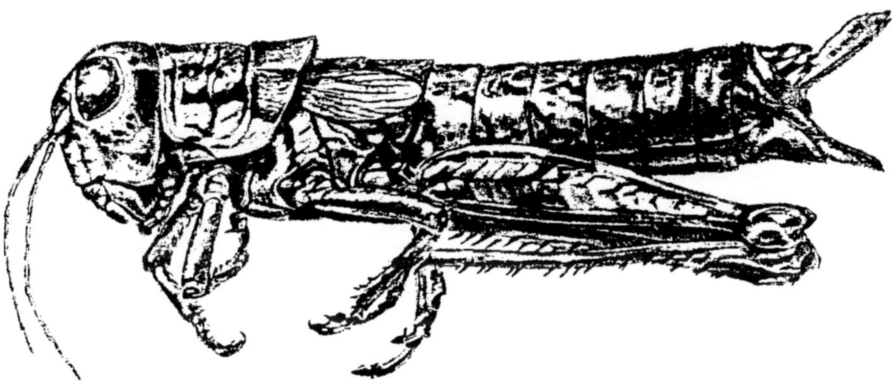

Fig. 623. *Appalachia arcana*, female allotype. After G. Eager *in* Hubbell and Cantrall (1938).

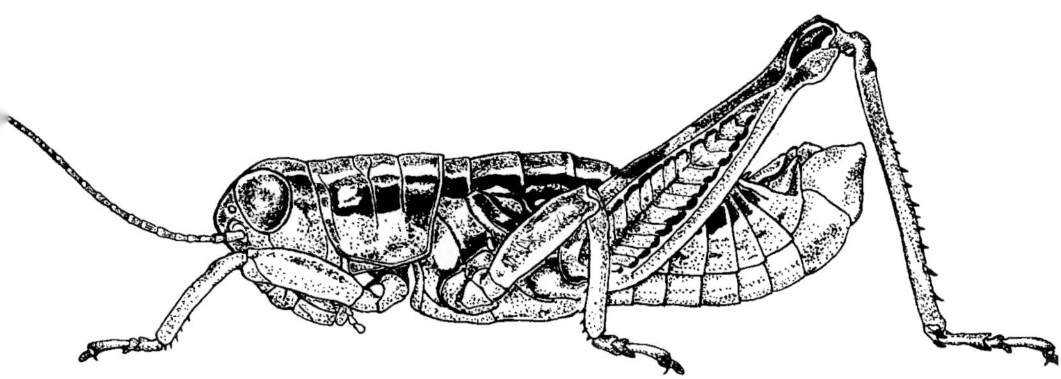

Fig. 624. *Buckellacris chilcotinae chilcotinae*, male.

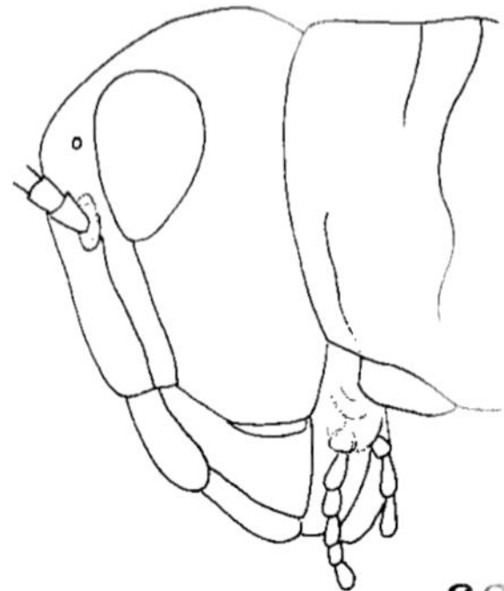

625

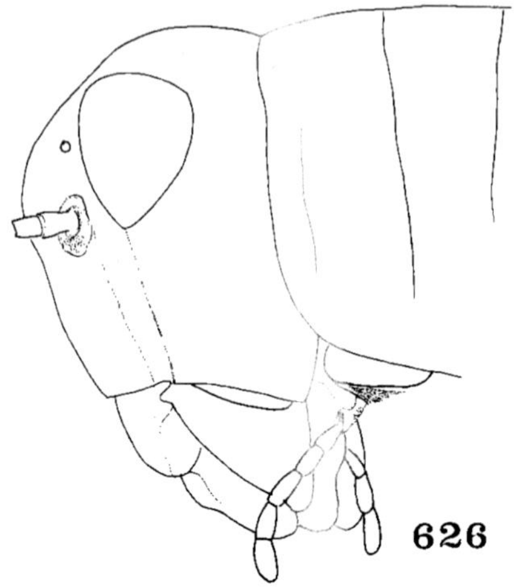

626

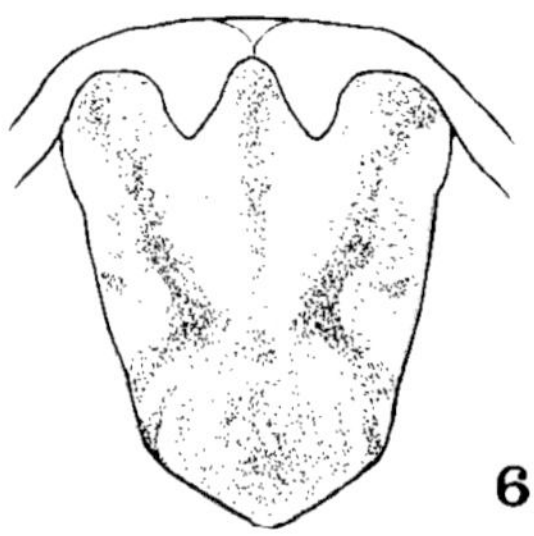

627

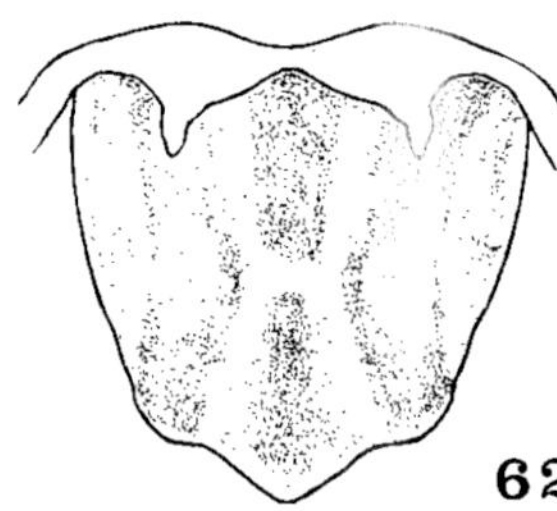

628

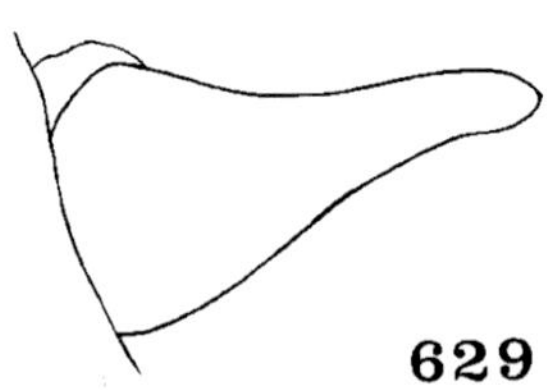

629

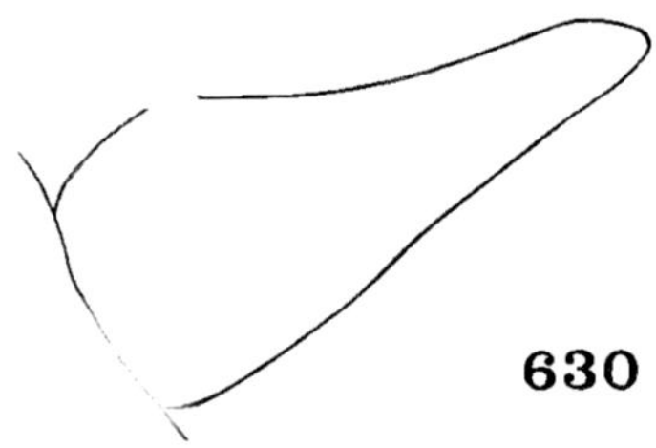

630

Figs. 625–630. *Buckellacris* spp. 625, *B. nuda nuda*, head and anterior part of pronotum, lateral; 626, *B. chilcotinae chilcotinae*, the same; 627, *B. nuda nuda*, male furculae and supra-anal plate, dorsal; 628, *B. chilcotinae chilcotinae*, the same; 629, *B. nuda nuda*, male cercus, from left; 630, *B. chilcotinae chilcotinae*, the same.

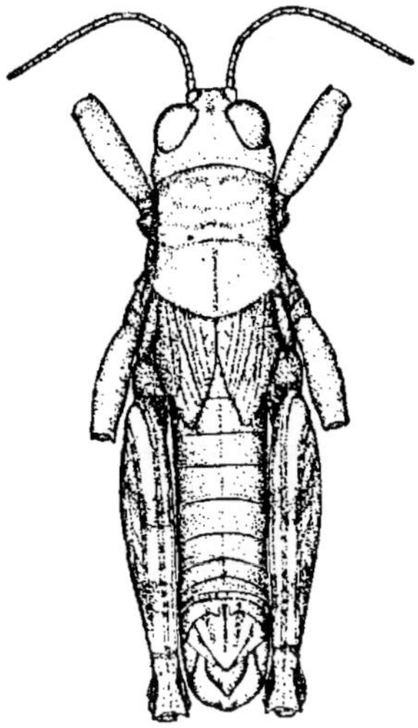

Fig. 631. *Argiacris rehni*, male. After Hebard (1918).

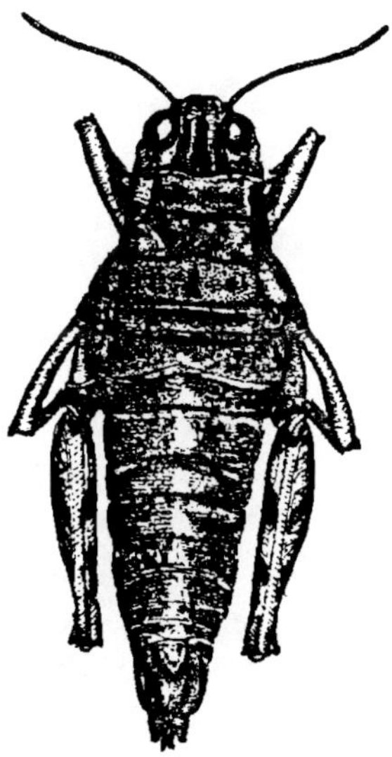

Fig. 632. *Bradynotes obesa obesa*, female. After Helfer (1963), courtesy of J.R. Helfer.

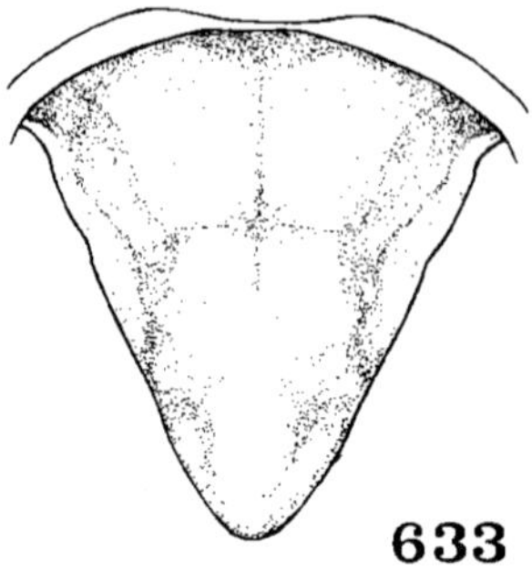

633

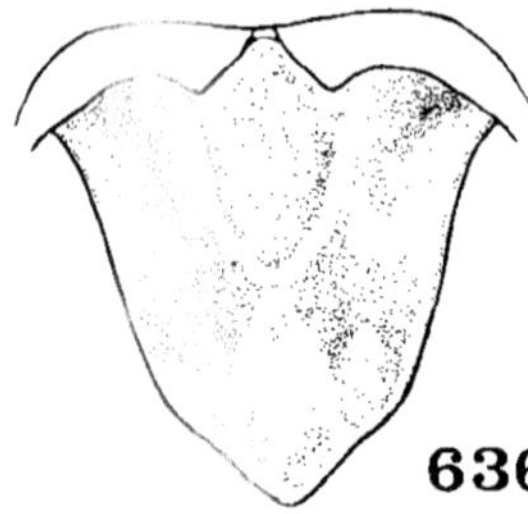

636

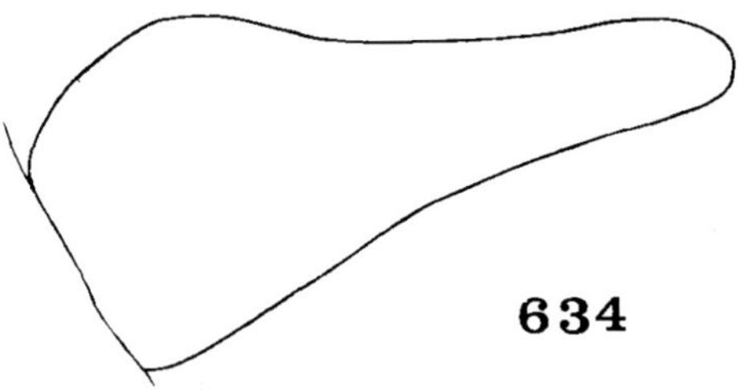

634

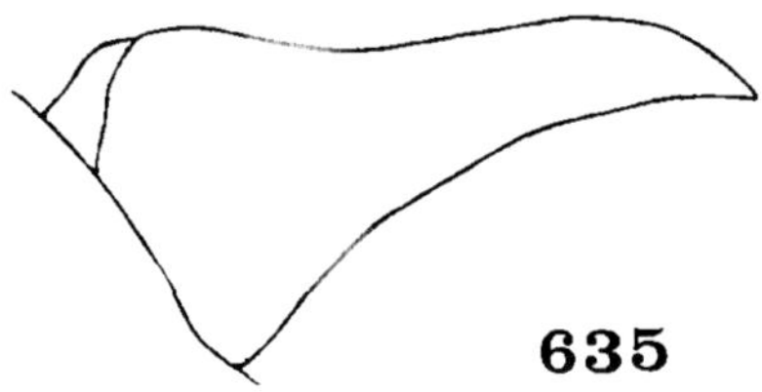

635

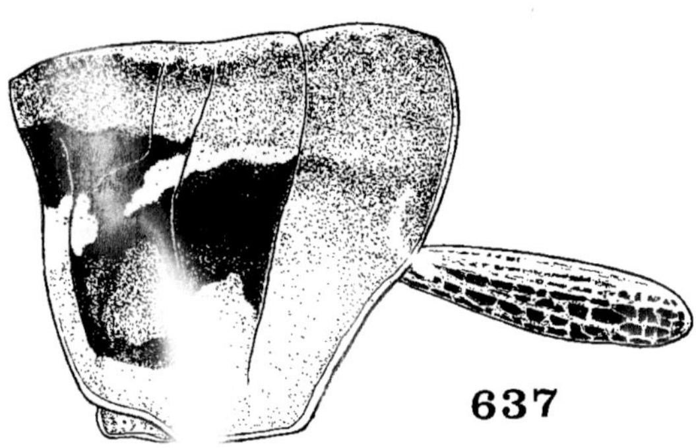

637

Figs. 633–637. *Bradynotes* and *Asemoplus* spp. 633, intermediate between *B. obesa obesa* and *B. obesa caurus*, from Osoyoos, B.C., male supra-anal plate, dorsal; 634, the same, from Rock Creek, B.C., male cercus, from left; 635, *A. montanus*, male cercus, from left; 636, the same, male supra-anal plate, dorsal; 637, the same, female pronotum and tegmen, from left.

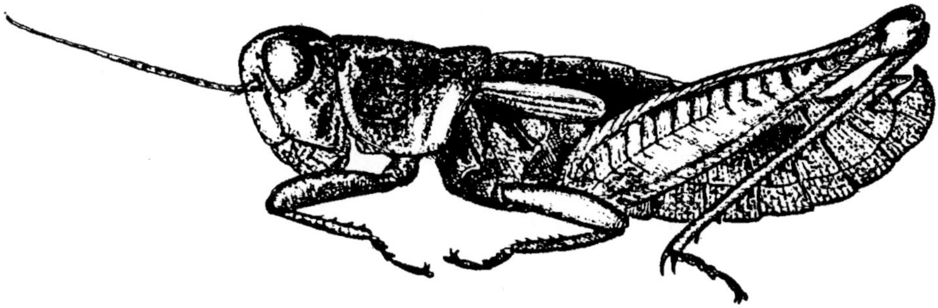

Fig. 638. *Asemoplus montanus*, male. Adapted from Helfer (1963), courtesy of J.R. Helfer.

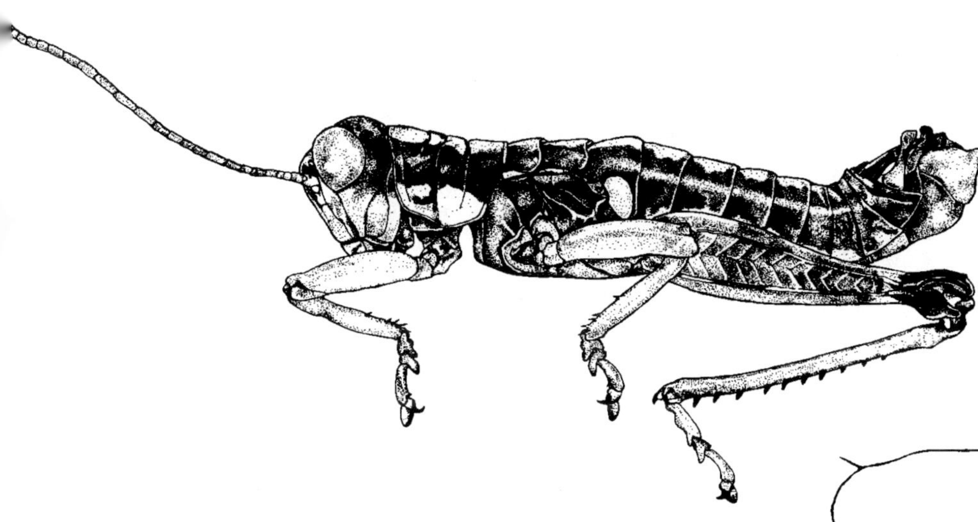

639

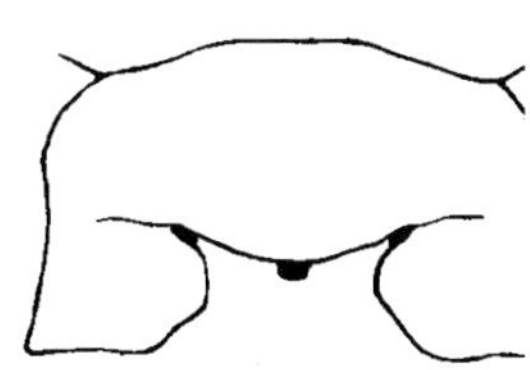

640

Figs. 639, 640. *Booneacris glacialis canadensis*. 639, male; 640, mesosternum, female.

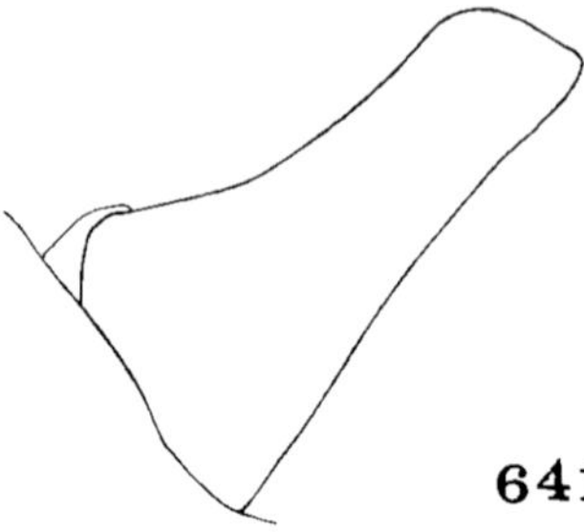

641

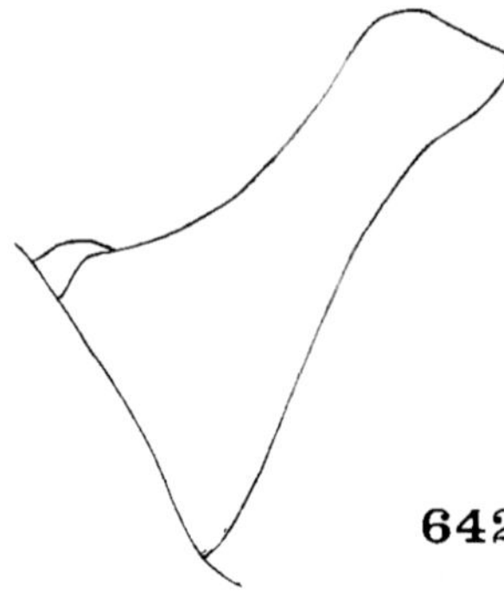

642

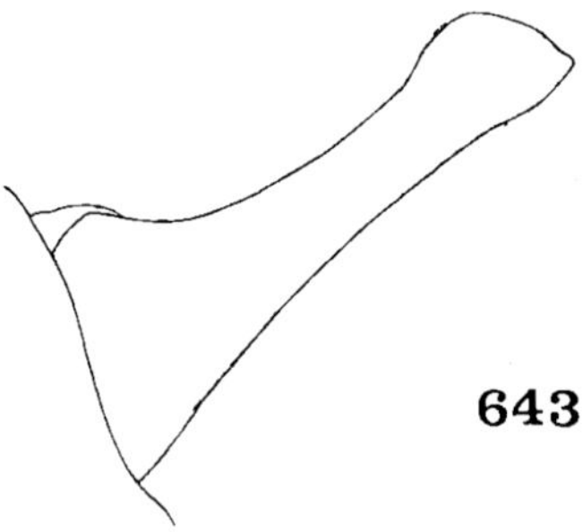

643

644

Figs. 641-644. *Booneacris* spp., male cerci, from left. 641, *B. glacialis glacialis*; 642, intermediate between *B. glacialis glacialis* and *B. glacialis canadensis*; 643, *B. glacialis canadensis*; 644, *B. variegata*. 641-643, after Vickery et al. (1974).

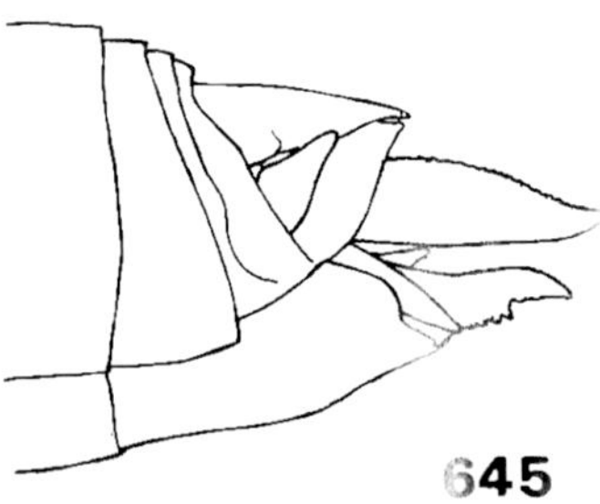

645

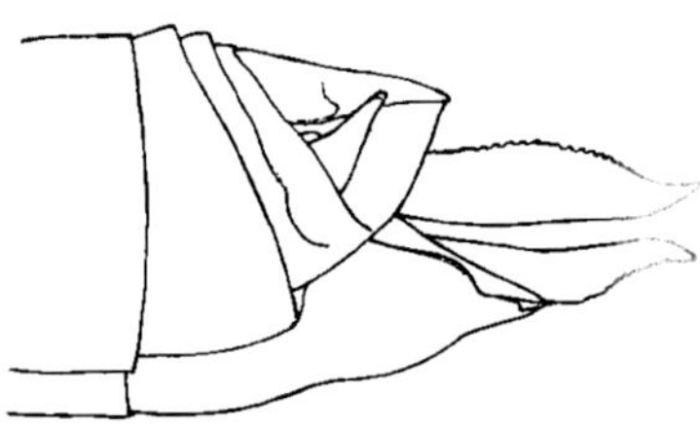

646

647

648

Figs. 645-648. *Booneacris* spp., female abdominal terminalia. 645, *B. glacialis glacialis*; 646, intermediate between *B. glacialis glacialis* and *B. glacialis canadensis*; 647, *B. glacialis canadensis*; 648, *B. variegata*. 645-647, after Vickery et al. (1974).

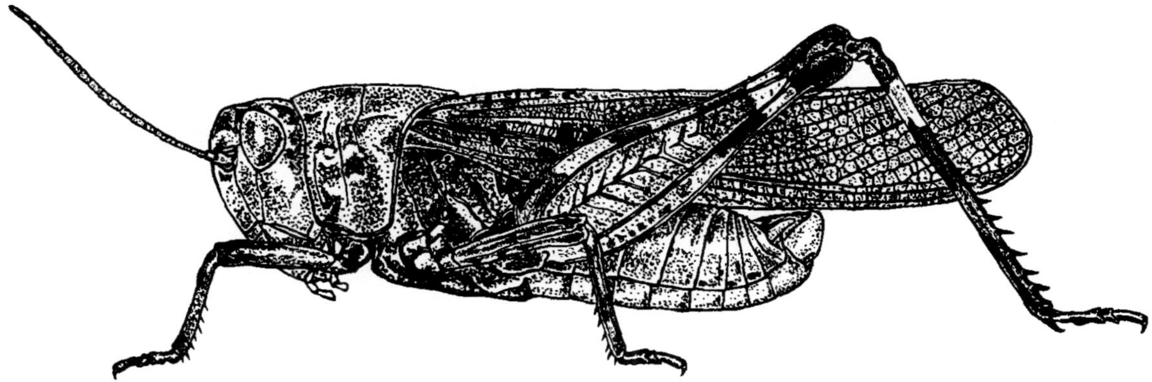

Fig. 649. *Arphia sulphurea*, male.

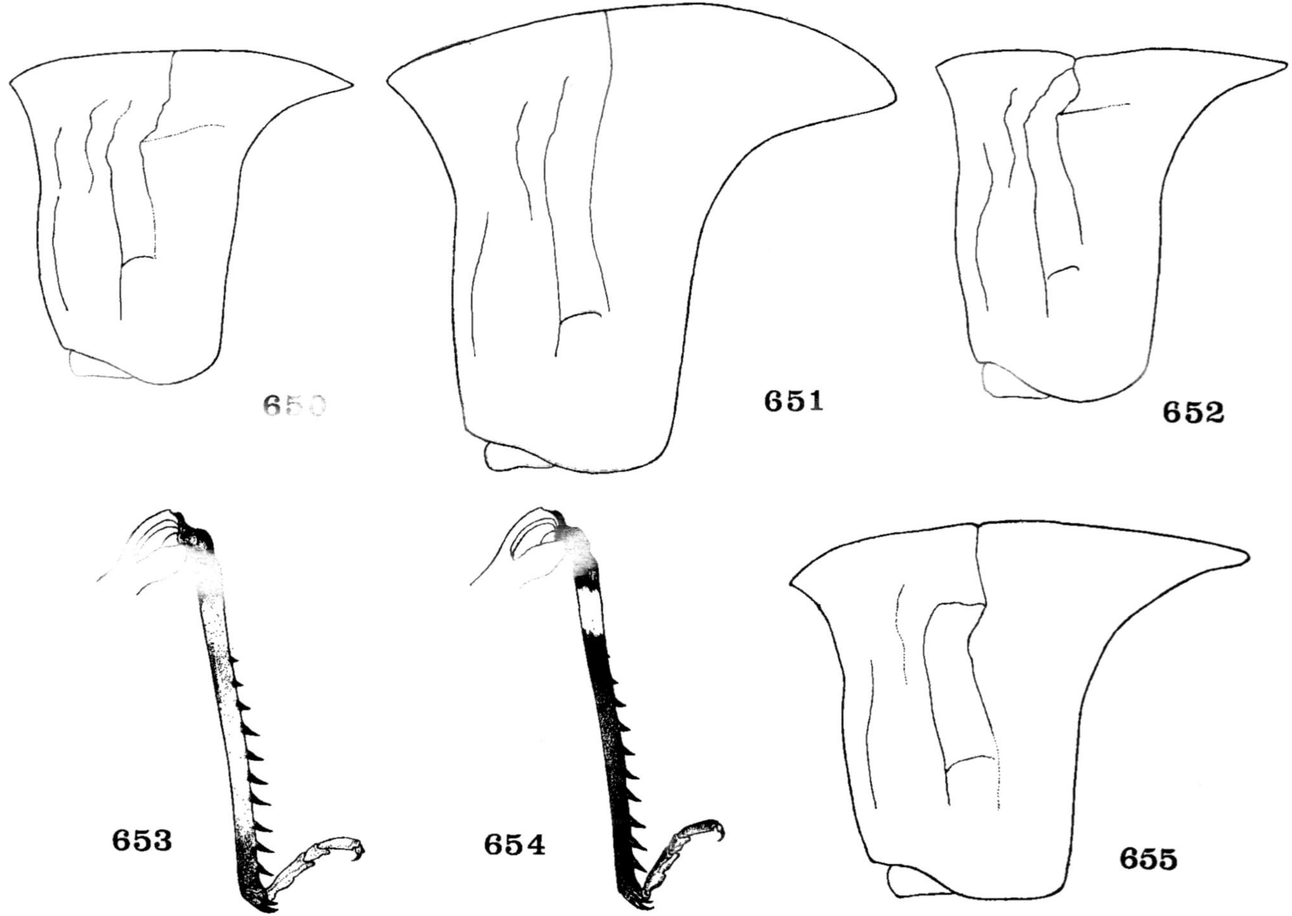

Figs. 650–655. *Arphia* and *Chortophaga* spp. 650–652, 655, pronota, from left; 653, 654, hind tibiae; 650, *A. sulphurea*; 651, *A. xanthoptera*; 652, *C. viridifasciata*; 653, *A. conspersa*; 654, *A. pseudonietana pseudonietana*; 655, *A. pseudonietana pseudonietana*.

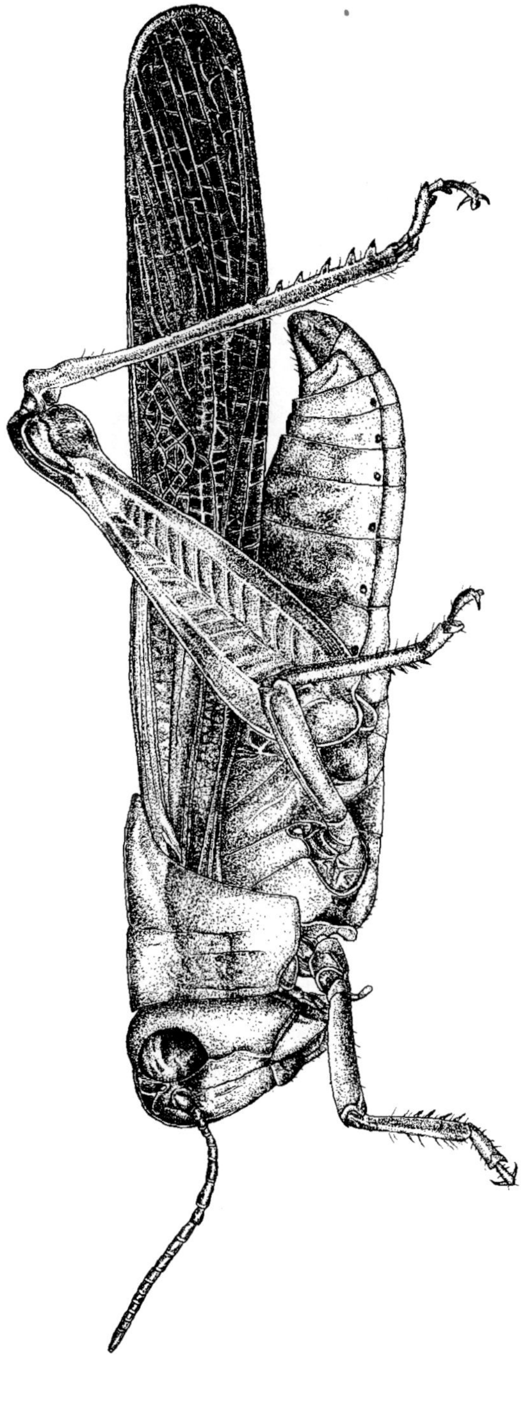

Fig. 656. *Chortophaga viridifasciata*, male.

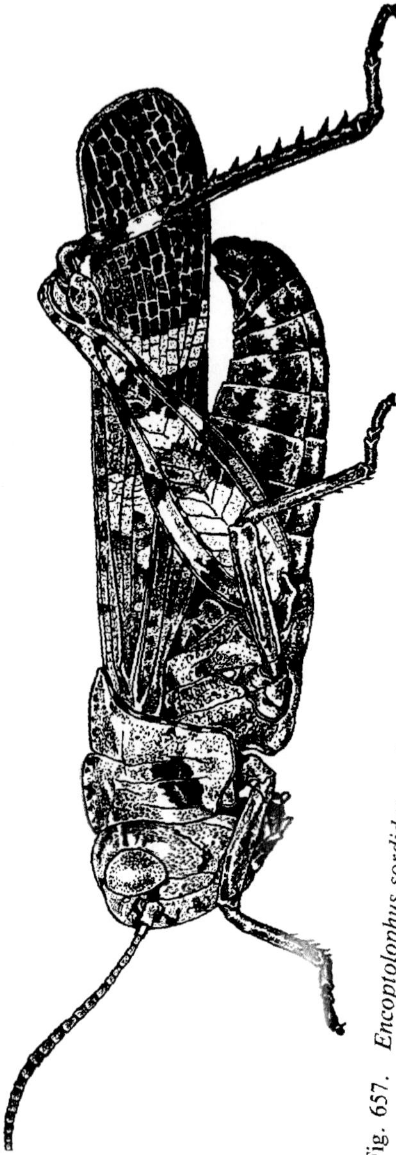

Fig. 657. *Encoptolophus sordidus*, male.

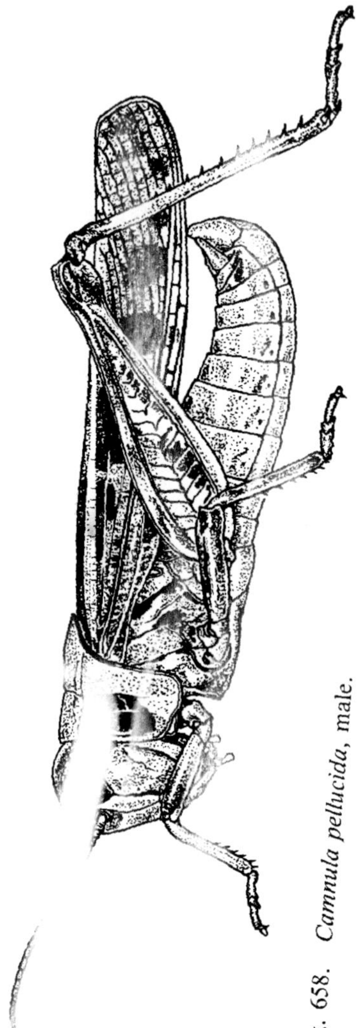

Fig. 658. *Camnula pellucida*, male.

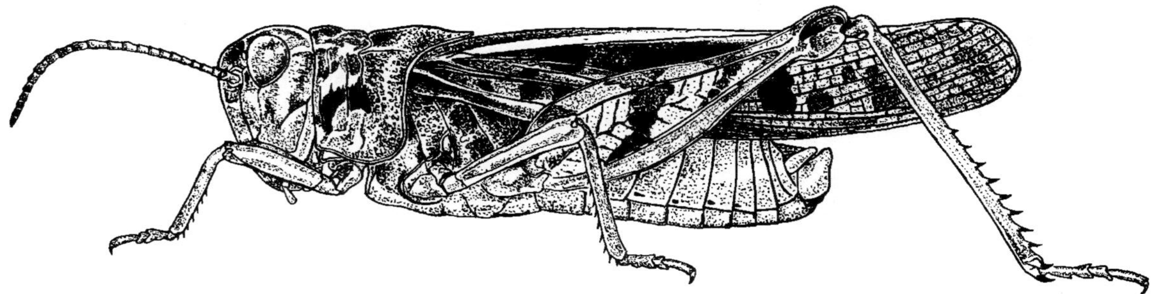

Fig. 659. *Pardalophora apiculata*, male.

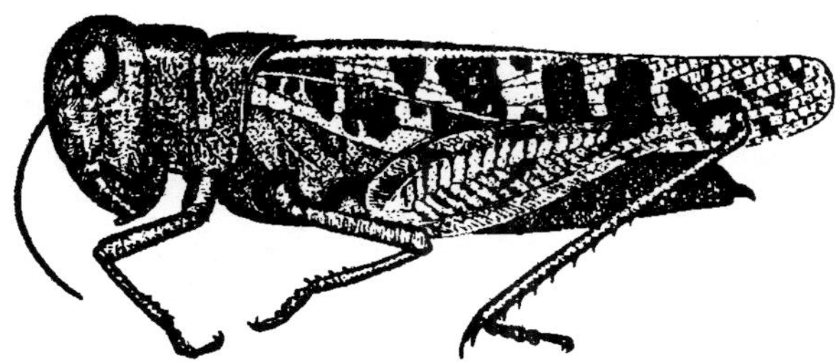

Fig. 660. *Hippiscus ocelote*, female. After Helfer (1963), courtesy of J. R. Helfer.

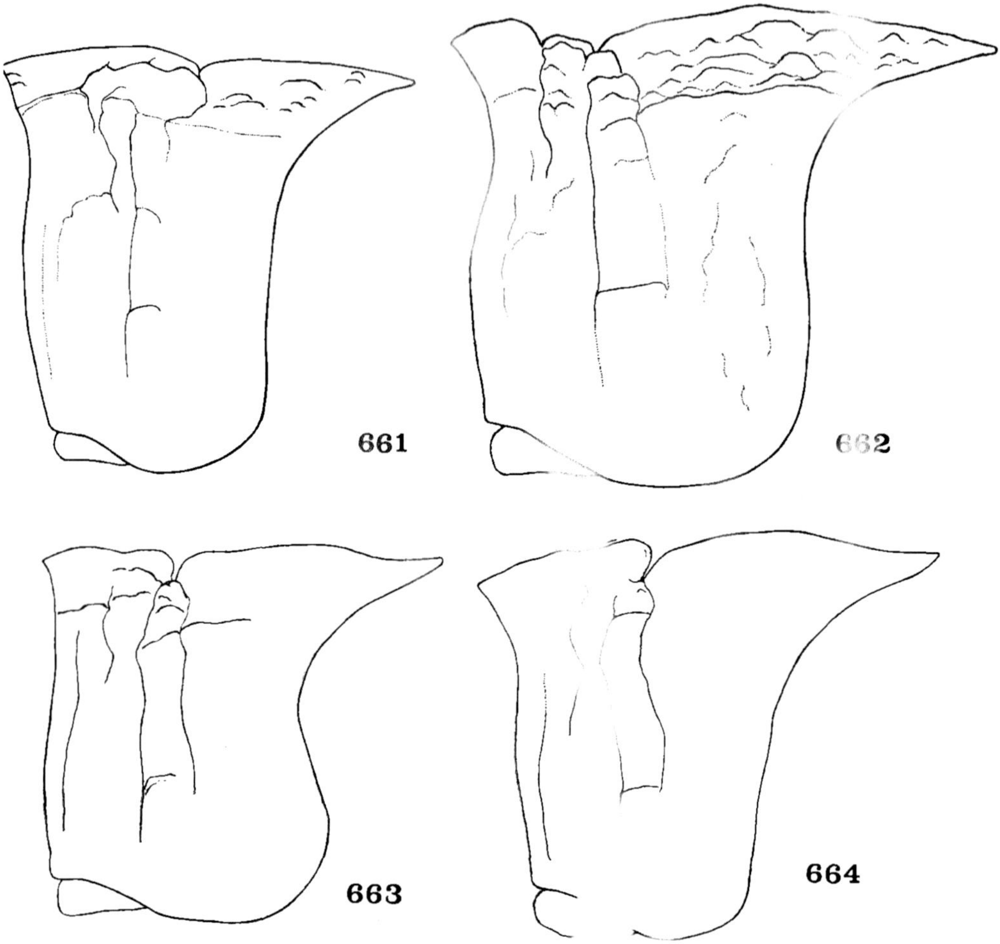

Figs. 661–664. Pronota, from left. 661, *Hi[illegible]us [illegible]elote*, male; 662, *Xanthippus corallipes latefasciatus*, female; 663, *Cratyp[illegible] ne[illegible]tus*, female; 664, *Dissosteira carolina*, male. 664, after Vickery et al. (19[illegible]

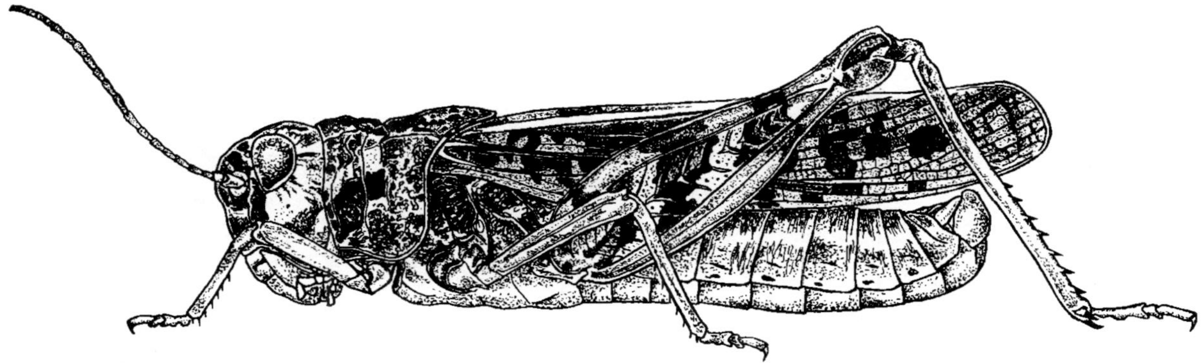

Fig. 665. *Xanthippus corallipes buckelli*, male.

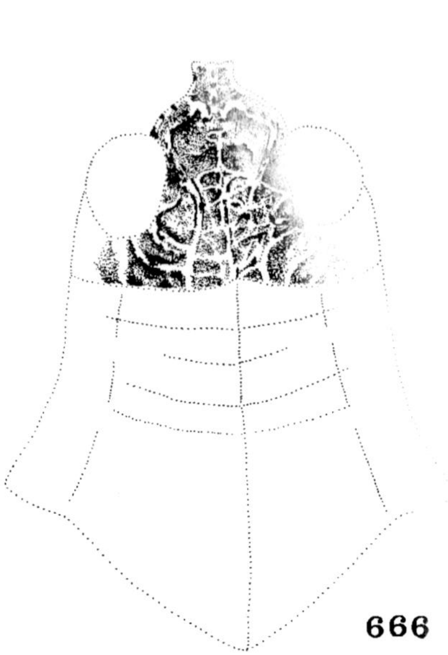

666

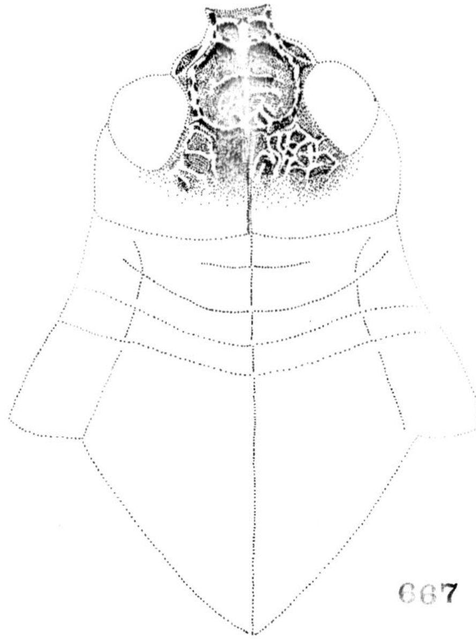

667

Figs. 666, 667. Vertices, dorsal. 666, *Xanthippus corallipes latefasciatus*; 667, *Cratypedes neglectus*.

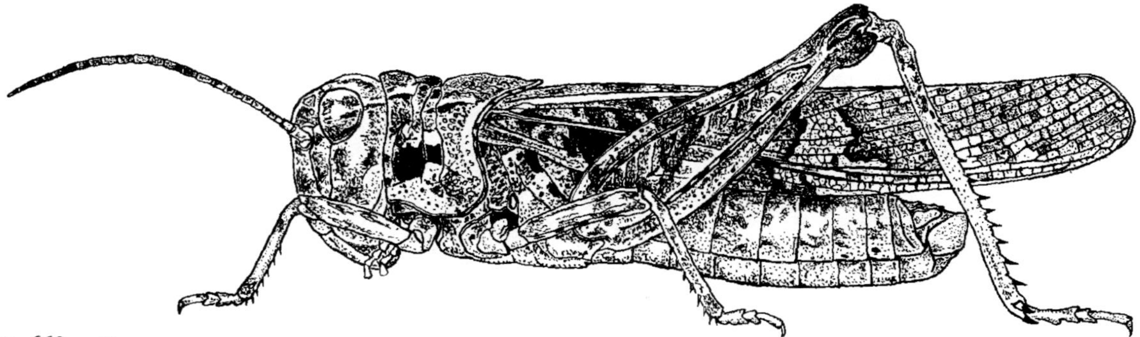

Fig. 668. *Cratypedes neglectus*, male.

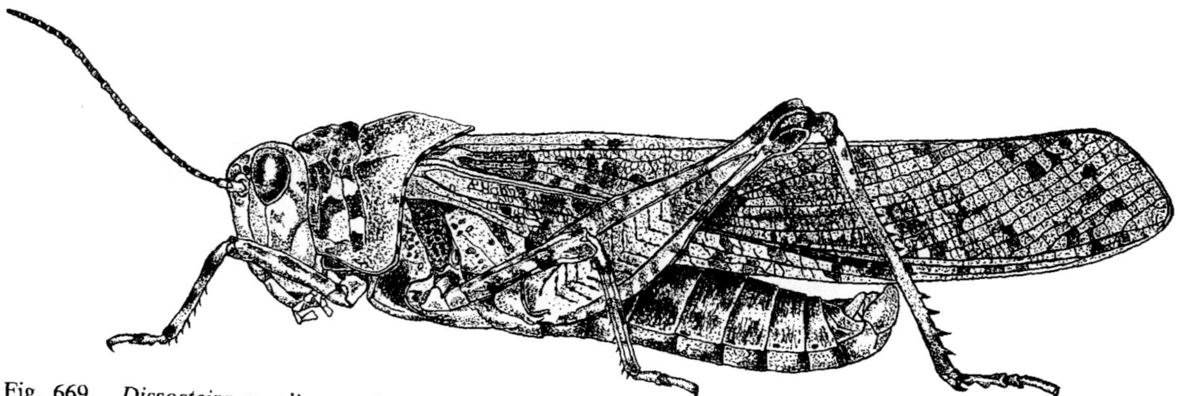

Fig. 669. *Dissosteira carolina*, male.

Fig. 670. *Gryllus carolinus*, male. Detail from color plate of Catesby (1743:2:pl. 89).

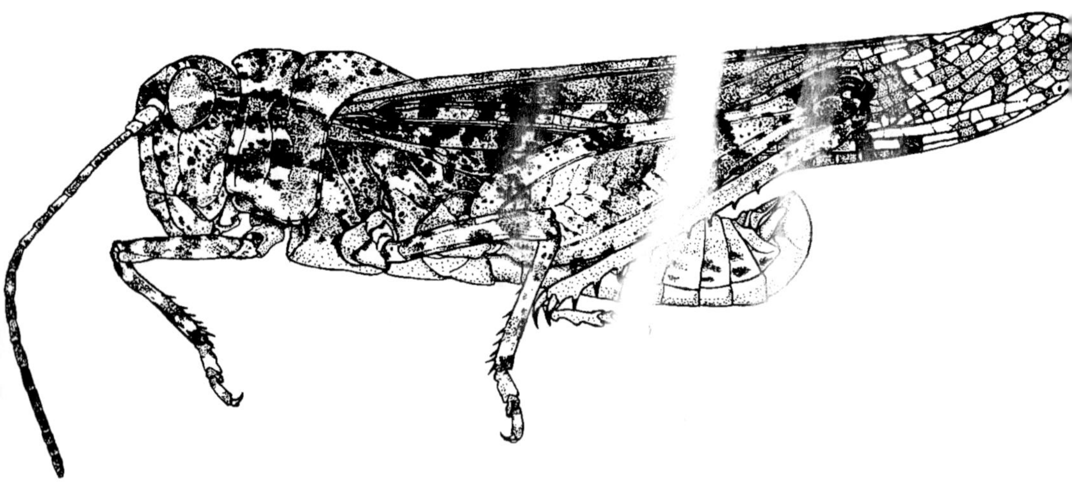

Fig. 671. *Spharagemon collare*, male.

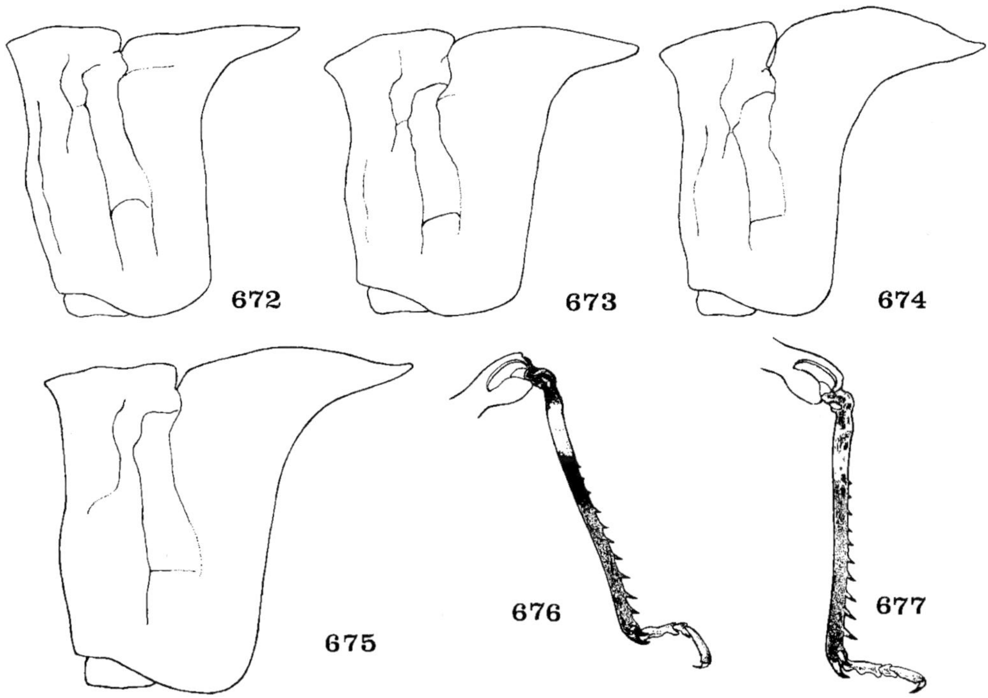

Figs. 672–677. *Spharagemon* spp. 672–675, pronota, from left; 676, 677, hind tibiae; 672, *S. equale*; 673, *S. bolli bolli*; 674, *S. collare*; 675, *S. saxatile saxatile*; 676, *S. bolli bolli*; 677, *S. collare*. 673, 674, after Vickery et al. (1974).

Fig. 678. *Scirtetica marmorata*, male.

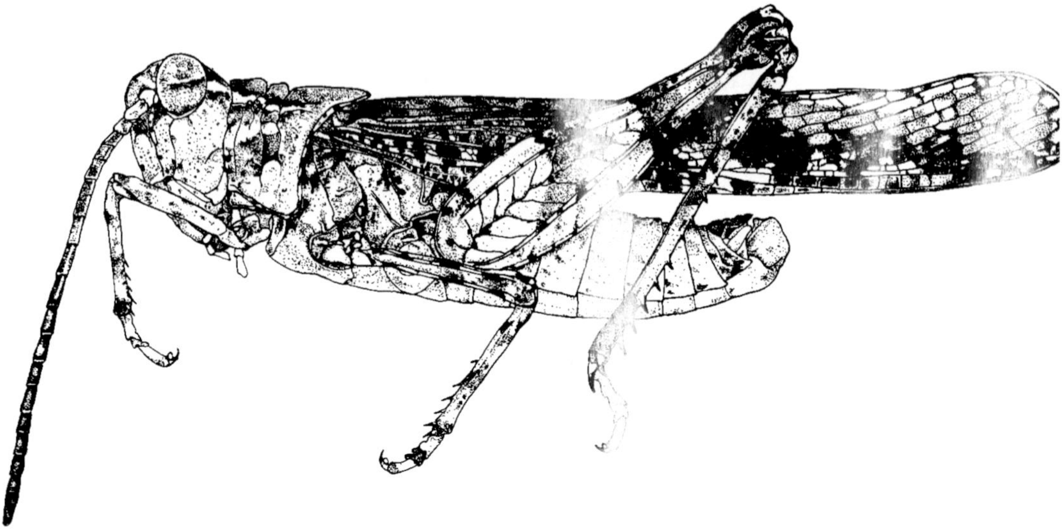

Fig. 679. *Psinidia fenestralis fenestralis*, male.

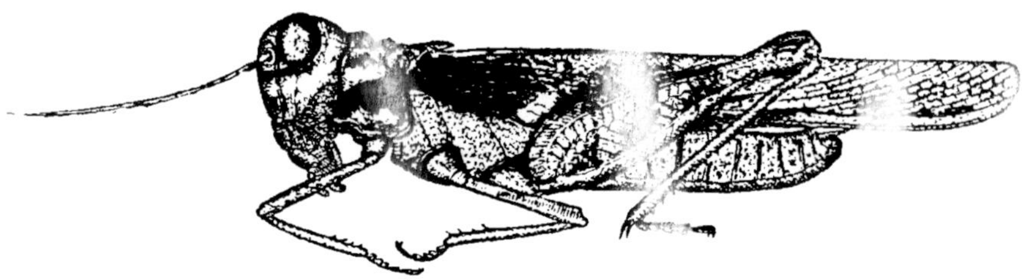

Fig. 680. *Mestrobregma plattei plattei*, male. Adapted from Helfer (1963), courtesy of J.R. Helfer.

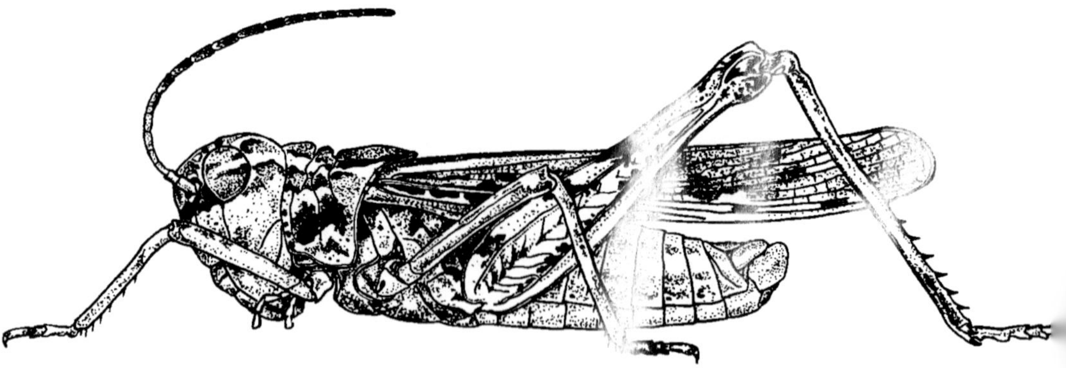

Fig. 681. *Metator nevadensis*, male.

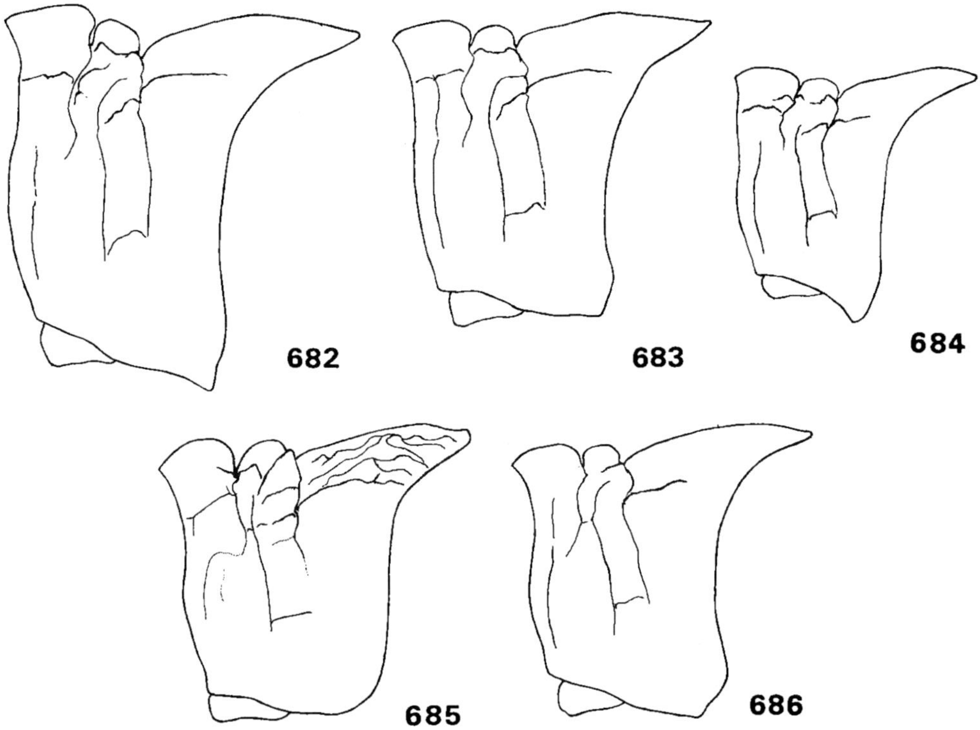

Figs. 682–686. *Metator, Trachyrhachys, Derotmema*, and *Conozoa* spp., pronota, from left. 682, *M. pardalinus*; 683, *M. nevadensis*; 684, *T. kiowa kiowa*; 685, *D. haydenii haydenii*; 686, *C. wallula*.

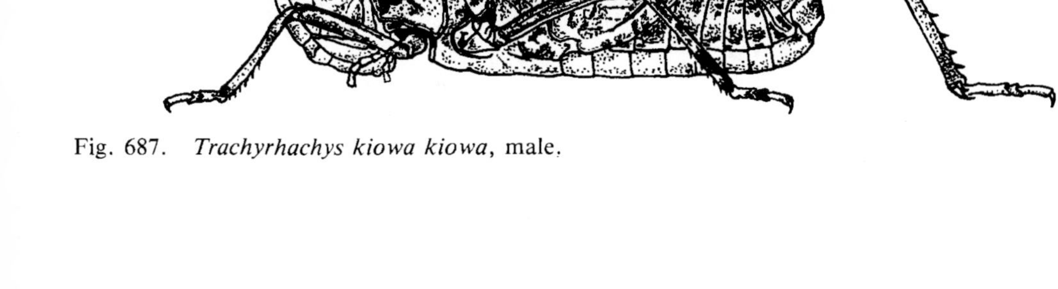

Fig. 687. *Trachyrhachys kiowa kiowa*, male.

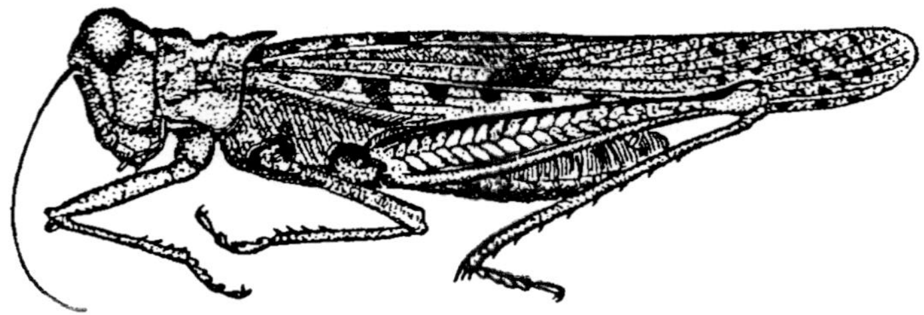

Fig. 688. *Derotmema haydenii haydenii*, male. After Helfer (1963), courtesy of J. R. Helfer.

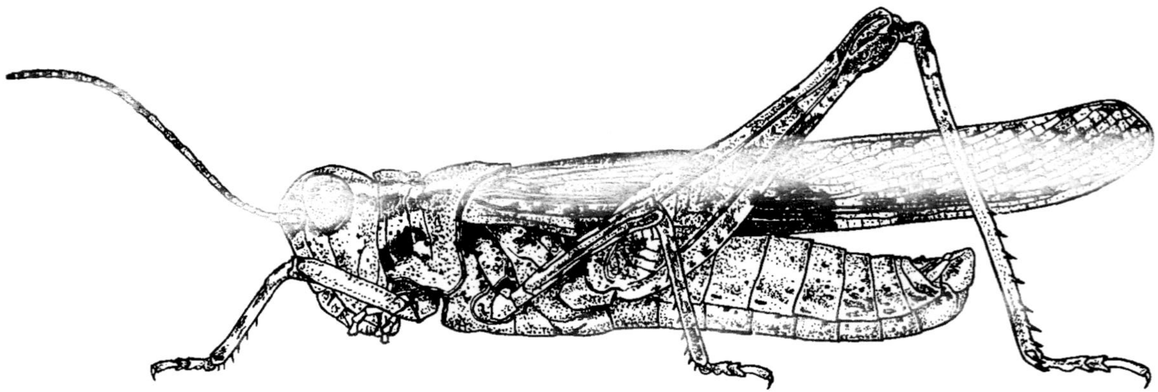

Fig. 689. *Conozoa wallula*, male.

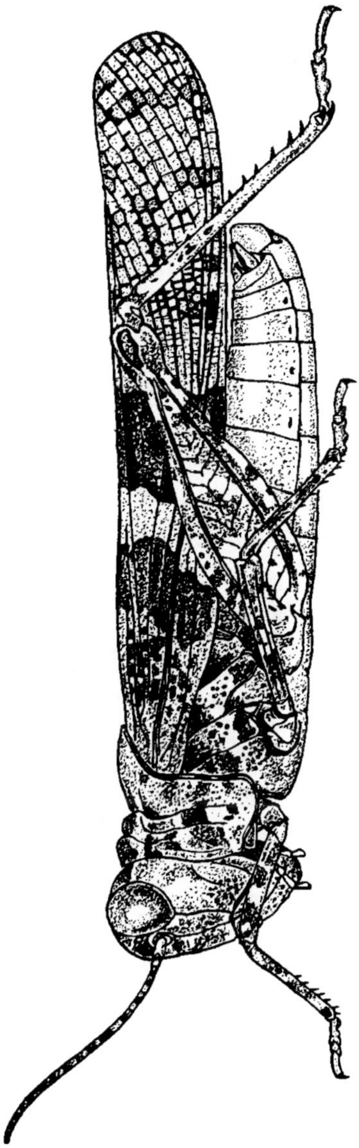

Fig. 690. *Trimerotropis pallidipennis pallidipennis*, male.

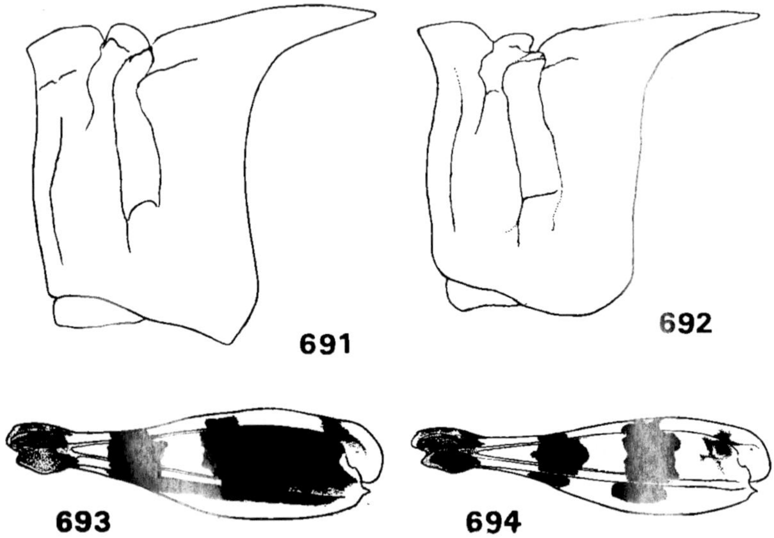

Figs. 691–694. *Trimerotropis* spp. 691, 692, pronota, from left; 693, 694, inner faces of hind femora. 691, *T. agrestis*; 692, *T. verruculata*; 693, *T. huroniana*; 694, *T. maritima interior*. 692, after Vickery et al. (1974).

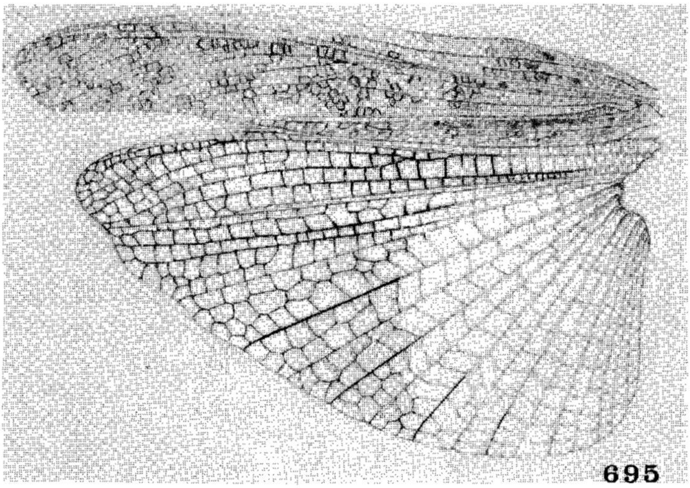

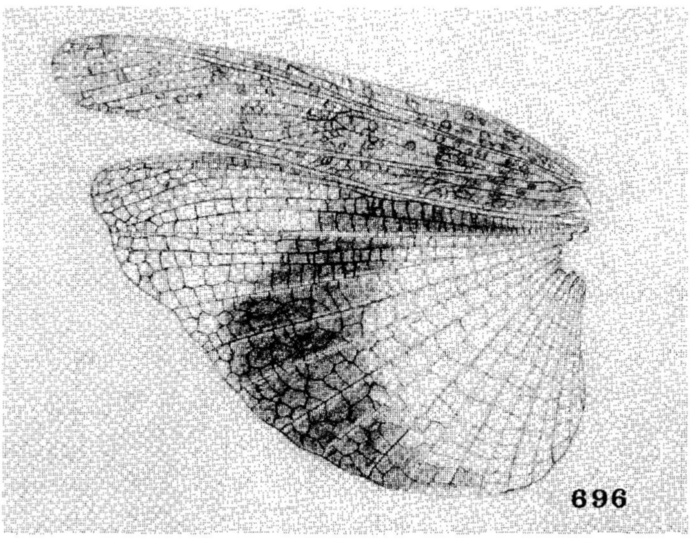

Figs. 695, 696. *Trimerotropis* spp., wing patterns. 695, *T. gracilis*; 696, *T. sordida*.

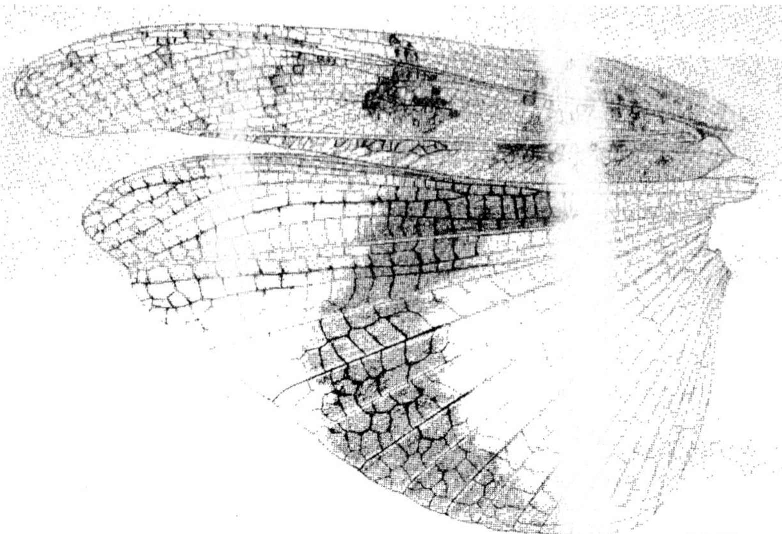

697

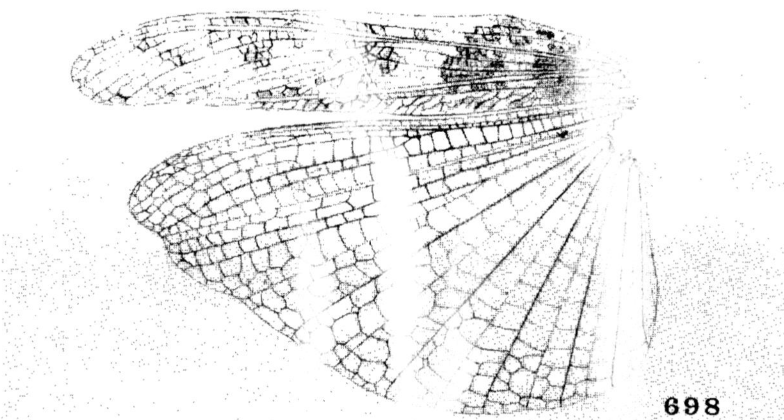

698

Figs. 697, 698. *Trimerotropis* spp., wing patterns. 697, *T. huroniana*; 698, *T. sparsa*.

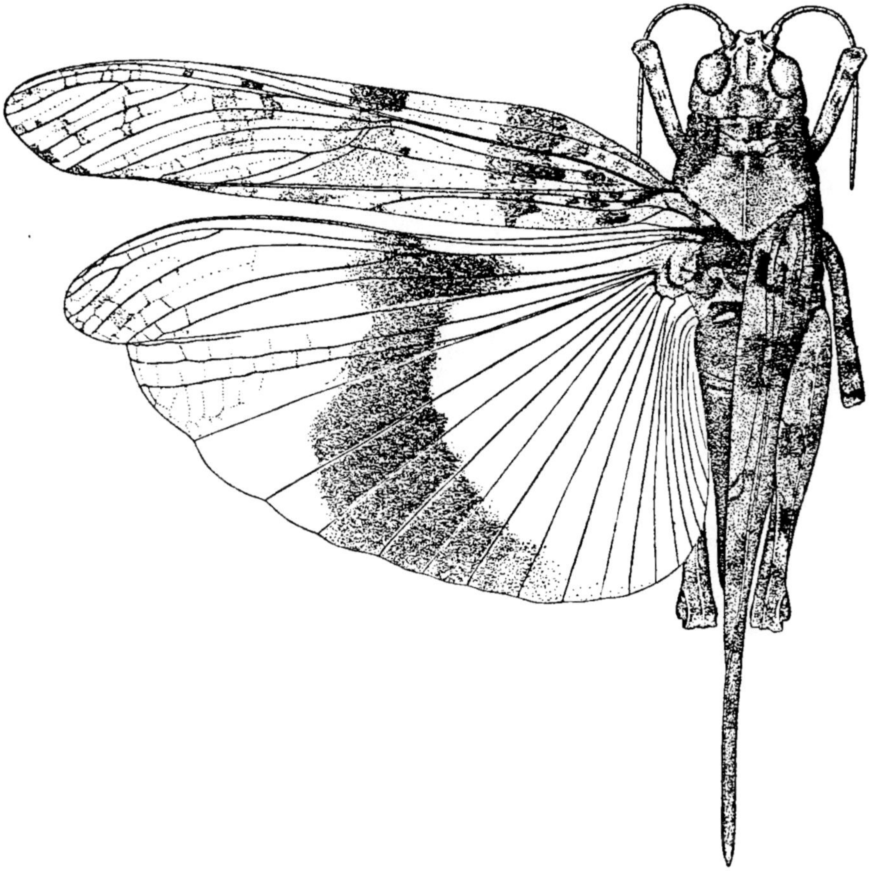

Fig. 699. *Trimerotropis diversellus*, male. After Hebard (1928).

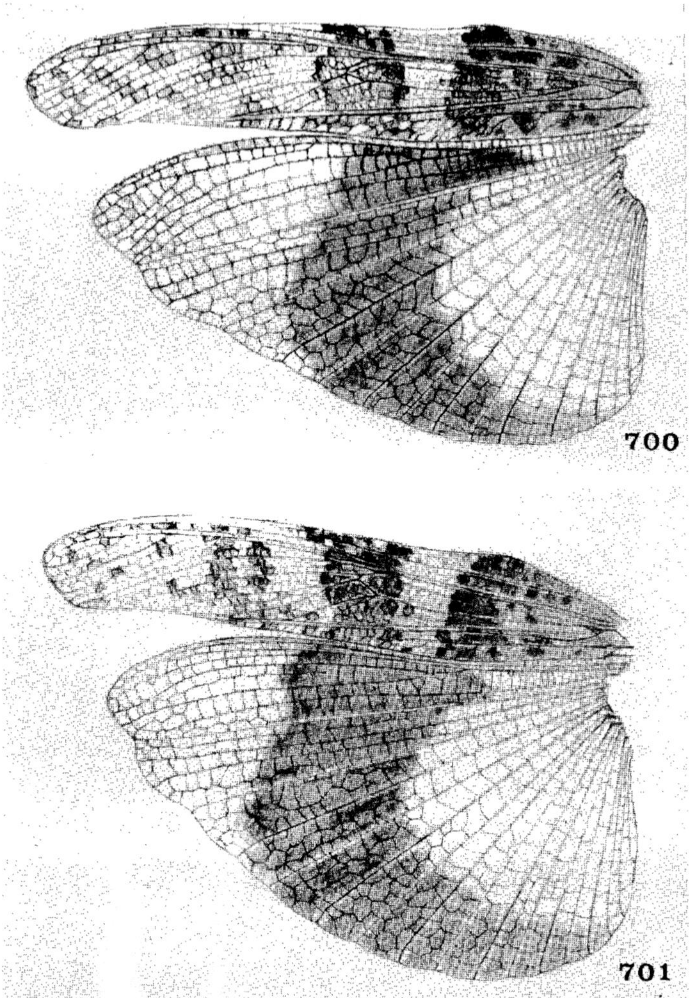

Figs. 700, 701. *Trimerotropis pallidipennis*, wing patterns. 700, *T. pallidipennis pallidipennis*; 701, *T. pallidipennis salina*.

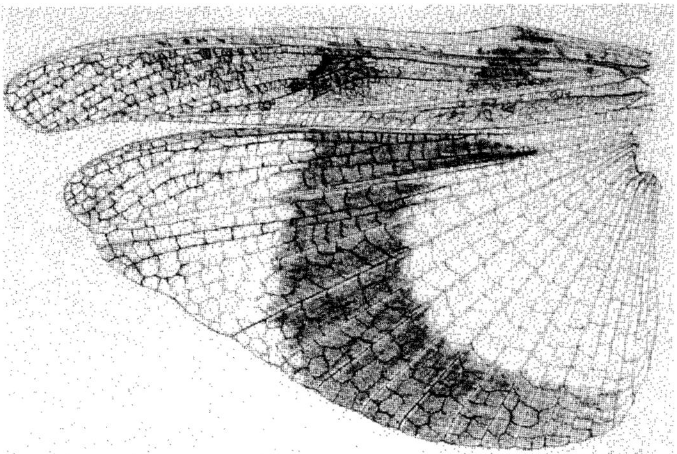

702

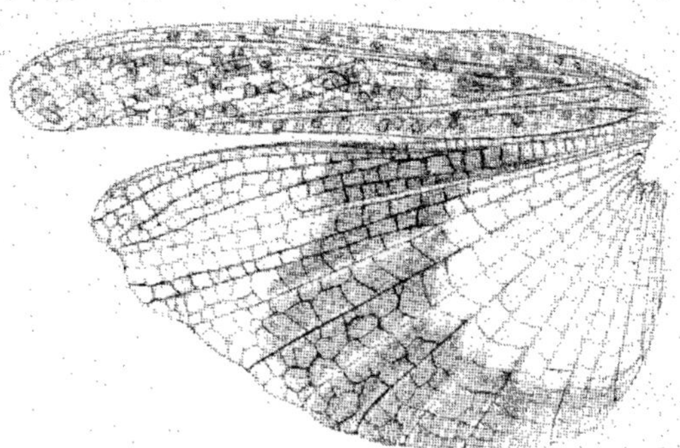

703

Figs. 702, 703. *Trimerotropis* spp., wing patterns. 702, *T. maritima interior*; 703, *T. agrestis*.

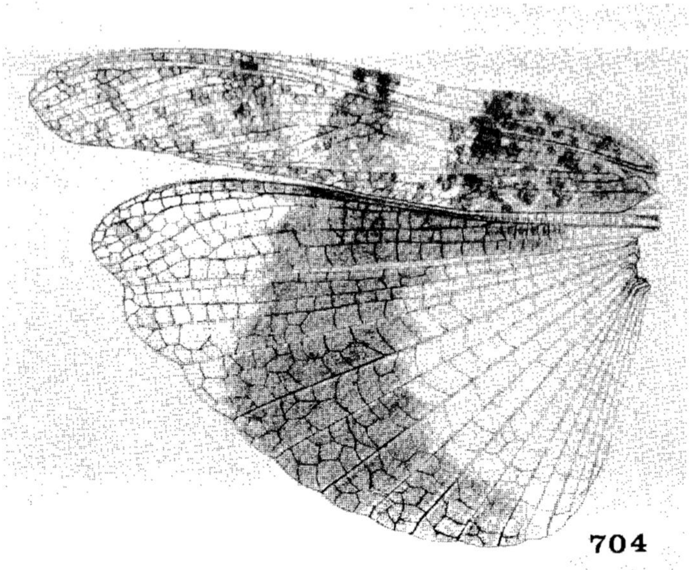

704

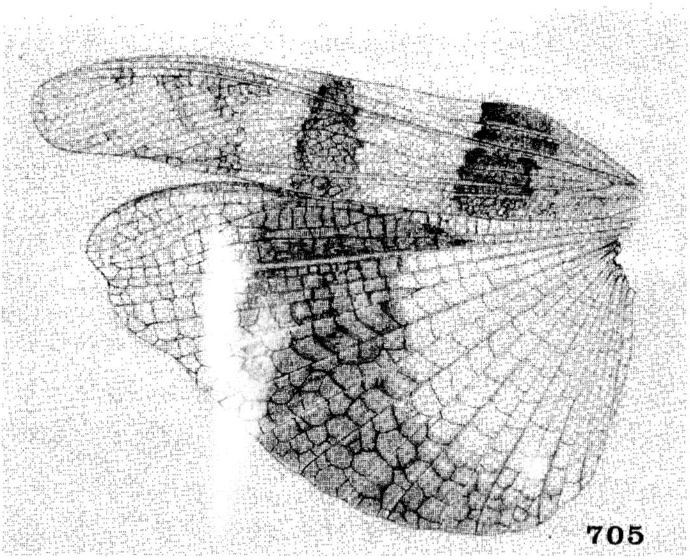

705

Figs. 704, 705. *Trimerotropis* spp., wing patterns. 704, *T. campestris*; 705, *T. pistrinaria*.

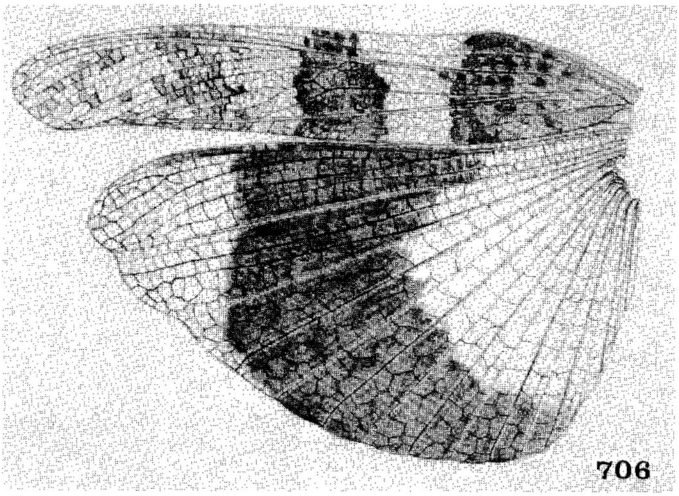

706

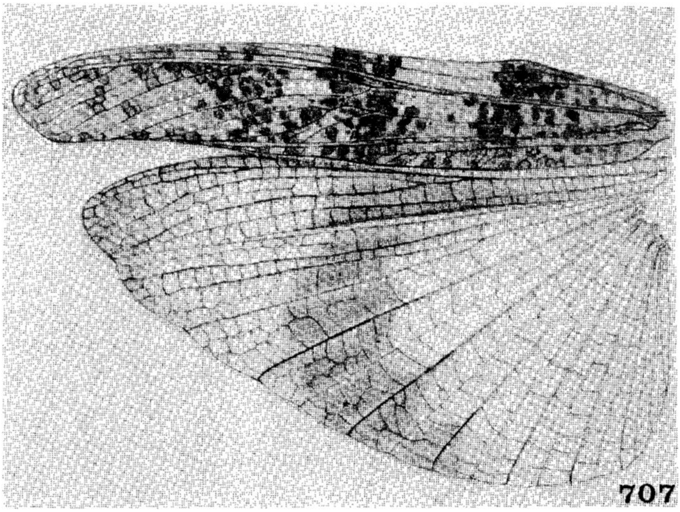

707

Figs. 706, 707. *Trimerotropis* spp., wing patterns. 706, *T. latifasciata*; 707, *T. fontana*.

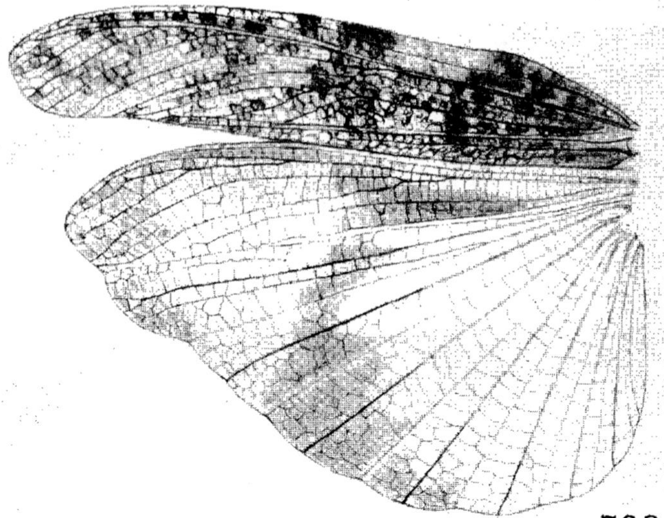

708

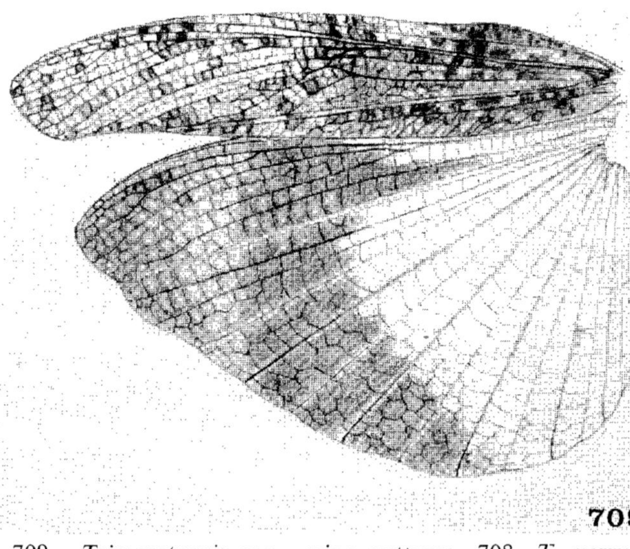

709

Figs. 708, 709. *Trimerotropis* spp., wing patterns. 708, *T. verruculata*; 709, *T. suffusa*.

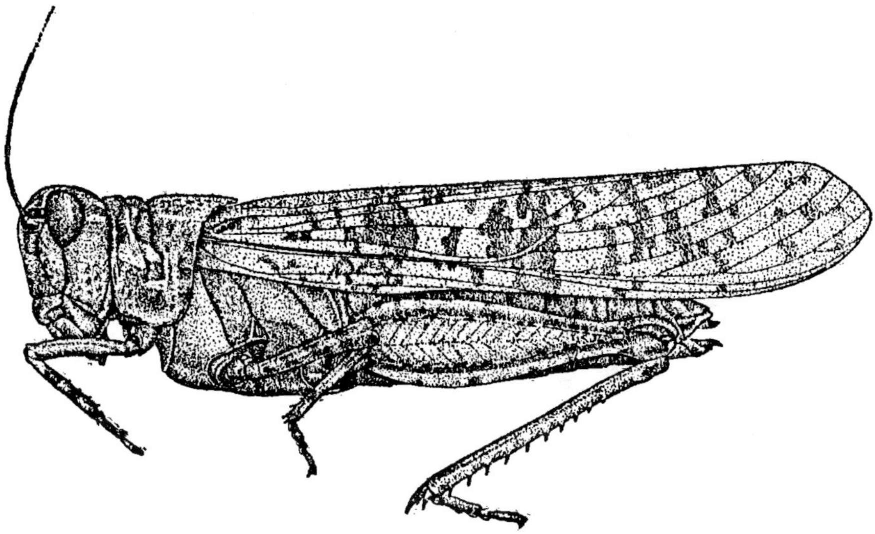

Fig. 710. *Circotettix rabula rabula*, female. After J. A. G. Rehn and Hebard (1906).

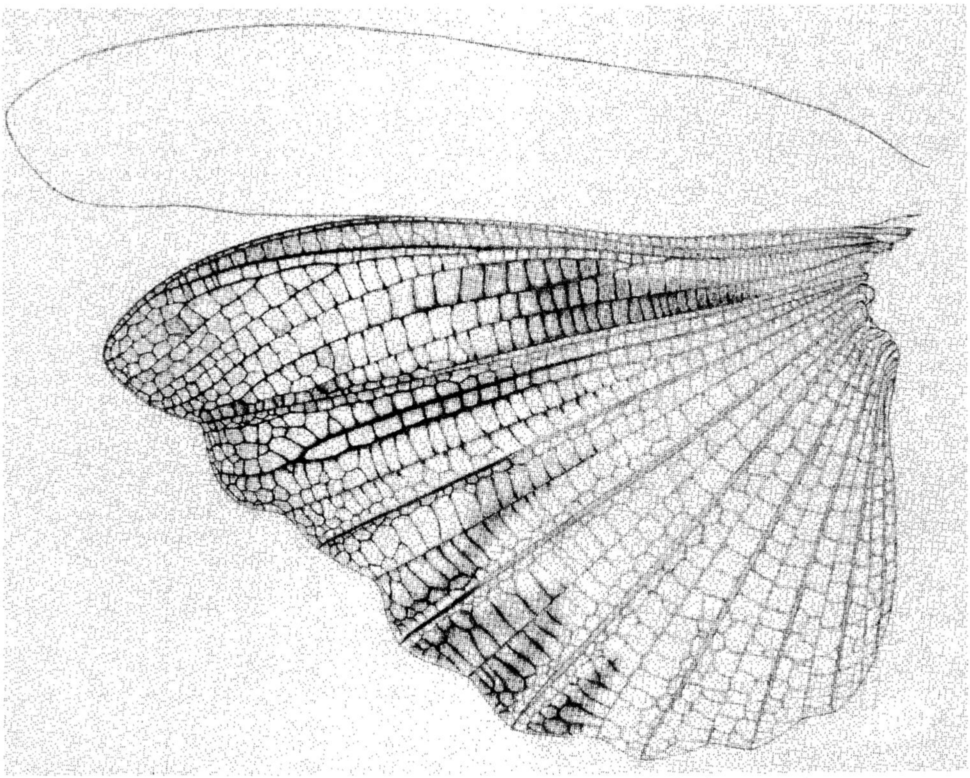

Fig. 711. *Circotettix rabula rabula*, tegmen outline and hind wing venation.

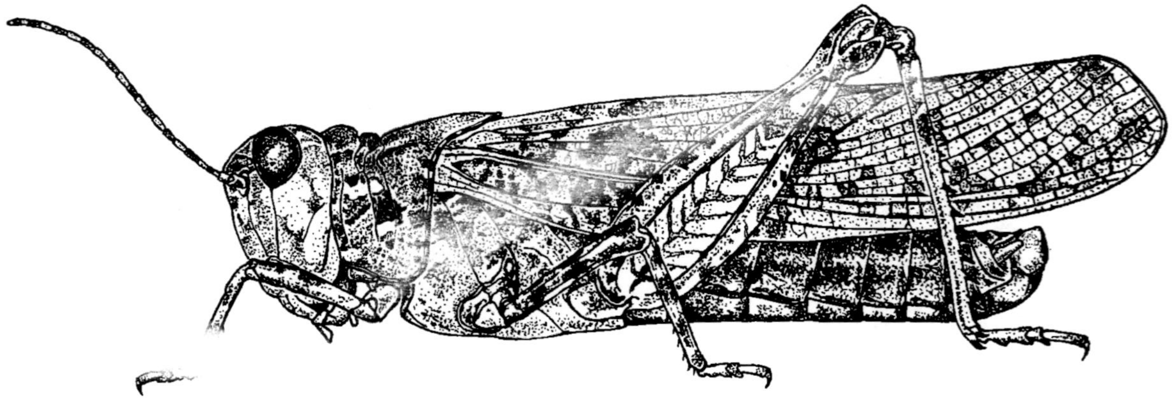

Fig. 712. *Aerochoreutes carlinianus strepitus*, male.

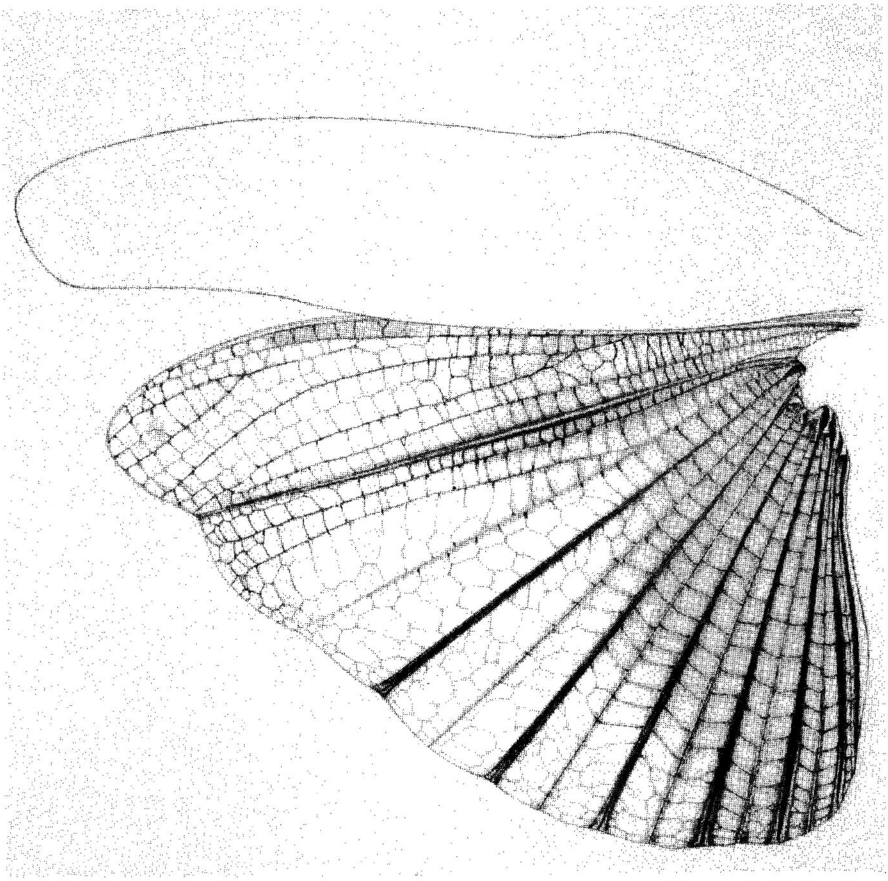

Fig. 713. *Aerochoreutes carlinianus carlinianus*, tegmen outline and hind wing venation.

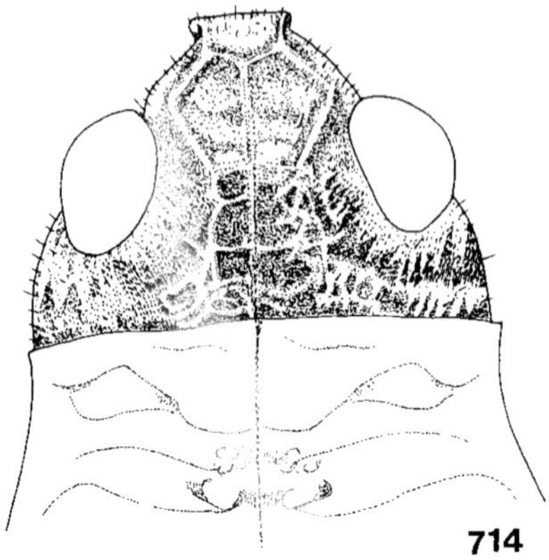

714

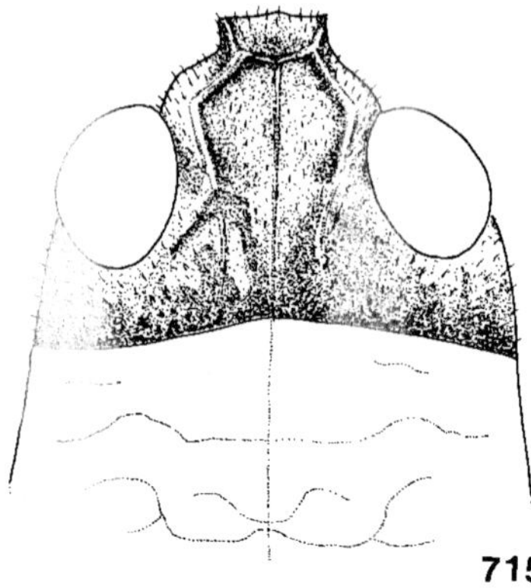

715

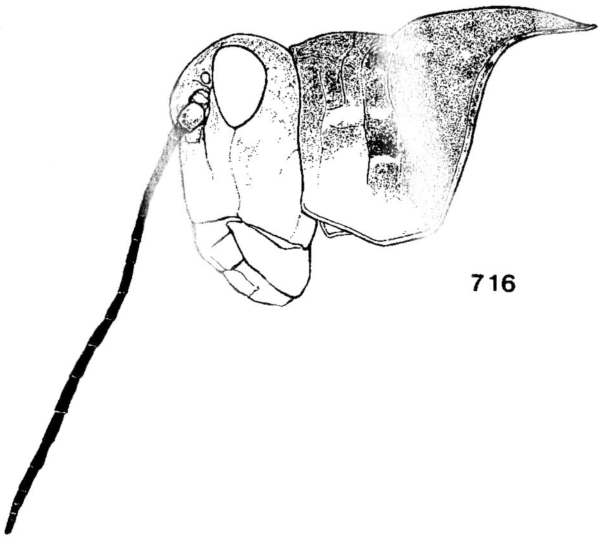

716

Figs. 714–716. *Aerochoreutes* and *Hadrotettix* spp. 714, *A. carlinianus carlinianus*, head, dorsal; 715, *A. carlinianus strepitus*, head, dorsal; 716, *H. trifasciatus*, head and pronotum, from left.

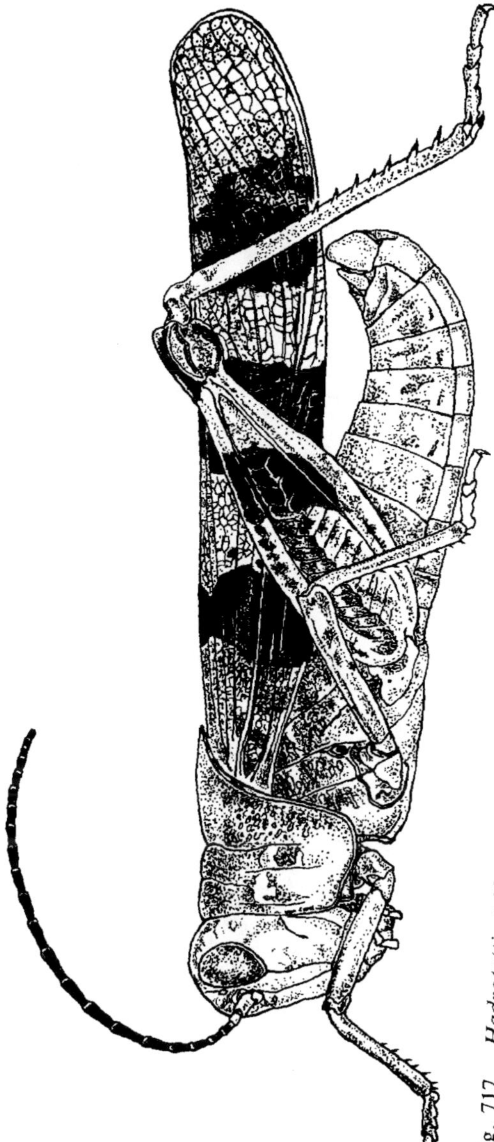

Fig. 717. *Hadrotettix trifasciatus*, male.

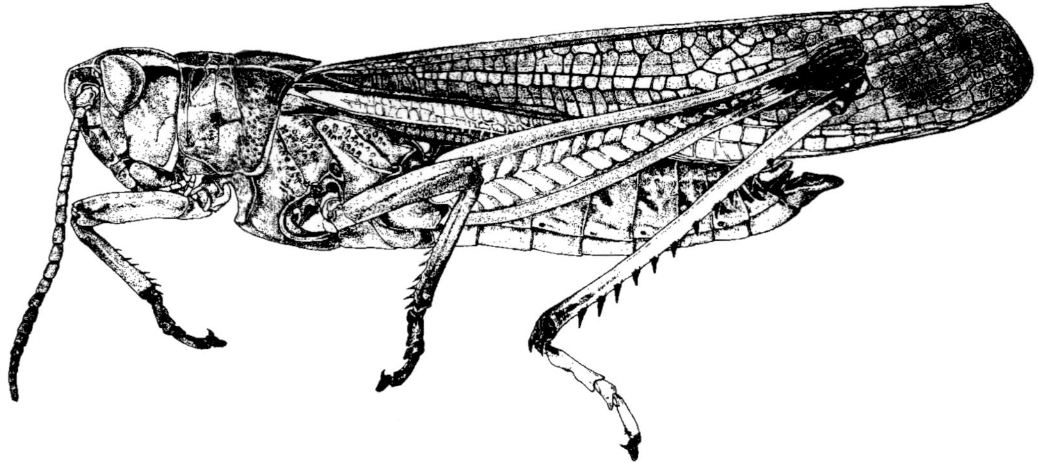

Fig. 718. *Stethophyma lineatum*, male.

Fig. 719. *Metaleptea brevicornis*, emblem of Pan American Acridological Society (Sociedad Panamericana de Acridiologia).

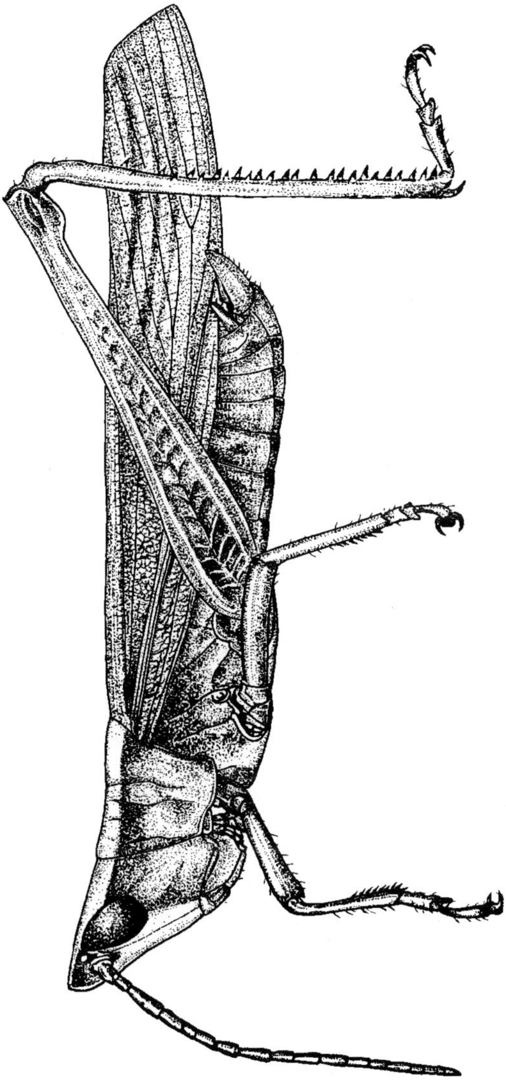

Fig. 720. *Metaleptea brevicornis brevicornis*, male.

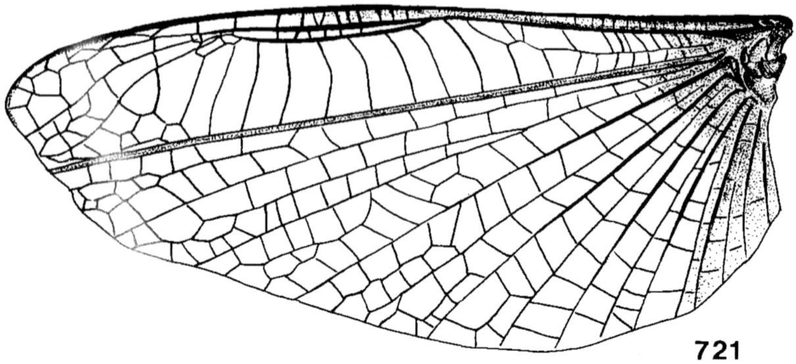

721

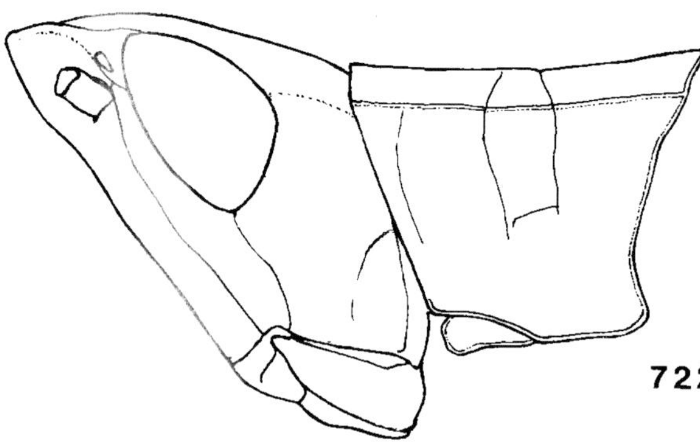

722

Figs. 721, 722. *Metaleptea brevicornis brevicornis*. 721, male, left hind wing; 722, head and pronotum, male, from left.

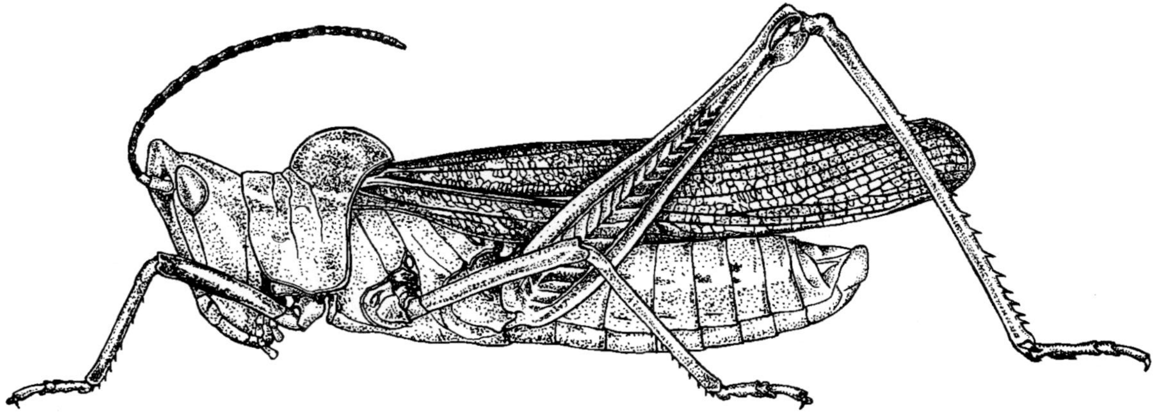

Fig. 723. *Acrolophitus hirtipes hirtipes*, head and pronotum from left.

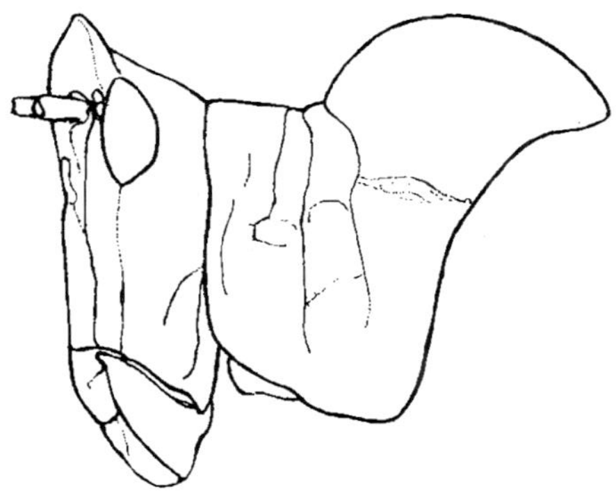

Fig. 724. *Acrolophitus hirtipes hirtipes*, male.

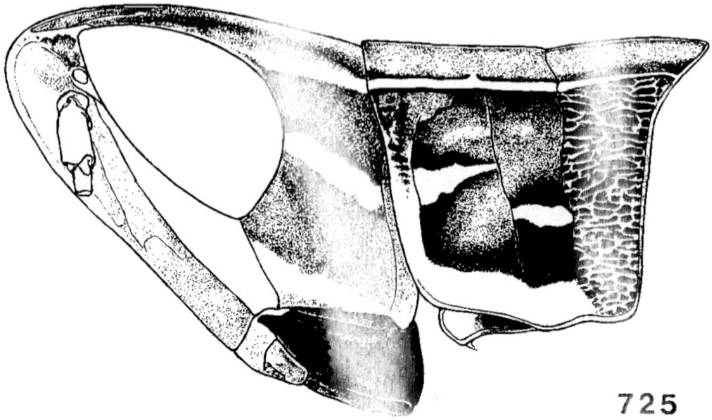

725

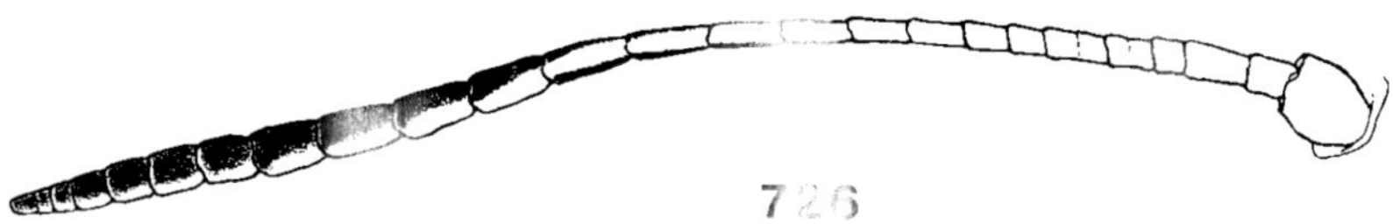

726

Figs. 725, 726. *Syrbula admirabilis*. 725, head and pronotum from left; 726, male antenna.

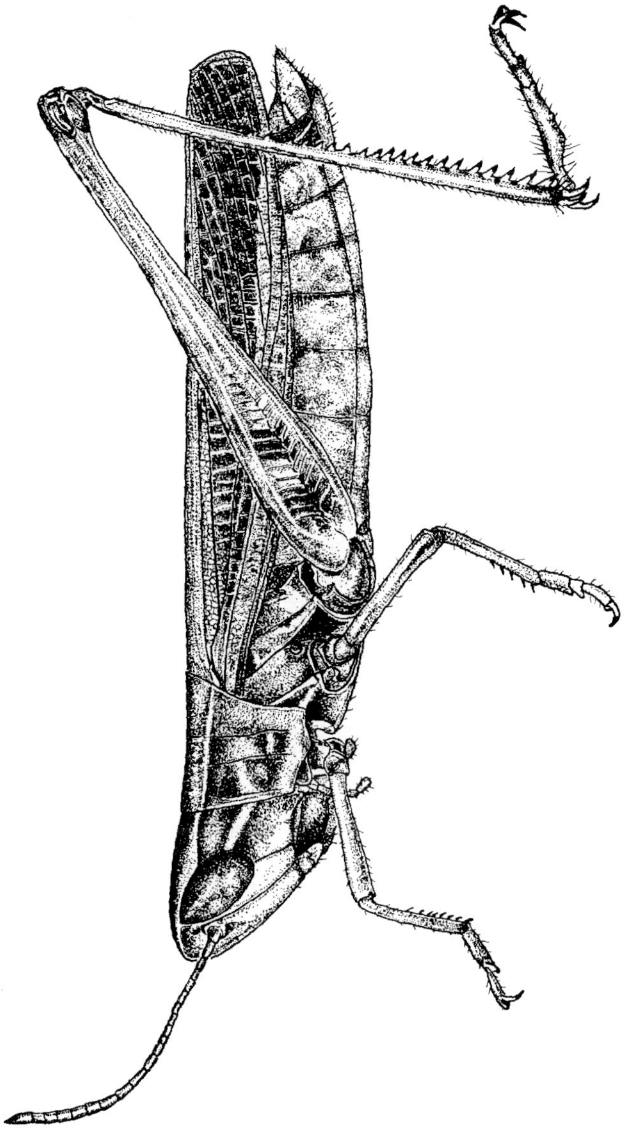

Fig. 727. *Syrbula admirabilis*, male.

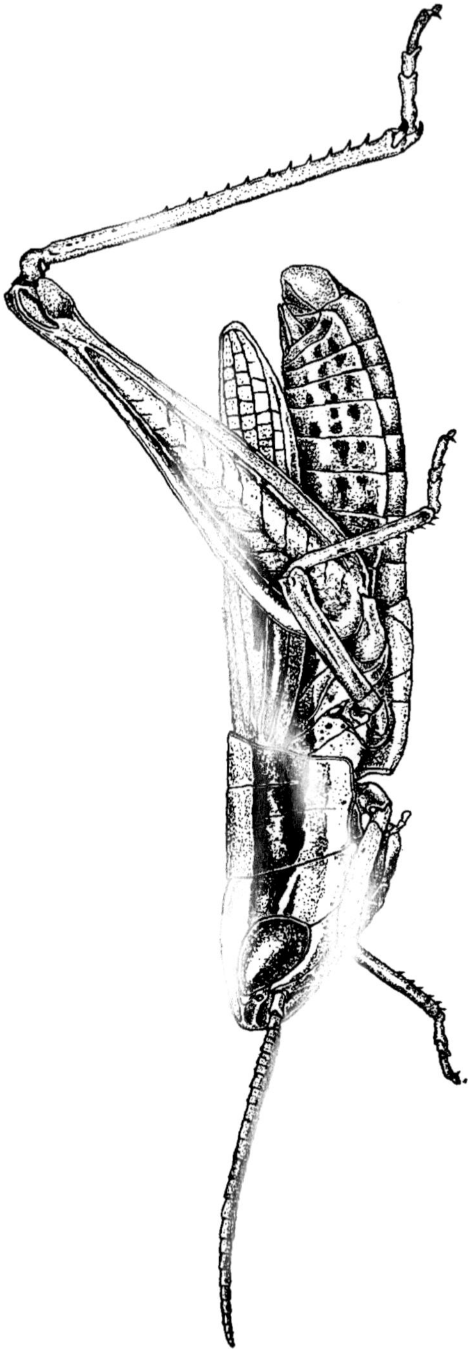

Fig. 728. *Opeia obscura*, male.

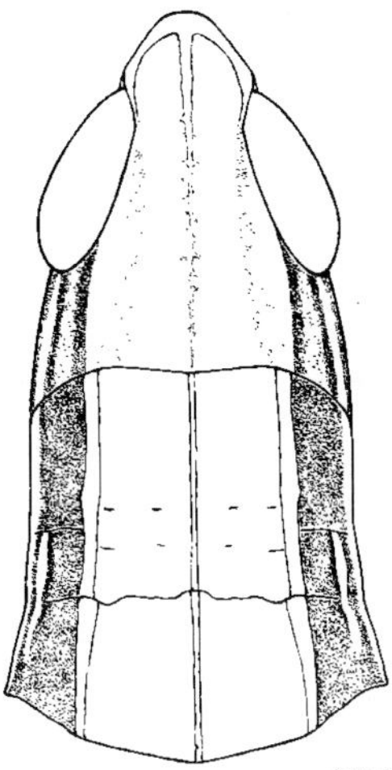

729

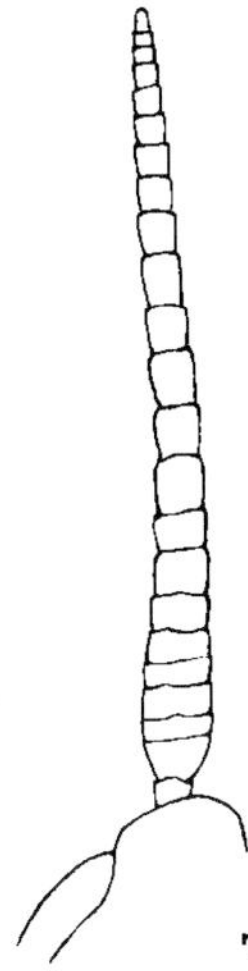

730

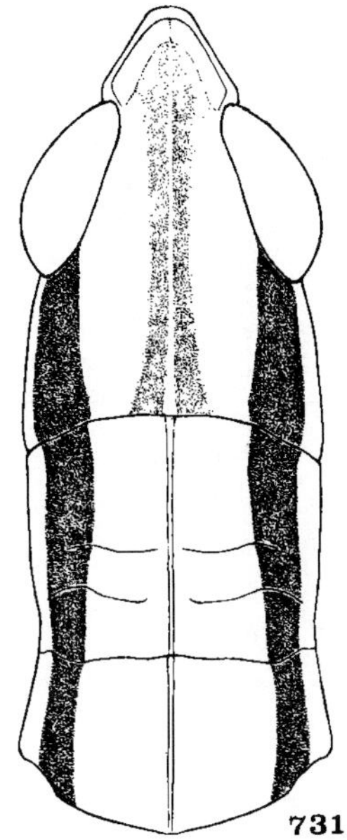

731

732

Figs. 729–732. *Opeia* and *Mermiria* spp. 729, *O. obscura*, head and pronotum, dorsal; 730, the same, male antenna; 731, *M. bivittata maculipennis* head and pronotum, dorsal; 732, the same, base of antenna.

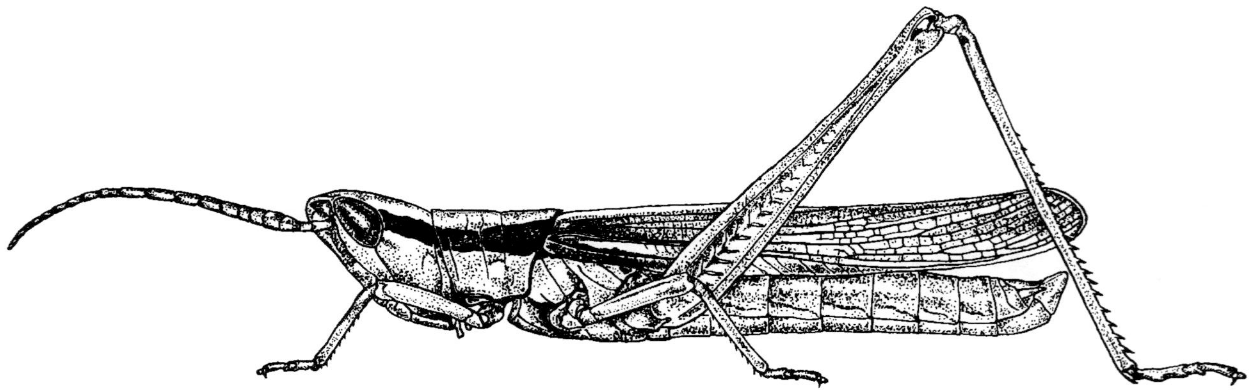

Fig. 733. *Mermiria bivittata maculipennis*, male.

Fig. 734. *Pseudopomala brachyptera*, male.

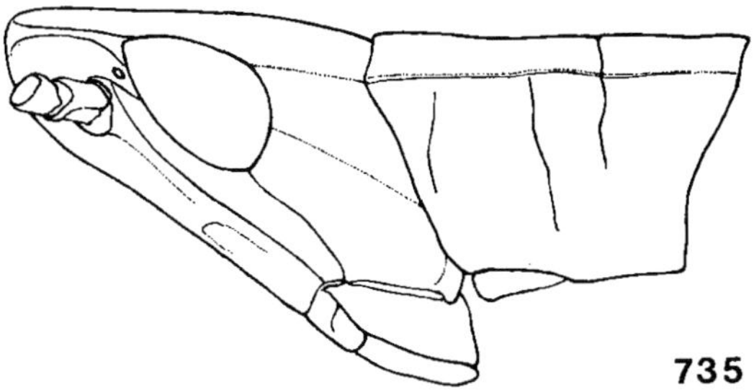

735

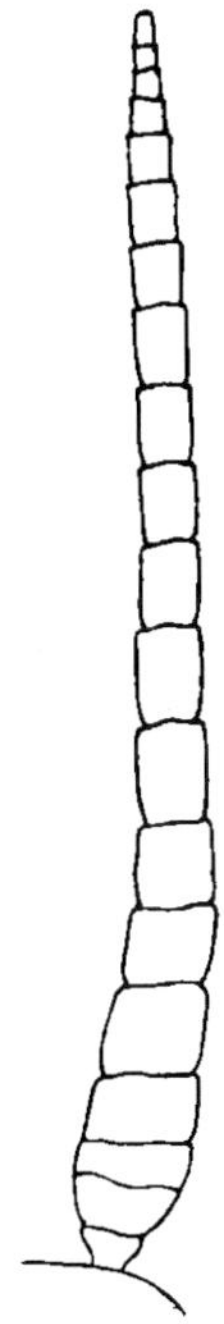

736

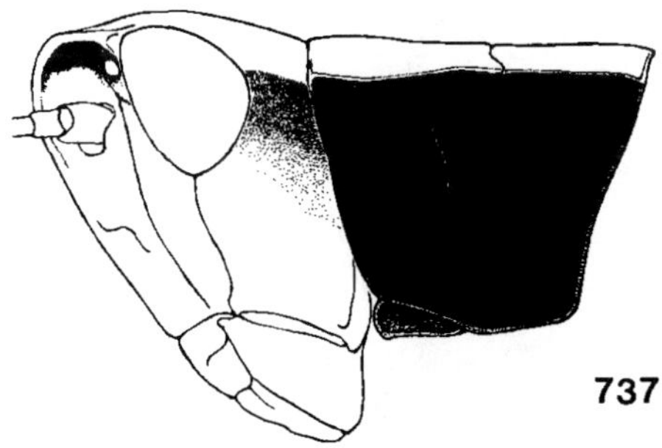

737

Figs. 735–737. *Pseudopomala* and *Chloealtis* spp. 735, *P. brachyptera*, head and pronotum, from left; 736, *P. brachyptera*, male antenna; 737, *C. conspersa*, head and pronotum, male, from left.

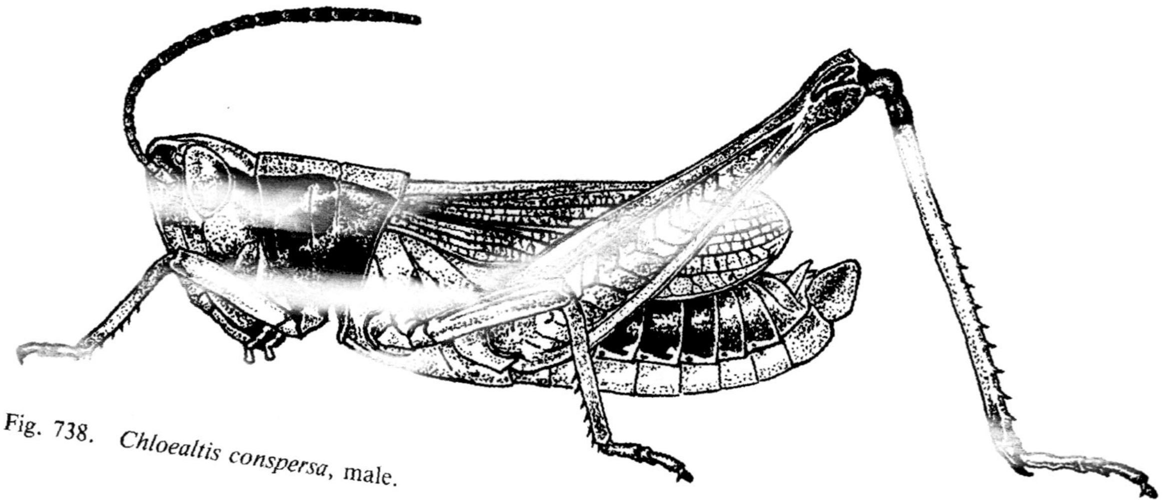

Fig. 738. *Chloealtis conspersa*, male.

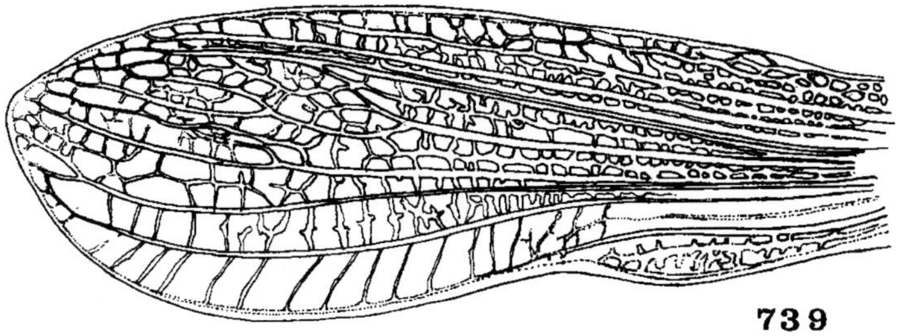

739

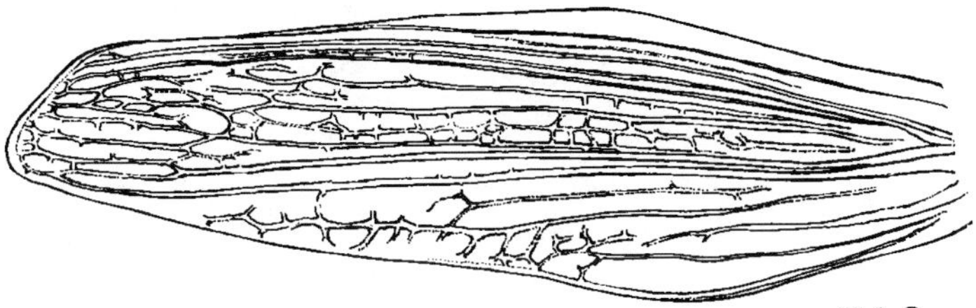

740

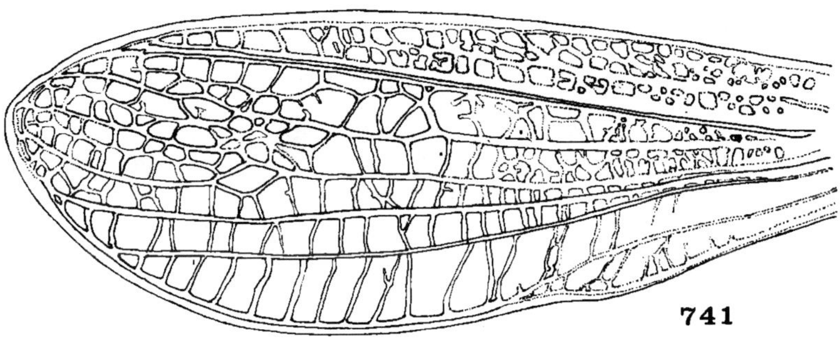

741

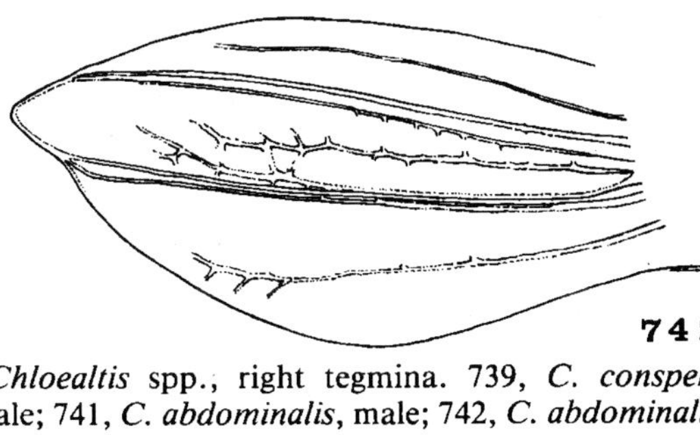

742

Figs. 739–742. *Chloealtis* spp., right tegmina. 739, *C. conspersa*, male; 740, *C. conspersa*, female; 741, *C. abdominalis*, male; 742, *C. abdominalis*, female. After Vickery et al. (1974).

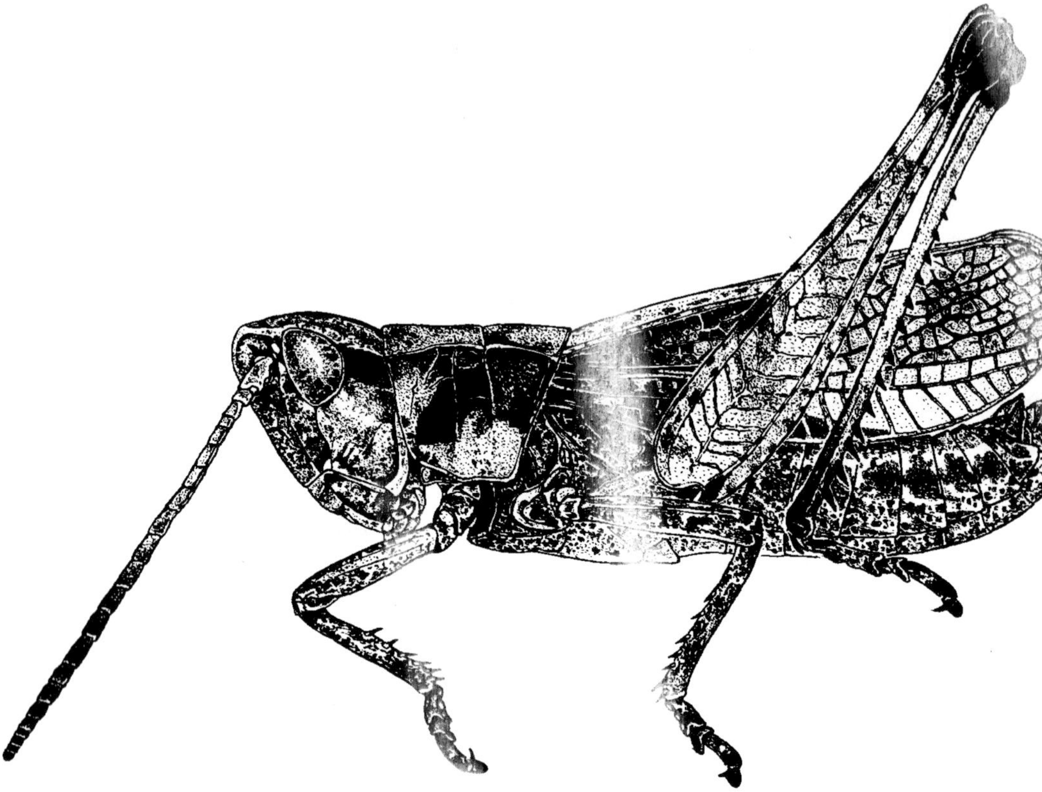

Fig. 743. *Chloealtis abdominalis*, male (tending toward melanic form, Mont Albert, Gaspé, Qué.).

Figs. 746–755. Gomphocerini and Orphulellini, lateral foveolae of vertices, from left. 746, *Chorthippus curtipennis curtipennis*, male; 747, *Bruneria brunnea*, male; 748, *Aeropedullus clavatus*, male; 749, *Psoloessa delicatula buckelli*, male; 750, *Aulocara elliotti*, male; 751, *A. femoratum*, male; 752, *Ageneotettix deorum*, male; 753, the same, female; 754, *Orphulella pelidna pelidna*, male; 755, the same, female.

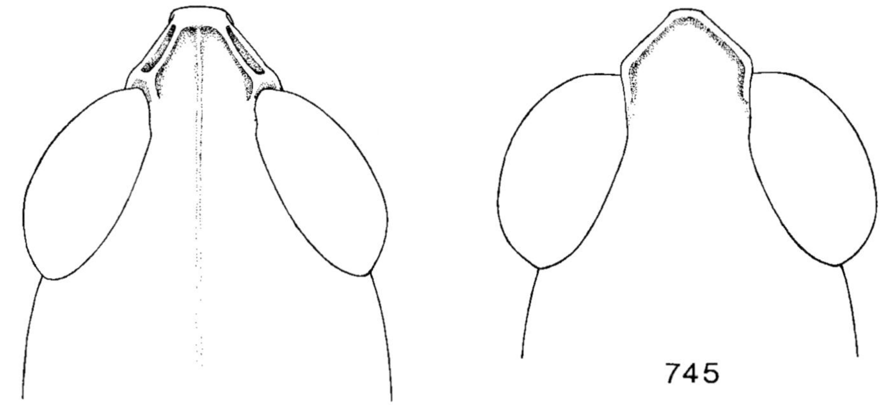

744 745

Figs. 744, 745. Gomphocerini and Orphulellini, head, dorsal, showing fastigia of vertices. 744, *Chorthippus curtipennis curtipennis*; 745, *Orphulella speciosa*. After Vickery et al. (1974).

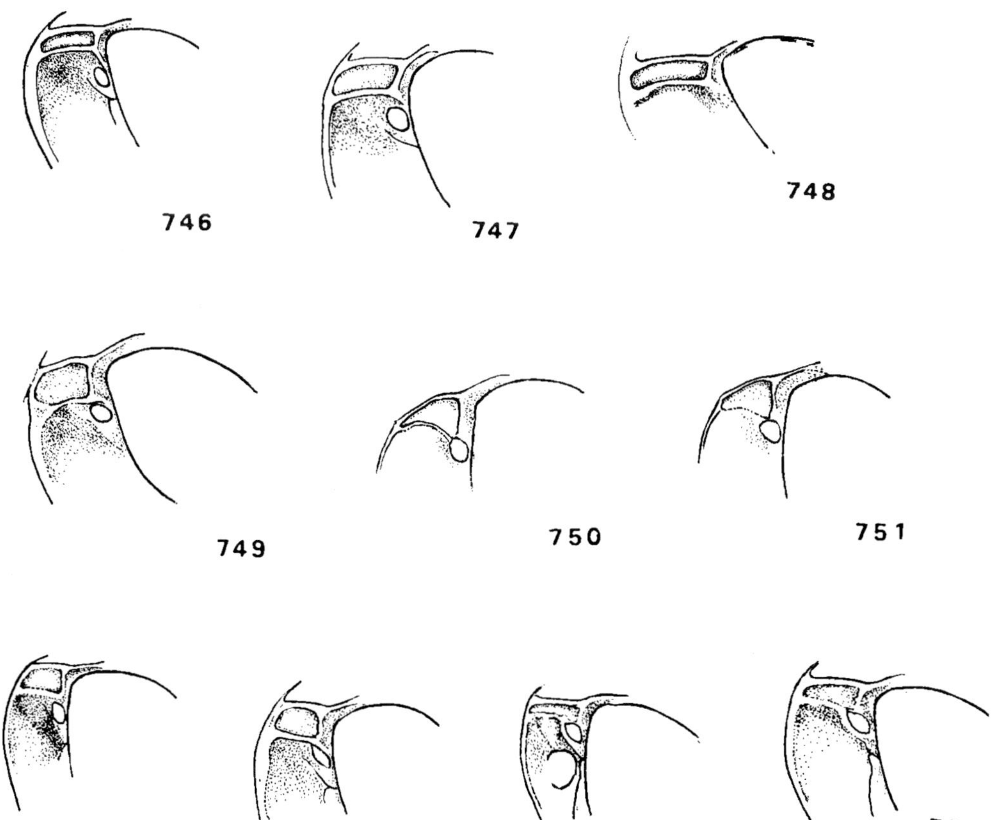

746 747 748

749 750 751

752 753 754 755

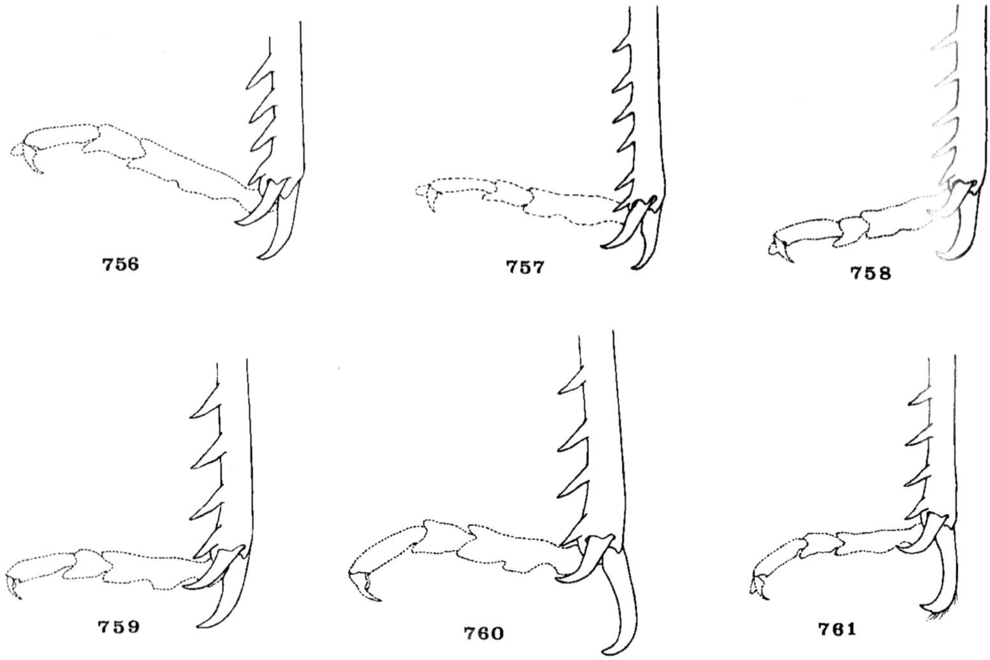

Figs. 756–761. Gomphocerini, apices of female right hind tibiae (hind tarsi in outline). 756, *Bruneria brunnea*; 757, *Amphitornus coloradus coloradus*; 758, *Eritettix simplex tricarinatus*; 759, *Aulocara elliotti*; 760, *Aulocara femoratum*; 761, *Ageneotettix deorum*.

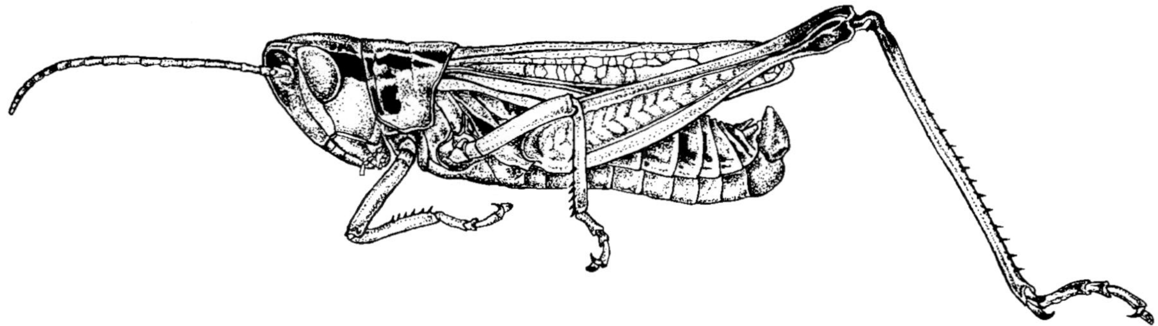

Fig. 762. *Chorthippus curtipennis curtipennis*, male.

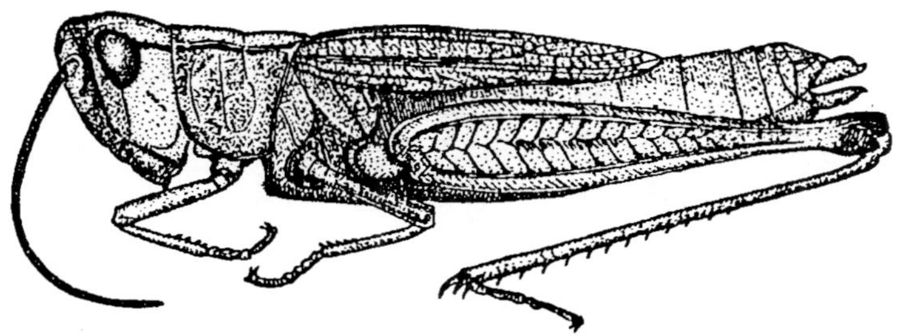

Fig. 763. *Chorthippus curtipennis curtipennis*, female with comparatively long tegmina. After Helfer (1963), courtesy of J. R. Helfer.

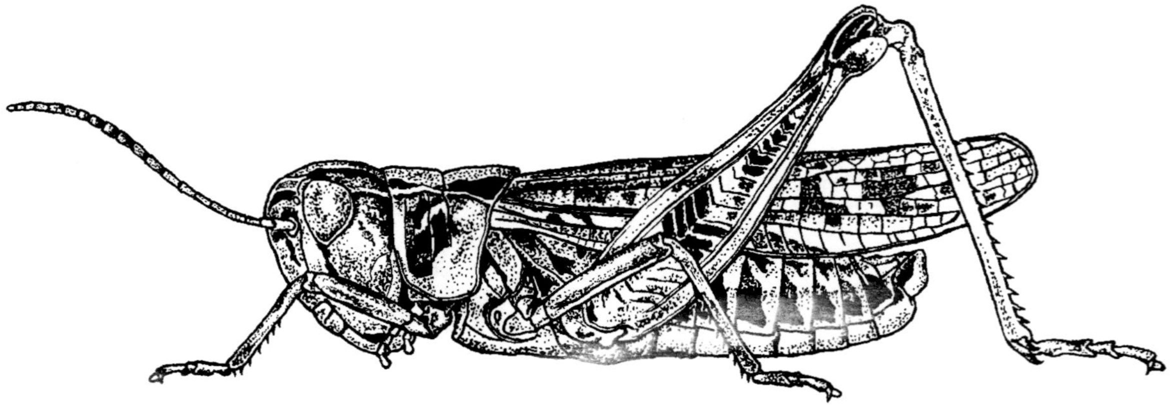

Fig. 764. *Bruneria brunnea*, male.

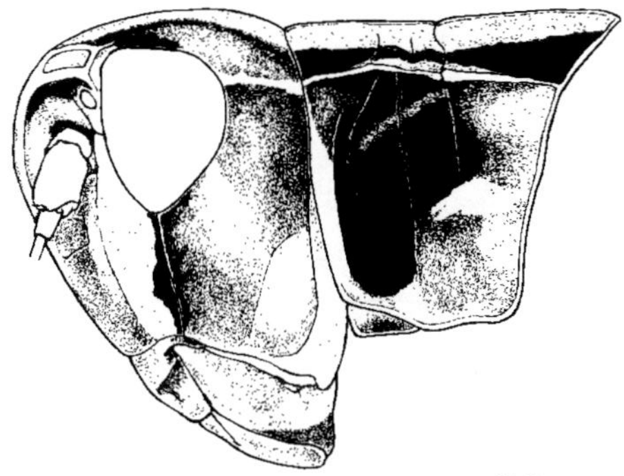

765

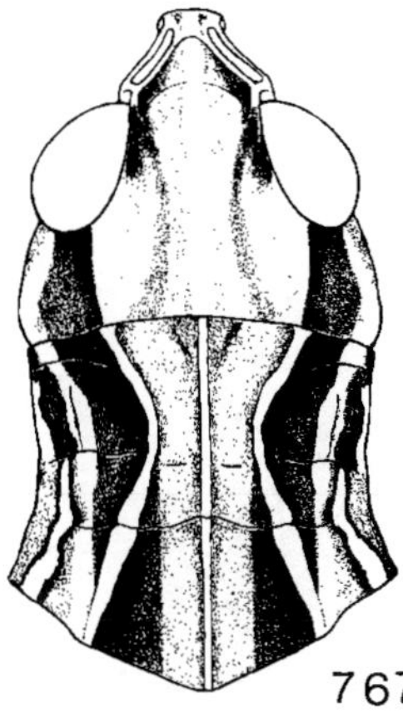

767

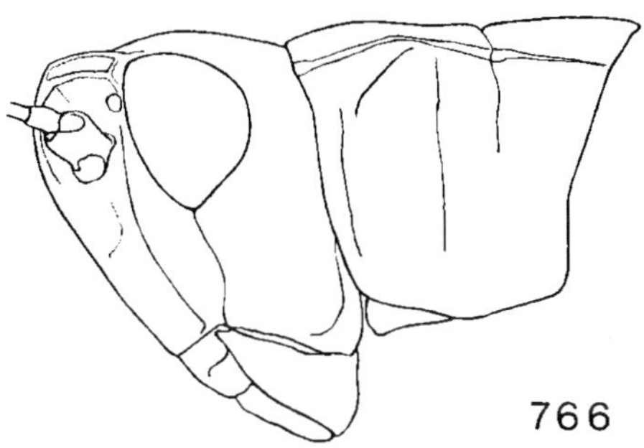

766

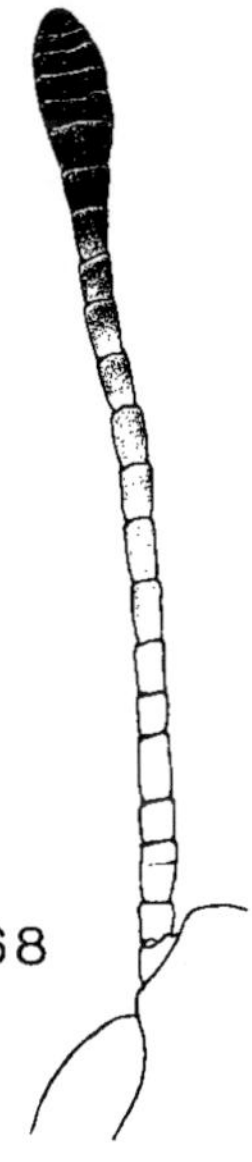

768

Figs. 765–768. *Bruneria* and *Aeropedellus* spp. 765, *B. brunnea*, head and pronotum, male, from left; 766, *A. clavatus*, male, head and pronotum, from left; 767, the same, female, dorsal; 768, the same, left antenna of male.

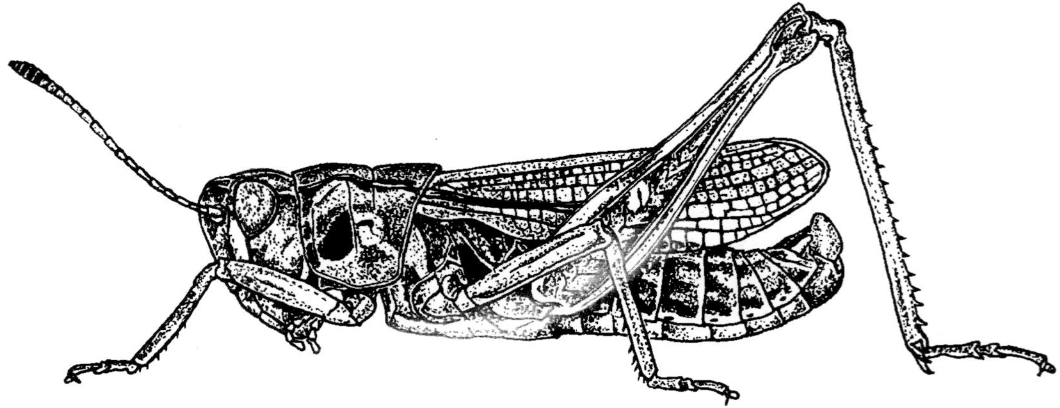

Fig. 769. [illegible]us *arcticus*, male.

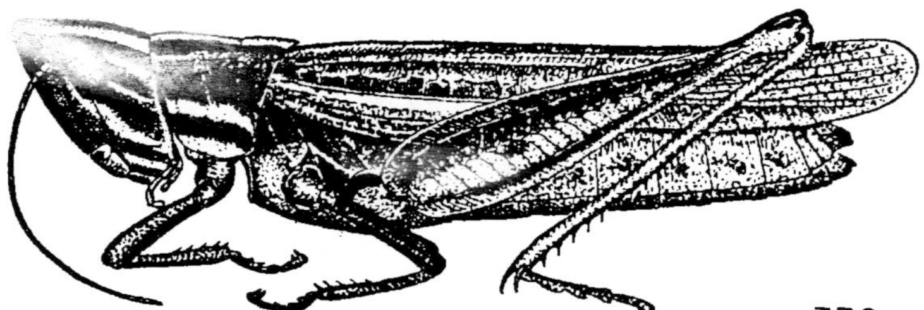

770

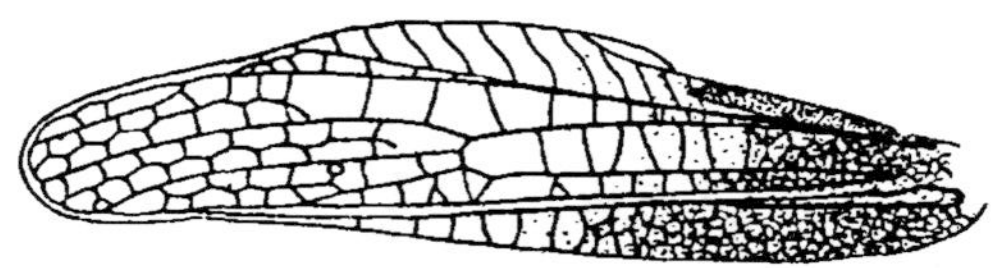

771

Figs. 770, 771. *Amphitornus coloradus ornatus*. 770, female; 771, left male tegmen. After Helfer (1963), courtesy of J. R. Helfer.

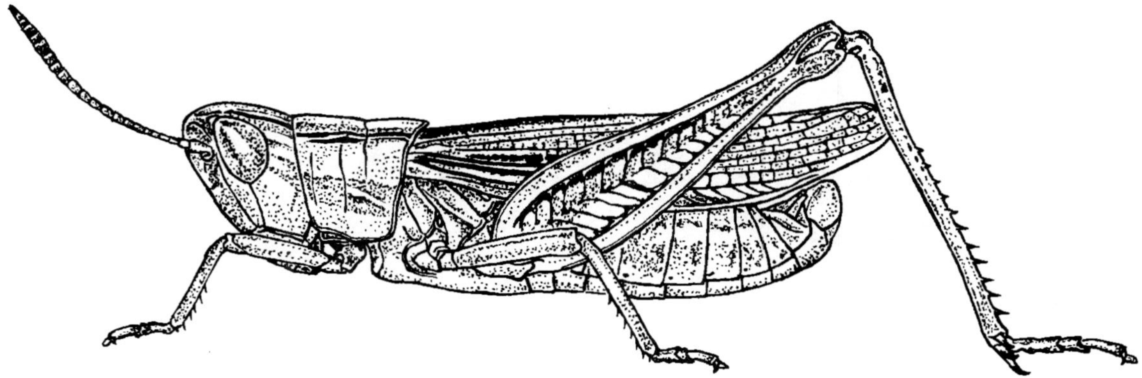

Fig. 772. *Eritettix simplex tricarinatus*, male.

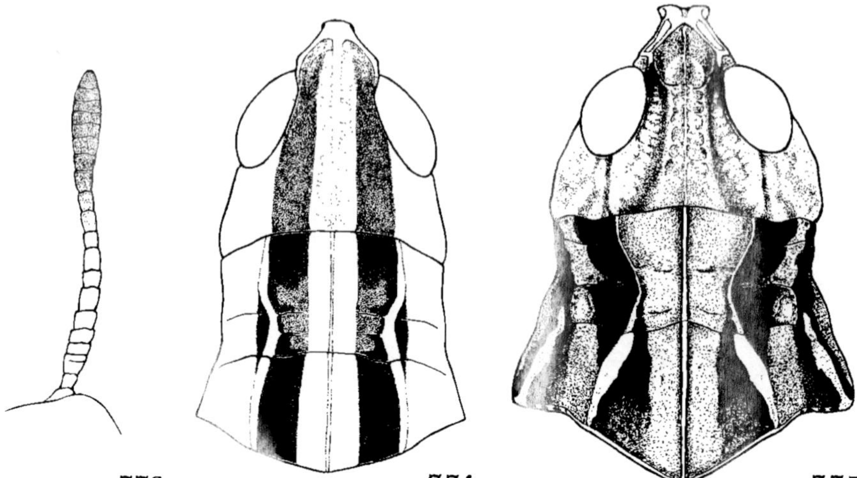

773 774 775

Figs. 773–775. *Eritettix* and *Psoloessa* spp. 773, *E. simplex tricarinatus*, left female antenna; 774, the same, head and pronotum, female, dorsal; 775, *P. delicatula buckelli*, head and pronotum, female, dorsal.

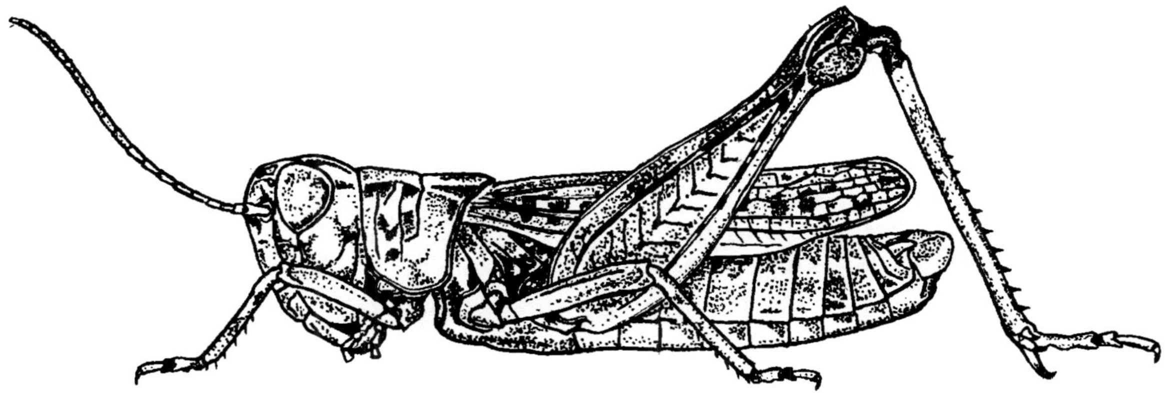

Fig. 776. *Psoloessa delicatula delicatula*, male.

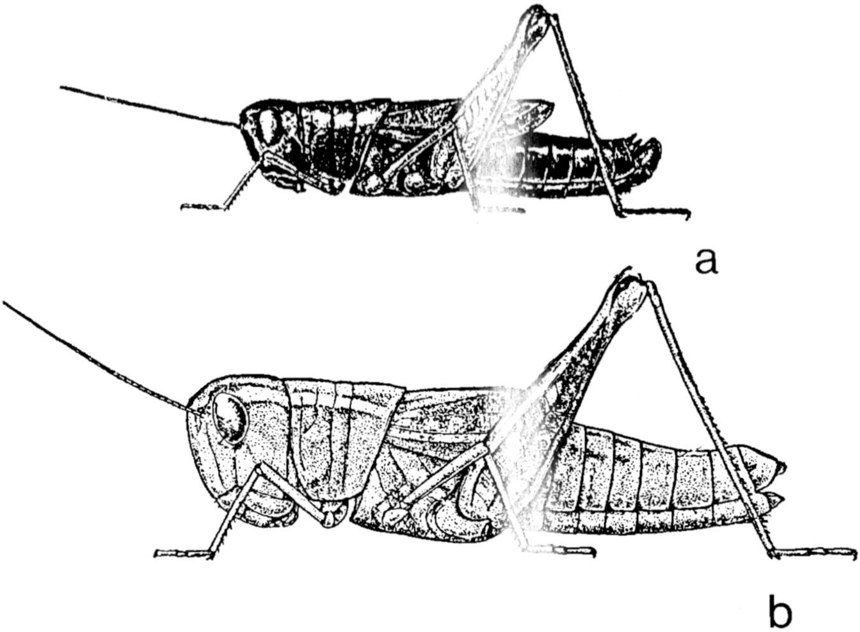

Fig. 777. *Boopedon nubilum*. *a*, male; *b*, female. Adapted from Ball et al. (1942).

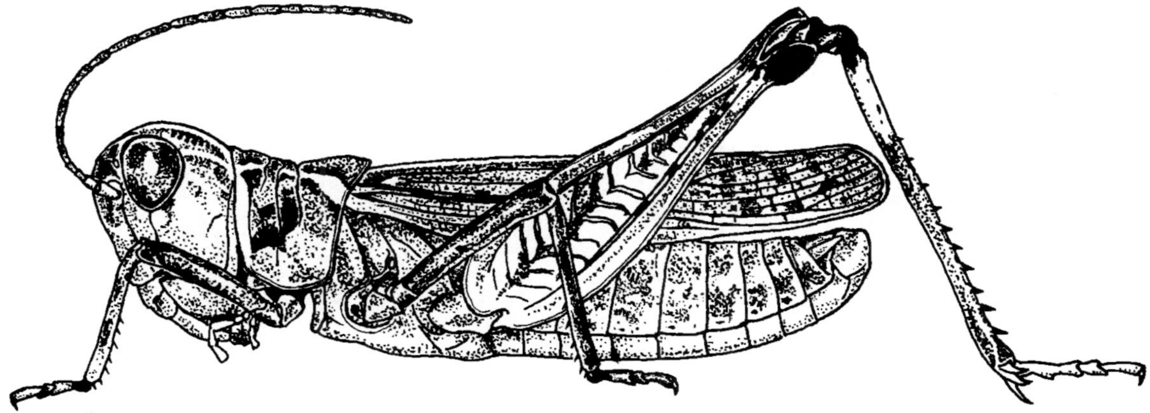

Fig. 778. *Aulocara elliotti*, male.

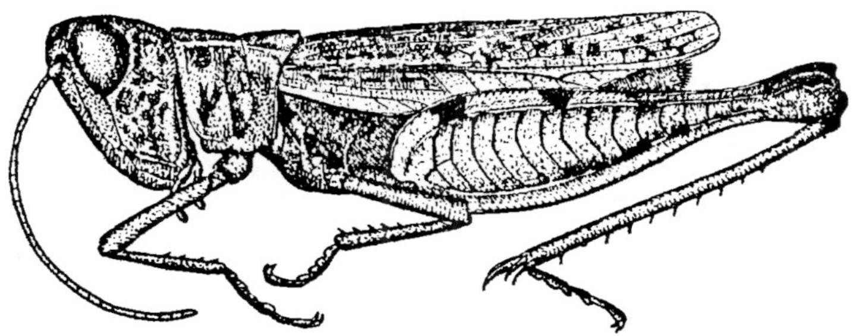

Fig. 779. *Ageneotettix deorum*, male. After Helfer (1963), courtesy of J. R. Helfer.

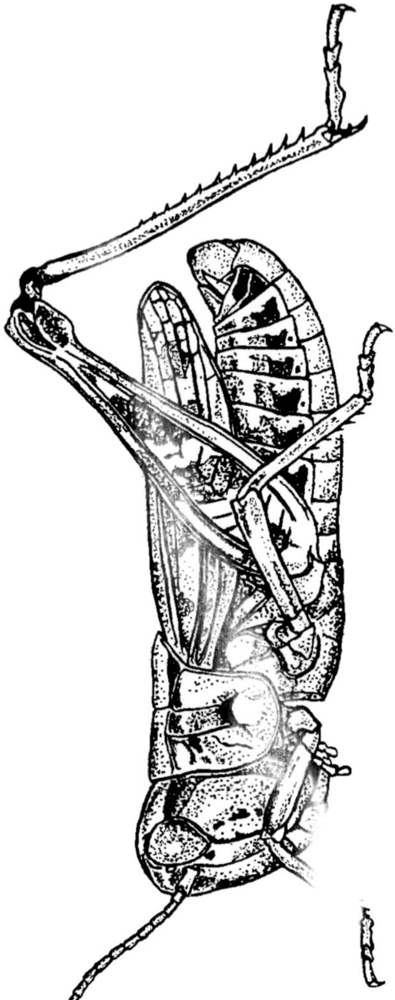

Fig. 780. *Phlibostroma quadrimaculatum*, male.

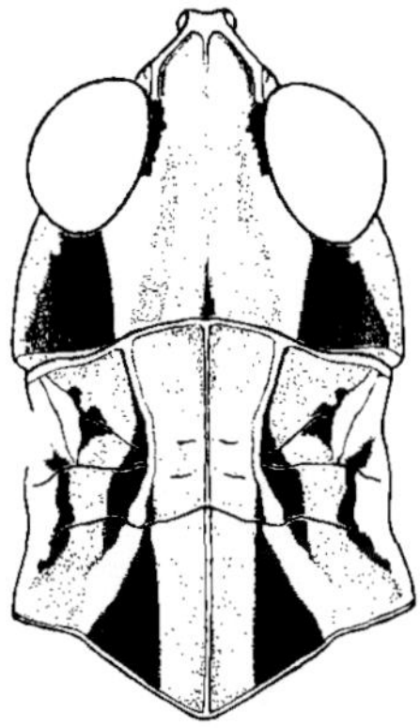

781

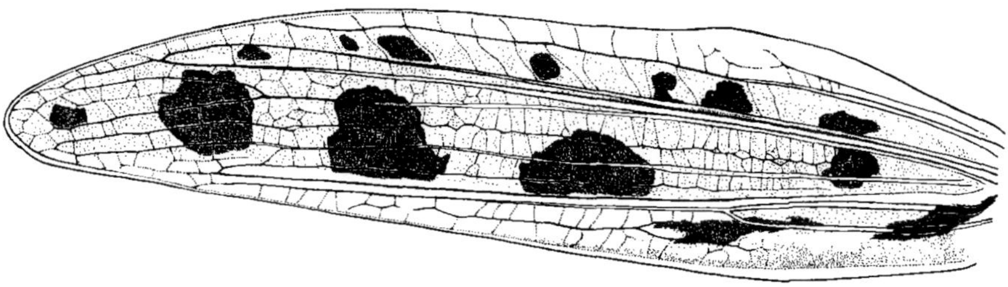

782

Figs. 781, 782. *Phlibostroma quadrimaculatum*. 781, head and pronotum, male, dorsal; 782, left tegmen, female.

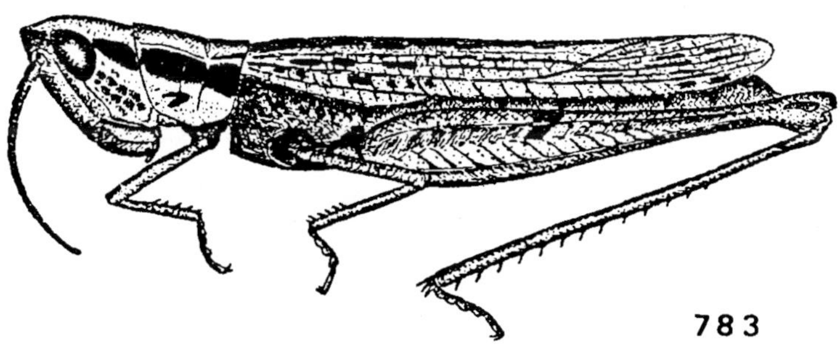

783

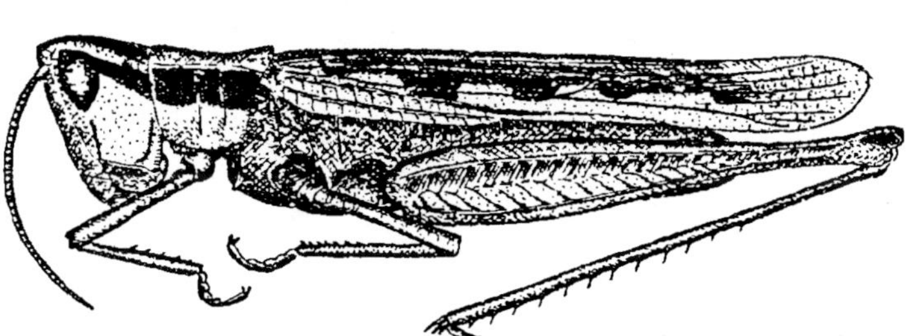

784

Figs. 783, 784. *Cordillacris* spp. 783, *C. occipitalis*, female; 784, *C. crenulata*, female. After Helfer (1963), courtesy of J. R. Helfer.

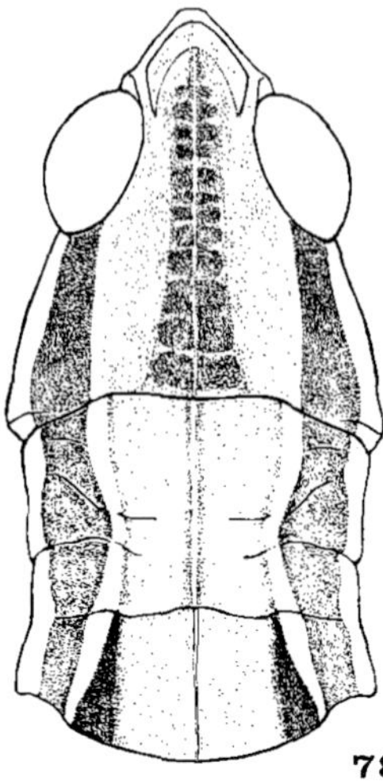

785

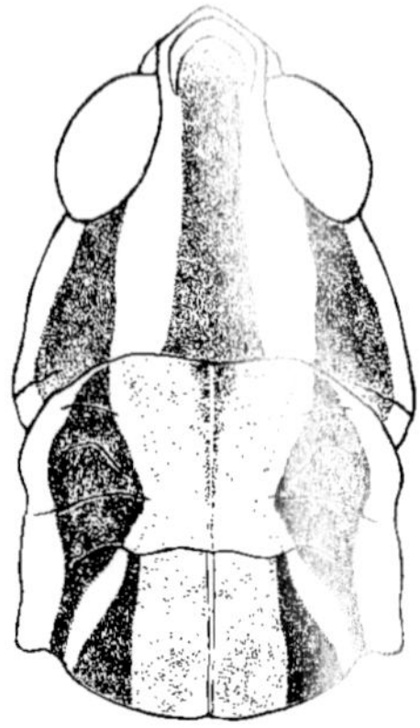

786

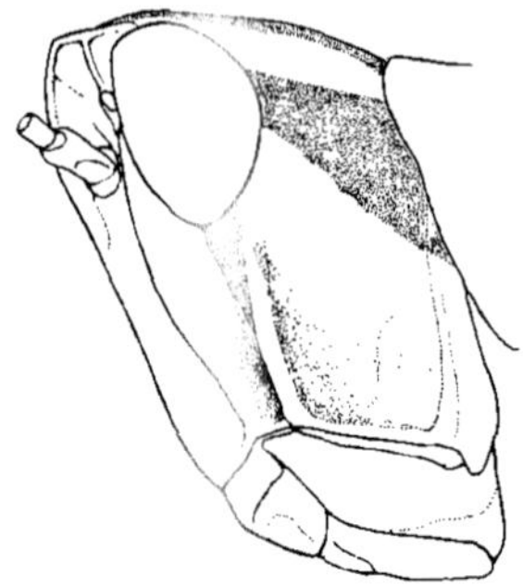

787

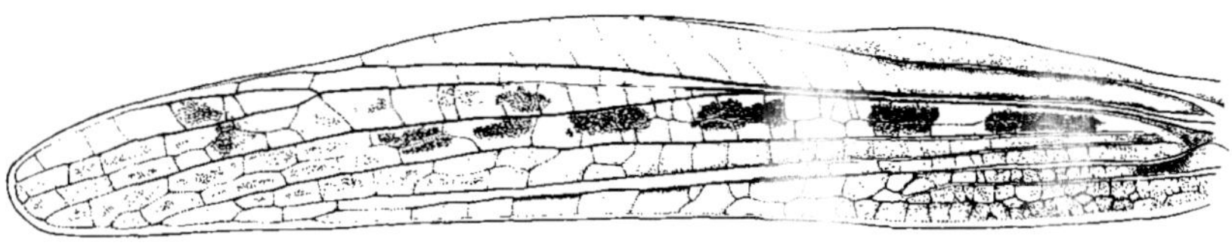

788

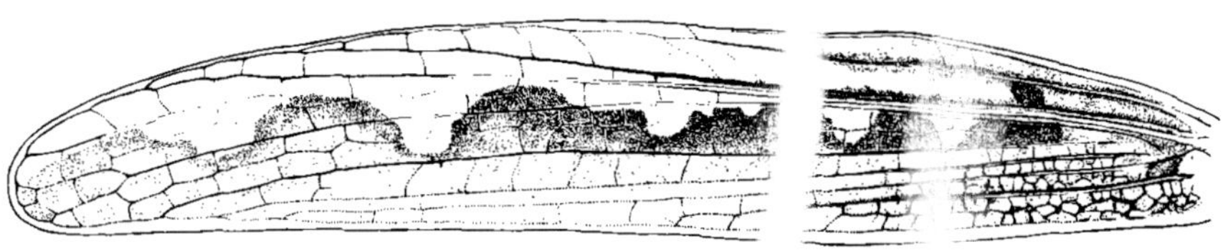

789

Figs. 785–789. *Cordillacris* spp. 785, *C. occipitalis*, head and pronotum, female, dorsal; 786, *C. crenulata*, head and pronotum, female, dorsal; 787, the same, head, female, from left; 788, *C. occipitalis*, left tegmen, male; 789, *C. crenulata*, left tegmen, male.

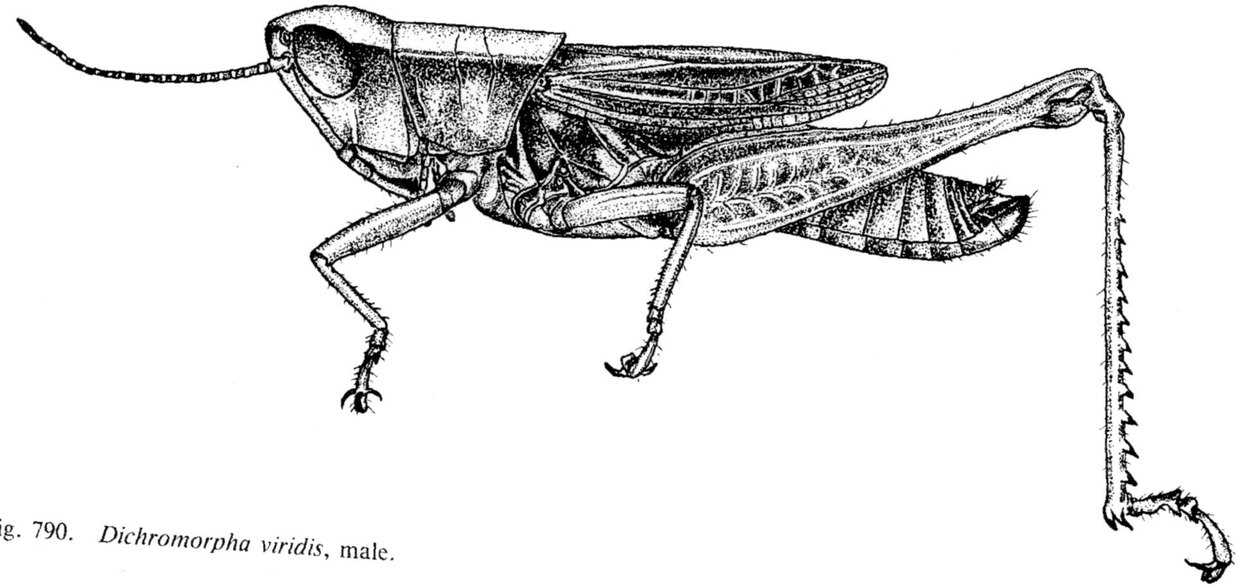

Fig. 790. *Dichromorpha viridis*, male.

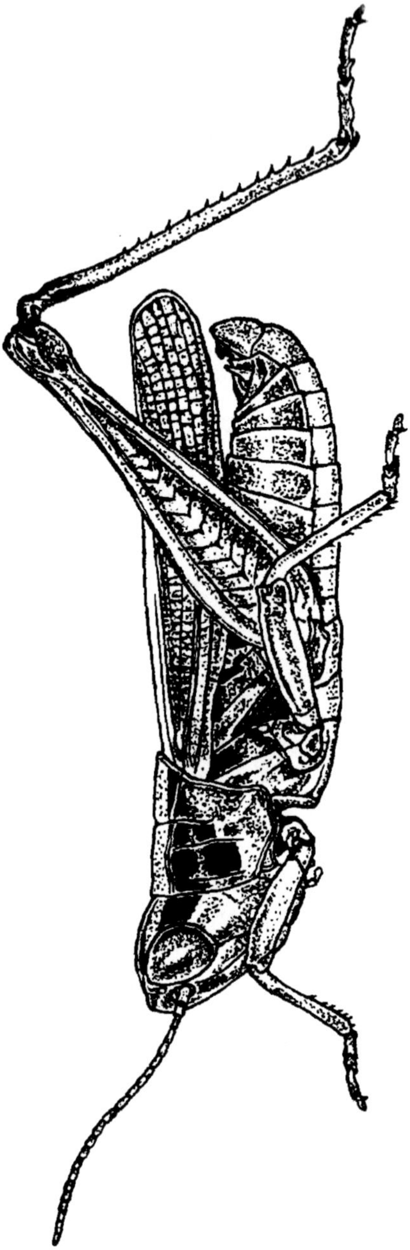

Orphulella speciosa, male.

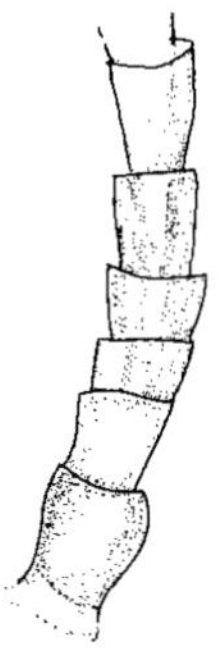

792

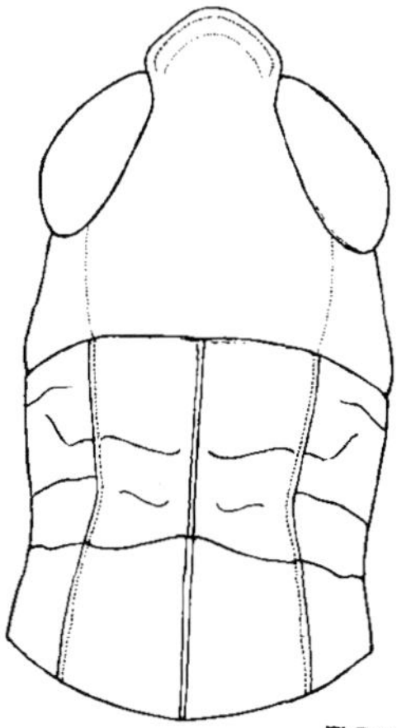

793

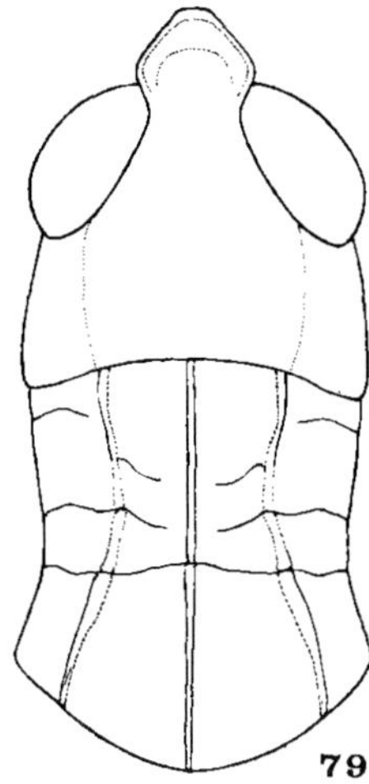

794

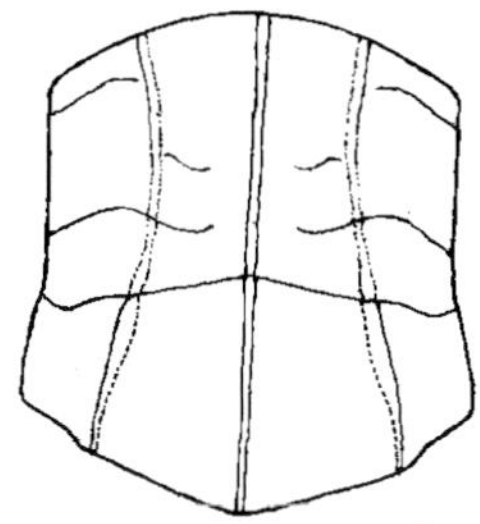

795

Figs. 792–795. *Orphulella* spp. 792, *O. speciosa*, base of antenna; 793, the same, head and pronotum, female, dorsal; 794, *O. pelidna pelidna*, head and pronotum, female, dorsal; 795, *O. pelidna desereta*, pronotum, female, dorsal.

Fig. 796. *Nomotettix cristatus cristatus* male.

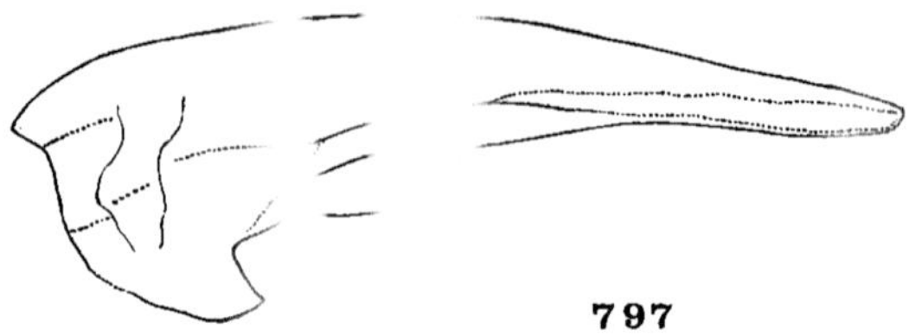

797

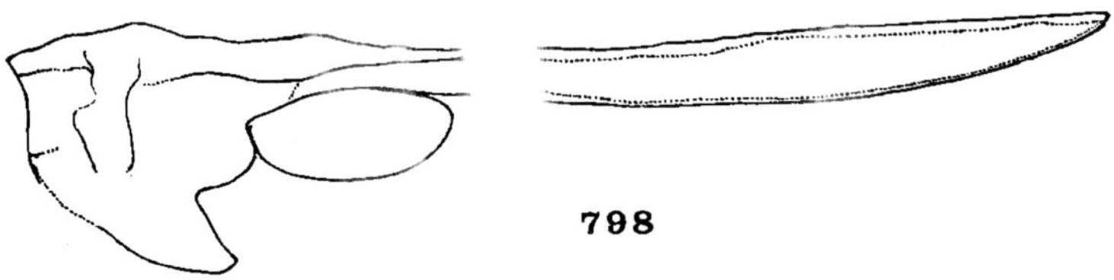

798

Figs. 797, 798. Tetrigini, pronota, from 797, *Nomotettix cristatus cristatus*; 798, *Tetrix arenosa angusta*. After Vicker al. (1974).

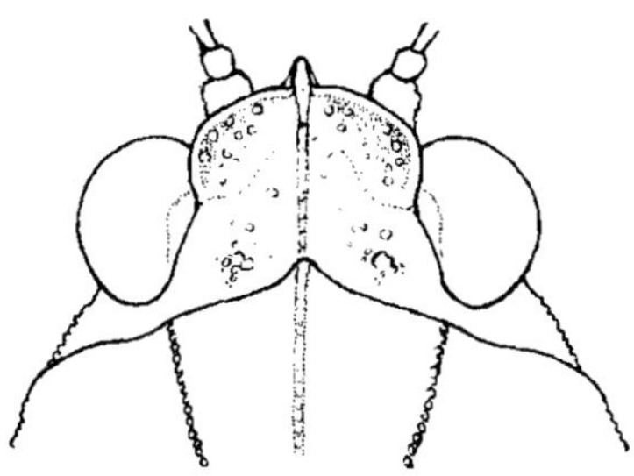

799

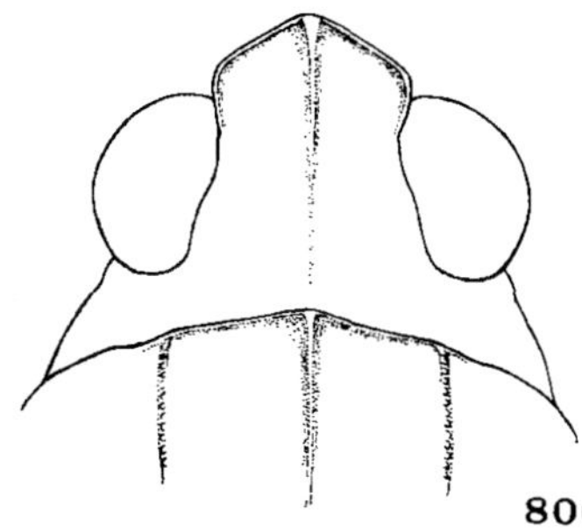

800

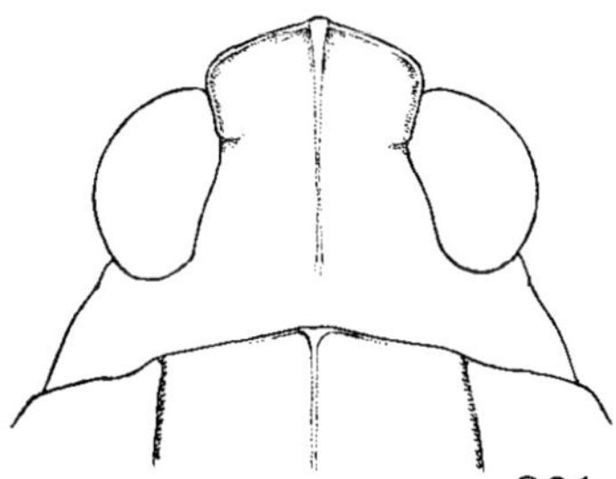

801

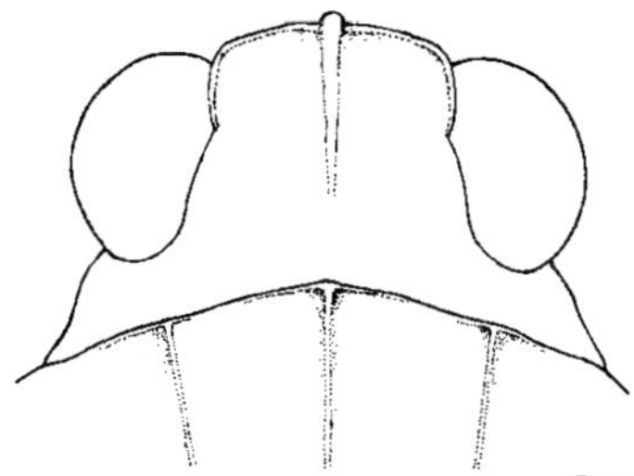

802

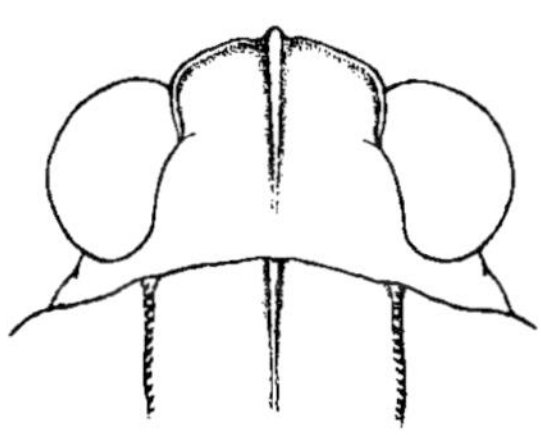

803

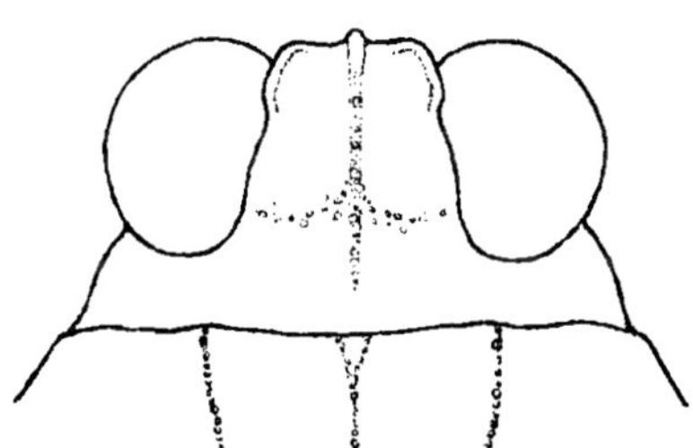

804

Figs. 799–804. Tetrigini, head and anterior margins of pronota, females, dorsal. 799, *Nomotettix cristatus cristatus*; 800, *Tetrix subulata*; 801, *T. brunnerii*; 802, *T. arenosa angusta*; 803, *T. ornata ornata*; 804, *Paratettix cucullatus*. 800, 802, 803, after Vickery et al. (1974).

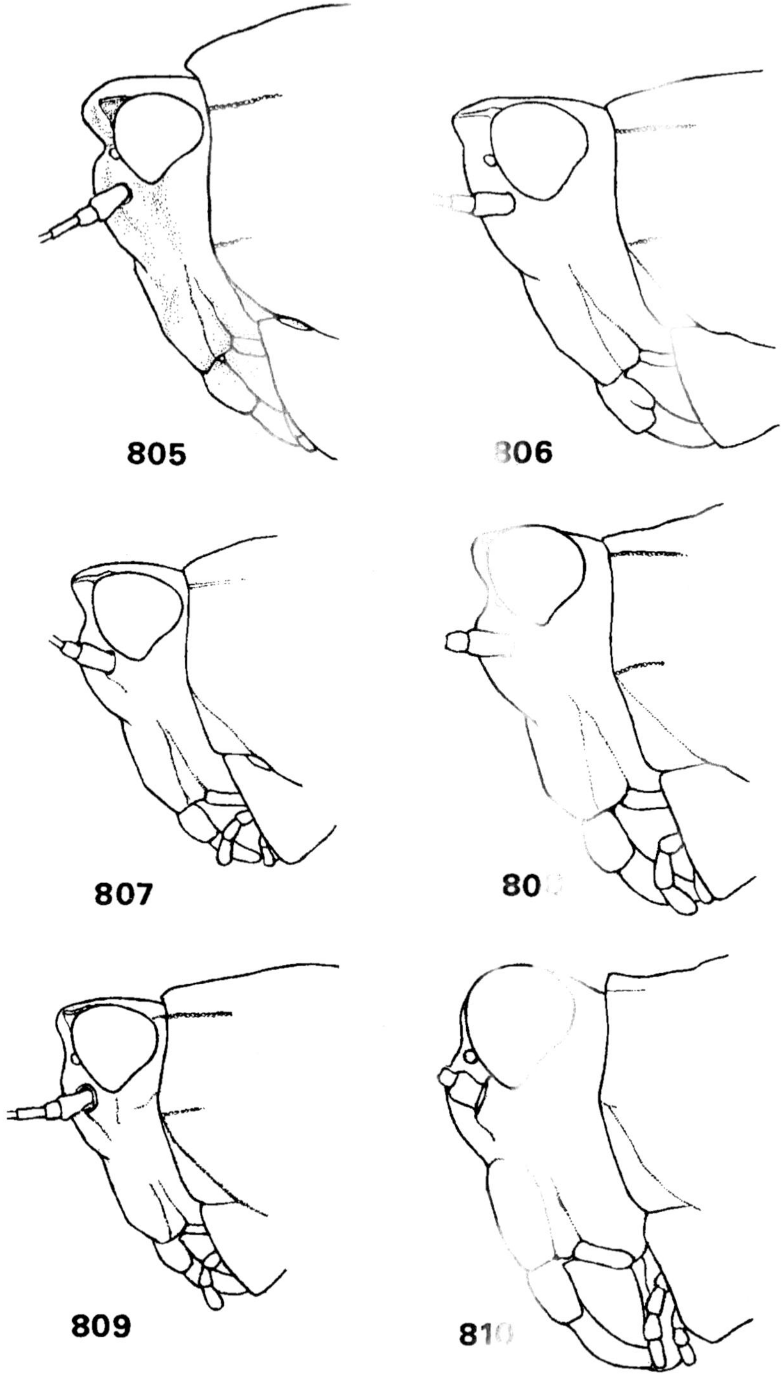

Figs. 805–810. Tetrigini, heads and anterior margins of pronota, from left. 805, *Nomotettix cristatus cristatus*, female; 806, *Tetrix subulata*, female; 807, *T. brunnerii*, male; 808, *T. arenosa angusta*, female; 809, *T. ornata ornata*, male; 810, *Paratettix cucullatus*, female.

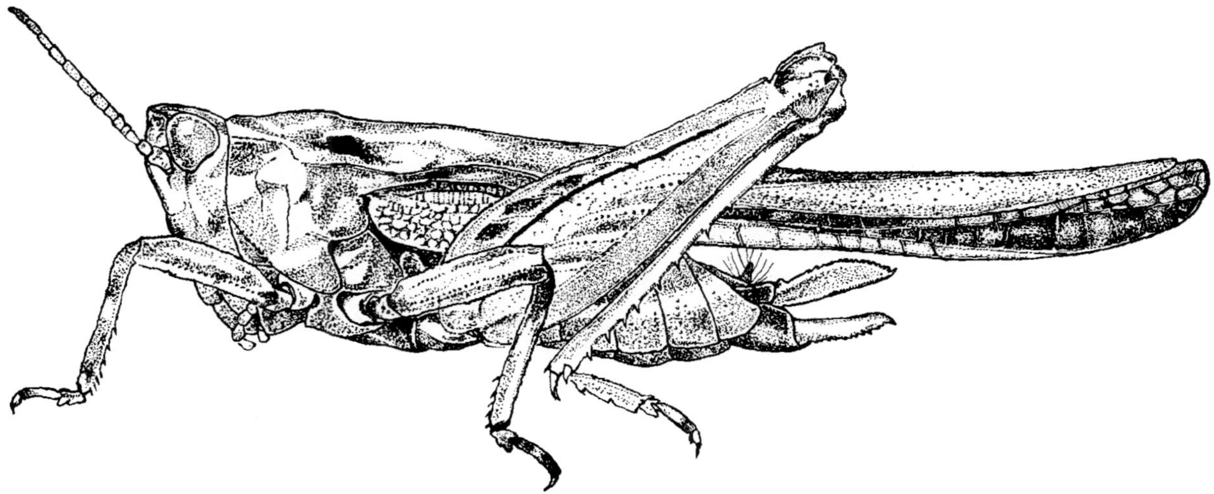

Fig. 811. *Tetrix subulata*, female.

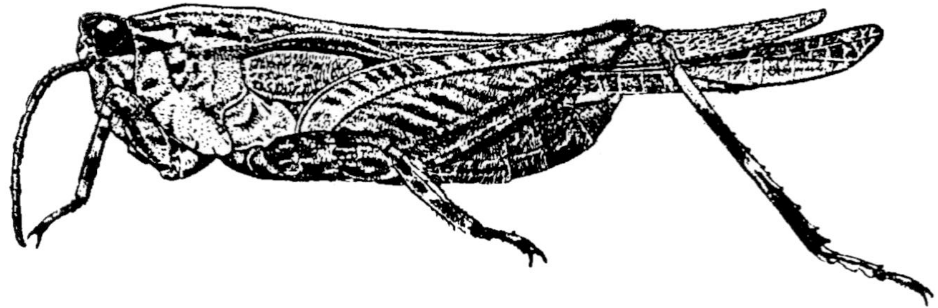

Fig. 812. *Tetrix ornata ornata*, male. Adapted from Hebard (1934*a*).

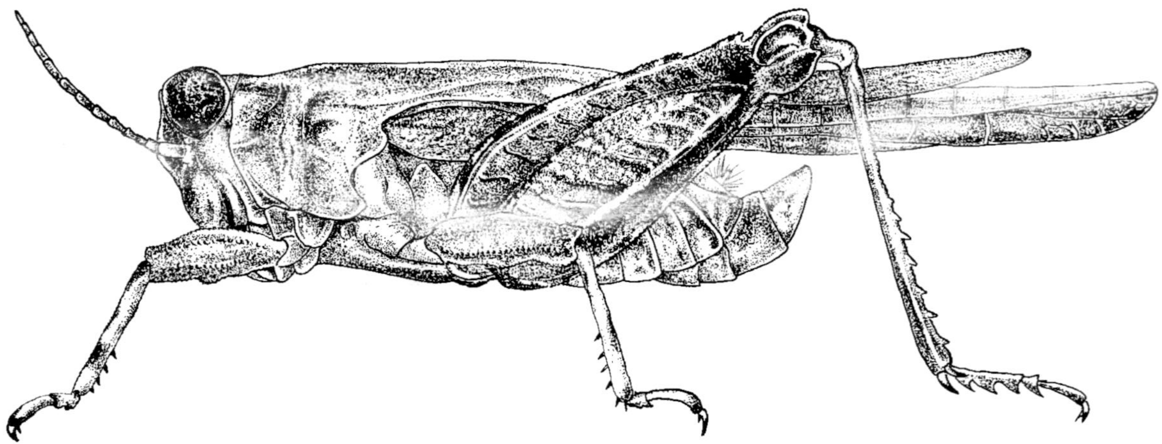

Fig. 813. *Paratettix cucullatus*, male.

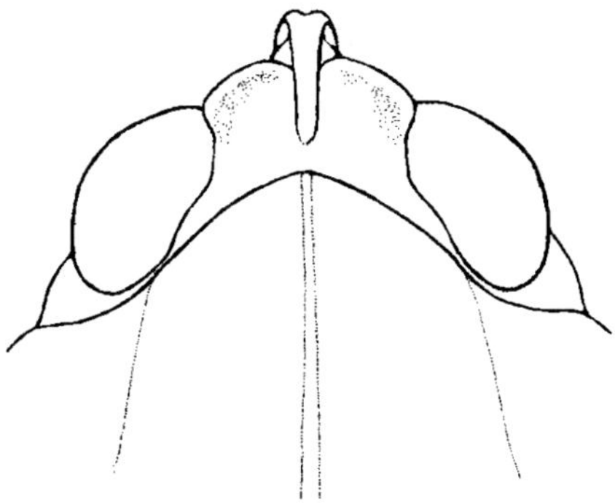

814

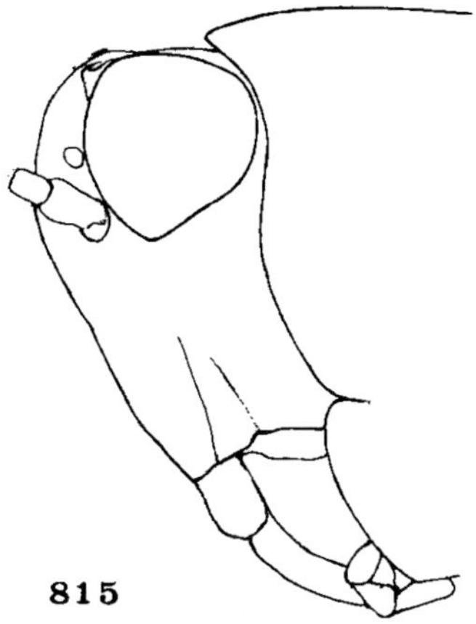

815

Figs. 814, 815. *Tettigidea lateralis*, intermediate between subspecies *lateralis* and *cazieri*, head and pronotum, female. 814, dorsal; 815, from left.

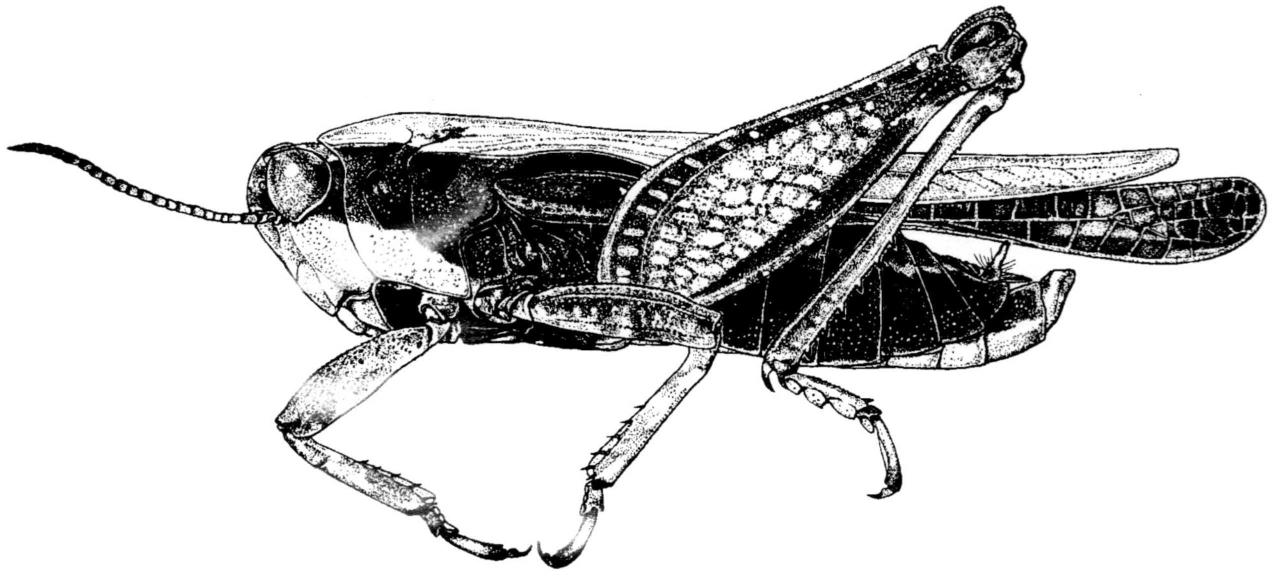

Fig. 816. *Tettigidea lateralis*, intermediate between subspecies *lateralis* and *cazieri*, male.

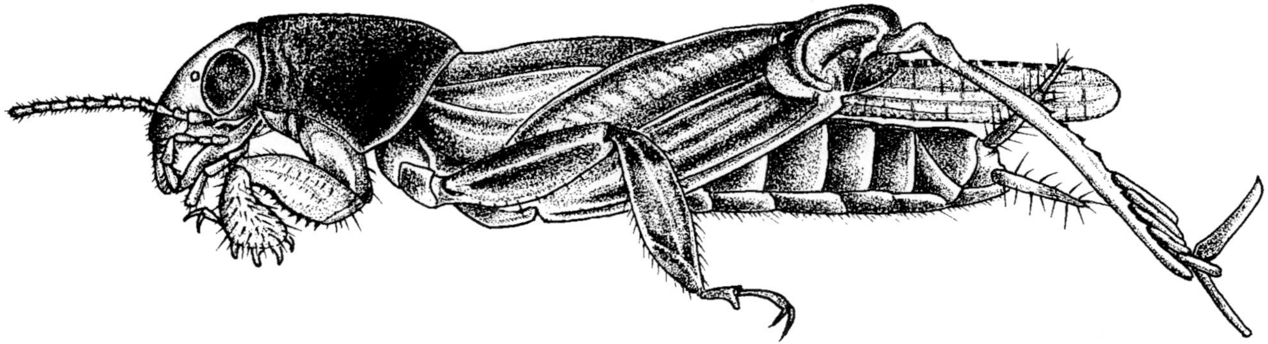

Fig. 817. *Neotridactylus apicialis*, male.

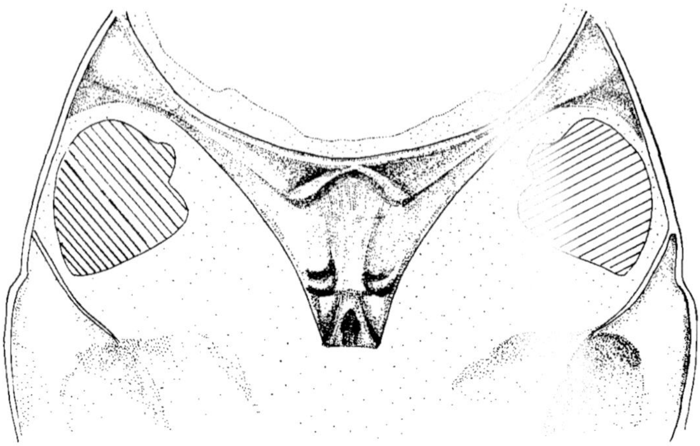

818

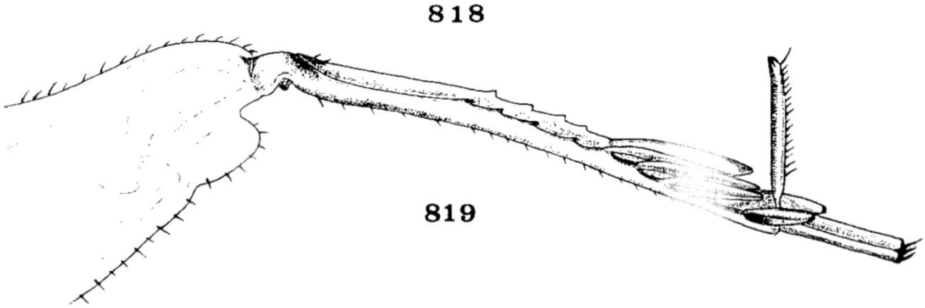

819

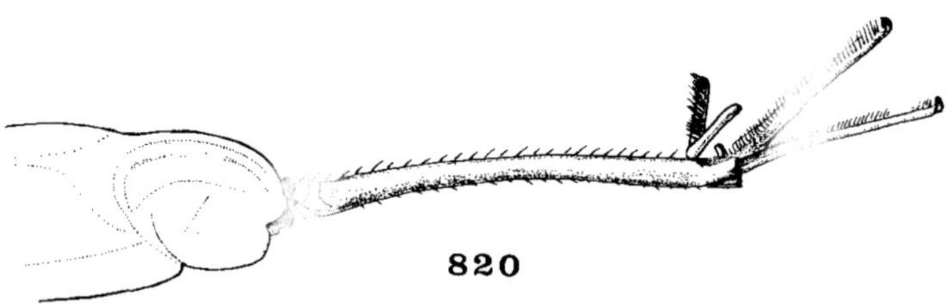

820

Figs. 818–820. Tridactylidae. 818, *Neotridactylus apicialis*, prosternum (crosshatched areas indicate positions of anterior coxae of legs); 819, the same, hind leg showing apex of hind femur, tibia, and tarsus; 820, *Ellipes minutus minutus*, hind leg showing apex of hind femur, tibia, and tarsus.

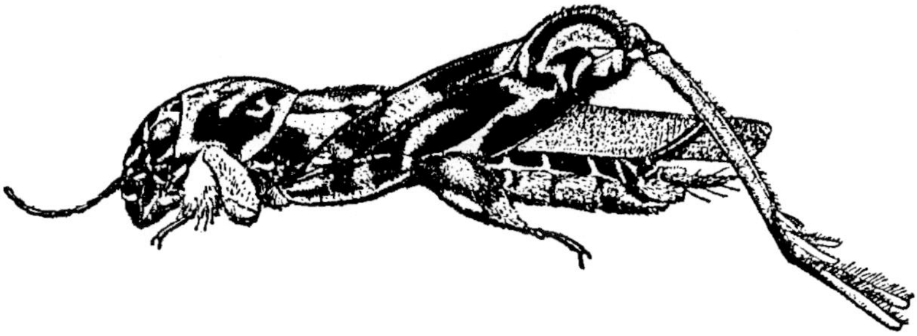

Fig. 821. *Ellipes minutus minutus*, male. Adapted from Hebard (1934*a*).

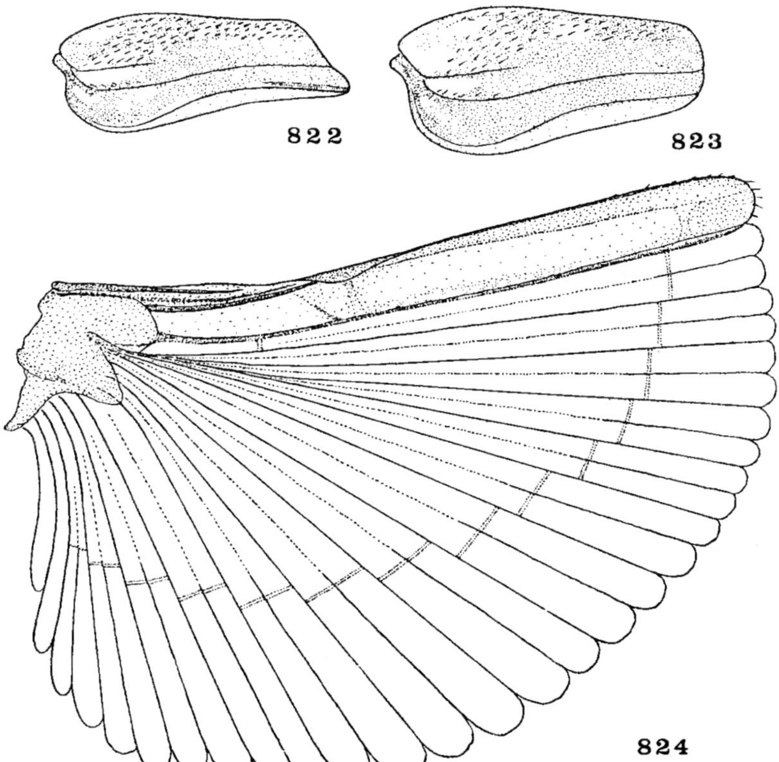

Figs. 822–824. *Ellipes minutus minutus*, left tegmina and hind wing, ventral. 822, male tegmen (showing row of stridulatory teeth); 823, female tegmen (without stridulatory teeth); 824, male hind wing. After K. K. Günther (1977).

List of species recorded as adventives, temporary introductions, or laboratory cultures

Order Dermaptera

Chelisochidae

Chelidurella acanthopygia (Gené). Fig. 99. Native to Europe. Adventive.

Forficulidae

Doru davisi Rehn & Hebard. Fig. 103. Native to Florida. Potential adventive.

Doru taeniatum (Dohrn). Figs. 104, 105. Native to southern United States, Mexico, and Central and South America. Adventive.

Order Dictuoptera

Mantidae

Stagmomantis carolina (Johannson). Fig. 77. Carolina mantid. Native to the United States (Utah to Indiana, south to Florida), Central America, and Cuba. Adventive.

Blattidae

Periplaneta fuliginosa (Audinet-Serville). Smokybrown cockroach. Adventive. Established in southern United States.

Eurycotis opaca (Brunner von Wattenwyl). Adventive. Intercepted on bananas. Vickery et al. (1974).

Eurycotis tibialis (Hebard). Adventive.

Eurycotis floridana (F. Walker). Fig. 24. Adventive.

Eurycotis caraibea (Bolívar). Adventive. Intercepted on fruit from Florida. Vickery et al. (1974).

Blaberidae

Blaberus giganteus (Linnaeus). Adventive. Intercepted on tropical fruit. Vickery et al. (1974). Cultured for laboratory study.

Blaberus discoidalis Audinet-Serville. Fig. 26. Temporary introduction in greenhouses in the United States. Cultured for laboratory study.

Blaberus craniifer Burmeister. Fig. 25. Native to Cuba, Mexico, and Central America. Introduced into southern Florida. Potential adventive.

Sibylloblatta panesthoides (F. Walker). Fig. 27. Apparently native to Jamaica. Potential adventive.

Nauphoetidae

Panchlora nivea (Linnaeus). Cuban cockroach. Fig. 28. Native to Caribbean, Central America, and southern United States. Adventive.

Panchlora latipennis Saussure & Zehntner. Native to Central America. Adventive.

Panchlora exoleta Burmeister. Native to tropical America. Adventive. Intercepted in bananas.

Nauphoeta cinerea (Olivier). Cinereous cockroach. Fig. 29. Native to Africa, now established in Bermuda, Florida, West Indies, Mexico, and Central and South America. Adventive. Vickery et al. (1974).

Rhyparobia maderae (Fabricius). Madeira cockroach. Fig. 30. Native to Africa, although described from Madeira Islands. Introduced into tropical America and into houses in New York City. Adventive in Canada. Cultured for laboratory study.

Pycnoscelus surinamensis (Linnaeus). Surinam cockroach. Fig. 31. Native to Far East, now cosmopolitan in warm countries. Adventive. Sometimes established in greenhouses.

Diplopteridae

Diploptera punctata (Eschscholtz). Native to Indo-Pacific region. Cultured for laboratory study.

Epilampridae

Epilampra maya maya J. A. G. Rehn. Fig. 32*b*. Native to Mexico and South America. Adventive.

Nauclidas nigra (Brunner von Wattenwyl). Native to West Indies. Adventive. Vickery et al. (1974).

Nyctiboridae

Nyctibora laevigata (Palisot de Beauvois). Native to West Indies. Adventive.

Nyctibora noctivaga J. A. G. Rehn. Native to tropical America. Adventive. Vickery et al. (1974).

Blattellidae

Neoblattella spp. Figs. 35, 36. Native to tropical America. Adventive. Possible species: *N. detersa* (Walker), *N. tridens* Rehn & Hebard, *N. carcina* Rehn & Hebard, *N. celeripes* Rehn & Hebard.

Symploce pallens (Stephens). Native to Florida, West Indies, and Mexico; more or less cosmopolitan in warm countries. Cultured for laboratory study.

Ectobius sylvestris (Poda). Native to Europe. Established at Geneva, New York.

Order Grylloptera

Phaneropteridae

Leptophyes punctatissima (Bose d'Antic). Fig. 189; Map 42. Native to Europe. Adventive.

Phaneroptera gracilis gracilis Burmeister. Fig. 190. Native to Africa and southern Asia. Adventive.

Pseudophyllidae

Nesoecia nigrispina (Stål). Fig. 221. Native to Central America. Adventive.

Conocephalidae

Neoconocephalus triops (Linnaeus). Native to warmer parts of the Americas. Adventive.

Belocephalus subapterus Scudder. Fig. 288. Native to Florida. Adventive.

Gryllidae

Gryllus assimilis (Fabricius). Native to West Indies and Central and South America. Established in California, Arizona, Texas, and Florida. Adventive. Cultured for fish food and for laboratory study.

Gryllodes supplicans (F. Walker). Fig. 335. Native to southern Asia. Established in Florida and the Bermudas. Potential adventive. Cultured for laboratory study.

Eneopteridae

Hapithus vagus Morse. Figs. 367, 368. Origin uncertain, but somewhere in tropical America. Once established in greenhouse in Massachusetts, but subsequently eradicated (Morse 1916, 1919*b*).

Order Orthoptera

Romaleidae

Romalea guttata (Houttuyn) (= *R. microptera* (Palisot de Beauvois); see Kevan (1980*a*) for nomenclature). Eastern lubber grasshopper. Fig. 389. Native to southern United States (Florida and the Gulf States). Adventive. Imported in preserved state for laboratory study.

Acrididae

Oxya hyla intricata (Stål). Native to Far East. Adventive. Vickery and Kevan (1967).

Schistocerca rubiginosa (Harris). Figs. 396, 401–404, 411. Native to southern United States (New Jersey to Florida and the Gulf States). Confused with various Canadian and northern United States species in the past (Hubbell 1960).

Schistocerca nitens nitens (Thunberg). Vagrant grasshopper. Native to southwestern United States, and Central and South America. Adventive.

Schistocerca gregaria gregaria (Forskål). Desert locust. Native to Africa and Asia. Cultured for laboratory study.

Glossary

aciculate Needlelike, or furnished with a sharp terminal point.
acuminate Tapering to a point.
acute Sharp, pointed; less than a right-angle.
alate Having wings; applied as a noun to the primary reproductive caste of a termite before it sheds its wings.
allopatric (allopatry) Occupying (the occupation of) different geographical areas, lacking contact with each other (see also sympatric and parapatric).
allotype A selected specimen of the sex opposite to that of the holotype (q.v.), and designated at the same time as the latter.
annulate Ringed, or banded.
annulus Ring, or band.
antepenultimate Two before the last.
ancorae Paired anterior processes on the epiproct.
anterior Toward the front. Note that what is called the anterior margin of the wing is directed downward when the wings are folded.
apex Terminal part of body.
apical Farthest from the base (see also distal).
apterous Entirely lacking wings.
arboreal Living in trees.
arbusticolous Living in small trees, bushes, or taller shrubs (see also thamnicolous).
arcuate Arched, or bowed.
arenicolous Living in (or among) sand.
armature Structures such as spines, spurs, and denticulations, usually found on the tibiae and femora, sometimes on parts of the thorax.
arolium (pl., **arolia)** A padlike structure between the tarsal claws.
attenuate Becoming slender; drawn out; tapering.
austral From the southern regions of the southern hemisphere; usually exclusive of the Antarctic and Subantarctic.

basal At or near the base; opposite of apical.
basivalvula (pl., basivalvulae) A very small lateral sclerite supporting the opening of the spermatheca, particularly in female cockroaches.
bisexual Both males and females occurring.
boreal From the northern regions of the northern hemisphere; usually applied to the broad zone between the subarctic and the deciduous forest or prairies.
brachypterous With wings short to very short, not functional for flight but readily visible. (If the wings are minute, the condition is known as micropterous.)

calcar (pl., **calcaria)** Large spine or spur; term usually refers to hind tibial spurs in Tridactylidae.
carina (pl., **carinae)** An elevated ridge.
carinula (pl., **carinulae)** A fine carina, usually applied to a longitudinal ridge on the vertex of the head; also used for the lateral raised margin of the frontal ridge, when applicable.

caudad Toward the hind end.
caudal At the hind end.
cephalad Toward the head end.
cephalic At the head end.
cercus (pl., **cerci**) One of the paired terminal or subterminal projections on the 10th abdominal segment.
cinereous Ash colored; gray tinged with black.
clavate Clubbed at apex; usually refers to antennae.
clypeus The sclerite immediately above the labrum.
compressed Pressed together; flattened as viewed from the side (contrast with depressed).
concave Hollowed out, even if very shallowly.
convex Bulging outward at center, even if only slightly.
cosmopolitan Widespread in the world; usually, but not always, referring to synanthropic species.
costa Literally, a rib; the vein at the anterior margin of a wing (including the tegmen); see also frontal costa.
costal Pertaining to the costa; thus "costal margin" of the wing is its anterior edge.
courtship Behavioral interactions between male and female following visual contact, and often leading to copulation.
coxa The most basal segment of a leg.
crenate Scalloped, with small blunt rounded teeth.
crenulate Having small scallops that are evenly rounded and rather strongly curved.
crepitation The production of snapping or crackling sounds made by the wings in flight.
crepuscular Active in subdued light, as in the evening twilight; sometimes used for the evening and the dawning generally.

dactyl One of the projections on the forelegs of mole crickets (see *Neocurtilla*).
dentate Having toothlike projections.
dentation Toothlike projection.
denticle A fine tooth.
denticulate Finely toothed.
denticulation Part or the whole of a series of denticles.
depressed Literally, pressed downward; flattened as viewed from above (contrast with compressed).
dimorphic Existing in two forms.
disc (of pronotum) Dorsal surface area (of pronotum). There is no reference to a circular form implied, despite the derivation of the word; discal may be used with reference to the principal areas of other broad structures, but this is rather unusual.
distad Toward the end.
distal Pertaining to the end farthest from the base (see also apical).
distoventral On the underside farthest from the anterior region of the body.
diurnal Active during hours of daylight; implying inactivity at night.
dorsad Toward the back, or dorsum.

dorsal On the back, or dorsum; upper.

dorsal field of tegmen, or fore wing The area of the tegmen that is parallel to the substrate when the wings are at rest; the posterior field of the tegmen.

dorsolateral At the sides of the dorsum.

dorsomedian At the upper and median surface.

dorsoventral From upper to lower surfaces.

dorsum Literally, the back; the upper part of the body when an animal is standing horizontally on the ground (whether or not this is a natural position).

egg pod The egg mass of Acridoidea and certain other Orthoptera together with the frothy material that binds it and acts as a plug above it in the soil.

elliptical Oblong-oval, with ends equally rounded; much the same as ovoid but generally of more regular and often more elongate form.

elytron (pl., **elytra**) The fore wing; in orthopteroids, only applicable to Dermaptera (although used synonymously with tegmen by many authors).

emarginate Scooped out at the edge; strongly concave or broadly notched (see also excised).

ensiform Sword-shaped, flattened, and tapering to apex; often refers to antennae (Acridodea), sometimes to ovipositor (Tettigoniodea).

epiproct The 11th, or last, abdominal tergal plate; that part of the supra-anal plate lying behind its transverse sulcus, or groove, when this is present; some authors use this term synonymously with supra-anal plate.

ergate A sterile worker-caste termite.

excised Cut out; can vary in meaning from simply concave to notched (see also emarginate).

exserted Protruding; not enclosed.

facet An individual unit of the surface of a compound eye.

fascia (pl., **fasciae**) A narrow band, usually of contrasting color; sometimes used for a contrasting dark mark of some other form.

fasciated Adorned with fasciae.

fastigial Pertaining to the fastigium.

fastigio-facial Pertaining to the fastigium and the face.

fastigio-frontal Pertaining to the fastigium and the frons.

fastigium The depressed, or flattened, area at front of vertex of head.

femur (pl., **femora**) Stout (and usually the longest) segment of the leg between the coxa (or trochanter) and the tibia.

fenestrate Windowlike, or having the appearance of being paned.

file The part of the stridulatory vein of Grylloptera that bears the stridulatory teeth.

filiform Threadlike (usually referring to antennae).

flagellum The distal part of the antenna; the basal segment is the scape.

fontanelle A small median ocelluslike depression on the frontal region of the head in some termites.

forb An herbaceous plant other than a grass or sedge; generally a non-woody broad-leaved (dicotyledonous) plant.

fossa (pl., **fossae**) A pit, usually referring to antennal socket.

foveola (pl., **foveolae**) A depression with well-defined edges (see lateral foveolae).

foveolar area That part of the head upon which lateral foveolae (q.v.) occur, whether these are present or not.

frons The upper part of front of head; the "face" above the clypeus.

frontal costa Raised area extending from anterior part of top of head vertically between eyes to near clypeus, often ridged at edges and grooved in center.

furcate Forked, or bearing a forklike structure.

furcula (pl., **furculae**) One of a pair of mediodorsal processes on the posterior margin of the ninth abdominal tergum in certain male Acrididae (notably Melanoplinae).

fuscous Dull dark brown.

gena Cheek.

genicula (pl., **geniculae**) Knee; joint between femur and tibia, or the distal end of femur, particularly of the hind leg (see also knee).

genicular area The crescentic area at the distal end of the hind femur (of Acridoidea in particular), often dark colored.

glabrous Smooth, hairless, and without punctures.

glaucous Whitish blue, tending to gray.

graminicolous Living among grasses.

graminivorous Feeding upon grasses.

granulate With small protuberances like grains of sand.

granulation A minute grainlike protuberance, or the state of possessing such protuberances.

griseous Grayish.

gula The underside of the head; the "throat."

habitat The place or type of situation that an insect inhabits.

haemolymph Insect blood.

holotype The single specimen representative of a species or subspecies, designated by the original author at time of publication of original description.

humeral sinus A notch in the posterior margin of the lateral pronotal lobe.

hyaline Transparent or partly so; glassy.

hydropyle A chorionic structure that conducts liquids to the developing embryo.

immaculate Without spots or other markings.

incised Cut into; deeply notched.

infumate Smoky, clouded.

infuscate Grayish brown to brownish, with blackish tinge, usually overlaying more general pigmentation.

intercalary Interposed; applied, in certain Acrididae, to a wing vein of secondary derivation lying between the normal main veins of the tegmen.

intercalated Interpolated; applied, in certain cockroaches, to the apical area of the hind wing lying beyond, but with its base between the branches of, the longest main veins.

interocular Situated between the compound eyes, usually dorsally.

junior synonym See synonym.

knee The thickened apex of the hind femur, usually together with the base of the hind tibia (see also genicula).

labial Pertaining to the labium.
labium The third set of mouthparts; the lower lip with its associated structures.
labrum The upper lip below the clypeus.
lanceolate Lance-shaped; elongated to an acute apex.
laterad Toward the side.
lateral At the side.
lateral carina (pl., **carinae**) One of a pair of lateral ridges, usually of the pronotal disc.
lateral foveolae Small depressions lying on either side of the fastigium of the vertex toward the upper part of the front of the head in certain Acrididae.
lateral frontal carina (pl., **carinae**) A ridge running toward the lateral part of the clypeus approximately from the base of the antenna.
lateral pronotal lobe A lateral expansion of the pronotum directed downward in the vertical plane.
laterosternite A small sclerite of the sternum of a segment, lying laterally to the median sternite, or sternellum. (Referred to in this work with respect to 10th abdominal segment of certain female cockroaches.)
lectotype A single specimen subsequently selected to represent a species or subspecies when it has no holotype; only a syntype qualifies for designation as a lectotype. However, if a holotype is lost or destroyed, a paratype shall serve (the *International Code of Zoological Nomenclature* is vague about this latter situation).
lophi Paired posterior processes on the epiphallus.

macropterous Having wings long, fully developed; usually implying ability to fly.
macula (pl., **maculae**) A spot, or blotch, of contrasting color.
maculate With one, or more, maculae.
maculation The condition when one, or more, maculae are present; also used more or less synonymously with macula.
mandible One of the pair of robust biting, or chewing, jaws constituting the first pair of mouthparts.
maxilla (pl., **maxillae**) The second pair of mouthparts.
maxillary Pertaining to the maxilla.
median Middle.
median carina A median longitudinal ridge running along the dorsal surface of head (the area usually termed median carinula), thorax (particularly the pronotum, upon which it may be interrupted by transverse sulci or be partly or wholly obsolete), and abdomen (where it is often obsolete); usually restricted to the pronotal region.
mesad Toward the middle.
mesal At the middle.
mesosternal Pertaining to the mesosternum.
mesosternal interspace The area between the mesosternal lobes occupied by the anterior part of the metasternum.

mesosternal lobes The lateral backward projections of the mesosternum.
mesosternum The underside of the second thoracic segment.
metasternal Pertaining to the metasternum.
metasternal interspace The area between the metasternal lobes.
metasternal lobes The lateral backward projections of the metasternum.
metasternum The underside of the third thoracic segment.
mesozona That part of the pronotal disc immediately in front of the principal, or typical, transverse sulcus and behind the weaker sulcus anterior to the latter; applicable to Acridoidea.
metapygidium The middle opisthomere of the epiproct.
metazona That area of the pronotal disc posterior to the principal, or typical, sulcus.
micropterous With wings so reduced that they are represented by mere scales or minute lobes.
mirror A large membranous area of the tegmen of male Grylloptera associated with resonance during stridulation.
moniliform Like a string of beads; referring to antennae.
monotypic A subgenus or genus that includes only one species.
morph One of two, or more, anatomical or color forms occurring within a single species or subspecies.
multiarticulate Composed of numerous segments.

neotype A single specimen selected to represent a species or subspecies when the holotype and all paratypes, or all syntypes are lost or destroyed.
nocturnal Active at night; implies inactivity by day.
nomen nudum (pl., **nomina nuda**) An invalid name published without a description or without any other indication of the genus, species, or subspecies to which it referred.
notum (pl., **nota**) An entire dorsal plate of a thoracic segment, i.e., a thoracic tergum.
nunatak An elevated region believed to have remained ice-free, projecting above a glacial ice sheet.
nymph An immature (preadult) active feeding stage.

obsolescent Present but becoming difficult to discern.
obsolete Lacking or barely distinguishable.
obtuse Blunt; more than a right angle.
occiput The posterior part of top of head where the head abuts on the pronotum.
ocellus A simple eye; typically three ocelli are present, one near the base of each antenna and one at the median frontal.
omnivorous Eating food of both plant and animal origin, more or less indiscriminately.
ootheca (pl., **oothecae**) The leathery, membranous, or spongy covering of the egg mass of many Dictuoptera; loosely applied to the eggs also.
opisthomere One of the three parts of the pygidium (q.v.); particularly applicable to Dermaptera.
oval Egg-shaped; loosely used for ovoid structures.
oviparous Laying eggs that hatch after an appreciable interval subsequent to deposition.

ovipositor The posterior, terminal, valvular structure of the female, used to deposit eggs.

ovoid Egg-shaped; generally meaning irregularly but broadly elliptical.

ovoviviparous Producing eggs that hatch immediately on deposition.

pair formation The initial association of male and female prior to courtship and mating; may be accidental or brought about by male stridulation or other means of attraction.

pallid Pale.

palp or **palpus** (pl., **palps** or **palpi**) A segmented, mainly sensory, adjunct to the maxilla or labium.

paralectotype A syntype other than one designated as a lectotype.

parapatric (parapatry) Occupying (the occupation of) non-overlapping geographical areas (see also allopatric and sympatric).

paratergite That part of a tergum (particularly of a cockroach) forming a definite lateral or marginal region, which is, however, not separated by a suture from the rest of the tergum.

paratype A specimen other than a holotype that was studied by the original author when preparing an original description of a species or subspecies and designated as such by the author at the time of publication.

paraproct One of a pair of lateral plates forming part of the abdominal terminalia and lying between supra-anal and subgenital plates.

parthenogenesis Parthenogenetic; the act of producing offspring without mating; the capacity for this.

pellucid Semitransparent, but not necessarily colorless; more or less translucent.

penis-lobe One of a pair of copulatory structures associated with the male pygidium (q.v.), one which is often suppressed or nonfunctional; particularly applicable to Dermaptera.

penultimate Next to the last.

persulcate With furrows or grooves.

phallic Pertaining to the phallus (q.v.).

phallomere One of the principal sclerites of the phallus.

phallus The entire sclerotized intromittant part of the male genitalic apparatus, together with the associated membranous structures.

pheromone A chemical that alters behavior in specific ways.

piceous Of a pitchy black color.

pilose Covered with long soft hairlike structures.

plantula (pl., **plantulae**) An exaggerated and often protruding pulvillus.

polymorphic Existing in more than two forms.

posterior Toward the rear; hind.

posterolateral At the side and toward the rear.

posteroventral At the posterior end and toward the venter.

postocular stripe A dark band extending from behind the compound eye, across the gena and the upper part of the lateral pronotal lobe, and sometimes on to the tegmen.

preapical Before but near the apex.

primary type A holotype, lectotype, or neotype, but not an allotype or a paratype.

principal sulcus The posterior transverse sulcus, or groove, on the pronotum; usually the deepest sulcus, cutting the median or lateral carinae or both; also called typical sulcus.

pronotum The dorsal plate of the first thoracic segment, comprising the disc and, in most forms, the lateral lobes.

prosternal Pertaining to the prosternum.

prosternal spine or tubercle A spine or projection extending downward between the first pair of legs; in some groups paired spines occur.

prosternum The underside of the first thoracic segment.

proximad Toward the base.

proximal At the base.

prozona The part of the pronotal disc anterior to the principal, or typical, posterior transverse sulcus. (If the mesozona is recognized, the prozona refers only to the part of the disc in front of it.)

pseudergate A sexually underdeveloped, wingless termite acting in the capacity of a worker but capable of becoming sexually functional under certain circumstances.

pubescent Covered with short soft fine hair.

pulvillus (pl., **pulvilli**) Padlike sole or cushion on tarsal segment.

puncturation A pattern of small pits or punctures on a surface.

punctured With minute depressions, or pits.

pygidium The 11th abdominal segment, which forms a small process between the bases of forceps; particularly applicable to Dermaptera.

quadrate Approximately as broad as long; more or less square.

raptorial Adapted for seizing and grasping; almost invariably applied to specialized forelegs, especially of mantids, but also of some tettigonioids.

relict Left behind; groups having a discontinuous world distribution are often regarded as relicts in that they have become extinct in intervening regions.

remigium Anterior part of hind wing.

rudimentary Not sufficiently developed to be functional; somewhat similar to vestigial (q.v.).

ruga (pl., **rugae**) An individual fold or wrinkle that forms a roughened surface.

rugose Rough, usually on account of the presence of numerous small irregular wrinkles.

rugosity The condition of being rough; a roughened area.

ruguso-punctate Rough and marked by pits or punctures.

saltatorial Habitually or capable of jumping (the hind legs being enlarged).

sand-basket A specialized arrangement of strong spines on the tibia of certain Ceuthophilinae (Rhaphidophorinae).

scape The basal segment of the antenna; the distal part of the antenna is the flagellum.

sclerite A hardened plate of the integument; not usually applied to appendages.

segment An individual ringlike or cylindrical part of the body or of an appendage.

senior synonym See synonym.
sensu lato In the broad sense.
sensus stricto In the narrow, or restricted, sense.
serrate(d) Saw-toothed.
serration One or a series of sawlike teeth; or the state of being saw-toothed.
serrulate(d) With minute sawlike teeth.
serrulation Same as serration but with individual teeth minute.
seta (pl., **setae**) A bristle or hair.
setaceous Slender, gradually tapering to an acute tip; bristlelike.
setose Bristly.
shaft of hind femur The slender distal part in the hind leg of saltatorial forms.
sinuate, sinuous Curved and recurved; winding in and out; S-shaped.
sinus A scooped-out cavity in the margin of a structure.
spatulate Spoon- or paddle-shaped; broad and flattened distally.
speculum A spoon-shaped structure.
spermatheca (pl., **spermathecae**) A small vesicular structure forming part of the female internal reproductive system in which the male sperms are received and stored after mating.
spine An immovable sharp or slender projection, particularly on a tibia.
spinose Armed with spines.
spur A movable articulated heavy spinelike projection on a leg, particularly on a hind tibia.
sternite An individual platelet forming part of a sternum.
sternum (pl., **sterna**) The entire ventral plate of a single body segment; collectively, the ventral aspect of the insect, particularly of the thoracic region.
stria (pl., **striae**) A longitudinal streak, slightly raised.
striate(d) Possessing striae.
stridulation Sound produced by rubbing one part of the body against another.
stridulatory pegs Small pegs, or teeth, usually in a row, that strike against a scraper ridge during stridulation; commonly located on the inner face of the hind femur or on the tegmina, depending upon the group, less frequently located elsewhere.
stridulatory ridge A ridge used in sound production, usually bearing stridulatory pegs, or teeth.
stridulatory teeth See stridulatory pegs.
strigil The scraping part of a stridulatory apparatus.
strumose Having numerous fine raised irregular ridges.
style A stiff rodlike structure, which may be slightly curved, particularly associated with the male genitalia of Dictuoptera.
sub- Somewhat below.
subelliptical Somewhat elliptical.
subequal Similar but not exactly equal in size to some other dimensional character.
subdistal Nearly terminal.
subgenital plate The last sternal plate of the abdomen, either male or female.

subocular sulcus A vertical groove, or depression, running from below the compound eye to the anterior base of the mandible.
subquadrate Almost as broad as long (or vice versa); shortly rectangular.
sulcus (pl., **sulci**) A groove, or depression, usually transverse on pronotal disc and vertical on lateral pronotal lobes.
supra-anal plate The 11th, or last, abdominal tergal plate; some authors use this term synonymously with supra-anal plate.
sylvan Living in woodland, usually on or near the ground.
sympatric (sympatry) Occupying (the occupation of) the same geographical area, at least in part (see also allopatric and parapatric).
synanthropic Found in or near the habitations of man; usually of alien origin.
synonym Each of two or more different names applied to the same taxon; a **senior synonym** is the earliest available name; a **junior synonym** is a name of later date.
syntype Any specimen of a series studied by the author at the time of description of a species or subspecies, provided that no holotype (q.v.) was designated.

tarsal Pertaining to a tarsus.
tarsus (pl., **tarsi**) The foot; the segmented part of a leg distal to the tibia (see Fig. 6, *ta*).
taxon A group of organisms included within a given systematic category, such as a species, a genus, or a family; often incorrectly applied to the category itself or to its name.
tectate Covered; rooflike or ridgelike.
tectiform Ridged in middle and sloping on each side, like a roof.
tegmen (pl., tegmina) The fore wing, when stiffened or leathery, as in most orthopteroids.
telson The embryonic 12th abdominal segment.
tegmino-alary A function involving both front and hind wings.
tergite An individual platelet forming part of a tergum.
tergum (pl., **terga**) The dorsal plate of a single body segment (Fig. 6, *te*); thoracic terga are usually called nota.
terminalia The distal extremity of the abdomen, more specifically the genitalia and associated structures.
terricolous Dwelling on the (bare) ground.
testaceous Dull yellowish brown, like old parchment.
thamnicolous Living on small shrubs and bushes (see also arbusticolous).
tomentose Covered with extremely fine depressed microscopic hairs; however, these hairs are insufficiently prominent to render the surface of the insect "matt" in appearance.
tomentum A covering of minute hairs; see tomentose.
trichobothrium (pl., **trichobothria**) A long erect or semierect sensory hair on the cercus.
tricholith A club-shaped sensory seta on the basal inner surface of a cercus, as found in all true crickets and in certain cockroaches.
trochanter The second segment of a leg from its base.
truncated Ending abruptly as if cut off.

tympanum Literally eardrum; an auditory organ; in most Orthoptera the tympanum is located laterally near the base of the abdomen, in Grylloptera it is located on the fore-tibiae.

type *The* type, loosely used, is synonymous with holotype (q.v.); *a* type is usually synonymous with syntype (q.v.).

type-genus A genus whose name is the basis of the names in the family-group taxa that include it; the name is the first to have been used in the derivation of that of a suprageneric taxon, not necessarily the earliest relevant generic name to have been published.

type locality The place from which the primary type of a species or subspecies originated.

type series The holotype and all paratypes, or else all syntypes, of a species or subspecies.

type-species The species that is deemed to represent a genus (or subgenus) and so designated either at the time of the original description of that genus (or subgenus), or subsequently; if only one species is mentioned at the time of publication, then that is the type-species (there are also a few special instances where a species is automatically the type-species in the absence of monotypy).

typical sulcus See principal sulcus.

valvifer A small terminal abdominal sclerite bearing an ovipositor valve.

vannus Posterior part of hind wing.

ventrad Toward the undersurface.

ventral Lower surface.

ventrolateral At the side but low down.

vertex The top of the head.

vestibular sclerite A small sclerite forming part of the floor of the vestibulum in a female cockroach.

vestibulum The posterior chamber in a female cockroach where the ootheca is formed.

vestigial Reduced from a functional to a functionless condition; similar to rudimentary (q.v.).

viviparous Producing young alive; eggs completing development within the body of the female.

References

Anonymous, 1962*a*. Secrets of city hall revealed. Winnipeg Tribune, 72 (40; 16 Feb., 1962):1.

Anonymous, 1962*b*. Old relics disclose history. Winnipeg Tribune, 72 (41; 17 Feb., 1962):23.

Adelung, N. 1902. Beitrag zur Kenntnis der paläarctischen Stenopelmatiden (Orthoptera, Locustodea). Ann. Mus. zool. Acad. Sci. St. Petersb. 7:55–75.

Alexander, G., and J. R. Hilliard. 1964. Life history of *Aeropedellus clavatus* (Orthoptera, Acrididae) in the alpine tundra of Colorado. Ann. ent. Soc. Am. 57:310–317.

Alexander, R. D. 1956. A comparative study of sound production in insects, with special reference to the singing Orthoptera and Cicadidae of the eastern United States. Columbus, OH: Ohio State Univ. Ph.D. Thesis. 529 pp.

Alexander, R. D. 1957*a*. The song relationships of four species of ground crickets (Orthoptera: Gryllidae: *Nemobius*). Ohio J. Sci. 57:153–163.

Alexander, R. D. 1957*b*. The taxonomy of the field crickets of the eastern United States (Orthoptera: Gryllidae: *Acheta*). Ann. ent. Soc. Am. 50:584–602.

Alexander, R. D. 1960. Sound communication in Orthoptera and Cicadidae. *In* W. E. Lanyov and W. N. Tavolga, eds. Animal sounds and communication. Publ. Am. Inst. biol. Sci. 7:38–92.

Alexander, R. D. 1963. Animal species, evolution and geographic isolation. Syst. Zool. 12:202–204.

Alexander, R. D. 1968. Life cycle origins, speciation, and related phenomena in crickets. Q. Rev. Biol. 43:1–41.

Alexander, R. D. 1975. Natural selection and specialized chorusing behavior in acoustical insects. Pages 35–37 *in* D. Pimentel, ed. Insects, science and society. Proc. Cornell Univ. Symp.

Alexander, R. D., and R. S. Bigelow. 1960. Allochronic speciation in field crickets, and a new species, *Acheta veletis*. Evolution 14:334–346.

Alexander, R. D., and D. Otte. 1967. The evolution of genitalia and mating behavior of crickets (Gryllidae) and other Orthoptera. Misc. Publs Mus. Zool. Univ. Mich. 113:5–62.

Alexander, R. D., A. E. Pace, and D. Otte. 1972. The singing insects of Michigan. Gt Lakes Ent. 5:33–69.

Alexander, R. D., and E. S. Thomas. 1959. Systematic and behavioral studies on the crickets of the *Nemobius fasciatus* group (Orthoptera: Gryllidae: Nemobiinae). Ann. ent. Soc. Am. 52:591–605.

Allard, H. A. 1911. The musical habits of some New England Orthoptera in September. Ent. News 22:28–39.

Allard, H. A. 1930. The chirping rates of the snowy tree cricket (*Oecanthus niveus*) as affected by external conditions. Can. Ent. 62:131–142.

Amedegnato, C. 1974. Les genres d'Acridiens néotropicaux, leur classification par familles, sous-familles et tribus. Acrida 3:193–204.

Ander, K. 1938. Ein abdominales stridulations Organ bei *Cyphoderris* (Prophalangopsidae) und über die systematische Einteilung der Ensiferen (Saltatoria). Opusc. ent. 3:32–38.

Ander, K. 1950. Orthoptera. *In* P. Brinck and K. G. Wingstrand. The mountain fauna of the Virihaure area in Swedish Lapland. II. Special account. Acta Univ. Lund. (Lunds Univ. Arsskrift) (N.F.) (2)46:135–137.

Ander, K. 1960. Dermaptera and Orthoptera from Newfoundland. Opusc. ent. 25:224–231.

Anderson, N. L. 1964. Some relationships between grasshoppers and vegetation. Ann. ent. Soc. Am. 57:736–742.

Anderson, N. L., and J. C. Wright. 1952. Grasshopper investigations on Montana range lands. Tech. Bull. Mont. agric. Exp. Stn 486:1–46.

Ando, H., ed. 1979. Studies on the Grylloblattodea [in Japanese except for English subtitles]. Tsukuba. University of Tsukuba. IV, 84 pp.

Audinet-Serville, J. G. 1831. Revue méthodique des Orthoptères. Annls Sci. nat. (Zool.) 22:28–65, 134–162, 262–292.

Audinet-Serville, J. G. 1838. Histoire naturelle des Insectes, Orthoptères. Coll. Suites à Buffon [7], Paris. Roret [dated 1839; published Dec. 1838]. XVIII, 786, 14 pl.

Back, E. A. 1936. *Gryllus domesticus* L. and city dumps. J. econ. Ent. 29:198–202.

Back, G. 1836. Narrative of the arctic land expedition to the mouth of the Great Fish River and along the shores of the Arctic Ocean, in the years 1833, 1834, and 1835. London, England: John Murray. X, 663 pp., 14 pl., 1 foldout (1 map). [Also published in Paris by A. & W. Galigniani & Co., 1836:1 VIII, 338 pp., and in Philadelphia by E. L. Corey & A. Hunt, 1836:IX, 456 pp.; reprinted 1970 by Charles E. Tuttle Co. Inc., with XVIII extra pages.]

Bailey, C. G., and P. W. Riegert. 1971. Food preferences of the dusky grasshopper, *Encoptolophus sordidus costalis* (Scudder) (Orthoptera: Acrididae). Can. J. Zool. 49:1271–1274.

Bailey, C. G., and P. W. Riegert. 1973. Energy dynamics of *Encoptolophus sordidus costalis* (Scudder) (Orthoptera: Acrididae) in a grassland ecosystem. Can. J. Zool. 51:91–100.

Bailey, V., and C. C. Sperry. 1929. Life history and habits of grasshopper mice, genus *Onychomys*. Tech. Bull. U.S. Dep. Agric. 145:1–19.

Baillie, J. L. 1929. William Couper — a pioneer Canadian naturalist. Can. Field-Nat. 43:169–176.

Baird, A. B., ed. 1938. Summary of insect parasites and predators liberated in Canada up to December 31, 1937. Can. Insect Pest Rev. 16:77–154.

Baird, A. B., ed. 1939. Summary of parasite liberations in Canada during the year 1938. Can. Insect Pest Rev. 17:102–128.

Baird, A. B., ed. 1940. Summary of insect parasites and predators liberated in Canada and Newfoundland 1939. Can. Insect Pest Rev. 18:94–126.

Baird, A. B., ed. 1941. Summary of parasite and predator liberations in Canada and Newfoundland in 1940. Can. Insect Pest Rev. 19:94–125.

Baird, A. B., ed. 1942. Summary of parasite and predator liberations in Canada in 1941. Can. Insect Pest Rev. 20:112–135.

Baird, A. B., ed. 1943. Summary of parasite and predator liberations in Canada in 1942. Can. Insect Pest Rev. 21:107–130.
Baird, A. B., ed. 1944. Summary of parasite and predator liberations in Canada in 1943. Can. Insect Pest Rev. 22:117–138.
Baird, A. B., ed. 1945. Summary of parasite and predator liberations in Canada in 1944. Can. Insect Pest Rev. 23:119–139.
Baird, A. B., ed. 1946. Summary of parasite and predator liberations in Canada in 1945. Can. Insect Pest Rev. 24:134–152.
Baker, C. F. 1905. Orthoptera: Second report on Pacific Slope Orthoptera. Invertebrata Pacifica 1:71–84.
Ball, E. D., E. R. Tinkham, R. Flock, and C. T. Vorhies. 1942. The grasshoppers and other Orthoptera of Arizona. Tech. Bull. Univ. Ariz. agric. Exp. Stn 93:257–353.
Banks, N. 1907. A new species of *Termes*. Ent. News 18:392–393.
Banks, N., and T. E. Snyder. 1920. A revision of the Nearctic termites with notes on biology and geographic distribution. Bull. U.S. natn. Mus. 108:1–228.
Barnum, A. H. 1959. The phallic complex in the Oedipodinae (Orthoptera, Acrididae). Ann Arbor, MI: Iowa State Coll. Available from: University Microfilms, Ann Arbor, MI; Publication no. Mic 59-3370. II, 220 pp. Ph.D. Dissertation.
Barnum, A. H. 1964. Orthoptera of the Nevada test site. Brigham Young Univ. Sci. Bull. (Biol.) 4:1–134.
Bate, J. 1969. On some aspects of the biology of *Acheta domesticus* (Insecta, Orthoptera) with reference to the infestation of refuse tips. Pedobiologia 9:300–322.
Bate, J. 1971. Life history of *Acheta domesticus* (Insecta, Orthoptera, Gryllidae). Pedobiologia 11:159–172.
Bate, J. 1972. Variation in fecundity of *Acheta domesticus* (Insecta, Orthoptera, Gryllidae) in relation to season and temperature. Pedobiologia 12:1–5.
Baumgartner, W. J. 1910. Observations on the Gryllidae. III. Notes on the classification and on some habits of certain crickets. Kans. Univ. Sci. Bull. 5:309-319.
Bazyluk, W. 1956. Karaczany—*Blattodea*. Pol. Tow. ent. Klucze Oznaczania Owadów Pol. 9 (Pol. Zwiaz. ent. 11):3–32.
Bazyluk, W. 1960. Rozprzestrzenienie geograficzne i zmienność *Mantis religiosa* (L.) (*Mantodea, Mantidae*) oraz opisy nowych podgatunków. Die geographische Verbreitung und Variabilität von *Mantis religiosa* (L.) (Mantodea, Mantidae), sowie Beschreibungen neuer Unterarten. Annls zool., Warsz. 18:231-270, pl. XII.
Beaudry, J. R. 1952. Additions à la faune orthoptérologiques de la Province de Québec ou du Canada. Annls ACFAS (Assoc. can. fr. Av. Sci.) 18:100–102.
Beaudry, J. R. 1954. A simplification of Hubbell's method for trapping and preserving specimens of *Ceuthophilus* (Orthoptera, Gryllacrididae). Can. Ent. 86:121–122.

Beaudry, J. R. 1973. Une analyse des compléments chromosomiques de certains Orthoptères du Québec et sa signification taxonomique et évolutionnaire. Can. J. Genet. Cytol. 15:155–170.

Beck, E. W., and J. L. Skinner. 1967. Screening of insecticides against the Changa and the southern mole cricket attacking seedling millet. J. econ. Ent. 60:1517–1519.

Begg, A., and W. P. Nursey. 1879. Ten years in Winnipeg. A narration of the principal events in the history of the city of Winnipeg from the year A.D. 1870 to the year A.D. 1879, inclusive. Winnipeg, Man.: Times Printing and Publishing House. IV, 226 pp.

Beĭ-Bienko, G. Ya. 1928. Opredelitel' lichinok glavneĭshikh zapadnosibirskikh saranchevȳkh. Trudȳ sib. Inst. sel'. Khoz. Lesov. 9:153–198.

Beĭ-[as Bey-]Bienko, G. [Ya.] 1932. Orthoptera Palaearctica critica XI. The group *Chrysochraontes* (Acrid.). Eos, Madr. 8:43–92.

Beĭ-Bienko [as Bey-Bienko], G. [Ya.] 1933. Records and descriptions of some Orthoptera from U.S.S.R. Boln Soc. esp. Hist. nat. 33:317–431.

Beier, M. 1929. Einige neue Mantiden aus der Sammlung des naturhistorischen Museums in Wien. Zool. Anz. 80:129–139.

Beier, M. 1955. Überordnung Orthopteroidea (Handlirsch 1903). Bronn's Kl. Ord. Tierr. Leipzig. Akademische Verlagsgesellschaft Geest & Portig K.-G. 5(3)[6(5)]:31–304.

Beier, M. 1957. Orthopteroidea: Ordnung Cheleutoptera Crampton 1915 (Phasmida Leach 1815). Bronn's Kl. Ord. Tierr. Leipzig. Akademische Verlagsgesellschaft Geest & Portig K.-G. 5(3)[6(2)]:305:454.

Beier, M. 1959. Ordnung: Dermaptera (Degeer 1773) Kirby 1813. Bronn's Kl. Ord. Tierr. Leipzig. Akademische Verlagsgesellschaft Geest & Portig K.-G. 5(3)[6(3)]:455–585.

Beier, M. 1960. Orthoptera Tettigoniidae (Pseudophyllinae II). Pages 1–396 *in* E. Mertens et al., eds. Das Tierreich. Berlin, West Germany: Walter de Gruyter & Co. Vol. 74. X, 396 pp.

Beier, M. 1961. Überordnung: Blattopteroidea Martynov 1938 (Blattaeformia Handlirsch 1903, part. Blattoidea Weber 1933). Bronn's Kl. Ord. Tierr. Leipzig. Akademische Verlagsgesellschaft Geest & Portig K.-G. 5(3)[6(3)]:587–848.

Beier, M. 1962. Orthoptera Tettigoniidae (Pseudophyllinae I). Pages 1–468 *in* E. Mertens et al., eds. Das Tierreich. Berlin, West Germany: Walter de Gruyter & Co. Vol. 73. XII, 468 pp. [Part II was published earlier, in 1960, (q.v.).]

Beier, M. 1963. Tettigoniidae: Subfam. Pseudophyllinae. Orthopterorum Catalogus. 's-Gravenhage. Dr. W. Junk. 5:1–240.

Beier, M. 1964. Blattopteroidea Mantodea. Bronn's Kl. Ord. Tierr. Leipsig. Akademische Verlagsgesellschaft Geest & Portig K.-G. 5(3)[5]:849–970.

Beier, M. 1968. Phasmida (Stab- oder Gespenstheuschrecken). Pages 1–56 *in* J.-G. Helmcke, D. Starck, and H. Wermuth, eds. Kükenthal's Handb. Zool. (Ed. 2). Berlin, West Germany: Walter de Gruyter & Co. Vol. 4(2)[2(10)].

Beier, M. 1972. Ordnung Saltatoria (Grillen und Heuschrecken). Pages 2–217 *in* J.-G. Helmcke, D. Starck, and H. Wermuth, eds. Kükenthal's Handb. Zool. (Ed. 2). Berlin and New York: Walter de Gruyter & Co. Vol. 4(2)[2(9)].

Beier, M. 1974. Ordnung Blattariae (Schaben). Pages 1–127 *in* J.-G. Helmcke, D. Starck, and H. Wermuth, eds. Kükenthal's Handb. Zool. (Ed. 2). Berlin and New York: Walter de Gruyter & Co. Vol. 4(2)[2(13)].

Beirne, B. P. 1972. Pest insects of annual crop plants in Canada. IV. Hemiptera–Homoptera, V. Orthoptera, VI. Other groups. Mem. ent. Soc. Can. 84:1–73.

Bell, P. D. 1979. Rearing the black-horned tree cricket, *Oecanthus nigricornis* (Orthoptera: Gryllidae). Can. Ent. 111:709–712.

Bell, P. D. 1980. Opportunistic feeding by the female tree cricket, *Oecanthus nigricornis* (Orthoptera: Gryllidae). Can. Ent. 112:431–432.

Bennett, C. B. 1904. Earwigs (*Anisolabia maritima* Bon.). Psyche, Camb. 11:47–53.

Bessey, C. E. 1877. A preliminary catalogue of the Orthoptera of Iowa. Bienn. Rep. Iowa St. Coll. 7(1876–77):205–210.

Bethune, C. J. S. 1875. Grasshoppers or locusts. Rep. ent. Soc. Ont. 5:29–42.

Bethune, C. J. S. 1898*a*. Some household insects. Rep. ent. Soc. Ont. 28:57–61.

Bethune, C. J. S. 1898*b*. The rise and progress of entomology in Canada. Trans. R. Soc. Can. 1898(4):155–165.

Beutenmüller, W. 1894*a*. Description of a new tree cricket. Jl N.Y. ent. Soc. 2:56.

Beutenmüller, W. 1894*b*. Notes on some species of North American Orthoptera, with descriptions of new species. Bull. Am. Mus. nat. Hist. 6:249–252.

Beutenmüller, W. 1894*c*. Descriptive catalogue of the Orthoptera found within fifty miles of New York City. Bull. Am. Mus. nat. Hist. 6: 253–316, pl. V–X.

Bharadwaj, R. K. 1966. Observations on the bionomics of *Euborellia annulipes* (Dermaptera: Labiduridae). Ann. ent. Soc. Am. 59:441–450.

Bigelow, R. S. 1958. Evolution in the field cricket, *Acheta assimilis* Fab. Can. J. Zool. 36:139–151.

Bigelow, R. S. 1960*a*. Interspecific hybrids and speciation in the genus *Acheta* (Orthoptera, Gryllidae). Can. J. Zool. 38:509–524.

Bigelow, R. S. 1960*b*. Development rates and diapause in *Acheta pennsylvanicus* (Burmeister) and *Acheta veletis* Alexander and Bigelow (Orthoptera: Gryllidae). Can. J. Zool. 38:973–988.

Bigelow, R. S. 1962. Factors affecting developmental rates and diapause in field crickets. Evolution 16:3[illegible]–406.

Bigsby, J. J. 1824. Notes on the geography and geology of Lake Huron. Trans. geol. Soc. Lond. for 1824:175–209, 6 pl., 1 map.

Bigsby, J. J. 1850. The shoe and canoe: Or pictures of travel in the Canadas, illustrative of their scenery and of colonial life. . . . London, England: Chapman & Hall. Vol. 2. VIII, 346 pp., 13 pl., 2 maps. [Vol. 1 (XVI, 352 pp., 7 pl., 2 maps) is irrelevant here.]

Binet, L. 1931. La vie de la mante religieuse. Paris, France: Vigot Frères. 93 pp.

Bird, R. D. 1961. Ecology of the aspen parkland of Western Canada in relation to land use. Can. Agric. Publ. 1066:(II) I, X, 155 pp., frontisp., 1 foldout (3 maps).

Blatchley, W. S. 1892*a*. Two new Orthoptera from Indiana. Can. Ent. 24:26–27.

Blatchley, W. S. 1892*b*. Some Indiana Acrididae.-II. Can. Ent. 24:28–34.

Blatchley, W. S. 1893*a*. The Locustidae of Indiana. Proc. Indiana Acad. nat. Sci. for 1892:92–153.

Blatchley, W. S. 1893*b*. The Blattidae of Indiana. Proc. Indiana Acad. nat. Sci. for 1892:153–165.

Blatchley, W. S. 1893*c*. Some new Locustidae from Indiana. Can. Ent. 25:89–93.

Blatchley, W. S. 1898*a*. Some Indiana Acrididae.-IV. Can. Ent. 30:54–64.

Blatchley, W. S. 1898*b*. Two new Melanopli from Les Cheneaux Islands, Michigan. Psyche, Camb. 8:195–197.

Blatchley, W. S. 1900. On the species of *Nemobius* known to occur in Indiana. Psyche, Camb. 9:51–54.

Blatchley, W. S. 1903. The Orthoptera of Indiana. An illustrated descriptive catalogue of the species known to occur in the State, with bibliography, synonymy and descriptions of new species. Rep. Indiana Dep. Geol. nat. Resour. 27(1902):123–471, 669–677.

Blatchley, W. S. 1920. Orthoptera of northeastern America with especial reference to the fauna of Indiana and Florida. Indianapolis, IN: Nature Publishing Co. 784 pp., 1 pl.

Blunck, H. 1942. *Leptophyes punctatissima* Bosc als Rosenschädling. Z. Pflanzenkr. Pflanzenschutz 52:192–204.

Boettger, C. R. 1950/1951. Die Gewächshausheuschrecke (*Tachycines asynamorus* Adelung)/the greenhouse stone cricket (*Tachycines asynamorus* Adelung). Abh. Braunschweig. wiss. Ges. 2:13–39; 3:19–20.

Bohart, R. M. 1947. Sod webworms and other lawn pests in California. Hilgardia 17:267–308.

Boheman, C. H. 1846. Två nya svenska *Gryllus*-arter, jemte några anmaerkningar rörande Orthopt. Öfvers. K. VetenskAkad. Förh. 1846:80–83.

Bolívar, I. 1887. Essai sur les acridiens de la tribu des Tettigidae. Annls Soc. ent. Belg. 31:175–313, pl. 4, 5.

Borror, D. J., and D. M. DeLong. 1971. An introduction to the study of insects. (3rd ed.). New York et al.: Holt, Rinehart & Winston. XIV, 812 pp.

Bosc d'Antic, L. A. G. 1792. Observations sur l'*Acheta sylvestris* et *Locusta punctatissima*. Acta Soc. Hist. nat. Paris 1:108–109, 110–111, pl. 1.

Boucher, P. 1664. Histoire véritable et naturelle des mœurs et productions du pays de la Nouvelle France, vulgairement dite le Canada. Paris, France: F. Lambert. (12), 168 pp.

Boudreaux, H. B. 1979. Arthropod phylogeny with special reference to insects. New York et al.: John Wiley and Sons, Inc. VIII, 320 pp.

Bradley, J. C., and B. S. Galil. 1977. The taxonomic arrangement of the Phasmatodea with keys to the subfamilies and tribes. Proc. ent. Soc. Wash. 79:176–208.

Briand, L. J., and I. Rivard. 1964. Observations sur *Mermis subnigrescens* Cobb (Mermithidae), nématode parasite des criquets au Québec. Phytoprotection 45:73–76.

Brindle, A. 1977. British earwigs (Dermaptera). Entomologist's Gaz. 28:29–37.

Britton, W. E. 1920. Check-list of the insects of Connecticut. Bull. Conn. St. geol. nat. Hist. Surv. 31:37–43.

Brodie, W., and J. E. White, eds. 1883. Label list of insects of the Dominion of Canada. Toronto, Ont.: Natural History Society of Toronto. II, 69 pp.

Brooks, A. R. 1958. Acridoidea of southern Alberta, Saskatchewan and Manitoba (Orthoptera). Can. Ent. Suppl. 9:3–92.

Brown, A. W. A. 1940. The Canadian forest insect survey in 1939. Rep. ent. Soc. Ont. 70[dated 1939]:95–114.

Bruner, L. 1876. New species of Nebraska Acrididae. Can. Ent. 8:123–125.

Bruner, L. 1877. List of Acrididae found in Nebraska. Can. Ent. 9:144–145.

Bruner, L. 1884*a*. Two new Myrmecophilae from the United States. Can. Ent. 16:41–43.

Bruner, L. 1884*b*. Observations on the Rocky Mountain locust during the summer of 1883. Bull. U.S. Dep. Agric. Div. Ent. 4:51–62.

Bruner, L. 1885*a*. First contribution to a knowledge of the Orthoptera of Kansas. Bull. Washburn Coll. Lab. nat. Hist. 1:125–139.

Bruner, L. 1885*b*. Contributions to the North Transcontinental Survey. Orthoptera. Can. Ent. 17:9–19.

Bruner, L. 1890. New North American Acrididae found north of the Mexican boundary. Proc. U.S. natn. Mus. 12(1889):47–82. [Issued 5.II.1890.]

Bruner, L. 1891*a*. Ten new species of Orthoptera from Nebraska—Notes on habits, wing varieties, etc. Can. Ent. 23:36–40, 56–59, 70–73.

Bruner, L. 1891*b*. On some destructive locusts of North America, together with notes on the occurrences in 1891. Can. Ent. 23:189–196.

Bruner, L. 1892. A change in the name of a recently described species of *Orchelimum*. Ent. News 3:264–265.

Bruner, L. 1893*a*. The more destructive locusts of America north of Mexico. Bull. U.S. Dep. Agric. Div. Ent. 28:1–40.

Bruner, L. 1893*b*. A list of Nebraska Orthoptera. Publs Neb. Acad. Sci. 3:19–33.

Bruner, L. 1897. The grasshoppers that occur in Nebraska. Rep. Neb. agric. Exp. Stn 1896:105–138.

Bruner, L. 1899. A new *Conocephalus*. Ent. News 10:38–39.

Bruner, L. 1904*a*. Some new Colorado Orthoptera. Bull. Colo. St. Univ. agric. Exp. Stn (tech. Ser.) 94:57–67.

Bruner, L. 1904*b*. Fam. Acrididae [Contd.]. Biol. Centr. Am. (Zool.) Ins. Orthoptera 2:33–72, pl. 1 (January); 73–104 (April).

Bruner, L. 1905. Fam. Acrididae [Contd.]. Biol. Centr. Am. (Zool.) Ins. Orthoptera 2:105–112 (January); 113–152 (March); 153–160 (June); 161–176, pl. 2 (October).

Bruner, L. 1906*a*. Fam. Acrididae [Contd.]. Biol. Centr. Am. (Zool.) Ins. Orthoptera, 2:177–208.

Bruner, L. 1906*b*. Synoptic list of Paraguayan Acrididae, or locusts, with descriptions of new forms. Proc. U.S. natn. Mus. 30:613–694.

Brunner de [von] Wattenwyl, C. 1865. Nouveau système des Blattaires. Vienne [Wien], G. Braumüller; Paris, A. Deyrolle; Leipsic [Leipzig], F. A. Brockhaus. XII, 426 pp., XIII pl.

Brunner von Wattenwyl, C. 1878. Monographie der Phaneropteriden. Wien. K. K. Zoologische-Botanische Gesellschaft. II, 401 pp., VIII, foldout pl.

Brunner von Wattenwyl, C. 1888. Monographie der Stenopelmatiden und Gryllacriden. Verh. zool.-bot. Ges. Wien 38:247–394, pl. V–IX.

Brunner de [von] Wattenwyl, C. 1893. Révision du système des Orthoptères et description des espèces rapportées par M. Leonardo Fea de Birmanie. Annali Mus. Stor. nat. Genova 33[= (2)13]:1–230, pl. I–VI.

Brunner von Wattenwyl, C. 1895*a*. *Tomonotus theresiae*, sp.n. Berl. ent. Z. 49:277.

Brunner von Wattenwyl, C. 1895*b*. Monographie der Pseudophylliden. Wien. K. K. Zoologische-Botanische Gesellschaft. IV, 282 pp., [X pl. issued separately].

Brünnich, M. T. 1763. Blatta transfuga. Pages 1–697 *in* E. Pontoppidan. Den Danske Atlas elle konge-riget Dannemark, med dets naturlige egenskaber, elementer, inbyggere, vaexter, dyr og andre affφdninger, dets gamle tildragelser og naervaerende omstaendigheder i alle provintzer, staeder; kirker, slotte og herregaarde . . . . Kiφbenhavn: A. H. Godiche. [5 vols. in 6,]. Pl. 29.

Bruton, F. A. 1930. Philip Henry Gosse's entomology of Newfoundland. Ent. News 41:34–38.

Buckell, E. R. 1920. Life history notes on some species of Acrididae (Orthoptera) found in British Columbia. Rep. ent. Soc. Ont. 50:53–61.

Buckell, E. R. 1922. A list of Orthoptera and Dermaptera recorded from British Columbia prior to the year 1922, with annotations. Proc. ent. Soc. Br. Columb. ([1] Syst.) 20:3–41.

Buckell, E. R. 1924. Additions and corrections to the list of British Columbia Orthoptera. Proc. ent. Soc. Br. Columb. (2)21:7–12.

Buckell, E. R. 1925. Notes on some British Columbia Grylloblattaria, Dermaptera and Orthoptera for the year 1925. Proc. ent. Soc. Br. Columb. (2)22:35–36.

Buckell, E. R. 1928. Entomology, British Columbia Orthoptera. Rep. prov. Mus. nat. Hist. Anthrop. Br. Columb. (1927):E15–E16.

Buckell, E. R. 1929. Notes on some Orthoptera from the Peace River District of British Columbia. Proc. ent. Soc. Br. Columb. (2)25(1928):10–14.

Buckell, E. R. 1930. The Dermaptera and Orthoptera of Vancouver Island, British Columbia. Proc. ent. Soc. Br. Columb. (2)27:17–51.

Buckell, E. R. 1937. List of Dermaptera & Orthoptera recorded from British Columbia. Located at: Dominion Entomological Laboratory, Vernon, B.C. 12 pp.

Buckell, E. R. 1941. Field crop and garden insects of the season 1940 in British Columbia. Can. Insect Pest Rev. 19:81–84.

Buckell, E. R. 1945. The grasshopper outbreak of 1944 in British Columbia. Can. Ent. 77:115–116.

Burmeister, H. [C. C.] 1838. Kaukerfe, *Gymnognatha* (Erste Hälfte; *vulgo* Orthoptera). Handbuch der Entomologie. Berlin. Theod. Chr. Friedr. Enslin. 2(2):I–VIII, 397–756.

Burr, M. 1897. British Orthoptera (earwigs, grasshoppers and crickets). Naturalist's J. & Guide, Huddersfield 6 (Suppl. 2):IV, 68 pp., 6 pl.

Bush, G. L. 1976. Modes of animal speciation. Ann. Rev. Ecol. System 6:339–364.

Caesar, L. 1941. Insects troublesome in the home. Bull. Ont. Dep. Agric. 416:1–52.

Caesar, L., and G. G. Dustan. 1939. Control of the house cricket. Rep. ent. Soc. Ont. 69 [dated 1938]:101–105.

Callan, E. McC. 1956. Observations on self-burial in *Acrotylus hirtus* Dirsh (Orth. Acrididae). Entomologist's mon. Mag. 92:116–117.

Campbell, M. G. 1949. Notes on Grylloblatta at Kamloops. Proc. ent. Soc. Br. Columb. 45:1–5.

Canadian Broadcasting Corporation. 1970. This Land [Television program.] (Broadcast January 1971.)

Cantrall, I. J. 1939–1940. Notes on collecting and studying *Orthoptera* [Part I]; Part II Notes. . . ; Part III Notes. . .; *Orthoptera* Part V [*sic*]; Orthoptera Part VI [*sic*]. Ward's Bull. 13(3):1–5; 13(4):1–6 [1939]; 13(5):4–5; 13(6):5–7; 13(7):4–5 [1940].

Cantrall, I. J. 1941. Compendium of entomological methods Part II: Notes on collecting and preserving Orthoptera. Rochester, NY: Ward's Natural Science Establishment, Inc. 28 pp.

Cantrall, I. J. 1943. The ecology of Orthoptera and Dermaptera of the George Reserve, Michigan. Misc. Publs Mus. Zool. Univ. Mich. 54:1–182.

Cantrall, I. J. 1968. An annotated list of the Dermaptera, Dictyoptera, Phasmatoptera and Orthoptera of Michigan. Mich. Ent. 1:299–346.

Cantrall, I. J. 1972. *Sago pedo* (Pallas) (Tettigoniidae: Saginae), an Old World katydid new to Michigan. Gt Lakes Ent. 5:103–106.

Cantrall, I. J., and F. N. Young. 1954. Contrasts in the Orthopteran faunas of grassland, forest and transitional areas in southern Indiana. Proc. Indiana Acad. Sci. 63(1953):157–162.

Carman, [W.] B. 1902. The pipes of Pan: Number One. From The book of myths. Boston, MA: L.C. Page & Co. 88 pp.

Castle, G. B. 1934. The damp-wood termite of western United States, genus *Zootermopsis* (formerly *Termopsis*). Pages 264–282 *in* C. F. Kofoid, ed. Termites and termite control. Berkeley, CA: University of California Press.

Catesby, M. 1731; 1743. The natural history of Carolina, Florida and the Bahama Islands: Containing the figures of birds, beasts, fishes, serpents, insects and plants; particularly the forest trees, shrubs and other plants not hitherto described, or very incorrectly figured by authors. Together with their descriptions in *English* and *French*. . . . London, England: Mark Catesby F.R.S.I. Vol. I: (I–V), 1–100, pl. 1–100; vol. 2:1–4, 1–100, pl. 1–100.

Caudell, A. N. 1903*a*. Some new or unrecorded Orthoptera from Arizona. Proc. ent. Soc. Wash. 5:162–166.

Caudell, A. N. 1903*b*. Notes on nomenclature of Blattidae. Proc. ent. Soc. Wash. 5:232–234.

Caudell, A. N. 1903*c*. Notes on the Orthoptera of Bermuda with the description of a new species. Proc. ent. Soc. Wash. 5:329–331.

Caudell, A. N. 1903*d*. Notes on Orthoptera from Colorado, New Mexico, Arizona and Texas, with descriptions of new species. Proc. U.S. natn. Mus. 26:775–809.
Caudell, A. N. 1904*a*. Notes on some Orthoptera from British Columbia. Ent. News 15:62–64.
Caudell, A. N. 1904*b*. The genus *Cyphoderris*. Jl N.Y. ent. Soc. 12:47–53.
Caudell, A. N. 1904*c*. Some Orthoptera taken at Moose Jaw, Assiniboia. Can. Ent. 36:248.
Caudell, A. N. 1904*d*. On a collection of non-saltatorial Orthoptera from Paraguay. Jl N.Y. ent. Soc. 12:179–188.
Caudell, A. N. 1904*e*. Orthoptera of the Expedition. Page 117 *in* T. Kincaid, ed. Insects Part I. Alaska [Harriman Alaska Expedition]. New York: Doubleday, Page & Co. Vol. 8.
Caudell, A. N. 1904*f*. Orthoptera from southwestern Texas. Collected by the Museum expedition of 1903, 1904. Sci. Bull. Brook. Inst. Arts Sci. 1:105–116, pl. VI, VII.
Caudell, A. N. 1904*g*. The synonymy of *Oedipoda cincta* Thomas. Proc. ent. Soc. Wash. 6:125.
Caudell, A. N. 1905*a*. A new species of the locustid genus *Amblycorypha* from Kansas. Jl N.Y. ent. Soc. 13:50.
Caudell, A. N. 1905*b*. Class I, Hexapoda. Order XI, Orthoptera. A new *Bacunculus* from Indiana. Jl N.Y. ent. Soc. 13:212.
Caudell, A. N. 1906. Class I, Hexapoda. Order XI, Orthoptera. The Cyrtophylli of the United States. Jl N.Y. ent. Soc. 14:32–45, pl. I.
Caudell, A. N. 1907*a*. Notes on United States Orthoptera, with the description of one new species. Proc. ent. Soc. Wash. 8:133–135.
Caudell, A. N. 1907*b*. The Decticinae (a group of Orthoptera) of North America. Proc. U.S. natn. Mus. 32:285–410.
Caudell, A. N. 1908. Notes on some western Orthoptera; with the description of one new species. Proc. U.S. natn. Mus. 34:71–81.
Caudell, A. N. 1915. Notes on some United States grasshoppers of the family Acrididae. Proc. U.S. natn. Mus. 49:25–31.
Caudell, A. N. 1916. The genera of the tettigoniid insects of the subfamily Rhaphidophorinae found in America north of Mexico. Proc. U.S. natn. Mus. 49:655–690.
Caulfield, F. B. 1886. List of Orthoptera taken in the vicinity of Montreal, P.Q. Can. Ent. 18:211–212.
Caulfield, F. B. 1887. List of Orthoptera taken in the vicinity of Montreal, P.Q. Can. Rec. Sci. 2:373–383; 393–404.
Caulfield, F. B. 1888. A sketch of Canadian Orthoptera. Rep. ent. Soc. Ont. 18:59–72.
Cawein, M. [J.]. 1906. 1891–1900 [1893]. Nature-notes and impressions, in prose and verse. New York: E. P. Dutton & Co. Pp. 79–134.
Chagnon, G. 1944*a*. Contribution à l'étude des Orthoptères et des Dermaptères du Québec. Naturaliste can. 71:15–39, 54–74, 127–128.
Chagnon, G. 1944*b*. Contribution à l'étude des Orthoptères et des Dermaptères du Québec. Contr. Inst. Biol. Univ. Montréal. Vol. 14. II, 68 pp.
Chagnon, G. 1947. The cockroaches of the Province of Quebec. Can. Ent. 79:57–58.

Chagnon, G. 1948. Les Blattidés du Québec (Ordre des Orthoptères). Rep. Québ. Soc. Prot. Pl. 30:129–130.

Charpentier, T. de. 1825. De Orthopteris Europaeis. Pages 61–181 *in* Horae Entomologicae (Neuroptera, Orthoptera, Coleoptera). Wratislawiae [Breslau/Bratislava]: Gosohorsky. Vol. 4.

Charpentier, T. de. 1843. Orthoptera descripta et depicta. Lipsiae [Leipzig]. Leop. Voss. 9:[99]-[110], pl. 49–54.

Children, J. C. 1836. Appendix No. III. Articulata. Catalogue of the Arachnida and insects collected by Mr. King, surgeon and naturalist of the expedition. Pages 532–542 *in* G. Back. Narrative of the Arctic Land Expedition to the mouth of the Great Fish River, and along the shores of the Arctic Ocean in the years 1833, 1834, and 1835. London: John Murray.

Chopard, L. 1922. Orthoptères et Dermaptères. Faune Fr. 3:VI, 212 pp.

Chopard, L. 1938. La biologie des Orthoptères. Encycl. ent. Ser. A 20:IV, 541 pp., 4 pl.

Chopard, L. 1949. Ordre des Dictyoptères Leach, 1818 [*sic*]. (= Blattaeformia Werner, 1906; = Oothecaria Karny, 1915); Ordre des Chéleutoptères Crampton 1915 (= Phasmoptères Jeannel, 1947); Ordre des Orthoptères Latreille 1793—Olivier 1789. Pages 355–407, 594–616, 617–722 *in* P.-P. Grassé, ed. Traité de Zoologie. Paris: Masson & Cie. Vol. 9. 1 pl. [Reprinted 1965.]

Chopard, L. 1951. Orthoptéroïdes. Faune Fr. 56:IV, 359 pp.

Chopard, L. 1967. Gryllides Fam. Gryllidae: Subfam. Gryllinae (Trib. Gymnogryllini, Gryllini, Gryllomorphini, Nemobiini). Pages 1–211 *in* M. Beier, ed. Orthopterorum Catalogus. Vol. 10.

Clark, J. T. 1975. A key to the subfamilies of Phasmida. Entomologist's Rec. J. Var. 87:104–107.

Clark, J. T. 1979. A key to eggs of stick and leaf insects (Phasmida). Syst. Ent. 4:325–331.

Clifford, C. W., R. M. Roe, and J. P. Woodring. 1977. Rearing methods for obtaining house crickets, *Acheta domesticus*, of known age, sex, and instar. Ann. ent. Soc. Am. 70:69–74.

Cockerell, T. D. A. 1889. On the variation of insects [5]. Entomologist 22:125–130.

Cockerell, T. D. A. 1917. Melanism in the Orthoptera. Entomologist's Rec. J. Var. 29:247.

Cohen, S. H., and L. M. Roth. 1970. Chromosome numbers of the Blattaria. Ann. ent. Soc. Am. 63:1520–1547.

Cohn, T. J., and I. J. Cantrall. 1974. Variation and speciation in the grasshoppers of the Conalcaeini (Orthoptera: Acrididae: Melanoplinae): The lowland forms of western Mexico, the genus *Barytettix*. Mem. San Diego Soc. nat. Hist. 6:3–131, 1 pl.

Comeau, N.-M. 1965. A glance at the history of entomology and entomological collections in Québec. Ann. ent. Soc. Québ. 10:85–90.

Connin, R. V. 1964. Notes on the biology of *Hesperotettix viridis*. J. econ. Ent. 57:606.

Cornwell, P. B. 1968. The cockroach. Volume I. Industrial Pest. London, England: Hutchinson. 391 pp., 1 pl.

Cornwell, P. B. 1976. The cockroach. Volume II. Insecticides and cockroach control. New York: St. Martin's Press. 557 pp.

Coues, E., ed. 1897. New light on the early history of the Greater Northwest. The manuscript journals of Alexander Henry . . . and of David Thompson . . . 1799–1814. . . . [3 vols.] New York: Francis P. Harper.

Cowan, F. T. 1929. Life history, habits and control of the Mormon cricket. Tech. Bull. U.S. Dep. Agric. 161:1–28.

Cowan, F. T. 1971. Field biology of the migratory phase of *Melanoplus rugglesi* (Orthoptera: Acrididae). Ann. ent. Soc. Am. 64:574–580.

Crampton, G. C. 1915. The thoracic sclerites and the systematic position of *Grylloblatta campodeiformis* Walker, a remarkable annectent, "Orthopteroid" insect. Ent. News 26:337–350, pl. XIII.

Crampton, G. C. 1919. Notes on the phylogeny of the Orthoptera. Ent. News 30:42–48.

Criddle, A. 1973. Criddle-de-diddle-ensis: A biographical history of the Criddles of Aweme, Manitoba pioneers of the 1880's. Winnipeg, Man.: Alma Criddle. VI, 288 pp.

Criddle, N. 1903. Mr. Criddle's Report. Pages 186–187 *in* J. Fletcher. Report of the Entomologist and Botanist 1902. Ottawa, Ont.: Exp. Fms Reps. (Appendix to Rep. Min. Agric.)

Criddle, N. 1913. Insect pests of southern Manitoba during 1912. Rep. ent. Soc. Ont. 43:97–100.

Criddle, N. 1917. Precipitation in relation to insect prevalence and distribution. Can. Ent. 49:77–80.

Criddle, N. 1918. The egg-laying habits of some of the Acrididae (Orthoptera). Can. Ent. 50:145–151.

Criddle, N. 1920. Locusts in Manitoba, with special reference to the outbreak of 1919. Rep. ent. Soc. Ont. 50:49–53.

Criddle, N. 1924. Notes on the early stages of grasshoppers. Can. Ent. 56:49–53.

Criddle, N. 1925. Field crickets in Manitoba. Can. Ent. 57:79–84.

Criddle, N. 1926*a*. The life history and habits of *Anabrus longipes* Caudell (Orthop.). Can. Ent. 58:261–265.

Criddle, N. 1926*b*. Studies of the immature stages of Manitoban Orthoptera. Trans. R. Soc. Can. (3)20(Sect. V):505–525, pl. 1–4.

Criddle, N. 1930*a*. Life-history of the cow grasshopper (Chrysochraon abdominalis Thom.) in Manitoba. Can. Ent. 62:25–28.

Criddle, N. 1930*b*. An outline of methods employed in rearing Orthoptera. Treesbank, Man. 6 pp.

Criddle, N. 1931. Grasshopper control in Canada east of the Rocky Mountains. Bull. Can. Dep. Agric. (N.S.) 143:1–18.

Criddle, N. 1932. The life-history of *Schistocerca lineata* Scud. Can. Ent. 64:98–102.

Criddle, N. 1935 (posthumous). Studies in the biology of North American Acrididae: Development and habits. Can. Soc. tech. Agriculturists 2:474–494.

Crozier, L. M. 1977. Identification of the immature stages of grasshoppers (Acrididae) of Québec. Montreal, Qué.: McGill Univ. M.Sc. Thesis. 136 pp., 77 pl.

Davis, A. C., and L. M. Smith. 1926. Notes on the genus *Stenopelmatus* with description of a new species (Orthoptera). Pan-Pacif. Ent. 2:174–181.

Davis, W. T. 1887. Notes on the Locustidae, with description of a new species. Can. Ent. 19:56–57.

Davis, W. T. 1888. . . . entomological notes of local interest. *In* Proceedings of Scientific Societies. Am. Nat. 22:1148.

Davis, W. T. 1893. The song of *Thyreonotus*. Can. Ent. 25:108–109.

Davis, W. T. 1907. A new tree cricket from Staten Island and New Jersey. Can. Ent. 39:173–174.

Davis, W. T. 1911. Miscellaneous notes on collecting in Georgia. Jl N.Y. ent. Soc. 19:216–219.

Davis, W. T. 1915*a*. A new species of *Atlanticus* from the mountains of Georgia and North Carolina. Bull. Brooklyn ent. Soc. 9(1914):104–106.

Davis, W. T. 1915*b*. *Dendrotettix quercus* Riley at Yaphank, Long Island, N.Y. Bull. Brooklyn ent. Soc. 10:33–34.

Davis, W. T. 1923. A new walking-stick insect from eastern North America. Jl N.Y. ent. Soc. 31:52–55, pl. X.

Davys, I. [= Davis, J.] 1589. The second voyage attempted by M. Iohn Dauis with others, for the discovery of the Northwest Passage, in Anno 1586. *In* R. Hakluyt, ed. The Principall Nauigacions, Voiages, Traffiques, & Discoueries of the English Nation. . . . London, England. 2nd ed. 3 vols.

Dawson, G. M., ed. 1875*a*. Report on the geology and resources of the region in the vicinity of the forty-ninth parallel from the Lake of the Woods to the Rocky Mountains with lists of plants and animals collected and notes on fossils. Montreal, Qué.: Dawson Bros. for British North American Boundary Commission.

Dawson, G. M. 1875*b*. Notes on the locust invasion of 1874 in Manitoba and the Northwest Territories. Can. Nat. (n.s.) 8:119–134.

Dawson, G. M. 1876. Notes on the appearance and migrations of the locust in Manitoba and the Northwest Territories, summer of 1875. Can. Nat. (n.s.) 8:207–226.

De Geer, C. 1773. Mémoires pour servir à l'Histoire des Insectes. Stockholm, Sweden: Pierre Hesselberg. Vol. 3. VIII, 696, 2 pp., 44 pl.

Deitz, L. L. 1979. Curatorial note on genitalia vials. N.Z. Ent. 7:10–11.

DeLong, D. M., and R. J. Keagy. 1949. Termite cultures in the laboratory. Turtox News 27:114–116.

Dirsh, V. M. 1961. A preliminary revision of the families and subfamilies of Acridoidea. Bull. Brit. Mus. (nat. Hist.) Ent. 10:350–419.

Dirsh, V. M. 1973. Genital organs in Acridomorphoidea (Insecta) as taxonomic character. Z. zool. Syst. EvolutsForsch. 11:133–154.

Dirsh, V. M. 1974. Genus *Schistocerca* (Acridomorpha, Insecta). The Hague, Netherlands: Dr. W. Junk (Ser. Ent. 10). VIII, 238 pp.

Dirsh, V. M. 1975. Classification of the acridomorphoid insects. Faringdon, England: E. W. Classey Ltd. VIII, 171 pp.

Dodge, G. M. 1876. New species of Acridini from Nebraska. Can. Ent. 8:9–12.

Dodge, G. M. 1877. New species of Orthoptera. Can. Ent. 9:111–113.

Dohrn, W. L. H. 1862. Die Dermapteren von Mexico. Stettin. ent. Ztg 23:225-232, pl. 1.

Doubleday, E. 1838. Communications on the natural history of North America [III]. Entomologist's mon. Mag. 5:269-300.

Doyon, D. 1962. Étude bio-écologique d'*Asclepias syriaca* L. Rep. Québ. Soc. Prot. Pl. 42(1960):25-53.

Drury, D. 1773. Illustrations of natural history, wherein are exhibited upward of two hundred and forty figures of exotic insects, according to their different genera; . . . London, England: White. Vol. 1, 2. 130 pp., 50 pl.

DuFour, L. 1841. Recherches anatomiques et physiologiques sur les Orthoptères, les Hyménoptères et les Nevroptères. Mém. Math. Sav. étrang. 7:265-647, pl. 1-13.

DuPorte, E. M. 1919. The propleura and the pronotal sulci of the Orthoptera. Can. Ent. 51:147-153.

Dwight, T. 1821 (posthumous). Travels in New England and New York. New Haven, CT. Vol. 2.

Eades, D. C. 1962. The identity of *Ceuthophilus guttulosus* and its subspecies (Orthoptera, Gryllacrididae, Rhaphidophorinae). Ent. News 73:147-152.

Eades, D. C. 1964. General biology and geographic variation of *Ceuthophilus guttulosus* Walker (Orthoptera: Gryllacrididae: Rhaphidophorinae). Trans. Am. ent. Soc. 90:73-110.

Ede, E. K. 1974. Some observations on *Metrioptera roeselii* (Grylloptera: Tettigoniidae) in Maine, New Hampshire and Massachusetts. Can. Ent. 106:1205-1209.

Edwards, G. A., and W. L. Nutting. 1950. The influence of temperature upon the respiration and heat activity of *Thermobia* and *Grylloblatta*. Psyche, Camb. 57:33-44.

Elliott, W. M., and B. N. Dhanvantari. 1973. Tree cricket (Orthoptera: Oecanthidae) damage to young peach trees. Can. Ent. 105:1035-1038.

Emerson, A. E. 1933. A revision of the genera of fossil and recent Termopsinae (Isoptera). Univ. Calif. Publs Ent. 6:165-196.

Esenther, G. R. 1969. Termites in Wisconsin. Ann. ent. Soc. Am. 62:1274-1284.

Essig, E. O. 1942. College entomology. New York: MacMillan Co. VIII, 900 pp.

Evans, H. E., and F. E. Kurczewski. 1966. Observation on the nesting behavior of some species of *Tachytes* (Hymenoptera: Sphecidae, Larrinae). J. Kans. ent. Soc. 39:323-332.

Fabricius, J. C. 1775. Systema entomologiae sistens Insectorum classes, ordines, genera, species, adjectis synonymis, locis, descriptionibus. Flensburgi & Lipsiae [Leipzig]:Korte. (30), 832 pp. [Appendix 816-832.]

Fabricius, J. C. 1781. Species insectorum exhibentes eorum differentias specificas, synonymas auctorum, loca natalia, metamorphosin adjectis observationibus descriptionibus. Hamburg & Kiel. Bohn. Vol. 1:VIII, 552 pp.; vol. 2:II, 517 pp.

Fabricius, J. C. 1793. Classis II. Ulonata. Pages 1-62 *in* Entomologia systematica emendata et aucta, secundum classes, ordines, genera, species, adjectis synonymis, locis, observationibus, descriptionibus. Hafniae [København]. Proft 2.

Fabricius, J. C. 1798. Supplementum entomologiae systematicae. Hafniae [Kϕbenhavn]. Proft & Storch. IV, 572 pp.
Farrow, R. A. 1963. A comparative study of the biology of *Tetrix ceperoi* I. Bolivar and *T. subulata* (L.). London, England: Imperial Coll., Univ. London. Diploma Dissertation. 188 pp.
Farrow, R. A. 1964*a*. The post-embryonic development of the external genitalia of *Tetrix* Latreille (Orthoptera: Tetrigidae). Ann. Mag. nat. Hist. 7:301–313.
Farrow, R. A. 1964*b*. The spermatophore of *Tetrix* Latreille (Orthoptera: Tetrigidae). Entomologist's mon. Mag. 99(1963):217–223.
Faull, J. H., ed. 1913. Chapter XXII. Insects and their allies by E. M. Walker, B.A., M.B. The natural history of the Toronto region, Ontario, Canada. Toronto, Ont.: Royal Canadian Institute. Pp. 295–403.
Faure, J. C. 1933. The phases of the Rocky Mountain locust *Melanoplus mexicanus* (Saussure). J. econ. Ent. 26:706–718, pl. 29.
Fernald, C. H. 1888. The Orthoptera of New England. Designed for the use of students in the Massachusetts Agricultural College and farmers of the State. Ann. Rep. Mass. agric. Coll. 25(1887) [= Mass. publ. Doc. 1888(31)]:85–145.
Fieber, F. X. 1853. Synopsis der europäischen Orthoptera mit besonderer Rücksicht auf die in Böhmen vorkommenden Arten als Auszug aus dem Drucke vorliegenden Werke, die europäischen Orthoptera. Lotos 3:115–129.
Fischer de [von] Waldheim, G. 1846*a*. Index Orthopterorum Societati traditorum. Byull. Mosk. Obshch. Ispȳt. Prir. (Bull. Soc. nat. Moscou) 19: 468–482, pl. XIII.
Fischer de [von] Waldheim, G. 1846*b*. Orthoptera Imperii Russici (Orthoptères de la Russie). Pages IV + 1–443 *in* Entomographia Imperii Russici (Entomographie de la Russie) . . . Moscou, USSR: Imprimerie d'Auguste Semen. 4 [= Nouv. Mém. Soc. nat. Moscou, 8]. XXXVII pl.
Fitch, A. 1847. List of noxious insects. Am. J. Agric. Sci. 6:145–152.
Fitch, A. 1856. Third report on the noxious and other insects of the state of New York. Trans. N.Y. State agric. Soc. 16:321–507.
Fletcher, J. 1892. The northern mole-cricket (*Gryllotalpa borealis* Burm.). Can. Ent. 24:23–25.
Fletcher, J. 1896. Insect injuries of the year 1895. Rep. ent. Soc. Ont. 26:31–36.
Fletcher, J. 1899. Report of the Entomologist and Botanist 1898. Ottawa, Ont.: Exp. Fms Reps. (Appendix to Rep. Min. Agric.) Pp. 167–212.
Fletcher, J. 1900. Report of the Entomologist and Botanist 1899. Ottawa, Ont.: Exp. Fms Reps. (Appendix to Rep. Min. Agric.) Pp. 159–204.
Fletcher, J. 1901. Report of the Entomologist and Botanist 1900. Ottawa, Ont.: Exp. Fms Reps. (Appendix to Rep. Min. Agric.) Pp. 220–228.
Fletcher, J. 1902*a*. Report of the Entomologist and Botanist 1901. Ottawa, Ont.: Exp. Fms Reps. (Appendix to Rep. Min. Agric.) Pp. 197–262.
Fletcher, J. 1902*b*. The painted lady butterfly (*Pyrameis Cardui*, L.). Rep. ent. Soc. Ont. 32:54–56.
Fletcher, J. 1903. Entomological Record, 1902. Rep. ent. Soc. Ont. 33:87–98.

Fletcher, J. 1906. Entomological Record, 1905. Rep. ent. Soc. Ont. 36:90–105.

Fletcher, J., and A. Gibson. 1908. Entomological Record, 1907. Rep. ent. Soc. Ont. 38:113–133.

Fletcher, J., and A. Gibson. 1909. Entomological Record, 1908. Rep. ent. Soc. Ont. 39:99–116, pl. Q, R [Orthoptera by E. M. Walker: 113–114].

Flint, O. S. 1951. A new cockroach record for the United States. Bull. Brooklyn ent. Soc. 46:53.

Fontana, P. G., and V. R. Vickery. 1973. Segregation-distortion in the B-chromosome system of *Tettigidea lateralis* (Say) (Orthoptera: Tetrigidae). Chromosoma 43:75–100.

Fontana, P. G., and V. R. Vickery. 1974. Heterochromatin content and chiasma distribution in the megameric chromosomes of *Stethophyma gracile* and *S. lineatum* (Orthoptera: Acrididae). Chromosoma 46:375–395.

Fontana, P. G., and V. R. Vickery. 1975. The B-chromosome system of *Tettigidea lateralis* (Say) II. New karyomorphs, patterns of pycnosity and giemsa banding. Chromosoma 50:371–391.

Fontana, P. G., and V. R. Vickery. 1976. Cytotaxonomic studies on the genus *Boonacris*. I. The "eastern" taxa and a comparison with the related genera *Dendrotettix* and *Appalachia* (Orthoptera: Catantopidae: Podismini). Can. J. Genet. Cytol. 18:625–652.

Ford, N. 1926. On the behavior of *Grylloblatta*. Can. Ent. 58:66–70.

Fox, C. J. S., and G. M. Stirrett. 1953. Annotated catalogue of insect and other invertebrate pests of tobacco in Canada. Rep. ent. Soc. Ont. 83:48–54.

Fox, H. 1912*a*. Two apparently hitherto undescribed species of *Xiphidium* from the salt marshes of the Atlantic coast of the United States (Orthop.). Ent. News 23:111–119, pl. VIII, IX.

Fox, H. 1912*b*. Types of *Xiphidium spartinae* and *nigropleuroides* (Orthop.). Ent. News 23:232.

Fox, H. 1917. Field notes on Virginia Orthoptera. Proc. U.S. natn. Mus. 52:199–234.

Freeman, P., ed. 1980. Common insect pests of stored food products: A guide to their identification. 6th ed. British Museum (Natural History), London. X, 69 pp.

Frings, H., and M. Frings. 1962. Simplified methods for laboratory maintenance of saltatory Orthoptera. J. econ. Ent. 55:1019–1020.

Frings, H., and M. Frings. 1971. CFAA: A fixing agent for insect tissues. Ent. News 82:157–159.

Froeschner, R. C. 1954. The grasshoppers and other Orthoptera of Iowa. Iowa St. Coll. J. Sci. 29:163–354.

Fulton, B. B. 1915. The tree crickets of New York: Life history and bionomics. Tech. Bull. N.Y. agric. Exp. Stn 42:1–47, pl. I–VI.

Fulton, B. B. 1925. Physiological variation in the snowy tree cricket, *Oecanthus niveus*, De Geer. Ann. ent. Soc. Am. 18:363–383.

Fulton, B. B. 1926. The tree crickets of Oregon. Bull. Oregon agric. Coll. Exp. Stn 223:1–20.

Fulton, B. B. 1928. The habitat of *Tropidischia xanthostoma* Scudder. Ent. News 39:8–11.

Fulton, B. B. 1930. Notes on Oregon Orthoptera with descriptions of new species and races. Ann. ent. Soc. Am. 23:611–641.

Fulton, B. B. 1931. A study of the genus *Nemobius* (Orthoptera: Gryllidae). Ann. ent. Soc. Am. 24:205–237.

Fulton, B. B. 1951. The seasonal succession of orthopteran stridulation near Raleigh, North Carolina. J. Elisha Mitchell Sci. Soc. 67:87–95.

Fulton, B. B. 1956. The genus *Anaxipha* in the United States (Orthoptera: Gryllidae). J. Elisha Mitchell Sci. Soc. 72:222–243.

Gage, S. H., M. K. Mukerji, and R. L. Randell. 1976. A predictive model for seasonal occurrence of three grasshopper species in Saskatchewan (Orthoptera: Acrididae). Can. Ent. 108:245–253.

Gallaway, H. E. 1948. *Melanoplus occidentalis occidentalis* a range species of grasshopper in Nevada. J. econ. Ent. 41:925–927.

Gallienne, G. 1969. Un pied d'ancre: Journal de Placide Vigneau: Trois quarts de siècle d'histoire sur la Côte Nord, le Labrador et les Îles de la Madeleine (1857–1926). Lévis, Qué.: Imprimerie Le Quotidien, Ltée. 311 pp.

Gangwere, S. K. 1960*a*. The feeding and culturing of Orthoptera in the laboratory. Ent. News 71:7–13, 37–45.

Gangwere, S. K. 1960*b*. The use of the mouthparts of Orthoptera during feeding. Ent. News 71:193–206.

Gangwere, S. K. 1961. A monograph on food selection in Orthoptera. Trans. Am. ent. Soc. 87:67–230.

Gangwere, S. K. 1965. Food selection in the oedipodine grasshopper, *Arphia sulphuria* (Fabricius). Am. Midl. Nat. 74:67–75.

Gangwere, S. K. 1966. The behavior of *Atlanticus testaceus* (Orthoptera: Tettigoniidae) on the E. S. George Reserve, Michigan. Mich. Ent. 1:95–100.

Gangwere, S. K. 1967. The feeding behavior of *Atlanticus testaceus* (Orthoptera: Tettigoniidae). Ann. ent. Soc. Am. 60:74–81.

Gaunitz, C. B. 1936. *Ectobius lapponicus* L. als Vorratschädling in Lappland, eine alte, sicher unrichtige Vermutung in neuer Beleuchtung. Konowia 15:162–166.

Gené, C. G. 1832. Saggio di una monografia delle Forficule indigine. Annali Sci. Regno lombardo-veneto, 2:215–228. [Also sep., Padova. 16 pp.]

Ghouri, A. S. K., and J. E. McFarlane. 1958*a*. Reproductive isolation in the house cricket (Orthoptera: Gryllidae). Psyche, Camb. 64(1957): 30–36.

Ghouri, A. S. K., and J. E. McFarlane. 1958*b*. Observations on the development of crickets. Can. Ent. 90:158–165.

Gibb, G. 1859. On the generation of sounds by Canadian insects. Can. Nat. Geol. 4:121–130.

Gibson, A. 1910*a*. Reports on insects of the year. Division No. 1, Ottawa District. Rep. ent. Soc. Ont. 40:9–16.

Gibson, A. 1910*b*. The Entomological Record, 1909. Rep. ent. Soc. Ont. 40:110–128.

Gibson, A. 1914. Reports on insects of the year [1913]. Division No. 1, Ottawa District. Rep. ent. Soc. Ont. 44:15–18.
Gibson, A. 1915*a*. Reports on insects for the year [1914]. Division No. 1, Ottawa District. Rep. ent. Soc. Ont. 45:13–16.
Gibson, A. 1915*b*. The Entomological Record, 1914. Rep. ent. Soc. Ont. 45:123–150.
Gibson, A. 1916*a*. Reports on insects of the year [1915]. Division No. 1, Ottawa District. Rep. ent. Soc. Ont. 46:11–14.
Gibson, A. 1916*b*. Locust control work with poisoned baits in eastern Canada in 1915. Rep. ent. Soc. Ont. 46:156–162.
Gibson, A. 1916*c*. The Entomological Record, 1915. Rep. ent. Soc. Ont. 46:194–230.
Gibson, A. 1918. Report on insects for the year [1917]. Division No. 1, Ottawa District. Rep. ent. Soc. Ont. 48:18–20.
Giese, R. L., and K. H. Knauer. 1977. Ecology of the walkingstick. Forest Sci. 23:45–63.
Gilbert, H. A. 1936. A note on grasshopper outbreaks in the counties of Renfrew and Hastings, Ontario, in 1935. Rep. ent. Soc. Ont. 66:60–62.
Gilbert, H. A., and R. W. Thompson. 1937. The grasshopper outbreak in Ontario in 1936. Rep. ent. Soc. Ont. 67:65–68.
Gillette, C. P. 1904. Annotated list of Colorado Orthoptera from material in the collections of the Colorado Agricultural College and Agricultural Experiment Station. Part I. Including families Forficulidae, Blattidae, Mantidae, Phasmidae and Acrididae. Tech. Bull. Colo. agric. Exp. Stn 94:17–56.
Gillette, C. P. 1905. The western cricket. Bull. Colo. St. agric. Coll. 101:2–10.
Girard, C. 1853. Orthopterous insects. *In* Appendix F. Zoology. Pages 257–261 *in* R. B. Marcy et al. Exploration of the Red River of Louisiana in the year 1852. Washington, DC: Robert Armstrong (Exec. Doc. 54, 32nd Congress, 2nd Session). Pl. 15.
Glaser, R. W. 1914. *Forficula auricularia* in Rhode Island. Psyche, Camb. 21:157–158.
Glen, R. 1956. Introduction; *and* Historical Review. *In* R. Glen, ed. Entomology in Canada up to 1956: A review of developments and accomplishments. Can. Ent. 88:291–292, 292–300.
Glover, T. 1872*a*. Entomological Record. Monogr. Rep. U.S. Dep. Agric. 1872(2):74–76.
Glover, T. 1872*b* [–1878]. Orthoptera. Illustration of North American entomology, (United States and Canada). Washington, DC: J. S. Tomlinson. Vol. 1. VIII, 11 pp., pl. I–XIII.
Gloyer, W. O., and B. B. Fulton. 1916. Tree crickets as carriers of *Leptosphaeria coniothyrium* (Fckl.) Sacc. and other fungi. Tech. Bull. N.Y. agric. Exp. Stn 50:1–22.
Gmelin, J. F., ed. 1789. Caroli a Linné, . . . Systema Naturae per Regna Tria Naturae, Secundum Classes, Ordines, Genera, Species; cum Characteribus, Differentiis, Synonymis, Locis. Edito decima tertia, aucta, reformata. Lund. J. B. Delamollière. Vol. 1(4). Pp. I–IV, 1517–2224.

Goble, H. W. 1966. Insects attacking agricultural crops and ornamental plants in Ontario in 1965. Proc. ent. Soc. Ont. 96:5–6.

Goellner, E. J. 1931. A new species of termite, *Reticulitermes arenicola* from the sand dunes of Indiana and Michigan, along the shores of Lake Michigan. Proc. ent. Soc. Wash. 33:227–234.

Gooderham, C. B. 1917. The Acrididae of Nova Scotia. Proc. ent. Soc. Nova Scotia 2(1916):21–30.

Gooderham, C. B. 1918. The Locustidae of Nova Scotia. Proc. ent. Soc. Nova Scotia 3(1917):23–36.

Gosse, P. H. 1840. The Canadian naturalist, a series of conversations on the natural history of Lower Canada. London. J. van Voorst. xii, 372 pp.

Gould, G. E., and H. O. Deay. 1940. The biology of six species of cockroaches which inhabit buildings. Bull. Purdue Univ. agric. Exp. Stn 451:1–31.

Gray, G. R. 1835. Synopsis of the species of insects belonging to the family of Phasmidae. London, England: Longman, Rees, Brown, Green & Longman. IV, 48 pp.

Gregson, J. D. 1938. *Grylloblatta campodeiformis* Walker—a new record. Can. Ent. 70:63–64.

Gregson, J. D. 1939. Notes on the occurrence of *Grylloblatta campodeiformis* Walker in the Kamloops district. Proc. ent. Soc. Br. Columb. 35:29–30.

Guérin-Méneville, F. E. 1844. Insectes. Iconographie du Règne Animal de G. Cuvier, ou représentations d'après nature de l'une des espèces les plus remarquables et souvent non encore figurées de chaque genre d'animaux; pouvant servir d'atlas à tous les traités de Zoologie. Paris, France: J. B. Baillière. Vol. 7. 576 pp., 104 pl.

Gunn, D. 1878. [Ravages in the Northwest in the early part of the century]. Pages [147]–[148] *in* C. V. Riley, A. S. Packard, and C. Thomas. First annual report of the United States Entomological Commission for the year 1877 relating to the Rocky Mountain Locust and the best methods of preventing its injuries. . . . Washington, D.C.: Government Printing Office.

Günther, K(laus). 1953. Über die taxonomische Gliederung und die geographische Verbreitung der Insektenordnung der Phasmatodea. Beitr. Ent. 3:541–563.

Günther, K., and K. Herter. 1974. Ordnung Dermaptera (Ohrwürmer). Pages 1–158 *in* J.-G. Helmcke, D. Starck, and H. Wermuth, eds. Kükenthal's Handbuch der Zoologie (Ed. 2). Berlin, New York. Walter de Gruyter. Vol. 4(2)[2(11)].

Günther, K(urt). K. 1972. Die Tridactyloidea von Uruguay (Orthopteroidea, Saltatoria). Dt. ent. Z. (N.F.) 19:211–236.

Günther, K. K. 1975. Das Genus *Neotridactylus* Günther, 1972 (Tridactylidae, Saltatoria, Insecta). Mitt. zool. Mus. Berl. 51:305–365.

Günther, K. K. 1977. Revision der Gattung *Ellipes* Scudder, 1902 (Saltatoria, Tridactylidae). Dt. ent. Z. (N.F.) 24:47–122.

Guppy, R. 1950. Biology of *Anisolabis maritima* (Gené) the seaside earwig, on Vancouver Island (Dermaptera, Labiduridae). Proc. ent. Soc. Br. Columb. 46:14–18.

Gurney, A. B. 1937*a*. Studies in certain genera of American Blattidae. Proc. ent. Soc. Wash. 39:101–112.

Gurney, A. B. 1937*b*. Synopsis of the Grylloblattidae with the description of a new species from Oregon. Pan.-Pacif. Ent. 13:160–170.

Gurney, A. B. 1939. Aids to the identification to the Mormon and Coulee crickets and their allies (Orthoptera: Tettigoniidae: Gryllacrididae). Mimeogr. Circ. U.S. Bur. Ent. Pl. Quarant. E. 479:1–11.

Gurney, A. B. 1940*a*. A revision of the grasshoppers of the genus *Orphulella* Giglio-Tos, from America north of Mexico (Orthoptera: Acrididae). Ent. Am. (n.s.) 20:85–157.

Gurney, A. B. 1940*b*. Notes on certain genera of North American grasshoppers of the subfamily Oedipodinae, with the description of a new genus and species (Orthoptera: Acrididae). Proc. ent. Soc. Wash. 42:1–12.

Gurney, A. B. 1941. Taxonomic and bionomic notes of the grasshopper *Melanoplus impudicus* Scudder (Orthoptera: Acrididae). Am. Midl. Nat. 26:558–569.

Gurney, A. B. 1948. The taxonomy and distribution of the Grylloblattidae. Proc. ent. Soc. Wash. 50:86–102.

Gurney, A. B. 1949. *Melanoplus rugglesi*, a migratory grasshopper from the Great Basin of North America. Proc. ent. Soc. Wash. 51:267–272.

Gurney, A. B. 1950*a*. An African earwig new to the United States, and a corrected list of the Nearctic Dermaptera. Proc. ent. Soc. Wash. 52:200–203.

Gurney, A. B. 1950*b*. The Linnaean subgeneric names of *Gryllus* (Orthoptera). J. Wash. Acad. Sci. 40:409–413.

Gurney, A. B. 1951*a*. Praying mantids of the United States, native and introduced. Rep. Smithson. Instn 1950:339–362.

Gurney, A. B. 1951*b*. The names of the field and house crickets. J. econ. Ent. 44:611.

Gurney, A. B. 1953*a*. Grasshopper glacier of Montana and its relation to long-distance flights of grasshoppers. Rep. Smithson. Instn 1952:305–325, pl. 1–8.

Gurney, A. B. 1953*b*. Recent advances in the taxonomy and distribution of *Grylloblatta* (Orthoptera: Grylloblattidae). J. Wash. Acad. Sci. 43:325–332.

Gurney, A. B. 1958. Which grasshoppers of temperate North America are migratory; notes on their migratory behavior and their names (Orthoptera: Acrididae). Proc. X Int. Congr. Ent. (Montreal, 1956) 3:463–464.

Gurney, A. B. 1961. Further advances in the taxonomy and distribution of the Grylloblattidae (Orthoptera). Proc. biol. Soc. Wash. 74:67–76.

Gurney, A. B. 1962. On the name of the migratory grasshopper of the United States and Canada, *Melanoplus sanguinipes* (F.). (Orthoptera: Acrididae). Proc. biol. Soc. Wash. 75:189–192.

Gurney, A. B. 1968. The spotted Mediterranean cockroach, *Ectobius pallidus* (Olivier) (Dictyoptera: Blattaria: Blattellidae), in the United States. Rep. U.S. Dep. Agric. coop. econ. Ins. 18:684–686.

Gurney, A. B. 1971. North American grasshoppers of the genus *Argiacris*, including two new species from Idaho (Orthoptera: Acrididae: Catantopinae). Proc. ent. Soc. Wash. 73:292–303.

Gurney, A. B., and A. R. Brooks. 1959. Grasshoppers of the *mexicanus* group, genus *Melanoplus* (Orthoptera: Acrididae). Proc. U.S. natn. Mus. 110:1–93.

Gurney, A. B., and G. M. Buxton. 1965. New California grasshoppers of the genus *Melanoplus* (Orthoptera: Acrididae). Occ. Pap. Calif. Dep. Agric. Bur. Ent. 7:I–II, 1–7, pl. 1–4.

Gurney, A. B., and G. M. Buxton. 1968. Grasshoppers of the genus *Karokia* and of the *saltator, immunis*, and *harperi* species groups of the genus *Melanoplus* in California and Oregon (Orthoptera: Acrididae). Occ. Pap. Calif. Dep. Agric. Bur. Ent. 13:1–56.

Gurney, A. B., J. P. Kramer, and G. C. Steyskal. 1964. Some techniques for the preparation, study, and storage in microvials of insect genitalia. Ann. ent. Soc. Am. 57:240–242.

Guthrie, D. M., and A. R. Tindall. 1968. The biology of the cockroach. London, England: Edward Arnold. VIII, 416 pp.

Gwynne, D. T. 1977. Mating behavior of *Neoconocephalus ensiger* (Orthoptera: Tettigoniidae) with notes on the calling song. Can. Ent. 109:237–242.

Haan, W. de. 1840–1844. Bijdragen tot de Kennis der Orthoptera. *In* K. J. Temminck, ed. Verhandlingen over de Natuurlijke Geschiedenis der Nederlandsche Overzeesche Bezittingers . . . Leiden, Netherlands: Natuurkuundige Commissie in Indie. Vols. (6; Zool. 3):pl. 10[1840]; (16; Zool. 6):45–124, pl. 11–20[1842]; (18; Zool. 7):125–164[1842]; (19/20; Zool. 8/9):165–228[1843]; (24; Zool. 10):229–248, pl. 21–23[1844].

Hagen, H. A. 1858. Monographie der Termiten. II. Linnaea ent. 12:1–342, pl. I–III.

Hagen, H. A. 1874. Report on the Pseudo-Neuroptera collected by Lieut. W. L. Carpenter in 1873 in Colorado. Rep. U.S. geol. Surv. Terr. 1873:571–606.

Hagenbach, J. J. 1822. Symbola Faunae Insectorum Helvetiae exhibentia vel species novas vel nondum depictas. Basileae [= Basel]. Neukirch. Vol. 1. 52 pp., 15 pl.

Haldeman, S. S. 1850. History of *Phalangopsis*, a genus of Orthoptera, with three new species, two of which form a subgenus. Proc. Am. Assoc. Advmt Sci. 2:346.

Haldeman, S. S. 1852. Appendix C. Insects. Pages 366–379 *in* H. Stansbury. Exploration and survey of the valley of the Great Salt Lake of Utah, including a reconnoissance of a new route through the Rocky Mountains. Washington, DC: United States Senate. Pl. 9, 10.

Haldeman, S. S. 1853. Description of some new species of insects, with observations on described species. Proc. Acad. nat. Sci. Philad. 6:361–365.

Hale, J. 1829. Observations upon crickets in Canada. Trans. lit. hist. Soc. Québ. 1:254–255.

Hancock, J. L. 1895. A new *Tettix*. Am. Nat. 29:761–762.

Hancock, J. L. 1896. On Illinois grouse locusts. Trans. Am. ent. Soc. 23:235–244, pl. VI–IX.
Hancock, J. L. 1899*a*. A new species of *Nomotettix* from Kansas. Ent. News 10:8.
Hancock, J. L. 1899*b*. Some Tettigian studies. Ent. News 10:275–282.
Hancock, J. L. 1902. The Tettigidae of North America. Chicago, IL: Mrs. Frank G. Logan. VIII, 188 pp., XI pl.
Hancock, J. L. 1906. Orthoptera Fam. Acridiidae Subfam. Tetriginae. Genera Insect. 48:I–II, 1–79, pl. 1–4.
Hancock, J. L. 1916. Pink katydids and the inheritance of pink coloration. Ent. News 27:70–82.
Handford, R. H. 1946. The identification of nymphs of the genus *Melanoplus* of Manitoba and adjacent areas. Sci. Agric. 26:147–180, pl. I–XII.
Handlirsch, A. 1906–1908. Die Fossilen Insekten und die Phylogenie der rezenten Formen: Ein Handbuch für Paläontologen und Zoologen. Leipzig, West Germany: Wilhelm Engelmann. Vol. 1:[II], X, 1430 pp., sep. diagr. VII, IX; vol. 2:(II), XL pp., XLI pl. [1906, 1(1–4):1–640; 2:I–XXVII, pl. I–XXXVI; 1907, 1(5–7):641–1122; 2:XXVII–XL, pl. XXXVII–XLI; 1908, the balance].
Hardman, J. M., and S. Smoliak. 1980. Potential economic impact of rangeland grasshoppers (Acrididae) in southeastern Alberta. Can. Ent. 112:277–284.
Hargrave, J. J. 1871. Red River. Montreal, Qué.: John Lovell. Pp. I–XVI, 17–506.
Harmon, D. W. 1820. A journal of voyages and travels in the interior of North America, between the 47th and 58th degrees of north latitude, extending from Montreal nearly to the Pacific Ocean. . . . Andover, MA: Flagg & Gould. XXIII, (25), 432 pp.
Harris, T. W. 1831. *Locusta* (*Tettigonia vitis*). Encycl. Am. Philad. 8:40–43.
Harris, T. W. 1833. A catalogue of the animals and plants in Massachusetts. VIII. Insects. Pages 566–595 *in* E. Hitchcock. Report on the geology of Massachusetts. Amherst, MA.
Harris, T. W. 1835. A catalogue of the animals and plants in Massachusetts. VIII. Insects. Pages 553–602 *in* E. Hitchcock. Report on the geology of Massachusetts (Ed. 2). Amherst, MA.
Harris, T. W. 1841. A report on the insects of Massachusetts injurious to vegetation. Cambridge, MA: Folsom, Wells & Thurton. VIII, 459 pp.
Harris, W. V. 1961. Termites: Their recognition and control. London, England: Longmans. XII, 187 pp., 26 pl., 1 foldout (table).
Harris, W. V. 1971. Termites: Their recognition and control (Ed. 2). London, England: Longmans. XIV, 186 pp., 26 pl., 1 foldout (table).
Harrison, R. G. 1977. Parallel variation at an enzyme locus in sibling species of field crickets. Nature, Lond. 266:168–170.
Harrison, R. J. 1980. Clé revisée des blattes adultes (Dictyoptera: Blattidae) du Canada. Ann. ent. Soc. Québ. 25:10–13.
Hartman, H. B., W. W. Walthall, L. P. Bennett, and R. R. Stewart. 1979. Giant interneurons mediating equilibrium reception in an insect. Science 205:503–505.

Harz, K. 1969. Die Orthopteren Europas/The Orthoptera of Europe, 1. The Hague, Netherlands: Dr. W. Junk (Series Entomologica 5). XX, 749 pp.
Hebard, M. 1913. A revision of the species of the genus *Nemobius* (Orthoptera: Gryllidae) found in North America north of the Isthmus of Panama. Proc. Acad. nat. Sci. Philad. 65:394–492.
Hebard, M. 1915*a*. Records of Orthoptera from Newfoundland. Ent. News 26:306.
Hebard, M. 1915*b*. Dermaptera and Orthoptera found in the vicinity of Miami, Florida in March 1915. Part I; *and* Part II. Ent. News 26: 397–408, 457–469.
Hebard, M. 1916*a*. A study of the species of the genus *Stenopelmatus* found in the United States. Jl N.Y. ent. Soc. 24:70–84.
Hebard, M. 1916*b*. Certain features found in the genus *Panchlora*, with other observations and the description of one new species (Orthoptera: Blattidae). Ent. News 27:217–222.
Hebard, M. 1917*a*. Notes on Mexican Melanopli. Proc. Acad. nat. Sci. Philad. 69:251–275.
Hebard, M. 1917*b*. Notes on the earwigs (Dermaptera) of North America, north of the Mexican Boundary. Ent. News 28:311–323.
Hebard, M. 1917*c*. The Blattidae of North America, north of the Mexican boundary. Mem. Am. ent. Soc. 2:1–284.
Hebard, M. 1918. New genera and species of Melanopli found within the United States (Orthoptera: Acrididae) [I]. Trans. Am. ent. Soc. 44: 141–169, pl. VIII.
Hebard, M. 1919*a*. Studies in the Dermaptera and Orthoptera of Colombia. First Paper: Dermaptera and Orthoptera, Families Blattidae, Mantidae and Phasmidae. Trans. Am. ent. Soc. 45:89–179, pl. XVI–XXIII.
Hebard, M. 1919*b*. New genera and species of Melanopli found within United States (Orthoptera: Acrididae) Part II. Trans Am. ent. Soc. 45:257–298, pl. XXIX–XXXI.
Hebard, M. 1920*a*. A revision of the North American species of the genus *Myrmecophila* (Orthoptera: Gryllidae: Myrmecophilinae). Trans. Am. ent. Soc. 46:91–111.
Hebard, M. 1920*b*. New genera and species of Melanopli found within the United States (Orthoptera: Acrididae) Part III. Trans. Am. ent. Soc. 46:355–403, pl. XVI–XVIII.
Hebard, M. 1920*c*. The Blattidae of Panama. Mem. Am. ent. Soc. 4(1919):(I–II), 1–148, I–VI, pl. I–VI.
Hebard, M. 1921. Mexican records of Blattidae (Orthoptera). Trans. Am. ent. Soc. 47:199–220.
Hebard, M. 1922. New genera and species of Melanopli found within the United States and Canada (Orthoptera: Acrididae). Part IV. Trans. Am. ent. Soc. 48:49–66, pl. II, III.
Hebard, M. 1923*a*. Expedition of the California Academy of Sciences to the Gulf of California in 1921. The Dermaptera and Orthoptera. Proc. Calif. Acad. Sci. (4)12:319–340.
Hebard, M. 1923*b*. Orthoptera. Page 141 *in* W. L. McAtee et al. A biological survey of the Pribilof Islands, Alaska. Part II. Insects, arachnids, and chilopods of the Pribilof Islands, Alaska. N. Am. Fauna. Vol. 46.

Hebard, M. 1925*a*. The Orthoptera of South Dakota. Proc. Acad. nat. Sci. Philad. 77:33–155.
Hebard, M. 1925*b*. Records of European Acrididae, Tettigoniidae and Gryllidae (Orthoptera). Trans. Am. ent. Soc. 51:35–55.
Hebard, M. 1926. A key to the North American genera of the Acridinae which occur north of Mexico (Orthoptera: Acrididae). Trans. Am. ent. Soc. 52:47–59.
Hebard, M. 1927. Fixation of the single types of species of Orthoptera described by Cyrus Thomas. Proc. Acad. nat. Sci. Philad. 79:1–11.
Hebard, M. 1928. The Orthoptera of Montana. Proc. Acad. nat. Sci. Philad. 80:211–306, pl. 24, 25.
Hebard, M. 1929*a*. Studies in Malayan Blattidae. Proc. Acad. nat. Sci. Philad. 81:1–109, pl. 1–6.
Hebard, M. 1929*b*. The Orthoptera of Colorado. Proc. Acad. nat. Sci. Philad. 81:303–425, pl. 11.
Hebard, M. 1931*a*. The Orthoptera of Alberta. Proc. Acad. nat. Sci. Philad. 82(1930):377–403.
Hebard, M. 1931*b*. The Orthoptera of Kansas. Proc. Acad. nat. Sci. Philad. 83:119–227.
Hebard, M. 1932*a*. The Orthoptera of Minnesota. Tech. Bull. Univ. Minn. agric. Exp. Stn 85:1–61.
Hebard, M. 1932*b*. New species and records of Mexican Orthoptera. Trans. Am. ent. Soc. 58:201–371, pl. XVII–XXI.
Hebard, M. 1932*c*. Notes on Montana Orthoptera. Proc. Acad. nat. Sci. Philad. 84:251–257.
Hebard, M. 1934*a*. The Dermaptera and Orthoptera of Illinois. Bull. Ill. nat. Hist. Surv. 20:125–279, 1 pl.
Hebard, M. 1934*b*. Studies in Orthoptera which occur in North America. II. *Cyphoderris*, a genus of katydid of southwestern Canada and the northwestern United States. Trans. Am. ent. Soc. 59:371–376, pl. XXII, figs. 7–12.
Hebard, M. 1935*a*. New genera and species of the Melanopli found within the United States and Canada (Orthoptera: Acrididae). Parts V–VI. Trans. Am. ent. Soc. 60:337–390, pl. XXVI.
Hebard, M. 1935*b*. Orthoptera of the upper Rio Grande Valley and the adjacent mountains in northern New Mexico. Proc. Acad. nat. Sci. Philad. 87:45–82.
Hebard, M. 1935*c*. Studies in the Orthoptera of Arizona, Part I. New genera, species and geographical races. Trans. Am. ent. Soc. 61:111–153, pl. IV–VIII.
Hebard, M. 1935*d*. Studies in the Orthoptera of Arizona, Part II. A list of the Dermaptera and Orthoptera of Arizona with new records and corrections of the literature subsequent to 1900. Trans. Am. ent. Soc. 61:269–316.
Hebard, M. 1935*e*. Notes on the group Gomphoceri and a key to its genera, including one new genus (Orthoptera: Acrididae: Acridinae). Ent. News 46:184–188, 204–208.
Hebard, M. 1936*a*. The Orthoptera of North Dakota. Tech. Bull. N. Dakota agric. Exp. Stn 284:1–69.

Hebard, M. 1936*b*. New genera and species of Melanopli found within the United States and Canada (Orthoptera: Acrididae) Parts VII, VIII and IX. Trans. Am. ent. Soc. 62:167–222, pl. XII–XVII.
Hebard, M. 1936*c*. Notes on North American Orthoptera of the Arctic-Alpine zone. Ent. News 47:13–15.
Hebard, M. 1937*a*. New genera and species of the Melanopli found within the United States and Canada (Orthoptera: Acrididae) Parts X–XIV. Trans. Am. ent. Soc. 63:147–173, pl. X, XI.
Hebard, M. 1937*b*. Studies in Orthoptera which occur in North America north of the Mexican boundary [VII–IX]. Trans. Am. ent. Soc. 63:347–379, pl. XXI–XXIII.
Hebard, M. 1937*c*/1938*a*. Where and when to find Orthoptera of Pennsylvania, with notes on the species which in distribution reach nearest this state. Ent. News 48(1937):219–225, 274–280; 49(1938):33–38, 97–103, 155–159.
Hebard, M. 1941. The group Pterophyllae as found in the United States (Tettigoniidae: Pseudophyllinae). Trans. Am. ent. Soc. 67:197–219, pl. XVIII, XIX.
Hebard, M. 1945. The Orthoptera of the Appalachian Mountains in the vicinity of Hot Springs, Virginia, and notes on other Appalachian species and recent extensions of the known range of still other southeastern species. Trans. Am. ent. Soc. 71:77–97.
Hegh, E. 1922. Les Termites. Brussels, Belgium: Ministère des Colonies. IV, 756 pp., 1 map.
Helfer, J. R. 1963. How to know the grasshoppers, cockroaches and their allies: Pictured-keys for identifying many of the grasshoppers, crickets, katydids, earwigs, termites, cockroaches, praying mantids, rock-crawlers, and walking sticks occurring in North America, north of Mexico. Dubuque, IA: Wm. C. Brown Co. VI, 354 pp.
Helwig, E. R. 1942. Unusual integrations of the chromatin in *Machaerocera* and other genera of the Acrididae (Orthoptera). J. Morph. 71:1–26, pl. 1–4.
Helwig, E. R. 1955. Spermatogenesis in hybrids between *Circotettix verruculatus* and *Trimerotropis suffusa* (Orthoptera: Oedipodidae). Univ. Colo. Stud. Ser. Biol. 3:47–64.
Henderson, W. W. 1931. Crickets and grasshoppers in Utah. Circ. Utah. agric. Exp. Stn 96:1–38.
Henriksen, K. L. 1939. A revised index of the insects of Grønland containing a supplement to the insect list in Kai L. Henriksen & Will. Lundbeck: Grønlands Landarthropoder (Medd. om Grønl. Bd. 2. 1917). Medded. Grønland. 119(10):1–112.
Henson, W. R. 1957. Temperature preference of *Grylloblatta campodeiformis* Walker. Nature, Lond. 179:637.
Hermann, O. 1874. Die Decticiden der Brunner von Wattenwyl'shen Sammlung. I. Genera. Verh. zool.-bot. Ges. Wien 24:191–210.
Hewitt, G. B. 1978. Reduction of western wheatgrass by the feeding of two rangeland grasshoppers, *Aulocara elliotti* and *Melanoplus infantilis*. J. econ. Ent. 71:419–421.

Hewitt, G. B., and W. F. Barr. 1967. The band-wing grasshoppers of Idaho (Orthoptera: Oedipodinae). Res. Bull. Idaho agric. Exp. Stn 72:1–64.

Heymons, R. 1921. Heuschrecken der Gattung *Leptophyes* und ihre Schädigungen an Pfirsichblättern. Z. angew. Ent. 7:453–456.

Hincks, W. D. 1954. Notes on Dermaptera. Proc. R. ent. Soc. Lond., Ser. B 23:159–163.

Hind, H. Y. 1857. Essay on the insects and diseases injurious to the wheat crops. Toronto, Ont.: Lovell & Gibson. VIII, 139 pp.

Hind, H. Y. 1859. Northwest Territory. Reports of progress; together with a preliminary and general report on the Assiniboine and Saskatchewan Exploring Expedition, made under instructions from the Provincial Secretary, Canada. Toronto, Ont.: J. Lovell, Queen's Printer. XII, 202 pp., 3 pl., 3 maps.

Holst, K. T. 1970. Kakerlakker Graeshopper og Ørentviste. Danmarks Fauna. København. G. E. C. Gads Forlag 79:1–221.

Howden, H. 1969. Effects of the Pleistocene on North American Insects. Am. Rev. Ent. 14:39–56.

Howse, P. E. 1968. On the division of labour in the primitive termite *Zootermopsis nevadensis* (Hagen). Insectes soc. 15:45–50.

Howse, P. E. 1970. Termites: A study in social behaviour. London, England: Hutchinson & Co. 150 pp., pl. I–IV.

Hubbell, T. H. 1922*a*. Notes on the Orthoptera of North Dakota. Occ. Pap. Mus. Zool. Univ. Mich. 113:1–56.

Hubbell, T. H. 1922*b*. The Dermaptera and Orthoptera of Berrien county, Michigan. Occ. Pap. Mus. Zool. Univ. Mich. 116:1–77.

Hubbell, T. H. 1925. A new species of *Pristoceuthophilus* from the Olympic Mountains, Washington (Orthoptera: Tettigoniidae). Pan-Pacif. Ent. 2:39–42.

Hubbell, T. H. 1926. Correction of records of *Ceuthophilus* (Orthoptera: Tettigoniidae). Fla Ent. 10:14.

Hubbell, T. H. 1936. A monographic revision of the genus *Ceuthophilus* (Orthoptera: Gryllacrididae: Rhapidophorinae). Univ. Fla Publ biol. Sci. Ser. 2:1–551, pl. I–XXXVIII.

Hubbell, T. H. 1956. A new collecting method: The oatmeal trail. Ent. News 67:49–51.

Hubbell, T. H. 1960. The sibling species of the *Alutacea* group of the bird-locust genus *Schistocerca* (Orthoptera: Acrididae: Cyrtacanthacridinae). Misc. Publs Mus. Zool. Univ. Mich. 116:1–91, pl. I–XXIII.

Hubbell, T. H., and I. J. Cantrall. 1938. A new species of *Appalachia* from Michigan (Orthoptera: Acrididae: Cyrtacanthacridinae). Occ. Pap. Mus. Zool. Univ. Mich. 389:1–22, pl. I.

Hull, G., and R. H. Davidson. 1958. The biology of the brown-banded cockroach and its relative susceptibility to five organic insecticides. J. econ. Ent. 51:606–610.

Hunt, P. 1978. Communicative behaviour of three grasshopper species of Quebec (Orthoptera: Acrididae: Gomphocerinae). Montréal, Qué.: McGill Univ. M.Sc. Thesis. [ii], IX, 115 pp.

Hunter-Jones, P. 1961. Rearing and breeding locusts in the laboratory. London. Anti-Locust Res. Cent. Rep. [now Cent. Overseas Pest Res. Misc. Rep.]: II, 12 pp., 4 pl.

Huntington, R. 1874. Sunrise on the Tusket. Pages 142–144 *in* T. H. Rand, ed. 1900. (2nd ed. 1904. Toronto, Ont.: Henry Frowde. XXIV, 412 pp.) A treasury of Canadian verse. Toronto, Ont.: William Briggs. Reprinted 1969.

Huyshe, G. L. 1871. The Red River expedition. London & New York: MacMillan & Co.; Toronto, Ont.: Adam, Stevenson & Co. XII, 276 pp., 1 pl., 3 foldouts (maps), 1 foldout (table).

Hyslop, J. A. 1934. Insect findings of recent years which are or may become of interest to nursery inspectors and plant quarantine officers. J. econ. Ent. 27:565.

Isely, F. B. 1938*a*. Survival value of acridian protective coloration. Ecology 19:370–389.

Isely, F. B. 1938*b*. The relations of Texas Acrididae to plants and soils. Ecol. Monogr. 8:551–604.

Jackson, W. B., and P. P. Maier. 1961. Additional studies of dispersion patterns of American cockroaches from sewer manholes in Phoenix, Arizona. Ohio J. Sci. 61:220–226.

Jacobs, W. 1953. Verhaltensbiologische Studien an Feldheuschrecken. Z. Tierpsychol., Beih. 1:I–VIII, 1–228.

Jago, N. D. 1969. A revision of the systematics and taxonomy of certain North American gomphocerine grasshoppers (Gomphocerinae: Acrididae: Orthoptera). Proc. Acad. nat. Sci. Philad. 121:229–333.

Jago, N. D. 1971. A review of the Gomphocerinae of the world with a key to the genera (Orthoptera: Acrididae). Proc. Acad. nat. Sci. Philad. 123:205–343.

Jago, N. D. 1977. The higher classification of the Acridoidea (*sensu* Dirsh, 1975). Mem. Lyman ent. Mus. Res. Lab. 4:47–52.

Jago, N. D., A. Antoniou, and P. Scott. 197[illegible] laboratory evidence showing the separate species status of *Schistoc*[illegible] *gregaria*, *americana* and *cancellata* (Acrididae: Crytacanthacridina[illegible]c]). Syst. Ent. 4:133–142.

James, H. G. 1959*a*. Egg development, hat[illegible]g, and prey taken by the European mantis, *Mantis religiosa* L., in [illegible]al habitats. Rep. ent. Soc. Ont. 89:50–55.

James, H. G. 1959*b*. New records of the Eur[illegible]n mantis, *Mantis religiosa* L. (Orthoptera: Mantidae), in Ontario [illegible]. ent. Soc. Ont. 89:70.

Johansson, B. 1763. Centuria Insectorum rari[illegible] Quam praeside D.D. Car. von Linne. Pages 384–415 *in* C. Linnaeus [illegible] Amoenitates Academicae Uppsalaiae. Vol. 6.

Johnstone, D. E. 1971. Additions to the orthopteran fauna of Canada. Can. Ent. 103:1193–1194.

Johnstone, D. E. 1973. An unusual color morph of *Melanoplus fasciatus* (F. Walker) (Orthoptera: Acrididae) from Mont Albert, Gaspésian Park, Québec. Ann. ent. Soc. Québ. 18:41–[illegible].

Josselyn, J. 1674. An account of two voyages to New-England, wherein you have the setting out of a ship with the charges . . .; A description of the country, natives and creatures; . . . London, England: G. Widdows. IV, 277, (3) pp.

Judd, W. W. 1955*a*. *Systellogaster ovivora* Gahan (Hymenoptera: Pteromalidae) reared from egg capsules of the woodroach, *Parcoblatta pennsylvanica* (De Geer), collected at Rondeau Park, Ontario. Can. Ent. 87:98–99.

Judd, W. W. 1955*b*. Grasshoppers (Acrididae) and caterpillars (Arctiidae) impaled on thorns by shrikes in the vicinity of London, Ontario. Ent. News 66:225–229.

Kalm, P. 1756*a*. Beskrifning på et slagts Gras-Hoppor, uti Norra America. Ofvers. K. VetenskAkad. Förh. Handl. 17:101–116.

Kalm, P. 1756*b*, 1761. En Resa til Norra America, på Kongl. Svenska Wetenskaps Akademiens befallning. Och. publici kostned. Stockholm, Sweden: Lars Salvius. Vol. 2(1756):526 pp.; vol 3(1761):538 pp. [Also German and English versions, 1754 and 1770.]

Kaltenbach, A. 1958. Eine neue Methode zur Trockenkonservierung von orthopteroiden Insekten. Zool. Anz. 160:116–118.

Kamp, J. W. 1963. New species of Grylloblattodea with an interpretation of their geographical distribution. Ann. ent. Soc. Am. 56:53–68.

Kamp, J. W. 1970. The cavernicolous Grylloblattodea of the western United States (I). Annls spéléol. 25:223–230.

Kamp, J. W. 1973*a*. Numerical classification of the orthopteroids, with special reference to Grylloblattodea. Can. Ent. 105:1235–1239.

Kamp, J. W. 1973*b*. Biosystematics of the Grylloblattodea. Vancouver, B.C.: Univ. Br. Columb. Ph.D. Thesis. 275 pp.

Kamp, J. W. 1979. Taxonomy, distribution, and zoogeographic evolution of *Grylloblatta* in Canada (Insecta: Notoptera). Can. Ent. 111:27–38.

Karny, H. [H]. 1907*a*. Ergebnisse der mit Subvention aus der Erbschaft Treitl unternommenen zoologischen Forschungsreise Dr. Franz Werner's nach dem ägyptischen Sudan und Nord-Uganda. IX. Die Orthopterenfauna des ägyptischen Sudans und von Nord-Uganda (Saltatoria, Gressoria, Dermaptera) mit besonderer Berücksichtigung der Acridoideengattung Cantantops. Sber. Akad. Wiss. Wien 116:267–378, pl. I–III [Feb].

Karny, H. [H]. 1907*b*. Bermerkungen zu dem Linné'schen Gattungsnamen "Tettigonia." Zool. Annln 2:202–208.

Karny, H. [H]. 1907*c*. Revisio Conocephalidarum. Abh. zool.-bot. Ges. Wien 4(3):1–114.

Karny, H. [H]. 1912*a*. Orthoptera Fam. Locustidae Subfam. Conocephalinae. Genera Insect. 135:I–II, 1–17, pl. 1–2.

Karny, H. [H]. 1912*b*. Orthoptera Fam. Locustidae Subfam. Copiphorinae. Genera Insect. 139:I–II, 1–50, pl. 1–7.

Karny, H. H. 1937. Orthoptera Fam. Gryllacrididae Subfamiliae Omnes. Genera Insect. 206:1–317, pl. 1–7.

Keasey, M. S. 1974. Arizona's six-legged rainbows. Insect Wld Dig. 1(6):7–8.

Kelleher, J. S., ed. 1979. Insect and related pests of households. Can. agric. Insect Pest Rev. 56(1978):47–50.

Kerr, G. E. 1974. Visual and acoustical communicative behaviour in *Dissosteira carolina* (Orthoptera: Acrididae). Can. Ent. 106:263–272.

Kerr, G. E. 1978. Uncertainty analyses of the behaviour of the Carolina locust, *Dissosteira carolina* (Orthoptera: Acrididae). Can. J. Zool. 56:201–214.

Kevan, D. K. McE. 1952*a*. A summary of the recorded distribution of British Orthopteroids. Trans. Soc. Brit. Ent. 11:165–180.

Kevan, D. K. McE. 1952*b*. A method of mounting genitalia, etc. for incorporation into collections of pinned insects. Ent. Rec. 64:195–197.

Kevan, D. K. McE. 1955. Méthodes inhabituelles de production de son chez les Orthoptères. Annls Épiphyt. Sér. C [= Annls Inst. natn. Rech. agron., Paris Sér. A.] (fasc. hors Série) 1954:103–141.

Kevan, D. K. McE. 1961*a*. A revised summary of the known distribution of British Orthopteroids. Trans. Soc. Br. Ent. 14:187–205.

Kevan, D. K. McE. 1961*b*. *Metrioptera* (*Roeseliana*) *roeseli* (Hagenbach, 1822) f. *diluta* (Charpentier, 1825) in the Montreal area (Orthoptera: Tettigoniidae). Can. Ent. 93:605–607.

Kevan, D. K. McE. 1962*a*. Gryllinae (Orthoptera) in the Linnaean Collection, London. Proc. Linn. Soc. Lond. 173:129–131, pl. 1.

Kevan, D. K. McE. 1962*b*. Soil animals. [Aspects of Zoology Series, 3.] London, England: H. F. & G. Witherby. XVI, 237 pp., pl. I–V. [Also 2nd ed., 1968:XII, 244 pp., pl. I–V.]

Kevan, D. K. McE. 1974*a*. Greetings from the Entomological Society of Canada and from the Entomological Society of Québec. Proc. ent. Soc. Ont. 104:52–59.

Kevan, D. K. McE. 1974*b*. Land of the grasshoppers, being some verses on grigs. . . . Mem. Lyman ent. Mus. Res. Lab. 2:IX, 326 pp.

Kevan, D. K. McE. 1976*a*. Classification of the Acridomorphoid Insects . . . by V. M. Dirsh. S. Afr. J. Sci. 72:316–319.

Kevan, D. K. McE. 1976*b*. Suprafamilial classification of "Orthopteroid" and related insects, applying the principles of symbolic logic—A draft scheme for discussion and consideration (XV. International Congress of Entomology, Washington, August, 1976). Notes Lyman ent. Mus. Res. Lab. 2:1–24.

Kevan, D. K. McE. 1977*a*. XV. International Congress of Entomology, Washington, D.C., U.S.A., 19–27 August, 1976. Section 1. Symposium: The higher classification of the orthopteroid insects. Mem. Lyman ent. Mus. Res. Lab. 4:IV, 52 pp.

Kevan, D. K. McE. 1977*b*. The higher classification of the Orthopteroid insects: A general view; *and* Appendix. Suprafamilial classification of "orthopteroid" and related insects; a draft scheme for discussion and consideration. Mem. Lyman ent. Mus. Res. Lab. 4:1–31; (1)–(26).

Kevan, D. K. McE. 1978*a*. Proposal to conserve *Blatta germanica* Linnaeus, 1767 and to designate it as type-species of *Blattella* Caudell, 1903 (Insecta Dictuoptera: Blattodea). Z.N.(S.) 680. Bull. zool. Nom. 35:34–39.

Kevan, D. K. McE. 1978*b*. The land of the locusts, being some further verses on grigs and cicadas. . . . Mem. Lyman ent. Mus. Res. Lab. 6:X, 530 pp.

Kevan, D. K. McE. 1979. 27. Dictuoptera; 28. Notoptera; 30. Grylloptera; 31. Orthoptera (*s.str.*); 32. Cheleutoptera. *In* H. V. Danks, ed. Canada and its insect fauna. Mem. ent. Soc. Can. 108:314–316, 316–317, 318–321, 321–323, 323–324.

Kevan, D. K. McE. 1980*a*. *Romalea guttata* (Houttuyn), name change for well-known "eastern lubber grasshopper" (Orthoptera: Romaleidae). Ent. News 91(4):139–140.

Kevan, D. K. McE. 1980*b*. Names involving the Madeira and Surinam cockroaches (Dictuoptera: Blattodea: Nauphoetidae). Ent. Rec. 92:77–82.

Kevan, D. K. McE. 1980*c*. The orthopteroid insects of the Bermudas. Mem. Lyman Ent. Mus. Res. Lab. 8 (Spec. Publ. 16). 182 pp.

Kevan, D. K. McE., E. J. LeRoux, and C. d'Ornellas. 1962. Further observations on *Metrioptera* (*Roeseliana*) *roeseli* (Hagenbach, 1822) in Quebec, with notes on the genus *Metrioptera* Wesmael, 1938 [*sic*] (Orthoptera: Tettigoniidae: Decticinae). Ann. ent. Soc. Québ. 7:70–86.

Kevan, D. K. McE., and V. R. Vickery. 1978. The orthopteroid insects of the Magdalen Islands with notes from adjacent regions. Ann. ent. Soc. Québ. 22:193–204.

Kevan, D. K. McE., and D. C. Wighton. 1981. Paleocene orthopteroids from south-central Alberta, Canada. Can. J. Earth Sci. 18:1824–1837.

King, R. L. 1933. Inheritance of melanism in the grasshopper, *Melanoplus differentialis*. Am. Nat. 67:80.

King, R. L. 1942. Inheritance of melanism in *Melanoplus differentialis*. *In* B. P. Kaufmann, ed. Abstracts of papers presented at the 1941 meetings of the Genetics Society of America Cold Spring Harbor, New York, August 27–29, 1941 Dallas, Texas, December 29–31, 1941. Genetics 27:151.

King, R. L., and E. H. Slifer. 1955. The inheritance of red and blue hind tibiae in the lesser migratory grasshopper, *Melanoplus mexicanus mexicanus* (Saussure). J. Hered. 46:302–304.

Kirby, C. S. 1963. Termites in Ontario, with particular reference to the Toronto region. Can. Dep. For., Forest Ent. Path. Branch Inf. Rep. (Forest Insect Lab., Sault Ste Marie, Ont.) 1963–1:II, 30 pp. [Mimeo.]

Kirby, W. 1824. An account of the animals seen by the later Northern Expedition whilst within the Arctic Circle. 4. Land invertebrate animals. Pages 183–310 *in* E. Sabine, W. Kirby, J. E. Gray, and R. Brown. A supplement to the appendix of Captain Parry's voyage for the discovery of a Northwest Passage, in the years 1819–20. Containing an account of the subjects of natural history. London, England: J. A. Murray. [1–5], pl. 1–6.

Kirby, W. 1837. Part the fourth and last. The insects. Pages XXX, 1–325 *in* J. Richardson, W. Swainson, and W. Kirby. Fauna Boreali Americana; or the zoology of the northern parts of British Columbia: Containing descriptions of the objects of natural history collected on the late Northern Land Expeditions, under the command of Captain Sir John Franklin, R.N. Vol. 4. Norwich, England: Josiah Fletcher. Errata slip, VIII pl.

Kirby, W. F. 1904–1910. A synonymic catalogue of the Orthoptera. London, England: British Museum. Vol. 1(1904):X, 501 pp.; vol. 2(1906):VIII, 562 pp.; vol. 3(1910):X, 674 pp.

Kofoid, C. F., ed. 1934. Termites and termite control. Berkeley, CA: University of California Press. XXVIII, 734 pp.

Kollar, V. 1837. Naturgeschichte der schädlichen Insecten. Verh. K. K. Landw-Ges. Wien (N.F.) 5:I–VIII, 1–421, (1)–(2).

Knipper, H., and D. K. McE. Kevan. 1954. Über Flügelfarbung und Sicheingraben von *Acrotylus junodi* Schulthess (Orth. Acrid. Oedipodinae). Veröff. Überseemus. Bremen (A)2:213–226, 1 pl.

Krauss, H. [A]. 1892. Systematisches Verzeichnis der canarischen Dermapteren und Orthopteren mit Diagnosen der neuen Gattungen und Arten. Zool. Anz. for 1892, pp. 163–171.

Kreasky, J. B. 1960. Extended diapause in eggs of high-altitude species of grasshoppers, and a note on food-plant preferences of *Melanoplus bruneri*. Ann. ent. Soc. Am. 53:436–438.

Krishna, K., and F. M. Weesner, eds. 1969. Biology of termites. New York & London: Academic Press. Vol. 1. XIV, 598 pp.

Krishna, K., and F. M. Weesner, eds. 1970. Biology of termites. New York & London: Academic Press. Vol. 2. XVI, 643 pp.

Kurczewski, F. E., and S. E. Ginsburg. 1971. Nesting behavior of *Tachytes* (*Tachyplena*) *validus*. J. Kansas ent. Soc. 44:113–131.

Laing, F. 1930. The cockroach: Its life-history and how to deal with it. Brit. Mus. (nat. Hist.) econ. Ser. 12 [Ed. 2]: I–IV, 1–23.

Laing, F. 1946. The cockroach: Its life-history and how to deal with it. Brit. Mus. (nat. Hist.) econ. Ser. 12 [Ed. 4]: I–IV, 1–28.

Lamb, R. J. 1975. Effects of dispersion, travel, and environmental heterogeneity on populations of the earwig *Forficula auricularia* L. Can. J. Zool. 53:1855–1867.

Lamb, R. J. 1976*a*. Polymorphisms among males of the European earwig, *Forficula auricularia* (Dermaptera: Forficulidae). Can. Ent. 108:69–75.

Lamb, R. J. 1976*b*. Dispersal by nesting earwigs, *Forficula auricularia* (Dermaptera: Forficulidae). Can. Ent. 108:213–216.

Lamb, R. J. 1979. 29. Dermaptera. *In* H. V. Danks, ed. Canada and its insect fauna. Mem. ent. Soc. Can. 108:317–318.

Lamb, R. J., and W. G. Wellington. 1975. Life history and population characteristics of the European earwig, *Forficula auricularia* (Dermaptera: Forficulidae), at Vancouver, British Columbia. Can. Ent. 107:819–824.

Lambert, J. 1810. Travels through Canada and the United States of North America, in the years 1806, 1807 & 1808. To which are added biographical notices and anecdotes of some of the leading characters in the United States. . . . London, England: Richard Phillips. 3 vols.

Lambert, J. 1813. Travels through Canada . . . etc. 2nd, revised ed. London & Edinburgh: C. Cradock & W. Joy; Doig & Stirling. 2 vols.

La Munyon, I. W. 1877. New Orthoptera. Proc. Nebraska Assoc. Adv. Sci. 1877 (8 March):1[sep.].

Langston, R. L., and J. A. Powell. 1975. The earwigs of California (Order Dermaptera). Bull. Calif. Insect Surv. 20:I–IV, 1–25.

Larochelle, A. 1978. Catalogue des Orthoptères du Québec. Cordulia, Suppl. 7:1–36.

Latreille, P.-A. [1804] ("An. XII"). Histoire naturelle, générale et particulière, des Crustacés et des Insectes. . . . Paris, France: F. Dufart. Vol. 12. IV, 424 pp., pl. 94–97.

Laurent, P., and H. de Saussure. 1898. A species of Orthoptera. Ent. News 9:144–145, pl. IX.

Lee, K. E., and T. G. Wood. 1971. Termites and soils. London, New York: Academic Press. X, 251 pp.

Lim, H.-C. 1971. Note on the chromosomes of Nemobiinae (Orthoptera: Gryllidae). Can. J. Zool. 49:391–395, pl. I.

Lim, H.-C., V. R. Vickery, and D. K. McE. Kevan. 1973. Cytogenetic studies in relation to taxonomy within the family Gryllidae. I. Subfamily Gryllinae. Can. J. Zool. 51:179–186, pl. I, II.

Linnaeus, C. 1758. Systema Naturae per regna tria naturae secundum classes, ordines, genera, species, cum characteribus, differentiis, synonymis, locis. (Ed. 10) Holmiae [= Stockholm], Sweden: Laur. Salvius. Vol. 1. III, 823 pp.

Linnaeus, C. 1761. Fauna Suecica sistens animalia Sueciae: Quadrupedia, aves, amphibia, pisces, insecta, vermes distributa per classes, et ordines, genera et species, cum differentiis specierum, synonymis autorum, nominibus incolarum, locis habitationum, descriptionibus insectorum. (Ed. 2) Stockholm, Sweden: Laur. Salvius. [Vol. 46], 578 pp., 2 pl.

Linnaeus, C. [Linné, C. a(= von)]. 1767. Systema Naturae per regna tria naturae secundum classes, ordines, genera, species, cum characteribus, differentiis, synonymis, locis. (Ed. 12) Holmiae [= Stockholm]. Laur. Salvius. 1(2):[I–II], 533–1327, (1)–(37).

Lochhead, W. 1898. A study of the Gryllidae (crickets). Rep. ent. Soc. Ont. 28:39–44.

Lord, J. K. 1866. The naturalist in Vancouver Island and British Columbia. London, England: Richard Bentley. Vol. 1: XVI, 358 pp., 7 pl; vol. 2:X, 375 pp., 4 pl.

Lucas, H. 1847. *Forficesila annulipes* Lucas. spec. nova. Annls Soc. ent. Fr. (2)5:(Bull.) LXXXIV–LXXXV.

Lugger, O. 1898. Grasshoppers, locusts, crickets, cockroaches, etc., of Minnesota. Bull. Minn. agric. Exp. Stn 55:91–386.

McCafferty, W. P., and J. L. Stern. 1976. Indiana Ensifera (Orthoptera). Gt Lakes Ent. 19:25–56.

McClung, C. E 1905. The chromosome complex of orthopteran spermatocytes. Biol. Bull. 9:304–340.

McClung, C. E. 1914. A comparative study of the chromosome in orthopteran spermatogenesis. J. Morph. 25:651–749.

McClung, C. E. 1917. The multiple chromosomes of *Hesperotettix* and *Mermiria* (Orthoptera). J. Morph. 29:519–605.

McKittrick, F. A. 1964. Evolutionary studies of cockroaches. Mem. Cornell Univ. agric. Exp. Stn 389:1–197.

McKittrick, F. A. 1965. A contribution to the understanding of cockroach-termite affinities. Ann. ent. Soc. Am. 58:18–22.

McLeod, J. H. 1962. Part I: Biological control of pests of crops, fruit trees, ornamentals, and weeds in Canada up to 1959. *In* J. H. McLeod, B. M. McGugan, and H. C. Coppel. A review of the biological control attempts against insects and weeds in Canada. Tech. Commun. Commonw. Inst. biol. Control 2:1–33.

MacNay, C. G. 1955*a*. Cockroaches and their control/Les Blattes (Coquerelles) comment s'en débarrasser. Publs Can. Dep. Agric. Sci. Serv. Ent. 924:1–8. [English and French Editions.]

MacNay, C. G. 1955*b*. Summary of important insect infestations, occurrences, and damage in Canada in 1954. Ann. Rep. ent. Soc. Ont. 85:61–91.

MacNay, C. G., ed. 1955*c*. A summary of parasite and predator liberations in Canada in 1954: Statement prepared by Biological Control Investigations Unit. Can. Insect Pest Rev. 32(1954):332–341.

MacNay, C. G. 1959*a*. Summary of important insect infestations, occurrences, and damage in agricultural areas of Canada in 1958. Ann. Rep. ent. Soc. Ont. 89:73–87.

MacNay, C. G. 1959*b*. January–May, 1959; *and* July, 1959; *and* Summary, 1959. Can. Insect Pest Rev. 32:123–145, 170–182, 2 maps, 266–297, 1 map.

MacNay, C. G. 1961*a*. Summary of important insect infestations, occurrences, and damage in agricultural areas of Canada in 1960. Proc. ent. Soc. Ont. 91:247–263.

MacNay, C. G. 1961*b*. January–May, 1961; *and* August, 1961. Can. Insect Pest Rev. 39:111–136, 2 maps, 179–206, 2 maps.

McNeill, J. 1891. A list of the Orthoptera of Illinois. Psyche, Camb. 6:3–9, 21–27, 62–66, 73–78.

McNeill, J. 1896/1897. Revision of the Truxalinae of North America. Proc. Davenport Acad. nat. Sci. 6:179–244 (Nov.–Dec. 1896), 245–274, pl. 1–6 (Jan. 1897).

McNeill, J. 1900*a*. The Orthopteran genus *Trimerotropis*. Psyche, Camb. 9:27–36.

McNeill, J. 1900*b*. *Orchelimum*, Serv. Can. Ent. 32:77–83.

McNeill, J. 1901. Revision of the orthopteran genus *Trimerotropis*. Proc. U.S. natn. Mus. 23:393–499, pl. XXI.

McNutt, D. M. 1976. Insec collecting in the tropics. London, England: Centre for Overseas Pe Research. IV, 68 pp.

Macoun, J. 1882. Manitoba the great Northwest: The field for investment; the home of the c rant, being a full and complete history of the country. . . . Guelp nt.: World Publishing Co. XXII, 687 pp., foldout maps, pls.

Mair, C. 1868. Dreamland an her poems. Montreal, Qué.: Dawson Bros.; London, England: S. L , Son & Marston. VI, 151 pp.

Martin, J. E. H., compiler. 77. The insects and arachnids of Canada. Part 1. Collecting, prepa g, and preserving insects, mites, and spiders. Agric. Can. Publ. 1643. tawa, Ont.: Ministry of Supply and Services Canada. 182 pp.

Matsuda, R. 1976. Morph gy and evolution of the insect abdomen with special reference to velopmental patterns and their bearing upon systematics. Pergam Int. Ser. pure & appl. Biol. (Zool.) London. Pergamon Press. Vol 56. VIII, 532 pp.

Maw, M. G. 1976. An ann ated list of insects associated with Canada thistle (*Cirsium arvense*) in Canada. Can. Ent. 108:235–244.

Mayr, E. 1963. Animal species and evolution. Cambridge, MA: Belknap Press, Harvard University. 797 pp.

Melander, A. L., and M. A. Yothers. 1917. The coulee cricket. Bull. Wash. St. Coll. agric. 137:1–56.

Meloche, F., J.-G. Pilon, G. Mailloux, and T. Vrain. 1980. Inventaire des problèmes entomologiques et nématologiques dans les plantations de tabac jaune au Québec. Ann. ent. Soc. Québ. 25:81–89.

Metcalf, C. L., and A. S. Colby. 1930. The meadow grasshopper, *Orchelimum vulgare* Harris, a new raspberry pest. J. econ. Ent. 23:97–108.

Mills, H. B., and J. H. Pepper. 1937. Observations on *Grylloblatta campodeiformis* Walker. Ann. ent. Soc. Am. 30:269–274, pl. I.

Mills, H. B., and J. H. Pepper. 1938. Key to the grasshoppers of Montana. Mimeogr. Circ. Montana St. Coll. 9:1–25, pl. 1–3.

Mishchenko, L. L. 1951. 1. Podsemeĭstvo Catantopinae (= Acridiidae Cyrtacanthacridinae). Pages 131–270 *in* G. Ya. Beĭ-Bienko, and L. L. Mishchenko. Saranchevȳe Faunȳ SSSR Chast. I. Opred. Faun. SSSR. Vol. 38.

Misra, S. D., and L. G. Putnam. 1966. The damage potential of the grasshopper, *Camnula pellucida* (Scudd.) (Orthoptera: Acrididae) on pastures and ranges in Canada. Indian J. Ent. 28:224–233.

Mitchell, J. E., and R. E. Pfadt. 1974. A role of grasshoppers in a shortgrass prairie ecosystem. Envir. Ent. 3:358–360.

Mitchell, P., and D. Wighton. 1979. Larval and adult insects from the Paleocene of Alberta, Canada. Can. Ent. 111:777–782.

Mitchener, A. V. 1954*a*. A history of grasshopper outbreaks and their control in Manitoba, 1799–1953. Rep. ent. Soc. Ont. 84:27–35.

Mitchener, A. V. 1954*b*. A history of grasshopper outbreaks and their control in Manitoba, 1799–1953. Proc. ent. Soc. Manitoba 9(1953):46–59.

Mitchener, A. V. 1956. A history of grasshopper control in Manitoba. Proc. ent. Soc. Manitoba 11(1955):40–48.

Moens, P. B. 1978. Kinetochores of grasshoppers with Robertsonian chromosome fusions. Chromosoma 67:41–54.

Morris, G. K. 1970. Sound analyses of *Metrioptera sphagnorum* (Orthoptera: Tettigoniidae). Can. Ent. 102:363–368.

Morris, G. K. 1971. Aggression in male conocephaline grasshoppers (Tettigoniidae). Anim. Behav. 19:132–137.

Morris, G. K. 1972. Phonotaxis of male meadow grasshoppers (Orthoptera: Tettigoniidae). Jl N.Y. ent. Soc. 80:5–6.

Morris, G. K., and D. T. Gwynne. 1979. Geographical distribution and biological observations of *Cyphoderris* (Orthoptera: Haglidae) with a description of a new species. Psyche, Camb. 85(1978):147–167.

Morris, G. K., G. E. Kerr, and J. H. Fullard. 1978. Phonotactic preferences of female meadow katydids (Orthoptera: Tettigoniidae: *Conocephalus nigropleurum*). Can. J. Zool. 56:1479–1487.

Morris, G. K., G. E. Kerr, and D. T. Gwynne. 1975*a*. Calling song function in the bog katydid, *Metrioptera sphagnorum* (F. Walker) (Orthoptera: Tettigoniidae): Female phonotaxis to normal and altered song. Z. Tierpsychol. 37:502–514.

Morris, G. K., G. E. Kerr, and D. T. Gwynne. 1975*b*. Ontogeny of phonotaxis in *Orchelimum gladiator* (Orthoptera: Tettigoniidae: Conocephaline). Can. J. Zool. 53:1127–1130.

Morris, G. K., and R. E. Pipher. 1967. Tegminal amplifiers and spectrum consistencies in *Conocephalus nigropleurum* (Bruner), Tettigoniidae. J. Ins. Physiol. 13:1075–1085.

Morris, G. K., and R. E. Pipher. 1972. The relation of song structure to tegminal movement in *Metrioptera sphagnorum* (Orthoptera: Tettigoniidae). Can. Ent. 104:977–985.

Morris, G. K., and T. J. Walker. 1976. Calling songs of *Orchelimum* meadow katydids (Tettigoniidae) I. Mechanism, terminology, and geographic distribution. Can. Ent. 108:785–800.

Morris, R. F. 1981. Note on the occurrence of the lubber grasshopper, *Brachystola magna* (Orthoptera: Romaleidae), in Newfoundland. Can. Ent. 113:659–660.

Morris, R. F., and H. G. Morry. 1978. Newfoundland. Can. Agric. Insect Pest Rev. 55(1977):1–9.

[Morris, R. F., and H. G. Morry.] 1979. Newfoundland. *In* J. S. Kelleher, ed. Insect and related pests of forage grasses and legumes. Can. agric. Insect Pest Rev. 56(1978):11.

Morrissey, R., and J. S. Edwards. 1979. Neural function in an alpine grylloblattid: A comparison with the house cricket, *Acheta domesticus*. Physiol. Ent. 4:241–250.

Morse, A. P. 1894*a*. Wing-length in some New England Acrididae—I; *and* II. Psyche, Camb. 7:13–14, 53–55.

Morse, A. P. 1894*b*. A preliminary list of the Acrididae of New England. Psyche, Camb. 7:102–107.

Morse, A. P. 1894*c*. *Spharagemon*: A study of the New England species. Proc. Boston Soc. nat. Hist. 26:220–240.

Morse, A. P. 1894*d*. Notes on the Acrididae of New England. I. Psyche, Camb. 7:147–154, 163–167, pl. 6.

Morse, A. P. 1894*e*. Notes on the Orthoptera of Penikese and Cuttyhunk. Psyche, Camb. 7:179–180.

Morse, A. P. 1895. New North American Tettiginae. I; *and* II. Jl N.Y. ent. Soc. 3:14–16, 107–109.

Morse, A. P. 1896*a*. Notes on the Acrididae of New England.—II. Tryxalinae. Psyche, Camb. 7:32[illegible]–327, 342–344, 382–384, 402–403, 407–411, 413–422, 433–445, pl. 7.

Morse, A. P. 1896*b*. Some notes on locust stridulation. Jl N.Y. ent. Soc. 4:16–20.

Morse, A. P. 1897. Notes on New England Acrididae.—III. Oedipodinae.—I–VII. Psyche, Camb. 8:6–8, 35–37, 50–51, 64–66, 80–82, 87–89, 111–114, pl. 2.

Morse, A. P. 1899. New North American Tettiginae.—III. Jl N.Y. ent. Soc. 7:198–201.

Morse, A. P. 1901. The *Xiphidiini* of the Pacific Coast. Can. Ent. 33:201–205.

Morse, A. P. 1906. New Acrididae from the southern states. Psyche, Camb. 13:119–122.

Morse, A. P. 1908. Tettigidean notes, and a new species. Psyche, Camb. 15:25.

Morse, A. P. 1909. *Melanoplus harrisii* n.sp. Psyche, Camb. 16:12.

Morse, A. P. 1916. A New England orthopteran adventive. Psyche, Camb. 23:178–180.

Morse, A. P. 1919*a*. New records of Orthoptera in New England. Psyche, Camb. 26:16–18.

Morse, A. P. 1919*b*. A list of the Orthoptera of New England. Psyche, Camb. 26:21–39.

Morse, A. P. 1920. Manual of Orthoptera of New England, including the locusts, grasshoppers, crickets and their allies. Proc. Boston Soc. nat. Hist. 35:197–556, pl. 10–29.

Morse, A. P. 1921. Orthoptera of Maine. Bull. Maine agric. Exp. Stn 296:II, 1–36.

Morton, A. S. 1937. Chapter VI. The fur-traders and their world. [5] Grasshoppers. Pages 138–144 *in* Under Western Skies, being a series of pen-pictures of the Canadian West in early fur-trade times. Toronto, Ont.: Thomas Nelson & Sons Ltd. [Whole work XIV, 232 pp., 9 pl.]

Morton, A. S. 1938. History of prairie settlement. Pages XI–XVIII, 1–186 *in* W. A. MacKintosh, and W. L. G. Joerg, eds. Canadian frontiers of settlement. Toronto, Ont.: MacMillan. Vol. 2(1).

Morton, A. S. 1939. A history of the Canadian west to 1870–71, being a history of Rupert's Land (the Hudson's Bay Company's territory) and of the Northwest Territory (including the Pacific Slope). London, Edinburgh, Paris, Melbourne, Toronto, New York: Thomas Nelson & Sons Ltd. XIV, 987 pp.

Mulkern, G. B., K. P. Preuss, H. Knutson, A. F. Hagen, J. B. Campbell, and J. D. Lambley. 1969. Food habits and preferences of grassland grasshoppers of the north central Great Plains. Bull. N. Dakota agric. Exp. Stn 481:1–32.

Nabours, R. K. 1947. The grouse locusts (Orthoptera: Acrididae: Tetriginae). J. Kansas ent. Soc. 20:127–141.

Neill, E. D. 1858. The history of Minnesota: From the earliest French explorations to the present time. Philadelphia, PA: J. B. Lippincott & Co. XLVIII, 49–628 pp., 4 maps.

Newton, R. C., and A. B. Gurney. 1956/1957. Distribution maps of range grasshoppers in the United States. U.S. co-op. econ. Ins. Rep. [CEIR] (U.S. Dep. Agric. agric. Res. Serv., Ent. Res. Br.) 6(1956):597–600, 710–712, 743–744, 776, 838–840, 883–884, 920, 938–940, 972, 987–988, 1002–1004, 1019–1020, 1036, 1050–1052, 1090–1092, 1122–1124, 1149–1950; 7(1957):50–52, 71–72, 89–90, 109–110, 129–130, 150, 188, 208, 225–226, 247–248, 263–264, 315–316, 368, 388, 409, 432, 455–456, 479–480.

Nicholls, C. F. 1963. Some entomological equipment. Inf. Bull. Res. Inst. Can. Dep. Agric. Belleville, Ont. 2:I–II, 1–85.

Nickle, D. A. 1976. Interspecific differences in frequency and other physical parameters of pair-forming sounds of bush katydids (Orthoptera: Tettigoniidae: Phaneropterinae). Ann. ent. Soc. Am. 69:1136–1144.

Nielsson, R. J., and M. H. Bass. 1967. Seasonal occurrence and number of instars of *Nemobius fasciatus*, a pest of white clover. J. econ. Ent. 60:699–701.

Nigam, L. N. 1933. The life-history of a common cockroach, *Periplaneta americana*. Indian J. agric. Sci. 3:530–543.

Olivier, A. G. 1789. Dictionnaire des Insectes 2. Introduction; A–Bom. Encyclopédie Méthodique, ou par ordre de matière, par une société de gens de lettres. Paris. Pankouke. Hist. Naturelle 4:1–44, 45–331, pls.

Onsager, J. A., and G. B. Mulkern. 1963. Identification of eggs and egg-pods of North Dakota grasshoppers (Orthoptera: Acrididae). Tech. Bull. N. Dakota St. Univ. agric. Exp. Stn 446:1–48.

Otte, D. 1970. A comparative study of communicative behavior in grasshoppers. Misc. Publs Mus. Zool. Univ. Mich. 141:1–168.

Otte, D. 1979*a*. Revision of the grasshopper tribe Orphulellini (Gomphocerinae: Acrididae). Proc. Acad. nat. Sci. Philad. 131:52–87.

Otte, D. 1979*b*. Descriptions of new North American gomphocerine grasshoppers (Gomphocerinae: Acrididae). Proc. Acad. nat. Sci. Philad. 131:231–243.

Otte, D. 1979*c*. Biogeographic patterns in flight capacity of Nearctic grasshoppers (Orthoptera: Acrididae). Ent. News 90:153–158.

Otte, D., and J. Loftus-Hills. 1979. Chorusing in *Syrbula* (Orthoptera: Acrididae). Cooperation, interference competition, or concealment? Ent. News 99:159–165.

Otte, D., and K. Williams. 1972. Environmentally induced color dimorphisms in grasshoppers. *Syrbula admirabilis, Dichromorpha viridis*, and *Chortophaga viridifasciata*. Ann. ent. Soc. Am. 65:1154–1161.

Packard, A. S. 1877. Report on the Rocky Mountain locust and other insects now injuring or likely to injure field and garden crops in the western states and territories. Rep. U.S. geol. geogr. Surv. Terr. 9(1875):589–815, pl. LXII–LXX, maps 1–5.

Packard, A. S. 1878. Chapter II. Chronological history. Pages 53–113 *in* C. V. Riley, A. S. Packard, and C. Thomas. First annual report of the United States Entomological Commission for the year 1877, relating to the Rocky Mountain locust and the best methods of preventing its injuries and of guarding against its invasions, in pursuance of an appropriation made by Congress for this purpose. Washington, DC: Government Printing Office.

Packard, A. S. 1881. Chapter VII. Summary of locust flights from 1877 to 1879; *and* Chapter VIII. The western cricket. Its habits and ravages. Pages 160–178 *in* C. V. Riley, A. S. Packard, and C. Thomas. Second report of the United States Entomological Commission for the years 1878 and 1879, relating to the Rocky Mountain locust and the western cricket and treating of the best means of subduing the locust in its permanent breeding grounds with a view of preventing its migrations into the more fertile portions of the Trans-Mississippi country, in pursuance of appropriations made by Congress for this purpose. Washington, DC: Government Printing Office. Maps 2–4. [Dated 1880.]

Packard, A. S. 1890. On insects injurious to forest and shade trees. Rep. U.S. ent. Commiss. 5:I–VIII, 1–958, pl. 1–40.

Packard, A. S., and C. V. Riley. 1881. Chapter I. Additions to the chronology of locust ravages. Pages 1–114 *in* C. V. Riley, A. S. Packard, and C. Thomas. Second report of the United States Entomological Commission for the years 1878 and 1879, relating to the Rocky Mountain locust and the western cricket and treating of the best means of subduing the locust in its permanent breeding grounds with a view of preventing its migrations into the more fertile portions of the Trans-Mississippi country, in pursuance of appropriations made by Congress for this purpose. Washington, DC: Government Printing Office. [Dated 1880.]

Painter, R. H. 1941. The grasshopper seed grain mortgages of 1876 in Manitoba. Can. Ent. 73:194.

Palisot de Beauvois, A.-M.-F.-J. 1807. Insectes recueillis en Afrique et en Amérique, dans les royaumes d'Oware et de Benin, à Sainte-Domingue et dans les États-Unis, pendant les années 1786–1797. Paris, France: Imprimerie de Fain et Cie. Parts 4, 5, pp. 57–88, 12 pl.

Pallas, P. S. 1771. Anhangen. Pages 453–504(467) *in* Reise durch verschieden Provinzen des Russichen Reichs [in den Jahren] 1768–1774. St. Petersburg, USSR: Kayserliche Akademie der Wissenschaften. Vol. 1(1771).

Parker, J. R. 1930. Some effects of temperature and moisture upon *Melanoplus mexicanus mexicanus* Saussure and *Camnula pellucida* Scudder (Orthoptera). Bull. Montana agric. Exp. Stn 223:1–132.

Parker, J. R. 1934. The grasshopper problem in Canada and the United States. Proc. 5th Pacif. Sci. Congr., 1933 5:3459–3471.

Parker, J. R., and R. V. Connin. 1964. Grasshoppers: Their habits and damage. U.S. Dep. Agric. Info. Bull. 287:1–28.

Parker, J. R., W. R. Walton, and R. L. Shotwell. 1932. How to control grasshoppers in cereal and forage crops. U.S. Dep. Agric. Fmrs Bull. 1691:1–13.

Parrott, P. J., and B. B. Fulton. 1914. Tree crickets injurious to orchard and garden fruits. Bull. N.Y. agric. Exp. Stn 388:417–461, pl. I–X.

Paul, L. C. 1941. Intersexuality in *Camnula pellucida* Scudder (Orthoptera). Can. Ent. 73:195–196.

Paul, L. C., and V. L. Berg. 1948. An outbreak of *Aeropedellus clavatus* (Thomas) (Orthoptera: Acrididae). Can. Ent. 80:174–175.

Perty, J. A. M. 1832. Delectus Animalium Articulatorum, quae in itinere per Brasiliam annis 1817–1820 jussu et auspiciis Maximiliani Josephi Bavariae regis augustissimi per acto, collegerunt Dr. J.B. Spix et Dr. C.F. Ph. de Martius; digessit, descripsit et pingenda curavit. Monachii [München]. Perty. Vol. 2. Pp. 61–124, pl. 13–24.

Petch, C. E. 1915. Insects injurious in southern Québec, 1914. Rep. ent. Soc. Ont. 45:70–71.

Peterson, A. 1964. A manual of entomological techniques. (10th ed.) Ann Arbor, MI: Edwards Brothers, Inc. 435 pp.

Pettit, R. H., and E. McDaniel. 1918. Key to Orthoptera of Michigan with annotations. Spec. Bull. Mich. agric. Coll. Exp. Stn 83:1–48.

Pickens, A. L. 1934. The biology and economic significance of far western subterranean termite, *Reticulitermes hesperus*. Pages 148–174 *in* C. F. Kofoid, ed. Termites and termite control. Berkeley, CA: University of California Press.

Pickford, R. 1953. A two-year life-cycle in grasshoppers (Orthoptera: Acrididae) overwintering as eggs and nymphs. Can. Ent. 85:9–14.

Pickford, R. 1963. Wheat crops and native prairie in relation to the nutritional ecology of *Camnula pellucida* (Scudder) (Orthoptera: Acrididae) in Saskatchewan. Can. Ent. 95:764–770.

Pickford, R. 1974. Reproductive behaviour of the clear-winged grasshopper, *Camnula pellucida* (Orthoptera: Acrididae). Can. Ent. 106:403–408.

Pickford, R., and M. E. Taylor. 1963. Note on oviposition in plant stems by grasshoppers (Orthoptera: Acrididae). Can. Ent. 95:556–557.

Piers, H. 1895. Notes on Nova Scotian zoology: No. 3. Trans. Nova Scotian Inst. Sci. 8:395–410.

Piers, H. 1896. Preliminary notes on the Orthoptera of Nova Scotia. Trans. Nova Scotian Inst. Sci. 9:208–218.

Piers, H. 1918. The Orthoptera (cockroaches, locusts, grasshoppers and crickets) of Nova Scotia; with descriptions of the species and notes on their occurrence and habits. Trans. Nova Scotian Inst. Sci. 14:201–356, pl. [I]–[IV].

Pipher, R. E., and G. K. Morris. 1974. Frequency modulation in *Conocephalus nigropleurum*, the black-sided meadow katydid (Orthoptera: Tettigoniidae). Can. Ent. 106:997–1001.

Popham, E. J. 1963. The geographical distribution of the Dermaptera. Entomologist 96:131–144.

Popham, E. J. 1965. A key to Dermapteran subfamilies. Entomologist 98:126–136.

Pratt, E. J. 1952. Towards the last spike: A verse panorama of a struggle to build the first Canadian Transcontinental from the time of the proposed terms of union with British Columbia (1870) to the hammering of the last spike in the Eagle Pass (1885). Toronto, Ont.: MacMillan. 53 pp.

Prest, V. K., J. Terasm[a]e, J. V. Matthews, and S. Lichti-Federovich. 1976. Late-quaternary history of Magdalen Islands, Québec. Marit. Sediments 12:39–59, pl. I–III.

Princis, K. 1960*a*. Zur Systematick der Blattaria. Eos, Madr. 36:427–449.

Princis, K. 1960*b*. Zur Kenntnis der Blattarien des Italienisch-Somalilands. Atti Soc. ital. Sci. nat. Mus. Stor. nat. Milano 99:187–194.

Pritchard, G., and P. Scholefield. 1978. Observations on the food, feeding behaviour, and associated sense organs of *Grylloblatta campodeiformis* (Grylloblattodea). Can. Ent. 110:205–212.

Pritchett, J. P. 1942. The Red River Valley 1811–1849: A regional study. New Haven, CT: Yale University Press; Toronto, Ont.: Ryerson Press. XXII, 295 pp.

Procter, W. 1946. Biological survey of the Mount Desert Region Incorporated. Part VII. The insect fauna with reference to methods of capture, food plants, the flora and other biological features. Philadelphia, PA: Wistar Institute of Anatomy and Biology. Pp. 36–41.

Provancher, L. 1876. Petite faune entomologique du Canada. Les Orthoptèpes [*sic*], Orthoptera. Naturaliste can. 8:13–26, 52–62, 72–81, 106–116, 134–143.

Provancher, L. 1877. Faune canadienne. Les Insectes–Orthoptères. Additions et corrections. Naturaliste can. 9:289–300.

Provancher, L. 1883. Deuxième Ordre. Les Orthoptères. Orthoptera. Pages 1–53 *in* Les Orthoptères, les Nevroptères et les Hymenoptères. Petite faune entomologique du Canada et particulièrement de la Province de Québec. Québec, Qué.: C. Darveau, Vol. 2.

Putnam, L. G. 1963*a*. Dispersal of nymphs of the grasshopper, *Camnula pellucida* (Scudd.). Can. Ent. 95:76–80.

Putnam, L. G. 1963*b*. The progress of nymphal development in pest grasshoppers (Acrididae) of western Canada. Can. Ent. 95:1210–1216.

Putnam, L. G., and R. H. Handford. 1958*a*. Control of grasshoppers in the Prairie Provinces. Publs Can. Dep. Agric. Ent. Div. 1036:1–9.

Putnam, L. G., and R. H. Handford. 1958*b*. Two-year and one-year life cycles in *Melanoplus bivittatus* (Say) (Orthoptera: Acrididae) in western Canada. Proc. X. Int. Congr. Ent. (Montreal, 1956) 2:651–656.

Rae, J. 1850. Narrative of an expedition to the shores of the Arctic Sea in 1846 and 1847. . . . London, England: T. & W. Boone. VIII, 247 pp., map.

Ragge, D. R. 1955. The wing-venation of the Orthoptera Saltatoria. London, England: British Museum (Natural History), London. VI, 159 pp.

Ragge, D. R. 1965. Grasshoppers, crickets and cockroaches of the British Isles. London, New York: Frederick Warne & Co. Ltd. XII, 299 pp., pl. I–XXII.

Ragge, D. R. 1973. Dictyoptera (cockroaches and praying mantids). Pages 399–403 *in* K. G. V. Smith, ed. Insects and other arthropods of medical importance. London, England: British Museum (Natural History), London.

Rakshpal, R. 1962*a*. Respiratory metabolism during embryogenesis of *Gryllus veletis* (Alexander and Bigelow) (Orthoptera: Gryllidae). Physiol. Zool. 35:47–51.

Rakshpal, R. 1962*b*. Effects of freezing temperatures on the eggs of *Gryllus pennsylvanicus* Burmeister and *Nemobius allardi* Alexander and Thomas (Orthoptera: Gryllidae). Indian J. Ent. 24:199–204.

Rakshpal, R. 1964*a*. Growth and respiration during post-embryonic development of a diapause species, *Gryllus veletis* (Alexander and Bigelow) (Orthoptera: Gryllidae). Indian J. Ent. 26:67–75.

Rakshpal, R. 1964*b*. Diapause in the eggs of *Nemobius allardi* Alexander and Thomas (Orthoptera: Gryllidae: Nemobiinae). Zool. Anz. 173:282–288.

Ramsay, A. J. (alias J. R.) 1859. The Canadian lyre. Hamilton, Ont.: C. W. Donnelley.

Randell, R. L. 1963. On the presence of concealed genitalic structures in female Caelifera (Insecta: Orthoptera). Trans. Am. ent. Soc. 88: 247–260, pl. XXII–XXX.

Randell, R. L. 1964. The male genitalia in Gryllinae (Orthoptera: Gryllidae) and a tribal revision. Can. Ent. 96:1565–1607.

Randell, R. L., and D. K. McE. Kevan. 1963. Research on field and house crickets (Orthoptera: Gryllidae: Gryllinae) at Macdonald College. Ann. ent. Soc. Québ. 8:102–107.

Redtenbacher, J. 1891. Monographie der Conocephaliden. Verh. zool.-bot. Ges. Wien 41:315–562, pl. III, IV.

Rees, N. E. 1973. Arthropod and nematode parasites, parasitoids, and predators of Acrididae in America north of Mexico. Tech. Bull. U.S. Dep. Agric. agric. Res. Serv. 1460:I–IV, 1–288.

Rehn, J. A. G. 1901*a*. Remarks on some Mexican Orthoptera, with descriptions of new species. Trans. Am. ent. Soc. 27:218–229.

Rehn, J. A. G. 1901*b*. Some necessary changes and corrections in names of Orthoptera. Can. Ent. 33:271–272.

Rehn, J. A. G. 1902*a*. A contribution to the knowledge of the Orthoptera of Mexico and Central America. Trans. Am. ent. Soc. 29:1–34.

Rehn, J. A. G. 1902*b*. Two new species of N. American Stenopelmatinae. Ent. News 13:240–241.

Rehn, J. A. G. 1904*a*. A new *Melanoplus* from New Jersey. Ent. News 15:85–87.

Rehn, J. A. G. 1904*b*. Notes on Orthoptera from northern and central Mexico. Proc. Acad. nat. Sci. Philad. 56:513–549.

Rehn, J. A. G. 1904*c*. Notes on the Orthoptera of the Keweenaw Bay Region of Baranga County, Michigan. Ent. News 15:229–236, 263–270.

Rehn, J. A. G. 1904*d*. Notes on Orthoptera from Arizona, New Mexico and Colorado. Proc. Acad. nat. Sci. Philad. 56:562–575.

Rehn, J. A. G. 1907. Notes on Orthoptera from southern Arizona, with descriptions of new species. Proc. Acad. nat. Sci. Philad. 59:24–81.

Rehn, J. A. G. 1910. Records of Orthoptera of western Canada. Ent. News 21:23–27.

Rehn, J. A. G. 1919*a*. A study of the genus *Mermiria* Stål. Proc. Acad. nat. Sci. Philad. 71:55–120, pl. V–VII.

Rehn, J. A. G. 1919*b*. Descriptions of new and critical notes upon previously known forms of North American Oedipodinae (Orthoptera: Acrididae). Trans. Am. ent. Soc. 45:229–255, pl. XXVI–XXVIII.

Rehn, J. A. G. 1921. Descriptions of new and critical notes upon previously known forms of North American Oedipodine (Orthoptera: Acrididae). Second Paper. Trans. Am. ent. Soc. 47:171–197, pl. XI, XII.

Rehn, J. A. G. 1927. On new and certain previously-known American genera of the Acridinae, with specific comments and descriptions (Orthoptera: Acrididae). Trans. Am. ent. Soc. 53:213–240, pl. XXII.

Rehn, J. A. G. 1931. On *Melanoplus borealis* in northern Labrador (Orthop.: Acrididae). Ent. News 42:33–35.

Rehn, J. A. G. 1937*a*. New or little known Neotropical Blattidae (Orthoptera). No. 4. Trans. Am. ent. Soc. 63:207–258.

Rehn, J. A. G. 1937*b*. A new subspecies of *Psoloessa delicatula* (Orthoptera: Acrididae). Trans. Am. ent. Soc. 63:325–332.

Rehn, J. A. G. 1942. On the locust genus *Psoloessa* (Orthoptera: Acrididae: Acridinae). Trans. Am. ent. Soc. 68:167–237.

Rehn, J. A. G. 1944*a*. The rhaphidophorid *Tachycines asynamorus* Adelung in America (Orthoptera: Gryllacrididae: Rhaphidophorinae). Ent. News 55:36–39.

Rehn, J. A. G. 1944*b*. A revision of the locusts of the group Hyalopteryges (Orthoptera: Acrididae: Acridinae). Trans. Am. ent. Soc. 70:181–234, pl. V.

Rehn, J. A. G. 1946*a*. On the *punctulatus* species-group of the genus *Melanoplus* (Orthoptera: Acrididae: Cyrtacanthacridinae), with the description of a new species from Kansas. Proc. Acad. nat. Sci. Philad. 98:241–269, pl. 1, 2.

Rehn, J. A. G. 1946*b*. The post-oak locust (*Dendrotettix quercus*) at Mount Misery, New Jersey, in 1944 (Orthoptera: Acrididae). Ent. News 37:147–148.

Rehn, J. A. G. 1964. A systematic and geographic analysis of the genus *Bradynotes* (Orthoptera: Acrididae: Cyrtacanthacridinae). Trans. Am. ent. Soc. 90:131–203, pl. 4–7.

Rehn, J. A. G., and H. J. Grant. 1955*a*. The North American tetrigid genus *Nometettix* (Orthoptera: Acridoidea: Tetrigidae). Proc. Acad. nat. Sci. Philad. 107:1–34, pl. I–III.

Rehn, J. A. G., and H. J. Grant. 1955*b*. *Tetrix subulata* (Orthoptera: Acridoidea: Tetrigidae) as occurring in North America. Proc. Acad. nat. Sci. Philad. 107:145–165.

Rehn, J. A. G., and H. J. Grant. 1956*a*. An analysis of *Tetrix arenosa* (Orthoptera: Acridoidea: Tetrigidae). Trans. Am. ent. Soc. 82:117–145.

Rehn, J. A. G., and H. J. Grant. 1956*b*. On *Tetrix brunneri* and a new species of *Tetrix* from California. (Orthoptera: Acridoidea: Tetrigidae). Proc. Acad. nat. Sci. Philad. 108:97–115.

Rehn, J. A. G., and H. J. Grant. 1956*c*. The components of *Tetrix ornata* (Orthoptera: Acridoidea: Tetrigidae). Proc. Acad. nat. Sci. Philad. 108:117–153.

Rehn, J. A. G., and H. J. Grant. 1958. The Batrachideinae (Orthoptera: Acridoidea: Tetrigidae) of North America. Trans. Am. ent. Soc. 85: 13–103, pl. I.

Rehn, J. A. G., and H. J. Grant. 1959. An analysis of the tribes of the Romaleinae with special reference to their internal genitalia (Orthoptera: Acrididae). Trans. Am. ent. Soc. 85:233–271.

Rehn, J. A. G., and H. J. Grant. 1961. A monograph of the Orthoptera of North America (north of Mexico) Vol. 1. Monogr. Acad. nat. Sci. Philad. 12:1–257, pl. I–VIII.

Rehn, J. A. G., and M. Hebard. 1905. A contribution to the knowledge of the Orthoptera of south and central Florida. Proc. Acad. nat. Sci. Philad. 57:29–55, pl. I.

Rehn, J. A. G., and M. Hebard. 1906. A contribution to the knowledge of the Orthoptera of Montana, Yellowstone Park, Utah, and Colorado. Proc. Acad. nat. Sci. Philad. 58:358–418.

Rehn, J. A. G., and M. Hebard. 1908. An orthopterological reconnaissance of the southwestern United States. Part I: Arizona. Proc. Acad. nat. Sci. Philad. 60:365–402.

Rehn, J. A. G., and M. Hebard. 1910*a*. A revision of the North American species of the genus *Ischnoptera* (Orthoptera). Proc. Acad. nat. Sci. Philad. 62:407–453.

Rehn, J. A. G., and M. Hebard. 1910*b*. Orthoptera found about Aweme, Manitoba. Ent. News 22(1911):5–10. [Mailed Dec. 31, 1910.]

Rehn, J. A. G., and M. Hebard. 1911*a*. Records of Georgia and Florida Orthoptera, with the descriptions of one new species and one new subspecies. Proc. Acad. nat. Sci. Philad. 62(1910):585–598.

Rehn, J. A. G., and M. Hebard. 1911*b*. Preliminary studies of North Carolina Orthoptera. Proc. Acad. nat. Sci. Philad. 62(1910):615–650.

Rehn, J. A. G., and M. Hebard. 1912*a*. Fixation of single type (lectotype) specimens of species of American Orthoptera. Section One. Proc. Acad. nat. Sci. Philad. 64:60–128.

Rehn, J. A. G., and M. Hebard. 1912*b*. On the Orthoptera found on the Florida Keys and in extreme southern Florida. I. Proc. Acad. nat. Sci. Philad. 64:235–276.

Rehn, J. A. G., and M. Hebard. 1914*a*. United States and Mexico records of species of the genus *Doru*. Jl N.Y. ent. Soc. 22:89–96.

Rehn, J. A. G., and M. Hebard. 1914*b*. Studies in American Tettigoniidae (Orthoptera) I. A synopsis of the species of the genus *Scudderia*. Trans. Am. ent. Soc. 40:271–314, 341–342, pl. XI-XXI.

Rehn, J. A. G., and M. Hebard. 1914*c*. Studies in American Tettigoniidae (Orthoptera) II. A synopsis of the species of the genus *Amblycorypha* found in America north of Mexico. Trans. Am. ent. Soc. 40:315–340, 342-344, pl. XI, XII.

Rehn, J. A. G., and M. Hebard. 1915*a*. Studies in American Tettigoniidae (Orthoptera) III. A synopsis of the species of the genus *Neoconocephalus* found in North America north of Mexico. Trans. Am. ent. Soc. 40:365–413, pl. XV, XVI.

Rehn, J. A. G., and M. Hebard. 1915*b*. Studies in American Tettigoniidae (Orthoptera) IV. A synopsis of the species of the genus *Orchelimum*. Trans. Am. ent. Soc. 41:11–83, pl. I–IV.

Rehn, J. A. G., and M. Hebard. 1915*c*. Studies in American Tettigoniidae (Orthoptera) V. A synopsis of the species of the genus *Conocephalus* found in North America north of Mexico. Trans. Am. ent. Soc. 41:115–224, pl. XV–XX.

Rehn, J. A. G., and M. Hebard. 1915*d*. The genus *Gryllus* (Orthoptera) as found in America. Proc. Acad. nat. Sci. Philad. 67:293–322, pl. IV.

Rehn, J. A. G., and M. Hebard. 1916*a*. Studies in American Tettigoniidae (Orthoptera) VII. A revision of the genus *Atlanticus* (Decticinae). Trans. Am. ent. Soc. 42:33–99, pl. VI–VIII.

Rehn, J. A. G., and M. Hebard. 1916*b*. Studies in the Dermaptera and Orthoptera of the coastal plain and Piedmont region of the southeastern United States. Proc. Acad. nat. Sci. Philad. 68:87–314, pl. XII–XIV.

Rehn, J. A. G., and M. Hebard. 1920. Descriptions of new genera and species of North American Decticinae (Orthoptera: Tettigoniidae). Trans. Am. ent. Soc. 46:225–265, pl. VIII–XI.

Rehn, J. A. G., and M. Hebard. 1927. The Orthoptera of the West Indies. Number 1. Blattidae. Bull. Am. Mus. nat. Hist. 54:1–320, pl. I–XXV.

Rehn, J. A. G., and R. L. Randell. 1962. *Boonacris*, a new generic component of the North American Melanoplini (Orthoptera: Acrididae: Cyrtacanthacridinae). Trans. Am. ent. Soc. 88:105–182, pl. V–VIII.

Rehn, J. A. G., and J. W. H. Rehn. 1938. The post-oak locust (*Dendrotettix quercus*) in the eastern United States, with notes on macropterism in the species (Orthoptera: Acrididae). Trans. Am. ent. Soc. 64:79-95.

Rehn, J. A. G., and J. W. H. Rehn. 1939. Studies of certain cyrtacanthacridoid genera (Orthoptera: Acrididae). Part I. The *Podisma* complex. Trans. Am. ent. Soc. 65:61-96, pl. VI, VII.

Rehn, J. A. G., and J. W. H. Rehn. 1945. Studies of certain cyrtacanthacridoid genera (Orthoptera: Acrididae). Part III. *Buckellacris*, another new North American genus of Holarctic type. Trans. Am. ent. Soc. 71: 1-45, pl. I, II.

Rehn, J. W. H. 1939*a*. Notes on Orthoptera of Nova Scotia and Newfoundland. Can. Ent. 71:175-178.

Rehn, J. W. H. 1939*b*. Dane faunistyczne dotyczace szarańczaków (*Orthoptera*) z Nowej Szkocji, Nowej Fundlandii i wysp francuskich Saint-Pierre et Miquelon. (Records of Orthoptera from Nova Scotia, Newfoundland and the French Islands Saint-Pierre et Miquelon). Fragm. faun. Mus. zool. polon. 4:259-265.

Rentz, D. C. 1962*a*. *Melanoplus alpinus* Scudder in California. Pan-Pacif. Ent. 38:167-168.

Rentz, D. C. 1962*b*. A technique useful for the dry preservation of soft-bodied Orthoptera. Wasmann J. Biol. 20:159-160.

Rentz, D. C., and J. D. Birchim. 1968. Revisionary studies in the Nearctic Decticinae. Mem. Pacif. Coast ent. Soc. 3:1-173.

Rentz, D. C., and D. C. Lightfoot. 1976. Notes on the distribution of Oregon shield-backed katydids with the description of a new species of *Idiostatus* (Orthoptera: Tettigoniidae: Decticinae). Ent. News 87:145-158.

Rentz, D. C. F., and D. B. Weissman. 1980. An annotated checklist of the grasshopper species of *Aerochoreutes, Circotettix*, and *Trimerotropis* (Orthoptera: Acrididae: Oeipodinae). Trans. Am. ent. Soc. 106:223-252.

Reynolds, W. J. 1980. A re-examination of the characters separating *Chorthippus montanus* and *C. parallelus* (Orthoptera: Acrididae). J. nat. Hist. 14:283-303.

Richardson, Sir J. 1851. Arctic searching expedition: A journal of a boat-voyage through Rupert's Land and the Arctic Sea, in search of the discovery ships under command of Sir John Franklin. With an appendix on the physical geography of North America (2 vols.). London, England: Longman, Brown, Green & Longman. Vol. 2. VIII, 426 pp. 1 pl.

Riegert, P. W. 1967*a*. Some observations on the biology and behaviour of *Camnula pellucida* (Orthoptera: Acrididae). Can. Ent. 99:952-971.

Riegert, P. W. 1967*b*. Association of subzero temperatures, snow cover, and winter mortality of grasshopper eggs in Saskatchewan. Can. Ent. 99:1000-1003.

Riegert, P. W. 1968. A history of grasshopper abundance surveys and forecasts of outbreaks in Saskatchewan. Mem. ent. Soc. Can. 52:1-99.

Riegert, P. W. 1980. From arsenic to DDT: A history of entomology in western Canada. Toronto, Ont.: University of Toronto Press. XII, 357 pp.

Riegert, P. W., R. Pickford, and L. G. Putnam. 1965. Outbreaks of *Camnula pellucida* (Scudder), (Orthoptera: Acrididae), in relation to native grasslands and cereal crops in Saskatchewan. Can. Ent. 97:508–514.

Riley, C. V. 1875. The Rocky Mountain locust—*Caloptenus spretus* Thomas [*sic*]. (Ord. Orthoptera: Fam. Acrididae). Ann. Rep. nox. benef. Ins. Missouri 7:VIII, 1–196.

Riley, C. V. 1876. The Rocky Mountain Locust—*Caloptenus spretus* Thomas [*sic*]. (Ord. Orthoptera: Fam. Acrididae). Ann. Rep. nox. benef. oth. Ins. Missouri 8:57–156. [Whole Report VIII, 185 pp.]

Riley, C. V. 1877*a*. The Rocky Mountain locust—*Caloptenus spretus* Thomas [*sic*]. (Ord. Orthoptera: Fam. Acrididae). Ann. Rep. nox. benef. oth. Ins. Missouri 9:56 (map), 57–124. [Whole Report XIII, 129 pp.]

Riley, C. V. 1877*b*. The locust plague in the United States: Being more particularly a treatise on the Rocky Mountain locust or so-called grasshopper, as it occurs east of the Rocky Mountains with practical recommendations for its destruction. Chicago, IL: Rand McNally Co. 236 pp.

Riley, C. V. 1881. General index and supplement to the nine reports on the insects of Missouri. Bull. U.S. ent. Commiss. 6:1–178.

Riley, C. V., A. S. Packard, and C. Thomas. 1878. First annual report of the United States Entomological Commission for the year 1877, relating to the Rocky Mountain locust and the best methods of preventing its injuries and of guarding against its invasions, in pursuance of an appropriation made by Congress for this purpose. Washington, DC: Government Printing Office. XVI, 477, [295] pp., pl. I–V, maps 1–3.

Ritchie, A. S. 1867. Notes on the "*Spectrum femoratum*." Can. Nat. Geol. (n.s.) 3:66–69.

Robert, A. 1960. Étude préliminaire des Orthoptères de la Vallée du Diable, parc du Mont Tremblant et du lac Mistassini, Nouveau-Québec. Ann. ent. Soc. Québ. 5(1959):44–48.

Roberts, C. G. D. 1893. Songs of the common day and Ave! An ode for the Shelly centenary. Toronto, Ont.: William Briggs; Montreal, Qué.: C. W. Coates; Halifax, N.S.: S. F. Huestis. X, 120 pp.

Roberts, H. R. 1941*a*. Nomenclature in the Orthoptera concerning genotype designations. Trans. Am. ent. Soc. 67:1–34.

Roberts, H. R. 1941*b*. A comparative study of the subfamilies of the Acrididae (Orthoptera) primarily on the basis of their phallic structures. Proc. Acad. nat. Sci. Philad. 93:201–246.

Roberts, H. R. 1942. Two subspecies of *Melanoplus differentialis* and related new species from Mexico with discussion of their variations (Orthoptera: Acrididae: Cyrtacanthacridinae). Trans. Am. ent. Soc. 68:151–166, pl. X, XI.

Roberts, H. R. 1947. Revision of the Mexican Melanoplini (Orthoptera: Acrididae: Cyrtacanthacridinae). Part I. Proc. Acad. nat. Sci. Philad. 99:201–230, pl. 6.

Roberts, R. A. 1937. Biology of the minor mantid, *Litaneutria minor* Scudder (Orthoptera: Mantidae). Ann. ent. Soc. Am. 30:111–121, pl. I, II.

Robinson, D. J. 1980. Acoustic communication between the sexes of the bush cricket, *Leptophyes punctatissima*. Physiol. Ent. 5:183–189.

Roeder, K. D. 1935. An experimental analysis of the sexual behavior of the praying mantis (*Mantis religiosa* L.). Biol. Bull. 69:203–220.

Rogers, C. E. 1975. Walkingsticks—insects or twigs? Insect Wld Dig. 2(2):1–4.

Roth, L. M. 1968. Oothecae of Blattariae. Ann. ent. Soc. Am. 61:83–111.

Roth, L. M. 1970. Evolution and taxonomic significance of reproduction in Blattaria. A. Rev. Ent. 15:75–96.

Roth, L. M., and H. B. Hartman. 1967. Sound production and its evolutionary significance in the Blattaria. Ann. ent. Soc. Am. 60:740–752.

Roth, L. M., and E. R. Willis. 1954. The reproduction of cockroaches. Smithson. misc. Collns 122(12):I–II, 1–49, pl. 1–12.

Roth, L. M., and E. R. Willis. 1956. Parthenogenesis in cockroaches. Ann. ent. Soc. Am. 49:195–204.

Roth, L. M., and E. R. Willis. 1957*a*. The medical and veterinary importance of cockroaches. Smithson. misc. Collns 134(10):I–IV, 1–147, pl. 1–7.

Roth, L. M., and E. R. Willis. 1957*b*. Observations on the biology of *Ectobius pallidus* (Olivier) (Blattaria: Blattidae). Trans. Am. ent. Soc. 83:31–37.

Roth, L. M., and E. R. Willis. 1960. The biotic associations of cockroaches. Smithson. misc. Collns 141:I–VI, 1–470, pl. 1–36.

Rubtsov, I. A. 1932. Opredelitel'lichinok saranchevȳkh vostochnoĭ Sibiri. [Key to immatures of grasshoppers of eastern Siberia.] Moskva, Irkutsk: Akademiya Nauk. 32 pp.

Sales, G., and D. Pye. 1974. Ultrasonic communication by animals. New York. John Wiley & Sons. 281 pp.

Saussure, H. de. 1859. Orthoptera nova americana (diagnoses praeliminaires) [II.] Locustidae. Rev. Mag. Zool. (2)11:201–212.

Saussure, H. de. 1861. Orthoptera Nova Americana (diagnoses praeliminaires) (Series II[e]) [1–4]. Rev. Mag. Zool. (2)13:126–130; 156–164, 313–324; 397–402.

Saussure, H. de. 1862. Orthoptera Nova Americana diagnoses praeliminaires. III Series. Rev. Mag. Zool. (2)14:163–171, 227–234.

Saussure, H. de. 1864*a*. Blattarum novarum species aliquot. Rev. Mag. Zool. (2)16:305–326, 341–349.

Saussure, H. de. 1864*b*. Orthoptères de l'Amérique Moyenne. [I.] Famille des Blattes. Pages 7–279 *in* Mémoires pour servir à l'histoire naturelle du Méxique, des Antilles et des États-Unis. Genève, Switzerland: Imprimerie Ramboz & Schuchardt. Vol. 1(3). Pl. I, II.

Saussure, H. de. 1871. Supplément au IIIme Fascicule Mantides. Mélanges Orthoptérologiques. Genève & Bale: H. Georg. Vol. 1 (3, Suppl):I–II, 363–460, pl. 7.

Saussure, H. de. 1874. Études sur les Insectes Orthoptères [IV]. Famille des Gryllides. Mission scientifique au Méxique et dans l'Amérique Centrale; Recherches Zoologiques pour servir à l'histoire de la faune de l'Amérique Central et du Méxique. Sixième Partie: Études sur les Myriapodes et les Insectes. Paris, France: Imprimerie Impériale. Pp. 296–519, pl. 7, 8.

Saussure, H. de. 1884. Prodromus Oedipodiorum Insectorum ex Ordine Orthopterorum. Mém. Soc. Phys. Hist. nat. Genève 28(9):1–254, pl. 1.

Saussure, H. de. 1888. Additimenta ad Prodromum Oedipodiorum. Mém. Soc. Phys. Hist. nat. Genève, 30(1):1–180, pl. 1.

Saussure, H. de. 1892. Orthoptera centrali americana. Soc. ent. [= Ent. Rdsch. (Suppl.)]7:121–124.

Saussure, H. de. 1894–1897. [Orthoptera Genuina.] Fam. Gryllidae. Biol. Centr. -Am. (Zool.) Insecta Orthoptera 1:198–200 (1894), 201–216, pl. 11 (1896), 217–284, pl. 12, 13 (1897).

Saussure, H. de. 1898. Analecta entomologica I. Orthopterologica. Revue suisse Zool. 5:183–250, pl. 9. [Appendix, ibid. 787–809 irrelevant here.]

Saussure, H. de, and A. Pictet. 1897–1899. [Orthoptera Genuina.] Fam. Locustidae. Biol. Centr. -Am. (Zool.) Ins. Orth. 1:285–344, pl. 14–16 (1897); 345–456, pl. 17–22 (1898); 457–458 (1899).

Saussure, H. de, and L. Zehntner. 1893–1894. Suborder Orthoptera Genuina. Fam. Blattidae; *and* Fam. Mantidae. Biol. Centr. -Am. (Zool.) Ins. Orth. 1:13–123, pl. 4, 5; 123–197, pl. 6–10 [pp. 13–112 publ. 1893; balance 1894].

Say, T. 1824*a*. American entomology, or descriptions of the insects of North America. Philadelphia Museum. Augustus Mitchell. Vol. 1. VIII, (106) pp., frontisp., pl. 1–18.

Say, T. 1824*b*. Appendix, Part 1, Natural history. 1. Zoology. Pages 254–378 *in* W. H. Keating. Narrative of an expedition to the source of St. Peter's River, Lake Winnepeek, Lake of the Woods, etc., etc., performed in the year 1823, by order of the Hon. John Caldwell Calhoun, Secretary of War, under the command of Stephen Harriman Long. . . . Philadelphia, PA: Government Printer. Vol. 2.

Say, T. 1825*a*. Descriptions of new hemipterous insects collected in the expedition to the Rocky Mountains, performed by order of Mr. Calhoun, Secretary of War, under command of Major Long. J. Acad. nat. Sci. Philad. 4:307–345.

Say, T. 1825*b*. American entomology, or descriptions of the insects of North America. Philadelphia Museum. Augustus Mitchell. Vol. 2. II, (125) pp., pl. 19–36.

Schennum, W. E., and R. B. Willey. 1979. A geographical analysis of quantitative morphological variation in the grasshopper *Arphia conspersa*. Evolution 33:64–84.

Schimmer, F. 1909. Beitrag zu einer Monographie der Gryllodeengattung Myrmecophila Latreille. Z. wiss. Zool. 93:409–534, pl. XXII–XXIV.

Schimmer, F. 1911. Eine neue *Myrmecophila*-Art aus den Vereinigten Staaten. (Orth.). Dt. ent. Z. 1911:443–448.

Schmidt, H., ed. 1955. Die Termiten ihre Erkennungsmerkmale und wirtschaftliche Bedeutung. Leipzig, Germany: Geest & Portig K.-G. 309 pp.

Scoggan, A. C., and M. A. Brusven. 1972. Differentiation and ecology of common immature Gomphocerinae and Oedipodinae (Orthoptera: Acrididae) of Idaho and adjacent areas. Melanderia 8:1–76.

Scudder, S. H. 1861. On the genus *Raphidophora*, Serville; with descriptions of four species from the caves of Kentucky, and from the Pacific Coast. Proc. Boston Soc. nat. Hist. 8:6–14.

Scudder, S. H. 1862. List of Orthoptera collected on a trip from Assiniboiâ to Cumberland. Can. Nat. Geol. 7:283–288.

Scudder, S. H. 1863*a*. Materials for a monograph of the North American Orthoptera, including a catalogue of the known New England species. J. Boston Soc. nat. Hist. 7 (Nov. 1862):409–480. [Publ. Jan. 1863.]

Scudder, S. H. 1863*b*. Remarks on some characteristics of the insect-fauna of the White Mountains, New Hampshire. J. Boston Soc. nat. Hist. 7:612–631, pl. 14, 15.

Scudder, S. H. 1868*a*. The insects of ancient America. Am. Nat. 1:625–631, pl. 16.

Scudder, S. H. 1868*b*. Notes on the stridulation of some New England Orthoptera. Proc. Boston Soc. nat. Hist. 11:306–313.

Scudder, S. H. 1868*c*. Catalogue of the Orthoptera of North America described previous to 1867. Smithson. misc. Collns 2(189):I–XX, 1–89.

Scudder, S. H. 1869*a*. Descriptions of new species of Orthoptera, in the collection of the American Entomological Society. Trans. Am. ent. Soc. 2:305–307.

Scudder, S. H. 1869*b*. Revision of the large, stylated, fossorial crickets. Mem. Peabody Acad. Sci. 1:1–29, pl. I.

Scudder, S. H. 1872. Notes on the Orthoptera collected by Dr. F. V. Hayden in Nebraska. Part III. Entomology. Pages 249–261 *in* F. V. Hayden. Final report of the United States Geological Survey of Nebraska and portions of the adjacent territories, made under the direction of the Commissioner of the General Land Office. Washington, DC: Government Printing Office.

Scudder, S. H. 1874. The distribution of insects in New Hampshire. Pages 331–380 *in* C. H. Hitchcock, ed. Final report upon the geology of New Hampshire. Concord, New Hampshire. E. A. Jenks, State Printer. Vol. 1, pl. A.

Scudder, S. H. 1875*a*. A century of Orthoptera. Decade II.—Locustariae. Proc. Boston Soc. nat. Hist. 17:454–462.

Scudder, S. H. 1875*b*. Appendix D. Notice of the butterflies and Orthoptera, collected by Mr. George M. Dawson, as naturalist of the B.N.A. Boundary Commission. Pages 341–345 *in* G. M. Dawson, ed. Rep. geol. Res. 49th Parallel. Montreal, Qué.

Scudder, S. H. 1875*c*. *Spharagemon*,—a genus of Oedipodidae; with a revision of the species. Proc. Boston Soc. nat. Hist. 17:467–471.

Scudder, S. H. 1875*d*. A century of Orthoptera. Decade III.—Acrydii (*Pezotettix: Caloptenus*). Proc. Boston Soc. nat. Hist. 17:472–478.

Scudder, S. H. 1875*e*. Revision of two American genera of Oedipodidae. Proc. Boston Soc. nat. Hist. 17:478–485.

Scudder, S. H. 1875*f*. A century of Orthoptera. Decade IV.—Acrydii [(cont'd.)]. Proc. Boston Soc. nat. Hist. 17:510–517 [Sep. 1879:22–29].

Scudder, S. H. 1876*a*. List of the Orthoptera collected by Dr. A. S. Packard in Colorado and the neighbouring territories, during the summer of 1875. Bull. U.S. geol. geogr. Surv. Terr. 2:261–267.

Scudder, S. H. 1876*b*. A century of Orthoptera. Decade VI. Forficulariae. Proc. Boston Soc. nat. Hist. 18:257–264.

Scudder, S. H. 1876*c*. Appendix H9. Report on the Orthoptera collected by United States geographical surveys west of the one hundredth meridian, under the direction of Lieutenant George M. Wheeler, during the season of 1875. Ann. Rep. Chief Engineer's U.S. geogr. Surv. W. of 100th Merid. 1876 (Appendix JJ):498–515. [Also as Ann. Rep. geogr. Surv. 100th Merid. 1876:278–295.]

Scudder, S. H. 1876*d*. Orthoptera from the Island of Guadalupe. Proc. Boston Soc. nat. Hist. 18:268–271.

Scudder, S. H. 1877*a*. New forms of saltatorial Orthoptera from the southern United States. Proc. Boston Soc. nat. Hist. 19:35–41.

Scudder, S. H. 1877*b*. The Florida Orthoptera collected by Mr. J. H. Comstock. Proc. Boston Soc. nat. Hist. 19:80–94.

Scudder, S. H. 1878*a*. Remarks on *Calliptenus* and *Melanoplus*, with a notice of the species found in New England. Proc. Boston Soc. nat. Hist. 19:281–286.

Scudder, S. H. 1878*b*. Brief notice of the American species of *Melanoplus* found west of the one hundred and seventeenth meridian. Proc. Boston Soc. nat. Hist. 19:286–290.

Scudder, S. H. 1879*a*. A century of Orthoptera. Decade VIII.—Acridii (*Melanoplus*). Proc. Boston Soc. nat. Hist. 20:63–75. [Sep. 57–69.]

Scudder, S. H. 1879*b*. A century of Orthoptera. Decade IX.—Acridii (*Pezotettix*). Proc. Boston Soc. nat. Hist. 20:75–86. [Sep. 64–75.]

Scudder, S. H. 1879*c*. A century of Orthoptera. Decade X.—Locustariae (*Conocephalus*). Proc. Boston Soc. nat. Hist. 20:87–95. [Sep. 76–84.]

Scudder, S. H. 1880. A few notes on N. American Acridii. Can. Ent. 12:75–76.

Scudder, S. H. 1881. Appendix II. List of the Orthoptera collected by Dr. A. S. Packard in the western United States in the summer of 1877. Pages (23)–(28) *in* C. V. Riley, A. S. Packard, and C. Thomas. Second report of the United States Entomological Commission for the years 1878 and 1879 relating to the Rocky Mountain locust and the western cricket. . . . Washington, DC: Government Printing Office. Pl. XVII.

Scudder, S. H. 1892. The orthopteran genus *Hippiscus*. Psyche, Camb. 6:265–274, 285–288, 301–304, 317–320, 333–336, 347–350, 359–363.

Scudder, S. H. 1894*a*. The North American Ceuthophili. Proc. Am. Acad. Arts Sci. 30:17–113.

Scudder, S. H. 1894*b*. A preliminary review of the North American Decticidae. Can. Ent. 26:177–184.

Scudder, S. H. 1896*a*. The North American species of *Nemobius*. Jl N.Y. ent. Soc. 4:99–108.

Scudder, S. H. 1896*b*. Index to the Mantidae of North America north of Mexico. Can. Ent. 28:207–215.

Scudder, S. H. 1896*c*. The Rocky Mountain locust and its allies in Canada. Ann. Rep. ent. Soc. Ont. 26:62–66.

Scudder, S. H. 1897*a*. The genera of North American Melanopli. Proc. Am. Acad. Arts Sci. 32:195–206.

Scudder, S. H. 1897*b*. Guide to the genera and classification of the North American Orthoptera found north of Mexico. Cambridge, MA: Edward W. Wheeler. 89 pp.

Scudder, S. H. 1897*c*. The species of the genus *Melanoplus*. Proc. Am. phil. Soc. 36:5–35.
Scudder, S. H. 1897*d*. Revision of the orthopteran group Melanopli (Acrididae) with species reference to North American forms. Proc. U.S. natn. Mus. 20:1–421, pl. I–XXVI.
Scudder, S. H. 1898*a*. The described species of *Xiphidium* in the United States and Canada. Can. Ent. 30:183–184.
Scudder, S. H. 1898*b*. The Orthopteran group Scudderiae. Proc. Am. Acad. Arts Sci. 33:271–290.
Scudder, S. H. 1898*c*. Supplement to a revision of the Melanopli. Proc. Davenport Acad. nat. Sci. 7:157–205, pl. VII–IX.
Scudder, S. H. 1898*d*. Brunner's genus *Metaleptea*. Psyche, Camb. 8:168.
Scudder, S. H. 1898*e*. The alpine Orthoptera of North America. Appalachia 8:299–319, pl. XLI–XLIV.
Scudder, S. H. 1899*a*. Stenopelmatinae of the Pacific Coast. Can. Ent. 31:113–121.
Scudder, S. H. 1899*b*. Catalogue of the described Orthoptera of the United States and Canada [1]. Proc. Davenport Acad. Sci. 8:1–48 (Oct.–Dec. 1899). [For continuation see Scudder 1900*a*.]
Scudder, S. H. 1899*c*. The North American species of *Orphulella*. Can. Ent. 31:178–188.
Scudder, S. H. 1899*d*. The orthopteran genus *Schistocerca*. Proc. Am. Acad. Arts Sci. 34:441–476.
Scudder, S. H. 1899*e*. The species of *Myrmecophila* in the United States. Psyche, Camb. 8:423–428.
Scudder, S. H. 1899*f*. Short studies of North American Tryxalinae. Proc. Am. Acad. Sci. 35:41–57.
Scudder, S. H. 1899*g*. Two genera of North American Decticinae. Proc. Am. Acad. Arts Sci. 35:83–93.
Scudder, S. H. 1900*a*. Catalogue of the described Orthoptera of the United States and Canada [2]. Proc. Davenport Acad. nat. Sci. 8:49–101 (Jan.–Feb., 1900), pl. I–III (1901); issued also complete as reprint dated 1900. [For first part see Scudder 1899*b*.]
Scudder, S. H. 1900*b*. A list of the Orthoptera of New England. Psyche, Camb. 9:99–106.
Scudder, S. H. 1901*a*. Alphabetical index to North American Orthoptera described in the eighteenth and nineteenth centuries. Occ. Pap. Boston Soc. nat. Hist. 6:I–VIII, 1–436.
Scudder, S. H. 1901*b*. The species of *Diapheromera* (Phasmidae) found in the United States and Canada. Psyche, Camb. 9:187–189.
Scudder, S. H. 1901*c*. *Cyphoderris monstrosa*. Can. Ent. 33:17–19.
Scudder, S. H. 1901*d*. Four new species of *Hippiscus*. Can. Ent. 33:88–92.
Scudder, S. H. 1902. On the United States Orthoptera which have been referred to the genus *Tridactylus*. Psyche, Camb. 9:308–310.
Scudder, S. H., and T. D. A. Cockerell. 1902. A first list of the Orthoptera of New Mexico. Proc. Davenport Acad. Sci. 9:1–62.
Seamans, H. L. 1956. Field crop and vegetable insects. *In* R. Glen, ed. Entomology in Canada up to 1956: A review of developments and accomplishments. Can. Ent. 88:322–331.

Severin, H. C. 1944. The grasshopper mite, *Eutrombidium trigonum* (Hermann): An important enemy of grasshoppers. Tech. Bull. S. Dakota St. Coll. agric. Exp. Stn 3:1–36.

Sharov, A. G. 1968. Filogeniya Ortopteroidnȳkh Nasekomȳkh [Phylogeny of Orthopteroid Insects]. Trudȳ paleont. Inst. Akad. Nauk SSSR 118: 1–217, pl. I–XII.

Sharov, A. G. 1971. Phylogeny of the Orthopteroidea [English translation, Jerusalem] VI, 251 pp., XII pl. Translated from the Russian by Israel Program for Scientific Translations, Jerusalem.

Sharp, D. 1895. Insecta Part I. Introduction, Aptera, Orthoptera, Neuroptera, and a portion of the Hymenoptera (Sessiliventres and Parasitica). Pages 81–565 *in* S. F. Harmer and A. E. Shipley, eds. The Cambridge Natural History. London, England: Macmillan. Vol. 5.

Shaw, K. C. 1968. An analysis of the phonoresponse of males of the true katydid, *Pterophylla camellifolia* (Fabricius) (Orthoptera: Tettigoniidae). Behaviour 31:203–206.

Shaw, K. C. 1975. Environmentally-induced modification of the chirp length of males of the true katydid, *Pterophylla camellifolia* (F.) (Orthoptera: Tettigoniidae). Ann. ent. Soc. Am. 68:245–250.

Shiraki, T. 1932. [Orthoptera.] Pages 2036–2114, pl. 16 *in* T. Eskai, et al. Nippon Konchû Zûkan. Iconographia Insectorum Japonicorum. Edito Prima. Tokyo, Japan: Hokuryan.

Silvestri, F. 1931. Notes on *Grylloblatta campodeiformis* and a description of a new variety (Grylloblattidae). Trans. Am. ent. Soc. 57:291–295, pl. XXV.

Slingerland, M. V. 1900. A new beneficial insect in America. The common European praying mantis. *Mantis religiosa* Linnaeus. Order Orthoptera; family Mantidae. Bull. Cornell Univ. agric. Exp. Stn 185:35–47, 1 pl.

Smith, D. S., and N. D. Holmes. 1977. The distribution and abundance of adult grasshoppers (Acrididae) in crops in Alberta, 1918–1975. Can. Ent. 109:575–592.

Smith, L. M. 1930. The snowy tree cricket and other insects injurious to raspberries. Bull. Calif. agric. Exp. Stn 505:1–38.

Smith, R. W. 1958. Parasites of nymphal and adult grasshoppers (Orthoptera: Acrididae) in western Canada. Can. J. Zool. 36:217–262.

Smith, S. I. 1868. The Orthoptera of Maine. Proc. Portland Soc. nat. Hist. 1:143–151.

Snodgrass, R. E. 1905. The coulee cricket of central Washington. (Peranabrus scabricollis Thomas.) Jl N.Y. ent. Soc. 13:74–82, pl. I, II.

Snyder, T. E. 1934. American subterranean termites other than those of the Pacific Coast. Pages 178–186 *in* C. F. Kofoid, ed. Termites and termite control. Berkeley, CA: University of California Press.

Snyder, T. E. 1935. Our enemy the termite. Ithaca, NY: Comstock Publishing Co. XIV, 196 pp.

Snyder, T. E. 1954. Order Isoptera: The termites of the United States and Canada. New York: National Pest Control Association. 64 pp.

Somes, M. P. 1914. Acridiidae of Minnesota. Bull. Univ. Minn. agric. Exp. Stn 141:1–98, pl. I–IV.

Spencer, G. J. 1945. On the incidence, density and decline of certain insects in British Columbia. Proc. ent. Soc. Br. Columb. 42:19-23.

Spencer, G. J. 1958. A record of a sand cricket, Stenopelmatinae, from the coastal wet belt of British Columbia. Proc. ent. Soc. Br. Columb. 55:25.

Spooner, J. D. 1964. The Texas bush katydid—its sounds and their significance. Anim. Behav. 12:235-244.

Spooner, J. D. 1968*a*. Collection of male phaneropterinae katydids by imitating sounds of the female. J. Georgia ent. Soc. 3:45-46.

Spooner, J. D. 1968*b*. Pair-forming acoustic systems of phaneropterine katydids (Orthoptera: Tettigoniidae). Anim. Behav. 16:197-212.

Spooner, J. D. 1973. Sound production in *Cyphoderris monstrosa* (Orthoptera: Prophalangopsidae). Ann. ent. Soc. Am. 66:4-5.

Stainer, J. E. R. 1975. The status of *Conocephalus fasciatus vicinus* (Morse, 1901) (Orthoptera: Conocephalidae). J. ent. Soc. Br. Columb. 72:31-34.

Stål, C. 1873*a*. Orthoptera nova descripsit. Öfvers. K. VetenskAkad. Förh. 30 (4):39-54.

Stål, C. 1873*b*. Recensio Orthopterorum. Revue critique des Orthoptères décrits par Linné, De Geer et Thunberg. Stockholm, Sweden: P.A. Norstedt & Söner. Vol. 1. IV, (20), 154 pp.

Stamplecoski, R. 1980. An invasion of grasshoppers? This week in the Madawaska Valley, Madawaska, Ontario 10(41):2.

Steinberg, J. B., and R. C. Conant. 1974. An informational analysis of the intermale behaviour of the grasshopper *Chortophaga viridifasciata*. Anim. Behav. 22:617-627.

Steinberg, J. B., and R. B. Willey. 1974. Visual and acoustical social displays by the grasshopper *Chortophaga viridifasciata* (Acrididae: Oedipodinae). Can. J. Zool. 52:1145-1154, pl. I-III.

Steinmann, H. 1971. The tetricids of the Nearctic subregion (Orthoptera). Acta zool. Acad. Sci. hungar. 17:381-385.

Steinmann, H. 1973. A zoological checklist of world Dermaptera. Fol. ent. hungar. (s.n.) 27:145-154.

Steinmann, H. 1975. Suprageneric classification of Dermaptera. Acta zool. Acad. Sci. hungar. 21:195-220.

Stephens, J. F. 1829*a*. Orthoptera. Pages 25-26 *in* The nomenclature of British insects being a compendious list of such species as are contained in the Systematic Catalogue of British Insects. London, England: Baldwin. [Whole work in 68 columns.]

Stephens, J. F. 1829*b*. Order Dermaptera; Order Orthoptera. A systematic catalogue of British insects, being an attempt to arrange all the hitherto discovered indigenous insects in accordance with their natural affinities; containing also the references to every English writer on entomology, and to the principal foreign authors; with all the published British genera to the present time. London, England: Baldwin. Vol. 1. Pp. 229, 300-304.

Stephens, J. F. 1835. Mandibulata [VI]: Order II.—Dermaptera, *DeGeer*.; Order III.—Orthoptera, *Olivier*. Illustrations of British entomology or a synopsis of indigenous insects: Containing their generic and specific distinctions, with an account of their metamorphoses, times of appearance, localities, food and economy, as far as practicable, with coloured figures (from Westwood) of the rarer and more interesting species. London, England: Baldwin & Craddock. Vol. 6. Pp. 2-9, 9-48, pl. XXVIII. [Whole vol. 240 pp., pl. XXVIII-XXXIV.]

Strohecker, H. F. 1952. Descriptions of new species and notes on North American Orthoptera. Am. Midl. Nat. 48:683-688.

Talbot, E. A. 1824. Letter XIV. Insects of various kinds—butterfly—grasshopper and locust—(etc.). Five years' residence in the Canadas: Including a tour through part of the United States of America, in the year 1823. London, England: Longman, Hurst, Rees, Orme, Brown & Green. (2 vols.) Vol. 1. Pp. 237-258 (238-240).

Taylor, A. S. 1859. An account of the grasshoppers and locusts in America. Smithson. Inst. ann. Rep. 1858:200-213.

Thomas, C. 1862. Note on *Tridactylus illinoiensis* n.sp. Proc. ent. Soc. Philad. 1:104.

Thomas, C. 1865. Insects injurious to vegetation in Illinois. Trans. Ill. St. agric. Soc. 5:401-468.

Thomas, C. 1870. Descriptions of grasshoppers, from Colorado. Proc. Acad. nat. Sci. Philad. 22:74-84.

Thomas, C. 1871*a*. A list and descriptions of new species of Orthoptera. Rep. U.S. geol. Surv. Terr. 2:265-284.

Thomas, C. 1871*b*. On a new grasshopper from Colorado. Can. Ent. 3:168.

Thomas, C. 1872. Notes on the saltatorial Orthoptera of the Rocky Mountain regions. Rep. Progr. U.S. geol. Surv. Montana adj. Terr. 5:423-466, pl. I, II.

Thomas, C. 1873. Synopsis of the Acrididae of North America. Rep. U.S. geol. geogr. Surv. Terr. 5:I-X, 1-262, 1 pl.

Thomas, C. 1874. Descriptions of some new Orthoptera, and notes on some species but little known. Bull. U.S. geol. geogr. Surv. Terr. 1:65-71.

Thomas, C. 1875. Report upon a collection of Orthoptera made in portions of Nevada, Utah, California, Colorado, New Mexico and Arizona, during the years 1871, 1872, 1873 and 1874. Rep. U.S. geol. Surv. W. 100th Merid. 5:843-908, pl. 42-45.

Thomas, C. 1876*a*. A list of Orthoptera, collected by J. Duncan Putnam, of Davenport, Iowa, during the summers of 1872-3-4 & 5, chiefly in Colorado, Utah and Wyoming Territories. Proc. Davenport Acad. nat. Sci. 1:249-264, pl. 36.

Thomas, C. 1876*b*. A list of the Orthoptera of Illinois. Bull. Ill. Mus. nat. Hist. 1:59-69.

Thomas, C. 1880*a*. Manual of economic entomology, Part III. The Acrididae of Illinois. Rep. St. Ent. Ill. 9:71-140.

Thomas, C. 1880*b*. Notes on Orthoptera. Can. Ent. 12:221-224.

Thomas, E. S., and R. D. Alexander. 1962. Systematic and behavioral studies on the meadow grasshoppers of the *Orchelimum concinnum* group (Orthoptera: Tettigoniidae). Occ. Pap. Mus. Zool. Univ. Mich. 626:1-31.

Thwaites, R. G., ed. 1896-1901. The Jesuit relations and allied documents: Travels and explorations of the Jesuit missionaries in New France, 1610-1791; the original French, Latin and Italian texts, with English translations and notes, illustrated by portraits, maps and facsimiles. Cleveland, OH: The Burrows Bros. Co. 73 vols.

Tinkham, E. R. 1939. Five new records and notes on Canadian Acrididae from the Higdon Ranch, southeastern Alberta. Can. Ent. 71:121–126.

Tinkham, E. R. 1947. New species, records and faunistic notes concerning Orthoptera in Arizona. Am. Midl. Nat. 38:127–149.

Tinkham, E. R. 1959. Notes on the self-burial habits of two Nearctic sand dune acridids (Orthoptera). Entomologist 92:185–188, pl. IX fig. [5].

Tinkham, E. R., and D. C. Rentz. 1969. Notes on the bionomics and distribution of the genus *Stenopelmatus* in central California with the description of a new species (Orthoptera: Gryllacrididae). Pan-Pacif. Ent. 45:4–14.

Treherne, R. C., and E. R. Buckell. 1924. The grasshoppers of British Columbia with particular reference to the influence of injurious species on the range lands of the province. Bull. Can. Dep. Agric. (n.s.) 39:1–47.

Tuck, J. B., and R. C. Smith. 1940. Identification of the eggs of mid-western grasshoppers by the chorionic sculpturing. Tech. Bull. Kansas agric. Exp. Stn 48:1–39.

Turley, L. W. 1901. *Cyphoderris monstrosa*. Can. Ent. 33:246–248.

Turton, W. 1802. 57. Blatta. Pages 527–532 *in* Insects. Order II. Hemiptera. A general system of nature, through the three Grand Kingdoms of animals, vegetables and minerals; systematically divided into their several classes, orders, genera, species and varieties, with their habitations, manners, economy, structure and peculiarities. London. Lockington, Allen & Co. Vol. 2.

Tyrkus, M., ed. 1983. Proceedings of the second triennial meeting of the Pan American Acridological Society, Bozeman, Montana, July 1979. Detroit, MI: Pan American Acridological Society.

Ueshima, N., and D. C. (F.) Rentz. 1979. Chromosome systems in the North American Decticinae with reference to Robertsonian changes (Orthoptera: Tettigoniidae). Cytologia 44:693–714.

Uhler, P. R. 1864. Orthopterological contributions. Proc. ent. Soc. Philad. 2:543–555.

Ulagaraj, S. M. 1975. Mole crickets: Ecology, behavior and dispersal flight. (Orthoptera: Gryllotalpidae: *Scapteriscus*). Envir. Ent. 4:265–273.

Ulagaraj, S. M., and T. J. Walker. 1973. Phonotaxis of crickets in flight: Attraction of male and female crickets to male calling songs. Science 182:1278–1279.

Ulagaraj, S. M., and T. J. Walker. 1975. Response of flying mole crickets to three parameters of synthetic songs broadcast outdoors. Nature, Lond. 253:530–532.

United Nations Educational, Scientific and Cultural Organization (UNESCO). 1962. Termites in the humid tropics: Proceedings of the New Delhi Symposium. Humid Tropics Research. Paris, France: UNESCO. [4] 259 pp., 1, (42 in), 16 pl.

Urquhart, F. A. 1937. Some notes on the sand cricket, (*Tidactylus apicalis* Say.). Can. Field-Nat. 51:28–29.

Urquhart, F. A. 1938. A new species of *Nemobius* from Ontario (Orthoptera). Can. Ent. 70:101–102.

Urquhart, F. A. 1940*a*. Notes on the Ontario species of *Scudderia* (Orthoptera: Ensifera). Can. Field-Nat. 54:102–104.

Urquhart, F. A. 1940*b*. Further notes on the sand cricket, *Tridactylus apicalis* Say. Can. Field-Nat. 54:106.

Urquhart, F. A. 1941*a*. A faunal investigation of Prince Edward County, Ontario. An annotated list of the crickets and grasshoppers (Orthoptera: Saltatoria) of Prince Edward County, Ontario. Univ. Toronto Stud. biol. Ser. 48:116–119.

Urquhart, F. A. 1941*b*. The Blattaria and Orthoptera of Essex County, Ontario. Contr. R. Ont. Mus. Zool. 20:1–32.

Urquhart, F. A. 1941*c*. An ecological study of the Saltatoria of Point Pelee, Ontario. Univ. Toronto Stud. biol. Ser. 50:1–91, pl. I–XXVII.

Urquhart, F. A. 1941*d*. The species of *Nemobius* (Orthoptera: Ensifera) in Ontario. Can. Field-Nat. 55:80–82, pl. I, II.

Urquhart, F. A. 1942. New records and notes of Saltatoria (Orthoptera) in Ontario. Can. Ent. 74:97–98.

Urquhart, F. A., and J. R. Beaudry. 1953. A recently introduced species of European grasshopper. Can. Ent. 85:78–79.

Uvarov, B. P. 1914. K'faunē pryamokrȳlȳkh Zaboĭkanvya. (Contributions à la faune des *Orthoptères* de la province de Transbaïkalie.) Ezheg. zool. Muz. Akad. Nauk S. Peterb. (Ann. Mus. zool. Acad. Sci. St. Petersb.) 19:167–172.

Uvarov, B. P. 1966*a*. Grasshoppers and locusts: A handbook of general acridology. Cambridge, England: Cambridge Univ. Press for Anti-Locust Research Centre, London. Vol. 1. XII, 481 pp., 1 pl.

Uvarov, B. P. 1966*b*. Hibernation of active stages of Acridoidea in temperate climates. Atti Accad. gioenia Sci. nat. (6)18:175–189.

Uvarov, B. P. 1977 (posthumous). Grasshoppers and locusts: A handbook of general acridology. London, England: Centre for Overseas Pest Research. Vol. 2. X, 613 pp., 1 pl.

Valek, D. A., and H. C. Coppel. 1972*a*. Bionomics of an oak defoliating grasshopper, *Dendrotettix quercus*, in Wisconsin. Ann. ent. Soc. Am. 65:310–319.

Valek, D. A., and H. C. Coppel. 1972*b*. Ovipositional site preferences of the oak defoliating grasshopper, *Dendrotettix quercus*, in Wisconsin. Proc. Wis. Acad. Sci. Arts Lett. 60:225–230.

Van Horn, D. 1965. Variations in size and phallic morphology among populations of *Melanoplus dodgei* (Thomas) in the Colorado Front Range (Orthoptera: Acrididae). Trans. Am. ent. Soc. 91:95–119.

Van Horn, S. N. 1965. The embryology of the stomodeal nervous system in *Aulocara elliotti* Thomas (Orthoptera: Acrididae). Proc. XII Int. Congr. Ent. (London, 1964) 152–153.

Van Horn, S. N. 1968. Changes in gland volume during the post-diapause development in embryos of *Aulocara elliotti* (Thomas) (Orthoptera: Acrididae). Proc. Montana Acad. Sci. 28:72–75.

Vickery, V. R. 1961. The Orthoptera of Nova Scotia. Proc. Trans. Nova Scotian Inst. Sci. 25:1–70.

Vickery, V. R. 1962. Additional notes on *Mantis religiosa* Linné in Québec. Ann. ent. Soc. Québ. 7:60–61.

Vickery, V. R. 1965. Factors governing the distribution and dispersal of the recently introduced grasshopper, *Metrioptera roeseli* (Hgb.) (Orthoptera: Ensifera). Ann. ent. Soc. Québ. 10:165–172.

Vickery, V. R. 1967*a*. The Orthoptera of Alaska, Yukon and the Mackenzie District of the Northwest Territories. Trans. Am. ent. Soc. 93:249–278.

Vickery, V. R. 1967*b*. Distribution and variation in North American *Chorthippus* (Orthoptera: Acrididae: Gomphocerinae). Ann. ent. Soc. Québ. 12:100–132.

Vickery, V. R. 1969*a*. Two species of *Pteronemobius* previously unreported in Québec (Orthoptera: Ensifera: Grylloidea: Nemobiinae). Ann. ent. Soc. Québ. 14:22–24.

Vickery, V. R. 1969*b*. Two new species of sub-arctic American Orthoptera. Ent. News 80:265–272.

Vickery, V. R. 1970. A new species of *Melanoplus* (Orthoptera: Acridoidea) from Québec. Ann. ent. Soc. Québ. 15:6–13.

Vickery, V. R. 1973. Notes on *Pteronemobius* and a new genus of the tribe Pteronemobiini (Orthoptera: Gryllidae: Nemobiinae). Can. Ent. 105:419–424.

Vickery, V. R. 1978. The value of cytology in taxonomy with particular reference to the Podismini (Acridoidea: Acrididae: Melanoplinae). Rev. Soc. ent. argent. 36 (1977):89–95.

Vickery, V. R. 1979. Notes on some Canadian Acrididae (Orthoptera). Can. Ent. 111:699–702.

Vickery, V. R. 1984. The Orthoproid insects of Yukon. Mem. Lyman ent. Mus. Res. Lab. 10:1–42.

Vickery, V. R., and D. E. Johnstone. 1970. Generic status of some Nemobiinae (Orthoptera: Gryllidae) in northern North America. Ann. ent. Soc. Am. 63:1740–1749.

Vickery, V. R., and D. E. Johnstone. 1973. The Nemobiinae (Orthoptera: Gryllidae) of Canada. Can. Ent. 105:623–645.

Vickery, V. R., D. E. Johnstone, and D. K. McE. Kevan. 1974. The orthopteroid insects of Québec and the Atlantic provinces of Canada. Mem. Lyman ent. Mus. Res. Lab. 1:I, 1–204, 1 pl.

Vickery, V. R., and G. E. Kerr. 1975. Additional records of Ensifera (Grylloptera) in Ontario. Proc. ent. Soc. Ont. 105 (1974):96–100.

Vickery, V. R., and D. K. McE. Kevan. 1964*a*. The Provancher types of Orthoptera. Can. Ent. 96:1549–1554.

Vickery, V. R., and D. K. McE. Kevan. 1964*b*. The genus *Schistocerca* (Orthoptera: Acrididae) in Canada. Can. Ent. 96:1555–1558.

Vickery, V. R., and D. K. McE. Kevan. 1967. Records of the orthopteroid insects in Ontario. Proc. ent. Soc. Ont. 97:13–68.

Vickery, V. R., and D. K. McE. Kevan. 1978. A new species of *Melanoplus* (Orthoptera: Acrididae: Melanoplini) from the Magdalen Islands, Québec. Ann. ent. Soc. Québ. 22:188–192.

Vickery, V. R., and B. Nagy. 1973. Ecological notes on Orthoptera (*s. str.*) in British Columbia. J. ent. Soc. Br. Columb. 70:27–33.

Vickery, V. R., L. M. Crozier, and M. O'c.Guibord. 1981. Immature grasshoppers of eastern Canada (Orthoptera: Acrididae). Mem. Lyman ent. Mus. Res. Lab. 9:V, 1–72.

Villiers, C. J. de. 1789. Caroli Linnaei Entomologia, faunae suecicae descriptionibus aucta; D. D. Scopoli, Geoffroy, De Geer, Fabricii, Schrank etc. speciebus vel in systemate non enumeratis, vel nuperrime detectis, vel speciebus Galliae australis locupletata, generum specierumque rariorum iconibus ornata. . . . Lungundi [= Leiden], Netherlands: Piestre & Demolliere. Vol. 1. Pp. XXIV, 1–765, pl. I–III.

Wakeland, C. 1959. Mormon crickets in North America. Tech. Bull. U.S. Dep. Agric. agric. Res. Serv. 1202:I–II, 1–77, 1 foldout (map).

Wakeland, C., and W. E. Shull. 1936. The Mormon cricket with suggestions for its control. Ext. Bull. Univ. Idaho Coll. Agric. 100:1–30.

Walden, B. H. 1911. Guide to the insects of Connecticut. Part II. The Euplexoptera and Orthoptera of Connecticut. Bull. Conn. geol. nat. Hist. Surv. 16:39–169, pl. III (part), VI–XI.

Walker, E. M. 1898*a*. A new grasshopper from Ontario. Can. Ent. 30:90–92.

Walker, E. M. 1898*b*. Notes on some Ontario Acrididae. Can. Ent. 30: 122–126, 258–263.

Walker, E. M. 1898*c*. A new alpine grasshopper from western Canada. Can. Ent. 30:197–199, pl. 6.

Walker, E. M. 1899. Notes on some Ontario Acrididae.—Part III [*sic*], [actually IV]. Can. Ent. 31:29–36.

Walker, E. M. 1901. Notes on some Ontario Acrididae.—Part IV [cont'd]. Can. Ent. 33:20–23.

Walker, E. M. 1902*a*. The Canadian species of *Trimerotropis*. Can. Ent. 34:1–11, pl. 1.

Walker, E. M. 1902*b*. A preliminary list of Acrididae of Ontario. Can. Ent. 34:251–258.

Walker, E. M. 1902*c*. A collecting trip in south-western Ontario. Rep. ent. Soc. Ont. 32:85–90.

Walker, E. M. 1902*d*. Entomological record [1901]: Orthoptera. Rep. ent. Soc. Ont. 32:108–109.

Walker, E. M. 1903*a*. The genus *Podisma* in eastern North America. Can. Ent. 35:295–302, pl. 6.

Walker, E. M. 1903*b*. Reports on the insects of the year [1902]. Division No. 3.—Toronto. Rep. ent. Soc. Ont. 33:39–41.

Walker, E. M. 1904*a*. The crickets of Ontario. Can. Ent. 36:142–144, 181–188, pl. 4, 249–255.

Walker, E. M. 1904*b*. Notes on the Locustidae of Ontario [I, II]. Can. Ent. 36:325–330, 337–341.

Walker, E. M. 1905*a*. Notes on the Locustidae of Ontario [III]. Can. Ent. 37:34–38.

Walker, E. M. 1905*b*. Notes on the Locustidae of Ontario [IV]. Can. Ent. 37:113–119, pl. IV, V.

Walker, E. M. 1906*a*. Records in Orthoptera from the Canadian Northwest. Can. Ent. 38:55–59.

Walker, E. M. 1906*b*. Orthoptera and Odonata from Algonquin Park, Ont. Rep. ent. Soc. Ont. 36:64–70.

Walker, E. M. 1909. On the Orthoptera of northern Ontario. Can. Ent. 41:137–144, 173–178, 205–212.

[Walker, E. M., ed.] 1910. The Orthoptera of western Canada. Can. Ent. 42:269-276, 293-300, 333-340, 351-356.

Walker, E. M. 1911. On the habits and stridulation of *Idionotus brevipes* Caudell, and other notes on Orthoptera. Can. Ent. 43:303-304.

Walker, E. M. 1914*a*. A new species of Orthoptera, forming a new genus and family. Can. Ent. 46:93-99.

[Walker, E. M.] 1914*b*. [Untitled note.] Can. Ent. 46:368.

Walker, E. M. 1915*a*. Field notes and questions. The occurrence of *Mantis religiosa* L. in Canada. Can. Ent. 47:135.

Walker, E. M. 1915*b*. Notes on a collection of Orthoptera from Prince Edward Island and the Magdalen Islands, Qué. Can. Ent. 47:339-344.

Walker, E. M. 1919*a*. The terminal structures of orthopteroid insects: A phylogenetic study. Ann. ent. Soc. Am. 12:267-316, pl. XX-XXXVIII.

Walker, E. M. 1919*b*. On the male and immature state of *Grylloblatta campodeiformis* Walker. Can. Ent. 51:131-139.

Walker, E. M. 1920. Orthoptera collected in the Canadian Arctic, 1913-1918. Report of the Canadian Arctic Expedition, 1913-18. Ottawa, Ont.: F. A. Acland, King's Printer. Vol. 3 (Insects) (J). Pp. 1-4.

Walker, E. M. 1937. Presidential address [of the Royal Society of Canada]: *Grylloblatta*, a living fossil. Trans. R. Soc. Can. (3) 31 [Sect. 5; title p. gives vol. no. erroneously as XXVI]:1-10.

Walker, E. M., and F. A. Urquhart. 1940. New records and notes of Orthoptera in Ontario. Can. Ent. 72:15-19.

Walker, F. 1853. Termitides. List of the specimens of neuropterous insects in the collection of the British Museum. London, England: British Museum (Natural History) Publ. III, pp. 501-529.

Walker, F. 1868. Catalogue of the specimens of Blattariae in the collection of the British Museum. London, England: British Museum (Natural History) Publ. IV, 239 pp.

Walker, F. 1869*a*. Catalogue of the specimens of Dermaptera Saltatoria [1] and supplement to the Blattariae in the collection of the British Museum. London, England: British Museum (Natural History) Publ. IV, 224 pp.

Walker, F. 1869*b*; 1870; 1871. Catalogue of the specimens of Dermaptera Saltatoria in the collection of the British Museum. London, England: British Museum. (Natural History) Publ. 2:I-IV, 225-423 (1869); 3:I-IV, 425-604 (1870); 4:I-IV, 605-809 (1870); 5:I-IV, 811-850 (1871). (Part 5 bound with "Supplement to the catalogue of Blattariae," pp. 1-40).

Walker, F. 1872. Hemiptera, Heteroptera and Dermaptera (Orthoptera) of America to the north of the United States. Can. Ent. 4:29-31.

Walker, T. J. 1962. The taxonomy and calling songs of United States tree crickets (Orthoptera: Gryllidae: Oecanthinae) I. The genus *Neoxabea* and the *niveus* and *varicornis* groups of the genus *Oecanthus*. Ann. ent. Soc. Am. 55:303-322.

Walker, T. J. 1963. The taxonomy and calling songs of United States tree crickets (Orthoptera: Gryllidae: Oecanthinae) II. The *nigricornis* group of the genus *Oecanthus*. Ann. ent. Soc. Am. 56:772-789.

Walker, T. J. 1964. Cryptic species among sound-producing ensiferan Orthoptera (Gryllidae and Tettigoniidae). Q. Rev. Biol. 39:345-355.

Walker, T. J. 1966. Annotated checklist of Oecanthinae (Orthoptera: Gryllidae) of the world. Fla Ent. 49:265–277.

Walker, T. J. 1969. Acoustic synchrony: Two mechanisms in the snowy tree cricket. Science 166:891–894.

Walker, T. J. 1975*a*. Effects of temperature on rates in poikilotherm nervous systems: Evidence from the calling songs of meadow katydids (Orthoptera: Tettigoniidae: *Orchelimum*) and reanalysis of published data. J. comp. Physiol. 101:57–69.

Walker, T. J. 1975*b*. Stridulatory movements in eight species of *Neoconocephalus* (Tettigoniidae). J. Insect Physiol. 21:595-603.

Walker, T. J., and A. B. Gurney. 1960. A new species of *Oecanthus* from the West Indies (Orthoptera: Gryllidae). Fla Ent. 43:9–13.

Walker, T. J., and A. B. Gurney. 1967. The metanotal gland as a taxonomic character in *Oecanthus* of the United States (Orthoptera: Gryllidae). Proc. ent. Soc. Wash. 69:157–161.

Walker, T. J., J. J. Whiteshell, and R. D. Alexander. 1973. The robust conehead: Two widespread sibling species (Orthoptera: Tettigoniidae: *Neoconocephalus "robustus"*). Ohio J. Sci. 73:321–330.

Wallace, H. S. 1955. Revision of the genus *Aeoloplides* (Orthoptera: Acrididae). Ann. ent. Soc. Am. 48:453–480.

Walsh, B. D. 1864. On phytophagic varieties and phytophagic species. Proc. ent. Soc. Philad. 3:403–430.

Walsh, B. D. 1866. Grasshoppers and locusts. Pract. Ent. 2:1–5.

Washburn, F. L. 1912*a*. Grasshopper work in Minnesota during the season 1911. J. econ. Ent. 5:111–121.

Washburn, F. L. 1912*b*. Grasshoppers and other injurious insects of 1911 and 1912. Rep. St. Ent. Minn. 14:XIV, 1–114, pl. 1–3.

Weesner, F. M. 1956. The biology of colony foundation in *Reticulitermes hesperus* Banks. Univ. Calif. Publs Zool. 61:253–314.

Weesner, F. M. 1965. The termites of the United States: A handbook. Elizabeth, NJ: National Pest Control Association. 71 pp.

Weesner, F. M. 1970. Termites of the Nearctic region. Pages 477–525 *in* K. Krishna and F. M. Weesner, eds. Biology of termites. New York & London: Academic Press. Vol. 2.

Weever, E. G., and J. A. Vernon. 1959. The auditory sensitivity of Orthoptera. Proc. natn. Acad. Sci. U.S.A. 45:413–419.

Weidner, H. 1970. Isoptera (Termiten). Pages 1–147 *in* J.-G. Helmcke, D. Starck, and H. Wermuth, eds. Kükenthal's Handbuch der Zoologie (ed. 2). Berlin, Germany: Walter de Gruyter & Co. Vol. 4(2) [2 (14)].

Weiss, H. B. 1914. Some facts about the egg nest of *Paratenodera sinensis* (Orth.). Ent. News 25:279-282.

Weiss, H. B. 1936. The pioneer century of American entomology. New Brunswick, NJ. II, 320 pp.

Weiss, H. B., and E. L. Dickerson. 1918. The European mole cricket, *Gryllotalpa gryllotalpa* L., an introduced insect pest. Jl N.Y. ent. Soc. 26:18–23, pl. II.

Weissman, D. B., and D. C. F. Rentz. 1977. Feral house crickets, *Acheta domesticus* (Orthoptera: Gryllidae) in southern California. Ent. News 88:246–248.

Weissman, D. B., and D. C. F. Rentz. 1980. Cytological, morphological, and crepitational characteristics of the trimerotropine (*Aerochoreutes, Circotettix* and *Trimerotropis*) grasshoppers (Orthoptera: Oedipodinae). Trans. Am. ent. Soc. 106:253–272.

Whitcomb, W. H. 1967. Field studies on predators of the second-instar bollworm, *Heliothis zea* (Boddie) (Lepidoptera: Noctuidae). J. Georgia ent. Soc. 2:113–118.

White, A. 1851. List of insects taken by Sir John Richardson and John Rae, Esq., in Arctic North America. Pages 351–363 *in* Sir J. Richardson, Arctic searching expedition: A journal of a boat-voyage through Rupert's Land and the Arctic Sea, in search of the discovery ships under command of Sir John Franklin. . . . London, England: Longman, Brown, Green & Longman. Vol. 2.

White, M. J. D. 1949. A cytological survey of wild populations of *Trimerotropis* and *Circotettix*. (Orthoptera: Acrididae). I. The chromosomes of twelve species. Genetics 34:537–563.

White, M. J. D. 1951*a*. A cytological survey of wild populations of *Trimerotropis* and *Circotettix* (Orthoptera: Acrididae). II. Racial differentiation in *T. sparsa*. Genetics 36:31–53.

White, M. J. D. 1951*b*. Structural heterozygosity in natural populations of the grasshopper *Trimerotropis sparsa*. Evolution 5:376–394.

White, M. J. D. 1951*c*. Cytogenetics of orthopteroid insects. Adv. Genet. 4:268–330.

White, M. J. D. 1954. Animal cytology & evolution (Second Edition). Cambridge, England: Cambridge University Press. XIV, 454 pp.

White, M. J. D. 1973. Animal cytology and evolution (Third Edition). Cambridge. England: Cambridge University Press. VIII, 961 pp.

White, R. M., and P. J. G. Rock. 1945. A contribution to the knowledge of the Acrididae of Alberta. Sci. Agric. 25:577–596.

Whitehead, F. E., and D. Miner. 1944. The biology and control of the camel cricket *Daihinia brevipes*. J. econ. Ent. 37:573–581.

Willemse, C. [J. M.] 1943. Het sjirpen van *Mecosthethus grossus* L. de groote weidesprinkhaan. Natuurhist. Maandbl. 32:7–8.

Willey, R. B. 1975. Slowed motion analysis of sound production in the grasshopper *Arphia sulphurea* (Acrididae: Oedipodinae). Psyche, Camb. 82:324–340.

Willey, R. B., and R. L. Willey. 1967. Barriers to gene flow in natural populations of grasshoppers. I. The Black Canyon of the Gunnison River and *Arphia conspersa*. Psyche, Camb. 74:42–57.

Willey, R. B., and R. L. Willey. 1969. Visual and acoustical social displays by the grasshopper *Arphia conspersa* (Orthoptera: Acrididae). Psyche, Camb. 76:280–305.

Willey, R. B., and R. L. Willey. 1971. Barriers to gene flow in natural populations of grasshoppers. II. Maintenance of narrow hybrid-zones between morphs of *Arphia conspersa* on Black Mesa, Colorado. Psyche, Camb. 78:330–349.

Williams, J. B. 1906. Reports on insects of the year [1905]. Division No. 3.—Toronto district. Rep. ent. Soc. Ont. 36:9–10.

Williamson, G. D. 1966. A summary of parasite and predator liberations in Canada in 1965. Can. Insect Pest Rev. 43 (1965):127–139.

Williamson, G. D., ed. 1973. Insect liberations in Canada: Parasites and predators 1972. Agric. Can. Liberation Bull. 36:I, II, 1–11.

Williamson, W. D. 1832. The history of the State of Maine; from its first discovery, A.D. 1602, to the separation, A.D. 1820, inclusive . . . (2 vols.). Hallowell, ME: Glazier, Masters & Co. Vol. 1. Pp. 102, 103, 172.

Willis, E. R., and N. Lewis. 1957. The longevity of starved cockroaches. J. econ. Ent. 50:438–440.

Willis, E. R., G. R. Riser, and L. M. Roth. 1958. Observations on reproduction and development in cockroaches. Ann. ent. Soc. Am. 51:53–69.

Wilson, M. V. H. 1977. New records of insect families from the freshwater middle Eocene of British Columbia. Can. J. Earth Sci. 14:1139–1155.

Wilson, T. 1915. The outbreak of locusts of 1914. Proc. ent. Soc. Br. Columb. [1] 7:41–42.

Wilson, W. A. 1971. Nematode occurrence in Ontario earwigs (Nematoda: Dermaptera). Can. Ent. 103:1045–1048.

Yakobson, G. G., and B. L. Bianki. 1902–1905. Pryamokrȳlȳya i Lozhnosêtchatokrȳlȳya Possiiskoi Imperĭi i Sopredêl'nȳkh Stran'. S. Peterburg, USSR: A. F. Devrien. Pp. I–V, 1–952, pl. I–XXV.

Yersin, A. 1860. Note sur quelques Orthoptères nouveaux ou peu connus. Annls Soc. ent. Fr. (3) 8:509–537, pl. X.

Zacher, F. 1928. Züchtung von Orthopteren. Pages 89–190 *in* E. Abderhalden, ed. Handbuch der biologischen Arbeitsmethoden. Berlin & Wein: Urban & Schwarzenberg. Vol. 9.

Zacher, F. 1949. Orthopteroidea, Geradflügler. Pages 228–253 *in* H. Blunck, ed. Tierische Schädlinge an Nutzpflanzen. Sorauer's Handbuch der Pflanzenkrankheiten (ed. 5). Berlin, Germany: Paul Parey. Vol. 4 (1).

Zeuner, F. E. 1941. The classification of Decticinae hitherto included in *Platycleis* Fieb. or *Metrioptera* Wesm. (Orthoptera, Saltatoria). Trans. R. ent. Soc. Lond. 91:1–50.

Index to names of ulonate taxa

(Page numbers of principal entries are in boldface; synonyms are in italic type.)

Index to plant hosts, parasites, and predators